BRITISH MEDICAL ASSOCIATION
1007683

AN ATLAS OF COMPARATIVE VERTEBRATE HISTOLOGY

AN ATLAS OF COMPARATIVE VERTEBRATE HISTOLOGY

DONALD B. MCMILLAN
Emeritus Professor, Department of Biology, Western University, London, ON, Canada

RICHARD J. HARRIS
Founding Manager (Retired)—Integrated Microscopy, The Biotron, University of Western Ontario, London, ON, Canada

Academic Press is an imprint of Elsevier
125 London Wall, London EC2Y 5AS, United Kingdom
525 B Street, Suite 1650, San Diego, CA 92101, United States
50 Hampshire Street, 5th Floor, Cambridge, MA 02139, United States
The Boulevard, Langford Lane, Kidlington, Oxford OX5 1GB, United Kingdom

British Library Cataloguing-in-Publication Data
A catalogue record for this book is available from the British Library

Library of Congress Cataloging-in-Publication Data
A catalog record for this book is available from the Library of Congress

ISBN: 978-0-12-410424-2

For Information on all Academic Press publications
visit our website at https://www.elsevier.com/books-and-journals

Publisher: Mica Hayley
Acquisition Editor: Tari Broderick
Editorial Project Manager: Tracy Tufaga
Production Project Manager: Mohana Natarajan
Designer: Christian Bilbow

Typeset by MPS Limited, Chennai, India

Dedication

We dedicate this book to our wives—Lone and June—for their support, patience, assistance, and understanding.

Contents

Introduction

Histology is the microscopic study of how cells and tissues collaborate to form organs. It is the meeting ground where genetics, physiology, anatomy, biochemistry, and evolution come together. The processes of life do not just happen in a sack of biological soup bounded by a membrane. On a subcellular level the microstructures and organelles in every cell are arranged in a deliberate and intricate fashion that facilitates the molecular interactions necessary for life—arrangements plainly visible in every electron micrograph. The cells themselves are ordered into precise patterns forming tissues and organs providing specific frameworks and environments for life's processes to happen. These frameworks are determined by the specific function of a tissue or organ—every function imposes a characteristic plan on the arrangement of the tissues and cells making up the organ.

The vertebrate body is a complex system of multiple organs each made up of millions of cells that vary in shape, size, and function. Yet all of these varying cells exist side by side in the same microenvironment carrying identical genetic information in their nuclei. Obviously there must be a tremendous evolutionary advantage to this redundant duplication in the body of every vertebrate. The diversity of cell shape, size, and function arises through a dual process of increasing specialization and loss of options to develop in other ways imposed on the pluripotent cells of the early embryo. Indeed, it is a fascinating line of inquiry—how and why cells differentiate—but that is the venue of molecular and developmental biology. Histology accepts cellular differentiation as a given and focuses on the organization, behavior, and interaction of these differentiated cells.

Vertebrates are a diverse and adaptable group, living in a wide variety of environments. Each species has evolved solutions in adapting to its environmental niche. The histology of the vertebrates reveals how species and groups have solved the problems of their environmental niches, adapting organs and tissues to meet the specific pressures imposed on them. This process varies substantially from one group to the next. For instance, in those vertebrate groups with no hollow bones (such as cartilaginous and bony fishes), hemopoietic tissue occurs in vascular "backwaters" such as the kidneys and sex organs. To deal with excess salt in the diet some marine birds have specialized "salt" glands in the head that excrete excess salt from the blood—a function typically relegated to the kidney. It is common in amphibians to use other organs, such as the skin, to aid lungs or gills in oxygen exchange.

Some structures go beyond their usual function such as the electric organ in certain fish, which is highly modified and specialized muscle tissue. Comparative histology allows us to understand the structural responses underlying the physiology unique to each vertebrate group.

Veterinary pathology and histology texts, while comparative, typically examine only a few species and typically only deal with mammals found in the practice such as dogs, cats, rodents, cattle, horses, and pigs. Other "comparative" texts deal with laboratory species such as rat, mouse, or primate, compared to human. This text looks at the histology of a wide range of animals, representative of the major vertebrate classes and families. The study of comparative histology in the vertebrates leads to an understanding of how various groups have addressed similar problems, opening doors to interesting research possibilities.

PREPARING AND STUDYING TISSUES FOR HISTOLOGY

Microscopes are a ubiquitous tool of science and, in one form or another, are used in laboratories from biomedical to materials engineering. The data produced by a microscope are not just images—but huge amounts of information presented in a visual format. We need some basic knowledge and understanding of the workings of the microscope, image formation, and the techniques used in order to understand and interpret that information we produce. In the study of cells and tissues we need to understand how the tissues or cells were prepared since these preparative processes have wide ranging effects at the molecular, cellular, and tissue levels.

Microscopic images are more than pictures. A micrograph is a two-dimensional representation of a three-dimensional structure; even when these images are of a thin slice of tissue. These slices have a finite thickness that can reveal important clues about the overlying three-dimensional organization of the specimen. For any organism this three-dimensional arrangement of the cells and tissue in its organs is important. Interpretation of a microscopic image requires consideration of the original three-dimensional nature of the cells and tissues represented in this two-dimensional image.

HOW BIG IS SMALL?

Before delving into the mysteries of specimen preparation and microscopy, consider sizes and dimensions. Eukaryotic cells tend to be tiny, and any units of measure used must to be appropriate to them. Consider an image of a typical slice (or section) through any piece of tissue viewed with an optical microscope at a nominal magnification between 10 and 1000 ×. Since each image shows hundreds to thousands of cells we need an appropriate "ruler" to comprehend the sizes of structures seen and markings on the "ruler" must be fine enough.

Microscope images and image analysis typically use standard metric measurements of micrometers (μm) and nanometers (nm) applicable to the level of magnification and resolution of the image:

$$1 \text{ mm} = 1000 \text{ (micrometers) } 1000\ \mu\text{m}$$
$$1\ \mu\text{m} = 1000 \text{ (nanometers) } 1000 \text{ nm}$$

A micrometer is often called micron, usually designated by μ. While the term micron is no longer sanctioned by the International System of Units, it is still commonly used in both sciences and industry in English speaking countries to differentiate the unit of measure from the micrometer, a measuring device. The human eye is capable of resolving structures as small as approximately 100 μm—about the diameter of a mature human ovum or that of an average human hair (Fig. 1a).

Mammalian cells range in diameter from 5to 25 μm. Of course there are exceptions: fertilized mammalian ova can be 40–150 μm. Cells from the snake-like amphibian *Amphiuma* are many times the volume of typical mammalian cells. In some specialized cells, such as avian and reptilian ova, the yolk is a massive single cell bounded by a membrane. Neurons, especially the peripheral nerves of the limbs, can be tens of centimeters in length in many species. But all cells no matter their size are bounded by a cell membrane. Red blood cells (RBCs) provide a convenient built in "ruler" since they are consistently sized within any particular vertebrate species, and are present in almost all tissues. RBCs are among the smallest cells of the body ranging in size from 6 to 8 μm in mammals to over 60 μm in *Amphiuma* (Figs. 1b–1e).

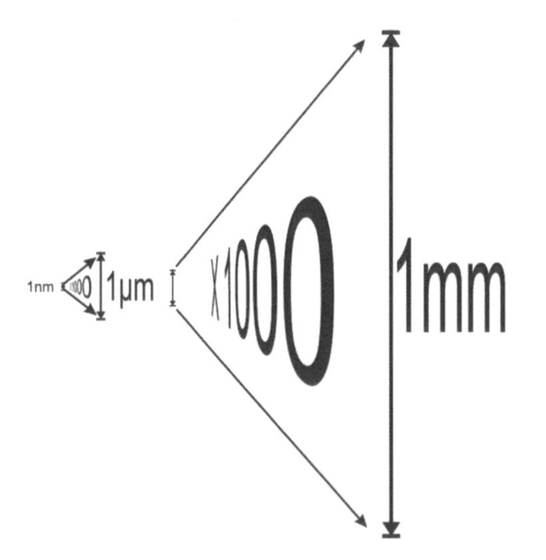

FIGURE 1a A graphical representation of metric scales typically used in microscopy

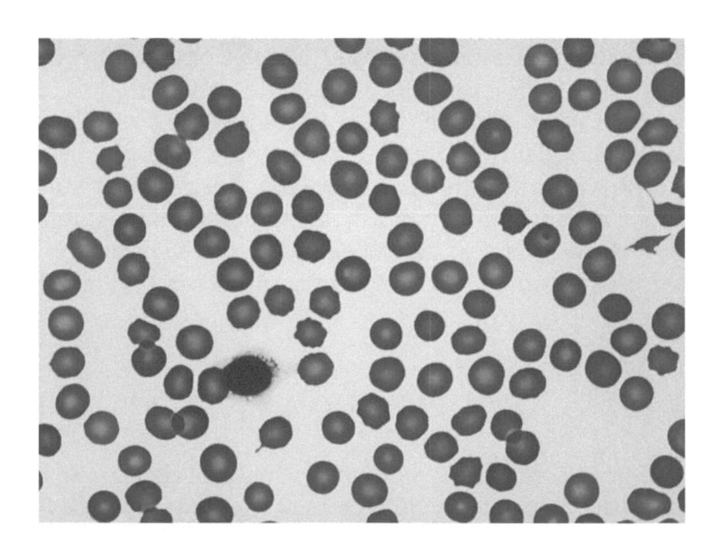

FIGURE 1b Mammalian blood cells are nonnucleated and biconvex in shape. The red blood cells of the Guinea pig are typical, approximately 7 μm in diameter—almost the same size as human red blood cells. 63 ×.

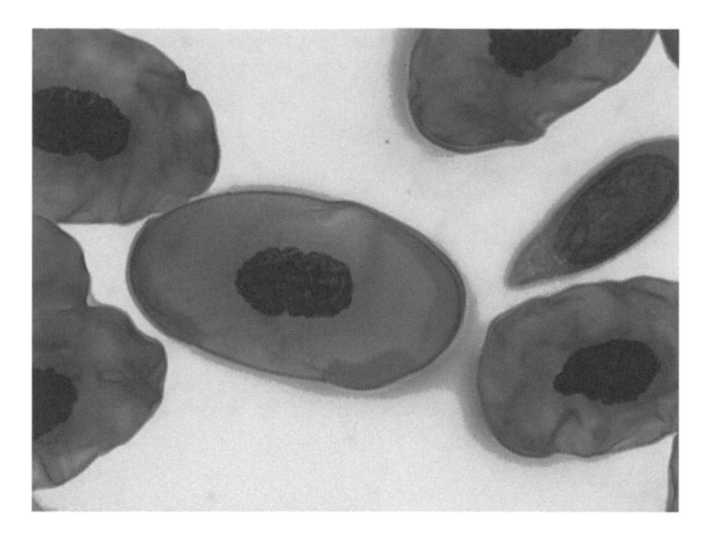

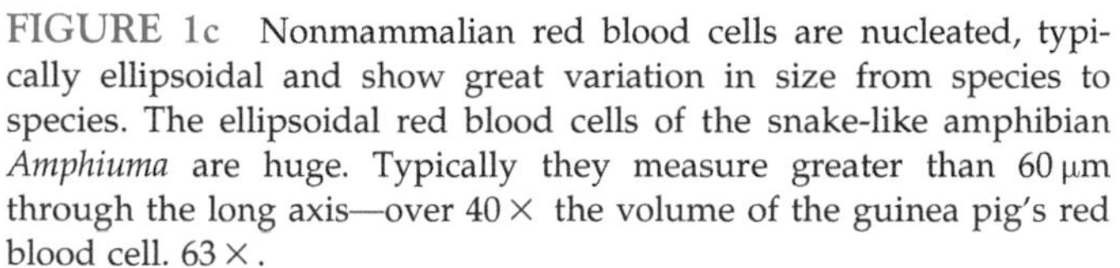

FIGURE 1c Nonmammalian red blood cells are nucleated, typically ellipsoidal and show great variation in size from species to species. The ellipsoidal red blood cells of the snake-like amphibian *Amphiuma* are huge. Typically they measure greater than 60 μm through the long axis—over 40 × the volume of the guinea pig's red blood cell. 63 ×.

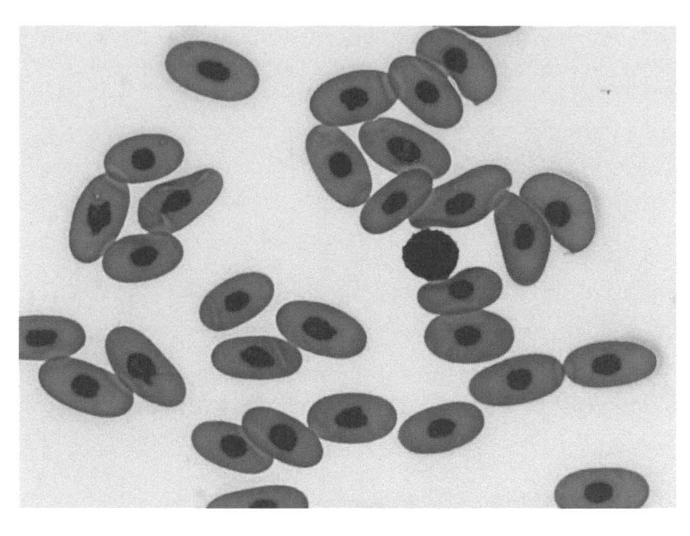

FIGURE 1d Red blood cells from the alligator are larger than those of the guinea pig but much smaller than those from *Amphiuma*. 63 ×.

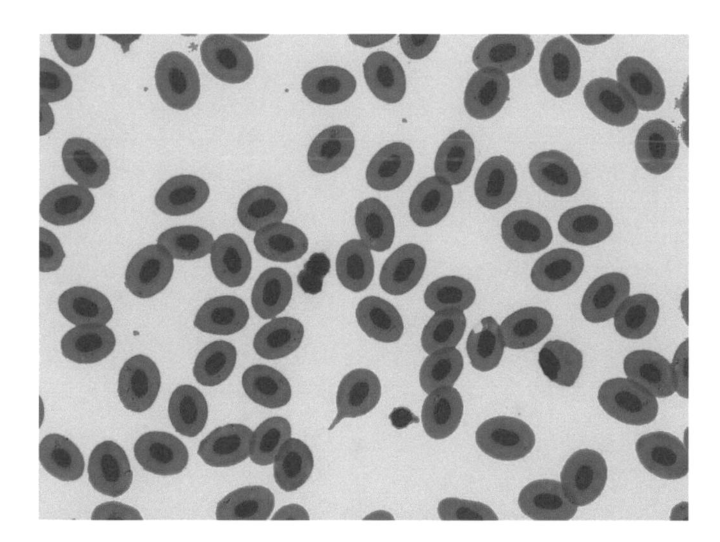

FIGURE 1e While similar in size to the guinea pig's red blood cells—those of the bass are nucleated. 63 ×.

Subcellular structures such as nuclei (Figs. 2a–2d), mitochondria, lipid membranes, and ribosomes require taking a further step down in scale (Figs. 2e–2h). These organelles and cellular components are usually only a small fraction of the volume of the cell, and are only clearly resolved using an electron microscope. Typically, subcellular components are measured in nanometers (nm)—there are 1,000,000,000 nm per *m* or 1000 nm in each *μm*. While a mitochondrion may be 1–2 μm in length, cristae inside are typically 100–200 nm in width. Ribosomes, the protein factories found in almost every cell, have varying sizes in the range of 25–30 nm in diameter. The lipid membrane bounding the cells of all eukaryotic organisms is a phospholipid bilayer approximately 70 nm in width (Fig. 2f). The dimensions of these subcellular structures are remarkably consistent across almost every vertebrate and provide us with a readymade "ruler" for electron micrographic images.

FIGURE 2(a–d) Light micrographs demonstrating the tremendous variation in nuclear size and shape seen in various tissues and animals. Nuclei and cells are bounded by a bilayered phospholipid membrane—but this membrane is far below the resolving power of the light microscope

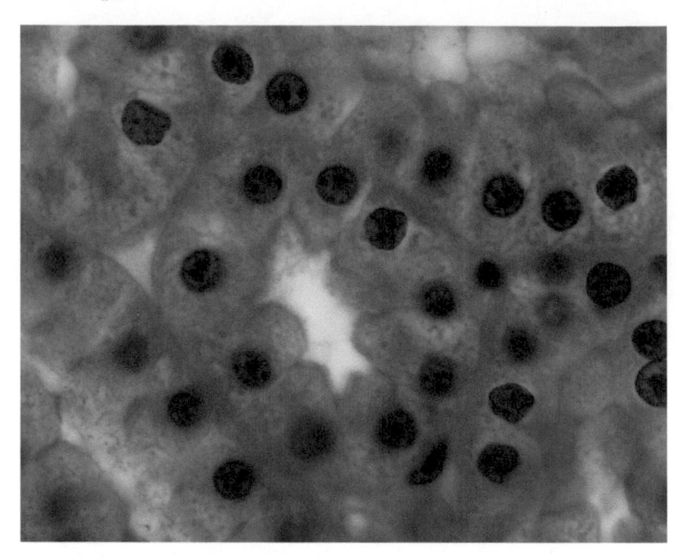

FIGURE 2a Kidney of a frog—the compact nuclei of the tubule cells. 63 ×.

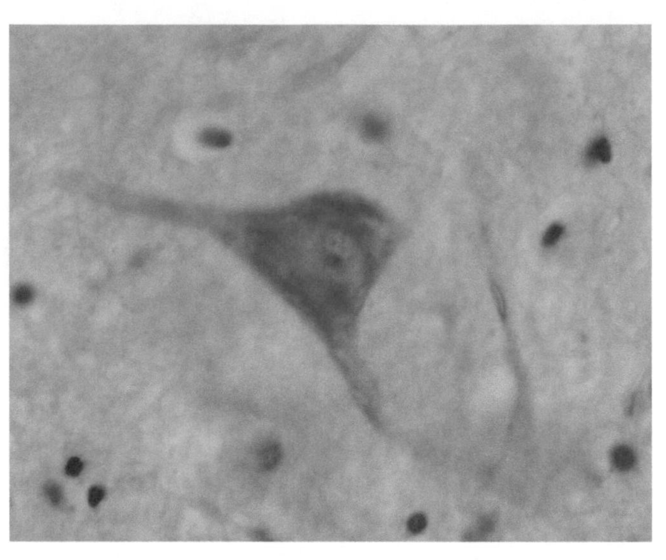

FIGURE 2b This section through a neuron in a mammalian spinal cord shows a large nucleus with a central nucleolus. 63 ×.

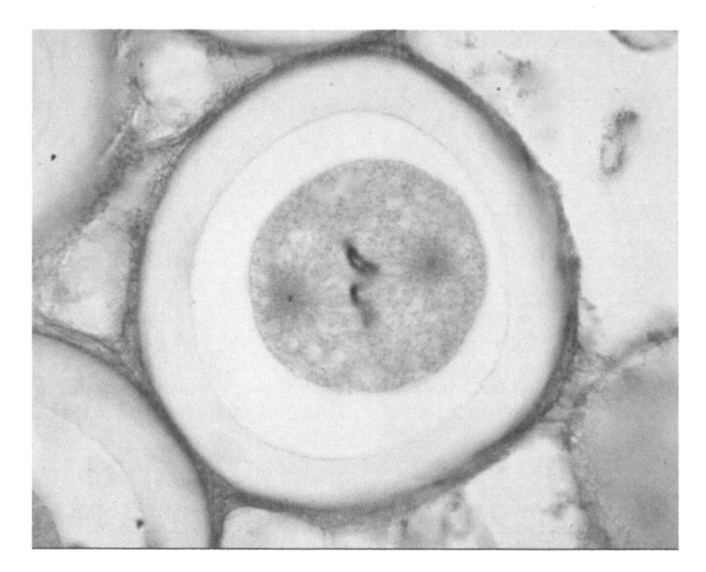

FIGURE 2c A section through the egg of the roundworm *Ascaris*, showing condensed chromosomes and a prominent mitotic spindle. 63 ×.

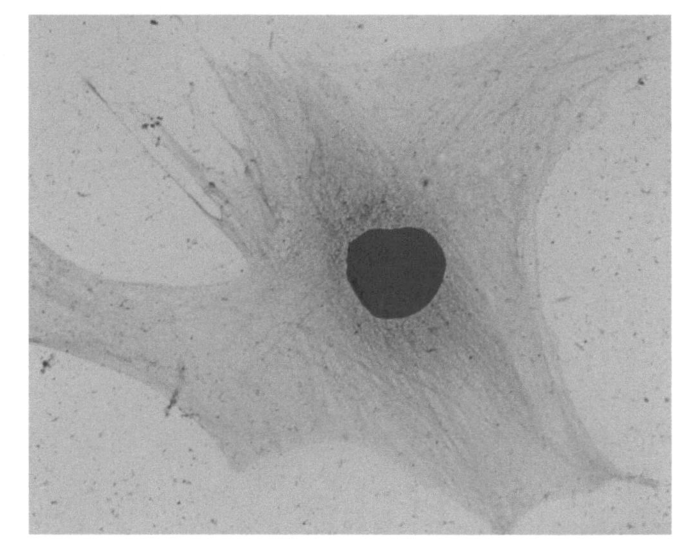

FIGURE 2d This cultured cell (BHK) has been stained to show its cytoskeleton with the nucleus enmeshed in the complex network of cytoskeletal fibers. 100 ×.

PREPARING TISSUE FOR MICROSCOPIC EXAMINATION

Almost every specimen examined in histology is a slice of dead, but carefully preserved, tissue. It is essentially a snapshot of the cells and tissues at the moment of death. Procedures are followed to preserve and maintain relationships between cells and the various components making up a tissue while also allowing illuminating radiation (light or electrons) to pass through the tissue. Light does not penetrate tissues to any great extent and most biological tissues are soft and squishy and optically dense. To examine the cellular nature of a tissue, specimens must be prepared in a way allowing the visualization of microscopic details.

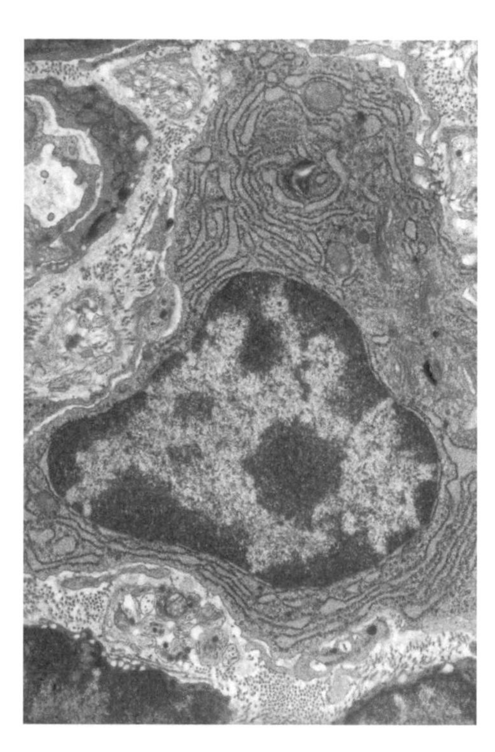

FIGURE 2e This electron micrograph of a plasmocyte in loose connective tissue of the rat reveals details in the nucleus and cytoplasm that are not clear in any light micrograph. The cytoplasm is rich in granular tubular structures (rough endoplasmic reticulum) surrounding the nucleus. The basophilia seen in the light microscope is due to the tremendous amounts of rough endoplasmic reticulum we see here. 29,000 ×.

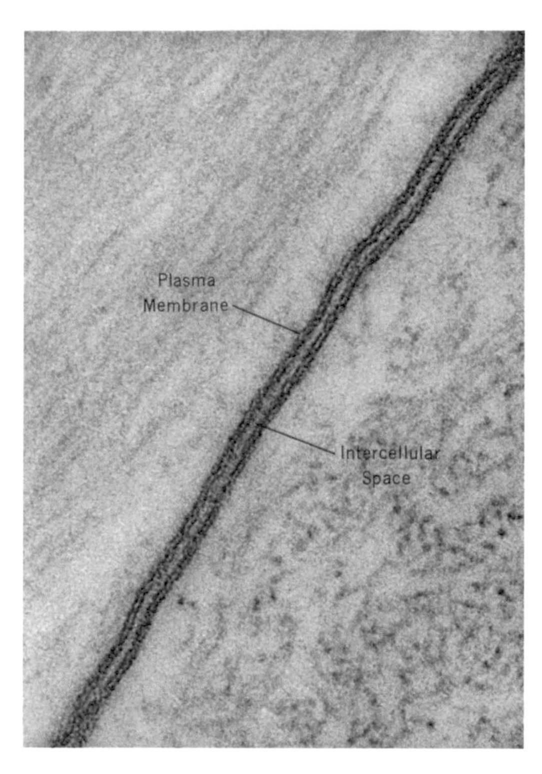

FIGURE 2f Electron microscopy is capable of showing us the fine structure of the plasma membrane in this section showing the apposition of two cells. The plasma membrane of each cell is 70 nm wide and has a trilaminar structure consisting of two dense lines and a less dense middle layer. These membranes run parallel to each other here and are separated by an intercellular space of uniform width. 260,000 ×.

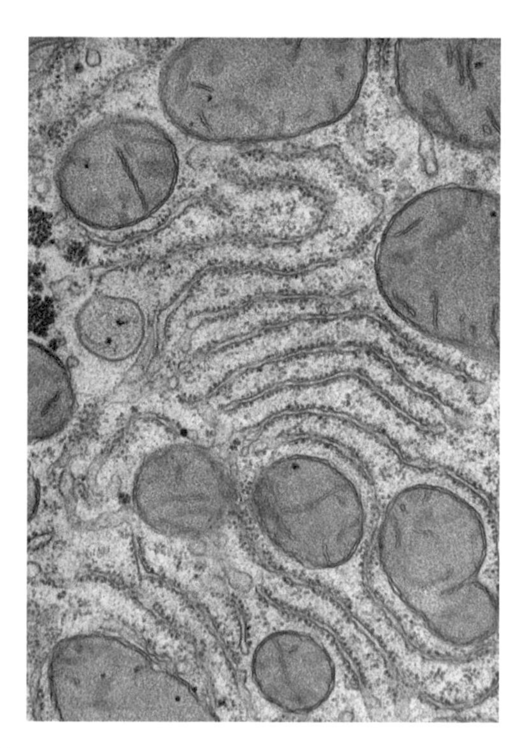

FIGURE 2g Ribosomes, which are particles of ribonucleoprotein, are attached to the surface of large, flattened sacs or CISTERNAE of the endoplasmic reticulum. We can see the fine structure of the mitochondria scattered throughout the field. 58,000 ×.

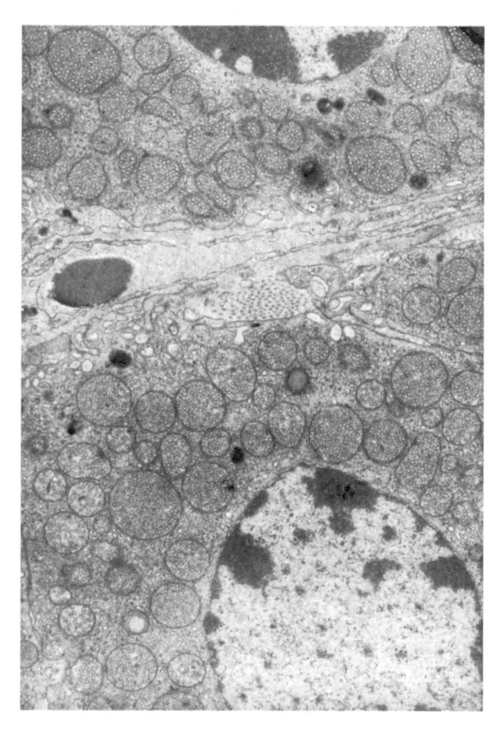

FIGURE 2h The cytoplasm of these cells is packed with agranular or SMOOTH ENDOPLASMIC RETICULUM (SER) and mitochondria with TUBULAR CRISTAE. A few ribosomes lie free in the cytoplasm. Portions of two nuclei (N) appear at the top and bottom. A few dark LYSOSOMES (Ly) appear near the top. 22,300 ×.

FIGURE 3a Microtomes used to section paraffin blocks use specially designed steel knives with a heavy wedge-shaped profile. The cutting bevel has a slightly wider angle than the wedge of the knife to facilitate free movement of the section during the sectioning process.

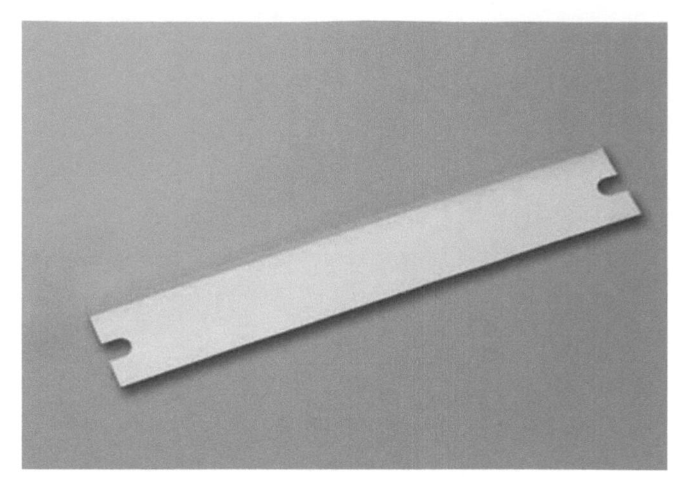

FIGURE 3b Disposable blades used with special holders have largely displaced the standard steel knives in paraffin microtomy. Disposable microtome blades avoid the care and maintenance and sharpening required with the steel knife.

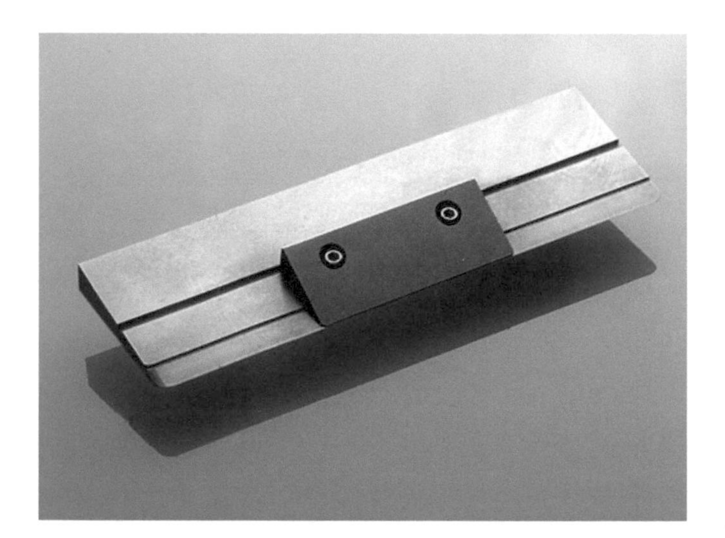

FIGURE 3c An example of a disposable blade holder.

FIGURE 3d Diamond knives are used in the ultramicrotome to cut sections epoxy-embedded tissue for electron microscopy. The diamond blade (the bit of light material at the top of the black vertical) is precision ground to produce an edge capable of reproducibly cutting 60–100 nm thick sections with little wear. Diamond knives have a long usage lifespan and can be reground several times.

Seeing the detail in preserved tissues usually requires cutting slices of the sample thin enough to permit illuminating radiation to pass through, allowing a detailed examination at high resolution and magnification. Color or contrast may be added to the sample to enhance visibility or emphasize particular features of the sample. Several methods are used to prepare tissues for examination with a microscope. On a typical microscope slide there is a thinly sliced section of tissue that looks colored. But the sliced tissues used for microscopy are thinner than any deli style sliced smoked meats. A typical slice (section) of tissue on a light microscope slide is between 2 and 20 μm thick; sections used in electron microscopy are an order of magnitude thinner: 60–100 nm. In order to slice tissues for histological examination, a process is followed that both preserves the tissue's structure and supports it for slicing. To cut a tissue sample thin enough for microscopic examination, specialized knives must be used (Figs. 3a–3d, 4e) on an instrument called a MICROTOME (Figs. 4a–4c). The tissue is made firmer by fixation to preserve structure, then infiltrated with a medium (such as wax or plastics) to support it (Figs. 5a and 5b) so that the tissues and cells can resist the distorting forces imposed when thinly sliced. Alternatively, fresh tissues

FIGURE 4a The microtome is a specialized cutting tool for paraffin sections. The manually driven rotary microtome is one of the most common types. The microtome maintains the orientation of the specimen block to the knife once it is set. With each pass the specimen is advanced a fixed distance, typically 5–20 μm.

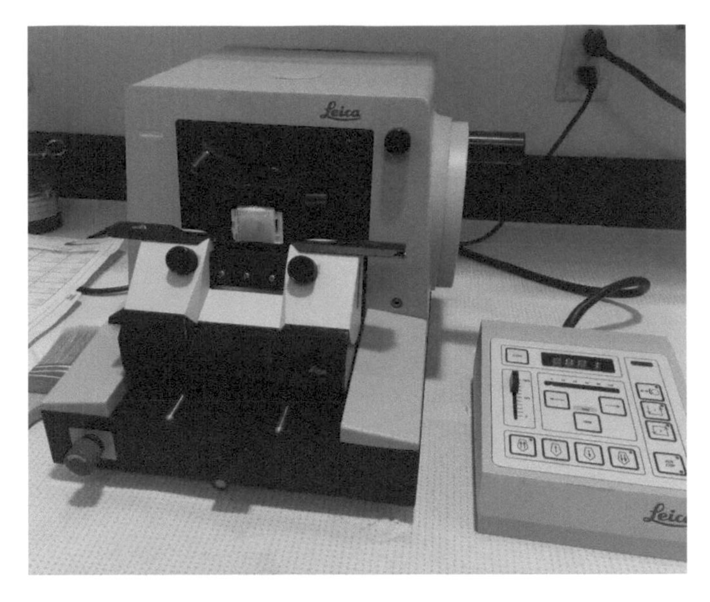

FIGURE 4b A motorized version of the manual rotary microtome seen in Fig. 4a.

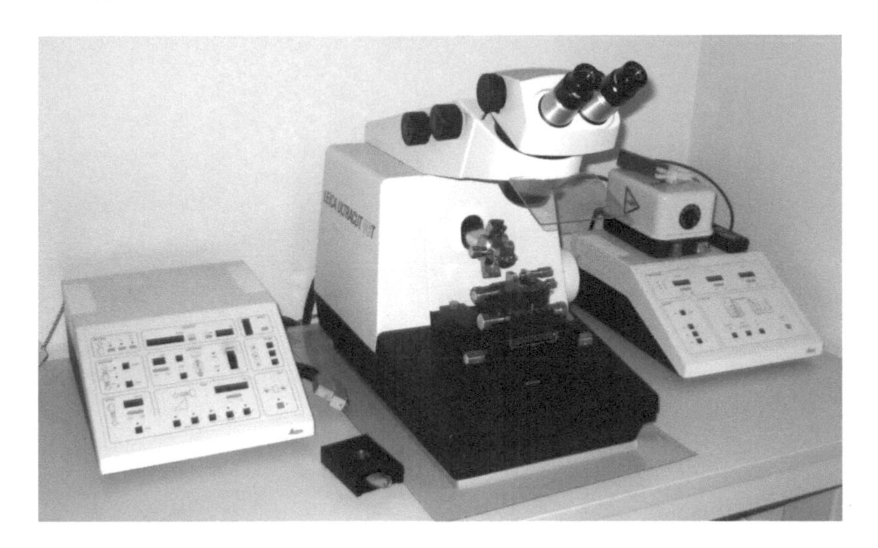

FIGURE 4c The ultramicrotome is a highly refined version of the rotary microtome used for paraffin sectioning. This instrument can reliably and reproducibly section plastic embedded tissue 60–100 nm at each pass. These sections are floated out on a water bath attached to back of a diamond or glass knife. Typically the specimen is advanced through the application of heat to the metal arm bearing the specimen. Since the heat is constant the arm will expand at a constant rate over time—thus the speed of the cutting stroke and the heat applied govern section thickness.

are sometimes rapidly frozen and then thinly sliced in a special microtome set-up in a precisely temperature controlled freezer called a CRYOSTAT MICROTOME (Fig. 4d). The frozen water in the sample supports the tissue allowing it to be sliced thinly. These sections are often used for specialized histochemical and immunological preparations and in surgical procedures where time is limited to guide the surgeon

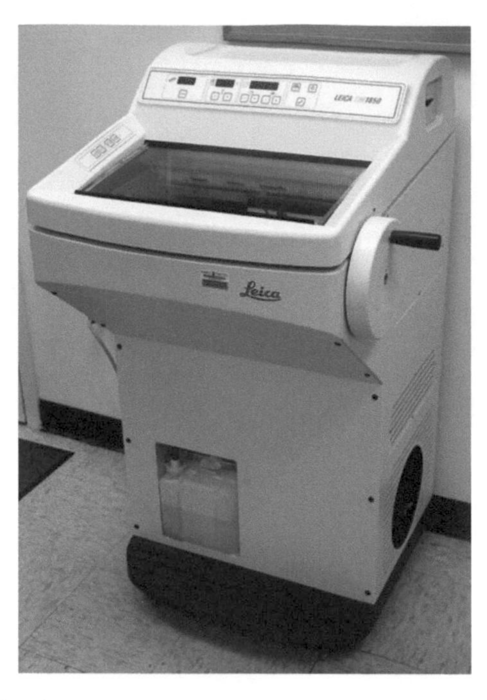

FIGURE 4d Cryostat microtomes are essentially rotary microtomes in a precision controlled freezer. The rotary drive wheel extends through the freezer cabinet.

FIGURE 4e Precisely scored and broken glass knives are often used in the ultramicrotome. The edges are smooth sharp and molecularly thin—but fragile. Glass knives are in fact sharper than diamond knives but not nearly as robust.

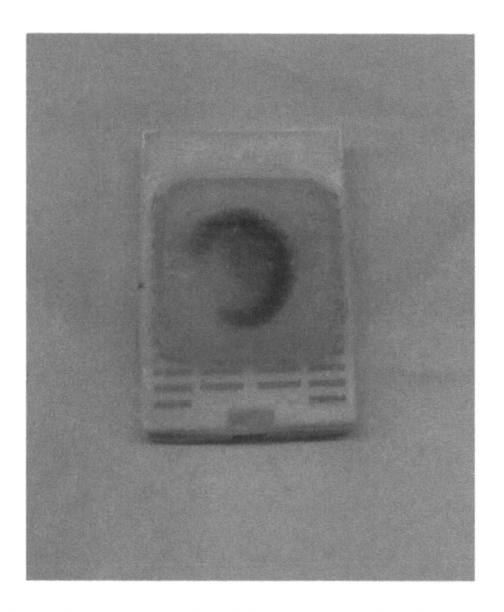

FIGURE 5a Tissue embedded in paraffin (the brownish material in the center) mounted and ready to be trimmed and sectioned.

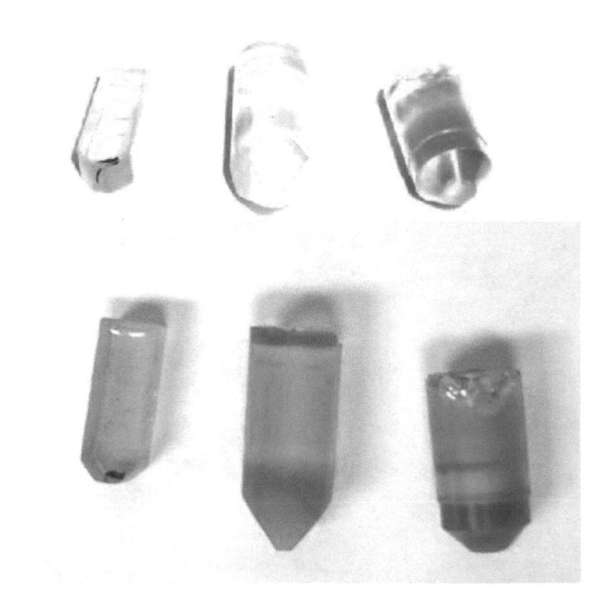

FIGURE 5b Tissues (the black material) embedded in various shapes of epoxy plastic blocks—ready to be trimmed and sectioned for electron microscopy.

FIXATION

Typically the standard process for preparing histological samples begins with careful removal of the tissue followed by rapid rinsing and washing in physiological saline, taking measures to ensure the tissue does not dry out, then quickly moving the tissue into a fixative bath. There are two basic types of fixatives: (1) *Coagulative* fixatives, such as alcohol and acetic acid, which cause proteins to precipitate out of solution—this process also happens when tissues are heated above 65°C for extended times.

(2) *Crosslinking* fixatives, such as formaldehyde or glutaraldehyde, create stabilizing chemical bonds within and between proteins, rendering them in a state similar to that found in the living tissues—without precipitating the proteins.

DEHYDRATION

Most histological specimens are embedded in paraffin or plastic wax prior to sectioning (Fig. 4a). The embedding medium (paraffin or plastic) supports the cellular structure of the tissue and allows thin slices to be cut while protecting delicate features of the cells and tissue from being crushed during the process. Biological tissues are about 80% water and waxes and plastics are not (usually) miscible with water. For plastic or wax to penetrate the cells and tissues, the water must be exchanged for something that is miscible with the embedding medium. A series of ever increasing concentrations of a dehydrating liquid, such as ethanol, up to a bath of absolute dehydrant is used to replace the water. At this point all the water in the tissue has been replaced with dehydrant and the tissue is ready to transfer into an embedding medium (wax or plastic).

Some plastics and all waxes are not miscible with alcohol and the tissue is passed through another series of washes that replaces the alcohol with a solvent miscible in both alcohol and the embedding medium. Again gradual replacement of dehydrant with a transitional solvent is employed—but larger steps can now be used since the dehydrant has stabilized the tissue. When the tissue is in 100% transitional solvent it can be moved into the embedding medium. For wax embedding the wax must be melted and kept at 62°C to avoid cooking the tissue. It usually takes two baths of hot, liquid wax to fully replace the transitional solvent and infuse the tissue completely. Many polymeric plastics used for histological embedding are soluble in alcohol and a few are soluble in water. With these plastics the dehydrant or water is gradually replaced with liquid, unpolymerized plastic through a series of ever increasing concentrations of resin and water or dehydrant mixes until the tissue is infused with 100% unpolymerized plastic resin. For those plastics not miscible with the dehydrant or water there are transitional steps through a solvent miscible with both the dehydrant and the liquid plastic resin similar those steps used with wax.

Tissues to be embedded in paraffin are typically processed using an automated tissue processor (Fig. 6). This machine can fix, dehydrate, and infiltrate specimens in paraffin, leaving only the embedding and trimming to be done manually.

FIGURE 6 Automated processors provide reliable and reproducible results. They take the sample from the fixative to paraffin infiltration without human intervention. The chemicals for each step are stored in the jugs below the processor's working chamber.

EMBEDDING

Following infiltration with wax or plastic the tissue is ready to be embedded in molds sized suitably to the tissue. For wax embedding a plastic ring is used to provide a strong grip to clamp the block into the chuck of the microtome. The tissue is oriented in the mold and the mold is filled with liquid wax and allowed to harden. This wax block with its piece of tissue is ready to be trimmed and cut on the microtome. For plastic embedding the process is similar but to harden the plastic it is mixed with a polymerizing agent and an accelerator. Again the tissue is placed in a mold, oriented, and the mold is filled with the complete plastic. The plastic is allowed to polymerize and harden (usually in an oven). Once the embedding medium is set the block can be unmolded and trimmed, ready for the microtome.

SECTIONING

A microtome is a specialized precision cutting instrument, which accurately and repeatedly slices sections from a block of embedded tissue. Different kinds of microtomes are used to section paraffin and plastic embedded tissues (Figs. 4a–4c) as well as the specialized microtomes used to section frozen tissues (Fig. 4d). In any microtome a sharp knife and the tissue block are held in a fixed relation to each other. With each pass of the tissue past the knife it advances the tissue block a preset amount—the section thickness. For frozen sections the section thickness typically ranges from 8 to 15 μm, for wax sections 4–10 μm, and for plastic histological sections 0.5–3 μm. In electron microscopy sections must be extremely thin, about 200× thinner than wax sections. Typically plastic sections used in transmission electron microscopy (TEM) are cut in the range 60–100 nm (Figs. 7a and 7b).

MOUNTING

As wax sections are sliced from the tissue block, sections stick to each other forming a ribbon (Figs. 7a and 7b) providing an automatic sequencing of sections, allowing alignment of sections in the order cut. For some studies (e.g., embryological) these "serial sections" permit a following a structure through the tissue. But in many cases, single sections are mounted on a slide. Typically paraffin wax sections are floated on a warm water bath (which relieves compression caused by sectioning), transferred to a slide, excess water is blotted away, and the slide and

FIGURE 7a Paraffin sections form "ribbons" during the sectioning process allow easy sequencing of section from first to last. The tissue profile is visible in the ribboned sections.

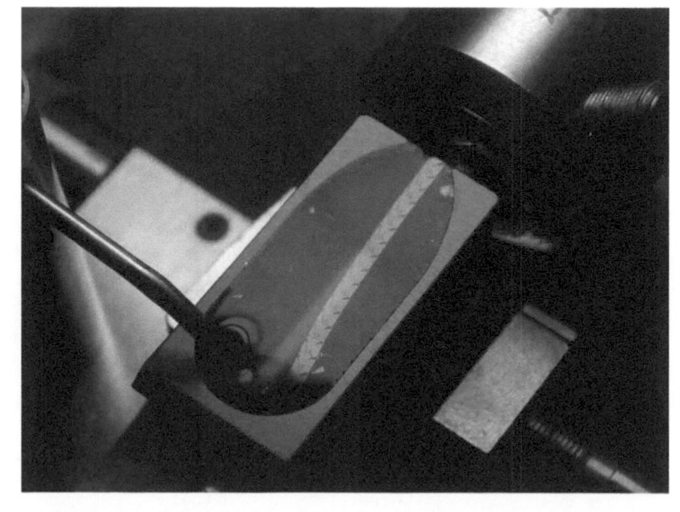

FIGURE 7b Plastic section cut on an ultramicrotome are floated onto the surface of a water bath behind the cutting edge; in this case a diamond knife. Plastic sections also form "ribbons."

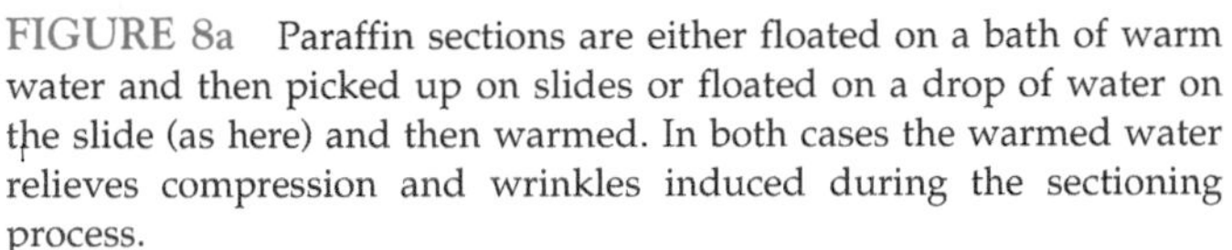
FIGURE 8a Paraffin sections are either floated on a bath of warm water and then picked up on slides or floated on a drop of water on the slide (as here) and then warmed. In both cases the warmed water relieves compression and wrinkles induced during the sectioning process.

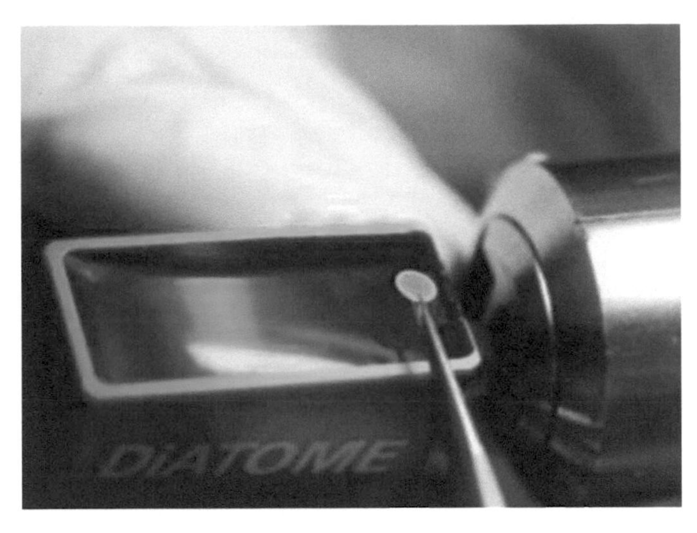

FIGURE 8b The tiny sections cut on the ultramicrotome are picked up directly on a copper grid and dried prior to viewing in the transmission electron microscopy (TEM).

section are allowed to dry on a warming plate for some hours before further processing (Fig. 8a). For histological plastic sections the process is similar but higher temperatures are applied to flatten the section and dry the slide. For TEM the sections are picked up on a copper grid (Fig. 8b).

STAINING AND COVERSLIPPING

Since most cytological and histological stains are water based, wax sections mounted on slides are not ready for staining, the wax must be removed from the sections and the tissue rehydrated before it can be stained. Common staining series begins with a bath of wax solvent followed by a series of baths of decreasing concentrations of alcohol to water. The tissues are then stained, rinsed and almost ready for the coverslips. But again there is the miscibility issue—most products used to mount coverslips to the slide are not water or alcohol soluble. Following staining the slides are run through a series of alcohol baths of increasing concentrations, through a transitional solvent bath (compatible with both the mounting medium and alcohol) and then the coverslips are applied. The mounting medium is allowed to set and the slides are ready for viewing. Staining plastic histological sections is simpler since the plastic is not removed; typical plastic histological sections are stained with water-based dyes, rinsed, and allowed to dry. The coverslips are affixed with a suitable mounting medium, and the slide is ready for viewing.

INTERPRETING MICROSCOPIC IMAGES

Interpreting the images seen under the microscope can be confusing, requiring three-dimensional thought while studying a two-dimensional image. Histological preparations are typically dead tissue, fixed, sectioned, and stained to retain and emphasize tissue structure and cellular relationships. In essence each section is a snapshot of the tissues and cells at the moment of death. The organ or tissue from which it was cut is a complex three-dimensional structure and the section is but a thin slice of that complex structure. Although a histological section is a thin slice, the microscope's focal plane is typically even narrower, and we can focus through the section gathering some three-dimensional information as we go. When examining a histological section (for all practical purposes) the microscope presents us with a series of two-dimensional images derived from the complex three-dimensional structure of the tissue section.

The same structures can take on different appearances depending upon the plane and angle of the slice; that is the orientation of the tissue was to the knife during sectioning. Orientation of the tissue in the section is indicated using typical anatomical terminology referring to the plane and direction of the cut (Figs. 9a and 9b). Any section

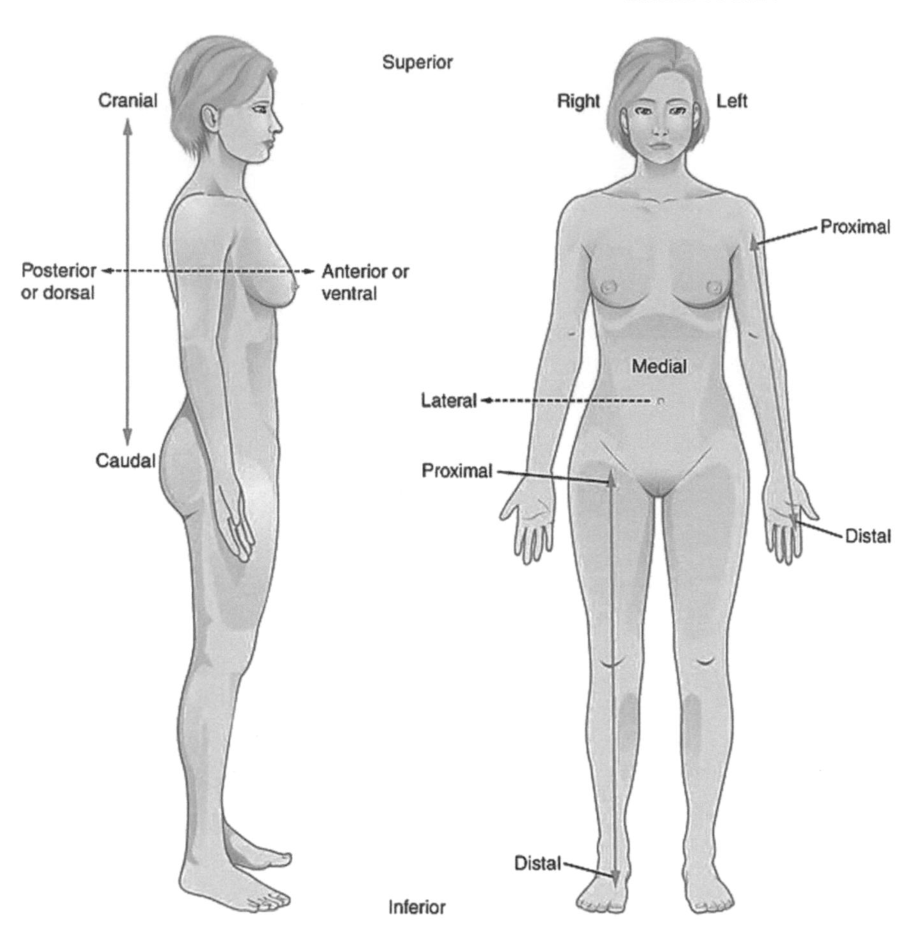

FIGURE 9a Histologists use many of the same terms to describe direction as anatomists use to describe anatomical directions.

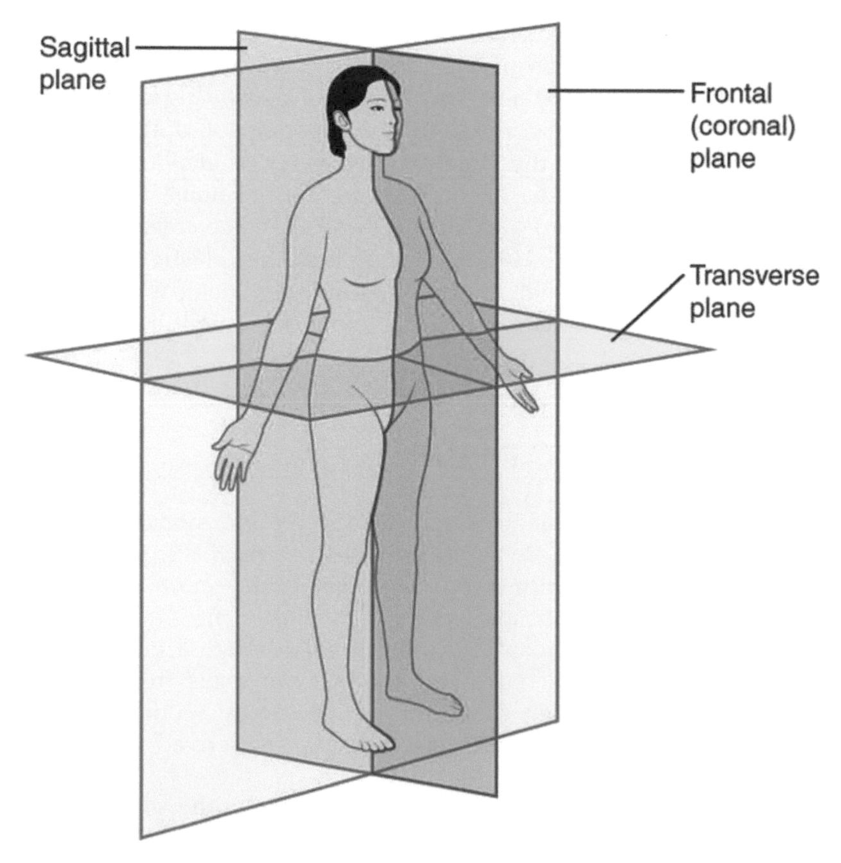

FIGURE 9b In histology, as in anatomy, the plane of the cut is an important clue in interpreting what we see. Histologists, in general, do not differentiate between sagittal and frontal planes—the term *longitudinal section* is used instead—meaning cut parallel to the long axis of an organ or structure. Transverse is replaced with *cross-section*—referring to cutting across the short axis of an organ or structure.

presents a bewildering array of profiles. Understanding and interpreting an image or micrograph of a microscope section requires visualizing it as part of a three-dimensional structure. It might be a useful tool to imagine the section as if it were a card taken from a three-dimensional "deck" of cards—the image on each card varying slightly from the one above it and the one below.

Simple structures such as spheres or tubes can take on different appearances; especially when sectioned in an atypical orientation. Even a simple sphere can present a different profile depending on the depth of the cut (Figs. 10). Tubular structures present an even more challenging set of section profiles depending on the angle and plane of the cut (Figs. 11a–11c).

As a model of a cell consider a glass of water with a small lime floating in it. Simplifying and rotating 180 degrees makes it more diagrammatic, easier to understand, and provides a basic cell model. Sections taken at various planes reveal profiles that might not look at all like the original object (Figs. 12a–12d). In some planes the lime will not be sectioned. In other sections it may be difficult to interpret the profile of the sectioned lime as a lime.

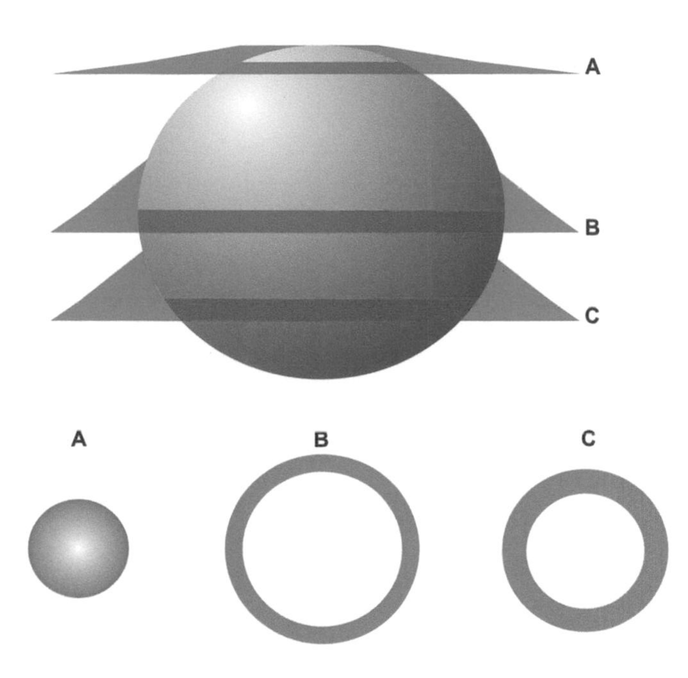

FIGURE 10 Any section through a sphere will produce a circular profile but the diameter of the circle will vary with the depth of the plane of cut. In a hollow sphere with a wall of finite thickness, section profiles will show a variable wall thickness on exiting and entering the sphere.

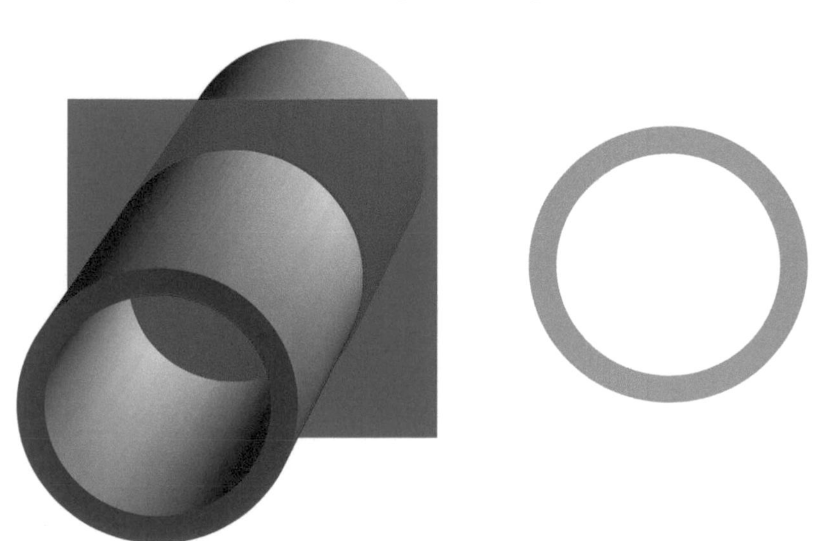

FIGURE 11a Straight tubes with wall of finite thickness sectioned at 90 degrees to the long axis will also produce circular profiles similar to the section profile of a sphere.

Sectioning a tube with a wall of finite thickness

Longitudinal

Section profiles produced cutting at the red planes - parallel to the long axis

Plane A

Plane B

Plane C

FIGURE 11b Straight tubes with wall of finite thickness sectioned parallel to the long axis will produce paired linear profiles as seen in the sections at Planes A and B. As the section plane enters (or exits) the tube near the edge, as in Plane C, a single linear profile will be produced.

Sectioning a tube with a wall of finite thickness

Oblique

Sections cut on planes between longitudinal and cross sections yield more complex profiles

Section profiles produced cutting at the red planes - oblique to the long axis

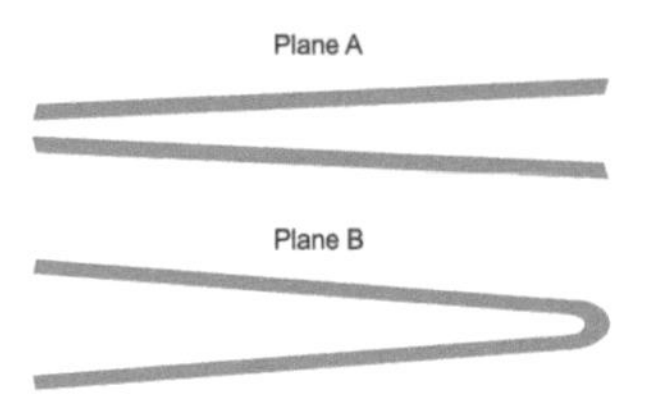

Section profiles produced cutting at the red planes - oblique to the short axis

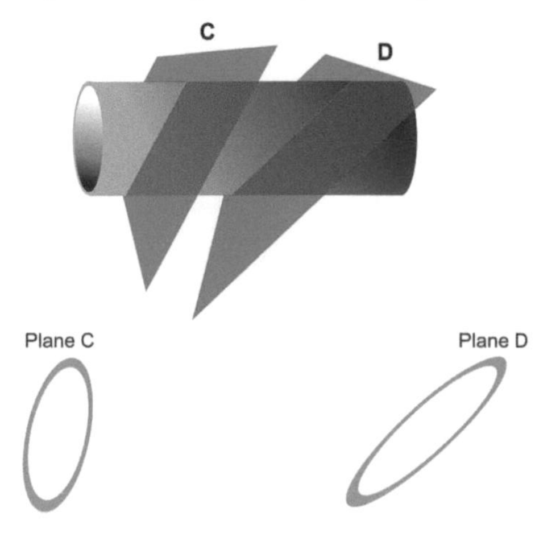

FIGURE 11c If the plane of section is not parallel to the long axis as in A and B the section profiles become variable and no longer have simple linear features. As the plane of section is skewed away from a 90 degree cross section, the circular profile is distorted into an oval shape.

FIGURE 12a Imagine a lime floating in an inverted glass full of water—A.
In order to make it easier to visualize section planes the image is rendered in black and white, making it more diagrammatic—B.
This image can be used to represent a simplified epithelial cell, the lime representing the nucleus (the lime also has an internal structure), the water represents the cytoplasm of the cell, and the glass the cell membrane.

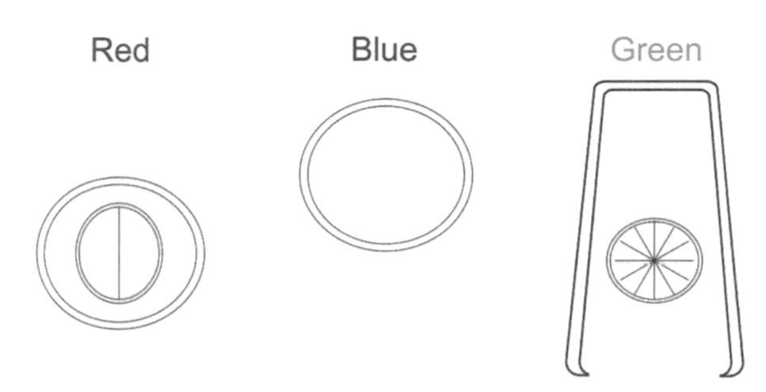

FIGURE 12b A cross-section through our cell model missing the nucleus (blue plane) produces a simple circular profile. The red plane passes through the nucleus and the profile becomes more complex as the internal structure of the lime is revealed. A longitudinal section along the mid-line (green plane) also reveals the internal structure of lime and it demonstrates the open end of the glass as well.

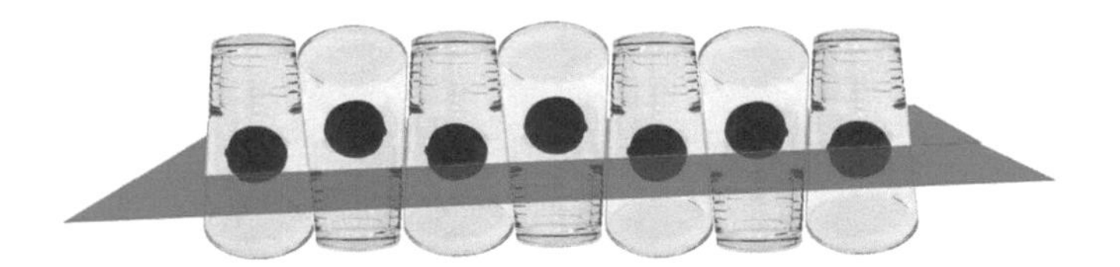

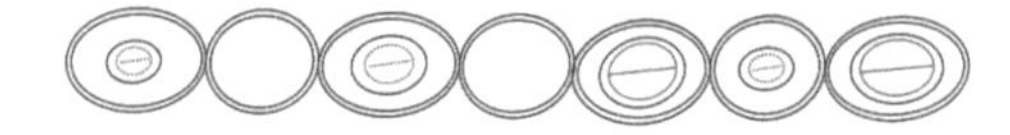

FIGURE 12c Lining up several of the model cells increases the complexity of the object being sectioned and produces an array of "cellular" section profiles—especially if the section plane is slightly oblique.

As the complexity of the object is increased to a line of "cells" (Fig. 12c), cut on a plane slightly askew from the long axis, the profiles presented can become a useful aid in understanding the three-dimensional structure of the model cell. Further increasing the complexity of the object as an array of "cells" laid out in an $X-Y$ grid we begin to see something that begins to resemble an epithelial sheet (Fig. 12d). The range of profiles present in a single slice through this "sheet" becomes ever vaster. Finally the model can attain an even greater level of complexity of the profiles seen in the sections if by visualizing the sheet thrown up into folds and ridges.

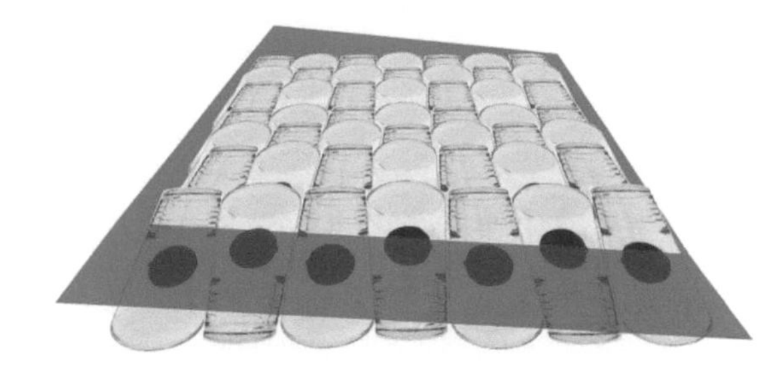

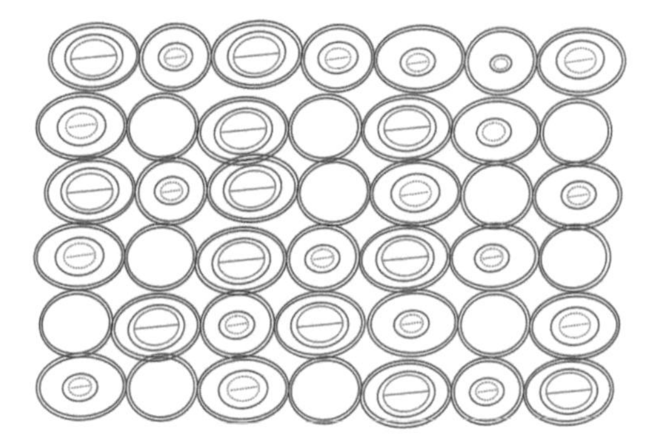

FIGURE 12d Here the model cells are arranged on an $X-Y$ grid, we can see the model resembles a sheet of epithelial cells. As the object becomes more complex the obvious outcome is a more complex section profile through any plane of cut. Try to visualize this array thrown up into folds—creating hills and valleys—further complicating the section profile.

HOW MICROSCOPES WORK?

What does a microscope do? Typically people answer this question is by saying a microscope makes things look bigger. But making things look bigger is only part of what a microscope does. What a microscope really does is improve visual acuity—that is the ability to discern and distinguish small things from one another. Making a specimen appear larger (magnification) is a by-product of increasing visual acuity using a microscope, or any other magnifying optical lens system.

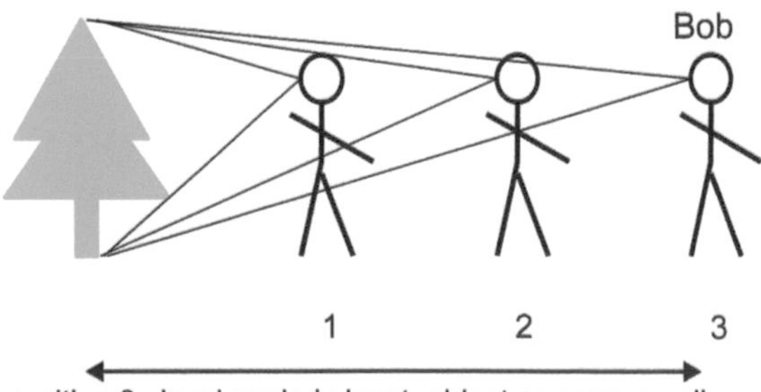

Consider friend Bob, above. In Position 3, Bob is looking at an evergreen tree from a distance. The tree subtends a small visual angle and appears small to Bob, with little apparent detail. As Bob moves closer to the tree, the visual angle the tree subtends increases—as do the details of the tree. At some point Bob can get no closer to the tree (or a part of it) without losing focus. At this point the tree, or at least some portion of it, occupies his entire field of view and has attained the greatest possible visual angle. Without some intervention Bob cannot increase the visual angle and examine a given portion of the tree in any greater detail. Optical lenses intervene and provide an apparent increase in the visual angle, making an image of the object appear larger and closer to the eye, revealing more details.

Lenses have been used to enhance human vision for at least 2000 years. Ancient Egyptians, Greek, Roman, Arabic, Indian, and Chinese writings make it clear these cultures were well aware of the magnifying and focusing power of convex lenses well before renaissance Europeans.

Many credit the Dutch optician father/son team of Hans and Zacharias Jansen, and Willem Boreel with the development of convex lens compound microscopes in the 1590s. Robert Hooke first used the word cell to describe the structures seen in cork in his 1665 book *Micrographia*, an amazing collection of micrographical drawings is based on his observations using a compound microscope. Anton van Leeuwenhoek's 1674 simple, single lens microscopes allowed him to resolve tiny "animicules" in water. Development of the compound microscope took another 150 years to equal the resolution achieved by van Leeuwenhoek's simple microscopes. Improvement in microscope design and function proceeded through trial and error methods until Ernst Abbé published a series of papers in the 1860s. Abbé's formulae and calculations, together with Carl Zeiss and Otto Schott, laid the ground work for modern lens and microscope design.

Two types of images are produced in a microscope: a REAL, inverted image at the focal plane of the objective lens and a VIRTUAL, noninverted image in the ocular lens.

The nature of photons allows us to consider them as having properties of both waves and particles. Without delving into quantum mechanics we can model the interaction of light in lens systems using simple geometric optical diagrams. The rays in geometric optic diagrams represent the direction of travel of the photonic wave fronts and can be used to model image formation in lenses.

A simple microscope consists of a single lens element as in Fig. 13 showing the creation of a real, inverted, and magnified image of an object.

In this case the object (op) is located just beyond the front focal point (f_o) the lens AB. By visualizing only three rays the location and size of the resultant real image ($o'p'$) can be determined. What makes a real image "real"? Simply put, a real image can be projected onto a screen. Real images are formed beyond the rear focal point by a lens only when the object is beyond the front focal point. In order to see a real image it must be projected onto a screen of some type.

Looking through a magnifying glass (CD) one sees a virtual image that is an image appearing to be projected in front of the lens being looked through. A virtual image is exactly that - an image existing only in the mind. The optics of the lens "fool" the eye and brain into seeing an virtual image (i.e., cannot be projected onto a screen). Fig. 14 shows how a virtual image is created when the specimen (in this case $o'p'$) sits between the lens and the front focal point (f'_o) and the lens of the eye is between the rear focal point (f'_i) and the magnifying glass. If we trace the ray paths we see that the magnified image projected onto the retina appears to originate in front of the lens CD—creating an extended visual angle greater than the eye could achieve on its own.

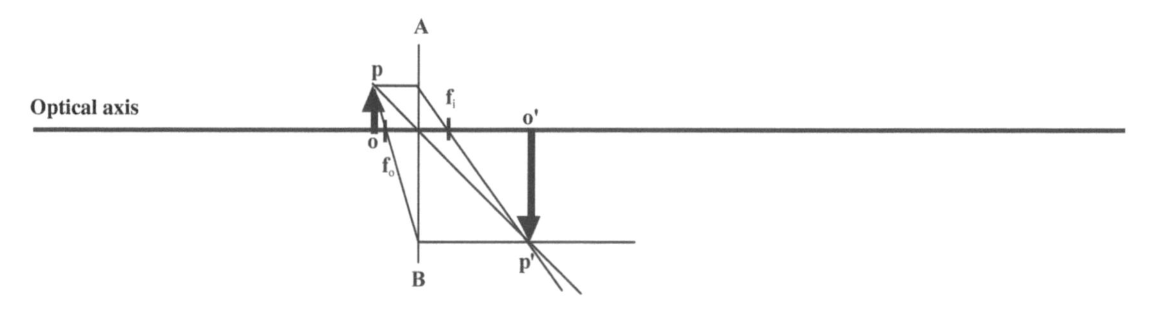

FIGURE 13 Formation of a real image at the rear focal of the objective.

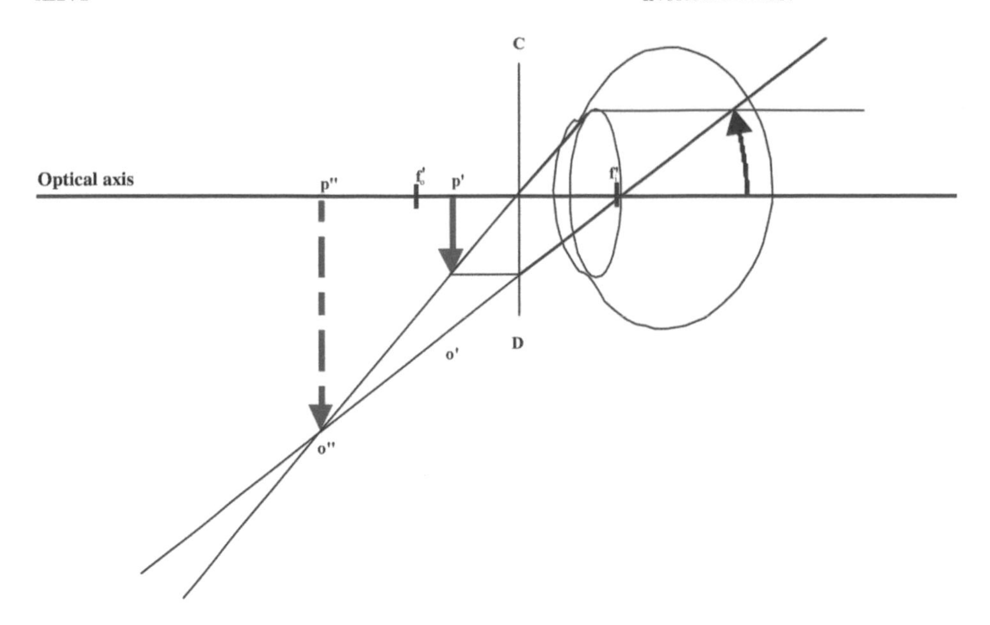

FIGURE 14 Formation of a virtual image at the ocular.

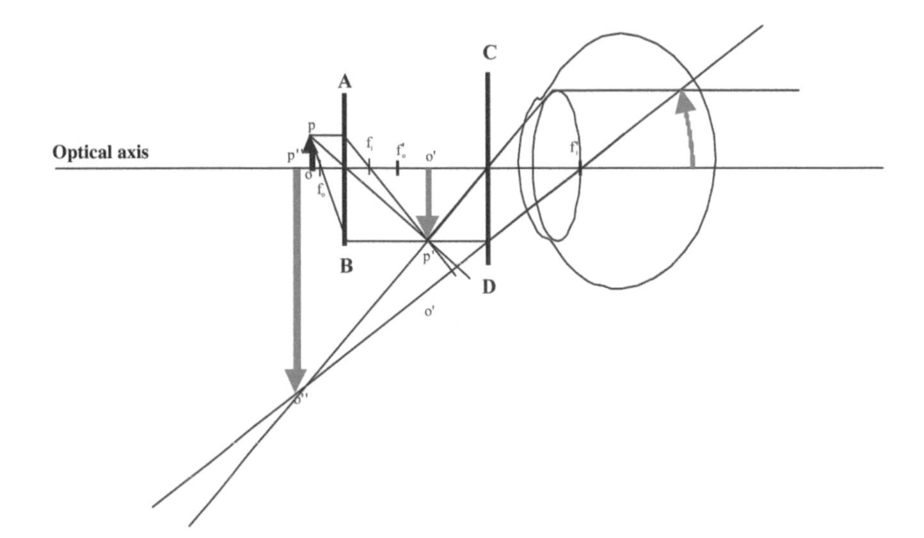

FIGURE 15 Combining diagrams Figs. 13 and 14 yields a simple microscope.

Fig. 15 shows the components of a simple compound microscope consisting of just two elements: the objective lens, which produces a real, magnified image of the object and the ocular that is used to capture the real image produced by the objective and then present a virtual, magnified image to the user.

Fig. 16 shows how additional components, such as a field lens in the ocular, may be added to the compound microscope to correct for imaging defects, increase the field of view or to otherwise enhance image quality as delivered to the eye.

Lens and Image Defects

Under perfect conditions using dimensionless, ideal lenses the previous systems would produce distortion free, high resolution images. But in the real world, neither lenses nor conditions are perfect, and lenses used in microscopes are subject to imperfections, much like ourselves.

Today the use of "synthetic" glasses, computer modeling, and laser guided lens grinding has given rise to microscope lenses capable of producing images virtually free of aberrations and imperfections. In fact the lenses on modern microscopes are capable of imaging at near the theoretical limits of resolution. Image imperfections due to spherical and chromatic aberrations have been abolished, for the most part, from the modern microscope.

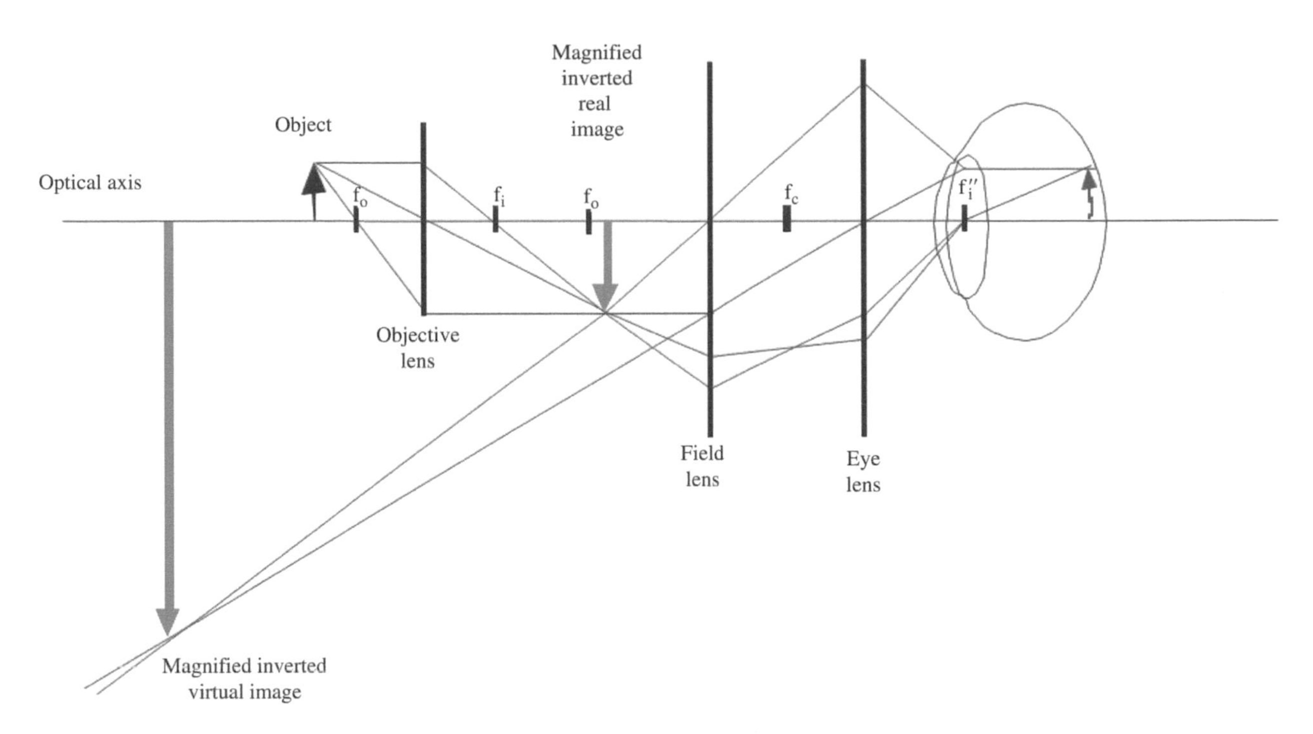

FIGURE 16 Addition of a field lens to the simple microscope increases field of view, sharpens the image, and reduces spherical aberration.

Resolution

Resolution (δ) is the ability of an imaging system to distinguish two closely associated points as separate objects. Resolution is a physical property of any optical system but is especially important to us as microscope users. The microscope should be set to optimize the balance between resolution and visibility of the specimen.

Resolution in an optical system is dependent upon several factors:

Wavelength (λ) of the incident light
Angular aperture of the lens (θ)
Refractive index of the medium through which the light travels (R.I.)

The wavelength of the light is important on many levels; human eyes are optimized to give the best visual acuity in the middle (green) portion of the visible spectrum. The theories and equations governing resolution show that shorter wavelengths (toward the ultraviolet) will yield greater resolution in an optical system. Unfortunately, as the wavelength of the light deceases, so does human visual acuity. As a result almost all microscopes are designed to operate best using light from the middle (green) part of the visible spectrum.

The angular aperture (θ) of a lens is a measure of a lens' ability to gather light. The angle θ of the cone of light entering the lens, as θ becomes wider the lens has greater light gathering power (see the diagram later). However, the light gathering power of a lens is usually expressed as the numerical aperture (N.A.), which describes the relationship between θ and the refractive index (R.I.) of the media through which the light travels to reach the lens. Mathematically this relationship is defined as:

$$\text{N.A.} = R.I. \times \sin\theta$$

In a microscope the *Practical* N.A. of a system will not exceed 1 if there is air between the condenser or the objective and the specimen. To achieve maximum resolution in a microscope there must be a high R.I. medium (usually immersion oil) between both the objective and the condenser and the specimen. Ernst Abbé, in the late 19th century, considered diffraction effects at both the lens and the specimen to be the major contributor to image formation and hence resolution. Abbé's mathematical consideration of diffraction and image formation led to the development of modern microscopes. Abbé's initial work considered diffraction effects at the lens. By his reasoning the limits of resolution of a lens system may be described as:

$$\text{Resolution}(\delta) = \frac{0.61 \times \text{Wavelength}(\lambda)}{\text{R.I.} \times \sin\theta}$$

$$\delta = \frac{0.61 \times \lambda}{\text{N.A.}}$$

When he considered diffraction effects at the specimen, his equation becomes slightly different:

$$\delta = \frac{0.5 \times \lambda}{\text{N.A.}}$$

In both cases, it is obvious resolution in an optical lens system can only be increased in three ways:

1. Increase the R.I. of the system (makes the denominator larger),
2. Increase the angular aperture (θ) of the system (makes the denominator larger),
3. Decrease the wavelength (λ) of the incident light (makes the numerator smaller).

From a practical point of view, for maximum resolution to be achieved we would work with oil immersion lenses (both objective and condenser) of high N.A. using short wavelength light.

ELECTRON MICROSCOPY

While the layout of components in the transmission electron microscope (TEM) is similar to that found in optical microscopes, the conditions and materials used are quite different. Optical microscopy relies on the nature of photons and their quantum interaction with glass lenses and the specimen while electron microscopy is dependent on wave-like properties associated with the negatively charged subatomic particles called electrons. To manipulate electrons we cannot use a glass lenses; instead electron optics relies on the interaction of electrons with magnetic fields of the annular electromagnets used as lenses in the electron microscope. The strength of the magnetic lens can be varied by changing the current flowing through the coils, increasing the current in the annular electromagnet increases the strength of its magnetic field. To further complicate matters the microscope must be operated under a high vacuum (typically 1×10^{-4} to 1×10^{-6} torr), allowing the illuminating electrons a relatively long free flight path (i.e., no air molecules in the way) to be manipulated by the annular magnetic lenses, to focus the beam of electrons on the specimen.

A TEM may look as if it has more lens components than an optical microscope—but that is just an illusion of the diagrams used here. Electron and optical lenses such as condensers, objectives, and projectors/oculars are, in reality complex systems, each made up of several components. Fig. 17 shows a geometrical optical path in simplified light and electron microscopes.

RESOLUTION IN TRANSMISSION ELECTRON MICROSCOPE

The TEM provides an increase in resolution and magnification of two or three orders of magnitude—giving a tool that allowing the visualization of cellular nanostructures. Electron microscopy takes advantage of the wave-like nature of electrons and their inherently short wavelength to make this leap in resolution. However, the complex relationship between the electrons, the accelerating voltage (the speed at which electrons are shot down the column) and the electron's interaction with the lens-shaped magnetic fields produced by the annular electromagnets used as lenses. The resolution equation for TEM is slightly more complex than the equation used for light microscopy because there are more variables.

The basic equation to resolution in electron microscopy is:

$$d = \frac{0.753}{\alpha \times \sqrt{V}}$$

Where *d*—resolution; α—half-opening angle of the objective (in radians); *V*—accelerating voltage.

Theoretically the resolution limit in the TEM is about 1 nm but the achieved maximum practical resolution for biological samples ranges from 2.5 to 5 nm, depending on the type of sample viewed.

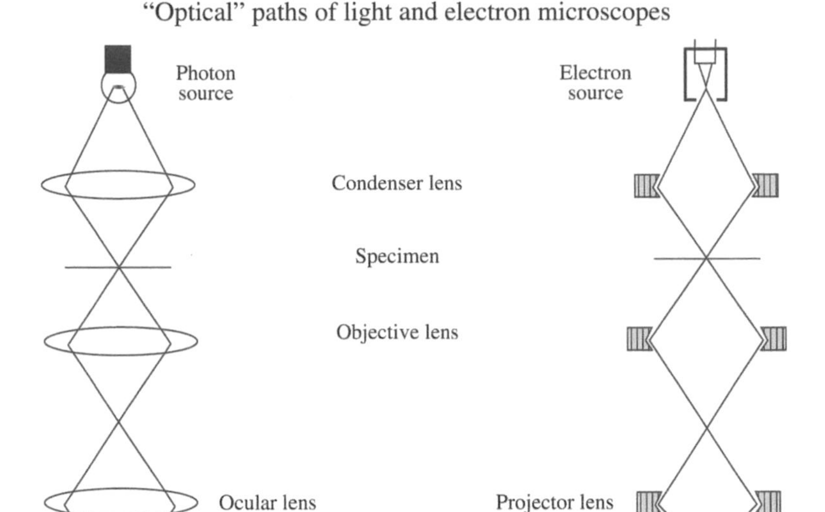

FIGURE 17 "Optical" paths of light and electron microscopes.

NOTES ABOUT MAGNIFICATIONS AND REPETITION OF IMAGES USED IN THIS BOOK

The magnifications indicated in this book depend on the way any particular image was sourced. For light micrographs taken specifically for this book—the magnification (e.g., 40 ×) refers to the magnification indicated on the particular objective lens used to capture that image. For images sourced from research papers, text books, and reference texts—the magnification indicated is that used by the authors. The final magnification of any image from this work will be variable and dependent on the physical size of output device used to view the image.

Examining a slide with a microscope typically requires repetitive examination using different powered objectives, differing focal points; scanning the slide at low power to find an area of interest, and then zooming in for a higher resolution look using objectives of greater power. The reader will see this in the way the images are presented from a lower to higher magnifications. This repetition is the key to a thorough study of each slide and we have tried to duplicate that experience here.

The reader may also notice that some slides have been used to demonstrate more than one type of tissue or cell. Organs are complex systems made up of many types of cells and tissues.

ACKNOWLEDGMENTS

The authors are grateful to the Department of Biology at Western University for allowing us to access and use of the wonderful comparative vertebrate slide collection housed in the department. Almost all of the light micrographs used in this work were generated from that collection.

We are grateful to Ian Craig—Digital Media Specialist, Department of Biology, Western University and Karen Nygard—Manager of Integrated Microscopy in the Biotron. Without their help this work would be a lesser thing.

PART I

CELLS

In 1838 and 1839 Schleiden and Schwann initiated the cell theory which states that all animals and plants are made up of cells and their products and that growth and reproduction are fundamentally due to the division of cells.

Cells are the smallest units of living matter that can lead an independent existence and reproduce their own kind.

C H A P T E R

A

The Animal Cell

INTRODUCTORY STUDY

"Typical" Animal Cell

An animal cell is a small mass of protoplasm externally limited by a CELL MEMBRANE (not a cell wall) and containing a roughly spherical NUCLEUS enclosed in its own envelope. Examination of an assortment of photomicrographs soon dispels the notion that most cells resemble the "croquet-ball-in-a-shoebox" depicted in many introductory textbooks. It is interesting to compare the depictions of typical cells over the years. Fig. A1 is a diagram from a classic textbook of 1928. The extravagant shapes of many cells reflect the myriad functions they perform in the animal body. Maximow and Bloom showed the great variation of cells in their "Textbook of Histology" of 1930 (Fig. A2). The complex structure of cells was revealed by the electron microscope and was summarized in a diagram by Jean Brachet in 1961 (Fig. A3). It is our purpose in this course to bring order to this apparent chaos and to explain why cells assume their various characteristics.

Examine the six micrographs for this exercise and notice the features that all the cells have in common. Identify the CELL MEMBRANE (PLASMALEMMA), CYTOPLASM, NUCLEAR ENVELOPE, NUCLEOPLASM, NUCLEOLUS, and CHROMATIN. Often the limiting membranes of cells are impossible to see with the light microscope. The cells in the first micrograph show a cross section of a tubule in the kidney of a frog (Fig. A4) most closely resemble the ideal "typical" cell: large, plump cells containing spherical nuclei with nucleoli. Cells similar in appearance are seen in the micrograph of the liver of an amphibian, *Amphiuma* (Fig. A5). Notice that some of the cells are packed with brown granules that almost obscure the nuclei. Most blood cells of another amphibian, *Necturus* (Fig. A6) contain a dense, dark nucleus, and the cytoplasm seems devoid of structure. One of the blood cells near the center, however, has taken on a different appearance, displaying an elaborate nucleus and granular cytoplasm. The communicating function of the nerve cells shown in the micrograph of a mammalian spinal cord (Fig. A7) is reflected in the long processes radiating from these cells. Although they are highly specialized, these cells still manifest their cellular nature, demonstrating well-developed cytoplasmic structures and a big, round nucleus with a nucleolus inside. Note that the tissue around the nerve cells is peppered with small nuclei; these represent packing cells whose outlines are indistinct. It is more difficult to discern the cells in the next two micrographs taken from the small intestine of a cat. The elongate cells in the first picture (Fig. A8) are muscle cells that produce the peristaltic movements of the gut. These cells have the shape of an archery bow with an ovoid nucleus in the position of the handgrip. The second picture from this slide (Fig. A9) shows an assortment of cells found in the wall of the gut. Some are secretory and others form packing tissues. Can you tell which is which? Can you draw any conclusions about the functions of these cells from their appearance?

Most of the slides used for these micrographs are thin (5–10 μm) sections of tissues stained to bring out cellular details. One of the most commonly used stains is a combination of hematoxylin (deep blue to black) and eosin (pink), the "H&E" stain. Hematoxylin behaves like a base in its affinity for various structures, and substances that stain deeply with it are BASOPHILIC. Conversely, structures that stain with the acid dye eosin are ACIDOPHILIC or EOSINOPHILIC. Because you are examining thin slices of a tissue, the cells will be cut in various planes and appear in a variety of ways; often a section will not include the nucleus. Try to reconstruct in your mind the three-dimensional form of the various structures being studied.

An Atlas of Comparative Vertebrate Histology.
DOI: https://doi.org/10.1016/B978-0-12-410424-2.00001-9

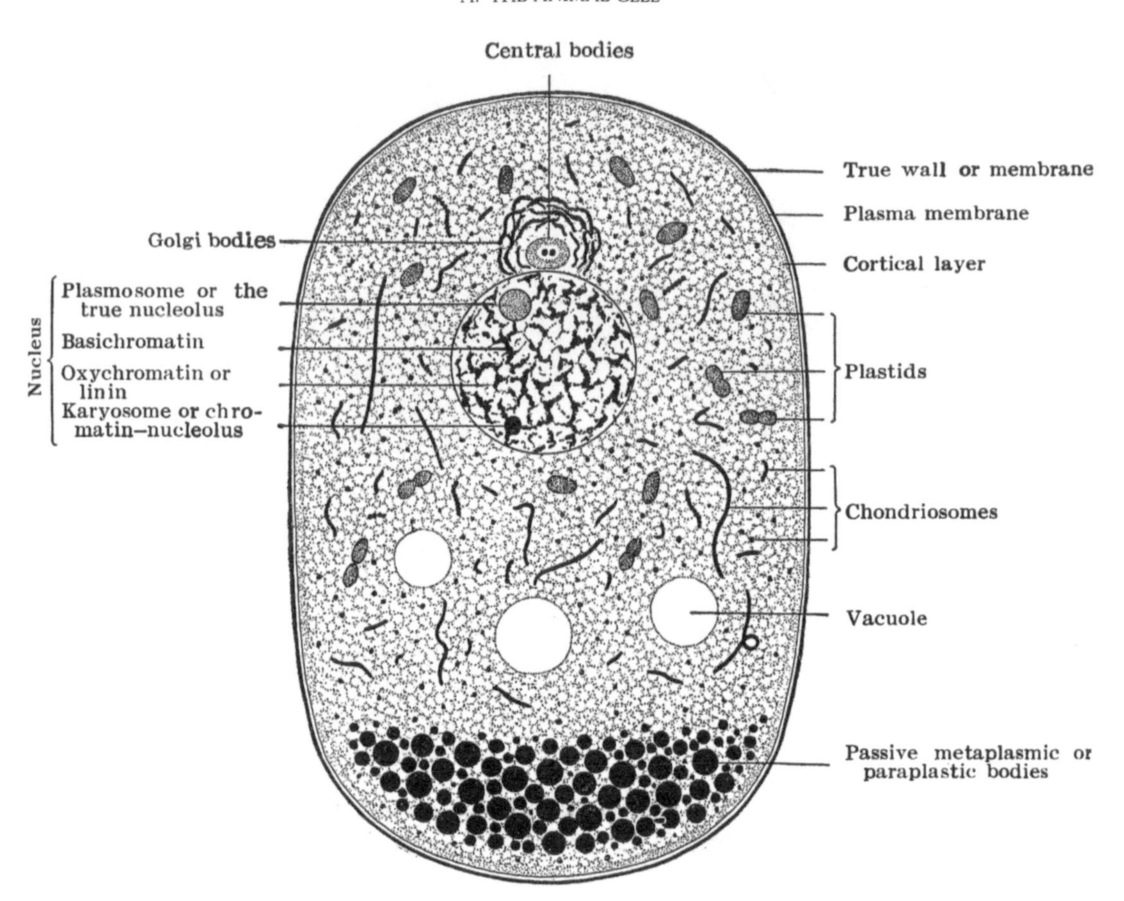

FIGURE A1 *General diagram of a cell.* "Its cytoplasmic basis is shown as a granular meshwork or framework in which are suspended various differentiated granules, fibrillae and other formed components."

STUDY OF CELLULAR STRUCTURE WITH THE LIGHT MICROSCOPE

Cytoplasmic Organelles

Cytoplasmic organelles are living protoplasmic structures having a definite form. They participate in the functional activity of the cell. This group includes the ergastoplasm, the mitochondria, the Golgi complex (lamellar apparatus), centrosomes, and fibrils.

1. *Ergastoplasm or Chromophil substance.* Scattered in the cytoplasm of some cells is material with a strong affinity for basic stains. It is especially obvious in cells that are manufacturing proteins and consists—as we shall soon see—of cytoplasmic masses rich in ribosomes, granular endoplasmic reticulum, and mitochondria. Look for these basophilic masses in sections of spinal cord stained with Nissl stain (Fig. A10). Ergastoplasm also shows to advantage in sections of pancreas where groups of cells unite to produce digestive enzymes (Figs. A11 and A12). Note that the outer regions of the cells are packed with deeply basophilic material while the central regions contain the acidophilic secretory product. Developing eggs cells in the ovary of a fish are deeply basophilic, indicating that they are actively producing the proteins of yolk (Fig. A13). Some of this yolk may be seen as acidophilic droplets in the larger, more mature, egg cells at the sides of the micrograph.
2. *Mitochondria* (Singular: mitochondrion). Although mitochondria are almost universal within cells, they are destroyed by routine procedures and are seen only in special preparations. The mitochondria have been preserved in this section of amphibian liver and appear as deep blue particles within the cytoplasm (Fig. A14). Mitochondria are also shown in the section of amphibian kidney (Fig. A15) where two types of tubules are present. One type is metabolically more active than the other and the tubules can be distinguished by the greater density of mitochondria in the cells of the more active tubules.

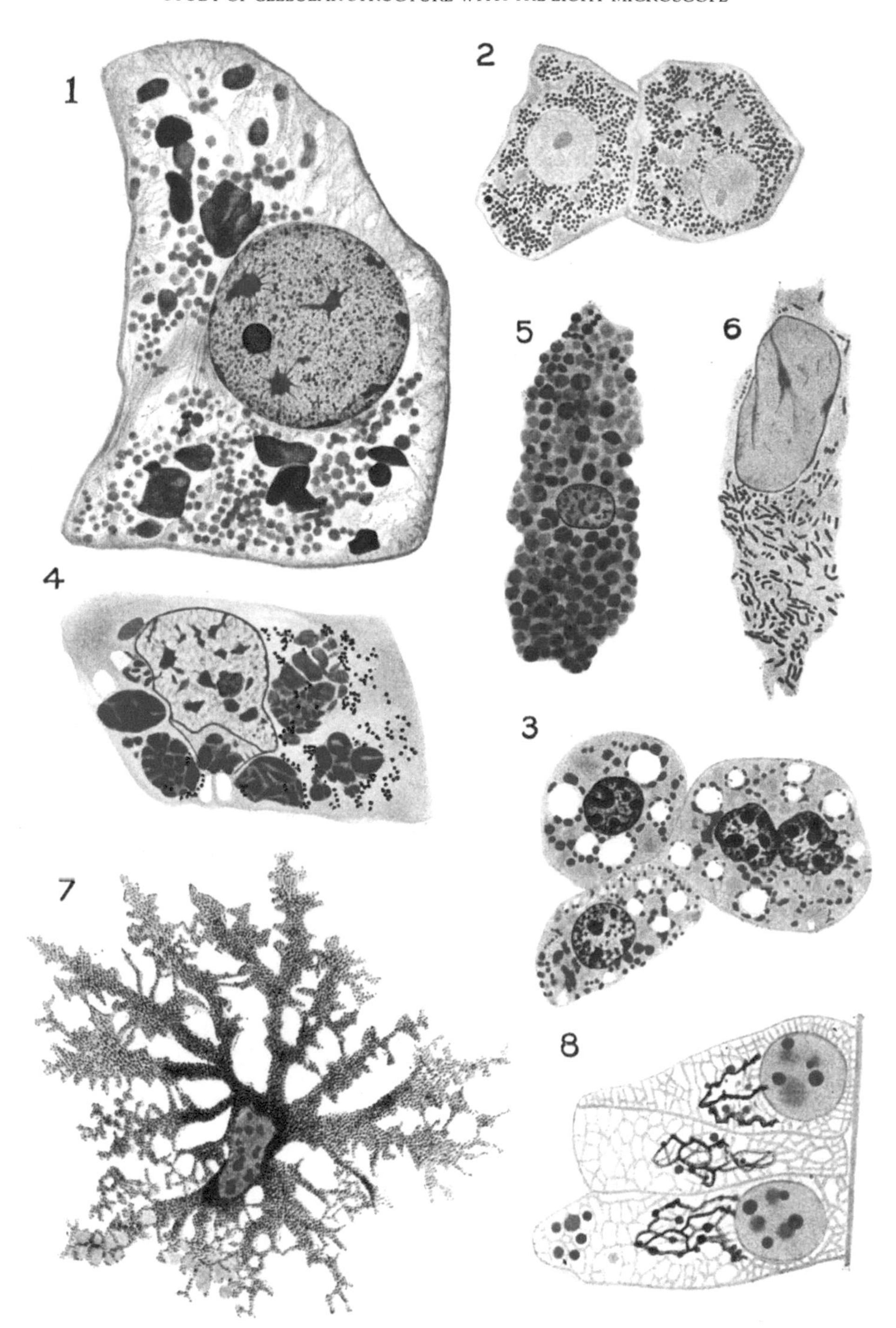

FIGURE A2 Cell types. (1) Liver cell of an axolotl. 1100 × ; (2) Liver cells of a rabbit. 750 × ; (3) Liver cells of a rat. One cell is binucleate. 800 × ; (4) Lining cell of the oral cavity of an embryo of axolotl. 1200 × ; (5) Connective tissue cell of a rat. 1200 × ; (6) Another cell of the connective tissue of a rat. 1200 × ; (7) Pigment cell from the embryo of an axolotl. 600 × ; (8) Secretory cells from the skin of the salamander *Triton taeniatus*.

3. *Golgi complex.* Again, special techniques are required to demonstrate the Golgi complex. In specially prepared micrographs note the black reticular structure near the nucleus. It is found in most animal cells but reaches its highest development in actively secreting cells. The Golgi complex has been stained black with silver in these cells from the intestine of *Amphiuma* (Fig. A16).

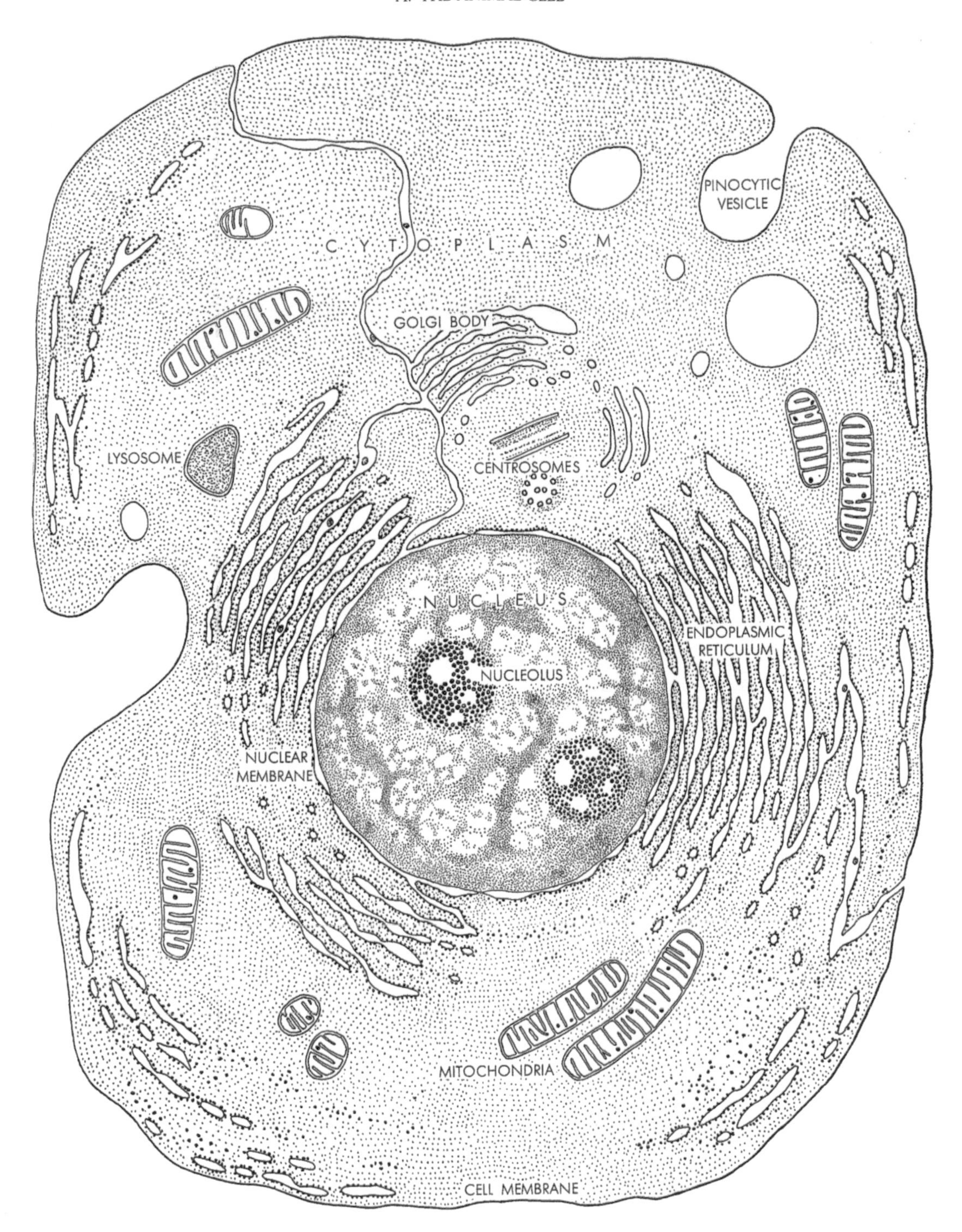

FIGURE A3 A more recent diagram of a typical cell is based on what is seen in electron micrographs. The mitochondria are the sites of the oxidative reactions that provide the cell with energy. The dots that line the endoplasmic reticulum are ribosomes: the sites of protein synthesis. In cell division the pair of centrosomes, one shown in longitudinal section (*rods*), other in cross section (circles), part to form poles of the spindle apparatus that separates two duplicate sets of centrosomes.

4. *Centrosomes and fibrils.* The term "fibrils" is an artificial grouping of a heterogeneous collection of filamentous structures within cells but, for convenience, the term has retained its place in the histological vocabulary. Examine cells undergoing mitosis, especially the cells of developing fish eggs, and notice the SPINDLE FIBERS arranged like iron filings around the poles of a magnet (Fig. A17). In some sections, the fibers are seen to

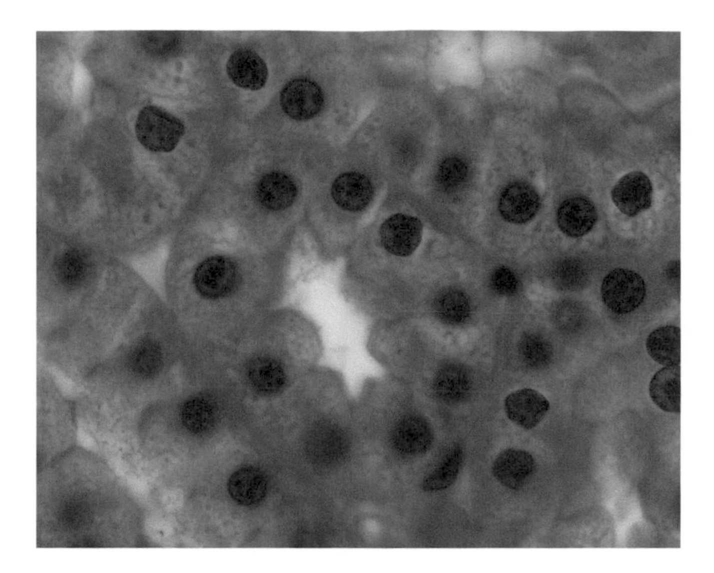

FIGURE A4 Section of the kidney of a frog. A cross section of a single tubule is shown at the center with about 10 wedge-shaped cells arranged around a single lumen or space. Often the plasma membrane between cells is difficult to discern; it is helpful of you *imagine* that there are membranes approximately halfway between each pair of nuclei. 100 ×.

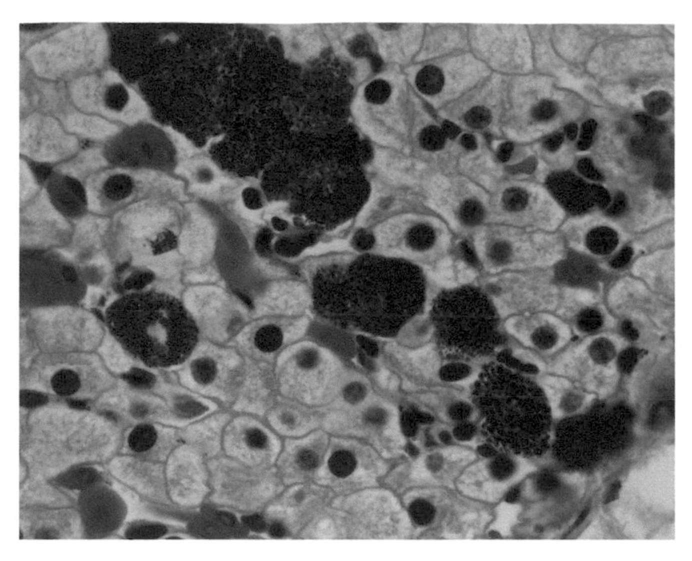

FIGURE A5 Section of the liver of *Amphiuma*. An amphibian whose cells are the exceptionally large. Two types of roughly cubical cells are shown: cells with pink cytoplasm (acidophilic) and spherical nuclei and cells packed with brown granules that obscure most of their nuclei. On closer inspection, a few elongate, deep red cells are seen wedged between the paler cells. These are blood cells contained within minute blood vessels. 40 ×.

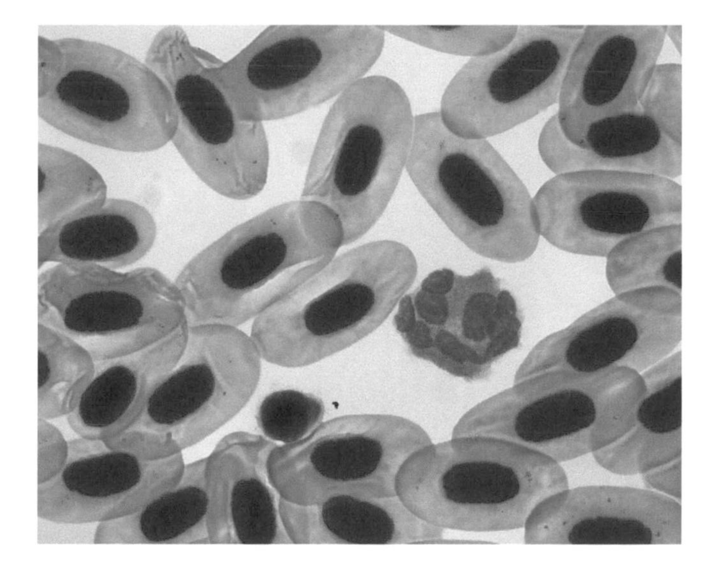

FIGURE A6 Blood smear of the mudpuppy, *Necturus*. The cytoplasm of most of the cells is acidophilic and characterless; their nuclei are ovoid and deeply stained. One cell near the center of the micrograph has pink, granular cytoplasm and an extremely elaborate nucleus that is made up of several lobes. 63 ×.

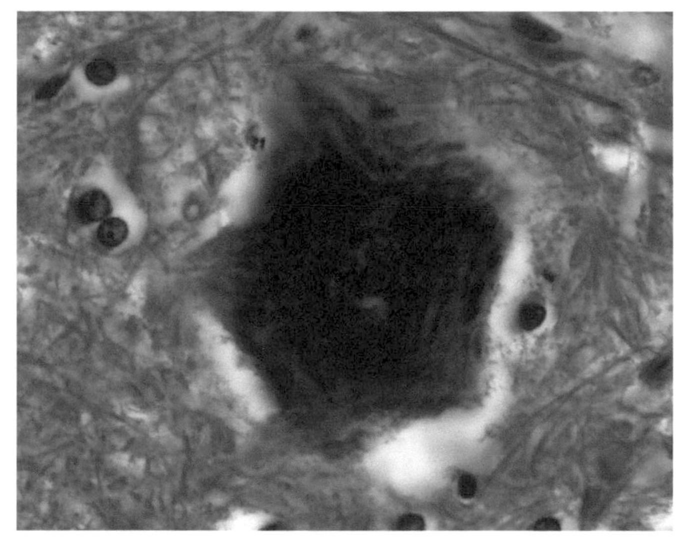

FIGURE A7 Cross section of the gray matter in the spinal cord of a small mammal. The nerve cell at the center sends out long processes, which are amputated in sectioning. Many processes form a mesh surrounding the nerve cells. The cells and processes are supported by tissue whose nuclei are scattered throughout. Some shrinkage has occurred around the nerve cell. 100 ×.

radiate from two small, deeply basophilic CENTROSOMES (Fig. A18). We will study these structures in greater detail in another exercise. Other fibrils are also seen in cells. Try to distinguish MYOFIBRILS in muscle cells (Fig. A19). Can you locate NEUROFIBRILS in sections of nerve cells (Fig. A20)? Cellular fibrils are well demonstrated in cultured cells where these delicate strands form a cellular framework or CYTOSKELETON (Fig. A21).

FIGURE A8 Section of the duodenum of a cat. Long, narrow cells of smooth muscle near at the right of the micrograph assume the shape of an archery bow when cut longitudinally; their elongate nuclei are in the position of the handgrip on the bow. 100 ×.

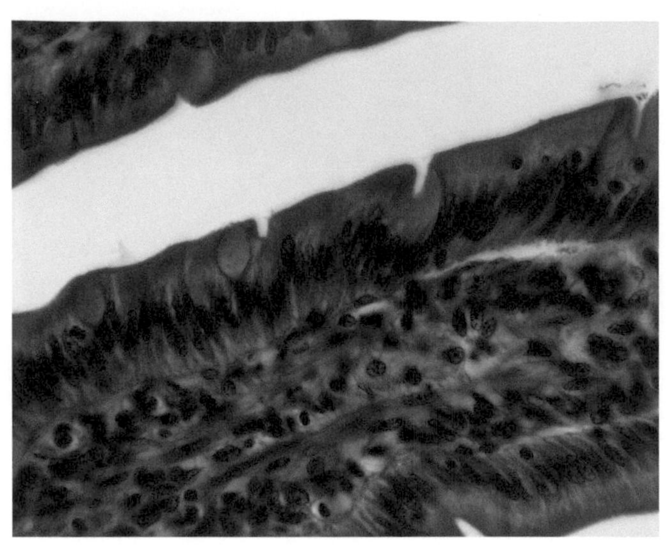

FIGURE A9 This micrograph is taken from another region of the slide shown in Fig. A8. It shows an assortment of cells, largely smooth muscle. 63 ×.

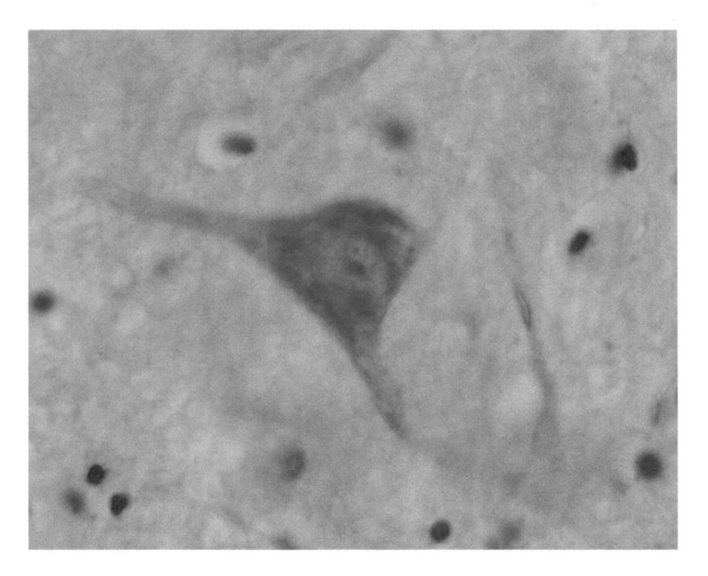

FIGURE A10 At the center is a multipolar nerve cell in a section from the gray matter of mammalian spinal cord. The deep blue basophilic flecks in the cytoplasm represent the ERGASTOPLASM. 100 ×.

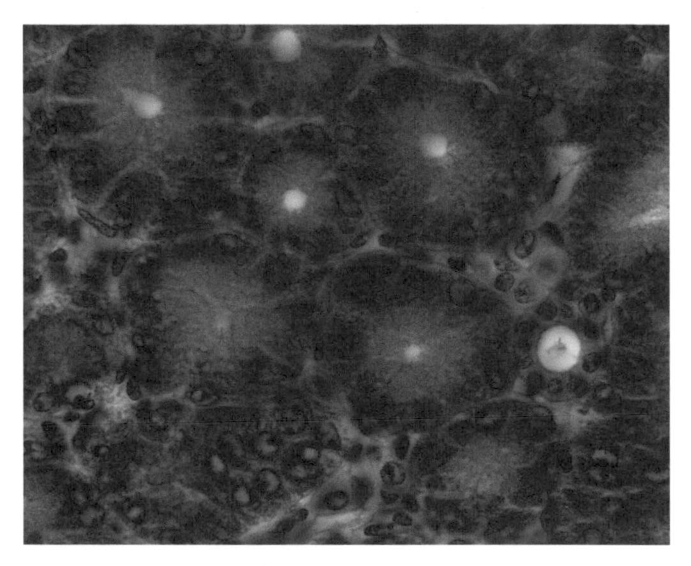

FIGURE A11 Section through the secretory units or ACINI (singular acinus) in the pancreas of a dogfish. Deep blue material around the periphery of each acinus is ergastoplasm, which is actively producing the acidophilic droplets of secretion at the center. 63 ×.

Cytoplasmic Inclusions

There are also nonprotoplasmic inclusions that are more or less temporary, nonliving, storage units of the cell. They include secretory granules, yolk granules, fat droplets, carbohydrates, pigments, and others. The organelles and inclusions are surrounded by the cytoplasmic GROUND SUBSTANCE, cytoplasmic MATRIX, or CYTOSOL.

1. *Secretory granules* have already been seen in sections of the pancreas where basophilic ergastoplasm produces droplets of proteinaceous secretion within grape-like ACINI (Figs. A11 and A22).
2. *Yolk granules.* Large, acidophilic yolk granules pack mature egg cells in the ovary of a lamprey (Figs. A23a and A23b).

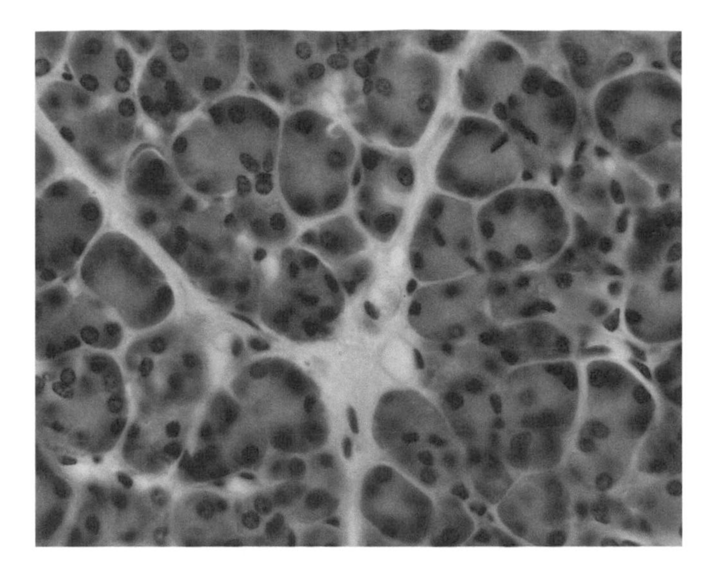

FIGURE A12 A similar situation is seen in the human pancreas where ergastoplasm around the periphery of an acinus produces acidophilic droplets of secretion that await release into the space or lumen near the center. 63 ×.

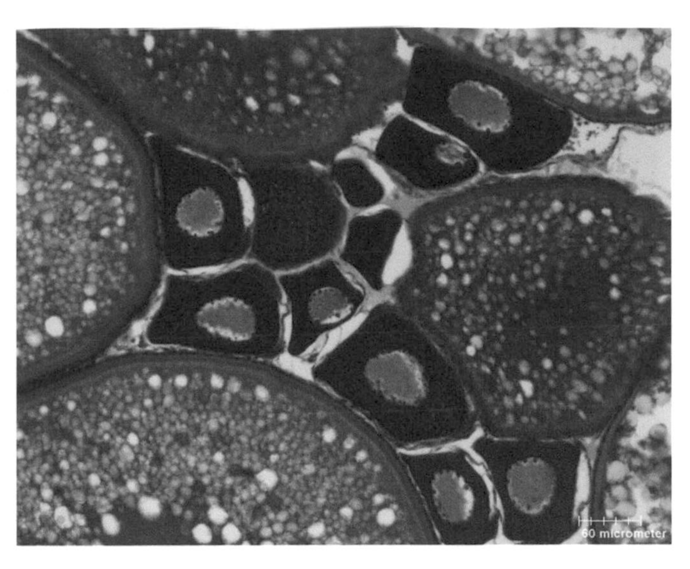

FIGURE A13 Section of the ovary of a rock bass. The cytoplasm of several deeply basophilic immature egg cells at the center contains dense masses of ERGASTOPLASM that produces droplets of acidophilic yolk as seen in the larger, more mature eggs at the periphery. 20 ×.

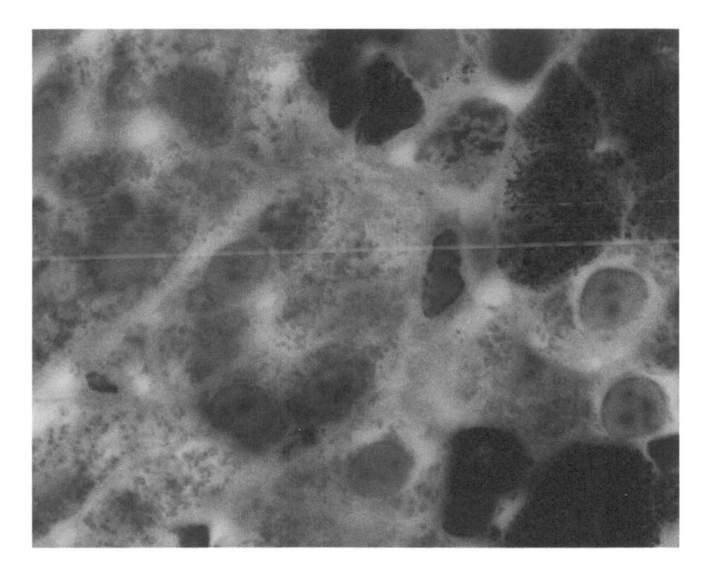

FIGURE A14 Section of the liver of an amphibian stained to preserve MITOCHONDRIA. The blue specks in these cells are mitochondria. Ignore the several cells near the edges of the micrograph that are packed with brown pigment granules. 100 ×.

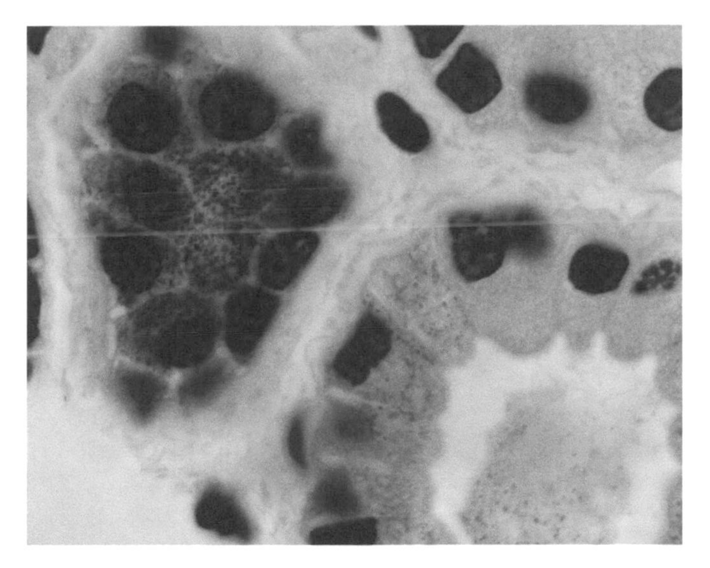

FIGURE A15 Cross section through tubules from an amphibian kidney (*Amphiuma*) stained to preserve mitochondria. Note that some of the tubules are more richly endowed with mitochondria than others. 100 ×.

3. *Fat* is usually removed in routine preparations, leaving a characteristic "chicken-wire" arrangement of the cellular remnants as seen in this section of the fat body of a toad (Fig. A24). Fat has a great affinity for the vapors from osmium tetroxide as shown in these four cells in a whole mount of the mesentery of a mammal where the fat has been preserved by special methods (Fig. A25). The fat has also been preserved in this similar whole mount of the mesentery and has been stained by the fat-soluble dye Sudan IV (Fig. A26).
4. *Carbohydrates.* Routine methods also remove carbohydrates, such as glycogen. In this section of a mammalian liver, the glycogen has been preserved and appears as deep red droplets in the cytoplasm of these cells (Fig. A27).
5. *Pigment* granules shield delicate cells from harmful radiations and provide color to the body. It is uncertain why pigments would be so abundantly distributed in the amphibian liver but it is well demonstrated in this

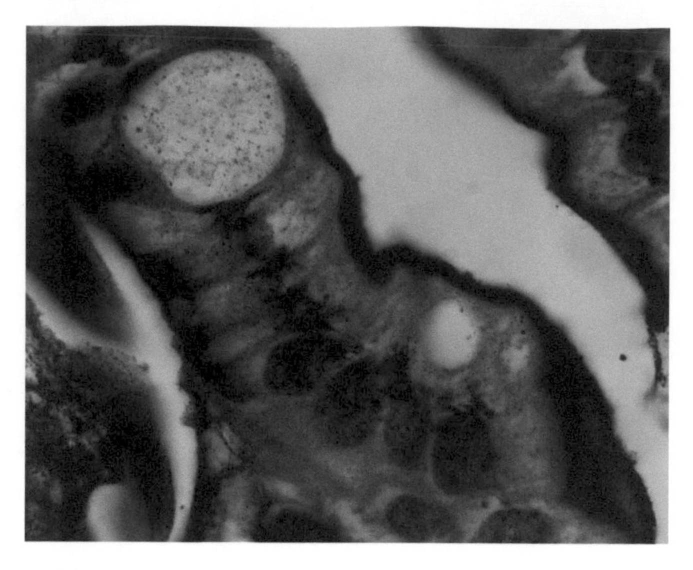

FIGURE A16 Section of absorptive cells from the intestine of an amphibian. Precautions were taken to preserve the GOLGI COMPLEX during preparation of the slide. The Golgi complexes are stained with silver and constitute the black skeins near the purple nuclei. 100×.

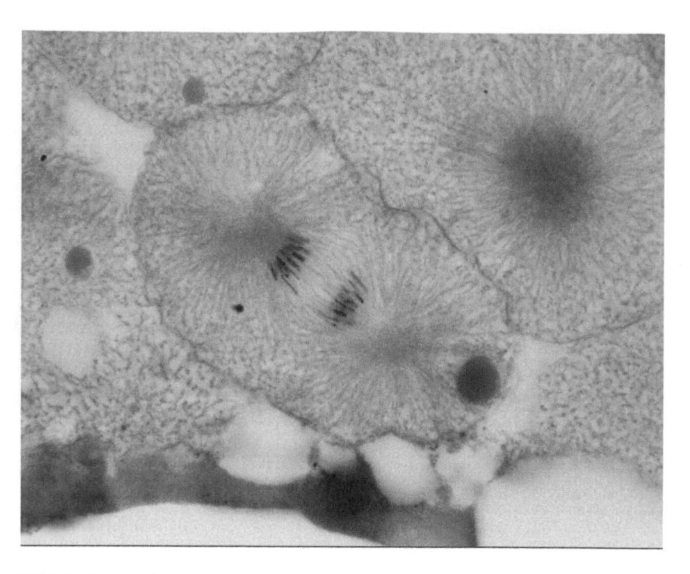

FIGURE A17 Section through an early stage of development (blastula) of a whitefish during active mitosis. Deeply staining chromosomes are pulling away from the equator of the cell. Delicate SPINDLE FIBERS form a pattern reminiscent of iron filings responding to a magnet. 100×.

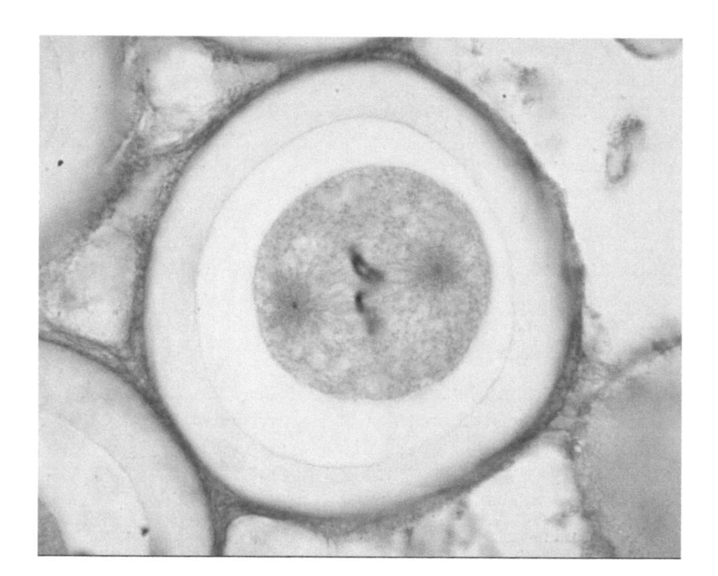

FIGURE A18 Section of the egg of a roundworm, *Ascaris*, showing the mitotic spindle. Mitosis is often studied in this worm because its cells contain only four pairs of chromosomes. Spindle fibers radiate from the two deeply basophilic CENTROSOMES. 100×.

FIGURE A19 Longitudinal section of striated muscle of a mammal. Contractile MYOFIBRILS run the length of these elongate cells. Striated muscle cells are so huge that individual cells may be seen with the naked eye. Such large cells contain many nuclei. (Contrast of the fibers has been enhanced by the use of differential interference contrast (DIC) microscopy.) 100×.

section (Fig. A28). Pigment cells in the skin of many vertebrates provide characteristic coloration. Note the pigment granules in these whole mounts of elaborate, star-shaped cells from the skin of a fish and an amphibian (Figs. A29 and A30). The granules are able to move back and forth in the extensions of the cell, thereby changing the color of the skin. Similar pigmented cells are seen in the skin of a rattlesnake (Fig. A31). Because this is a section, many of the extensions have been cut off and the extravagant shape of the cells is less obvious. Pigment cells improve visual acuity by minimizing scattering of light in the retina of the eye (Fig. A32).

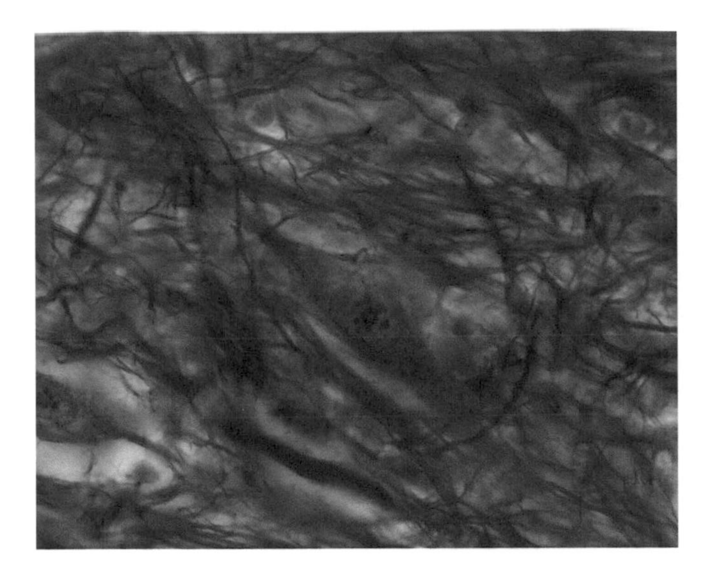

FIGURE A20 Section of the spinal cord of a mammal. NEUROFIBRILS, stained black with silver, course throughout the multipolar nerve cell at the center and in the surrounding processes. 100 ×.

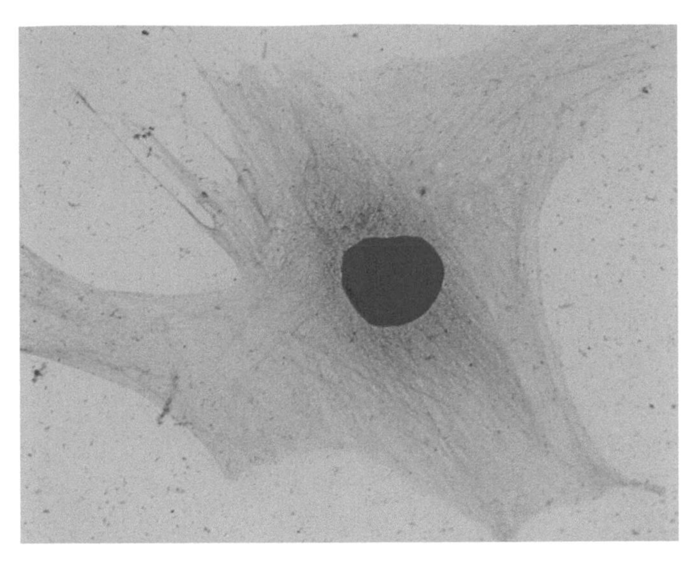

FIGURE A21 This is an entire cultured cell spread out on a microscope slide and stained. Delicate fibrils form a cellular framework or CYTOSKELETON for the cell. 100 ×.

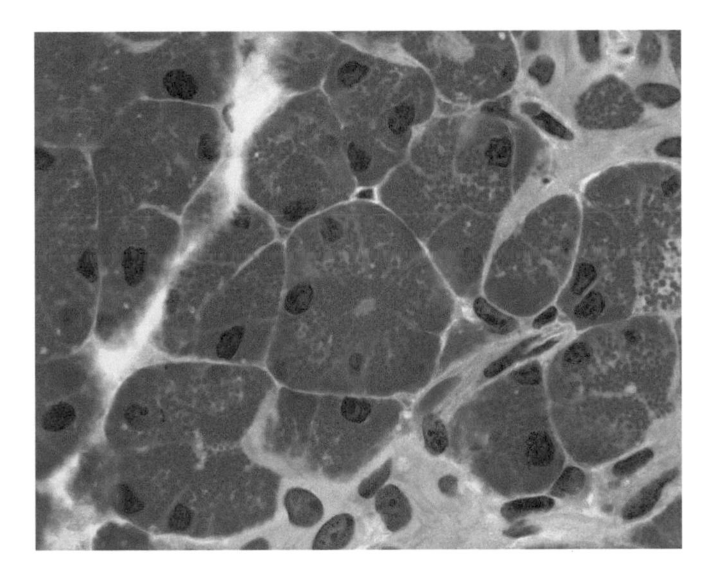

FIGURE A22 Thin section through the pancreas of a monkey. Secretory droplets are packed in the acinus at the center of the micrograph. These droplets find their way into the lumen at the center of the acinus and from there these are transported into the intestine. 100 ×.

Nucleus

Review the micrographs you have examined so far, noting the variations in size, shape, and position of the nuclei. Usually nuclei are spherical or ovoid but sometimes they assume more elaborate shapes as seen in the blood cell of *Necturus* near the left of Fig. A33. Is there a similar degree of variation in nuclear content? In each case observe the NUCLEOPLASM, CHROMATIN, and the NUCLEOLUS. Nucleoli are well developed in cells that are actively producing proteins. Several hundred nucleoli are seen in the developing egg cells of some vertebrates (Fig. A34).

Chromatin may occur as deeply staining clumps of condensed HETEROCHROMATIN, which stains with basic dyes such as hematoxylin, and extended EUCHROMATIN in the clear areas between. Metabolically active cells display relatively larger amounts of euchromatin. The chromatin in most of the blood cells in Fig. A33 is heterochromatic.

The presence of the SEX CHROMATIN mass against the nuclear membrane (BARR BODY) in stained smears of cells from the lining of the human cheek has been used to distinguish between male and female nuclei. A small mass of sex chromatin abuts the nucleolus of the nerve cell in Fig. A35a, appearing much like a tennis ball beside a basketball. Sex chromatin is not seen in all nuclei in a smear of female cells so that a large number of nuclei

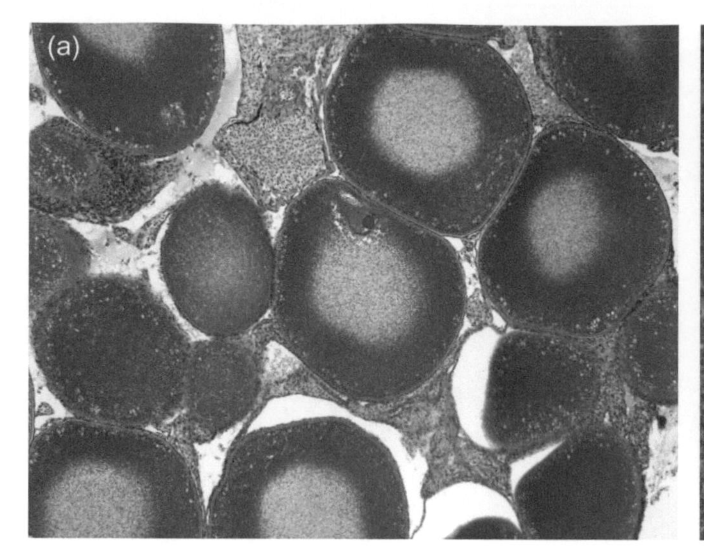

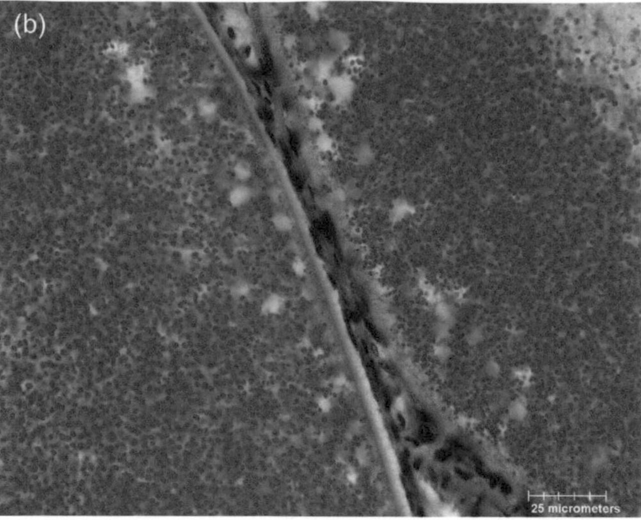

FIGURE A23 (a and b) Sections through egg cells in the ovary of a lamprey at low (a) and high power (b). Egg cells are suspended in a vascularized mesh of connective tissue. They are packed with yolk droplets that appear denser at the periphery of the cells. Acidophilic yolk droplets in some of these egg cells are densely packed below the cell membrane. 10 × and 63 ×.

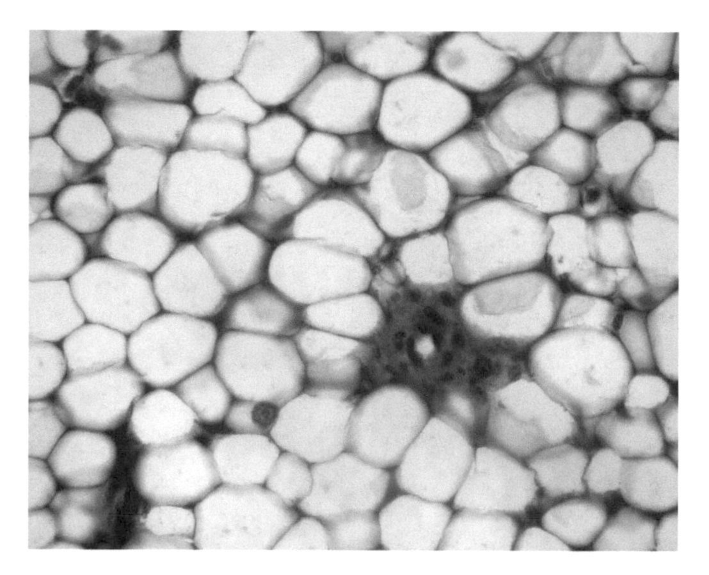

FIGURE A24 FAT CELLS in section of the fat body of a toad. In life, each of these cells would have contained a large droplet of fat. Since fat solvents are used in routine histological preparations, these droplets are not seen in most sections. All that is left of the fat cells in this preparation is the thin rim of cytoplasm that once enclosed the fat droplets. A small artery is shown near the center. The tissue is supported by delicate strands of connective tissue. 40 ×.

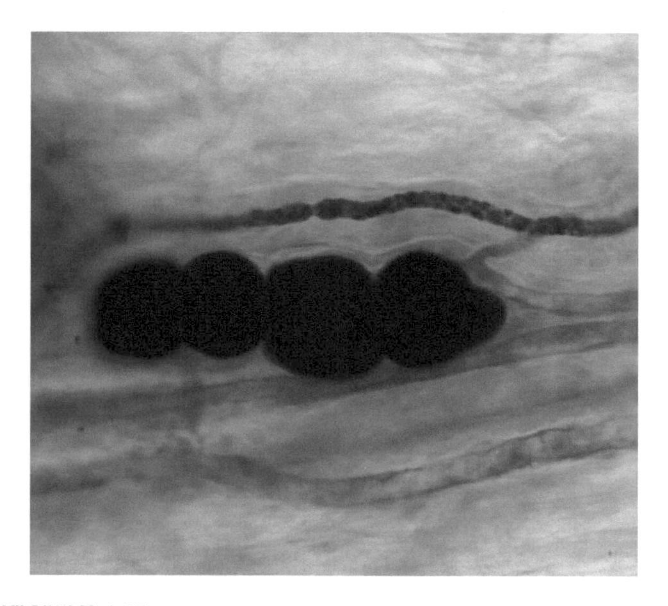

FIGURE A25 No fat solvents were used in this whole mount of a few fat cells in the mesentery of a mammal. The fat was stained by exposing the cells to the vapors from a solution of osmium tetroxide (OsO_4). The surrounding thin rim of cytoplasm of the fat cells is not visible; a few blood vessels course across the section. 25 ×.

should be examined before a diagnosis is made. Is the sex chromatin present in cells of the female and absent from cells of the male throughout the animal kingdom?

STUDY OF CELLULAR STRUCTURE WITH THE ELECTRON MICROSCOPE

Throughout these studies we will be investigating ultrastructure using ELECTRON MICROGRAPHS. Most of the material we will be studying is of tissues fixed in solutions of glutaraldehyde and osmium tetroxide, embedded in epoxy resin, sectioned on an ULTRAMICROTOME to thicknesses of about 50–100 nm, and "stained" with compounds of lead and uranium. The heavy metals accumulate selectively in certain parts of the cell and enhance the contrast of electron micrographs by scattering the electron beam, thereby producing a shadow.

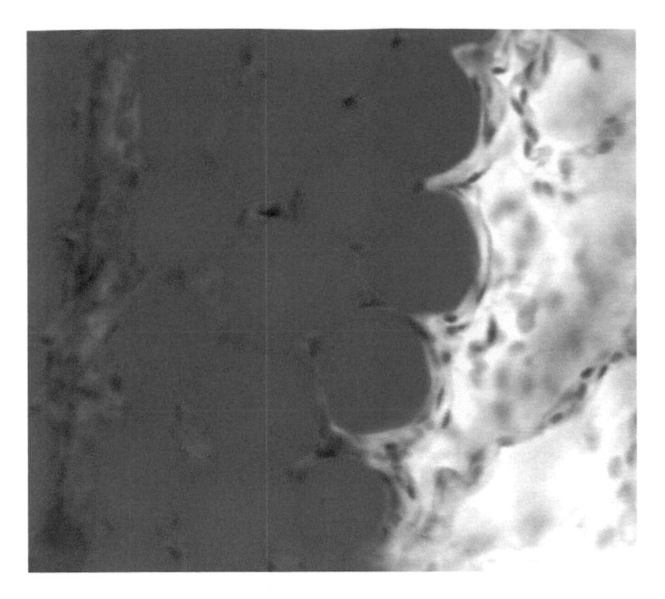

FIGURE A26 Like Fig. A25, no fat solvents were used in preparing this whole mount of fat cells in the mesentery of a mammal. The fat was stained by immersing the tissue in an aqueous solution of the dye Sudan IV, which has a greater affinity for lipids than for water. The surrounding thin rim of cytoplasm of the fat cells has been stained with a blue dye. 25 ×.

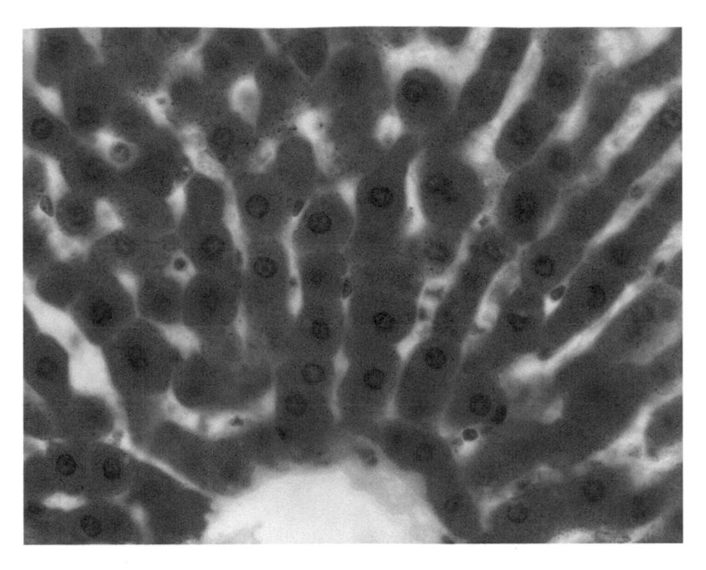

FIGURE A27 Section of the liver of a mammal stained by the periodic acid-Schiff (PAS) method to show GLYCOGEN. Although it is abundant in many cells, glycogen is not seen unless precautions are taken to avoid its leaching out. Deep red droplets of glycogen are richly packed within the cells of the liver of a mammal. Nuclei are stained with a basophilic dye. Blood spaces occur between the liver cells. 63 ×.

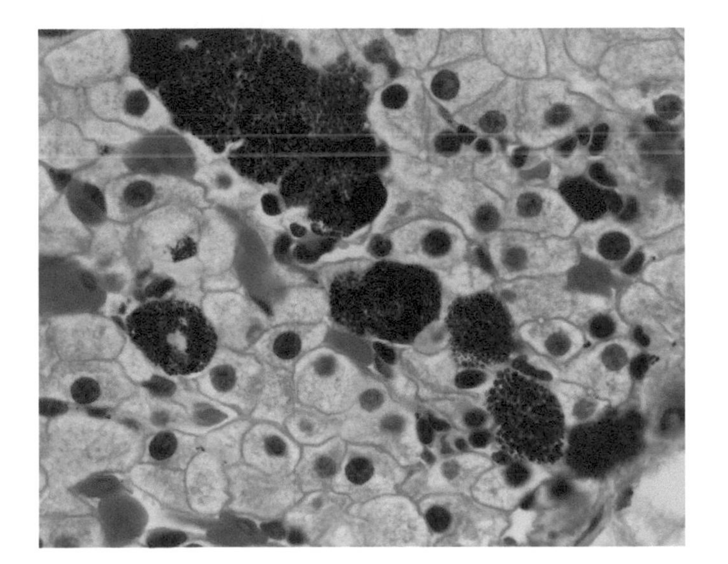

FIGURE A28 Section of the liver of *Amphiuma* stained with hematoxylin and eosin. The pale liver cells are slightly acidophilic with basophilic nuclei. The deep red blobs are blood cells in small blood vessels that permeate the liver. Scattered throughout are deep brown pigment cells packed with brown granules that display their natural color. 40 ×.

FIGURE A29 Whole mount of the skin of a shiner (bony fish). The tissue was preserved in a fixative containing yellow picric acid; otherwise no stains have been used. Elaborate star-shaped PIGMENT CELLS darken the skin in certain areas. The cells are packed with black pigment granules, which roll in and out of the processes, lightening and darkening the skin. 10 ×.

Osmium has an affinity for the abundant trilaminar UNIT MEMBRANES of the cell: the plasmalemma and many of the internal membranes. Many organelles and inclusions are surrounded by unit membranes and these form chambered labyrinths within the cytoplasm. The membranes greatly increase the surfaces on which metabolic reactions occur and also enclose compartments where materials may be segregated.

In the assortment of electron micrographs that follows, identify the cellular features noted earlier. Often the magnifications are so great that only a small portion of a cell is visible in a single picture. In addition, because of the thinness of the sections, many structures may not appear in a particular cell and the mental gymnastics required to reconstruct three-dimensional forms are even more complex than with the light microscope.

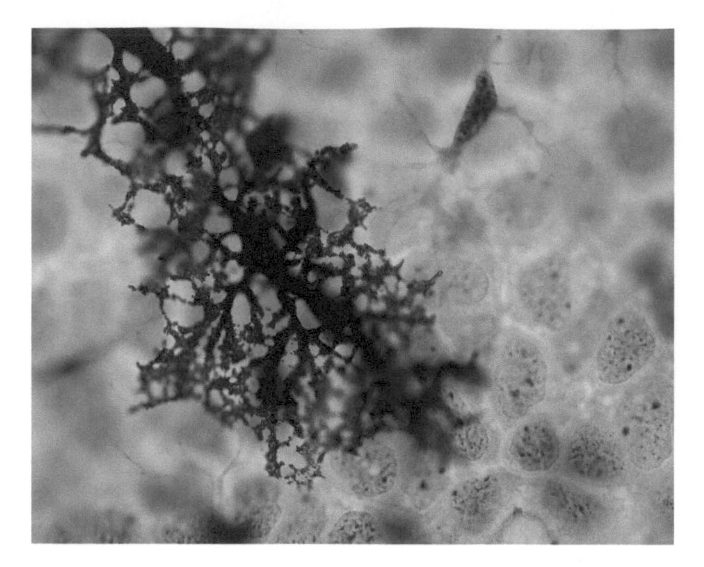

FIGURE A30 Whole mount of skin from a salamander, *Amphiuma*. Brown pigment granules are dispersed throughout the elaborate processes of this cell. 63×.

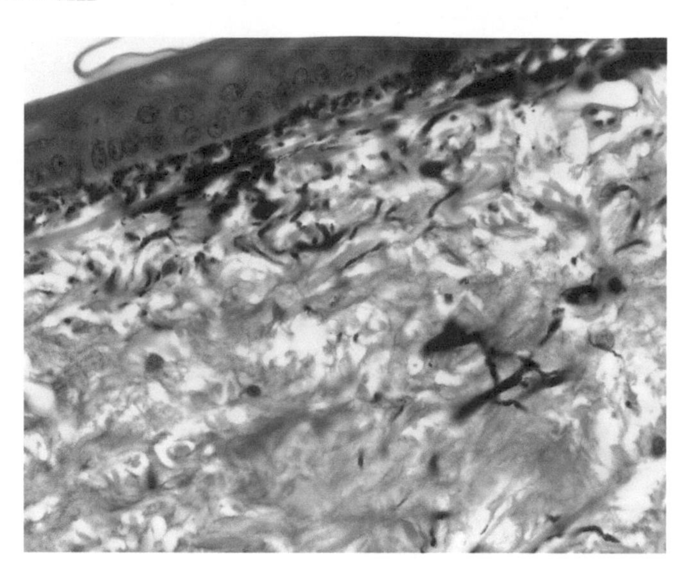

FIGURE A31 Section through the skin of a rattlesnake. Pigment cells lurk under the tough outer layers of the skin (top). Many of the long processes of these delicate cells have been lopped off in sectioning. A few pigment cells are dispersed in the deeper connective tissue of the skin. 63×.

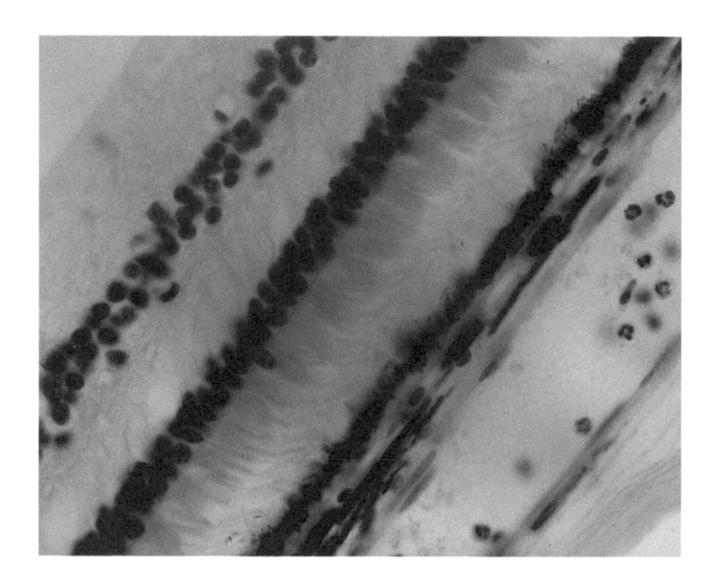

FIGURE A32 Section through the retina in the eye of a monkey. Light enters at the upper left and passes through several layers of cells to stimulate sensory cells at the center. Immediately to the right of these cells is a brown layer of pigment cells, which minimizes scattering of light within the retina. 63×.

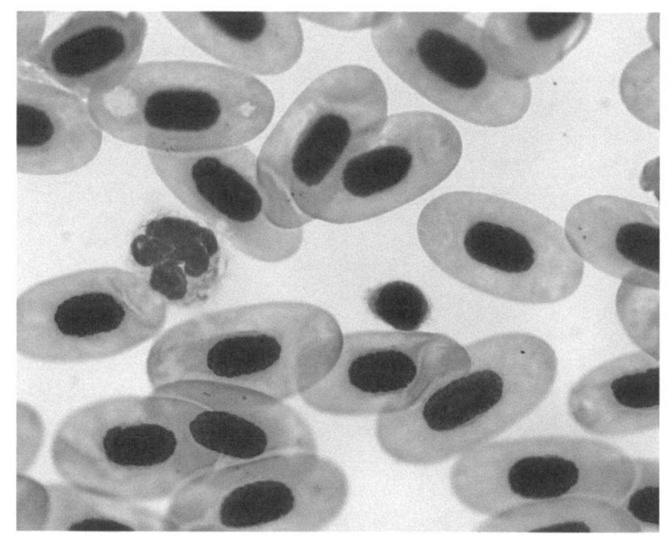

FIGURE A33 Stained smear of the blood of a mudpuppy *Necturus*. Most of the cells are red blood cells with dense, ovoid nuclei. A white blood cell at the left contains a complex lobulated nucleus. 63×.

Cell Membrane (Plasmalemma)

The cell membrane is a thin (8–10 nm) osmiophilic (osmium loving) line, which appears double in the best micrographs (Figs. A35b and A36). On its outer surface it has a CELL COAT of variable thickness consisting of glycoproteins and glycolipids (Fig. A37). Vesicles, indicative of the processes of endocytosis and exocytosis, may be associated with the cell membrane.

Endoplasmic Reticulum and Ribosomes

The cytoplasm of nearly all cells contains a network of membrane-bound, fluid-filled channels, the ENDOPLASMIC RETICULUM, which consists of intercommunicating tubules, cisternae (singular: cisterna), and isolated

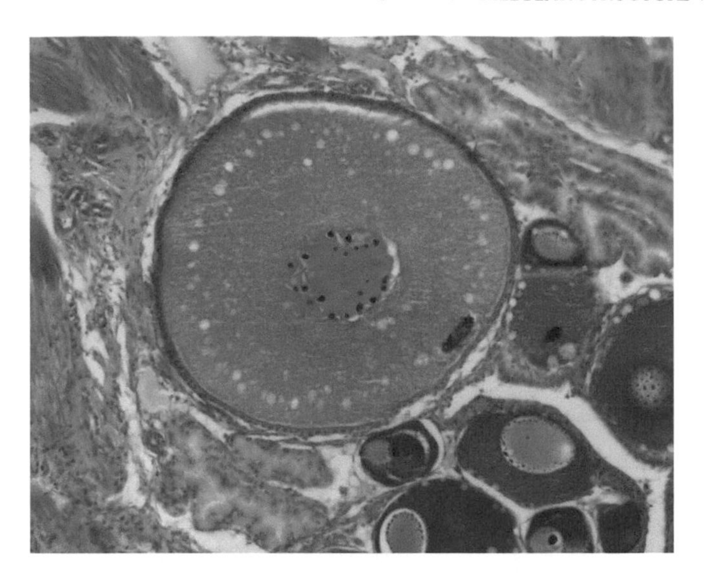

FIGURE A34 Section of the ovary of a pike. The nuclei of these large cells contain several nucleoli just inside the nuclear membrane. 20 ×.

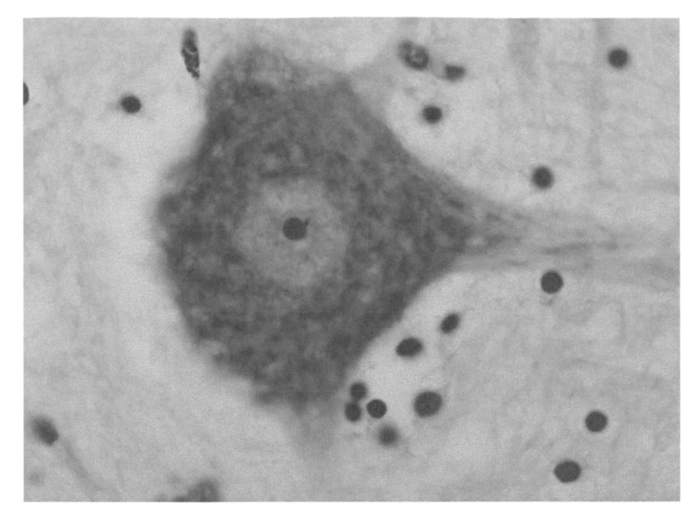

FIGURE A35a Section of the spinal cord of a female mammal. The sex chromatin, characteristic of female mammals, is nestled beside the nucleolus, a tennis ball beside a basketball. 63 ×.

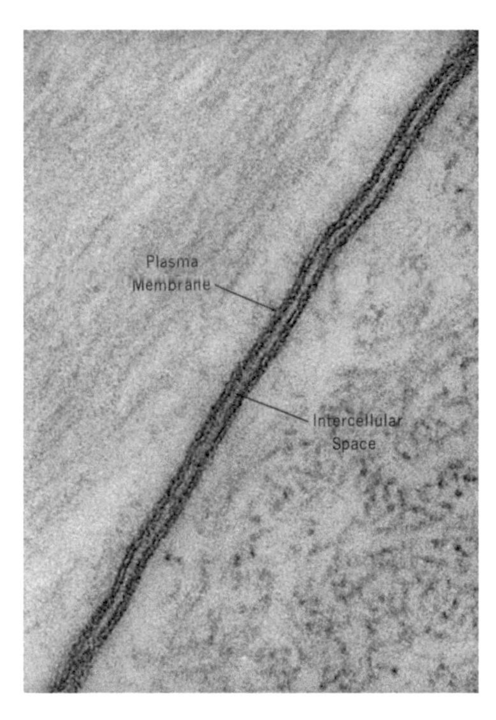

FIGURE A35b This micrograph is of a thin section showing the apposition of two cells. The plasma membrane of each appears as a trilaminar structure consisting of two dense lines and a less dense intermediate layer. The unit membranes of these cells run parallel and are separated by an intercellular space of uniform width. 260,000 ×.

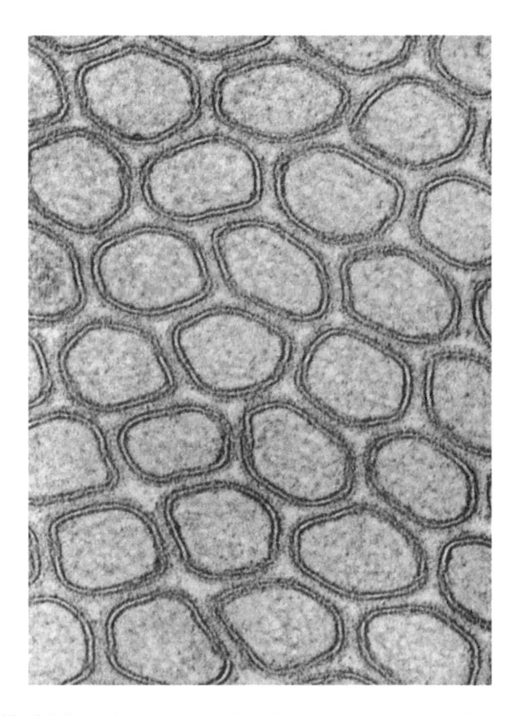

FIGURE A36 This section has been cut tangentially through a field of microvilli—consider a lawnmower passing over a freshly cut lawn. The trilaminar plasmalemma encloses the contents of each microvillus. 230,000 ×.

saccules (Fig. A38). The degree of development of the endoplasmic reticulum varies greatly between different types of cells and during different functional phases of the same cell.

The GRANULAR or ROUGH ENDOPLASMIC RETICULUM (rER) is associated with small particles containing ribonucleic acid, the RIBOSOMES, and constitutes the ergastoplasm seen with the light microscope (Figs. A38–A41). It is abundant in cells actively secreting proteins such as the pancreatic acinar cell and the plasmocyte. The basophilia resides in the large numbers of ribosomes adhering to the outer surfaces of the membranes. These ATTACHED RIBOSOMES synthesize the proteins secreted by the cell. FREE RIBOSOMES in the cytoplasmic matrix are considered to

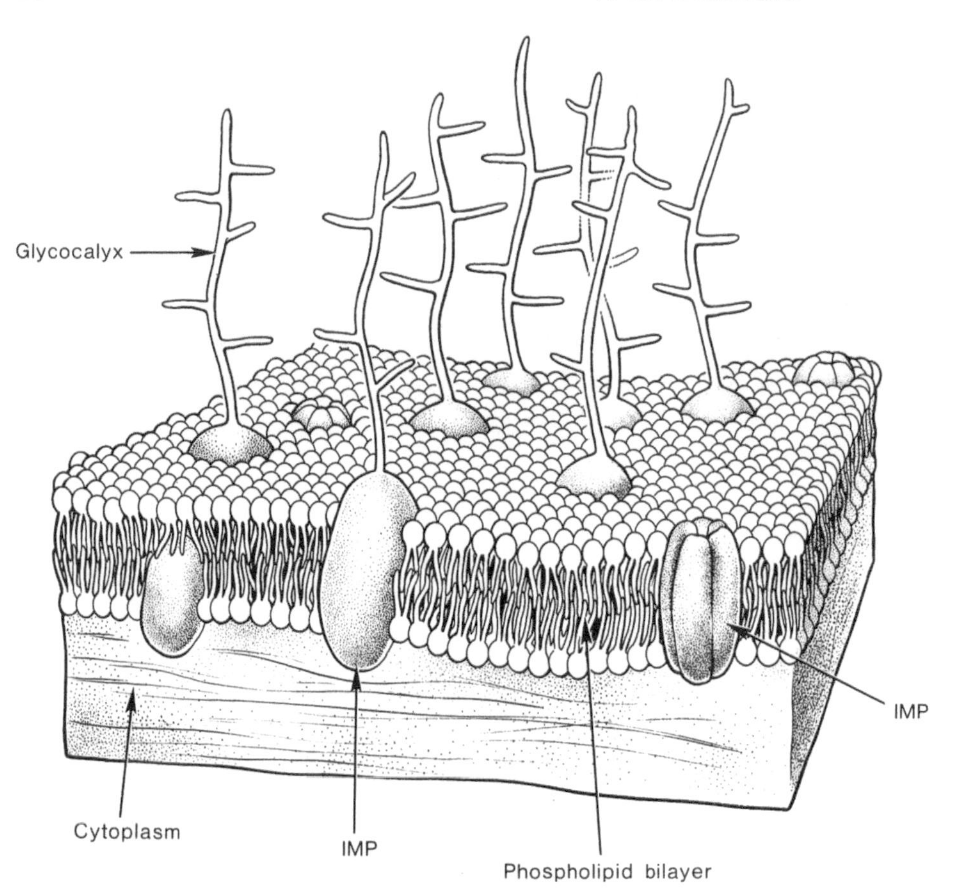

FIGURE A37 Diagram to illustrate the trilaminar nature of the PLASMALEMMA. The membrane consists of two layers of phospholipid molecules which have a double nature: a water-soluble phosphate group at one end and a fatty acid, which is insoluble in water, at the other. In water, the lipid tails of the molecules form a double hydrophobic sheet sandwiched between the phosphate groups, which are exposed to the aqueous environment. The lipid bilayer is a barrier to the free diffusion of dissolved molecules and ions and passage through the membranes is provided by intramembranous particles (IMP), largely proteins, which are embedded in them; presumably the central part of each particle is hydrophobic and is held in position by the central lipid core of the membrane. The outer surface of the membrane may have a felt-like coating or GLYCOCALYX, which consists of molecules with a protein or lipid base attached to complex, branching sugars. The glycocalyx is anchored to the cell membrane by the protein or lipid part of the molecule.

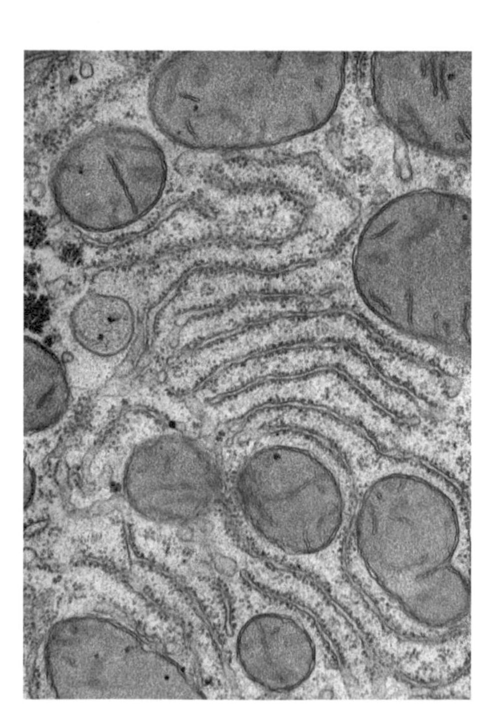

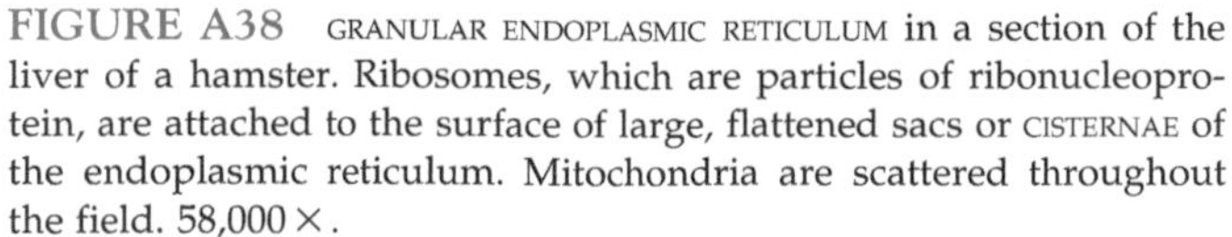
FIGURE A38 GRANULAR ENDOPLASMIC RETICULUM in a section of the liver of a hamster. Ribosomes, which are particles of ribonucleoprotein, are attached to the surface of large, flattened sacs or CISTERNAE of the endoplasmic reticulum. Mitochondria are scattered throughout the field. 58,000 ×.

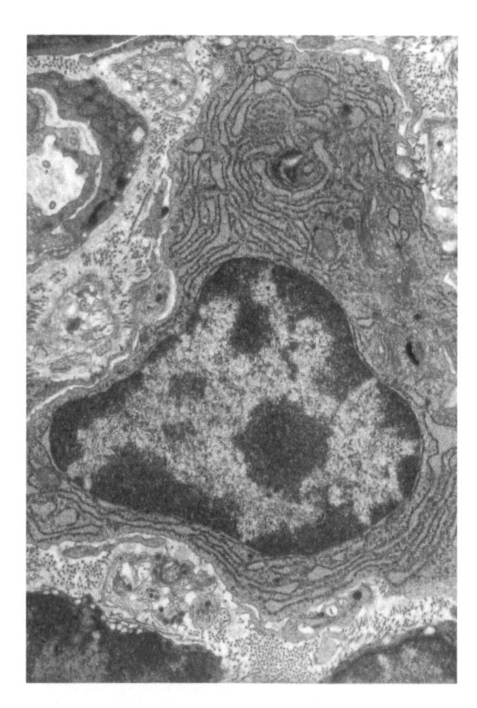

FIGURE A39 Plasmocyte in loose connective tissue of the rat. Cytoplasm that is rich in granular endoplasmic reticulum surrounds the nucleus of this plasmocyte—the presence of large amounts of granular endoplasmic reticulum accounts for the deep basophilia of plasmocytes in light microscope preparations. The granular endoplasmic reticulum consists of large, flattened cisternae studded on their outer surfaces with ribosomes; it is the ribosomes that produce the deep basophilia of these cells. 29,000 ×.

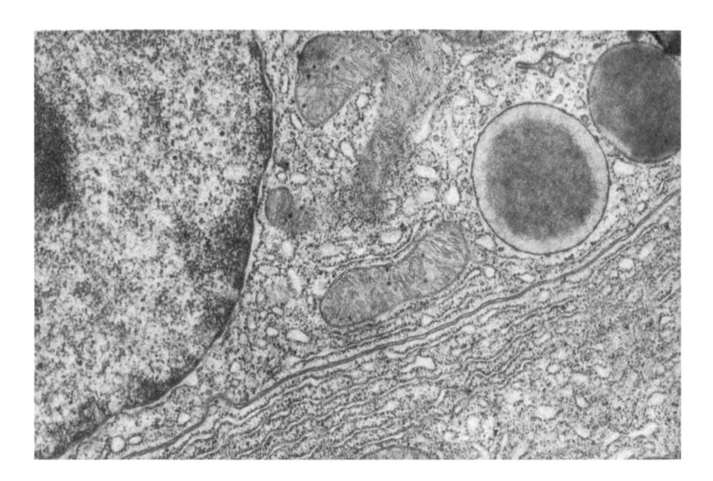

FIGURE A40 Acinar cell in the pancreas of a frog. A portion of the nucleus is shown at the left. The nucleus is enclosed by the NUCLEAR ENVELOPE, which consists of two trilaminar membranes with a space between; a few PORES penetrate the nuclear envelope at the lower left and near the top—note that the outer and inner laminae of the nuclear envelope are continuous at the site of the pores. Cisternae of granular endoplasmic reticulum, studded with ribosomes, abound in the cytoplasm of this cell. There are two DROPLETS OF SECRETION at the upper right. A portion of an adjacent cell is shown at the lower right of the micrograph—presumably the two cells are held about 20 nm apart by their glycocalyces. Sections of a few mitochondria are scattered throughout the granular endoplasmic reticulum. 43,000 ×.

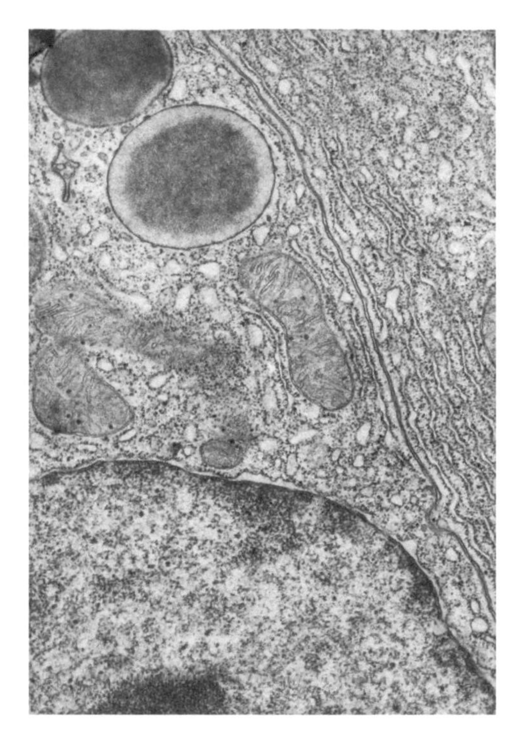

FIGURE A41 Pancreatic acinar cells from a bat. Granular endoplasmic reticulum of many parallel cisternae packs the cytoplasm at the base of these cells (right). Droplets of proteinaceous secretion, ready to be discharged, are seen at the top. Part of a nucleus is seen at the left as well as a few mitochondria enmeshed in the endoplasmic reticulum. 13,500 ×.

be the sites of the protein synthesis necessary to sustain cellular proliferation. Whether free or attached, ribosomes often occur as POLYRIBOSOMES in clusters of 3 to more than 30 held together by a slender filament.

SMOOTH or AGRANULAR ENDOPLASMIC RETICULUM (sER) consists of closely meshed three-dimensional tubules that lack ribosomes; flat cisternae are seldom found (Figs. A42–A44). Several diverse functions have been shown: in muscle it releases and captures calcium ions during contraction and relaxation; in the endocrine glands it is engaged in the biosynthesis of steroids; and in the liver, where both granular and agranular endoplasmic reticulum occur in almost equal amounts, it is involved in lipid and cholesterol metabolism and plays a role in the detoxification of drugs. The agranular endoplasmic reticulum is also involved in lipid absorption, the metabolism and synthesis of glycoprotein, glycogen metabolism, and membrane formation.

Mitochondria

Mitochondria provide energy for numerous chemical reactions and active transport mechanisms within the cell (Figs. A41 and A42). They also contain enzymes for protein synthesis and lipid metabolism. They are enclosed by two unit membranes, 5–6 nm in thickness, between which is a narrow compartment, the MEMBRANE SPACE. The outer membrane is smooth, but the inner forms narrow folds, the CRISTAE (singular: crista) that project into the INTERCRISTAL SPACE, which is filled with MITOCHONDRIAL MATRIX. Dense MITOCHONDRIAL GRANULES in the matrix are cation-binding sites and function in the regulation of the ionic concentration of the cytoplasmic matrix. Mitochondria of cells involved in lipid metabolism often have tubular cristae and are associated with agranular endoplasmic reticulum (Figs. A44 and A45).

Golgi Complex

The Golgi complex is active in the concentration and packaging of secretory products of glandular cells and is usually positioned between the nucleus and the secreting end of the cell (Figs. A46–A48). It consists of

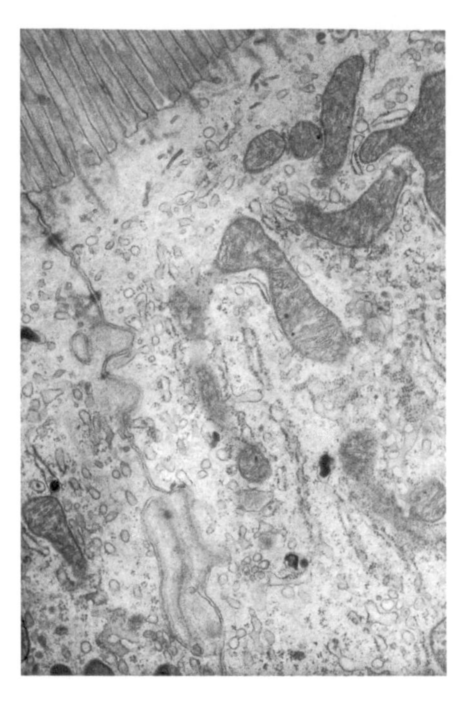

FIGURE A42 Section of two lining cells from the jejunum of the rat. This micrograph is limited to the apical portion of these two cells. The area of the absorptive surface of the intestine (upper left) is greatly enhanced by the formation of finger-like extensions, the MICROVILLI. The plasma membranes of the two cells are apposed at the left—JUNCTIONAL COMPLEXES within these membranes, strengthen the apposition of these cells. MITOCHONDRIA, with distinctive shelf-like CRISTAE, are abundant in this apical cytoplasm. The endoplasmic reticulum near the center of the micrograph appears denuded of its ribosomes—this is AGRANULAR ENDOPLASMIC RETICULUM where triglycerides are synthesized. 42,000 ×.

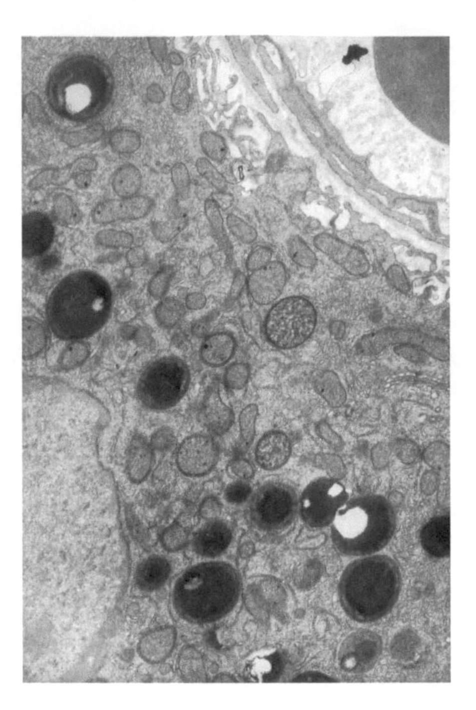

FIGURE A43 Testosterone-secreting interstitial cell from the testis of a mouse. The cell is packed with agranular endoplasmic reticulum (SER), mitochondria, and spherical droplets of secretion; a few fragments of granular endoplasmic reticulum are seen. Its nucleus is at the lower left. 29,000 ×.

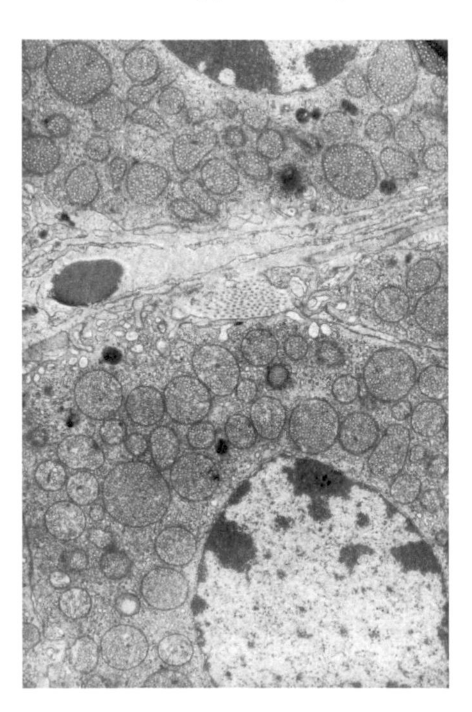

FIGURE A44 Adrenal, rat. Some cells of the adrenal cortex secrete steroid hormones. The cytoplasm of these cells is packed with AGRANULAR ENDOPLASMIC RETICULUM (SER) and mitochondria with TUBULAR CRISTAE. A few ribosomes lie free in the cytoplasm. Portions of two nuclei (N) appear at the top and bottom. A few dark LYSOSOMES (Ly) appear near the top. 22,300 ×.

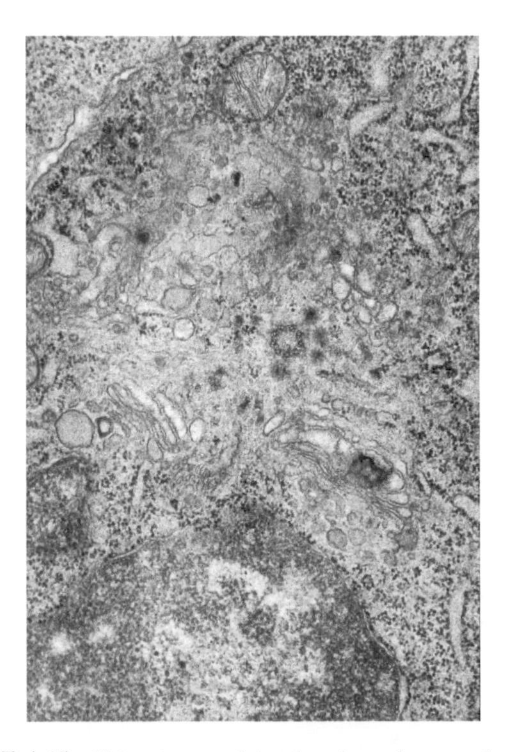

FIGURE A45 This micrograph is taken from the central region of a plasmocyte in loose connective tissue of the rat. Part of the nucleus (N) is shown at the bottom. Above the nucleus and a little to the right are flattened cisternae of a GOLGI COMPLEX (G). With careful examination, you may be able to discern more Golgi complexes. Outside this central region are abundant ribosomes and cisternae of endoplasmic reticulum. 67,000 ×.

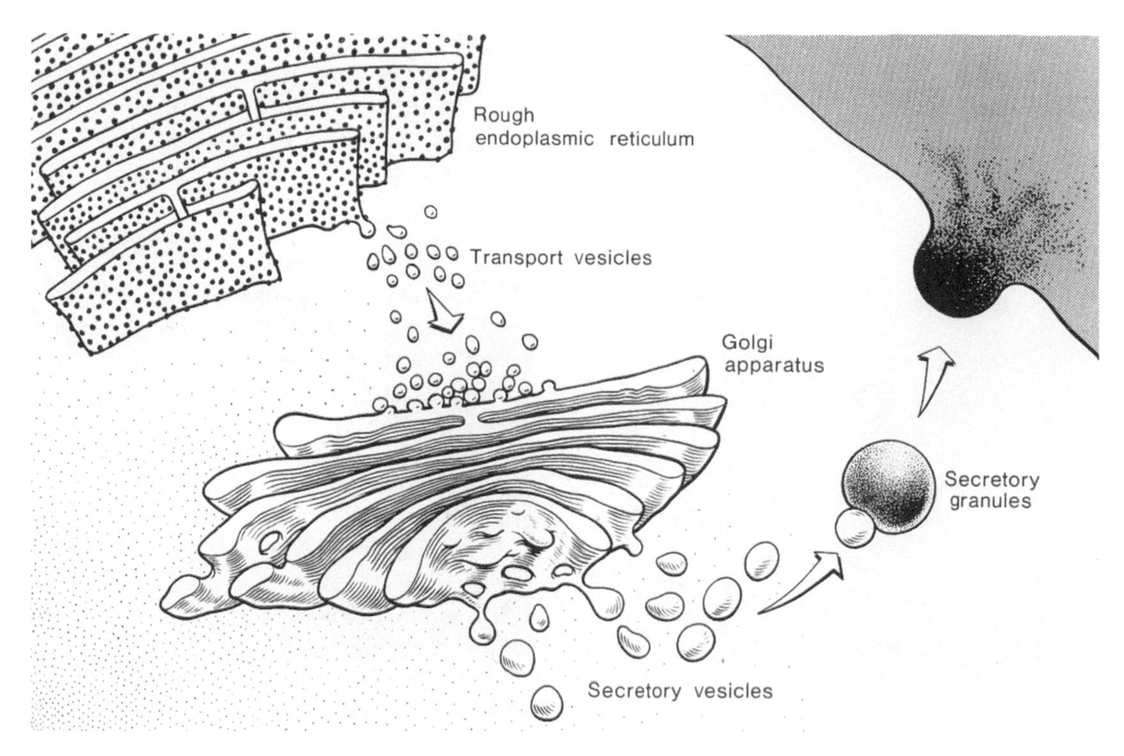

FIGURE A46 Diagram illustrating the involvement of the GOLGI COMPLEX in the secretions of protein. Dilute solutions of proteins produced within the granular endoplasmic reticulum are sequestered within TRANSFER VESICLES and passed to the convex FORMING FACE of the Golgi complex where they are incorporated and their contents concentrated and sugars added. CONDENSING VESICLES are released at the convex MATURING FACE. These condense and coalesce to produce SECRETORY DROPLETS that are released from the cell.

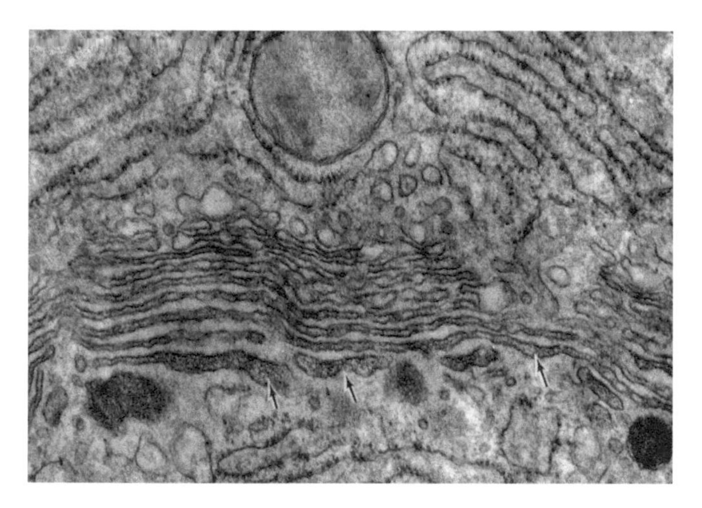

FIGURE A47 GOLGI COMPLEX in the submucosal gland of the intestine of a mouse. The convex FORMING FACE of the Golgi complex is at the top of the stack, the concave MATURING FACE at the bottom. Some granular endoplasmic reticulum may be seen above and below the complex. 54,000×.

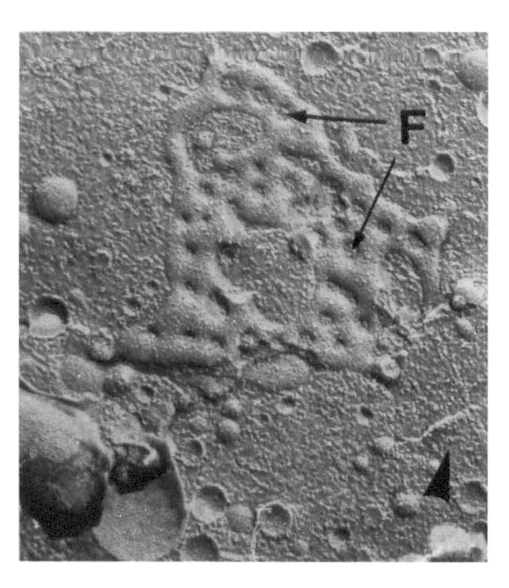

FIGURE A48 A freeze-fractured Golgi complex from a kidney tubule of the snake; the fracture plane is roughly parallel to the surface, Fenestrae (F) in the saccule have a vesicular appearance. 39,000×.

characteristic stacks of curved, parallel, flattened saccules, or cisternae that are often expanded at their ends. TRANSFER VESICLES of proteinaceous material from the granular endoplasmic reticulum present themselves at the convex inner surface or FORMING FACE of the stack and, in an active cell, there is a constant coalescence of the membranes and contents of these vesicles to form new saccules every few minutes. At the same time, CONDENSING VACUOLES are released at the opposite surface, the MATURE FACE, and the Golgi complex may be said to be in a steady state between uptake and discharge of vesicles. These vesicles condense and coalesce to produce the SECRETORY

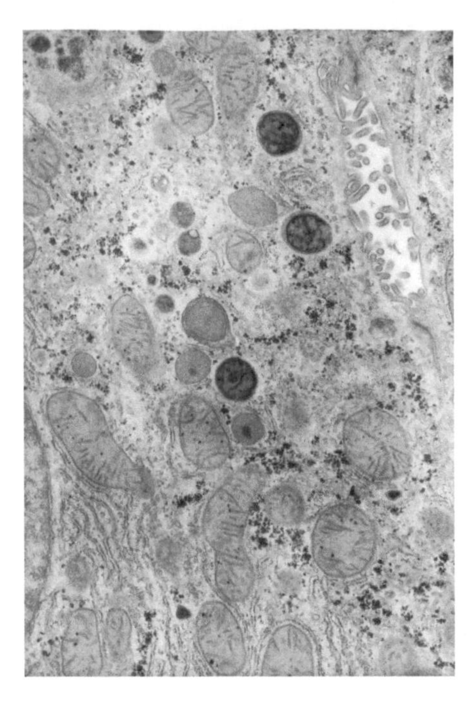

FIGURE A49 Liver cell of a rat. The large dense bodies in this micrograph are LYSOSOMES; the small dense bodies peppered throughout are granules of GLYCOGEN. Both granular and agranular endoplasmic reticulum and mitochondria may be recognized. There is a Golgi complex in the upper left corner. A small amount of the nucleus lurks at the lower left. Lysosomes contain portions of cytoplasmic components such as glycogen, mitochondria, or cisternae of the endoplasmic reticulum. Hydrolytic enzymes (phosphatases and proteases) provide intracellular digestion of worn-out cellular organelles and materials taken into the cell by endocytosis. 30,000 ×.

GRANULES or, more properly, the SECRETORY DROPLETS, which are released from the cell. Although it plays little part in the synthesis of proteins, the Golgi complex may actively participate in the synthesis of polysaccharides.

Lysosomes

Lysosomes are dense bodies in the cytoplasm, which were originally defined biochemically as being limited by a membrane and containing acid hydrolases (hydrolytic enzymes that function in slightly acid conditions) (Figs. A49 and A50). The structures identified in electron micrographs that fulfill these criteria are diverse and may be spherical, ovoid, or irregular in outline, with a pale to dense matrix of varying degrees of homogeneity. Identification of lysosomes in electron micrographs is unreliable and should be verified by other means such as their uptake of certain fluorescent dyes (e.g., acridine orange) as observed under ultraviolet light and by their histochemical reactions for acid hydrolases.

Lysosomes assist in the intracellular digestion of worn-out cellular organelles and materials taken into the cell by endocytosis. The indigestible residues accumulate in the cell as LIPOFUSCIN GRANULES, the "age pigments" or "wear and tear pigments" (Fig. A51). The hydrolytic enzymes of the lysosomes are normally safely contained within the membrane but may be released and cause breakdown of injured cells.

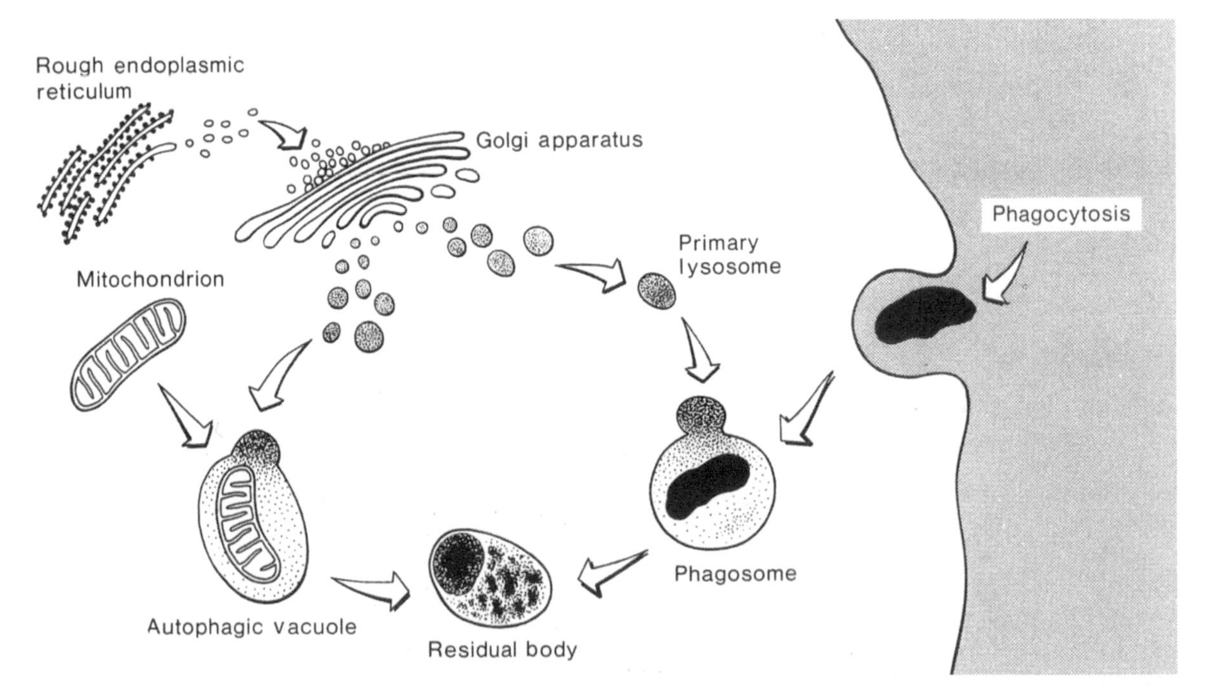

FIGURE A50 The synthesis and fate of LYSOSOMES. PRIMARY LYSOSOMES are formed from the Golgi sacs. When they fuse with a substance to be digested they become SECONDARY LYSOSOMES. They may digest materials absorbed from outside the cell by phagocytosis and become PHAGOSOMES. They may absorb worn-out organelles within the cell and become AUTOPHAGIC VACUOLES. RESIDUAL BODIES are lysosomes containing undigested material.

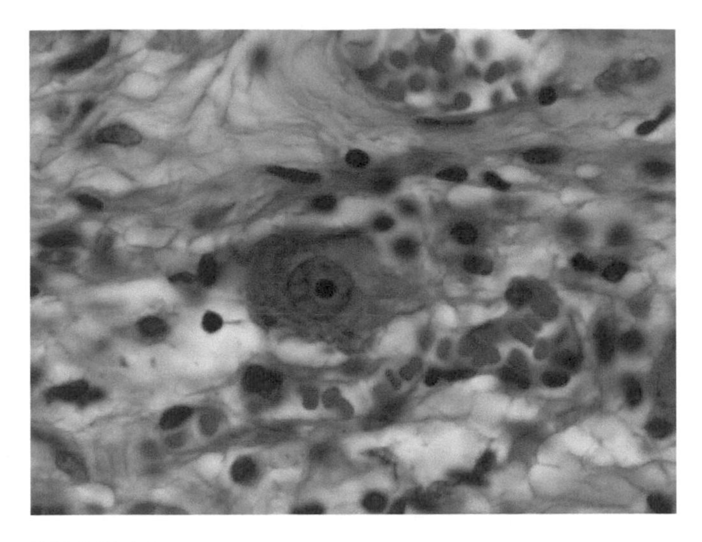

FIGURE A51 Granules of the brownish pigment LIPOFUSCIN have accumulated in the cytoplasm of a neuron in this mammalian spinal ganglion. 63×.

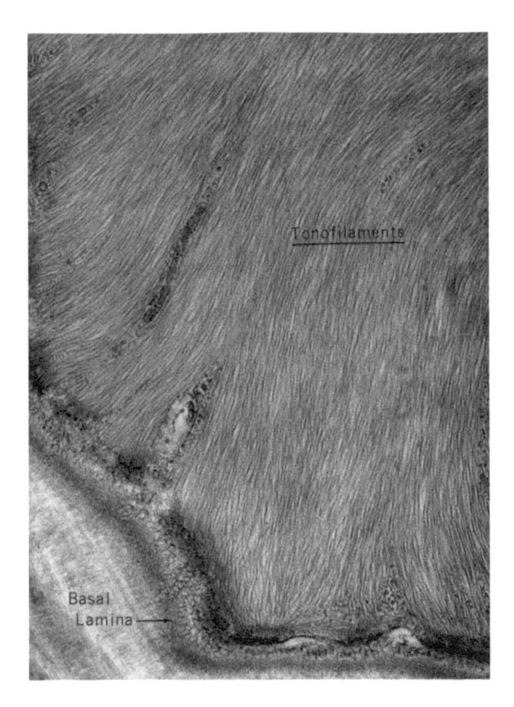

FIGURE A52 Microfilaments pass in all directions through a nerve cell in the mammalian spinal cord. 100×.

Lysosomes are formed from the Golgi sacs (Fig. A50). Newly formed lysosomes are PRIMARY LYSOSOMES. When they fuse with substances to be digested they become SECONDARY LYSOSOMES; these may be further subdivided on the basis of the material being digested as PHAGOSOMES, DIGESTIVE VACUOLES, or AUTOPHAGIC VACUOLES. Lysosomes are formed from the inner Golgi sacs which, in turn, are derived from the rough endoplasmic reticulum. These three components are sometimes referred to as the CERL.

Centrosomes and Fibrils

The electron microscope has demonstrated the artificiality of grouping the diverse fibrous structures as fibrils. These include the scattered actin MICROFILAMENTS, which may be more or less organized into bundles in the cytoplasmic matrix of most cells (Fig. A52), depending upon the mechanical stresses on the cell, and are associated with cellular movements (Figs. A52 and A53). INTERMEDIATE FILAMENTS are a heterogeneous group of structural elements 8–10 nm in diameter that occur in a variety of cells and participate in the formation of the cytoskeleton. In nerve and glial cells they are called NEUROFILAMENTS (Figs. A51 and A54) and GLIAL FILAMENTS (Fig. A55), respectively. The MYOFILAMENTS of muscle cells are actually overlapping rodlets of the fibrous proteins, ACTIN and MYOSIN (Figs. A56 and A57).

MICROTUBULES, 20–27 nm in diameter, are an important part of the cytoskeleton and are found in almost all cells scattered singly throughout the cytoplasm and in bundles (Figs. A58 and A59). They converge upon the centrosome and constitute the fibrils of the spindle apparatus. The CENTROSOME consists of two CENTRIOLES in a zone of dense cytoplasm and lies close to the center of the cell. Each centriole is a cylindrical bundle of nine sets of microtubular triplets. The two centrioles lie at right angles to each other. The centrosomes replicate before mitosis occurs and migrate to opposite poles of the cell. During the formation of cilia, centrioles replicate and migrate to the surface of cells on which cilia will develop. Each centriole organizes the assembly of microtubules in the cilium and remains as the BASAL BODY. Microtubules in nerve cells are called NEUROTUBULES (Fig. A60).

Cytoplasmic Inclusions

Study electron micrographs showing various cytoplasmic inclusions and note whether or not each is bounded by a membrane.

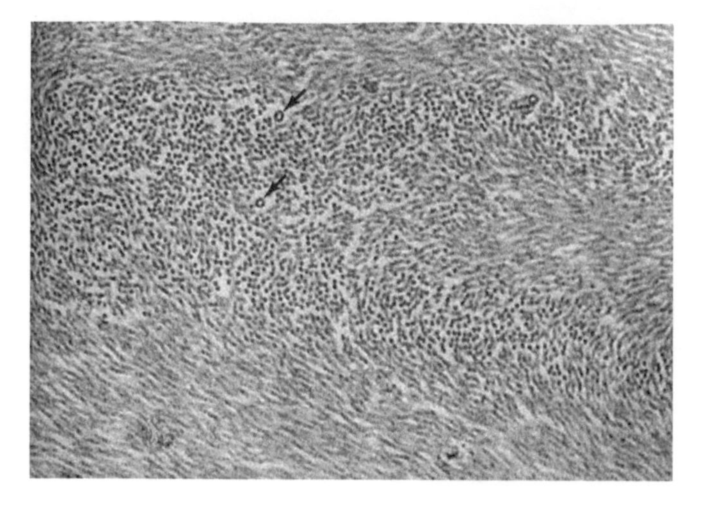

FIGURE A53 An extreme concentration of MICROFILAMENTS strengthening a cell exposed to great tensile stress in the skin of an ammocoete. 80,000 ×.

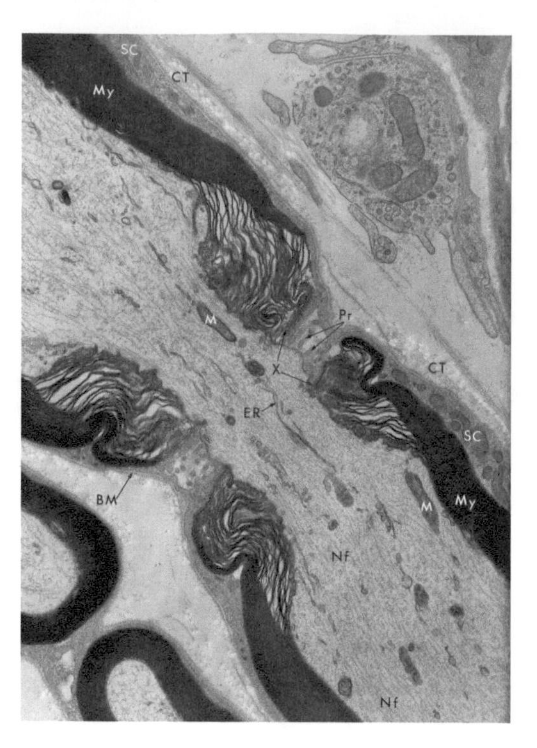

FIGURE A54 Here neurofilaments reinforce a delicate nerve fiber that is subjected to the strains produced by bodily movements. BM - basement membrane; My - myelin sheath; M - mitochondira; SC - Schwann (neurilemma) Cell; Nf neurofilaments; ER - endoplamic reticulum; X - neurilemma folds 22,500 ×.

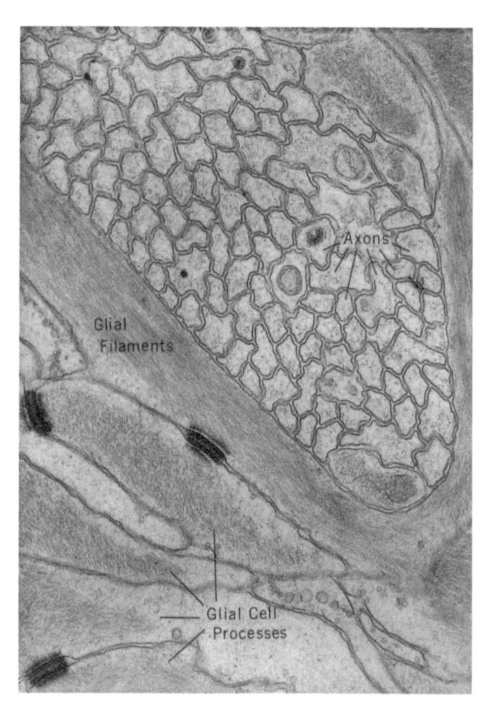

FIGURE A55 The supportive or glial cells of nervous tissue are also reinforced by filaments, the GLIAL FILAMENTS. 40,000 ×

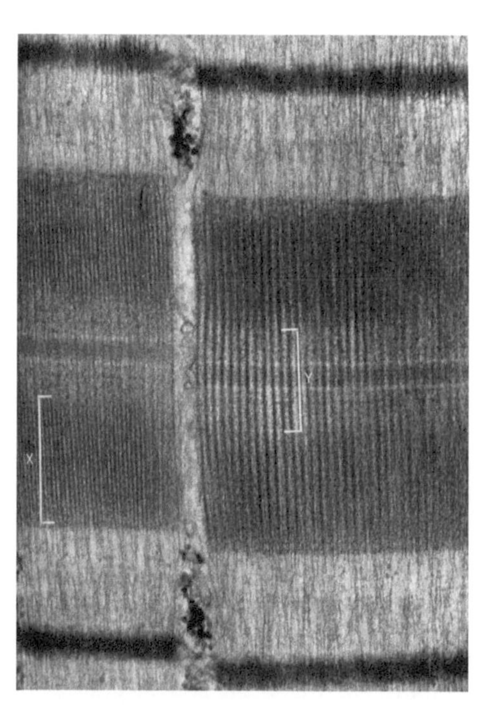

FIGURE A56 Myofilaments of muscle cells are precisely overlapping rodlets of the fibrous proteins, ACTIN and MYOSIN. X indicates the portion of the A-band where actin and myosin are both present. Y indicates the H-band where only myosin fibers are present. 78,000 ×.

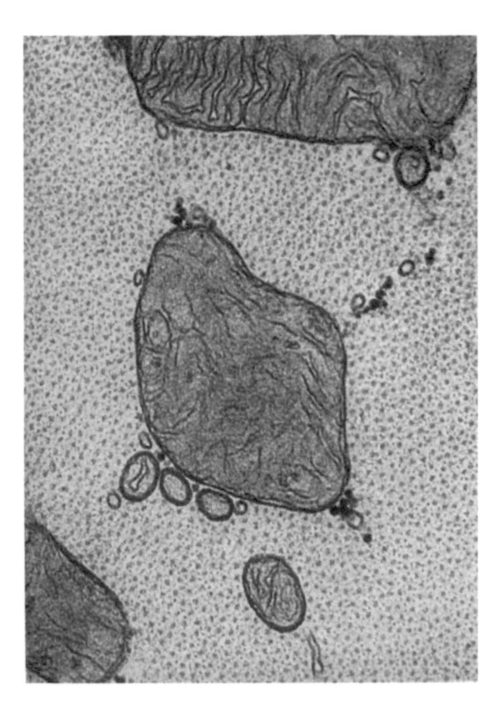

FIGURE A57 The precise arrangement of the filaments of ACTIN and MYOSIN is seen in this cross section of a muscle cell. 117,000 ×.

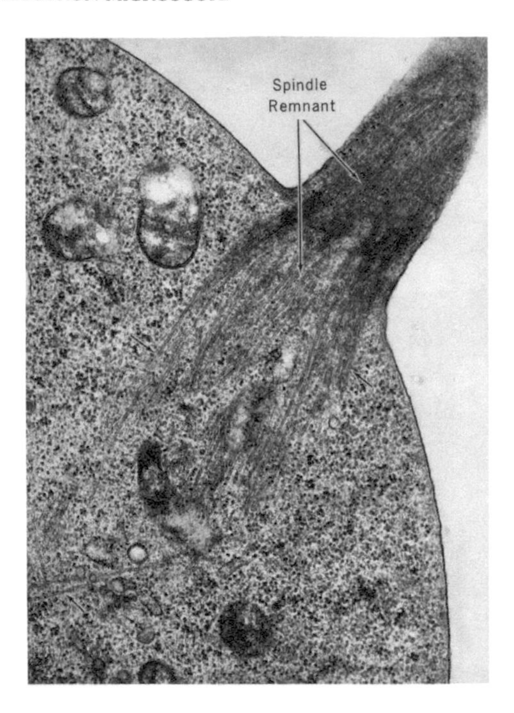

FIGURE A58 MICROTUBULES form the spindle fibers that are seen during mitotic division. This is a newly formed blood cell in the bone marrow of a guinea pig where remnants of the spindle fibers remain after division has occurred. 44,000 ×.

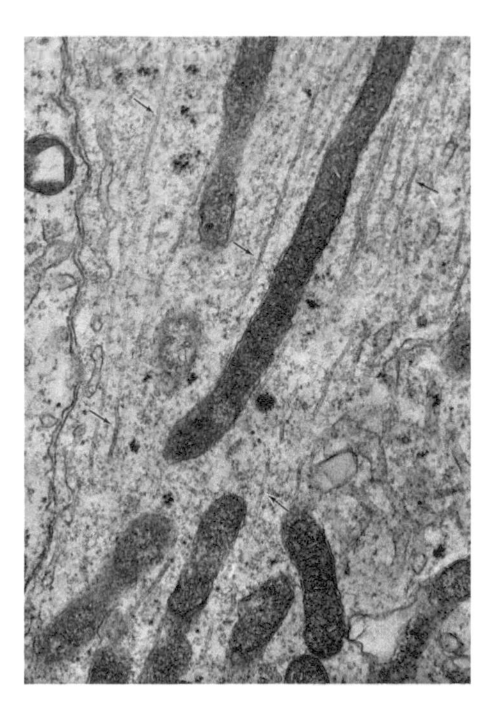

FIGURE A59 Microtubules sometimes appear apart from cell division: they provide support in this hormone-secreting cell of the testis of a guinea pig. 50,000 ×.

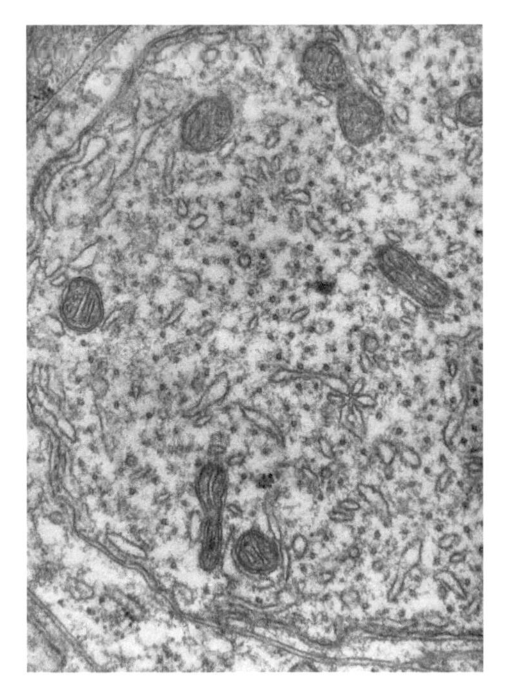

FIGURE A60 Cross sections of microtubules seen in the process of a nerve cell. These strengthen the long, delicate process against the pulls and yanks of everyday movement. A few mitochondria are scattered throughout. 72,000 ×.

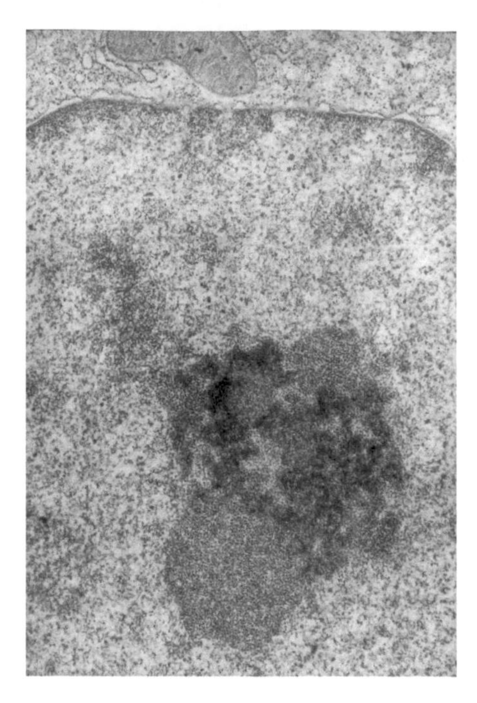

FIGURE A61 This is a section of a cell from the pancreas of a frog. The NUCLEAR ENVELOPE, consisting of two layers with the PERINUCLEAR SPACE between them, passes across the micrograph near the top. You may be able to distinguish NUCLEAR PORES that are spanned by a fine diaphragm. Most of the micrograph is occupied by the NUCLEOPLASM, which contains fine granules of CHROMATIN. The dark blob is the NUCLEOLUS; it consists of two parts, the PARS GRANULOSA and a dense, central area, the PARS FIBROSA of tightly packed filaments of ribonucleic acid. 40,000 ×.

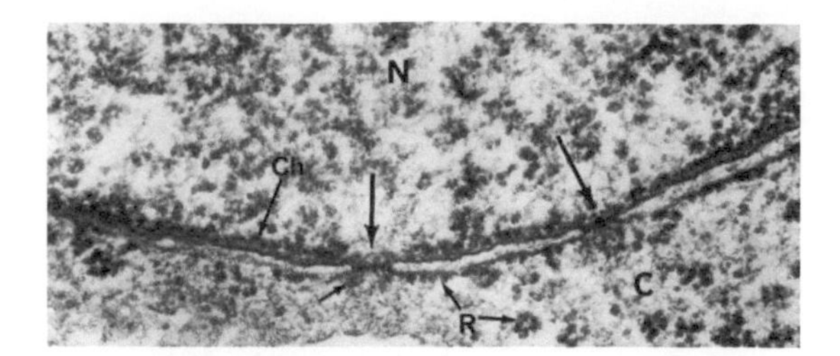

FIGURE A62 Thin DIAPHRAGMS span the NUCLEAR PORES in the NUCLEAR ENVELOPE. The outer and inner membranes of the nuclear envelope are continuous at the periphery of the nuclear pores. 90,000 ×.

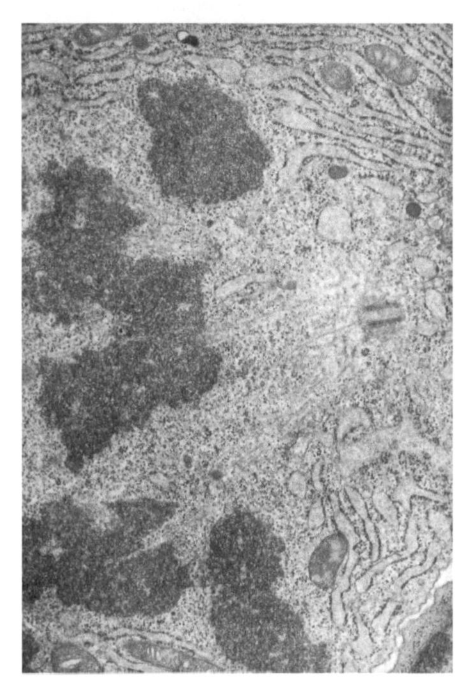

FIGURE A63 Cell of the connective tissue from a rat intestine. When mitosis begins, the chromatin aggregates into larger and larger masses, the CHROMOSOMES. Note MICROTUBULES of the spindle apparatus radiating from the CENTRIOLE, seen halfway up on the right. GRANULAR ENDOPLASMIC RETICULUM is scattered throughout the micrograph, especially on the right side. 40,000 ×.

Nucleus

The NUCLEAR ENVELOPE is porous and consists of two layers, which enclose the PERINUCLEAR SPACE between them (Fig. A61). High magnifications show that each layer displays the trilaminar structure of the unit membrane. The round or octagonal NUCLEAR PORES are about 70 nm in diameter and are spanned by a fine diaphragm (Fig. A62). The NUCLEOPLASM is a homogeneous matrix composed mainly of proteins (including various enzymes), soluble ribonucleic acid, and ribosomes. While the NUCLEOLUS appears as a solid body with the light microscope it is a nonmembranous structure contained within the nucleus of animal cells and consists of granular material, the PARS GRANULOSA of ribonucleic acid granules and a dense central area, and the filamentous and the PARS FIBROSA, of tightly packed filaments of ribonucleic acid. The pars granulosa and pars fibrosa form a network called the NUCLEOLONEMA.

During interphase, when the chromosomes are dispersed in the nuclear sap, fine CHROMATIN GRANULES disposed on a filamentous mesh are seen throughout the nucleus. When mitosis begins, the chromatin aggregates into larger and larger masses, the CHROMOSOMES (Fig. A63).

FREEZE FRACTURE

Much has been learned about the structure of cells from the technique of FREEZE FRACTURE where a piece of tissue is frozen in liquid nitrogen, transferred to a high vacuum, and then split by a sharp swipe of a blade, much the way firewood is split with an axe (Figs. A64 and A65). Often the tissue splits along the hydrophobic portions of membranes, revealing their surface structure. Carbon replicas of the surface, "shadowed" with platinum, and viewed in the transmission electron microscope, reveal beautiful images of irregularities in the surface (Fig. A66).

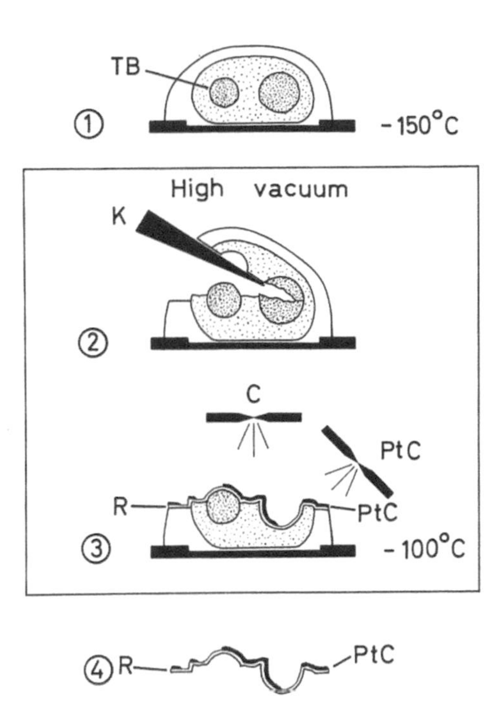

FIGURE A64 FREEZE FRACTURE. (1) A block of tissue (TB) is quickly frozen in liquid nitrogen and transferred to a high vacuum. (2) The specimen is fractured with a sharp swipe of a blade, much the way firewood is split with an axe. (3) A thin film of carbon (C) is deposited from above to produce a replica (R) of the contours of the tissue and the replica is shadowed at an angle with platinum–carbon (PtC). (4) The specimen is warmed and the tissue removed from the replica with strong acid. The replica is placed on a grid for examination with the transmission electron microscope.

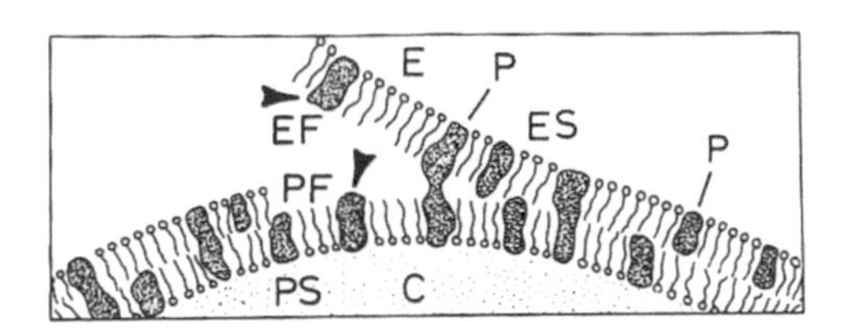

EF=extraplasmatic fracture face
ES=extraplasmatic surface
PF=plasmatic fracture face
PS=plasmatic surface

FIGURE A65 The plane of fracture may be of two kinds: some occur randomly and do not follow natural boundary lines, exposing such structures as the cytoplasm, nucleoplasm, or extracellular space. In others, the tissue fractures along the line of least resistance: the hydrophobic portion of the trilaminar unit cell membrane of the cell, the nucleus, or cytoplasmic organelles. The cell coat will adhere to the external surface (ES) of the external leaflet (EF) of the membrane and integral proteins (P) will adhere to both the internal (PF) and external leaflets. Thus four surfaces of a unit membrane are exposed:

ES: The extraplasmatic surface, the outer surface of the plasmalemma, which is in contact with the external milieu.

EF: The inner hydrophobic fracture face of this outer lamina—it faces in the direction of the cytoplasm;

PF: The outer hydrophobic fracture face of the inner lamina—it faces away from the cytoplasm;

PS: The inner surface of the inner lamina that is in contact with the cytoplasm.

Particles of intramembranous PROTEINS (P) are entrapped in both of the laminae—more particles are entrapped in the plasmatic lamina than in the extraplasmatic lamina.

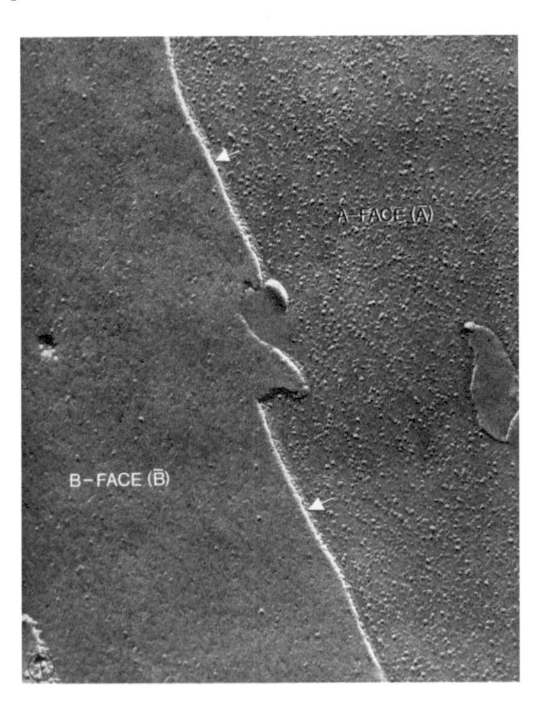

FIGURE A66 This is an image by transmission electron microscope of a replica of the cell membrane of an endocrine cell from the pancreas of a rat. The outer surface of the inner leaflet—the PLASMATIC FRACTURE FACE—is shown at the right (A). It is studded with many INTRAMEMBRANOUS PARTICLES of protein. The fracturing jumps from one leaflet to the other in the diagonal line across the micrograph and the left is occupied by the EXTRAPLASMATIC SURFACE (B) that exhibits few protein particles.

The arrow at the bottom of the micrograph indicates the direction of shadowing with platinum–carbon. Without this arrow, depressions would be indistinguishable from elevations—turn the picture upside-down and notice the effect. It is a convention that freeze-fracture micrographs are always displayed with the direction of shadowing going from the bottom to the top: i.e., the arrow should always point up. 163,000 ×.

C H A P T E R

B

Cell Division

All cells arise from preexisting cells by a process of cell division. Many organs of the mature animal, however, show few cells in division and it is sometimes said that there are none in the central nervous system. In many tissues the average period between divisions is measured in years but frequent divisions occur in the skin, the lining of the gut, and in the gonads. Rapidly growing tissues are the most favorable for the study of cell division. A typical cell division consists of a division of nuclear material (KARYOKINESIS) followed by a division of the cell body (CYTOKINESIS), each of the daughter cells receiving one of the two daughter nuclei. In certain cells, however, karyokinesis may occur without cytokinesis, resulting in binucleate or multinucleate cells. Karyokinesis is accomplished in most cells by MITOSIS.

MITOSIS

Mitosis is the shortest part of the CELL CYCLE: the M PHASE, when mitosis and cytokinesis divide the nucleus and cytoplasm (Fig. B1a). Between successive mitotic divisions is an INTERPHASE when the cells grow and copy their chromosomes in preparation for cell division. There are three phases of interphase when cells grow by synthesizing proteins and producing cytoplasmic organelles: the G_1 PHASE, the S PHASE, and the G_2 PHASE. The CHROMOSOMES, which carry the genetic material of the cell, are replicated only in the S phase (S for synthesis of DNA). There is considerable variation in the length of the phases in different cell types but the length of the arrows in Fig. B1a represents the duration of each phase in a typical cell where the entire cycle lasted about 24 hours.

In mitosis there is an almost equal qualitative and quantitative division of genetic material. Extremely rarely, division of nuclear material is not perfectly equal and mutations occur, thereby producing changes in genetic inheritance. Mutations are the raw material that is essential for evolution to occur. Mitosis is a continuous process but, for convenience in description, the mitotic cycle is divided into five arbitrary stages PROPHASE, PROMETAPHASE, METAPHASE, ANAPHASE, and TELOPHASE. The basic process has been understood for over 100 years and the illustration from Maximow and Bloom's first edition of 1930 is still useful in the study of these stages (Fig. B1b). The photomicrographs of dividing cells are taken from three widely differing sources: dividing cells during cleavage in the egg of a whitefish (Figs. B2a–B2l), cleavage in egg cells of a parasitic roundworm, *Ascaris megalocephala* (Figs. B3a–B3k), and dividing cells within the epidermis of a salamander, *Ambystoma* sp. (Figs. B4a–B4j). The fish egg was chosen because fine cytoplasmic particles within the dividing cells help to delineate the fibers of the mitotic spindle; the mitotic minuet is easy to follow in the cells of *Ascaris* because there are only four pairs of chromosomes taking part; and the epidermal cells of the salamander present mitosis as it appears in actively dividing cells of a growing animal. The developing egg cells are shown in sections while the epidermal cells are seen in a whole mount of the skin of a salamander. The process of mitosis is similar in all dividing cells of animals. In all of the preparations, the chromosomes have been stained a blue/black color with iron hematoxylin.

Interphase

Any cell not in the process of division is in INTERPHASE. Look for cells with ordinary spherical nuclei with a nuclear membrane. In the cytoplasm outside the nuclear membrane the centrosome (or MICROTUBULE ORGANIZING CENTER) splits during early interphase and microtubules radiate from the two resulting centrosomes to form arrays called ASTERS. In the nucleoplasm, the deeply stained CHROMATIN is in the form of fine, tangled threads

An Atlas of Comparative Vertebrate Histology.
DOI: https://doi.org/10.1016/B978-0-12-410424-2.00002-0

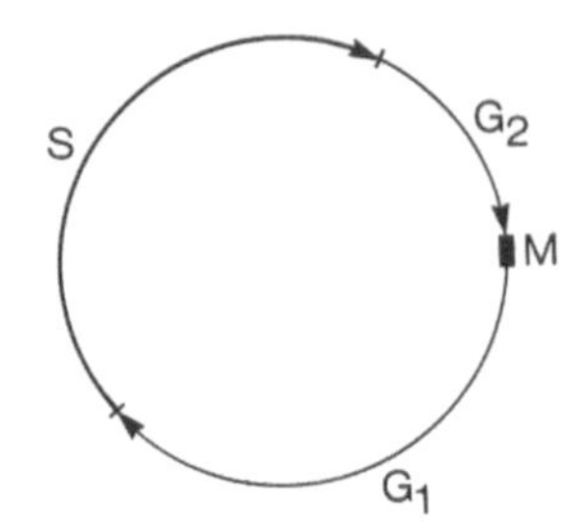

FIGURE B1a The cell cycle is made up of two parts: MITOSIS (M) and the much longer INTERPHASE that occurs in three phases, G_1, S, and G_2. There is considerable variation in the length of the phases in different cell types; the length of the arrows in the diagram represents the duration of each phase in a typical cell whose entire cycle lasted about 24 hours. During all three phases of interphase, the cell grows by synthesizing proteins and producing cytoplasmic organelles. Chromosomes are duplicated during the S phase (*S* for synthesis of DNA).

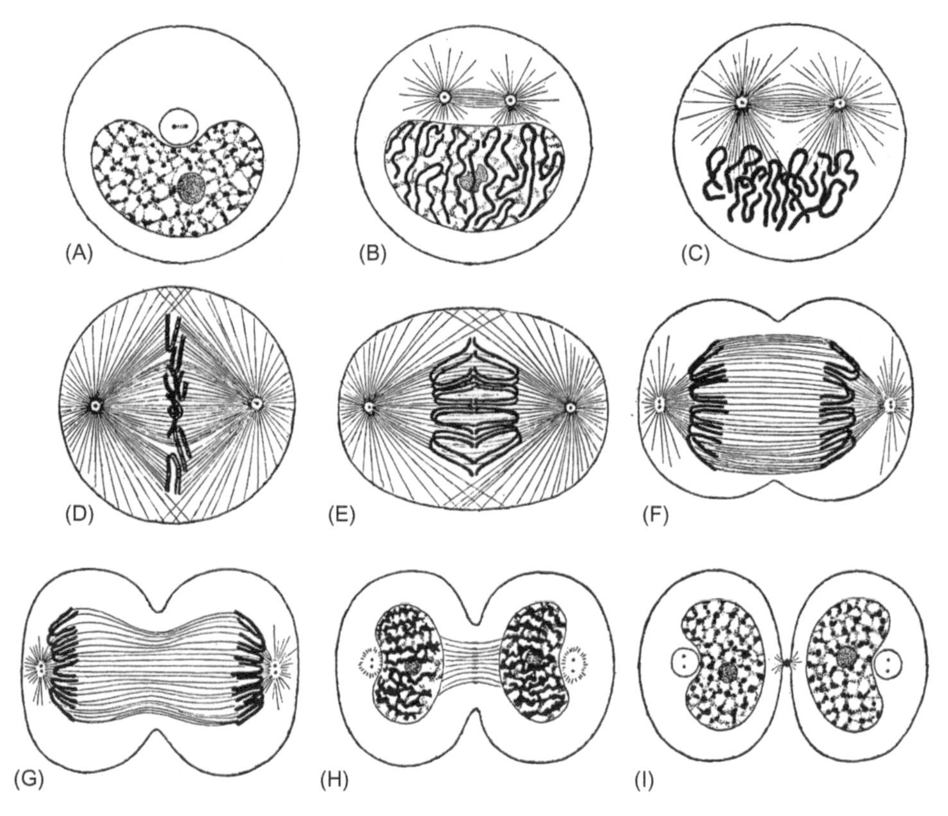

FIGURE B1b Diagram from the first edition of Maximow and Bloom's *Text-Book of Histology*, 1930, showing the stages of mitosis. (A) RESTING CELL; (B) PROPHASE; (C) transition from prophase to metaphase; (D) METAPHASE with longitudinal splitting of chromosomes; (E–G) ANAPHASE; (E) beginning of separation of the split chromosomes; (F) movement toward the poles; (G) chromosomes approaching the centrioles; (H) TELOPHASE with reconstruction of the daughter nuclei; (I) daughter cells connected only by an INTERMEDIATE BODY.

distributed in the pale HYALOPLASM. This meshwork gives the erroneous impression of a lack of organization although permanent CHROMOSOMES are present in an extremely extended condition. The chromosomes have already replicated during the S phase of the cell cycle. Each chromosome consists of a pair of loosely coiled strands, the CHROMATIDS. Portions of the strands may be tightly coiled and stain deeply with basic dyes and are readily visible with the light microscope as HETEROCHROMATIN, or CONDENSED CHROMATIN, and are in an inactive phase. The loose coils, invisible with the light microscope, constitute the EUCHROMATIN or EXTENDED CHROMATIN and direct the transcription of DNA through the formation of messenger RNA (mRNA); in this way euchromatin controls the production of proteins. Varying amounts and regions of chromatids will be active in different types of cells so that some nuclei will appear darker than others although, in a given species, all nuclei contain the same

FIGURE B2(a–l) Mitosis in the blastodisc of a whitefish. The following are micrographs of sections of rapidly dividing cells in the blastodisc of a whitefish, stained with iron hematoxylin, showing various stages of mitosis. These specimens were chosen because fine cytoplasmic particles within the dividing cells help to delineate the fibers of the mitotic spindle. 100 ×.

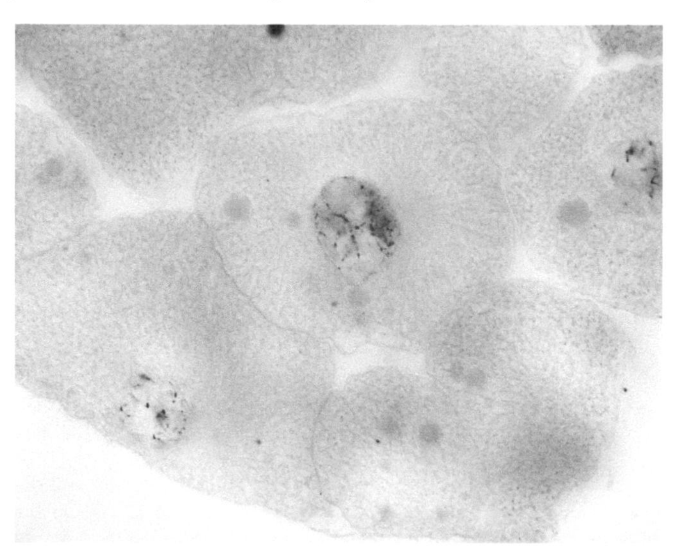

FIGURE B2a Section of the blastula of a whitefish showing an INTERPHASE nucleus near the center. The nuclear membrane is visible.

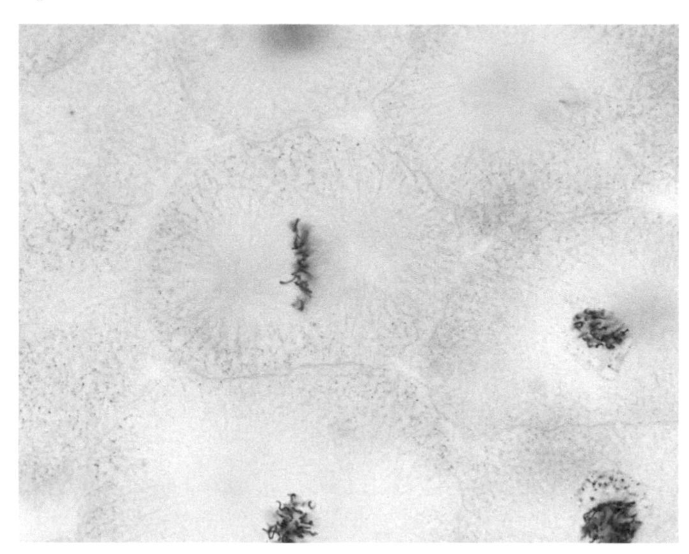

FIGURE B2b At the center is a cell in METAPHASE. Spindle fibers extend from the centrosomes at the poles (not visible) to the chromosomes lined up at the METAPHASE PLATE. Three cells in PROPHASE are shown at the lower right.

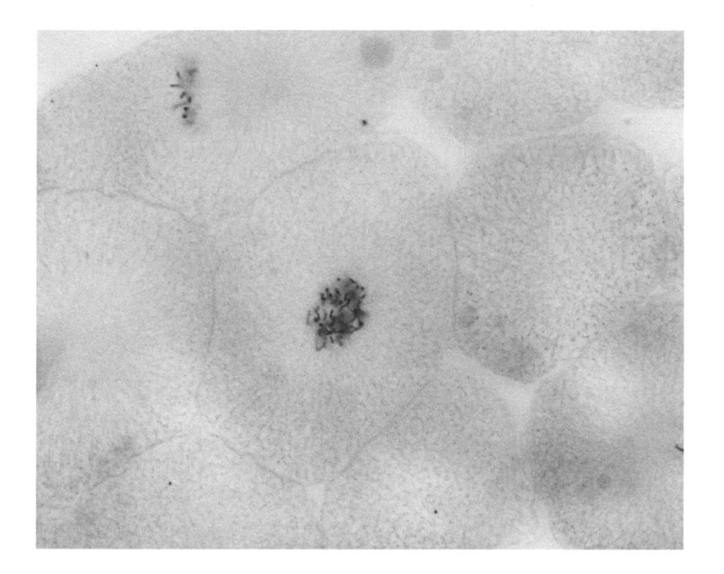

FIGURE B2c A PROPHASE nucleus is shown at the center. Chromosomes are visible inside the nucleus. Chromosomes have lined up at the metaphase plate in the cell at the top left.

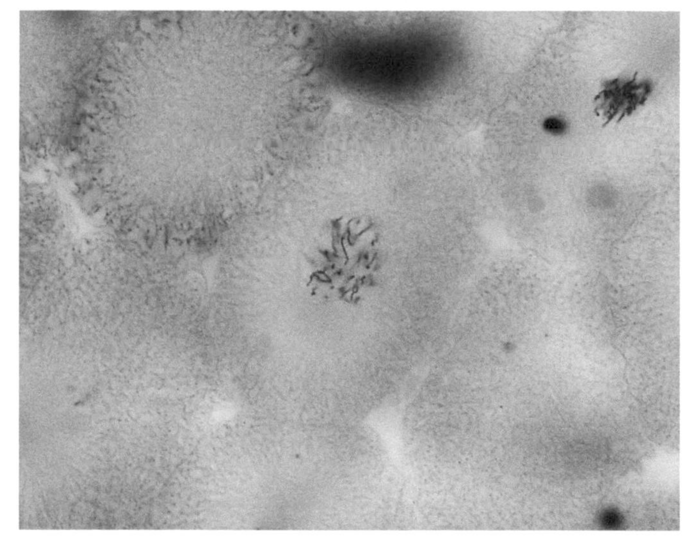

FIGURE B2d A PROPHASE nucleus is at the center. The nuclear membrane is faint. A later stage of prophase is seen at the upper right.

total amount of chromatin. The greater the proportion of active chromatin that is present in a nucleus, the paler the nucleus will appear.

Only one X-chromosome may function in a cell and, when there are two X-chromosomes present (as in cells of the mammalian female), one of them remains condensed and inactive throughout interphase and constitutes the SEX CHROMATIN or BARR BODY (Fig. B5).

Prophase

During prophase the nucleoli disappear and the chromatin fibers thicken and shorten to form discrete chromosomes visible with the light microscope. Each replicated chromosome appears as two identical chromatids joined at the CENTROMERE.

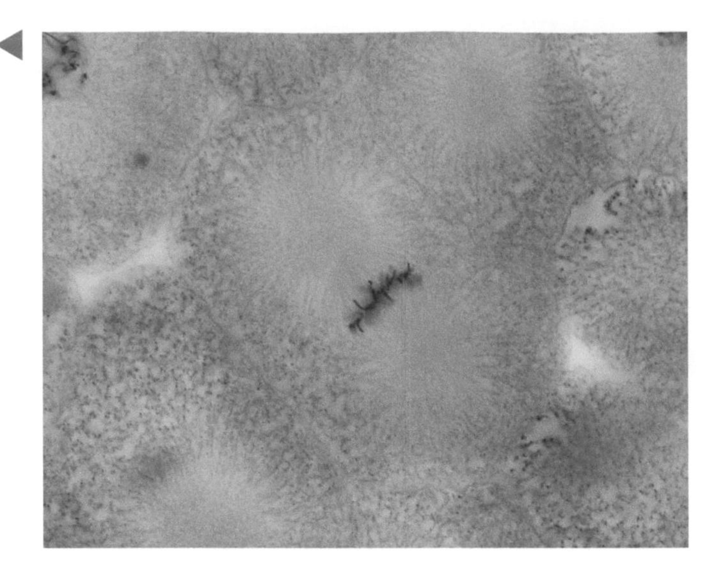

FIGURE B2e METAPHASE. Spindle fibers extend from the centrosomes at the poles (not visible) to the chromosomes lined up at the metaphase plate.

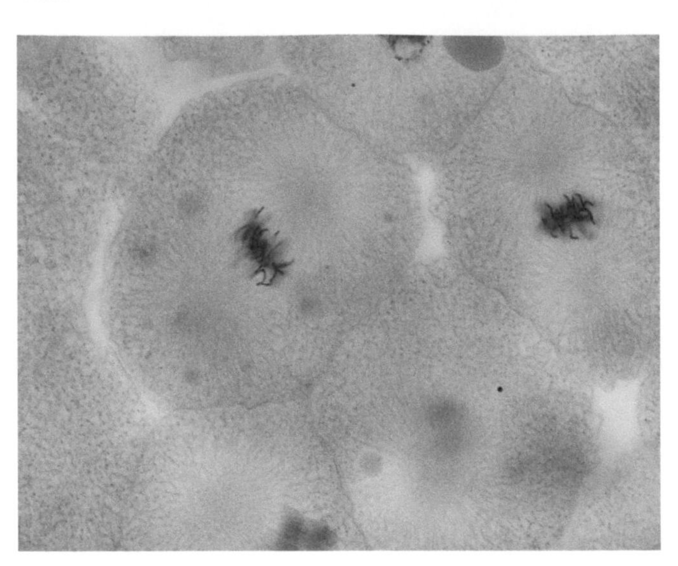

FIGURE B2f Two cells in METAPHASE are shown. In both, spindle fibers extend from the centrosomes at the poles (not visible) to the chromosomes lined up at the metaphase plate.

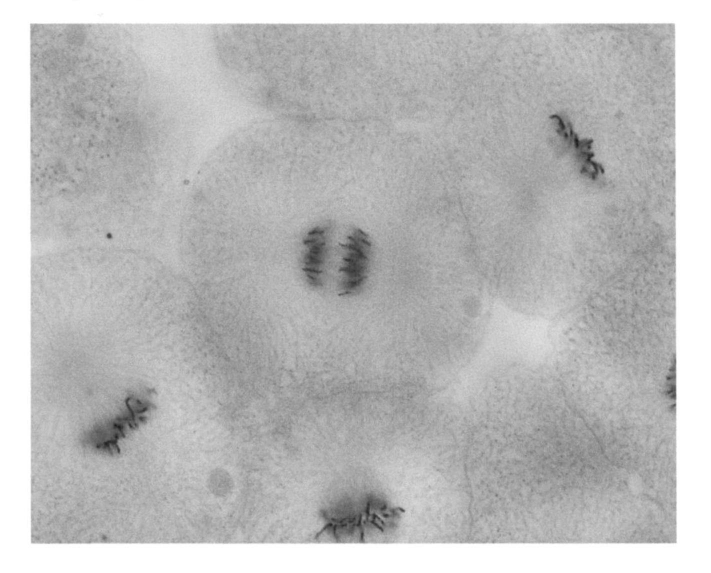

FIGURE B2g The chromosomes of the ANAPHASE nucleus in the center cell are pulling apart. Three cells with chromosomes lined up at the metaphase plate are seen at the edges of the micrograph.

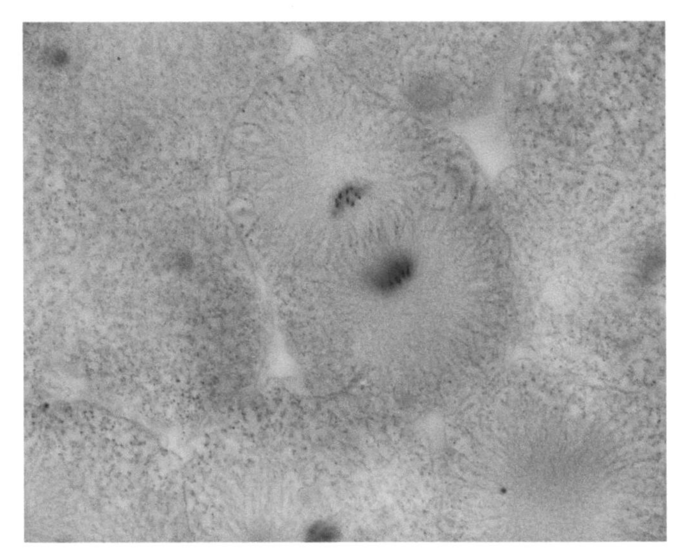

FIGURE B2h TELOPHASE. It is possible to imagine that the cell membrane is contracting at the equator of the cell (especially at the left) at the beginning of CYTOKINESIS.

The chromatids shorten and thicken and become more tightly coiled; the individuality of the separate chromosomes becomes clear. As prophase continues it can be seen that each chromosome consists of two parallel, touching chromatids held together at the centromere. In the adjacent cytoplasm the small, dark centrosome appears surrounded by a dense, "corona" of radiating fibrils, the ASTER. The centrosome divides and the daughter centrosomes begin their migration to the opposite poles of the cell, apparently propelled by the lengthening bundles of microtubules between them. Each daughter centriole is surrounded by an aster of radially arranged fibrils and some of these fibrils extend between the centrioles to form the SPINDLE. (The appearance of the spindle resembles the distribution of iron filings between the poles of a magnet.) The fibrils are seen to good advantage in the blastoderm of the fish where they are outlined by the presence of cytoplasmic granules. The fibrils are seen with the electron microscope to be microtubules and microtubule-associated proteins.

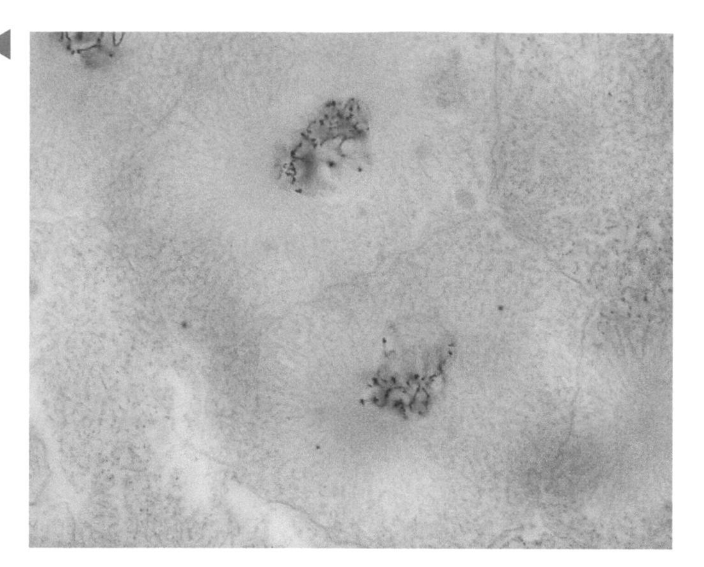

FIGURE B2i LATE TELOPHASE. Cleavage is complete and plasma membranes separate the daughter cells. Nuclear membranes are beginning to form around the separate masses of chromosomes.

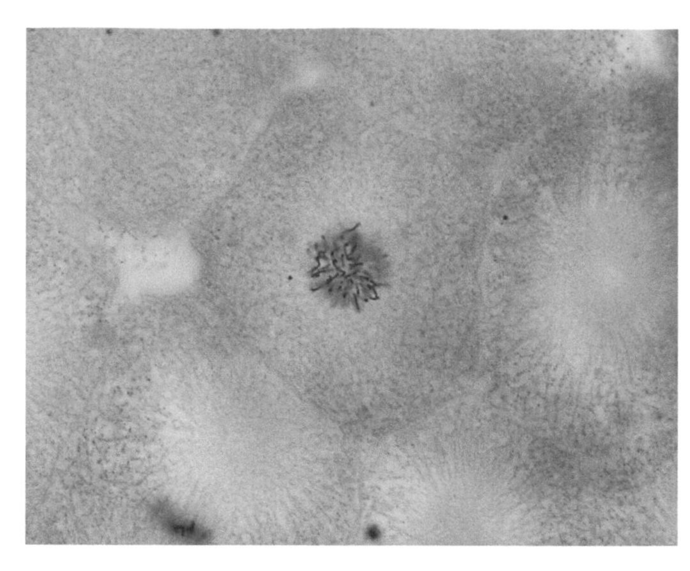

FIGURE B2j PROPHASE. Chromosomes are visible inside the nucleus.

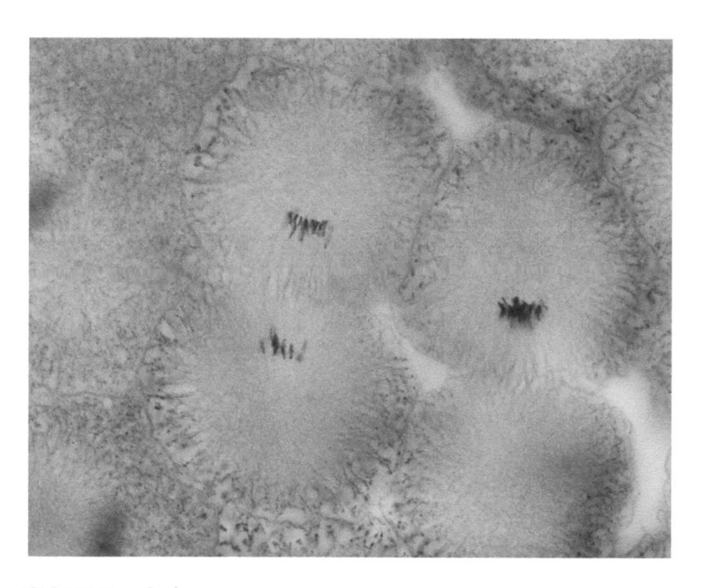

FIGURE B2k TELOPHASE. Cytokinesis is well underway. Cytokinesis is more advanced in the cell at the right; in this oblique section of the two daughter cells, the chromosomes are seen in only one of the cells.

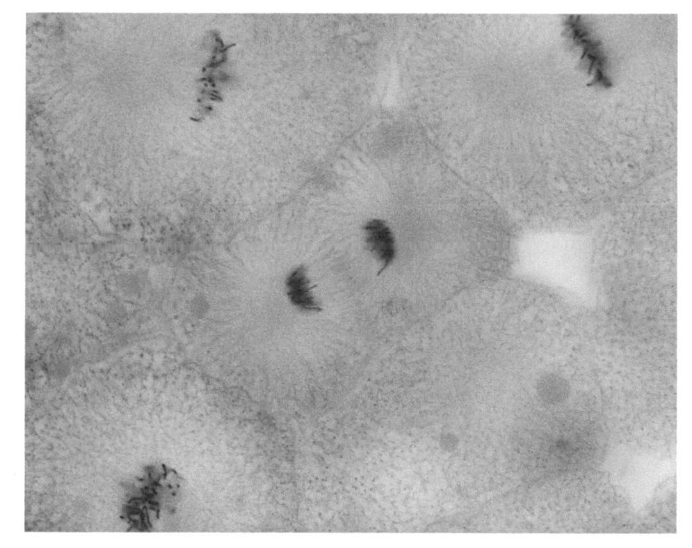

FIGURE B2l ANAPHASE at the center. Two cells in METAPHASE are shown at the top. In both, spindle fibers extends from the centrosomes at the poles (not visible) to the chromosomes lined up at the metaphase plate. A cell in PROPHASE is at the lower left.

Prometaphase

The nuclear membrane fragments during prometaphase and the MITOTIC SPINDLE, consisting of microtubules and associated proteins, forms between the two centrioles. Bundles of microtubules from the centrioles form SPINDLE FIBERS that extend between the poles of the cell and interact with KINETOCHORES that are located at the centromere region of each chromosome (Fig. B6a). Some spindle fibers attach to the kinetochores and are the KINETOCHORE MICROTUBULES of the mitotic spindle; the NONKINETOCHORE MICROTUBULES extend between the poles of the cell without attaching to the chromosomes.

In electron micrographs, CENTRIOLES appear as cylindrical structures which occur in pairs lying at right angles to each other (Figs. B6b and B7). Each cylinder is closed at one end and consists of nine parallel, overlapping "blades," each blade a simple "pipe of Pan" composed of three fused microtubules. The centrioles divide before

FIGURE B3 (a–k) Mitosis during cleavage in the eggs of the parasitic roundworm, *Ascaris megalocephala*.

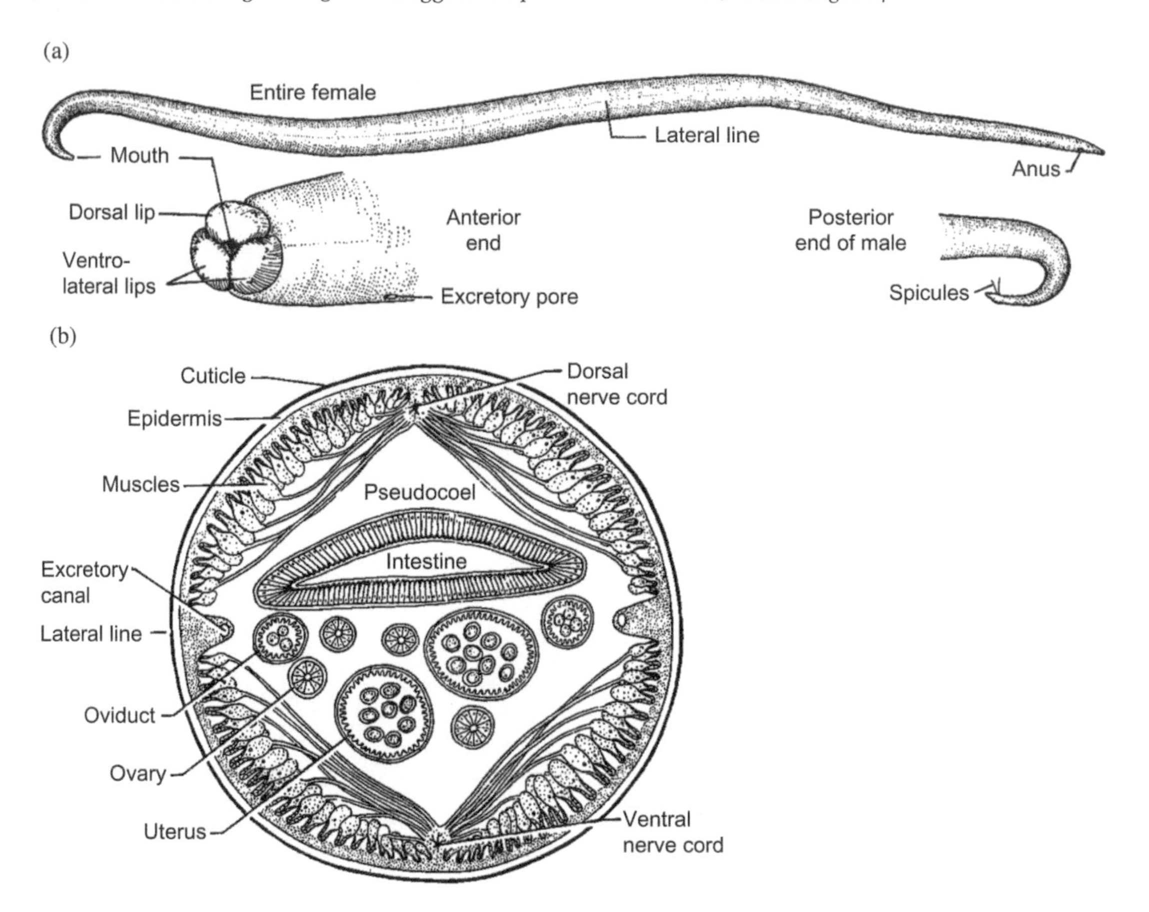

FIGURE B3a and B3b The following are micrographs of early cleavage in eggs in the ovary of a parasitic roundworm, *Ascaris megalocephala*, a parasitic worm of the invertebrate Class Nematoda; although not a vertebrate, it is useful for this study because the presence of only four pairs of chromosomes, which simplifies observation of their separation. The dividing egg cells are still in the uterus of this worm. 100 ×.

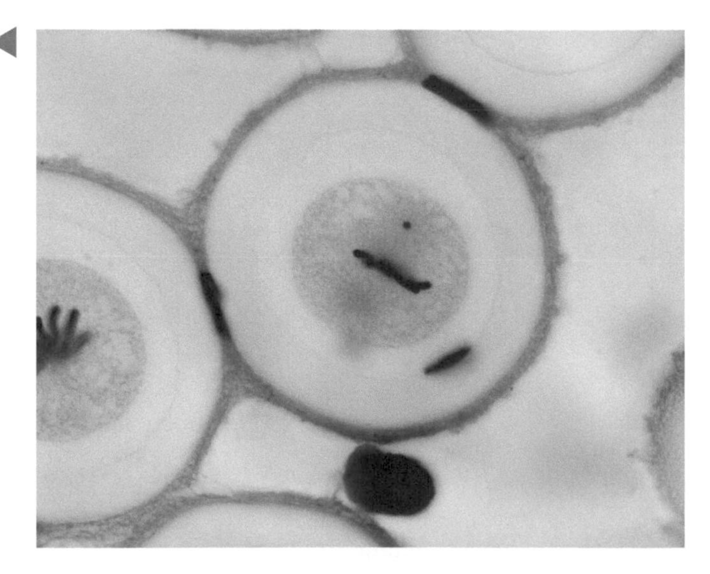

FIGURE B3c METAPHASE. The chromosomes are lined up at the METAPHASE PLATE at the equator of the cell. One CENTRIOLE is visible at the upper right of the cell. Spindle fibers are vaguely visible.

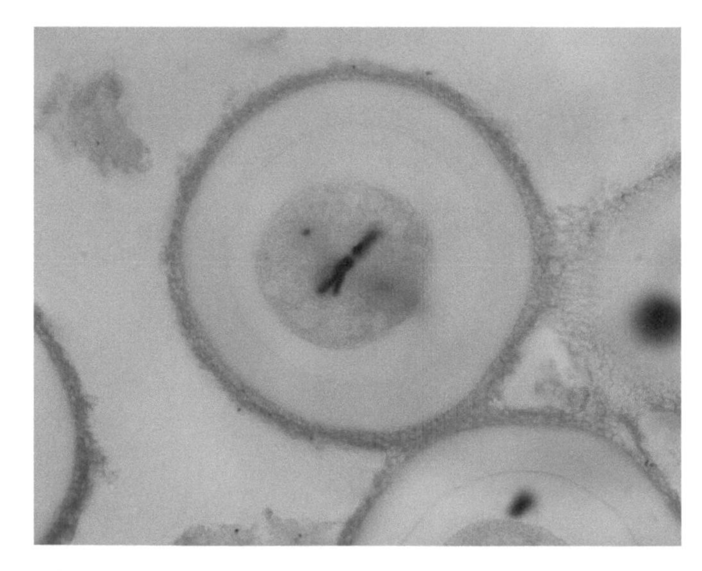

FIGURE B3d METAPHASE. The chromosomes are lined up at the METAPHASE PLATE at the equator of the cell. One CENTRIOLE is visible at the upper left of the cell. Spindle fibers are vaguely visible.

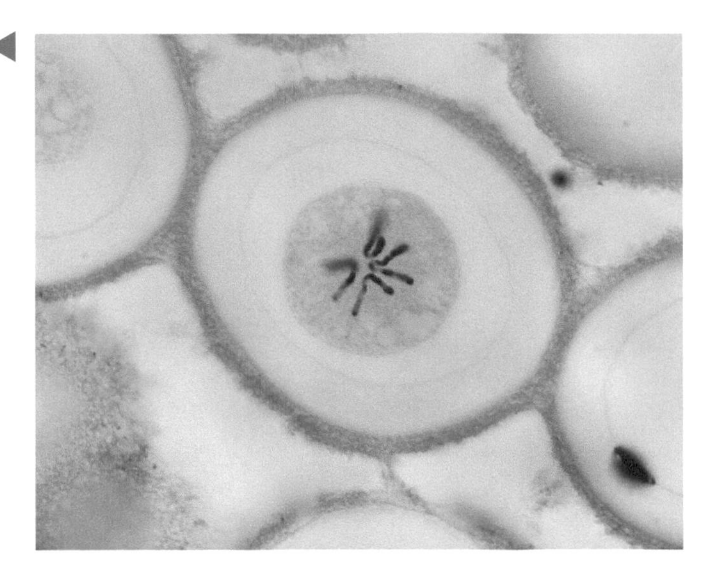

FIGURE B3e METAPHASE. Polar view of the cell with chromosomes lined up at its equator—this is the METAPHASE PLATE.

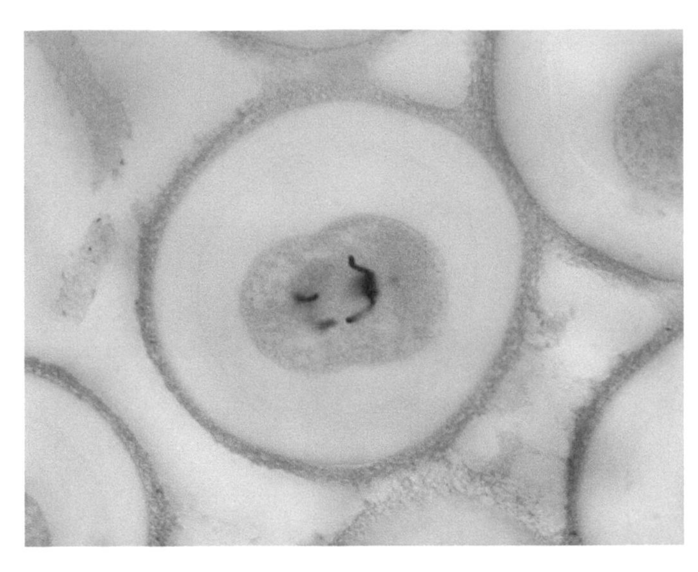

FIGURE B3f Beginning of ANAPHASE. The chromosomes are pulling apart.

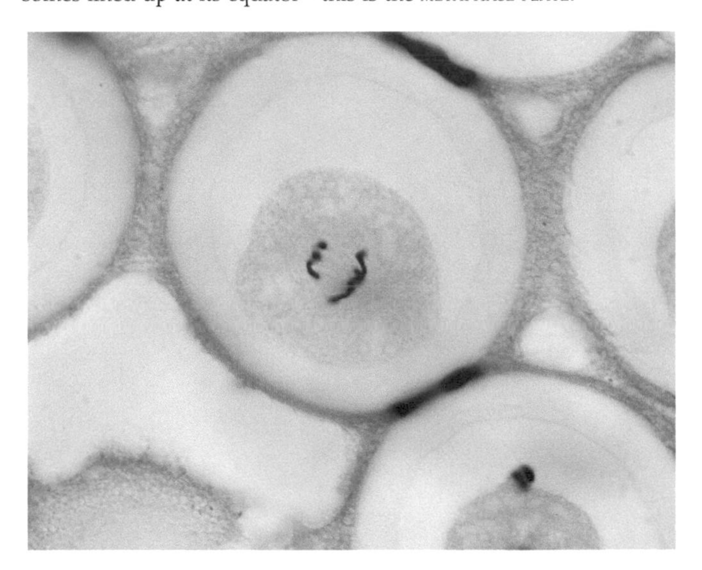

FIGURE B3g ANAPHASE. Chromosomes are pulling apart to the poles of the cell.

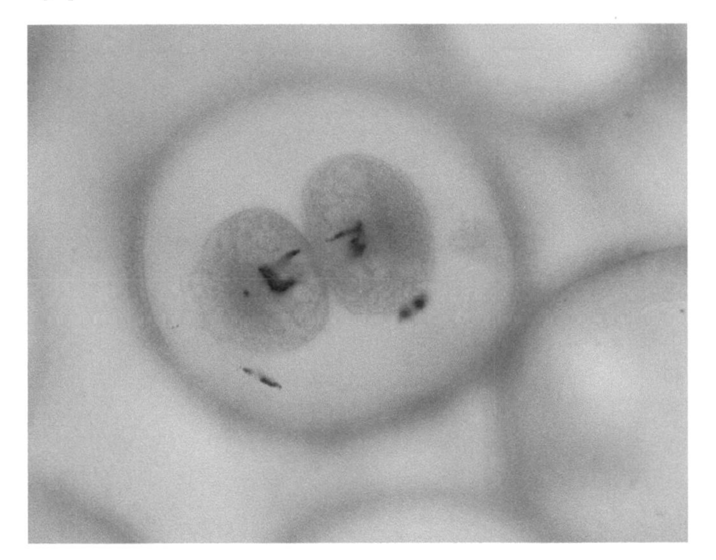

FIGURE B3h TELOPHASE; CHROMOSOMES are separating to the poles—a CENTRIOLE is visible in the left cell. CYTOKINESIS is complete and two cells have formed. The two black blobs at the lower right are POLAR BODIES remaining from the meiotic division that produced the ovum.

prophase and, at the beginning of prophase, four centrioles may be seen together. The daughter centriole is always arranged at right angles to the parent.

Metaphase

The chromosomes arrange themselves on the equatorial plane of the cell, thereby forming the EQUATORIAL PLATE. The centromeres of all the chromosomes align at the equator. The centromeres of all the chromosomes are aligned at the metaphase plate so that the sister pairs of chromosomes face opposite poles of the cell. Thus identical chromatids of each chromosome migrate to opposite poles of the parent cell. Microtubules of the spindle interact with the chromosomes. Bundles of SPINDLE FIBERS, extend from the poles toward the chromosomes at the equator where some attach to KINETOCHORES in the centromere region.

At this time, the chromosomes continue to condense and become even shorter and thicker. In an EQUATORIAL VIEW, the chromosomes appear as a dense line traversing the center of the spindle; in the POLAR VIEW, the

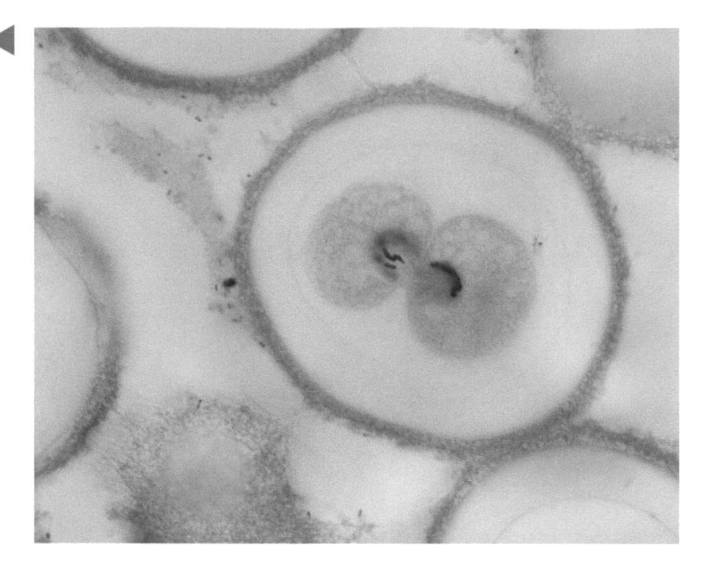

FIGURE B3i TELOPHASE. The chromosomes are approaching the poles and the cells have separated in CYTOKINESIS.

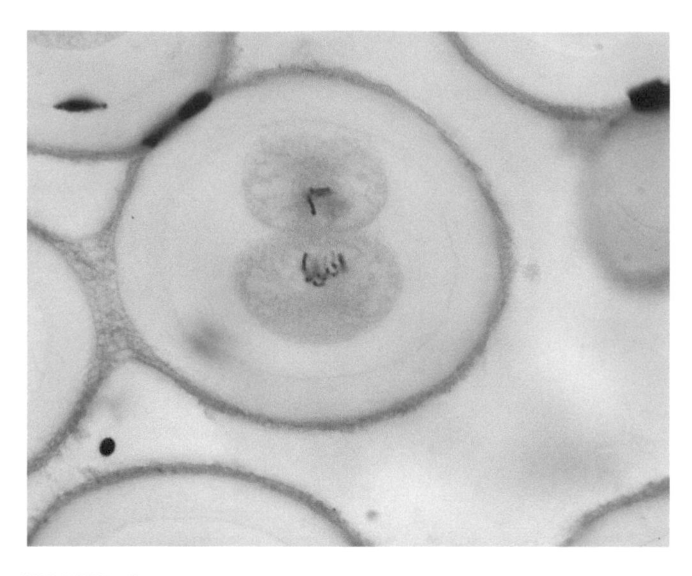

FIGURE B3j TELOPHASE. CYTOKINESIS is complete and two cells have formed.

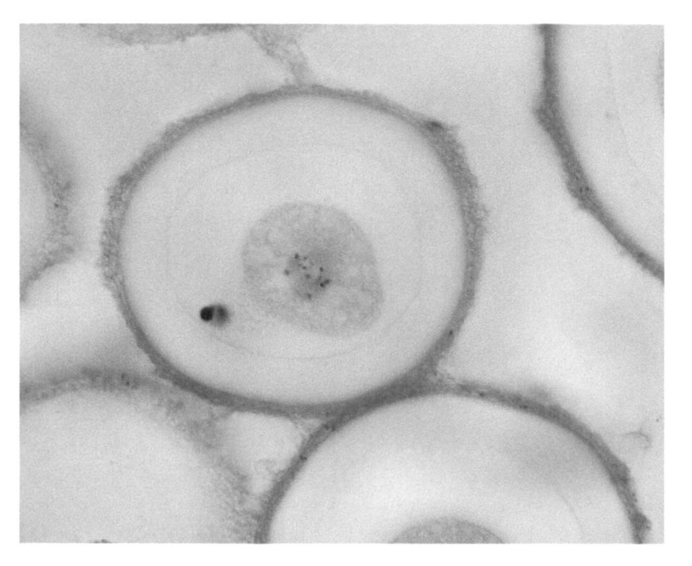

FIGURE B3k INTERPHASE. There has been no condensation to form chromosomes. Two POLAR BODIES are visible at the lower left; the polar bodies are contained within the egg membrane but they are outside the dividing cell. They will disappear.

individual chromosomes are arranged within a circle. Each chromosome develops a longitudinal split along its center or, more correctly, the split between the chromatids becomes visible. This split may be seen to advantage in the whole mounts of the larval skin of *Ambystoma*. The asters are at the poles of the cell and the spindle extends between them. For each chromosome, the kinetochores face opposite poles of the cell.

With the electron microscope, the chromosomes are seen to consist of dark granules associated with a lighter material (Figs. B8 and B9). The dark "granules" are actually sections of strands containing DNA. They resemble the condensed chromatin of the interphase nucleus.

Anaphase

The centromeres divide at the beginning of ANAPHASE so that the two chromatids of each chromosome become separated from each other and are now considered to be chromosomes. The sister chromosomes move toward opposite poles of the cell with their centromeres in the lead. The movements of the chromatids appear to result from two activities within the spindle: the kinetochore fibrils pull the separated chromosomes to the opposite

FIGURE B4 (a–j) Mitosis in the skin of a salamander, Ambystoma.

FIGURE B4a Mitosis is active in the skin of a larval salamander, *Ambydstoma*. (The figure is of an adult.) Whole mounts of the skin have been flattened on a slide, fixed, and stained with iron hematoxylin. 100×.

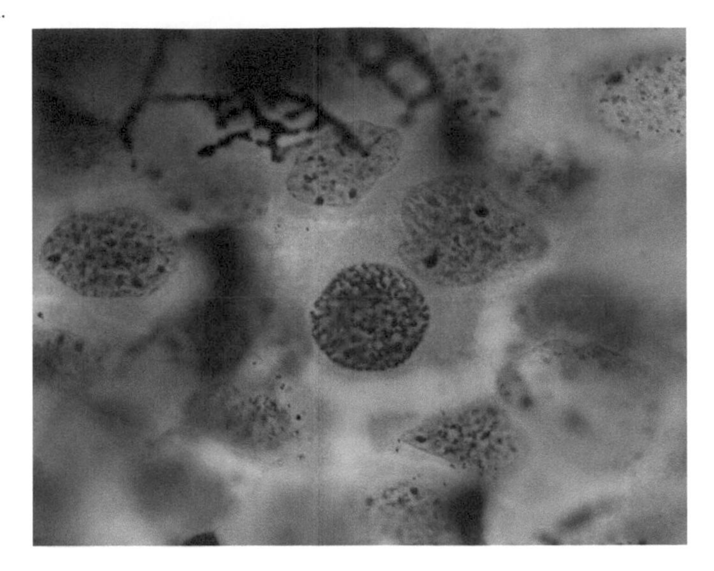

FIGURE B4b Condensation of chromatin has begun in this cell in the PROPHASE of mitosis. Compare its nucleus with those of the surrounding cells. Processes of a pigment cell, packed with pigment granules, can be seen at the top.

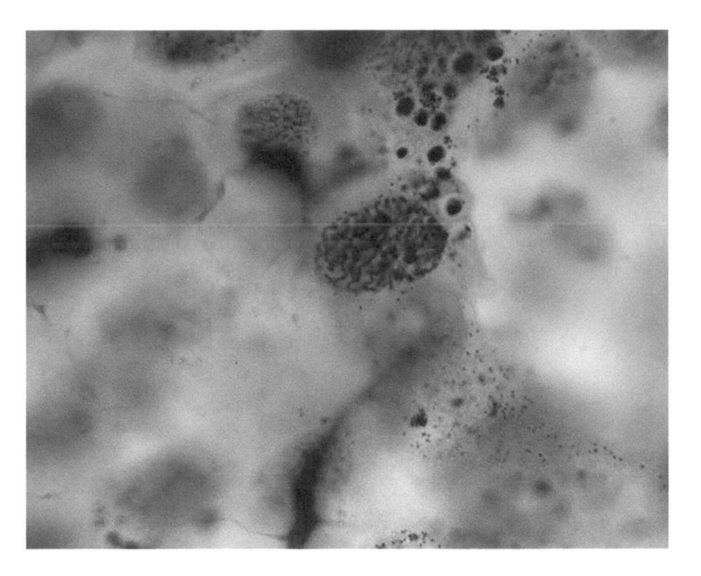

FIGURE B4c Condensation of chromatin has begun in this cell in the PROPHASE of mitosis.

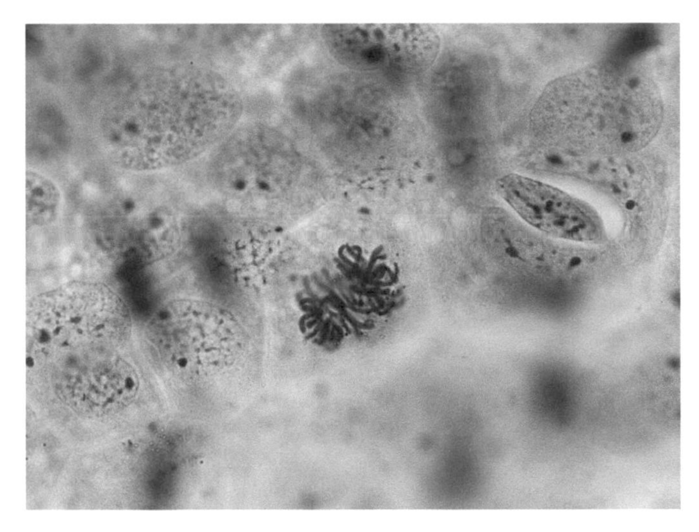

FIGURE B4d Chromosomes have gathered at the metaphase plate.

poles of the cell through their attachment to the centromeres and the elongation and sliding of the polar fibrils pushes the two poles farther apart. By the end of anaphase, the two poles of the cell have equivalent and complete collections of chromosomes.

Telophase and Cytokinesis

The processes of prophase occur in reverse to TELOPHASE. The chromosomes again become diffuse and the spiral threads partly unwind. The nucleolus and the nuclear membrane reappear and the nucleus revert to the interphasic condition with the chromatin appearing as delicate threads. The nuclear membrane reforms from fragments of the parent cell's nuclear envelope as well as other portions of the inracellular membranes. The CLEAVAGE FURROW across the equator is complete and the future CELL MEMBRANE appears across the cell through the center of the spindle. Two daughter cells similar to the original one have been produced. The mitochondria and the Golgi

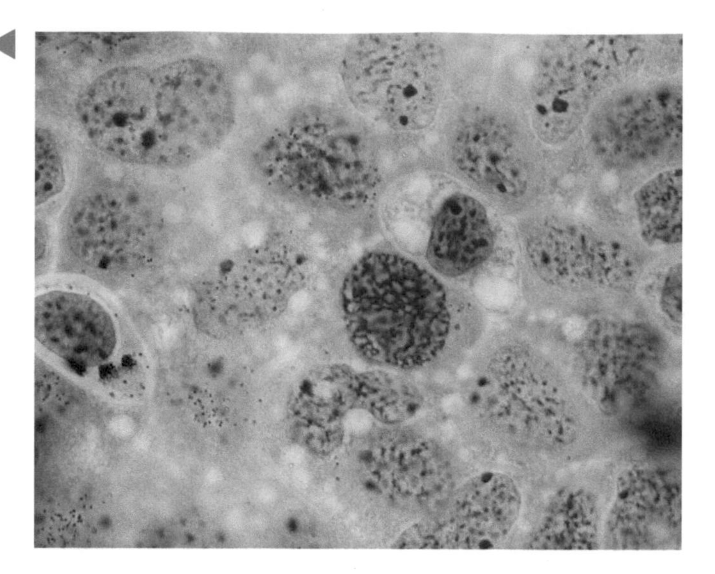

FIGURE B4e Most of the nuclei in this area are in INTERPHASE of mitosis, but chromatin material has begun to condense in the cell at THE CENTER AND PROPHASE HAS BEGUN.

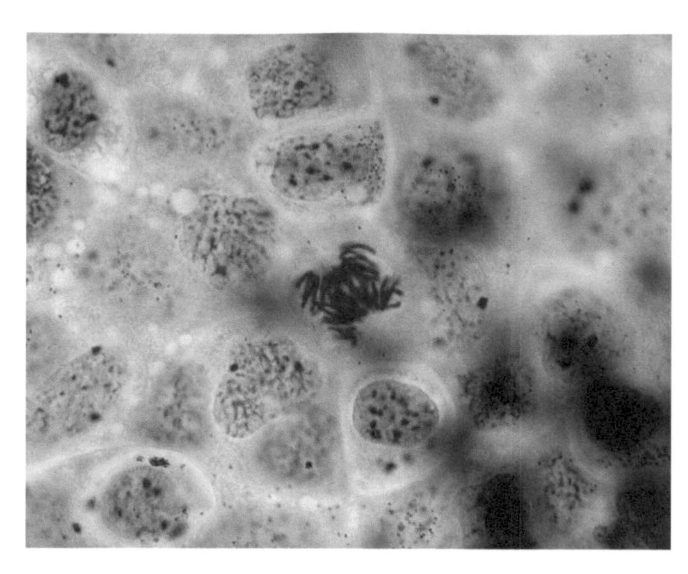

FIGURE B4f Chromosomes are pulling toward the poles in the center of the field in ANAPHASE of mitosis.

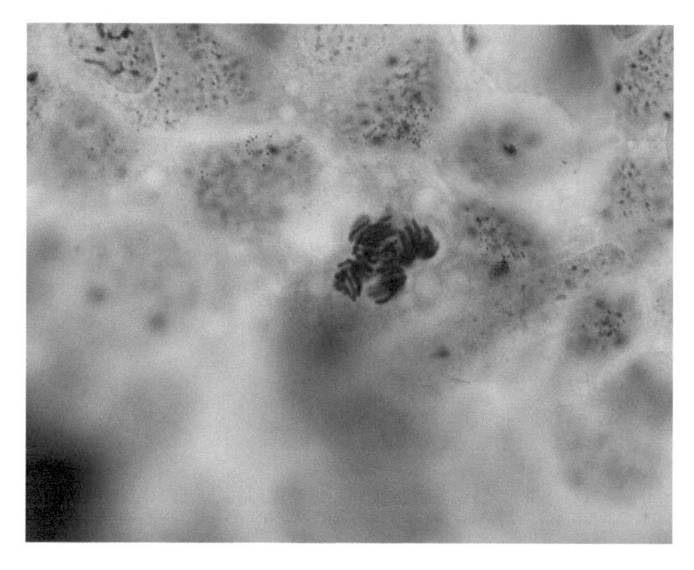

FIGURE B4g The chromosomes have just begun to pull apart and ANAPHASE is beginning in this cell.

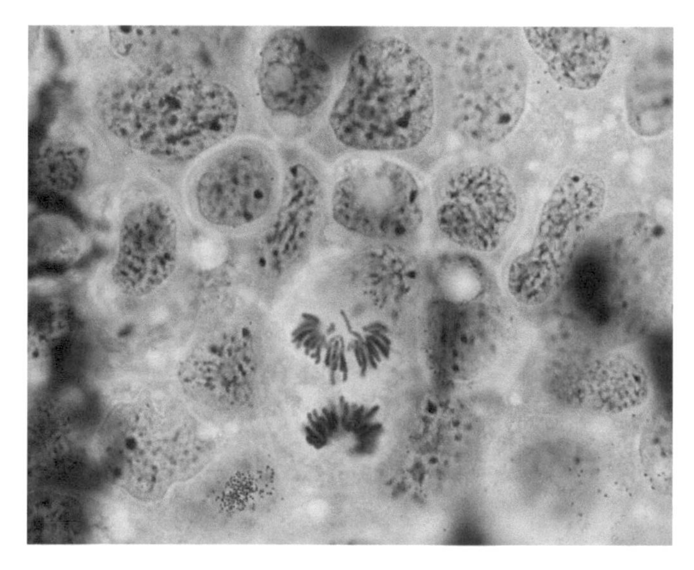

FIGURE B4h LATE ANAPHASE. Chromosomes have migrated to the opposite poles of the cell. A pigment cell intrudes at the far left.

complex are distributed to the daughter cells in approximately equal amounts. The centriole–aster–spindle complex ceases to function.

COLCHICINE, MITOSIS, AND KARYOTYPES

The alkaloid drug, colchicine, extracted from the corm of the autumn crocus *(Colchicum autumnale)*, arrests mitosis in metaphase by interfering with the formation of spindle fibrils, thereby retarding the division of the centromeres and preventing division of the centrioles. Cells begin mitosis but the chromosomes cannot separate. If colchicine is given to an animal or added to a culture of cells, all the cells beginning mitosis after the treatment will be arrested in the metaphase stage. This provides a useful procedure for determining mitotic rate, especially in tissues where this rate is low.

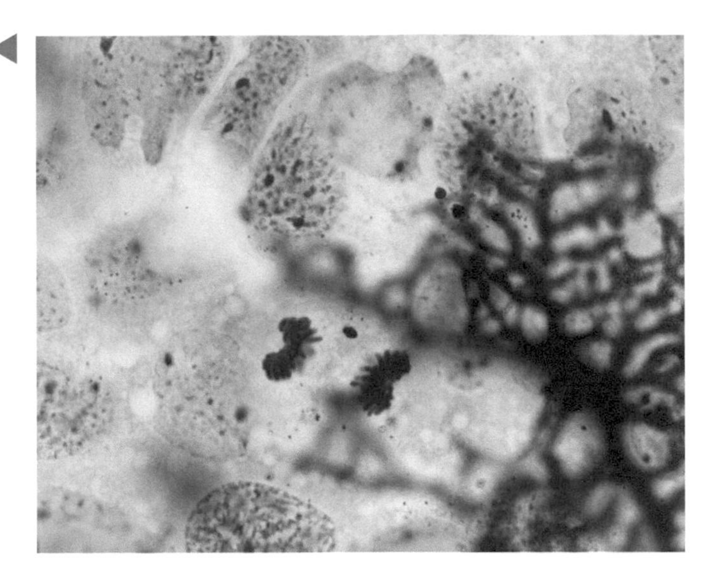

FIGURE B4i LATE ANAPHASE. Chromosomes have migrated to the opposite poles of the cell. A pigment cell with complex processes appears at the right.

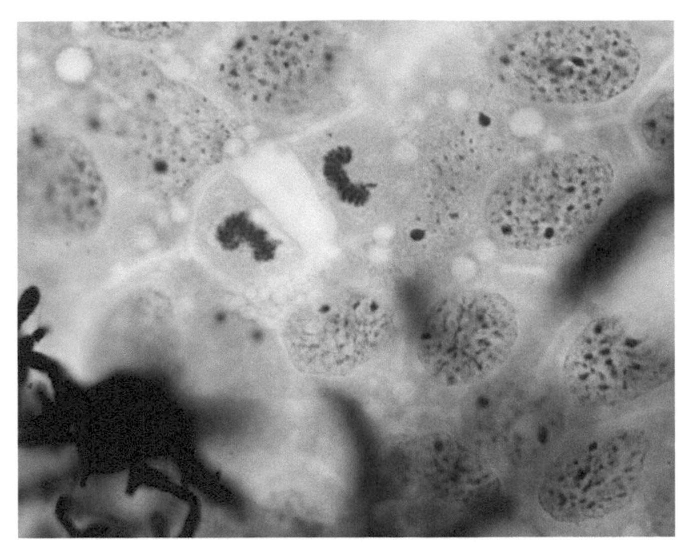

FIGURE B4j TELOPHASE. Cleavage is complete and two cells have been formed but no nuclear membrane is visible around the two masses of chromosomes. A pigment cell lurks in the lower left corner.

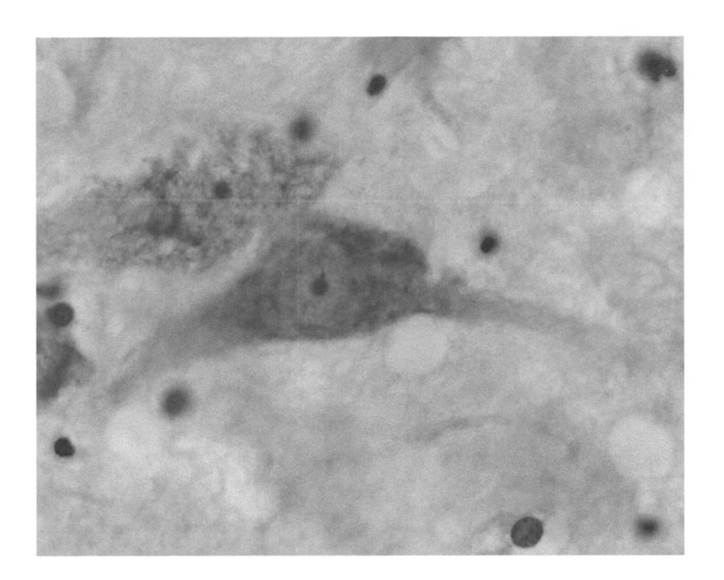

FIGURE B5 SEX CHROMATIN (Barr body) in the nucleus of a motor neuron from the spinal cord of a female mammal. The Barr body resembles a ping-pong ball alongside a basketball, which is the nucleolus.

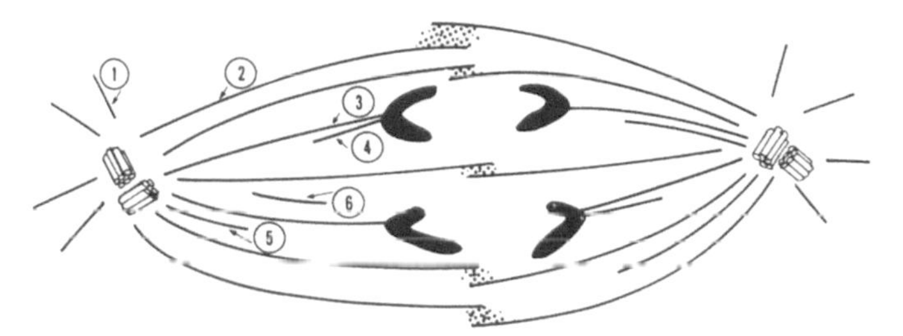

FIGURE B6a Diagram of the microtubules in a mitotic spindle. *Legend*: (1) microtubules of the aster; (2) Nonkinetochore microtubules; (3) kinetochore microtubules attached to the kinetochores on the chromosomes; (4–6) incomplete microtubules.

Colchicine treatment is also useful for the study of metaphase chromosomes. Cells are cultured in a medium containing colchicine for a period of time, fixed, stained, and the coverslips subjected to several kilograms of pressure. The cells are crushed and the chromosomes of the cells that were arrested in metaphase are spread and clearly delineated (Fig. B10a). KARYOTYPES are prepared from photomicrographs of such squashes by cutting out the chromosomes and pasting them in rows (Fig. B10b). Each chromosome appears double because the centromere has not split and holds the chromatids together (Fig. B11). When the centromere is approximately central, the chromosome is METACENTRIC. If it is about halfway between the middle and the end of the chromosome, the chromosome is SUBMETACENTRIC, near the end, ACROCENTRIC and at the end, TELOCENTRIC. In making a karyotype, cutout

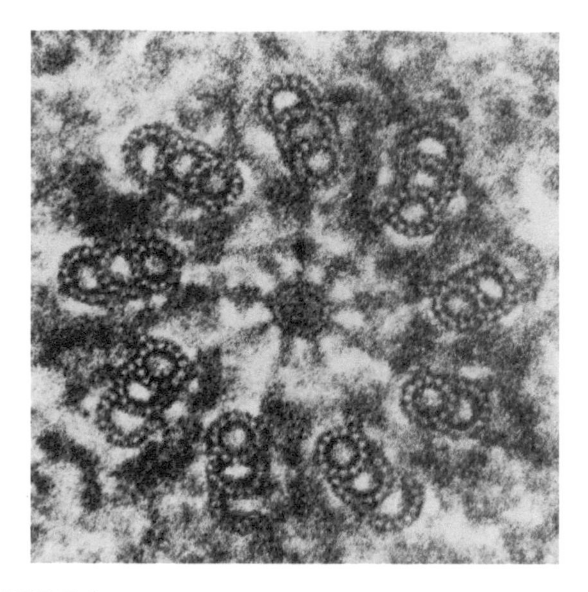

FIGURE B6b Electron micrograph of a cross section of a CENTRIOLE in a cell of the trachea of a chick. The centriole is a cylindrical structure that consists of nine parallel, overlapping "blades," each blade a simple "pipe of Pan" composed of three fused microtubules. 500,000 ×.

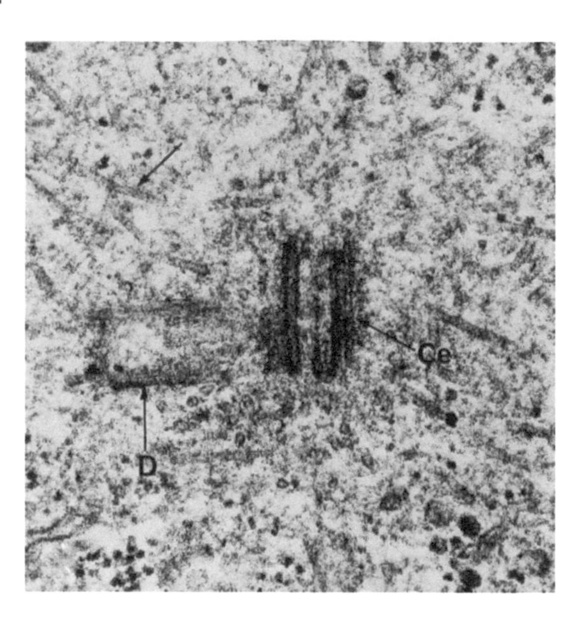

FIGURE B7 The cylindrical pairs of centrioles lie at right angles to each other in this cell from metaphase in the nervous tissue of a rat. *Ce* is the original centriole, *D* is its daughter.

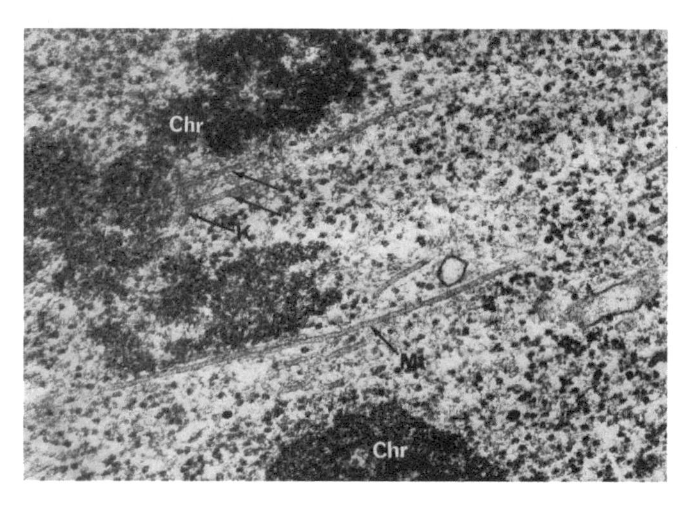

FIGURE B8 Electron micrograph of part of a cell in metaphase. Microtubules (*Mt*) course between chromosomes (*Chr*). The chromosomes appear as dark granules associated with a lighter material. The "granules" are actually sections of strands containing DNA. 50,000 ×.

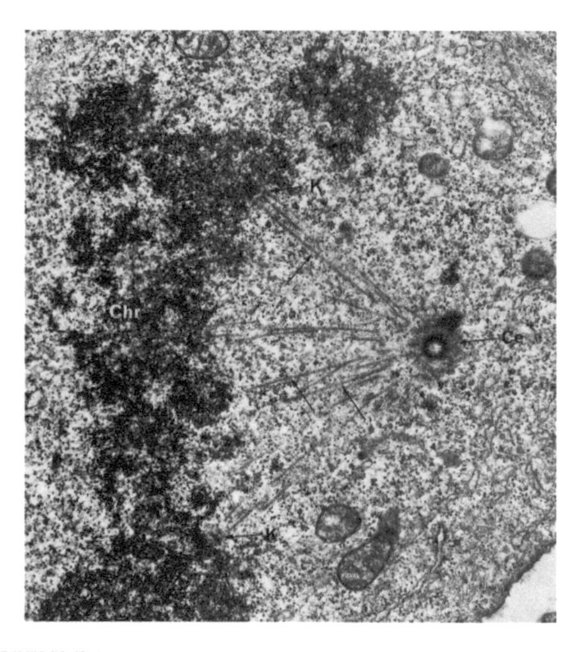

FIGURE B9 Electron micrograph of part of a cell in anaphase from the nervous tissue of a rat embryo. A centriole (*Ce*) is at the pole of the mitotic spindle. Spindle fibers radiate toward the dark masses of chromosomes (*Chr*). 22,000 ×.

chromosomes from a micrograph are arranged in order of decreasing length with their shorter arms extended upward. Then the cutouts are secondarily arranged according to the position of the centromere. In mammals, males are the heterogametic sex with unpaired sex chromosomes; the situation is reversed in birds. Karyotypes are characteristic of each species and have been used to distinguish taxonomic groups. They have also been used as a diagnostic aid in medicine where congenital chromosomal abnormalities may be detected. The first such condition discovered was the triplication of Chromosome 21 (TRISOMY 21), which produces DOWN SYNDROME.

FIGURE B10a Human chromosomes arrested in metaphase by COLCHICINE; from a squash preparation stained with acetoorcein. The dark blob is an interphase nucleus. 2400 ×.

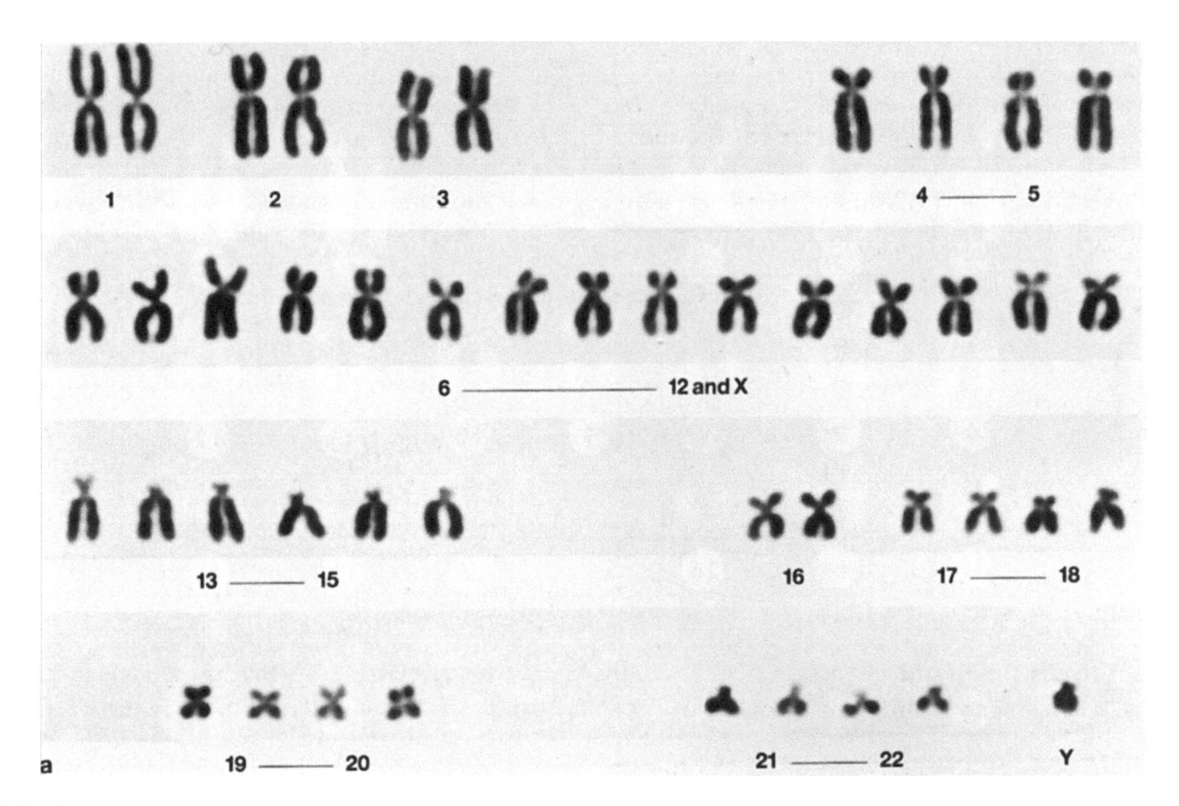

FIGURE B10b This human KARYOTYPE was prepared from the squash shown in Fig. B10a. Each chromosome appears double because the CENTROMERE has not split and holds the CHROMATIDS together. For convenience, the chromosomes are numbered from the longest to the shortest. 1400 ×.

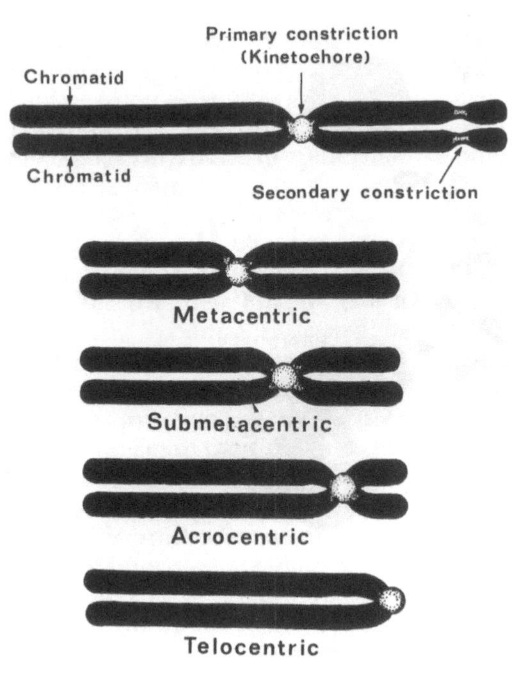

FIGURE B11 When the centromere is approximately central, the chromosome is METACENTRIC. If it is about halfway between the middle and the end of the chromosome, the chromosome is SUBMETACENTRIC, near the end, ACROCENTRIC, and at the end, TELOCENTRIC.

DURATION OF THE MITOTIC CYCLE

The total period varies with animal types, different organs of the same individual, and with physiological state and age. The following are examples of timing of living cells as observed in tissue culture.

	Embryonic mesenchyme (min)	Fibroblasts (min)
Prophase	10–15	30–60
Metaphase	2–3	2–10
Anaphase	3–4	2–3
Telophase	7–15	3–12
Interphase	37–97	30–120

AMITOSIS

In contrast to the precise mechanisms of mitosis, there have been suggestions that there is a direct method of cell division called AMITOSIS. Here a groove appears on the surface of the cytoplasm and is said to deepen and cut the nucleus and cytoplasm in two more or less equal parts. It may occur in transient or highly differentiated cells, such as certain cells of the placenta or of the blood, where a precise replication of chromatin is not essential. Notice this constriction in stained smears of urodele blood (Fig. B12).

MEIOSIS

This type of division, where the chromosome number is precisely halved, occurs during the formation of sex cells and will be discussed in the section on the testis (Fig. B13).

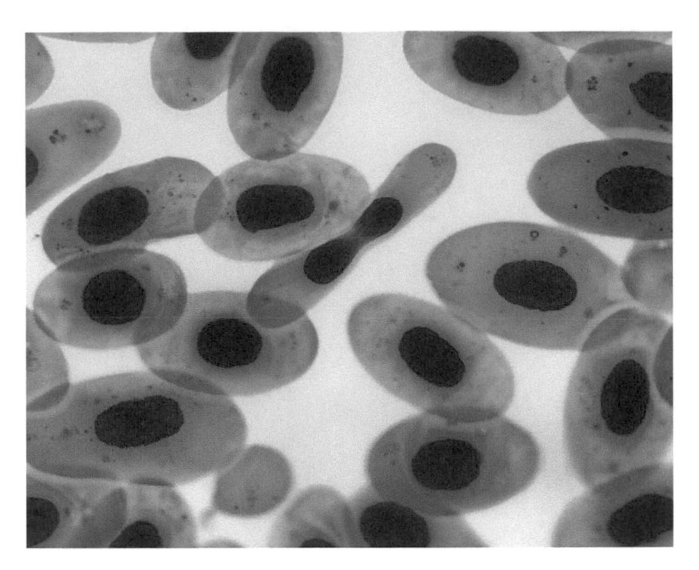

FIGURE B12 Blood smear of the mudpuppy, *Necturus*, stained with Giemsa stain. The red blood cell near the center appears to be separating into two cells. A skeptic might argue that this is the result of the smearing action in preparation. 63 × .

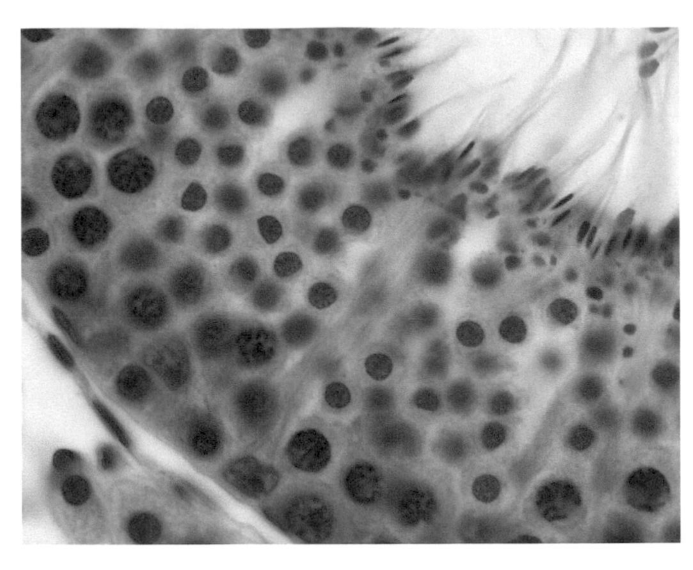

FIGURE B13 Section of the wall of a seminiferous tubule of the testis of a dog. MEIOTIC division occurs as the cells progress from DIPLOID cells at the periphery of the tubule to HAPLOID sperm in the lumen.

PART II

TISSUES

In the strict sense, histology is the study of tissues. However, the term is most commonly used to include the study of cells and organs. A tissue is a group of similar cells and their products, specialized in a common direction, and performing a common function. Tissues are the building materials from which organs are fashioned. There are five functionally and structurally distinct primary tissue types:

- *Epithelial*
- *Connective*
- *Muscular*
- *Nervous*
- *Vascular*

CHAPTER

C

Epithelial Tissues

An epithelium is a layer of cells that covers an internal or external surface and may be involved in protection, secretion, excretion, and the senses. Its function is reflected in its structure and may be deduced by microscopic study. The surface may be an expansive sheet or a microscopic tubule. Beneath the epithelium and attached to it is a noncellular BASEMENT MEMBRANE that almost always rests on a vascularized bed of connective tissue (Fig. C1). There is little material between the cells. Many epithelial cells are strengthened by an internal meshwork of intermediate filaments of keratin, the TONOFILAMENTS, that may aggregate into bundles, the TONOFIBRILS (Fig. C2). The filaments are concentrated just inside the cell membrane and comprise the CELL WEB. The cell web is often denser just below the free surface of epithelial cells, forming a brush-heap of tonofilaments from which organelles are excluded, and constitutes the TERMINAL WEB.

Epithelia have no blood capillaries and obtain their nutrients by diffusion from tissue fluids of the underlying connective tissues. Because substances entering and leaving the body must cross an epithelium in a direction perpendicular to the surface, the cells are functionally and structurally polarized: the BASAL portion of the cell, which is directed toward the basement membrane, usually differs from the APICAL portion, which is directed toward the free surface. In pictures of epithelia it is a convention that the apical surface is shown at the top. When freeze-fracture images of epithelia are shown, however, this convention gives way to the convention of orienting the direction of shadowing from the bottom to the top.

During your study of the various epithelia described later, note the BASEMENT MEMBRANE. It appears as an indistinct line in hematoxylin and eosin (H&E) preparations but is clearly delineated with the periodic acid-Schiff (PAS) (Fig. C3) or silver techniques. With the electron microscope, it is seen to be composed of two layers: an osmiophilic BASAL LAMINA, 20–100 nm thick, consisting of 3- to 4-nm fibrils[1] embedded in an amorphous matrix, and the RETICULAR LAMINA of condensed ground substance containing reticular microfibrils that are firmly rooted in the underlying connective tissue (Fig. C4). The basal lamina is generally separated from the epithelial cells by a clear LAMINA LUCIDA through which run fine microfilaments that attach the basal lamina to the epithelial cell membranes.

CLASSIFICATION OF EPITHELIA

Epithelia are classified on the basis of the shape, stratification, and specialization of the cells (Fig. C5).

Shape of Epithelial Cells

1. *Squamous epithelial cells.* The form and appearance of isolated squamous cells can be observed in desquamated cells from the superficial layer of the lining of the mouth (Figs. C6a and C6b). In the preparation shown here, little detail is obtained using an ordinary bright field microscope (Fig. C6a); the use of differential interference contrast (DIC) (Fig. C6b) enhances the image. Note the thin, scale-like appearance of the cells, their nuclei, cytoplasm, cell membrane, and the relative size and location of the nuclei. Fig. C7 is a whole mount of the layer of flattened cells that constitutes the peritoneal lining of the coelom. The tissue is composed of flat, polygonal cells fitted closely together in the manner of a pan-full of fried eggs; in this preparation, the intercellular cement has been blackened with a silver precipitate and the nuclei are stained pale violet.

[1]A nanometer (nm) is one thousandth of a micrometer (μm) and one millionth of a millimeter (mm).

An Atlas of Comparative Vertebrate Histology.
DOI: https://doi.org/10.1016/B978-0-12-410424-2.00003-2

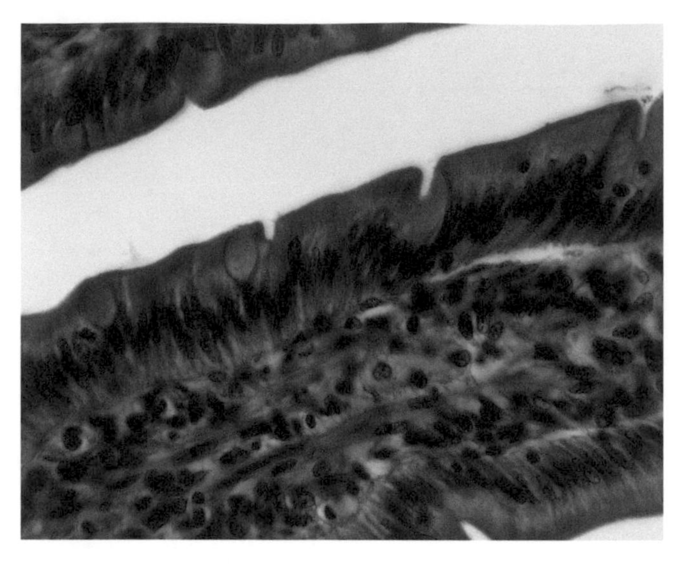

FIGURE C1 Section through the absorptive region of the small intestine of a cat. Absorptive epithelial cells at the top and bottom of the structure are subtended by vascular connective tissue. Two cells in the epithelium have the appearance of wine goblets—they are GOBLET CELLS and produce mucus. 63 ×.

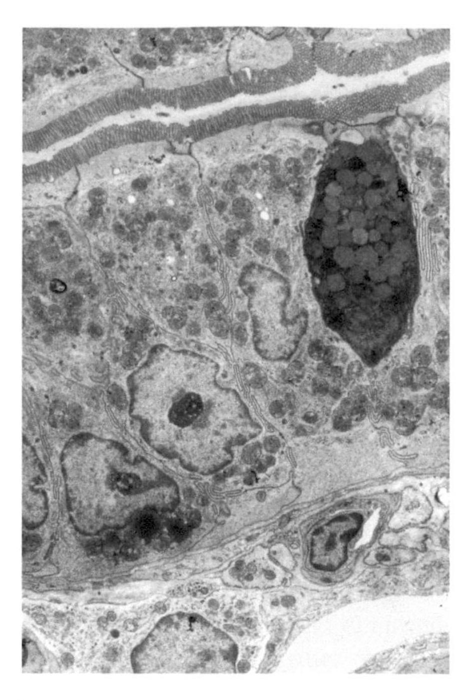

FIGURE C2 Electron micrograph of a similar epithelium in the intestine of a rat. Notice the large goblet cell packed with droplets of mucus. The upper portion of the epithelial cells is devoid of organelles; this is the TERMINAL WEB and contains a meshwork of TONOFILAMENTS. There is little space between the cells; they are separated from a layer of vascular connective tissue by a BASEMENT MEMBRANE. 7500 ×.

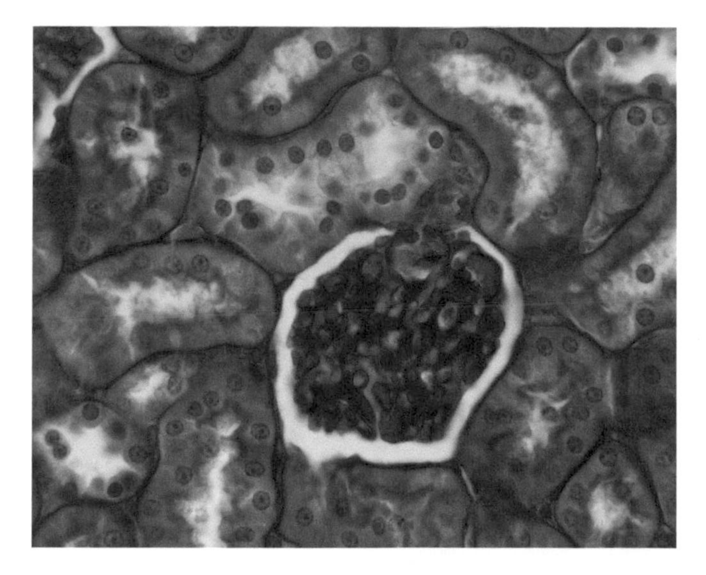

FIGURE C3 Section, stained with periodic acid-Schiff (PAS), of the kidney of a gerbil. A knot of blood vessels at the center, the NEPHIC CAPSULE, is surrounded by kidney tubules. The fuchsin dye has been avidly absorbed by the BASEMENT MEMBRANES of the kidney tubules. Nuclei of the epithelial cells of the tubules are stained deep green. 63 ×.

FIGURE C4 The basement membrane is clearly defined in this electron micrograph of a tubule in the kidney of a mouse. It separates the epithelial cells (above) from a capillary in the connective tissue (lower right). 21,000 ×.

In Fig. C8 these cells are shown in a section of this tissue. Fig. C9 shows squamous cells as they appear in a section of the nephric capsule of the kidney. The "fried eggs" may be rolled up to form tubes—blood capillaries—that constitute the smallest components of the vascular system; exchange of nutrients and wastes occurs across these thin cells (Figs. C10a and C10b).

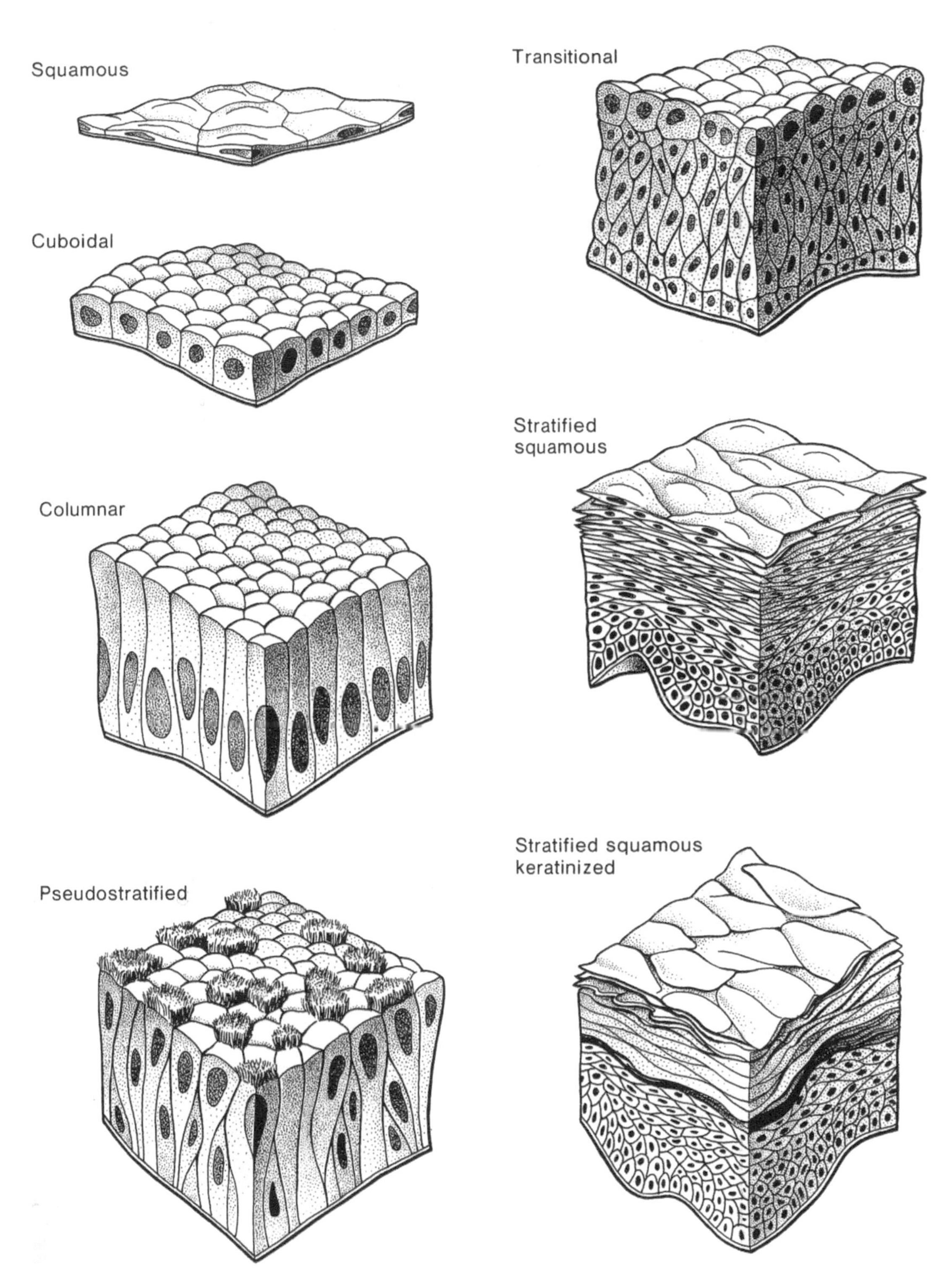

FIGURE C5 Epithelia are classified on the basis of shape, stratification, and specialization of the cells. Epithelial cells that are flat, like a fried egg, are SQUAMOUS. Those that are about as high as they are wide, are CUBOIDAL, and those that are tall and slender are COLUMNAR. Note that when they are pressed together to form an epithelium, they are approximately octagonal in surface view. These cells sit on a thin, noncellular membrane, the basement membrane. If there is only one layer of cells, the epithelium is SIMPLE; if the cells are piled on top of each other, the epithelium is STRATIFIED. Although an assortment of shapes may be found in a single stratified epithelium, the cell type that forms the superficial layer is used in the classification and thus human skin is classified as a stratified squamous epithelium even though the basal layers are cuboidal.

FIGURE C6 Fresh squamous epithelial cells scraped from the inside of the human cheek. 63 ×.

FIGURE C6a Little detail is seen with the ordinary bright field microscope.

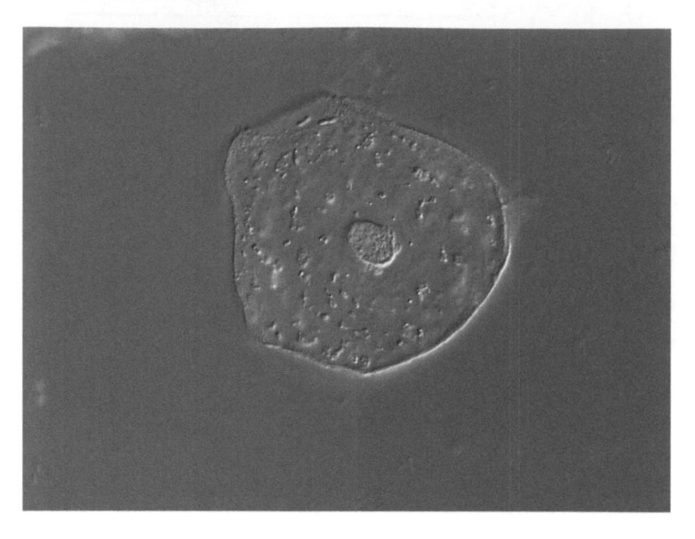

FIGURE C6b Detail is enhanced by the use of differential interference microscopy.

FIGURE C7 This is a surface view of the peritoneum that lines the body cavity of a vertebrate. The cementing substance between individual cells is stained black with silver. Like a pan of fried eggs, mutual pressure of the cells of this simple squamous epithelium causes them to assume a roughly hexagonal shape. The faint violet nuclei of some of the cells may be discerned. 40 ×.

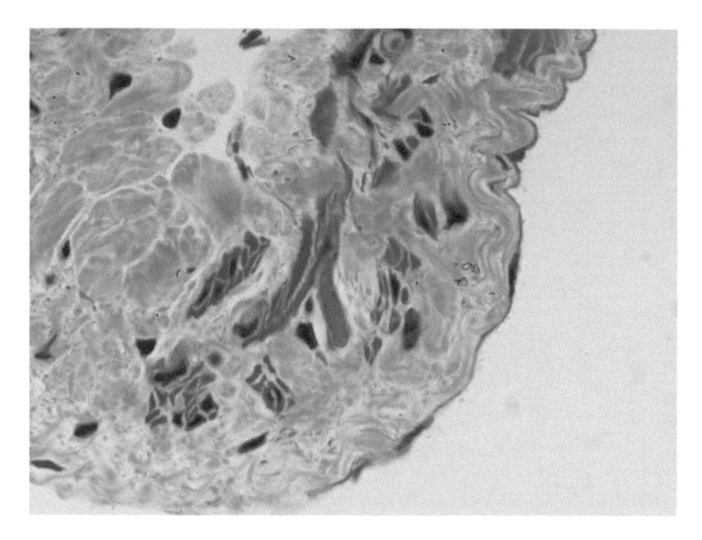

FIGURE C8 The "fried eggs" of a SIMPLE SQUAMOUS EPITHELIUM are sliced and seen in side view at the right this section of the peritoneum of a mammal. They rest on vascular connective tissue. 63 ×.

2. *Cuboidal epithelial cells.* In this section of the thyroid gland, note the shape of the cells forming the follicles as well as the shape and position of the nuclei (Fig. C11).
3. *Columnar epithelial cells.* Columnar epithelial cells range from those whose height is only slightly greater than their other two dimensions to those that are greatly elongated. The nuclei are usually found toward the base. Identify the columnar epithelium in sections of vertebrate gallbladder and intestine (Figs. C12 and C13). Note the shape and location of the nuclei.

Stratification of Epithelial Cells

All of the cells that you have studied so far occur in a single layer resting on a BASEMENT MEMBRANE; these are SIMPLE EPITHELIA. STRATIFIED EPITHELIA consists of two or more layers of cells with only the BASAL LAYER touching the

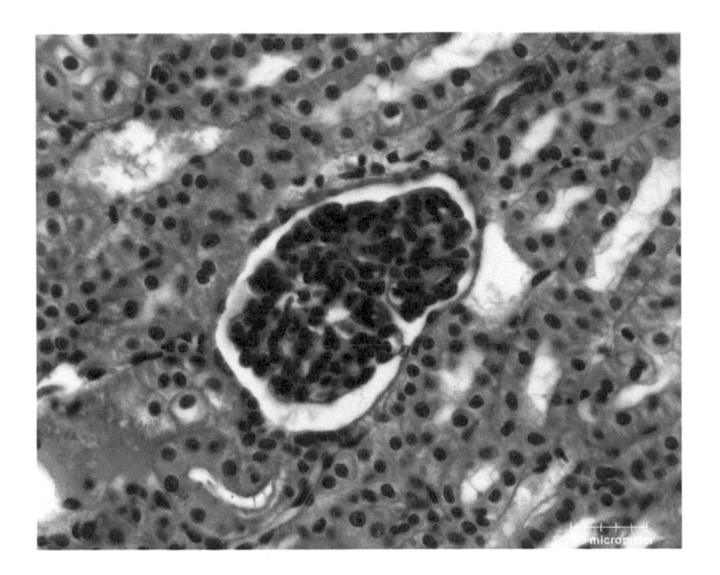

FIGURE C9 A SIMPLE SQUAMOUS EPITHELIUM lines the urinary space in the kidney of a frog. Most of the micrograph is occupied by sections through tubules of the kidney. 40 ×.

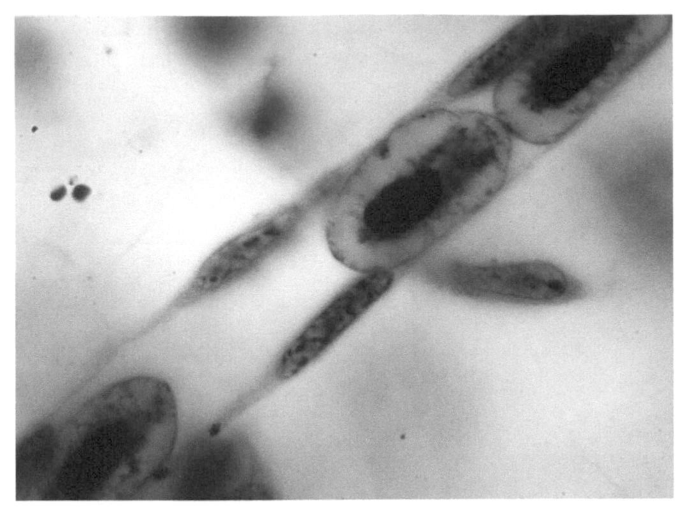

FIGURE C10a Simple "fried eggs" may be rolled up to form a thin-walled tube, as in this whole mount of the skin of *Ambystoma* showing a thin-walled blood vessel, a CAPILLARY containing oval nucleated blood cells. 100 ×.

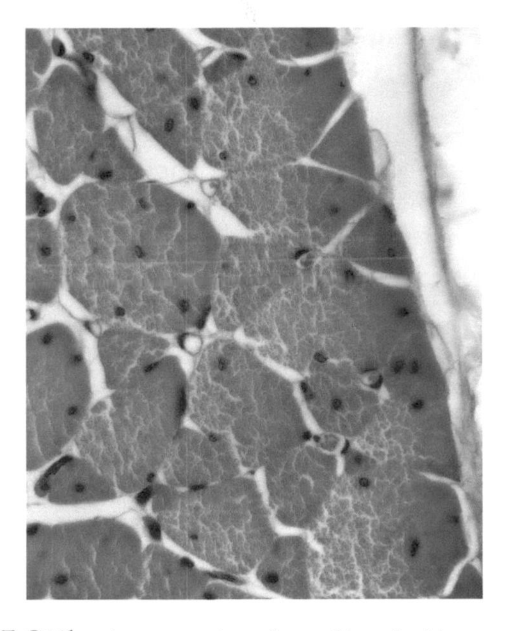

FIGURE C10b A cross section of a capillary, looking much like a signet ring, is at the center is of this section of the foot of a toad. Most of the section is occupied by cross sections of muscle cells. Several other sectioned capillaries are scattered throughout. 63 ×.

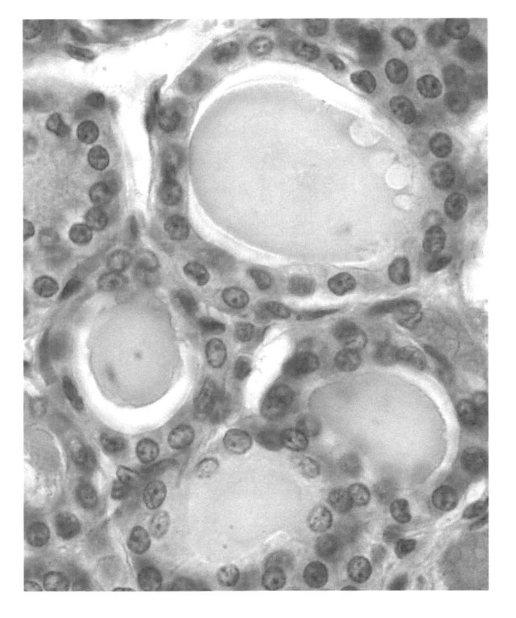

FIGURE C11 A SIMPLE CUBOIDAL EPITHELIUM surrounds yellowish colloidal material in follicles of the thyroid gland of a mammal. Often cell membranes are not visible in sections. If you imagine cell membranes between each pair of nuclei, the cells become simple cuboidal in your mind. 100 ×.

basement membrane. An assortment of shapes of cells may be found in a single stratified epithelium, but the cell type that forms the superficial layer is used in the classification. For example, although cells forming the deepest layer of human skin are cuboidal, the superficial layers are squamous, and this tissue is classified as a STRATIFIED SQUAMOUS EPITHELIUM.

Since the cell boundaries are usually poorly stained, it is often necessary to observe the shape and position of the nuclei to determine the shape of the cells. If nuclei of the outer layer are round and occupy the center of a mass of cytoplasm, the cells are probably CUBOIDAL. If the nuclei are oval or elongated with their long axes perpendicular to the basal line and located so that most of the cytoplasm is peripheral to the nucleus, the epithelium is probably COLUMNAR. SQUAMOUS cells appear flattened with the long axis of both the cytoplasm and the nucleus parallel to the surface. To identify an epithelium in which the cell membranes are indistinct it is helpful to form an imaginary picture of the outlines of each cell by assuming that there is one nucleus per cell and that the cell membranes pass about halfway between each pair of nuclei.

Thus an epithelium which looks like this:

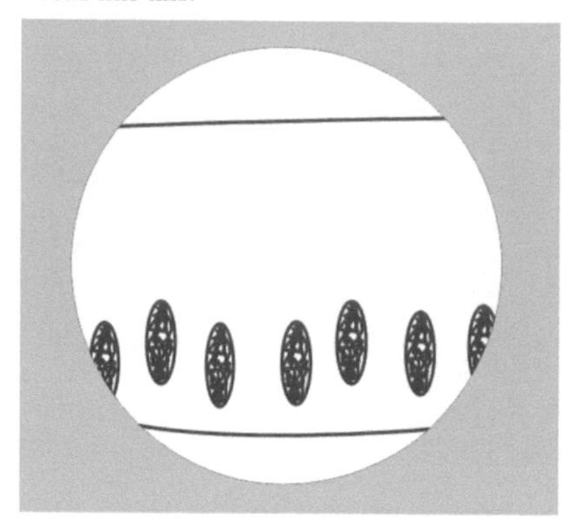

becomes like this in imagination:

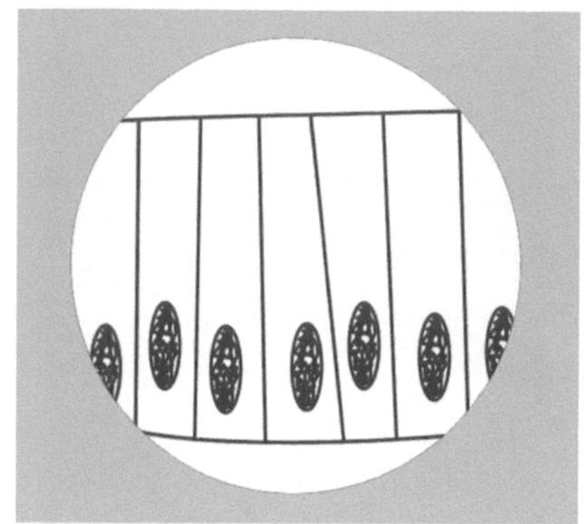

and is correctly identified as SIMPLE COLUMNAR.

1. *Stratified squamous epithelium.* You have already studied the surface view of a stratified squamous epithelium in the cells sloughed from the lining of the cheek. Study the appearance of these cells in sections of stratified squamous epithelia, noting the superficial flattened cells, the deeper layers of polyhedral cells, and the basal cuboidal (or low columnar) cells (Figs. C14–C16). These examples have all shown tough, protective stratified squamous epithelia that resist abrasions and the other rigors of daily life; Fig. C17 is the stratified squamous epithelium from a more protected region of the human body.
2. *Stratified columnar cells.* In this epithelium lining the urethra in the human penis, the superficial cells are low columnar; beneath is a jumble of polyhedral cells (Fig. C18). The basement membrane is not evident.
3. *Transitional epithelium.* This specialized epithelium lines the walls of hollow organs that are subject to distension (e.g., mammalian ureter and bladder) (Fig. C19). The cells are capable of flattening and sliding over one another, thus accommodating the stretching of the organ. In the contracted condition, the epithelium is many layered but a layer of as few as two cells is found when the organ is distended. The superficial cells have a dome-shaped peripheral margin while their lower surfaces are concave and fit on the cells of the intermediate zone. In the intermediate zone, the cells are irregularly polyhedral while the basal cells are cuboidal to low columnar.

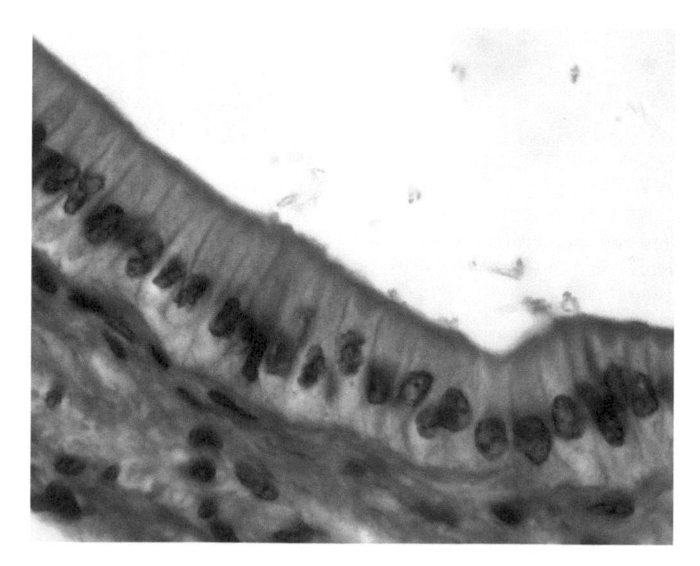

FIGURE C12 The gallbladder of a toad is lined with a SIMPLE COLUMNAR EPITHELIUM. 100 ×.

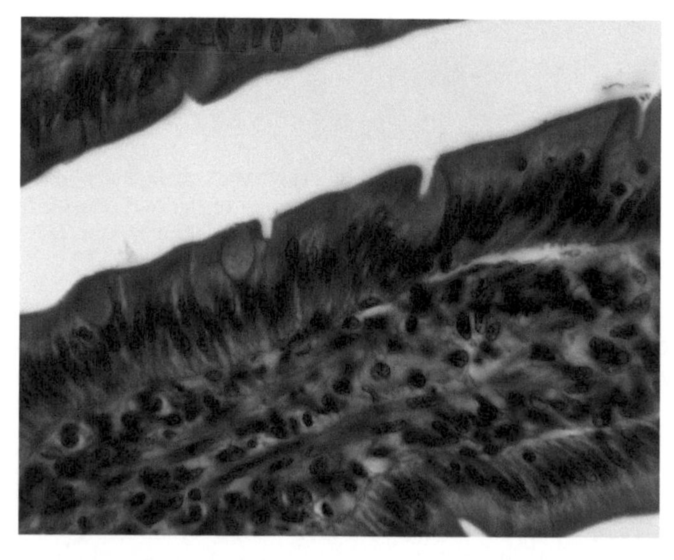

FIGURE C13 An absorptive SIMPLE, COLUMNAR EPITHELIUM lines the intestine of a cat. Note the two cells at the left that resemble stemmed goblets; they secrete mucus. 63 ×.

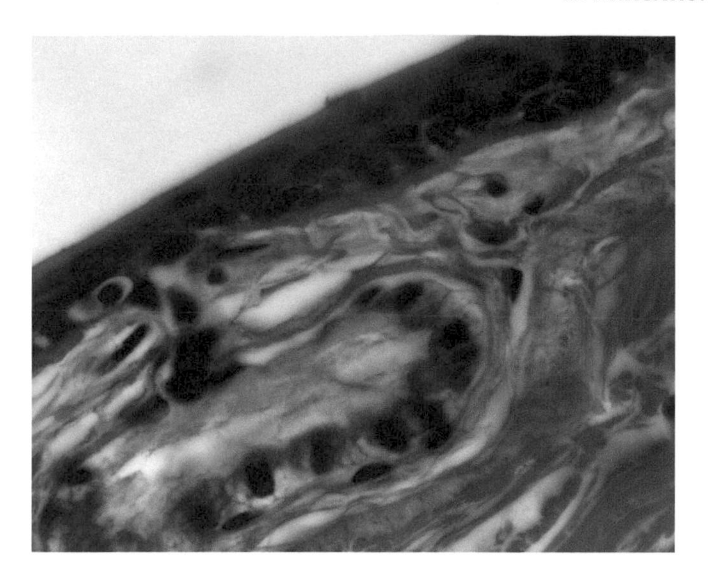

FIGURE C14 This section shows a STRATIFIED SQUAMOUS EPITHELIUM in the skin of a frog. A few flattened cells cover the surface; beneath them are irregular cuboidal cells. The basement membrane that separates the epithelium from the vascular connective tissue below is not seen. The squashed balloon at the lower left is a gland that secretes mucus; part of its duct to the surface can be seen. 100 ×.

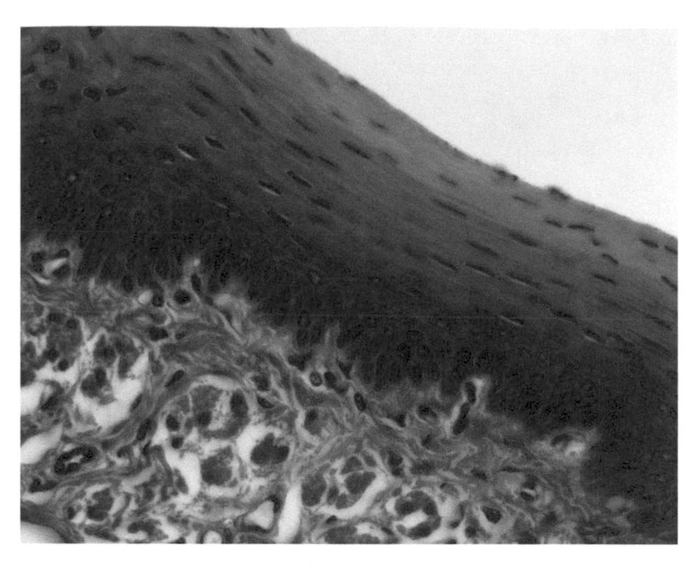

FIGURE C15 This STRATIFIED SQUAMOUS EPITHELIUM lining the esophagus of a rabbit resists the abrasions of the coarse food that is eaten. The basal cells of the epithelium are low columnar and cuboidal and actively divide by mitosis. As the cells move to the surface, they become flattened and die, thereby forming an "armor plate" over the surface. The lower portion of the micrograph is occupied by the tough leather-like vascular connective tissue of the wall of the esophagus. 63 ×.

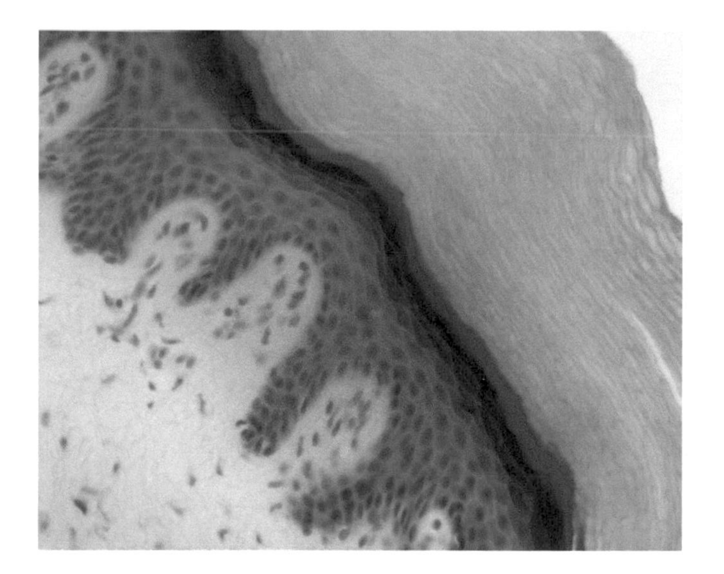

FIGURE C16 The skin of the human fingertip is an extreme example of a STRATIFIED SQUAMOUS EPITHELIUM. The pale area at the right consists of dead, squamous cells. They are separated from living epithelial cells by the bright red line. Cells along the wavy edge divide by mitosis to produce new cells that move toward the surface, die, and eventually wear away. The pale area at the lower left is vascular connective tissue. 40 ×.

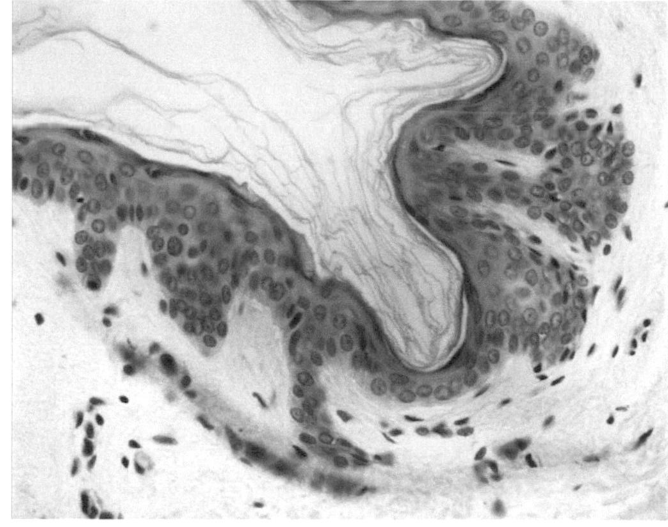

FIGURE C17 Skin from protected areas of the human body is covered with a thinner STRATIFIED SQUAMOUS EPITHELIUM. Divisions still occur in the basal level of epithelial cells at the wavy junction with the vascular connective tissue below. The cells still migrate toward the surface, die, and form a protective layer. Note that some of the dead squamous cells that have sloughed from the surface, have remained. 40 ×.

4. *Pseudostratified columnar epithelium.* Pseudostratified columnar epithelium is actually a simple epithelium whose cells are all in contact with the basement membrane but not all reach the surface. It appears to be stratified because the form and position of the cells have been altered by compression and the nuclei appear to be stratified. In this section of amphibian intestine, the nuclei are arranged in about three layers (Fig. C20). Identify the low triangular BASAL CELLS with spherical nuclei. PYRIFORM CELLS are of medium height with nuclei midway in the epithelium. COLUMNAR CELLS extend from the basement membrane to the free surface. Their nuclei are oval and form the outer nuclear layer.

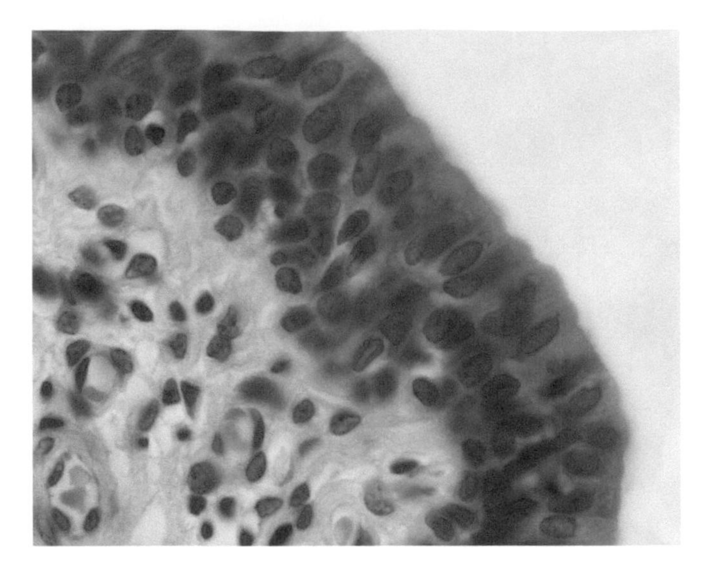

FIGURE C18 The lumen of the urinary duct (urethra) of the human penis is at the right. It is lined with a STRATIFIED COLUMNAR EPITHELIUM. The basal cells of the epithelium appear to be cuboidal, but those at the surface are clearly columnar. Vascular connective tissue is at the lower left. 100 ×.

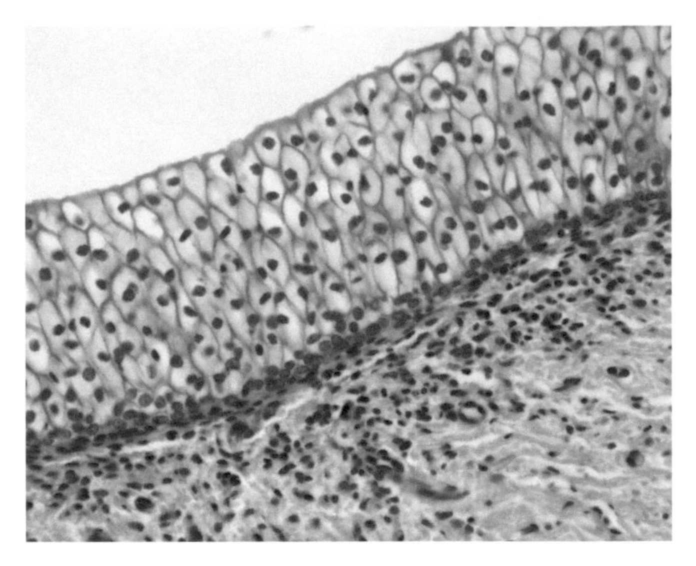

FIGURE C19 This TRANSITIONAL EPITHELIUM lining the ureter of a horse—the urinary duct—accommodates the stretching of the urinary tract between each urination. The surface cells slide across each other and tend to flatten as the bladder and ureter fill. This example is in the contracted state. By piling on top of each other, the epithelium contracts without forming wrinkles that could accommodate the proliferation of disease organisms. The vascular connective tissue below is elastic.

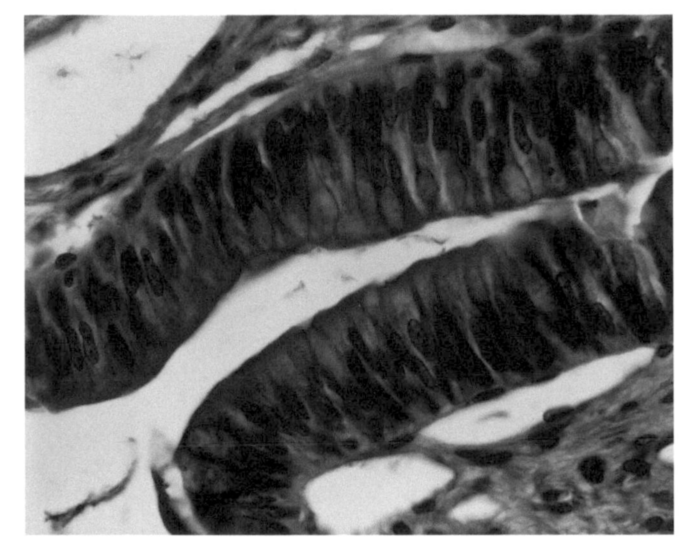

FIGURE C20 PSEUDOSTRATITIFIED COLUMNAR EPITHELIUM lining two sides of a furrow in a section of the intestine of a toad. It is difficult to identify the cells of the epithelium with certainty, but it can be seen that not all of them reach the surface. Try to identify the low, triangular BASAL CELLS with spherical nuclei; PYRIFORM CELLS of medium height with nuclei midway in the cells, and the more superficial nuclei of COLUMNAR CELLS, which extend from the basement membrane to the free surface. Blobs of mucus can be seen in GOBLET CELLS that extend from the basement membrane to the free surface. Many of the open spaces in the connective tissue below the epithelium are LYMPH SPACES that are lined with a simple squamous epithelium. 63 ×.

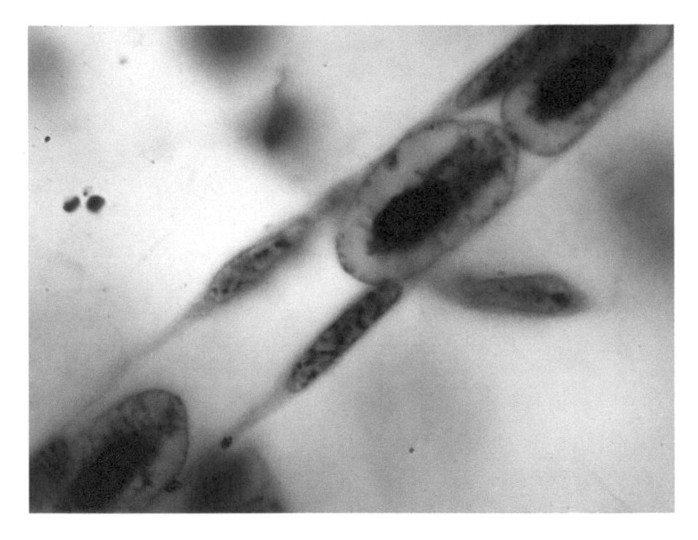

FIGURE C21 Capillary in a whole mount of the skin of a larval amphibian, *Ambystoma*. Squamous ENDOTHELIAL CELLS are rolled up to form the capillary. Flattened nuclei are seen in one of the endothelial cells. 100 ×.

ENDOTHELIUM AND MESOTHELIUM

Special names are employed for two simple squamous epithelia of wide occurrence. ENDOTHELIUM is the lining of the blood vessels and may be seen in the capillaries in a whole mount of the skin of a larval amphibian, *Ambystoma* (imagine a tube formed by rolling up a fried egg) (Fig. C21). MESOTHELIUM lines several of the body

FIGURE C22 MESOTHELIAL CELLS are the squamous cells that line the body cavities. Here a silver stain has been used to show the intercellular substance thereby revealing shape of the cells. Like a pan of fried eggs, mutual pressure imposes a roughly hexagonal shape on the cells. 40 ×.

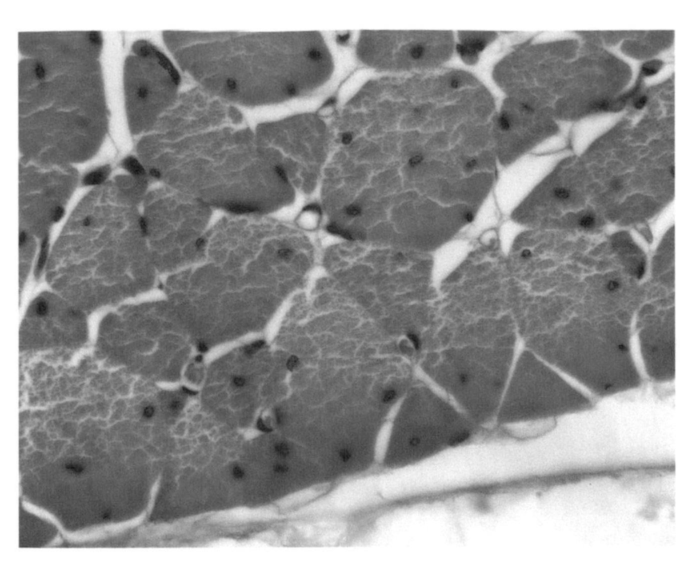

FIGURE C23 A capillary is seen in a section of muscle from the foot of a toad: it is a simple tube formed of rolled-up ENDOTHELIAL CELLS. All the blood vessels, including the heart, are lined with squamous endothelial cells. 63 ×.

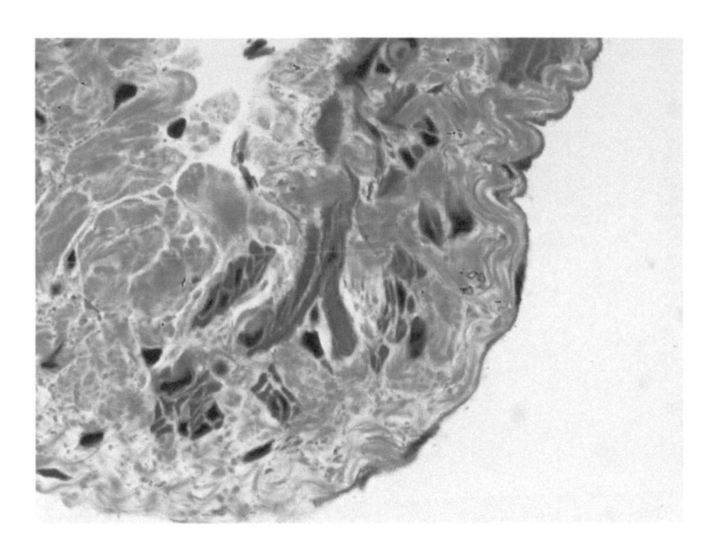

FIGURE C24 Body cavities are lined with a simple squamous epithelium, the MESOTHELIUM. This section is from the peritoneal sling, the MESENTERY, that suspends the gut. Mesentery is formed of a sheet of tough vascular connective tissue that is covered on both sides by squamous mesothelial cells. 63 ×.

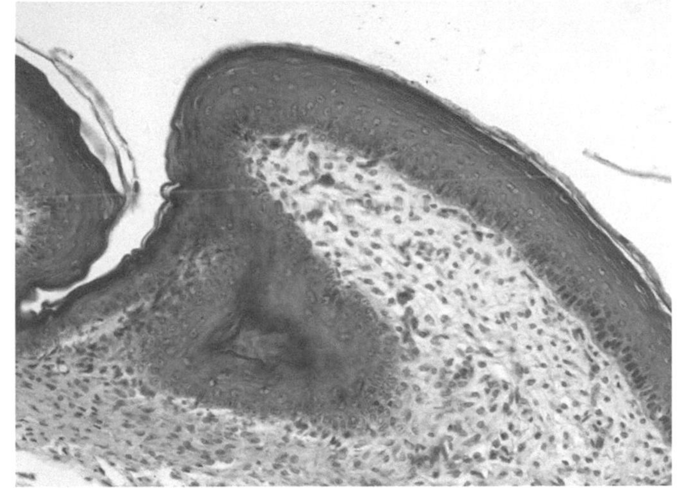

FIGURE C25 This section is from the tough, shiny epithelium that covers the leg of a chick. The outer layers of cells of the stratified squamous epithelium have become CORNIFIED and resist abrasion and dehydration. The ridges in the section are the bumps that give the leg its scaly appearance. 20 ×.

cavities and may be seen in whole mounts of spread mesentery (Fig. C22). Note also the appearance of these delicate cells in sections (Figs. C23 and C24).

Specialization of Epithelia

1. *Cornification or keratinization.* The superficial cells of epithelia covering dry surfaces usually have transformed into a tough, cornified layer of dead cells that are firmly attached to the living epithelial cells below. These provide a tough, flexible, impermeable armor that minimizes dehydration in land animals (Figure C25). Note these cells in human skin; cornification is especially well developed in mammalian plantar pad (Figs. C26 and C27). Cornification also occurs in wet, internal surfaces where severe abrasion occurs as in the esophagus of many mammals (Fig. C28).

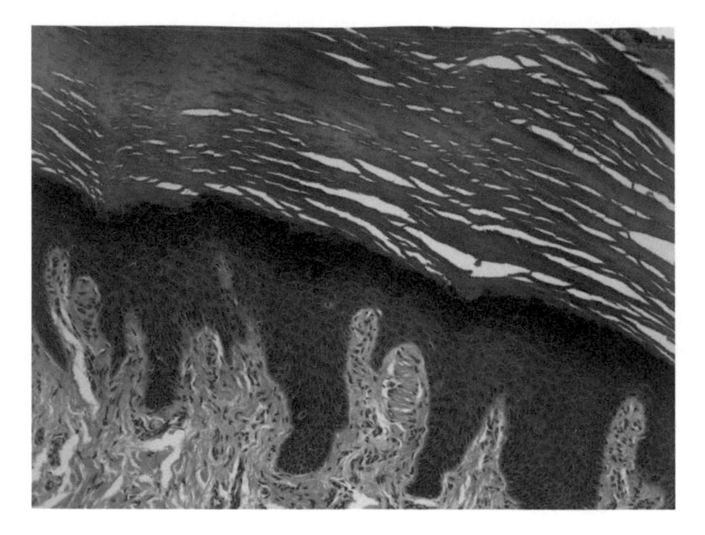

FIGURE C26 This is a section of the fingertip of a graduate student who got too close to the blade of a paper trimmer while preparing his master's thesis. Actively dividing basal cells are well nourished by blood vessels in the tough connective tissue below. Note that the junction between these two layers is corrugated, thereby maximizing the surface area for the exchange of materials. As the cells move toward the surface they produce the precursors of the tough and impermeable KERATIN that constitutes the dead plates on the surface. 10 ×.

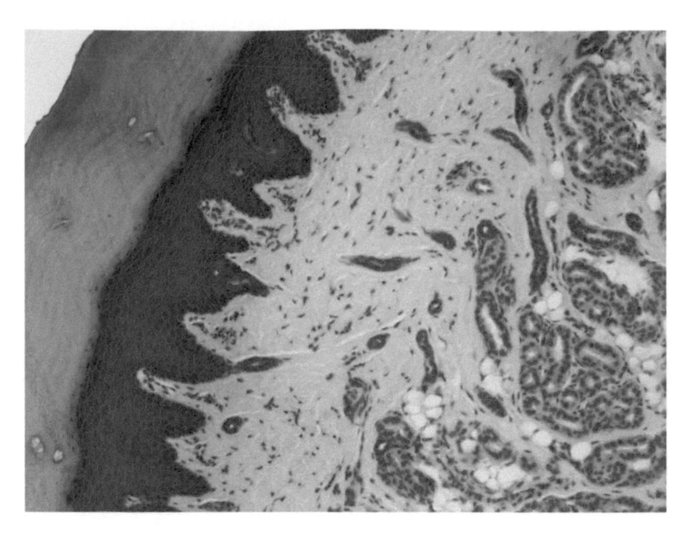

FIGURE C27 The footpad of a squirrel, in addition to being extremely tough, provides traction for these agile animals. In the connective tissue below the epithelium note the numerous glands that provide moisture to enhance grip. 10 ×.

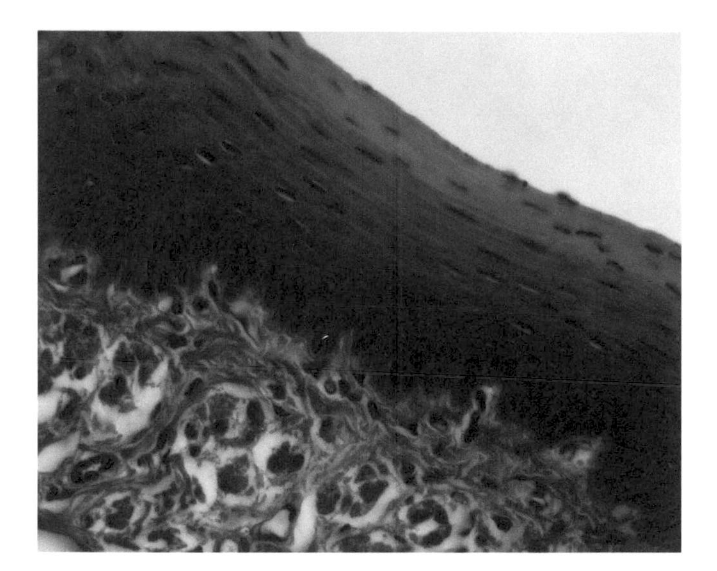

FIGURE C28 This STRATIFIED SQUAMOUS EPITHELIUM is from the moist lining of the esophagus of a rabbit. Cells from the surface that are worn away during ingestion of abrasive food are replaced by cells rising from below. 63 ×.

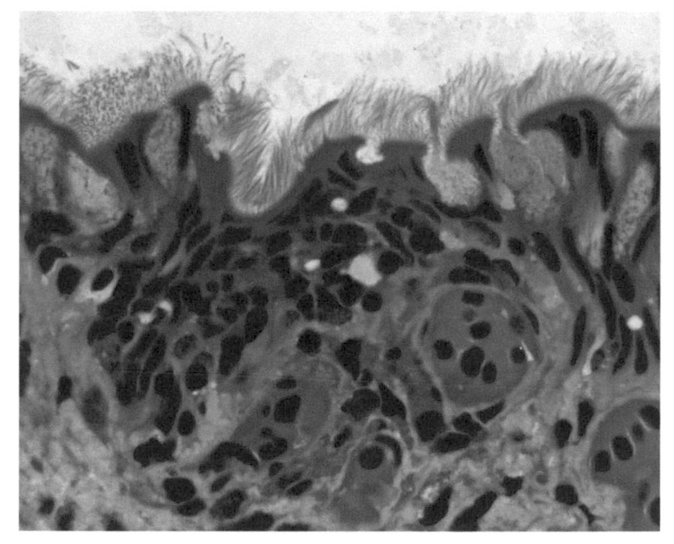

FIGURE C29 The epithelium covering the upper surface of the tongue of a frog is CILIATED. Cilia are cytoplasmic extensions of the cell that are covered by the cell membrane. Cilia are motile: they beat rhythmically, moving materials across the surface. When the frog closes its mouth, cilia on the tongue and lining the oral cavity carry morsels of food to the esophagus where other fields of cilia carry them to the stomach. 100 ×.

2. *Ciliated epithelia.* Sometimes the exposed surfaces of epithelial cells are provided with delicate, lashing extensions of protoplasm, the CILIA. Examine sections through the ciliated epithelium of the tongue and esophagus of a frog (Figs. C29 and C30). Cilia arise from a row of deeply staining BASAL BODIES just below the cell membrane; in the micrographs these appear as a darkened band at the bases of the cilia. Examine electron micrographs of ciliated epithelial cells and note the internal structure of a cilium (Figs. C31 and C32).

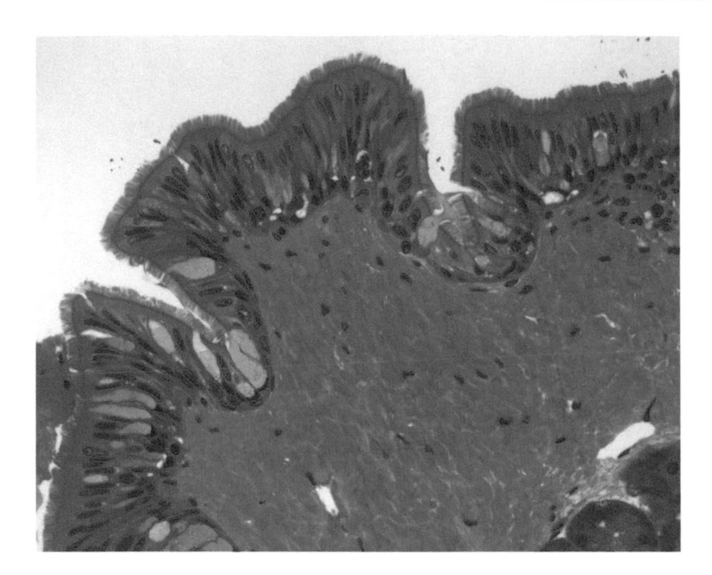

FIGURE C30 Lashing cilia in the epithelium lining the esophagus of a frog carry morsels of food to the stomach. The pale, granular GOBLET CELLS produce mucus that captures these materials. 40 ×.

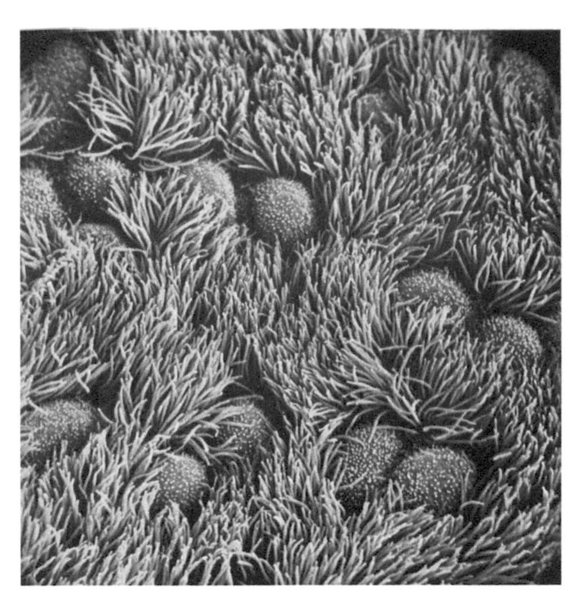

FIGURE C31 The surface view of the lining the trachea of a rat as seen with the scanning electron microscope. The cilia give the appearance of a field of waving grain; the spheres among the cilia are blobs of mucus in goblet cells. 4200 ×.

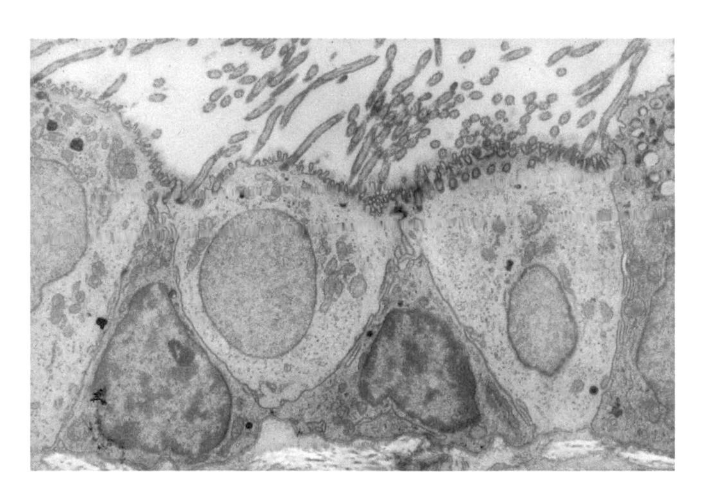

FIGURE C32 A transmission electron micrograph of the CILIATED EPITHELIUM lining the trachea of a bat. The cell membranes covering the cilia are continuous with the membranes enclosing the cells. A few MICROVILLI are seen between the stumps of cilia. 13,000 ×.

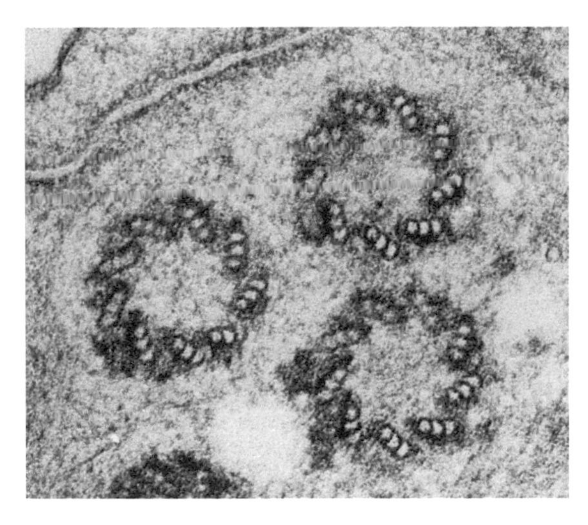

FIGURE C33a Electron micrograph of a cross section of a single CILIUM from an epithelial cell in the trachea of a rat. The core of a cilium consists of a formal arrangement of microtubules—nine double microtubules surrounding a central pair—that is consistent in all cilia and flagella of animals and plants. The nine doublets of microtubules surrounding a pair of single microtubules at the center constitutes the AXONEME. The membrane around the cilium is continuous with the plasma membrane of the cell. 220,000 ×.

Each cilium is a cytoplasmic extension of the cell and is covered by the cell membrane. At its base, the basal body resembles a centriole and is composed of a cylindrical bundle of nine triplet rodlets (Fig. C33a). Microtubules run lengthwise in the core of the cilium and provide the motile force; these constitute the AXONEME and consist of nine doublets surrounding two single microtubules at the center (Fig. C33b). In the doublets, note the relationship of the microtubules to each other. These microtubules insert on the basal body and the central microtubules terminate at the BASAL PLATE. The nine outer doublets are continuous with the nine triplets of the basal body. The basal body is firmly anchored in the terminal web by filamentous ROOTLETS of the BASAL FOOT (Figs. C34a and C34b). A single cilium may be present on individual cells of some epithelia; these cilia are thought to have a sensory function.

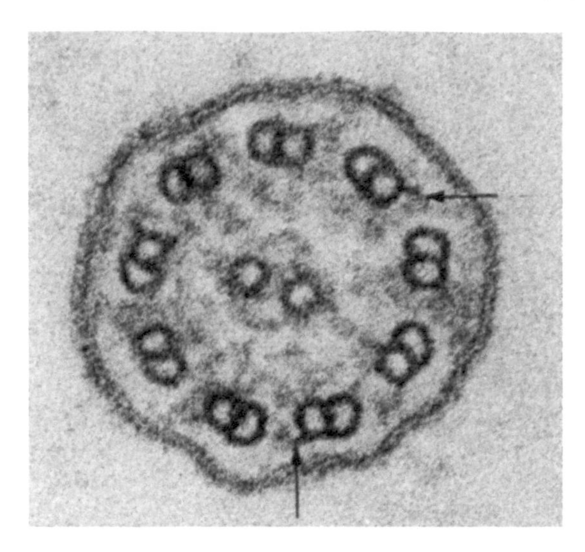

FIGURE C33b Electron micrograph of a cross section of BASAL BODIES of cilia in the trachea of a rat. Each basal body is a cylindrical bundle of nine triplet rodlets. 74,000 ×.

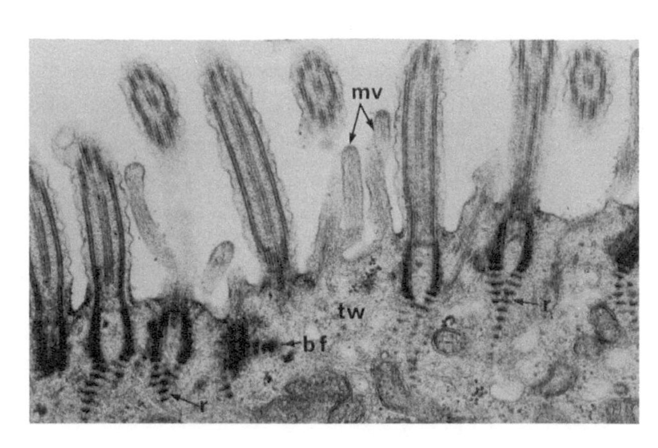

FIGURE C34a Electron micrograph of cilia in longitudinal and oblique sections from the lining of the trachea of a duck. Their basal bodies are firmly rooted in the apical terminal web (tw) of the cell by ciliary rootlets (r) and basal feet (bf). A few microvilli (mv) project between the cilia.

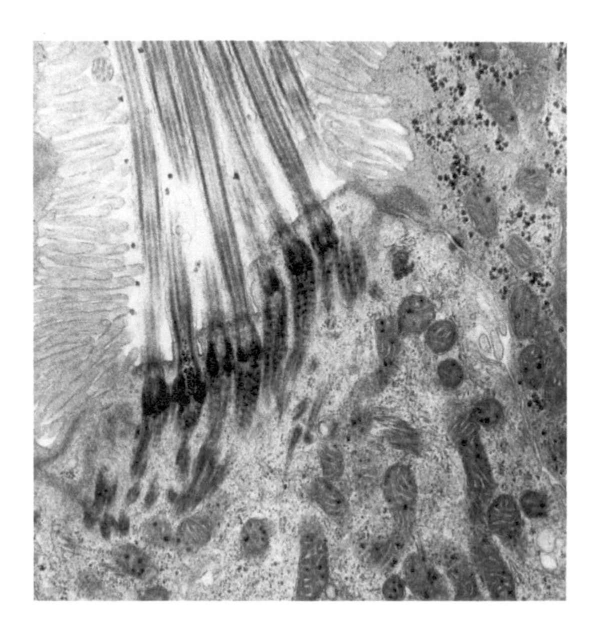

FIGURE C34b Electron micrograph of cilia from the kidney of a garter snake. Cilia arise from BASAL BODIES on the apical portion of these epithelial cells. The basal bodies are firmly anchored in the cytoplasm of these cells by long CILIARY ROOTLETS with a distinct cross-banded appearance. Note the MICROVILLI at the left and the large number of MITOCHONDRIA. c.27,500 ×.

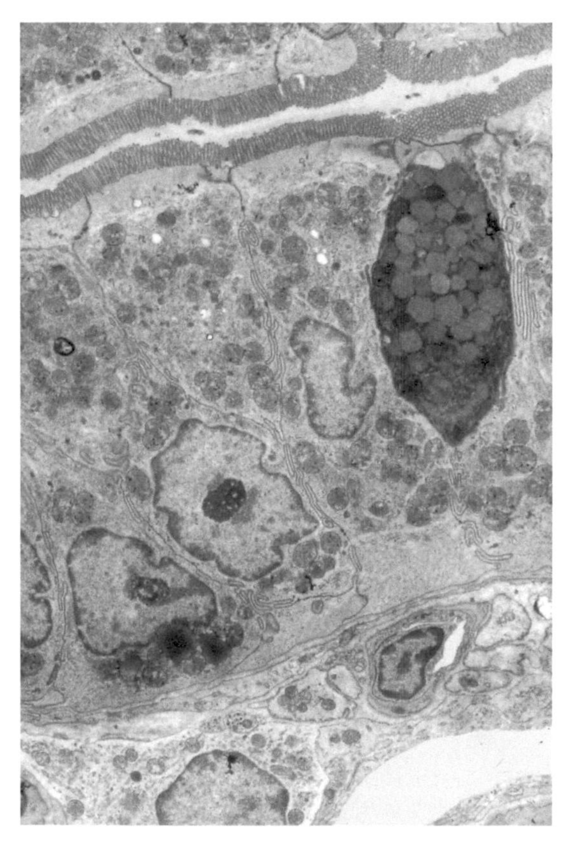

FIGURE C35 Electron micrograph of a few columnar cells lining the duodenum of a bat. The MICROVILLI are arrayed with military precision in a STRIATED BORDER. These filamentous cytoplasmic projections increase the area of the surface of the cell, enhancing the exchange of materials. A goblet cell appears among the columnar epithelial cells. 7500 ×.

3. *Brush border and striated border: Microvilli.* The electron microscope has shown that the free surfaces of certain absorptive cells (e.g., intestinal epithelium, kidney tubules) are covered with finger-like processes, the MICROVILLI (Figs. C35 and C36). Each microvillus is enclosed in an extension of the cell membrane and is filled with homogeneous, finely granular cytoplasm. Inside is a core consisting of a bundle of actin microfilaments

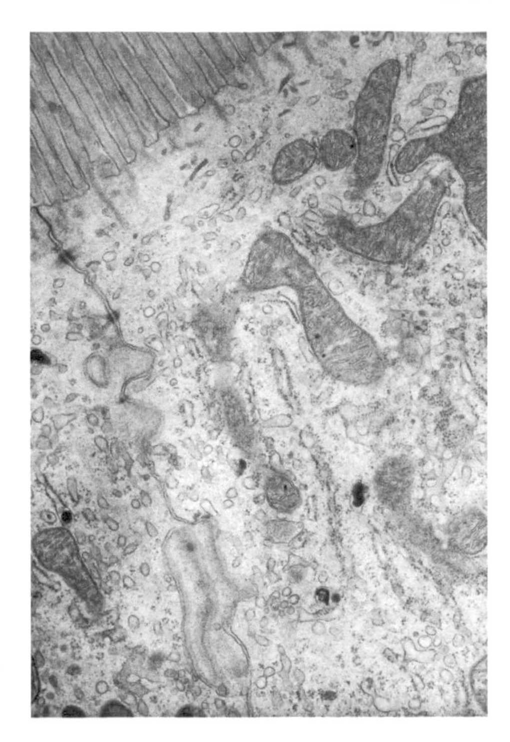

FIGURE C36 The precise arrangement of the MICROVILLI is apparent in this electron micrograph showing the striated border on the epithelial cells in the intestine of a rat. Each microvillus is enclosed in an extension of the cell membrane and is filled with homogeneous, finely granular cytoplasm. Inside is a core consisting of a bundle of actin microfilaments that insert on the terminal web. 42,000 ×.

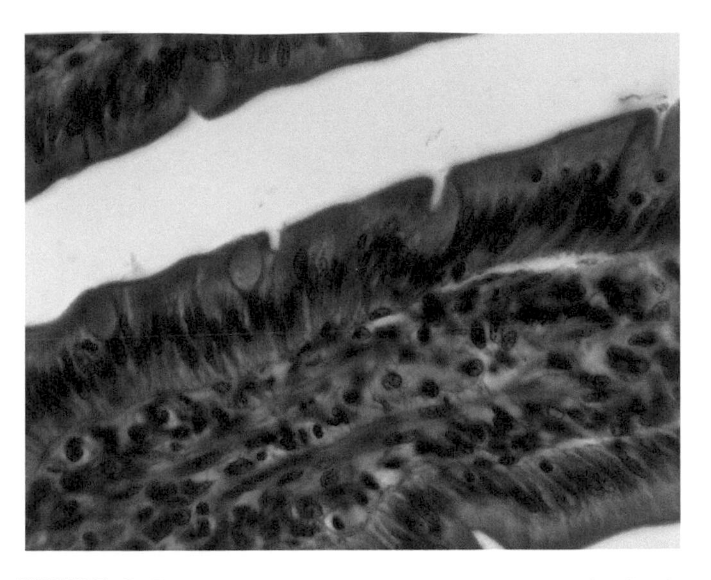

FIGURE C37 Photomicrograph of ABSORPTIVE CELLS in the duodenum of a cat. A few GOBLET CELLS provide a protective coat of mucus. The STRIATED BORDER of the absorptive cells is indistinct—the microvilli are below the resolving power of the microscope. 63 ×.

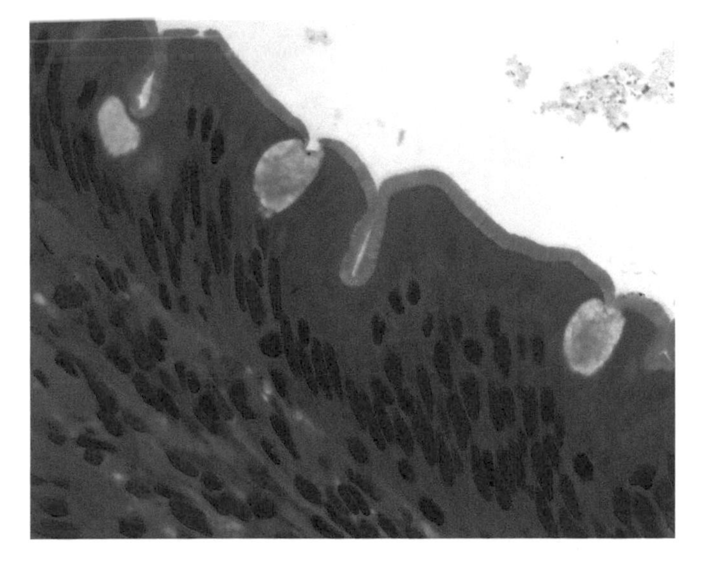

FIGURE C38 Greater detail is obtained when tissues are embedded in epoxy resin and sectioned on an ULTRAMICROTOME to produce SEMITHIN SECTIONS—this section is 1.5 μm thick. In this photomicrograph of a section of the duodenum of a monkey, the filamentous structure of microvilli in the STRIATED BORDER is just visible. A few goblet cells are seen. 100 ×.

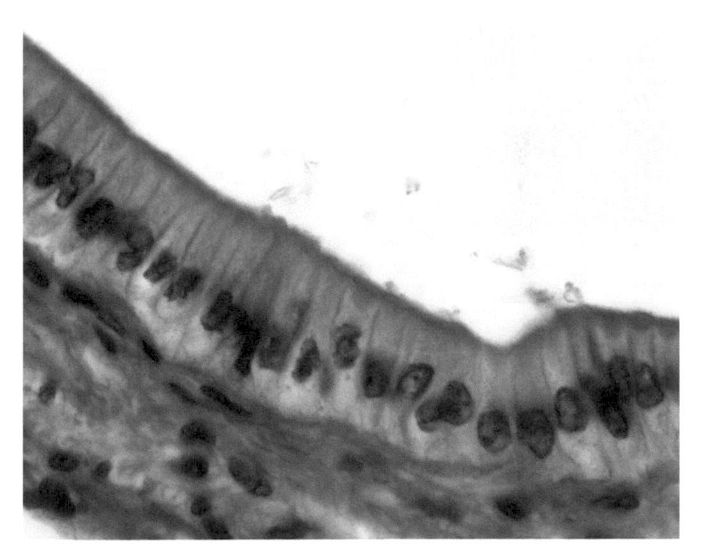

FIGURE C39 The microvillous border appears as a dark smudge on the apical surface of the epithelial cells in this paraffin section of the simple columnar epithelium from the gallbladder of a toad. 100 ×.

that insert on the terminal web. Microvilli vary from small, irregular projections to tall, closely packed uniform extensions of the cell. Observe microvillus borders with the light microscope in sections of the absorptive epithelia of the intestine (Figs. C37 and C38), gallbladder (Fig. C39), and excretory tubules (Figs. C40 and C41); they appear as a darkened shadow at the apical border of the cells. Microvillus borders

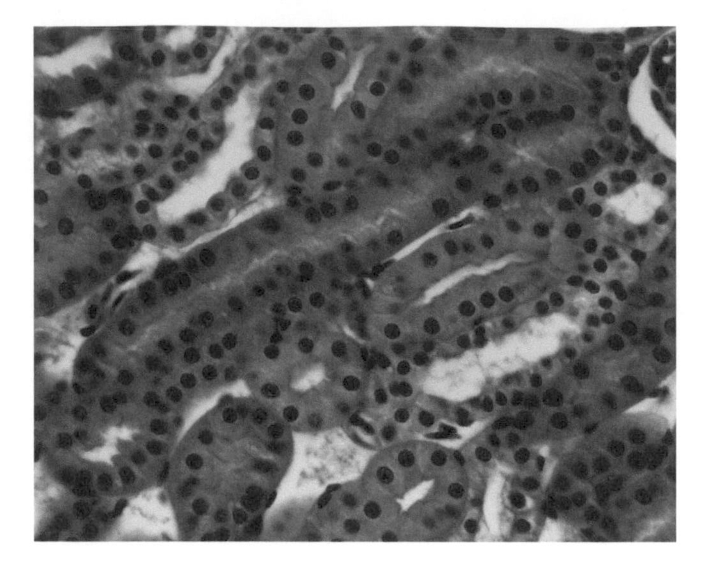

FIGURE C40 One gets the impression of a ragged BRUSH BORDER of microvilli lining the tubules in this section of the kidney of a bullfrog. The kidney consists largely of a tangle of tubules, many of which are seen in cross section in this photomicrograph. Ovoid nucleated red blood cells lurk in various capillaries. 40 ×.

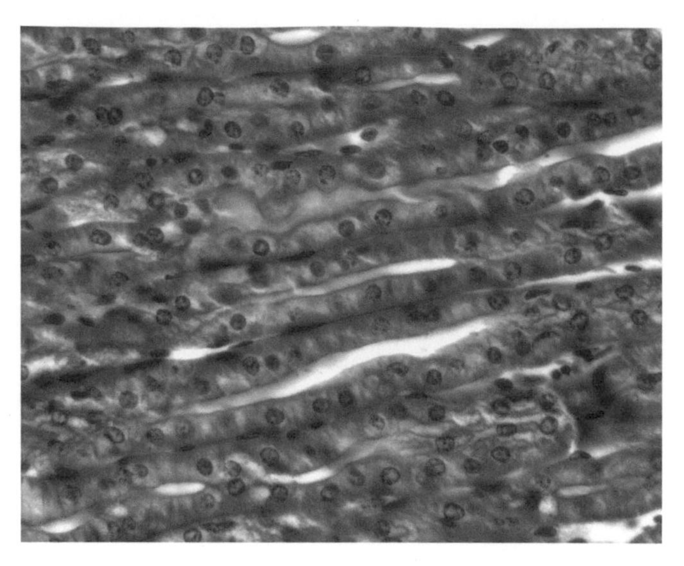

FIGURE C41 This is a photomicrograph of a section of the rectal gland of a dogfish, an outpouching of the gut that functions in maintaining the salt balance of the animal. The elaborate tubules in this section are cut longitudinally. A dark band can be discerned just under the free surface of the cells: this is the MICROVILLUS BORDER. Note the CAPILLARIES formed of squamous endothelial cells between the tubules. 40 ×.

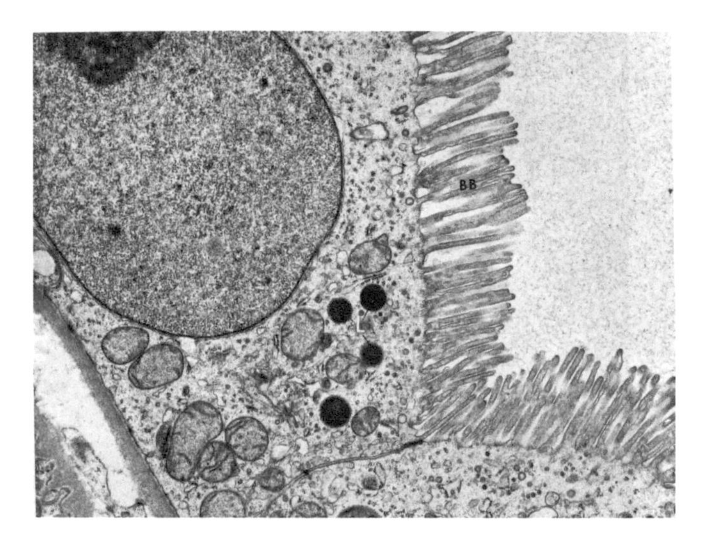

FIGURE C42 This electron micrograph shows a ragged brush border (BB) in a section of human kidney. 16,000 ×.

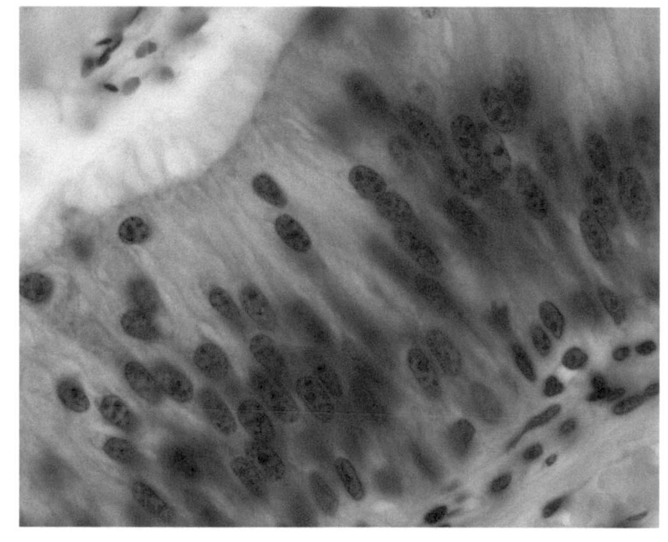

FIGURE C43 Photomicrograph of a section of the epididymis of a dog. STEREOCILIA—in reality extra-long microvilli—may enhance the environment in the epididymis for the storage of sperm. 100 ×.

have been referred to as STRIATED BORDERS in the intestine, where the microvilli are aligned with military precision, and BRUSH BORDERS in the kidney tubules, where their appearance is more ragged (Fig. C42). Although microvilli occur on many epithelial cells, their presence is often overlooked because they cannot easily be resolved with the light microscope.

4. *Stereocilia.* Identify long, hair-like processes on the epithelial cells of the vas deferens and the epididymis (Figs. C43 and C44). These resemble cilia but are nonmotile. They are often clumped together like the bristles of a wet brush and droplets of secretion can often be seen oozing at the tip. In electron micrographs, stereocilia are seen to be long, sinuous microvilli that lack filamentous cores (Fig. C45).
5. *Sensory "Hairs."* Certain sensory epithelial cells have hair-like receptor processes. Locate these in sections of olfactory epithelium (Figs. C46 and C47), taste buds (Figs. C48–C50), and the inner ear (Figs. C51a–C51c).

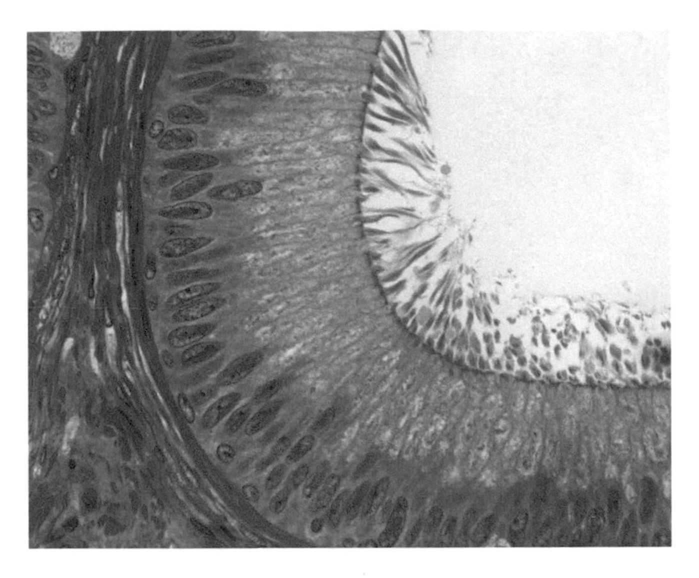

FIGURE C44 STEREOCILIA are clearly visible in this semithin section of epididymis. 63 ×.

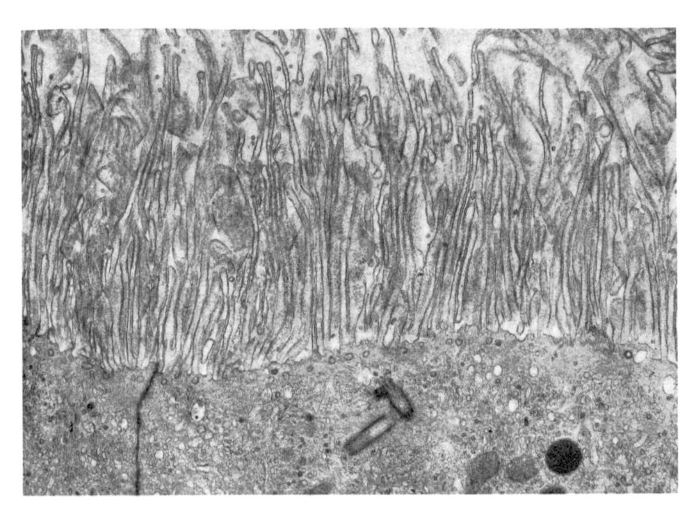

FIGURE C45 Electron micrograph of a section of the epididymis of a rabbit. The long, flexible STEREOCILIA are simply elongated microvilli. 16,000 ×.

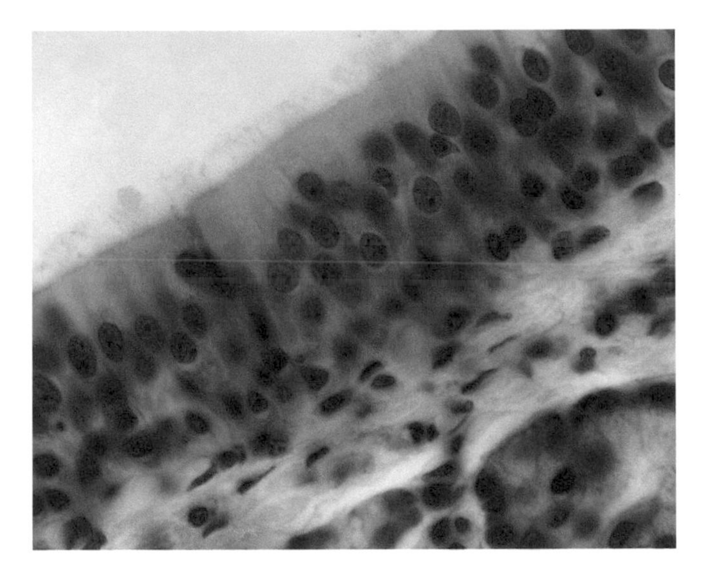

FIGURE C46 Section of the pseudostratified OLFACTORY EPITHELIUM from the nasal cavity of a mammal where respired air is sampled as it passes. The filamentous nature of the "sensory hairs" protruding into the space is just discernible. 100 ×.

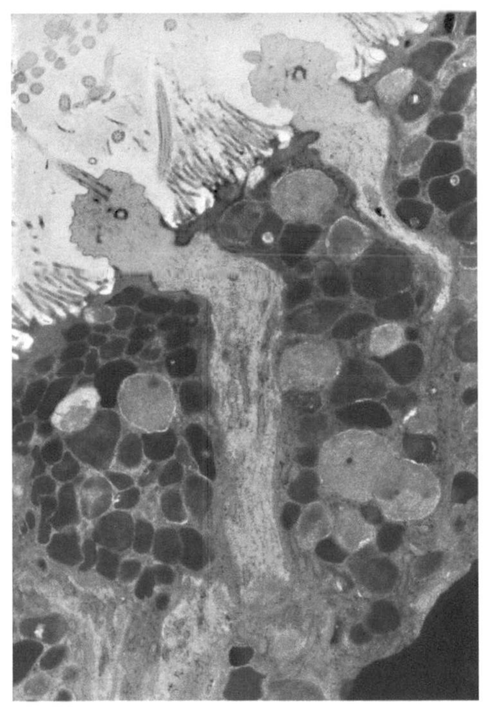

FIGURE C47 Electron micrograph of the olfactory epithelium of a leopard frog. OLFACTORY HAIRS are elongated cilia that extend into the nasal cavity from the pseudostratified olfactory epithelium. Some of the epithelial cells contain droplets of mucus. 20,000 ×.

Sensory hairs are a heterogeneous group of cytoplasmic processes, many of which resemble either cilia or stereocilia. Their ultrastructure will be studied when the sense organs are considered.

6. *Surface folds.* Examine electron micrographs of transporting epithelia showing the elaborate folding of the lateral and basal borders of the epithelial cells (Figs. C52a–C52c). These folds often interdigitate with those of neighboring cells. They create elaborate extracellular channels and greatly facilitate the transport of ions. Sometimes these channels are relatively simple (Fig. C53a), sometimes they are complex (Fig. C53b). They may take the form of intracellular tubules (Fig. C54). It appears that the more complex the channels, the greater the degree of concentration of ions. Note the distribution of mitochondria relative to these surface channels.

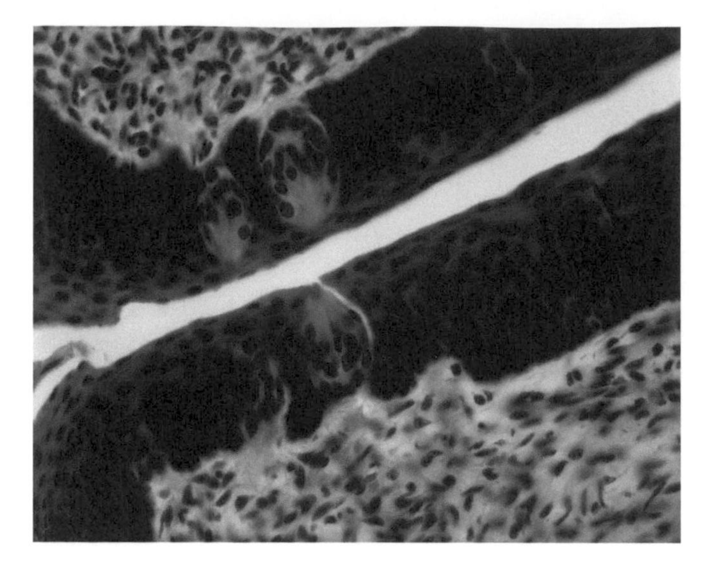

FIGURE C48 Taste cells form into TASTE BUDS in the tongue of mammals. Taste buds occupy protected trenches and spaces where they are not damaged by abrasive food but are bathed by the fluids resulting from mastication. Taste cells are roughly the shape of a banana with a "TASTE HAIR" protruding from one end. The taste hairs crowd into a small pore at the surface of the epithelium where they are able to sample the fluids that come their way. 40 ×.

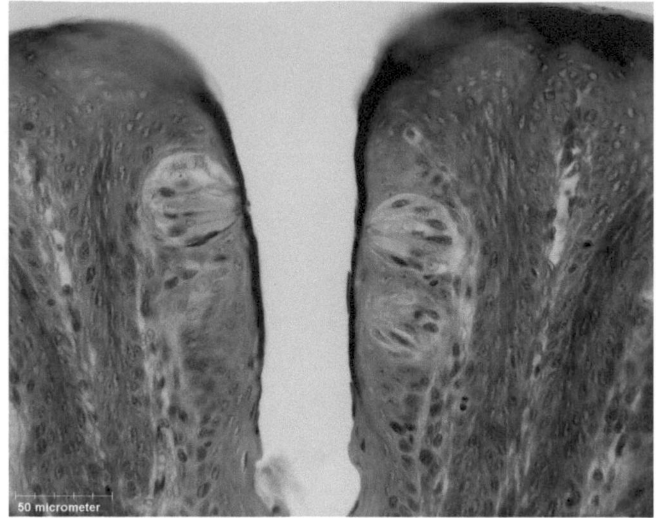

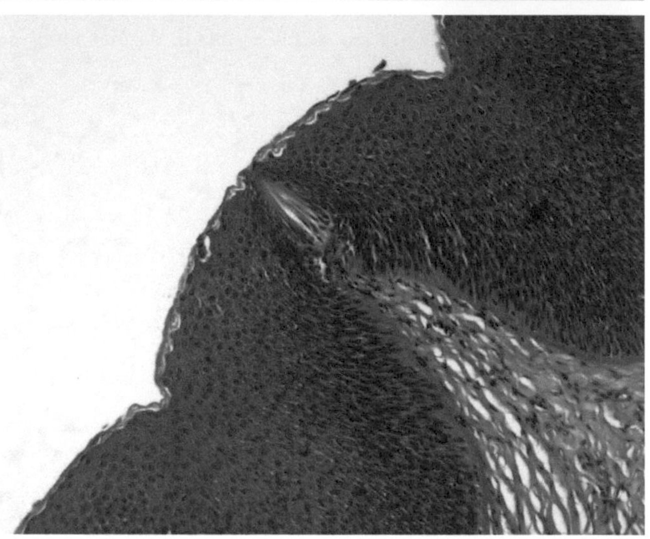

FIGURE C49 (a) Taste buds on the foliate papillae of rabbit tongue. This section is stained with iron hematoxylin. 20 ×. (b) "Taste bud" from the barbel (whisker) of a carp. These sense organs sample the ambient water and warn the fish against entering an inhospitable environment. 20 ×.

INTERCELLULAR SPECIALIZATIONS OF EPITHELIA

Desmosomes and Intercellular Bridges

Intercellular connections occur in many epithelia but they are most conspicuous in the lower layers of stratified epithelia, especially those subjected to mechanical stress. Note the strands between the cells of the deeply stained "prickle cells" in sections of mammalian epidermis (Fig. C55). Processes or "prickles" from adjacent cells meet and are firmly attached to each other by a DESMOSOME, which stains as a dense dot. Because of shrinkage during preparation of the tissue, the cells pull away from one another while the desmosomes hold, thereby producing a spiny appearance. (Visualize buttons straining on a fat man's shirt.) The intercellular connections, which have been artificially stretched, have long been known as INTERCELLULAR BRIDGES although there is no cytoplasmic continuity between adjacent cells.

Electron micrographs show that desmosomes consist of two dense ATTACHMENT PLAQUES fused with the inner leaflet of the cell membrane of adjacent cells, so that each cell contributes one half of the desmosome (Figs. C56 and C57). Each half desmosome is anchored to its cell web by 10-nm intermediate filaments, the TONOFILAMENTS. The membranes of adjacent cells do not fuse but are held apart by the material of the cell coat. At the center of this space is a thin dark line, the INTERMEDIATE LINE, that may have cementing properties. Single HALF DESMOSOMES are seen in some locations where epithelial cells abut on the basement membrane overlying the connective tissue (Fig. C58).

Terminal Bars and Junctional Complexes

In columnar or cuboidal epithelia, smoothly outlined rods of dense material have been described between the apical lateral margins of the cell. Identify these TERMINAL BARS in sections of intestine stained with iron hematoxylin (Fig. C59). The electron microscope has shown, however, that instead of crevices between cells being filled with calking, as had been originally supposed, the membranes of adjacent cells are joined near their free surfaces by JUNCTIONAL COMPLEXES (Figs. C60a, C60b, C61). The three components of a junctional complex are

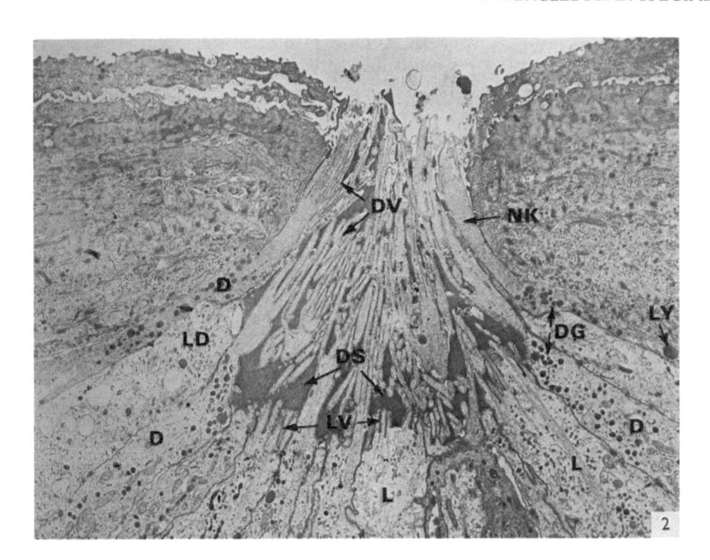

FIGURE C50 Electron micrograph of the apical ends of cells in the taste bud of a rabbit. Long microvilli from these cells crowd into the taste pore. *D*, Dark cells; *DG*, Dense granules; *NK*, Long necks; *DV*, Long microvilli; *L*, Light cells; *LV*, Short microvilli; *DS*, Dense substance; *LD*, Light cell; *LY*, Lysosome. 6700 ×.

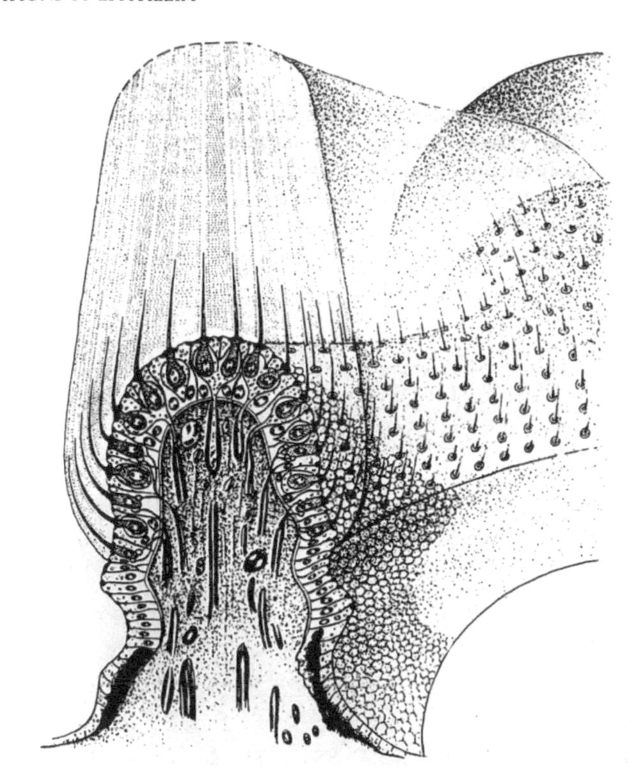

FIGURE C51a Sensory cells in the inner ear of a mammal detect movement. When the animal moves, fluids flow past this structure, stimulating the sensory "hairs" of these cells. The sensory hairs are embedded in a gelatinous material.

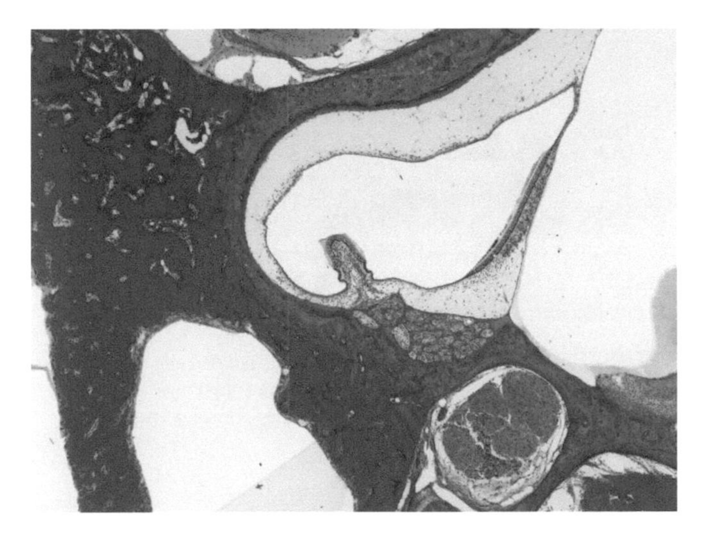

FIGURE C51b The sensory organ for movement is the CRISTA AMPULLARIS located in the fluid-filled labyrinth of the inner ear. 2.5 ×.

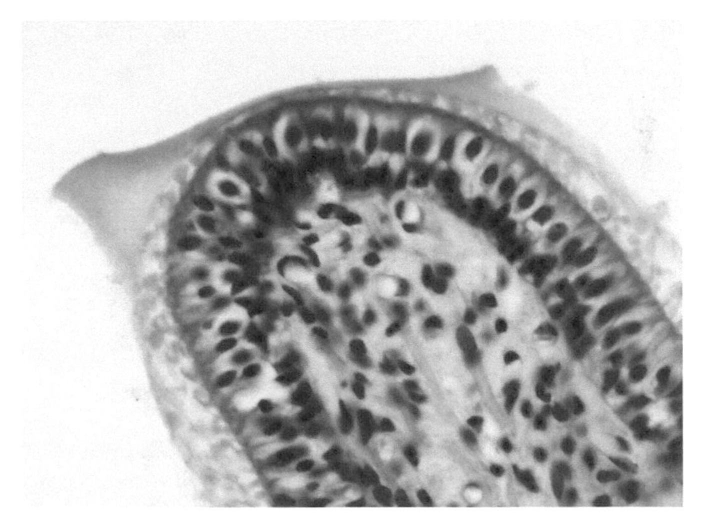

FIGURE C51c The CRISTA AMPULLARIS is a ridge bearing sensory cells at its peak. These cells extend sensory hairs into a cap of gelatinous material, the CUPULA. Movement of the body causes a flow of fluid past the crista, thereby deflecting the cupula and providing the sensation of movement. 40 ×.

shown diagrammatically in Fig. C62. In the superficial ZONULA OCCLUDENS or TIGHT JUNCTION, the outer leaflets of the membranes of adjacent cells come together at close intervals resulting in multiple sites of membrane fusion. These lines of contact is how well in freeze-fracture preparations of epithelial cells (Fig. C63). The zonula occludens appears as a series of strands that provide a seal for the junction. These strands consist of apposing, fused INTRAMEMBRANOUS PARTICLES (Fig. C64). These freeze-fracture images show that some tight junctions are tighter than others—Fig. C65 is an example of a "leakier" tight junction than that shown in Fig. C63. Below the zonula

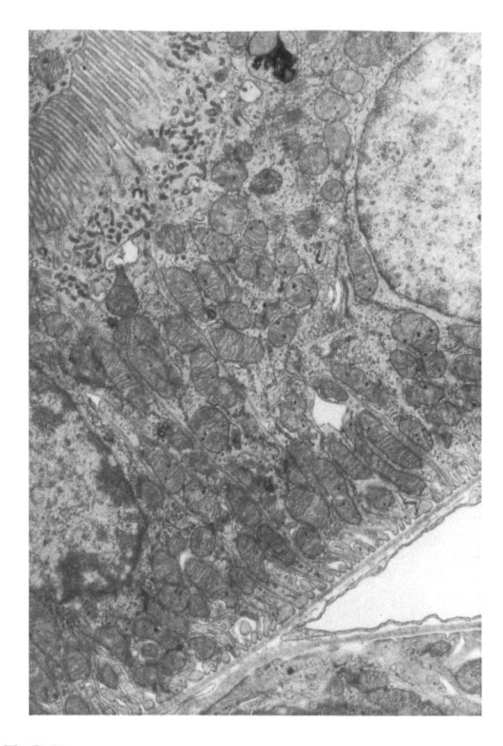

FIGURE C52a The transporting epithelium of a tubule in the kidney of a mouse bristles with MICROVILLI on its free apical surface (upper left). Note that the cell membrane at the bases of the cells is deeply folded, forming little hollows resembling the spaces of an egg carton. Inside these spaces—much like eggs in their carton—are MITOCHONDRIA. 21,000 ×.

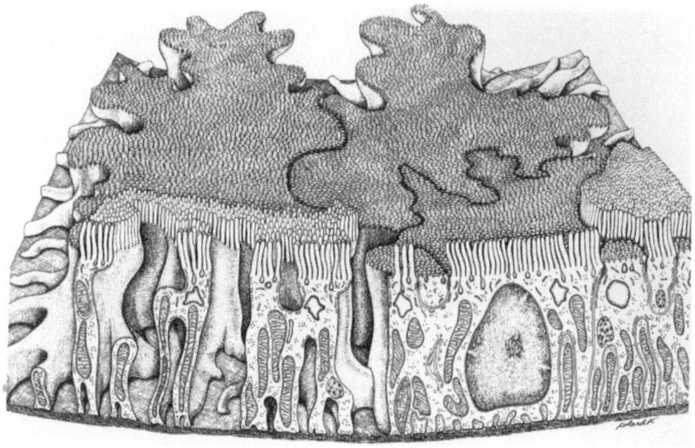

FIGURE C52b Diagram of a cell from the kidney tubule showing its basal infoldings that interdigitate with similar extensions of adjacent cells.

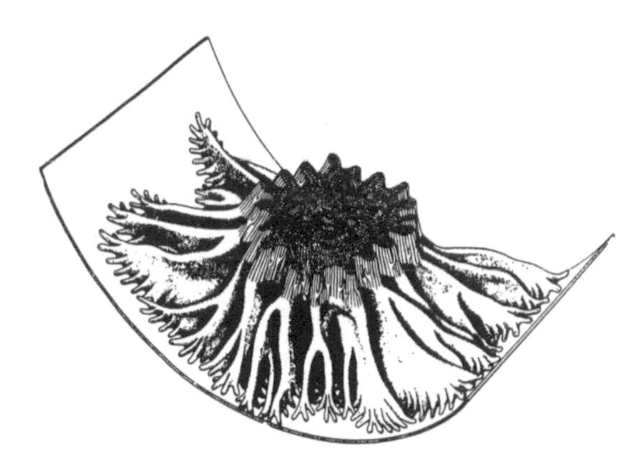

FIGURE C52c Diagram summarizing the many ways that a cell of the kidney tubule enhances its surface area as well as the channels typical of a transporting epithelium: microvilli, intercellular channels, and basal infoldings. Compare this diagram to the electron micrograph in Fig. C52a.

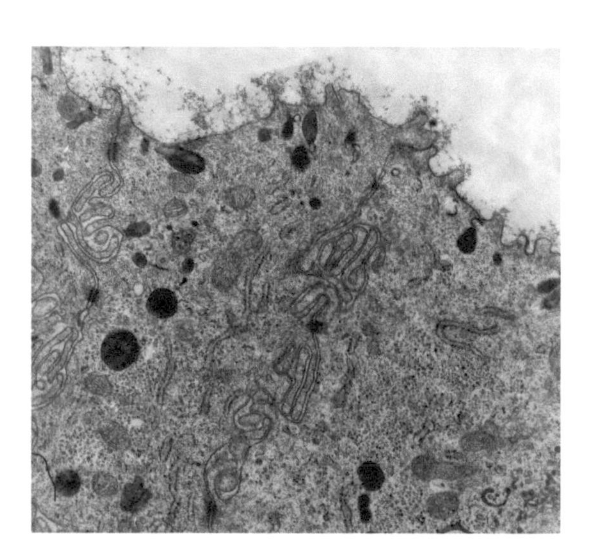

FIGURE C53a The membranes of epithelial cells lining the urinary bladder of a marine toad (*Bufo marinus*) are elaborately folded and interdigitated thereby creating complex extracellular channels.

occludens, the membranes are separated by a space of about 20–30 nm filled with material of low electron density; this is the ZONULA ADHAERENS. A moderately electron-dense material is associated with the cytoplasmic leaflet of the plasmalemma, and 6-nm actin microfilaments appear to anchor this electron-dense material to the terminal web. (Accumulations of stain within the zonula adhaerens give the appearance of the terminal bars as seen with the light microscope.) The two zonulae actually encircle each cell forming a seal between the lumen and the intercellular space. Additional strength is applied to the junction by the deeper presence of DESMOSOMES or MACULAE ADHAERENTES (*Singular*: Macula adhaerens), which exist as the spots already described. Junctional complexes are firmly rooted in the terminal webs of epithelial cells, thereby providing great tensile strength across the surface.

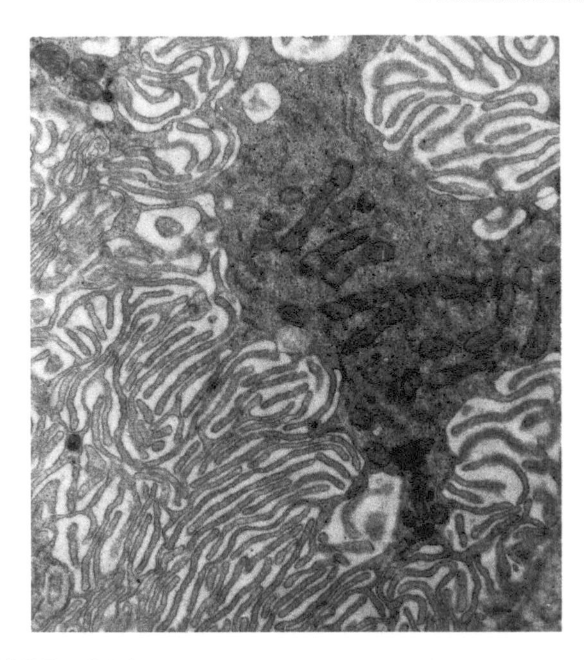

FIGURE C53b These extracellular channels may become extremely complex as in this section of cells from a kidney tubule of a snake. Note the aggregation of MITOCHONDRIA in the cytoplasmic island at the center. 24,000 ×.

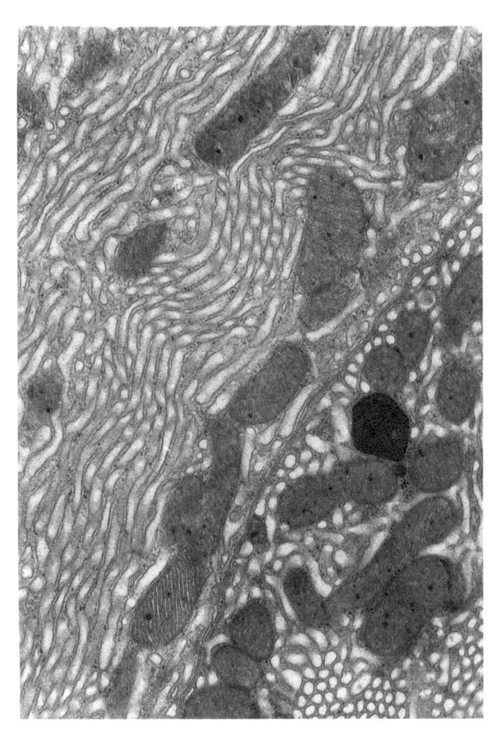

FIGURE C54 Extracellular channels may take the form of tubules. Some of the cells in the gill of the lamprey are actively pumping salt out of the body. These CHLORIDE CELLS are packed with smooth tubules that are continuous with the plasma membranes of the cells and their space is continuous with the extracellular space. Note the presence of mitochondria. 43,000 ×.

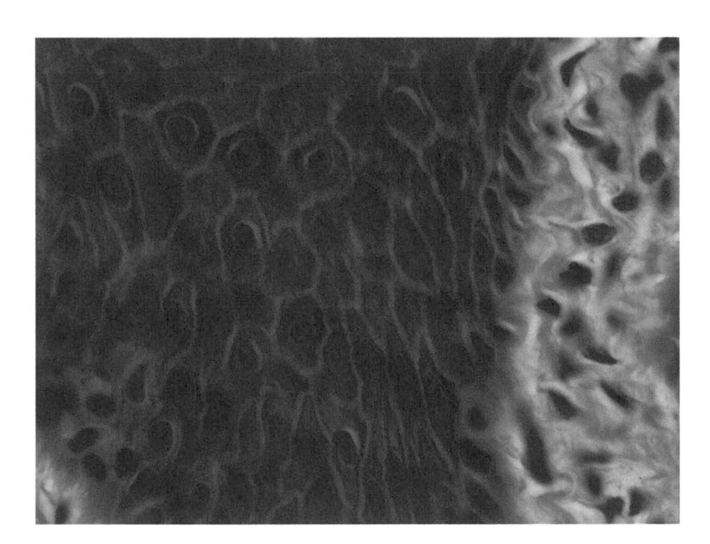

FIGURE C55 Photomicrograph of a section of human skin showing "prickle cells" in the stratified squamous epithelium. The cells are held together by attachment plaques, the DESMOSOMES, that hold when the cells shrink during fixation. INTERCELLULAR BRIDGES between adjacent cells are an artifact produced by shrinkage of the cells; there is no cytoplasmic continuity between adjacent cells. 63 ×.

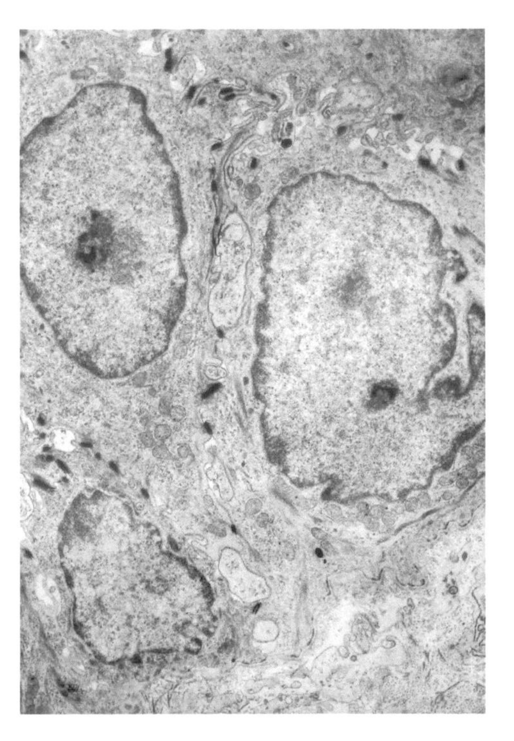

FIGURE C56 The DESMOSOMES (*D*) in this electron micrograph of a section of the skin of a newborn rat appear as black structures between adjacent cells. Each desmosome consists of two HEMIDESMOSOMES, one from each of the adjacent cells. 12,500 × *Inset*: 124,00 ×.

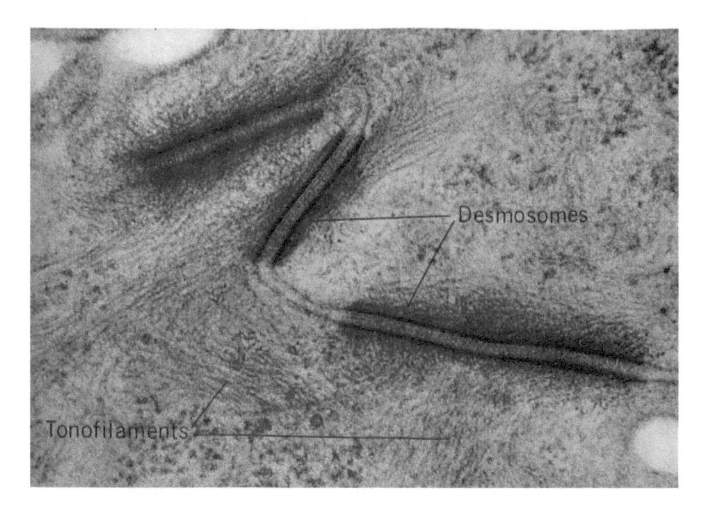

FIGURE C57 Electron micrograph of DESMOSOMES in the stratified squamous epithelium of the cheek of a hamster. Each half desmosome is anchored to its cell web by 10-nm intermediate filaments, the TONOFILAMENTS. The membranes of adjacent cells do not fuse but are held apart by material of the cell coat. At the center of this space is a dark line, the INTERMEDIATE LINE, that may have cementing properties. 70,000 ×.

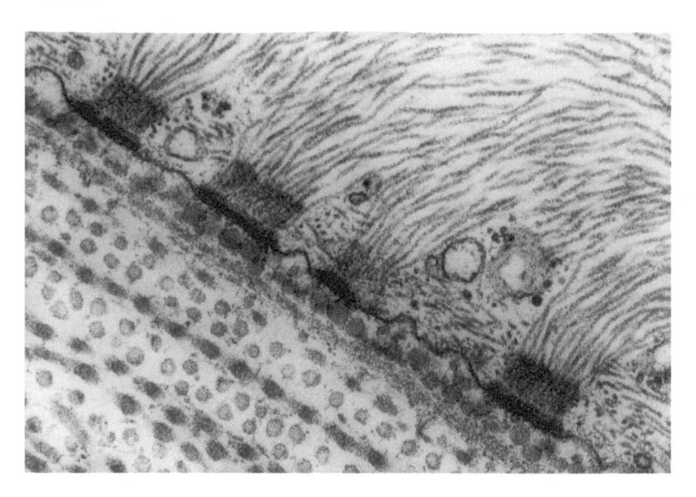

FIGURE C58 In locations where a hemidesmosome's attentions are unrequited, such as the junction of an epithelial cell with its basement membrane, hemidesmosomes appear alone. In this electron micrograph of the base of epidermis of an amphibian, tonofilaments converge on the hemidesmosomes, which appear firmly affixed to the basement membrane. The material at the lower left is the connective tissue layer of the skin.

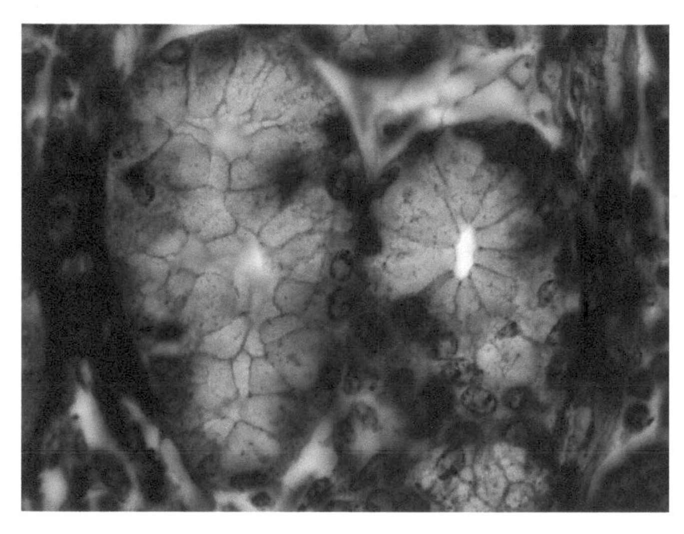

FIGURE C59 TERMINAL BARS between epithelial cells in the small intestine of a piglet. It is thought that the black hematoxylin accumulates in the zonula adhaerens of the junctional complex. Glancing sections of the junctions appear like chicken wire; cut in section, the terminal bars appear as dots. 63 ×.

Gap Junctions

Gap junctions provide communicating channels that bridge a 2- to 3-nm gap between adjacent cells and allow small molecules and ions to pass. Note their appearance in freeze-fracture preparations (Fig. C66). They consist of a variable number of 11-nm intramembranous particles that are packed together more closely than the somewhat smaller particles associated with the rest of the membrane (Fig. C67). Note the appearance of gap junctions in electron micrographs of sections (Fig. C68).

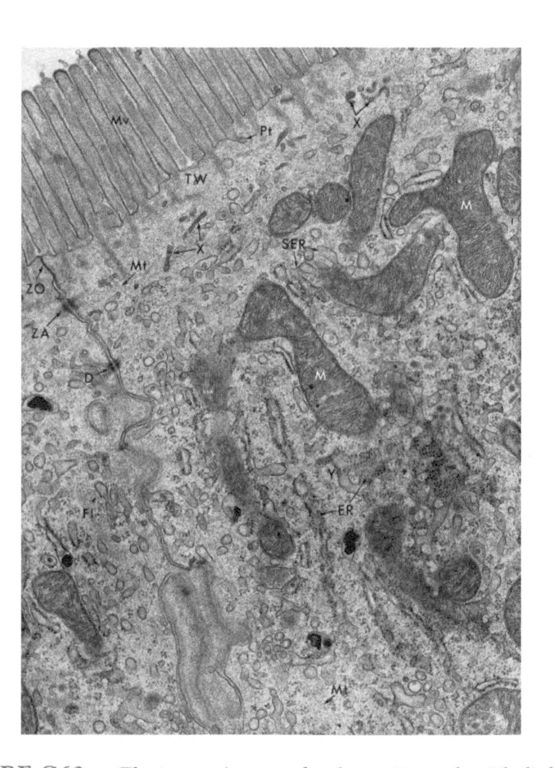

FIGURE C60a Electron micrograph of a section of epithelial cells in the intestine of a rat. A JUNCTIONAL COMPLEX at the extreme left seals the intercellular spaces of the epithelial cells from the lumen at the top left. In life, the junctional complex encircles the cell like a belt. It consists of three parts: the ZONULA OCCLUDENS (*ZO*) or TIGHT JUNCTION at the bases of the microvilli where the adjacent plasma membranes are fused; the ZONULA ADHAERENS (*ZA*) or TERMINAL BAR below; and the MACULA ADHAERENS or DESMOSOME (*D*). As we saw earlier, the DESMOSOME is a small, disc-shaped area where adjacent plasma membranes adhere. (*Zonula* is Latin for a little belt, *macula* is a spot). 42,000 ×.

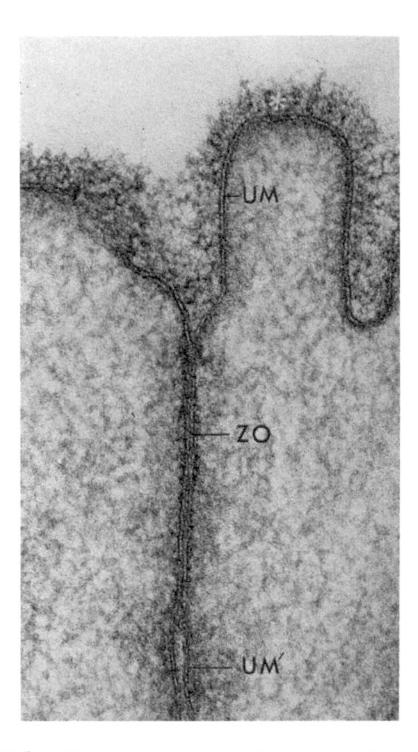

FIGURE C60b This is an electron micrograph of a section through the epithelial lining the stomach of a bat. The surfaces of the cells consist of a trilaminar unit membrane (*UM*) on their free and lateral surfaces. These membranes are covered by a fuzzy SURFACE COAT of polysaccharide material. At the opposing surfaces, the outer dense lamellae of the plasma membranes are united to form a TIGHT JUNCTION or ZONULA OCCLUDENS (ZO). This structure is believed to act as a seal preventing the diffusion of even small molecules between the cells. 180,000 ×.

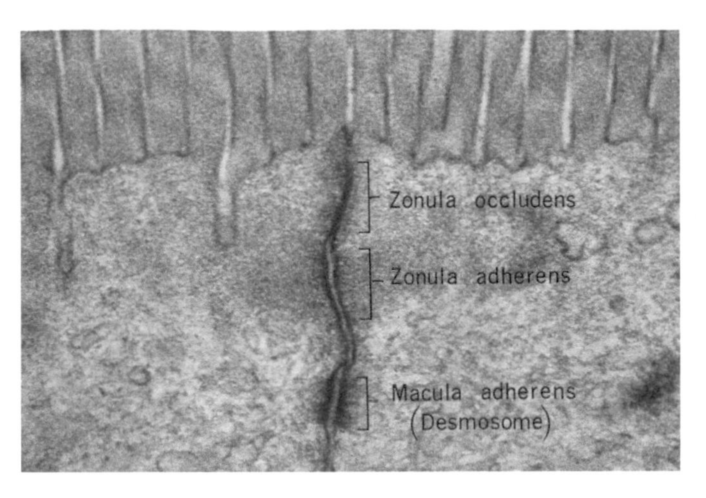

FIGURE C61 JUNCTIONAL COMPLEX from the intestinal epithelium of the hamster. The adjacent cell membranes appear fused in the ZONULA OCCLUDENS. There is an intercellular space of about 20 nm between the membranes in the ZONULA ADHAERENS or TERMINAL BAR; the thickenings of the membrane appear to be anchored by filamentous material in the adjacent cytoplasm. The MACULA ADHAERENS or DESMOSOME consists of two disc-shaped spots on opposing membranes. Fine TONOFILAMENTS extend into the adjacent cytoplasm. 70,000 ×.

GLANDULAR EPITHELIA

Secretion is one of the major functions of epithelia. Some secretion is carried on by a single cell in an epithelium but in many cases a GLAND is formed by an invagination of the epithelium that grows into the supporting connective tissue. When a duct persists to connect the gland to the outer surface, the gland is EXOCRINE, but when the cellular connection to the surface is lost, the gland secretes directly into the blood stream and is ENDOCRINE (Fig. C69). Since the endocrine cells no longer have a free border, they are sometimes referred to as EPITHELIOID CELLS. The endocrine glands will be considered in another exercise. Exocrine glands may be classified on the basis of several criteria (Fig. C70):

- Number of cells (unicellular or multicellular)
- Shape of the secretory units (tubular or alveolar)
- Type of secretion (mucous or serous)
- Multiplicity of secretory units (simple or compound)

Unicellular Gland: Goblet Cell

Goblet cells produce a mucoid secretion and are present in many organs. In sections of the intestine, note the clear spaces, oval in outline, among the columnar epithelial cells (Figs. C71a and C71b). The clear spaces are droplets of mucus surrounded by a thin layer of cytoplasm. Locate the deeply staining basal nucleus surrounded by granular cytoplasm. Observe several goblet cells in various phases of secretion (Figs. C72 and C73).

Multicellular Glands (Fig. C74)

1. *Simple Tubular Glands.* Note the tubular invaginations that extend into the connective tissue in the lining of the colon of a dog (Fig. C75). Identify the narrow distal NECK, the long expanded BODY, and the blind end, the FUNDUS. The central cavity of the gland is the LUMEN. Note the large numbers of goblet cells lining the tubular glands of the colon. This micrograph shows an almost ideal situation where many of the glands are cut from

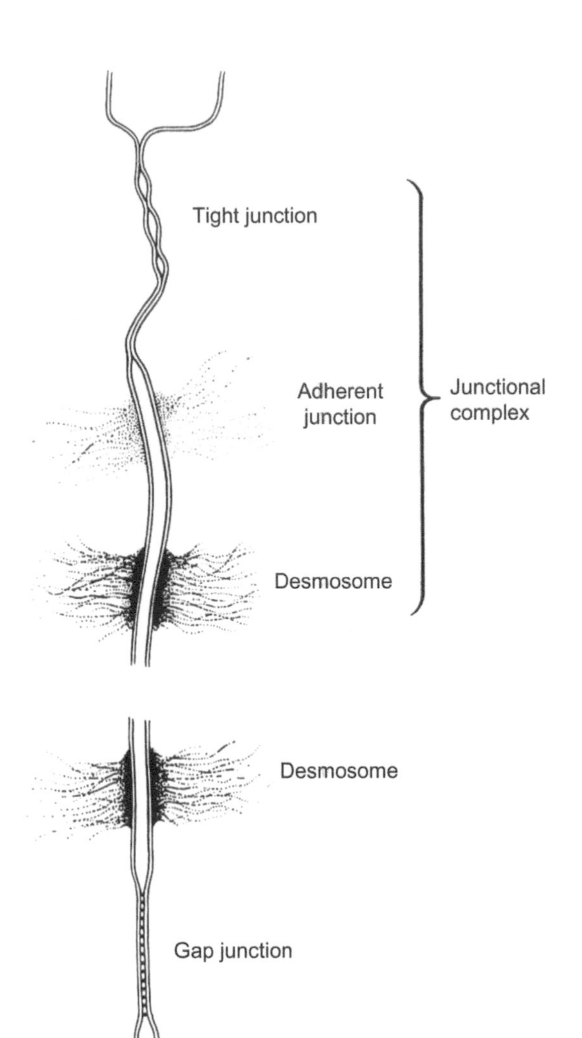

FIGURE C62 The three components of a tight junction (as well as other junctions) are shown in this drawing.

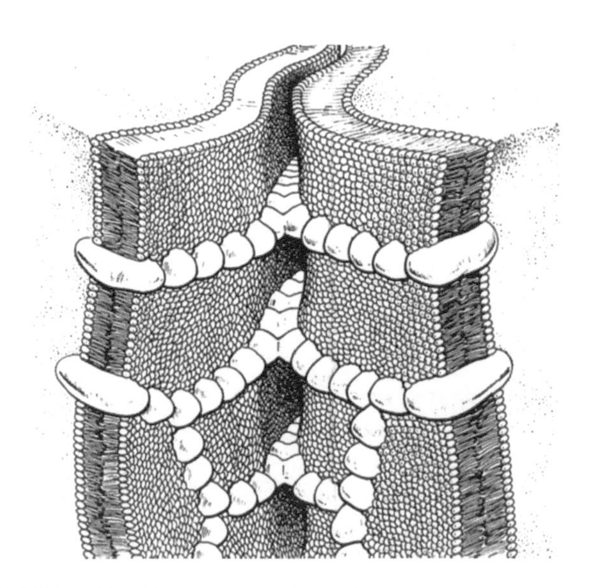

FIGURE C64 Diagrammatic view of a TIGHT JUNCTION or ZONULA OCCLUDENS. Two trilaminar plasma membranes are pulled apart in freeze fracture, revealing the intramembranous particles that form the zonula occludens. These particles fuse and hold the membranes tightly together. Note that the fracture in Figs. C63 and C65 has occurred *within* one of the trilaminar membranes, splitting the protoplasmic face (P), from the extraplasmatic (E) face.

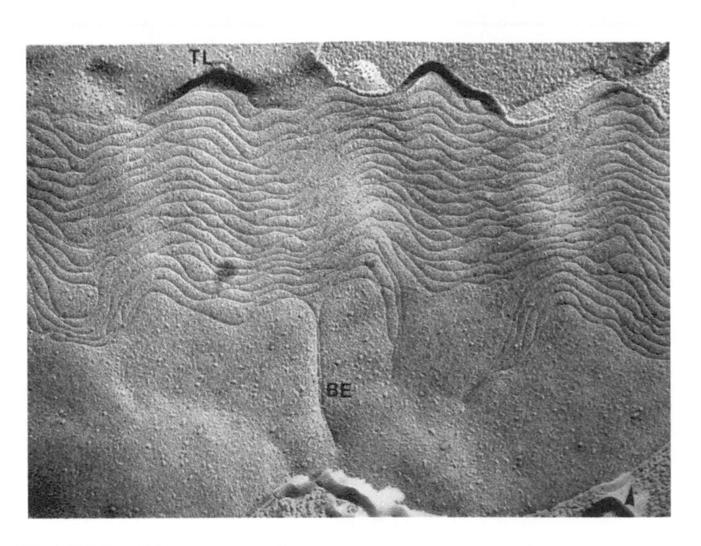

FIGURE C63 Freeze-fracture preparation of the ZONULA OCCLUDENS from the distal convoluted tubule of the kidney of a snake. The lumen of the tubule is at the top. The grooves on the E-face (extraplasmatic fracture face) form a complex elongate meshwork. The arrowhead indicates the direction of shadowing. *TL*, tubular lumen; *BE*, basal extensions. 102,000 ×.

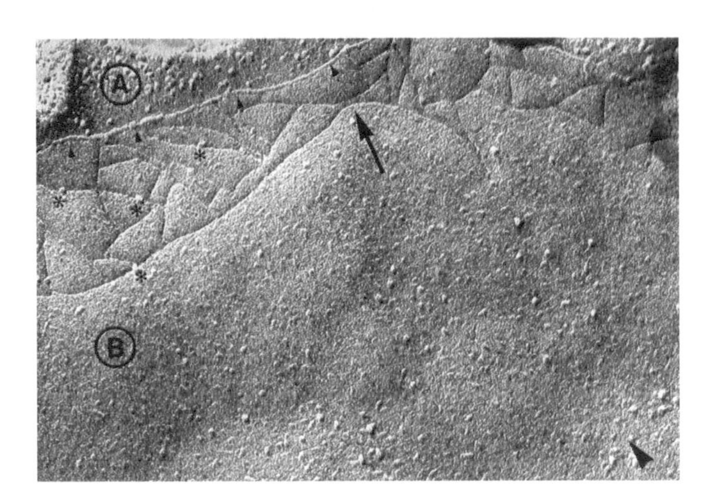

FIGURE C65 Some tight junctions are tighter than others. Compare this tight junction from the proximal convoluted tubule of a snake, with that shown in Fig. C63. This junction is much less complex and forms a less effective seal between two cells. Again, the ZONULA OCCLUDENS consists of interconnected grooves on the B-face. *A*, A-face; *B*, B-face; *, B-face particles in grooves except at large arrow; *Small arrowheads*, B-face ledge; *Large arrowhead*, direction of shadowing. 125,000 ×.

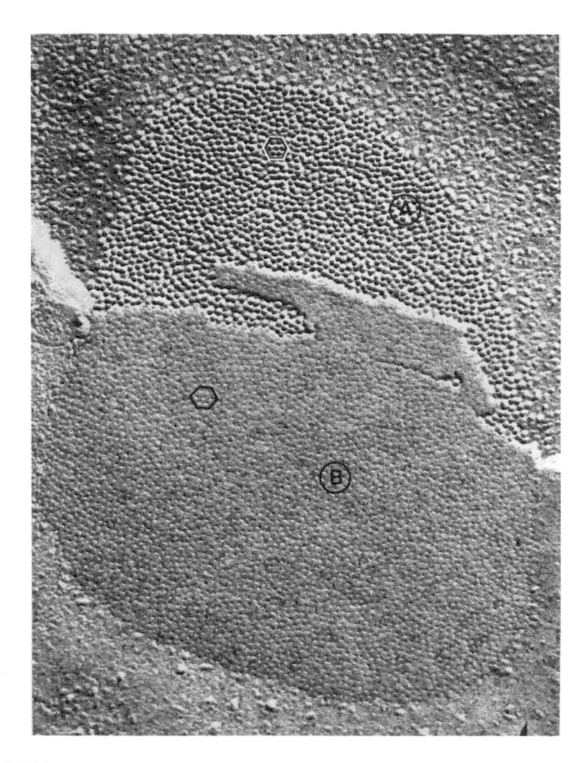

FIGURE C66 Freeze-fracture preparation of a gap junction in the proximal convoluted tubule of the snake kidney. INTRAMEMBRANOUS PARTICLES (A) are shown on the protoplasmic or P face at the top. Pits in the E-face are shown at the bottom (B). The arrays are arranged in a roughly hexagonal distribution. 203,000 ×.

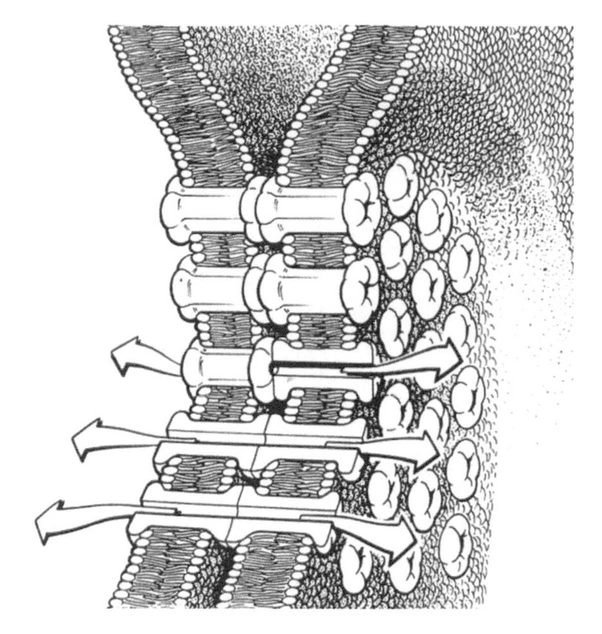

FIGURE C67 Intramembranous particles of gap junctions are visualized as small spools that provide electrical continuity between cells.

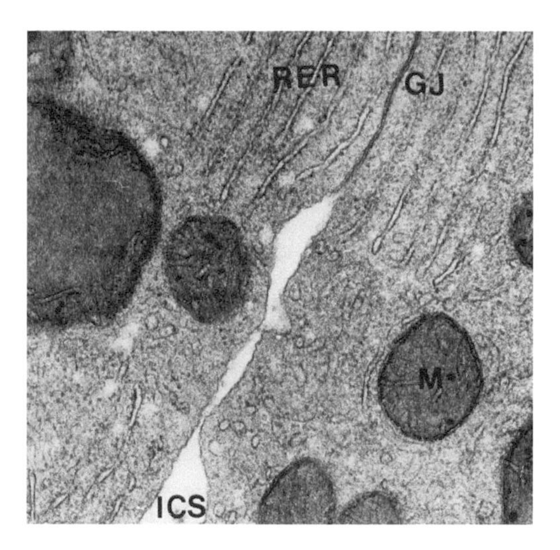

FIGURE C68 Electron micrograph of a section of the proximal convoluted tubule of the snake kidney showing a gap junction (*GJ*) between two cells. *ICS*, intercellular space; *M*, mitochondrion; *RER*, granular endoplasmic reticulum. 39,000 ×.

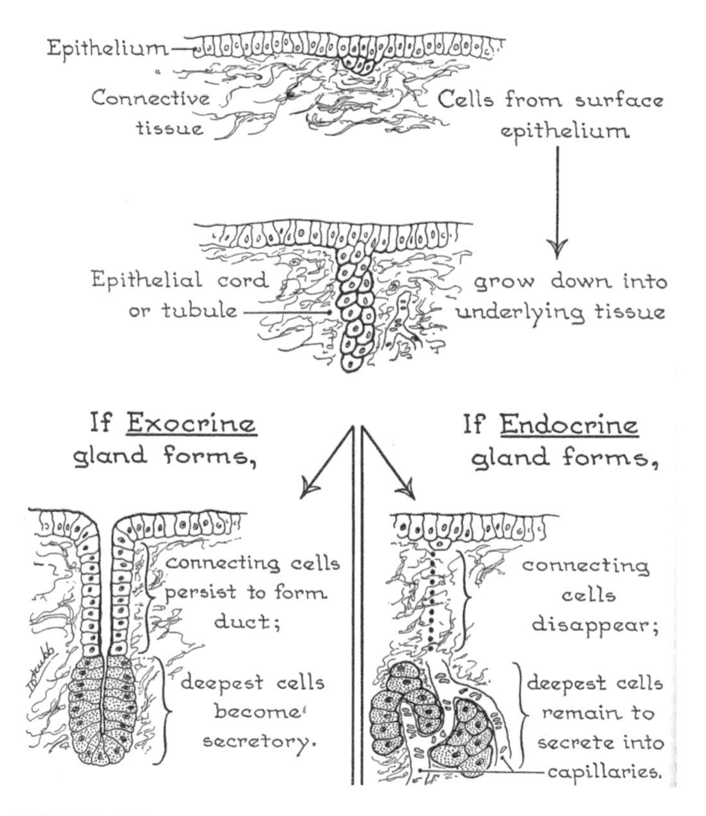

FIGURE C69 Glands develop from an epithelium, which sends a stalk of cells into the vascular connective tissue below. This stalk may become hollow and retain its connection to the surface. The deeper cells become secretory and pour their products onto the surface through the hollow stalk of cells. This is an EXOCRINE GLAND. Alternatively, the stalk of cells may disappear so that the connection to the surface is lost. The secretory cells remain in the vascular connective tissue and pour their products—HORMONES—into the blood stream. This is an ENDOCRINE GLAND.

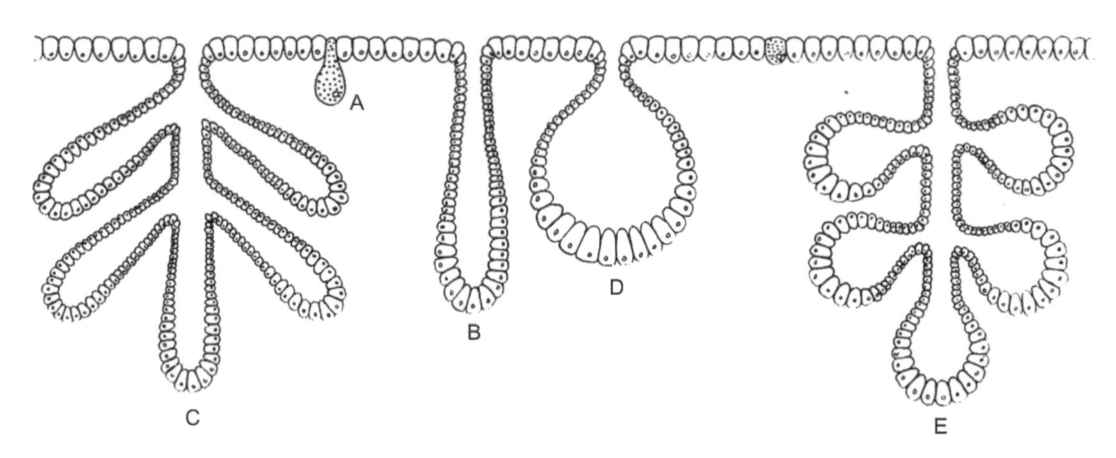

FIGURE C70 Diagram of UNICELLULAR and MULTICELLULAR GLANDS. (*A*) Unicellular gland; (*B*) simple tubular gland; (*C*) compound tubular gland; (*D*) simple alveolar gland; (*E*) compound alveolar gland.

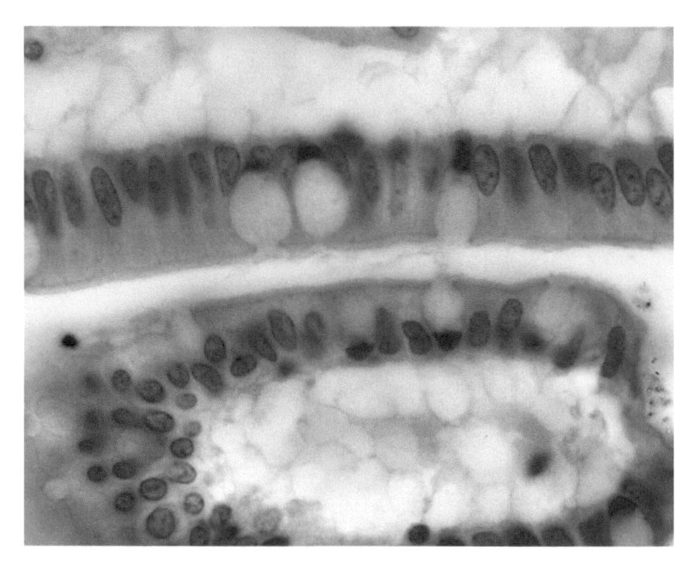

FIGURE C71a Section of human intestine (jejunum). Goblet cells abound in the simple columnar epithelium. Note the striated border on the epithelial cells.

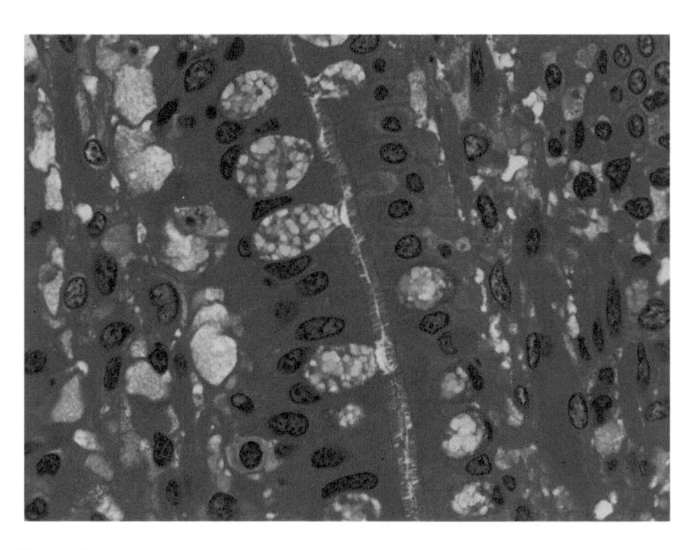

FIGURE C71b Semithin section of the jejunum of a monkey showing abundant goblet cells containing granular mucus. Note the microvillus border on the epithelial cells and the various stages of secretion. 63 ×.

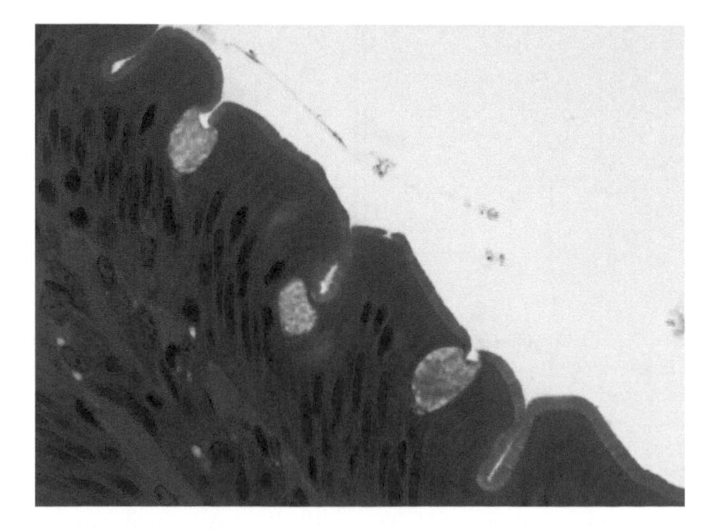

FIGURE C72 Semithin section of the intestine (duodenum) of a monkey showing several goblet cells and a striated border. 63 ×.

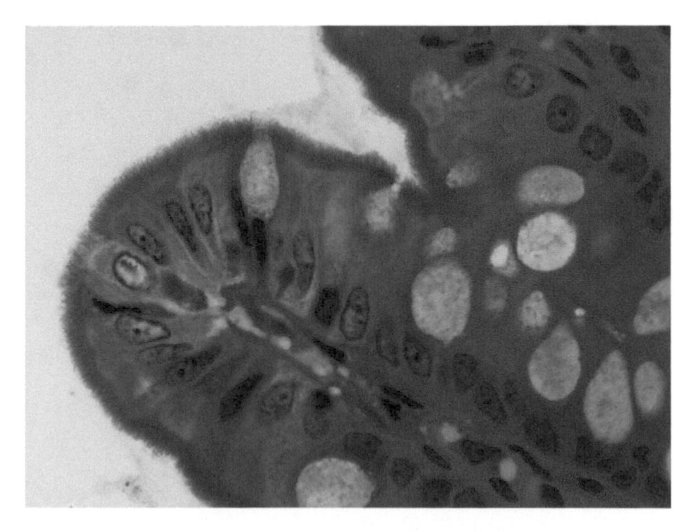

FIGURE C73 Semithin section of the large intestine (colon) of a monkey. 63 ×.

If secretory portion is:

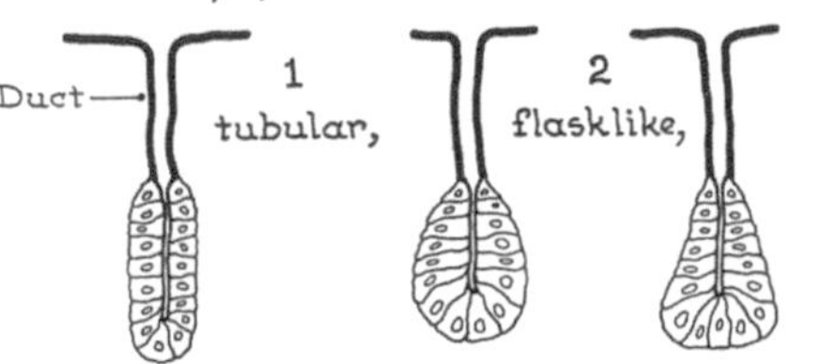

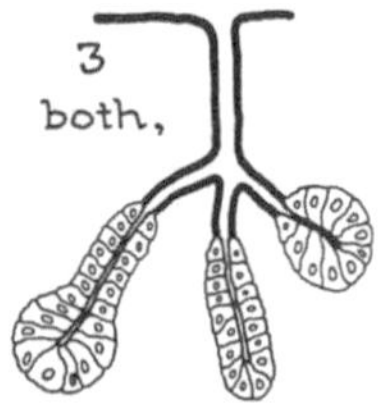

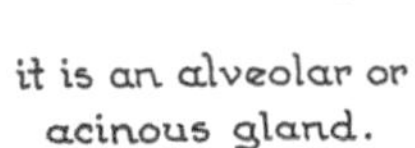

If duct doesn't branch:

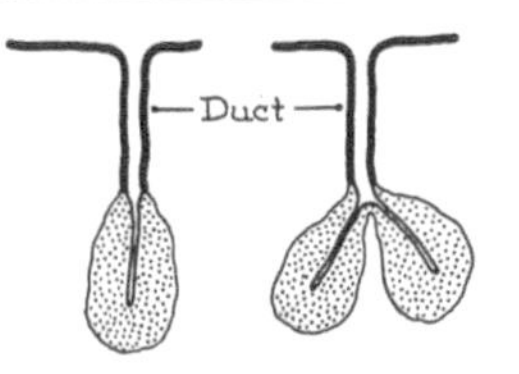

If duct branches:

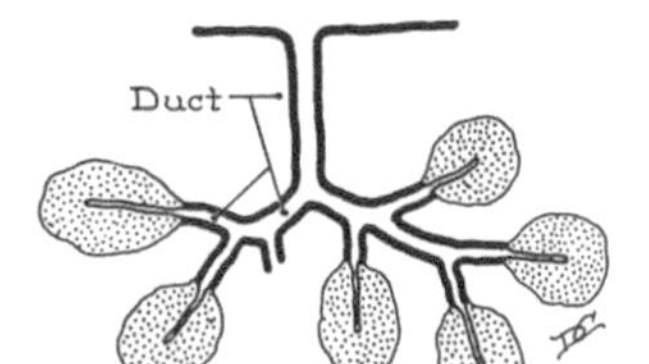

FIGURE C74 Multicellular glands are defined on the basis of the shape and multiplicity of their secretory units.

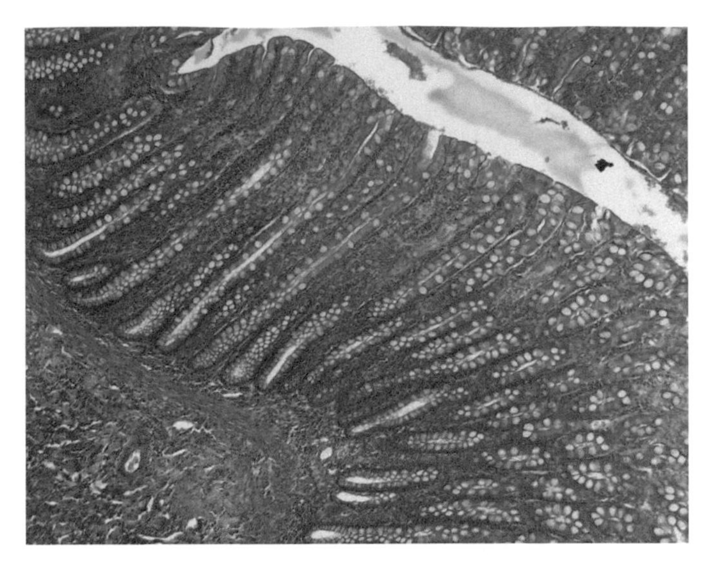

FIGURE C75 Section through the lining of the COLON of a dog. The epithelium is elaborated into myriad SIMPLE TUBULAR GLANDS, abundantly provided with mucus-secreting goblet cells that rest on vascular connective tissue in the lower left corner. The narrow distal NECK, is clear in a few of the glands at the upper left. The BODY, is long and leads to a blind end, the FUNDUS. The central space is the LUMEN. This micrograph shows an almost ideal situation where many of the glands are cut from top to bottom, clearly demonstrating their tubular nature. Often, however, the glands are cut obliquely, as shown at the lower right, and their tubular nature is more difficult to discern. 10 ×.

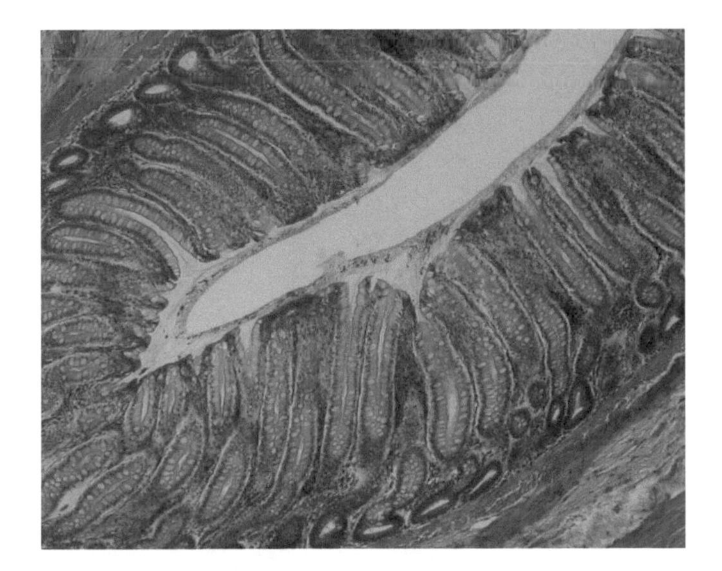

FIGURE C76 Section through the lining of the RECTUM of a dog. The simple tubular glands secrete large amounts of mucus that assists hardening waste materials to slide along the gut. Note that the section is not perfectly parallel to the line of some of the glands, and their connections to the FUNDUS is lost. 10 ×.

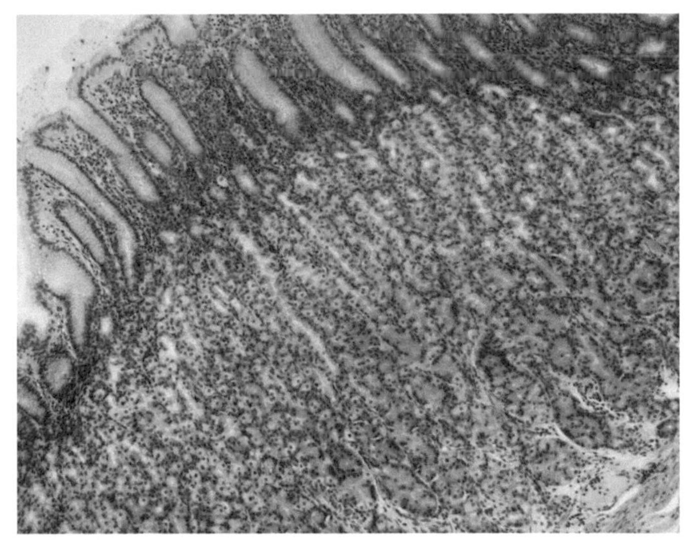

FIGURE C77 The simple tubular nature of the glands shown in this section of the lining of human STOMACH is difficult to discern. The NECKS of the glands at the upper left are richly provided with mucous cells (the mucus helps protect the lining of the stomach from self-digesting). The BODIES of the glands are long and twisted, thereby producing the confusing picture at the lower two-thirds of the micrograph. Can you identify the FUNDUS of any of these glands? 10 ×.

top to bottom, clearly demonstrating their tubular nature. Often, however, the glands are cut obliquely, as shown at the lower right, and their tubular nature is more difficult to discern. Tubular glands are characteristic of the large intestine and stomach (Figs. C76 and C77). The usual formation of the vertebrate stomach is exploited in an amazing way in the GIZZARD of birds (Figs. C78a and C78b) where the simple tubular glands

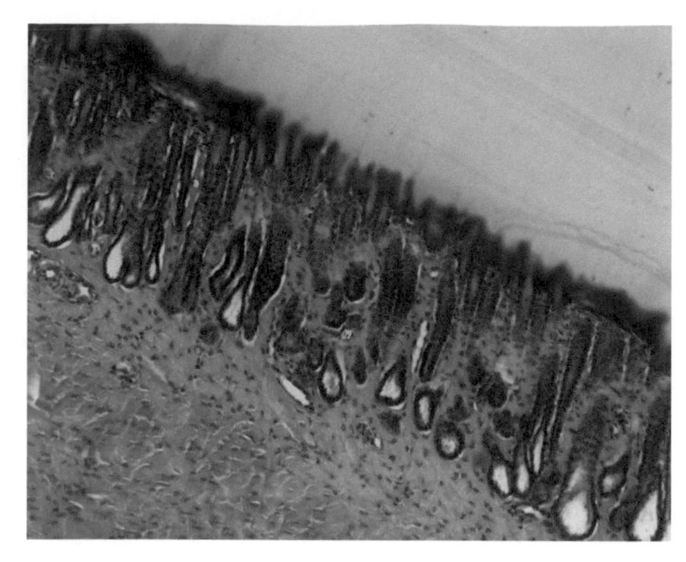

FIGURE C78a An unusual function of simple tubular glands is seen in this section of the GIZZARD of a pigeon. The glands secrete a viscous substance that hardens on arrival in the lumen of the gizzard. This material issues from the glands in the form of "pencils" that form the abrasive surface that grinds the food before it enters the stomach proper. The pale pink material at the upper right is this lining. Most of the section consists of a thick layer of muscle. 20 ×.

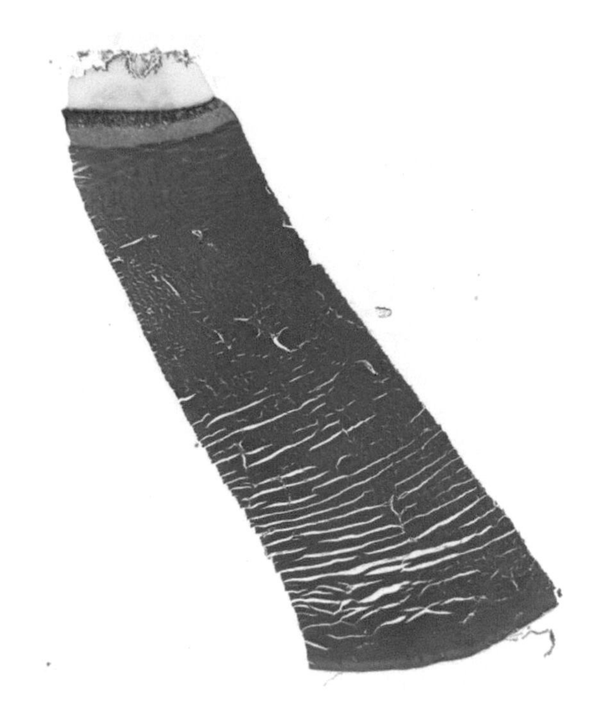

FIGURE C78b The adaptation of the avian stomach for the grinding function of the GIZZARD is well shown in this micrograph taken at a low power. The translucent grinding surface is seen at the top. Note that the secretion of the "pencils" has produced a rough, abrasive surface. Below this lining is the layer of simple tubular glands that produce the lining. The thick, red area that occupies most of the micrograph is a thick layer of muscle that grinds the food. 1 ×.

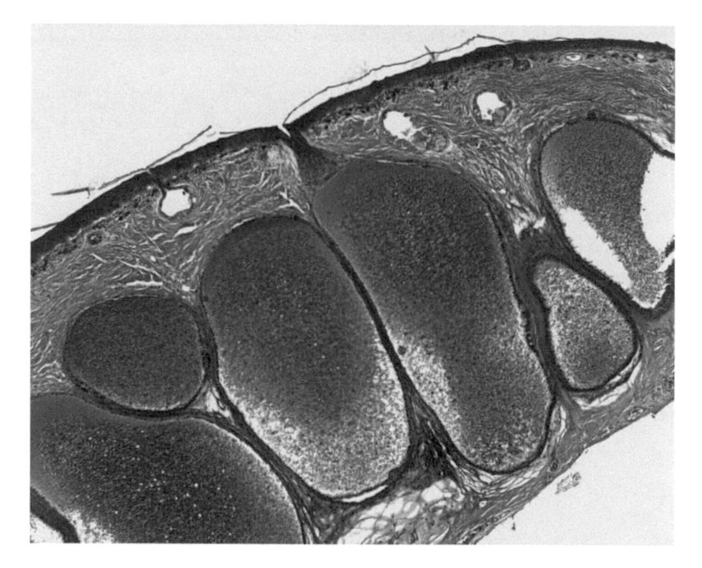

FIGURES C79a Section of a large "wart" in the skin of the head of a toad. The wart consists of several simple ALVEOLAR GLANDS in a cluster. These glands have pushed below the epithelium (from which they are derived) and lie in the vascular connective tissue below. The large glands with the dark secretion are POISON GLANDS. Note the much smaller MUCUS-SECRETING alveolar glands in the upper part of the section. The section has passed through the ducts of one of the mucous glands and one of the poison glands—these pour their secretions onto the surface of the skin. 5 ×.

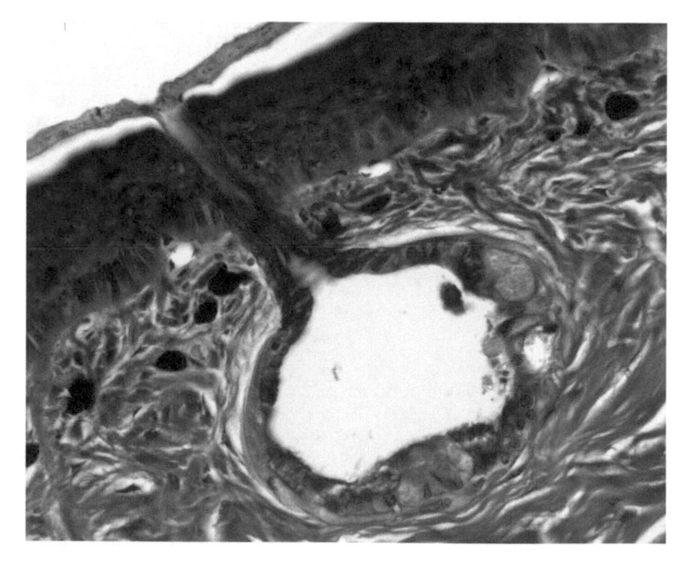

FIGURE C79b The duct of this ALVEOLAR MUCOUS GLAND in the skin of the toad carries secretions to the surface of the skin where they help to prevent dehydration. 40 ×.

secrete a viscous material that hardens to form a tough lining in the stomach; the muscle layers of the wall are greatly enhanced to form an effective grinding machine.

2. *Simple alveolar glands* are glandular invaginations that terminate in flask-shaped structures. In sections of the skin of the toad identify the DUCT and the flask-shaped ALVEOLUS composed of secreting cells (Figs. C79a and C79b). Examples are given of simple alveolar glands in the skin of amphibians (Figs. C80 and C81).

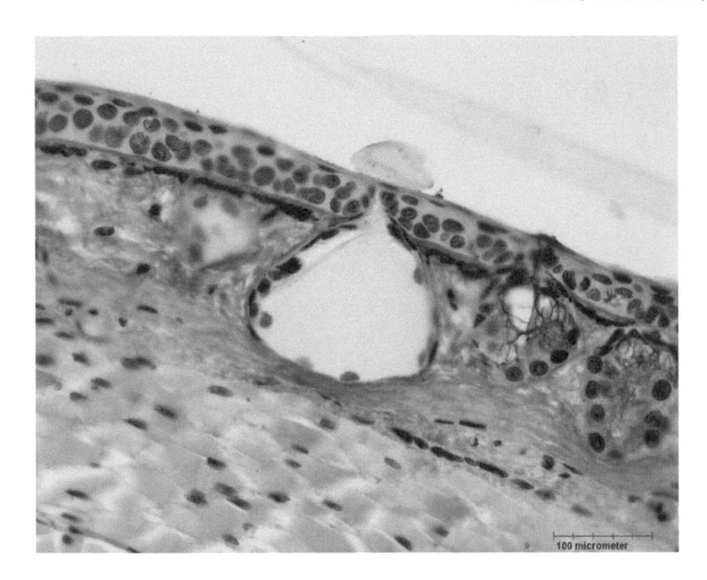

FIGURE C80 Mucus issues from this simple alveolar gland in the skin of *Amphiuma*. 20 ×.

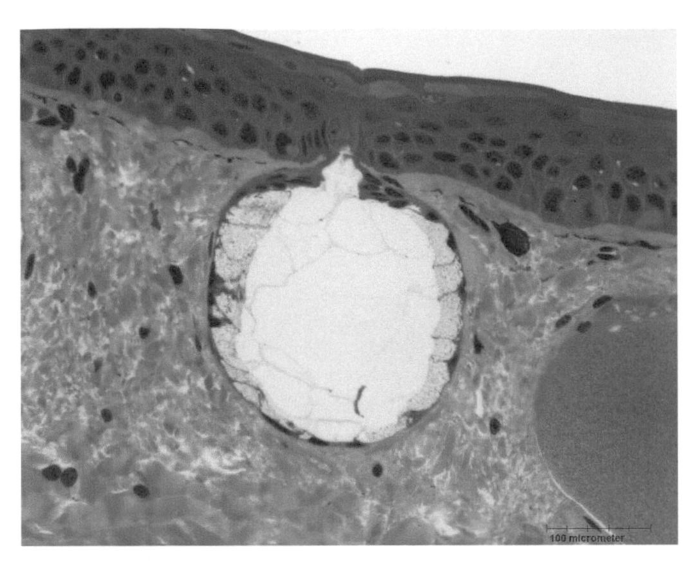

FIGURE C81 Semithin section of a simple alveolar mucus gland in the skin of *Amphiuma*. A portion of a poison gland is at the right; its size is similar to that of the mucous gland. 63 ×.

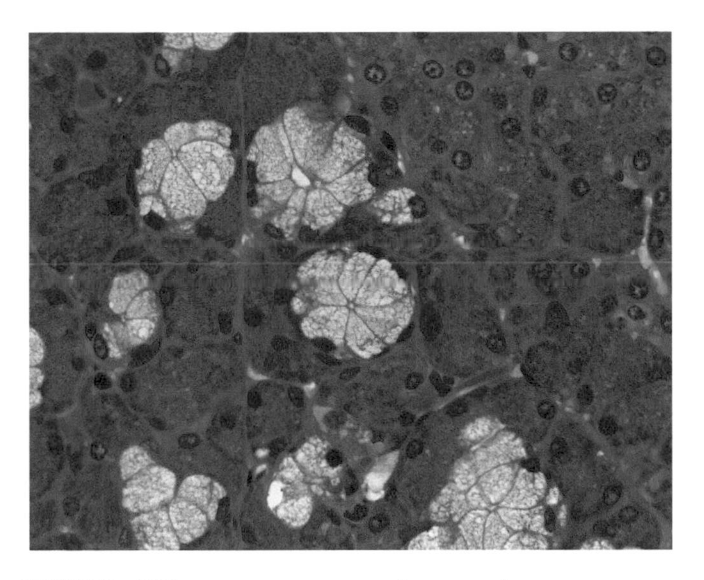

FIGURE C82 SALIVARY GLANDS of mammals are COMPOUND ALVEOLAR GLANDS. The SUBMAXILLARY GLANDS are often MIXED and consist of mucous and serous cells. In this semithin section of the submaxillary gland of a monkey, MUCUS-SECRETING CELLS are almost white while SEROUS CELLS are packed with purplish granules. Note the tiny LUMEN at the center of some of the mucous alveoli. 63 ×.

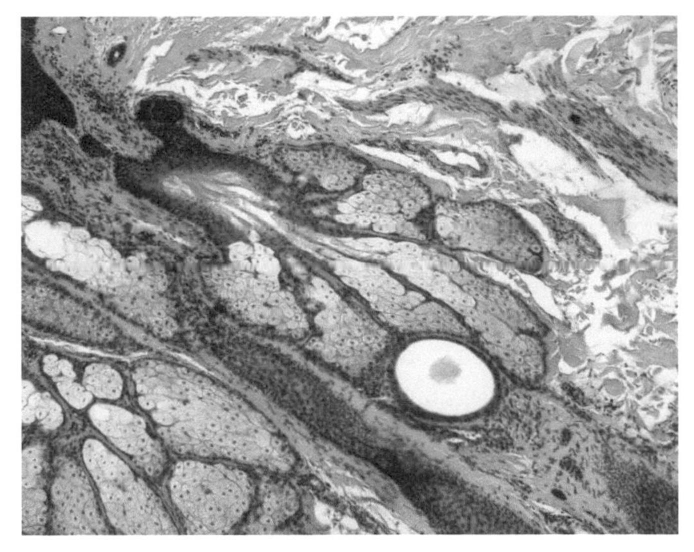

FIGURE C83a Section of sebaceous glands from human scalp. A gland pours its oily secretion from the right onto a hair follicle at the left (not seen). 10 ×.

3. *Compound alveolar glands.* Examine individual alveoli of the submaxillary gland of a mammal and note the presence of peripheral glandular cells that possess a densely granular cytoplasm (SEROUS CELLS) and central cells in which the cytoplasm is clear and nongranular (MUCUS CELLS) (Fig. C82). Observe the position and the shape of the nuclei. This gland is composed of many ALVEOLI that pour their secretion into a common duct. It is a COMPOUND ALVEOLAR GLAND with a dual function: that of producing both mucus and serous secretions.

HOLOCRINE, MEROCRINE, AND APOCRINE GLANDS

A cell of a HOLOCRINE GLAND accumulates secretory products in its cytoplasm and then dies and disintegrates. The entire dead cells are discharged and constitute the secretion. Constant reproduction of less highly differentiated basal cells provides a steady supply of cells that differentiate into secretory cells. Observe this gradual

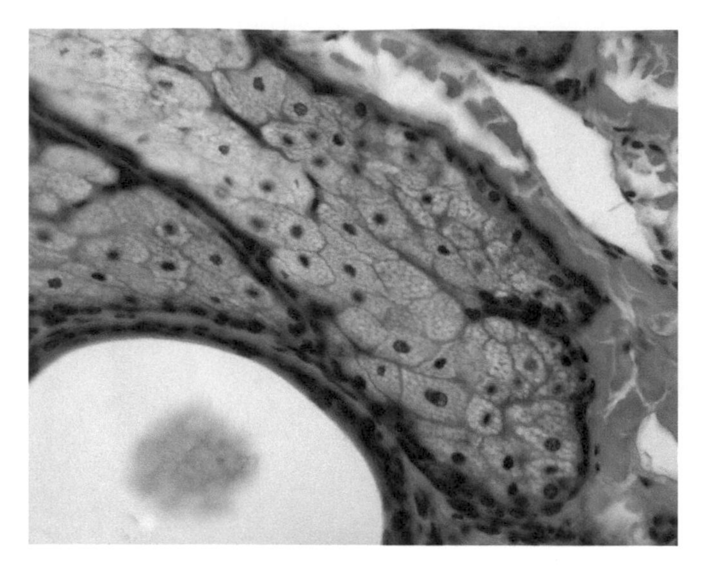

FIGURE C83b Cells reproduce at the fundus of the sebaceous gland at the lower right. As they move upwards (to the left) they accumulate oily droplets and die, pouring their secretions and cellular débris into the hair follicle. This débris probably forms a component of the dreaded curse of shampoo advertisements: dandruff. 40 ×.

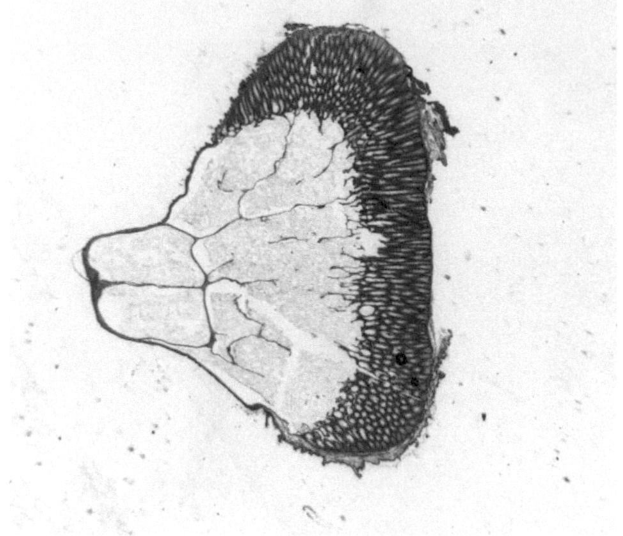

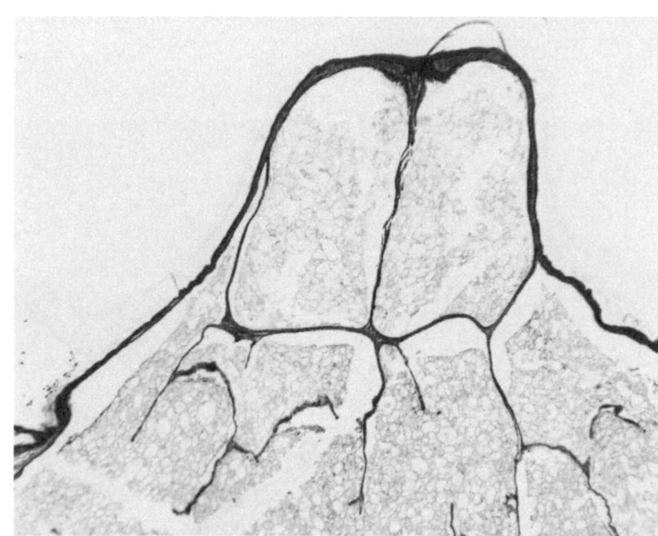

FIGURES C84(a and b) (a) Section through the preen gland of a cowbird. Tubular HOLOCRINE glands at the base of the gland produce cells laden with fatty droplets. These cells move toward the surface where they die, leaving an oily secretion that is squeezed out of the nipple at the top by the bird's beak. 1 ×. (b) The oily secretion contains débris that remains from the dead cells. The NIPPLE is at the top. 4 ×.

breakdown in the SEBACEOUS GLAND of mammals (Figs. C83a and C83b). A spectacular example of a holocrine gland is seen in the PREEN GLAND of a bird (Figs. C84a and b, C85a–c). It seems as if all of the sebaceous glands of the bird's skin are gathered together in one place at the base of the tail where they produce an oily secretion that the bird gathers in its mouth and spreads onto the feathers in the familiar preening action.

MEROCRINE GLANDS secrete without loss of any part of the cell. Since secretory droplets are cytoplasmic inclusions they are not actually part of the cytoplasm although they are manufactured by it. These droplets are passed through the cell membrane by exocytosis with no loss of cytoplasm in the process. Most glands are merocrine.

In APOCRINE GLANDS, the process is much the same as in merocrine glands except that a little cytoplasm is lost during secretion (Figs. C86 and C87). The secretory product is often lipid and, since the cell cannot form a lipid-containing membrane around a lipid droplet, a thin film of aqueous cytoplasmic matrix intervenes between the droplet and its membrane; this small amount of cytoplasm is lost (Figs. C88a and C88b). Mammary glands and some specialized sweat glands are the examples of apocrine glands usually given.

FIGURES C85 These sections are taken from the PREEN GLAND of a junco.

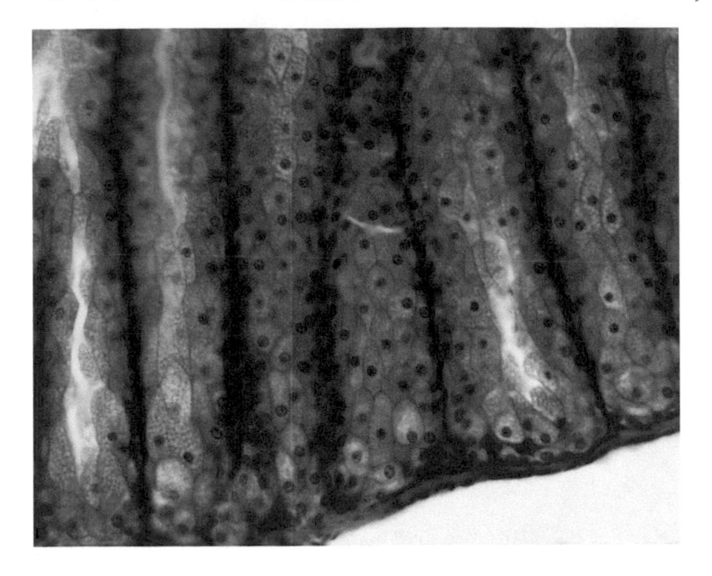

FIGURES C85a Cells are actively dividing by mitosis at their funduses. As the cells move toward the surface and die, oily droplets accumulate in their cytoplasm. 40 × .

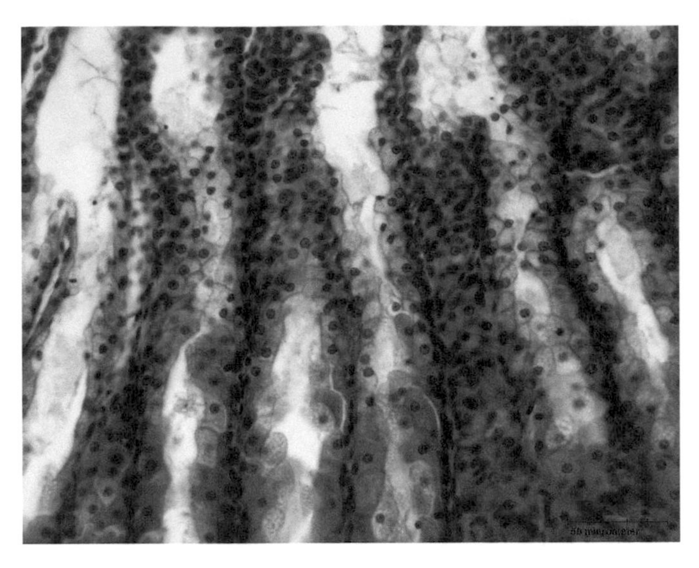

FIGURES C85b Partway to the surface it is seen that the cells have accumulated oily droplets and are beginning to die. 40 × .

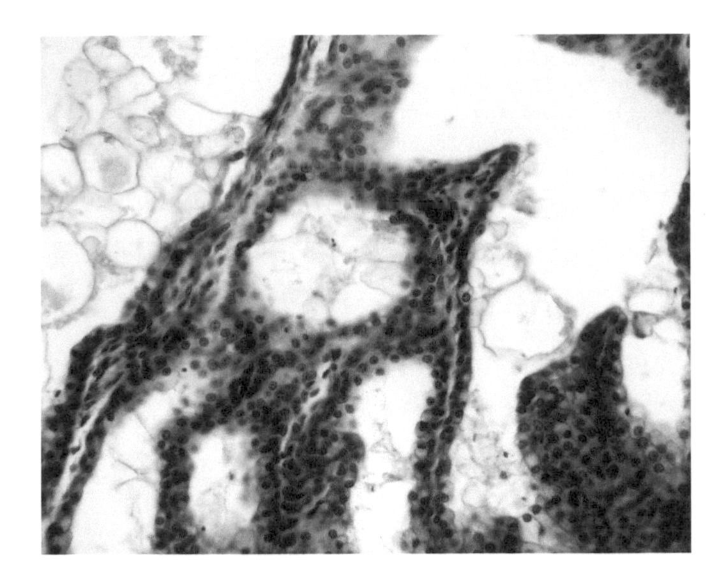

FIGURES C85c The oily secretion is stored just below the nipple. Accumulated débris can be seen in this section taken just below the nipple. 40 × .

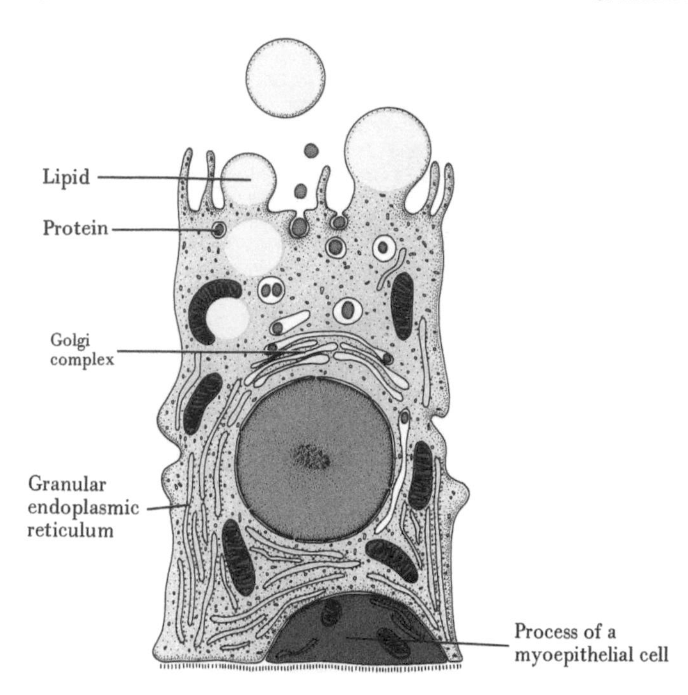

FIGURE C86 The mammary gland produces milk that contains a lipid component and a water-soluble component. The water-soluble component is secreted with no loss to the cell: MEROCRINE SECRETION. Lipid droplets, however, cannot be bounded by a lipid-soluble membrane and the lipid droplets are surrounded by a thin rim of aqueous cytoplasm that permits a lipid membrane to be formed round them. These droplets are then secreted with the loss of a small amount of this cytoplasm: APOCRINE SECRETION.

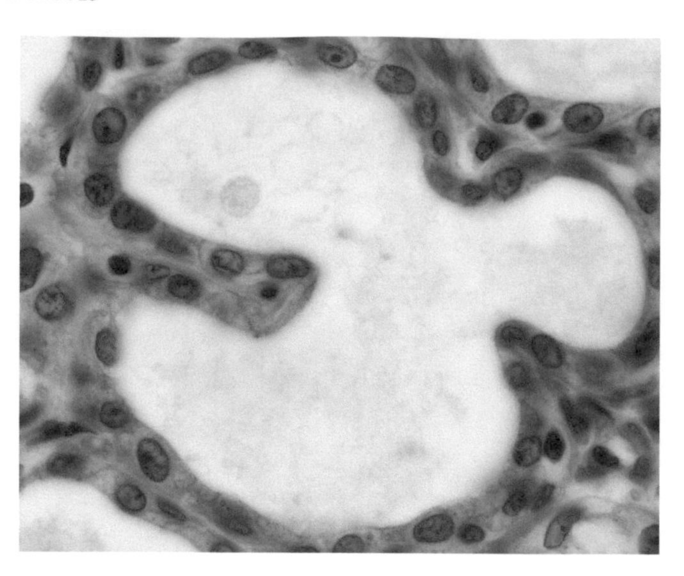

FIGURE C87 This "double" nature if apocrine secretion is not apparent in this section of the active human mammary gland. A small amount of débris is seen in the lumen. 100 ×.

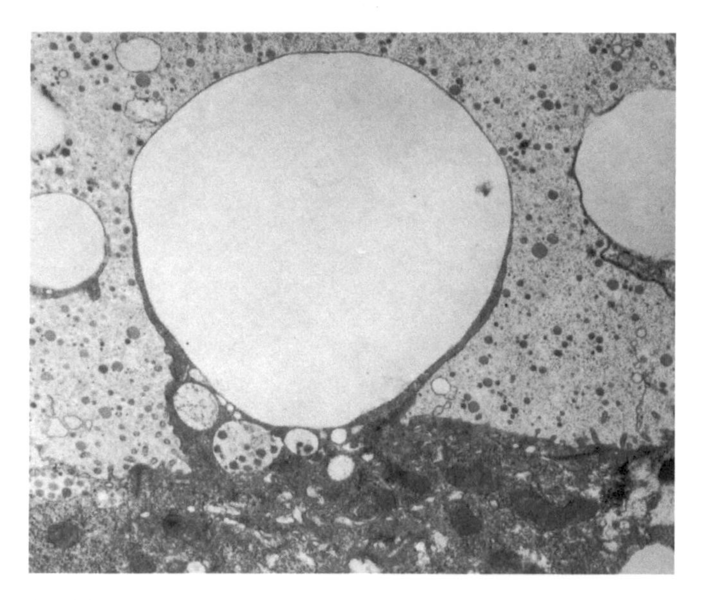

FIGURE C88a Section of the mammary gland of a lactating mouse. A large apocrine process protrudes into the lumen of the gland, awaiting its release. The large droplet of fat is surrounded by a thin film of aqueous cytoplasm which, in turn, is surrounded by the plasma membrane. A few proteinaceous granules are seen in the stalk of the process. 9000 ×.

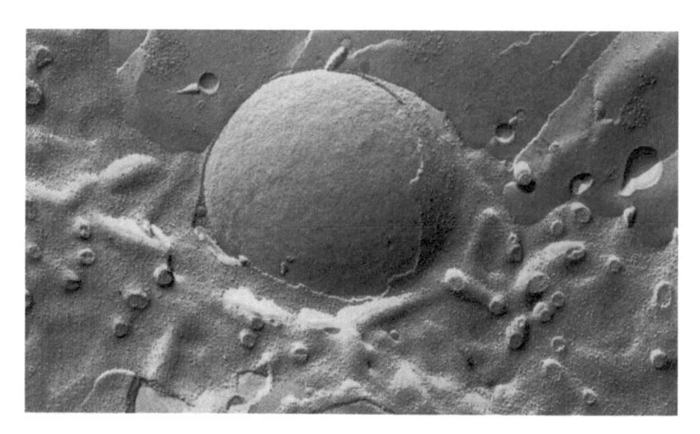

FIGURE C88b Freeze-fracture image of apocrine secretion in the mammary gland of a lactating mouse. Part of the cytoplasmic layer coating a lipid droplet has been torn away leaving the droplet with a smooth surface. 32,000 ×.

CHAPTER

D

Connective Tissues

Connective tissues are the binding or supporting elements of the animal body. Their morphology is variable but in each case the cells are relatively inconspicuous while the nonliving intercellular material is abundant, giving the tissue its particular character and providing its strength. Connective tissues are classified primarily on the basis of the nature and arrangement of the intercellular material, including the fibers contained within it.

EMBRYONIC CONNECTIVE TISSUE

Mesenchyme

Mesenchyme is the typical, unspecialized packing tissue of a developing embryo and its cells enter into the formation of specialized tissues. In sections of vertebrate embryos, note the loose tissue under the skin (Figs. D1 and D2). Look for stellate (star-shaped) cells with numerous cytoplasmic processes making contact with each other. The nuclei are large while the cytoplasm is scanty. Note the clear intercellular spaces that, in the living condition, are occupied by a fluid that constitutes the GROUND SUBSTANCE or MATRIX of the tissue. The matrix is permeated with delicate RETICULAR FIBERS that cannot be seen without special stains. In older embryos, the cells and fibers aggregate within the intercellular spaces, indicating the transformation of the mesenchyme into adult connective tissue (Fig. D3). With the electron microscope the clear, amorphous cytoplasm of mesenchyme cells shows little sign of metabolic activity; the fibers show clearly (Figs. D4a and D4b). Young mesenchyme is a multipotential tissue whose further development produces supporting tissues, vascular tissues, blood, and smooth muscle. A small number of mesenchymal cells remain in the adult as reserve elements with the ability to develop into a variety of tissues.

Mucous Tissue

As an embryo develops, the intercellular matrix of the mesenchyme thickens to become a mucoid jelly containing fine fibers, and the tissue transforms imperceptibly into mucous tissue. At this time the electron microscope shows the appearance of extensive granular endoplasmic reticulum and free ribosomes within the mesenchymal cells, indicating active protein production. Mucous tissue occurs temporarily in the normal development of supporting tissues. It may be found in abundance as WHARTON'S JELLY of the umbilical cord where its development does not progress further (Fig. D5). Note the CELLS and scanty FIBERS distributed throughout the pale, gelatinous MATRIX. The cells, called FIBROBLASTS, have long processes that communicate with those of other cells. Fine fibrils aggregate into bundles, the COLLAGENOUS FIBERS.

COMPONENTS OF ADULT CONNECTIVE TISSUES

Before discussing specific types of adult connective tissues we will describe separately the cells, fibers, and intercellular substances that enter into the formation of these tissues. More of these components can be seen in flattened whole mounts of AREOLAR CONNECTIVE TISSUE than in any other type of preparation. Areolar tissue is the loose connective tissue that is abundant throughout the body, binding and supporting most organs. In the sample shown here, the tissue was obtained by tearing away the skin from an anesthetized small mammal and removing the tissue that connected it to the muscle or other tissue below. The tissue was spread on a slide, stained with hematoxylin and eosin, and coverslipped.

An Atlas of Comparative Vertebrate Histology.
DOI: https://doi.org/10.1016/B978-0-12-410424-2.00004-4

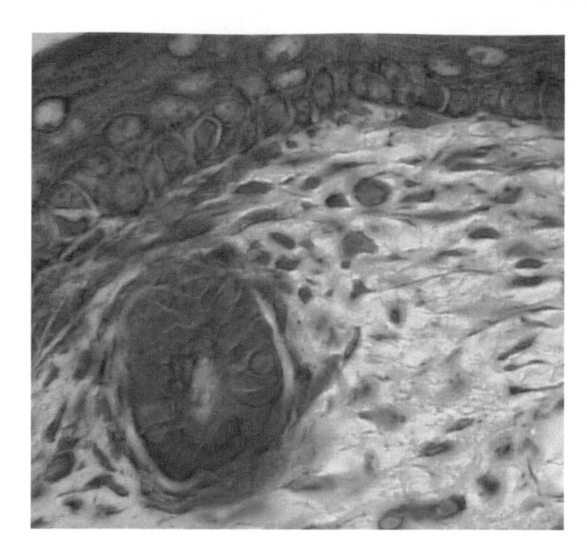

FIGURE D1 Mesenchyme in the developing skull of a fetal mouse is the loose tissue in the lower right. 100 ×.

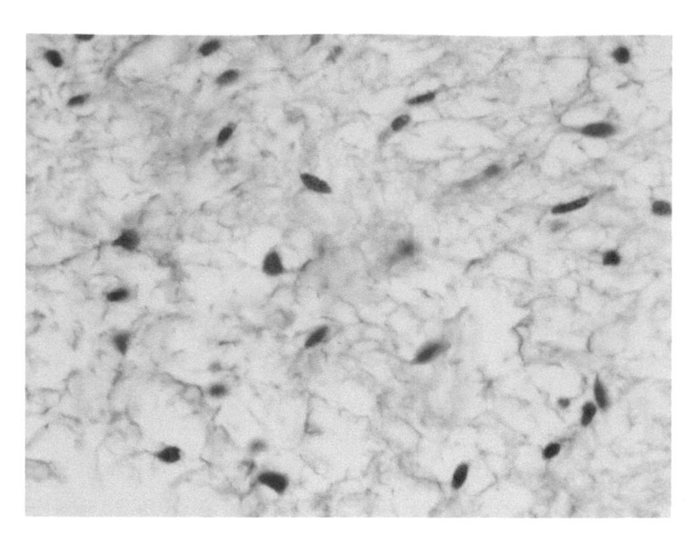

FIGURE D2 Mesenchyme under the skin of a developing dogfish. Most of mesenchyme consists of the gelatinous matrix between the cells. 63 ×.

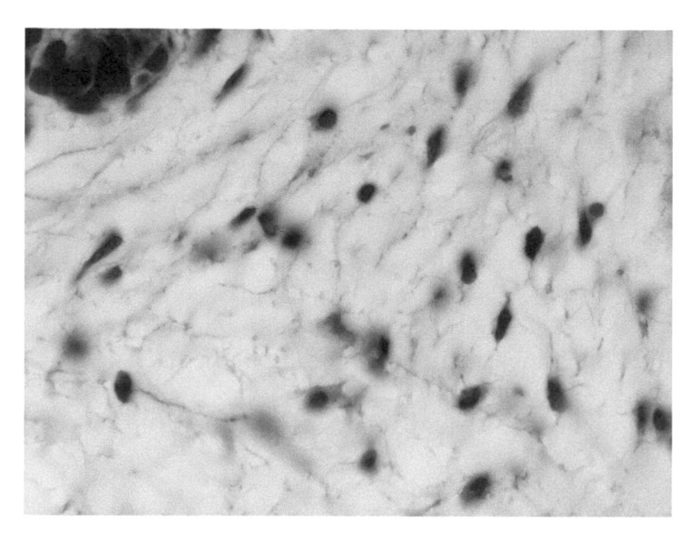

FIGURE D3 Mesenchyme in a region of developing membrane bone of a newborn mouse. As an embryo ages, the mesenchyme concentrates in specific areas. 63 ×.

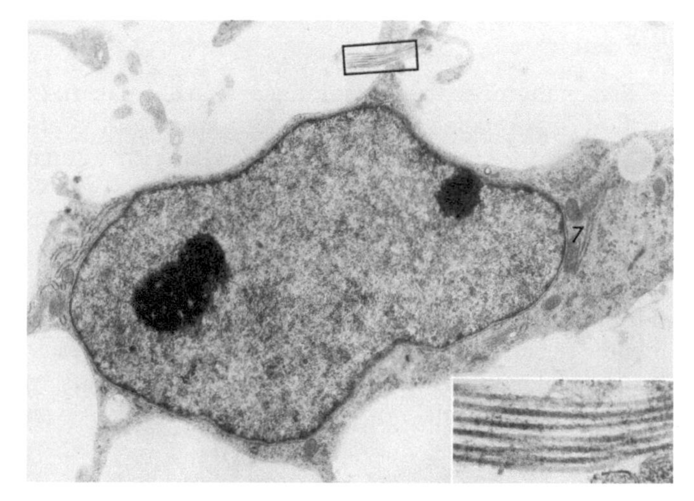

FIGURE D4a Electron micrograph of a section of a MESENCHYME CELL from the fetus of a rat (9000 ×). The scant cytoplasm contains granular endoplasmic reticulum, mitochondria, and a Golgi complex. The box at the top center shows a bundle of delicate reticular fibers shown in higher power at the lower right. 62000 ×.

Connective Tissue Cells

1. *Fibroblasts and Fibrocytes* (Fig. D6). FIBROBLASTS are responsible for the formation of the intercellular fibers and amorphous material of the matrix. In mature tissues they are almost immobile but, following injury, they become active and form new fibers. Look for the extended fibroblasts in the stained preparation of areolar connective tissue. Note the large, oval, nuclei and pale, homogeneous cytoplasm. Fibroblasts are the commonest cells of areolar tissue and the only cells found in tendon, where they are flattened between the fibers.

   Although the term *fibroblast* is firmly entrenched in the histological literature for all stages in the life of the fiber-producing cell, it should probably be reserved for young cells that are engaged in the production of the proteins of the fibrils and matrix. Ultrastructural evidence of protein synthesis is apparent in the fibroblast: well-developed Golgi complex and abundant granular endoplasmic reticulum (Fig. D7). In addition, the cell is surrounded by extracellular fibers and its outer surface is irregular, presumably reflecting a high level of surface activity, such as movement and protein secretion.

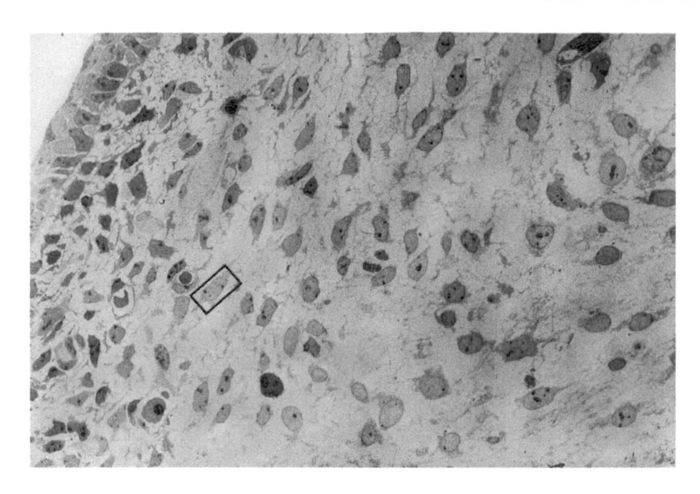

FIGURE D4b Subcutaneous mesenchymal tissue from the fetus of a rat showing the cell of Fig. D4a in context (rectangle). Epidermis is at the left. The mesenchymal cells are immersed in a gelatinous GROUND SUBSTANCE.

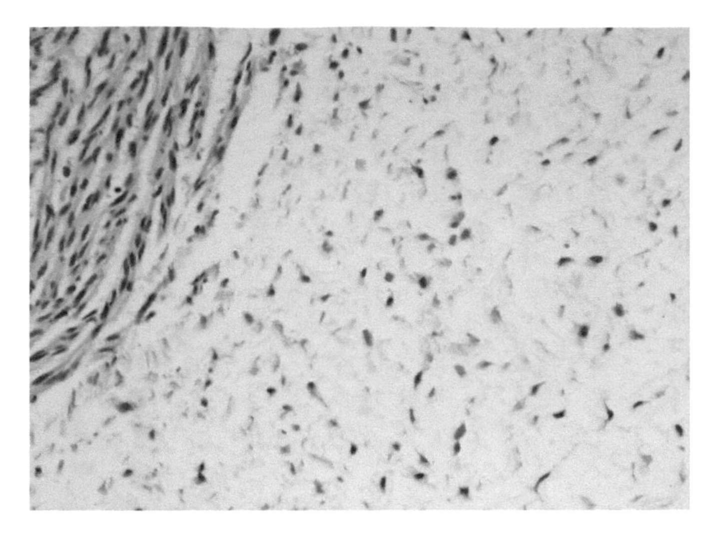

FIGURE D5 Cross section of mammalian umbilical cord showing MUCOUS CONNECTIVE TISSUE (Wharton's jelly) on the right. This tissue consists of fibroblasts scattered irregularly in a pale, gelatinous matrix. At the left, the fibroblasts are organizing themselves to form the wall of a blood vessel. Since the umbilical cord of a newborn has ceased to function in the nourishment of the fetus, this will never come to pass. 20 ×.

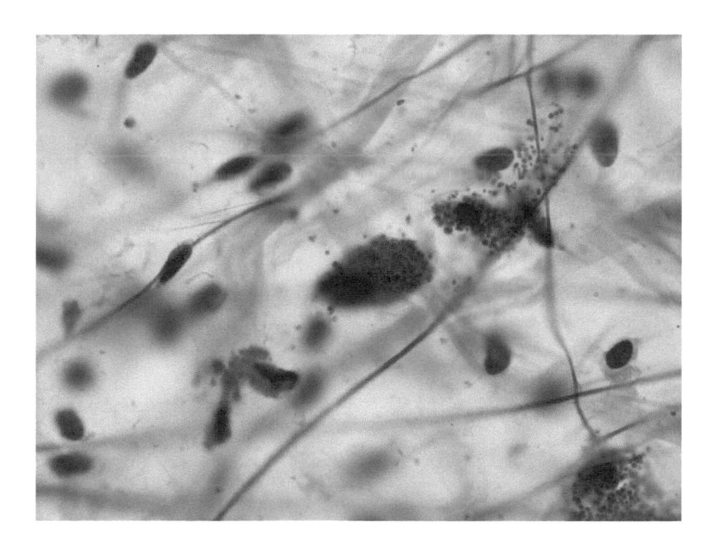

FIGURE D6 Whole mount of mammalian AREOLAR CONNECTIVE TISSUE, the loose fibrous connective tissue that binds skin to the layers of muscle below. Pink and black fibers crisscross the field. FIBROBLASTS are the graceful, elongate cells occurring throughout. A MAST CELL at the center is packed with deeply basophilic granules. Another mast cell nearby has burst and released its granules. You may be able to identify two macrophages in the lower right quadrant and elsewhere. These are wandering cells that are larger than fibroblasts and have a rounder nucleus. 63 ×.

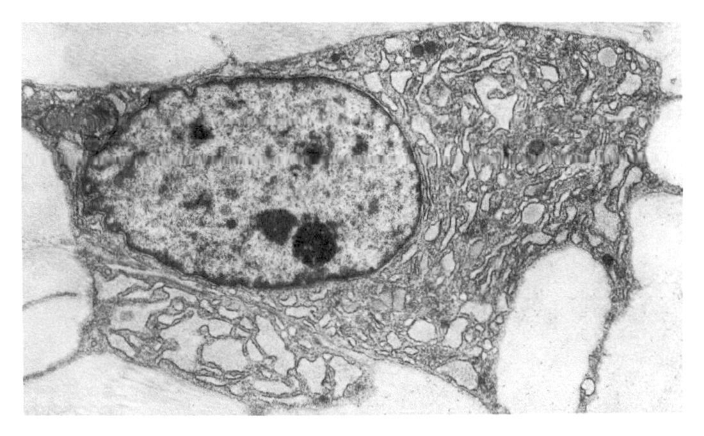

FIGURE D7 Electron micrograph of a section of a FIBROBLAST from the tendon in the tail of a rat. Abundant granular endoplasmic reticulum fills the cytoplasm of this cell. There is a small Golgi complex and a few mitochondria. It is difficult to see the fibers surrounding the cell—they are most apparent at the top left. 9500 ×.

Following this period of active synthesis, the cell matures into a FIBROCYTE. Its scant cytoplasm is weakly basophilic but the appearance of the cell with the light microscope is only slightly different from that of the fibroblast. With the electron microscope, however, the granular endoplasmic reticulum and Golgi complex are seen to have become inconspicuous and the outer surface smooth and regular (Figs. D8 and D9).

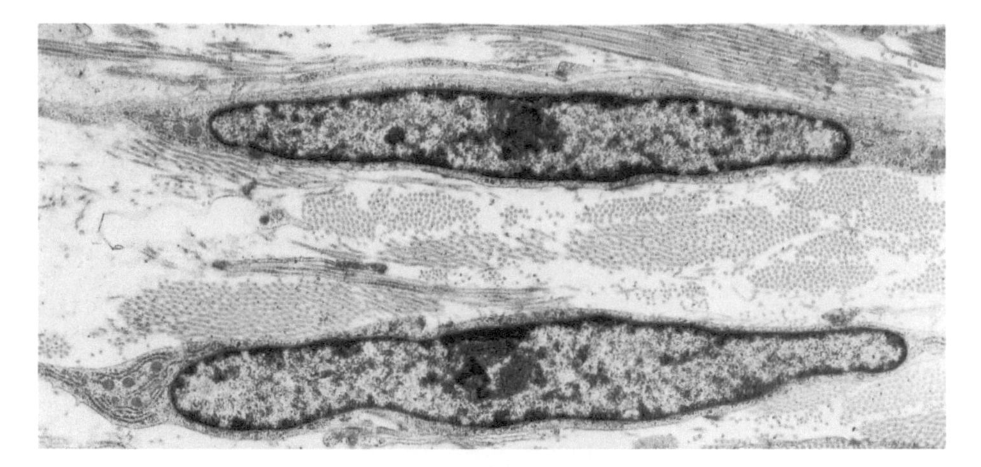

FIGURE D8 Fibroblasts mature into FIBROCYTES where signs of activity are greatly reduced and there is scant cytoplasm. These are electron micrographs of cells from a section of connective tissue of a cat. Abundant bundles of connective tissue fibers surround the cells. 9500 ×.

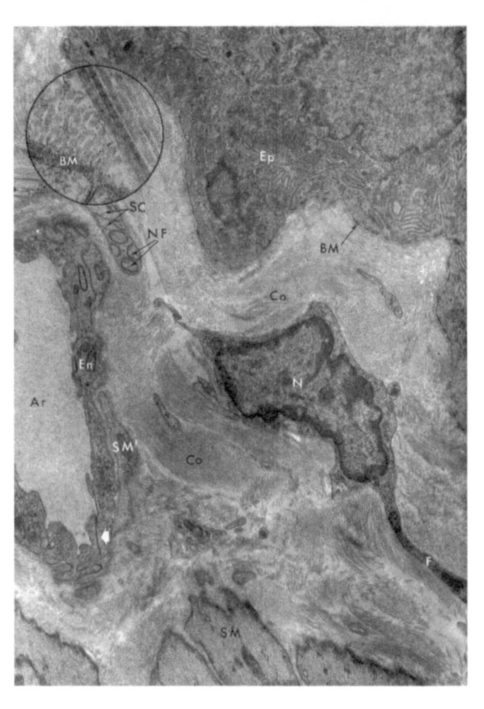

FIGURE D9 At the center of this electron micrograph is a quiescent FIBROCYTE from connective tissue in the esophagus of a bat. It is surrounded by tough fibers of connective tissue. There is an epithelial cell at the upper right and a small blood vessel at the lower left. *Ep*, esophageal epithelium; *BM*, basement membrane; *SM*, smooth muscle; *N*, fibrocyte nucleus; *F*, fibrocyte process; *Co*, collagen fibrils; *Ar*, arteriole; *En*, endothelial cell; *NF*, nerve fibers; *SC*, Schwann (neurilemma) cell. 15,000 ×, *Inset*: 77,000 ×.

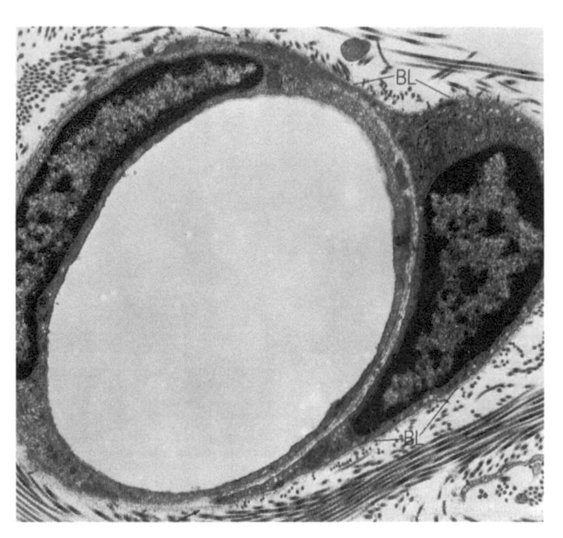

FIGURE D10 In this electron micrograph, an endothelial cell (nucleus on the left) forms a capillary and a PERICYTE (right) clings to one side. Pericytes are found throughout the body associated with capillaries. They are in perfect positions to serve as a reserve of mesenchymal cells that assist in tissue growth and repair. Note that the basal lamina that surrounds the endothelium splits to surround the pericyte.

2. *Mesenchymal cells*. Many authorities consider that cells of this "embryonic type" persist in the adult and, when stimulated, are able to differentiate into various cell types (Figs. D1–D3). Enigmatic cells, called PERICYTES, lurk within the basement membrane of capillaries throughout the body and may constitute outposts of reserves of mesenchymal cells (Fig. D10). Fibroblasts have begun differentiation and cannot transform into different cell types.
3. *Macrophages* have an irregular outline with short, blunt processes (Fig. D6). They are normally immobile (FIXED MACROPHAGES) but in inflammation they become amoeboid (WANDERING MACROPHAGES). The nucleus is smaller than that of fibroblasts, oval, and often indented. Macrophages readily ingest particulate matter. Quiescent macrophages are difficult to identify unless they display evidence of phagocytic activity (Figs. D11 and D12). They can be identified with the light and electron microscopes, however, by their positive reaction to stains that demonstrate acid phosphatase activity.

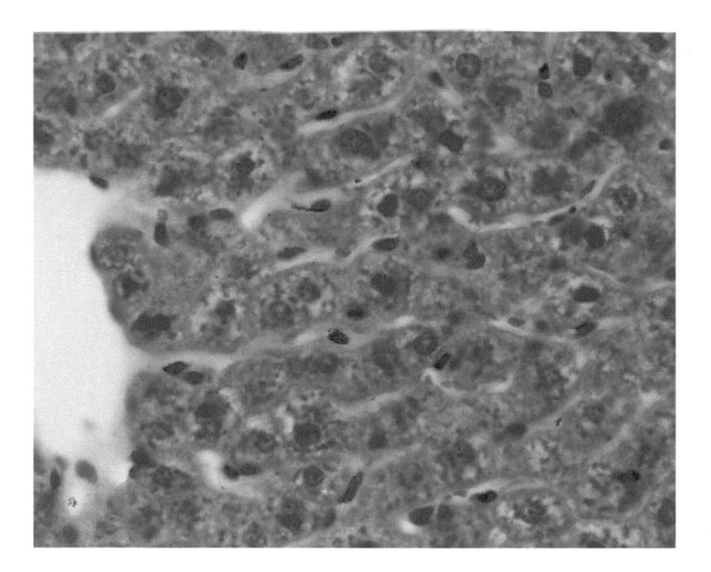

FIGURE D11 Photomicrograph of a section of the liver of a mammal that had been injected some time previously with a particulate stain, trypan blue. MACROPHAGES lining the clear blood spaces of the liver have ingested the stain. 63 ×.

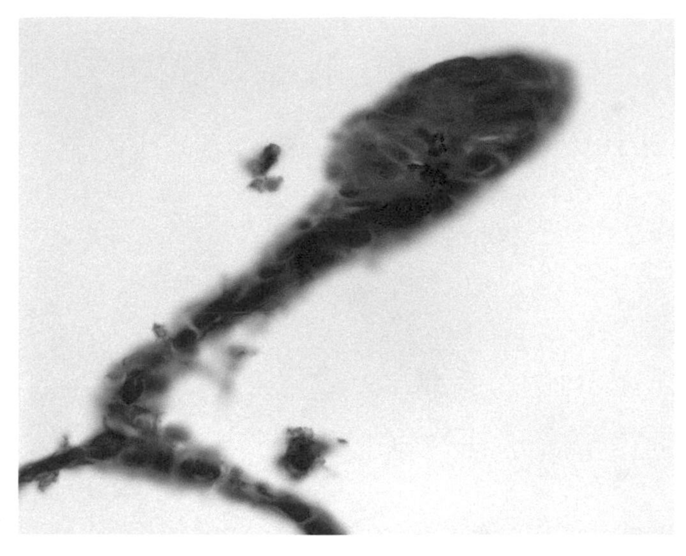

FIGURE D12 This is a section of the delicate respiratory surface in the lung of a smoker. MACROPHAGES within the lung tissue have absorbed black carbon particles. 63 ×.

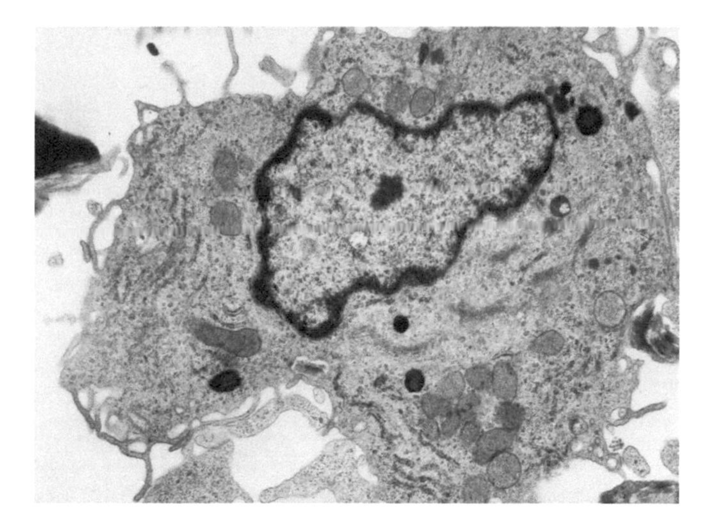

FIGURE D13 Section of the spleen of a rat showing a free MACROPHAGE. Note irregular folds and processes of the cell membrane, which are indicators of phagocytic activity. 19,000 ×.

The most conspicuous features seen in electron micrographs of active macrophages are the presence of phagocytosed material in vacuoles within the cytoplasm and irregular folds and finger-like projections that cover the surface, indicating phagocytic activity and movement (Fig. D13). Also present are a large Golgi complex, granular and agranular endoplasmic reticulum, mitochondria, and lysosomes (Fig. D14). In addition, there may be endocytotic vacuoles and secondary lysosomes. Note the large numbers of long, delicate processes in scanning electron micrographs of macrophages (Fig. D15).

4. *Fat cells (adipocytes)* are normal components of areolar tissue and occur singly or in clusters arranged along blood vessels. They do not form intercellular fibers or matrix but store fats within their cytoplasm. The cells are derived from mesenchyme but lose their protoplasmic processes early and become rounded (Fig. D16). Fat is deposited in small droplets within the cytoplasm; these gradually fuse into a single, large globule that pushes the nucleus to one side of the cell. This is a UNILOCULAR fat cell; it can be 100 μm or more in diameter. A Golgi complex may be seen and mitochondria are numerous around the nucleus (Fig. D17). Is the lipid globule enclosed by a unit membrane? Fat cells are reinforced by a mesh bag of delicate reticular fibers formed by fibroblasts that lurk in the interstitial areas (Figs. D18a and D18b).

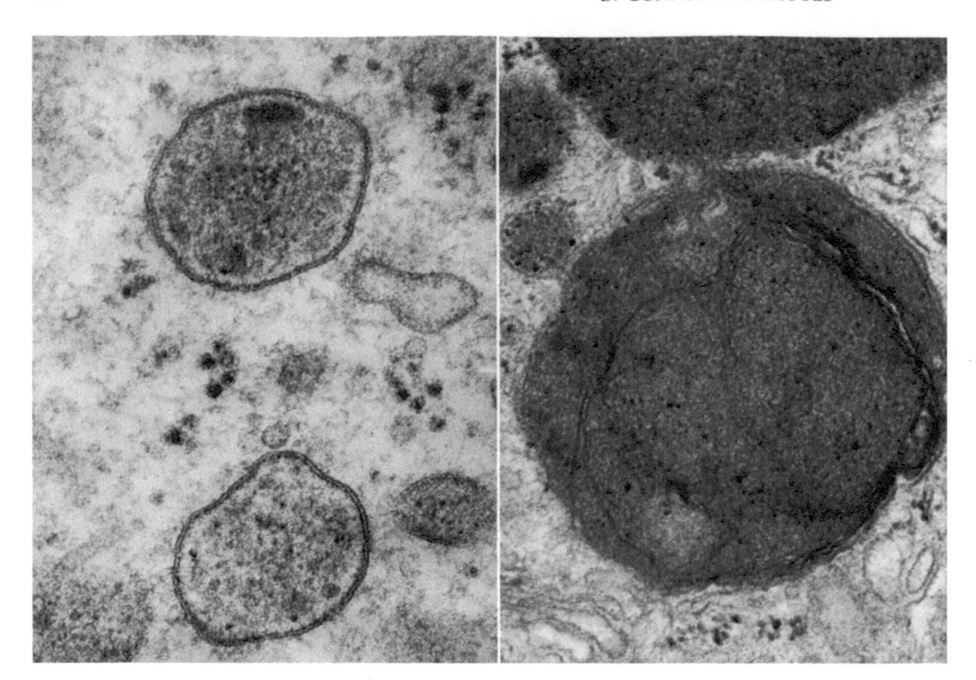

FIGURE D14 LYSOSOMES from macrophages. (Left) Primary lysosomes 120,000 ×; (Right) Secondary lysosomes. 60,000 ×.

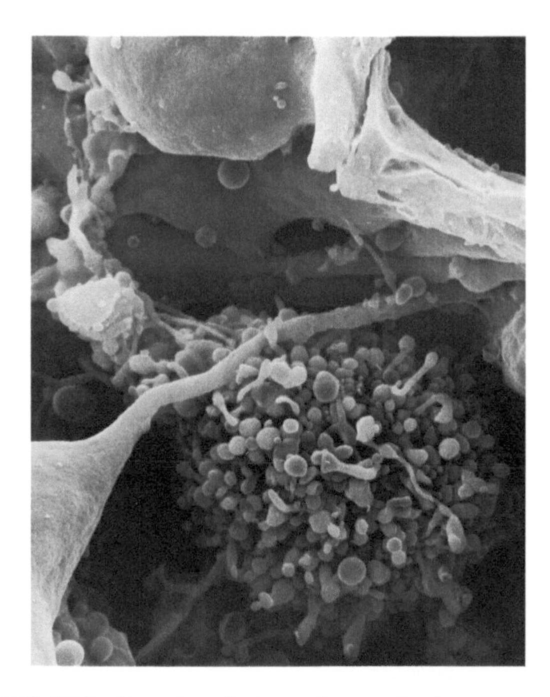

FIGURE D15 Scanning electron micrograph of a MACROPHAGE in the human spleen. Movement and phagocytic activity are indicated by irregular folds and finger-like processes that cover the surface. 10,000 ×.

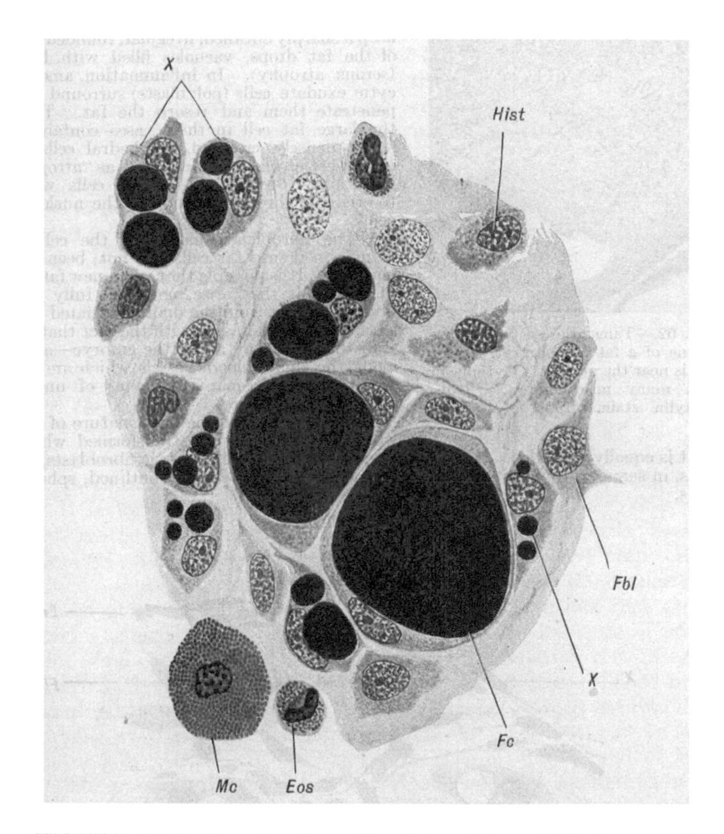

FIGURE D16 Drawing from the first edition of Maximow and Bloom's *Text-Book of Histology* of 1930 showing cells in the subcutaneous connective tissue of the rat. Fat droplets in the fat cells have been blackened with osmium tetroxide. Several droplets of fat are stored initially within cells that resemble fibroblasts (*X*). As these cells accumulate greater quantities of fat, the droplets coalesce into a single globule and a UNILOCULAR FAT CELL (*Fc*) is formed. *Eos*, eosinophilic leukocyte; *Fbl*, fibroblast; *Hist*, macrophage; *Mc*, mast cell. The fat cells are in different stages of differentiation. About 1000 ×.

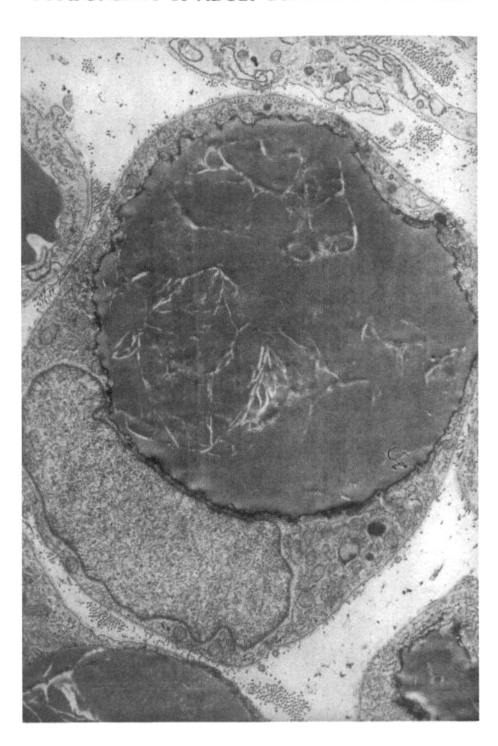

FIGURE D17 Fat cell from the subcutaneous connective tissue of a newborn rat. Fat is deposited in small droplets within the cytoplasm; these gradually fuse into a single, large globule that pushes the nucleus to one side of the cell; this is a UNILOCULAR fat cell. The nucleus is contained in a mass of cytoplasm at one side that also contains granular endoplasmic reticulum, a Golgi complex (at the left), and a few mitochondria. Fat cells are reinforced by a mesh bag of delicate reticular fibers formed by fibroblasts within the interstitial areas; sections of these fibers may be seen in the tissue around the fat cell. The appearance of cracking within the fat droplet is an artifact of preparation. 7200 ×.

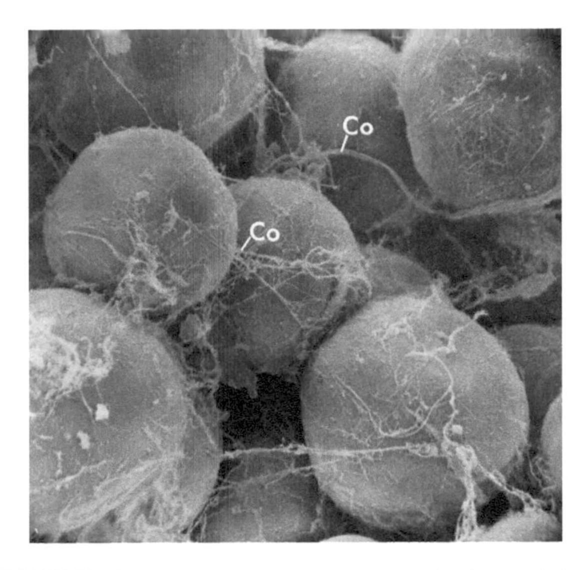

FIGURES D18a Scanning electron micrograph of several fat cells from a rat showing their reinforcing reticular fibers. Cells of this size would be easily squashed were it not for this reinforcement. *Co*, connective tissue fibers. 300 ×.

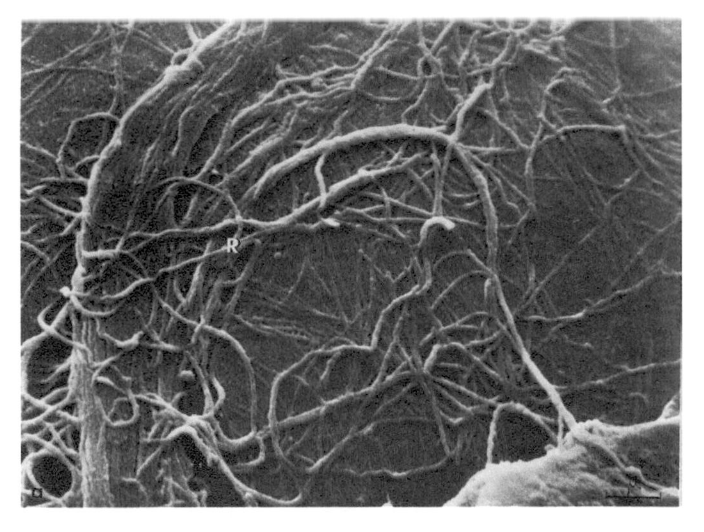

FIGURE D18b A scanning electron micrograph at a higher magnification showing the basket of reticular fibers that reinforce a fat cell of the mouse. 11,800 ×.

Fat cells may be visualized in whole mounts stained with Sudan IV or osmium tetroxide when no fat solvents have been used (Figs. D19a and D19b). Compare these with sections of adipose tissue stained by routine methods (Figs. D20a and D20b) where the outlines of the cells are modified by shrinkage resulting from the action of fat-dissolving reagents used in the preparation of the tissue.

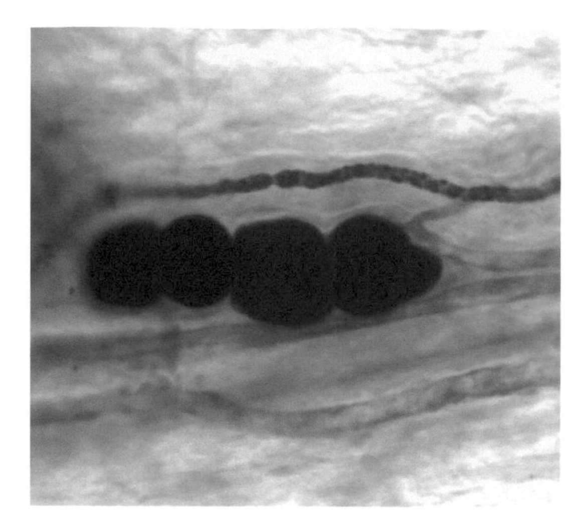

FIGURES D19a Whole mount of mammalian mesenteric fat that has been exposed to vapors containing osmium tetroxide. Osmium is more soluble in lipids than in aqueous media and the unilocular globules of fat have been stained black. Note several blood vessels coursing through the area. 25 ×.

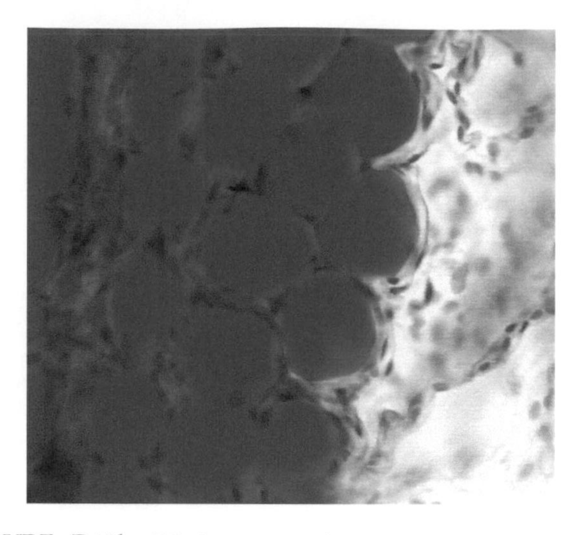

FIGURE D19b Whole mount of mammalian mesenteric fat stained with Sudan IV, a dye with a great affinity for lipids. A thin rim of cytoplasm surrounds the single globules of fat. A few capillaries course through the area. 25 ×.

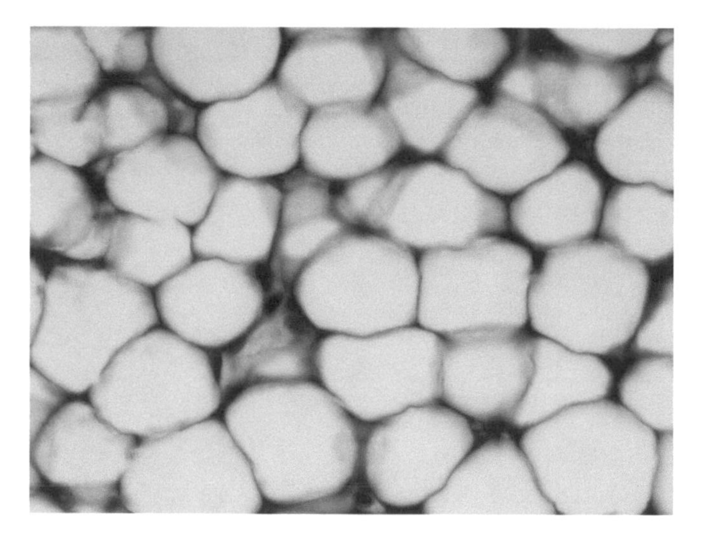

FIGURES D20a Section of the fat body of a toad stained by routine methods. Fat was removed during preparation. 40 ×.

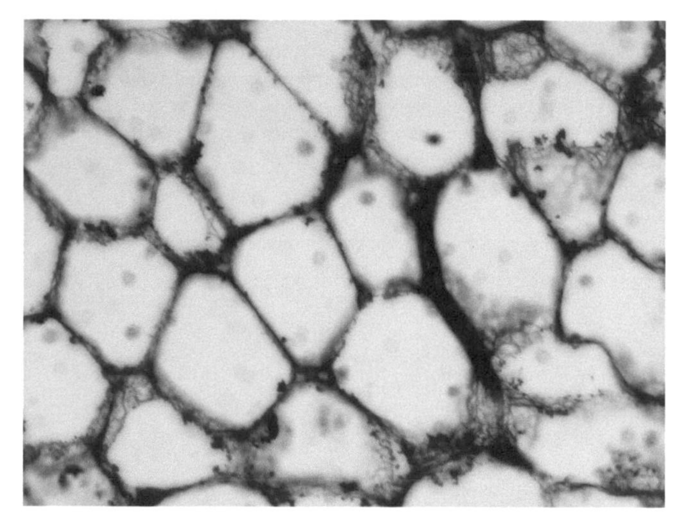

FIGURE D20b Section stained with silver of adipose tissue from a mammal showing the reticular mesh surrounding individual fat cells. 40 ×.

5. *Brown fat.* A special form of adipose tissue, brown fat, accumulates in many hibernating mammals, particularly rodents. It is also found in newborn humans and diminishes with age. The fat droplets do not coalesce into one large globule but remain discrete within the cytoplasm and the cell is MULTILOCULAR (Fig. D21). The cytoplasm is packed with large mitochondria, indicating a high level of metabolic activity (Figs. D22a–D22c). Elementary particles, which contain the enzymes for ATP production, are lacking in these mitochondria and energy is dissipated as heat rather than being stored as ATP (Fig. D23). In addition, fatty acids released into the blood from these cells provide fuel for increased metabolism and heat production elsewhere in the body.
6. *Leukocytes.* Lymphocytes and monocytes may outwander from the blood into the connective tissue and some of these may arise in the connective tissue itself. Eosinophils and neutrophils readily leave the blood stream and enter the connective tissue. These cells will be considered in a later period.

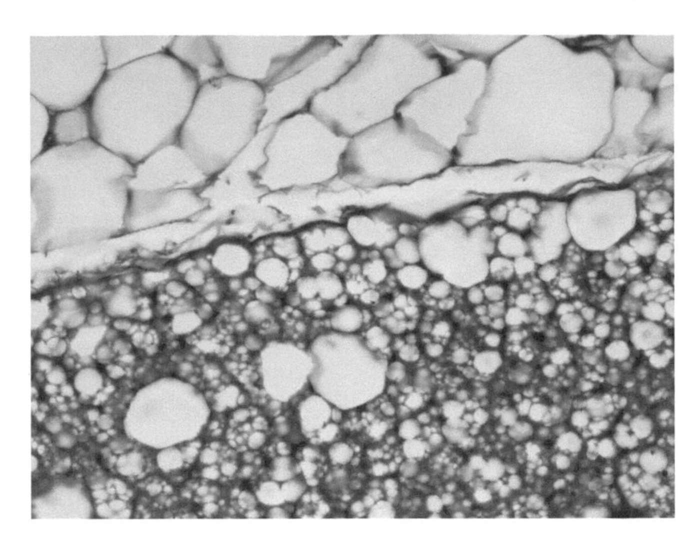

FIGURE D21 Section stained by routine methods through fat of a gerbil. Unilocular fat cells of white fat are located at the top and multilocular fat cells of brown fat at the bottom. The fat droplets have been leeched out in preparation of the slide and the cells appear shrunken. Cytoplasm is much more abundant in the brown fat cells. 20×.

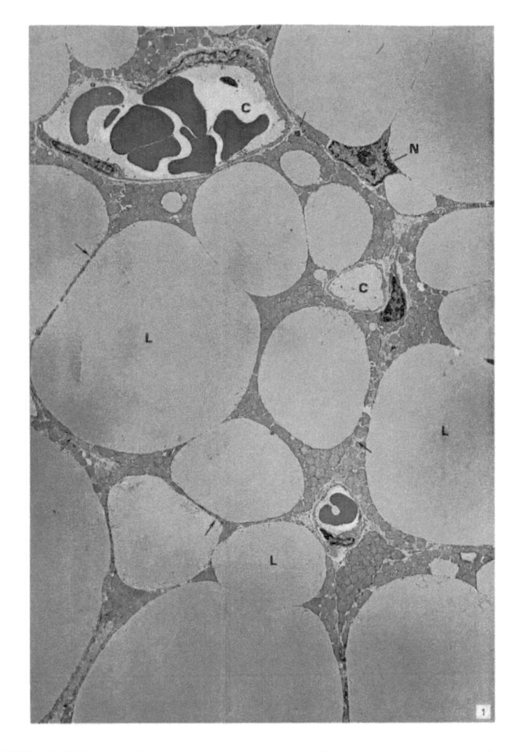

FIGURES D22a Electron micrograph of a section of BROWN FAT from a rat. The borders of the ADIPOCYTE at the center are marked by arrows. There is more than one droplet of fat (L) in this cell and the intervening cytoplasm is packed with mitochondria. A few capillaries (C) are shown. 3200×.

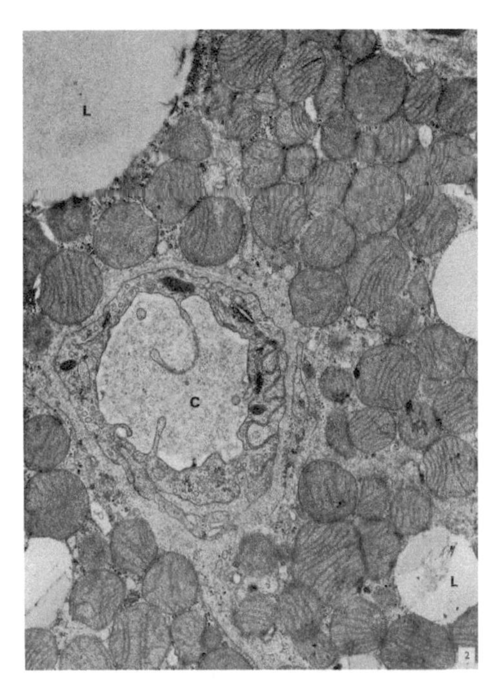

FIGURE D22b Detail from the Fig. D22a micrograph showing a capillary (C) and the cytoplasm of a brown fat cell packed with mitochondria, an indication of a high level of metabolic activity. A few lipid droplets (L) are seen. 16,000×.

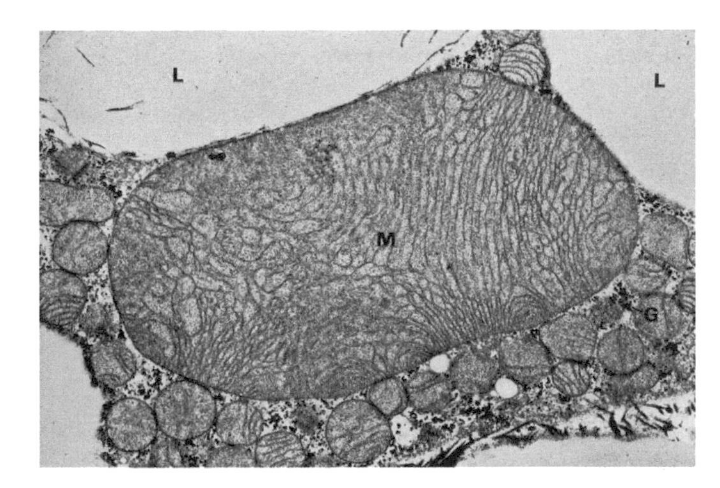

FIGURE D22c Electron micrograph of a section of an adipocyte from the brown fat of a rat. A giant mitochondrion (M) dominates the picture. 14,000×. *C*, capillary; *db*, dense bodies; *G*, glycogen granules; *L*, lipid droplets; *v*, vesicle.

7. *Plasmocytes* or plasma cells are rare generally but are plentiful in areas of chronic inflammation. They are instrumental in the production of antibodies. The round nucleus is eccentrically placed in the basophilic cytoplasm of these oval cells (Fig. D24). It may display large clumps of peripheral heterochromatin giving it the appearance of a cartwheel; this configuration has been termed a RADKERN (German for "wheel nucleus"). Near the nucleus is usually a clearing in the cytoplasm, the JUXTANUCLEAR SPACE or CELL CENTER, that contains

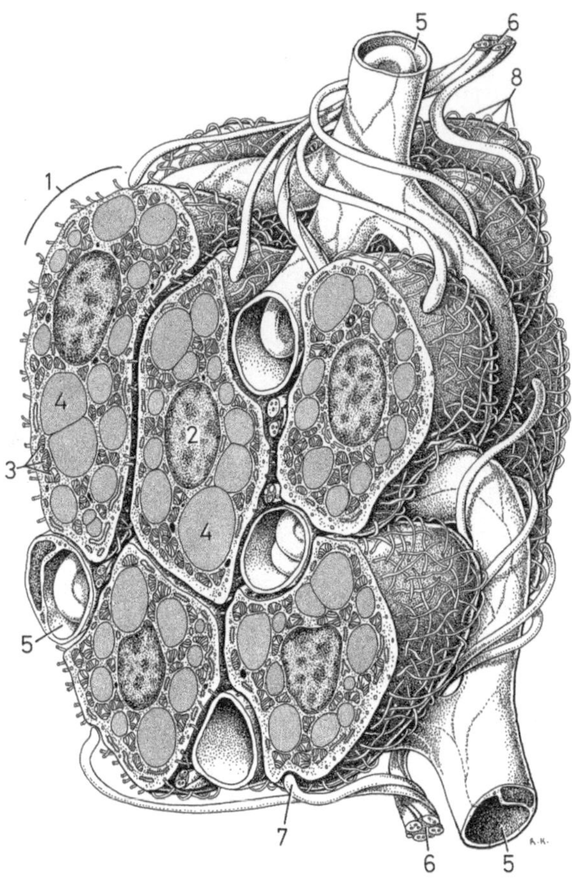

FIGURE D23 Three-dimensional drawing of a group of brown fat cells from a rat. The multilocular fat cells (1) are reinforced by baskets of collagenous fibers (8). The fat cells are packed with mitochondria (3), which are integral in the production of heat that is carried away to the rest of the body by abundant capillaries (5). 1 - brown fat cell; 2 - nucleus; 3 - mitochondria; 4 - fat droplets; 6 - unmyelinated nerve fiber; 7 - nerve ending 3200 ×.

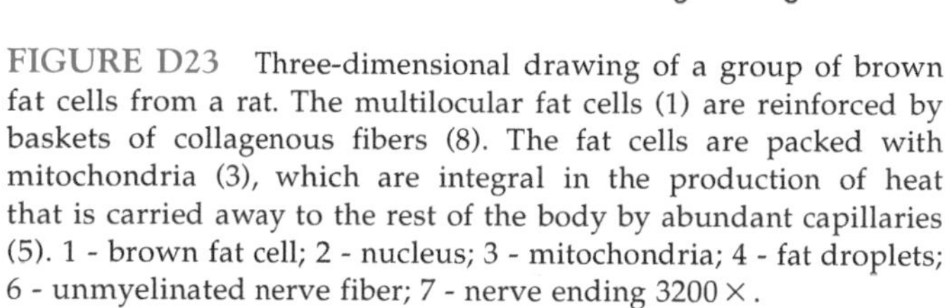

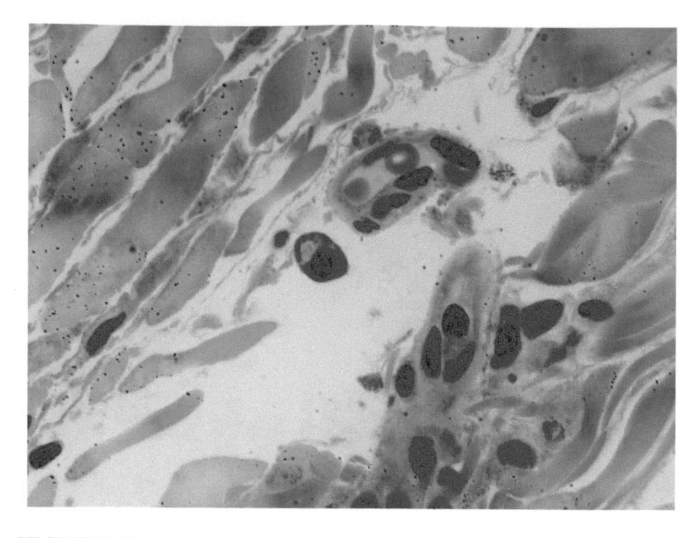

FIGURE D24 A plasmocyte from a semithin section of the aorta of a monkey appears near the center. The round nucleus shows (with the aid of some imagination) the appearance of a wheel (*Radkern*). The basophilic cytoplasm proclaims protein production; the JUXTANUCLEAR CLEARING is at the left. The dark spots throughout are artifacts of preparation. 63 ×.

the centriole and Golgi complex. In electron micrographs note the well-developed granular endoplasmic reticulum studded with ribosomes, some of which form spiral chains, the POLYRIBOSOMES. Locate the structures of the cell center (Figs. D25a and D25b).

8. *Mast cells.* These "tissue basophils" are distinct from the basophils of the blood (Fig. D26). Mast cells are large ovoid cells with small, pale nuclei. The cytoplasm is crowded with coarse, deeply basophilic granules. The granules are METACHROMATIC with certain basic aniline dyes; this means that the color of the dye is altered when taken up by the granules so that blue dyes such as toluidine blue or methylene blue become purple in the granules. This metachromatic reaction indicates the presence of sulfate groups. The cell membrane ruptures readily during fixation and liberates the granules. The granules contain HEPARIN (an anticoagulant), HISTAMINE (which increases capillary permeability) and, in some

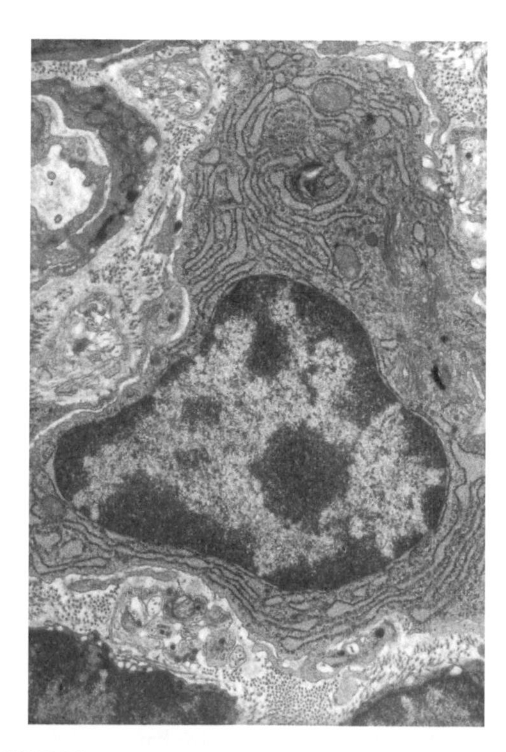

FIGURES D25a Electron micrograph of a plasmocyte from a section of connective tissue of a rat. Its cytoplasm is packed with granular endoplasmic reticulum and the nucleus shows the typical "CARTWHEEL" appearance (*Radkern*). The juxtanuclear clearing or cell center at the right contains aa Golgi complex. 29,000 ×.

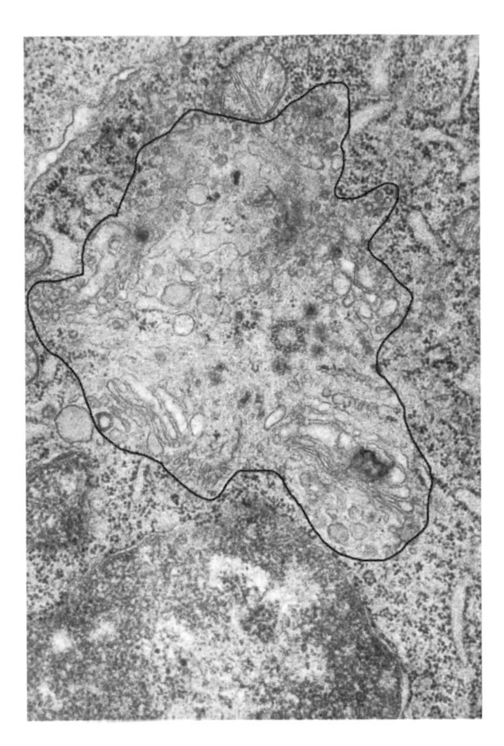

FIGURE D25b This electron micrograph of a section of the cell center of a plasmocyte contains several Golgi complexes—the granular endoplasmic reticulum appears to have been forced out of this area. That this cell is capable of division is indicated by the presence of a centriole near the center of the micrograph. The cell center is outlined. 67,000 ×.

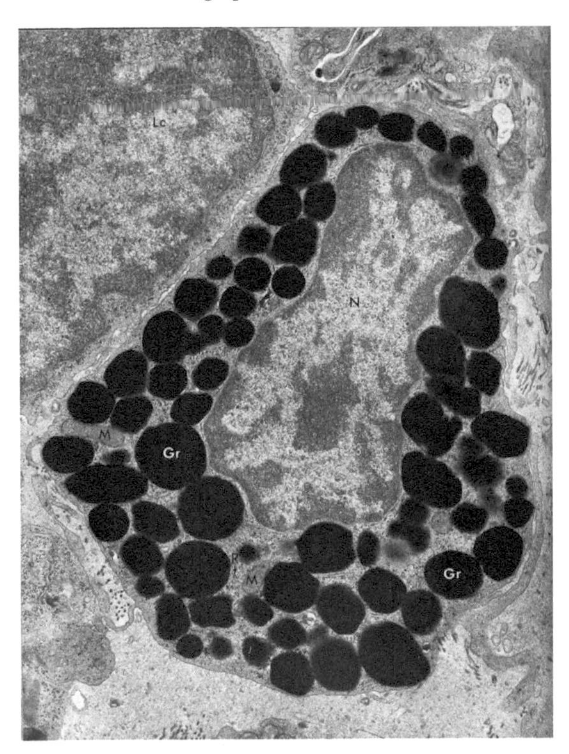

FIGURE D27 Electron micrograph of a section of a mast cell from the intestinal connective tissue of a rat. It presents the appearance of a differentiated cell where little synthesis or metabolic activity is occurring—the cytoplasm is packed with electron-dense, membrane-bound granules leaving little space for an occasional mitochondrion, a few ribosomes, and some membranes of both granular and agranular endoplasmic reticulum. 27,400 ×.

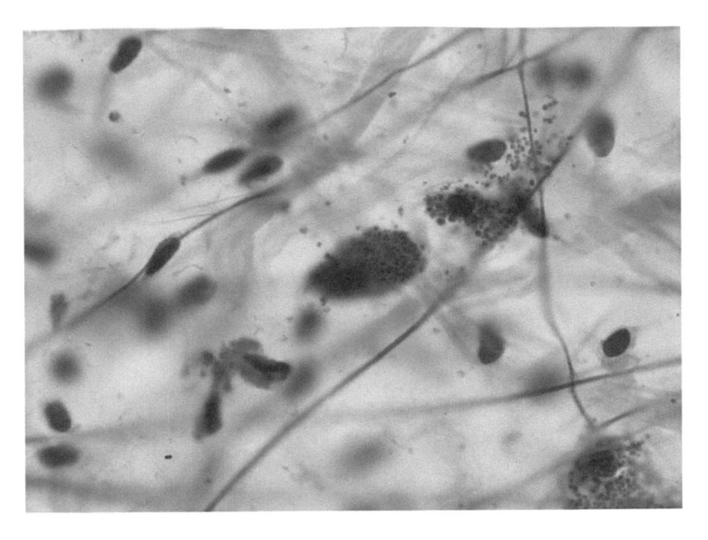

FIGURE D26 mast cells in areolar connective tissue. The mast cell at the center is intact, but the one to its right has burst, releasing its content of metachromatic granules. 63 ×.

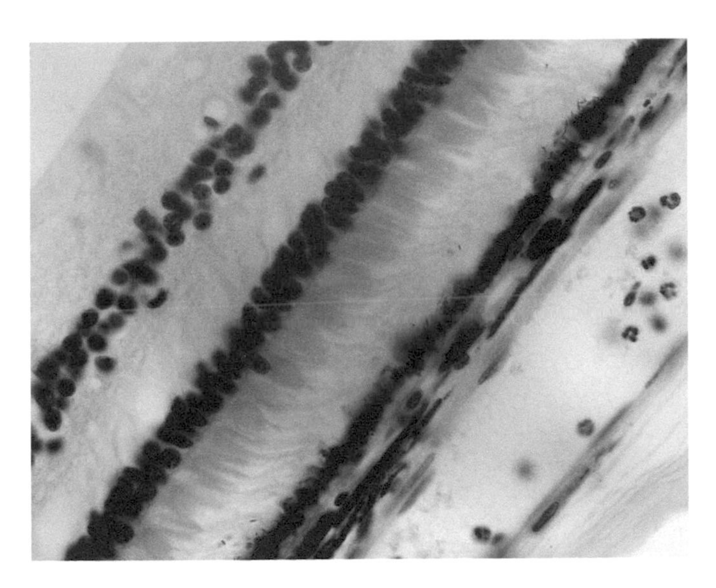

FIGURE D28 chromatophores are distributed in a brown layer at the right of this micrograph of a mammalian retina. 63 ×.

species, serotonin (a vasoconstrictor). The electron microscope shows that the cytoplasm is packed with electron-dense, membrane-bound granules leaving little space for an occasional mitochondrion, a few ribosomes, and some membranes of both granular and agranular endoplasmic reticulum (Fig. D27). The content of the granules varies from species to species, appearing homogeneous, fibrous, and even crystalline. The mast cell presents the appearance of a differentiated cell where little synthesis or metabolic activity is occurring.

9. *Chromatophores* or pigment cells are rare in the loose connective tissue of mammals; they occur more frequently in dense connective tissue of the skin, pia mater, and choroid coat of the eye (Fig. D28). They may be seen to good advantage in whole mounts of

FIGURE D29 CHROMATOPHORES in a whole mount of skin of a fish. Pigment granules extend into the farthest reaches of the elaborate processes of these cells. 10×.

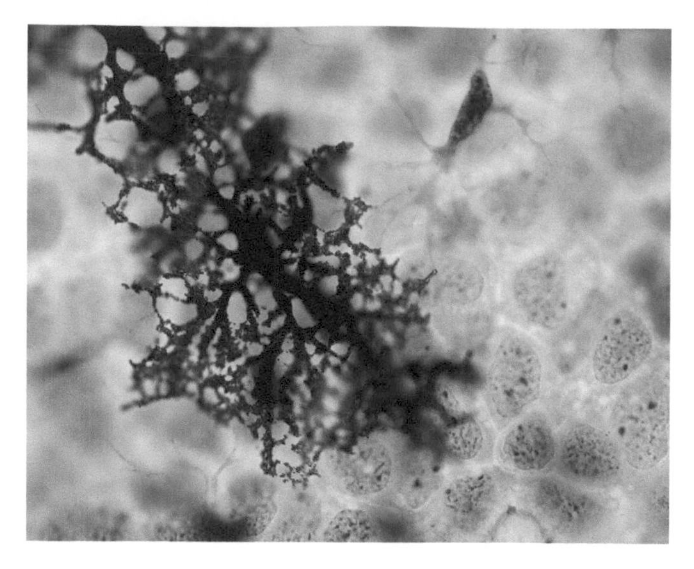

FIGURE D30 CHROMATOPHORE in a whole mount of the skin of a tailed amphibian, *Ambystoma*. These animals can darken by rolling pigment granules into these elaborate processes and lighten by retracting them. 63×.

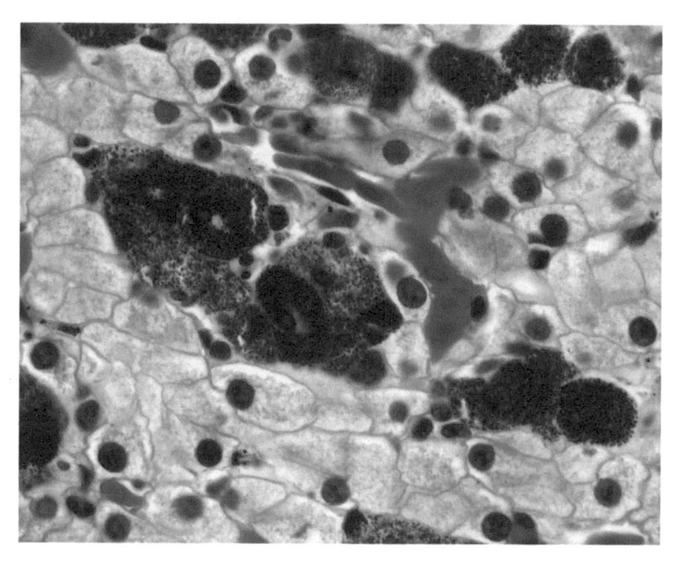

FIGURE D31 It is a puzzle as to why CHROMATOPHORES are scattered among liver cells in this section of the liver of another tailed amphibian, *Amphiuma*. 40×.

fish (Fig. D29) and amphibian skin (Fig. D30). It is a puzzle as to why pigment cells are scattered among the liver cells of amphibians (Fig. D31). The cytoplasm of chromatophores contains small granules of pigment of different colors and the cells have irregular processes of variable length. Some species of animals are able to alter their color by the migration of pigment granules in and out of the processes (Figs. D32a, D32b, D33). More rarely, the chromatophores themselves may change position. The term "chromatophore" is a generalized name for a pigmented cell, whereas MELANOPHORE (brown), XANTHOPHORE (yellow), etc., designate the particular color bearers (Fig. D34). Only the brown pigment MELANIN is found in mammals. (The melanocytes that color human skin produce melanin but are epithelial cells.) Compare light and electron micrographs of chromatophores showing the pigment both concentrated around the nucleus and extended in the processes (Figs. D35 and D36). Note the relationship of the pigment granules to the cytoskeleton.

10. *Cells of cartilage and bone.* The matrix of cartilage and bone is produced by cells derived from mesenchyme, the CHONDROBLASTS and OSTEOBLASTS, respectively. Eventually these cells become embedded in the matrix and are no longer actively secretory; these mature cells are called CHONDROCYTES and OSTEOCYTES. The cells of cartilage and bone will be studied more fully later in this exercise.

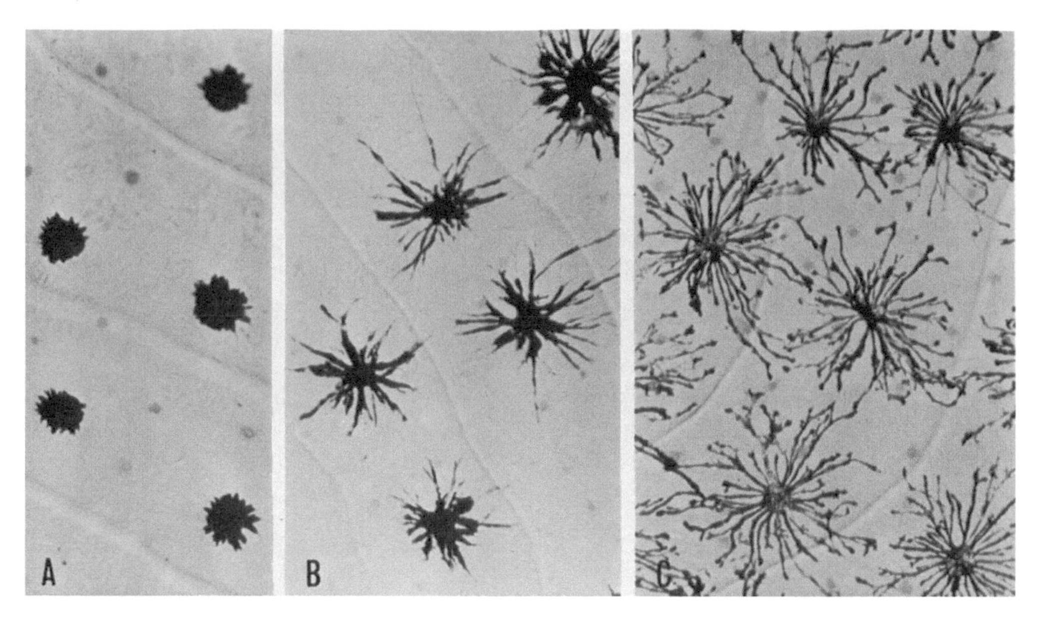

FIGURES D32a Photomicrographs of living MELANOPHORES in a whole mount of the skin of the mummichog, *Fundulus heteroclitus*, in gradual stages of pigment dispersion that darkens the skin. (A) Total aggregation; (B and C) Partial and total dispersion. 135×.

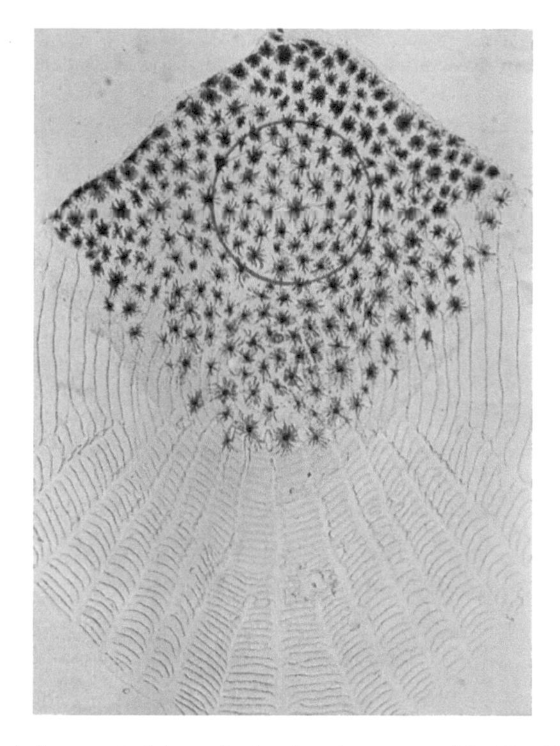

FIGURE D32b Photomicrograph of a whole mount of the scale showing the area used for Fig. 33a (circled).

Connective Tissue Fibers

Three kinds of fibers are found in adult connective tissues. All are represented, to varying degrees, in each type of connective tissue.

1. *Collagenous fibers (White fibers)*. Collagenous fibers are bundles about 1–12 μm in diameter of fine filaments 0.3–0.5 μm in diameter held together by a cementing substance (Fig. D37). The individual filaments do not

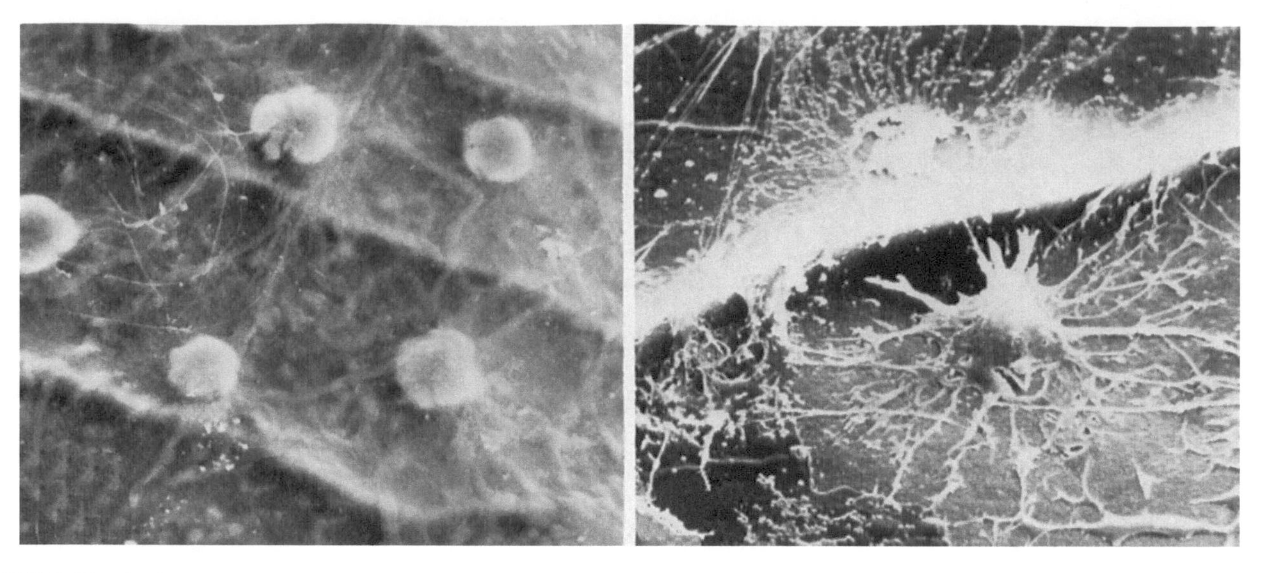

FIGURE D33 (Left) Scanning electron micrograph of melanophores in whole mounts of the skin of *Fundulus heteroclitus* with granules aggregated around the nucleus. 1500 ×. (Right) Similar preparation showing pigment granules dispersed within the elaborate processes of a melanophore. 1500 ×.

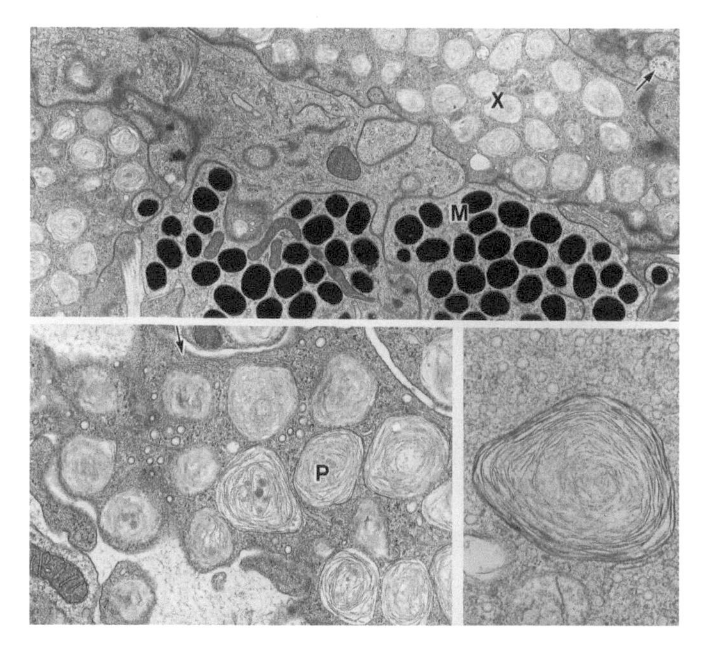

FIGURE D34 Electron micrographs of PIGMENT GRANULES of different colors from scales of the medaka, *Oryzias latipes*. (Top) *M* is a MELANOCYTE, MELANOSOMES are the brown granules within; *X* is a XANTHOPHORE, with yellow granules. Bar = 2.0 μm. 12,600 ×. (Bottom left) *P* are yellow to red PTERINOSOMES in a xanthophore. Bar = 1.0 μm. 24,700 ×. (Bottom right) Mature PTERINOSOME with a trilaminar limiting membrane. Bar = 0.5 μm. 47,100 ×.

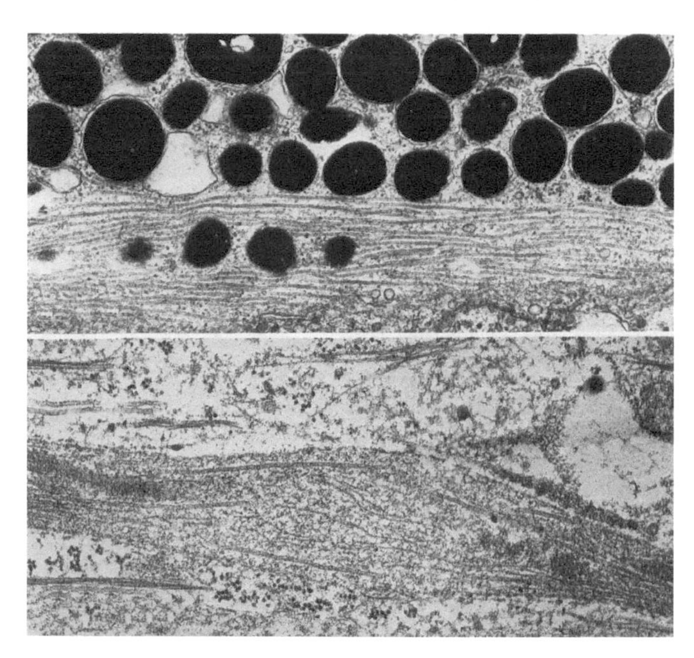

FIGURE D35 (Top) Electron micrograph of a horizontal section of a process of a melanophore showing dispersed melanosomes and microtubules aligned parallel to the long axis of the process. 25,000 ×. (Bottom) Process of a melanophore in which the melanosomes have departed to aggregate around the nucleus. 42,200 ×.

branch or unite with other filaments but the bundles frequently branch and recombine forming a network. The electron microscope shows that the fine filaments actually consist of even finer threads, the COLLAGEN FIBRILS, that vary in diameter from 20 to as much as 200 nm and show a banded appearance with a periodicity of 64 nm (Fig. D38a and b). There are several bands within each period.

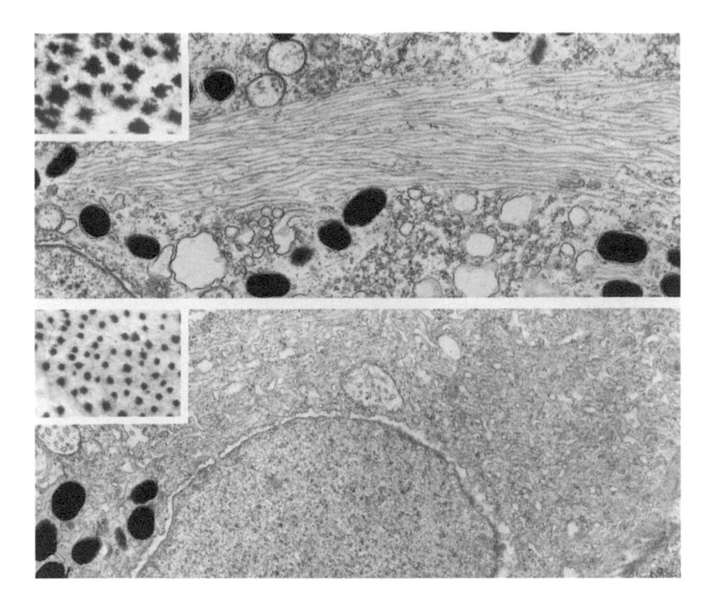

FIGURE D36 Intracellular migration of pigment granules depends upon the presence of cytoplasmic microtubules. These electron micrographs are of melanophores in an Antarctic teleost, *Pagothenia borchrgrevinki*. The insets are photomicrographs of cells with dispersed (Top) and aggregated (Bottom) pigment. 100 ×. (Top) Portion of a MELANOPHORE with dispersed pigment. A large bundle of microtubules runs across the field. 21,400 ×. (Bottom) Portion of a melanophore with aggregated pigment. Microtubules and membranous components are packed in the pigment-free area. 16,700 ×.

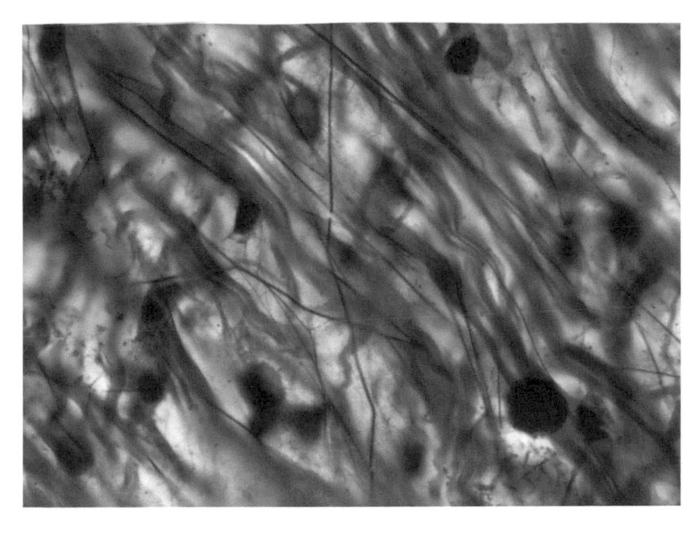

FIGURE D37 Collagenous fibers are the coarse pink fibers in this whole mount of areolar connective tissue. A different preparation was used for Fig. D6 where the eosin is paler because of the resistance to staining of collagenous fibers. 63 ×.

FIGURES D38 (a and b) These figures show negative staining with phosphotungstic acid of the band patterns of COLLAGEN FIBERS from the skin of amphibians. (1000 Å in the magnification scale is equal to 100 nm.) Collagen fibrils vary in diameter from 20 to as much as 200 nm and show a banded appearance with a periodicity of 64 nm. There are several bands within each period.

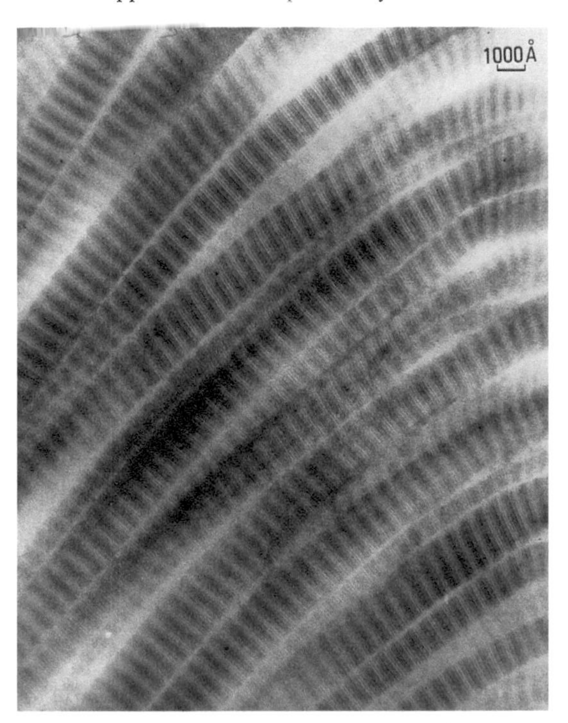

FIGURES D38a Collagen fibers from the skin of a frog (*Rana nigromaculata*).

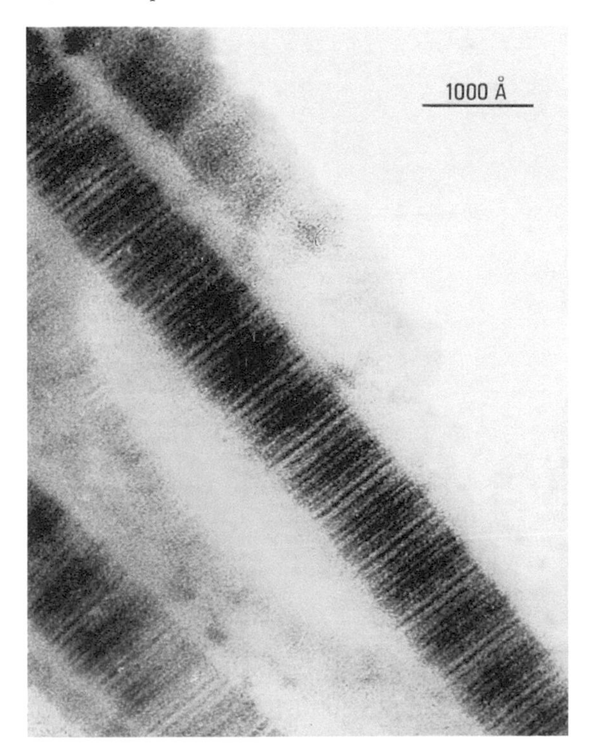

FIGURES D38b Collagen fibers from the skin of a newt (*Triturus pyrrhogaster*).

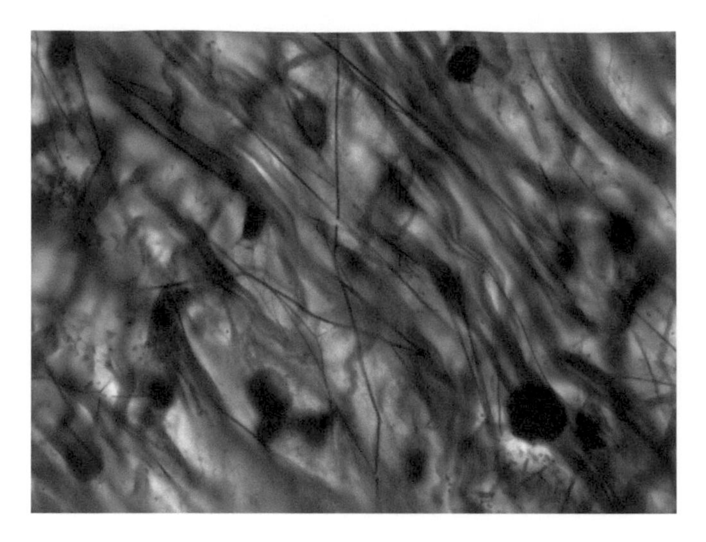

FIGURE D39 Whole mount of areolar connective tissue stained with hematoxylin, eosin, and orcein. The collagenous fibers have been stained heavily by liberal amounts of eosin. Elastic fibers are much finer, are branching, and stain brown/black with orcein. 63 ×.

FIGURES D40 Electron micrographs of sections of the connective tissue of human skin.

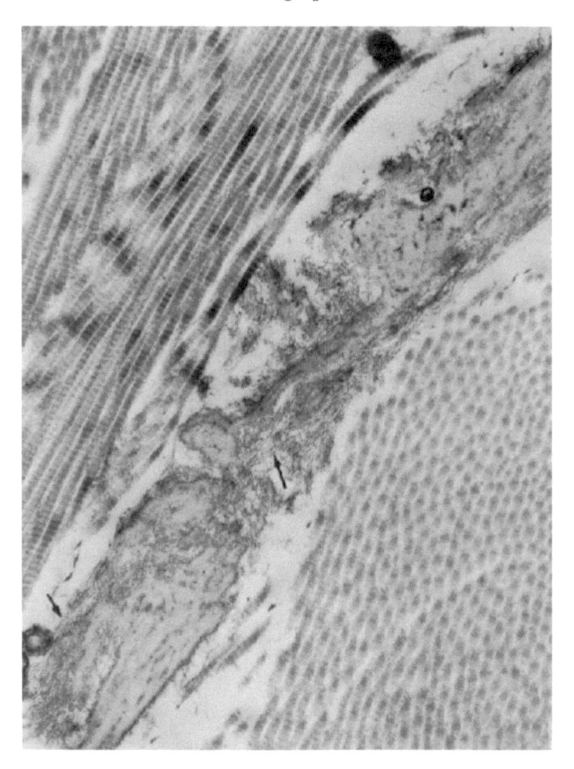

FIGURES D40a Skin of a young man. At the upper left are collagenous fibers with their characteristic periodic banding. An ELASTIC FIBER showing its two components of MICROFIBRILS (arrows) and the amorphous ELASTIN runs diagonally across the middle. Collagen is sectioned obliquely at the lower right. 21,000 ×.

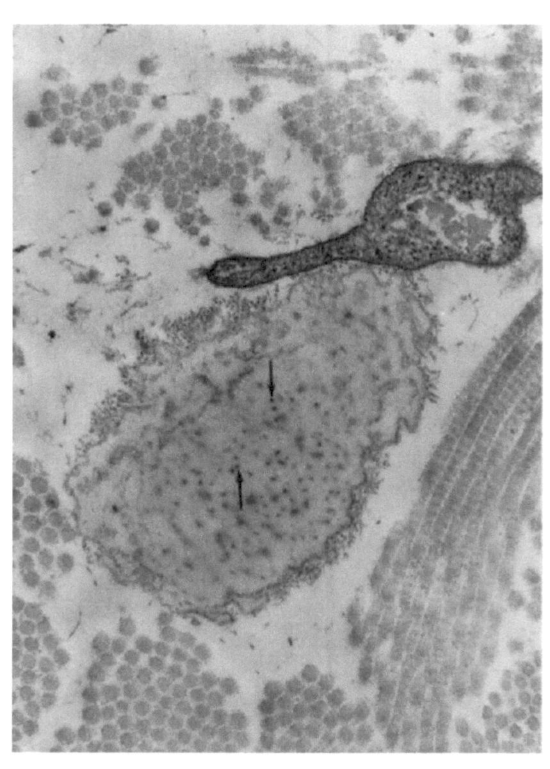

FIGURES D40b Longitudinal and cross sections of collagenous fibers surround an ELASTIC FIBER at the center of this micrograph from an adult. Amorphous ELASTIN is ensheathed by peripheral MICROFIBRILS. Many microfibrils (arrows) are embedded within the interstices of the elastic fiber. Above the fiber is a small segment of a cell, presumably a fibroblast. 32,000 ×.

2. *Elastic fibers (Yellow fibers)* are always solitary, never occurring in bundles (Fig. D39). They branch and anastomose freely thereby forming a network. In life they are stretched and under tension. They may be selectively stained with the dyes orcein and resorcin fuchsin. With the electron microscope they are seen to be composed of 12-nm MICROFIBRILS embedded in amorphous ELASTIN (Figs. D40a and b).
An assortment of fibers is presented in histological sections of skin from various vertebrates (Figs. D41a–D41g).

FIGURE D41(a–g) This series of photomicrographs illustrates the various forms taken by COLLAGENOUS FIBERS in providing strength and flexibility to the skin of vertebrates. All of the slides were stained with hematoxylin and eosin; stains to show elastic fibers were not used with the result that, although present—and even abundant in some cases—elastic fibers are not visible.

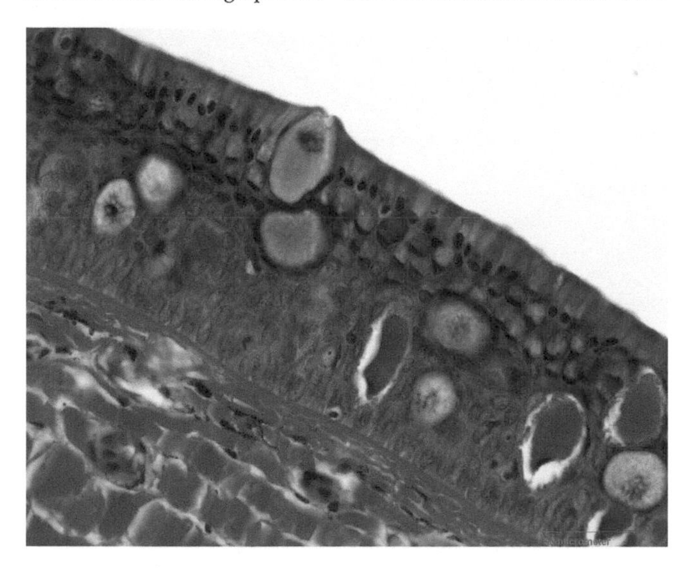

FIGURE D41a Collagenous fibers form regular bands under the epithelium of the skin of a HAGFISH. Several interesting secretory cells rise toward the surface of the epithelium. 40 ×.

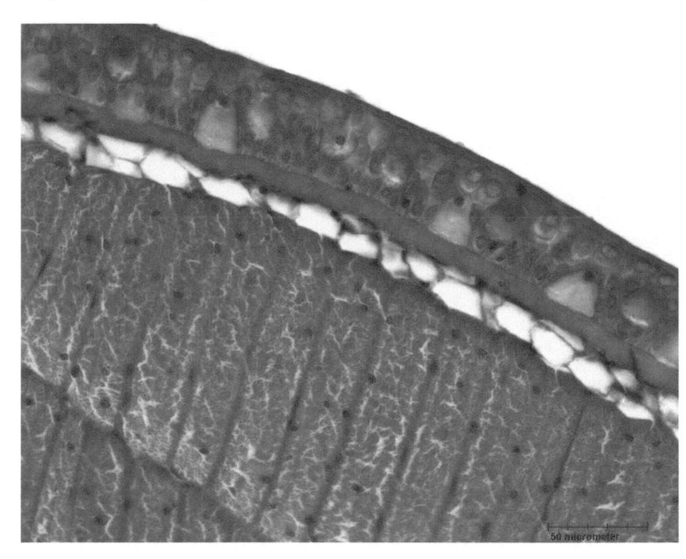

FIGURE D41b Collagenous fibers separate the epithelium from a layer of adipose cells in the skin of a larval lamprey (ammocoete). 40 ×. Note again all of the interesting cells in the epithelium. A few adipose cells separate the epithelium from the layer of connective tissue. Most of the body wall consists of muscle. 40 ×.

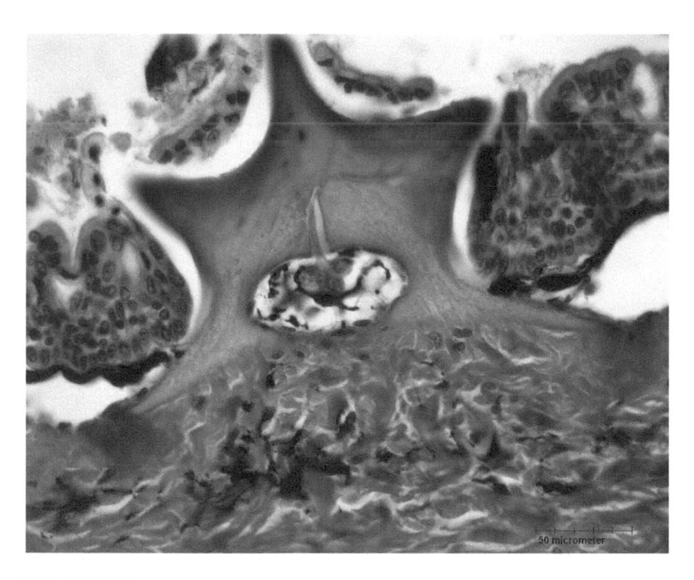

FIGURE D41c This is a section of the skin of a DOGFISH SHARK; a large scale is lodged in the outer layer of epithelium. Below is a tangled mass of collagenous fibers of the connective layer of the skin. Note the elaborate pigment cells in the connective tissue. 40 ×.

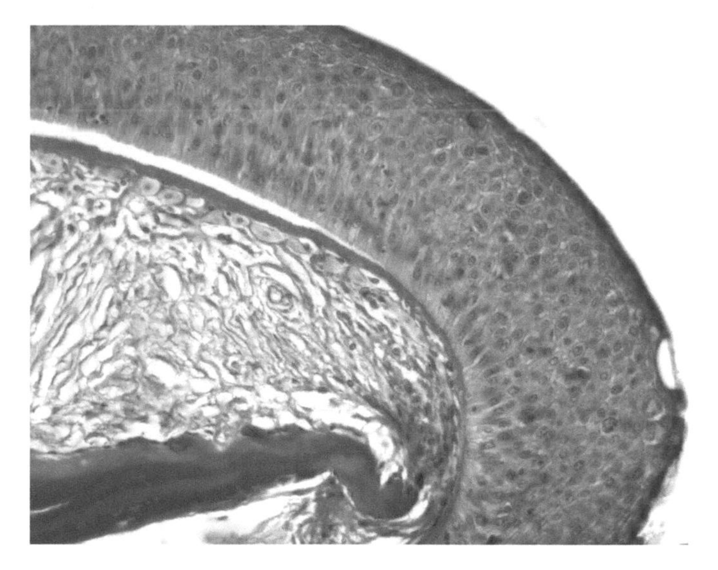

FIGURE D41d Collagenous fibers form a woven mass in the connective tissue of a FLOUNDER (an ocean-going flatfish). A deep pink scale within its "pocket" is seen at the bottom. 40 ×.

3. *Reticular fibers*. Reticular fibers are the first connective tissue fibers to appear in development and, although they provide a delicate support for most cells in mature vertebrates, they do not stain readily and are not apparent in ordinary preparations. Study the meshwork of reticular fibers in sections of lymph node and kidney impregnated with silver nitrate and compare with a similar region stained by routine methods (Figs. D42a and D42b and D43a–D43c). Reticular fibers have many of the characteristics of collagen fibrils and show the 64-nm periodicity with the electron microscope. They are always about 20 nm in diameter and do not form the coarse

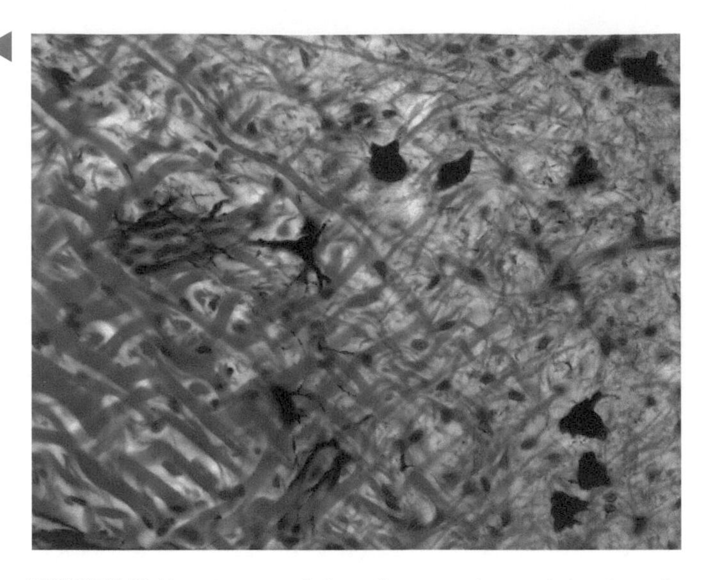

FIGURE D41e Section of the collagenous layer of the skin of a CAIMAN (a reptile that resembles a small crocodile). Collagenous fibers form a tightly woven mesh. A few chromatophores are seen. 40×.

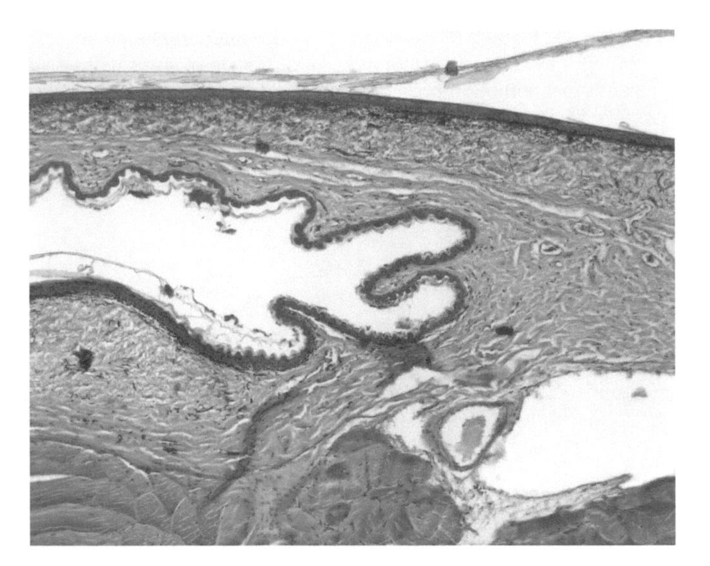

FIGURE D41f This is a section through the skin of a RATTLE SNAKE. The skin forms flaps that appear as scales and give the animal its incredibly sinuous flexibility. Note that dead cells are sloughing off from the stratified squamous, cornified epithelium. Remember that the snake tends to lose these cells all at once as it sheds its old skin. There are pigment cells in the upper layers of the connective tissue. Collagenous fibers in the skin form a fairly tangled, but tough "leather layer." 10×.

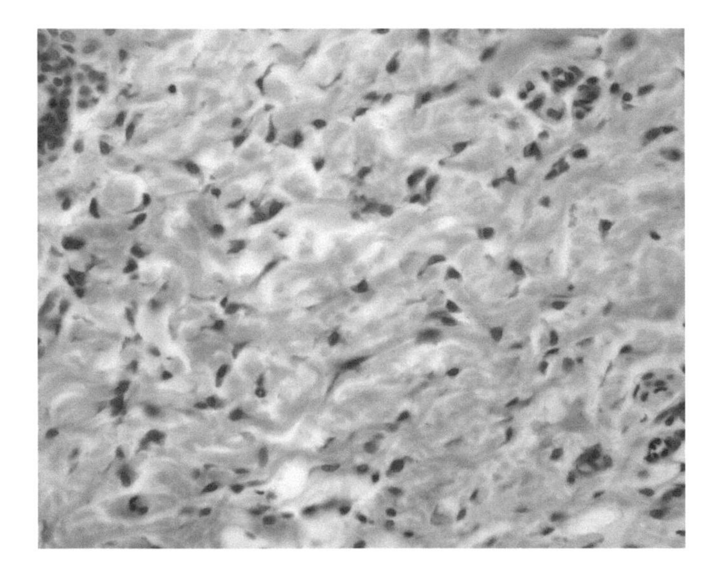

FIGURE D41g The "leather layer" of MAMMALIAN skin is formed of a tight mesh of collagenous fibers. Fibroblasts abound. 40×.

bundles seen in collagenous tissue (Figs. D44a–D44c). The branching fibers unite in an irregular, fine meshwork, the RETICULUM. In many locations they become continuous with collagenous fibers. Reticular fibers are unlike collagenous or elastic fibers in that they are ARGYROPHILIC, i.e., they are blackened by silver nitrate.

There is little difference between collagenous and reticular fibers when viewed with the electron microscope and all are called collagenous (Fig. D45). The distinction is valid for light microscopy where collagenous fibers are much coarser than reticular and possess different staining properties. The affinity for silver of the reticular fibers may be due to their coating of glycoprotein. This material probably is squeezed out when many fibrils are packed together, thereby preventing their staining.

FIGURES D42 A comparison of paraffin sections of lymph nodes stained with H&E and a sliver stain to show the reticulum.

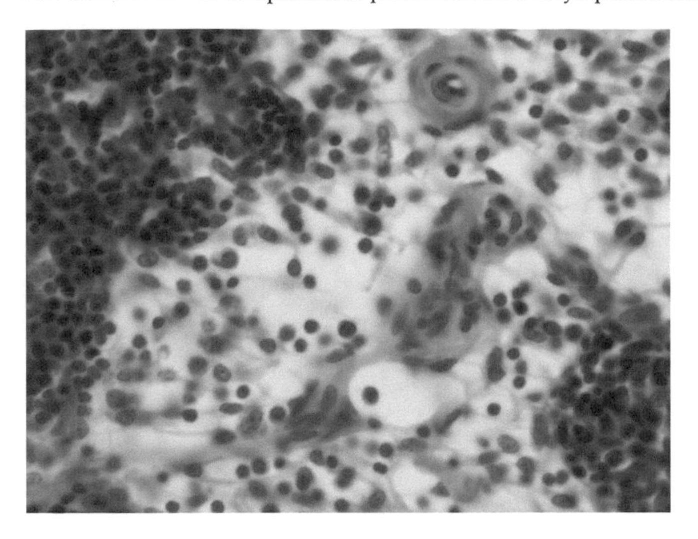

FIGURE D42a Routine section of a lymph space in a lymph node stained with hematoxylin and eosin. 40 ×.

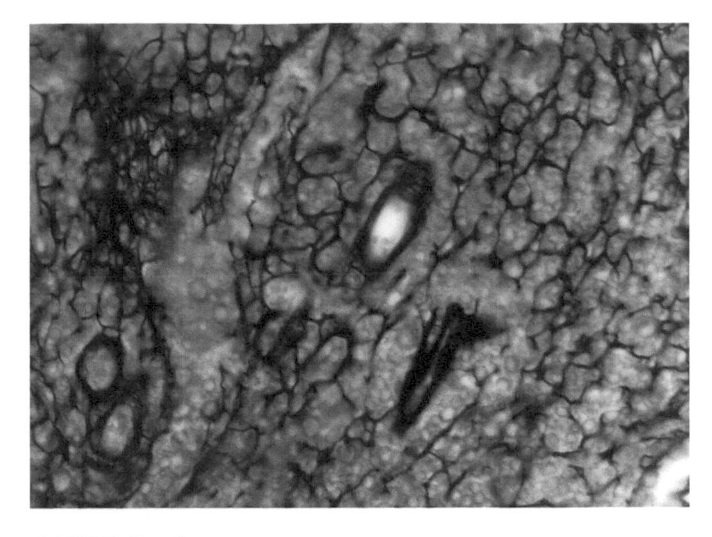

FIGURE D42b Section of a similar part of the lymph node stained with silver to show reticular fibers. 40 ×.

Amorphous Intercellular Substances

1. *Ground substance.* Connective tissue cells and fibers are embedded in an amorphous ground substance that is optically homogeneous and transparent when fresh; it is produced by fibroblasts and consists of proteoglycans and glycosaminoglycans of various types. It is extracted by ordinary fixatives and is not seen in most preparations. Proteoglycans contain sulfate groups and impart a metachromatic[1] property to ground substance when present in high concentrations. The ground substance of cartilage is composed principally of long proteoglycan–glycosaminoglycan chains associated with collagen fibrils; this provides rigidity to the tissue. The ground substance of bone is impregnated with crystals of hydroxyapatite $[Ca_{10}(PO_4)_6(OH)_2]$.
2. *Tissue fluid.* Connective tissue contains a transudate that seeps from blood plasma and constitutes one-third of the total body fluid (Fig. D46). Materials diffuse back and forth between the blood and cells of the connective tissues in this tissue fluid.

LOOSE CONNECTIVE TISSUES

Areolar Connective Tissue

Areolar tissue is a loosely arranged, fibroelastic connective tissue; it is the most widespread of all connective tissues and is encountered in sections taken from any region of the body; indeed, it enmeshes all blood vessels, thereby penetrating the farthest reaches of the body. This tissue connects the skin to the underlying tissue, appears as intermuscular connective tissue, and underlies the epithelium of the gut.

You have already studied a stained stretch preparation of areolar tissue, noting the COLLAGENOUS and ELASTIC FIBERS (Figs. D6 and D37). The two commonest cells are FIBROBLASTS and MACROPHAGES. They occur in spaces between the fibers, the AREOLAE. Note the small blood vessels that permeate the tissue. The MATRIX is the finely granular material between the fibers.

[1]Metachromasia: Staining a different color from that of the stain used; e.g., metachromatic ground substance stains purple when the dye methylene blue is used.

FIGURES D43 (a–c) A comparison of photomicrographs of kidney tubules of a gerbil stained by three different methods.

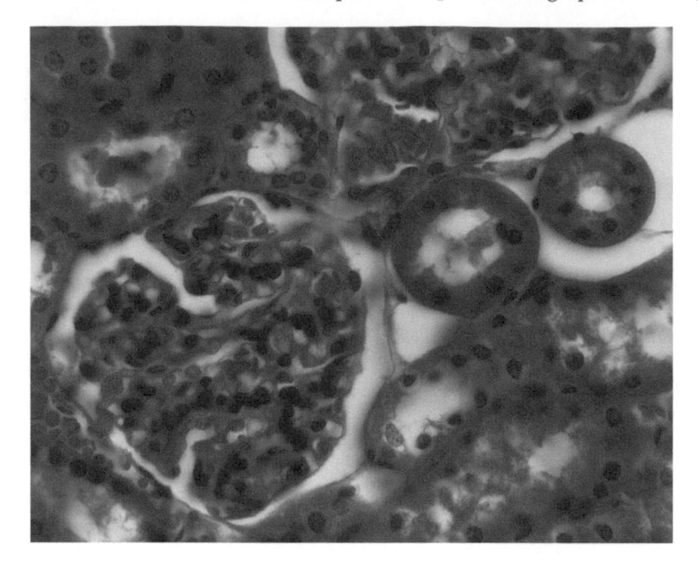

FIGURES D43a Routine section stained with hematoxylin and eosin. Little detail can be seen of the fibers supporting the various structures. 63 ×.

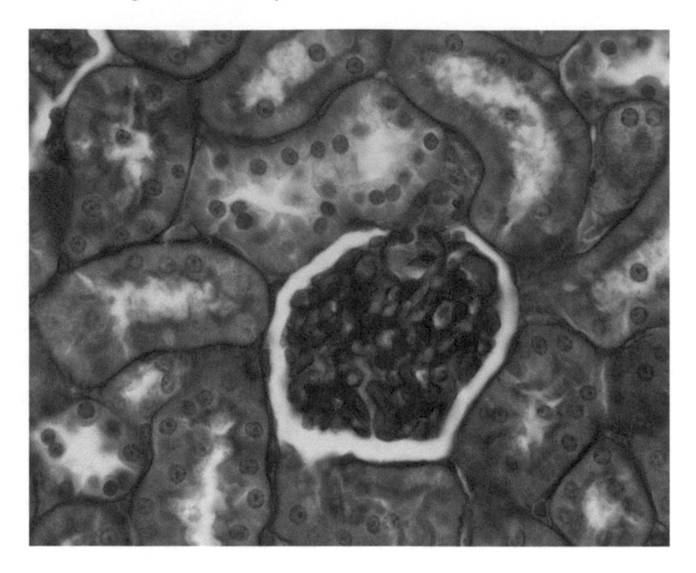

FIGURE D43b A combination of stains incorporating fast green, orange G, and the periodic acid-Schiff (PAS) stain highlights the basement membranes of the tubules and other structures in the kidney; nuclei are green and red blood cells are orange. (Note that the deep red of the fuchsin in the PAS has colored the brush borders of the tubules.) 63 ×.

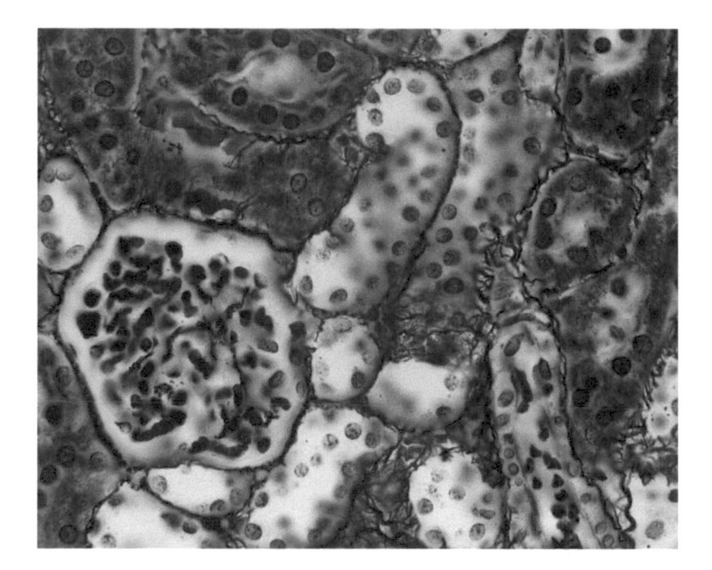

FIGURE D43c The elegant reticular mesh supporting the tubules of the kidney is stained black by Wilder's silver impregnation method. 63 ×.

Adipose Tissue

Fat is an atypical connective tissue in that the cell contents rather than the interstitial substance dominate the scene. In the micrograph of WHITE ADIPOSE TISSUE stained with Sudan IV (where no fat solvents have been used), the cytoplasm of the unilocular fat cells is a thin shell surrounding the fat globule and is somewhat more abundant about the flattened nucleus; the cytoplasm and nucleus have a signet-ring appearance (Fig. D19b). See similar preparations treated with osmium tetroxide (Fig. D19a). A fine network of reticular fibers envelops each cell and the tissue is bound together by richly vascularized areolar tissue (Figs. D18a and D18b). Figs. D20a and D20b show groups of fat cells in a section of adipose tissue stained by routine methods. The outlines of the cells are

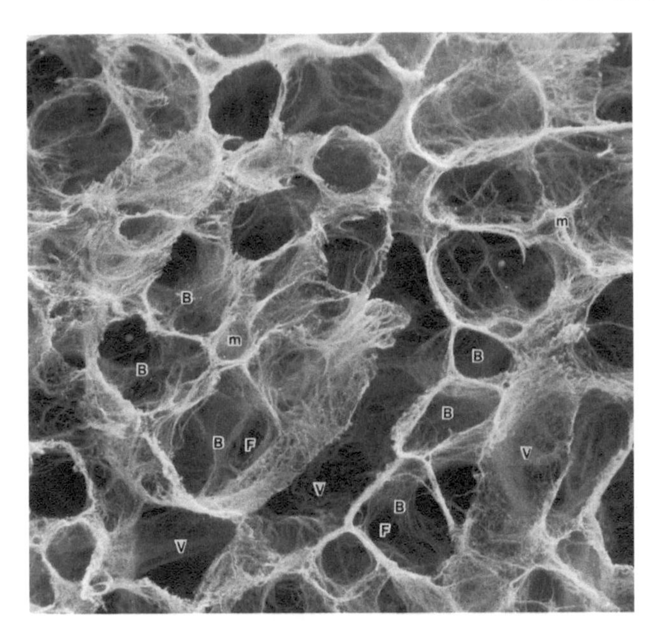

FIGURE D44a Scanning electron micrograph of the reticular framework of delicate fibers supporting cells of the adrenal gland of a rat. Cells and other components of the gland were removed by an alkali–water maceration process. The fibers support a vein (V), clusters of cells (B), and capillaries (m). Open spaces (F) in the baskets allow for communication within the gland. 900 ×.

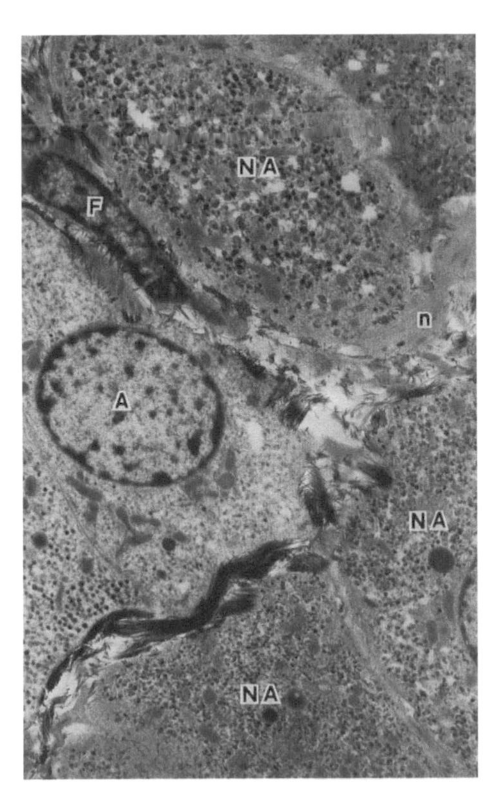

FIGURE D44c A view with the transmission electron microscope of a section of adrenal gland stained with tannic acid to show the plentiful reticular fibers supporting the cells of the adrenal gland (A, NA). A fibroblast (F) and a nerve fiber (n) lurk in the spaces between the cells. 5500 ×.

FIGURE D44b A closer view of the wall of the basket-like compartment interposed between the cluster of cells and the vein. 5000 ×. (Lower right) Reticular fibers support nervous elements (N and arrows) in the adrenal gland. Open spaces (F) in the baskets allow for communication within the gland. 1200 ×.

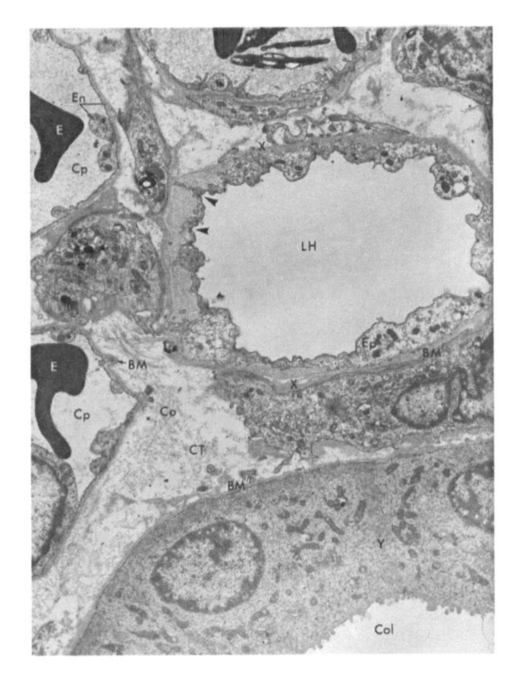

FIGURE D45 Transmission electron micrograph of a section of the kidney of a rat. Delicate reticular fibers in the connective tissue around the tubules provide support for the various structures. E, erythrocyte; LH, lumen of the Loop of Henle; BM, basement membrane; Co, collagen fibrils; CT, connective tissue fibrils; Cp, capillary; En, endothelium; X, lamellar basement membrane; arrowheads - junction complexes. 11,000 ×.

modified by shrinkage resulting from the action of the fat-dissolving reagents used to prepare the tissue. The fat cells form a chicken-wire appearance; the cytoplasm is a thin, peripheral rim around a vacuole that was originally occupied by fat. The nucleus is compressed to the edge of the cell. Fig. D17 is an electron micrograph of a section of a fat cell.

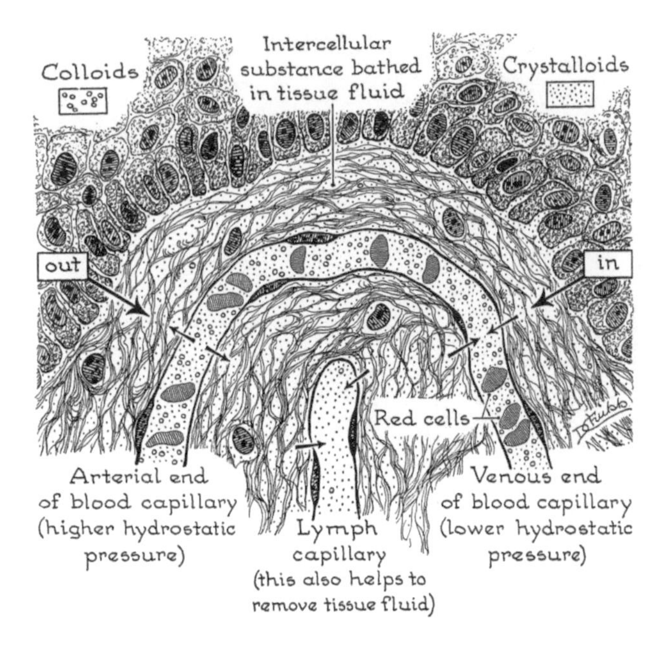

FIGURE D46 TISSUE FLUID seeps back and forth between the blood plasma within blood capillaries and the surrounding tissues where it constitutes about one-third of the total body fluid. Excess fluids are drained by LYMPHATICS that invade the tissues. HYDROSTATIC PRESSURE of the blood drives fluids out of capillaries; OSMOTIC PRESSURE, produced by the presence of COLLOIDS in the plasma within capillaries, but not in the tissue fluids, has an opposite effect on the fluids and draws them back into the capillaries. Under normal circumstances, a balance is affected between these two influences. If this balance is disturbed, e.g., by an obstruction of venous or lymphatic drainage, or by an increase in the accumulation of colloids in the tissues, the tissue may swell and become EDEMATOUS. CRYSTALLOIDS, (largely sodium chloride) and dissolved gases are present in similar amounts in the plasma and tissue fluid and have little effect on fluid exchange.

Accumulations of multilocular BROWN FAT cells in certain areas of the body constitute the "hibernating glands" of some mammals (Fig. D21). In electron micrographs of brown fat note the abundant mitochondria within the multilocular cells, and the rich blood supply and abundant nerve fibers permeating the interstitial tissue (Figs. D22a–D22c, D23).

Reticular Tissue

Reticular tissue provides delicate support for many of the cells of the body. It constitutes at least part of the framework of most organs (Figs. D44a and D44b) and of the mucous membranes of the gut and respiratory tract. Examine an area of the image of a lymph node stained with hematoxylin and eosin in Fig. D42a where the free cells are sparse. RETICULAR CELLS resemble mesenchymal cells with delicate, anastomosing processes forming a network. The nuclei are small. Note the fine mesh of RETICULAR FIBERS in the silver stained sections of the same tissue (Fig. D42b). The fibers lie close to the cells and form the true reticulum. The MATRIX is an abundant fluid in which the free cells of the organ float.

DENSE CONNECTIVE TISSUES

Tendon

Tendon is composed of thick, closely packed, parallel collagenous bundles, similar in structure to those of areolar tissue (Fig. D47a). Fibrocytes are the only cells present and are squeezed between the fibers in rows. In the micrograph of a longitudinal section of tendon, note the groups of fibrocytes squeezed between wavy collagenous fibers. In the cross section, note the fibrocytes between the fibers and that several fibers aggregate to form BUNDLES (Fig. D47b). Fine elastic networks have been described between the collagenous bundles but are not visible in these preparations.

Irregular Dense Collagenous Tissues

In many locations there are sheets of densely packed collagenous fibers similar to tendons except that the fibers form an irregular feltwork instead of parallel bundles. These include the "leather layer" of the skin (Fig. D48), capsules of many organs; the sheaths of muscles, bones, and cartilages; the membranes covering the central nervous system (Fig. D49)—dura mater; and the walls of some of the blood vessels (Fig. D50). Irregular dense collagenous tissue constitutes a large part of the wall of the swim bladder of many fish, providing great strength to resist the huge pressures developed by the gases within (Fig. D51). The pliable shell of the snapping turtle consists of irregular dense collagenous tissue reinforced by deposits of calcium salts between the fibers (Fig. D52).

FIGURE D47 (a and b) Cross and longitudinal sections of tendon. 10×.

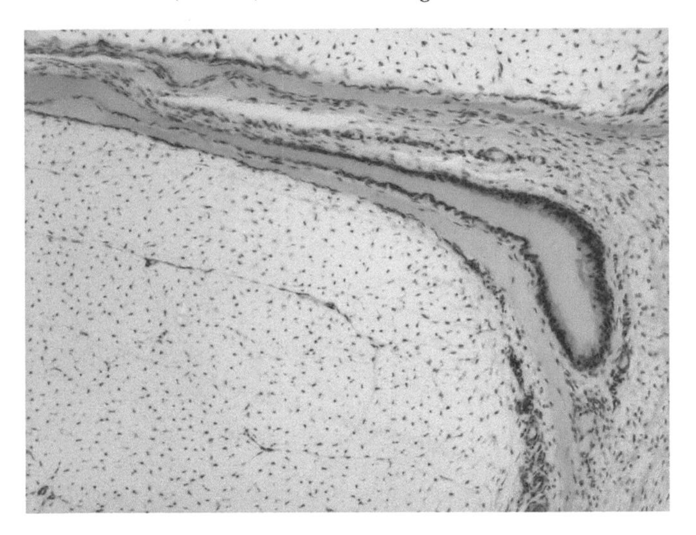

FIGURE D47a Longitudinal section. Bundles of collagenous fibers constitute most of the substance of the tendon. Elongate fibrocytes are squeezed between the fibers. 10×.

FIGURE D47b Cross section of tendon. Masses of strong collagenous fibers form the cables that connect bones to muscles. Fibrocytes are the dark dots between the fibers. The TENDON SHEATH of connective tissue crosses the top and right of the field. 10×.

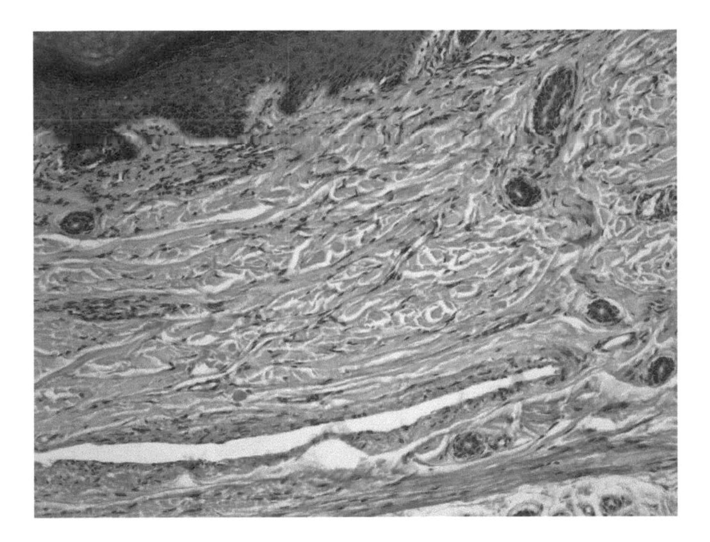

FIGURE D48 Cornified human skin showing the densely packed, tough dermal layer of irregular collagen fibers and sections of a few sweat glands. The cornified stratified squamous epithelium is at the top. 10×.

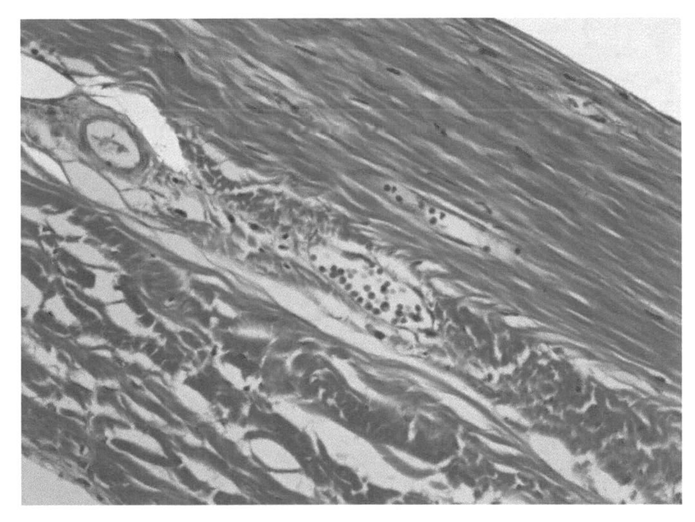

FIGURE D49 The dura mater is the tough sheath covering the brain. Irregular collagenous fibers permeated with blood vessels form this protective later. 20×.

Dense Elastic Tissue

This is a specialized tissue in which elastic fibers predominate and give it a yellow color. In longitudinal and transverse sections of the ligament of the neck of quadrupeds, the LIGAMENTUM NUCHAE, note the coarse elastic fibers with fine collagenous fibers between, and the rows of small, inconspicuous fibrocytes (Fig. D53a and b). Compare them with tendon cells. The elastic fibers of ligaments are usually arranged in parallel bundles but in other regions, such as the walls of the large arteries, their arrangement is less regular (Fig. D54).

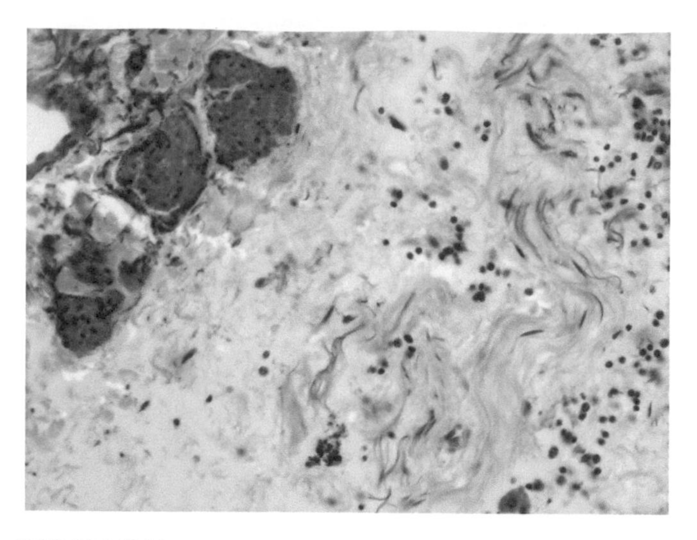

FIGURE D50 Veins are subjected to high internal pressures as blood is squeezed back to the heart by activity of the body. Their walls are reinforced by a thick layer of irregular collagenous fibers; a few bundles of smooth muscle are seen at the upper left. 20×.

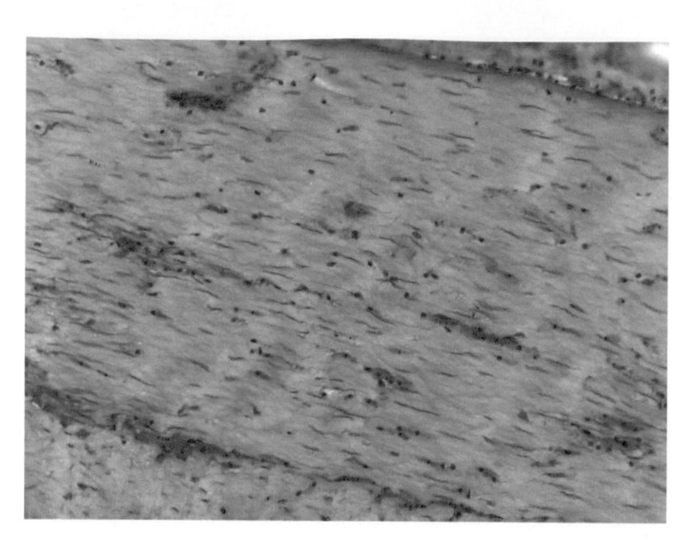

FIGURE D51 The swim bladder of fish must withstand great pressures as the animal swims up from great depths. This section of the wall of the swim bladder of the Atlantic sheepshead, *Archosargus probatocephalus*, is reinforced by reticular fibers. 20×.

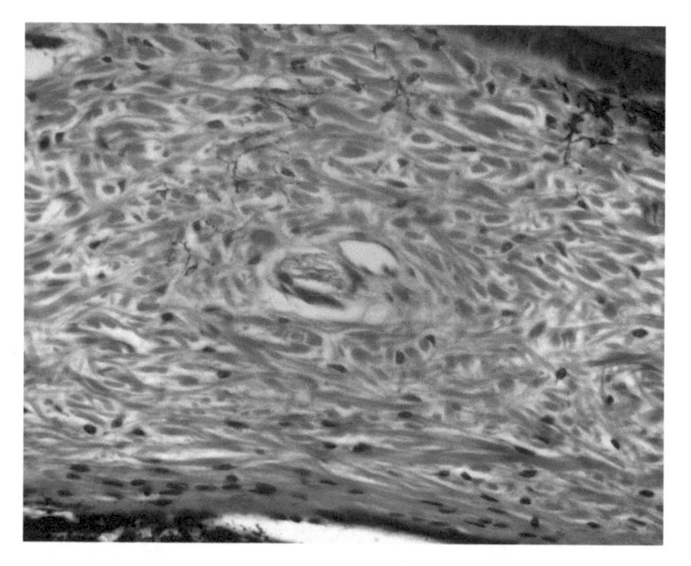

FIGURE D52 The pliable shell of the snapping turtle consists of irregular dense collagenous tissue reinforced by deposits of calcium salts between the fibers. Note the abundant fibroblasts throughout. Blood vessels pass through at the center. 40×.

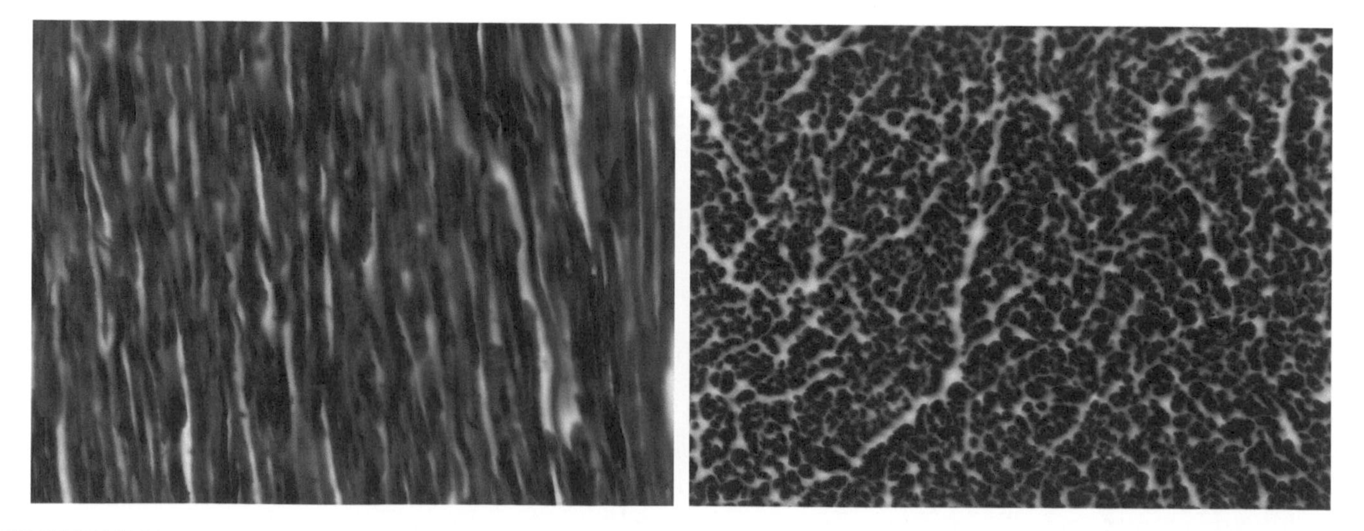

FIGURES D53 Longitudinal (a) and cross sections (b) of the ligament from the neck of a horse (ligamentum nuchae). The elastic fibers are stained with orcein. These ligaments are strong, elastic cables that connect bones and provide a measure of shock absorption. Ligaments in other locations contain more collagenous fibers and fewer elastic fibers. Fibrocytes are squashed between the fibers and are difficult to see. Both 20×.

FIGURE D54 Deeply stained elastic fibers in the wall of an artery provide strength and elasticity to the wall of this pulsating vessel. Between the elastic fibers are collagenous fibers and strands of smooth muscle. 20 ×.

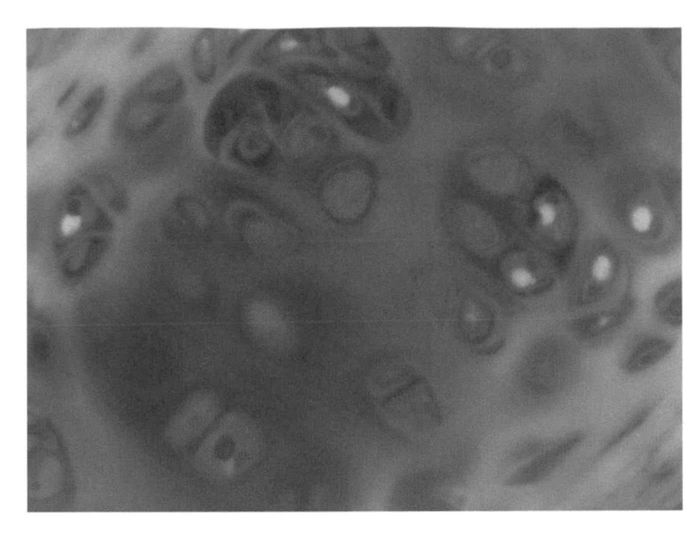

FIGURE D55 Section of hyaline cartilage from the trachea of a monkey. Clusters of chondrocytes are embedded in a hyaline matrix. The matrix surrounding the lacunae is more refractive than the rest of the matrix and is called the CAPSULE. 40 ×.

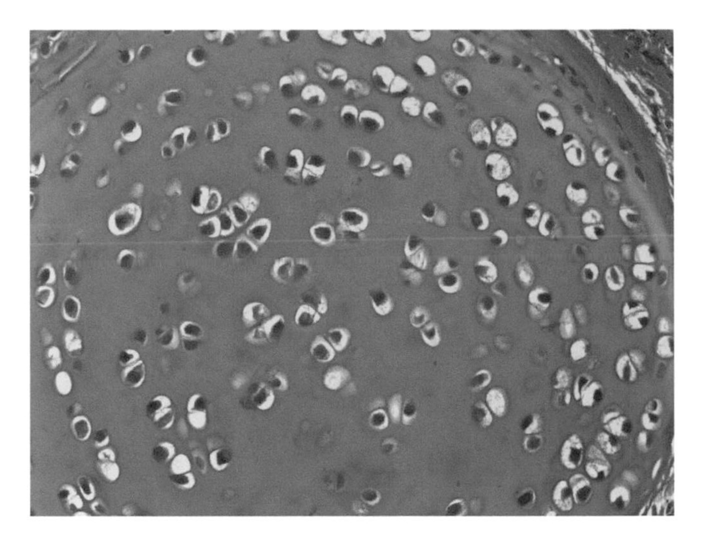

FIGURE D56a Paraffin section of hyaline cartilage of a lamprey shows considerable shrinkage of the chondrocytes within their lacunae. 10 ×.

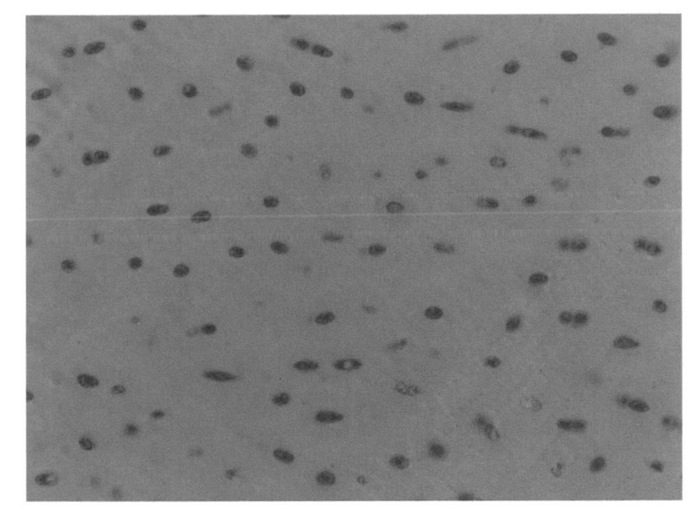

FIGURE D56b Hyaline cartilage from a shark. A hint of the collagenous supporting fibers in the cartilagenous matrix may be seen. 20 ×.

Cartilage

1. *Hyaline cartilage*. Fresh cartilage is firm and resilient, bluish-white in color, and has a pearly luster. The cells (CHONDROCYTES) lie in spaces or LACUNAE within the homogeneous MATRIX (Fig. D55). Although they often cannot be seen in the micrograph, COLLAGENOUS FIBERS are embedded in the matrix, greatly strengthening it. The matrix immediately surrounding the lacunae is more refractive than the rest of the matrix and is called the CAPSULE. Note the characteristic shapes of single cells and groups of cells. Since the matrix is avascular, how are the cells nourished? Cartilage is enclosed in a tough sheath of dense collagenous tissue, the PERICHONDRIUM (Figs. D56a–D56c). New chondrocytes are formed beneath this covering so that, in a growing animal, the lacunae are smaller and closer together than in the deeper parts of the matrix. During synthesis of matrix material, electron micrographs of immature cartilage cells or CHONDROBLASTS show the familiar characteristics of protein production: ragged cell borders with well-developed granular endoplasmic reticulum and Golgi complex (Fig. D57). As the cells mature into CHONDROCYTES these features diminish while glycogen and lipid material accumulate in the cytoplasm. Note collagenous fibers in electron micrographs of the matrix.

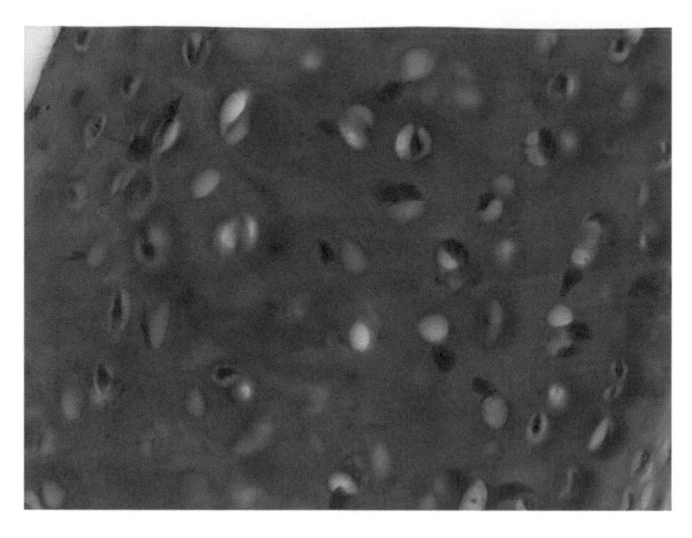

FIGURE D56c Hyaline cartilage from the trachea of a caiman. 40 ×.

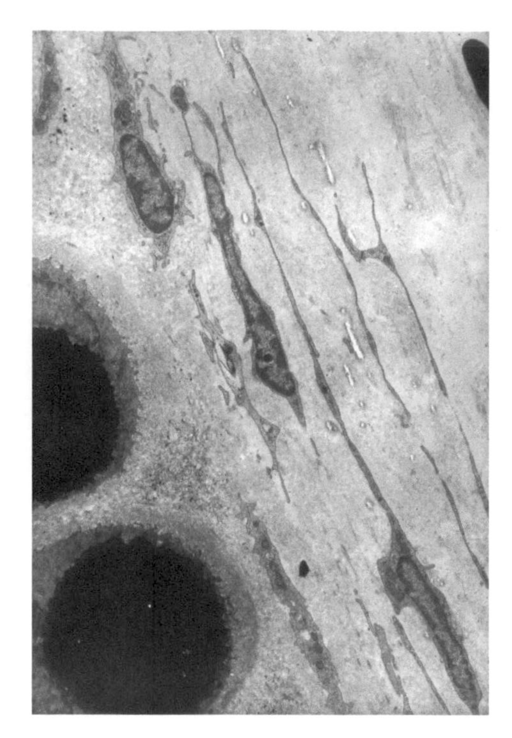

FIGURE D57 Transmission electron micrograph of tracheal hyaline cartilage of a bat. Two CHONDROCYTES at the left are embedded in their LACUNAE in the CARTILAGINOUS MATRIX. These chondrocytes contain large lipid droplets. The matrix is reinforced by an irregular three-dimensional lattice of delicate collagenous fibers. MICROVILLI, projecting from the surface of the chondrocytes probably enhance the exchange of metabolites with the matrix. A layer of tough fibrous connective tissue of the PERICHONDRIUM (right), encloses the cartilage; the perichondrium contains a few FIBROCYTES squashed between its fibers. 11,000 ×.

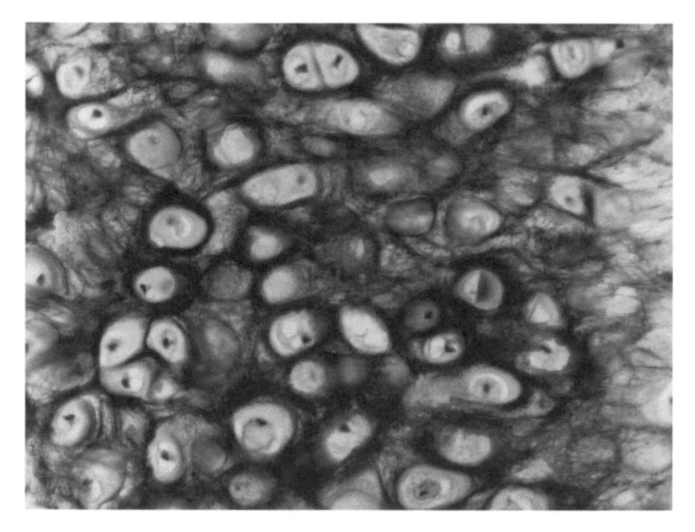

FIGURE D58a The matrix of this mammalian elastic cartilage is reinforced by deeply stained elastic fibers, which confer flexibility and strength to the structure. 20 ×.

2. *Elastic cartilage* has a yellowish tinge in the fresh condition because of numerous elastic fibers in the matrix. It is found in sections of ear pinna or epiglottis that have been stained to show elastic tissue (Figs. D58a and D58b).
3. *Fibrous cartilage.* Fibrocartilage never occurs alone but merges imperceptibly with neighboring hyaline cartilage or fibrous tissue. It resists compression and occurs in places where a tough support is required as in the intervertebral discs and the pubic symphysis (Fig. D59). Heavy bundles of collagenous fibers pass irregularly through the transparent hyaline matrix. Although hyaline cartilage also contains collagenous fibers in the matrix it differs from fibrocartilage largely in the amounts present.

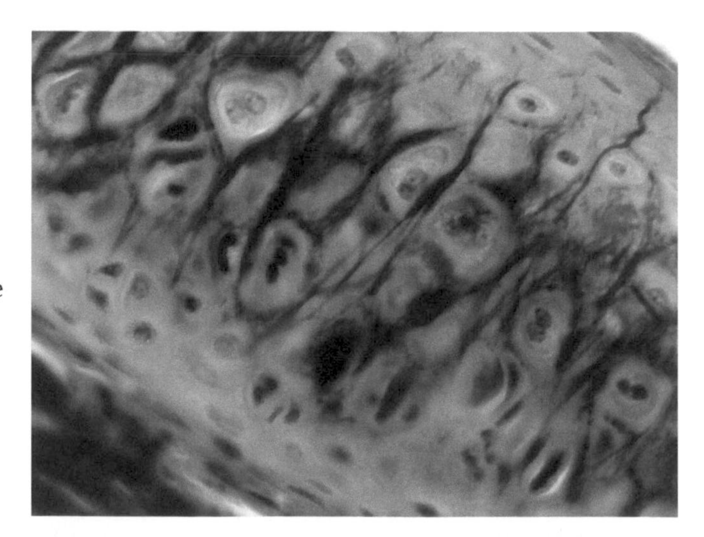

FIGURE D58b Elastic cartilage from the external ear of a cat; the elastic fibers coursing through the matrix are stained a deep blue-black with iron hematoxylin. A small portion of the perichondrium is seen at the lower left. 40 ×.

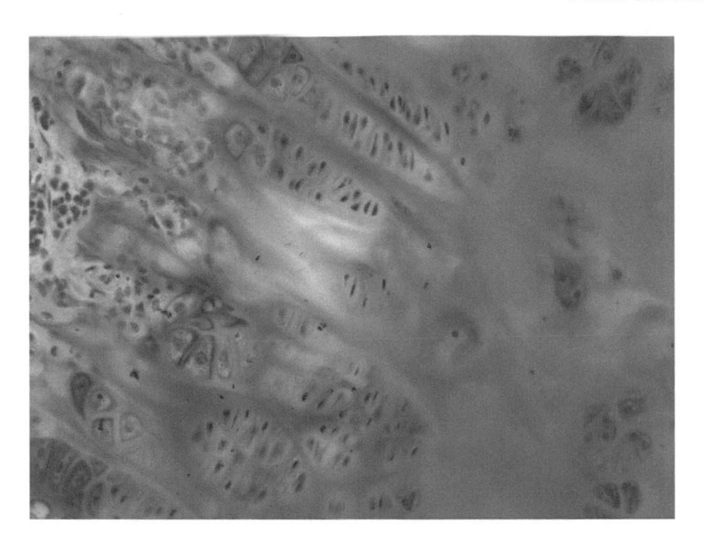

FIGURE D59 Hyaline cartilage on the right merges with fibrocartilage on the left where prominent collagenous fibers strengthen the matrix between aggregations of chondrocytes. 20 ×.

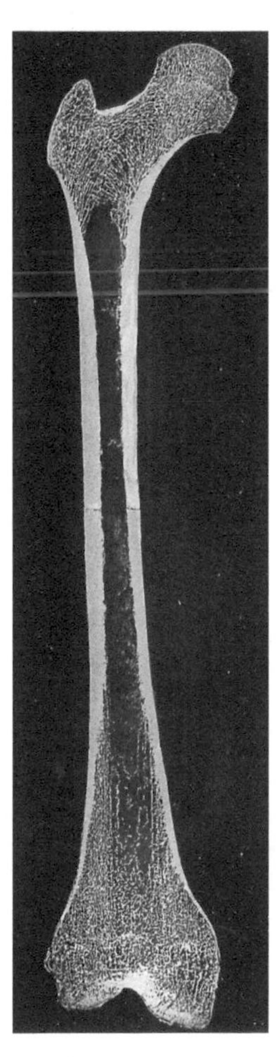

FIGURE D60 Frontal longitudinal mid-section of the left human femur. In the region of the shaft or DIAPHYSIS, the bone surrounding the marrow cavity is compact. The compact bone at the ends (EPIPHYSES) encloses spongy bone that is composed of a lattice-like network of irregular bony TRABECULAE.

Bone

The matrix of cartilage is relatively permeable to nutrients and wastes so that internal vascularization is unnecessary. Bony matrix, on the other hand, is impermeable and the structure of bone results, in large part, from the necessity of providing small channels in the matrix for the passage of materials.

1. *Gross structure.* Fig. D60 is a photograph of a dried human femur cut in halves longitudinally. Lightness is achieved by hollowing out the bones, and strengthening them internally by a scaffolding of delicate TRABECULAE. In the region of the shaft or DIAPHYSIS the bone surrounding the MARROW CAVITY is COMPACT. The bone at the ends (EPIPHYSES) of the long bone is SPONGY and is composed of a lattice-like network of irregular bony TRABECULAE that are in turn enclosed by an outer layer of compact bone (Figs. D61a and D61b). At first glance it might appear that the trabeculae of spongy bone are arranged in a random fashion. Note, however, that they are arranged along the lines of mechanical stress (Figs. D62a and D62b), in the form of graceful, strong gothic arches (Figs. D63a and D63b).

   A fresh bone is covered by a tough sheath of fibrous connective tissue, the PERIOSTEUM, to which the muscles attach (Figs. D64 and D65). This sheath is firmly affixed to the bony matrix by PERFORATING FIBERS (Sharpey's fibers), coarse collagenous fibers of the periosteum that plunge inward to penetrate the outer lamellae of bone. The marrow cavity is lined by a similar, but more delicate, sheath, the ENDOSTEUM.
2. *Thin compact bone.* Living bone cells or OSTEOCYTES are enclosed within small chambers, the LACUNAE, in the BONY MATRIX. Adjacent lacunae are connected by small CANALICULI that contain PROCESSES from the osteocytes. Scanning electron micrographs of decalcified bone show this intricate connection between osteocytes (Figs. D66a and D66b). There is an exchange of metabolites and wastes through these processes.

   The organic part of the matrix is the OSTEOID and consists of a dense feltwork of collagenous fibers bathed in a mucopolysaccharide ground substance. This becomes impregnated and hardened with densely packed crystals of HYDROXYAPATITE. In contrast to hyaline cartilage, the fibers constitute the greater proportion of the matrix. The mineral component of the matrix may be dissolved by soaking the bone in acid; this decalcified bone is rubbery (Fig. D67a). Alternatively, the organic components of the matrix may be burned away, leaving only the mineral components; such a bone retains the appearance of an intact bone but is brittle and may be easily crumbled between the fingers (Fig. D67b).

FIGURE D61(a and b) Scanning electron micrographs of dried bone from the human sternum.

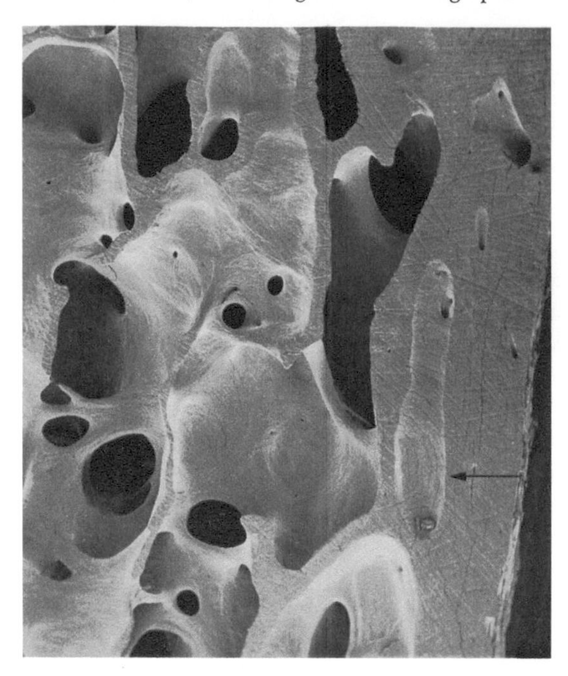

FIGURE D61a The junction of compact bone (right) and spongy bone. 30×.

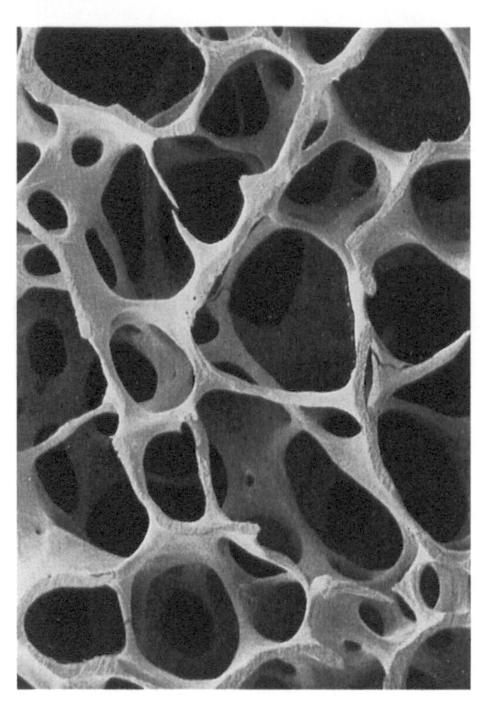

FIGURE D61b Delicate trabeculae from a medial portion of the sternum. 20×.

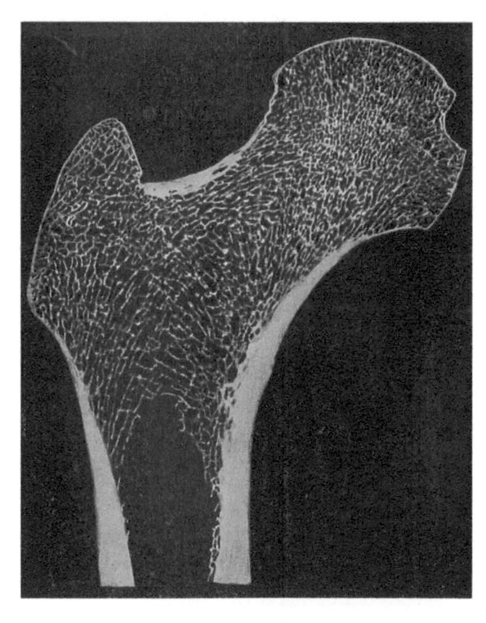

FIGURE D62a Frontal longitudinal mid-section of the upper left human femur. COMPACT BONE surrounds the structure; the remainder consists of TRABECULAE of spongy bone and the MARROW CAVITY at the bottom.

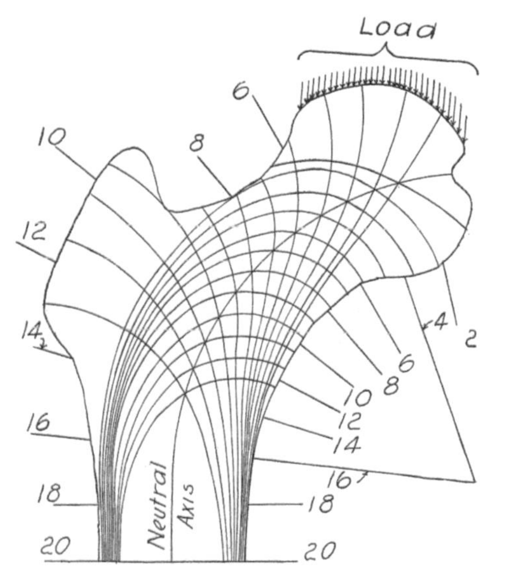

FIGURE D62b Diagram of the lines of stress (numbered from 1 through 20) in the upper femur based on a mathematical analysis. It is apparent that the trabeculae shown in Fig. D62a form Gothic arches that are arranged in the directions in which the bone is stressed.

In electron micrographs of bone, note the mineralized and fibrous components of the matrix (Fig. D68). Note that each lacuna is lined by a dense feltwork of unmineralized bony matrix, the OSTEOID. The osteocytes display the appearance of cells actively producing proteins.

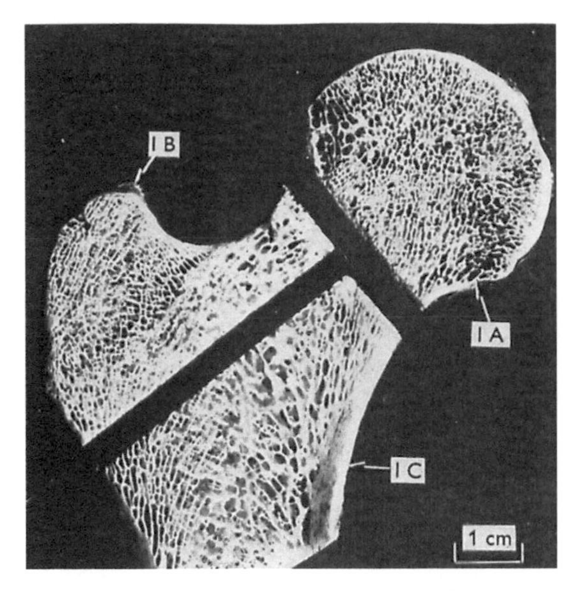

FIGURE D63a Three sections from the proximal epiphysis of a human femur. The trabeculae are arranged in stress-induced Gothic arches as seen in Fig, D62b. The numbers indicate load points.

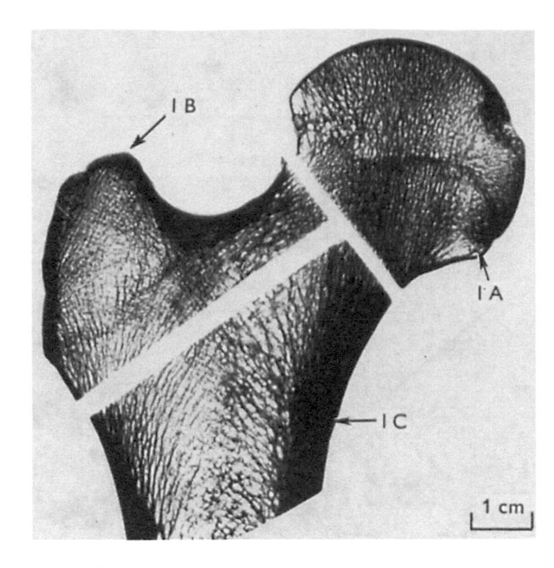

FIGURE D63b The arched arrangement of trabeculae is more apparent in this contact radiograph of the same three sections. The numbers indicate load points.

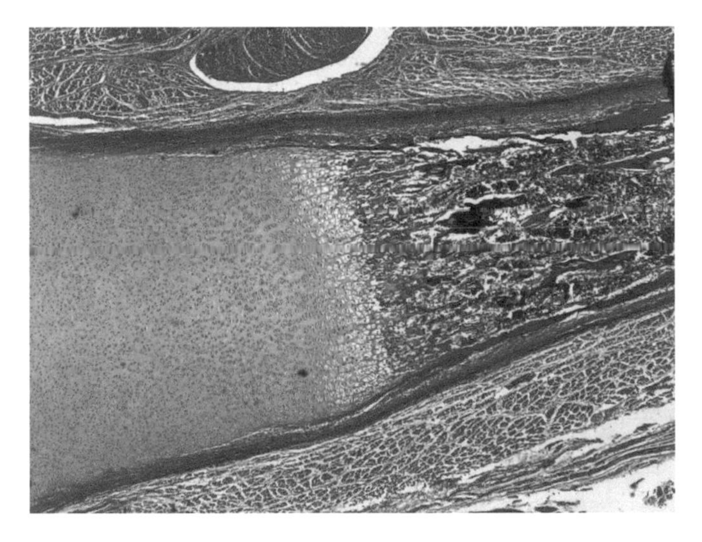

FIGURE D64 Longitudinal section of a developing mammalian long bone. The bright red structure enclosing the marrow cavity is compact bone. The periosteum is the pink layer around it. 2.5 ×.

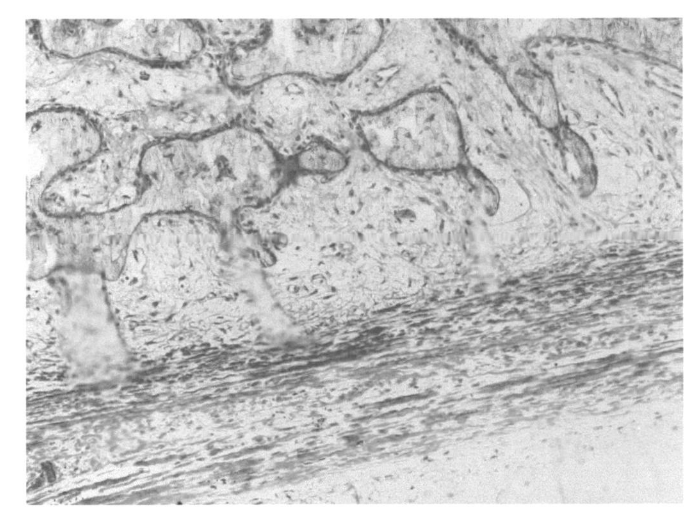

FIGURE D65 Section of undecalcified human bone cut on a special microtome with a very strong knife. The section has been stained with toluidine blue and is mounted on a slide in immersion oil. Cancellous bone at the top contains marrow. The periosteum of dense collagenous tissue, is at the bottom. 10 ×.

3. *Ground sections of dense bone*. Because of the hardness of the matrix, the microscopic structure of dense bone is often studied in thin sections prepared by grinding the tissue on a polishing wheel (Fig. D71). No cells remain and the spaces are made visible by their content of polishing compound and entrapped air. In Fig. D69a the image has been enhanced by a red dye. In this cross section of a long bone, note the OSTEONS (Haversian systems), each consisting of a series of concentric lamellae of bony matrix enclosing an OSTEON CANAL (Haversian canal) and the LACUNAE. Collagenous fibers in successive lamellae of an osteon alternate in direction, like the sheets of veneer in plywood, thereby providing an extremely strong structure (Fig. D70).

Filling the spaces between the circular osteons are the irregular INTERSTITIAL LAMELLAE with their associated lacunae and canaliculi. In life, the osteon canal contained blood vessels and nerves and the lacunae contained osteocytes. The osteon canals of two osteons are connected by a PERFORATING CANAL (Volkmann's canal) (Fig. D69b); perforating canals are not surrounded by concentric lamellae. In a longitudinal section of ground bone identify the osteons, the bony lamellae, lacunae, canaliculi, etc. (Fig. D71).

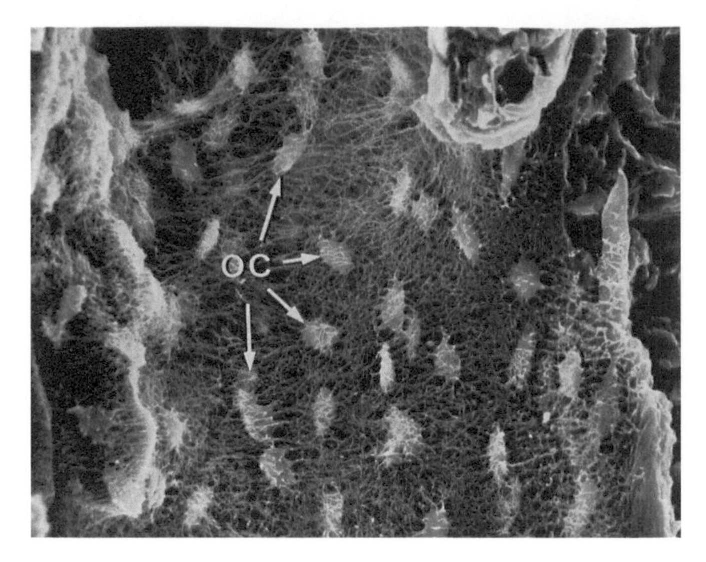

FIGURE D66a Scanning electron micrograph of decalcified (in hydrochloric acid) bone of a rat. The intricate cellular network formed by many OSTEOCYTES (OC) and their processes is shown. In life, osteocytes reside in snug LACUNAE in the bony matrix and the processes reach out to each other in fine tunnels, the CANALICULI, that penetrate the matrix. 1200×.

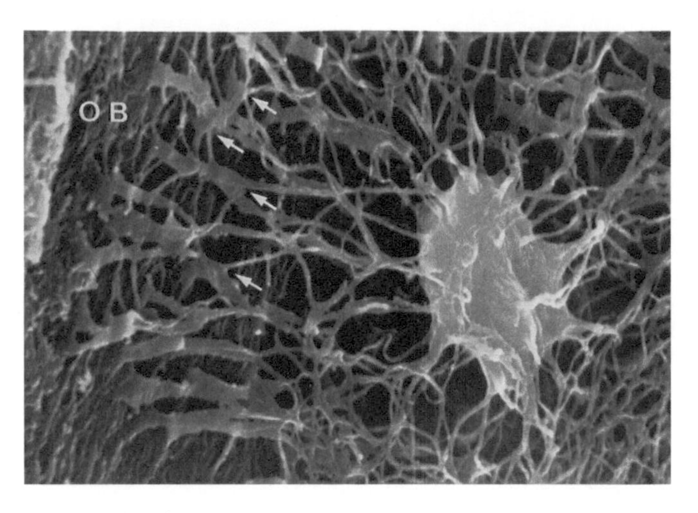

FIGURE D66b In the preparation seen in Fig. D66a, an osteocyte is connected with an osteoblast (OB)—a young bone cell—by many delicate processes that inhabit the canaliculi in life. Arrows indicate branching of the processes. 7100×.

FIGURE D67 (a and b) Bony matrix resembles reinforced concrete where steel rods, represented by collagenous fibers, reinforce the concrete as represented by the mineral component. The strength derived from two components can be shown by a simple demonstration.

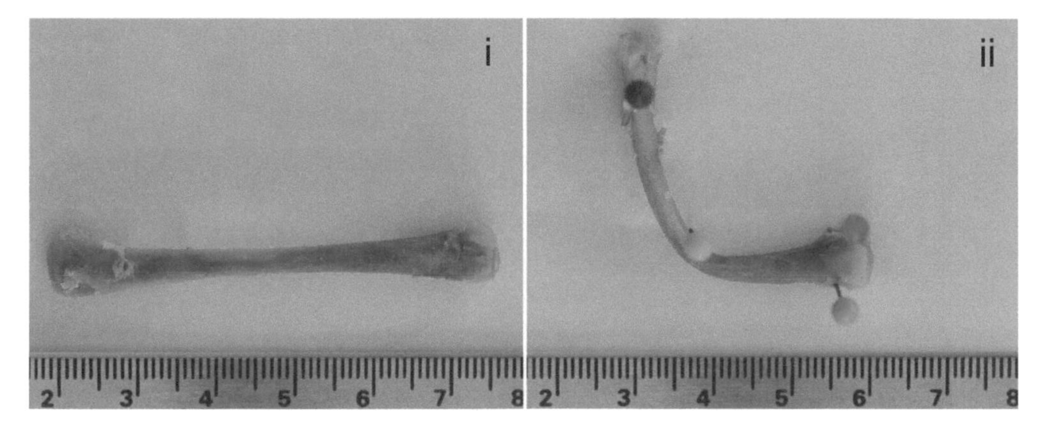

FIGURE D67a Demineralized bone from a chicken. The mineral content of the bone has been removed by soaking in a vinegar solution and the bone has a flexible, rubbery nature.

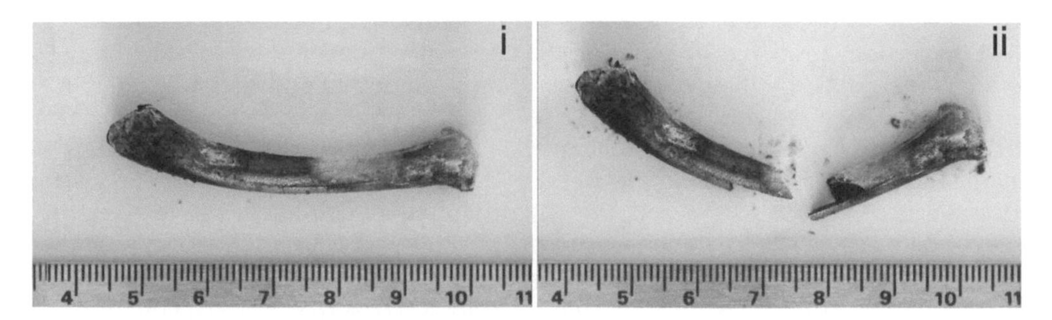

FIGURE D67b On the other hand the organic content of this bone was removed by burning (i) and in (ii) the remaining bone crumbles easily in the fingers.

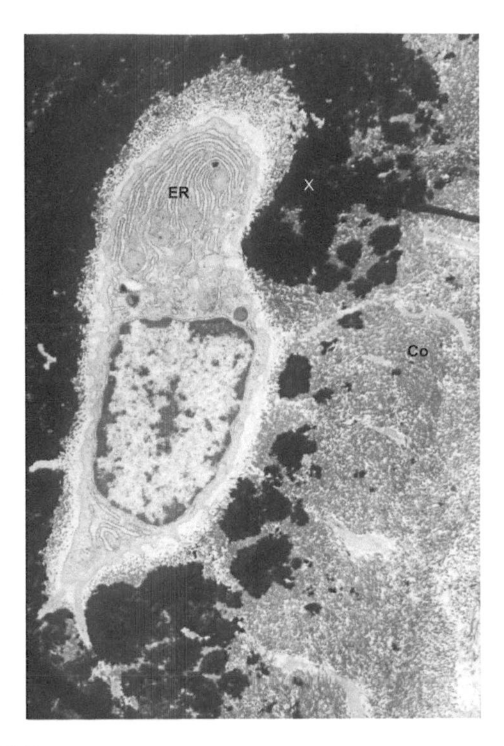

FIGURE D68 Transmission electron micrograph of an OSTEOCYTE embedded in its BONY MATRIX. From the fibula of a mouse. The bony matrix (X) is black. The organic matrix consists of a dense feltwork of collagenous fibers (Co) embedded in an amorphous ground substance. The lacuna is lined by a dense feltwork of unmineralized bony matrix, the OSTEOID. The osteocytes display the appearance of cells actively producing proteins: the GRANULAR ENDOPLASMIC RETICULUM is well developed (ER). 15,500 ×.

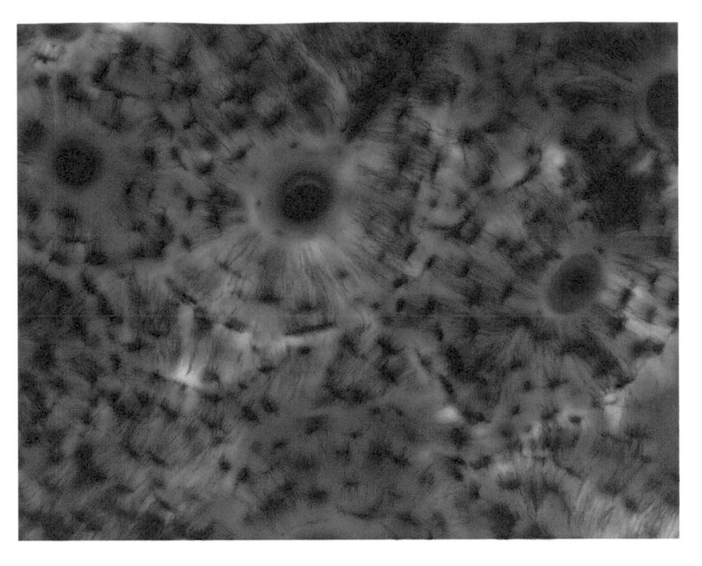

FIGURE D69a In this stained cross section of a long bone, note the OSTEONS (Haversian systems), each consisting of a series of concentric lamellae of bony matrix enclosing an OSTEON CANAL (Haversian canal) and the LACUNAE. Filling the spaces between the circular OSTEONS are the irregular INTERSTITIAL LAMELLAE with their associated lacunae and canaliculi. In life the osteon canals contained blood vessels and nerves and the lacunae contained osteocytes. 20 ×.

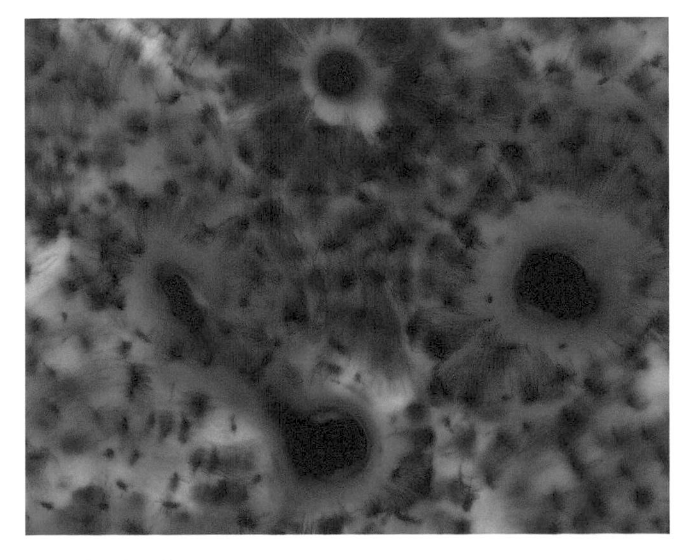

FIGURE D69b The osteon canals of two osteons are connected in this micrograph by a PERFORATING CANAL (Volkmann's canal); perforating canals are not surrounded by concentric lamellae. 20 ×.

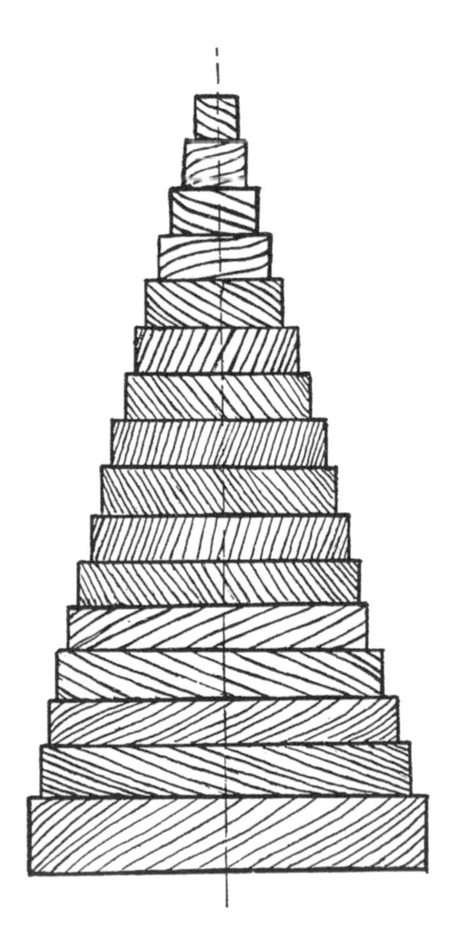

FIGURE D70 The collagenous fibers in successive lamellae of an osteon alternate in direction, like the sheets of veneer in plywood, thereby providing an extremely strong structure.

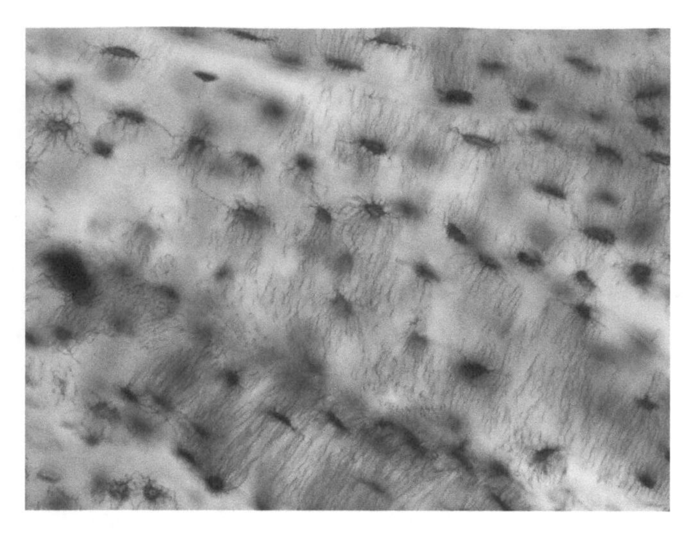

FIGURE D71 Longitudinal section of ground bone—unstained.

FIGURE D72a Section of cancellous bone with fatty and hemopoietic marrow. 2.5 ×.

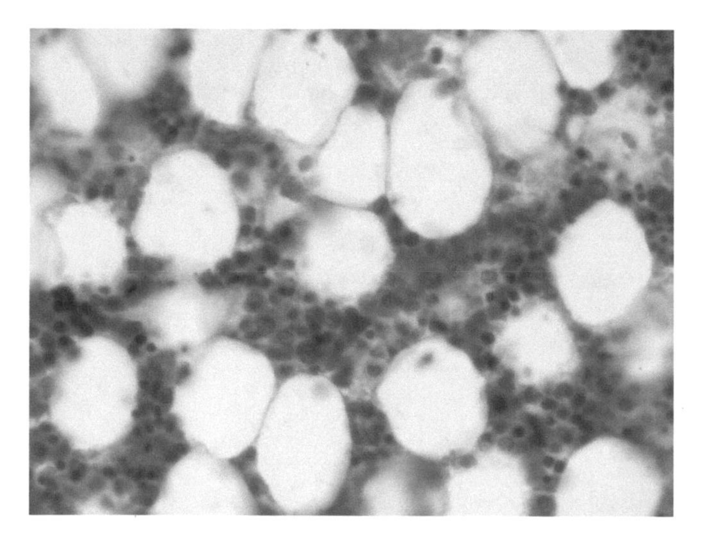

FIGURE D72b Section of mammalian bone marrow. Large adipose cells dominate the scene. Between them, all stages of granulocyte development may be seen. 20 ×.

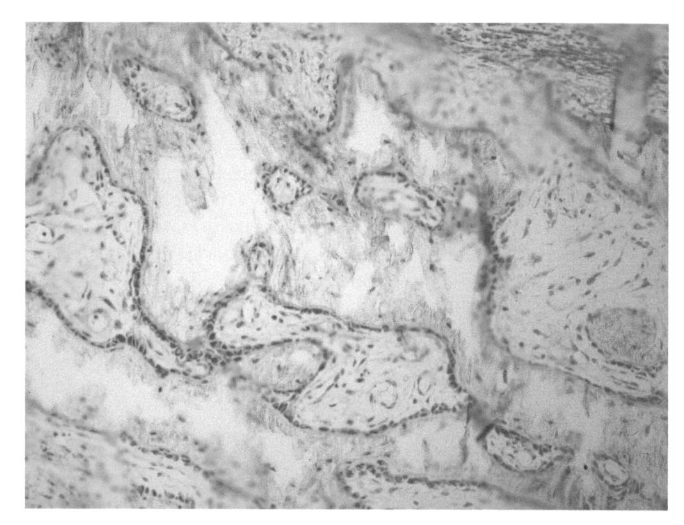

FIGURE D73a Trabecular bone, nondecalcified, from a biopsy of the human iliac crest. The irregular trabeculae of bone share the space with hemopoietic marrow. 10 ×.

4. *Spongy or cancellous bone.* Spongy bone is similar to compact bone except that the mineralized tissue is in the form of irregular partitions, the TRABECULAE, which enclose marrow-filled cavities (Figs. D61a, D61b, D72a, D72b, D73a, D73b). Since the marrow spaces between the trabeculae are richly vascular, there is no need for the elaborate vascularization seen in the Haversian systems; the matrix, however, is lamellated.
5. *Bone development.* The study of the development and adult structure of bone is complicated by the fact that there are two types of ossification and two kinds of arrangement of the tissue in its fully formed state. Differences in development arise from the fact that, in embryos, some of the bones are laid down directly in undifferentiated mesenchyme whereas in other parts of the body a temporary supporting system of cartilage precedes bone formation. The first type of ossification (INTRAMEMBRANOUS) is comparatively simple. In the second type (CARTILAGE REPLACEMENT or ENDOCHONDRAL), stages of cartilage erosion and of bone formation are intimately associated and form a more confusing picture. The essential processes by which the bony matrix is formed are the same in both; the differences lie in the tissue that precedes each. The result is the same in both cases, namely the formation of a mass of irregular trabeculae of spongy bone penetrated by blood vessels and connective tissue. In both cases the newly developed spongy bone undergoes secondary changes consisting of EROSION and REBUILDING. Differences in the manner and extent of rebuilding parts of the bone result in the

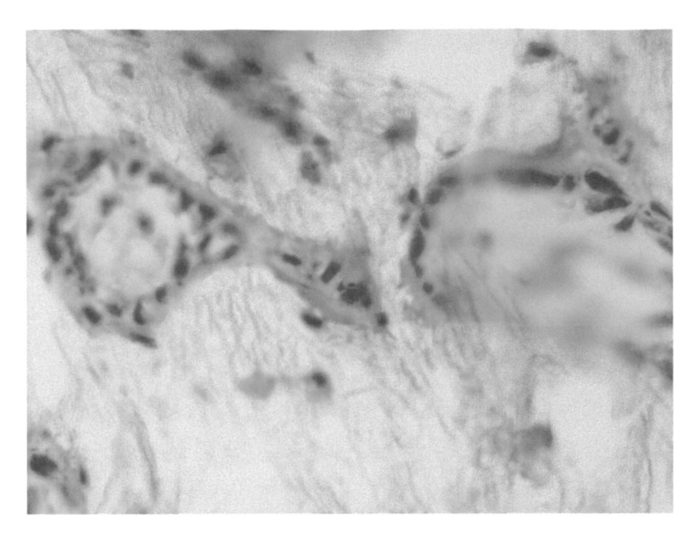

FIGURE D73b At a higher magnification the large basophilic osteoblasts are seen to be cooperating in the formation of bony spicules. Reticular fibers cross the space between the bony fragments. 40 ×.

development of two types of adult structure. In some regions the bone is eroded and rebuilt in its original SPONGY form. In others rebuilding follows a new pattern and is more extensive so that the tissue has the arrangement of COMPACT BONE. Compact and spongy bone are alike in their essential elements but differ in their arrangement and relative amounts of the matrix, blood vessels, and marrow spaces. Both types were seen in the long bone for example in Fig. D60.

- *Intramembranous ossification.* Some bones arise directly from the differentiating mesenchyme cells and are the DERMAL or MEMBRANE BONES (Figs. D74a–D74c). Most of the bones constituting the roof of the skull are formed in this manner. The micrographs show irregular spicules of bone developing under the dorsal skin in a section through the head of a newborn mouse. The vascular mesenchyme in the region of the developing bone is the PRIMITIVE MARROW and provides the cells, fibers, and blood vessels for bone formation (Fig. D65). OSTEOCOLLAGENOUS FIBERS are seen coursing through this mesenchymal tissue and numerous cuboidal cells, the OSTEOBLASTS (bone formers), line up along them and lay down slivers of bony matrix, the BONE SPICULES. As the spicules enlarge, osteoblasts huddle around on all sides. As noted earlier, osteoblasts have delicate processes that connect with those of adjacent osteoblasts (Figs. D66a and D66b) and as deposition of the matrix proceeds many osteoblasts become imprisoned within it and become OSTEOCYTES (bone cells) located within individual LACUNAE. The cytoplasmic connections remain between the cells during this "entombment" and account for the formation of the CANALICULI between the lacunae in the bony matrix. After an osteoblast becomes an osteocyte its matrix-forming activities cease except for the formation of the LACUNAR CAPSULE.

  Osteoblasts are derived from mesenchymal cells and give rise to osteocytes (Fig. D75); these are three phases in the life of one cell. Osteoblasts have abundant basophilic cytoplasm and an extensive pale Golgi area. Their borders show a prickly outline, the prickles representing the stumps of their many processes that cannot be seen in ordinary sections. Osteoblasts lay down the fibers and ground substance and they bring about the precipitation of mineral salts in the matrix.

  Bone spicules increase in size and eventually fuse with other spicules to form larger, irregular plates of bony matrix, the TRABECULAE. Eventually many trabeculae fuse to form the spongy bone studied earlier. In many positions this spongy bone is eroded and replaced by compact, lamellar bone. Even in regions where spongy bone remains there is a constant erosion of the matrix as the bone is modeled and remodeled—bone is not a static tissue but is constantly changing in response to the stresses and strains it experiences. This erosion is under the control of large, acidophilic, multinucleate OSTEOCLASTS (bone breakers) that often lie in little bays in the bony matrix, the LACUNAE OF EROSION (Howship's lacunae) that they have hollowed out for themselves (Fig. D76). Osteoclasts are the product of the fusion of monocytes from the blood.

- *Endochondral ossification.* The major portion of the skeleton of most vertebrates is outlined as hyaline cartilage that is later replaced by bone (Fig. D77). This indirect method of bone formation is ENDOCHONDRAL OSSIFICATION. The bones of the limb are outlined in cartilage that stains light blue with hematoxylin and eosin Figs. D78a–D78e. In the central (DIAPHYSIAL) region of the cartilage there may be evidence of the disintegration or dissolution of the cartilaginous matrix.

FIGURES D74(a–c) Developing membrane bone in the roof of the skull of a newborn mouse. 63×.

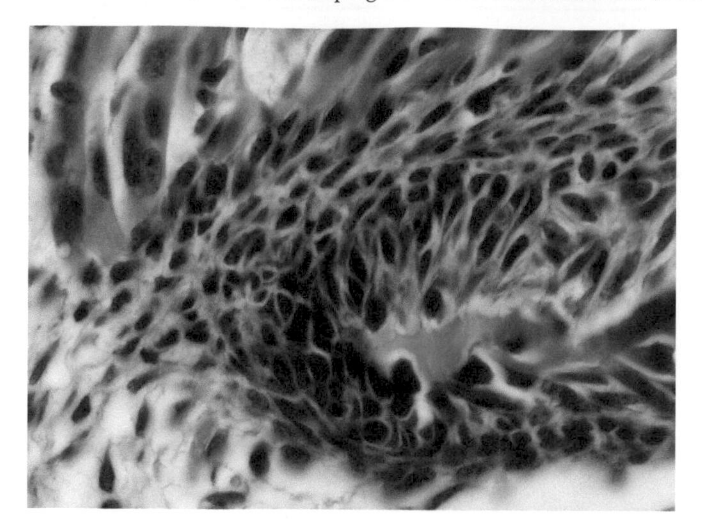

FIGURES D74a A spicule of bone is developing within mesenchymatous tissue. Some of the mesenchyme cells have transformed into osteoblasts and have huddled around the spicule, enlarging it.

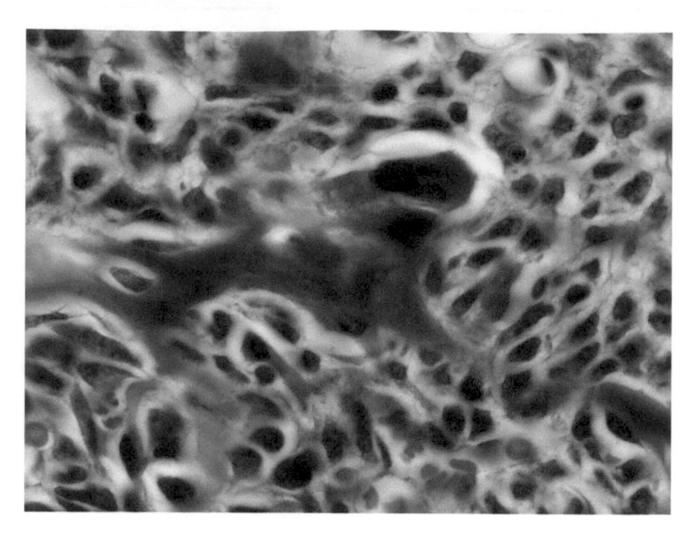

FIGURE D74b A few osteocytes have become trapped within lacunae in the developing bone of this spicule. A capillary invades the mesenchyme at the bottom center.

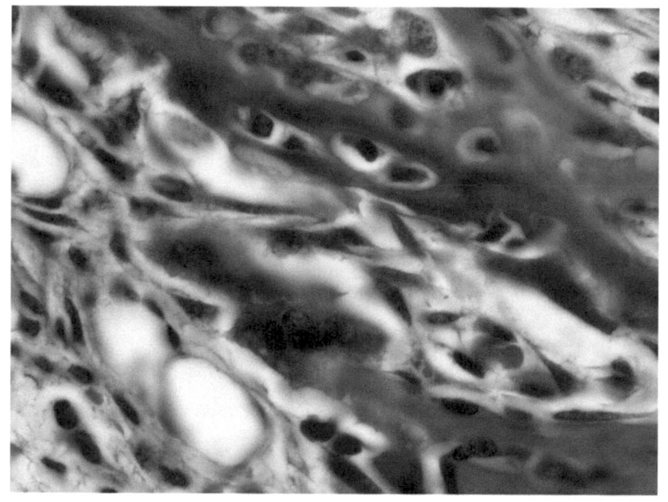

FIGURE D74c Several bony spicules have united at the top right, to form a TRABECULA of spongy bone. Vascular mesenchyme surrounds the trabecula. Slightly to the left of center, a grayish, multinucleate osteoclast is carving away the bony tissue.

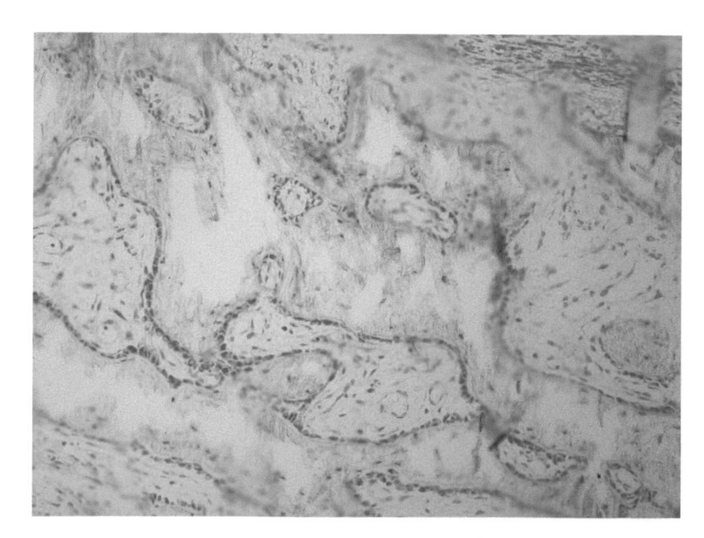

FIGURE D75 Section of undecalcified bone from a human biopsy lightly stained with toluidine blue. Osteoblasts arrayed around the periphery of the bony spicules lay down the bony matrix; eventually they become entrapped in the matrix as osteocytes. Large multinucleate osteoclasts, hollow out small "bays" in the matrix, the lacunae of erosion (Howship's lacunae). 10×.

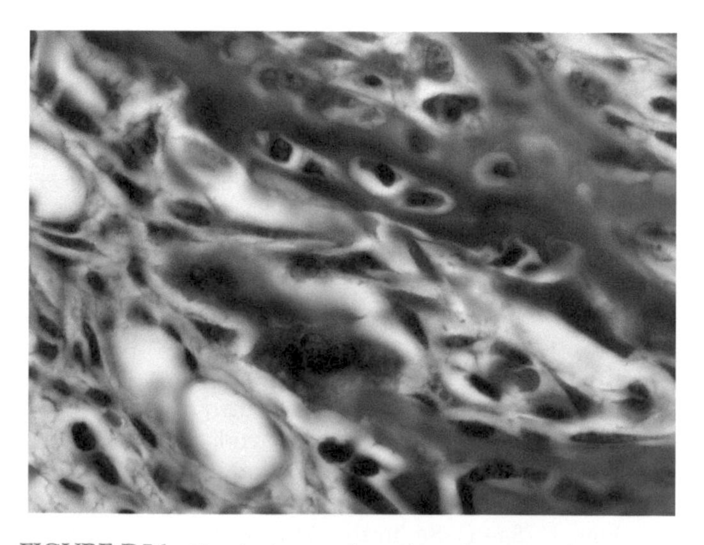

FIGURE D76 Developing membrane bone from a newborn mouse. Trabeculae of developing membrane bone are surrounded by vascular mesenchyme. A few osteoblasts at the sides of the trabeculae create new bony matrix; some become entrapped in the matrix as osteocytes. Large multinucleate osteoclasts (slightly to the left of center) dissolve the matrix in certain areas, thereby remodeling the bone. 63×.

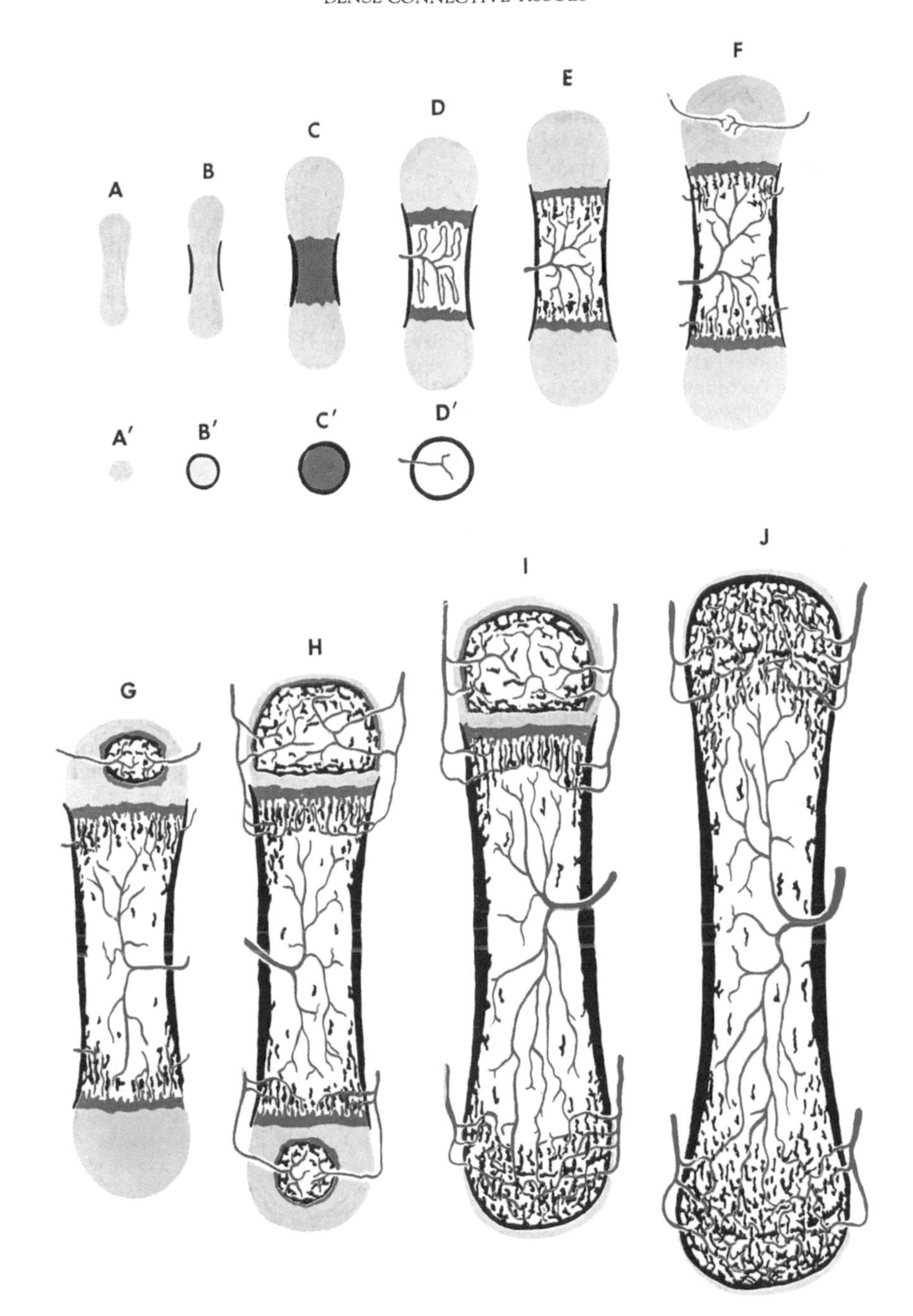

FIGURE D77 Diagrams of the development of a long bone in longitudinal sections (A–J) and in cross sections (A′, B′, C′, and D′) through the centers of A, B, C, and D. Pale blue = cartilage; purple = calcified cartilage, black = bone; red = arteries. (A) The original cartilaginous model of the bone; (B) a PERIOSTEAL collar of bone forms around the diaphysis; (C) Cartilage begins to calcify; (D) vascular mesenchyme of the PERIOSTEAL BUD enters the calcified cartilage; (E) osteoblasts lay down bone on the remaining spicules of calcified cartilage; (F, G) blood vessels and mesenchyme invade the epiphyseal cartilage and the EPIPHYSEAL CENTER OF OSSIFICATION develops; (H) a similar center of ossification develops in the lower epiphyseal cartilage; (I) the lower epiphyseal plate disappears as the bone ceases to grow in length (J).

The cartilaginous precursor of the bone is invaded by a mass of capillaries and cells, the OSTEOGENIC BUD (periosteal bud), that melts away cartilage as it advances. The cells include fibroblasts and osteoblasts; also present are CHONDROCLASTS that break down cartilage. The site of this activity is the PRIMARY OSSIFICATION CENTER and, in long bones, occurs in the shaft or diaphysis. Later, further invasions may occur at SECONDARY CENTERS. The newly formed cavity contains many cells and fibers. Notice that the cartilage cells on the margins of this area are undergoing cytolysis, have assumed a linear arrangement, and that their lacunae have enlarged. The CARTILAGINOUS SPIKES are the irregularly shaped remnants of cartilage that have not been resorbed. The spaces between the spikes contain the cells, fibers, and blood vessels of the PRIMARY MARROW. The spikes of cartilage become calcified and

FIGURE D78 (a–e) Series to show endochondral ossification in a long bone of a mammal.

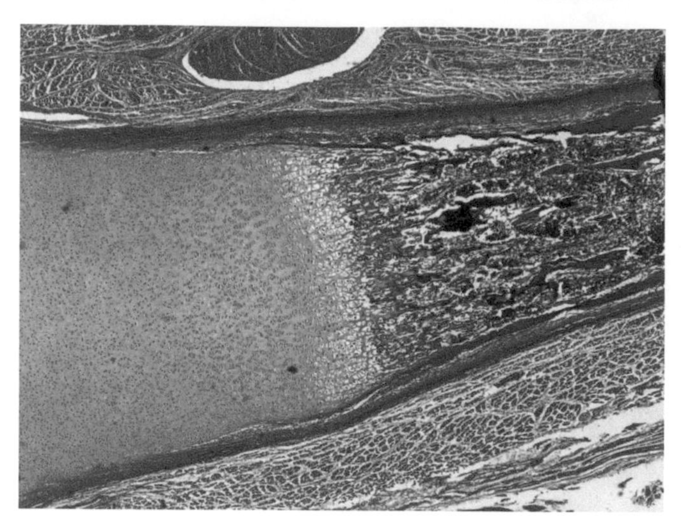

FIGURE D78a Overall view of the replacement of cartilage on the left and the forces of reconstruction advancing from the right. The cartilage on the left is distant from the scene of action and its lacunae are distributed more or less at random. As the advancing line of reconstruction approaches, the lacunae aggregate into rows, and eventually become empty. Bony spicules are laid down in the avascular mesenchyme at the right. The developing bone is enclosed in perichondrium to which muscles are attached. 2.5 ×.

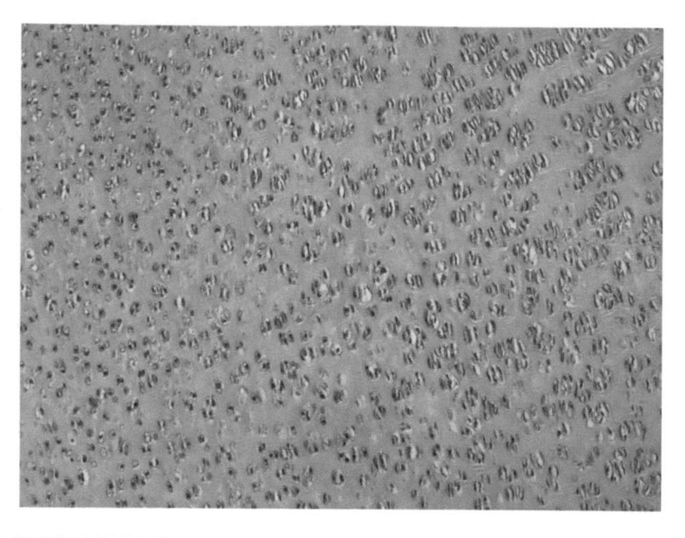

FIGURE D78b In cartilage of the epiphysis there is a realignment of lacunae from a random distribution at the left to the stacked arrangement at the right. 10 ×.

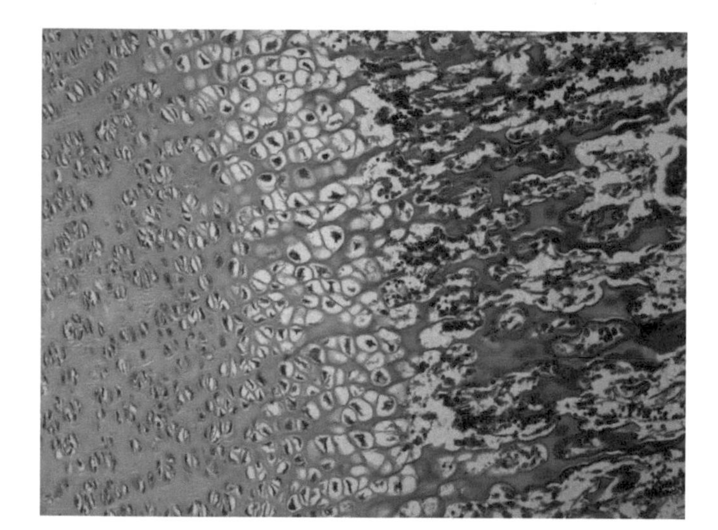

FIGURE D78c The "floors" of the lacunae tend to disappear, leaving irregular walls. The irregular compartments become invaded by the advancing forces of vascular mesenchyme, which includes osteoblasts that lay down bone on the remnants of cartilage, thereby laying down a framework for the development of osteons. 10 ×.

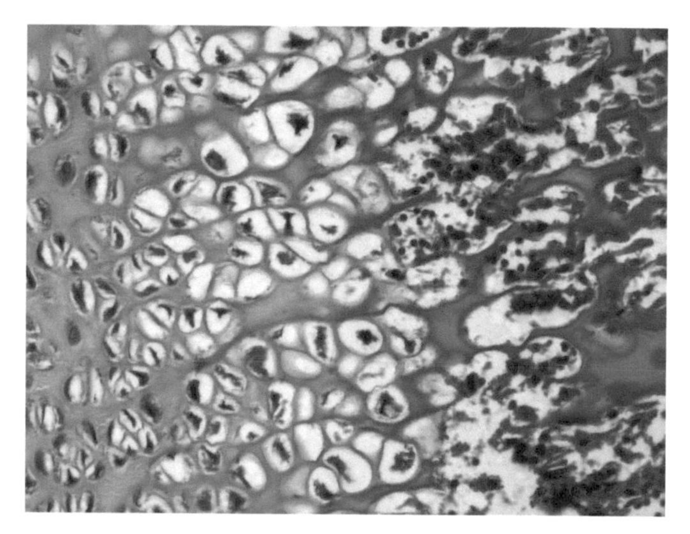

FIGURE D78d A higher powered view of the "battle line" between the losing forces of cartilage at the left and the advancing forces of bone formation at the right. Note that the chondrocytes have degenerated, leaving their lacunae almost empty. Vascular mesenchyme advancing from the right is laying down bone on the remnants of cartilage that remain. Bone is laid down in layers within the empty shafts left by the erosion of the cartilage; it surrounds capillaries within the shafts, layer upon layer, thereby forming the concentric lamellae of the osteons. 20 ×.

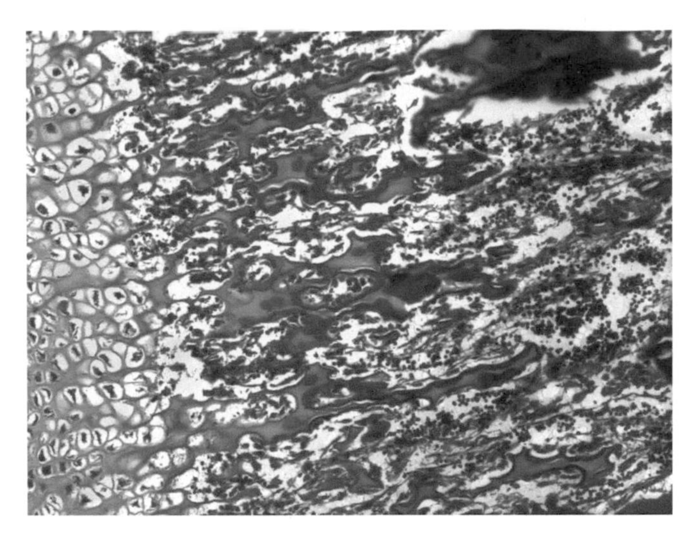

FIGURE D78e Another view of the advancing forces of primary ossification in the epiphysis of a mammalian long bone. Cartilage at the left is being invaded by chondroclasts, fibroblasts, osteoblasts, capillaries—from the right. Chondrocytes are undergoing cytolysis. Bony matrix is laid down on the remaining spicules of cartilage. 10×.

OSTEOBLASTS line up around them using them as a scaffold, which they encrust with bone. Since the calcified cartilage was in the form of an irregular sponge, the encrusting bone is spongy as well. Its trabeculae are characterized by having cores of calcified cartilage (Fig. D78e). OSTEOCLASTS may be seen in the primary marrow.

Discs of cartilage persist for some time as the EPIPHYSEAL DISCS providing for elongation of the bone (Fig. D77). They disappear when growth ceases.

The outer sheath of dense fibrous connective tissue, the PERIOSTEUM, has an important role in bone formation (Fig. D78a). Osteoblasts actively lay down bone beneath this membrane and this INTRAMEMBRANOUS OSSIFICATION increases the girth of the bone. Increases in length occur by the growth of cartilage that is then replaced by bone; the cartilage is never completely replaced until growth ceases. In long bones this growth occurs in the region between the primary and secondary centers of ossification, the EPIPHYSEAL LINE.

With the electron microscope, osteoblasts show a prominent granular endoplasmic reticulum and well-developed Golgi complex, both characteristics of synthetic activity (Fig. D79). These cells are surrounded by the products of their efforts: collagenous fibers, mucopolysaccharide ground substance, and scattered, electron-dense crystals of hydroxyapatite that fuse to form the hard matrix material. When its synthetic work is done and the osteoblast is solidly embedded in the matrix, it becomes an osteocyte. Its cytoplasm is reduced but some granular endoplasmic reticulum is retained.

Osteoclasts show a ruffled border of complex folds and projections apposed to the surface of the matrix where breakdown is occurring (Figs. D80a, D80b, D81). There are few organelles in this region although some lysosomes are present. Elsewhere in the cell may be seen several Golgi complexes, numerous large mitochondria, sparse endoplasmic reticulum, and a few free ribosomes.

Notochord

The rigid supporting rod that appears in the development of all chordates constitutes an atypical connective tissue where there is no intercellular matrix. It is a longitudinal flexible rod lying just below the spinal cord and above the gut (Fig. D82). It consists of vacuolated cells stuffed into a tough sheath of collagenous tissue: imagine a cylindrical canvas bag filled with inflated balloons that forms a rigid, but flexible, rod (Figs. D83–D87). The notochord is unusual among supportive structures in that it maintains its strength, not from extracellular material—e.g., cartilage—but by TURGOR PRESSURE within the cells as seen in plants like celery.[2] In higher vertebrates, it is progressively replaced by the centra of the vertebrae—the somites surrounding the notochord grow inward to constrict it segmentally, often completely, like a string of sausages so that, in tetrapods, the notochord persists in the adult only as the gelatinous NUCLEI PULPOSI (*singular:* nucleus pulposus) within the intervertebral discs that provide soft cushions between the vertebrae (Fig. D87).

[2]Fat is the only other vertebrate tissue that maintains its shape by its content of intracellular fluid.

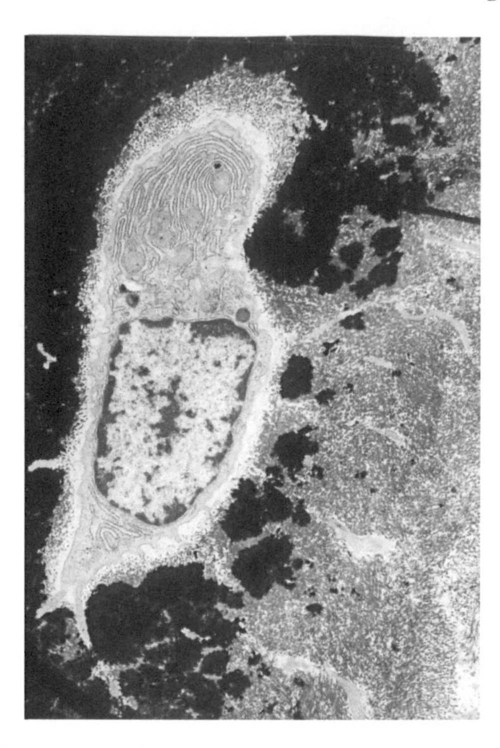

FIGURE D79 Transmission electron micrograph of an osteocyte from the fibula of a mouse. The presence of abundant granular endoplasmic reticulum indicates the secretory activity of the cell. Some of the products of this endeavor—collagenous fibrils—are shown at the lower right. Mineralization of the matrix by hydroxyapatite (dense black) has only partially engulfed the osteocyte. Osteocytes also produce the mucopolysaccharides of the amorphous ground substance. 15,500 ×.

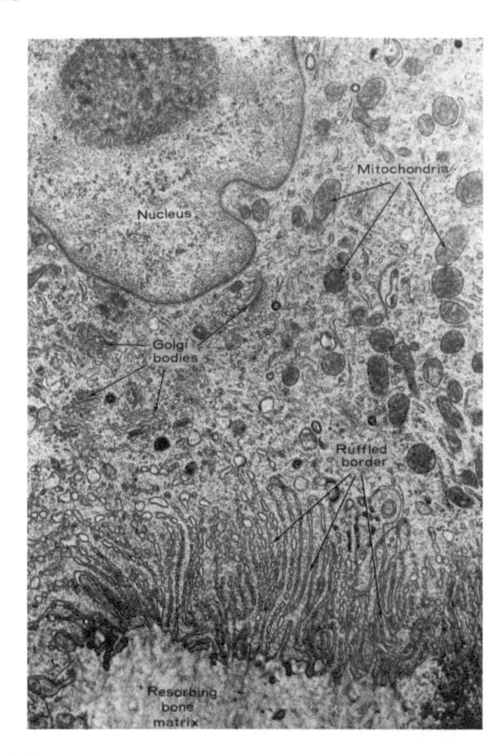

FIGURE D80a Electron micrograph showing the ruffled border of an osteoclast and its nucleus. The ruffled border is closely applied to an area of resorbing bony matrix.

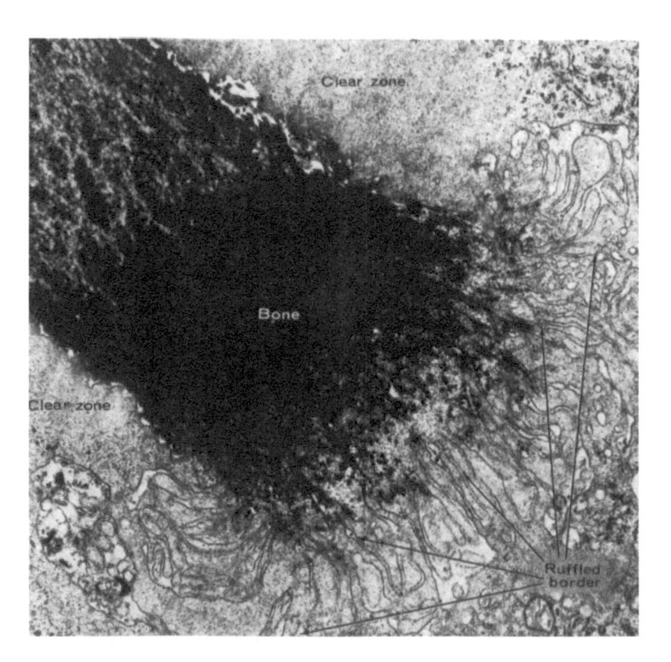

FIGURE D80b Electron micrograph of a portion of an osteoclast at the end of a spicule of bone. The bone is undergoing resorption at the ruffled border.

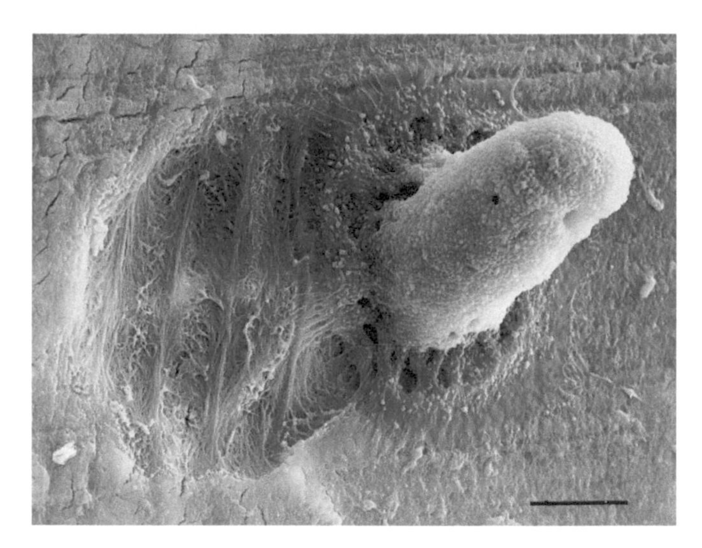

FIGURE D81 Scanning electron micrograph of an osteoclast adjacent to its resorption pit. Bar = 10 μm.

The nucleus pulposus is semifluid gel composed of fine collagenous and elastic fibers as well as a few round mesenchymal cells (at least in the young) that forms the central cushioning part of the intervertebral disc between the centra of vertebrae (Fig. D88). It is surrounded by tough fibrocartilage, the ANNULUS FIBROSUS (Fig. D89). The cells of the nuclei pulposi accommodate huge, fluid-filled spaces, both between the cells and within them (Fig. D90).

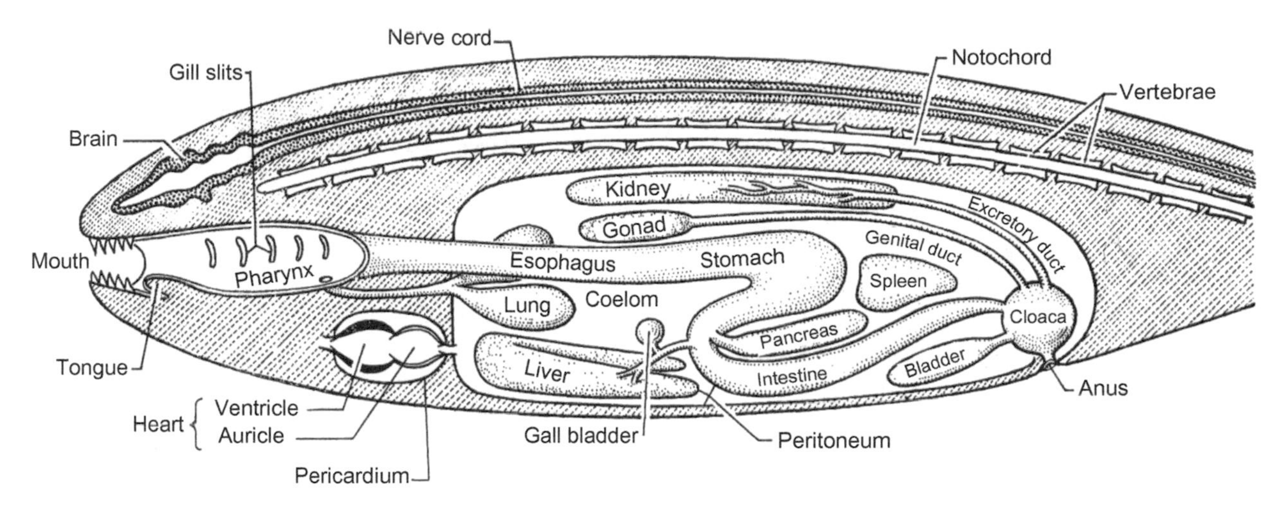

FIGURE D82 The notochord of a hypothetical vertebrate is a flexible rod lying just below the spinal cord and above the gut. Note that the notochord is surrounded by the centra of the vertebrae.

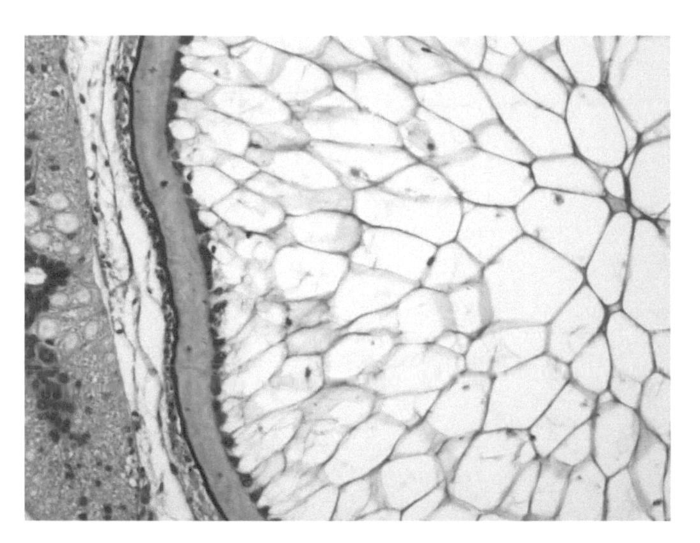

FIGURE D83 Cross section of an ammocoete showing the notochord containing fluid-filled cells stuffed within a tough collagenous sheath thereby creating turgor pressure within the flexible notochord that supports the body. 20 ×.

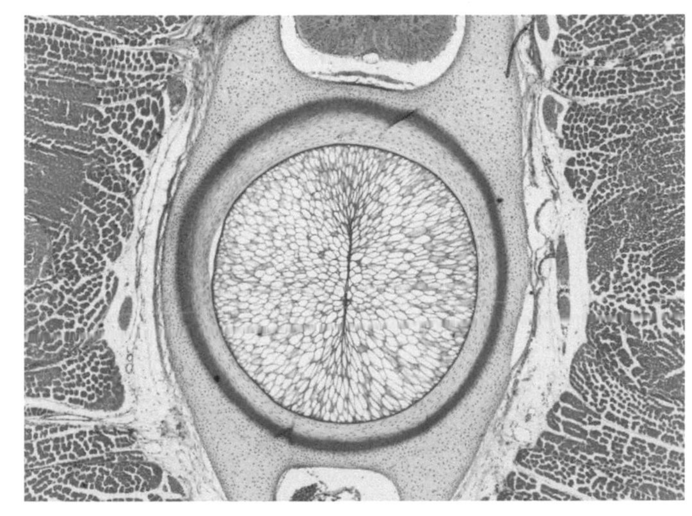

FIGURES D84a Cross section of a developing shark. Vacuolated cells are contained within a tough collagenous sheath to form the rigid notochord that supports the body. 2.5 ×

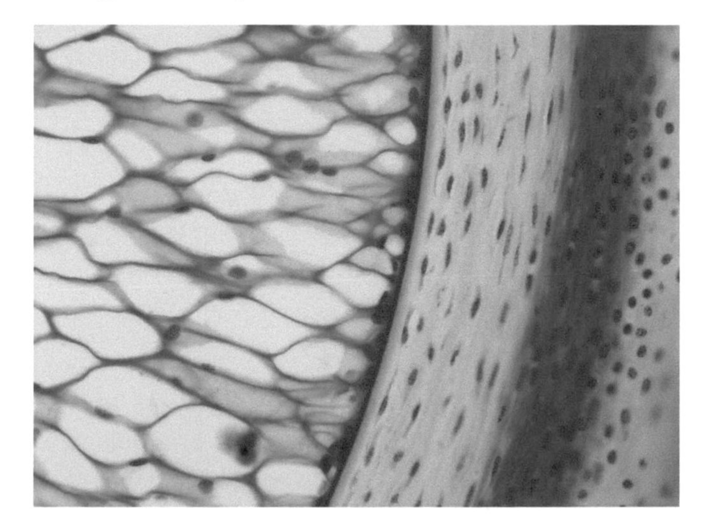

FIGURES D84b Cross section of a developing shark. At higher magnification the vacuolated cells are seen to be contained within a tough laminated collagenous sheath to form the rigid notochord that supports the body. 20 ×.

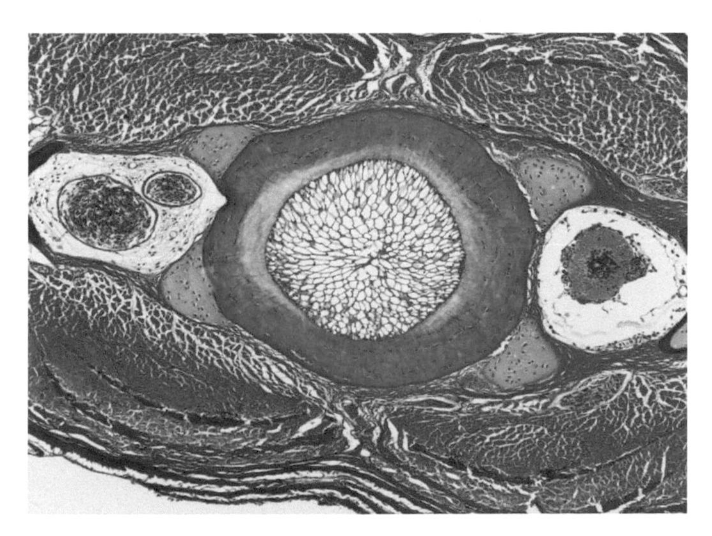

FIGURES D85a The same sort of structure is seen in the supporting notochord of a bony fish. 2.5 ×.

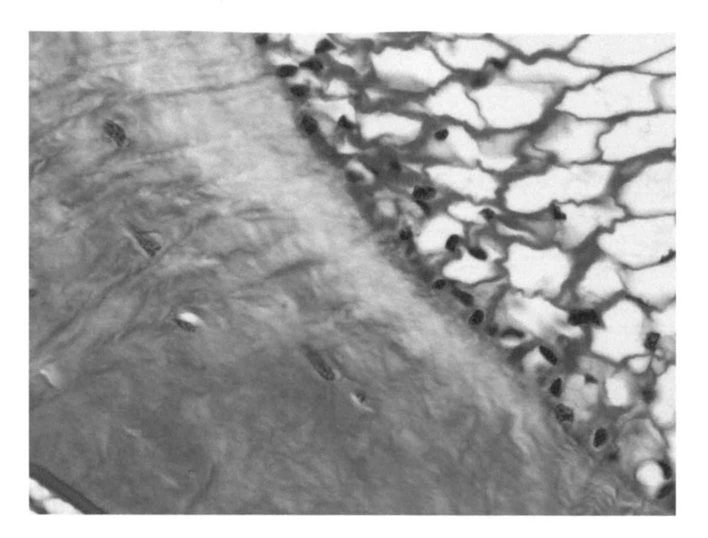

FIGURES D85b At higher maginifcation the tough sheath containing the supporting notochord of a bony fish. 20 × .

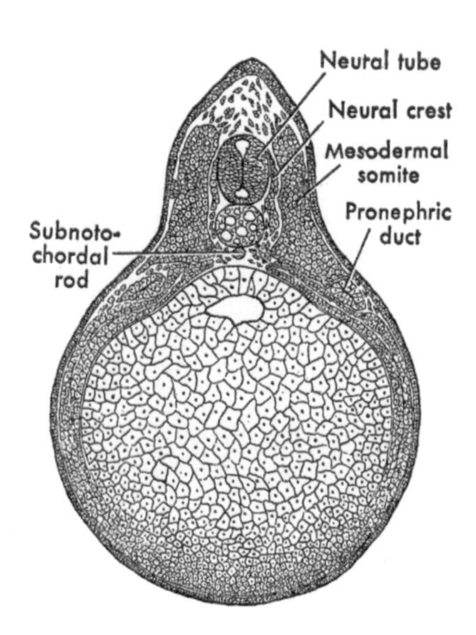

FIGURE D86 In this drawing of a cross section of a 5.5 mm tadpole, the notochord is located immediately below the neural tube.

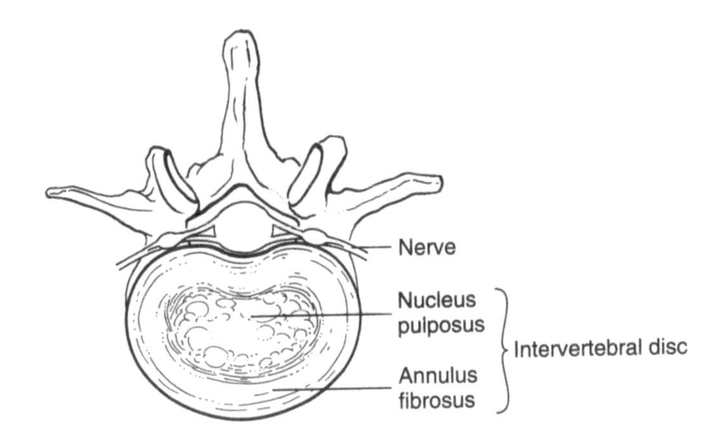

FIGURE D87 Fragments of the notochord persist in adult mammals as the NULCLEI PULPOSI (*Singular:* nucleus pulposus) that contributes to the intervertebral discs, the cushioning pads between the vertebrae. They form the semifluid masses at the center of the intervertebral disc and consist of fine white and elastic fibers and large fluid-filled cells.

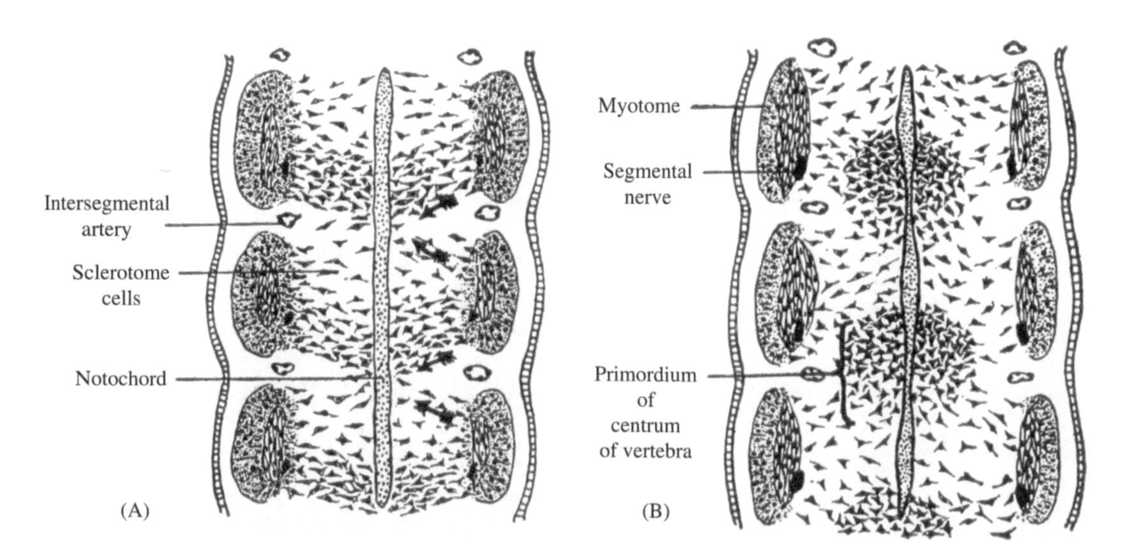

FIGURE D88 Segmental masses of mesenchymal cells, the sclerotomes, surround the notochord, constricting and segmenting it (A). These masses form the vertebrae and the segmented notochordal fragments contribute to the nuclei pulposi of the intervertebral discs (B).

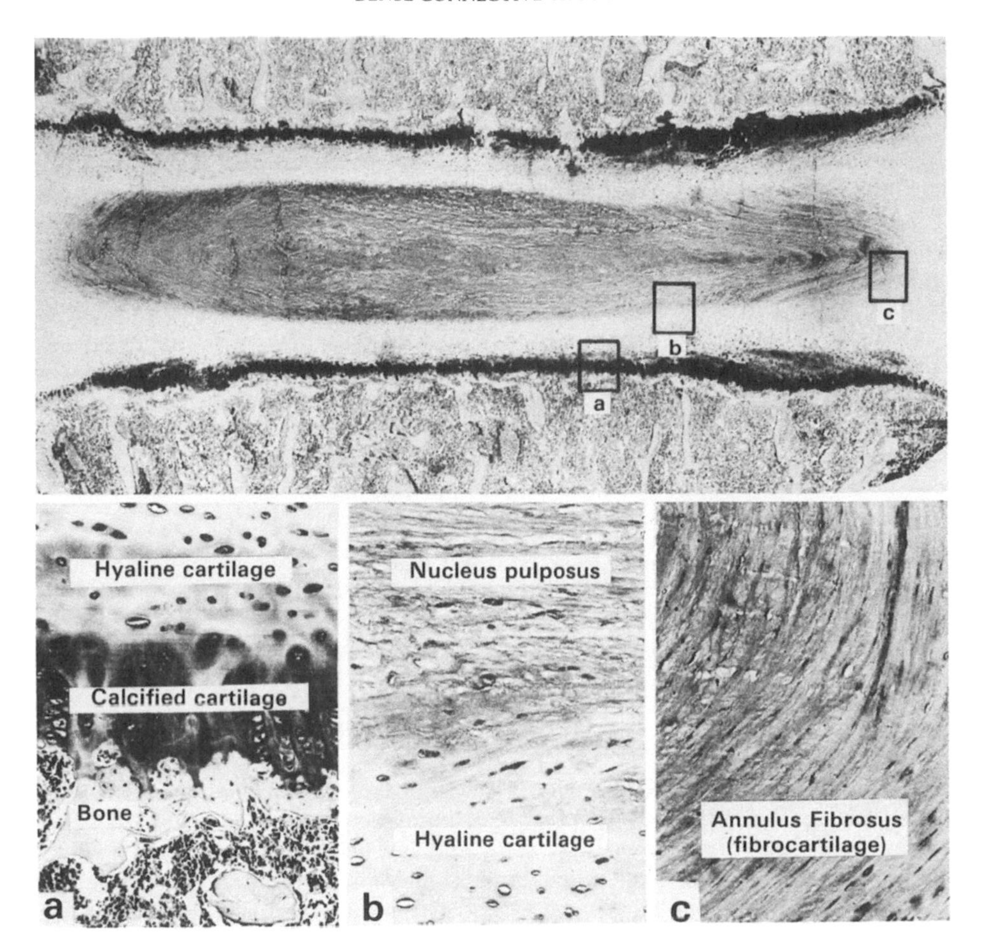

FIGURE D89 Lateral view of the intervertebral disc between two vertebrae of a young child. The three areas within the boxes are shown at a higher power below.

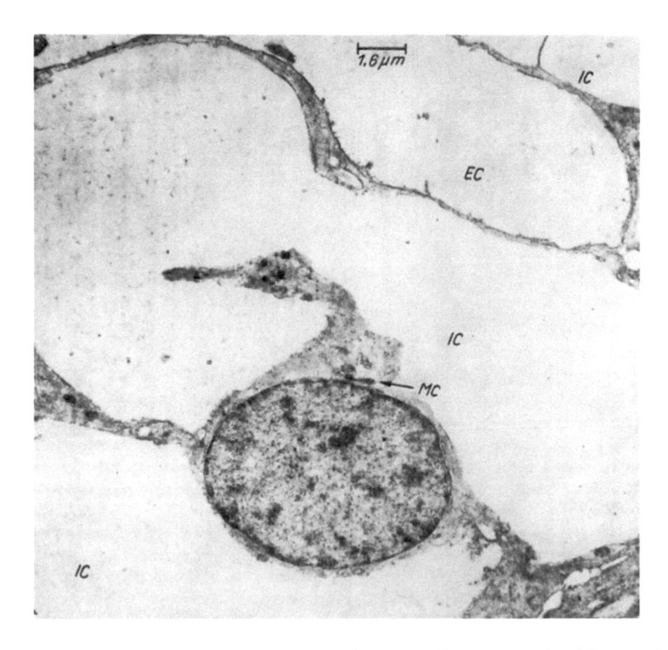

FIGURE D90 Electron micrograph of a section of the nucleus pulposus of a 7-month-old cat showing part of a highly vacuolated notochordal cell. *EC*, extracellular space; *IC*, intracellular space; MC, mitochondria. 6000 ×.

FIGURE D91a Lateral view of a whole mount of amphioxus showing the notochord (brown) as a stack of fluid-filled cells. The mouth is at the left and the gills occupy most of the bottom half of the picture. 2.5 ×.

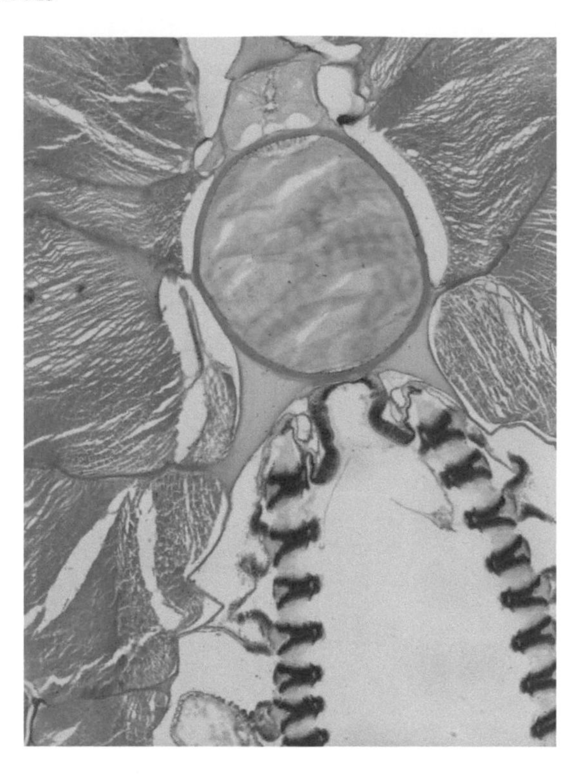

FIGURE D91b Cross section of amphioxus. The notochord is at the top center between the nerve cord (above) and the gills (below). The notochord is wrapped with a tough connective sheath and its cells are nondescript. 7.5 ×.

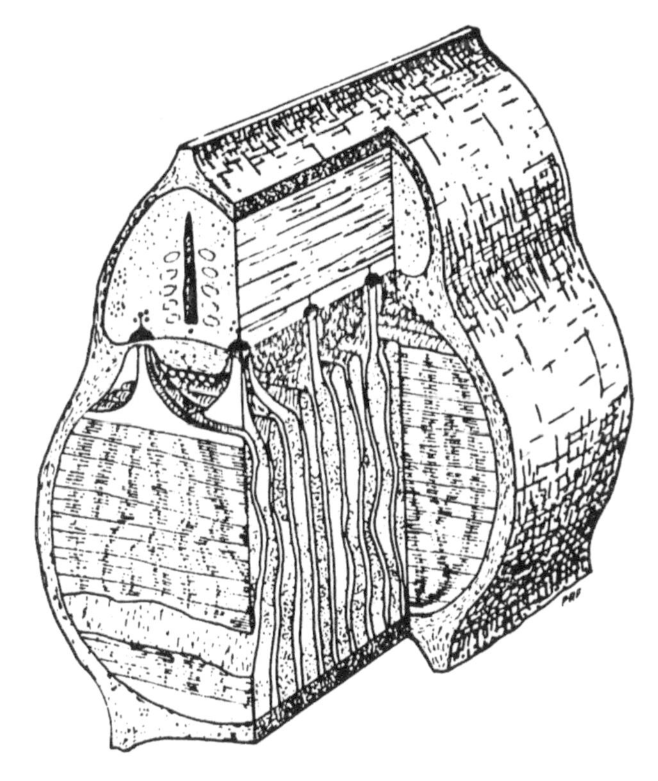

FIGURE D92 It has been suggested that the notochord of amphioxus bears a close relationship to the nerve cord: each lamella of the notochord appears to be contractile and has a process that extends to the nerve cord.

Although the notochord of amphioxus is homologous to that of other chordates, it is quite different (Fig. D91a). It consists of a stack of fluid-filled discoidal cells or LAMELLAE, contained, like a roll of coins within a tough collagenous wrapper (Fig. D91b). In cross sections these flat cells are cut horizontally and reveal little of their internal structure. It has been shown that the lamellae are innervated from the spinal cord and contain horizontally oriented contractile fibers (Figs. D92–D94). Fluid-filled spaces occur within the lamellae and spaces filled with extracellular fluid occur between the lamellae. Scattered among the lamellae are Müller cells, which are said to form the extracellular fluid as well as new lamellae. "It is perhaps most highly specialized hydroskeleton ever built and should not be considered as a degenerate organ."

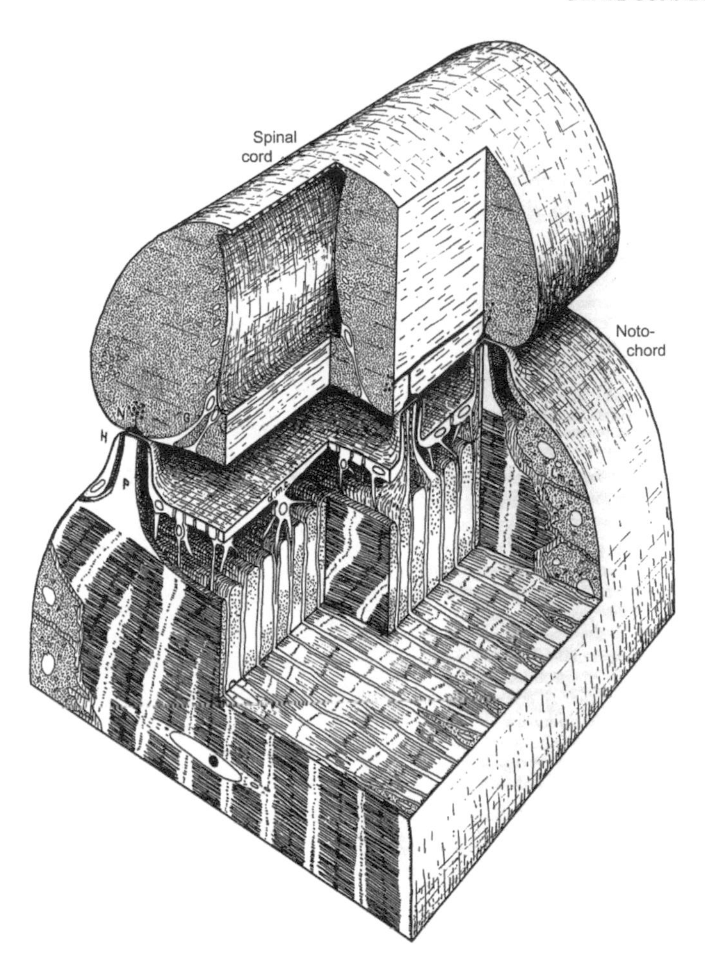

FIGURE D93 Diagram of the relationship between the spinal cord of amphioxus and the notochord. Cytoplasmic processes from the muscular notochordal lamellae (P) reach the spinal cord. Müller cells (dmc, lmc) are said to form the extracellular fluid as well as new lamellae.

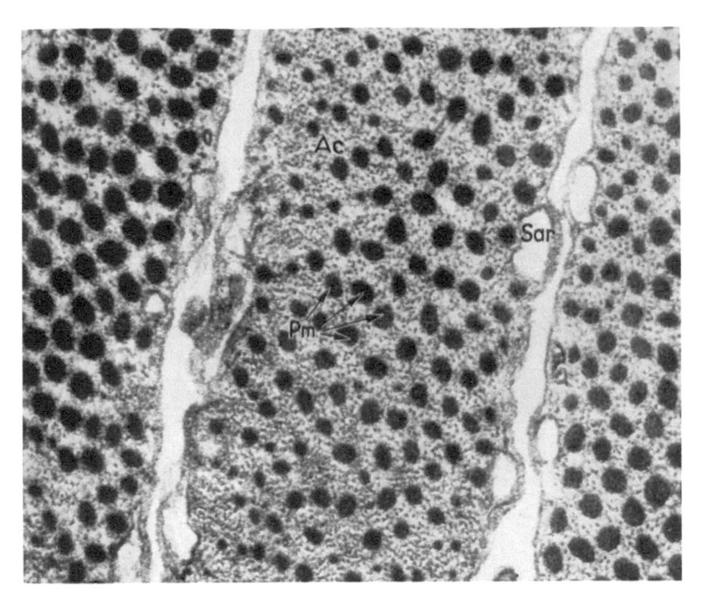

FIGURE D94 Electron micrograph of a sagittal section through the notochord of amphioxus. The regular array of filaments resembles that of muscle of some invertebrates Pm, paramyosin filaments; Ac, actin microfilaments; Sar, sarcoplasmic reticulum. 42,000 ×.

C H A P T E R

E

Muscle

Muscle tissues are the chief motor elements of the body, moving either parts of the skeleton or portions of the walls of the visceral organs. Muscle cells are elongated forming FIBERS; this use of the term "fiber" to designate an individual cell is unique to muscle. While muscle fibers possess the contractile properties of simple protoplasm, their contraction is limited to one direction. Usually the fibers are parallel so that the line of contraction of the muscle or muscle layer is the same as that of each of its fibers. The result of contraction in both the whole muscle and in its individual fibers is a shortening of the longitudinal axis and an increase in the transverse axis. Contractile filaments, the MYOFILAMENTS, occupy the bulk of each muscle fiber. Special terminology is applied to the parts of a muscle cell: the cytoplasm is the SARCOPLASM, the cell membrane is the SARCOLEMMA, the mitochondria are sometimes called SARCOSOMES, and the endoplasmic reticulum of some muscle cells is the SARCOPLASMIC RETICULUM. There are three types of muscle tissues in the body: SMOOTH MUSCLE, STRIATED MUSCLE, and CARDIAC MUSCLE.

SMOOTH MUSCLE

This is the least specialized type. It occurs as multicellular sheets in the visceral organs and blood vessels and as isolated bundles of fibers in the skin. It is under the control of the autonomic nervous system. Fig. E1a is a photomicrograph of individual smooth muscle cells that have been teased apart after soaking the muscular wall of the gut in a macerating fluid, such as 30% ethanol, for a few days. One cell is a fiber. Note the spindle shape of the cell, the sausage-shaped nucleus, and the cytoplasm. Try to identify MYOFIBRILS at the end of the broken fiber (Fig. E1b). Myofibrils are thought to be artifacts produced by clumping of finer myofilaments that are visible only with the electron microscope. The myofibrils are contained in the fluid sarcoplasm.

Study smooth muscle in cross sections of gut (Figs. E2a–E2c). At the outside of each section is a layer of muscle fibers extending longitudinally along the gut; these fibers are cut transversely giving this region a granular appearance when viewed under low power. The fibers are arranged in bundles imperfectly outlined by septa of connective tissue. Inside the layer of longitudinal muscle is a thicker layer of circular muscle. The circular muscle fibers encircle the gut and in these cross sections where the individual fibers are cut longitudinally, appearing spindle shaped. It is difficult to distinguish the outlines of these cells because the fibers overlap and the cell membranes are indistinct. Sections almost never include the full length of a fiber. In places where the fibers are separated, delicate strands of connective tissue may be seen between the cells. Identify the nucleus, sarcoplasm, and (if possible) the myofibrils in these cells. RETICULAR FIBERS are invisible in routine preparations but show as an elaborate net with silver stains (Fig. E3). What would be the result of the contraction of each of these muscle layers?

Electron micrographs of smooth muscle cells (Figs. E4, E5a, E5b, E6) show that most of the cytoplasm is packed with fine parallel filaments of ACTIN that run the length of the cell. Thick filaments, presumably MYOSIN are visible in appropriately prepared specimens (Fig. E7). In Fig. E6 look for electron-dense areas of α-ACTININ on the inner surface of the sarcolemma and interspersed among the myofilaments; these DENSE BODIES represent points of attachment of the contractile filaments. Cytoplasmic organelles are few and gather near the ends of the nucleus (Fig. E6): a few sarcosomes (mitochondria), sparse sarcoplasmic reticulum (agranular endoplasmic reticulum), clusters of free RIBOSOMES, and a small GOLGI COMPLEX. PINOCYTOTIC VESICLES associated with the sarcolemma and sarcoplasmic reticulum are characteristic of smooth muscle cells and have been shown to sequester calcium (Figs. E5a and E5b). Since not all smooth muscle cells are innervated, GAP JUNCTIONS between adjacent smooth muscle cells are thought to connect the fibers into functional groups.

An Atlas of Comparative Vertebrate Histology.
DOI: https://doi.org/10.1016/B978-0-12-410424-2.00005-6

FIGURE E1a Photomicrograph of smooth muscle cells that have been isolated by maceration of gut wall in 30% ethanol.

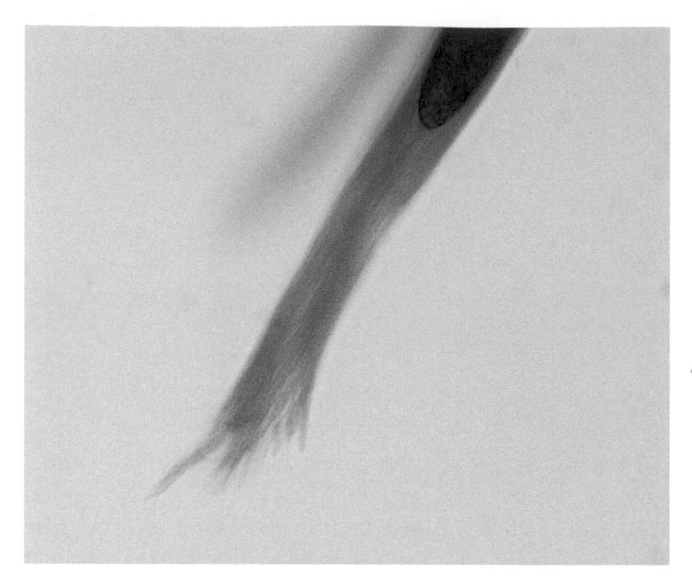

FIGURE E1b Photomicrograph of a maceration preparation of smooth muscle cells showing myofibrils in the broken end of a fiber.

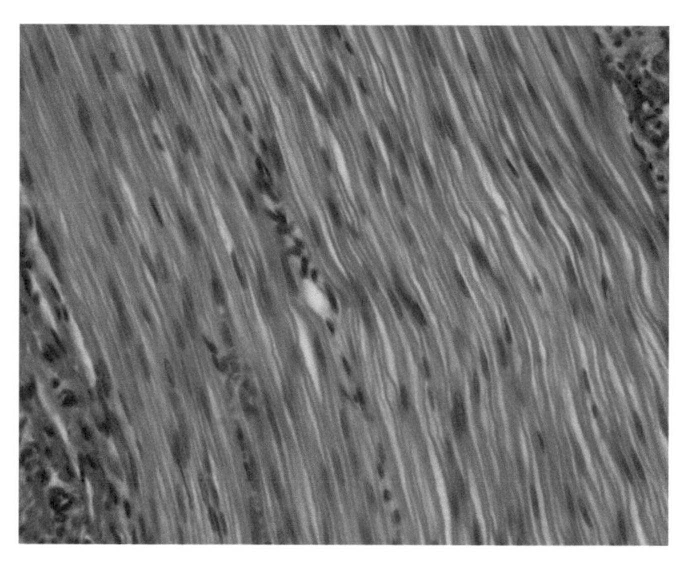

FIGURE E2a Smooth muscle in a cross section of the duodenum of a cat. At the right are longitudinal sections of individual muscle cells; cross sections of individual muscle cells are seen at the left bottom corner. 100 ×.

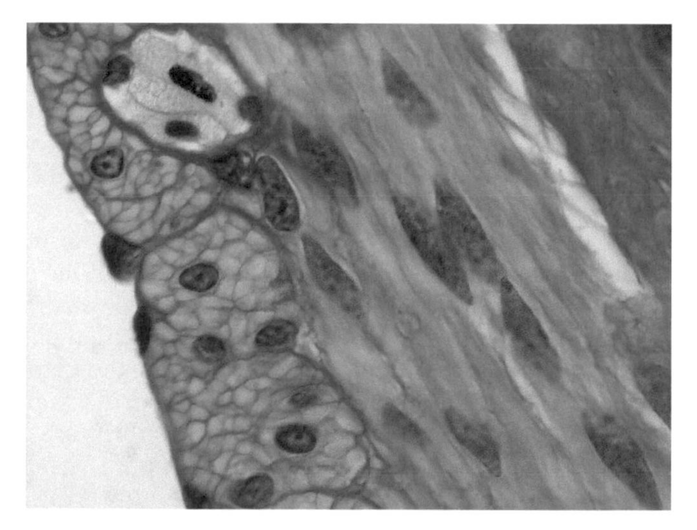

FIGURE E2b Semithin section of the intestine of an amphibian, *Necturus*. At the left are cross sections of smooth muscle cells that run the length of the intestine; at the right are longitudinal sections of cells that encircle the intestine. 40 ×.

STRIATED OR SKELETAL MUSCLE

Striated muscle is connected directly to the central nervous system and is under the control of the will. In most cases it is attached to the skeleton. The muscle fibers are much larger than those of smooth muscle: an individual fiber may be from 1 to 40 mm long and 10 to 100 μm in diameter—the size of a piece of thread—so that many are visible with the naked eye (Fig. E8a). This large cell is a SYNCYTIUM, a multinucleate mass of cytoplasm enclosed within a single cell membrane (Figs. E8b and E8c). There are approximately 35 nuclei per millimeter of length.

The fibers are unbranched, except in the tongue. Each fiber is enclosed in a thin sarcolemma (Figs. E9a–E9e). Compare these with the micrograph of striated muscle from a frog (Fig. E9f). The sarcolemma is coated with an

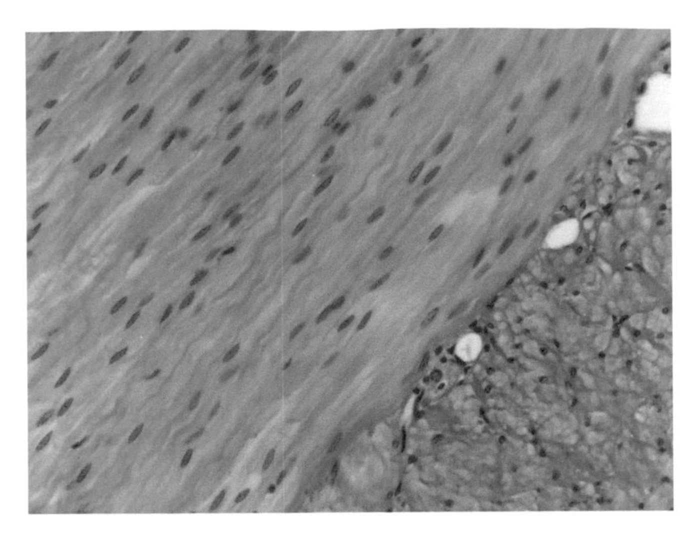

FIGURE E2c Smooth muscle in the wall of the duodenum of a caiman, a reptile. Most of the field is made up of longitudinal sections of smooth muscle cells; the lower right corner consists of cross sections of smooth muscle cells. 20 ×.

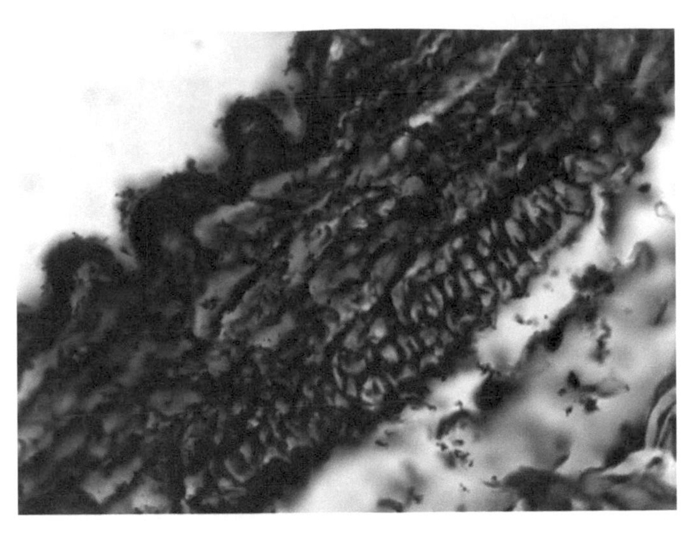

FIGURE E3 Smooth muscle in the wall of a small artery from a lymph node. The silver stain blackens the harness of reticular fibers. 63 ×.

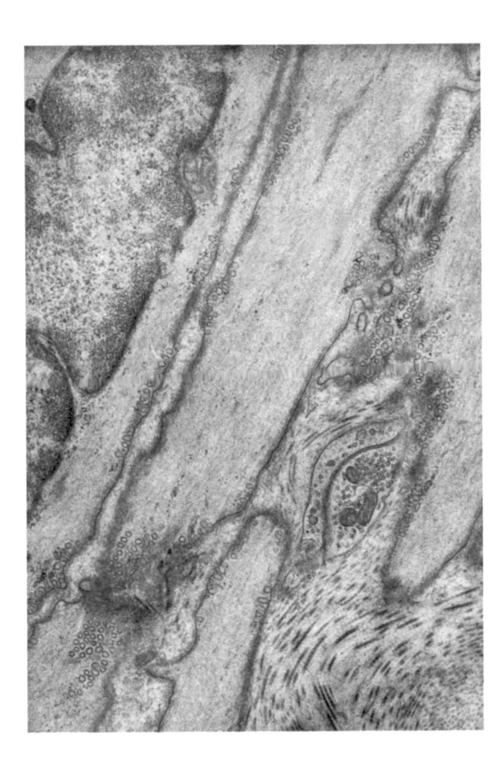

FIGURE E4 Longitudinal section of smooth muscle cells from the esophagus of a bat. Delicate filaments of ACTIN extend the length of these muscle fibers. It is assumed that coarser filaments of MYOSIN are present, perhaps in a disaggregated form, but their identity is unclear. A mitochondrion is shown huddled next to the nucleus of a smooth muscle cell on the left. Small pits on the surface of the smooth muscle cells at the upper left suggest that pinocytosis is characteristic of smooth muscle cells. Fine feltworks of moderately dense material lie on opposite sides of the plasma membrane slightly below center. The extracellular component is closely interwoven with the matrix and fibrils of the surrounding connective tissue and it is possible that the feltwork is homologous with the basement membrane of epithelia, binding together the reticular fibers of the "harness" (lower right) that permits the contracting cells to perform useful work. Actin filaments appear to be organized into bundles that transmit their contraction to the plasma membrane at DENSE BODIES (the "streamer" at the upper right). A cross section of a small autonomic nerve fiber containing vesicles and mitochondria is seen at the lower right; a cell apposed to the nerve fiber partially ensheaths it. 38,500 ×.

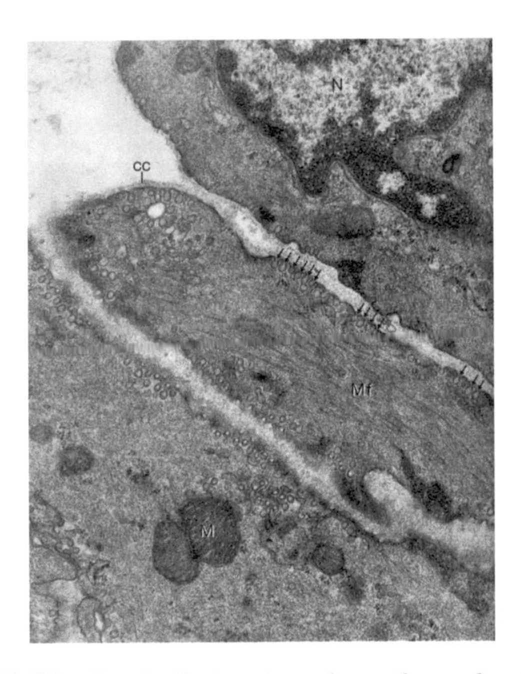

FIGURE E5a Longitudinal sections of smooth muscle cells from the wall of an artery in the pancreas of a rat. The cells are packed with delicate filaments of ACTIN (Mf). The plasma membrane abounds with the characteristic pits (arrows) that are seen in smooth muscle cells; these pits resemble endocytotic vesicles and appear to occur in delimited areas of the cell membrane. Few vesicles are seen free in the cytoplasm M - mitochondria; N - nucleus; cc - cell coat. 43,000 ×.

amorphous basal lamina in which collagenous fibers are embedded (Fig. E9g). The sarcoplasm contains fine longitudinal myofibrils that extend throughout. Electron micrographs (Figs. E10a–E10c, E11a, E11b) shows that each myofibril is composed of two types of short overlapping myofilaments that give the muscle the appearance of transverse banding: the STRIATIONS. Identify the striations in the micrographs. They are:

- A OR ANISOTROPIC BAND, the relatively broad, dark, doubly refracting or birefringent band;
- I OR ISOTROPIC BAND, the pale staining, singly refracting band;
- H BAND (H for "hell," bright), the light band bisecting the A band;
- M BAND, a thin, dark line, seen only in the best preparations, crossing the H band;
- Z BAND (Z for "Zwischenscheibe," intermediate disc), the dark line bisecting the I band.

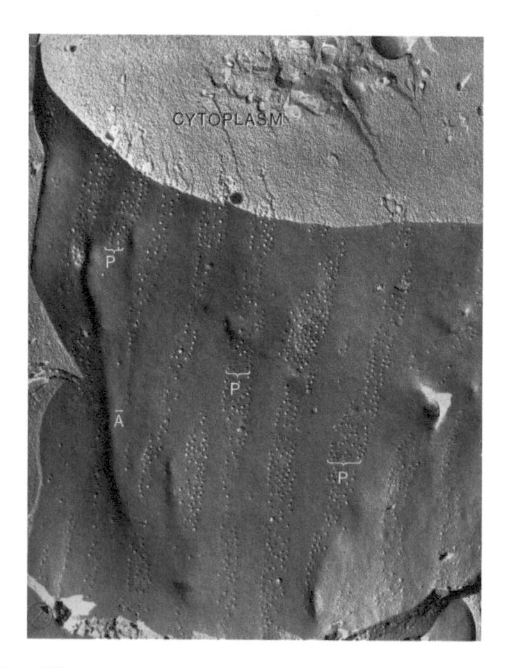

FIGURE E5b Electron micrograph of a freeze-fracture preparation of smooth muscle cells from the intestine of a rat. A large area of the PF face membrane (A) of the smooth muscle cell is shown in the lower portion of the picture. Note the nonrandom distribution of the pits (P) the arrow indicates the direction of shadowing. 25,000 ×.

The interval between successive Z bands is a SARCOMERE. Note the nuclei beneath the sarcolemma (Fig. E8a). Myofibrils occur in bundles, the SARCOSTYLES, separated by more open sarcoplasm (Figs. E9a and E10b). In sections of striated muscle nuclei, sarcolemma, and sarcoplasm are readily identified (Figs. E9a, E9b and E9c). The reader should take note of the appearance of groups of myofibrils, the sarcostyles.

In some cross sections of striated muscle you may notice that some fibers stain more intensely than others (Fig. E12a). These smaller RED FIBERS are richer in mitochondria than the larger WHITE FIBERS (Compare Figs. E12b and E12c). Red fibers contract more slowly than white and are almost impossible to fatigue. White or fast-twitch fibers function mainly anaerobically producing rapid bursts of activity; they have a meager blood supply and fatigue rapidly. INTERMEDIATE FIBERS, with characteristics between the two, are also found. "Dark meat" of fowl consists largely of red fibers, "white meat" of white fibers.

Locate the framework of fibrous connective tissue in transverse sections of entire body muscles (e.g., gastrocnemius, biceps brachii, etc.) (Fig. E13). Individual muscle fibers are surrounded by fibrous connective tissue, the ENDOMYSIUM, which carries capillaries,

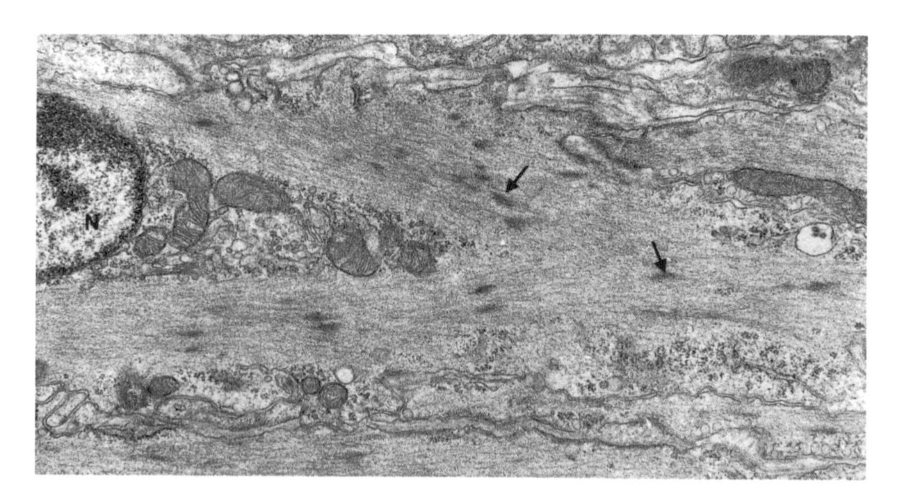

FIGURE E6 Electron micrograph of a longitudinal section of a smooth muscle fiber from the intestine of a 13-day-old rat. Mitochondria, a Golgi complex, and ribosomes, displaced by the thin filaments of actin, huddle at the end of the nucleus (N). The filaments appear to insert on the plasma membrane at dense bodies (arrows). 17,220 ×.

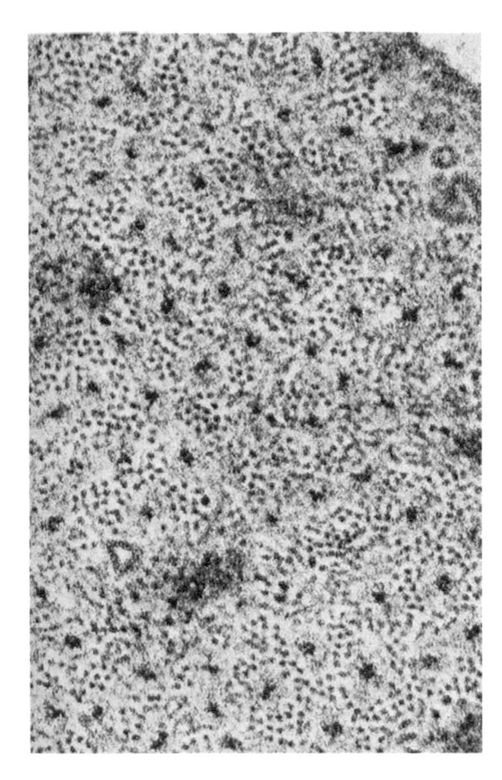

FIGURE E7 Electron micrograph of a cross section of a smooth muscle fiber from the vein of a rabbit. Thick and thin filaments are intermingled, apparently randomly. 153,000 ×.

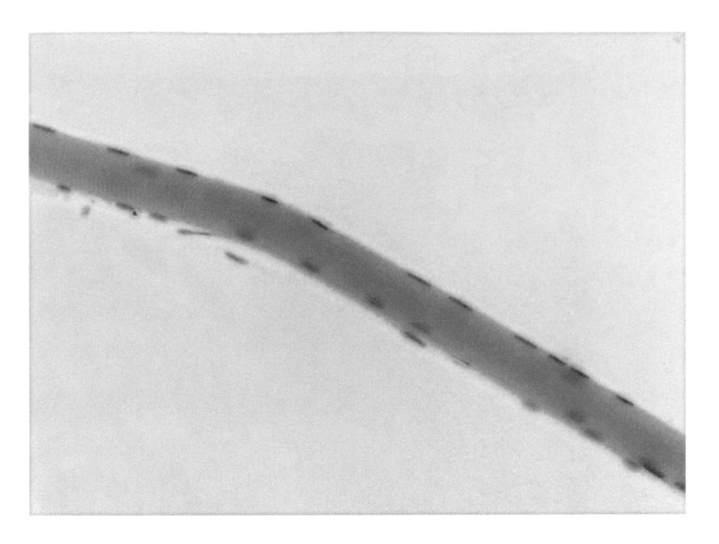

FIGURE E8a A teased striated muscle fiber is visible with the naked eye—a single multinucleate cell about the size of a piece of thread 3 cm long. 20 ×.

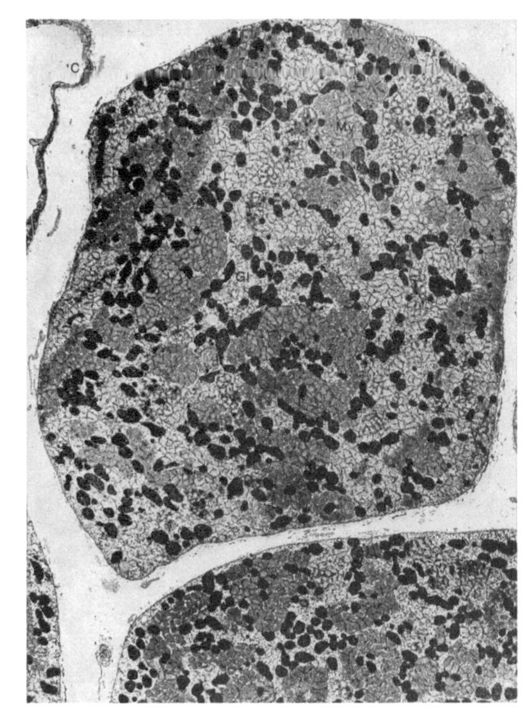

FIGURE E8b Low-power electron micrographs of section of the striated shaker muscle of a rattlesnake. Cross section. Columns of myofibrils (My) form SARCOSTYLES that are separated by sarcoplasm containing glycogen (Gl) and mitochondria. No nucleus appears in this section. Collagenous fibers form a harness around the sarcolemma. C - capillary. 5040 ×.

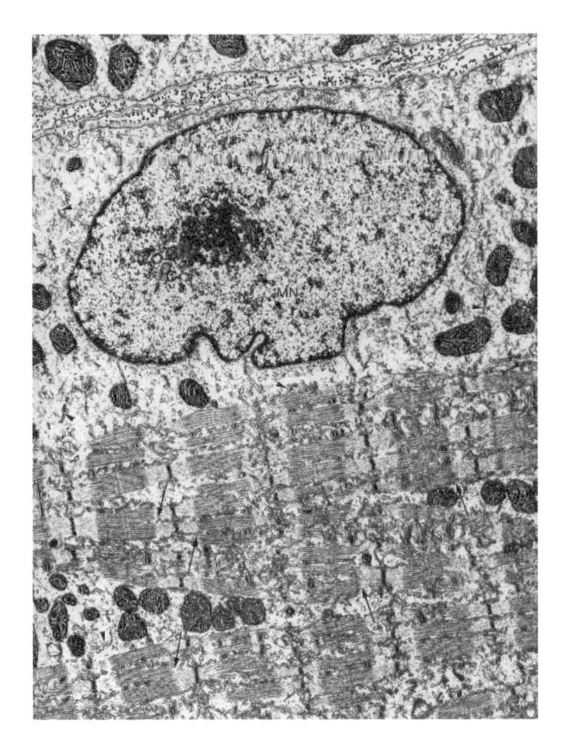

FIGURE E8c Low-power electron micrographs of section of the striated shaker muscle of a rattlesnake. Longitudinal section. A harness of collagenous fibers at the top surrounds the sarcolemma; note the collagenous fibers adjacent to the sarcolemma of the wedge of another muscle fiber at the extreme top. The nucleus (MN) of the muscle cell is surrounded by sarcoplasm richly provided with glycogen and mitochondria. Columns of myofibrils at the bottom are separated by more sarcoplasm. Arrowheads indicate elements of the sarcoplasmic reticulum or T-system. 14,600 ×.

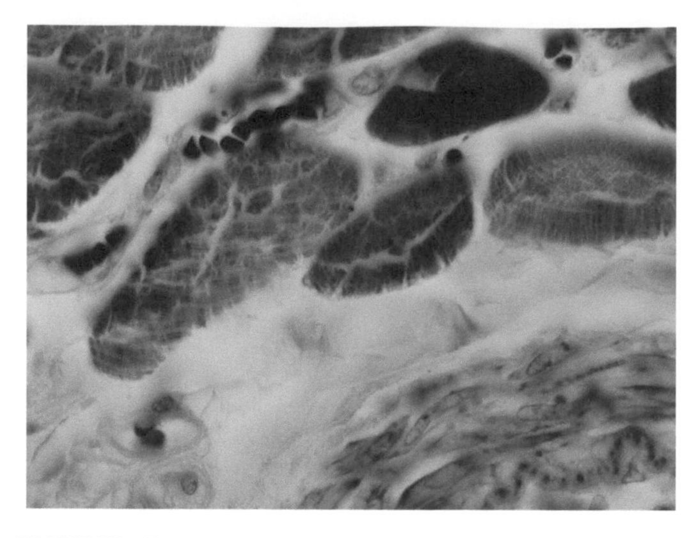

FIGURES E9a Striated muscle from mammalian diaphragm. Cross section. Groups of muscle cells form sarcostyles.

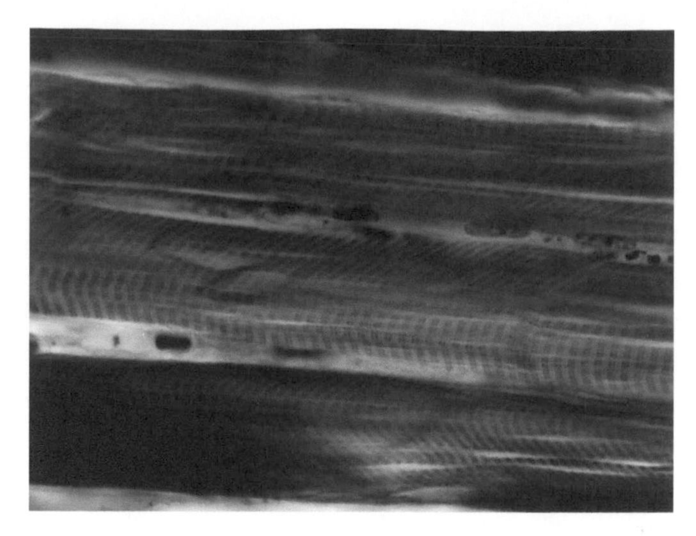

FIGURE E9b Striated muscle from mammalian diaphragm. Longitudinal section. Striations are apparent within the muscle cells.

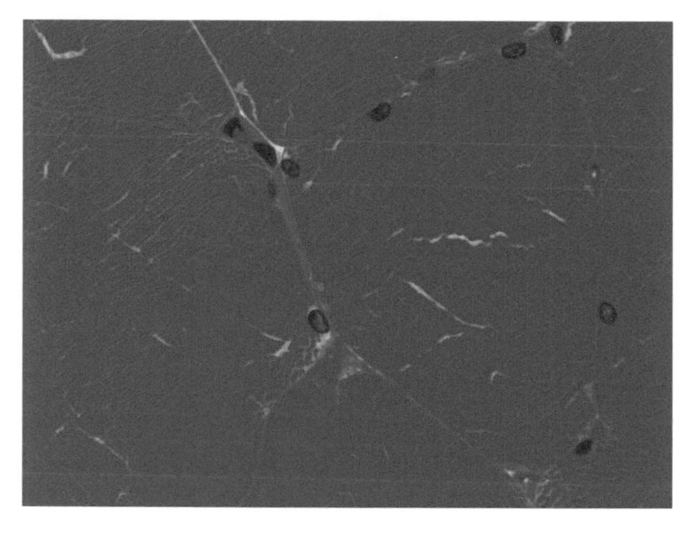

FIGURE E9c Semithin cross section of striated muscle of a mammal. Several nuclei are distributed immediately below the sarcolemma of the fibers. Contractile fibrils occur in bunches or sarcostyles within the fibers. Endomysium between the fibers harnesses their contractions. 63 ×.

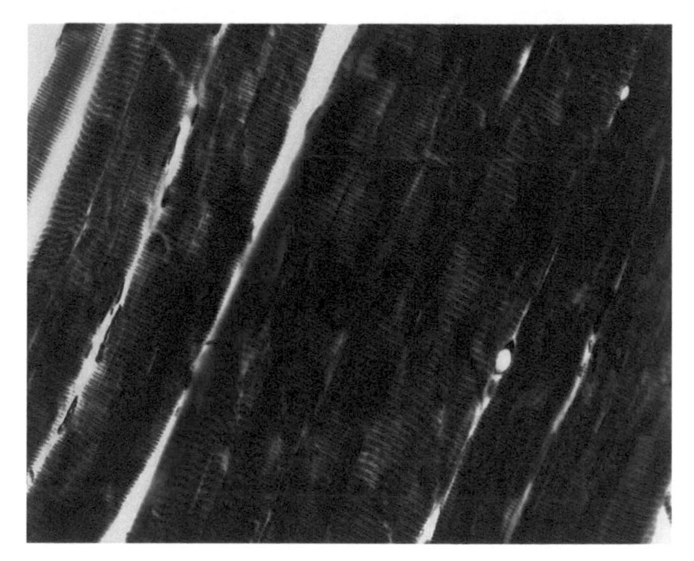

FIGURE E9d Longitudinal section of striated muscle of a cat. The striations are well developed. 63 ×.

lymph vessels, and nerves. Groups of fibers are bound together into bundles, the FASCICULI, by PERIMYSIUM. The entire muscle is sheathed in tough fibrous connective tissue, the EPIMYSIUM. (Fig. E14).

Study of sections of the junctions between striated muscle and tendons (Fig. E15) are unsatisfactory, partly due to the fact that collagenous fibers are resistant to staining. The scanning electron microscope shows that the end of each muscle fiber in the junction has a jagged dome shape (Fig. E16). This jagged contour increases the surface area of contact between the muscle and tendon, thereby increasing its strength (Fig. E17). Electron micrographs also show that the sarcolemma seen with the light microscope consists of the plasma membrane as well as an extracellular basal lamina permeated with a network of reticular fibers that are continuous with the collagenous fibers of the endomysium (Figs. E8b, E8c, E9f). The collagenous fibers form a hardness, which continues as fibers of the tendon. It is unclear how the collagenous fibers are attached to the sarcolemma and whether there is a physical connection between the Z lines of the contractile apparatus and the collagenous fibers. It is presumed that the collagenous fibers of the epimysium, perimysium, and endomysium continue into the tendon.

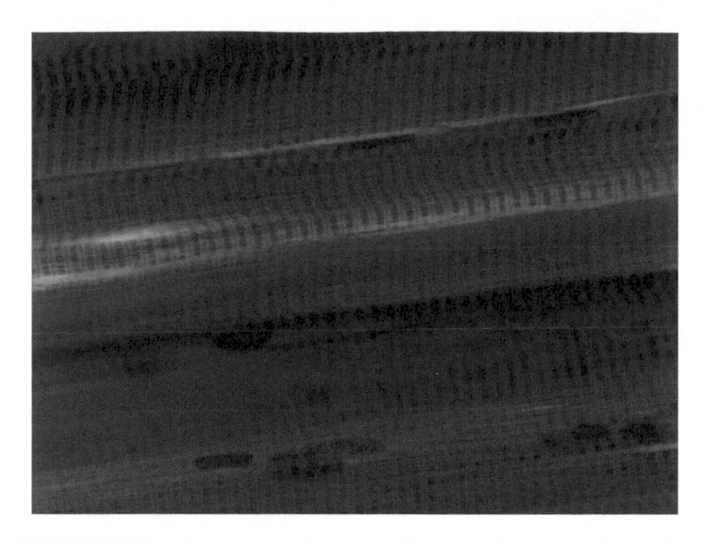

FIGURE E9e Longitudinal section of mammalian striated muscle. The nuclei of these multinuclear cells are distributed beneath the sarcolemma of the fibers. The striations are well defined. 63×.

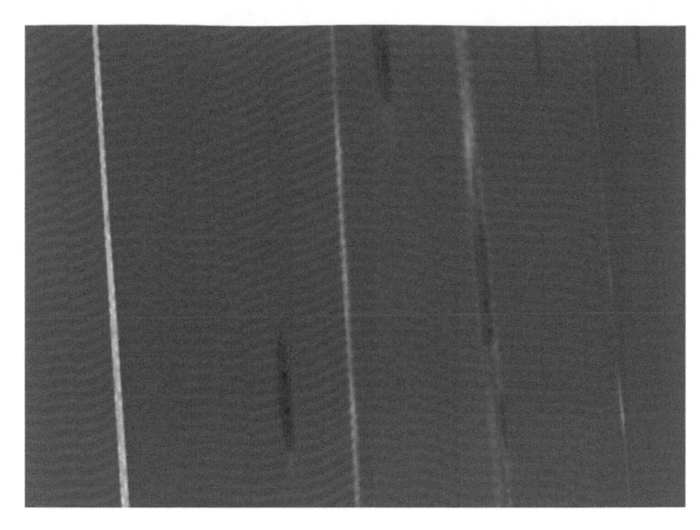

FIGURE E9f Longitudinal section of striated muscle from a frog. The characteristic banding pattern is clearly demonstrated. 63×.

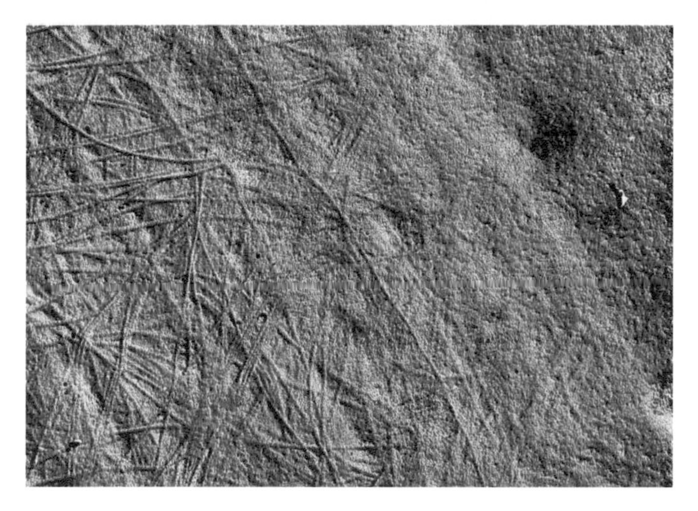

FIGURE E9g Freeze-fracture replica of the sarcolemma of a skeletal muscle fiber of a frog. The sarcolemma is coated with an amorphous substance in which collagenous fibers are embedded. 14,000×.

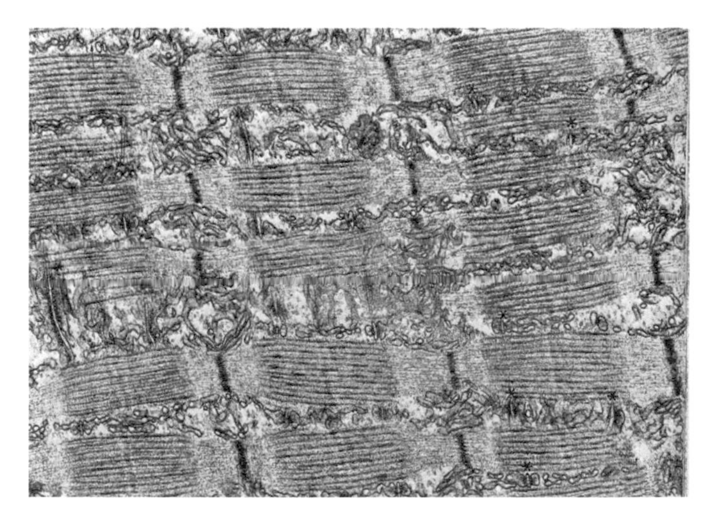

FIGURE E10a Electron micrograph through the shaker muscle of a rattlesnake. Longitudinal section through a column of myofibrils. An abundant sarcoplasmic reticulum penetrates between the myofibrils. 28,800×.

The blood vessels of skeletal muscle may be abundant and capillaries extend the length of the fibers to form a dense mesh. Figs. E18a and E18b show this vascularization in sections of striated muscle in which the blood vessels have been injected with colored material

Electron micrographs of longitudinal sections of striated muscle fibers show that the myofibrils consist of bundles of myofilaments whose precise arrangement of overlapping thin and thick filaments of fibrous proteins produces the striated appearance seen with the light microscope (Figs. E10a, E11a, E11b). The thick filaments are approximately 10 nm wide and 1.5 μm long and consist of MYOSIN. They run the length of the A band. The thin filaments are largely ACTIN, 5 nm wide and 2.0 μm long, and extend from the Z band to the edge of the H band. The rodlets of actin attach to the Z band keeping them in alignment and transmitting their contractile force to the sarcolemma; the Z band consists in part of α-ACTININ. The dark portion of the A band is produced by the overlapping of both types of filaments. In contraction the actin filaments slide between the myosin, reducing the breadth of the I and H bands. In micrographs of the highest power the myosin filaments show small lateral projections at one end that connect with and move the actin filaments during contraction. A diagram of the process of contraction is shown in Fig. E19.

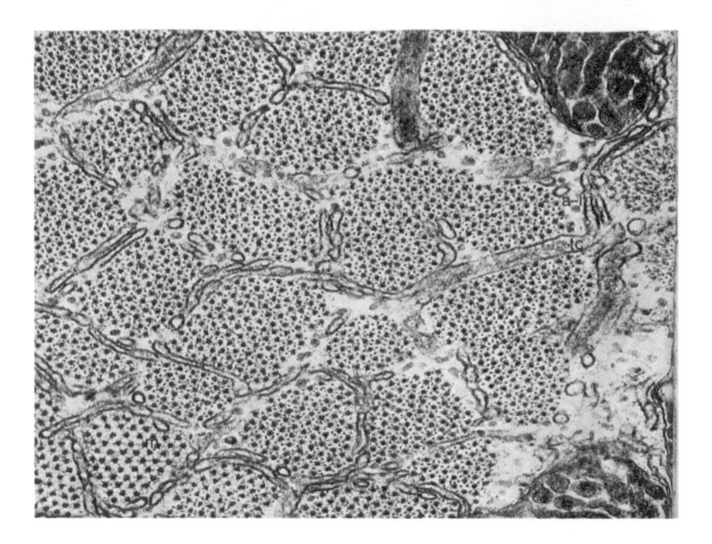

FIGURE E10b Electron micrograph through the shaker muscle of a rattlesnake. Cross section through a column of myofibrils. The sarcoplasmic reticulum pervades between the myofibrils. 64,400 ×.

FIGURE E11a Electron micrograph of a longitudinal section of the tail muscle of a tadpole from a leopard frog. Several myofibrils, consisting of a large number of myofilaments, extend across the micrograph. Alternating dark and light bands in the fibrils are formed from overlapping thick and thin filaments, which are in precise register. Each repeating sequence of striations constitutes a SARCOMERE. Each sarcomere is limited by a Z band. The isotropic I band is bisected by the Z. The anisotropic A is denser and is bisected by a narrow light band, the H. A faint M band is sometimes discerned and appears along the middle of the H band. The thick filaments are composed of the protein MYOSIN and the thin filaments consist of ACTIN. During contraction, the actin filaments slide between the myosin filaments and the I and H bands disappear. The complex sarcoplasmic reticulum penetrates between the myofibrils. In the region of the Z band, the tubules of the sarcoplasmic reticulum converge as dilated sacs (SR). These sacs are closely apposed to tubular invaginations of the sarcolemma, the T-tubules (TS) of the T-system. These T-tubules transmit an electrical impulse from the sarcolemma that encloses the cell to the sarcoplasmic reticulum thereby stimulating the contractile fibrils so that all myofibrils contract simultaneously. 29,000 ×.

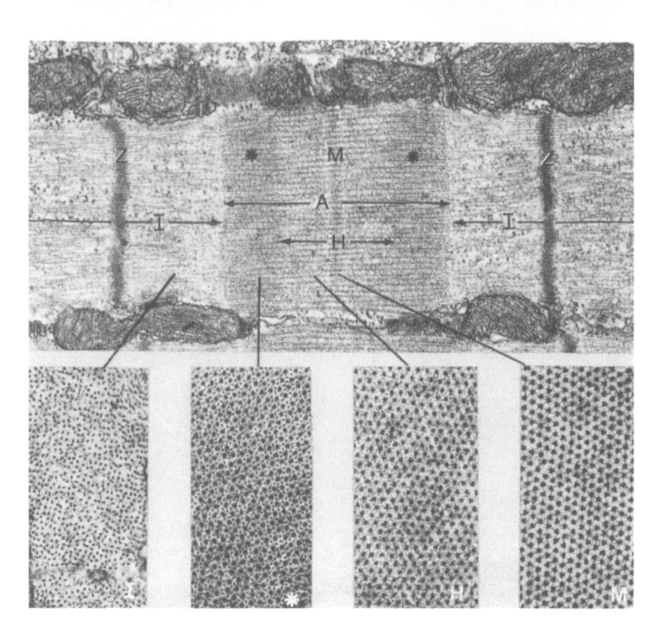

FIGURE E10c Longitudinal section (upper) and cross sections (below) of a sarcomere (from one Z line to the next) of striated muscle. The cross sections illustrate the amazing precision in the spacing of the filaments in each of the bands. At the *left*, in the I band, only thin filaments are seen; the *second* micrograph (*) shows thin filaments interdigitating with the thick filaments in the A band; the *third* shows thick filaments in the H zone; and the *fourth* shows thick filaments at the level of the M line where interconnections link them together.

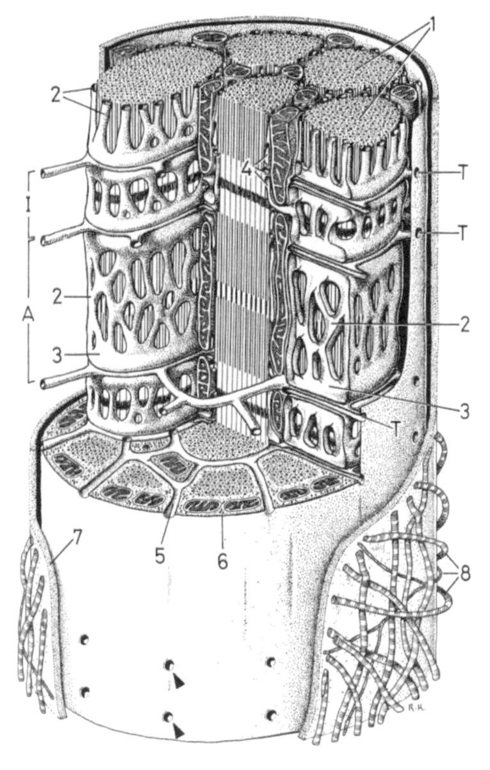

FIGURE E11b The three-dimensional relationships of the sarcoplasmic reticulum (2 & 3), the T-system (4,5 & T), and contractile filaments (1) are shown diagramaticly. 7 - basal lamina; 8 - reticular fibers.

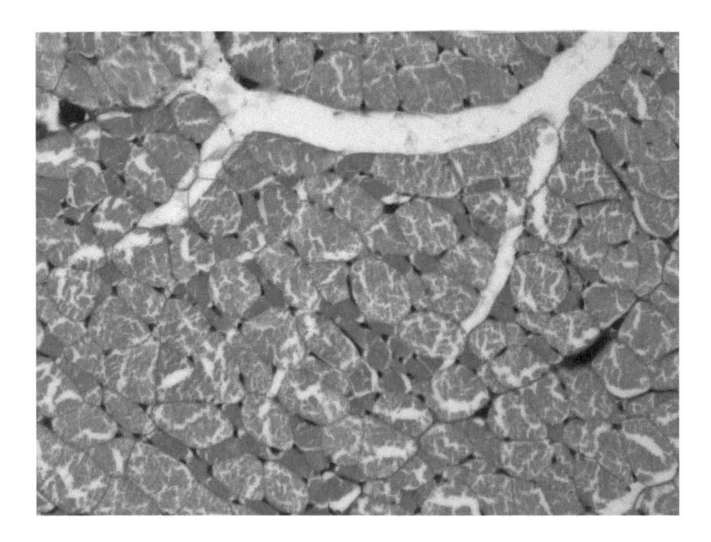

FIGURE E12a Cross section of striated muscle from a mammal. The dark "red" tonic fibers are smaller than the paler "white" twitch fibers. 20 ×.

The precise hexagonal arrangement of the rodlets of actin and myosin also may be seen in micrographs of transverse sections of striated muscle fibers (Fig. E10b). Note that both types of filaments will be found only in cross sections of the A band; only myosin filaments will appear in sections through the H, and only actin in sections through the I. A summary view of this arrangement may be seen in Fig. E10c.

Between the myofibrils the sarcoplasm consists of a typical cytoplasmic matrix containing organelles and inclusions (Figs. E8b and E8c). A Golgi complex huddles near one pole of each nucleus and rows of numerous large sarcosomes (mitochondria) with closely packed cristae fill the spaces between the myofibrils and are concentrated at the ends of the nuclei. Scattered throughout the sarcoplasm are small masses of lipid droplets and granules, 20–40 nm in diameter, that are probably glycogen.

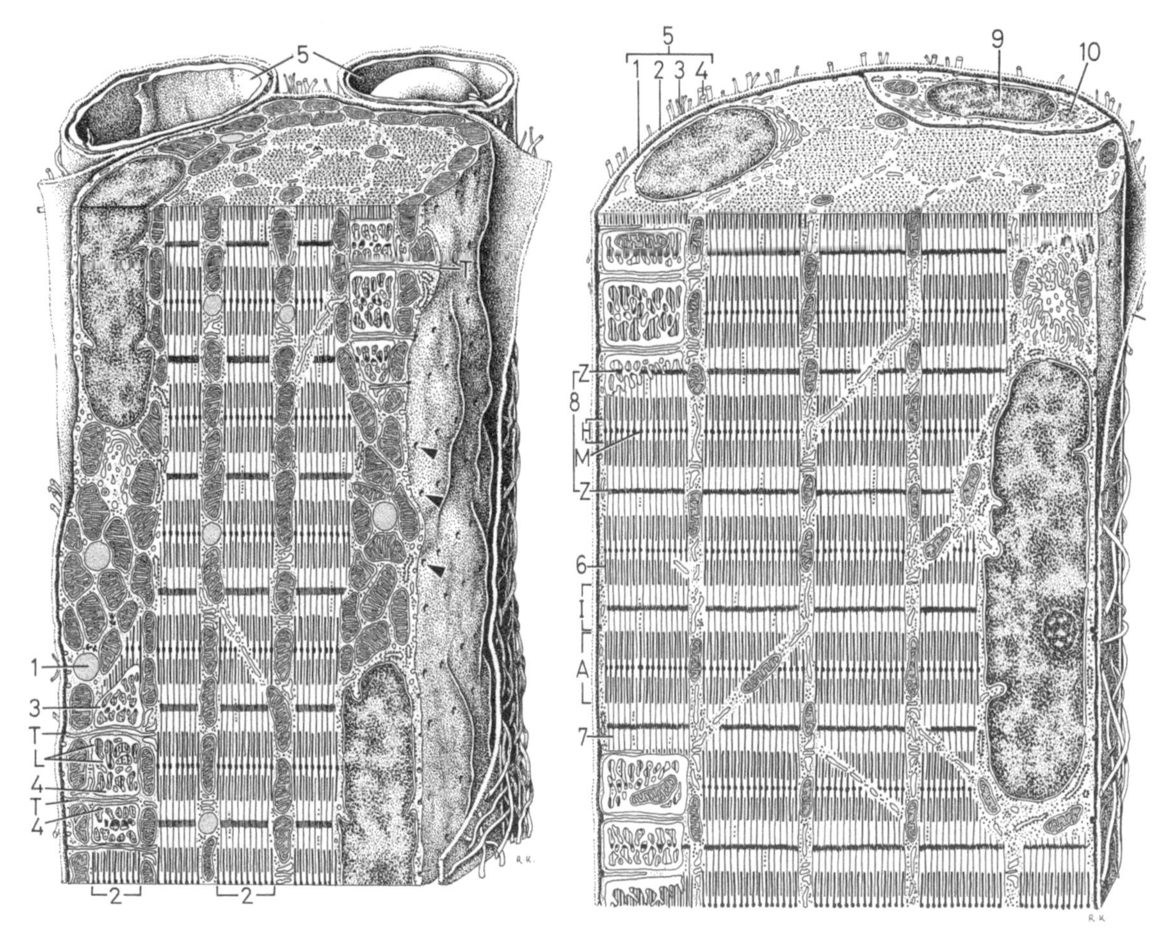

FIGURES E12b and E12c Compare these drawings of electron micrographs of red (tonic or slow fibers) striated muscle fibers (Fig. E12b) and white (twitch or fast fibers from a mammal). The tonic fibers are smaller in diameter, and contain more sarcoplasm, myoglobin, mitochondria, and lipid droplets than the twitch fibers. In addition, they have a richer blood supply. Their myofibrils are thinner and the Z lines thicker. Tonic fibers are slower to contract and slower to fatigue that twitch fibers and come into play for prolonged contractions such as the maintenance of posture. Twitch fibers are used in sudden actions as in jumping.

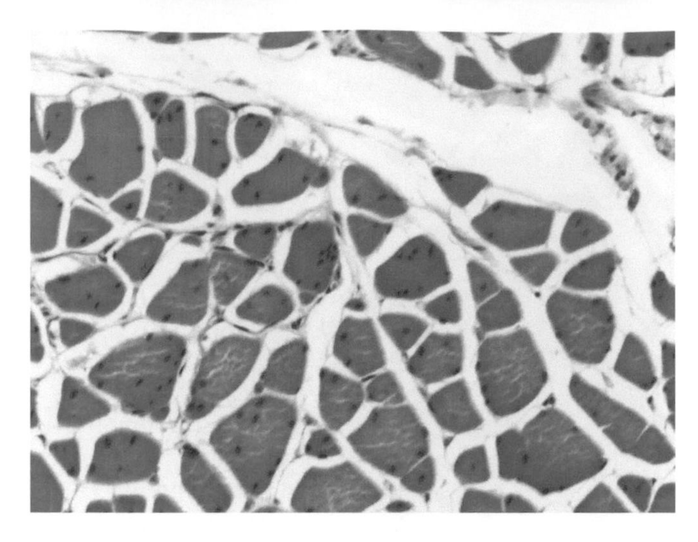

FIGURE E13 Cross section of the foot of a toad. A small amount of shrinkage serves to reveal the various levels of connective tissue within the muscle of the leg. An entire muscle, such as the gastrocnemius, is bound by a sheath of tough connective tissue, the EPIMYSIUM. Within the muscle itself, PERIMYSIUM binds the groups of muscle fibers into FASCICULI. Individual fibers are bound by ENDOMYSIUM. All these levels of connective tissue serve to transfer the contractions of the muscle fibers, to the skeleton. 20×.

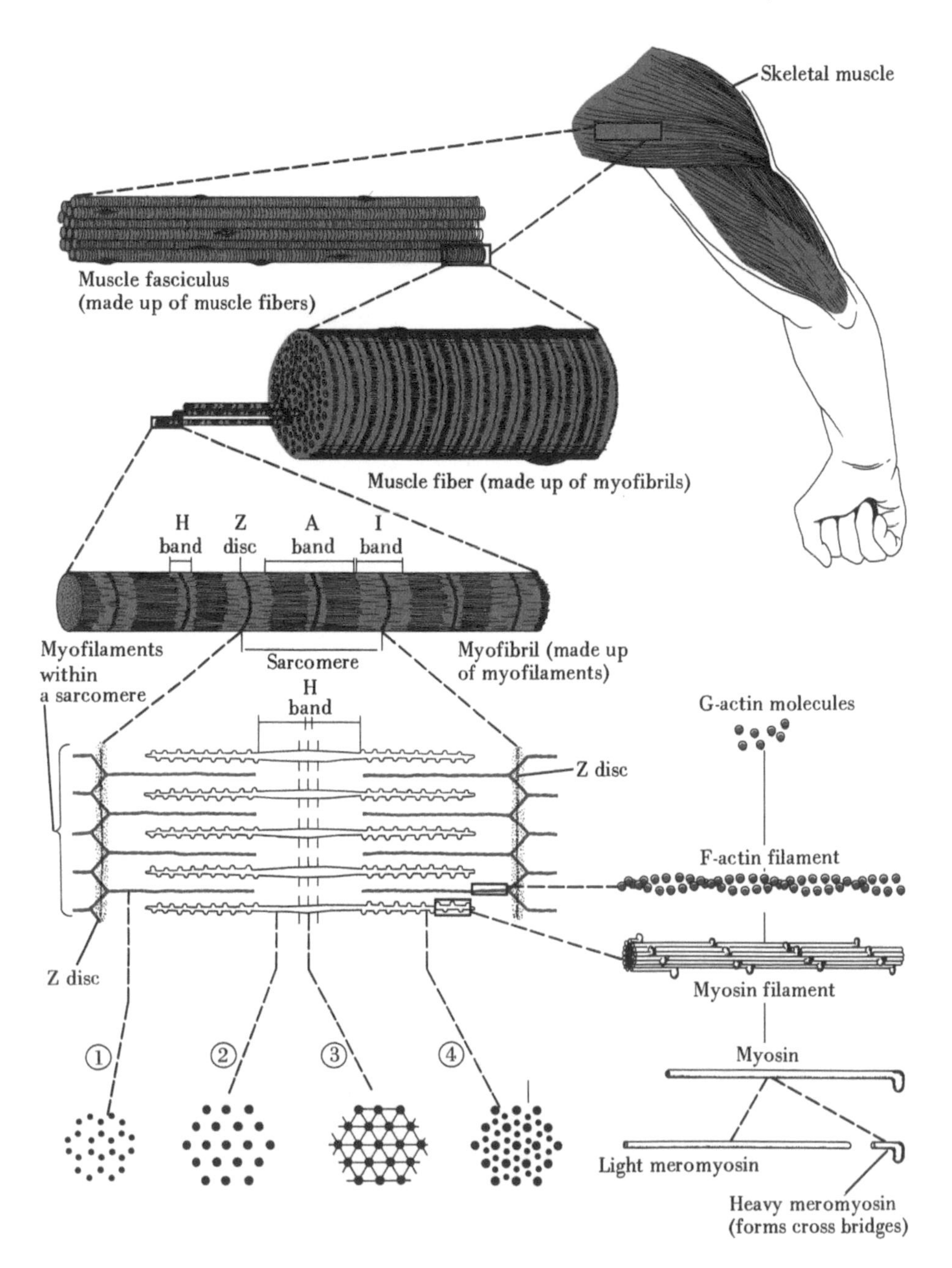

FIGURE E14 Organization of skeletal muscle. An entire muscle is sheathed in tough fibrous connective tissue, the EPIMYSIUM. It consists of groups of muscle fibers (cells) which are bound together into bundles, the FASCICULI, (sing. fasciculus) by PERIMYSIUM. Within the fiber are bundles of MYOFIBRILS. The myofibrils consist of bundles of MYOFILAMENTS whose precise arrangement of overlapping thin and thick filaments of fibrous protein produces the striated appearance seen with the light microscope. The thick filaments are approximately 10 nm wide and 1.5 μm long and consist of MYOSIN. They run the length of the A band. The thin filaments are largely ACTIN, 5 nm wide and 2.0 μm long, and extend from the Z band to the edge of the H band. The rodlets of actin attach to the Z band, which keeps them in alignment and transmits their contractile force to the sarcolemma. The dark portion of the A band is produced by the overlapping of both types of filaments. In contraction the actin filaments slide between the myosin, reducing the breadth of the I and H bands. In micrographs of the highest power, the myosin filaments show small lateral projections at one end that connect with and move the actin filaments during contraction. Sections through different portions of the sarcomere, 1, 2, 3, and 4 illustrate the three-dimensional arrangement of these myofilaments.

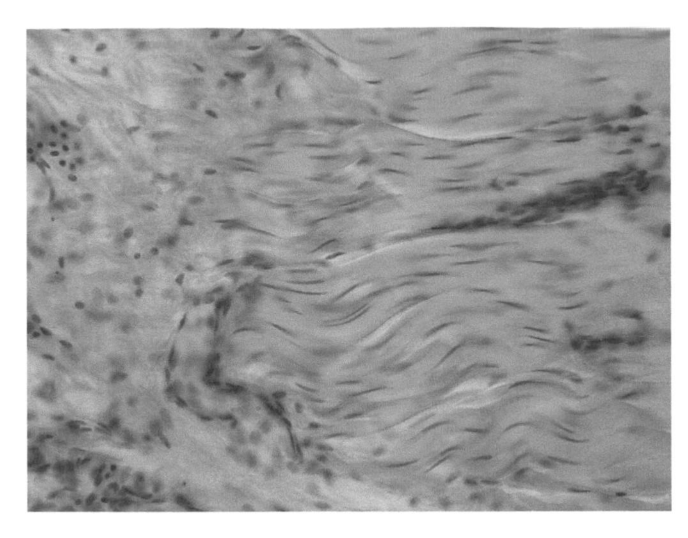

FIGURE E15 It is difficult to ascertain the relationship between muscle fibers and the collagenous fibers of a tendon in this photomicrograph of a section through their junction. Muscle cells are pink, collagenous fibers pale. 20 ×.

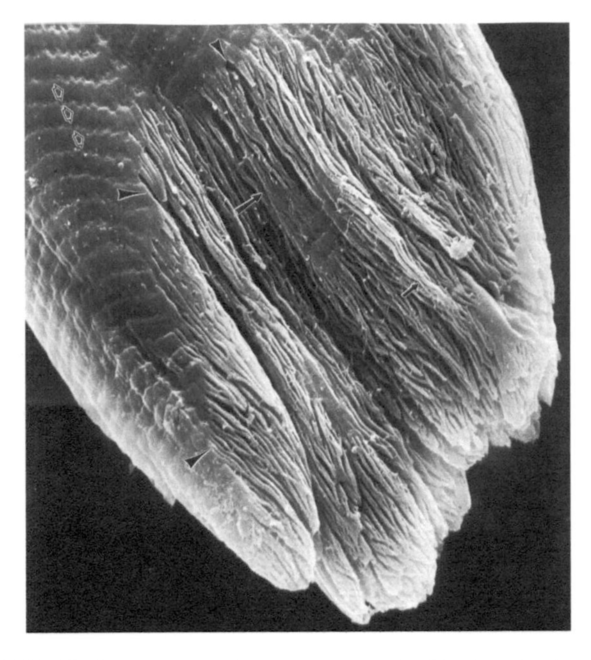

FIGURE E16 Scanning electron micrograph of the junctional surface of a single fiber from the plantaris muscle of a mouse. The junctional surface of the muscle fiber is enhanced by microvillus folding (open arrowheads and small arrows). Large arrows indicate non-junctional surfaces. 4600 ×.

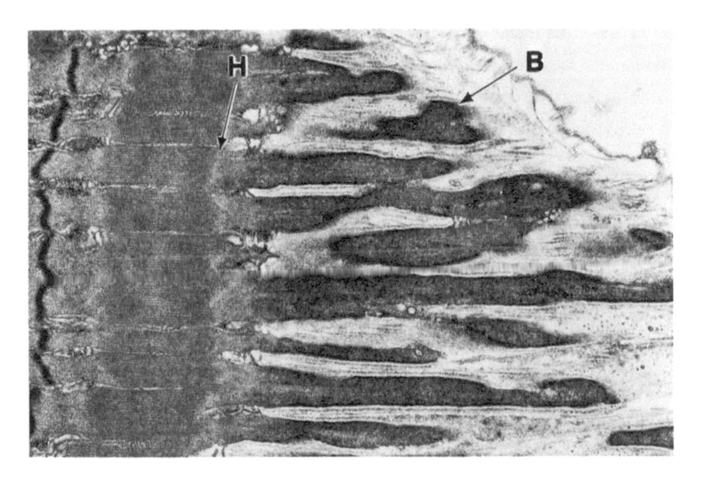

FIGURE E17 Electron micrograph of a section through the myotendinous junction in the plantaris muscle of a mouse. The junctional area is greatly enhanced by the interdigitation of cytoplasmic projections of the muscle cell and the collagenous fibers of the tendon. A basal lamina (B) coats the sarcolemma of the muscle cell. Actin filaments of the terminal sarcomeres (H) extend into the projections. 20,000 ×.

The SARCOPLASMIC RETICULUM is an extremely elaborate agranular endoplasmic reticulum that encloses each myofibril in a basket of SARCOTUBULES (Figs. E10a, E10b, E11a, and E11b). The reticulum shows a repeating pattern of tubules and cisternae whose arrangement parallels the banding pattern of the myofibrils (Fig. E20). The pattern varies in different species, but is basically the same in all; the following description applies to amphibian skeletal muscle (Fig. E21). Encircling the myofibril on each side of the Z band is a tire-shaped dilated sac, the TERMINAL CISTERNA. Extending the length of one sarcomere is a lacy network of sarcotubules that enclose the myofibril and are confluent with the terminal cisternae at both ends of the sarcomere. Between adjacent terminal cisternae and exactly parallel to the Z band is an inward extension of the sarcolemma, the TRANSVERSE TUBULE or T-TUBULE. The two terminal cisternae together with the T-tubule constitute a TRIAD; the system of T-tubules makes up the T-SYSTEM. The T-system is not a part of the sarcoplasmic reticulum and its lumen is continuous with the extracellular space. Stimuli for contraction reach the interior of the cell by way of the T-system and are transmitted to the myofibrils by the sarcoplasmic reticulum.

Electric Organs

During their evolution, several groups of elasmobranch and teleost fish have independently developed electric organs by accentuating the natural ability of muscle cells to generate an electric current. These organs arise either from presumptive or formed striated muscle masses and some are capable of producing potentials of several hundred volts (Fig. E22). Commonly a skeletal muscle cell widens and shortens to form a wafer-shaped ELECTROCYTE oriented at right angles to the original fiber (Fig. E23). These electrocytes are stacked in columns surrounded by perimysium and are embedded in a jelly-like material contained within a compartment of connective tissue (Figs. E24a–E24c).

FIGURE E18 (a and b) Sections of striated muscle that has been injected with a colored substance. The capillary network wraps around individual fibers bringing oxygen and nourishment and carrying away wastes. 20 ×.

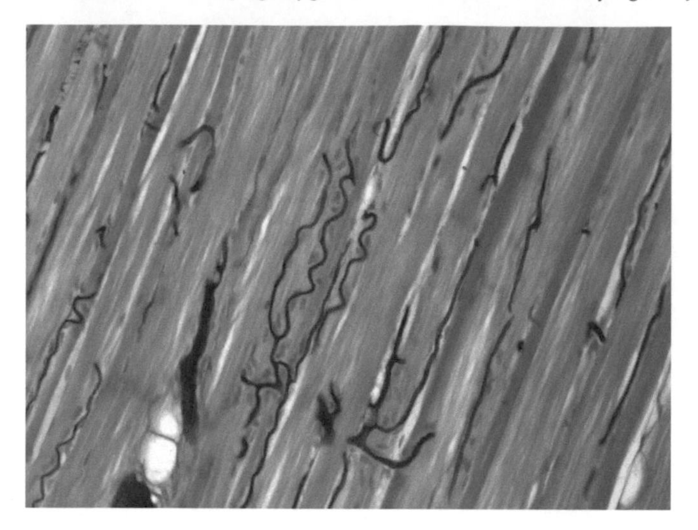

FIGURE E18a Longitudinal section.

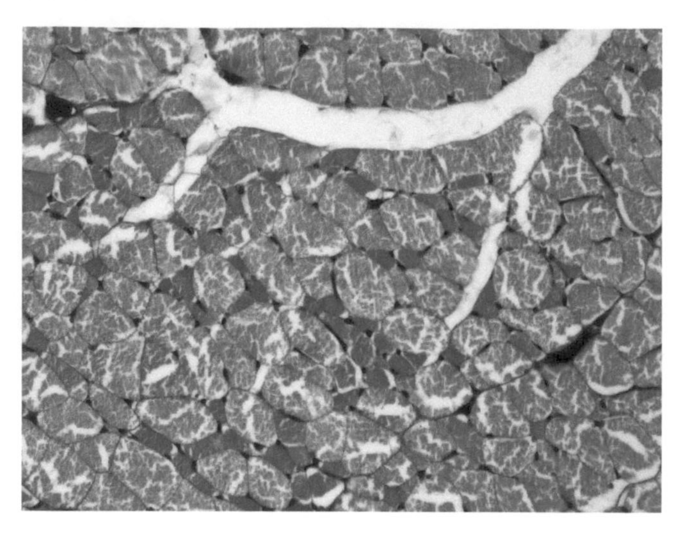

FIGURE E18b Cross section. This section illustrates the distinction between red and white muscle fibers.

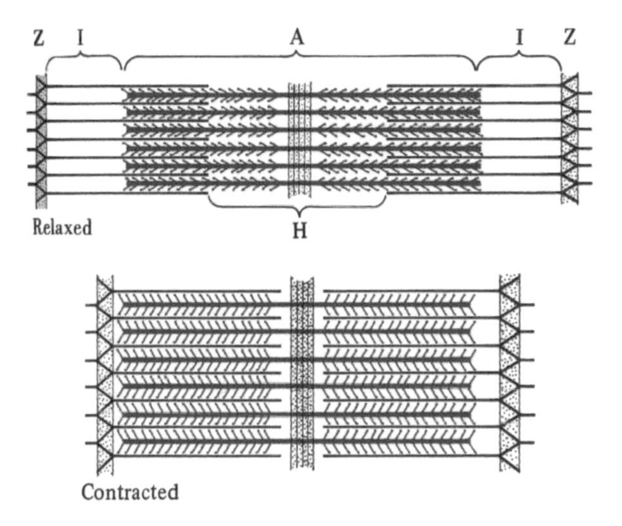

FIGURE E19 These diagrams illustrate relaxed and contracted sarcomeres during the contraction of striated muscle. In the relaxed state, interdigitation of actin and myosin filaments is incomplete and the H and I bands are wide. During contraction, the bands slide together, shortening the H and I bands but the length of the A band remains the same.

The electrocytes retain some of the characteristics of striated muscle cells. They are multinucleate and most of the nuclei are peripherally arranged (Fig. E25). The myofibrils are widely variable from those that show almost typical striations to those that are quite irregular (Figs. E26 and E27). Mitochondria and other organelles are present (Fig. E28). Large nerves run along the electric organ sending a branch to one surface of each electrocyte (Figs. E29a and E29b). This side of the electrocyte is relatively smooth while the opposite side is deeply convoluted and is nutritive. The same side of each electrocyte is innervated, either the posterior face if the current flows from the head to the tail, or the anterior if it flows in the opposite direction. Stimulation by the nerve causes an inrush of sodium ions on one side so that the interior of one side of the cell becomes positively charged. A current flows across the cell and skips across the cell membrane and the intercellular fluid to the next cell where a similar phenomenon is occurring. The pile of electrocytes operate in series so that each adds to the total voltage produced by the organ.

CARDIAC MUSCLE

The branched fibers of heart muscle form a fairly open, interconnecting network (Figs. E30a, E31a, E31b) Cardiac muscle fibers, although not under the control of the will, are striated. Identify the striations, sarcoplasm, myofibrils, and nuclei. Characteristic of cardiac muscle are the INTERCALATED DISCS that cross the fibers at intervals and are not visible in all preparations (Figs. E30a, E31a, E31b). These represent the junctions between individual cells that are joined end-to-end to make up a fiber. Thus, unlike skeletal muscle, cardiac muscle consists of many mononuclear cells aligned end-to-end (Fig. E32). Where fibers are cut transversely, note the central nucleus surrounded by granular sarcoplasm, and the arrangement of sarcostyles and myofibrils (Fig. E33). The delicate interstitial connective tissue between the bundles of fibers contains a rich network of capillaries and lymph vessels (Fig. E30b).

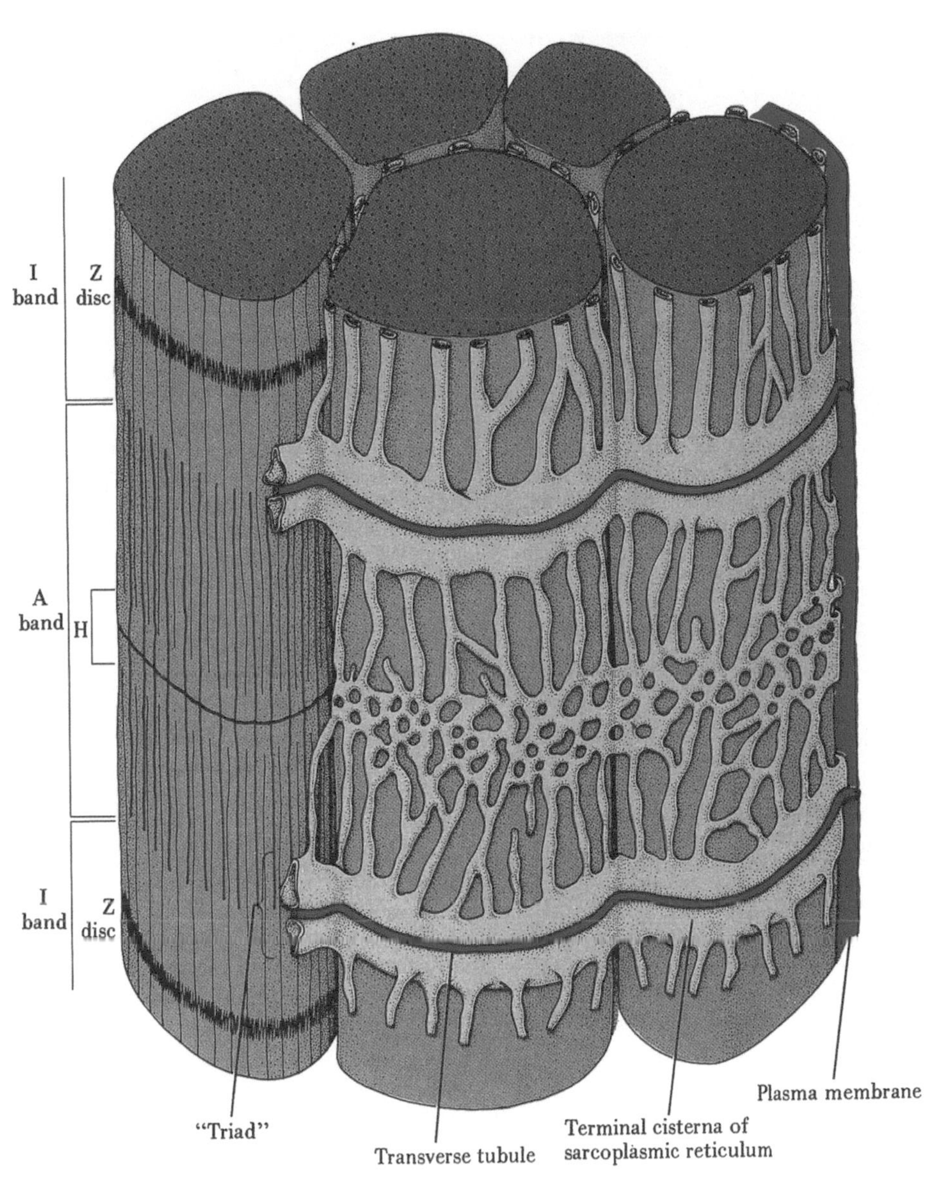

FIGURE E20 Diagram of part of a mammalian a striated muscle fiber illustrating the relationships of the sarcoplasmic reticulum to the myofibrils and the T-tubules, which are invaginations of the sarcolemma. Meshworks of sarcoplasmic tubules enclose the myofibrils. In the mammal, two transverse (T) tubules supply a sarcomere. Each T-tubule is located close to the junction between an A and an I band, where it is associated with two terminal cisternae of sarcoplasmic reticulum, on either side of it. Each terminal cisterna connects with meshwork of sarcoplasmic tubules located around the A band and extending across the H band. The two cisternae as well as the T-tubule constitute a TRIAD.

Note that differences between cardiac and skeletal muscles are largely in details (Compare Figs. E11a and E11b and E34). Perhaps the most obvious feature of cardiac muscle is the presence of large sarcosomes aligned in columns between the myofibrils. Myofibrils may branch and are not as clearly defined as in skeletal muscle. The sarcoplasmic reticulum encloses the myofibrils in a similar lacy basket of sarcotubules that extend the length of one sarcomere (Figs. E20 and E35). There are no large terminal cisternae but the tubules form small TERMINAL SACS that are in contact with the large T-tubule. The intercalated disc is an elaborate interdigitating junction between the ends of adjacent cells; it occurs at the level of the Z lines and encloses an intercellular space of 20 nm (Fig. E34). Here the myofilaments of adjacent I bands terminate on desmosomes and desmosome-like FASCIAE ADHAERENTES (*Singular*: fascia adhaerens) that firmly join the cells. At irregular points in the intercalated discs, gap junctions connect adjacent plasma membranes and assist in transmitting the stimulus for contraction.

The histological structure of cardiac muscle is basically the same in all classes of vertebrates. Compare Figs. E11a and E34. The hearts of many lower vertebrates is spongy with no clearly defined lumina as are seen in the hearts of mammals, birds, and reptiles. The blood oozes within a sponge of muscle fibers, being pumped along in a leisurely fashion while nourishing the cardiac muscle cells as it passes (Figs. E36–E39).

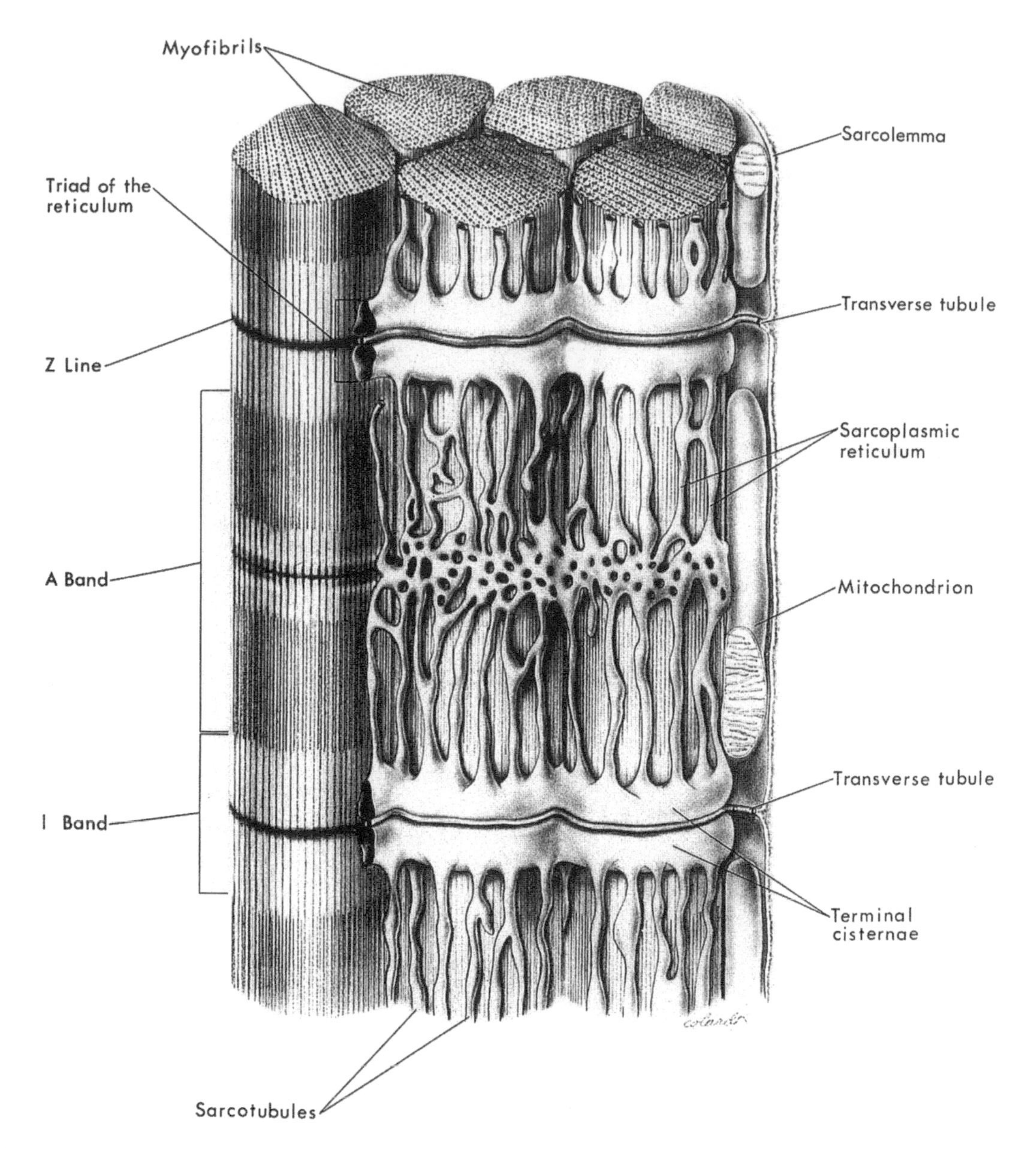

FIGURE E21 Schematic drawing showing the distribution of the sarcoplasmic reticulum around myofibrils of amphibian skeletal muscle. The longitudinal sarcotubules are confluent with transverse terminal cisternae. A slender transverse T-tubule, extending inward from the sarcolemma is flanked by two terminal cisternae to form the TRIADS of the reticulum. In this amphibian muscle the triads are at the Z lines. There are two to each sarcomere in mammalian muscle, located at the A–I junction.

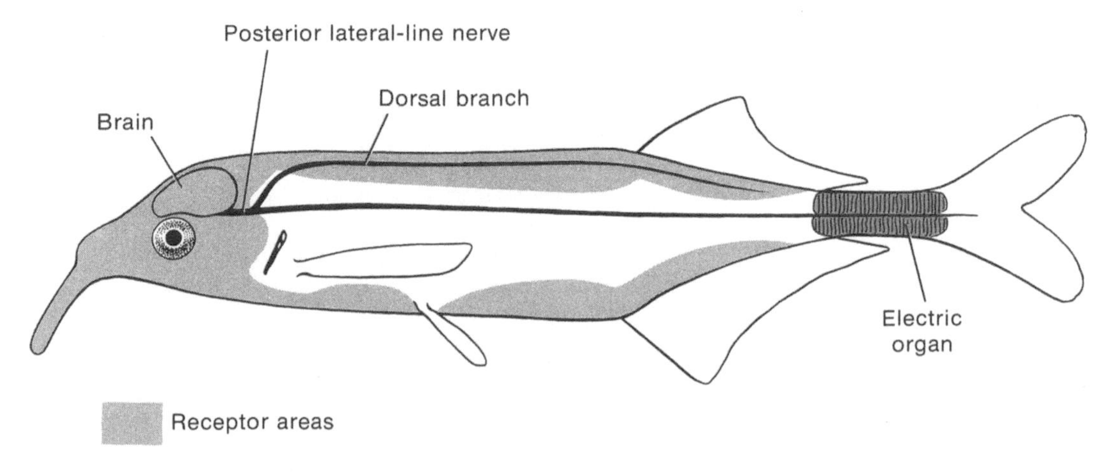

FIGURE E22 Positions of the electric organ and the lateral-line nerve trunk in the weakly electric fish, *Gnathonemus petersii*.

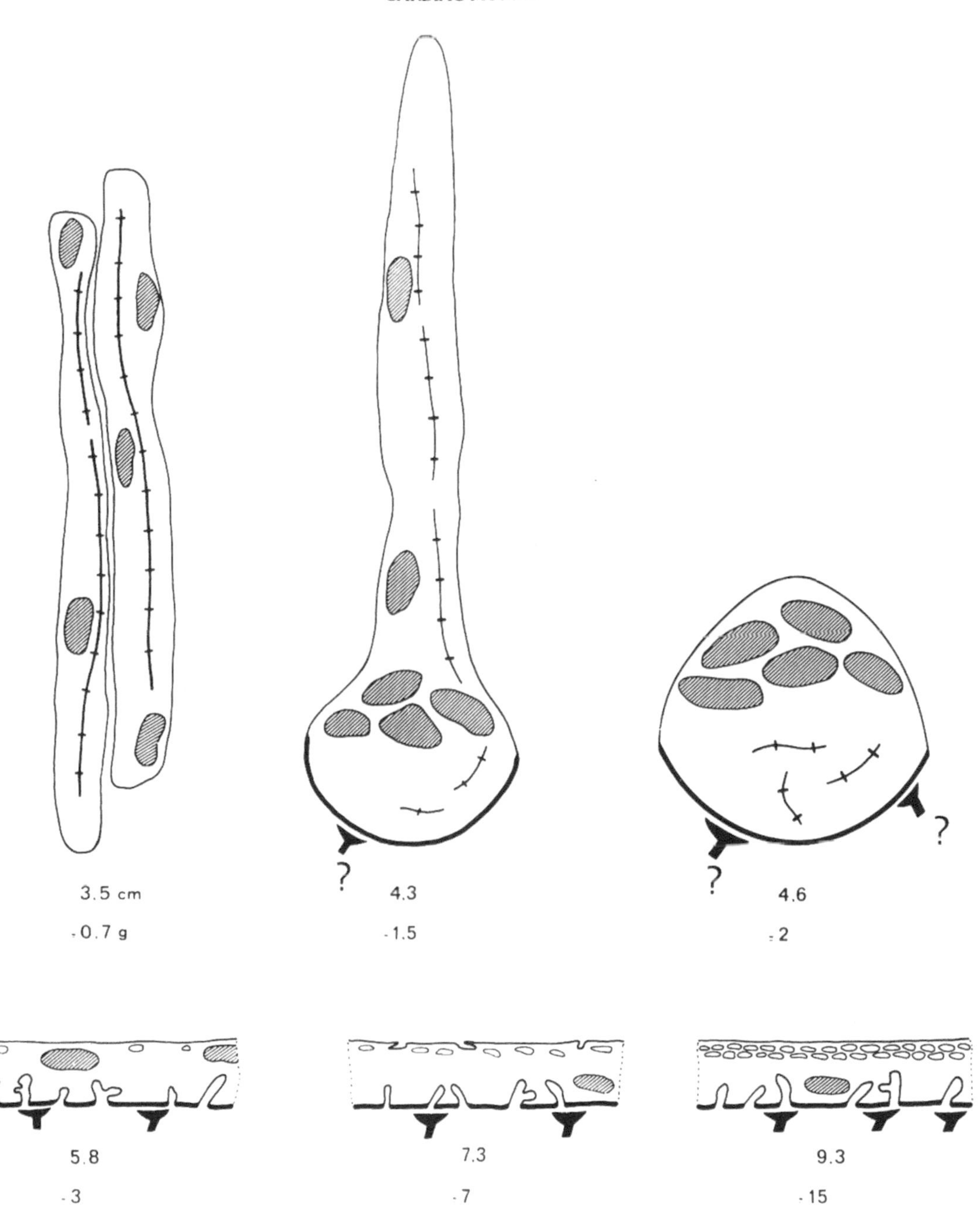

FIGURE E23 Schematic drawings of six stages in the development of an electrocyte in embryos of the electric ray, *Torpedo marmorata*. Lengths and weights are shown of the embryos examined. The natural orientation of the structures is indicated; shaded areas are nuclei; black lines with cross bars are myofibrils; and the heavy black lines represent postsynaptic membranes. The small drawings at the bottom represent the last three stages and indicate the development of the endoplasmic reticulum on the dorsal sides.

Conducting System of the Heart

A network of atypical muscle fibers, the CONDUCTING MYOFIBERS (Purkinje system), occurs under the endocardium that lines the internal surface of the mammalian heart (Fig. E40). Large amounts of sarcoplasm are arranged around the central nuclei and the myofibrils are mainly in the peripheral portions of the fibers (Fig. E41). These fibers form a mechanism that conducts the stimuli for contraction of the ordinary heart muscle. The cardiac impulse is transmitted from the sinoatrial node into the atria, then into the atrioventricular node, and finally through the conducting myofibers to all parts of the ventricles.

FIGURE E24(a–c) Anatomy of the electric organ of an 84-day old, mormyrid fish, *Pollimyrus isidori*.

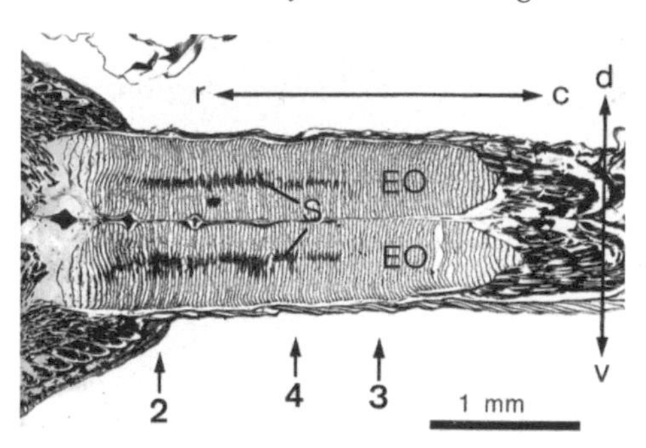

FIGURE E24a Parasagittal section showing the left half of the electric organ (EO). The electrocytes of the dorsal and ventral columns are juxtaposed. In the middle of the columns the innervated stalks (S) are visible. The levels of the cross section (indicated by the numbers) (Fig. E24b) and the sagittal section (Fig. E24c) are shown d, dorsal; v, ventral; r, rostral; c, caudal.

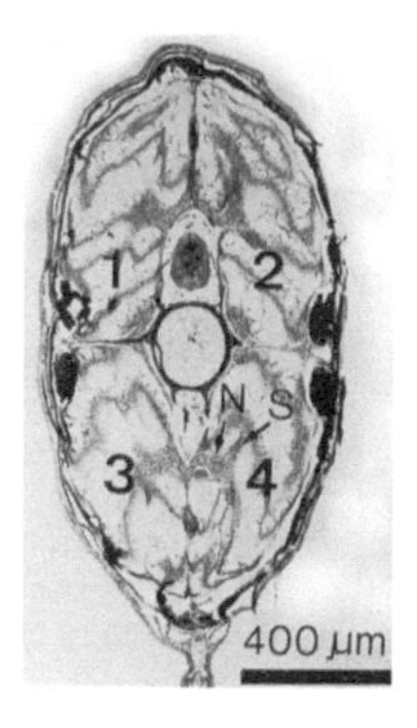

FIGURE E24b Cross section of the caudal peduncle showing the four columns (1–4) of electrocytes filling nearly the whole volume. The innervated stalks of the electrocyte (S) and the electric nerve (N) are indicated.

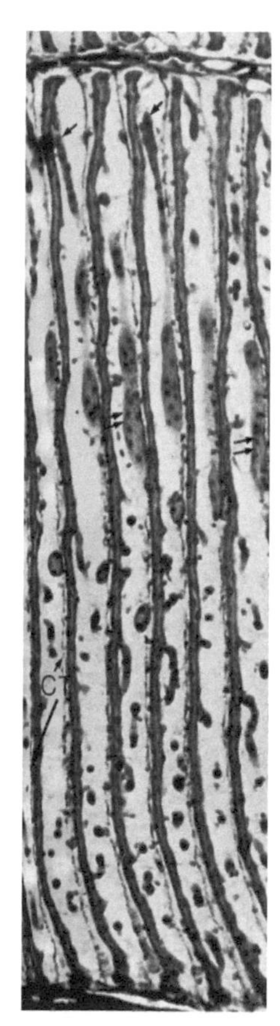

FIGURE E24c Sagittal section of a stack of electrocytes (E) from region 3 in Fig. E24a. Small stalks originate caudally (arrows) and join together at the center of the electrocytes to form the large stalk (double arrows), which receives the nerve terminals. Each electrocyte is surrounded by a connective tissue sheath (CT).

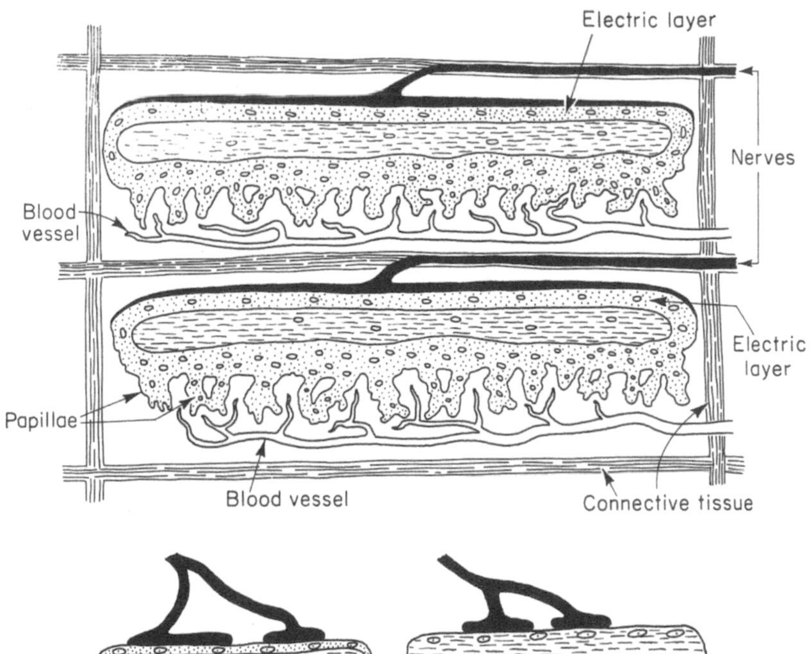

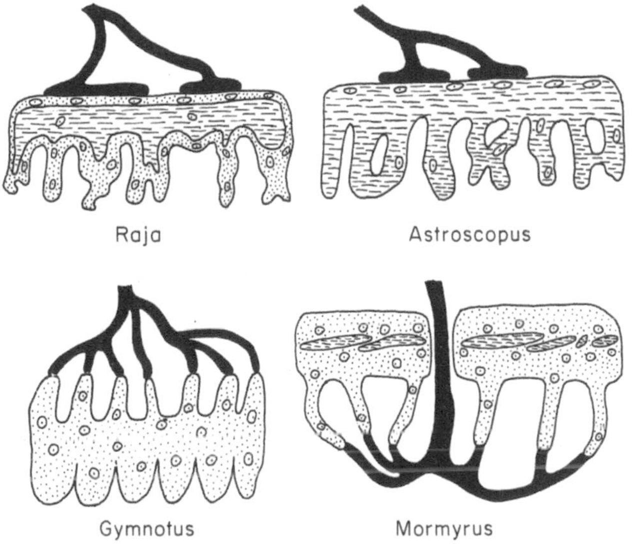

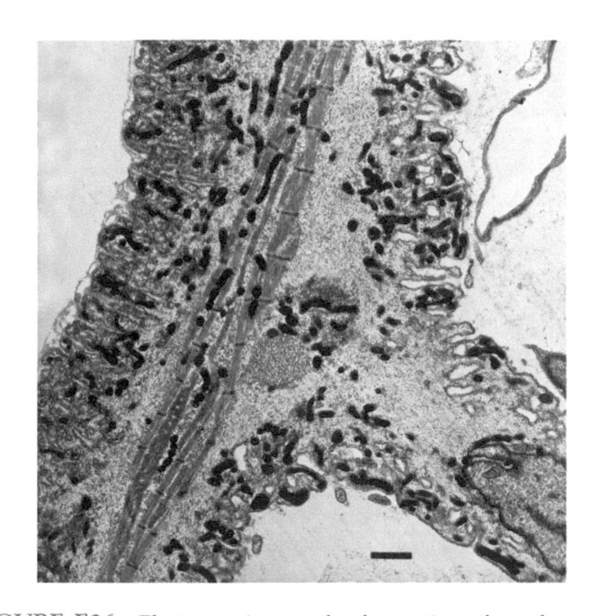

FIGURE E25 Structure of electrocytes. Two neighboring electrocytes are shown at the top. Four kinds of electrocytes are shown at the bottom. Cross-hatching indicates modified muscle, black is nervous tissue.

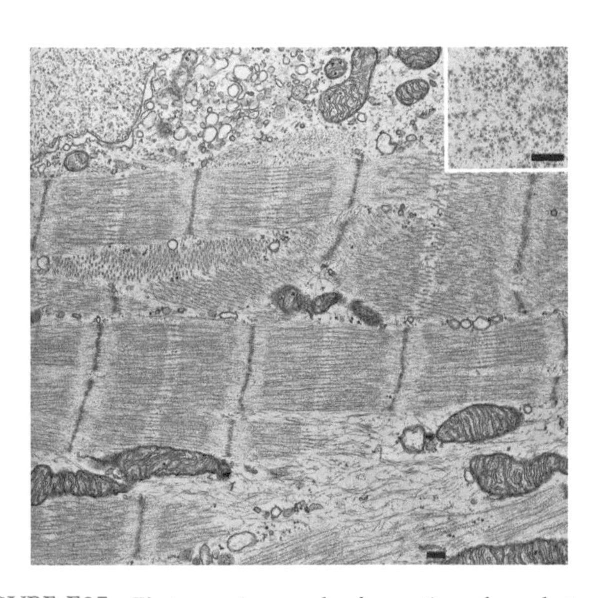

FIGURE E26 Electron micrograph of a section of an electrocyte of the mormyrid fish *Gnathonemus*. At the right a stalk projects from the innervated face; at the left the nonstalk face is slightly more proliferated than the stalk face. Mitochondria are interdigitated between the invaginations from both surfaces. Organized filaments with prominent *Z* bands run in the cytoplasm of the central region but do not extend into the stalk. 19,800 ×.

FIGURE E27 Electron micrograph of a section of an electrocyte of the mormyrid fish, *Gnathonemus*. The central region contains organized filaments possessing the band pattern of normal striated muscle although the *M* band is not clearly defined. The fibers run in all directions so that some are cut obliquely. Mitochondria, profiles of tubules (possibly a remnant of the sarcoplasmic reticulum), glycogen particles, and disordered filaments occur between the ordered filaments. 22,500 ×. The inset shows cross sections of thin and thick fibers in a rough hexagonal arrangement. 44,000 ×.

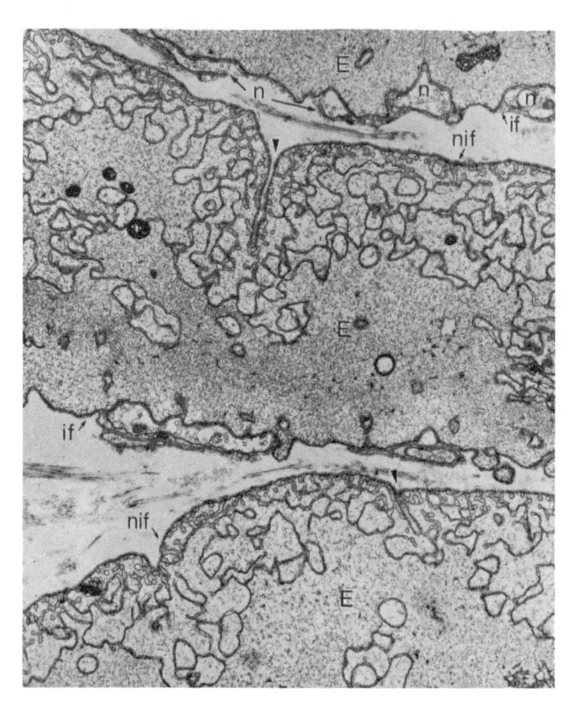

FIGURE E28 Electron micrograph of a section through electrocytes from the electric organs of the electric ray, *Discopyge tschudii*. The three cells are separated by connective tissue. The dorsal, noninnervated face (nif) exhibits narrow infoldings, which penetrate deep into the cell (arrowheads); nerve endings are applied to the surface of the ventral innervated face (if). 10,500 ×.

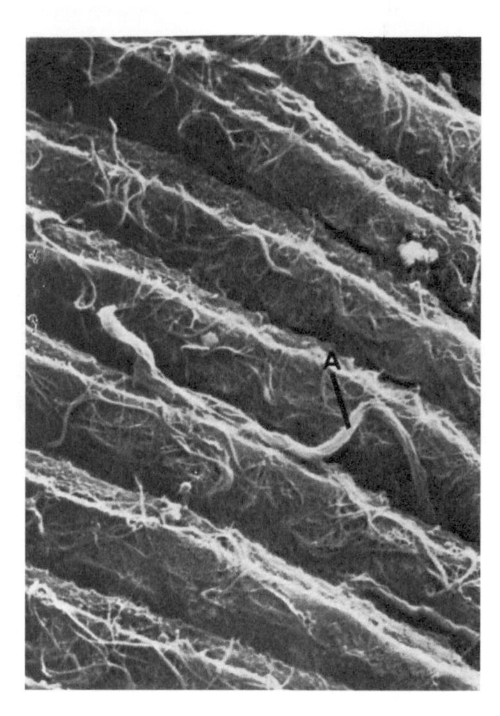

FIGURE E29a Scanning electron micrographs of the innervated ventral surface of stacked electrocytes of the ray, *Narcine brasiliensis*. *A*, large axon. 2,000 ×.

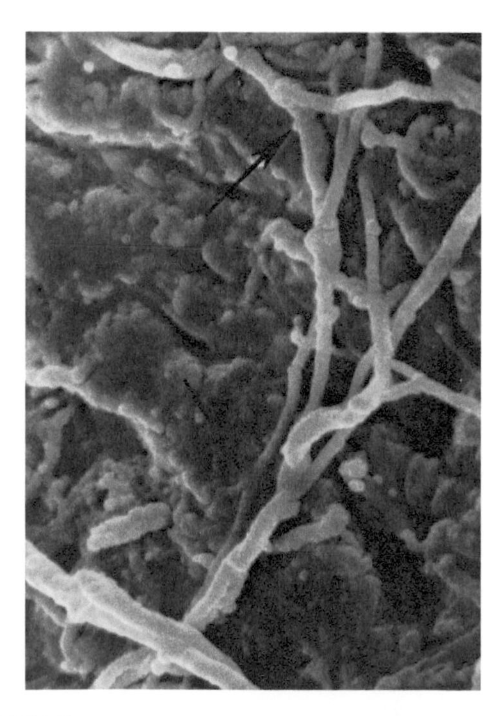

FIGURE E29b Detail of the innervated ventral surface of electrocytes demonstrating troughs and projections of the postsynaptic membrane and branching nerves (arrows) in close contact with the surface. 10,000 ×.

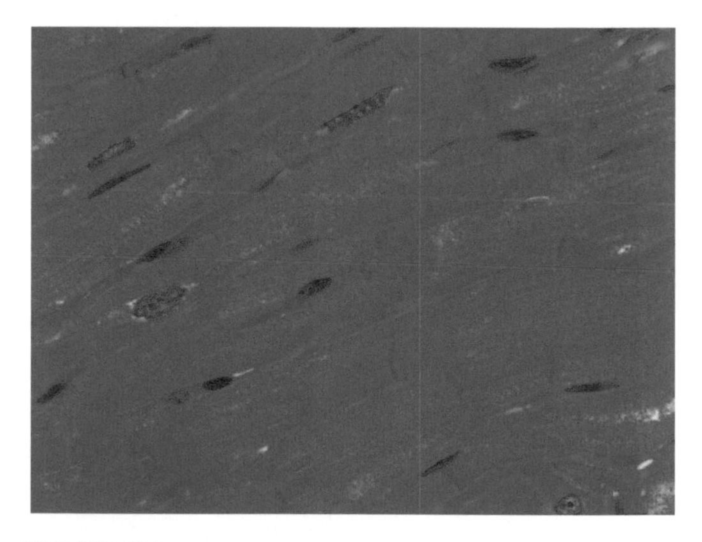

FIGURE E30a Semithin section of cardiac muscle of a monkey. Close examination reveals striations on the muscle fibers and the delicate interstitial connective tissue between the cells. A few intercalated discs join individual muscle cells end-to-end. Nuclei are central in the fibers. Branching of the fibers can be seen at the upper right. 63 ×.

FIGURE E30b Semithin section of a cross section of mammalian cardiac muscle fibers. The outlines of the cells are apparent. The nuclei of the endomysium lie between individual fibers. 63 ×.

FIGURE E31(a and b) Mammalian cardiac muscle.

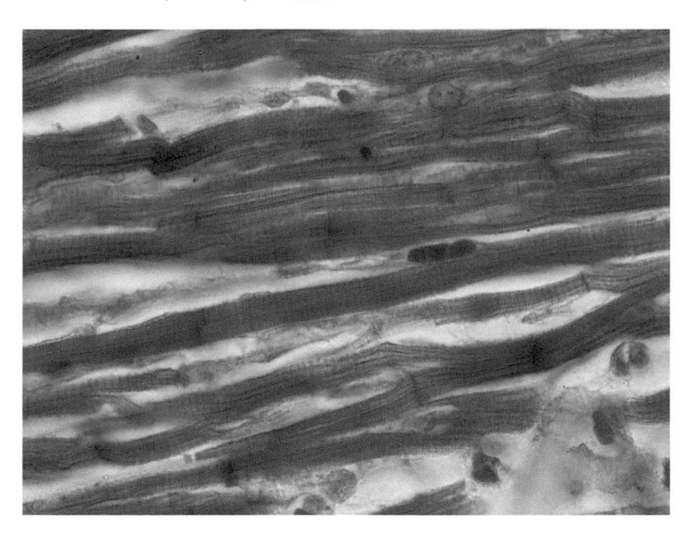

FIGURE E31a Paraffin section of mammalian cardiac muscle. Branching of the fibers can be seen more readily in this section. Striations and intercalated discs are apparent. Epimysium as well as small blood and lymph vessels occupy the spaces between the muscle fibers. 63 ×.

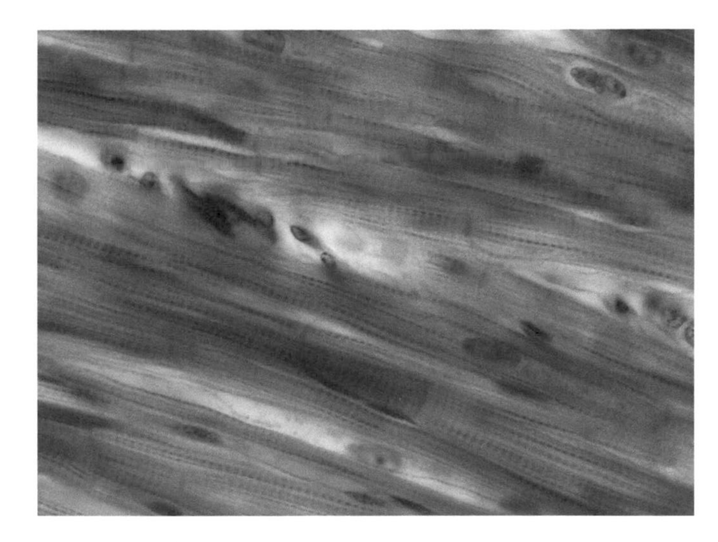

FIGURE E31b Section of cardiac muscle from a sheep, showing branching fibers, well-developed striations, intercalated discs as well as epimysium and blood vessel between the fibers. 63 ×.

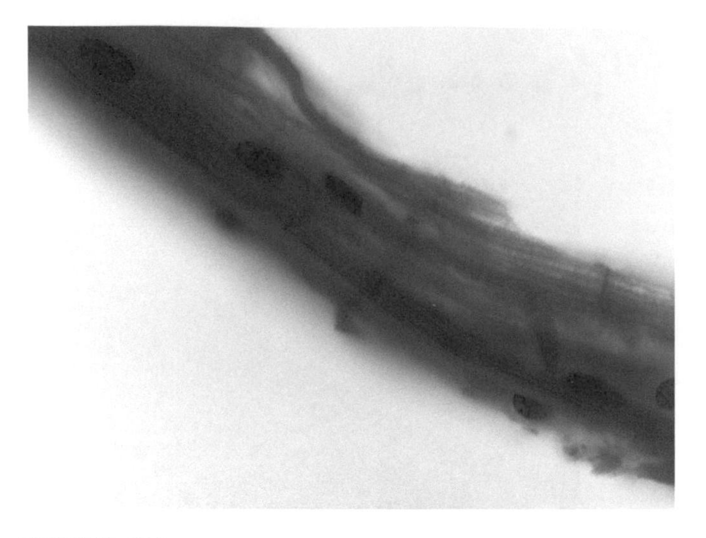

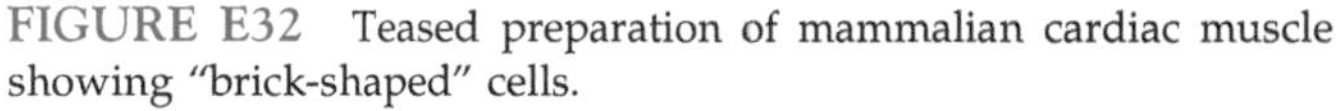

FIGURE E32 Teased preparation of mammalian cardiac muscle showing "brick-shaped" cells.

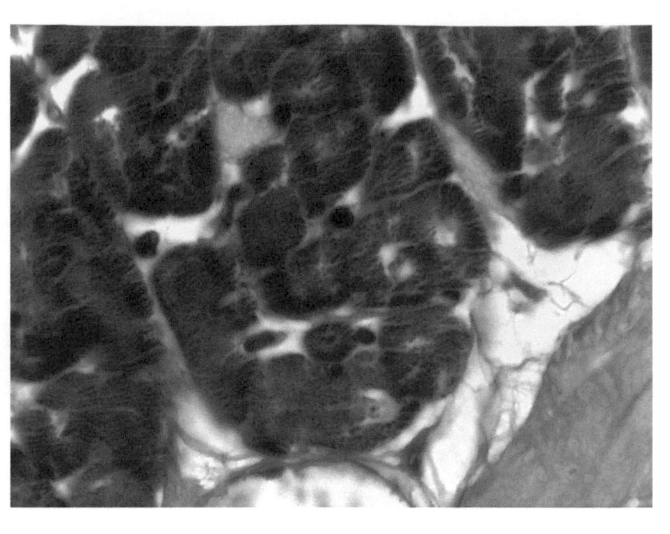

FIGURE E33 Paraffin section of mammalian cardiac muscle showing fibers in cross section.

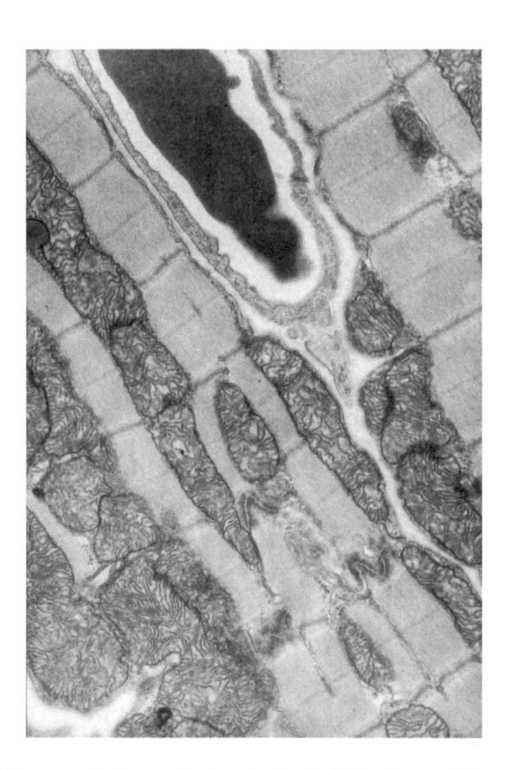

FIGURE E34 Electron micrograph of cardiac muscle from the heart of a bat. Portions of the cytoplasm of several cardiac muscle cells are shown. The cell at the upper right is separated from the lateral surfaces of the column of cells at the left by a thin layer of fine collagenous fibrils, which enclose a capillary containing an erythrocyte. Myofibrils run parallel to the long axis of the cell but, unlike the fibrils of skeletal muscle, these fibrils may branch. The sarcomeres show the same pattern of banding as skeletal muscle. The Z lines delimiting the sarcomere are prominent. The muscle is contracted and the I band regions have nearly disappeared and the A band occupies more of the length of the sarcomere. In contraction a dark band normally appears at the position of the H band. There are a large number of large mitochondria or sarcosomes aligned in columns between the myofibrils, presumably satisfying the cell's great need for ATP. Often lipid droplets lie alongside the sarcosomes (upper left). Unlike skeletal muscle, cardiac muscle is not a syncytium; the intercalated discs constitute strong barrier between cells laid end-to-end. Two cells are joined by intercalated discs at the level of the Z lines, although it does not always occur at Z lines that are in register. 30,500×.

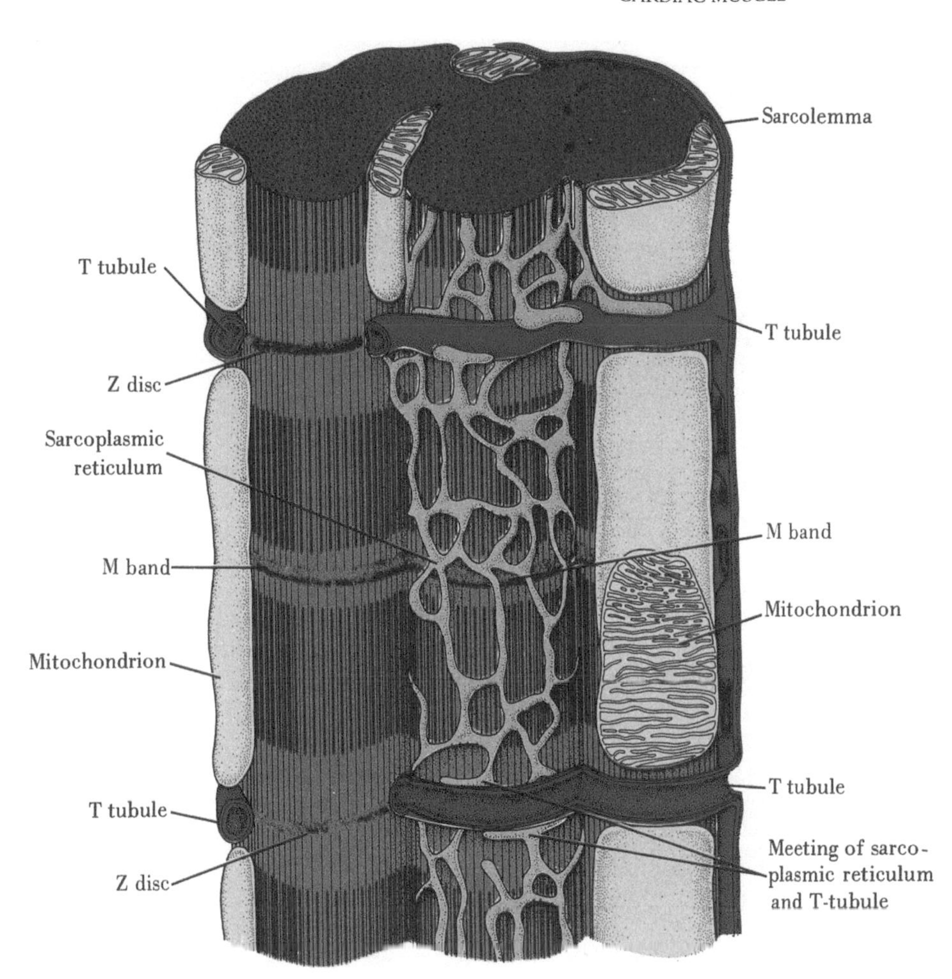

FIGURE E35 Diagram showing the organization of a mammalian cardiac muscle fiber. The transverse tubules are larger than those of skeletal muscle and carry an investment of basal lamina into the cell. They also differ in that they are located at the level of the Z disc. The portion of the sarcoplasmic reticulum adjacent to the T-tubule is not in the form of an expanded cisterna but is organized as an anastomosing network.

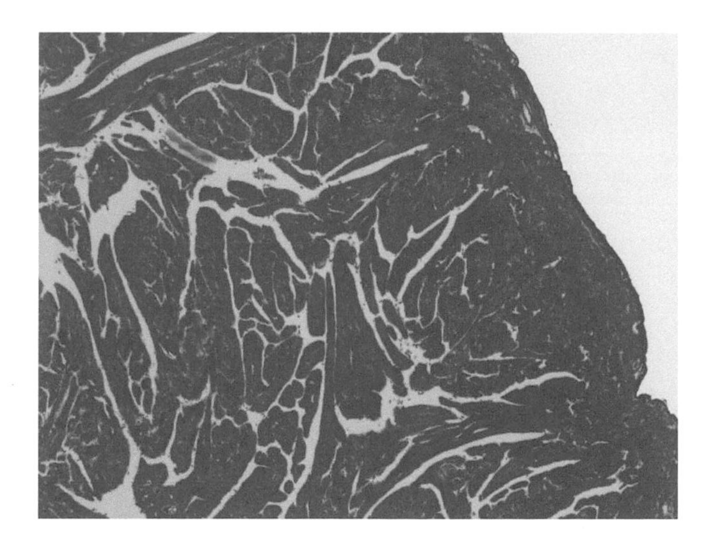

FIGURE E36 Section of dogfish heart. In both sharks and bony fish the heart resembles a sponge; the blood passing through nourishes the muscle and the squeezing of the muscle propels the blood forward to the ventral aorta. The blood spaces are lined with endothelium and contain a few nucleated red blood cells. 2.5 ×.

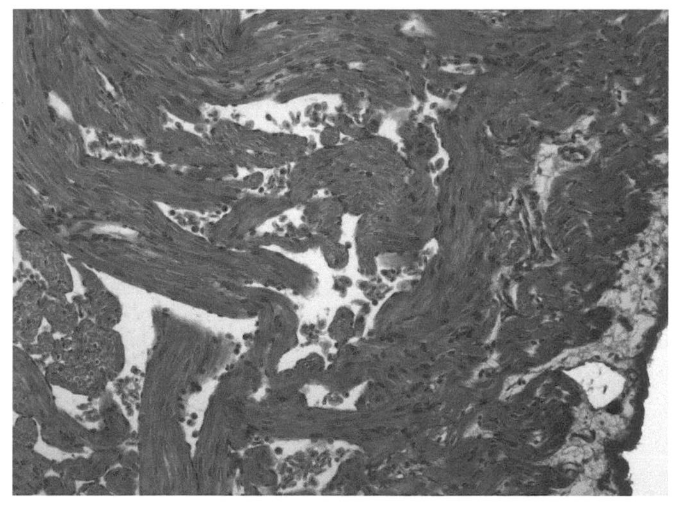

FIGURE E37 Section of heart of a garpike. As in sharks the heart of teleost fish resembles a sponge; the blood passing through nourishes the muscle and the squeezing of the muscle propels the blood forward to the ventral aorta. The blood spaces are lined with endothelium and contain a few nucleated red blood cells. 10 ×.

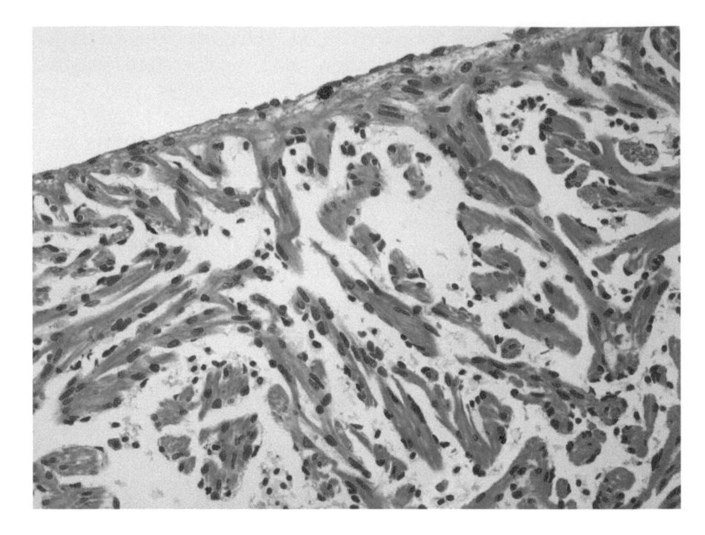

FIGURE E38 Section of cardiac muscle of the lungfish *Protopterus* sp. Bands of cardiac muscle are permeated by blood spaces. Contractions of this muscle forces blood through the arteries of the fish. 10 ×.

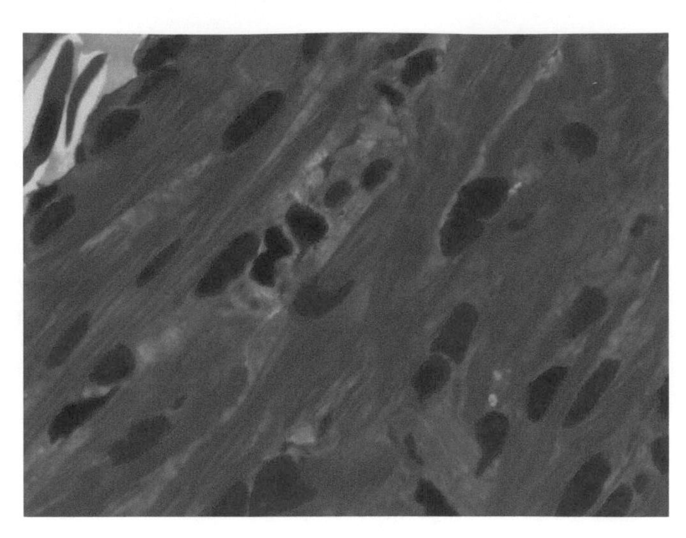

FIGURE E39 Cardiac muscle cells of the amphibian, *Amphiuma*, are similar to those of all vertebrates. 40 ×.

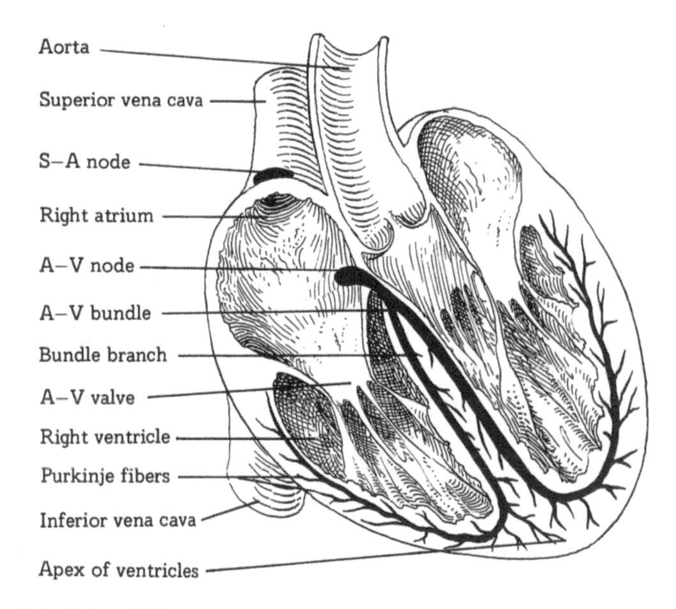

FIGURE E40 Diagram of the impulse-conducting system (Purkinje system) of the human heart. This heart has been cut open in the coronal plane to expose its interior and the main parts of its impulse-conducting system (indicated in black).

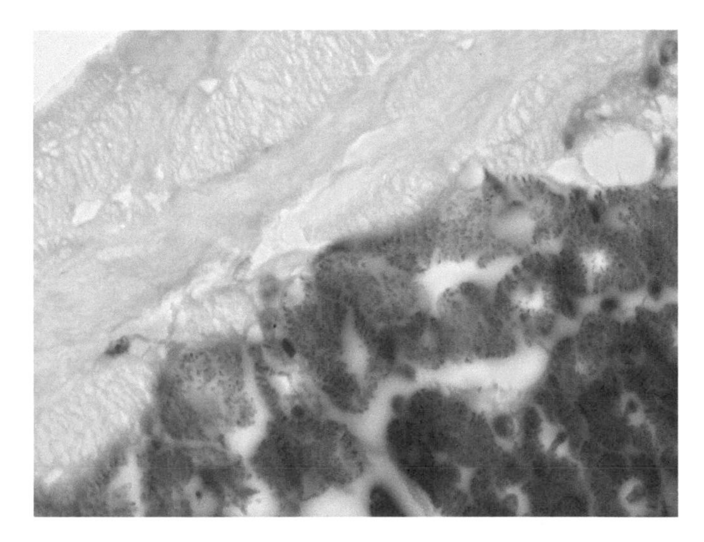

FIGURE E41 Conducting (Purkinje) fibers in beef heart. The sarcoplasm in the large spaces between the myofibrils is rich in glycogen. 63 ×.

CHAPTER

F

Nervous Tissue

Nervous tissue is massed in or near the central nervous system and sends out branches to innervate nearly all parts of the body. The CENTRAL NERVOUS SYSTEM (Fig. F1a) is composed largely of nerve cells, the NEURONS, and their cytoplasmic processes together with a specialized type of supporting tissue, the NEUROGLIA. The nerve trunks of the PERIPHERAL NERVOUS SYSTEM are composed of nerve cell processes enclosed in a sheath formed of neuroglial cells and bound together with delicate connective tissue. The GANGLIA (*Singular*: ganglion) that appear at various locations along nerve trunks are aggregations of nerve cell bodies and neuroglial cells. All neurons are enclosed by sheaths of neuroglial cells and are isolated from other neurons except at small points of contact, the SYNAPSES.

THE NEURON

A neuron consists of a nerve CELL BODY or PERIKARYON and its processes (Fig. F1b). It arises from a single embryonic cell, the NEUROBLAST, and retains its physical independence throughout life, connecting with other neurons only by contact at the synapses. From the central perikaryon extend two kinds of processes: DENDRITES which may be both numerous and much branched, and AXONS that are always single and generally unbranched except at their ends. Impulses are received at synapses on the dendrites as well as the surface of the cell body; the axon generates an ACTION POTENTIAL and conducts the impulse along its length. In many nerve cells, dendrites conduct toward the perikaryon while axons conduct away, but this is not always so.

Perikaryon

The perikaryon or nerve cell body may be seen in sections of spinal cord stained by the Nissl technique (Fig. F2a). In this survey view, notice the darkly staining neurons in the denser, central region of gray matter. At a higher magnification (Fig. F2b) note that the cell body is apparent but the processes have been cut off in sectioning so that only their stumps remain. Nerve cells may be viewed in their entirety in smears of nervous tissue (Fig. F3a). The single nucleus is EUCHROMATIC—i.e., it contains large amounts of pale-staining EUCHROMATIN, the extruded, genetically active form of chromatin. Its prominent nucleolus is visible. Sometimes SEX CHROMATIN (Barr body) is visible within the nucleus (Fig. F3b)—this feature of female mammalian nuclei appears in this section as a golf ball beside the basket ball–sized nucleolus. Basophilic clumps of CHROMOPHIL SUBSTANCE (Nissl substance) are scattered throughout the cytoplasm (NEUROPLASM) and are characteristic of neurons; note that they occur in the bases of the dendrites but avoid the axon, leaving an area of pale-staining cytoplasm, the AXON HILLOCK. Delicate NEUROFIBRILS, 0.5–3 μm in diameter, extend through the cell body and its processes but can be seen only when special stains are used (Fig. F4).

Compare the appearance of nerve cells stained with hematoxylin and eosin (Fig. F5) with those where special stains have been used (Figs. F4, F6, F7a, F7b) and try to decide the special purpose of each stain.[1] You may encounter the brownish pigment, LIPOFUSCIN, that accumulates in neurons with advancing age (Fig. F8). It probably represents insoluble remnants of metabolism and may occupy more than half of the perikaryon of some human brain cells.

[1]Impregnations of silver (and other metals), using methods similar to those of photography, are often used to delineate special features of nerve cells.

An Atlas of Comparative Vertebrate Histology.
DOI: https://doi.org/10.1016/B978-0-12-410424-2.00006-8

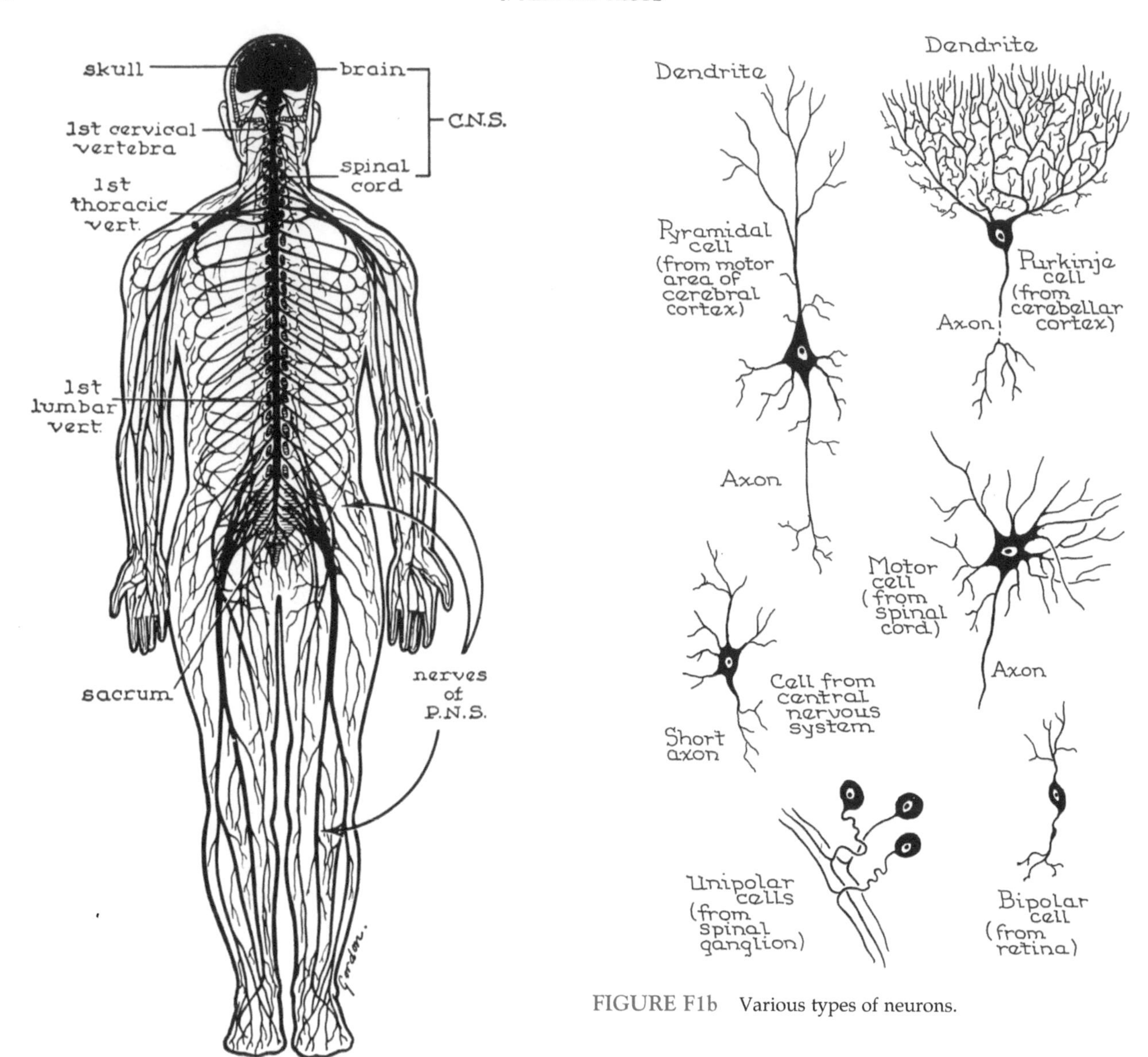

FIGURE F1b Various types of neurons.

FIGURE F1a Diagram of the human central (CNS) and peripheral (PNS) nervous systems.

Electron micrographs show that the perikaryon is densely packed with organelles (Fig. F9). The chromophil substance consists of dense masses of granular endoplasmic reticulum and clusters of free ribosomes. Fine NEUROFILAMENTS and microtubules (NEUROTUBULES) constitute the network of neurofibrils that extends throughout the neuron. Numerous mitochondria, fragments of the Golgi complex, and lysosomes are present. The neuron displays the characteristics of cells actively producing proteins. What are these proteins?

Nerve Cell Processes or Fibers

Nerve cell processes may be long or short, branched or unbranched (Fig. F1), and all are ensheathed by flattened NEURILEMMA CELLS that may or may not form an elaborate MYELIN SHEATH (Fig. F10).

1. *Myelinated nerve fibers.* Many axons possess a myelinated sheath throughout most of their length and are MYELINATED or MEDULLATED NERVES. Only the axon hillock and the terminal ramifications lack a myelin sheath. Myelin is responsible for the white appearance of the white matter of the brain and spinal cord.

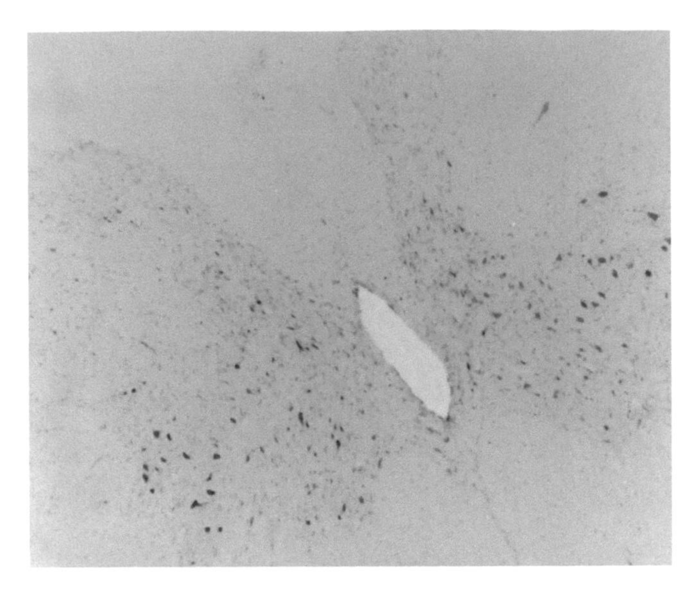

FIGURE F2a Cross section of mammalian spinal cord with Nissl stain. Neurons are deeply stained and are confined to the GRAY MATTER at the center of the cord. Outside the gray matter is the WHITE MATTER and at its center is the CENTRAL CANAL. 2.5 ×.

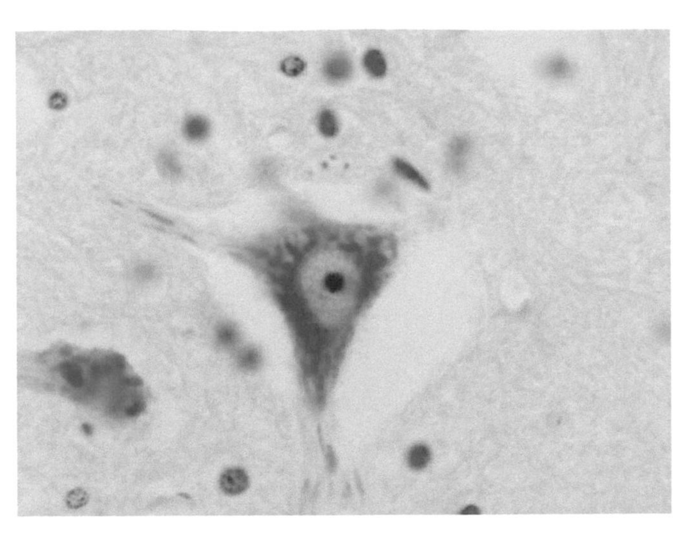

FIGURE F2b A neuron from the same slide as Fig. F2a. Clumps of basophilic material (Nissl substance) fill the cytoplasm. Processes extend in various directions from the cell body but are inevitably cut off during sectioning. 63 ×.

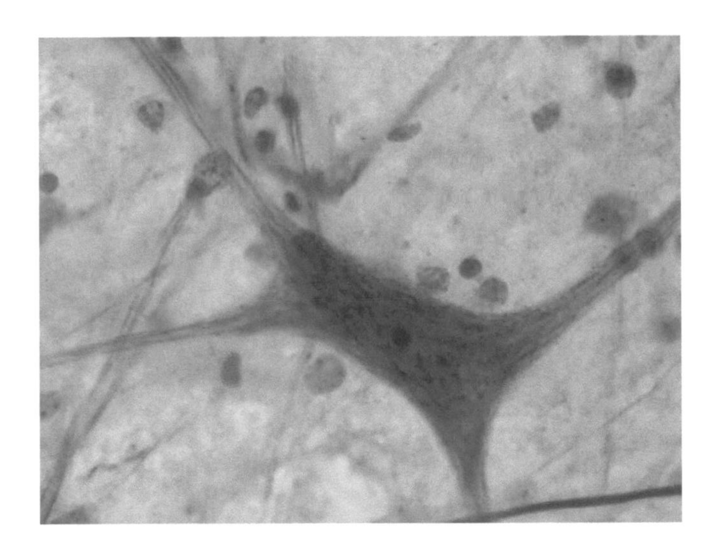

FIGURE F3a Neuron from a smear preparation of the gray matter from the spinal cord of an ox; the processes have not been cut off in sectioning. 63 ×.

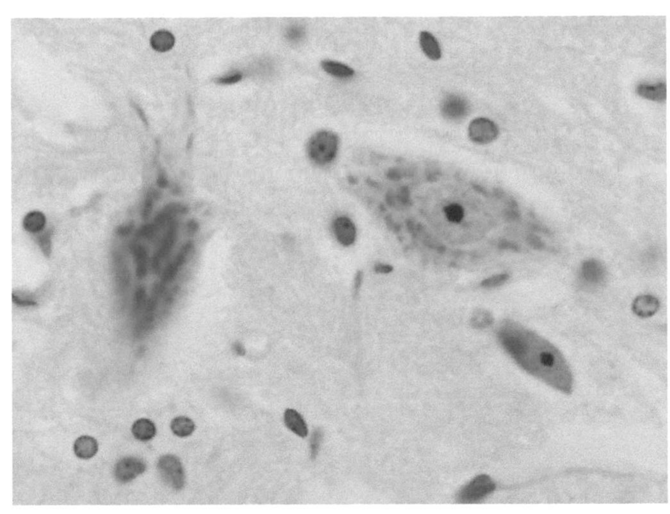

FIGURE F3b Section of a neuron from the gray matter of a female mammal. At one side of the nucleolus is a small mass of SEX CHROMATIN (the Barr body). 63 ×.

Individual myelinated fibers from the sciatic nerve trunk of a rat has been teased apart in the preparation of Fig. F11 and the myelin sheath has been stained black with osmium tetroxide. A nerve trunk is composed of many cylindrical fibers bound together by delicate connective tissue. Note the central axis cylinder and the thick surrounding myelin sheath; the axis cylinder is a protoplasmic extension of the perikaryon. The plasma membrane of the axon is called the AXOLEMMA, and its cytoplasm is the AXOPLASM. The myelin sheath represents the inner portions of the neurilemma cells (Schwann cells) whose nuclei and cytoplasmic mass occur at the periphery of the fiber. The myelin sheath is constricted at intervals of 0.08–0.1 mm, the NODES (of Ranvier),

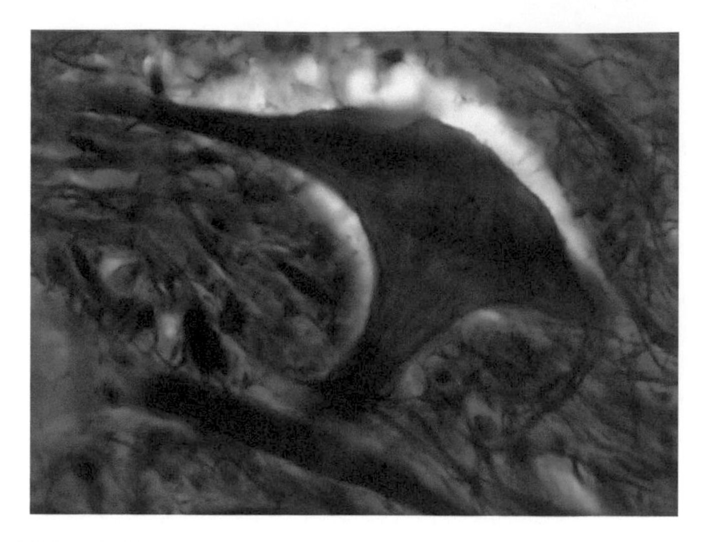

FIGURE F4 Section of mammalian gray matter blackened with a silver stain. 100 ×.

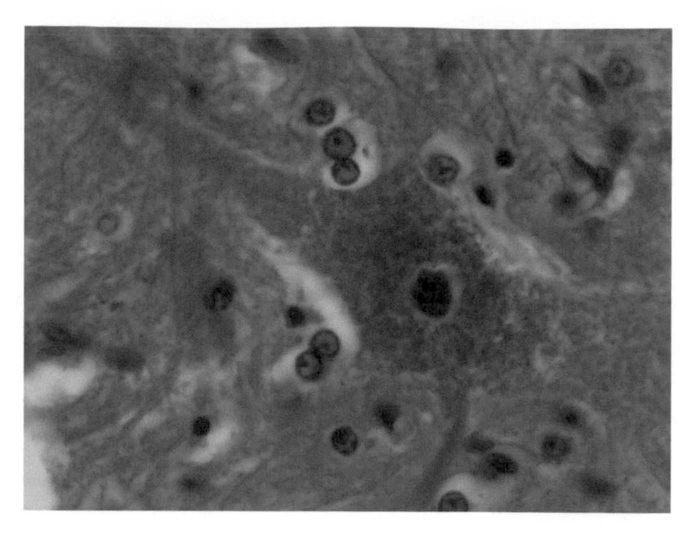

FIGURE F5 Section of mammalian gray matter stained with hematoxylin and eosin. 63 ×.

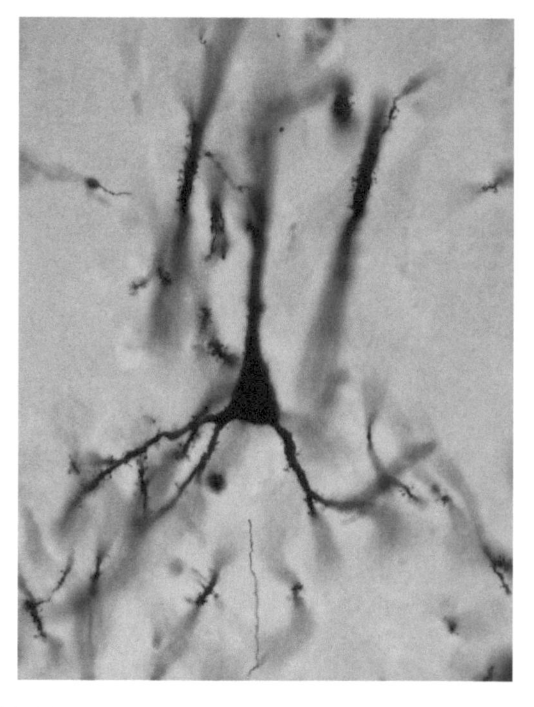

FIGURE F6 Pyramidal neuron from a section of the cerebrum of a cat. Silver stain. 40 ×.

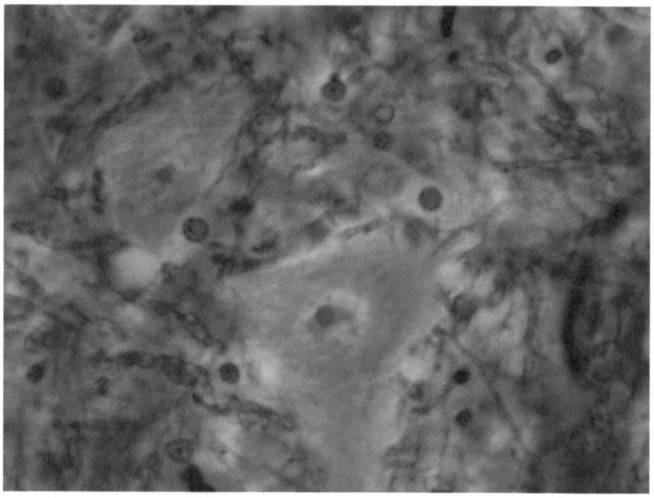

FIGURE F7a Multipolar neurons in a section of mammalian spinal cord stained with iron hematoxylin and fuchsin. 63 ×.

and the outer layer of the neurilemma dips in to touch the axis cylinder. Only one neurilemma cell ensheaths the INTERNODE between two nodes. Oblique interruptions, the INCISURES (clefts of Schmidt-Lantermann) occur between the nodes.

Electron micrographs of cross sections of a myelinated nerve trunk (Figs. F12 and F13) presents the striking appearance of the end of a tightly wound scroll. It has been determined that the nerve fiber becomes completely enveloped by a neurilemma cell and is suspended in a sling of the plasma membrane of the neurilemma cell, the MESAXON (Fig. F14a). As development continues the neurilemma cell elaborates sheets of plasma membrane that wind round and round the fiber forming concentric layers (Fig. F14b). The cytoplasm between the membranes is squeezed out and the inner leaflets fuse forming a dark line, the MAJOR DENSE LINE (Fig. F12). The outer leaflets of adjacent membranes also fuse forming the less distinct INTRAPERIOD LINE.

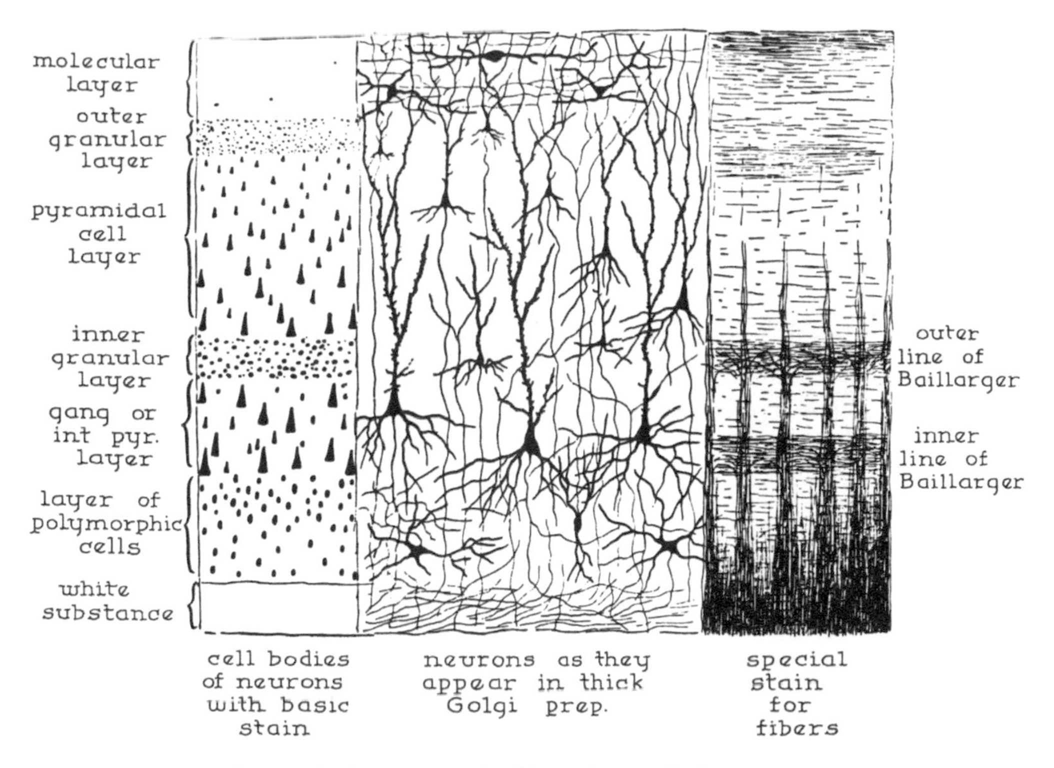

FIGURE F7b Diagram of the layers of the cerebral cortex as stained by various methods.

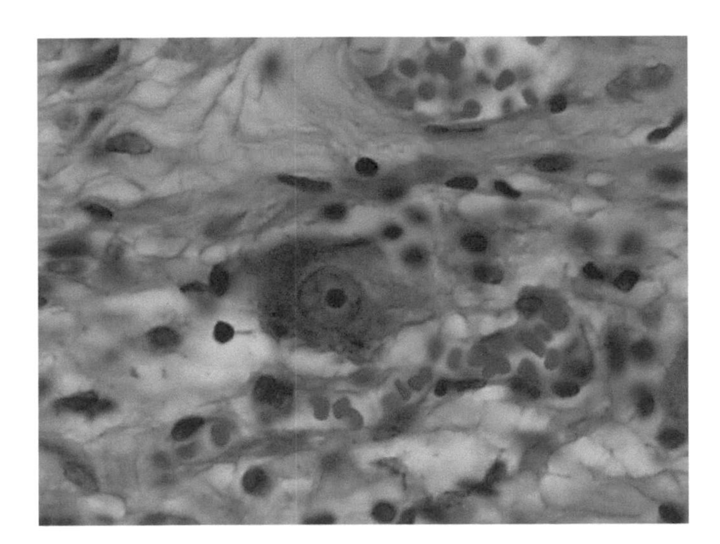

FIGURE F8 Section stained with hematoxylin and eosin of dorsal root ganglion from a mammal. The cell contains an abundance of brownish pigment, LIPOFUSCIN, the "age pigment."

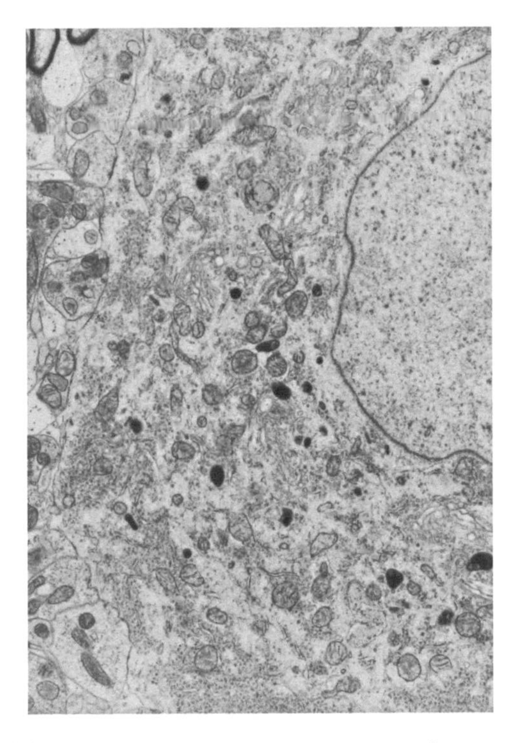

FIGURE F9 Electron micrograph of a section of a motor neuron from the spinal cord of a bat. The nucleus of this large cell is at the upper right. In the region outside the nucleus, the PERIKARYON, the clumps of CHROMOPHIL SUBSTANCE of light microscopy, are seen to be dense masses of granular endoplasmic reticulum and clusters of FREE RIBOSOMES. The GOLGI REGION is at the lower right, immediately below the nucleus. A few small MITOCHONDRIA are scattered here and there. The black blobs distributed throughout are LYSOSOMES. 16,000 ×.

Because all other material has been excluded by this fusion, the myelin sheath consists only of the material from the plasma membrane of the neurilemma cell. The nucleus and cell body of the neurilemma cell remain at the periphery of the sheath where scattered mitochondria and elements of the Golgi complex may be observed (Fig. F15). The neurilemma is completely covered by a basal lamina. In electron micrographs of longitudinal sections of myelinated nerves note that

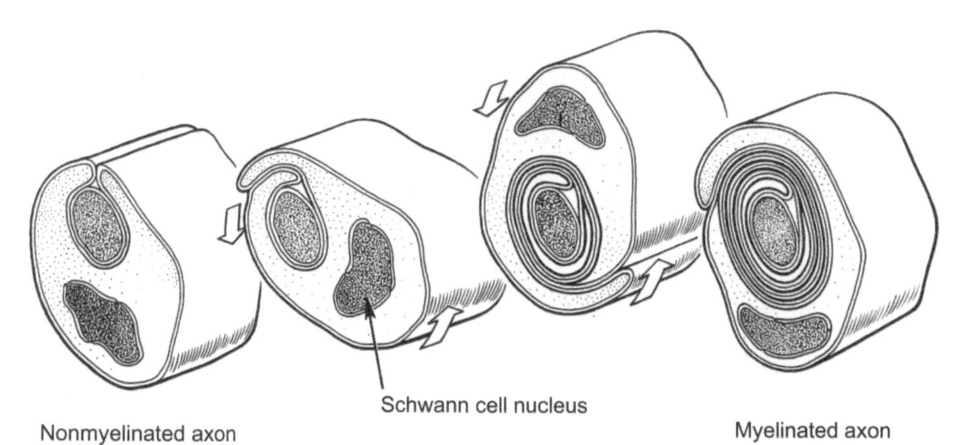

FIGURE F10 Diagrams to illustrate the formation of myelin around an axon. A NEURILEMMA CELL (Schwann cell) engulfs an axon. The axon retains its contact with the outside world by a suspensory sling, the MESAXON. The membranes of the mesaxon proliferate and fuse, forming the future INTRAPERIOD LINE of myelin at the line of fusion. The membranes continue to proliferate, winding round and round the axon. Eventually the layers of membrane constitute the MYELIN SHEATH of the nerve process. When a NONMYELINATED NERVE is formed, a neurilemma cell engulfs the process but the winding of membrane does not occur.

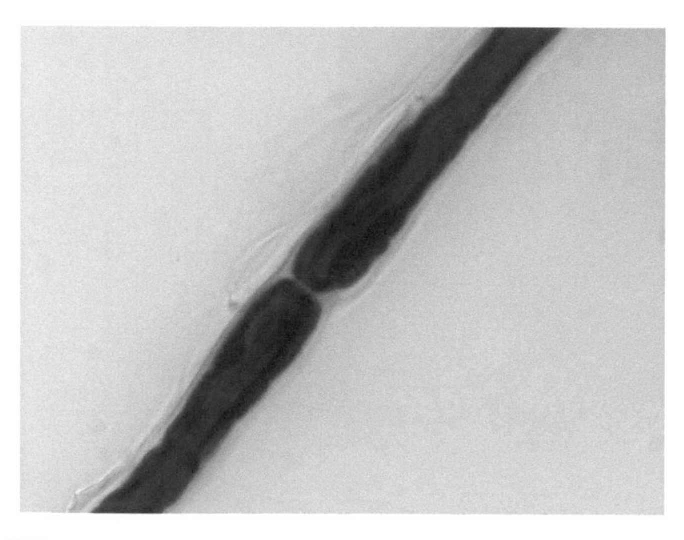

FIGURE F11 Whole mount of a single nerve fiber from the sciatic nerve trunk of a rat. The myelin sheath has been stained black by the vapors of osmium tetroxide. A NODE (of Ranvier) intrudes on the myelin sheath about halfway along the nerve fiber. The central AXIS CYLINDER is seen. The nucleus of one of the neurilemma cell presses on the myelin sheath about one-third of the distance from the bottom. 63×.

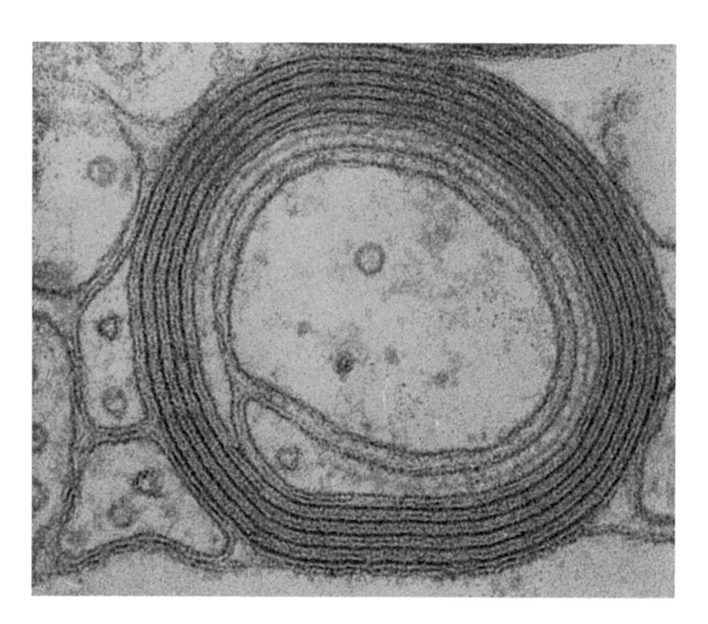

FIGURE F12 Electron micrograph of a cross section of a myelinated nerve fiber from the brain of a rat. In the brain, myelin sheaths are formed by neuroglial cells. Because of shrinkage in ordinary preparations, the axis cylinder appears thinner than is shown here. The growing lip of cell cytoplasm at the lower left is winding around the axis cylinder. As it winds, it intrudes itself into the space between the plasma membrane of the axon and the thin layer of cytoplasm left behind by the growing lip during its previous turn. This cytoplasm disappears as the inner leaflets of its plasma membrane fuse to form the MAJOR DENSE LINE. The outer leaflets of the plasma membrane surrounding the lip fuse on the next turn and form the less dense INTRAPERIOD LINE of the sheath. 166,000×.

the myelin sheath does not extend over the nodes; these are covered by a single thin layer formed by an extension of the cytoplasmic margins of the neurilemma cells (Figs. F16a and F16b). In these figures note also the presence of neurotubules, neurofilaments, endoplasmic reticulum, and mitochondria in the axoplasm.

Fig. F17 is a diagram of an imaginary unrolled neurilemma cell. Note that the fusion of the plasma membranes is not complete in the flattened area. Note the outer CELL BODY containing the nucleus and cytoplasm. On each side of the flattened area are the PERINODAL CYTOPLASMIC CHANNELS that form the loops seen in electron micrographs at each side of the node. The INCISURES (of Schmidt-Lanterman) are channels that extend from the cell body to the inner collar of cytoplasm that engulfs the axon (Fig. F18). Although there are no neurilemma cells in the central nervous system, the myelin sheaths are formed by neuroglial cells where one neuroglial cell forms several myelin sheaths whose structure is similar to that of the peripheral nerves (Fig. F19).

Micrographs of transverse and longitudinal sections of a nerve trunk show hundreds of fibers, each composed of a nerve process, some wrapped in a myelin sheath, some not (Figs. F20 and F21). The myelin sheath, being fatty in nature, is dissolved out of sections prepared by many methods and appears as a clear space around the axis cylinder. The axis cylinder is a protoplasmic extension of the perikaryon and in life is several times thicker than the myelin sheath; many fixatives, however, shrink it to a thin thread. In the lower powered cross section observe the grouping of fibers into FASCICULI or bundles bound together by loose, vascular fibroelastic connective tissue. The connective tissue immediately around each fasciculus is the PERINEURIUM, around all the fasciculi the EPINEURIUM, and between individual fibers the ENDONEURIUM. (Note that this terminology parallels that used for muscle fibers.)

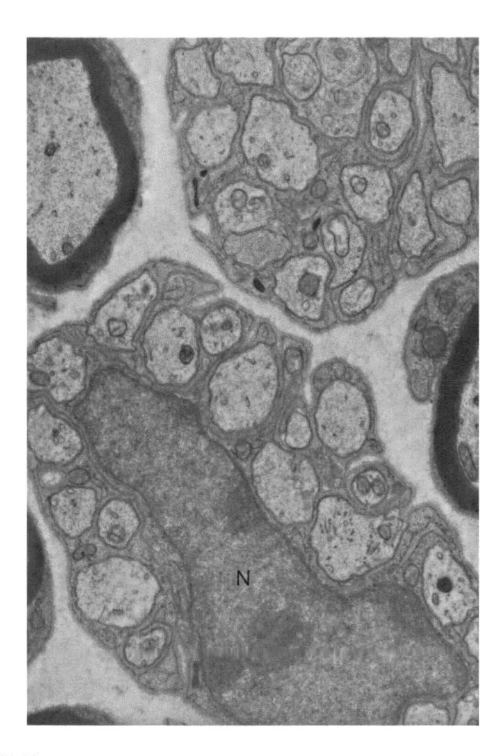

FIGURE F13 Electron micrograph of a section through peripheral nerve fibers in the skin of a mouse. Both myelinated and nonmyelinated fibers are contained within the same nerve trunk. A single neurilemma cell, with its large nucleus, engulfs many slow-conducting fibers in the lower half of this micrograph but each myelinated nerve (upper left, mid right) is enclosed by spiral laminae of its own neurilemma cell. Mesaxons connect nonmyelinated fibers with the surface of the nerve trunk. The enclosing lip of the neurilemma cell is seen on the myelinated nerve on the right. Each neurilemma cell is enclosed by a basement membrane that separates it from the connective tissue of the endoneurium. 27,500 ×.

2. *Nonmyelinated fibers*. Peripheral nerve trunks contain both myelinated and nonmyelinated fibers. Nonmyelinated nerves are found also in the motor nerves of the visceral smooth muscles and the sensory nerves of pain, taste, and smell. Electron micrographs of cross sections of nonmyelinated fibers show that several axons may be enveloped by a single long neurilemma cell and that they occupy separate troughs invaginated along its length (Fig. F22). The lips of the trough form a MESAXON but fail to fuse so that a tiny channel connects one side of the cavity to the exterior. Nonmyelinated fibers of the gray matter of the brain and spinal cord are similarly ensheathed by neuroglial cells.

   Nonmyelinated fibers can be seen readily with the light microscope in sections of the brain and spinal cord (Figs. F23a and F23b) but those of the peripheral nervous system are more difficult to locate. Note the star-shaped sympathetic ganglion in the micrograph of a silver-stained whole mount of the muscle layers of the gut (Fig. F24). The appearance of the nonmyelinated fibers is disappointing in sections of various organs and the inexperienced histologist will have trouble finding them. Ganglia may be more easily identified between the muscle layers of the wall of the gut (Fig. F25) and in the connective tissue around the secretory portions of glands. They stain paler than muscle or connective tissue and are delimited from the surrounding tissue by a definite sheath.

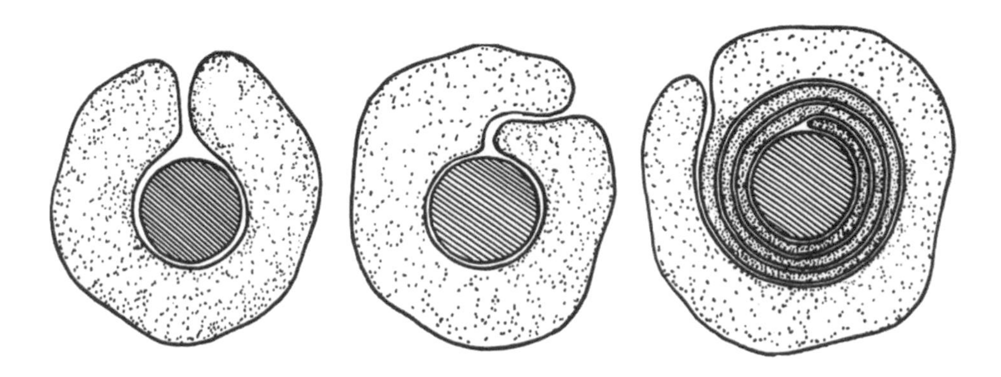

FIGURE F14a Diagram to illustrate the early stages in the formation of a myelin sheath by a neurilemma cell in the peripheral nervous system. The nerve fiber is engulfed by a neurilemma cell and is connected to the surface by the mesaxon. The mesaxon continues to produce more membrane resulting in a many-layered sheath around the axis cylinder.

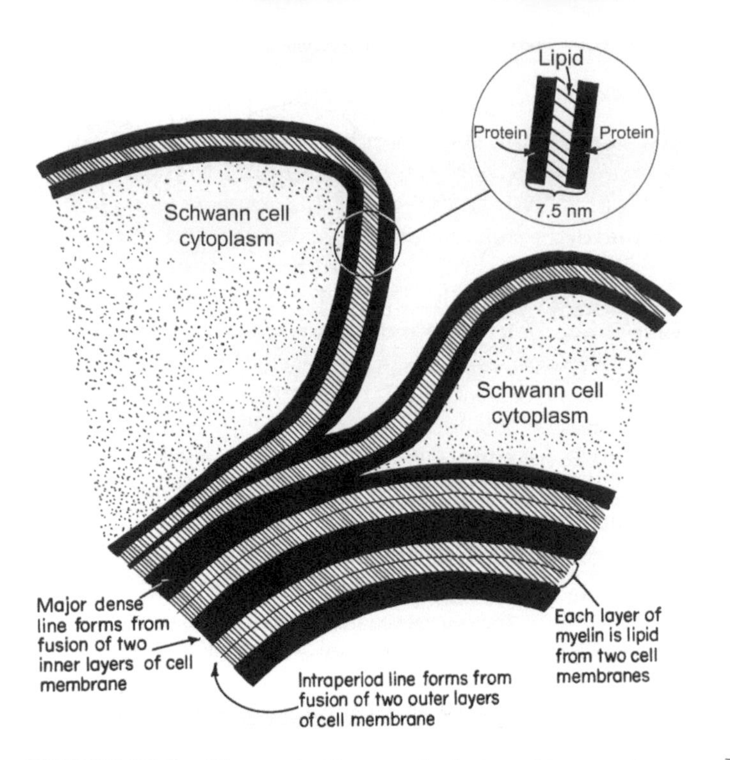

FIGURE F14b Diagram to illustrate the fusion of the membranes of a neurilemma cell to become a myelin sheath. The myelin sheath is pure membrane in a "jelly-roll" configuration.

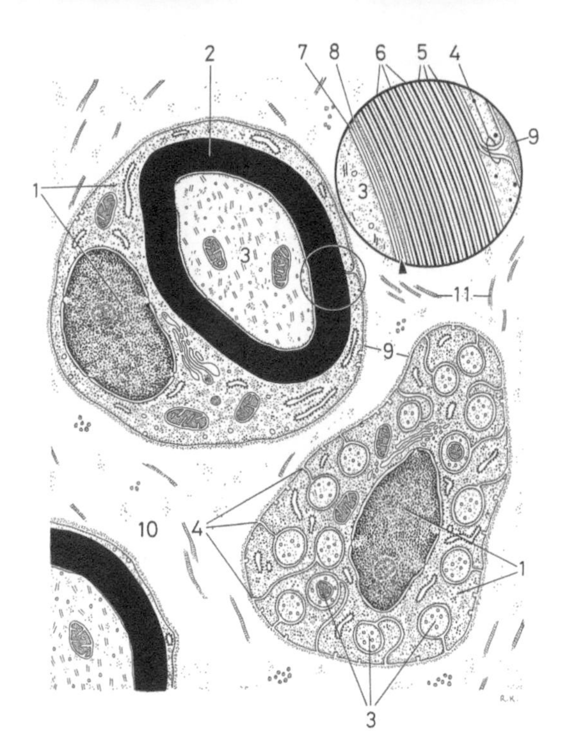

FIGURE F15 Drawings to review our observations on the structure of nerve trunks in vertebrates. A MYELINATED NERVE at the upper left is engulfed by a neurilemma cell (9), which encloses it with a many-layered myelin sheath (2) whose layers are not resolved at this magnification. The formation of the major dense line and the intraperiod line by proliferation of membrane from the mesaxon is shown at a higher magnification at the upper right (5-8). The neurilemma cell at the lower right engulfs several NONMYELINATED FIBERS (3) each of which is provided with a mesaxon (4).

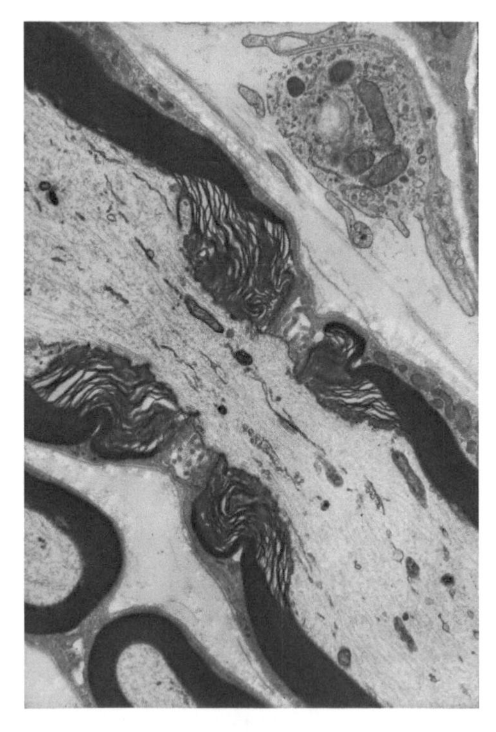

FIGURE F16a Electron micrograph of a longitudinal section of the sciatic nerve of a mouse. A single neurilemma cell encloses a limited length of a nerve fiber and a nerve trunk of any length is enclosed by a succession of several of these cells. The periodic interruptions between successive neurilemma cells are the NODES (nodes of Ranvier). On each side of a node some cytoplasm remains in the extended margins of the sheath layers forming folds that enclose the fiber. At the node itself, finger-like processes from the neurilemma cells interdigitate to cover the nodal area. 22,500 ×.

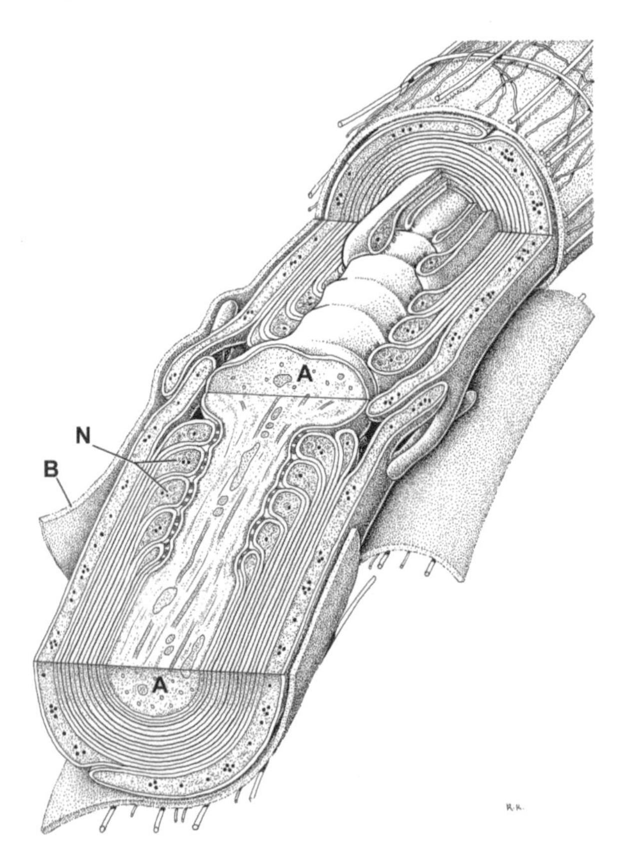

FIGURE F16b Idealized drawing of a longitudinal section of a node from a myelinated nerve. The nerve is enclosed by a basal lamina and endoneurium (B). At the node the axonal process is thickened (A) and is covered by finger-like processes from the adjacent neurilemma cells. Thickened borders (N) of the neurilemma cells enclose the adjacent axon.

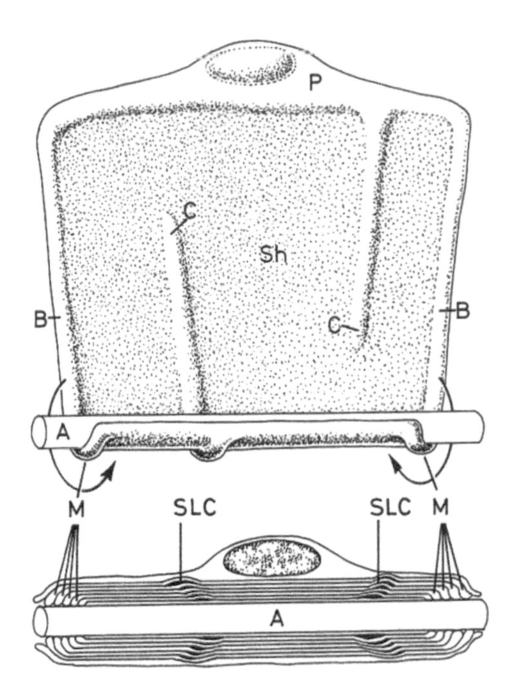

FIGURE F17 This is a fanciful drawing of an unrolled neurilemma cell that appears as a thin trapezoidal sheet (Sh). In the course of wrapping around an axon, all cytoplasm is extruded from most of the cell leaving cytoplasm in thickened portions: the elongated perikaryon (P), the thickened borders (B) at each side of the node, and the cytoplasmic channels (C) that become incisures (SLC). M, mesaxons.

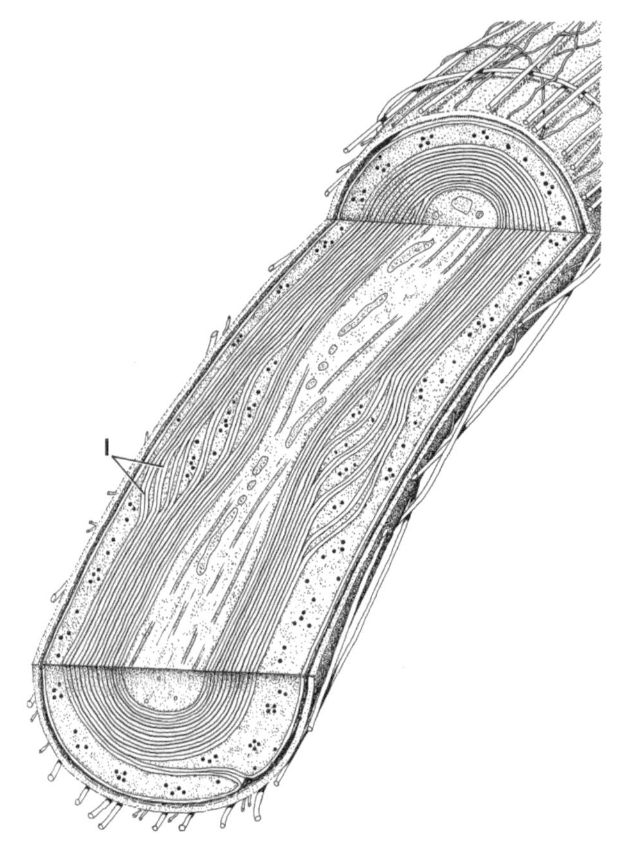

FIGURE F18 Idealized drawing of a longitudinal section of a myelinated nerve showing incisures (of Schmidt-Lanterman). Separations in the myelin at the level of the major dense line (I) produce the slanting "cracks" in the myelin sheath that are seen with the light microscope.

Cell Shape

The shape of nerve cells is dependent upon the number and arrangement of the cell processes (Fig. F1b).

1. *Pseudounipolar (unipolar) cell.* Note the large sensory pseudounipolar cells in sections of dorsal root ganglia (Fig. F26). A single process represents the union of two processes, one bringing impulses to the perikaryon, the other carrying them away. (Fig. F27; Since these are sections, it is unlikely that the process will be visible.) The nucleus is large and vesicular. Between the cells note numerous myelinated fibers and supporting cells. Each cell body is surrounded by a double layer of cells, an outer one composed of fibrocytes (CAPSULE CELLS) and connective tissue fibers continuous with the endoneurium, and an inner layer of flat SATELLITE CELLS lying directly on the sensory cell body.
2. *Bipolar cell.* This is a spindle-shaped cell with a process at each end (Fig. F28). Bipolar cells are found in the acoustic ganglion, in the retina of the eye, and in olfactory neurons, and will be considered later.
3. *Multipolar cells.* Multipolar cells are the commonest type (Figs. F29 and F30). They have several processes and have been seen already in the motor neurons of the spinal cord that have an axon and relatively few simple dendrites. More elaborate multipolar cells can be seen in the brain; the following are two examples:
   a. *Pyramidal cells* (Fig. F31). In sections of the cerebral cortex prepared by silver impregnation, note the pyramidal cells with their apices pointed toward the surface (Fig. F32a). One dendrite emerges from the cell at its apex and four or more extend outward from the base; the axon originates at the base and passes into the white matter. Note that the surfaces of many dendrites are covered with minute spines or GEMMULES resulting in a test-tube brush appearance (Fig. F32b). These spines serve as SYNAPTIC ORGANS: they are points of contact with the processes of other nerve cells. One of the characteristics placing mammals above all other animals and man as the highest mammal is the degree of elaboration of the pyramidal cells.

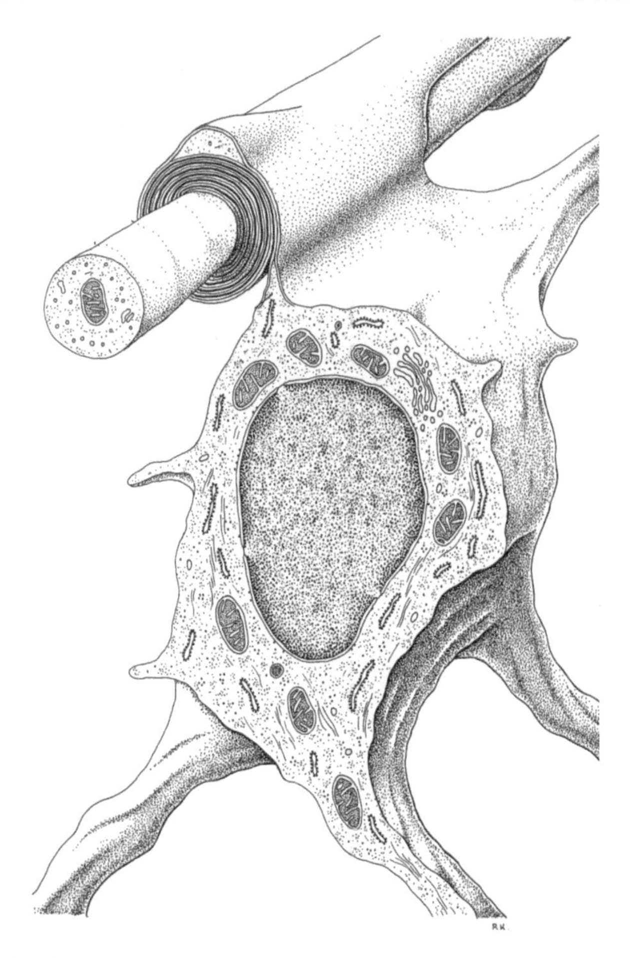

FIGURE F19 The myelin sheath formed in the central nervous system is the same as that of the peripheral nervous system but its method of formation is slightly different: it is produced by oligodendroglial cells, which can isolate several processes simultaneously. In this three-dimensional representation a process from an oligodendrocyte has wrapped around an axon, forming a myelin sheath. Other processes on this cell are able to form similar sheaths.

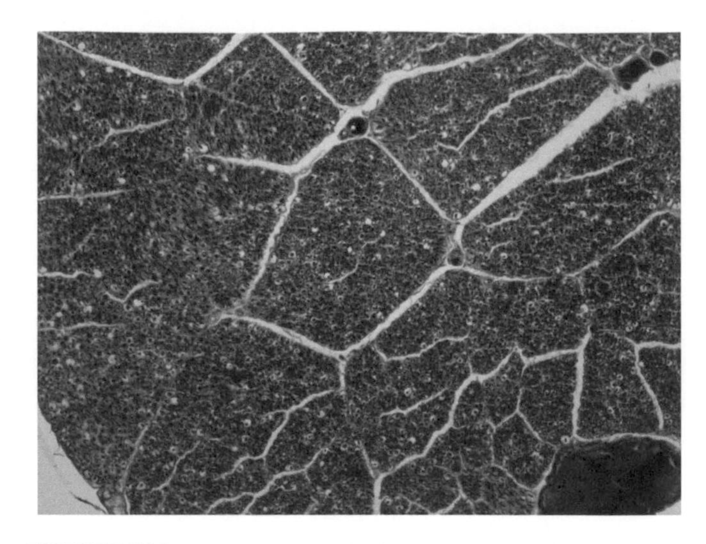

FIGURE F20 Light micrograph of a cross section of a mammalian nerve trunk containing hundreds of fibers. Many of the nerves are myelinated and their dark axis cylinders can be seen within the white myelin sheath. Within the nerve trunk, individual fibers are bound together by ENDONEURIUM. Large numbers of nerve fibers are bound into FASCICULI by PERINEURIUM, and the entire nerve trunk enclosed within EPINEURIUM. 10 ×.

FIGURE F21 Longitudinal section of a mammalian nerve trunk. The dark axis cylinders are visible in many of the nerve fibers. The lighter purple is largely myelin. 63 ×.

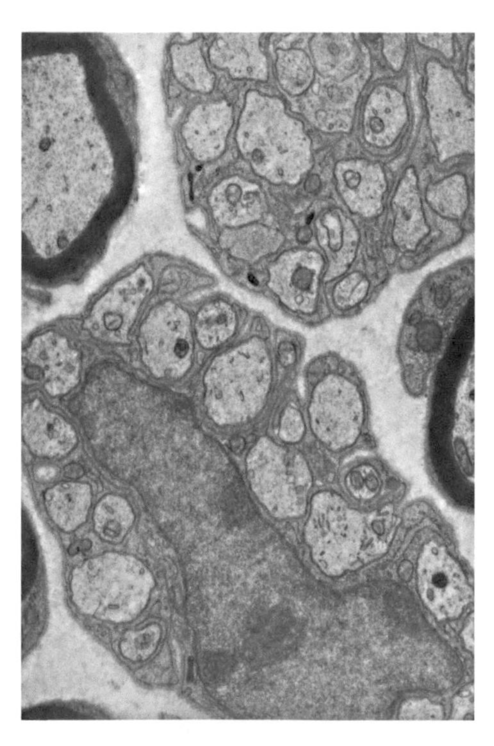

FIGURE F22 Electron micrograph of a cross section of a peripheral nerve trunk from the skin of a mouse. Most of the field is occupied by two bundles of slow-acting nonmyelinated neurons held together by separate neurilemma cells. Single myelinated fibers at the edges of the field are surrounded by individual neurilemma cells that formed the myelin sheaths. 27,500 ×.

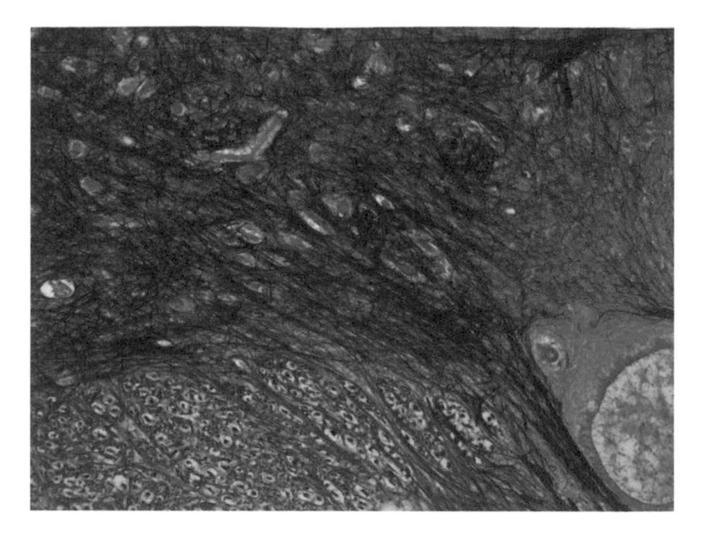

FIGURE F23a Section of mammalian spinal cord with a silver stain. Gray matter showing many cross sections of myelinated nerve fibers, is at the bottom; a few black nonmyelinated fibers course between these fibers. White matter is at the top and consists largely of black nonmyelinated fibers. A few neurons occupy spaces between them. 10 ×.

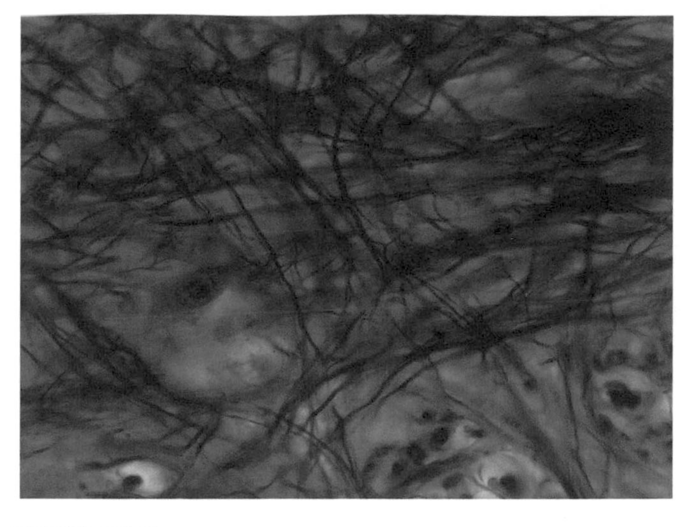

FIGURE F23b In this high-powered micrograph of the gray matter of the same spinal cord stained with silver, nonmyelinated fibers crisscross the field. A few neuronal cell bodies are seen between the fibers. 63 ×.

FIGURE F24 Whole mount of the smooth muscle layers in the gut of a small mammal. Black, nonmyelinated sympathetic nerve fibers run in straight lines between the muscle fibers. Many of these fibers emanate from a sympathetic ganglion at the left. 10 ×.

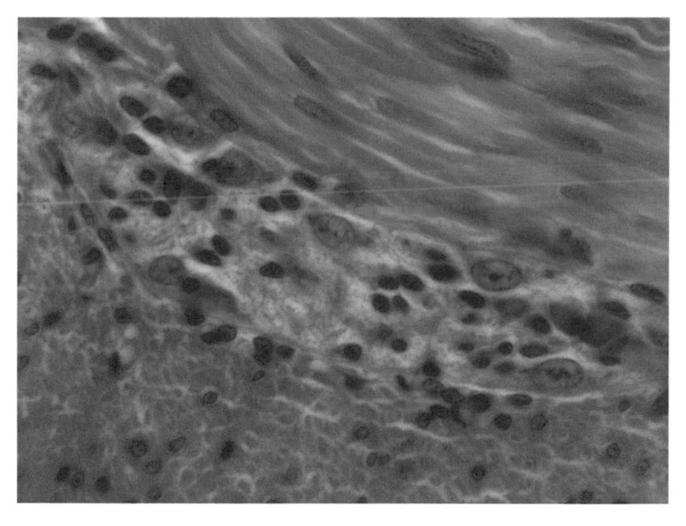

FIGURE F25 Photomicrograph of a cross section of the duodenum of a cat showing a sympathetic ganglion lodged between the longitudinal layer of smooth muscle at the bottom left corner and the layer of circular smooth muscle at the upper right. 63 ×.

b. *Piriform (Purkinje) cells* (Fig. F33). Identify the conspicuous piriform cells in a section of cerebellar cortex (Fig. F34). The body is large and pear shaped (Fig. F35). A thick dendrite proceeding toward the free surface divides into two branches, each of which in turn branches repeatedly forming a fan-shaped mass looking like a tumbleweed that has been flattened by a steamroller. The many spiny gemmules on these processes are evidence of the large numbers of synapses with other nerve processes. A single axon passes downward from the deep surface of the cell.

Nerve Endings

1. *Motor endings.* Figs. F36a and F36b are from a teased preparation of intact whole striated muscle cells stained with gold chloride. Note the myelinated nerve approaching a muscle fiber; the nerve branches are repeated until single fibers come into contact with single muscle cells at a MYONEURAL JUNCTION or MOTOR END-PLATE.

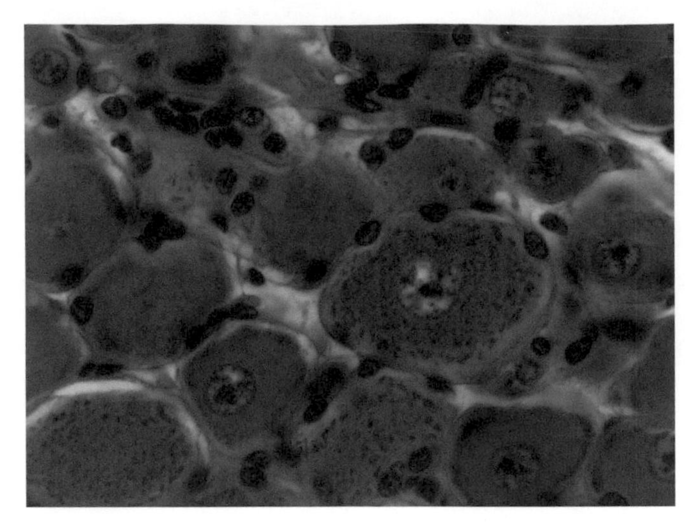

FIGURE F26 Photomicrograph of pseudounipolar neurons clustered in the dorsal root ganglion of a small mammal. Each neuron is enclosed within a thin layer of capsule cells. Connective tissue of the endoneurium fills in the spaces between the neurons. 63 ×.

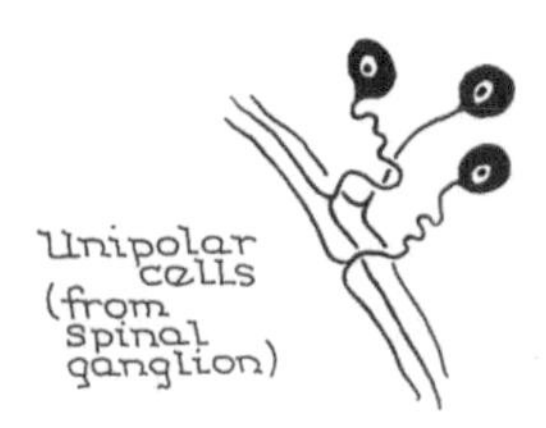

FIGURE F27 Diagrammatic representation of a PSEUDOUNIPOLAR NERVE CELL from a spinal ganglion. In early development of these cells the two processes of a bipolar neuron migrate together and fuse.

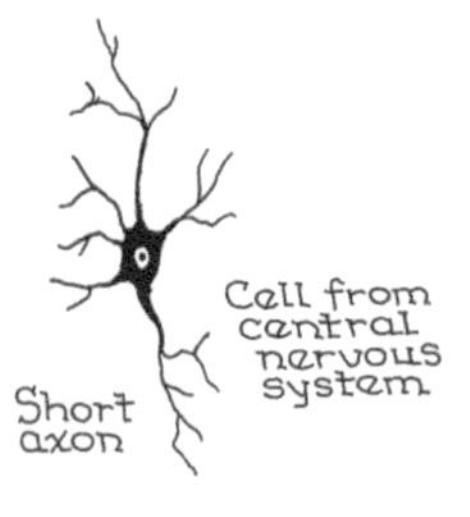

FIGURE F29 Diagrammatic representation of a MULTIPOLAR NEURON with short processes from the central nervous system.

FIGURE F28 Diagrammatic representation of a BIPOLAR NEURON from the retina.

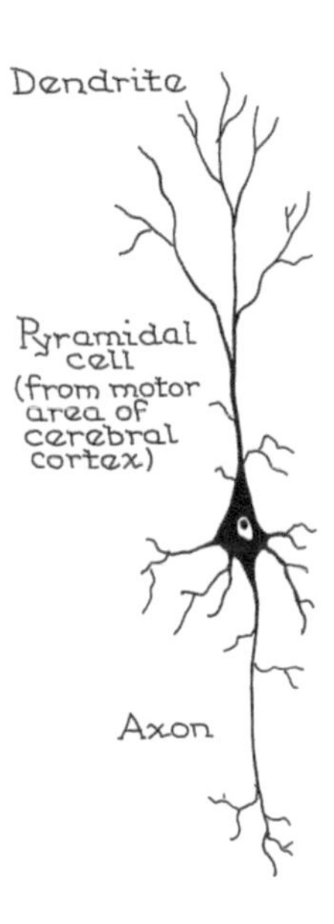

FIGURE F31 Diagrammatic representation of a PYRAMIDAL CELL, a multipolar neuron, from the cerebrum.

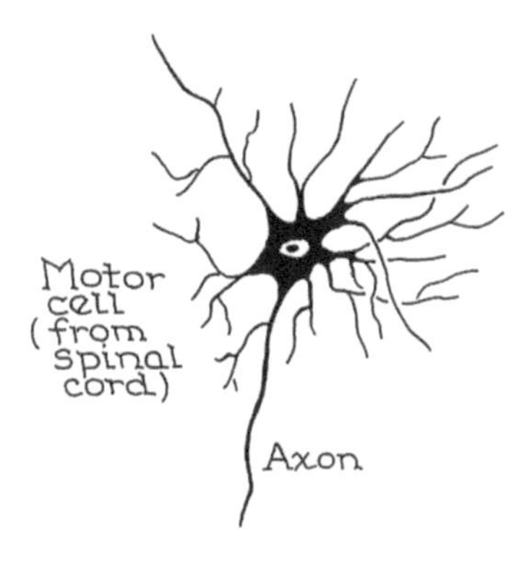

FIGURE F30 Diagrammatic representation of a MULTIPOLAR NEURON from the central nervous system; it may have processes of almost 1 m length.

The nerve branch loses its myelin sheath but is still covered by neurilemma cells and ends in a "crows foot" pattern directly on the sarcolemma which it invaginates. Several nuclei and mitochondria (sarcosomes) of the muscle cell aggregate in the sarcoplasm of this area giving the appearance of a shoe sole; for this reason the myoneural junction is often called a SOLE PLATE. Branches from one nerve may innervate many muscle fibers in most instances.

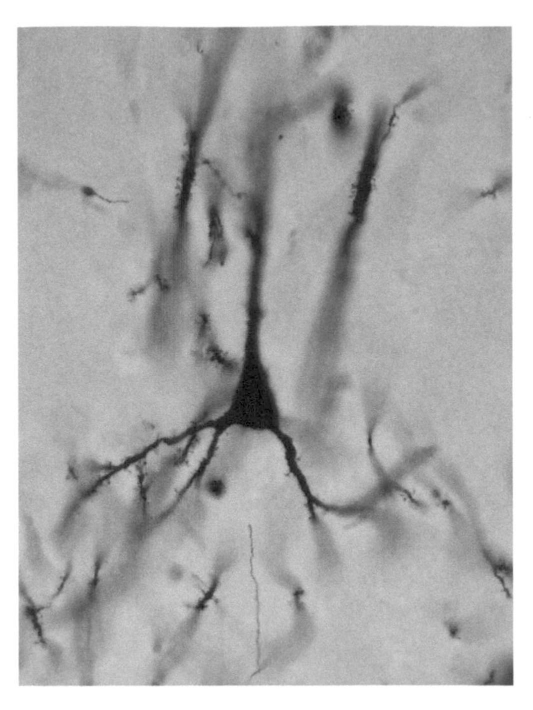

FIGURE F32a Photomicrograph of a section of the cerebrum of a cat with a pyramidal cell at the center. Note the large numbers of synaptic GEMMULES not only in the processes but on the cell body of the neuron as well. Silver impregnation. 40 ×.

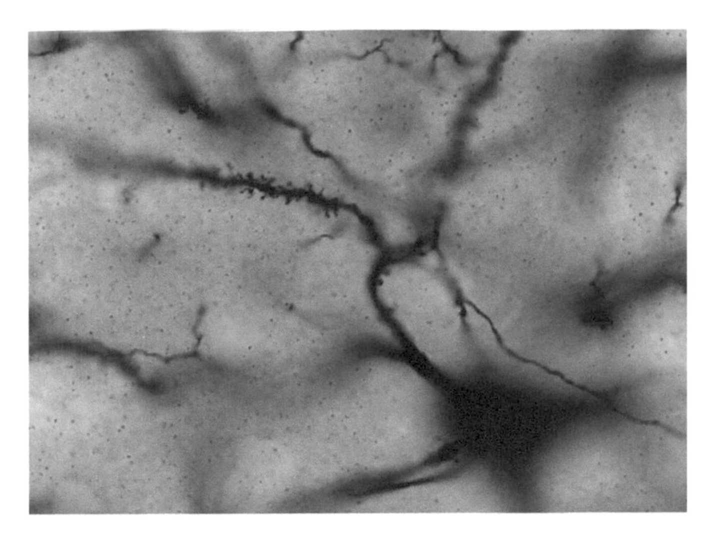

FIGURE F32b Photomicrograph of a section of a silver impregnation of mammalian cerebrum showing well defined GEMMULES on the processes. 63 ×.

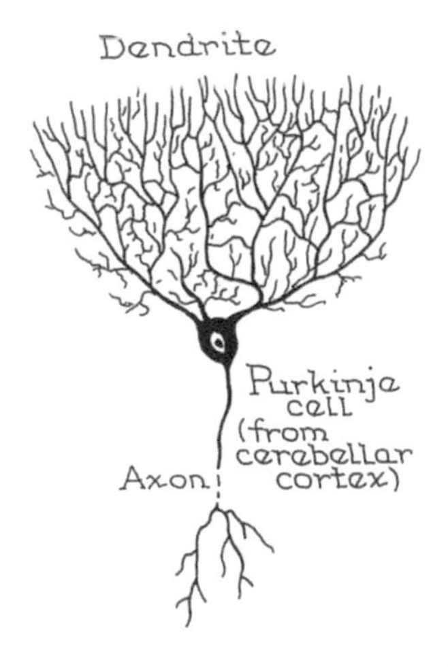

FIGURE F33 Diagrammatic representation of a PIRIFORM CELL (Purkinje cell), a multipolar neuron in the cerebellum.

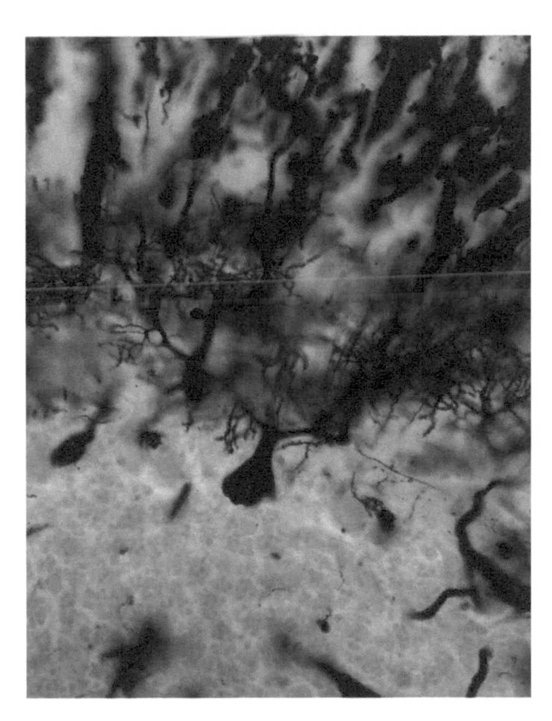

FIGURE F34 Photomicrograph of a silver impregnated section of mammalian cerebellum. Synaptic GREMMULES may be seen on the complex dendrites of the piriform cell near the center. 20 ×.

Electron micrographs show that its sheath of neurilemma cells ends as the nerve fiber approaches the muscle cell and that the endoneurium becomes continuous with the endomysium (Fig. F37). The nerve ends on the surface of the muscle cell in a number of finger-like projections that contain numerous mitochondria and synaptic vesicles in their cytoplasm (AXOPLASM) (Fig. F38). The nerve endings indent the surface of the muscle cell to form SUBNEURAL CLEFTS (Figs. F39 and F40). The sarcolemma lining the clefts forms elaborate folds, the JUNCTIONAL FOLDS, that invade the muscle cell. A sheath of basal lamina covers the entire muscle cell, lining the clefts and the junctional folds, thus separating the nerve endings from the muscle fiber.

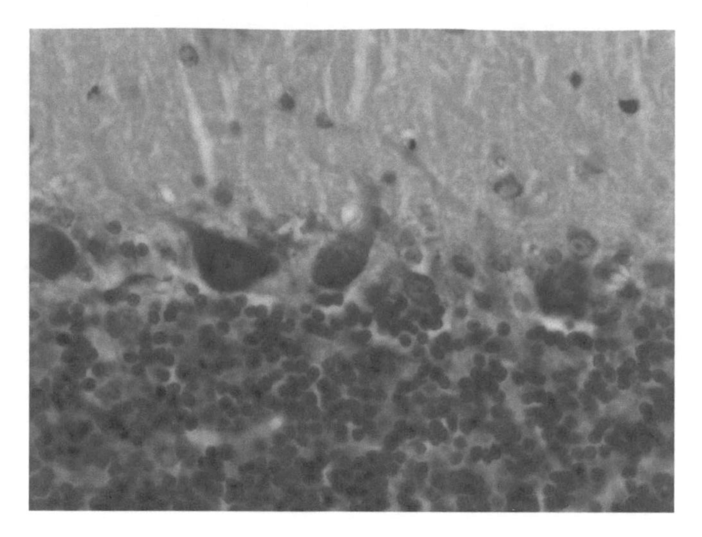

FIGURE F35 Photomicrograph of a section stained with hematoxylin and eosin of the cerebellum from a cat. Processes emanate from the pear-shaped piriform cells near the center. 40 ×.

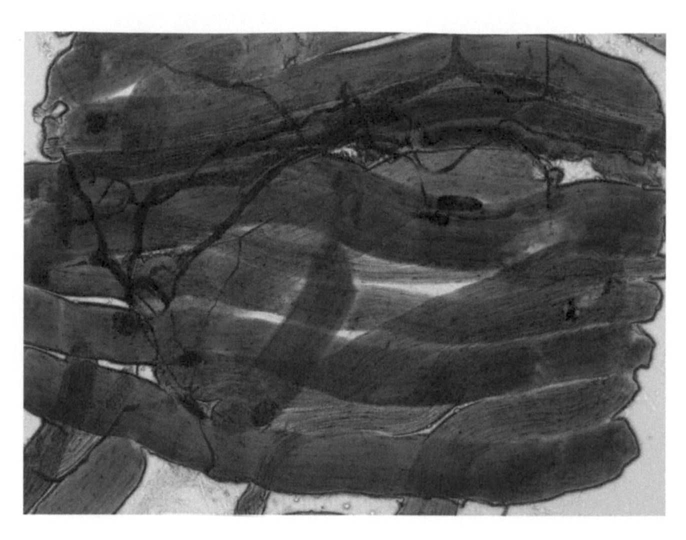

FIGURE F36a Teased preparation of mammalian striated muscle impregnated with silver to show the branching sensory neurons that innervate the tissue. 10 ×.

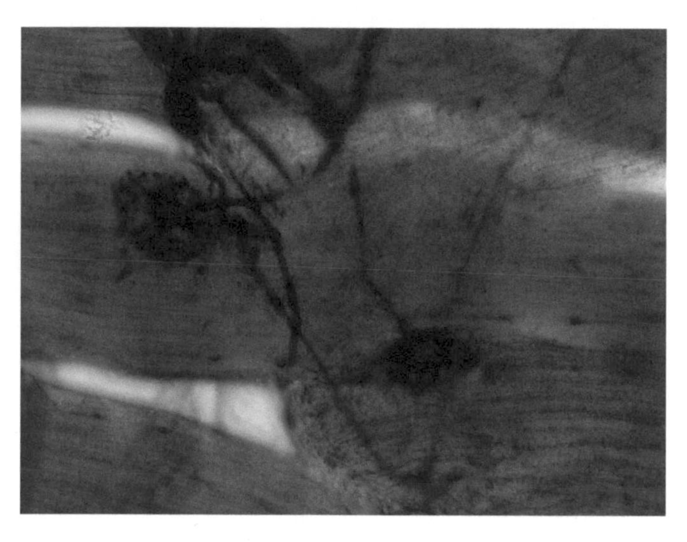

FIGURE F36b Higher powered view of the tissue shown here. There is a MYONEURAL JUNCTION or MOTOR END-PLATE at the upper left. 40 ×.

The innervation of striated muscle may be compared with that of the electric cell or ELECTROCYTE in the electric organ of the mormyrid fish (*Pollimyrus isidori*) (Figs. F41 and F42). Electrocytes are derived from muscle tissue and their nerves show many of the characteristics of motor endings.

2. *Proprioceptors.* These are the nerve endings of muscle sense providing information of movements and position of the body. NEUROMUSCULAR SPINDLES may be found in teased preparations of striated muscle stained with gold chloride (similar to those seen in the previous exercise) (Fig. F43). Note the terminal fibers of the nerve that wind around groups of 3–12 specialized muscle fibers, the INTRAFUSAL FIBERS, contained within a fluid-filled capsule of connective tissue (Figs. F44a and F44b). The muscle fibers are smaller than regular fibers but they receive motor neurons and are attached to the tendons. They contain many nuclei and few myofibrils in the area of the proprioceptive nerve ending. These endings are stimulated by changes in length or tension of the muscle fibers. Muscle spindles inform the brain of muscle action but cannot be considered sensory because we are not aware of their functioning. Of similar appearance and function are TENDON ORGANS, encapsulated proprioceptive endings that spiral around muscle tendons and are sensitive to sudden changes in tension (Figs. F45a, F45b, F46).
3. *Sensory endings.* A few simple examples of sensory endings will be studied in this section. The more elaborate sense organs will be considered in detail later.
   a. *Tactile corpuscles* (Meissner's Corpuscles). Pressure-sensitive tactile corpuscles are shown in the connective tissue of large dermal papillae in sections of skin from the human fingertip (Fig. F47). They appear as elliptical to cylindrical units with a connective tissue capsule. The cavity of the corpuscle contains a stack of wedge-shaped TACTILE CELLS that are thought to be modified neurilemma cells (Fig. F48). One or more unmyelinated nerve fibers enter the base of each corpuscle to produce a net of processes that penetrate between the tactile cells.
   b. *Lamellar corpuscles* (Pacinian corpuscles) are large and readily seen, even with the naked eye. They are regular, egg-shaped, white bodies that occur in deep subcutaneous connective tissue, near joints, in mesenteries, and in the interstitial tissue of the pancreas. In a whole mount of mesentery the corpuscle

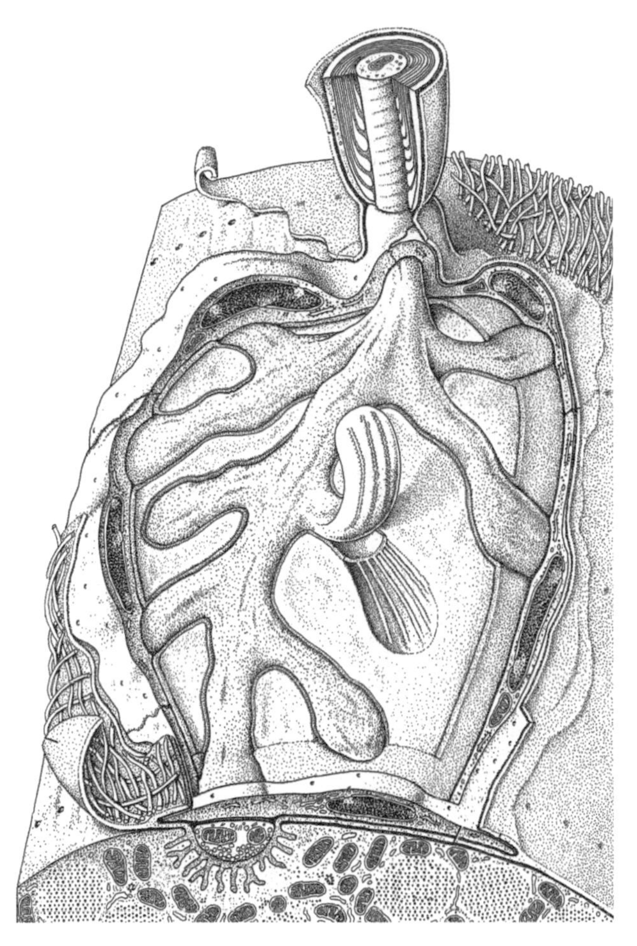

FIGURE F37 Diagram of a MOTOR END-PLATE in striated muscle. This is the synaptic contact between a motor nerve fiber and a striated muscle cell. As the motor axon approaches the muscle cell, it loses its myelin sheath and the neurilemma cells form a "lid" over the myoneural junction. The motor axon forms several branches, which indent the sarcolemma to form SUBNEURAL CLEFTS. The sarcolemma lining the clefts forms elaborate folds, JUNCTIONAL FOLDS, that invade the muscle cell. The endoneurium of the motor nerve is continuous with the endomysium of the muscle cell so that the motor ending and adjacent structures are enclosed in connective tissue. A sheath of basal lamina covers the entire muscle cell, lining the clefts and the junctional folds, thus separating the nerve endings from the muscle fiber.

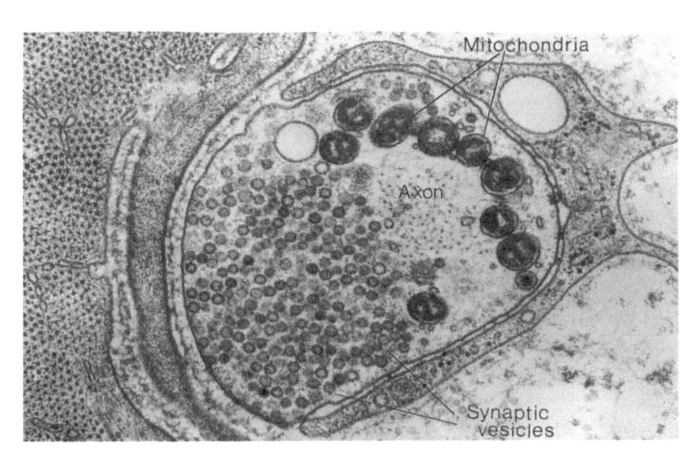

FIGURE F38 Electron micrograph of the nerve ending at a myoneural junction. The axon terminal contains a number of mitochondria and synaptic vesicles.

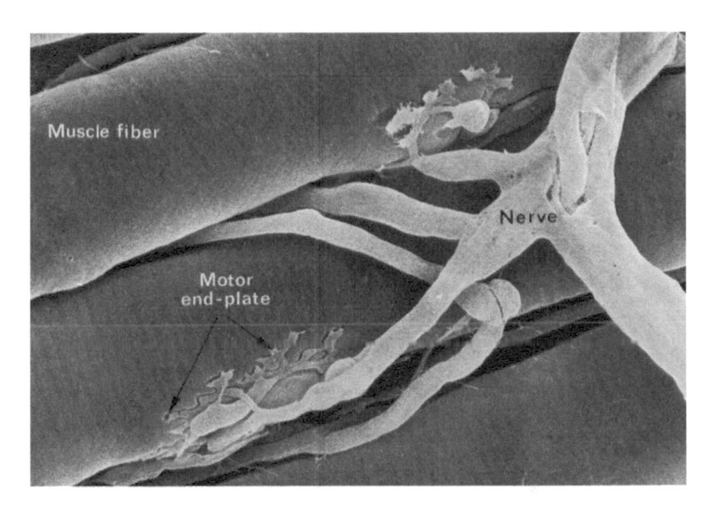

FIGURE F39 Scanning electron micrograph of a motor nerve giving rise to two motor end-plates on adjacent striated muscle fibers.

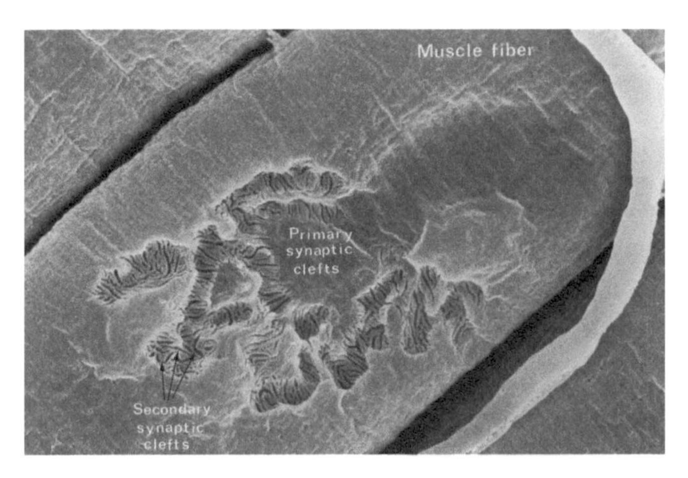

FIGURE F40 Scanning electron micrograph of a striated muscle fiber in the region of a myoneural junction where the nerve terminal has been pulled away, revealing the subneural clefts and junctional folds.

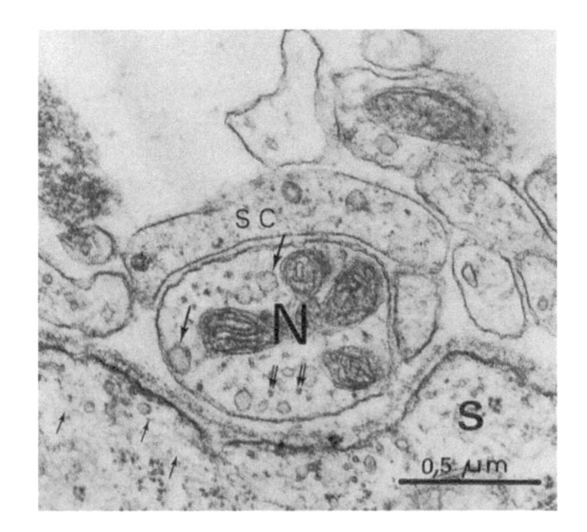

FIGURE F41 Electron micrograph showing contact between a nerve ending (N) and an electrocyte (S) of the electric organ of a mormyrid fish. The nerve fiber is enveloped by a neurilemma (Schwann) cell (SC) and contains large vesicles (big arrows) and possibly microtubules (double arrows).

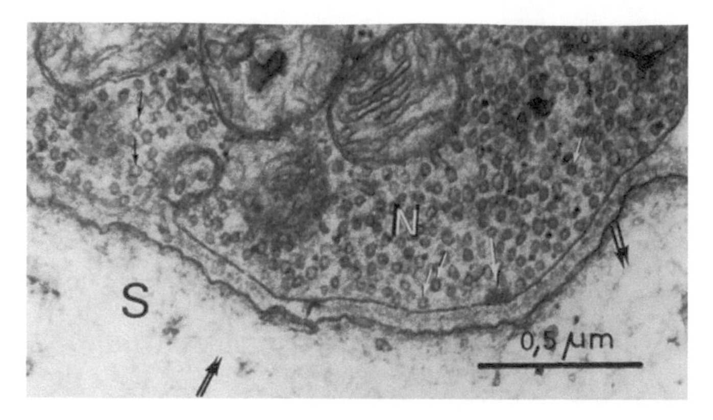

FIGURE F42 Detail of a nerve terminal (N) on an electrocyte (S) of a mormyrid fish. Mitochondria and many synaptic vesicles (small arrows) are present in the nerve endings. The plasma membranes of these cells are separated by a narrow synaptic cleft.

FIGURE F43 Photomicrograph of a teased preparation of striated muscle impregnated with gold chloride showing two muscle spindles, a small, indistinct one at the top, and a larger one near the bottom. Nerve endings wrapped around an INTRAFUSAL MUSCLE FIBER, smaller than the usual muscle fiber, and keep the brain informed of muscle tensions and movements. 20 ×.

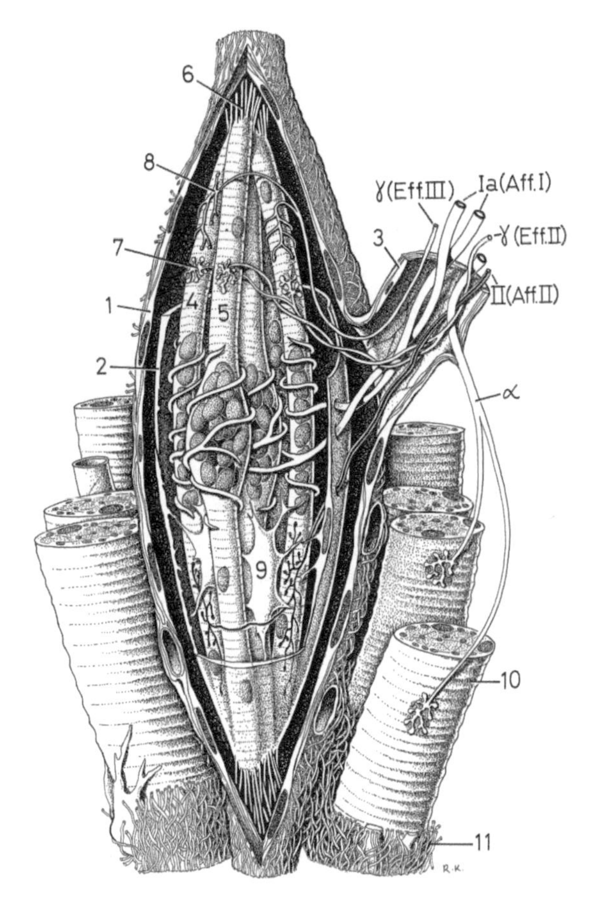

FIGURE F44a Diagram of a NEUROMUSCLE SPINDLE in striated muscle. Terminal fibers of an afferent nerve (Aff) wind around groups of several special muscle fibers, the INTRAFUSAL FIBERS (4, 5), contained within a fluid-filled capsule of connective tissue (1, 2) that is continuous with the perineurium (3) of the nerve fiber and is attached to the endomysium (11) of the regular muscle fibers. The intrafusal fibers are smaller than regular muscle fibers (10) but they receive motor neurons and are attached to tendons (6). They contain many nuclei and a few myofibrils in the area of the proprioceptive nerve ending (7, 8). These endings are stimulated by changes in the length or tension of the muscle fibers and inform the brain of muscle action. *9*, fibrocyte.

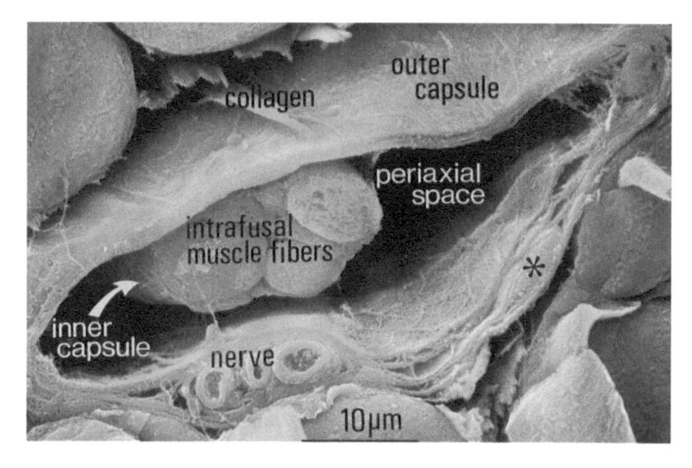

FIGURE F44b Scanning electron micrograph of a muscle spindle of a hamster. Three intrafusal muscle fibers are enveloped by an inner capsule. The inner and outer surfaces of the outer capsule are covered by a network of connective tissue fibers. Three nerve fibers and a capillary (*) are embedded within the capsular wall. 3300 ×.

resembles an onion. It is made up of many dense LAMELLAE or layers of squamous fibrocytes (continuous with the endoneurium) with fluid between (Figs. F49a–F51). The INNER BULB at the core of the corpuscle is a slender protoplasmic cylinder of granular semifluid consistency containing a single nerve fiber (Figs. F52 and F53). Only the initial part of the nerve fiber retains its myelin sheath; the terminal part is unmyelinated. The connective tissue CAPSULE of the corpuscle is continuous with the perineurium of the nerve. Lamellar corpuscles are sensitive to grosser changes in pressure than the tactile corpuscles.

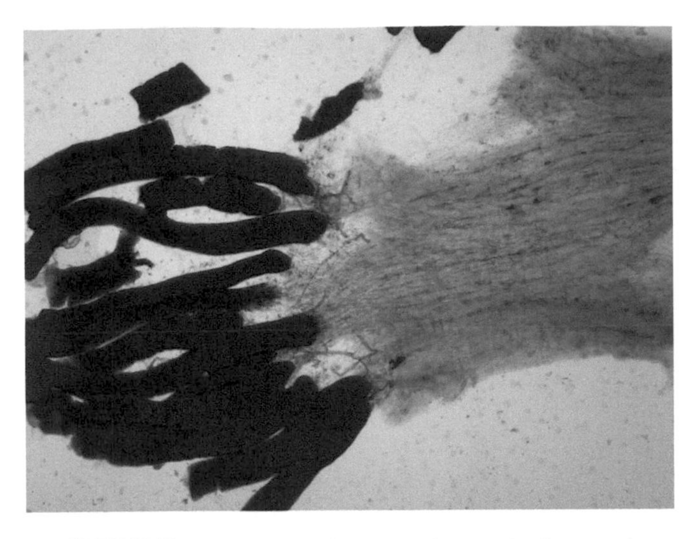

FIGURES F45a Receptors of grosser changes in the muscle are located in TENDON ORGANS as seen in this photomicrograph of a gold chloride impregnation of a tendon. Deeply stained muscle fibers are at the left. 10 ×.

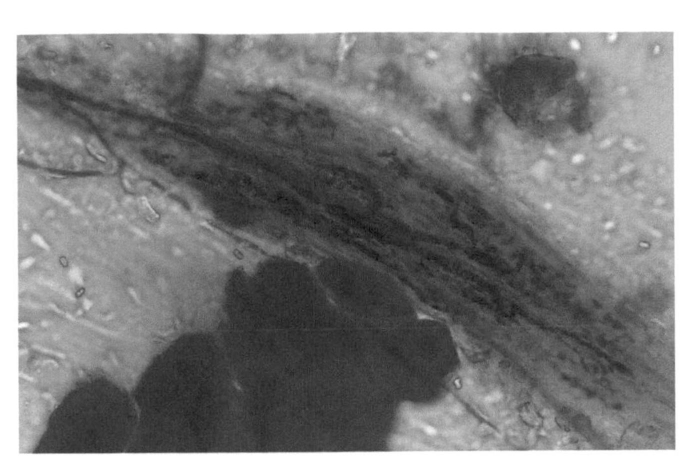

FIGURES F45b Sensory nerves in the tendon organ can be seen in this high-powered image. 20 ×.

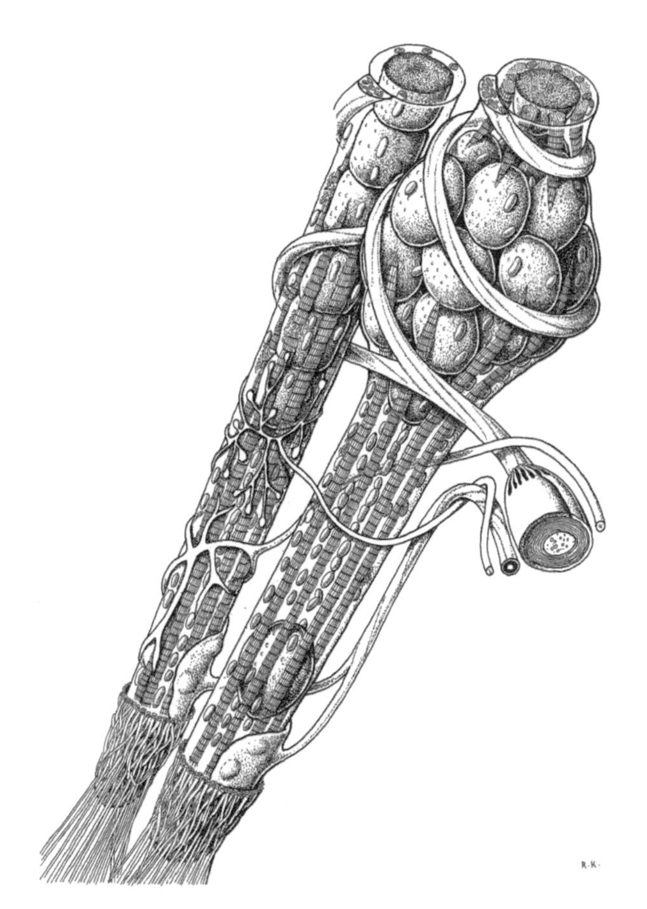

FIGURE F46 Diagram of a proprioceptive TENDON ORGAN from the tendon of striated muscle. Several myelinated nerve fibers lose their myelin sheaths and entwine the collagenous fibers of a tendon. A capsule around the organ is continuous with the perineurium. Contraction of the muscle compresses the naked nerve endings thereby exciting them.

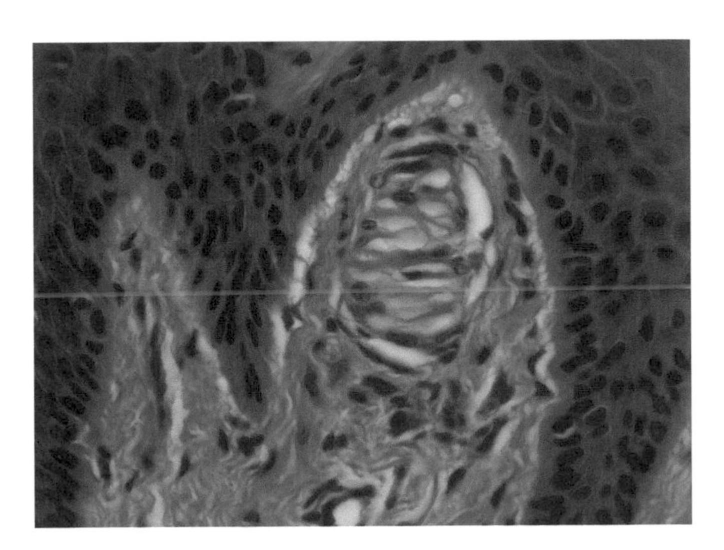

FIGURE F47 Photomicrograph of a TACTILE CORPUSCLE (Meissner's corpuscle) in a section of human skin stained with hematoxylin, eosin, and orange G. These touch receptors lie in the connective layer of the skin immediately beneath the epidermis. 40 ×.

c. *Corpuscles of Grandry and Herbst*. Similar sensory endings have been noted in the palate, skin of the bill, in the mucous membranes of the tongue, and joint capsules of some birds. The CORPUSCLES OF GRANDRY (Fig. F54) occur in subepithelial connective tissue; they are enclosed by a capsule of collagenous connective tissue and consist of two or more large hemispherical cells enclosing a nerve terminal (Figs. F55 and F56). CORPUSCLES OF HERBST resemble lamellated corpuscles and consist of several concentric lamellae surrounding an afferent nerve ending and enclosed within a capsule of collagenous tissue (Fig. F57). Several outer lamellae enclose a core of 60–70 sheets of cytoplasmic extensions (Figs. F58–F62).

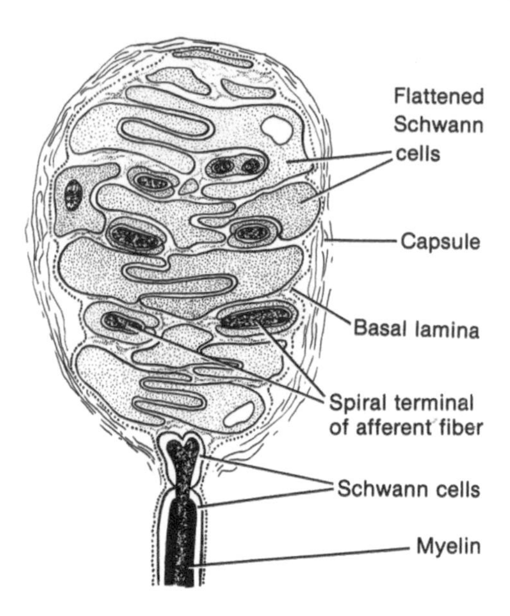

FIGURE F48 Diagram of a tactile corpuscle from a dermal papilla in mammalian skin. An unmyelinated nerve penetrates a stack of wedge-shaped TACTILE CELLS that are enclosed in a CAPSULE of connective tissue. The tactile cells may be derived from modified neurilemma cells.

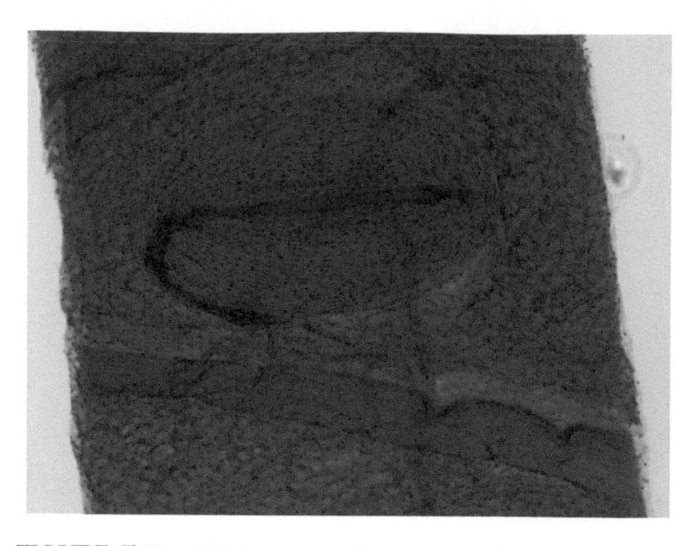

FIGURE F49a Whole mount of mesentery of a small mammal stained to show a LAMELLAR CORPUSCLE (Pacinian corpuscle). 5 ×.

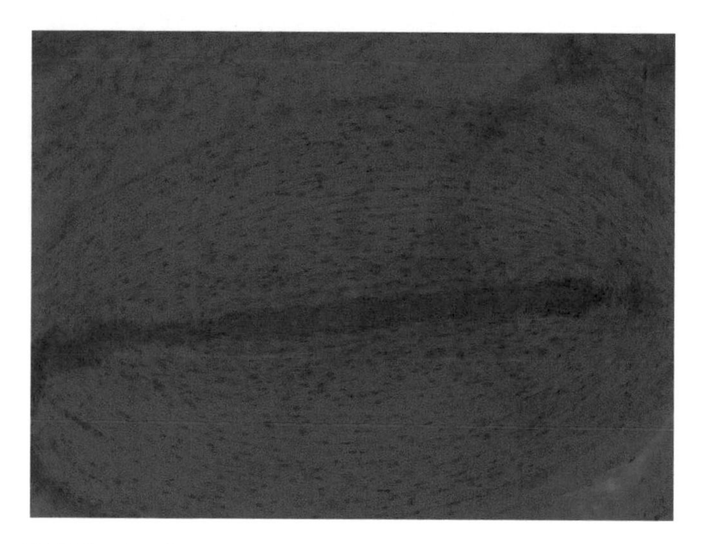

FIGURE F49b A high-powered view of the whole mount of a LAMELLAR CORPUSCLE shown. The layers of flattened cells are well defined. 10 ×.

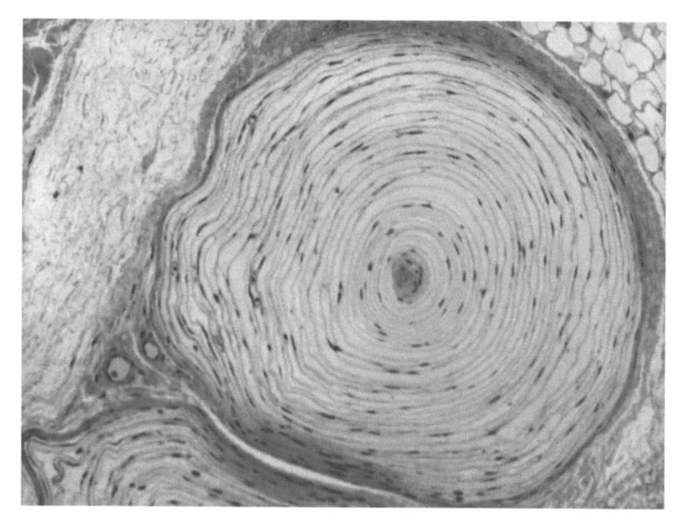

FIGURE F50a Cross section of a lamellar corpuscle with an unmyelinated nerve fiber at its center surrounded by concentric layers of flattened cells. 10 ×.

4. *Synapses*. The point of contact between an axon and another neuron occurs not only on the dendrites (AXODENDRITIC SYNAPSE) but on the cell body (AXOSOMATIC SYNAPSE), and even on another axon (AXOAXONIC SYNAPSE) (Fig. F63). The latter are believed to be inhibitory. As many as several thousand axonal terminals may bring impulses to one neuron. Commonly the axonal endings swell into tiny button-like END BULBS or BOUTONS TERMINAUX. Other types are brushes or baskets that adhere to a second neuron or clasp it. In every instance the point of association is intimate but of contact only. You have already seen synapses—the spiny gemmules—in sections of the central nervous system stained to show nerve fibers (Figs. F64 and F65).

In electron micrographs (Figs. F66 and F67) the end bulb of synapses is seen to contain numerous mitochondria and SYNAPTIC VESICLES and is separated by a narrow SYNAPTIC CLEFT from the dendrite, perikaryon, or axon with which it communicates. The pre- and postsynaptic membranes are thickened and may be joined by INTERSYNAPTIC FILAMENTS across the cleft. A network of filaments forms a SUBSYNAPTIC WEB in the cytoplasm immediately beneath the postsynaptic membrane.

FIGURE F50b Higher powered view of the cross section of the tactile corpuscle seen above. Fluid fills the spaces between the flattened cells. 63 ×.

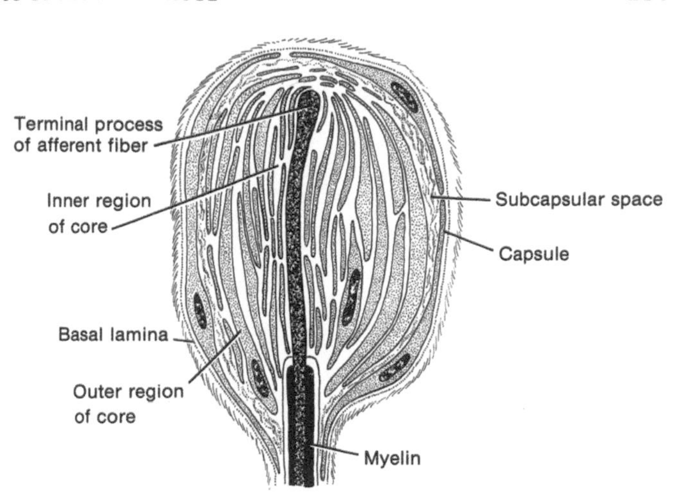

FIGURE F51 An onion-like lamellar corpuscle is composed of concentric lamellae of flattened fibrocytes that surround an afferent nerve fiber. The cells are immersed in a lymph-like fluid enclosed in a connective tissue capsule that is continuous with the perineurium of the nerve fiber.

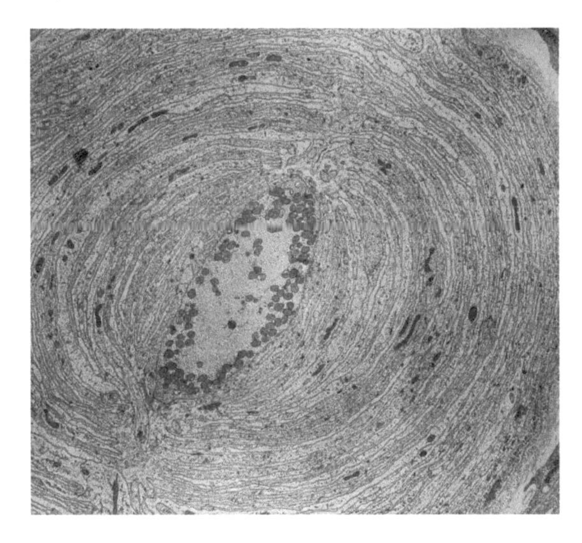

FIGURE F52 Electron micrograph of a cross section through the central core of a LAMELLAR CORPUSCLE (Pacinian corpuscle). The central axon lost its myelin sheath on entering the corpuscle and is surrounded by a neurilemma cell and concentrically arranged flattened cells. 8800 ×.

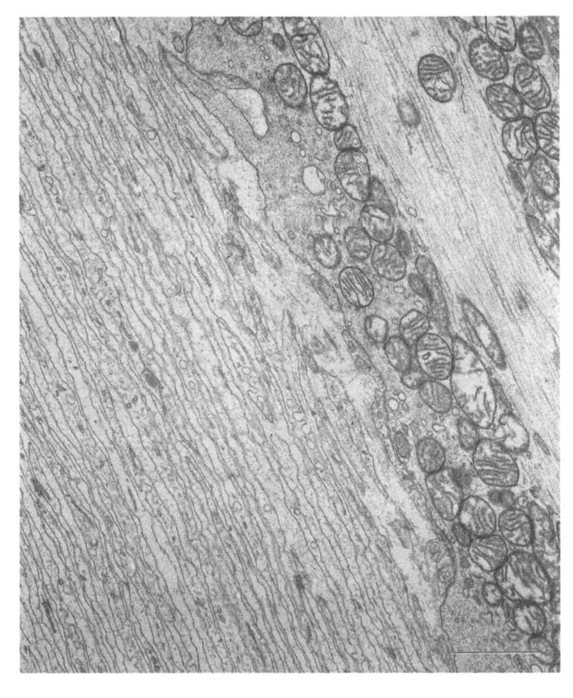

FIGURE F53 Electron micrograph of the inner core of the LAMELLAR CORPUSCLE. The axon terminal at the right contains synaptic vesicles and a large number of mitochondria. The inner core consists of centrically arranged flattened cells with a gel-like fluid between. 30,000 ×.

NONNERVOUS COMPONENTS OF NERVOUS TISSUE

Neurilemma and Satellite Cells

These ensheathing elements of the peripheral neurons and ganglia are ectodermal in origin, being derived from the neural crest. They were studied earlier in sections "Nerve Cell Processes or Fibers" and "Cell Shape (Pseudounipolar (unipolar) Cell)."

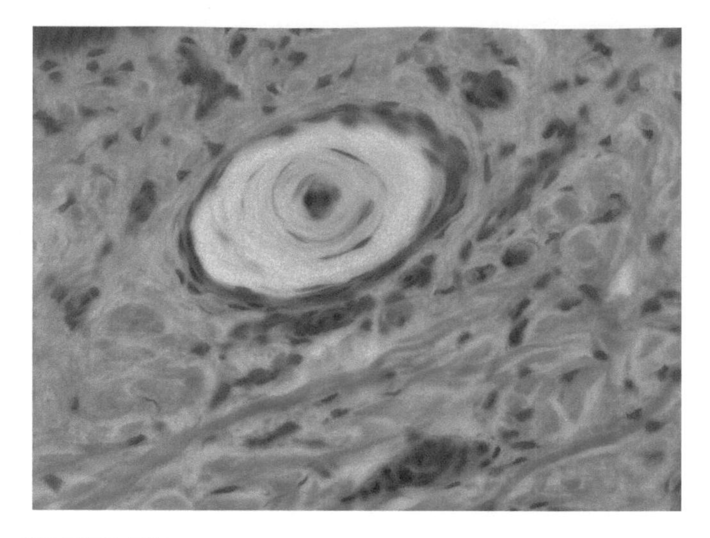

FIGURE F54 The CORPUSCLE OF GRANDRY occurs in subepithelial connective tissue of birds and resembles a simplified lamellar corpuscle. There is a central terminal axon sandwiched between two hemispherical cells. The corpuscle is surrounded by a dense formation of collagenous fibers. 40 ×.

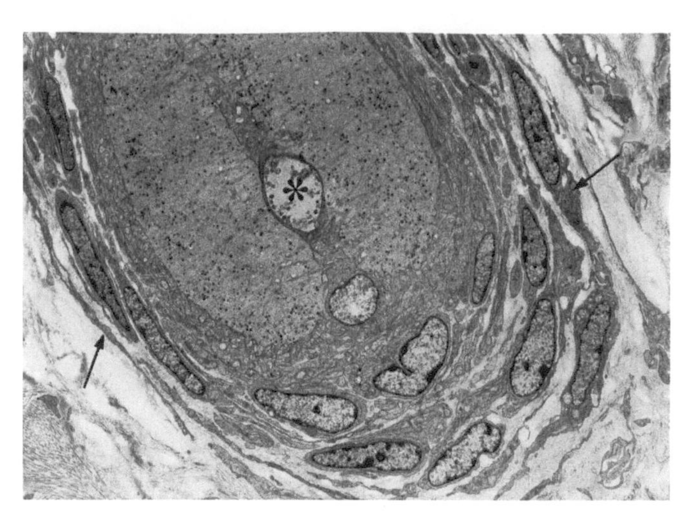

FIGURE F55 Transmission electron micrograph of a section through the CORPUSCLE OF GRANDRY of the oral lining of the duck. The terminal axon (*) is located in the space between the hemispherical Grandry cells. Several lamellae of capsular elements are noted (arrows). 2100 ×.

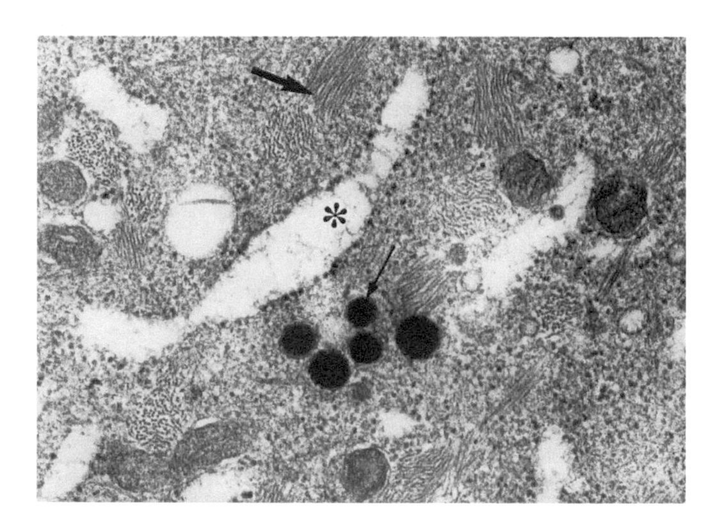

FIGURE F56 Transmission electron micrograph of a part of the cytoplasm of a GRANDRY CELL. Large arrow, filaments; small arrow, electron-dense granules of 50–100 nm; *, granular endoplasmic reticulum. 36,000 ×.

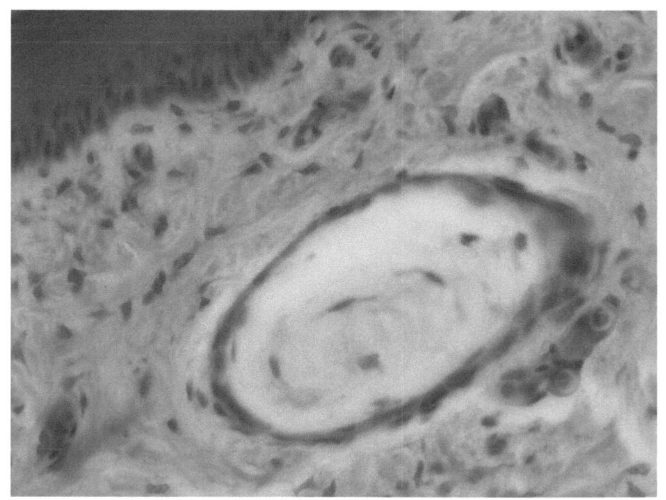

FIGURE F57 Light micrograph of a Herbst corpuscle from the oral lining of a duck. The terminal axon is enclosed by lamellae of capsule cells. The corpuscle is surrounded by dense collagenous tissue. 40 ×.

Neuroglia

Neuroglial cells have elaborate processes and serve supportive, metabolic, and other functions in the wall of the brain and spinal cord. They are responsible for isolating neurons from all other tissues of the body (including other neurones) except at the points of synapse. There are three types of neuroglial cells: ASTROCYTES, OLIGODENDROCYTES, and MICROGLIA. Neuroglial cells are not seen readily in routine preparations of nervous tissue and, with hematoxylin and eosin, only their nuclei are visible. The micrographs shown are from silver/gold impregnations of the central nervous system stained to show neuroglial cells (Fig. F68).

1. *Astrocytes.* These star-shaped cells are the largest of the glial cells and have a supportive and insulating function. They hold together the nervous tissue and anchor it to the blood vessels that course through it (Fig. F69). Hand-like expansions of the processes of astrocytes enclose blood vessels and other nerve cell

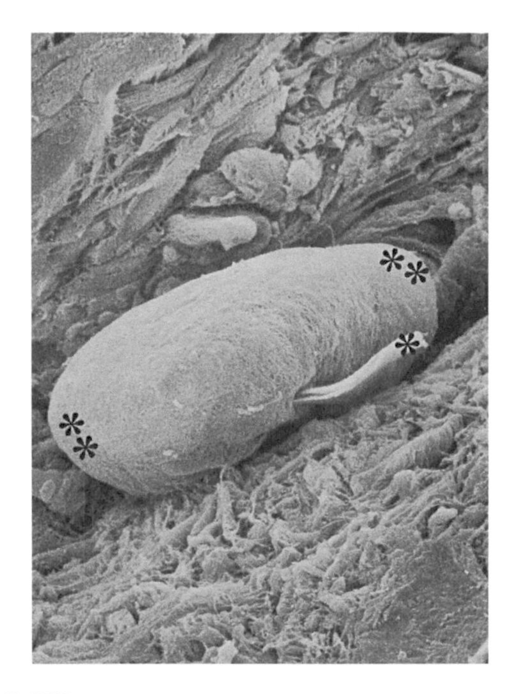

FIGURE F58 Scanning electron micrograph of a freeze-cracked preparation of a CORPUSCLE OF HERBST from the oral lining of a duck. *, Afferent nerve fiber; **, the two poles of the corpuscle. 750 ×.

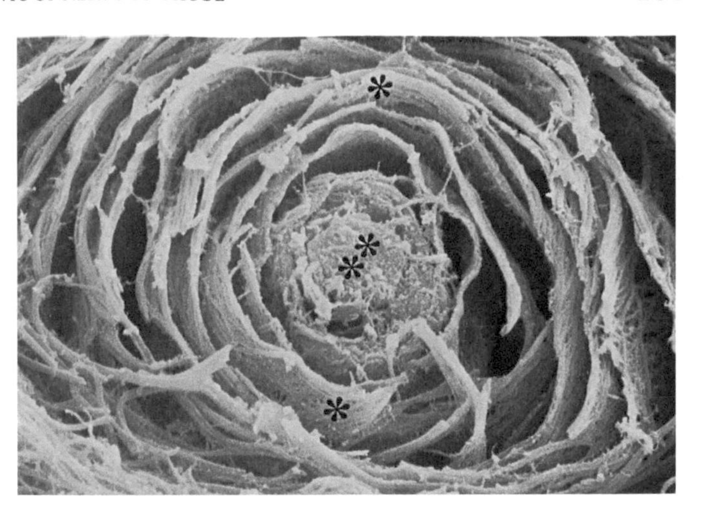

FIGURE F59 Scanning electron micrograph of a freeze-cracked preparation showing the interior of a CORPUSCLE OF HERBST from the oral lining of a duck. The transversely fractured image shows the lamellated outer capsule (*) and compact inner core (**). 3000 ×.

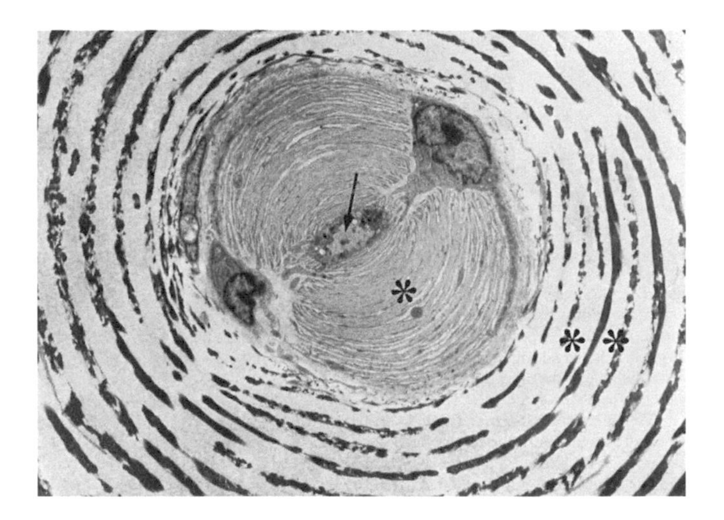

FIGURE F60 Low power transmission electron micrograph of a cross section of a CORPUSCLE OF HERBST from the oral lining of a duck. The arrow indicates the nerve ending. *, Compact inner core; **, outer capsule. 3000 ×.

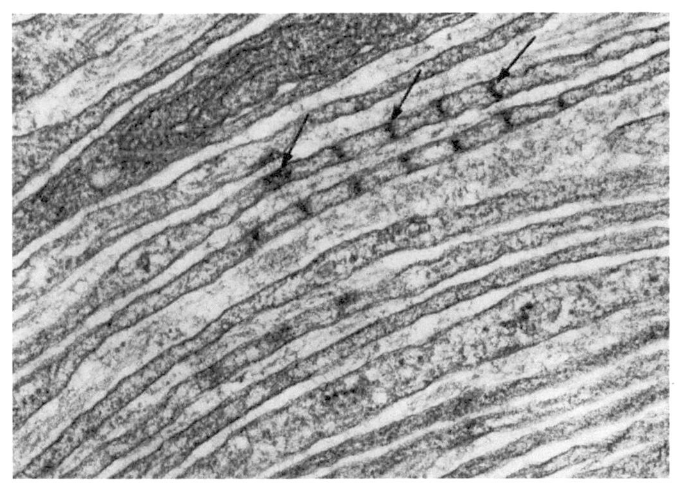

FIGURE F61 Transmission electron micrograph of a section of the inner core of a CORPUSCLE OF HERBST from the oral lining of a duck. The arrows indicate a flattened lamellar cell; a gel-like fluid fills the space between the lamellar cells. 33,000 ×.

bodies as well as forming a LIMITING MEMBRANE enclosing the nervous tissue on the surface of the brain and spinal cord. There are two types of these cells:

a. *Protoplasmic astrocytes* or *mossy cells* occur chiefly in the gray matter of the brain and spinal cord. They have thick, branching processes, and granular cytoplasm (Fig. F70). Some processes attach to blood vessels by expanded PERIVASCULAR FEET.

b. *Fibrous astrocytes* are found mainly in the white matter of nervous tissue and have fewer but longer processes that are thinner, straighter, and less branching than those of the protoplasmic astrocytes (Fig. F71). Some attach to blood vessels. Long, unbranched fibrils develop in the cytoplasm.

2. *Oligodendrocytes.* These cells are the commonest of the neuroglia and are found throughout the wall of the central nervous system forming rows between the nerve fibers in the white matter and clustering around neurons as SATELLITES in the gray matter (Fig. F72). Their short, beaded processes are little branched. Processes of oligodendrocytes form the myelin sheaths of the nerves of the central nervous system.
3. *Microglia.* Microglia are found throughout the brain and spinal cord occurring near the nerve cells and blood vessels (Figs. F73a and F73b). They are migratory, highly phagocytic macrophages of the central nervous system. They are the smallest glial cells and have relatively small nuclei.

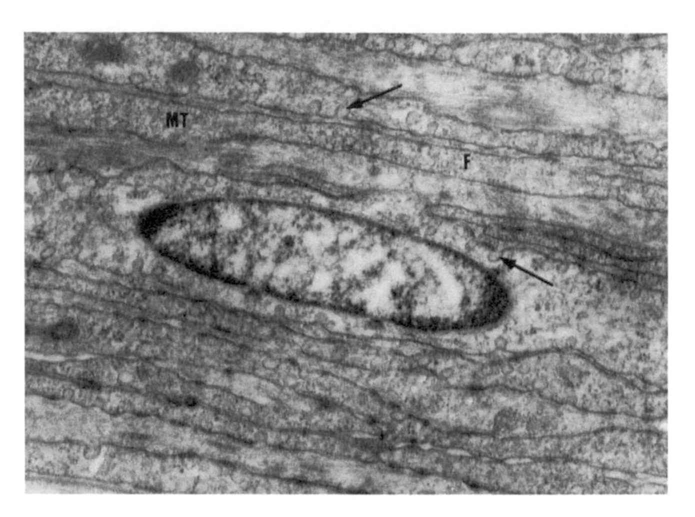

FIGURE F62 Transmission electron micrograph of a section of the capsular cells of a CORPUSCLE OF HERBST from the oral lining of a duck. The nuclear area of one lamellar cell is shown—these cells are probably derived from the perineurial cells of the peripheral nerve. Arrows indicate caveolae in the lamellar cells. *F*, filaments; *MT*, microtubules. 19,000 ×.

Ependymal Cells

The nonnervous ependymal cells are derived from the primitive neural ectoderm and form the lining of the cavities of the brain and spinal cord. They are often classified with neuroglia. Embryonic ependymal cells are ciliated and some retain their cilia permanently. Observe ependymal cells in sections of vertebrates (Figs. F74a–l). As development proceeds these cells become columnar and develop long cytoplasmic processes that extend out to the periphery of the neural tube (Fig. F75). Mature ependymal cells form an epithelial lining for the central canal of the spinal cord and the ventricles of the brain and vary from columnar to low cuboidal cells. They have a supportive as well as a lining function.

Meninges

The brain and spinal cord are enclosed in a jacket of connective tissue consisting of three layers, the MENINGES, all containing various arrangements of

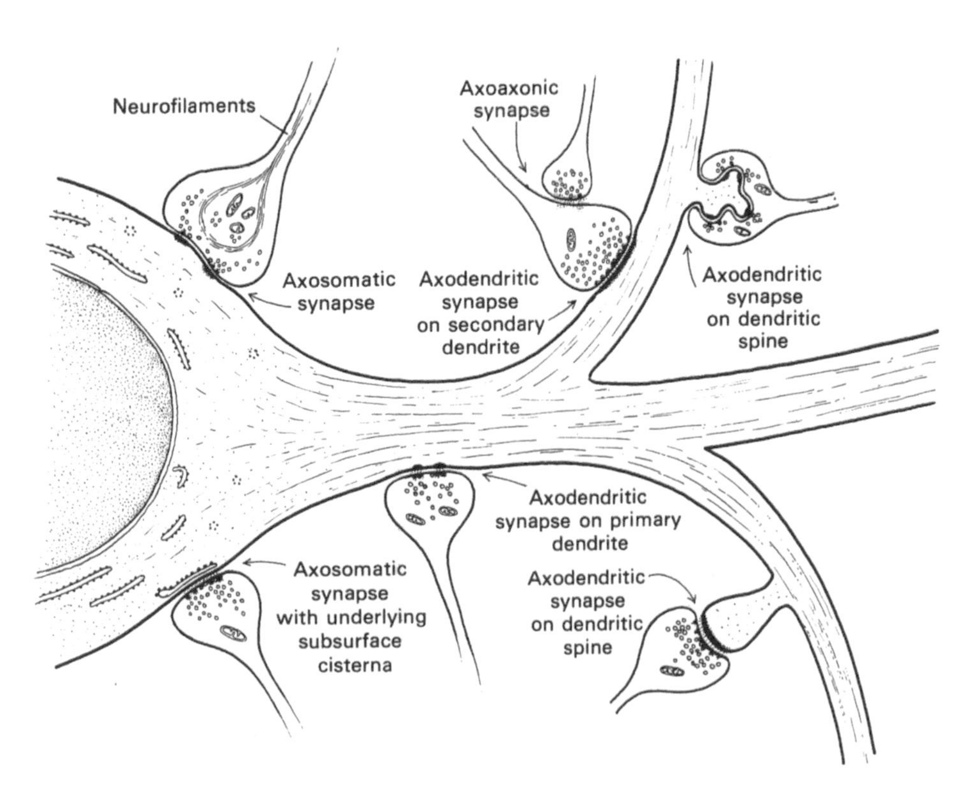

FIGURE F63 Synapses are named according to the regions of the neurons where the attachment occurs. The degree of membrane "thickening" varies in different types of synapses.

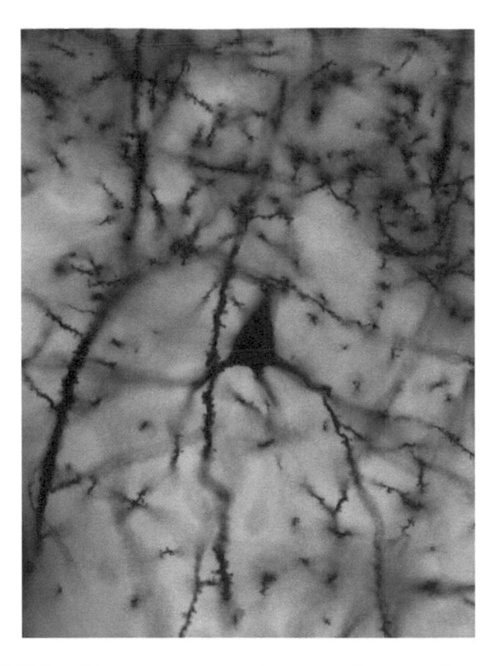

FIGURE F64 Silver impregnated section of mammalian cerebrum. A pyramidal cell is near the center. Processes of nerve cells bristle with gemmules: synaptic endings. 40 ×.

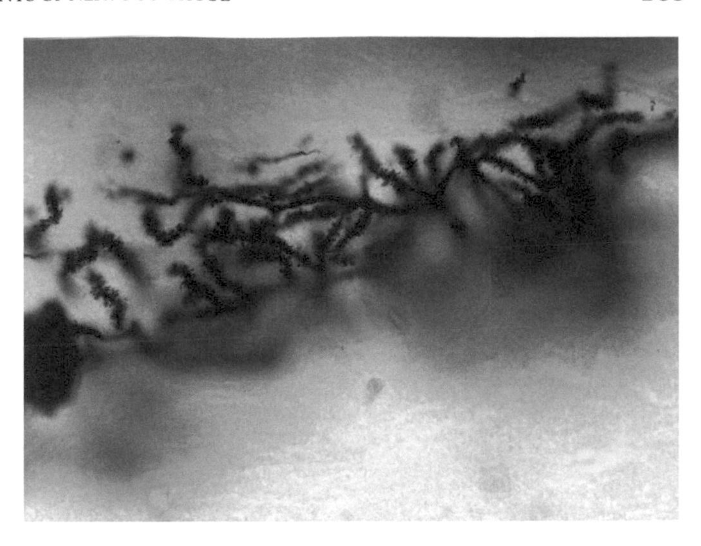

FIGURE F65 Silver impregnated section of mammalian cerebellum showing a dendritic branch of a piriform cell bristling with gemmules. 63 ×.

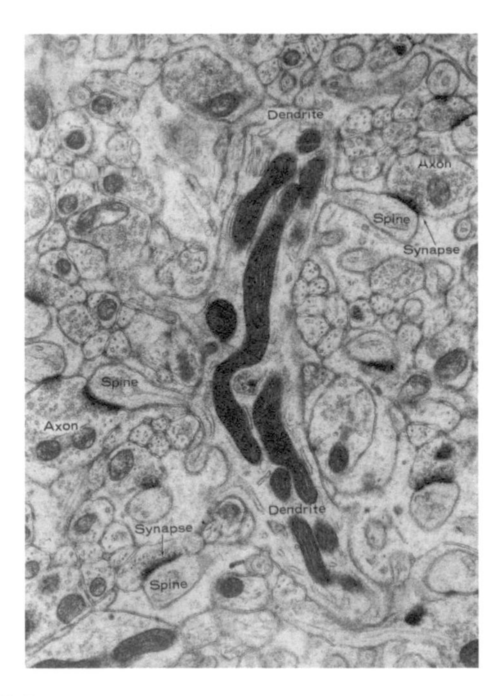

FIGURE F66 Electron micrograph of a section through the cerebellum. A branch of the dendritic tree of a piriform cell (Purkinje cell) running vertically through the field contains several mitochondria. Projecting laterally from the dendrite are "spines" or "thorns" with bulbous tips and narrow stalks. Axons of granule cells form synapses with the dendrite of the piriform cell.

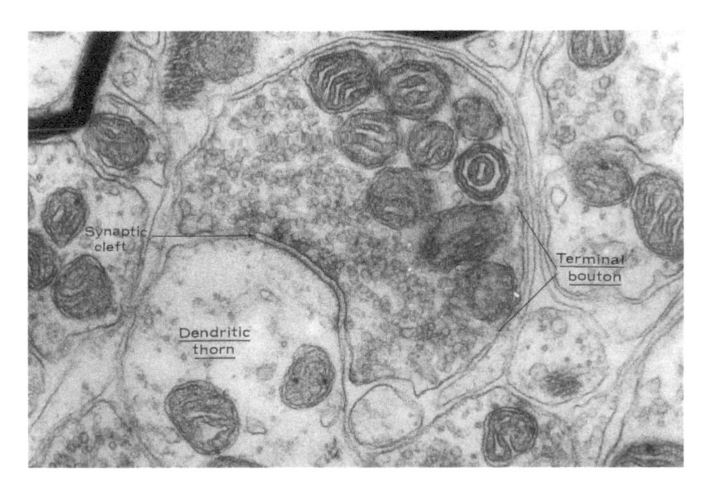

FIGURE F67 Electron micrograph of a section of a synapse from the ventral horn of the spinal cord of a rat. A dendritic thorn is capped by a terminal bouton. The typical features of synapses—mitochondria, clustered synaptic vesicles, and the cleft—are well shown. The terminal is enclosed within a thin astrocytic process.

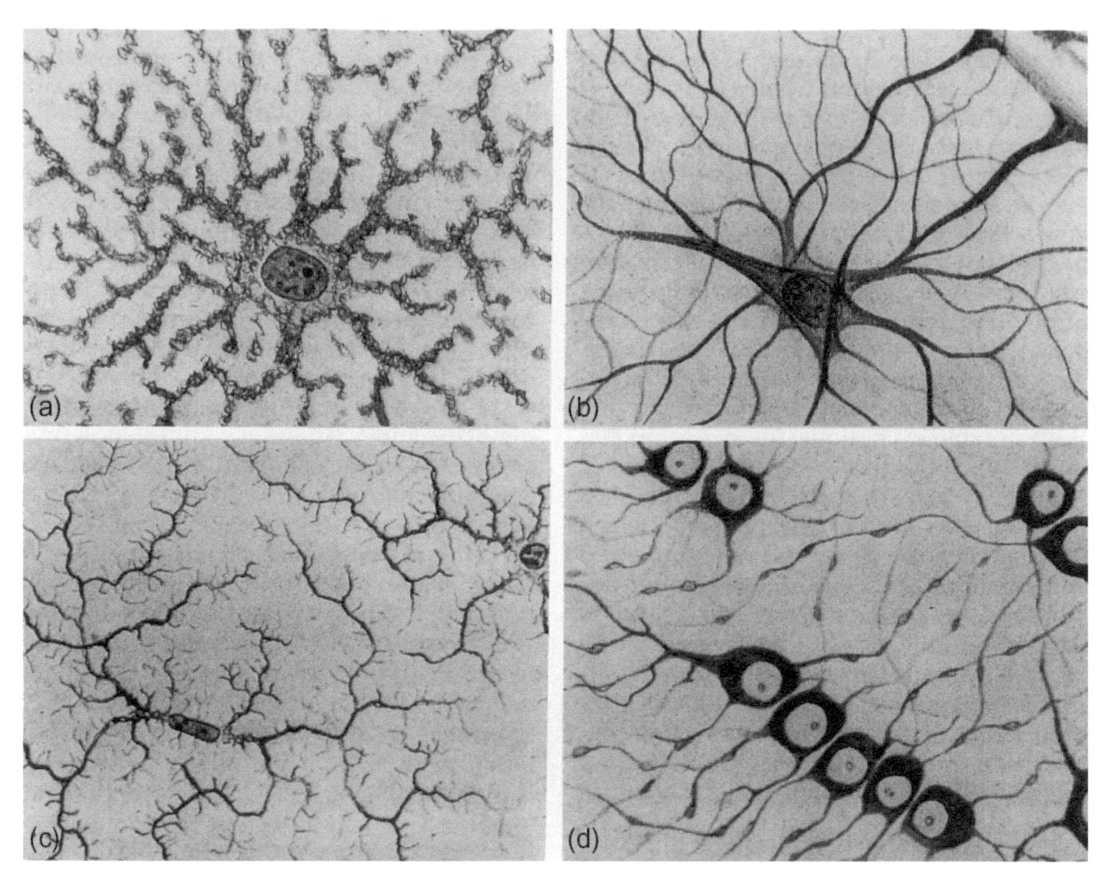

FIGURE F68 Neuroglial cells of the central nervous system. (a) Protoplasmic astrocyte; (b) fibrous astrocyte; (c) microglia; (d) oligodendroglia.

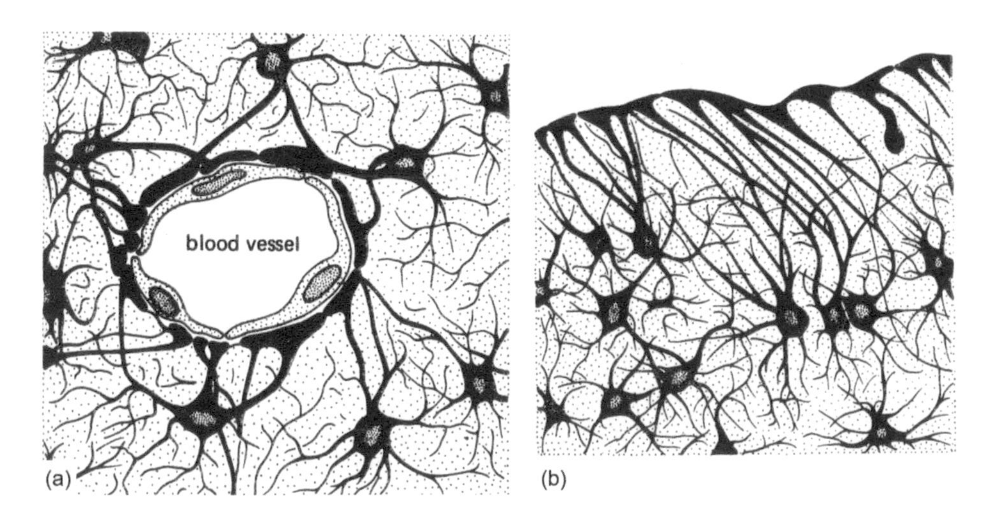

FIGURE F69 Drawing of neuroglial cells in mammals. (a) Perivascular feet of astrocytes attached to a blood vessel. (b) Perivascular feet of astrocytes forming the surface layer of the central nervous system.

collagenous fibers interspersed with some elastic fibers. The outer DURA MATER is a tough collagenous sheath (Fig. F76). The innermost layer is the more delicate PIA MATER, which is directly apposed to the nervous tissues. A few macrophages and fibroblasts roam throughout the pia and blood vessels are distributed in the pia over the brain (Fig. F77). Between the dura mater and the pia is the spidery ARACHNOID that is permeated with blood vessels and where the connective tissue fibers assume a cobweb-like conformation (Figs. F78 and F79). They also form a membranous covering, the pia. The spaces of this meshwork are filled with cerebrospinal fluid.

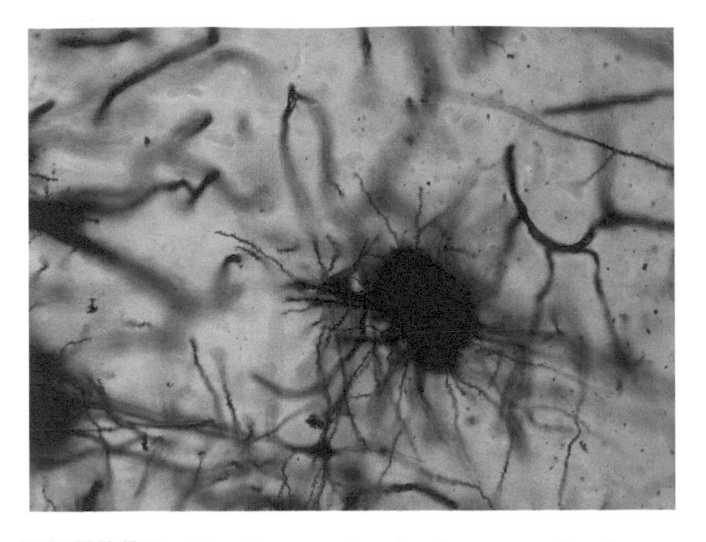

FIGURE F70 Silver impregnation showing a protoplasmic astrocyte from gray matter. 20 ×.

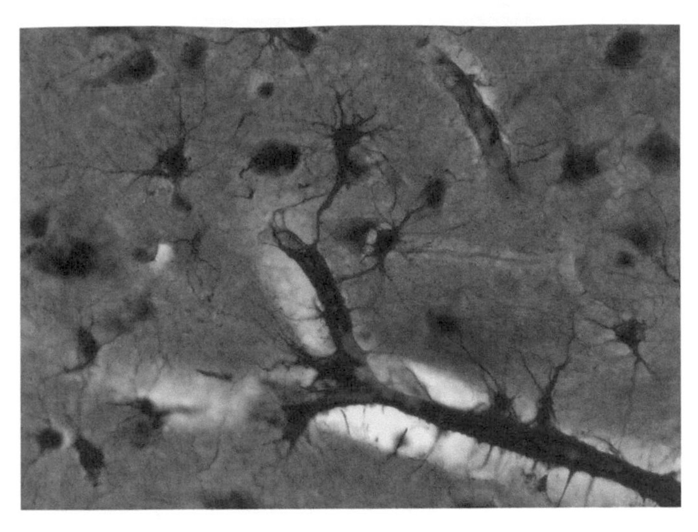

FIGURE F71 The long process of fibrous astrocyte in white matter are attached to the walls of capillaries and appear to be to be anchoring blood vessels within the delicate nervous tissue. 40 ×.

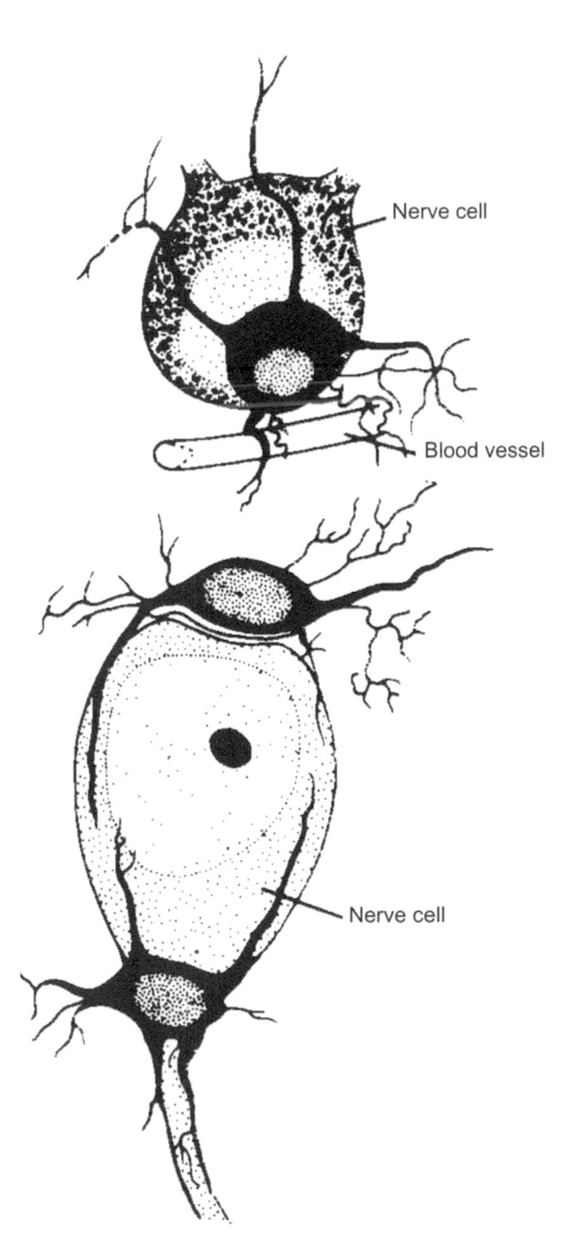

FIGURE F72 Oligodendrocytes as satellite cells embracing neuronal bodies.

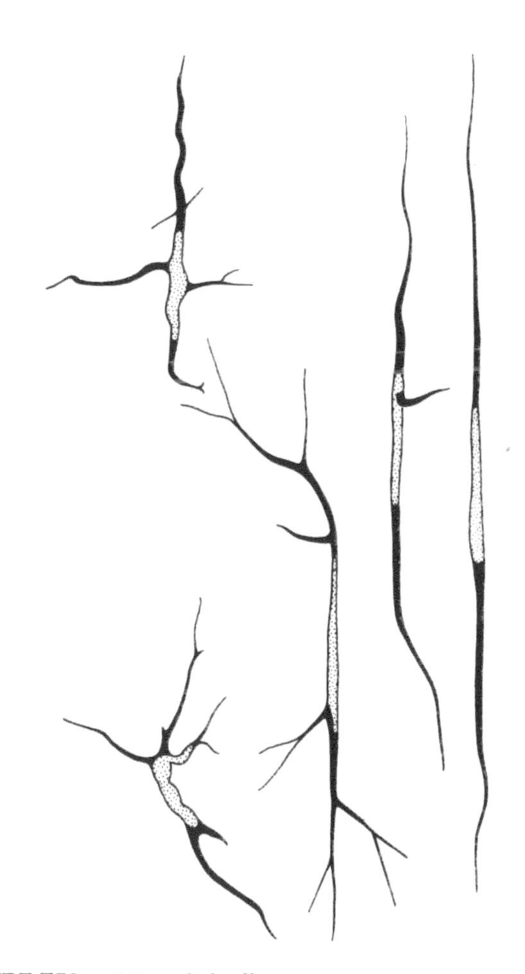

FIGURE F73a Microglial cells.

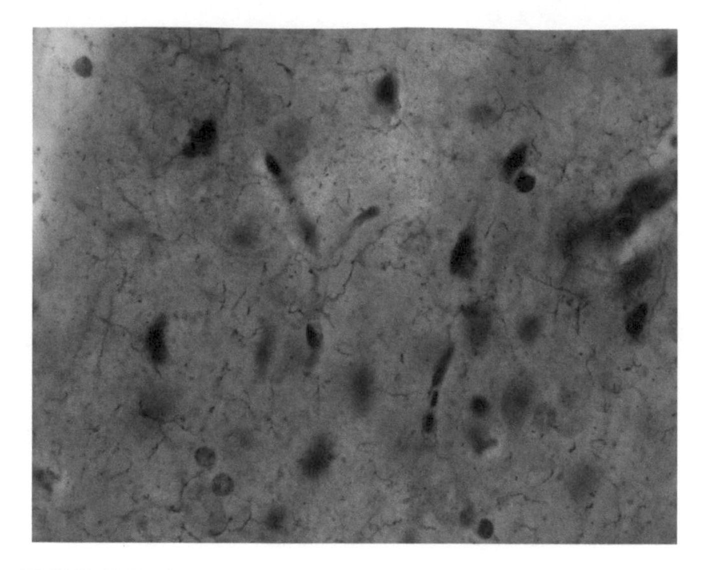

FIGURE F73b Microglia are migratory macrophages of the central nervous system. 40 ×.

Choroid Plexuses

Invading the third and fourth ventricles of the brain are tufts of nonnervous tissue the CHOROID PLEXUSES (Figs. F80a and F80b). These are richly vascular and produce most of the cerebrospinal fluid that is contained within the ventricles of the brain and the canal of the spinal cord as well as that contained within the layers of the meninges that surround the central nervous system. They are complex branching structures derived from the ependyma. They contain delicate fibrous connective tissue invaded by arterioles that open into a capillary bed and are covered with an epithelium. The free surface of the epithelium is covered with microvilli.

It is important to note that the brain and spinal cord are protected from shocks not only by the bony skull and vertebrae but they are also afloat in cerebral spinal fluid contained within the meninges. This tough sheath is not only protective but also nutritive bringing blood vessels to the surface of the central nervous system.

FIGURE F74(a–l) Cross sections of an assortment of vertebrates showing ependymal cells lining the central canal of the spinal cord.

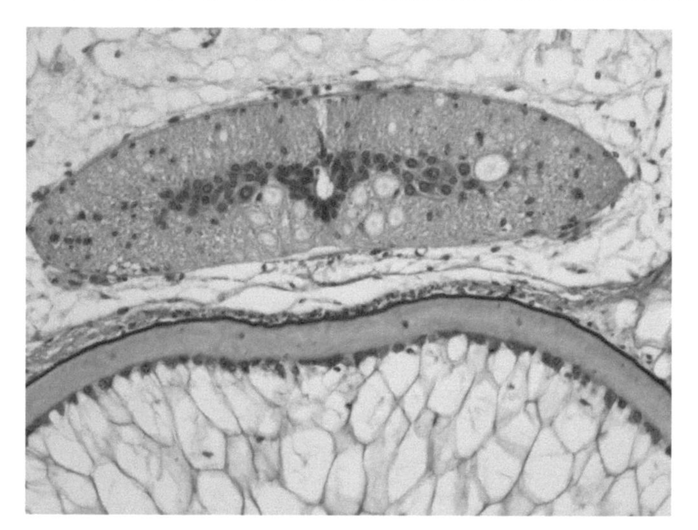

FIGURE F74a Ammocoete (larval lamprey). 20 ×.

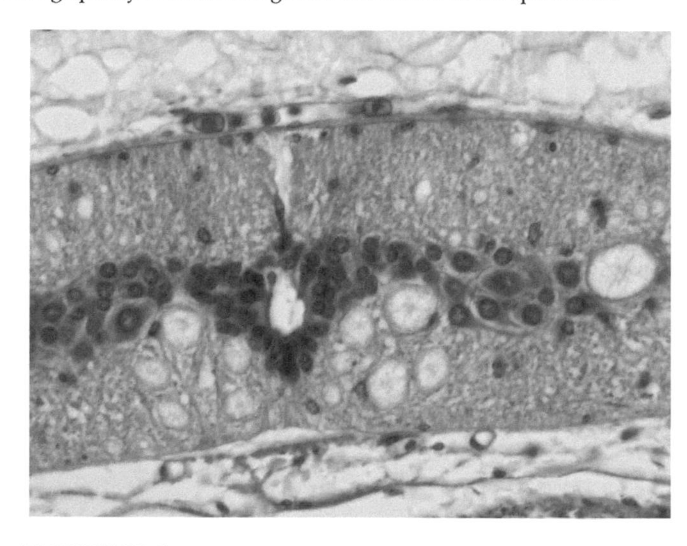

FIGURE F74b Ammocoete (larval lamprey). 40 ×.

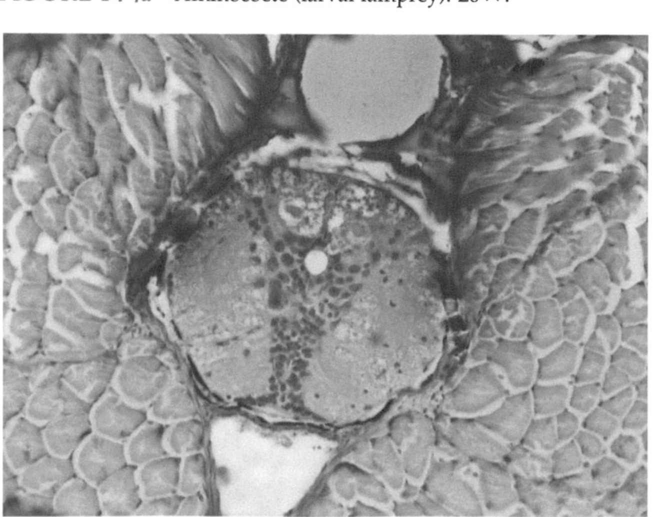

FIGURE F74c Stickleback (teleost fish) (19-392). 20 ×.

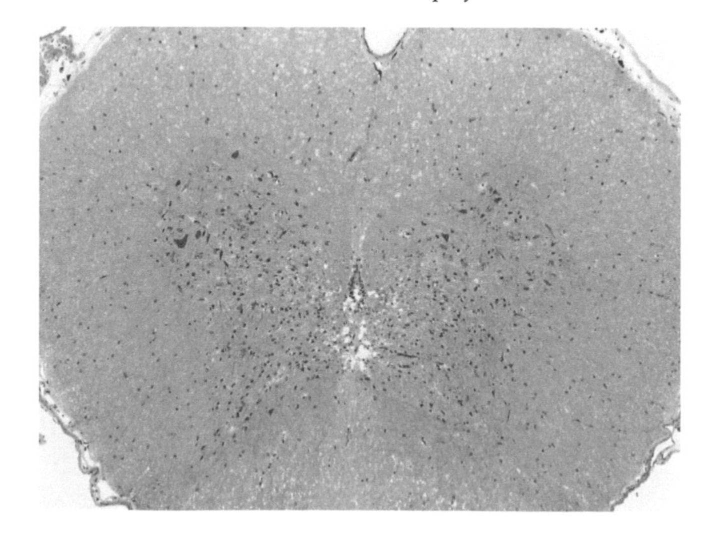

FIGURE F74d Frog. 5 ×.

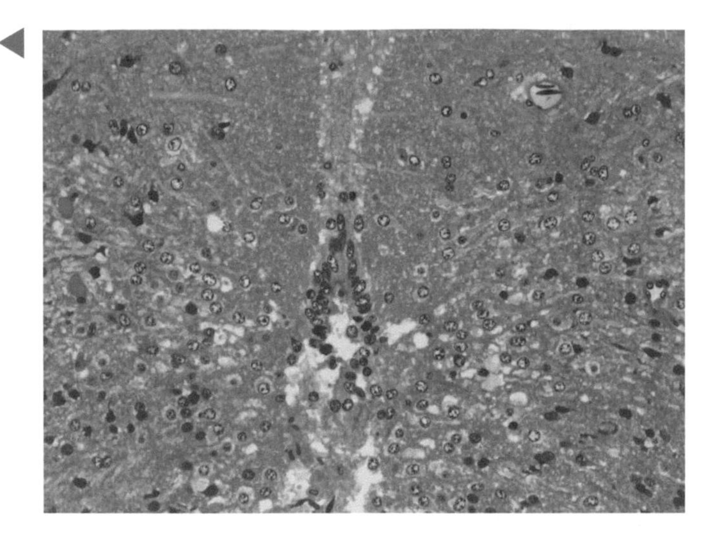

FIGURE F74e Frog. 20×.

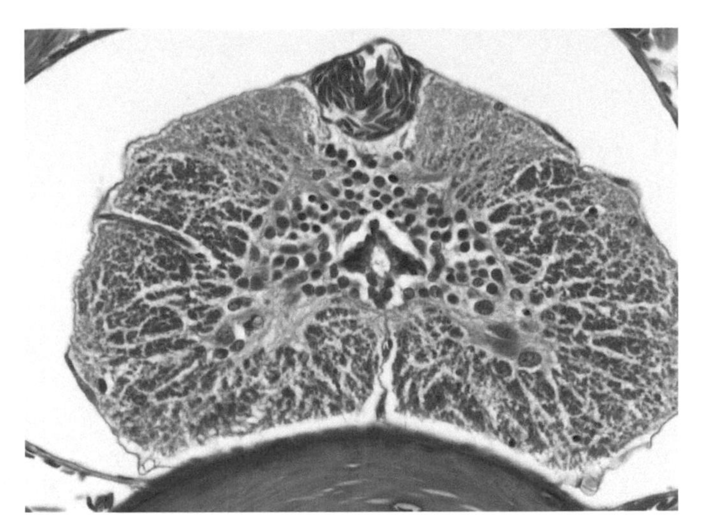

FIGURE F74f Apodan, a legless amphibian from Tanzania. This animal was collected in Africa some time ago, preserved in alcohol, sat on the shelf in a museum for over 20 years, was embedded in paraffin, sectioned, and stained in hematoxylin and eosin; some shrinkage inevitably has occurred.

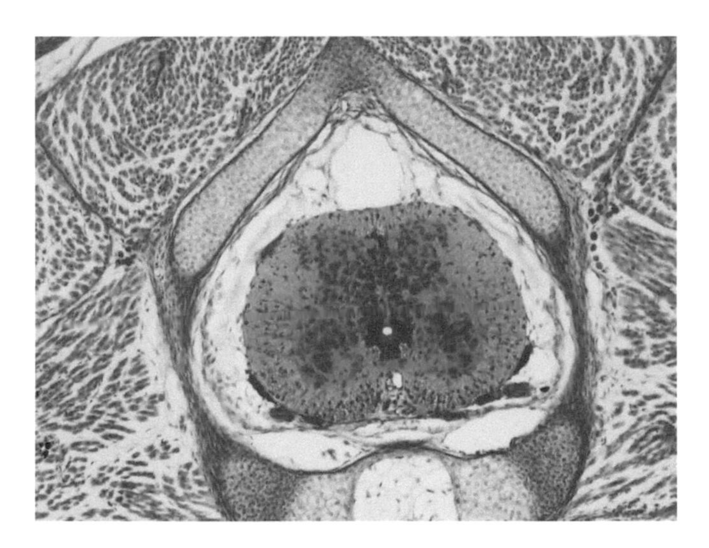

FIGURE F74g Lizard. 10×.

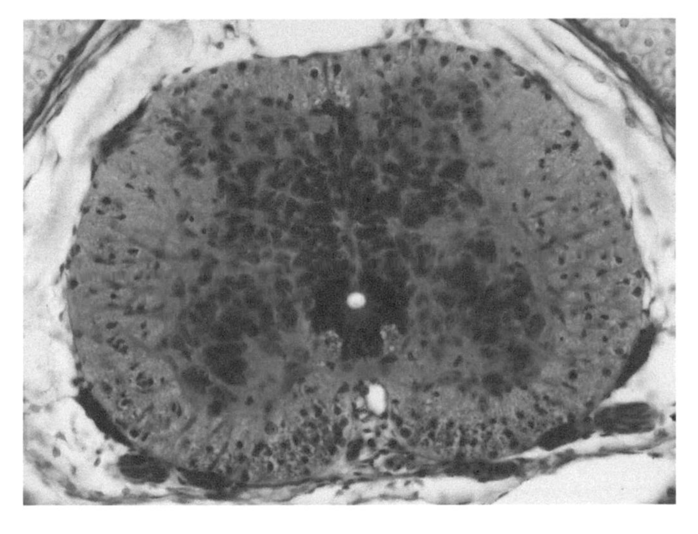

FIGURE F74h Lizard. 20×.

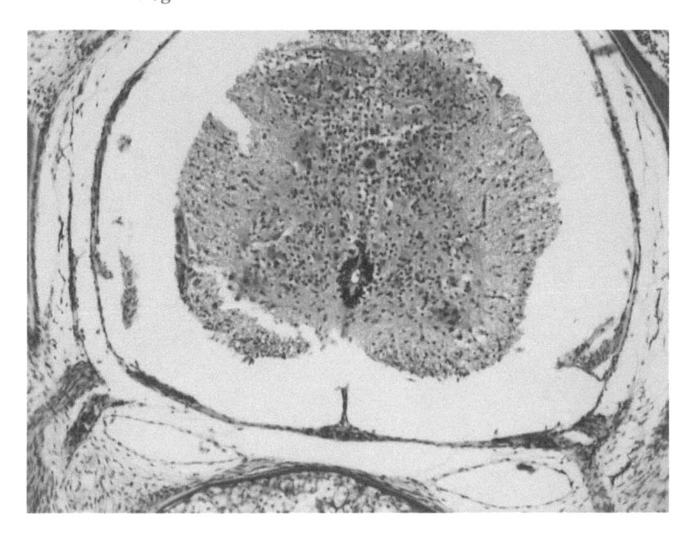

FIGURE F74i Snake. 10×.

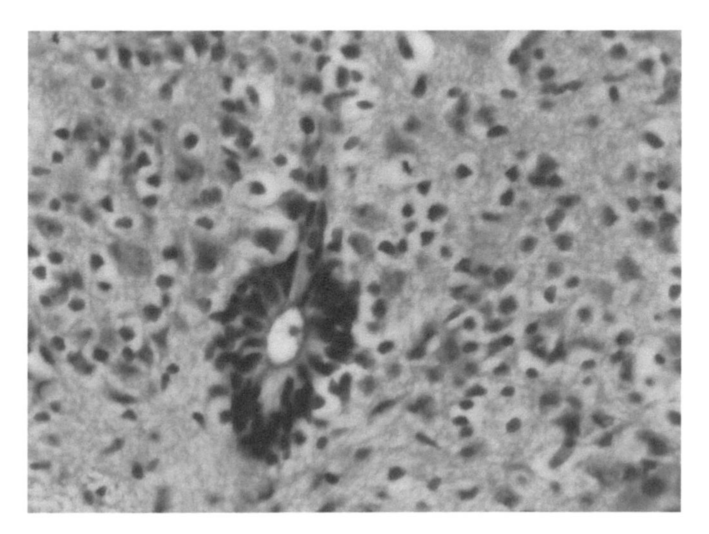

FIGURE F74j Snake. 40×.

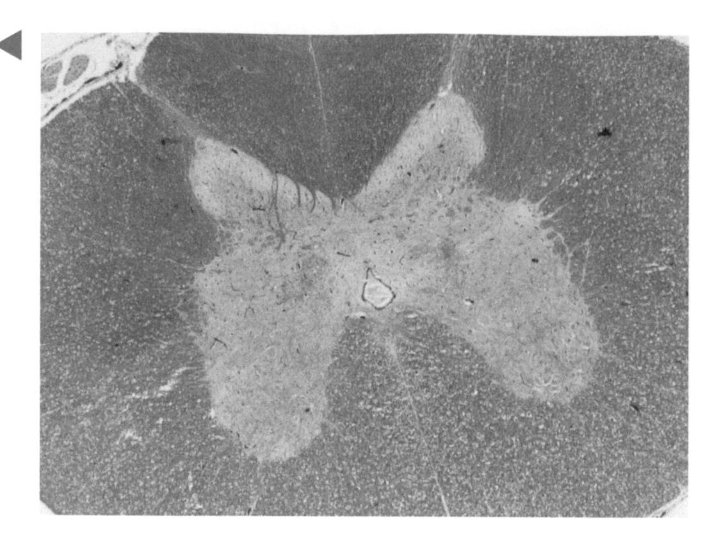

FIGURE F74k Mammal, Weigert stain. 2.5 ×.

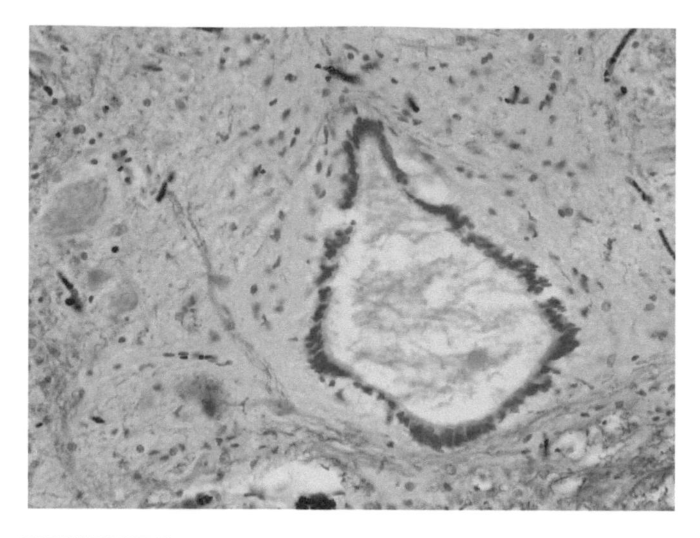

FIGURE F74l Mammal, Weigert stain. 20 ×.

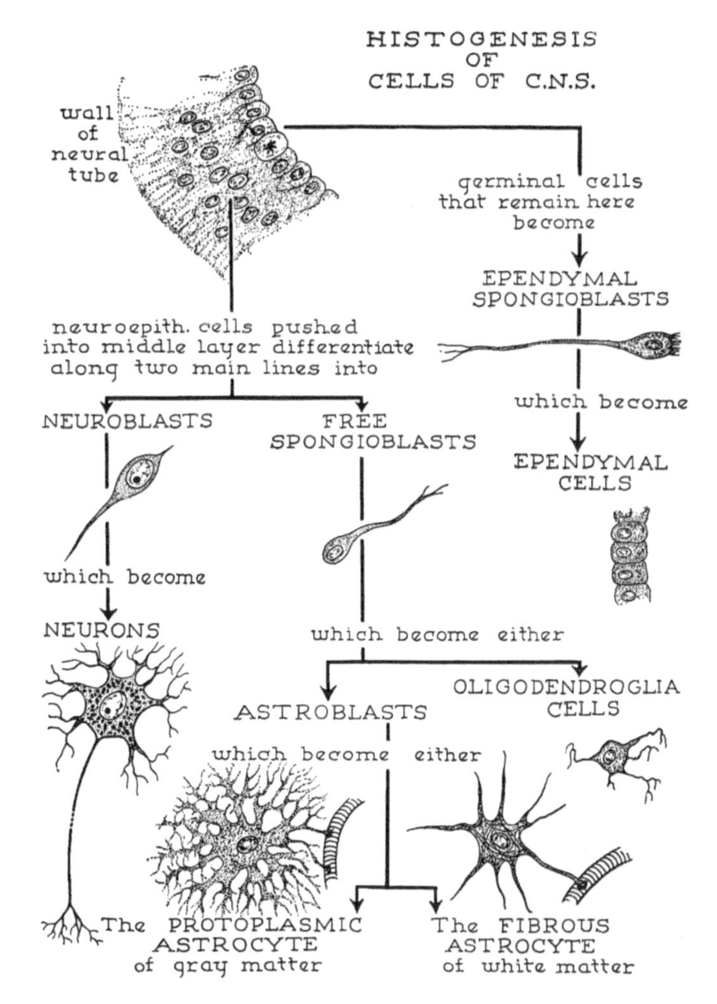

FIGURE F75 Diagram illustrating the main lines along which neuroectodermal cells of the neural tube differentiate in forming the central nervous system.

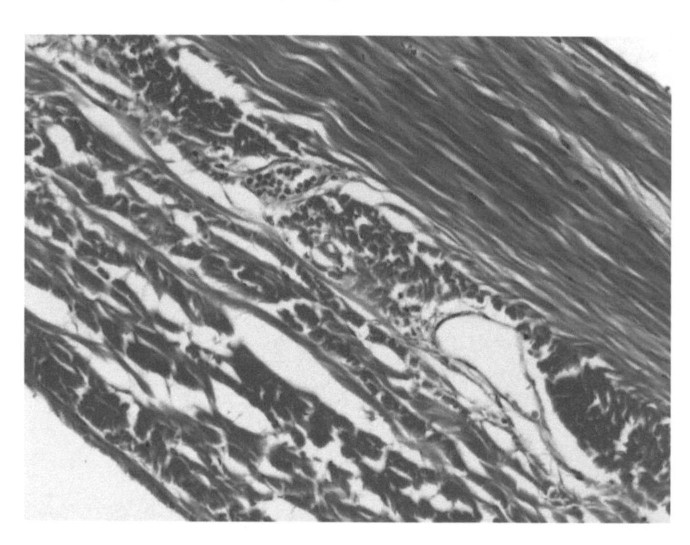

FIGURE F76 Section through the meninges of a mammal. The tough DURA MATER, composed largely of collagenous fibers, is at the top. At the bottom is the thin, delicate ARACHNOID LAYER. In between are the spaces of the ARACHNOID permeated by a few blood vessels. 20 ×.

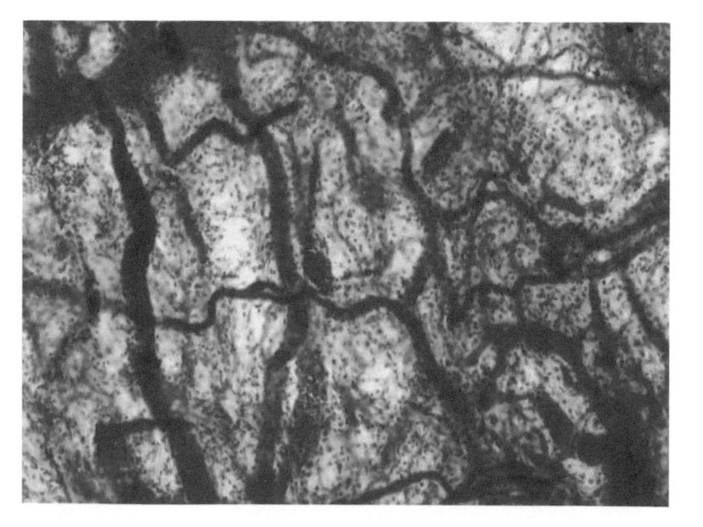

FIGURE F77 Whole mount of a fragment of the meninges of a mammal showing the multiplicity of blood vessels coursing through the arachnoid layer. 10 ×.

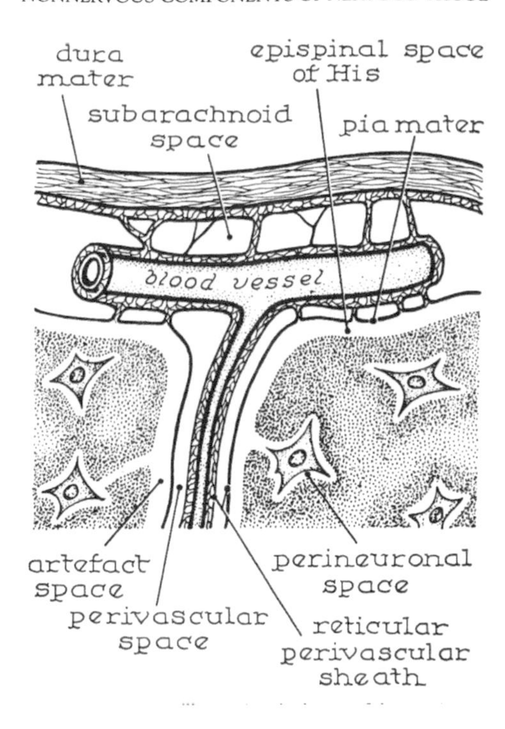

FIGURE F78 Diagram showing the three layers of meninges and the arachnoid space. Blood vessels permeate the meninges.

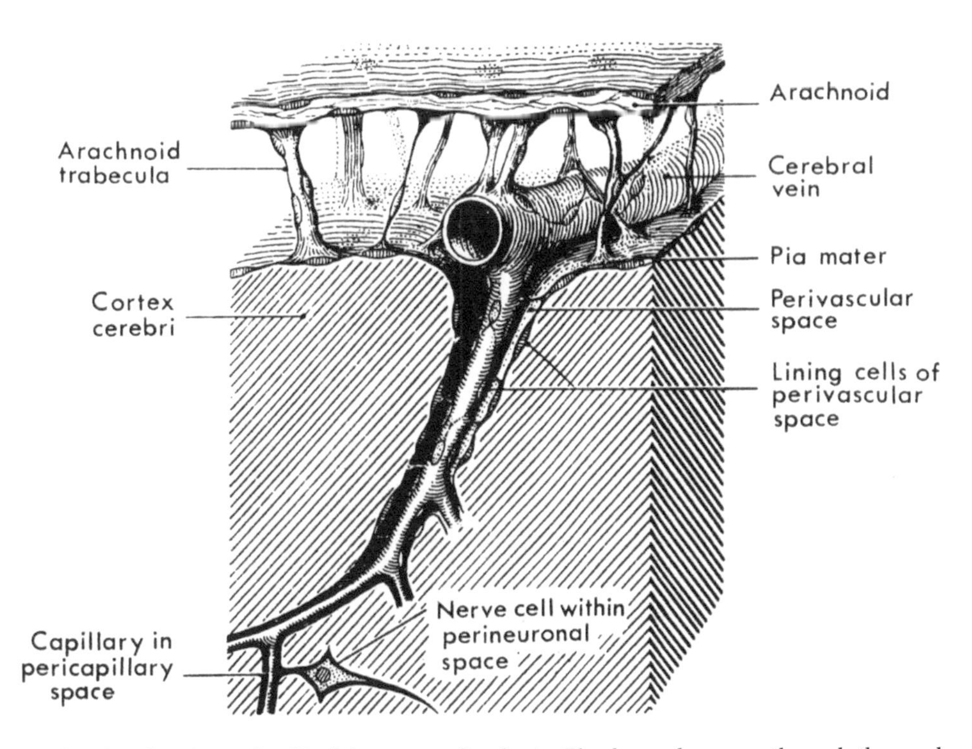

FIGURE F79 Diagram showing the pia-arachnoid of the mammalian brain. Blood vessels course through the arachnoid space. Those permeating the brain tissue are surrounded by fluid-filled perivascular spaces.

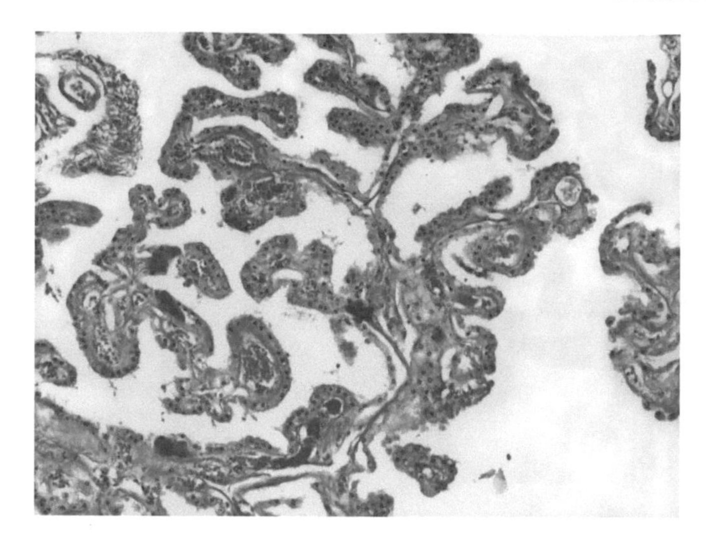

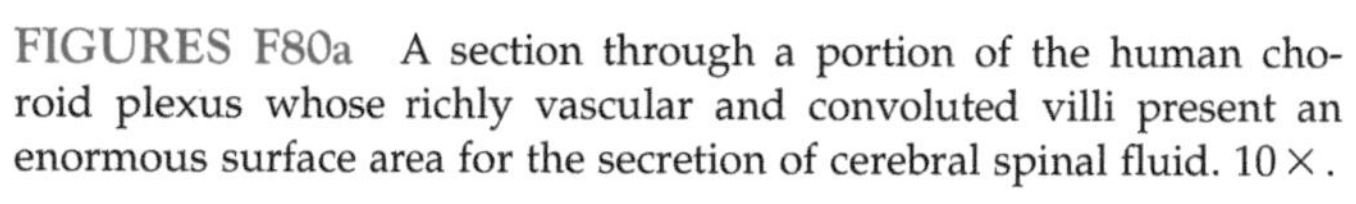

FIGURES F80a A section through a portion of the human choroid plexus whose richly vascular and convoluted villi present an enormous surface area for the secretion of cerebral spinal fluid. 10 ×.

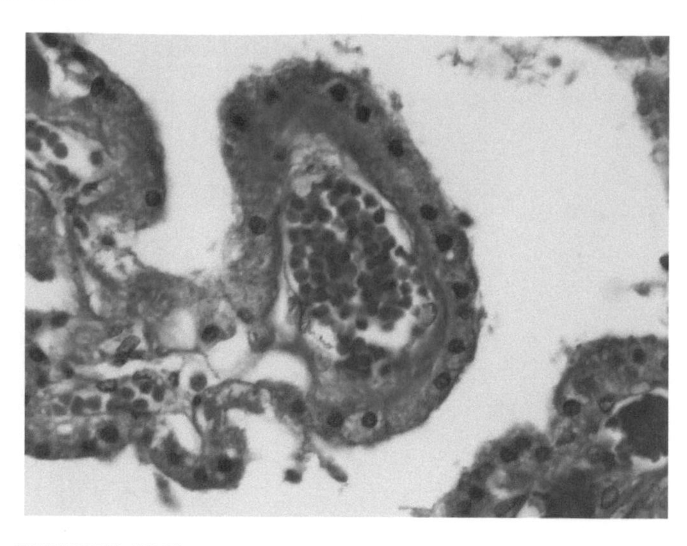

FIGURES F80b A higher magnification view of the section in Fig. F80a sectioned through a portion of the human choroid plexus 40 ×.

CHAPTER

G

Blood and Lymph

BLOOD

Blood is a tissue whose intercellular matrix, the PLASMA, is a liquid. The cellular components are ERYTHROCYTES and LEUKOCYTES, the former outnumbering the latter in humans 600 to 1 (Fig. G1). Erythrocytes, or red blood corpuscles, contain the respiratory pigment HEMOGLOBIN and have an orange tint when seen under the microscope; leukocytes, or white cells, are colorless. Blood is usually enclosed in vessels that are lined by endothelium. Blood, together with the endothelium, lymph, and blood-forming tissues, constitute the vascular tissues.

Blood is often studied in SMEARS where a drop of blood is placed on a cleaned glass slide and drawn across the slide by a second slide (Fig. G2). The cells may be stained and examined under a microscope with or without mounting a coverslip. The stain used is often Wright's stain, one of a group, ROMANOWSKY STAINS, that are used for blood smears. They are mixtures of the pink acid dye EOSIN with METHYLENE BLUE (basic). The methylene blue is in part oxidized into basic AZURES that are various shades between violet and blue. In addition to these components there is an eosin–azure–methylene blue complex that is neutral in reaction and lilac in color. Blood cells stain in a range of colors with these mixtures and their classification is based largely upon their staining reactions (Fig. G1).

Erythrocytes

In a stained smear of human blood, the erythrocytes measure 7.6 μm in diameter but the living cell is 8.5 μm. Since their size remains constant they can be used as built-in scales for estimating the size of other cells. Mammalian erythrocytes are typically biconcave discs where the center is thinner than the periphery and appears lighter in color than the edges (Figs. G3 and G4). Erythrocytes contain the pigment HEMOGLOBIN that increases their capacity for carrying oxygen. Mammalian erythrocytes have a marked tendency to adhere to one another by their broad surfaces and to assemble in long, curved columns resembling stacks of coins (Fig. G5). These columns are called ROULEAUX and are most prevalent in smears that have not been prepared quickly. Compare the shape and size of human erythrocytes with those of other mammals (including members of the camel family)(Figs. G5, G6a–G6g). All mammalian erythrocytes are enucleate; most are biconcave discs except for those of the camel family which are ovoid, bulging at the center. The erythrocytes of other vertebrates: birds, reptiles, amphibian, fish, and agnathans (Figs. G7a–G7i) are nucleated and ovoid. The blood of some Antarctic fish may contain no erythrocytes nor hemoglobin. Their tissues receive adequate supplies of oxygen, however, because oxygen concentrations are high in cold water while metabolism of the fish is low.

Seen with the electron microscope (Fig. G8a), the mammalian erythrocyte is a characteristically dense and homogeneous mass enclosed by a plasma membrane. A few ribosomes and vestiges of organelles may persist from an earlier nucleated stage. The cytoplasm of nucleated erythrocytes of nonmammals is similar and may be recognized easily in sections (Figs. G8b–G8e). Marginal bands of microtubules help to maintain the characteristic ovoid shape of these cells (Fig. G8c).

An Atlas of Comparative Vertebrate Histology.
DOI: https://doi.org/10.1016/B978-0-12-410424-2.00007-X

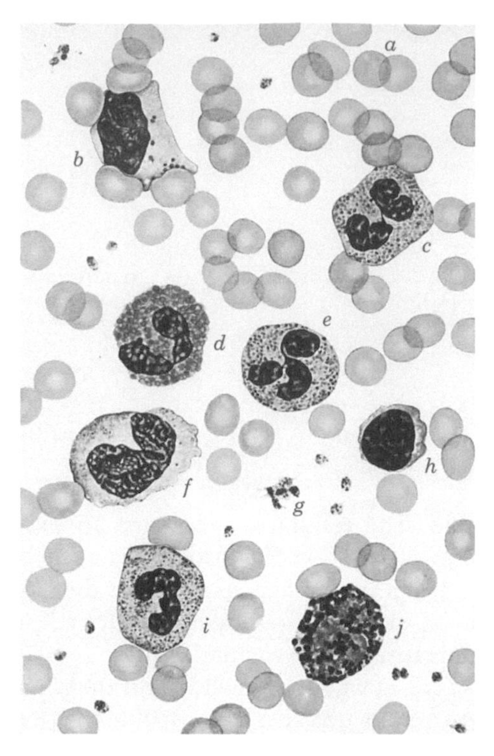

FIGURE G1 Cell types found in smears of peripheral human blood stained with Wright's stain. (a) Erythrocytes, (b) Large lymphocyte, (c) Neutrophil with segmented nucleus, (d) Eosinophil, (e) Neutrophil with segmented nucleus, (f) Monocyte, (g) Platelets, (h) Lymphocyte, (i) Neutrophil (band), (j) Basophil.

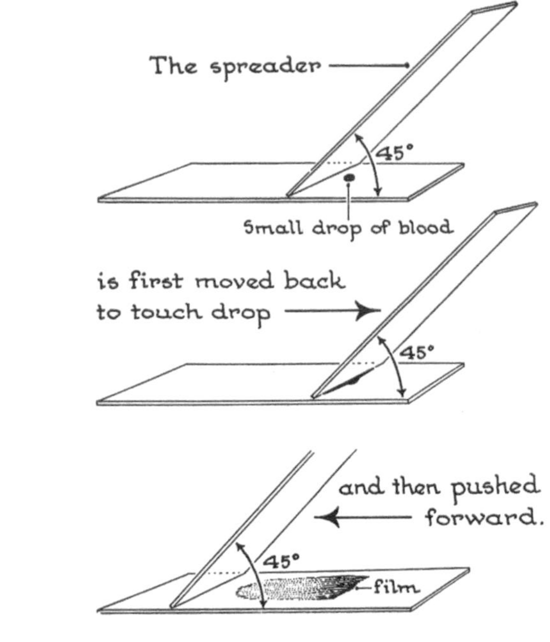

FIGURE G2 Preparation of a blood smear.

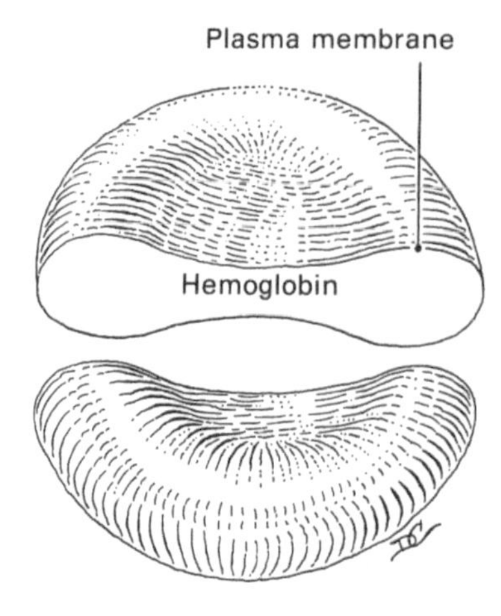

FIGURE G3 Erythrocytes of most mammals are *biconcave discs*—i.e., they resemble a tire on a wheel: narrower at the center, wider at the periphery.

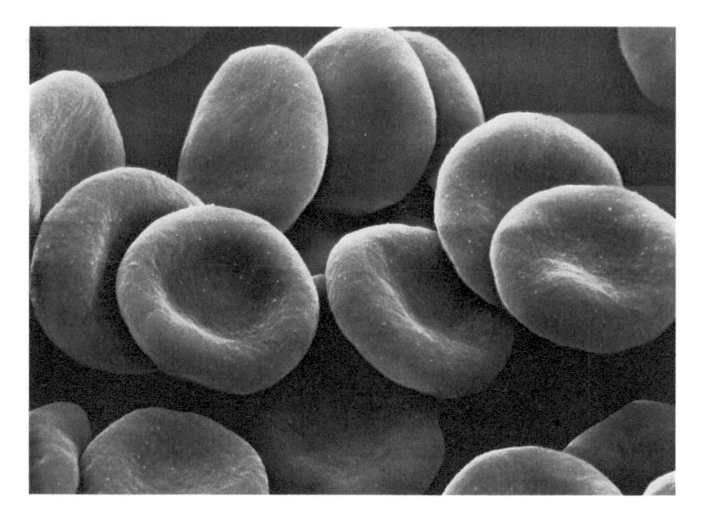

FIGURE G4 Human erythrocytes as seen with the scanning electron microscope. 7800 ×.

Leukocytes

Leukocytes contain no pigment and are colorless in unstained preparations; their designation is based on their staining reaction with a Romanowsky stain (Fig. G9). They usually do not have a constant shape because of their capacity for amoeboid movement. They are differentiated from erythrocytes: they are larger, they possess nuclei in all vertebrates, hemoglobin is not present in their cytoplasm, and they are capable of migration into other tissues of the body. Leukocytes may be divided into two main groups: GRANULOCYTES that have granules (in life these are membrane-bound semifluid droplets) in their cytoplasm, and AGRANULOCYTES that do not.

Granulocytes

Granulocytes are further subdivided on the basis of the staining characteristics of their granules.

a. *Neutrophils* constitute 65%–75% of leukocytes in human blood. Neutrophils are active phagocytes and destroy small organisms, especially bacteria. Numerous small, closely packed granules of indefinite color are present in the cytoplasm. In a neutrophil the nucleus is elongated, crescentic, or fragmented and usually consists of three to five lobes (polymorphic nucleus). The degree of lobulation increases progressively with the age of the cell (Fig. G10). Some neutrophils of females show a small projection from the nucleus, the DRUMSTICK, that represents the sex chromatin (Figs. G11a and G11b). Its presence is used in the diagnosis of chromosomal sex.

In electron micrographs, most vertebrate leukocytes often show fine cytoplasmic processes (Figs. G12a–G12f). Neutrophils are actively motile and phagocytic (Fig. G12g). They usually have a polymorphic nucleus that may appear fragmented in thin sections. Aside from the granules, the cytoplasm contains diminished numbers of the usual organelles. Two kinds of granules are found in neutrophils: AZUROPHILIC and SPECIFIC GRANULES (Figs. G13ba and G13b). The dense-cored azurophilic granules are lysosomes; the specific granules are smaller, more numerous, and have an antibacterial function.

b. *Eosinophils or acidophils* constitute 2%–5% of leukocytes in human blood (Fig. G14). Conspicuous acidophilic granules are present in the cytoplasm. The nucleus may appear bean shaped or lobulated. The background cytoplasm is usually lightly stained. Eosinophils help destroy parasites and modulate allergic inflammatory responses. The granules of eosinophils (Figs. G15a and G15b) are membrane limited and are lysosomes. They may be over 1 μm in diameter and usually take the form of biconvex discs with a characteristic dense crystalline core (Figs. G16a–G16d).

Neutrophils of many vertebrates (e.g., rabbit, chicken) are intensely eosinophilic and are often referred to as HETEROPHILS. The granules vary in shape from rod-like to spherical, in both

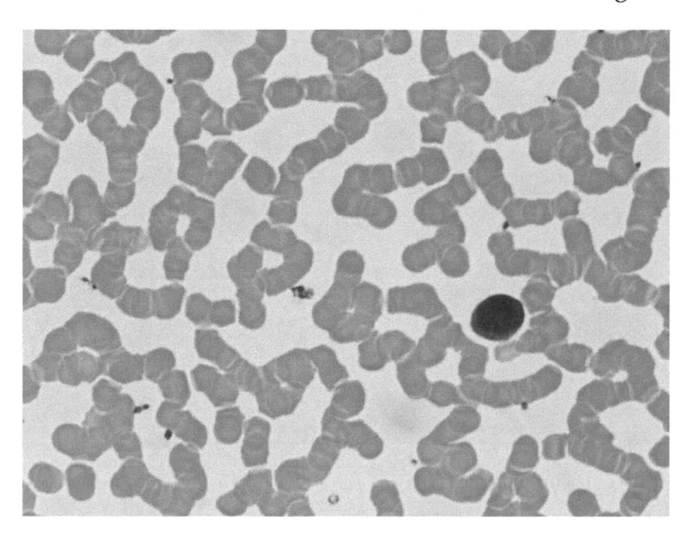

FIGURE G5 Armadillo. The erythrocytes in this preparation have linked together like stacks of coins. These are *rouleaux* and indicate delays in the preparation of the smear. A few platelets and a lymphocyte are seen. (This figure shows blood smears stained with Romanowsky stain). 63x.

FIGURE G6 (a–g) These figures show samples of mammalian blood smears stained with Romanowsky stain. All Photomicrographs were taken with a 63x objective lens.

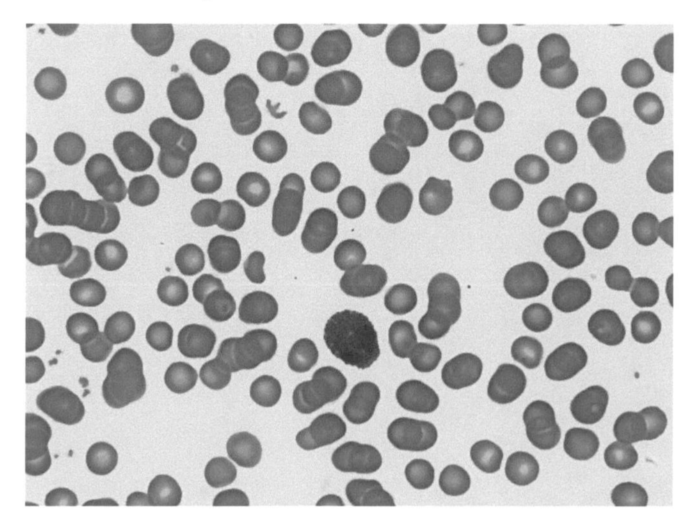

FIGURE G6a Human. There is an eosinophil near the center. 63x.

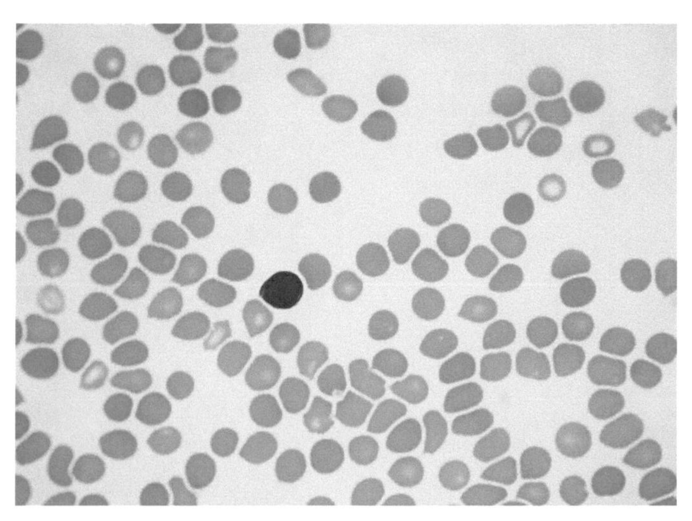

FIGURE G6b Rabbit. There is a small lymphocyte near the center. 63x.

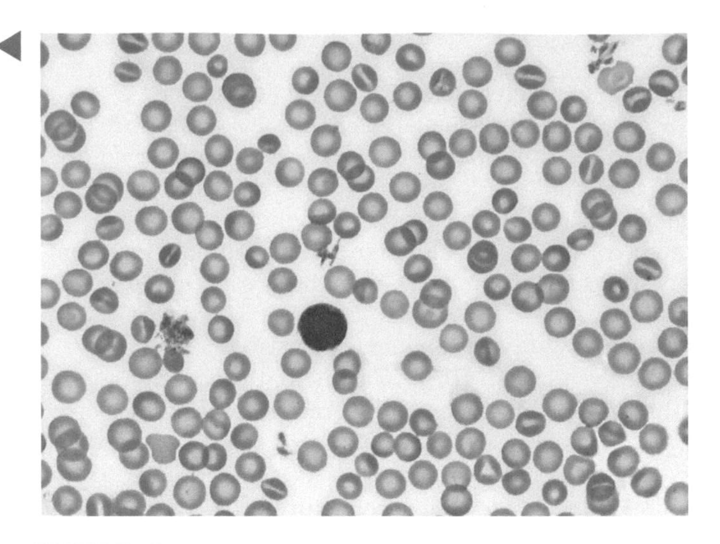

FIGURE G6c Squirrel. Another small lymphocyte among the erythrocytes. 63x.

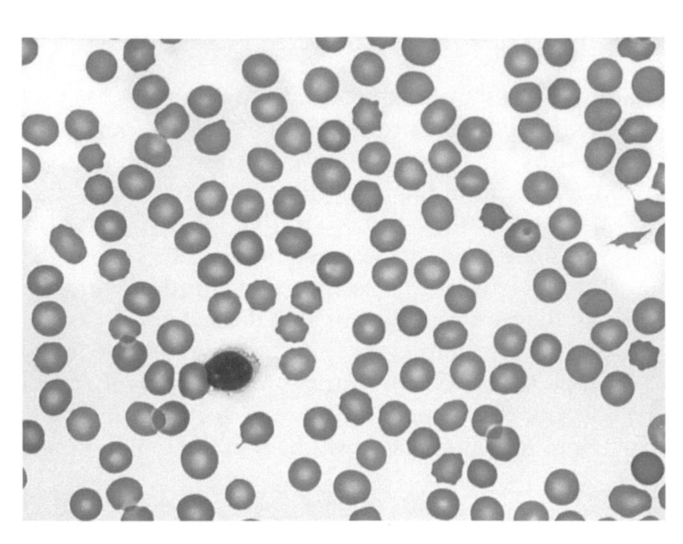

FIGURE G6d Guinea pig. This may be a medium-sized lymphocyte, but it is hard to tell. 63x.

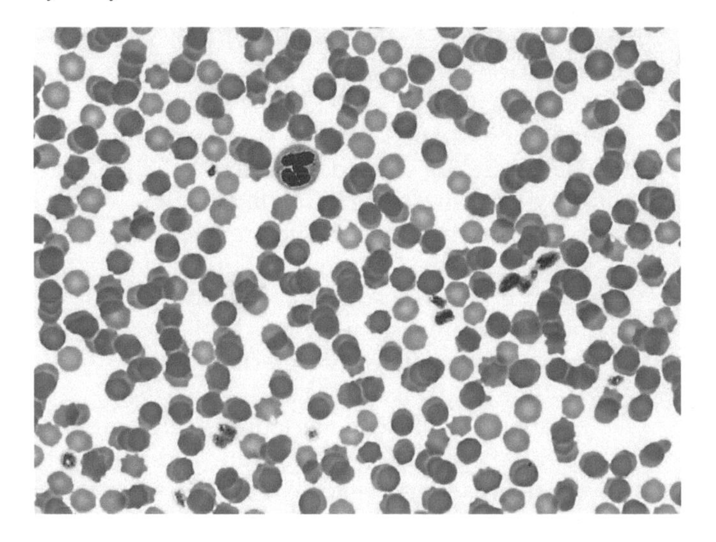

FIGURE G6e Cat. A polymorphonuclear neutrophil and a few platelets show up in the blood of a cat. 63x.

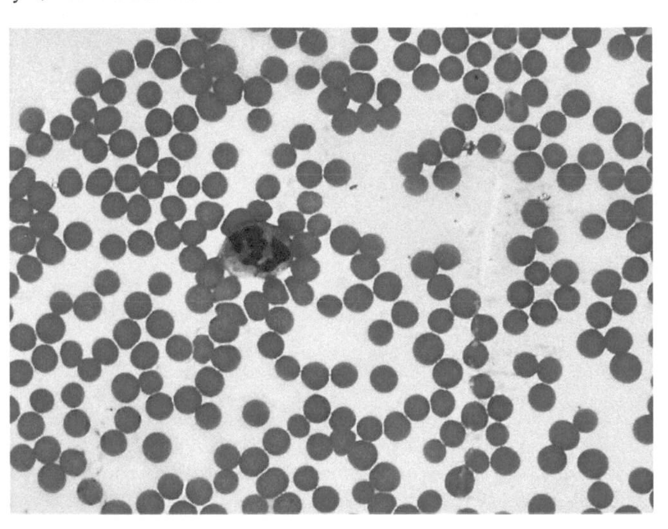

FIGURE G6f Horse. Although the horse is a large mammal, its blood cells are of comparable size to those seen in the other slides. There is a polymorphonuclear neutrophil near the center. 63x.

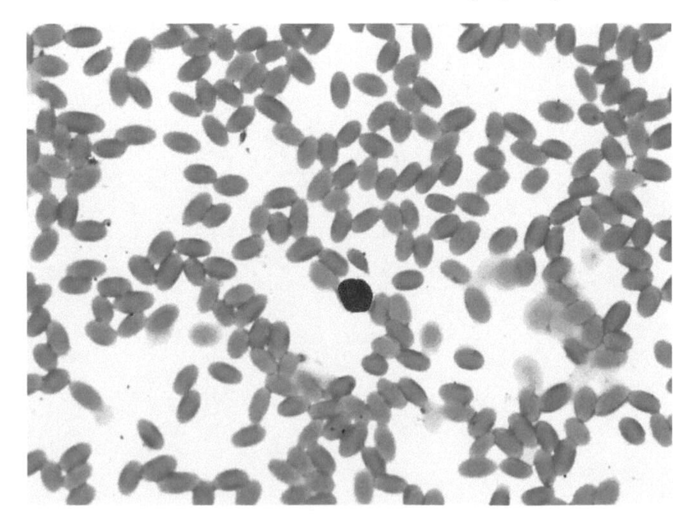

FIGURE G6g Camel. Erythrocytes of camels and their relatives are unlike the biconcave discs of most mammals but are ovoids that bulge at the center. There is a small lymphocyte near the center. 63x.

FIGURE G7 (a–i) These figures show blood smears from a variety of non-mammalian animals stained with a Romanowsky stain. All photomicrograhs were taken with a 63x objective lens.

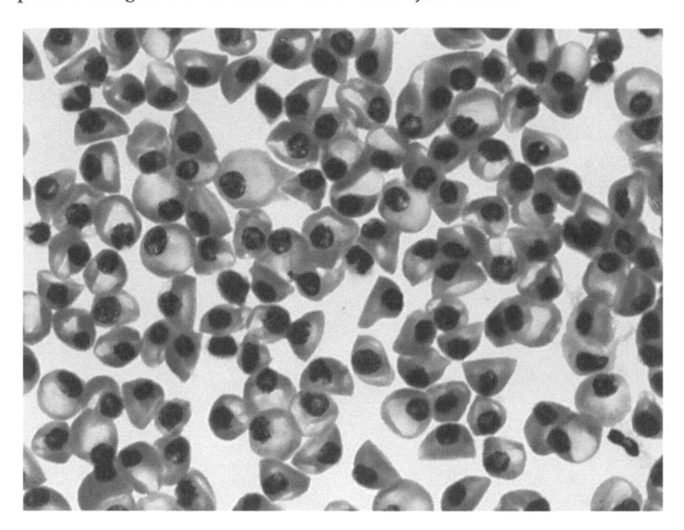

FIGURE G7a Lamprey. Roughly speaking these crumpled cells are ovoid. Their nuclei are apparent. 63x.

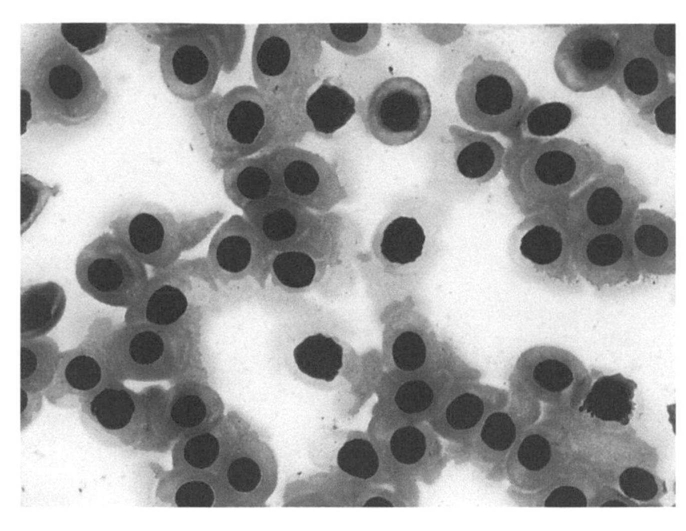

FIGURE G7b Shark. Several unidentified purple "smudges" are seen. The erythrocyte with slightly basophilic cytoplasm at the upper center is a developing erythrocyte; while all the cells in mammalian blood are mature, immature cells are released into the blood of lower vertebrates and maturation takes place in the circulating blood. 63x.

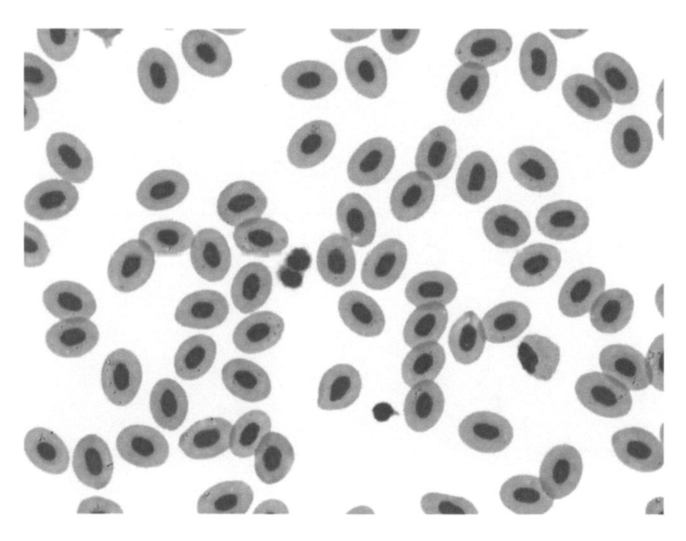

FIGURE G7c Bass. There are two thrombocytes clinging together near the center. 63x.

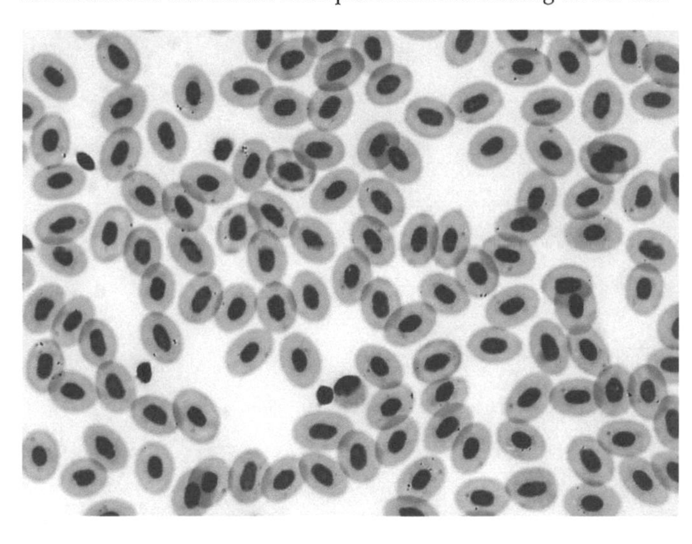

FIGURE G7d Carp. Near the upper center is an erythrocyte with bluish cytoplasm indicating an immature cell. 63x.

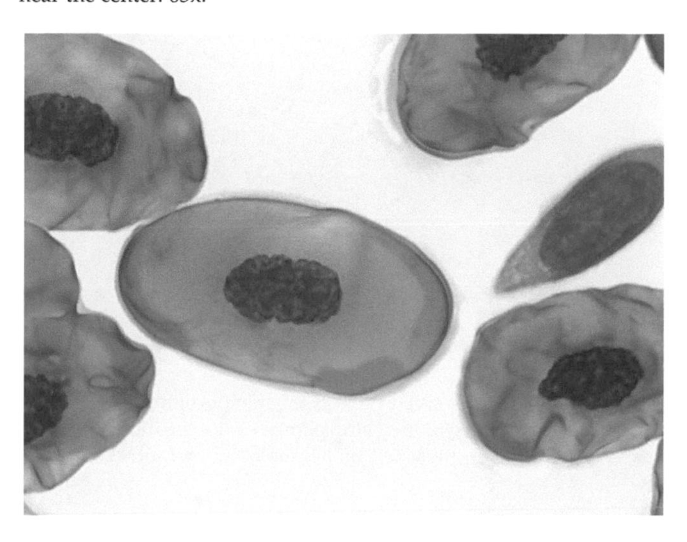

FIGURE G7e *Amphiuma*. The cells of *Amphiuma* are enormous; it is hard to believe that these cells were photographed at the same magnification as the other blood cells shown here. 63x.

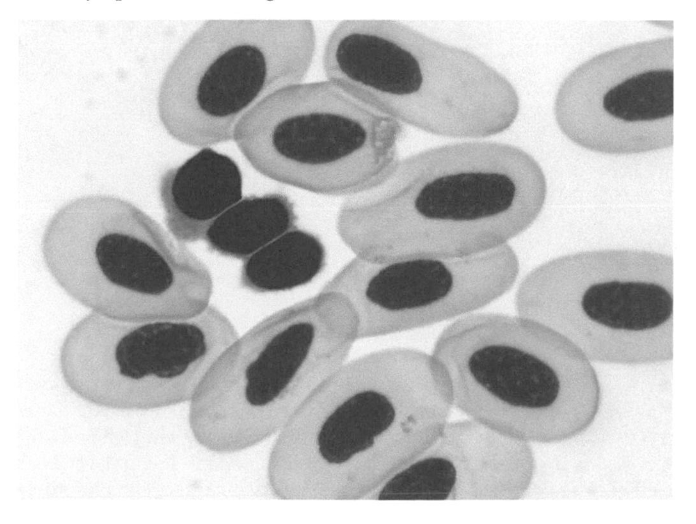

FIGURE G7f *Cryptobranchus*. The three cells holding together at the left of center are thrombocytes. 63x.

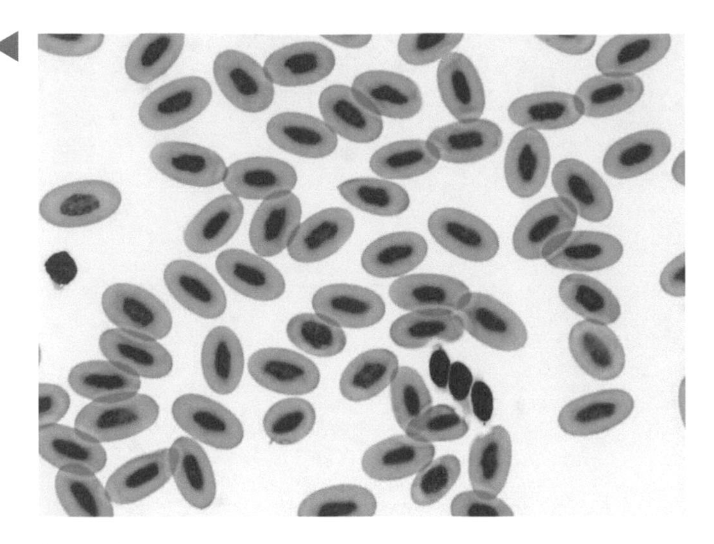

FIGURE G7g Lizard. Thrombocytes are characteristic of non-mammals and three are shown here at the bottom right. 63x.

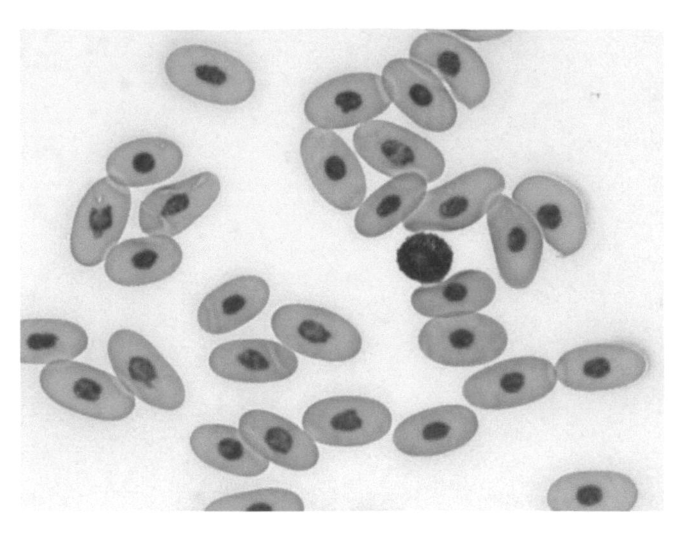

FIGURE G7h Alligator. The cell at the right of center seems to be a small lymphocyte. 63x.

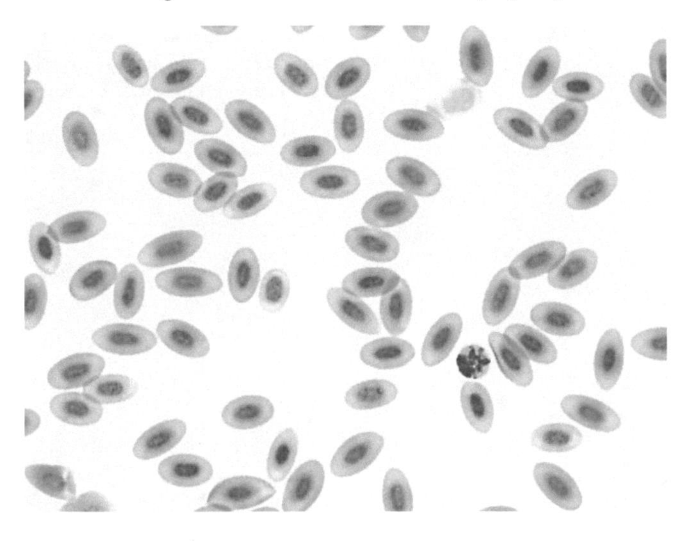

FIGURE G7i Chick. The cell at the lower right is a polymorphonuclear neutrophil. 63x.

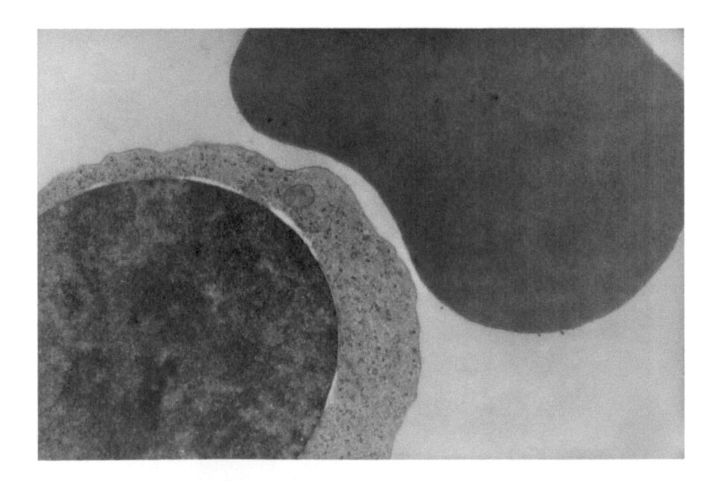

FIGURE G8a Transmission electron micrograph of an erythrocyte (*upper right*) and an erythroblast. Mature mammalian erythrocytes have shed their nuclei during development and are enucleate. At the lower *left* is a developing erythrocyte—its nucleus has not yet been expelled. The cytoplasm of this cell contains several organelles; these will be absorbed during maturation of the cell. 30,000 ×.

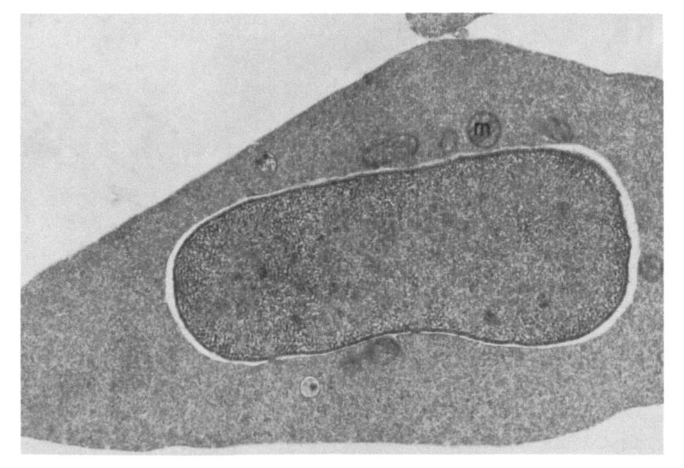

FIGURE G8b Transmission electron micrograph of an avian erythrocyte. The density of the cytoplasm is characteristic of erythrocytes and reflects the presence of hemoglobin. *m*, mitochondrion. 18,000 ×.

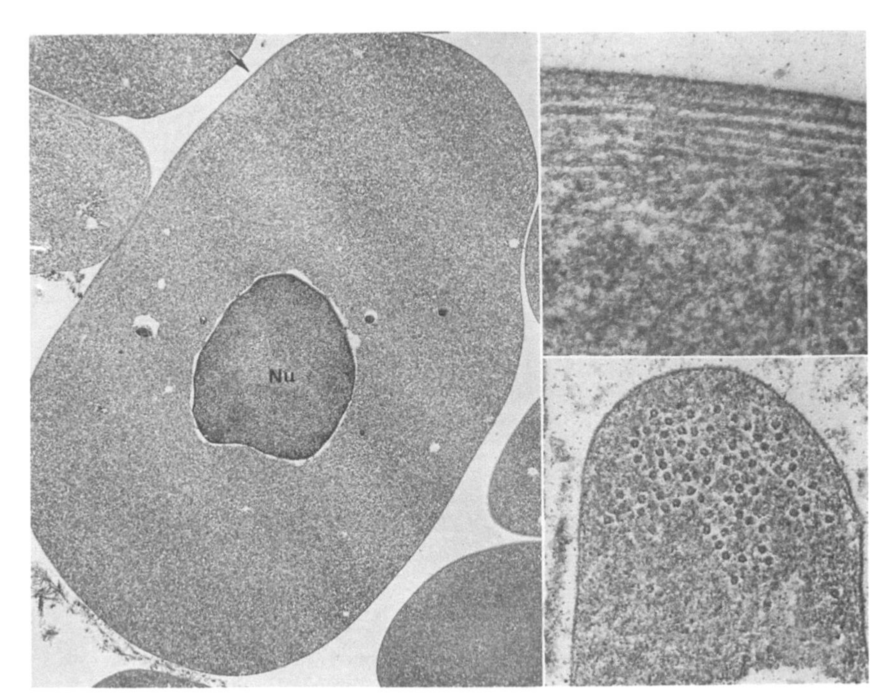

FIGURE G8c Transmission electron micrograph of a longitudinally sectioned mature erythrocyte of a tuatara (reptile). A marginal band of microtubules is shown at a higher magnification in the insets at the right. 68,000 × *top*; 66,000 × *bottom*.

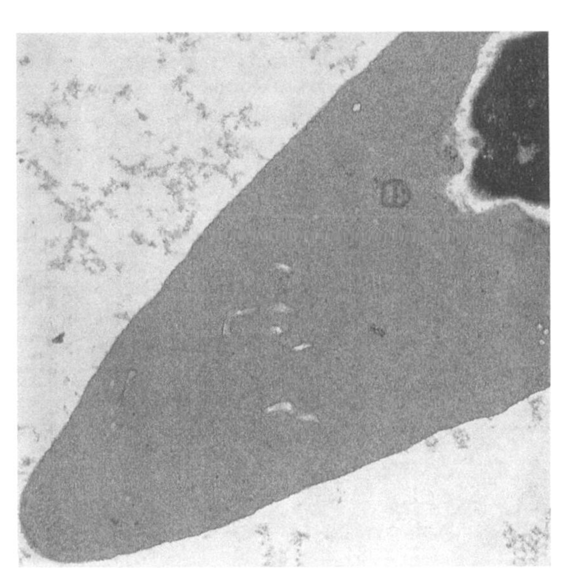

FIGURE G8d Transmission electron micrograph of an erythrocyte of a mature tadpole. A few ribosomes are seen. A marginal band of microtubules is seen at the lower left corner. 32,000 × .

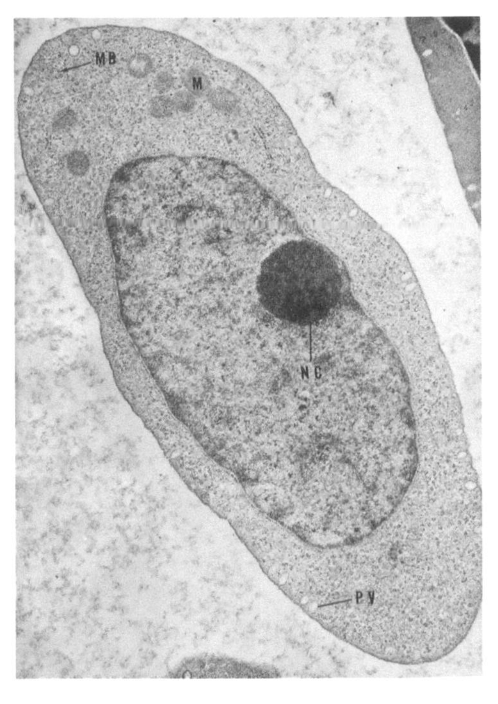

FIGURE G8e Transmission electron micrograph of a young erythrocyte of a hagfish. Many ribosomes and a few mitochondria (M) are seen in the cytoplasm as well as numerous pinocytotic vesicles (PV) just below the surface. A marginal band (MB) of microtubules encircles the cell. There is a well-developed nucleolus (NC). 22,400 × .

eosinophils and neutrophils. In many lower vertebrates it is difficult to distinguish between eosinophils and heterophils (neutrophils) and many cells designated as eosinophils are probably more closely related to mammalian neutrophils than to mammalian eosinophils. In view of these difficulties, heterophils (or neutrophils) probably should be defined functionally as the most active scavengers of the granulocytes. Eosinophils have been found in most species of vertebrates and in many fishes they constitute the most

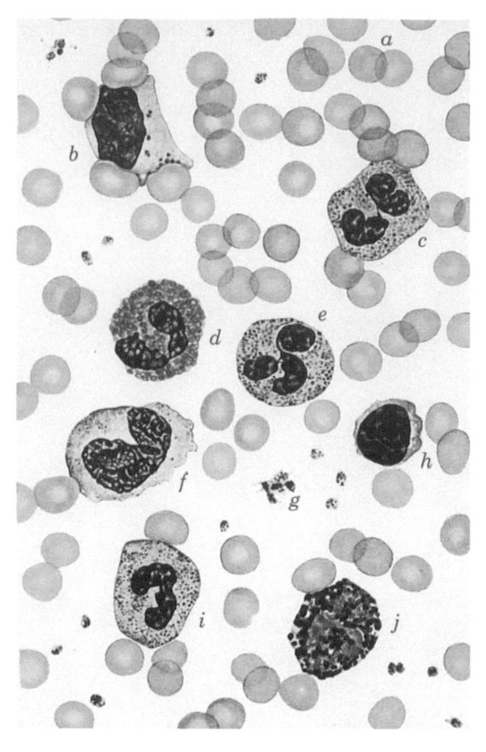

FIGURE G9 Cells found in a smear of normal human blood. (a) erythrocytes, (b) large lymphocyte, (c) neutrophil with segmented nucleus, (d) eosinophil, (e) neutrophil with segmented nucleus, (f) monocyte, (g) platelets, (h) lymphocyte, (i) neutrophil with band nucleus, (j) basophil.

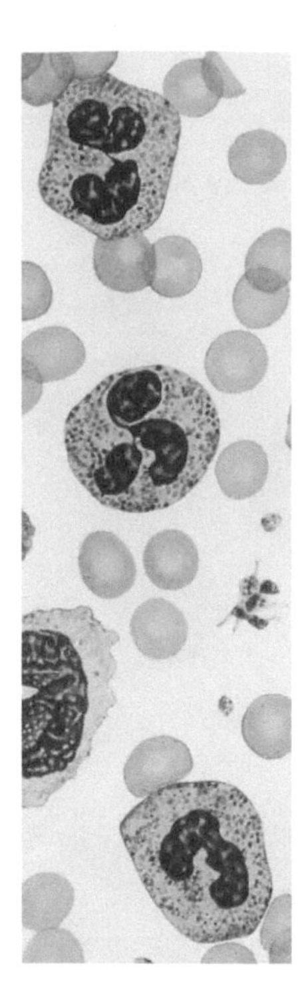

FIGURE G10 Neutrophils from a smear of normal human blood showing various degrees of nuclear segmentation. At the top and center, the nuclei are segmented. At the bottom the nucleus is in the band form. A few platelets and a portion of a monocyte also appear.

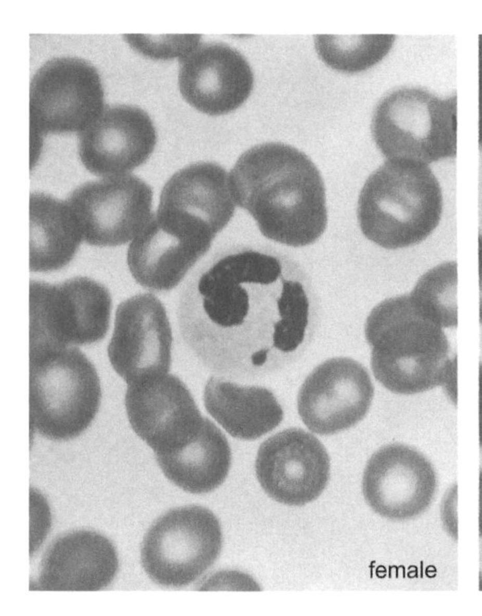

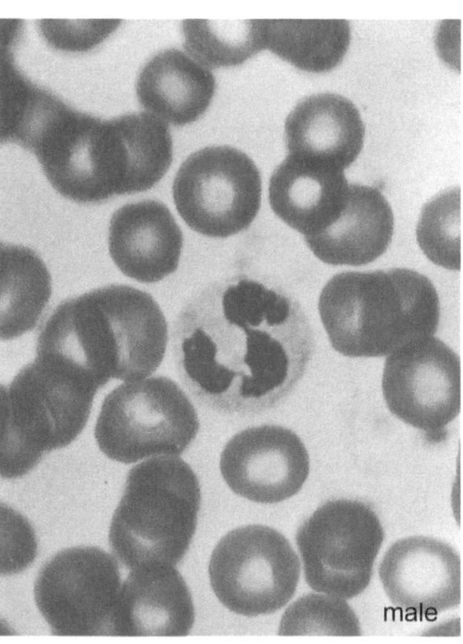

FIGURE G11a Photomicrographs of smears of human blood showing neutrophils at the center. The neutrophil on the left displays the "drumstick" of female blood while this is absent from the blood of a male (*right*).

abundant form of granulocyte. Much of the development of blood cells of lower vertebrates takes place in the circulating blood so that a great variety of cells, both immature and mature, may be observed. Many of the immature cells resemble agranulocytes.

Basophils. Only 0.5%–1.0% of human leukocytes are basophils (Fig. G17). These cells are packed with large, basophilic granules. The nucleus may have a crescentic shape or be lobulated and may be partially obscured by deeper staining granules that overlie it. Granules of basophils are membrane limited (Figs. G18a and G18b); their

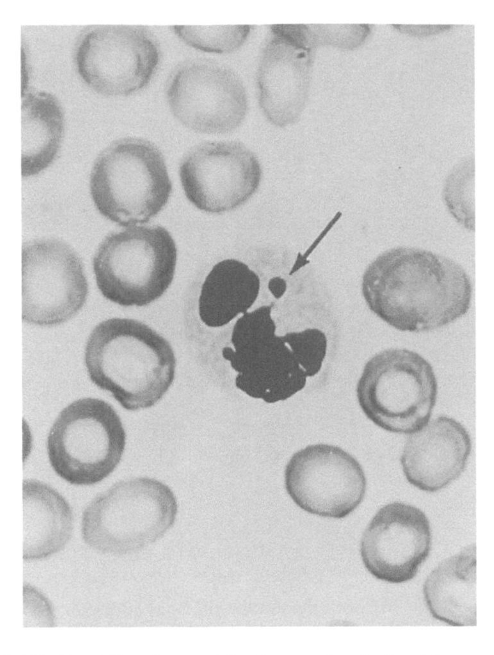

FIGURE G11b The drumstick shows clearly in this neutrophil of a human female.

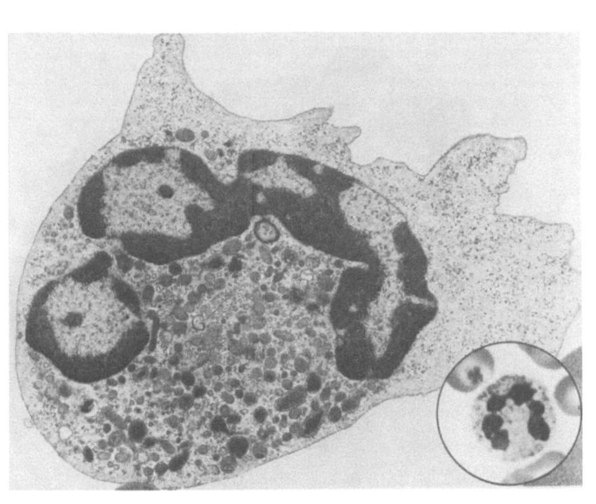

FIGURE G12a Transmission electron micrograph of a human mature neutrophil. The nucleus is multilobed; its heterochromatin is peripheral. There is a small Golgi complex (*G*); other organelles are sparse. Specific granules are less dense and more rounded than azurophilic granules. A photomicrograph of a comparable neutrophil in a smear is shown at the lower right.

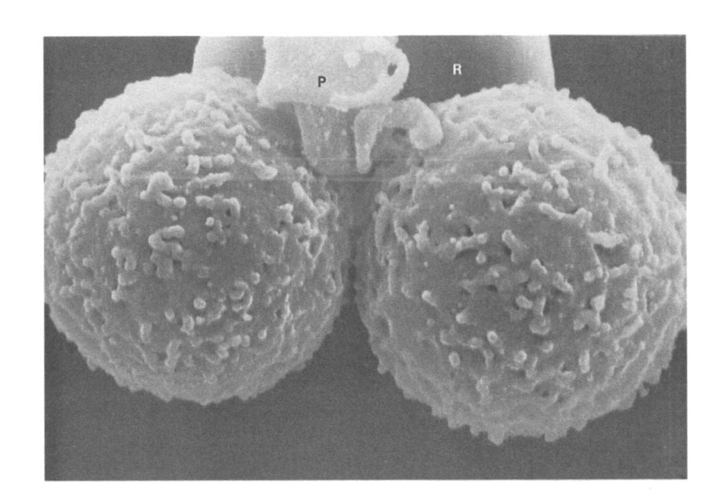

FIGURE G12b Scanning electron micrograph of two neutrophils in normal human blood. These cells bristle with short, twisted microvilli. *P*, platelets; *R*, erythrocyte. 16,000 ×.

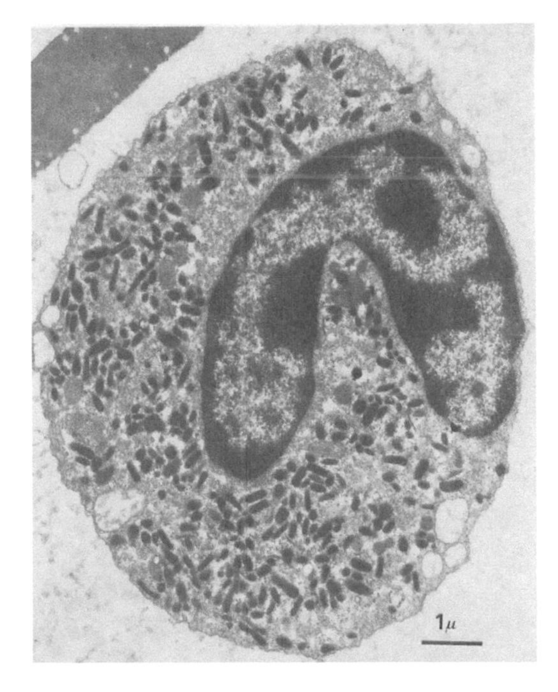

FIGURE G12c Transmission electron micrograph of a granulocyte from a hagfish. Granules and glycogen-like particles abound. 12,700 ×.

internal structure varies from species to species and may be fibrous, crystalline, or uniformly dense (Figs. G19a–G19e). Basophils are distinct from the mast cells or "tissue basophils" of connective tissue. Basophils secrete histamine that increases capillary permeability with a consequent drop in blood pressure. Serotonin is also released in some species.

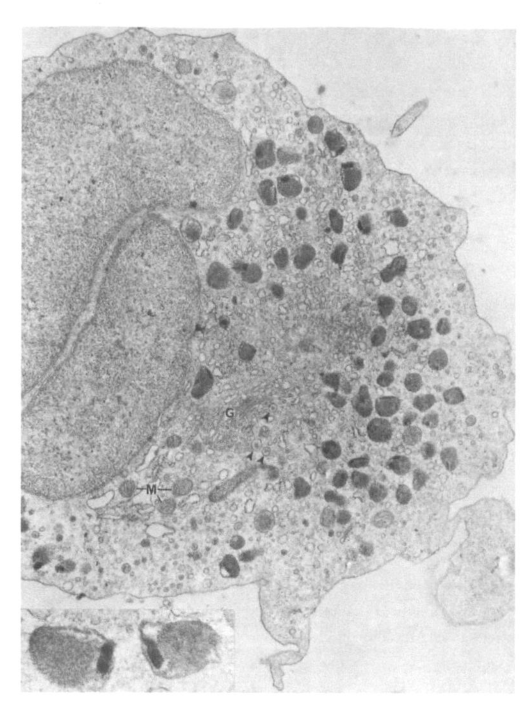

FIGURE G12d Transmission electron micrograph of a neutrophil from a white sucker (*Catostomus commersoni*). A bilobed nucleus, spherical mitochondria(M), and a Golgi complex (G) are surrounded by ribosome-studded cytoplasm. Two neutrophilic granules are shown at the lower left. 63,000 ×.

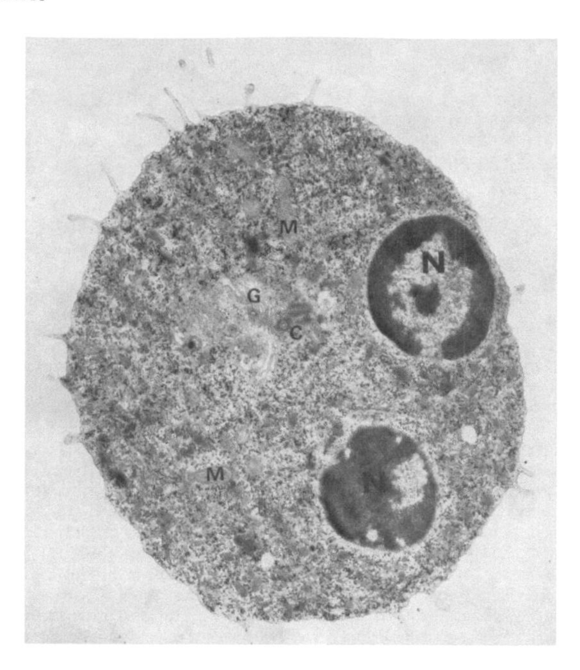

FIGURE G12e Transmission electron micrograph of a neutrophil from a toad (*Bufo vulgaris japonicus*) showing glycogen granules packed in the cytoplasm. *G*, Golgi complex; *M*, mitochondria; *N*, nucleus; *C*, centromere. 11,800 ×.

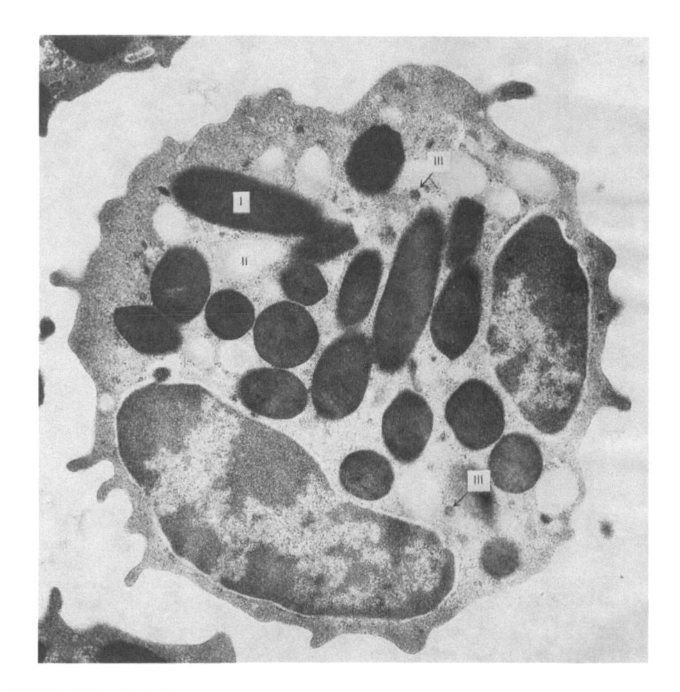

FIGURE G12f Transmission electron micrograph of a mature heterophil of a white leghorn chicken. There are three types of granules (I, II, and III). 24,000 ×.

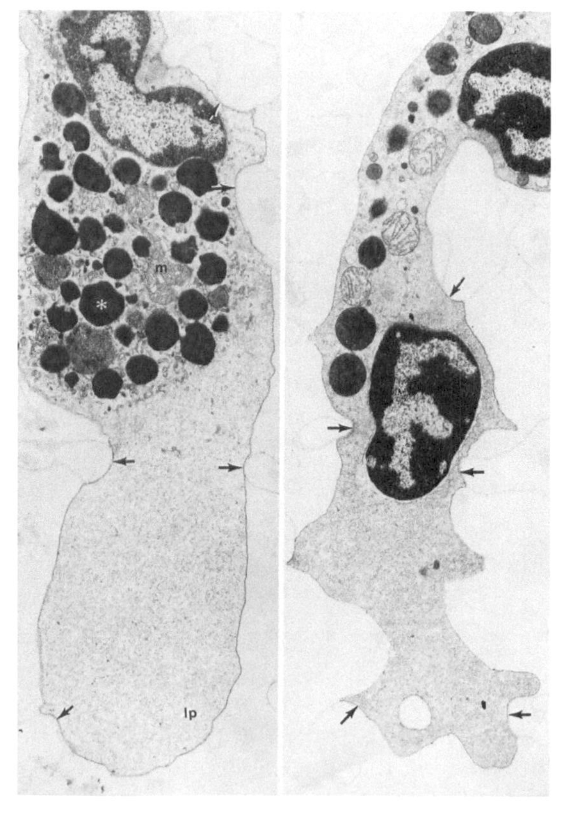

FIGURE G12g Transmission electron micrographs of sections through granulocytes undergoing migration in a nurse shark (*Ginglymostoma cirrhatum*). Movement follows the tongue-shaped lamellipodium (*lp*). Arrows indicate points of attachment to an external surface. *, granules; *m*, mitochondrion; 15,000 ×.

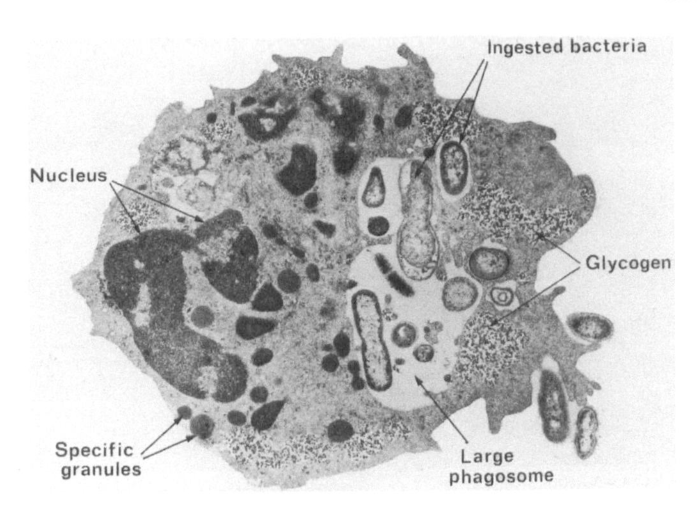

FIGURE G13a Transmission electron micrograph of a section of a neutrophil that has ingested several bacteria. The bacteria in the large vacuole show signs of beginning degeneration.

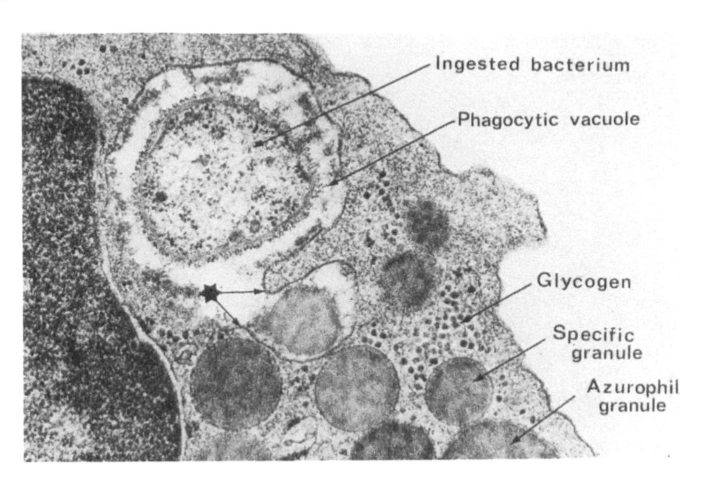

FIGURE G13b Transmission electron micrograph of a portion of a neutrophil that has recently ingested a bacterium. Arrows leading from the asterisk (*) indicate sites of continuity of the membrane of the phagosome with that of a granule about to discharge its contents into the phagosome.

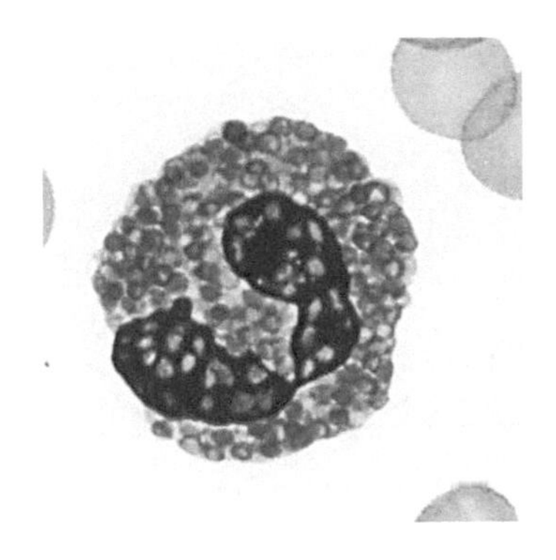

FIGURE G14 Drawing of an eosinophil from a smear of normal human blood.

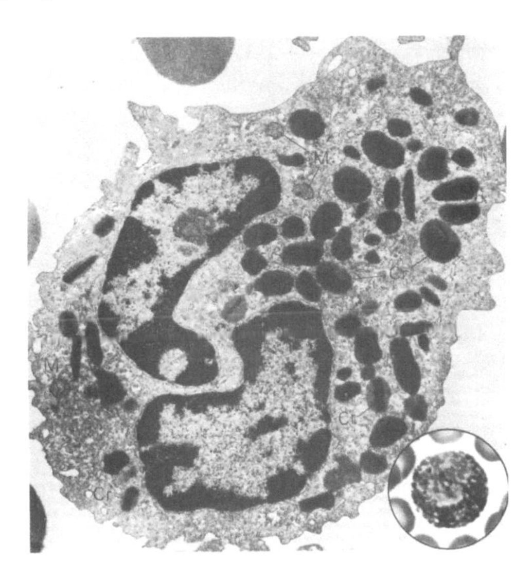

FIGURE G15a Transmission electron micrograph of a section of a human eosinophil. Some granules contain a crystalline body (Cr). A few mitochondria with rod-shaped crystals are seen. *Inset:* Photomicrograph of an eosinophil from a smear.

Agranulocytes

There are two types of agranulocytes in mammals: LYMPHOCYTES and MONOCYTES (Figs. G20a and G20b). A third type of agranulocyte is found in nonmammals: the THROMBOCYTE (Fig. G20c). The nuclear chromatin of lymphocytes and monocytes is highly condensed (Fig. G21). The cells are actively motile (Fig. G22). The cytoplasm contains the usual organelles but is especially rich in free ribosomes. Monocytes, which are active phagocytes, are characterized by the presence of abundant vesicles and lysosomes in the cytoplasm (Fig. G23).

Lymphocytes constitute 20%–25% of the leukocytes in human blood. Their deeply staining, large, spherical nuclei contain large masses of chromatin (Figs. G24a and G24b). The cytoplasm, which is basophilic, is variable in amount. This latter characteristic is the basis for the classification of lymphocytes (Fig. G25). SMALL LYMPHOCYTES of humans are 6–8 μm in diameter; the cytoplasm forms a thin film around the large, deeply stained nucleus. MEDIUM-SIZED LYMPHOCYTES have a relatively greater amount of cytoplasm than small lymphocytes and are 8–10 μm in diameter. The nucleus is more or less central. LARGE LYMPHOCYTES normally appear only in the blood-forming organs but may be found in the circulating blood in pathological conditions. They are 10–12 μm in diameter. The

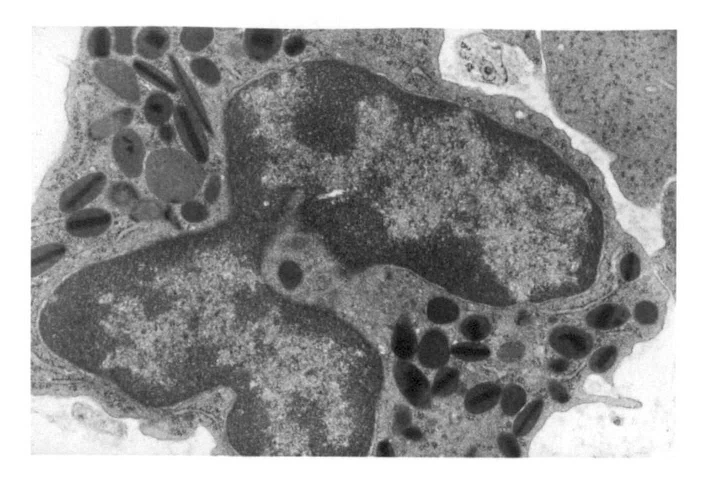

FIGURE G15b Transmission electron micrograph of a maturing eosinophil in a section of bone marrow of a rat. The nucleus is bilobed. A few cisternae of endoplasmic reticulum are seen at the lower right. The characteristic granules of eosinophils are biconcave discs surrounded by a membrane and containing rods of dense crystalline material cells. 37,000 ×.

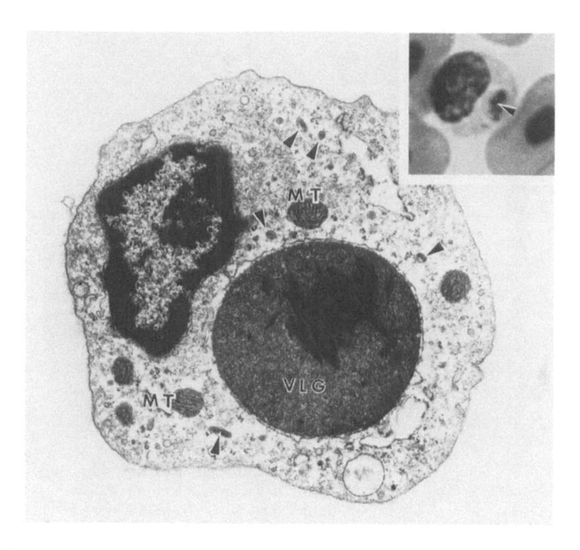

FIGURE G16a Transmission electron micrograph a section of a mature eosinophil from the blood of a loach (*Misgurnus anguillicaudatus*). *CC*, electron-dense crystalline core; *MT*, mitochondria; *VLG*, large dense granule; *arrowheads* indicate small dense granules. 16,500 ×. *Inset:* Photomicrograph of blood smear with a mature eosinophil. 2000 ×.

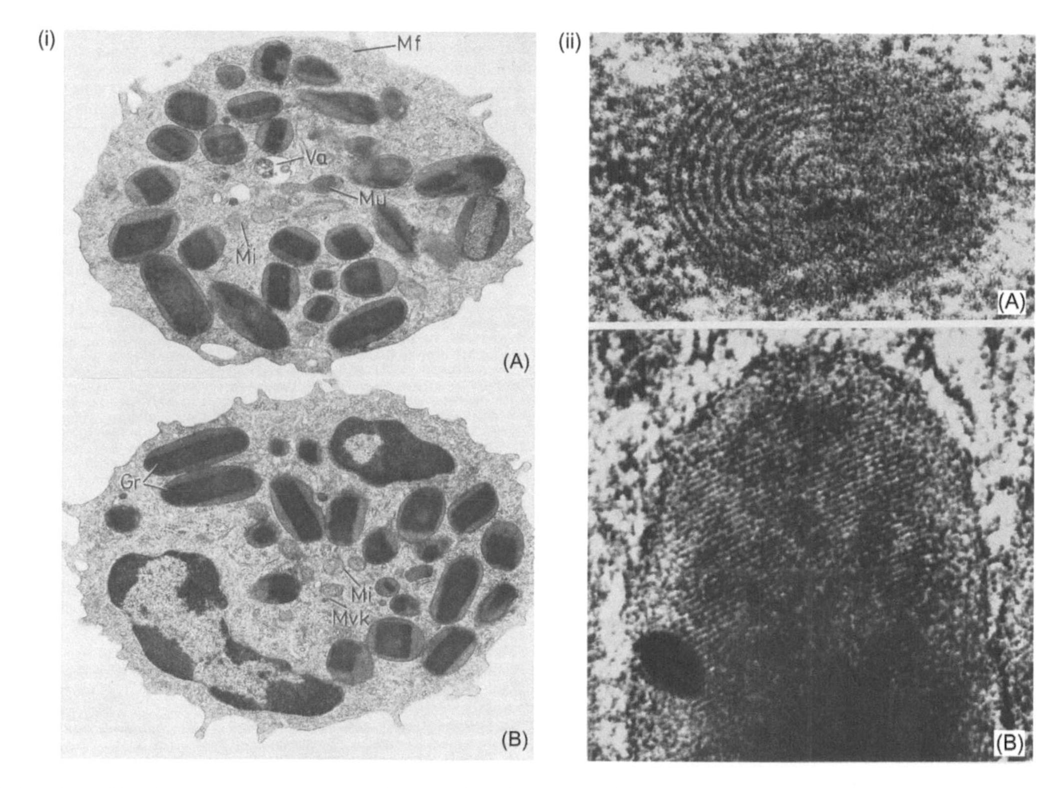

FIGURES G16b (i) and (ii) Transmission electron micrographs of sections of eosinophils from a domestic goose (*Anser anser*). The granules (*Gr*) in these cells contain characteristic rod-shaped crystals. High-powered views of the crystals are shown. *Mf*, microfibers; *Mi*, mitochondrion; *Mu*, transforming mitochndion; *MvK*, multivesicular body; *Va*, vacuole. (i) 17,500 ×; (ii) 205,000 ×.

oval or round nucleus is surrounded by a generous amount of cytoplasm. The cytoplasm, which may be slightly vacuolar, may contain a few azurophilic granules. Lymphocytes in the blood of all classes of vertebrates show little variation described in Figs. G26a–G26e.

While erythrocytes function in carrying oxygen to the farthest regions of the body, leukocytes are agents of the immune system protecting the body from infectious agents or ANTIGENS: bacteria, viruses, funguses, or parasites. Specialized cells may respond to the presence of foreign antigens either by a CELL-MEDIATED IMMUNE RESPONSE or by the

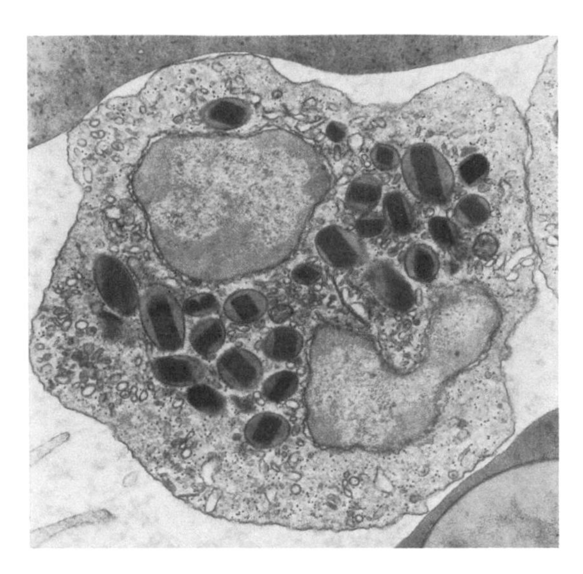

FIGURE G16c Transmission electron micrograph of a section of an eosinophil from the goose. Dark crystalline cores appear within the granules. 20,520 ×.

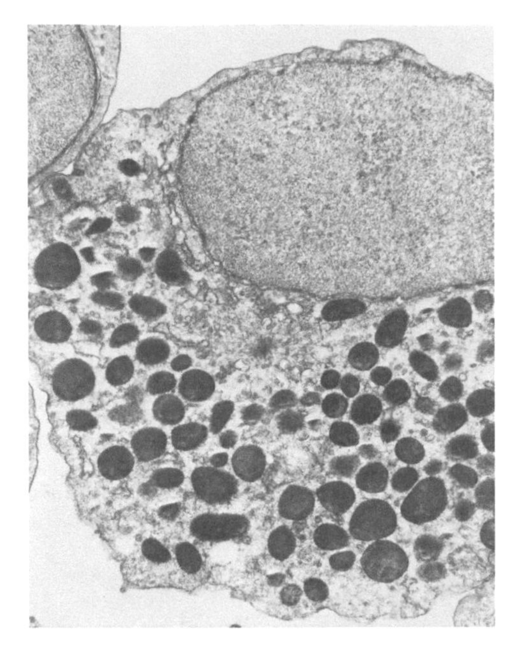

FIGURE G16d Transmission electron micrograph of a section through an eosinophil of the sucker (*Catostomus commersoni*) showing electron-dense granules in the cytoplasm. 21,000 ×.

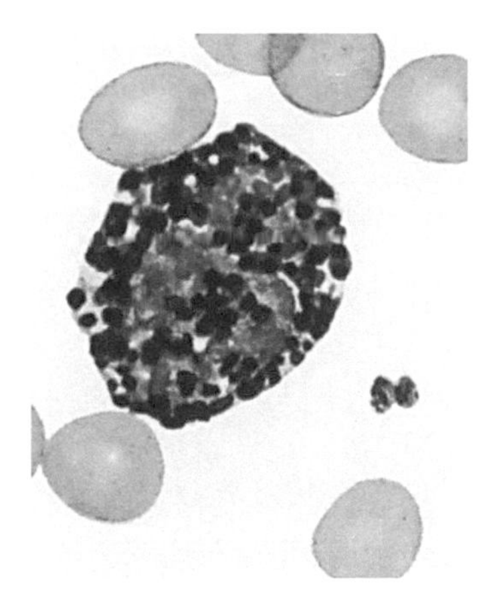

FIGURE G17 Basophil from a smear of normal human blood.

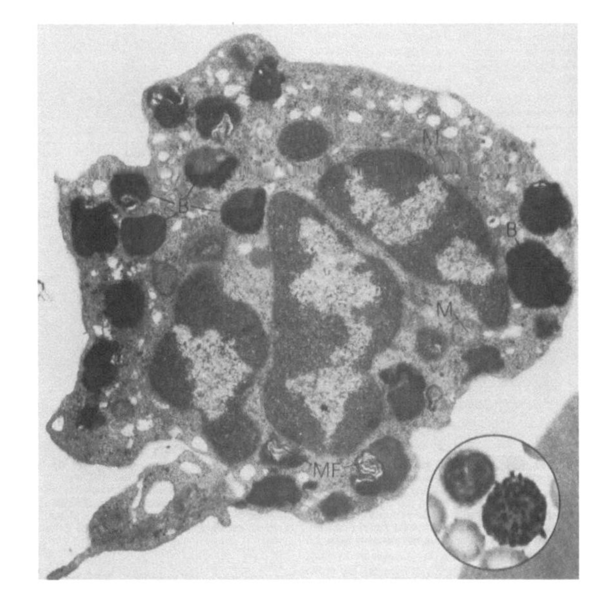

FIGURE G18a Transmission electron micrograph of a section through a human basophil. Although connected, the nucleus appears as three separate bodies; the connecting strands are not within the plane of section. The granules are large with an irregular shape. *M*, mitochondrion; *Mf*, myelin figure. *Inset*: Photomicrograph of a blood smear showing a comparable basophil.

production of ANTIBODIES. These cells are the lymphocytes, which occur in staggering numbers—about 2×10^{12} in the human body. Although they are indistinguishable morphologically, there are two major types of lymphocytes.

B LYMPHOCYTES, which are produced from stem cells in the bone marrow (and fetal liver of mammals) produce antibody; T LYMPHOCYTES arise in the thymus from primary stem cells and are responsible for cell-mediated immune responses. B and T lymphocytes can be distinguished only after stimulation by an antigen that results in differences in the glycoproteins of their plasma membranes. In the presence of antigen, B lymphocytes proliferate further and develop into PLASMA CELLS (Figs. G27a–G27c), which are antibody secreting cells containing abundant granular endoplasmic reticulum.

Activated T cells contain little endoplasmic reticulum and do not secrete antibodies. It is largely in the secondary lymphoid organs: lymph nodes, spleen, and gut-associated lymphoid tissue that T and B lymphocytes react with foreign antigens. The immune system responds to millions of different foreign antigens and it is proposed

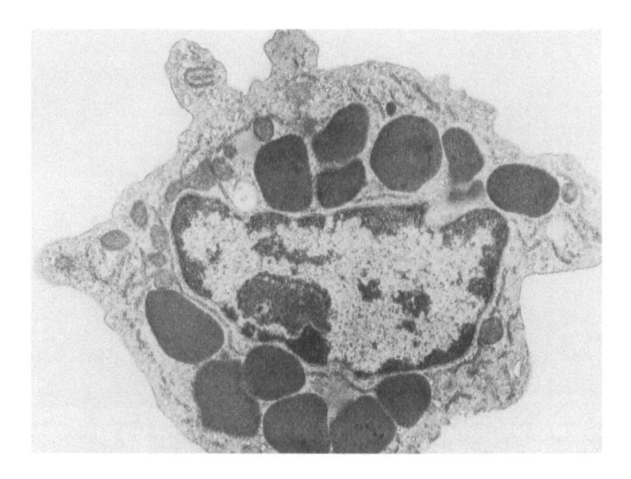

FIGURE G18b Transmission electron micrograph of a section through a mammalian basophil. The granules are large and an irregular shape.

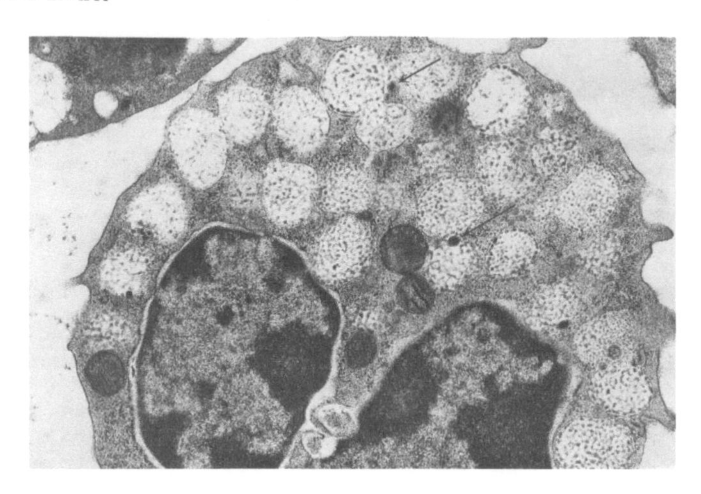

FIGURE G19a Transmission electron micrograph of a mature basophil of a white Leghorn chicken. The granules are roughly spherical, electron lucent, and with particulate contents. 24,000 ×.

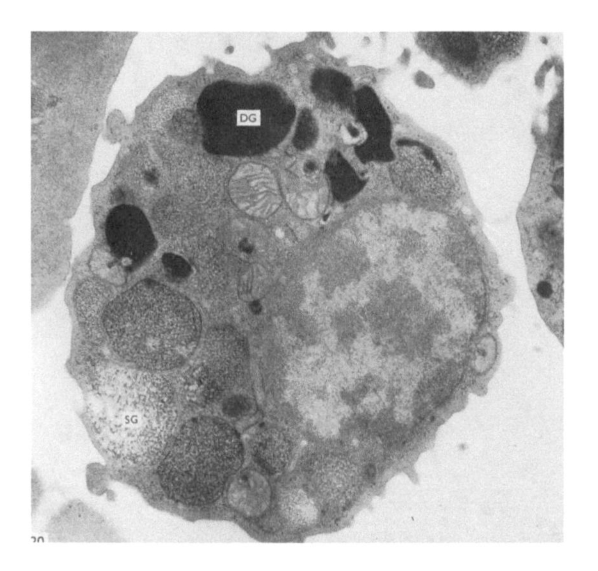

FIGURE G19b Transmission electron micrograph of a section of a basophil of the Guinea fowl (*Numida meleagris*). Some irregular dense granules (DG) and many stippled granules of various shapes are seen. 25,000 ×.

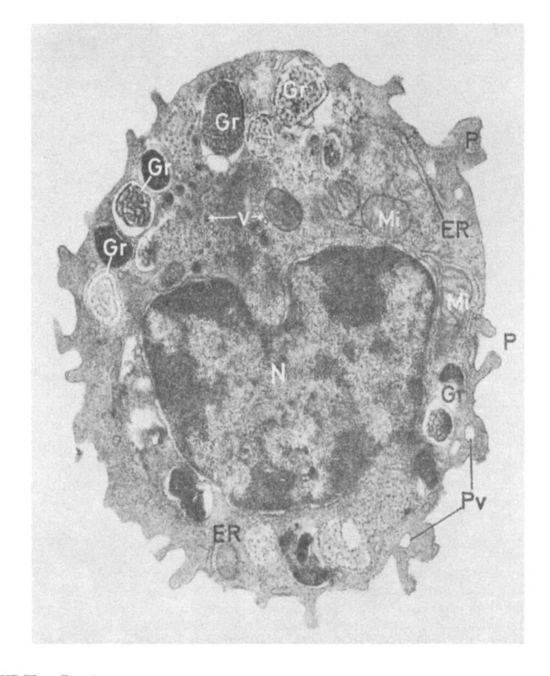

FIGURE G19c Transmission electron micrograph of a section through a basophilic granulocyte of a domestic hen. *ER*, endoplasmic reticulum; *Gr*, granules of varied shapes and content; *Mi*, mitochondria; *N*, nucleus; *P*, podocytes; *Pv*, pinocytotic vesicles; *V*, vesicles. 24,500 ×.

that, during development, each lymphocyte is programmed to react with a particular antigen by causing it to proliferate and mature. It is also proposed that, during development, each lymphocyte is programmed to react with a particular antigen, which causes it to proliferate and mature. Millions of different clones of cells, each consisting of T or B lymphocytes, descended from a common ancestor that is committed to make one particular antigen-specific receptor protein. The immune system therefore is composed of millions of families—CLONES—of cells each consisting of T or B lymphocytes descended from a common ancestor: CLONAL SELECTION. Lymphocytes are programmed to respond to a particular antigen by receptors on their surfaces that specifically bind to the antigen. Most T cells also play a role in immunity as HELPER CELLS and SUPPRESSOR T CELLS regulating the responses of leukocytes. Other T cells are CYTOTOXIC and kill virus-infected cells. Both cytotoxic T cells and B lymphocytes protect against infection and are called EFFECTOR CELLS. On a second exposure to the antigen—even years later—the response will be swifter, more powerful, and sustained for a longer period than the PRIMARY IMMUNE RESPONSE.

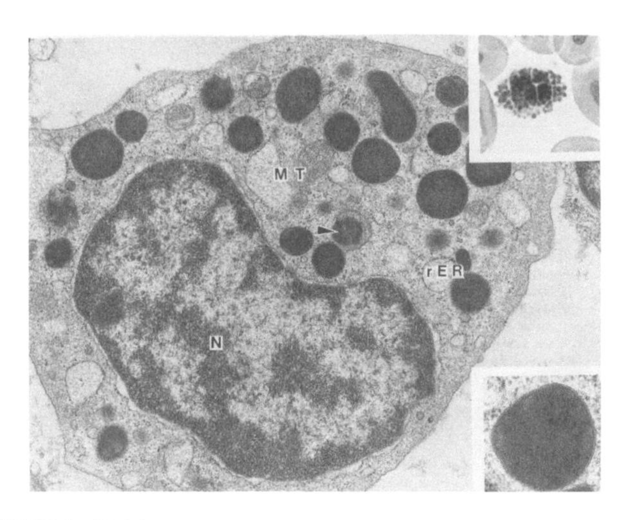

FIGURE G19d Transmission electron micrograph of a section through a basophil from the intertubular space of the kidney of a loach (*Misgurnus anguillicaudatus*). *MT*, mitochondria; *N*, nucleus; *rER*, granular endoplasmic reticulum. 20,400 ×. *Upper inset*: Photomicrograph of a mature basophil from a blood smear 1500 ×; *lower inset:* Transmission electron micrograph of a mature granule with an electron-dense, homogeneous matrix. 46,000 ×.

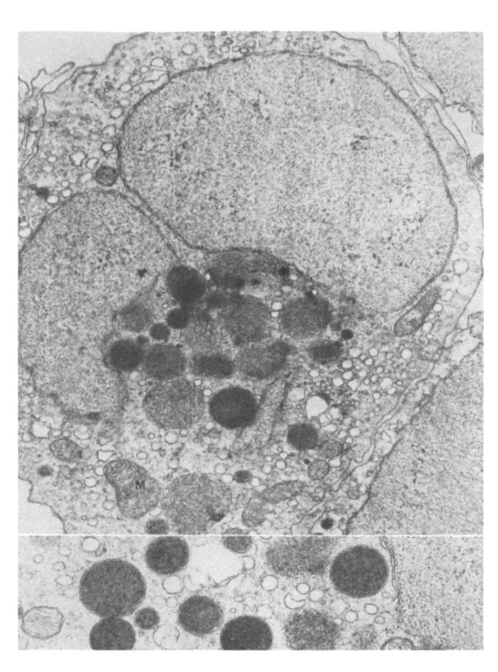

FIGURE G19e Transmission electron micrographs through sections of basophils of a sucker (*Catostomus commersoni*). *Upper*: 21,000 ×; *Lower*: 30,000 ×.

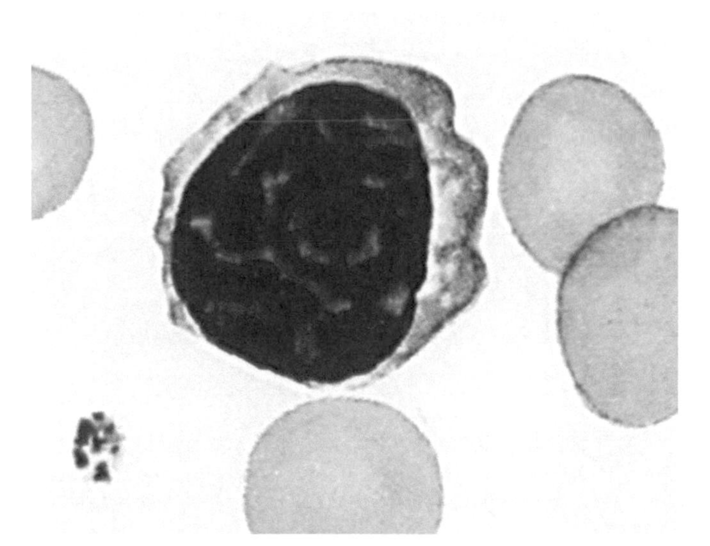

FIGURE G20a Drawing of a lymphocyte from a smear of normal human blood. A few platelets appear in the lower left corner.

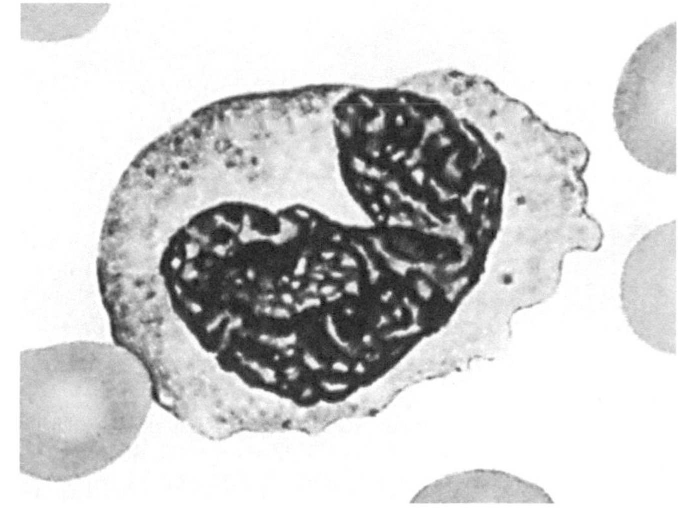

FIGURE G20b Drawing of a monocyte from a smear of normal human blood.

This more decisive SECONDARY IMMUNE RESPONSE indicates that the immune system has "remembered" its original encounter with the antigen.

Monocytes are the largest cells in the normal bloodstream and constitute 3%–6% of the leukocytes (Figs. G28 and G29a). They are between 12 and 15 μm in diameter in humans. The nucleus is usually eccentric and oval to bean shaped with pale staining, finely granular chromatin. The cytoplasm is faintly basophilic, may appear vacuolated, and is usually more granular than that of the lymphocyte. Monocytes leave the bloodstream to become macrophages throughout the body. They are active phagocytes and are characterized by the presence of abundant vesicles and lysosomes in their cytoplasm. Their appearance is similar in all vertebrates (Figs. G29b–G29d).

Thrombocytes are ovoid or spindle-shaped cells that resemble lymphocytes in their staining properties (Figs. G30a–G30e). These are agranular cells that are seen only in the blood of nonmammalian vertebrates.

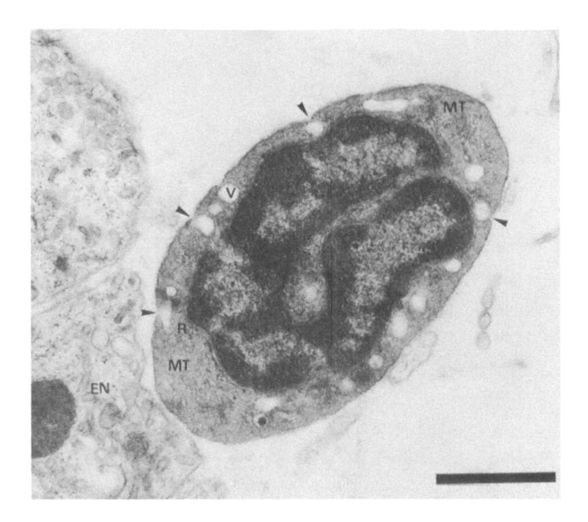

FIGURE G20c Transmission electron micrograph of a section of a thrombocyte from the medaka (*Oryzias latipes*). Note the bundles of microtubules (MT) vacuoles (V), and indentations of the plasma membrane (*arrowheads*). Bar = 1 μm.

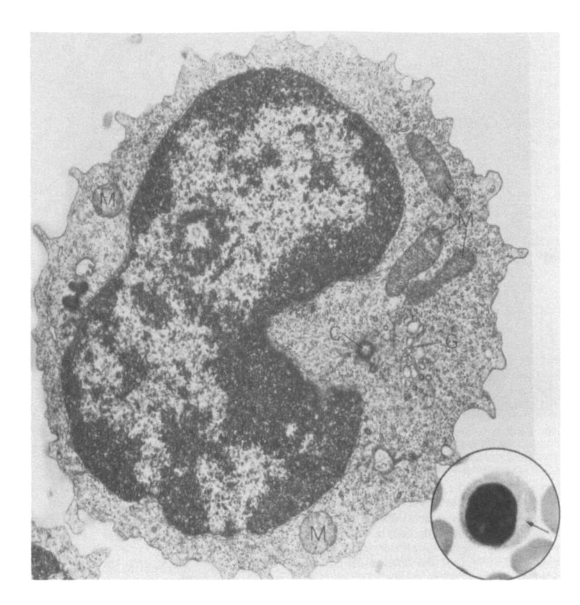

FIGURE G21 Transmission electron micrograph of a section through a medium-sized lymphocyte from a mammal. Numerous free ribosomes are present in the cytoplasm. *C*, centriole; *G*, Golgi complex; *M*, mitochondria. *Inset*: Photomicrograph of a medium-sized lymphocyte from a blood smear. The *arrow* indicates the Golgi-containing centrosphere region.

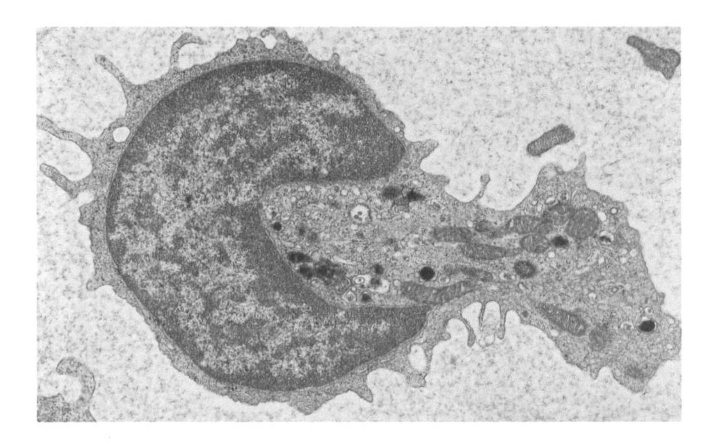

FIGURE G22 Transmission electron micrograph of a section of a human lymphocyte moving to the left, trailing its UROPOD behind. Microvilli are active on the surface of the cell. 9,200 ×.

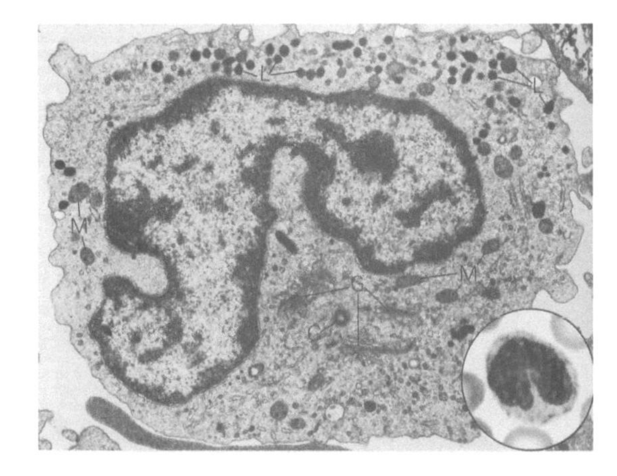

FIGURE G23 Transmission electron micrograph of a section of a human monocyte. The small dark granules are lysosomes (*L*). *C*, centriole, *G*, Golgi complexes. *Inset*: Photomicrograph of a monocyte in a blood smear.

A canalicular system, open to the surface, pervades the cytoplasm of thrombocytes (Figs. G31a and G31b). It is suggested that the increase in surface area provided by the canaliculi, amplifies the surface area for the exchange of metabolites. Thrombocytes are responsible for clot formation; although their appearance is similar to that of lymphocytes, they are considered to be unrelated.

Blood Platelets[1]

Blood platelets are small, irregularly shaped, enucleate masses of protoplasm found only in the blood of mammals (Figs. G32 and G33). They are formed by the fragmentation of the cytoplasm of megakaryocytes—think of a

[1]The term *thrombocyte* implies the designation of intact cells. It is often used, however, to designate cellular fragments—the platelets. We will reserve the term *thrombocyte* to describe the ovoid cells in the blood of nonmammals that are involved in the clotting of blood and use the term *platelets* for the cellular fragments in the blood of mammals that also participate in this function.

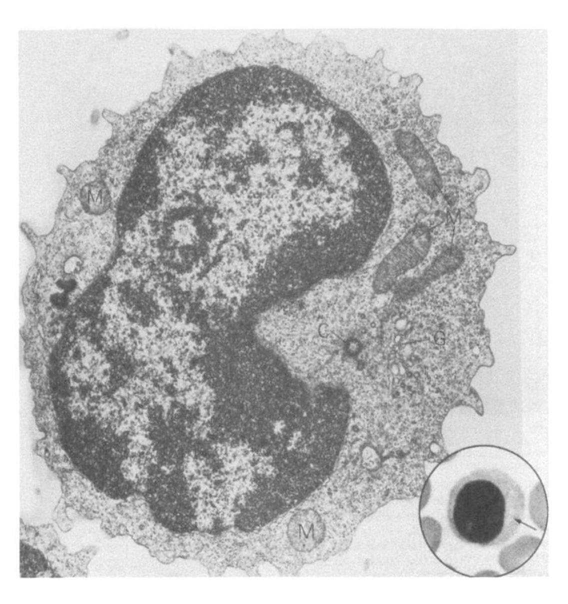

FIGURE G24a Transmission electron micrograph of a section of a medium-sized lymphocyte. *C*, centriole; *G*, Golgi complex. *M*, mitochondria. *Inset*: Photomicrograph of a medium-sized lymphocyte in a blood smear.

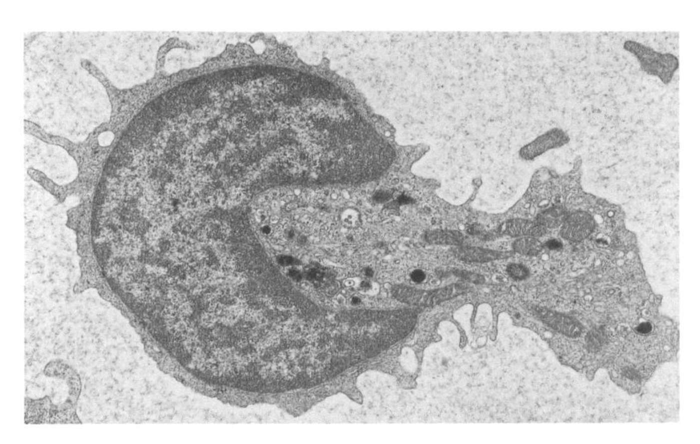

FIGURE G24b Transmission electron micrograph of a section of a human lymphocyte moving to the left, trailing its UROPOD behind. Microvilli are active on the surface of the cell. 9200 ×.

FIGURE G25 Drawings of small, medium-sized, and large lymphocyte from a smear of human blood.

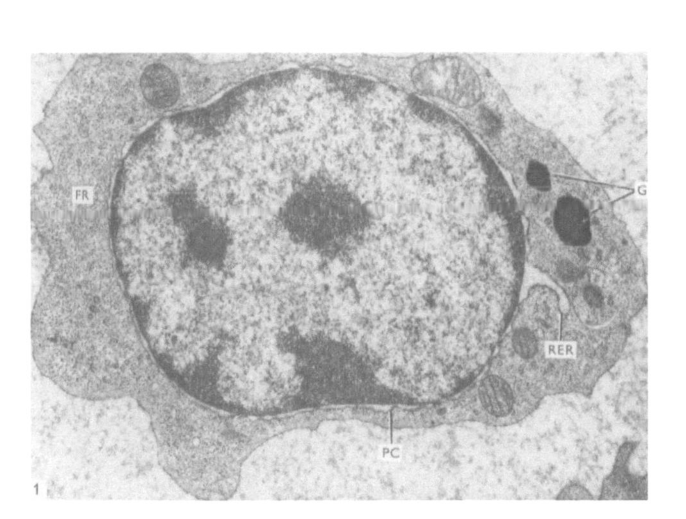

FIGURE G26a Transmission electron micrograph of a section of a medium-sized lymphocyte of a domestic duck (*Anas platyrhynchos*). *FR*, free ribosomes; *G*, dense granules; *PC*, perinuclear cisternae; *RER*, granular endoplasmic reticulum. 31,920 ×.

perforated sheet of stamps (Figs. G34 and G35) and are responsible for blood clotting. The tendency of platelets to clump together and adhere to any surface presented to them as soon as the blood is drawn creates technical difficulties with regard to their enumeration and study. To keep them apart so that they may be seen as individuals it is necessary to mix the blood with an anticoagulant as soon as it is taken from the body. Their contents may be differentiated into the peripheral hyaline HYALOMERE, and the central granular CHROMOMERE. The electron microscope shows that each platelet encloses within its plasma membrane cytoplasmic components such as

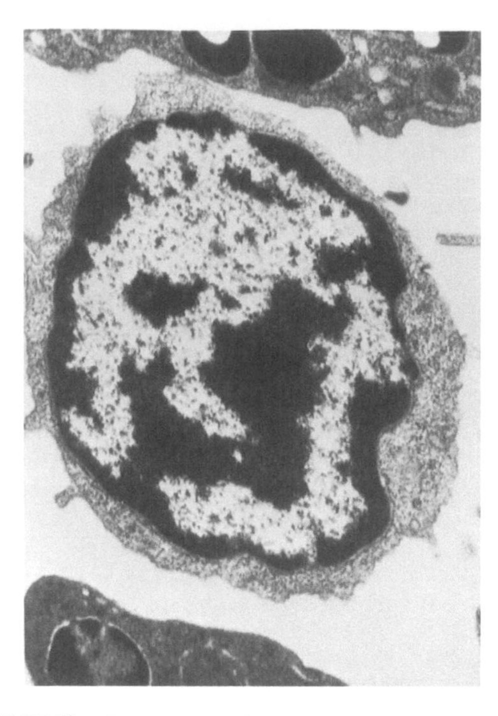

FIGURE G26b Transmission electron micrograph of a section of a small lymphocyte from a lizard (*Lacerta hispanica*). *R*, free ribosomes. 16,500 ×.

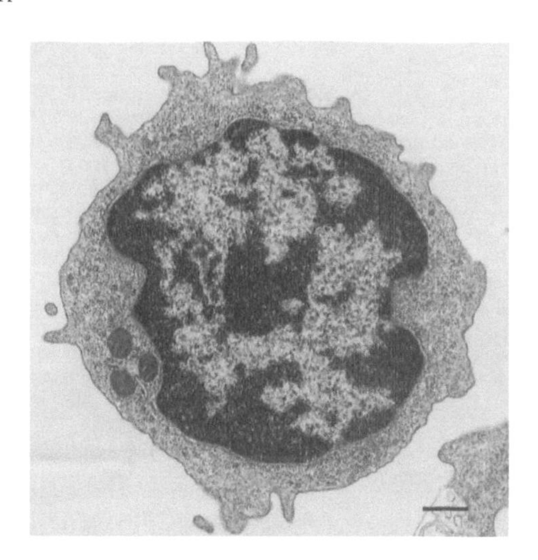

FIGURE G26c Transmission electron micrograph of a section of a small lymphocyte from the peripheral blood of a snake (*Elaphe obsoleta quadrivitatta*). Bar = 0.6 μm.

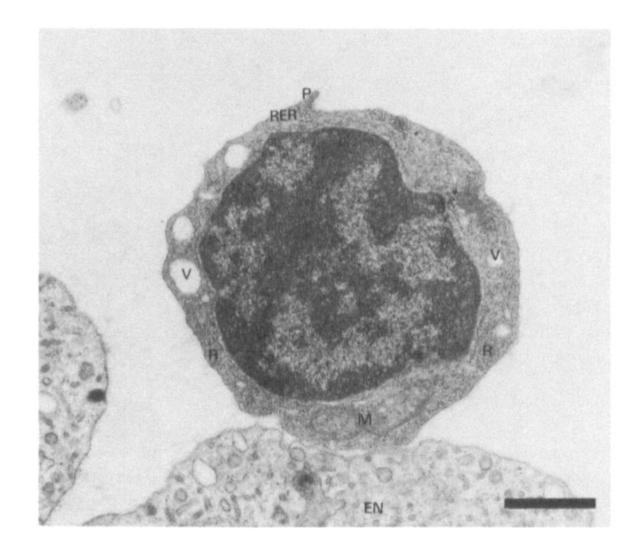

FIGURE G26d Transmission electron micrograph of a section of a small lymphocyte from the peripheral blood of a medaka (*Oryzias latipes*). *EN*, endothelial cell; *M*, mitochondria; *P*, pseudopodium; *R*, free ribosomes; *RER*, granular endoplasmic reticulum; V, vacuole. Bar = 1 μm.

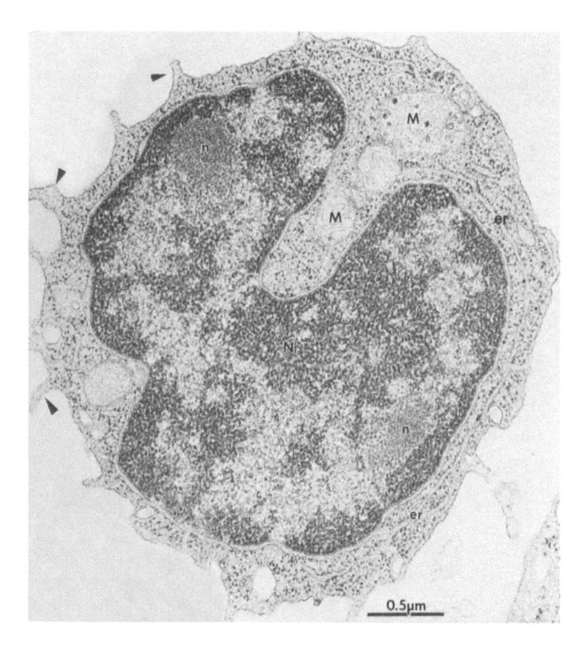

FIGURE G26e Transmission electron micrograph of a section of a small lymphocyte from the catfish (*Ictalurus punctatus*). *er*, granular endoplasmic reticulum; *M*, mitochondria; *N*, nucleus; *arrowheads*, pseudopodia. Bar = 0.5 μm.

mitochondria, ribosomes, vesicles of the endoplasmic reticulum, and two types of granules: ALPHA LYSOSOMAL GRANULES and VERY DENSE GRANULES (which contain serotonin) (Fig. G36). Also found are two systems of tubules, one connecting to the surface and another containing an electron-dense material. Cytoskeletal elements include microtubules and filaments. Mammalian platelets are sometimes referred to as THROMBOCYTES; we will reserve this term for the nucleated cells of nonmammals.

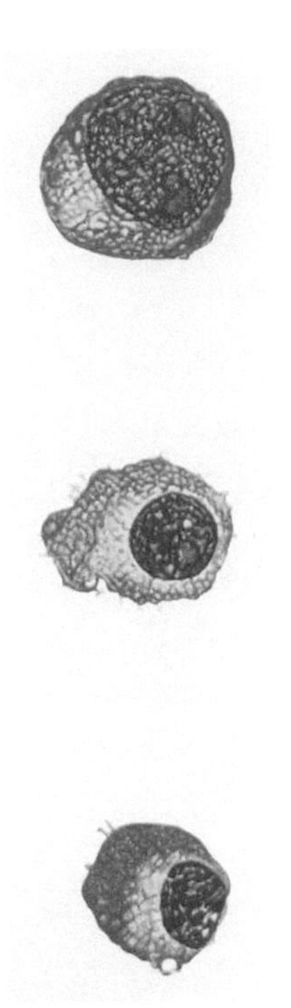

FIGURE G27a Plasmocytes are not found in circulating blood of normal individuals but in loose connective tissue. This series shows its development from a plasmoblast (top), to proplasmocyte (middle), and plasmocyte (bottom) as seen in photomicrographs.

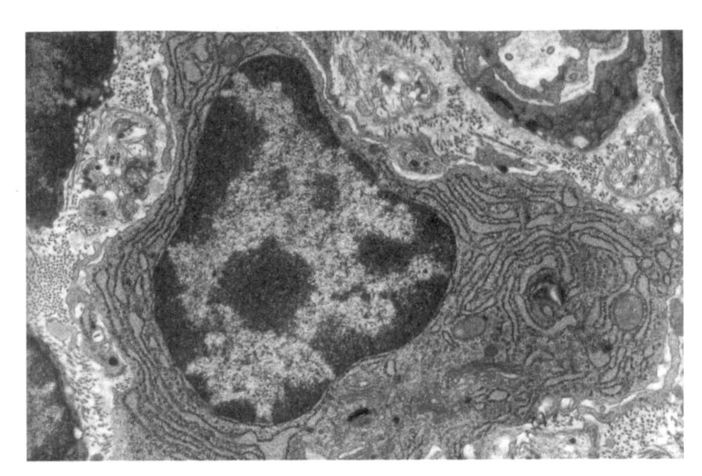

FIGURE G27b Transmission electron micrograph of a section of a plasmocyte from the connective tissue of a rat (*Mus norvegicus*). Characteristically the chromatin has clumped in "cartwheel" formation around the inside of the nuclear envelope and the cytoplasm is packed with the flattened cisternae of granular endoplasmic reticulum that are studded with ribosomes. Flattened sacs of the Golgi complex lurk near the nucleus at the lower right. A few mitochondria are seen. 29,000 ×.

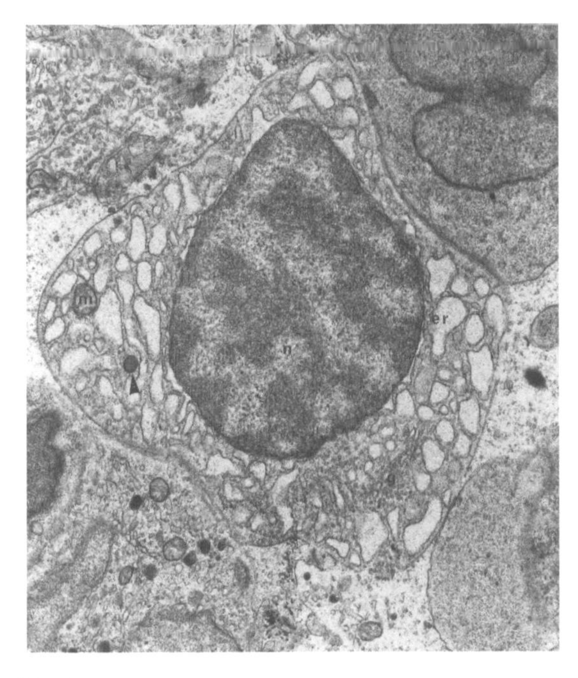

FIGURE G27c Transmission electron micrograph of a section of a plasmocyte from the peritubular loose connective tissue in the kidney of the masou salmon (*Oncorhynchus masou*). The cytoplasm is packed with the dilated cisternae of the granular endoplasmic reticulum (*er*). *g*, Golgi complex; *m*, mitochondria are sparse; *n*, nucleus showing characteristic clumping of chromatin. 13,000 ×.

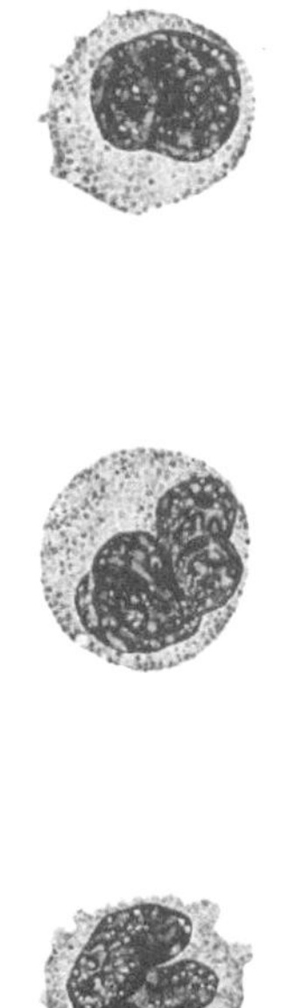

FIGURE G28 Drawings of human monocytes showing various nuclear conformations.

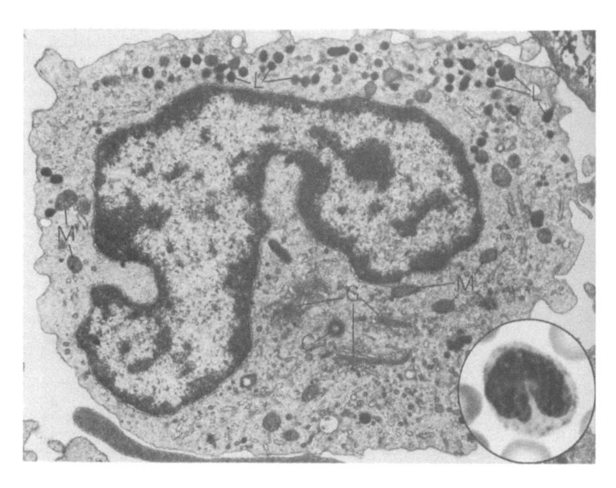

FIGURE G29a Transmission electron micrograph of a section of a monocyte from the peripheral blood.

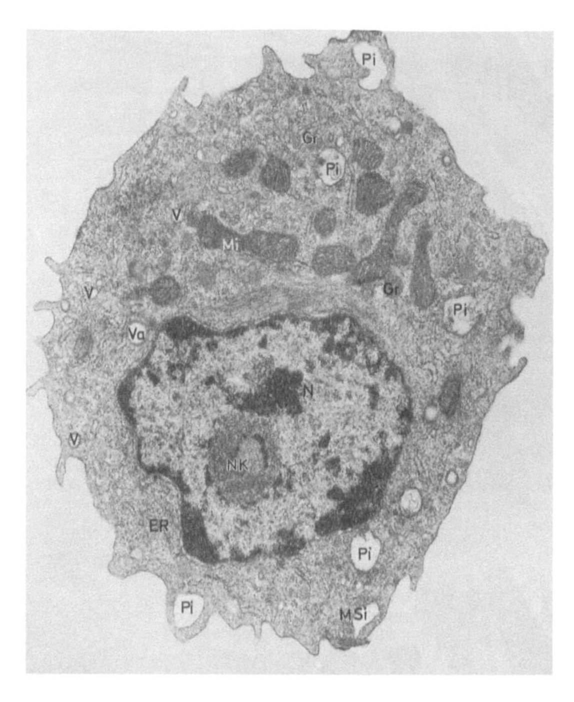

FIGURE G29b Transmission electron micrograph of a section of a monocyte from the peripheral blood of a domestic hen. *ER*, endoplasmic reticulum; *Mi*, mitochondrion; *MSi*, microtubules; *N*, nucleus; *NK*, nucleolus; *Pi*, pinocytotic vesicles; *V*, vesicle; *Va*, lipid vacuole. 17,000 ×.

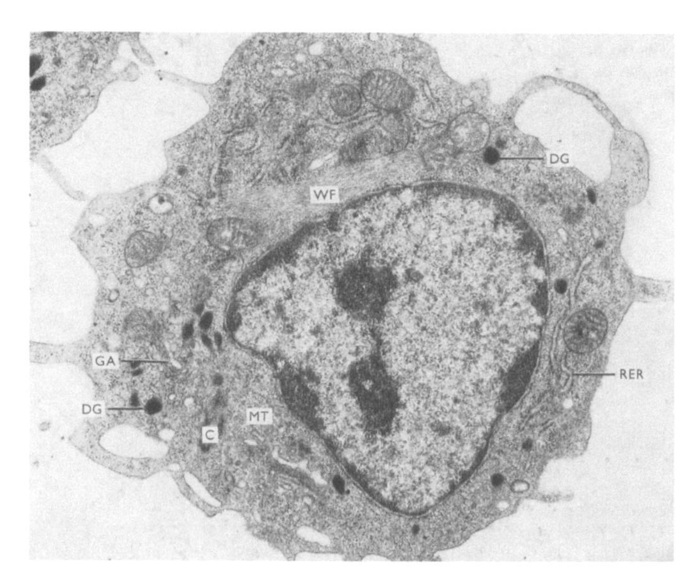

FIGURE G29c Transmission electron micrograph of a section of a monocyte from the peripheral blood of a domestic duck (*Anas platyrhynchos*). *C*, centrioles; *DG*, dense granules; *GA*, Golgi complex; *MT*, microtubules; *RER*, granular endoplasmic reticulum; *WF*, bundle of fine wavy filaments. 31,920 ×.

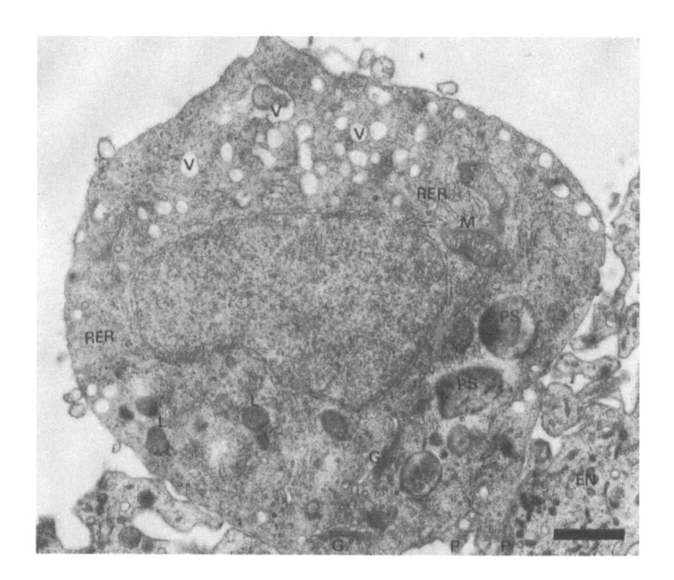

FIGURE G29d Transmission electron micrograph of a section of a monocyte from the peripheral blood of a medaka (*Oryzias latipes*). *EN*, endothelial cell of the heart; *L*, lysosome; *P*, pseudopodia; *PS*, phagosomes; *RER*, granular endoplasmic reticulum. Bar = 1 μm.

LYMPH

Lymph is a fluid that, having leaked into the tissues from blood in the capillaries, is being returned to the bloodstream by way of lymph vessels (Fig. G37). Cells are added when it passes through the lymph nodes; these cells are chiefly small lymphocytes.

HEMOPOIESIS

Hemopoiesis refers to the process of formation of blood cells. In the early embryo the blood cells differentiate in the mesenchyme in condensations known as BLOOD ISLANDS. Later in fetal life formation of blood cells occurs in the connective tissue of the liver, spleen, and lymph nodes. Following the appearance of hollow bones in the

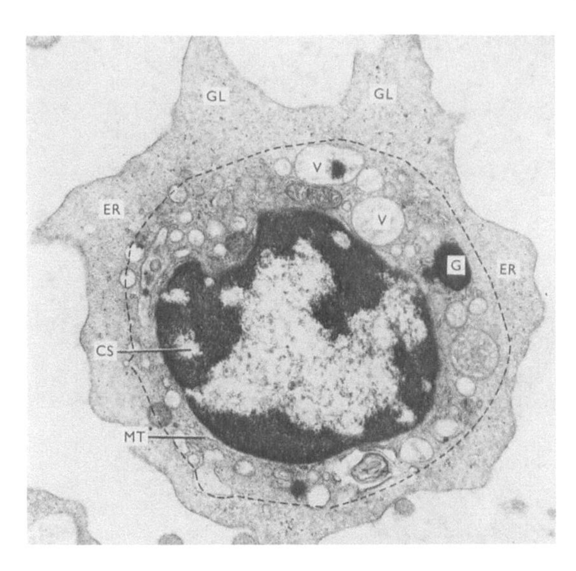

FIGURE G30a Transmission electron micrograph of a section of a thrombocyte from a domestic duck (*Anas platytrhynchos*). Organelles are distributed close to the nucleus. *CS*, clear spaces; *ER*, ectoplasmic ring; *G*, granules; *GL*, glycogen granules; *MT*, microtubules; *V*, vacuole. 39,920 ×.

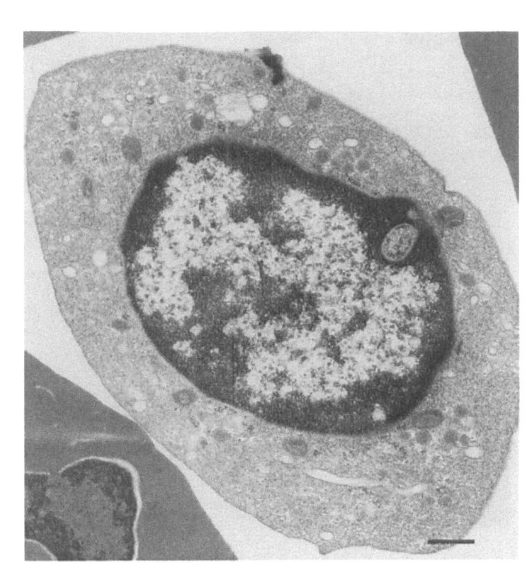

FIGURE G30b Transmission electron micrograph of a section of a thrombocyte from the peripheral blood of a yellow rat snake (*Elaphe obsoleta quadrivitatta*). Bar = 0.5 μm

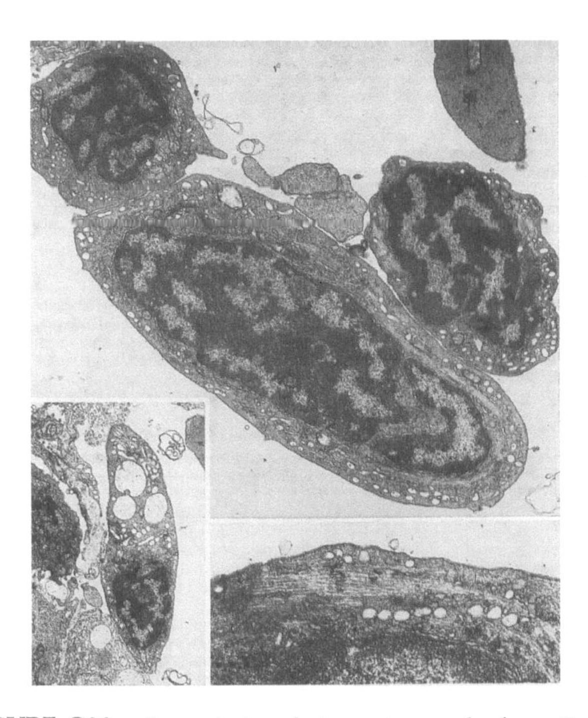

FIGURE G30c Transmission electron micrograph of a section of thrombocytes from the peripheral blood of a salmon (Top), group of three thrombocytes in long and cross section 10,000 × ; (left), thrombocyte with large vacuoles, perhaps phagocytic vesicles (5000 ×); (bottom), note vacuoles and marginal band of microtubules 20,000 ×.

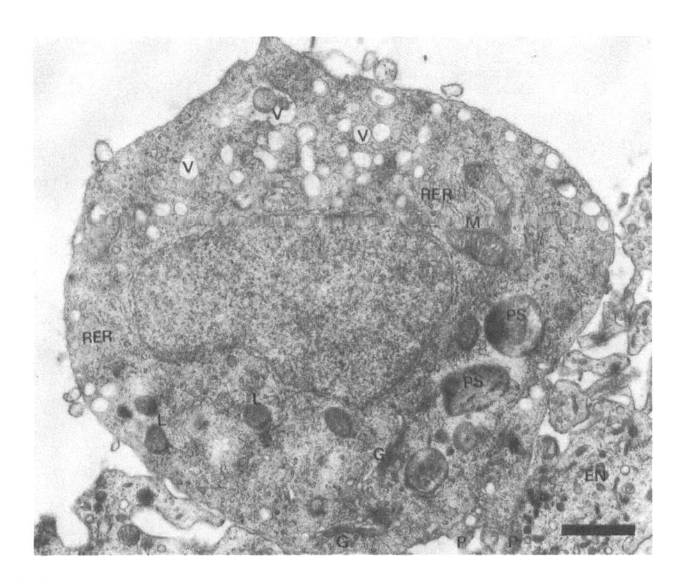

FIGURE G30d Transmission electron micrograph of a section of a thrombocyte from the heart of a medaka (*Oryzias latipes*). *Arrowheads* indicate indentations of the plasma membrane; *EN*, endothelial cells of the heart; *MT*, bundle of microtubules; *R*, free ribosomes; *V*, vacuole; *RER*, rough endoplasmic reticulum; *G*, Golgi; *M*, mitochondrion. Bar = 1 μm.

higher vertebrates, the site of hemopoiesis shifts to the bone marrow, although the spleen and other lymphoid tissues remain as the sole centers for the origin of the agranulocytes in the adult. Thus there are two kinds of hemopoietic tissue in the higher vertebrates: MYELOID and LYMPHOID. Hemopoiesis may occur in a wide variety of organs, wherever there is a loose stroma of connective tissue with a fairly sluggish blood supply: connective tissue of the gut, kidney, spleen, and even the ovary may contain active centers. In animals lacking hollow bones,

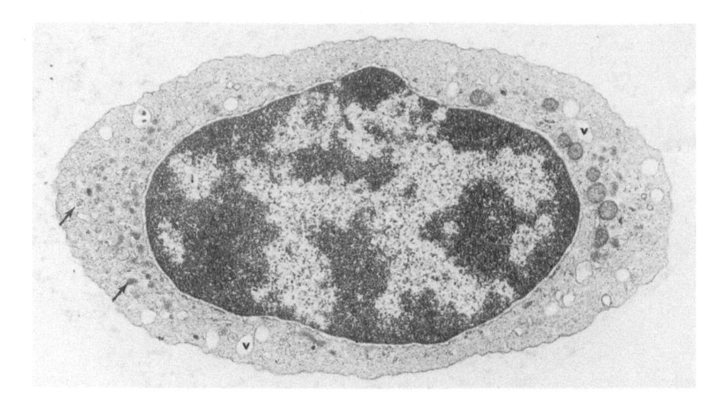

FIGURE G30e Transmission electron micrograph of a section of a thrombocyte from the peripheral blood of a nurse shark (*Ginglymostoma cirratum*). *v*, small vesicle; *arrows*, electron dense vesicles. 13,000 ×.

all types of blood cells must be derived from these extramedullary sites, which may be designated as LYMPHOMYELOID TISSUES.

Pluripotent Stem Cell

All blood cells are derived from pluripotent STEM CELLS, early cells with no specific identifying features. They are extremely rare and indistinguishable from small lymphocytes so that they are difficult to study (Fig. G38). The nucleus is round and relatively large with well-defined nucleoli and finely reticulated chromatin that takes a reddish stain. The intensity of basophilia of the cytoplasm varies widely. As a rule the cytoplasm has a mottled or foamy nature and may be less densely stained than the nucleus. The size of the stem cell varies widely from 10 to 40 μm. The margins of these cells are usually jagged.

Electron micrographs of presumed stem cells show abundance of free ribosomes occurring singly or in clusters as polysomes (polyribosomes). Otherwise the complement of organelles is not unusual: centrioles and a Golgi complex near the nucleus, granular endoplasmic reticulum, and a few scattered mitochondria. This appearance is characteristic of an undifferentiated cell. Stem cells divide slowly and cell replacement is largely achieved by the proliferation of more differentiated cells.

Nomenclature

The study of the development of cells of the blood has been seriously hampered in the past by confusing and conflicting terminology of the cells involved. In the description that follows the terms used are based on those recommended in *Nomina Histologica;* these are followed, in brackets, by the most generally accepted of the other terms.

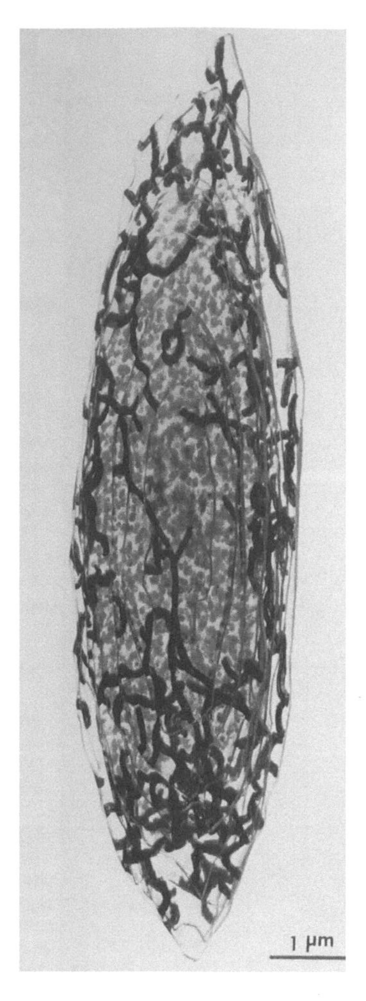

FIGURE G31a Plastic reconstruction of thick sections showing the three-dimensional arrangement of canaliculi of the surface-connected canalicular system in thrombocytes of the carp (*Cyprinus carpio*). 15,000 ×.

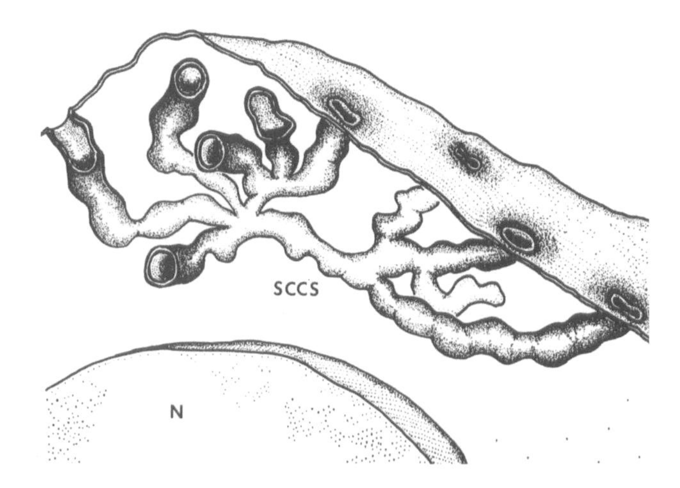

FIGURE G31b Drawing of the canaliculi of the surface-connected canalicular system of the carp.

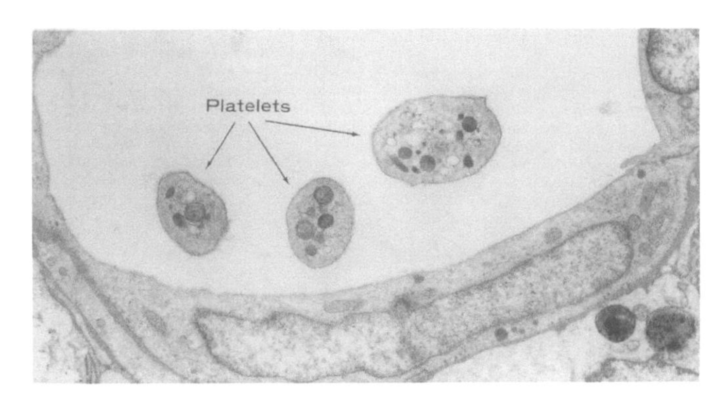

FIGURE G32 Transmission electron micrograph showing a section of platelets within the endothelium of a capillary.

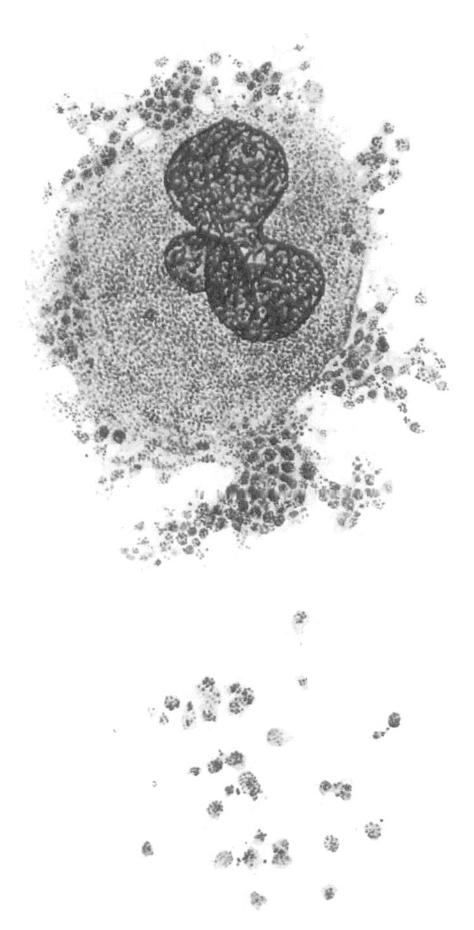

FIGURE G34 Drawing of a megakaryocyte from human bone marrow; its cytoplasm contains many platelets at its periphery. Fragments of this cytoplasm are released to become the platelets of the blood.

FIGURE G33 Transmission electron micrograph of a section of a lymphocyte from the blood of a guinea pig. (A) Transmission electron micrograph of a platelet sectioned transversely showing its dense secretory granules and an aggregation of glycogen particles. The arrows show cross sections of the circumferential bundles. (B) Platelet sectioned parallel to its broad dimension; its microtubules are sectioned longitudinally.

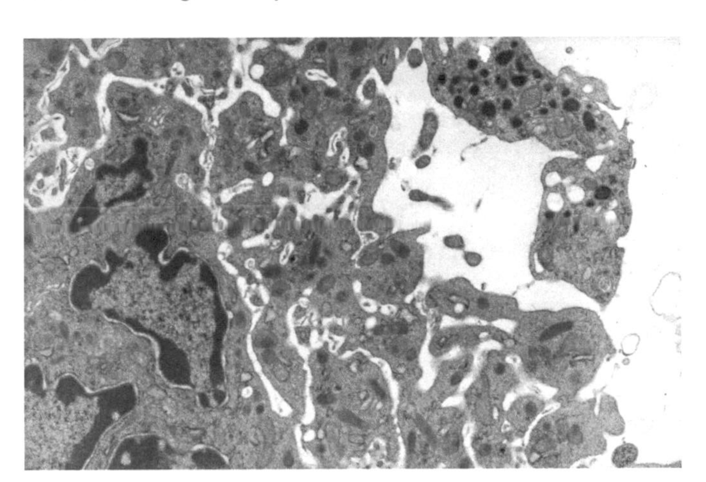

FIGURE G35 Transmission electron micrograph of a section of the bone marrow of a rat showing the periphery of a megakaryocyte. The cytoplasm of this huge cell is riddled with membranous partitions that outline future platelets. A portion of the multilobed nucleus is seen slightly to the left of center of the micrograph. 30,000 ×.

General Considerations

All types of blood cells reveal comparable changes as they develop from immature cells to mature, definitive types. These changes have been summarized for the maturation of blood cells in human bone marrow stained with a Romanowsky stain (Fig. G39a).

- As cells mature they become smaller as do their nuclei.
- The ratio of nucleus to cytoplasm decreases with maturity.
  - A young nucleus is more or less spherical; as it matures it tends to become indented and lobulated.
  - The stained nucleus changes from reddish-purple to pure blue.
  - The young nucleus is paler than the cytoplasm; with maturity its color becomes deeper.

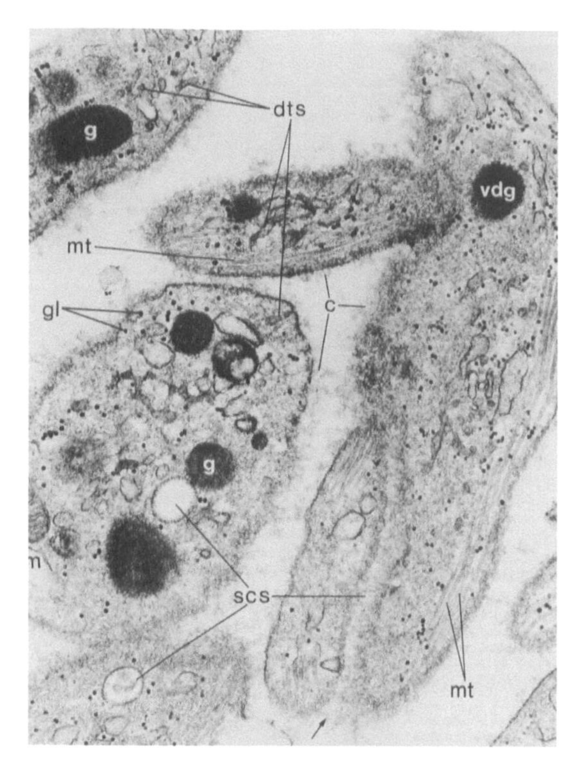

FIGURE G36 Transmission electron micrograph of a section of a group of platelets. Canalicular systems (scs) connect to openings on the surface (bottom center). These channels display the same kind of carbohydrate-containing coat as is seen on the plasma membrane of the platelets. A second system of tubules contains an electron-dense material (*dts*). *c*, cell coats; *g*, α granules; *gl*, glycogen particles; *m*, mitochondrion; *mt*, bundles of microtubules that encircle the platelets; *vdg*, very dense granule. 21,000 ×.

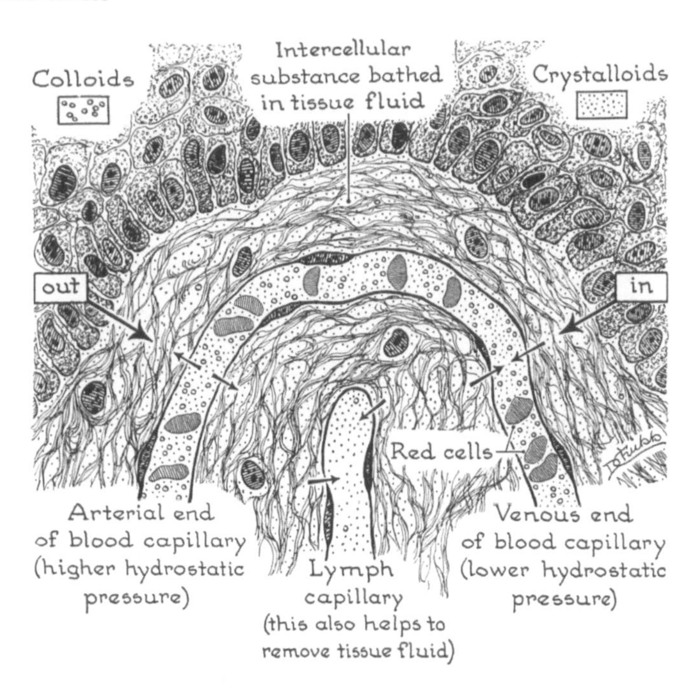

FIGURE G37 Diagram to illustrate the formation of lymph. Blood capillaries, at the arterial end, leak fluid—a plasma ultrafiltrate—into the tissues; this is the TISSUE FLUID. Protein molecules (colloids) are retained within the blood but crystalloids pass through. The fluid is resorbed to the blood capillaries at the venous end and by lymph vessels.

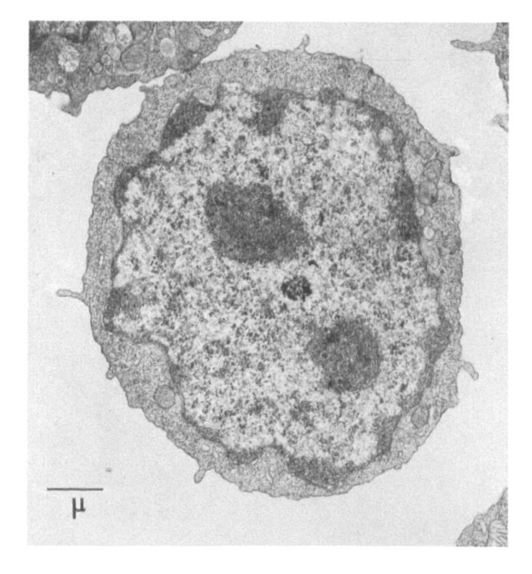

FIGURE G38 The extremely rare, presumed pluripotential, stem cell of mammalian hematopoietic tissue is indistinguishable from small lymphocytes. Free ribosomes abound in the cytoplasm. Bar = 1 μm 15,130 ×.

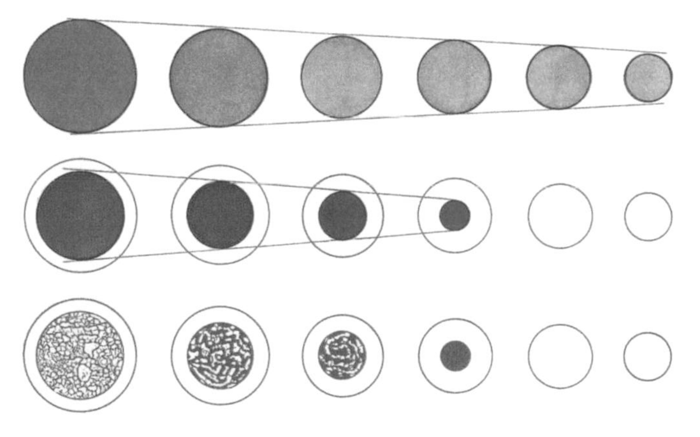

FIGURE G39a Sequence in the maturation of vertebrate blood cells as seen in a Romanowsky-stained smear. At the *top*, the changes and cell size and cytoplasmic color are shown as development proceeds; *center* are seen the changes in nuclear size and color; and, at the *bottom* the changes in nuclear chromatin are shown.

- Nucleoli disappear.
- The intense cytoplasmic basophilia of youth tends to become more acidophilic.
- Aging cells develop cytoplasmic granules.

An application of these generalizations is shown in Fig. G39b for the development of a human erythrocyte in bone marrow.

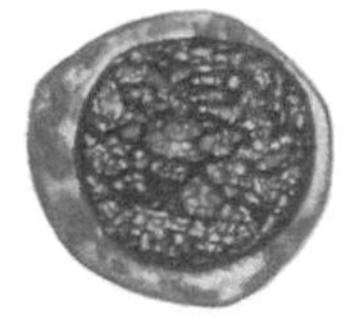 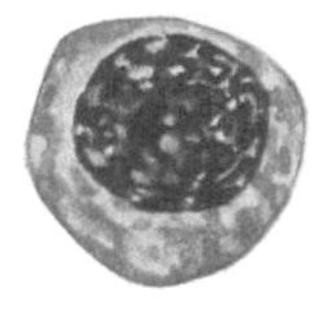 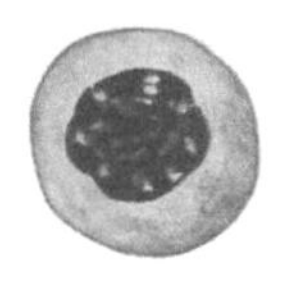 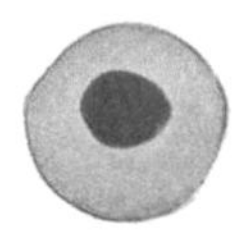  

FIGURE G39b A composite of these changes as seen in the development of a mammalian erythrocyte. These cells may be identified from the left to right: proerythroblast, basophilic erythroblast, polychromatophilic erythroblast, orthochromatic erythroblast, and mature erythrocyte.

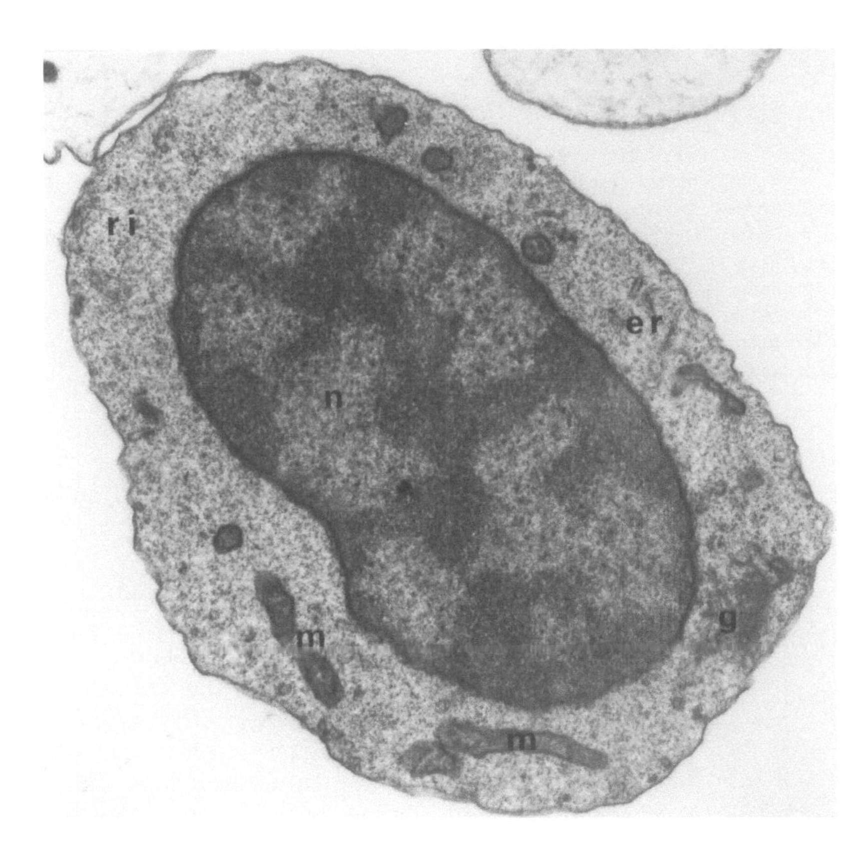

FIGURE G40 Transmission electron micrograph of prorubricyte in the kidney of a masou salmon (*Oncorhynchus masou*). The nucleus has been extruded. *er*, granular endoplasmic reticulum; *g*, Golgi complex; *m*, mitochondria; *n*, nucleus; *ri*, free ribosomes. 14,300 ×.

Erythrocytic Series: Erythropoiesis

The development of the mammalian erythrocyte involves numerous generations of cells that may be classified into four groups on the basis of several characteristics (Fig. G39b).

1. *Proerythroblast* (rubriblast). In the earliest forms the cytoplasm stains light blue but in later, more prevalent forms, in association with the beginning formation of hemoglobin, there is a superimposed reddish tint that gives the cytoplasm a distinctive, dark, royal blue color. Rubriblasts constitute less than 1% of nucleated cells of normal mammalian marrow.
2. *Basophilic erythroblast* (prorubricyte, basophilic normoblast, intermediate erythroblast, intermediate normoblast). An example of this cell from the blood of the salmon is shown in Fig. G40. This cell is slightly smaller than the proerythroblast and the nuclear structure is less delicate. Nucleoli are indefinite. The cytoplasm is still predominantly basophilic. These cells constitute 1%–4% of nucleated marrow cells—cell numbers are maintained by their proliferation.
3. *Polychromatophilic erythroblast* (rubricyte, polychromatophilic normoblast, intermediate erythroblast, intermediate normoblast) (Fig. G41). The nucleus of the polychromatophilic erythroblast resembles that of a mature lymphocyte. Nucleoli can no longer be seen. The cytoplasm is POLYCHROMATOPHILIC, where there is a mixing of the blueness of the cytoplasmic basophilia with the redness of the emerging hemoglobin. Normally these constitute 5%–10% of nucleated cells of the marrow.

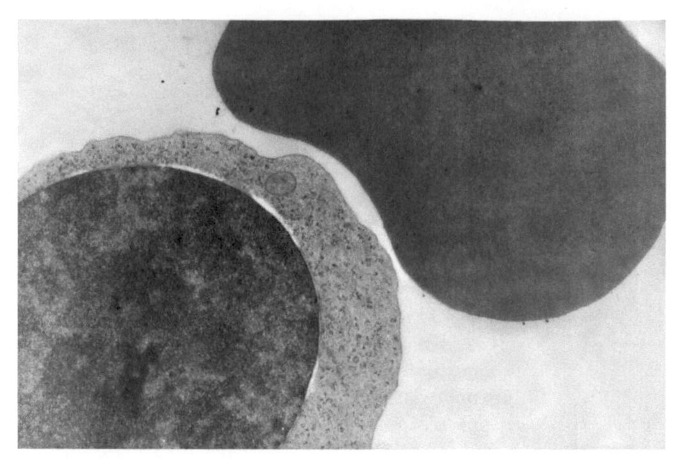

FIGURE G41 Transmission electron micrograph of an erythroblast (lower left) and an erythrocyte (upper right) from the liver of a rat embryo. The contents of the erythrocyte are electron dense, typical of material containing hemoglobin; occasional flecks (probably ribosomes) are scattered throughout. The erythroblast is no longer is able to divide but still has its nucleus. Also present are a few mitochondria and ribosomes. Occasional homogeneous masses of hemoglobin are beginning to appear in the cytoplasm. 30,000 ×.

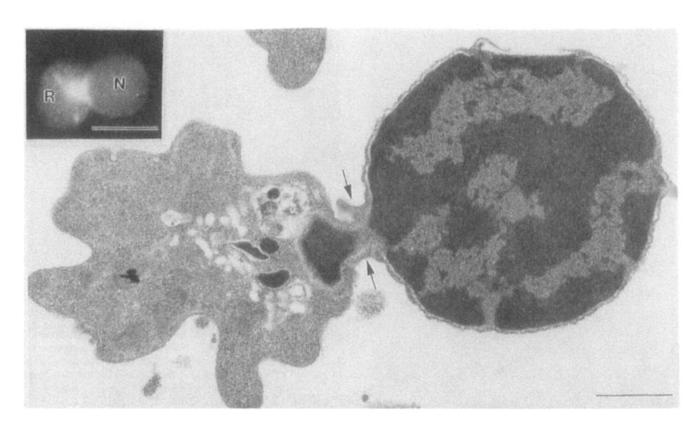

FIGURE G42 Transmission electron micrograph of a section of an enucleating erythroblast in cultured splenic tissue of the mouse. The arrows indicate the developing constriction between the nucleus and the incipient reticulocyte. Bar = 1 μm. *Inset*: Photomicrograph of a similar cell undergoing the process of separation. Bar = 5 μm.

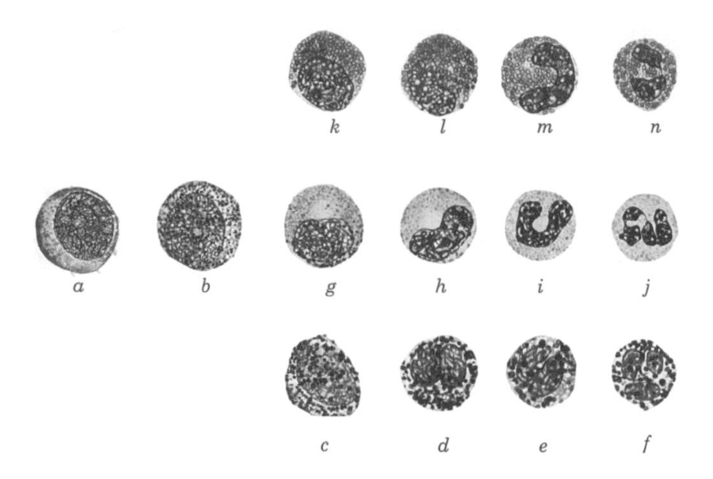

FIGURE G43 Developmental sequence of granulocytes. The cells are (a), myeloblast; (b), promyelocyte; (c, g, and k), myelocytes, basophilic, neutrophilic, and eosinophilic (d, h, and l), metamyelocytes, basophil, neutrophilic, and eosinophilic; (e, i, and m), granulocytes with band nuclei, basophilic, neutrophilic and eosinophilic (f, j, and n), mature basophil, neutrophil, and eosinophil.

4. *Orthochromatic erythroblast* (metarubricyte, normoblast, orthochromatic normoblast, acidophilic or eosinophilic erythroblast or normoblast). This cell has ceased to divide and its cytoplasm is predominantly red but usually there is a slight residue of basophilia due to the presence of a few polysomes and fragments of other organelles. In some cells the cytoplasm stains in much the same manner as the normal erythrocytes in the same field thus forming the basis for the prefixes "normo" and "ortho" that have been applied to these cells. The chromatin is more condensed than in earlier stages and the nucleus appears as a solid, almost blue/black sphere. When the metarubricyte differentiates into an erythrocyte, the nucleus is expelled (Fig. G42). Following nuclear extrusion, a few polysomes and fragments of organelles remain in the newly minted erythrocyte. These bits of basophilia persist for a short time in young erythrocytes, which are called RETICULOCYTES.

Granulocytic Series: Granulopoiesis

All granulocytes undergo similar sequences of development (Fig. G43 l and r). Later stages of the three types are distinguished by the presence of their specific granules. Immature stages are always larger than the mature elements. The developmental stages are as follows:

1. *Myeloblast*. The earliest cell of the granulocytic series has the morphological characteristics of all immature blood cells. There are no granules. These cells make up less than 1% of nucleated cells of the marrow.
2. *Promyelocyte* (progranulocyte). The nuclear chromatin of these cells is coarser than that of a myeloblast, the nucleoli are less well defined, the cytoplasm is less basophilic and, in relation to the nucleus, more abundant. Nonspecific azurophilic granules are present in varying numbers; the granules are dark and usually

basophilic but in some cells they appear reddish or purple. Promyelocytes constitute 1%–5% of nucleated cells of the marrow.

3. *Myelocytes.* At this level, specific granules have developed and three types of myelocytes are recognized: neutrophilic, eosinophilic, and basophilic. The nucleus is round or bean shaped, the chromatin coarser than in the promyelocyte, and nucleoli are no longer visible. Neutrophilic myelocytes constitute 8% of the nucleated cells of normal marrow.
4. *Metamyelocytes.* Metamyelocytes resemble myelocytes but have a slightly indented nucleus. The nuclear characteristics are intermediate between a myelocyte and the polymorphous mature types. The cytoplasm is more acidophilic and smaller in amount than in the myelocyte. The three types of metamyelocytes can be distinguished on the basis of their specific granules: neutrophilic, eosinophilic, and basophilic.

Electron micrographs of developing granulocytes (Figs. G44a–d) show that as the stem cell transforms into a promyelocyte the Golgi complex becomes more extensive and vacuoles containing a core of electron-dense material can be seen pinching off from the ends of its lamellae. These vacuoles fuse and their contents become denser forming the nonspecific AZUROPHILIC GRANULES. During the myelocyte stage, production of azurophilic granules ceases and specific granules (neutrophilic, eosinophilic, or basophilic) are manufactured. Azurophilic granules are produced at the concave surface of the Golgi complex but specific granules are produced at the convex surface. As the cell divides by mitosis, the relative numbers of azurophilic granules become progressively less although a few remain to appear in the mature granulocytes. The organelles diminish with maturation. Although the Golgi complex is reduced it is still involved in the formation of specific granules, pinching off small vacuoles of dense material. These vacuoles coalesce to form larger, membrane-bound specific granules.

Nonspecific azurophilic granules are homogeneous, electron-dense droplets. Neutrophilic granules are less dense and finely granular; eosinophilic granules are ovoid, less dense than the azurophilic, and consist of flocculent material with a crystalline core. Some basophilic granules are homogeneous and electron-dense and are similar to azurophilic granules although they are much larger; others, however, may have a distinctive crystalline, fibrous, or lamellar structure that often is characteristic of the species being studied or method of fixation.

Megakaryocyte or Giant Cell

The megakaryocyte of mammals is readily recognized because of its huge size and large multilobed nucleus (Fig. G45); these cells are often more than 100 μm in diameter. Megakaryocytes originate in the bone marrow from stem cells. There is DNA replication by ENDOMITOSIS (incomplete mitosis without nuclear division) and the nucleus becomes polyploid; it has an intermediate or mature chromatin pattern and several nucleoli. The

FIGURE G44(a–d) Transmission electron micrographs of sections of developing granulocytes showing the relationship of the Golgi complex with the formation of azurophilic and specific granules.

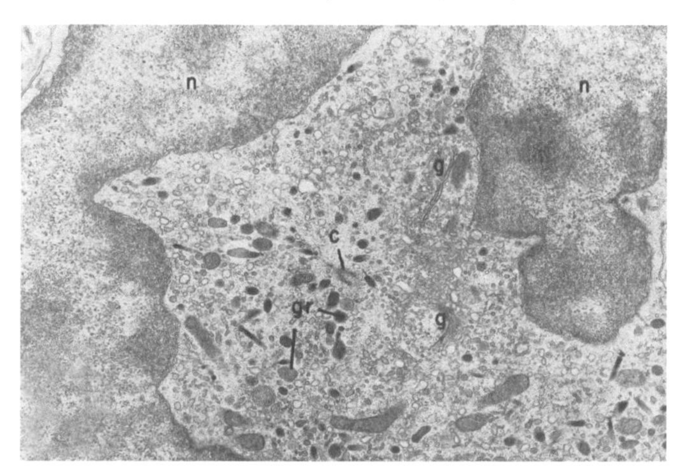

FIGURE G44a Heterophil myelocyte in the slender salamander (*Batrachoseps attenuatus*), showing granules (gr), Golgi elements (g), a centriole (c), and the nucleus (n). 9000 ×.

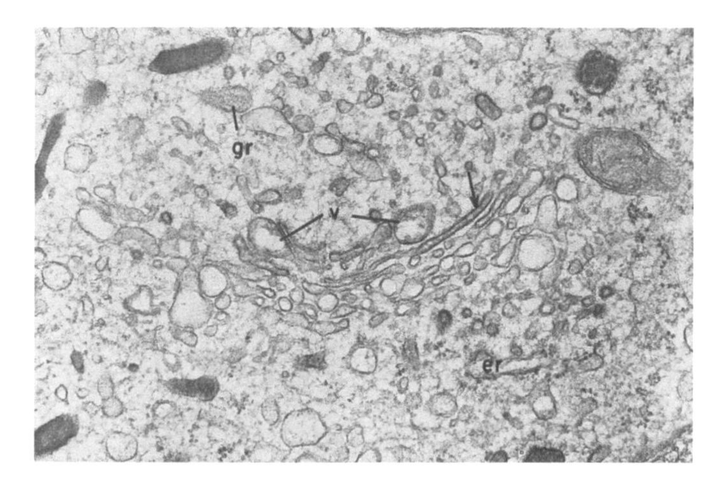

FIGURE G44b Golgi element of the heterophil myelocyte showing precipitated material in the vesicles (v) and cisterna (arrow). *er*, granular endoplasmic reticulum; *gr*, granule. 38,000 ×.

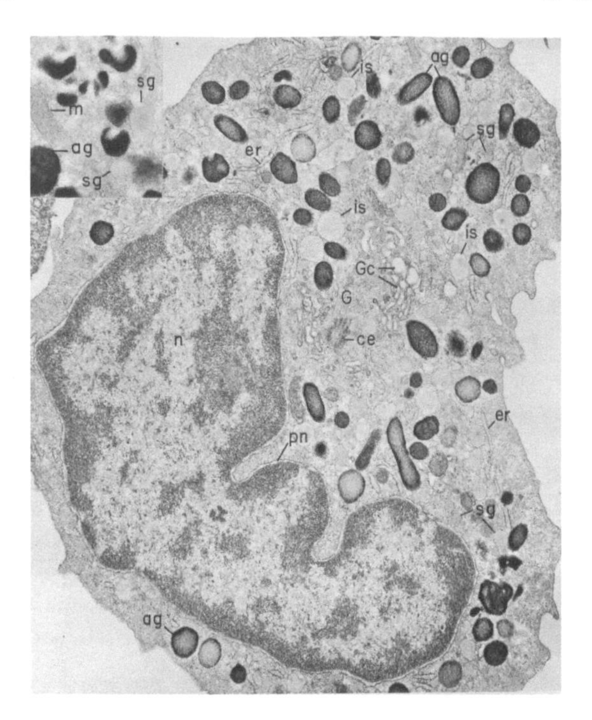

FIGURE G44c Human neutrophilic myelocyte. The peroxidase reaction used to differentiate the granules. Specific granules (sg) do not stain for peroxidase; *ag*, peroxidase-positive azurophils; *ce*, centriole; *er*, granular endoplasmic reticulum; is, immature specific granules; *Gc*, Golgi cisterna; *m*, mitochondrion; *n*, nucleus; *pn*, perinuclear cisterna. 41,000 ×.

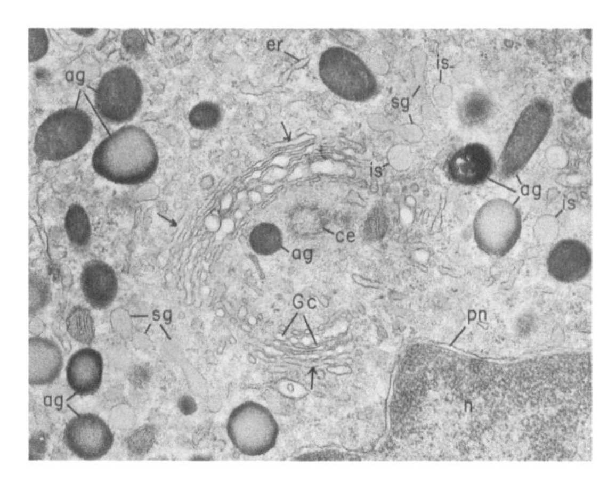

FIGURE G44d Higher power of the Golgi region in a cell similar to Fig. G44c. The peroxidase reaction is seen in the azurophils (ag) but not in specific granules (sg). Golgi cisternae (Gc) are oriented around the centriole (ce); the outer cisternae (arrows) have a content of medium density that is similar to the content of the specific granules. 33,000 ×.

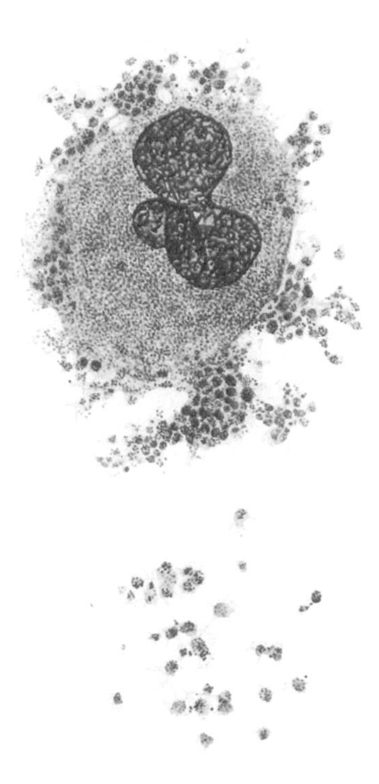

FIGURE G45 Photomicrograph of a megakaryocyte from the human bone marrow. The cytoplasm of these huge cells fragments into platelets (below) that contain a random assortment of organelles.

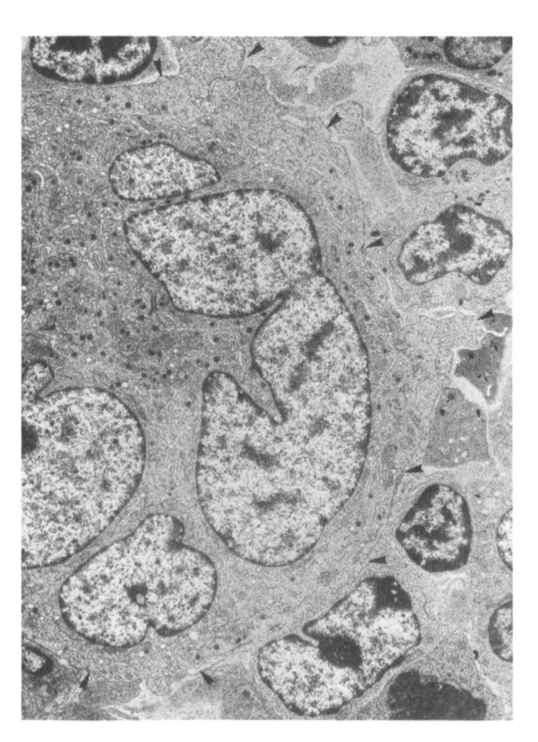

FIGURE G46a Transmission electron micrograph of the bone marrow of a rat with a huge megakaryocyte occupying most of the field; arrowheads indicate its outer limits. The multilobed nucleus is polyploid; compare its size with the nuclei of other cells at the right.

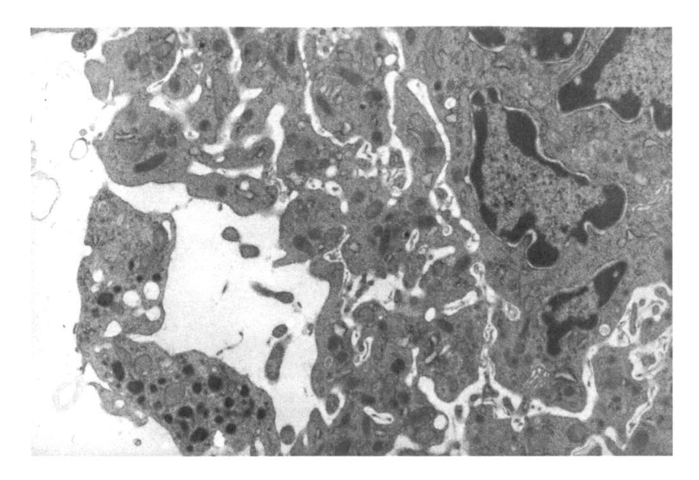

FIGURE G46b The fragmented cytoplasm of a megakaryocyte is shown in this electron micrograph of a section of the bone marrow of a rat. Portions of the polyploid, multilobed nucleus are shown at the bottom right. A few granules, Golgi complexes, mitochondria, and ribosomes may be noted. Flattened cisternae permeate the cytoplasm of these huge cells and eventually become the demarcation membranes where the cytoplasmic fragments tear apart. 30,000 ×.

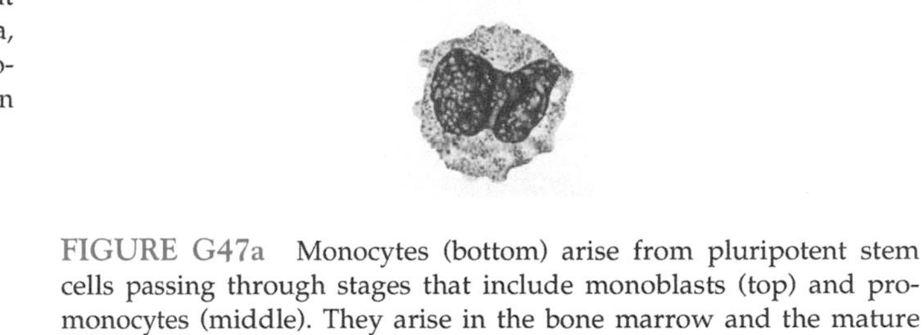

FIGURE G47a Monocytes (bottom) arise from pluripotent stem cells passing through stages that include monoblasts (top) and promonocytes (middle). They arise in the bone marrow and the mature cells migrate to the bone marrow.

cytoplasm is plentiful and is characterized by relatively coarse granules. Fragmentation of the mature cells produces blood platelets (Figs. G46a and G46b). About 0.2%–0.5% of nucleated cells of the marrow are megakaryocytes.

The tattered appearance of mature megakaryocytes is their most conspicuous feature in electron micrographs. Portions of the cell are partitioned off by the DEMARCATION SYSTEM OF MEMBRANES, chains of vesicles that coalesce, eventually isolating membrane-bound fragments of cytoplasm, the PLATELETS, from the main body of the cell. No cytoplasm is lost to the surrounding medium during this process. Lobes of the nucleus are scattered throughout the cell. Small mitochondria, small Golgi complexes, sparse granular endoplasmic reticulum, and abundant free ribosomes are present in the cytoplasm as well as membrane-bound granules of moderate electron density. The individual platelets are devoid of nuclei and contain mitochondria, ribosomes, vesicles of the endoplasmic reticulum, and membrane-bound granules.

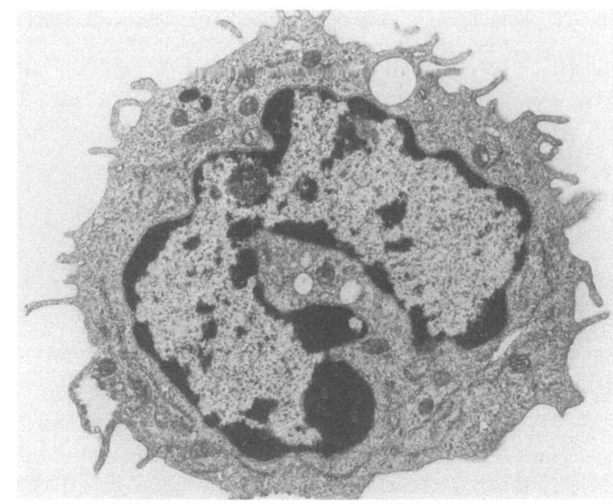

FIGURE G47b Transmission electron micrograph of a bovine monocyte.

Monocytic Series

Monocytes arise from pluripotent stem cells in the bone marrow. They pass through stages of decreasing basophilia of their cytoplasm and increasing lobulation of the nucleus: Monoblast, promonocyte, and monocyte (Figs. G47a and G47b).

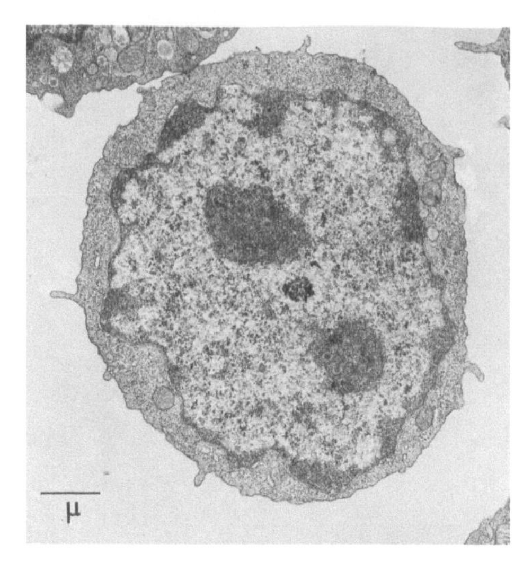

FIGURE G48 Transmission electron micrograph of a section of the extremely rare stem cell of all the blood cells. This pluripotential cell is undistinguished with a thin rim of basophilic cytoplasm.

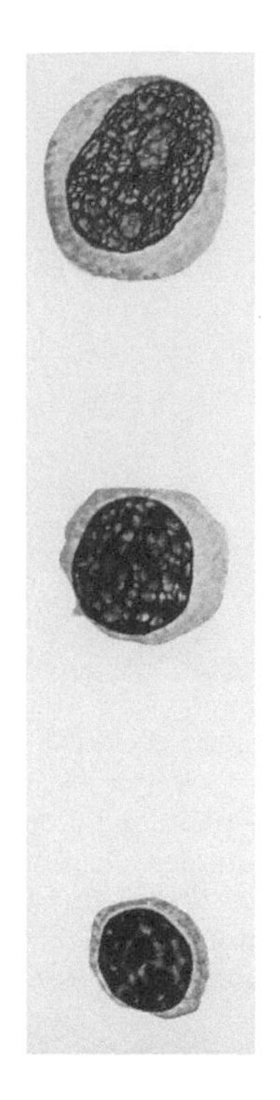

FIGURE G49 The development of the lymphocyte is shown in these photomicrographs of Romanowsky-stained cells. The earliest stage shown is the lymphoblast (top); the cell becomes smaller (middle), the cytoplasm becomes paler and the cell is classified as a prolymphocyte. The mature lymphocyte is smaller, its nucleus is more heterochromatic, and the cytoplasm forms a thin rim around the nucleus. Morphologically, one cannot differentiate B and T lymphocytes.

HEMOPOIESIS IN LYMPHOID TISSUE

The developing cells of lymphoid tissue are often studied in IMPRINT preparations where the cut surface of the organ (e.g., thymus, spleen, or lymph node) is gently pressed against a clean slide, fixed in methanol or the vapor from crystals of iodine, or simply dried in air. The cells are then stained with a Romanowsky stain. The stages in the development of lymphocytes, plasmocytes, and monocytes may be followed.

Stem Cell

These early cells may have no apparent cytoplasm and may be round or exceedingly elongated (Fig. G48). Pale blue cytoplasm may be visible in some of the rounded cells.

Lymphocytic Series

The morphology of the lymphoblast is similar to that of other early cells (e.g., myeloblast, rubriblast) (Fig. G49). This is a large basophilic cell with a large, pale nucleus. The sequence through prolymphocyte to small lymphocyte involves increasing amounts of heterochromatin in the nucleus and increasing basophilia in the diminishing cytoplasm. B and T lymphocytes cannot be differentiated in cells stained with the Romanowsky stains. Small, medium-sized, and large lymphocytes are present in great numbers. The amount of cytoplasm in the small lymphocyte may be so slight as to be not apparent. These cells often assume shapes other than spherical.

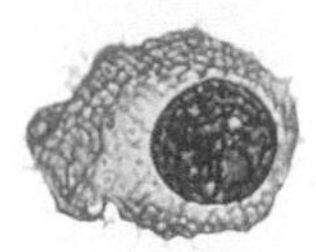

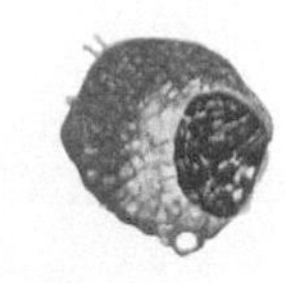

FIGURE G50 Three stages in the development of the plasmocyte are shown: plasmoblast, proplasmocyte, and mature plasma cell. Plasmocytes are distinguished by the intense basophilia of their cytoplasm and the eccentric positioning of the nucleus. The pale area in the cytoplasm is the juxtanuclear clearing that contains the centrioles and Golgi complex.

FIGURE G51 This section of the connective tissue surrounding the intestine of a rat shows the cell center of a plasmocyte surrounded by granular endoplasmic reticulum. It lies near the nucleus and contains a centriole as well as a Golgi complex. 67,000×.

Plasmocytic Series

Plasmocytes arise from plasmoblasts, cells with a rim of conspicuous basophilic cytoplasm; they resemble lymphoblasts but are more basophilic (Fig. G50). They mature through proplasmocytes to plasmocytes (plasma cells). Plasmocytes have a large amount of basophilic, agranular cytoplasm and a relatively small, eccentric nucleus. The cytoplasm adjacent to the nucleus is often more lightly stained than the remainder of the cytoplasm; this is the JUXTANUCLEAR CLEARING. It contains centriole and Golgi complex (Fig. G51).

PART III

ORGANS AND ORGAN SYSTEMS

Different tissues come together in specific and characteristic ways forming the structural and functional units that make up organs. In such an aggregation one tissue is of primary functional importance; other tissues involved perform structural, secondary, and auxiliary roles supporting the primary function of the organ. Groups of organs collaborate in carrying out related functions and constitute the organ systems.

CHAPTER

H

The Circulatory System

The circulatory system consists of the arteries and veins, microvessels, lymph vessels, and the heart. All are lined by simple squamous epithelium, the ENDOTHELIUM. ARTERIES carry blood *from* the heart, VEINS return it *to* the heart (Fig. H1). Arteries leaving the heart branch repeatedly, becoming smaller as they go. Eventually they reach the MICROVESSELS that invade the organs and are the small connections between arteries and veins. All blood vessels with diameters smaller than about 100 μm constitute the MICROCIRCULATION; these include ARTERIOLES, VENULES, CAPILLARIES, and other vessels (Fig. H2). LYMPH, a filtrate from the blood serum leaks into the tissue spaces and is returned to the blood stream in blind-ending LYMPHATICS or lymph vessels (Fig. H3).

The pattern of blood flow in vertebrates varies depending on whether the animal breathes with gills or lungs (Fig. H4). *The cyclostomes, cartilaginous fishes, and bony fishes* breathe by gills and have a two-chambered heart consisting of one ATRIUM and one VENTRICLE. Blood is pumped from the ventricle to the gills where it picks up oxygen and gets rid of wastes. It relies on residual pressure to carry it from the gills, through the veins, to the farthest reaches of the body, and back to the heart. In *amphibians,* the heart has three chambers. Blood from the body's tissues—depleted of oxygen—is carried to the right atrium and ventricle to be pumped to the lungs where it picks up oxygen to be carried throughout the body. It then returns to the heart to be pumped to the lungs to repeat the circuit. Since there is only one ventricle, some mixing of oxygenated and deoxygenated blood occurs, although baffles cut down on this inefficiency. In *birds* and *mammals,* there are two atria and two ventricles creating separate PULMONARY and SYSTEMIC CIRCULATIONS thereby avoiding mixing of the blood. The basic components of the parts of the circulatory system are similar regardless of the complexity of the heart. In spite of differences in function, the parts of the vascular systems of different vertebrates have many features in common.

FUNDAMENTAL PLAN OF THE VASCULAR SYSTEM

A fundamental plan is seen in all but the smallest vessels—even in the wall of the heart. The wall of each is usually composed of three concentric TUNICS (Fig. H5a)(DM diagram of generalized xs). Note that certain features of this common plan may be emphasized, reduced, or omitted and try to relate these variations from region to region with the functions of those parts (Fig. H5b). Although blood pressure attains similar values in all mammals, large and small, the complexity of the walls of the vessels increases with increasing body size.

The internal layer of the vessels is the TUNICA INTIMA. It is composed of elements usually arranged in three layers. ENDOTHELIUM is the simple squamous epithelium lining the vessel as well as its fine microfibrillar basal lamina. Endothelium lines *all* the blood vessels and the walls of some of the vessels in the microcirculation consist solely of endothelium. SUBENDOTHELIAL CONNECTIVE TISSUE is a thin layer of areolar connective tissue upon which the endothelium rests and the INTERNAL ELASTIC MEMBRANE is a layer of elastic fibrous connective tissue that may be fenestrated (Fig. H5c). In fixed preparations it often presents a wavy appearance because of contraction of the elastic components.

The TUNICA MEDIA is composed of smooth muscle and fibrous connective tissue. The smooth muscle fibers are usually longitudinally and circularly arranged, often as a low-pitched spiral, and are connected by gap junctions; elastic and collagenous fibers form the intermuscular connective tissue.

The elements of the external TUNICA ADVENTITIA are usually longitudinally arranged. Elastic fibers commonly concentrate nearest to the tunica media as an EXTERNAL ELASTIC MEMBRANE. The remainder of the coat is composed of fibroelastic tissue that grades into the adjacent areolar tissue to guy and support the vessel.

An Atlas of Comparative Vertebrate Histology.
DOI: https://doi.org/10.1016/B978-0-12-410424-2.00008-1

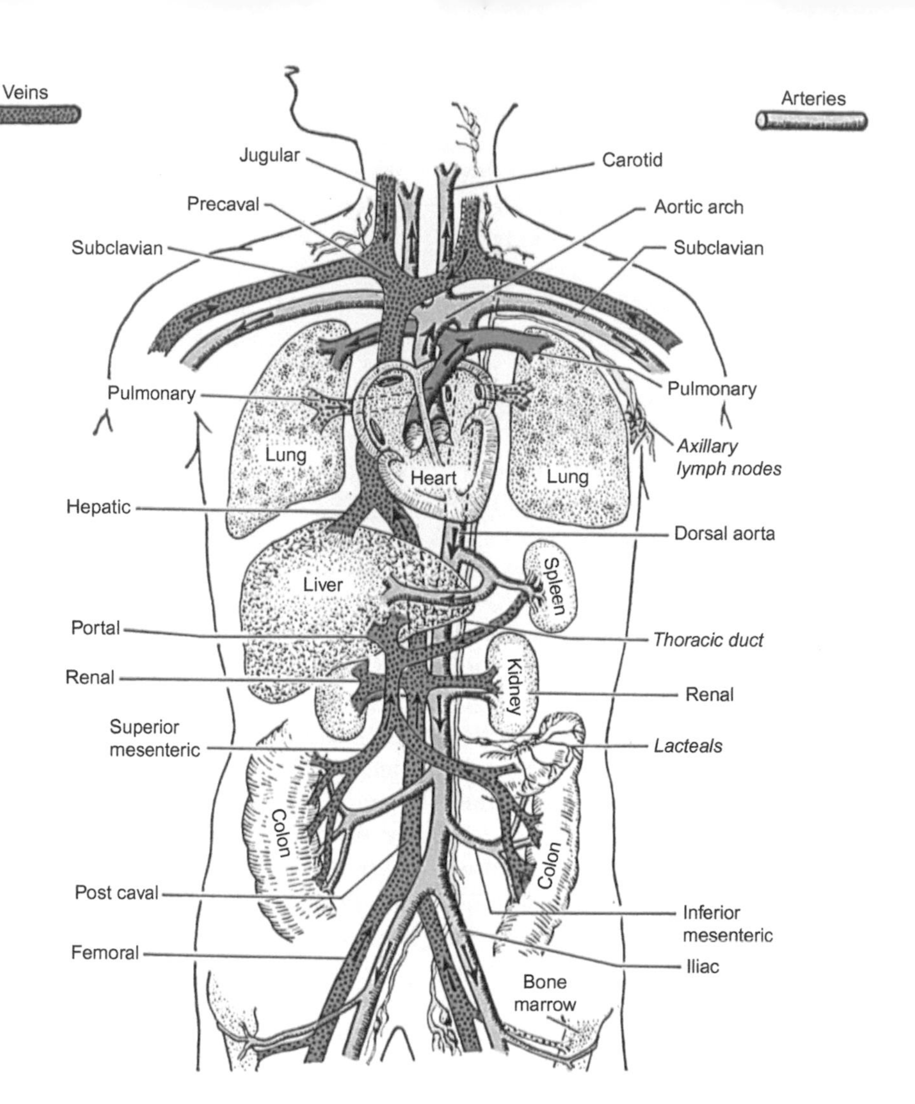

FIGURE H1 Principal blood vessels of the human circulatory system are shown in their relationship with the internal organs. Arrows indicate the paths of blood flow. *Blue*: blood low in oxygen. The thoracic duct of the lymphatic system and a few lymph nodes are shown in yellow and labeled in italics.

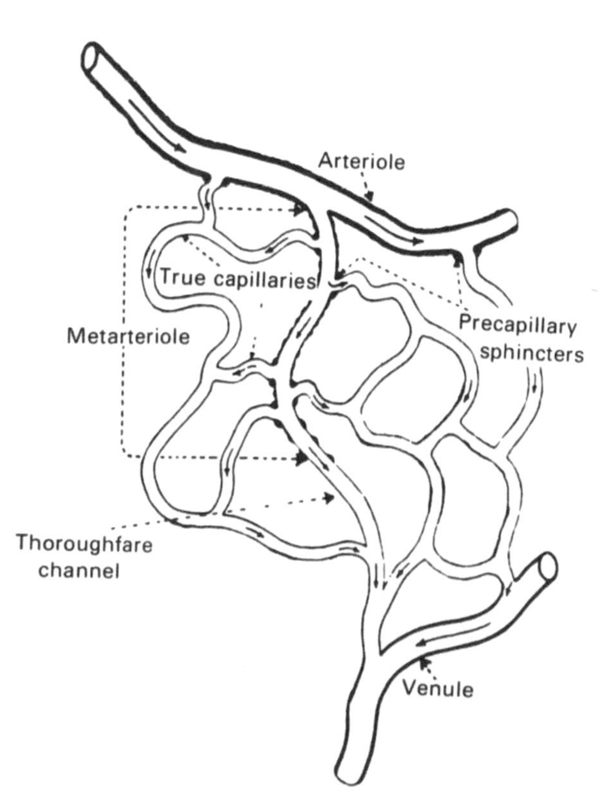

FIGURE H2 Schematic diagram of the vessels that constitute the terminal vascular bed in vertebrates: the MICROCIRCULATION.

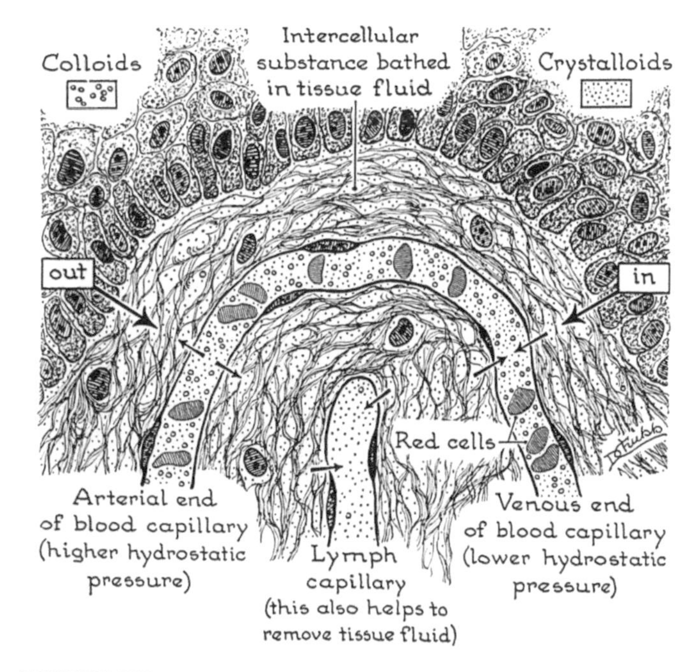

FIGURE H3 Diagram illustrating the formation of tissue fluid by capillaries and its absorption by capillaries and lymphatics. The proteins in the blood are represented by small circles and the crystalloids of blood and lymph by dots. Normally little colloid escapes from the capillaries; the small amount that does escape is returned to the circulation by the lymphatics.

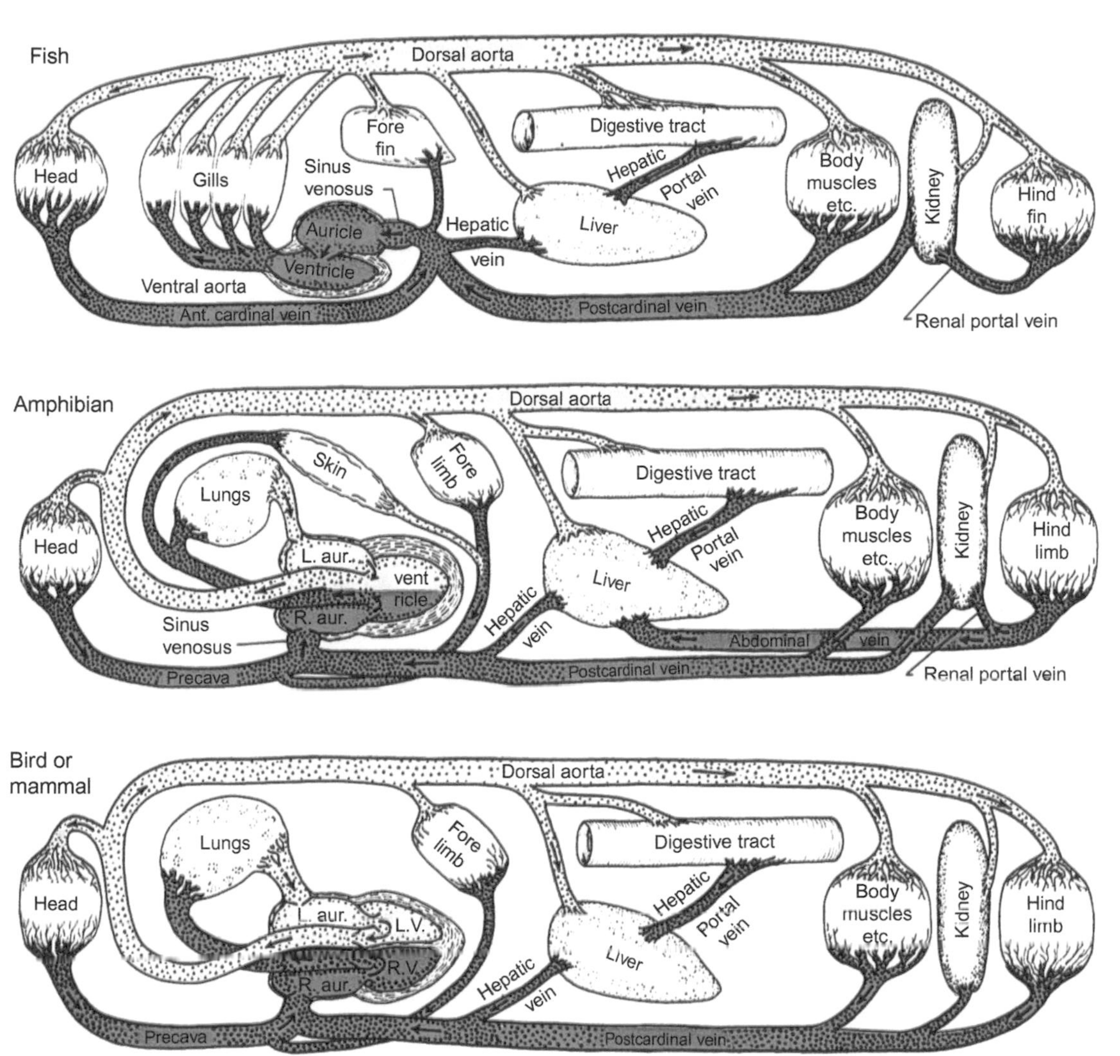

FIGURE H4 Paths of blood circulation in vertebrates. The hearts of fish have one atrium or auricle and one ventricle; amphibians have two atria and a single ventricle; birds and mammals have two atria and two ventricles. The arrows indicate the paths of blood flow. *Blue*: blood low in oxygen; *red*: oxygenated blood.

Blood vessels more than a millimeter in diameter have nutrient vessels, the VASA VASORUM or "vessels of the vessels" (*Singular*: VAS VASIS). Because of the high pressures inside the lumina of arteries, the vasa vasorum penetrate the wall only as far as the tunica adventitia; they extend through the tunica media of veins where the internal pressure is low. Lymph vessels are present in the walls of the largest blood vessels. The NERVI VASORUM (*Singular*: NERVUS VASIS) consists of branching unmyelinated vasomotor fibers in the tunica adventitia that terminate on the smooth muscle of the tunica media as well as branching myelinated afferent fibers that extend inward from the tunica adventitia as far as the tunica intima.

ARTERIES

Compare sections of large ELASTIC ARTERIES, (Figs. H6a and H6b) such as the aorta, with sections of small MUSCULAR ARTERIES (Fig. H6c). Squamous endothelial cells of the TUNICA INTIMA form a continuous pavement surrounding the lumen. In electron micrographs note that these cells rest on a basal lamina and contain the usual organelles; they are distinguished by an abundance of pinocytotic pits and vesicles (Figs. H7a and H7b). The cells are joined by tight junctions (zonulae occludentes) and gap junctions (Fig. H7c). The internal elastic membrane is apparent in all but the large arteries where similar membranes are distributed through the tunica media. This membrane contracts after death throwing the tunica intima into folds and causing the endothelial cells to project into the lumen.

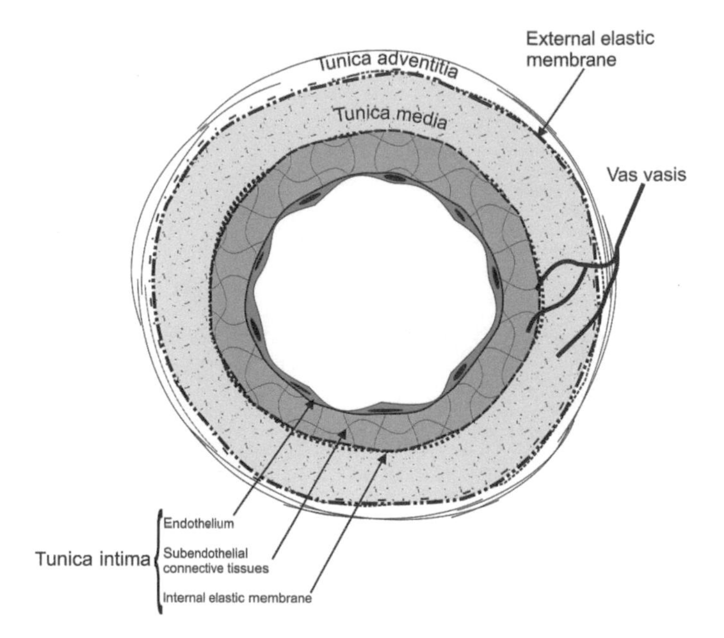

FIGURE H5a Diagram of a cross section of a generalized blood vessel. Three layers consisting of largely variable amounts of elastic connective tissue, smooth muscle, and collagenous tissue, constitutes the wall of the larger blood vessels. All blood vessels are lined with a simple squamous epithelium: the ENDOTHELIUM. The tissues in the three layers vary according to the functions they perform. The vessel walls, being living tissue, require a blood supply. This is brought to them by small vessels, the VASA VASORUM. This becomes a problem in the large arteries where the internal pressure is high.

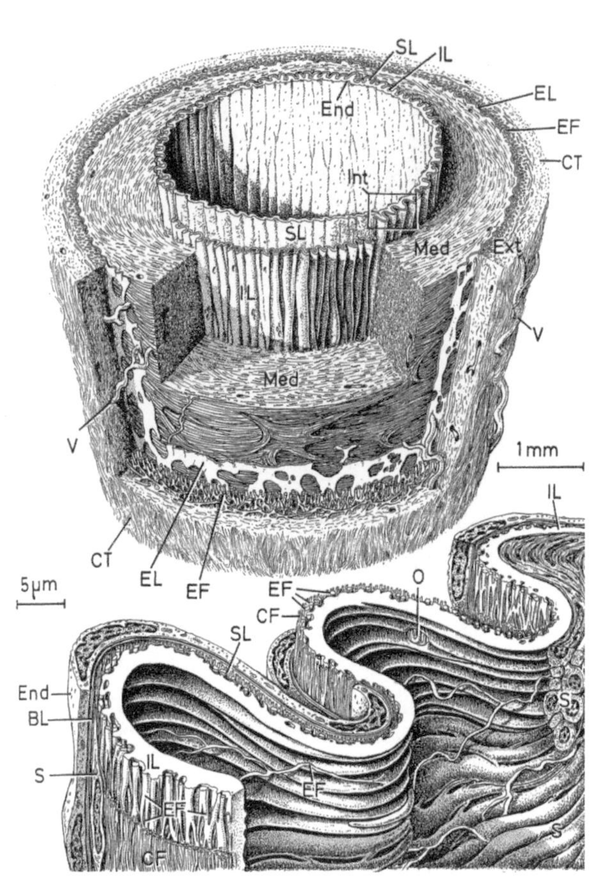

FIGURE H5b (Top) Drawing of a muscular artery of a human. The vessel is lined by the TUNICA INTIMA (TI), which consists of the inner simple squamous ENDOTHELIUM (End), the SUBENDOTHELIAL CONNECTIVE TISSUE (SL), and the INTERNAL ELASTIC MEMBRANE (IL). A thick layer of smooth muscle of the TUNICA MEDIA (Med) encircles the tunica intima. (Bottom) The basal lamina (BL) of the endothelial cells of the tunica intima is shown. Collagen fibers (CF), elastic fibers (EF), the elastic membrane (EL), and smooth muscle cells (S) form the internal elastic membrane of the tunica intima, small blood vessels (V) supply nutrients. Some may open (O) from the lumen of the larger vessel. The vessel is contained in a sheath (Ext) of connective tissue fibers (CT). *Note*: The wavy nature of the tunica intima is an artifact produced after death.

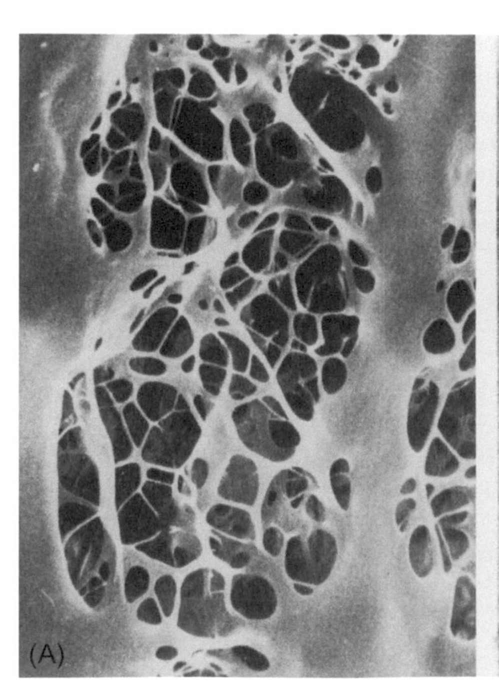

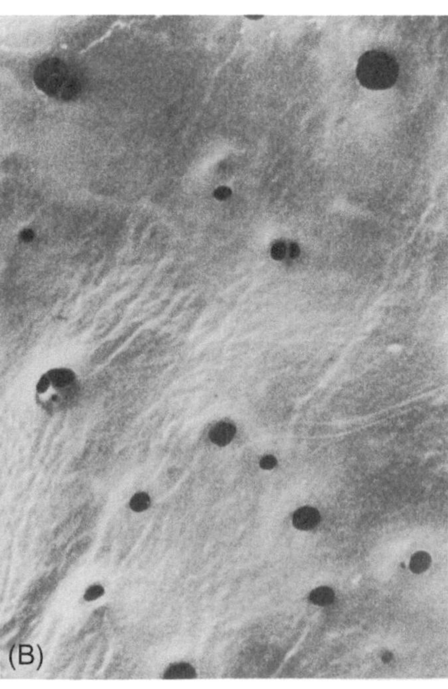

FIGURE H5c Scanning electron micrograph showing a surface view of the internal elastic lamina of the aorta of a rat. Large fenestrations are traversed by an irregular meshwork of thin strands of elastin (A). The internal elastic membrane of its femoral artery shows small, round fenestrations (B).

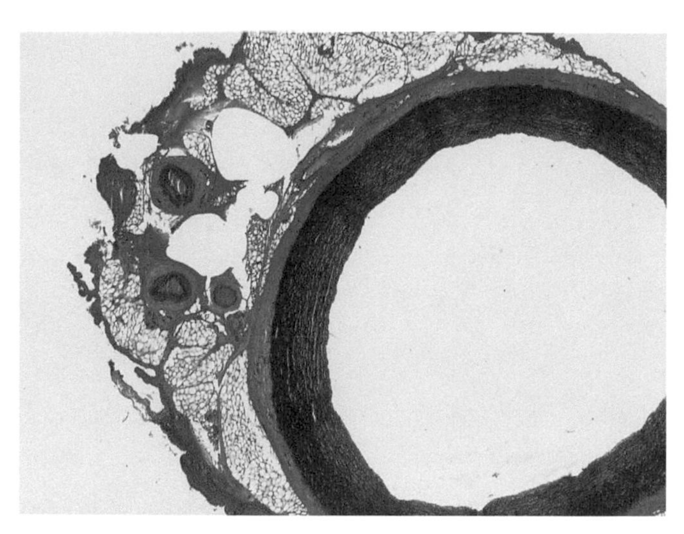

FIGURE H6a Photomicrograph of a cross section of a large, elastic artery of a mammal. Elastic fibers in the thick tunica media have been blackened by an elastic tissue stain. A few small branches from the artery are seen in the surrounding tissues. Smooth muscle fibers mingle with the elastic fibers to encircle the artery. 2.5 ×.

FIGURE H6b Photomicrograph of a cross section of a large, elastic artery of a mammal at a higher magnification. Elastic fibers in the tunica media blackened by an elastic tissue stain are visible individually. Smooth muscle fibers mingle with the elastic fibers encircling the artery. 20 ×.

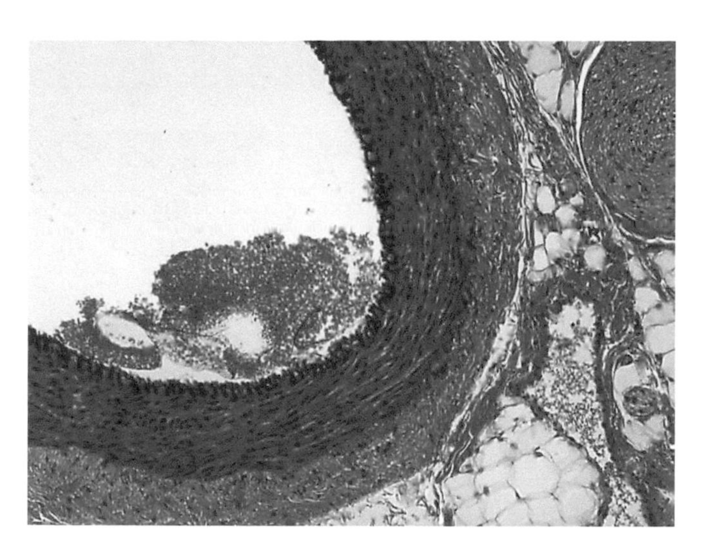

FIGURE H6c Cross section of a smaller muscular artery. Although an elastic tissue stain has been used, none is seen in the tunica media. The internal elastic membrane of the tunica intima has taken up the stain. 20 ×.

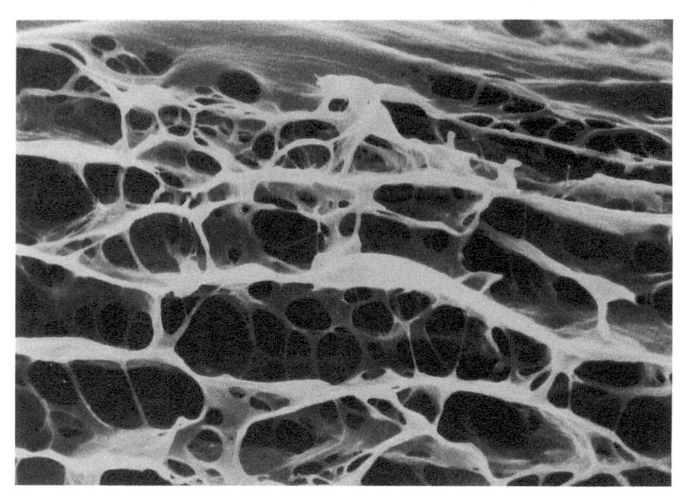

FIGURE H6d Scanning electron micrograph of the elastin in the wall of the rat aorta. The elastic layer was extracted with hot formic acid to remove all other tissue components. Multiple concentric elastic lamellae are interconnected by radially oriented strands and fenestrated septa. In the intact aorta, smooth muscle cells occupied the spaces between the elastic lamellae.

The thickest coat of an artery is the TUNICA MEDIA, which consists of muscle interspersed with elastic tissue. Compare the relative amounts of elastic and muscular tissue in large and small arteries. The elastic tissue in the wall of large arteries absorbs the pulsations of the blood arising from the heart so that a fairly even flow reaches the tissues. The muscle in the wall serves to direct the flow of blood to the various organs as needs dictate. (Figs. H6b–H6d). In arteries the TUNICA ADVENTITIA is thinner than the tunica media. It contains the outer elastic membrane and there are seldom any muscle fibers in it. Vasa and nervi vasorum are present in the tunica adventitia.

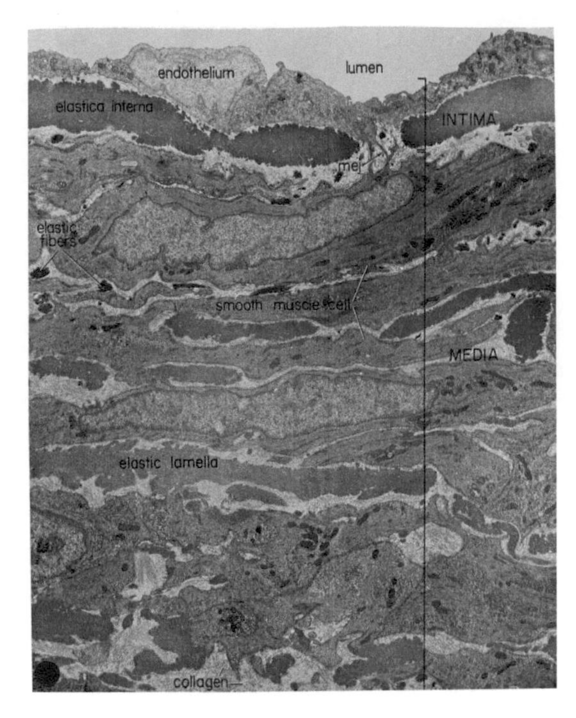

FIGURE H7a Electron micrograph of a cross section of a muscular artery. Processes of the endothelial cells penetrate fenestrations in the internal elastic lamina to form myoendothelial junctions (mej) with the smooth muscle cells of the tunica media. 6000 ×.

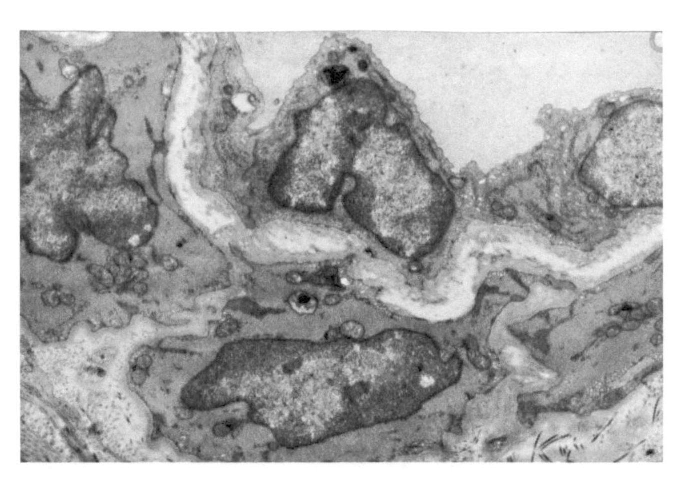

FIGURE H7b Electron micrograph of a cross section of an arteriole from the epidermis of a hamster. Cells of its endothelial lining protrude into the lumen of the vessel. A layer of elastic connective tissue separates the endothelial cells from the smooth muscle cells of the tunica media. The smooth muscle of the tunica media contracts and relaxes to regulate the flow of blood into the tissues. Collagenous fibers of the tunica adventitia are at the bottom. 21,500 ×.

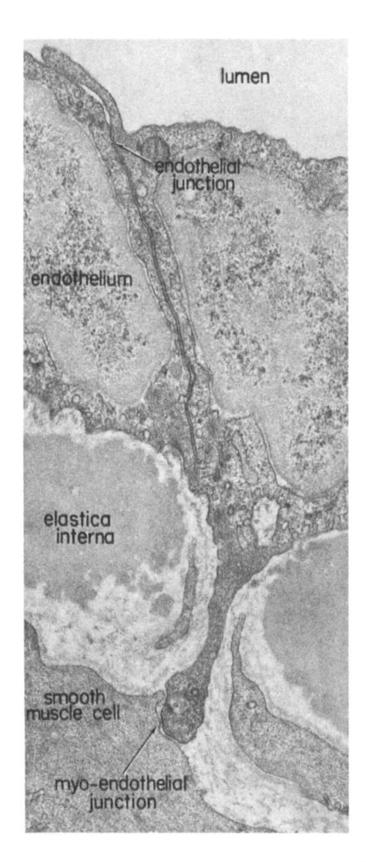

FIGURE H7c Electron micrograph of an endothelial junction and a myoendothelial junction in a muscular artery. A tongue-like extension of the endothelial cell penetrates a fenestra of the internal elastic membrane to reach the smooth muscle cell. 25,000 ×.

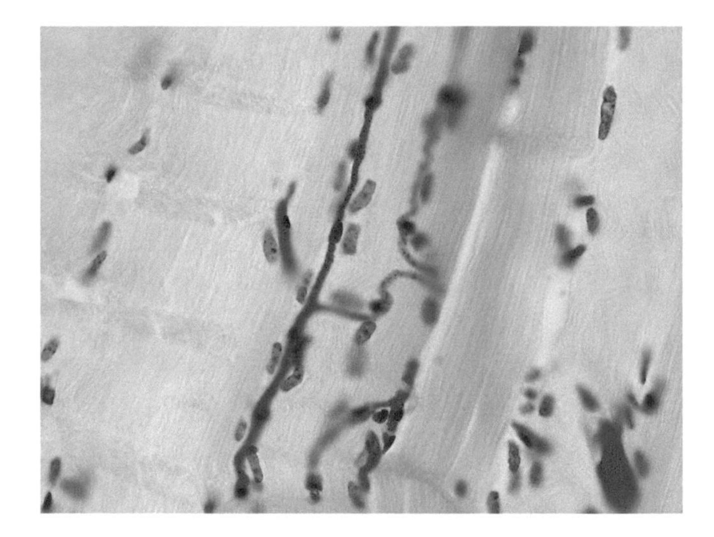

FIGURE H8a Photomicrograph of a longitudinal section of mammalian striated muscle; a colored mass has been injected into the blood vessels and shows the abundance of the blood supply to the muscle cells. 40 ×.

THE MICROCIRCULATION

There are several routes that blood can take when it leaves arterioles to make its way to venules (Fig. H8a). If these channels are smaller than 100 μm in diameter, they are considered to be the MICROCIRCULATION. Networks of capillaries are the most familiar of these but there are other channels that perform specialized functions and their roles are reflected in their structure. Blood may be

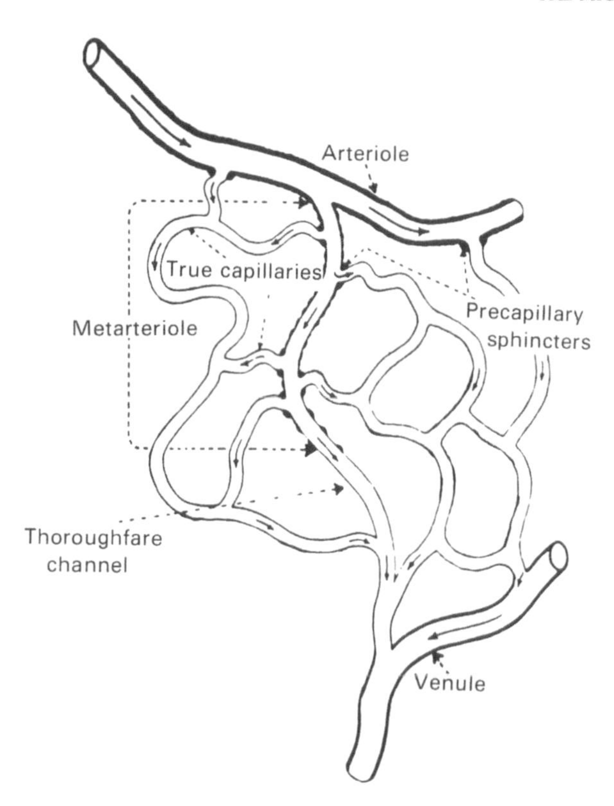

FIGURE H8b Diagram of the some of the components that constitute the microcirculation. The usual flow of blood is from arterioles through a capillary bed to venules and back to the heart. While in the capillaries the blood provides oxygen to the tissues and removes wastes. In certain areas, however, the blood flows through ARTERIOVENOUS ANASTOMOSES or AV SHUNTS, which provide a direct channel from the arteriole to the venule. These channels may be utilized to deliver heat to areas exposed to the cold, such as the skin of fingertips, nose, lips, or they may have a hydraulic function as in the erectile tissue of the penis and clitoris. Blood flow through the various parts of the microcirculation is controlled by PRECAPILLARY SPHINCTERS.

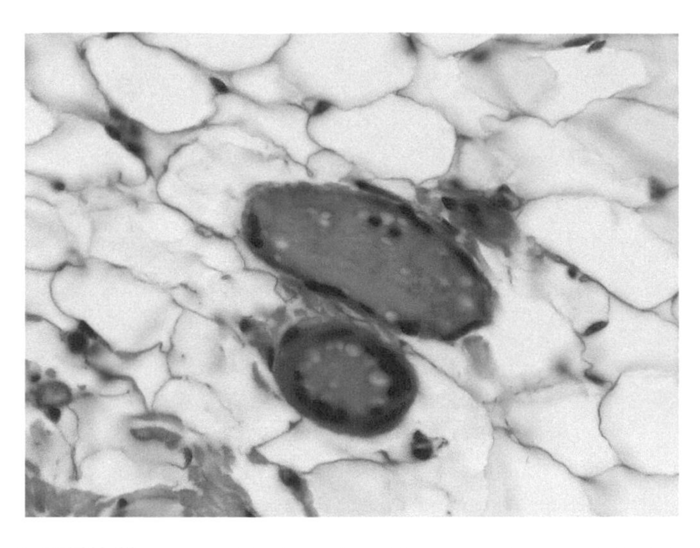

FIGURE H9a Arteriole (lower) and its companion venule in fatty tissue of the cat. There is a single layer of smooth muscle cells in the wall of the arteriole. 40 ×.

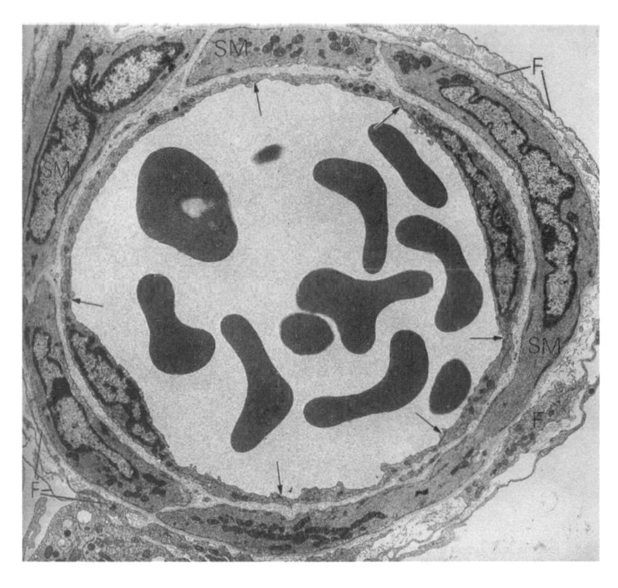

FIGURE H9c Electron micrograph of a cross section of a small mammalian arteriole. The tunica intima consists of squamous endothelial cells, their basement lamina, and a thin layer of subendothelial connective tissue. The *arrows* indicate junctions between adjoining endothelial cells. The tunica media is a single layer of smooth muscle cells (SM). The tunica adventitia is composed of collagen fibrils and attenuated fibroblasts (F).

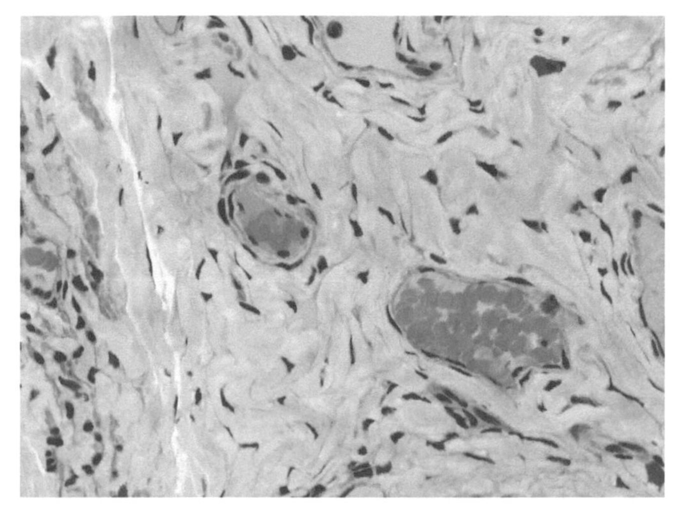

FIGURE H9b Photomicrograph of a semithin section of the mesentery of a monkey showing a cross section of an arteriole (upper) and its companion venule. 40 ×.

routed directly to venules by ARTERIOVENOUS ANASTOMOSES or pass through PREFERRED CHANNELS (Fig. H8b). As usual, all channels are lined by endothelial cells.

Arterioles

Two arterioles are shown in Figs. H9a and H9b. As is often the case they are accompanied by venules. In electron micrographs (Figs. H9c and H9d) the tunica media is seen to be reduced to a single or double layer of smooth muscle. The tunica intima consists of a lining of

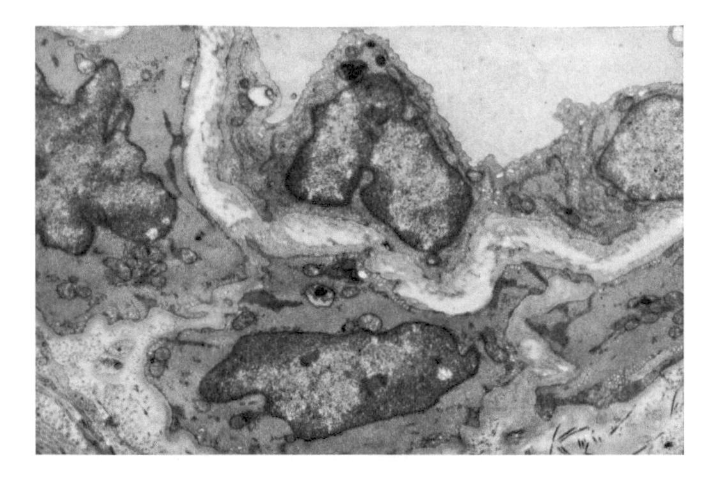

FIGURE H9d Transmission electron micrograph of a section of an arteriole from the epididymis of a hamster. Two endothelial cells at the top of the micrograph line the lumen of the arteriole and rest on a layer of elastic connective tissue. Below is the single layer of circumferentially arranged smooth muscle cells cut in cross section. Pinocytotic vesicles are plentiful under the plasma membranes of the smooth muscle cells. They are in the contracted state as evidenced by the irregular folds in their nuclear membranes. A basement lamina and a few collagenous and elastic fibers surround the smooth muscle cells on their inner and outer surfaces. 21,500 ×.

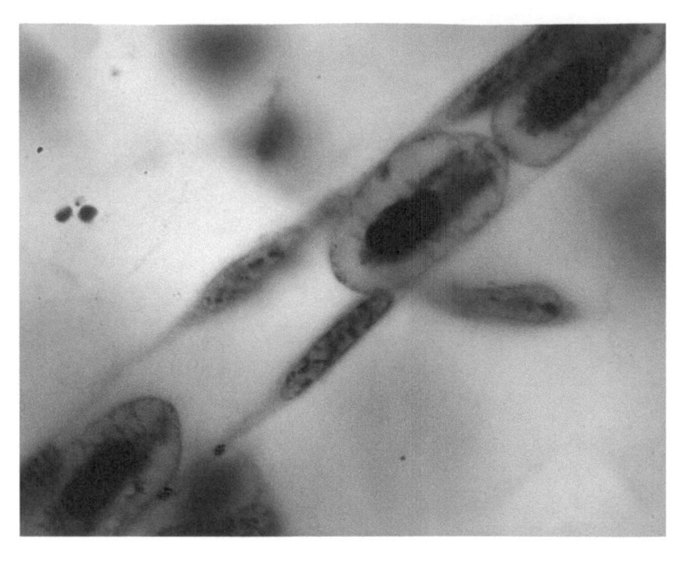

FIGURE H10a Capillary from a whole mount of the skin of a salamander (*Ambystoma*) stained with an iron haematoxilyn, showing the nuclei of the simple squamous endothelial cells. The lumen contains several nucleated erythrocytes. 100 ×.

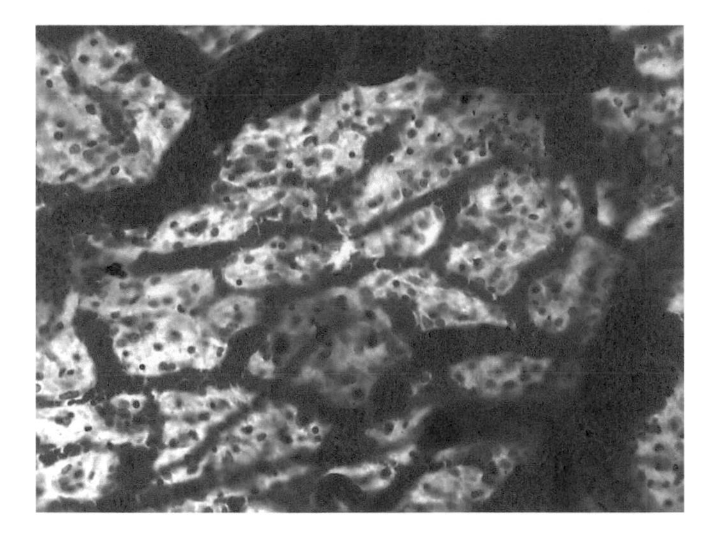

FIGURE H10b Whole mount of mammalian pia mater (a membrane covering the brain) showing the complexity of the capillary circulation. 20 ×.

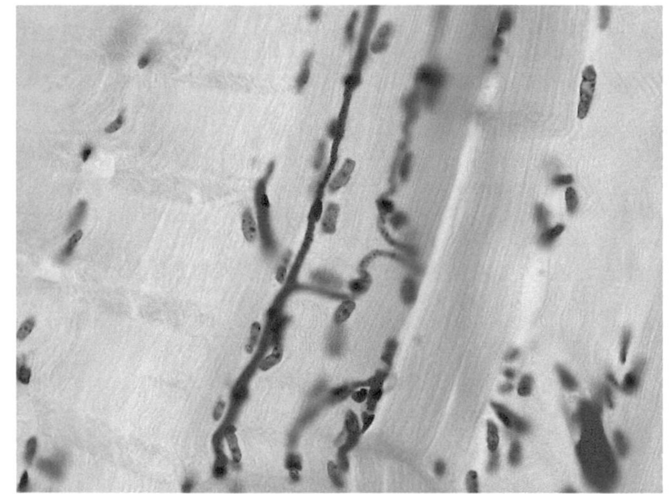

FIGURE H10c Longitudinal section of mammalian striated muscle. Blood vessels in this tissue were injected with a red gelatinous mass to show the complexity of the capillary circulation. 40 ×.

endothelial cells linked by occluding and gap junctions; it rests on a basal lamina and is subtended by loose connective tissue. Basal processes of the endothelial cells penetrate fenestrae in the internal elastic membrane to form contacts with the muscle cells in the tunica media to form, the MYOENDOTHELIAL JUNCTIONS.

Capillaries

Capillaries are small endothelial tubes, one cell in thickness, that connect arterioles and venules (imagine several fried eggs rolled up to form a tube) (Fig. H10a). The endothelial cells may be continuous or they may be FENSTRATED. They form a vascular network throughout the body, bringing oxygen to the tissues and carrying away wastes (Figs. H10b–H10e). They are supported by a basement lamina and fine reticular fibers (Fig. H10f). The capillary is clasped at intervals by branching PERICYTES interposed between leaflets of the basement lamina.

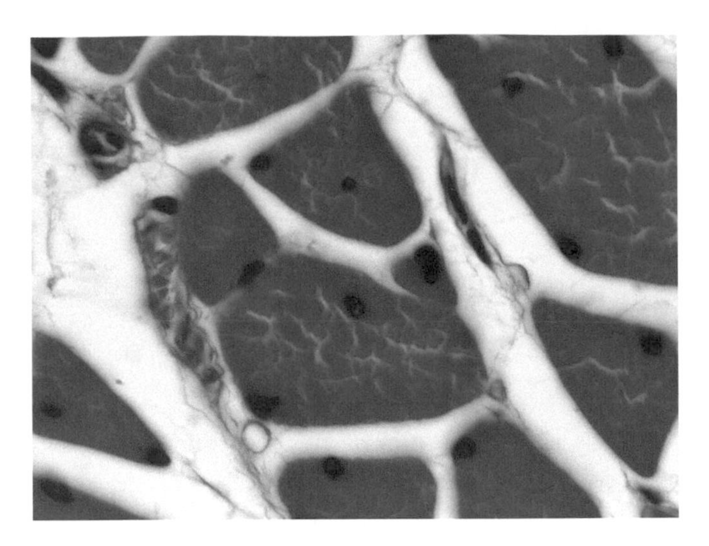

FIGURE H10d Cross section of striated muscle in a section of the foot of a toad; the slight shrinkage helps to visualize the structures between the muscle cells. Cross sections of capillaries resemble signet rings; a longitudinal section of a capillary is at the left. 63 ×.

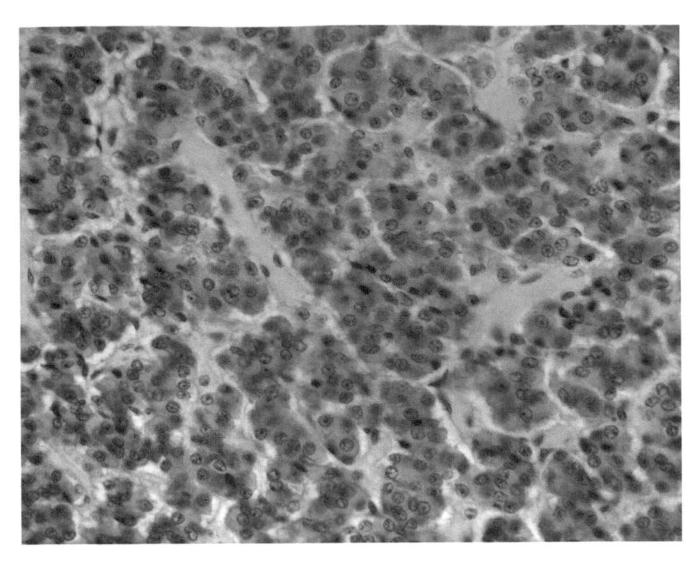

FIGURE H10e Photomicrograph of a section of mammalian pituitary. A longitudinal section of a capillary crosses the field near the center. 40 ×.

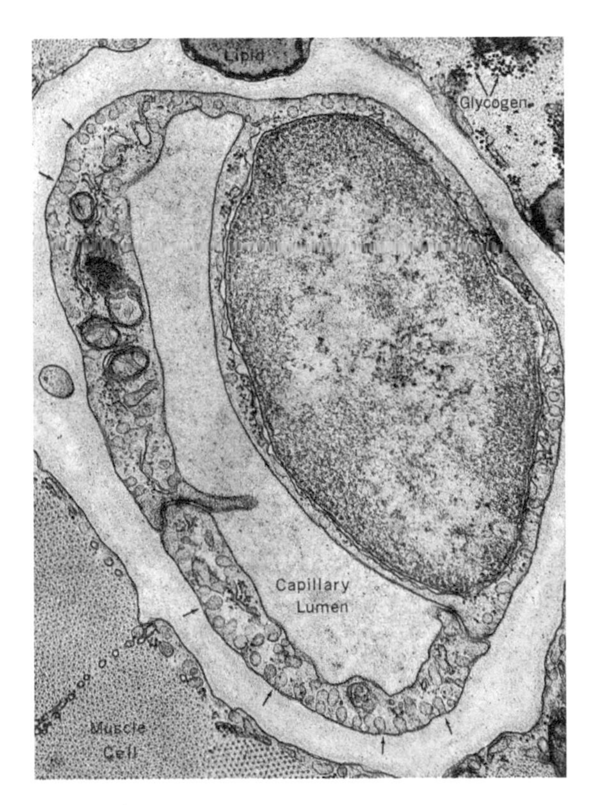

FIGURE H10f Electron micrograph of a cross section of a capillary in the myocardium of a cat. Micropinocytotic vesicles are abundant in the endothelial cells of the capillary. Similar vesicles open into the surrounding space (*arrows*); others are found free in the cytoplasm of the endothelial cells. At this point the capillary consists of parts of two endothelial cells fastened to each other by tight junctions. Marginal folds at the point of contact project into the lumen of the capillary. 41,000 ×.

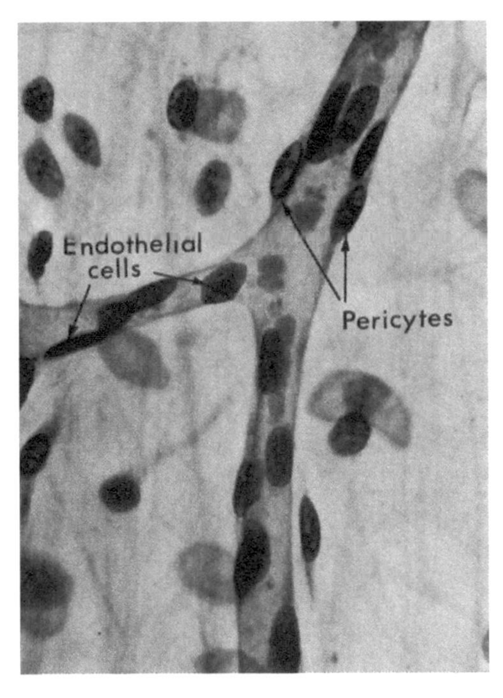

FIGURE H10g Photomicrograph of a whole mount of rat mesentery showing intact capillaries embraced by pericytes.

(Figs. H10g and H10h). The diameter of the lumen of capillaries is roughly the same as that of the erythrocytes of the species being examined and some capillaries are so small that only one erythrocyte at a time may squeeze through. Capillaries are defined as vessels consisting solely of endothelium, its basal lamina, and pericytes(Figs. H10i and H10j).

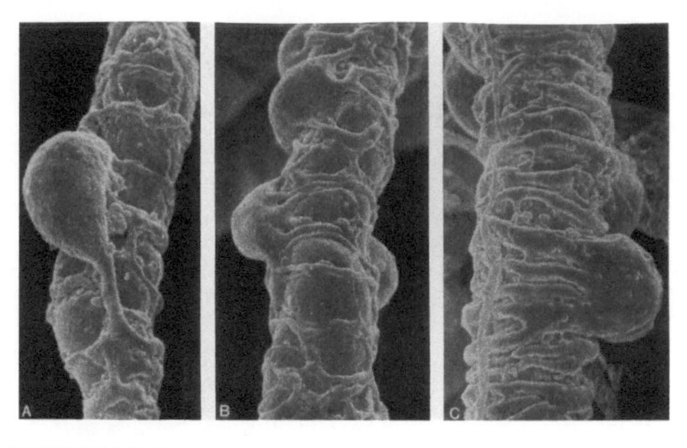

FIGURE H10h Scanning electron micrographs showing pericytes embracing small vessels: (A) capillary; (B) capillary; and (C) an arteriole.

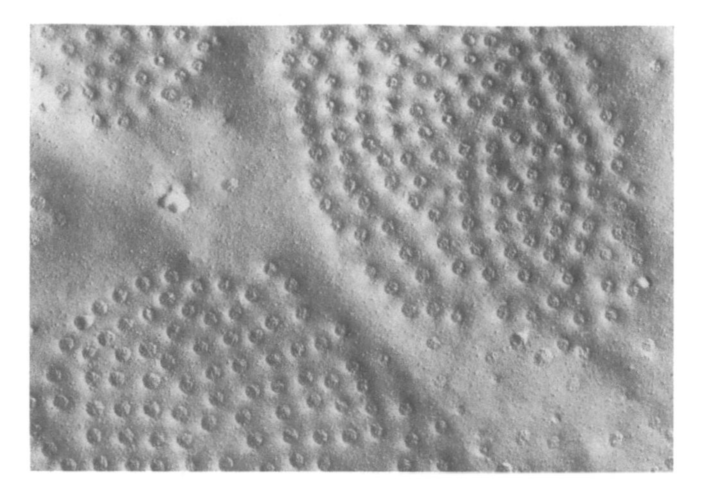

FIGURE H10i Electron micrograph of a freeze-fracture replica of a fenestrated capillary. Fenestrated areas are separated by nonfenestrated areas.

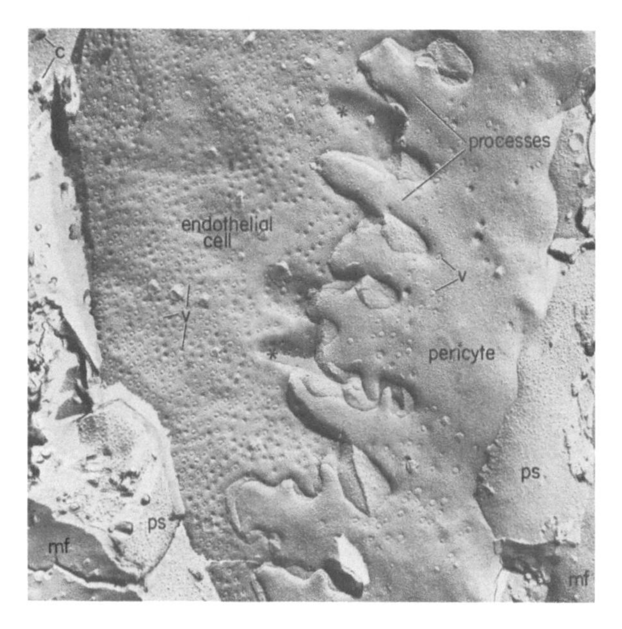

FIGURE H10j Electron micrograph of a freeze-fracture replica of a blood capillary from the diaphragm of a rat. The endothelial cell is partially covered by a pericyte. Some of the processes of the pericyte have been removed exposing complementary depressions (*) on the endothelium. There are more vesicular openings (*v*) on the endothelial cell than on the pericyte. *c*, collagen; *mf*, muscle fiber; *ps*, pericapillary space. 27,000 ×.

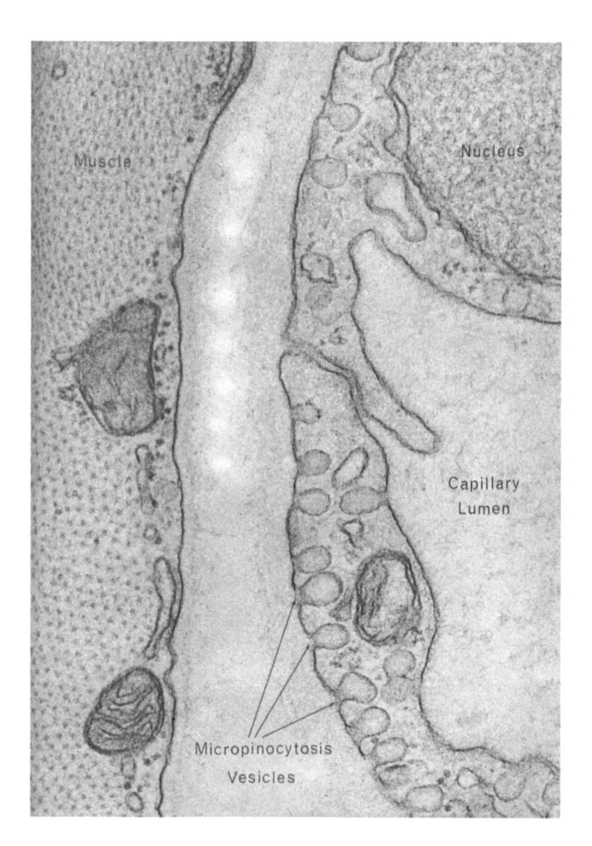

FIGURE H11a Electron micrograph of a section of mammalian heart muscle showing a continuous capillary containing many micropinocytotic vesicles opening either into the lumen of the vessel or into the extravascular space. 95,000 ×.

Electron micrographs of endothelial cells show abundant pinocytotic pits and vesicles throughout the cytoplasm (Fig. H11a). Note the zonulae occludentes in freeze-fracture preparations (Figs. H11b and H11c). Two types of capillaries may be recognized on the basis of the integrity of the wall (Fig. H11d). CONTINUOUS CAPILLARIES are seen in micrographs of muscle, lung, skin, etc., where the margins of the endothelial cells are closely apposed (Fig. H11e). Tight junctions in the form of plaques (FASCIAE OCCLUDENTES) occur between adjacent cells (Fig. H11c); elsewhere a thin layer of intercellular material intervenes. In the brain the tight junctions are zonulae occludentes and completely seal the intercellular spaces. The FENESTRATED CAPILLARIES of the kidney, some endocrine glands, exocrine pancreas, and intestinal villi consist of attenuated endothelial cells perforated by pores that may be spanned by a thin diaphragm. The basement lamina is continuous over the pores (Figs. H10i & H11f).

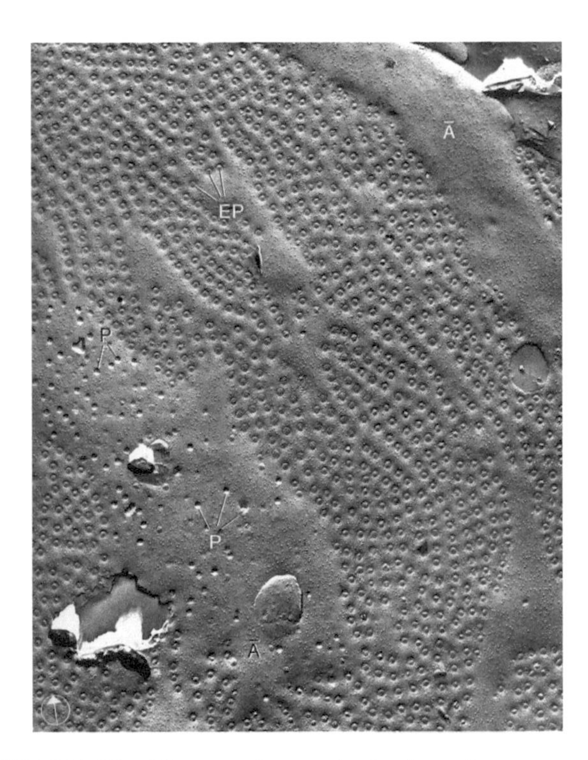

FIGURE H11b Electron micrograph of a freeze-fracture preparation of an fenestrated endothelial cell from the kidney of a spiny mouse. There are many shallow depressions, characteristic of endothelial pores (EP) and a few endocytotic pits (P). 39,000 ×.

Sinusoids

Sinusoids are structurally different from capillaries in that they enclose irregular tortuous spaces of variable diameter. They occur in sections of liver, bone marrow, spleen, and certain endocrine glands. Their walls are molded on the underlying epithelial cells and contain large gaps between the endothelial cells (Figs. H12a and H12b).

In electron micrographs the endothelial cells of sinusoids may appear much thicker than other endothelial cells. They contain abundant pinocytotic pits and vesicles (Fig. H12c). The basement lamina is incomplete or lacking and no diaphragms are seen covering the fenestrations (Fig. H12d).

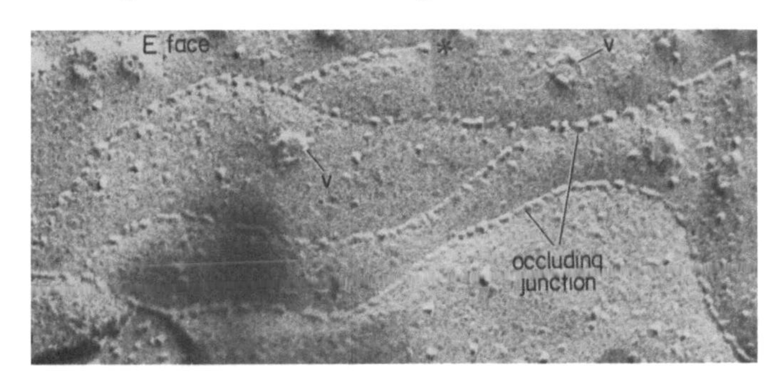

FIGURE H11c Electron micrograph of a freeze-fracture preparation of an occluding junction from the endothelial cells of a capillary. *V*, vesicle opening. 160,000 ×.

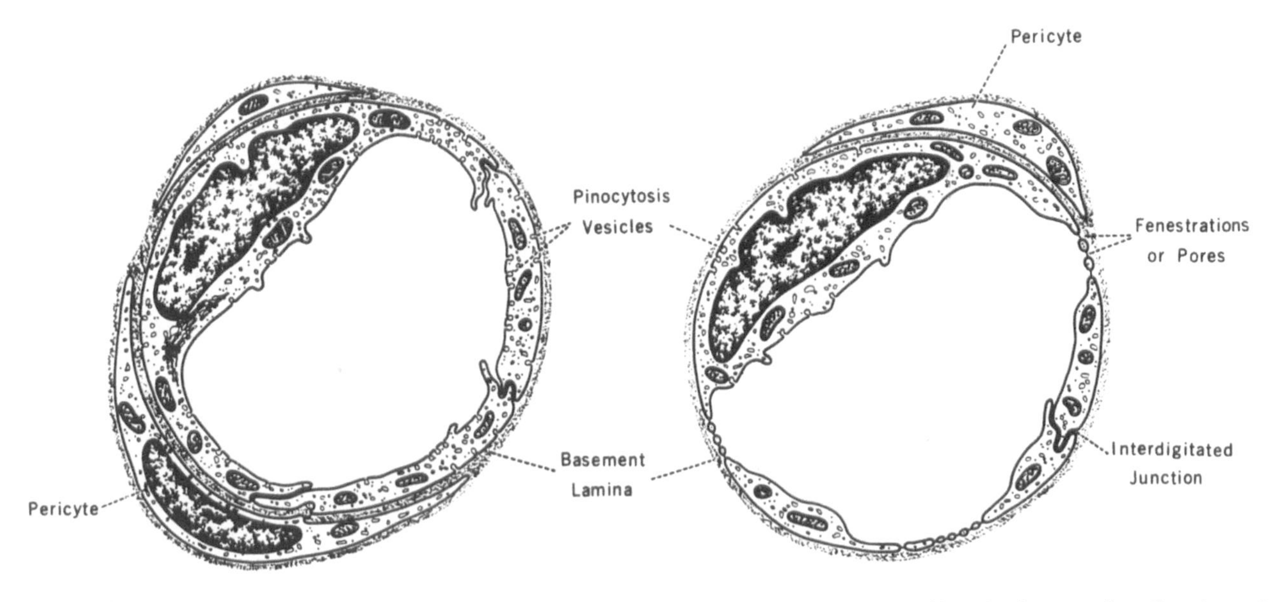

FIGURE H11d Diagrams comparing the structure of a *continuous capillary* from striated muscle (left) and a *fenestrated capillary* from the viscera (right).

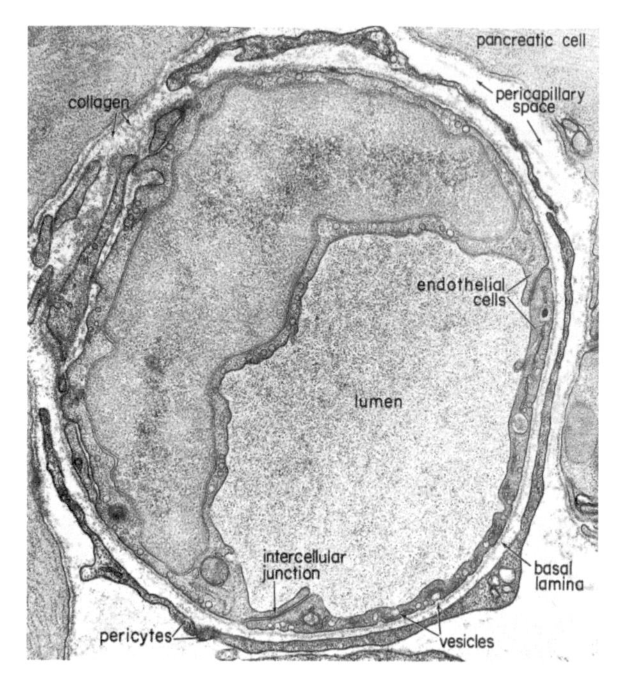

FIGURE H11e Cross section of a continuous capillary from the pancreas of a rat. The large endothelial nucleus at the left partially obliterates the lumen. Two endothelial cells, joined at intercellular junctions, constitute the lining of the capillary. A pericyte embraces the capillary. 40,000 ×.

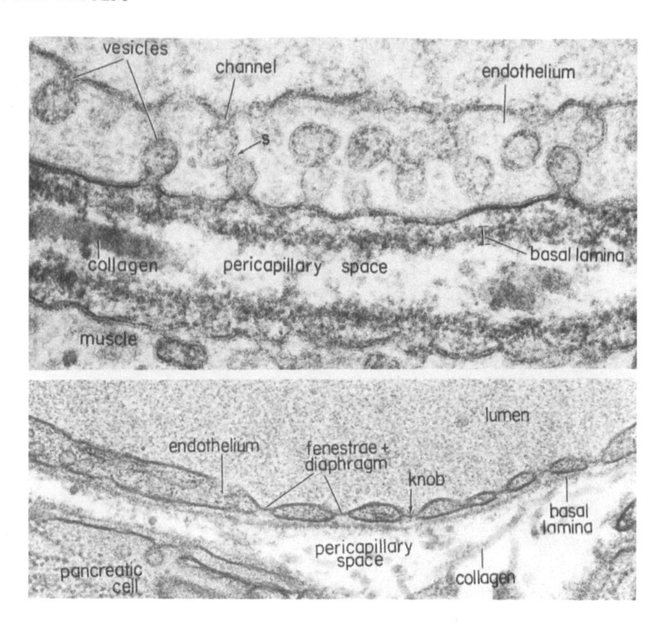

FIGURE H11f (Top) Electron micrograph of a section of a continuous capillary from the diaphragm of a rat. 120,000 ×. (Bottom) Electron micrograph of a fenestrated capillary from the pancreas of a rat. 70,000 ×.

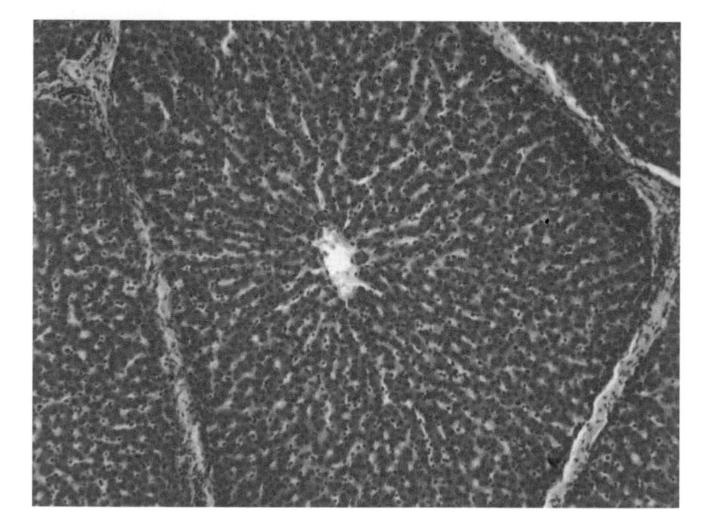

FIGURE H12a Photomicrographs of a section of a liver lobule from a pig. Blood enters the lobule by arterioles at its periphery, passes through sinusoids to the center of the lobule, and thence to the hepatic vein. 10 ×.

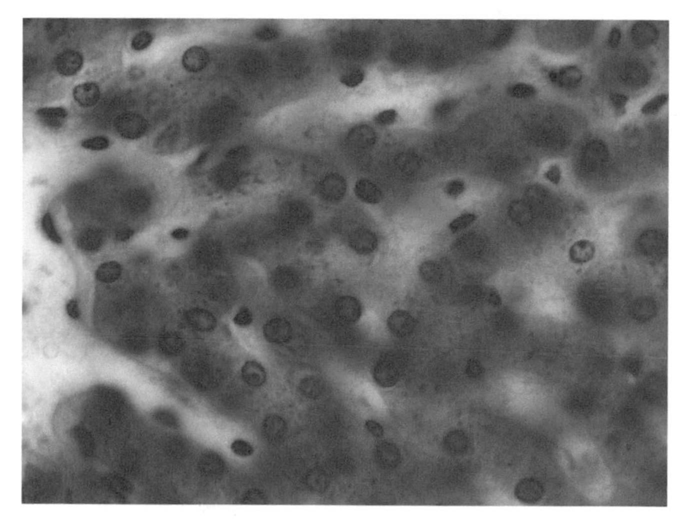

FIGURE H12b Photomicrographs from the same section of a liver lobule from a pig at a higher magnification. The sinusoids are of irregular shape. Their endothelial cells are stained pale blue. 63 ×.

Rete Mirabile

This "marvelous network" occurs in parts of the body where arterioles or venules break up into capillary-like vessels and recombine to form larger vessels of the same type, i.e., arteriole to arteriole or venule to venule. The only familiar example in mammals is in the glomerulus of the kidney (Figs. H13a–H13c). Retia mirabilia (*note plural*) form the capillary bed of gills (Fig. H13d) and, in certain fish, they are found in the red body of the swim bladder (Fig. H13e). Histologically their structure is similar to that of capillaries.

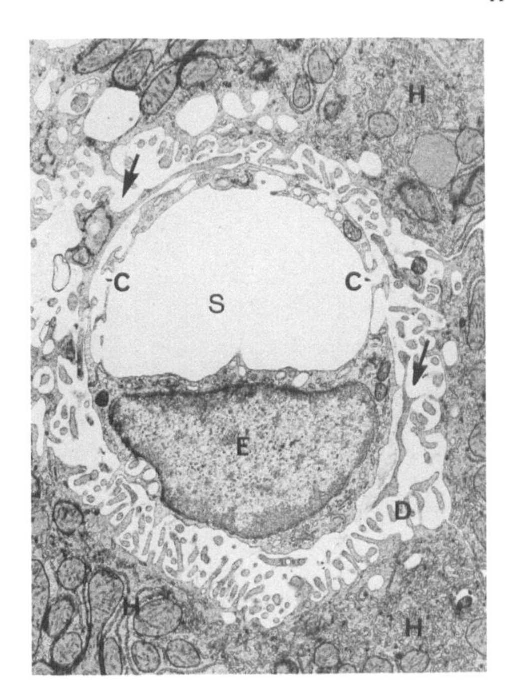

FIGURE H12c Electron micrograph of a cross section of a sinusoid from the liver of a rat. The endothelial cells (E) are surrounded by a forest of microvilli (D) from the cells of the liver cords (H). Large fenestrations in the endothelial cells (C) permit communication between the lumen of the sinusoid (S) and the surrounding space. 14,300 ×.

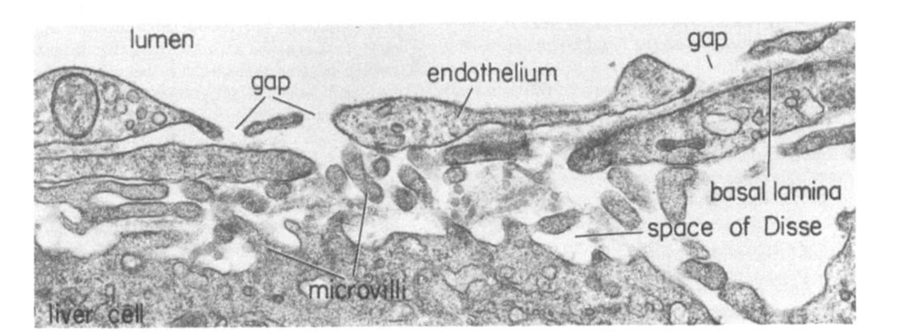

FIGURE H12d Electron micrograph of the discontinuous endothelial lining of a sinusoid from the liver of a rat. 30,000 ×.

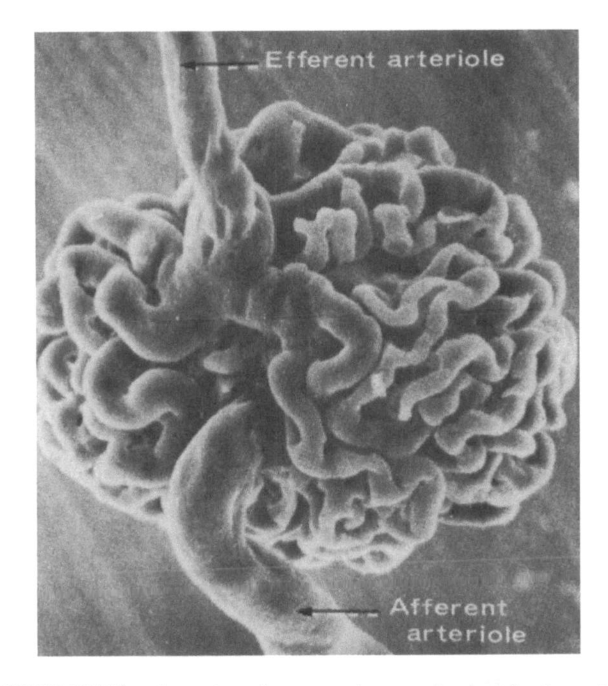

FIGURE H13b Scanning electron micrograph of a plastic cast of a glomerulus showing an afferent arteriole bringing blood *to* a knot of capillaries, the glomerulus, and an efferent arteriole carrying it away.

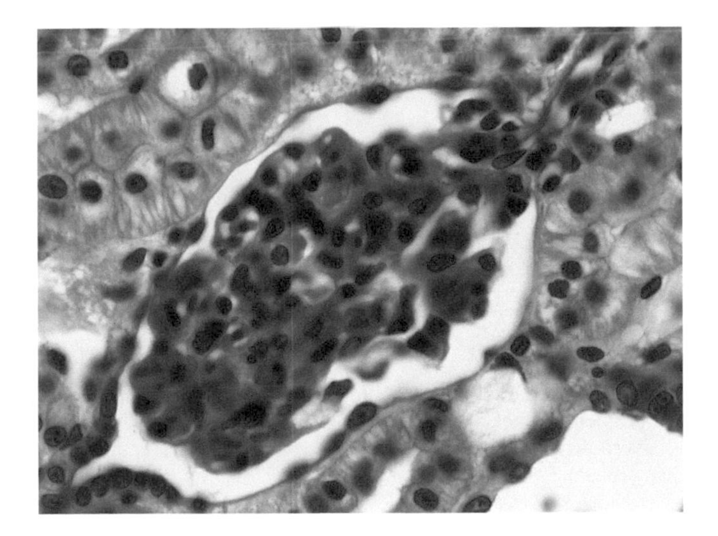

FIGURE H13a Photomicrograph of a section through a renal corpuscle of a monkey. The glomerulus at the center of the corpuscle is a RETE MIRABILE, a knot of capillaries that carries blood from one arteriole to another arteriole. 40 ×.

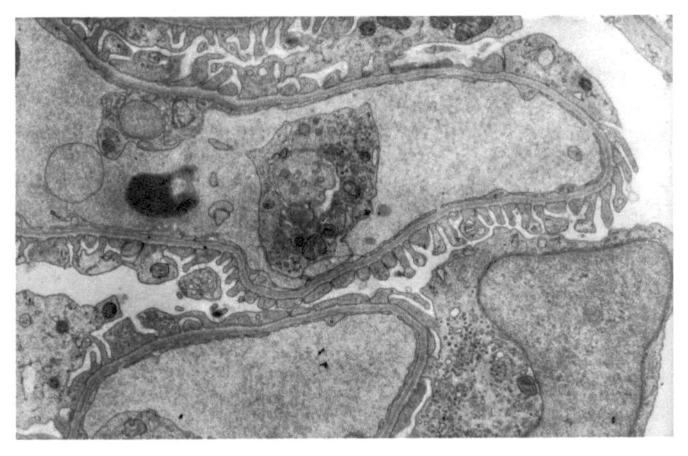

FIGURE H13c Transmission electron micrograph of a section of a glomerulus in the kidney of a mouse. These *fenestrated capillaries* are ultrafilters permitting water and molecules below a molecular weight of about 45,000 to pass into the urinary space; larger molecules are retained in the circulatory system. The capillary wall consists of thin endothelial cells surrounded by a basement lamina, which is supported by foot-like cells, the podocytes. 29,500 ×.

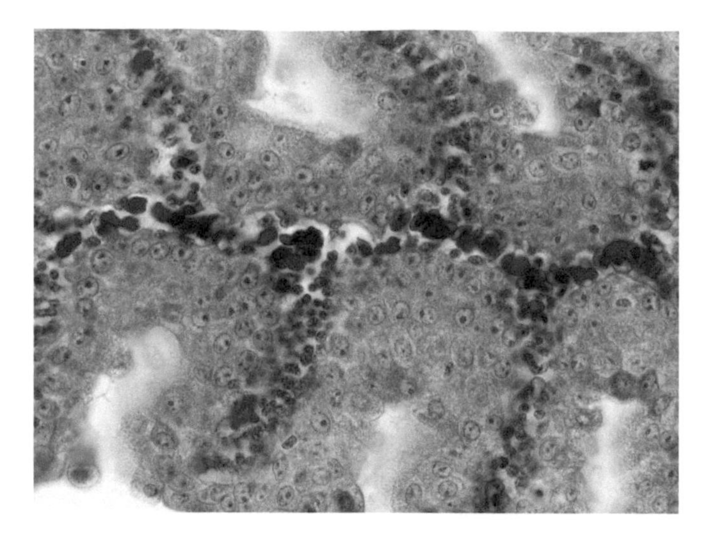

FIGURE H13d Cross section through the gill of an ammocoete (larval lamprey). The gills transmit blood from an afferent arteriole to an efferent arteriole and constitute a rete mirabile. 40 ×.

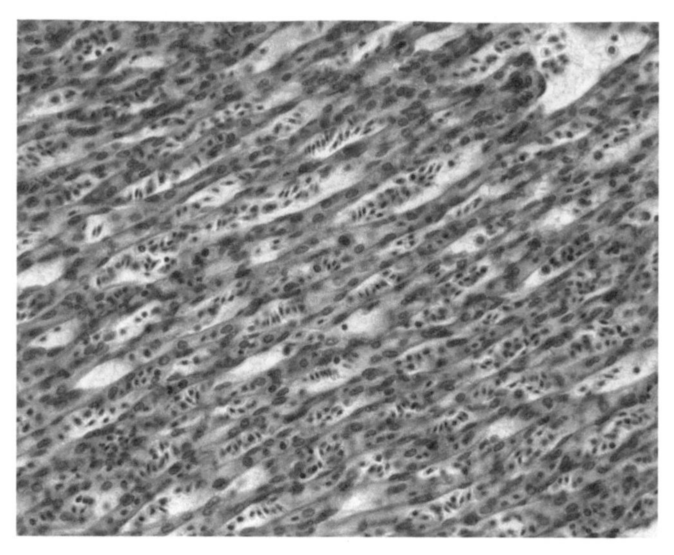

FIGURE H13e Section of a double RETE MIRABILE in the red body of the swim bladder of a fish. Blood passing through these complex vessels release gasses into the swim bladder and thereby control the degree of flotation of the fish.

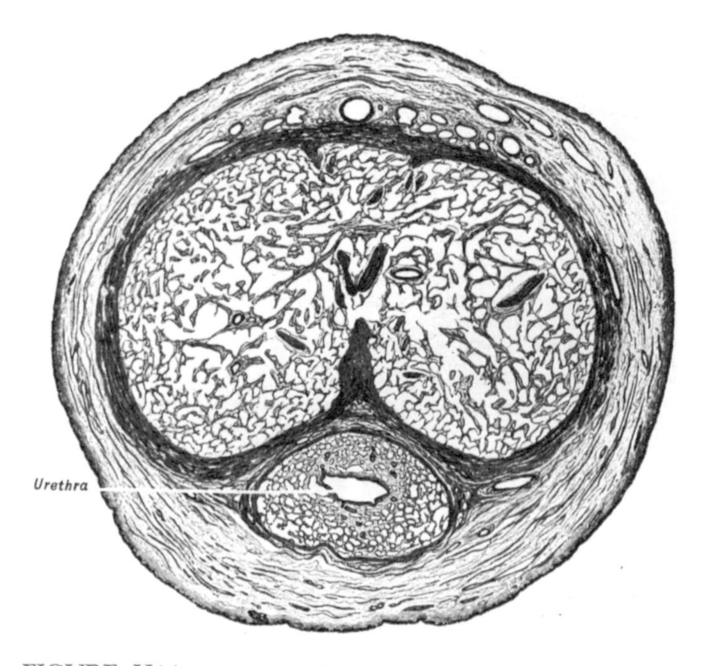

FIGURE H14a Drawing of a cross section of the penis of a 21-year-old man. TWO CORPORA CAVERNOSA are seen at the top of the figure; the urethra at the bottom is surrounded by a smaller mass of distensible tissue.

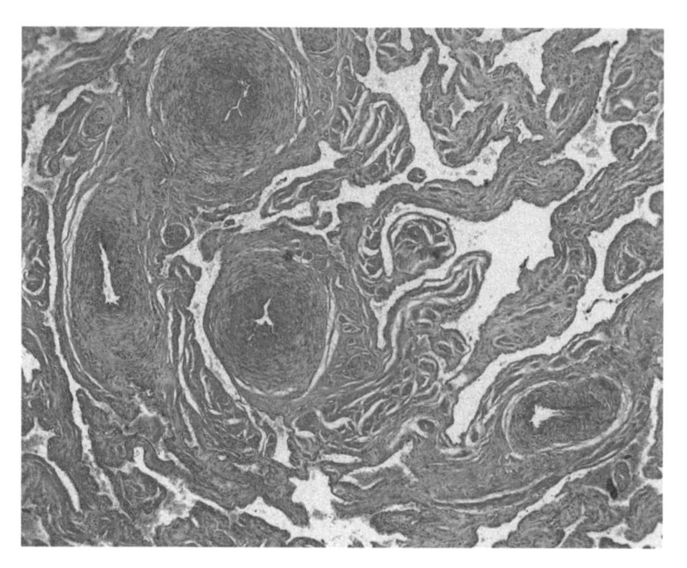

FIGURE H14b Photomicrograph of a portion of a cross section of a human penis. Thick-walled muscular arteries (left) bring blood to the cavernous bodies (right), which become distended at times of sexual arousal. 5 ×.

Cavernous Bodies

Cavernous or erectile tissue is found in the mammalian penis or clitoris and consists of large, irregular vascular spaces interposed between arteries and veins (Fig. H14a). It is supplied directly by muscular arteries that enter the central portion of the mass (Fig. H14b). The veins are peripheral and leave the blood spaces at an oblique angle. Sexual excitement causes an increased flow of blood into the tissue. The central spaces fill first and their distension compresses the peripheral spaces, obstructing the flow of blood through the angular openings into the veins, thus producing rigidity. The spaces of cavernous bodies of mammalian penis are lined by endothelium firmly reinforced by trabeculae of stout collagenous tissue (Fig. H14c).

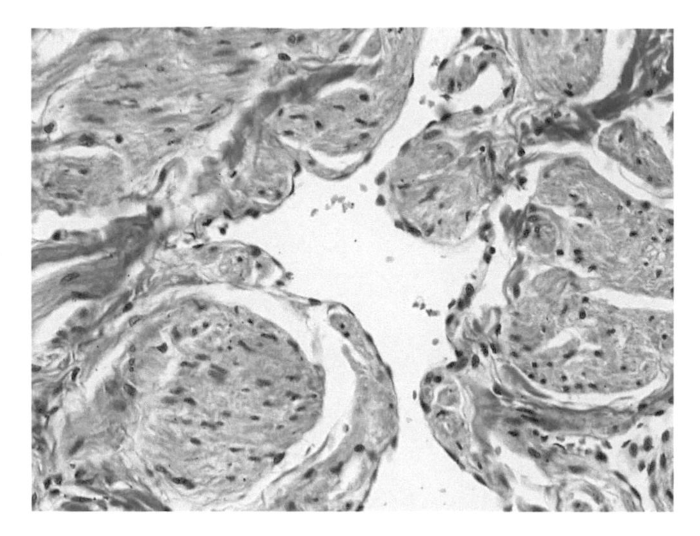

FIGURE H14c Photomicrograph of a cavernous body in a section of human penis. The blood spaces are lined with squamous endothelial cells. Tough collagenous tissue surrounding the corpora must withstand great pressure of blood during erection. 20 ×.

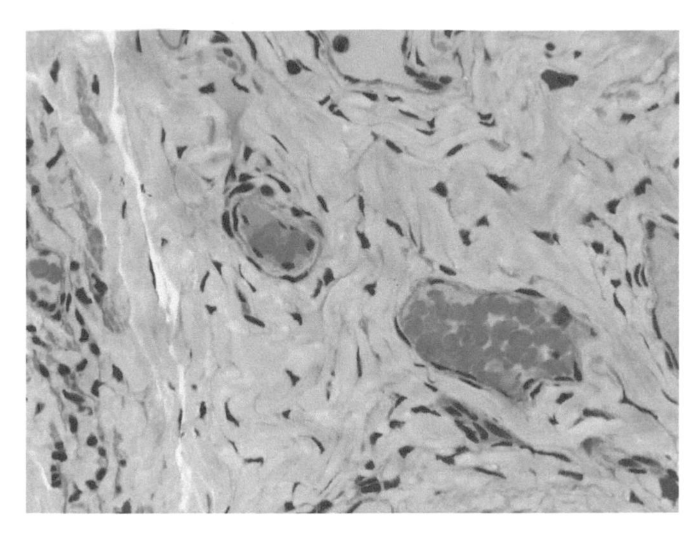

FIGURE H15a Photomicrograph through a semithin section of mesothelium showing cross sections through an arteriole and its (larger) accompanying venule. 40 ×.

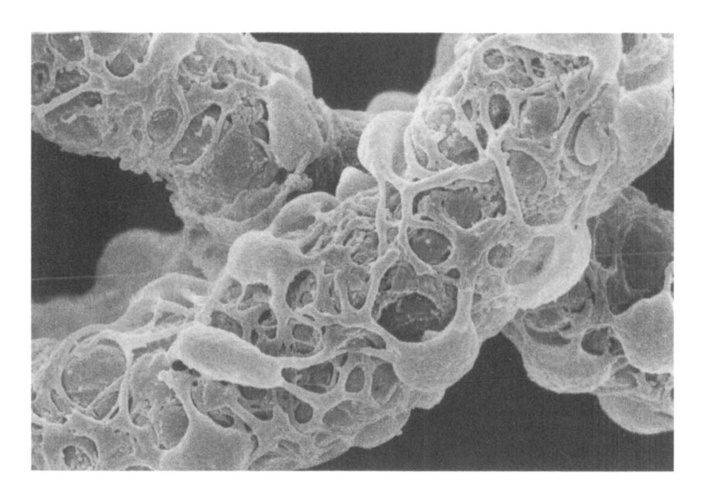

FIGURE H15b Scanning electron micrograph of pericytes clasping postcapillary venules of the mammary gland of a cat.

Venules

Blood is carried in venules from the capillaries back to the heart (Fig. H15a). Electron micrographs show that the small venules draining capillaries are embraced by rich populations of pericytes interposed in the thin basal lamina of the endothelium (Fig. H15b). The loosest endothelial junctions of the vascular system are found in venules and gap junctions are absent (Fig. H15c). Venules constitute the most permeable region of the vascular system to large molecules and cellular migration (Fig. H15c). A tunica media is lacking and there is a loose tunica adventitia of fibrous connective tissue. Larger venules accompany arterioles but can be distinguished by their thinner walls and collapsed lumina. The tunica intima is continuous throughout. The tunica

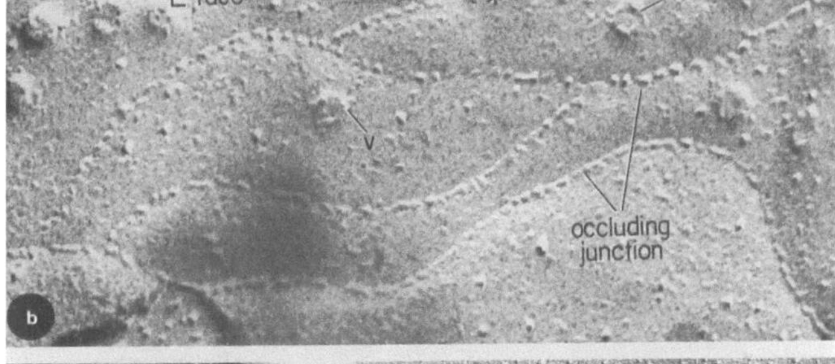

FIGURE H15c Electron micrographs of freeze-fracture preparations of endothelial junctions in three parts of the microvasculature. The junctions in the venule are most permeable (A) arteriole, 140,000 ×; (B) capillary, 160,000 ×; (C) venule. v, *vessicle;* E, *e-face junction component;* P,*p-face junction component*, *, *e-face groove*. 145,000 ×.

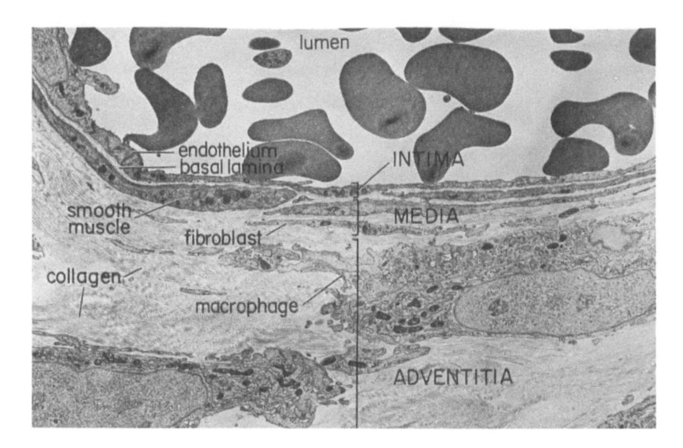

FIGURE H15d Electron micrograph of a cross section through a muscular venule. 4600 ×.

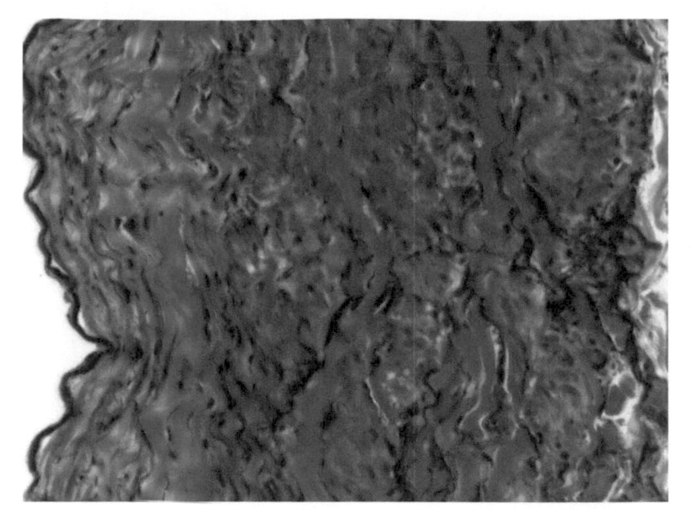

FIGURE H16a Photomicrograph of a section of mammalian vena cava, the lumen is to the left. Note the band of elastic fibers immediately under the endothelium of the tunica intima. Although the tunica intima is fairly clearly defined at the left of this photomicrograph, there is little distinction between the tunica media and the tunica adventia. Pink collagenous fibers are plentiful. A few elastic fibers (stained black), and scattered smooth muscle cells are present. 20 ×.

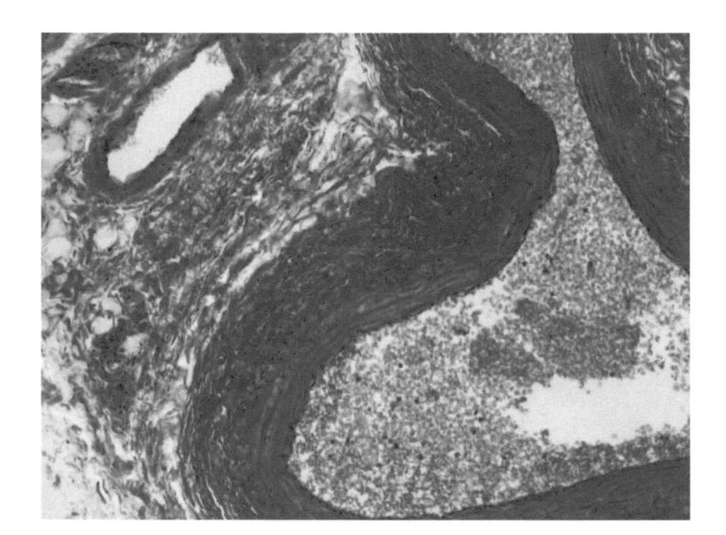

FIGURE H16b Photomicrograph of a medium-sized mammalian artery (upper left) and its corresponding vein (lower right). Arteries are often partnered with veins and carry equivalent amounts of blood; since the blood in veins travels at a much lower speed and pressure, the vein is considerably larger than the artery. 10 ×.

media of these venules consists of one or two layers of smooth muscle cells (Fig. H15d). There is a relatively thick tunica adventitia of fibrous connective tissue.

VEINS

Veins have thinner walls and larger lumina than the accompanying arteries and show much greater variation in structure. Compare the three tunics of the wall in sections of large and small veins (Figs. H16a and H16b). The boundaries of the layers are often indistinct and some consider the tunica media and the tunica adventitia to be one layer.

The endothelium of veins remains smooth (Fig. H17). The internal elastic membrane of the TUNICA INTIMA is often absent, especially in small veins and, when present, is

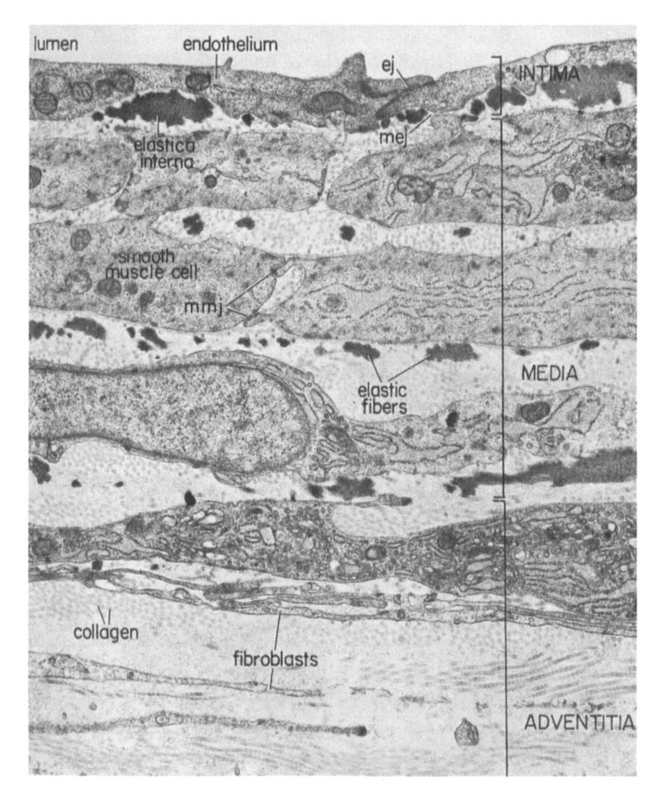

FIGURE H17 Electron micrograph of a cross section of the wall of a large vein of a rat. The internal elastic membrane is fenestrated and permits the penetration of myoendothelial junctions. *ej*, endothelial junction; *mmj*, junction between smooth muscle cells, *mej*, endothelial muscle cell junction. 15,000 ×.

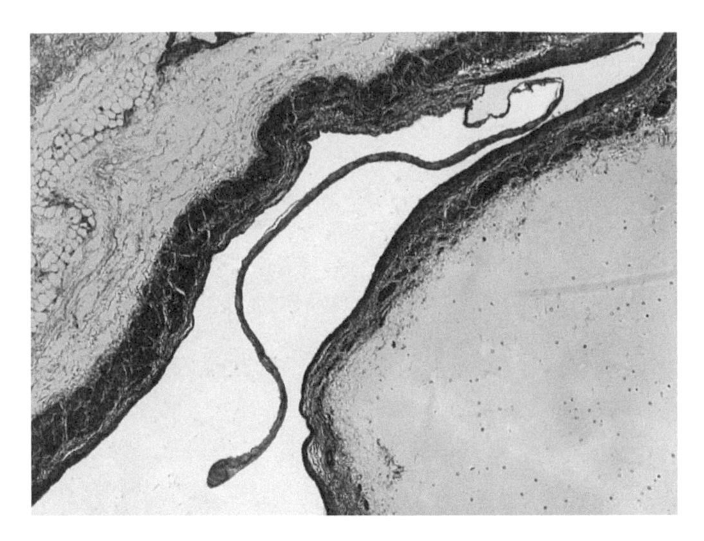

FIGURE H18a Photomicrograph of a longitudinal section of a large mammalian vein. This section includes a valve 2.5×.

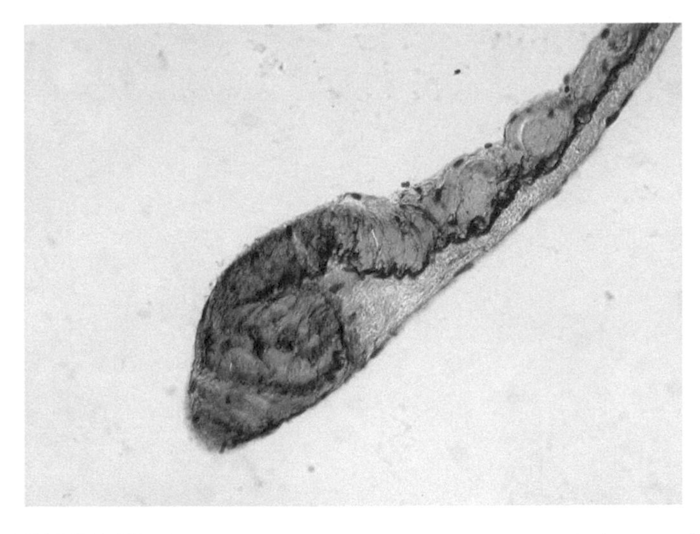

FIGURE H18b Higher powered photomicrograph of the tip of the venous valve shown in Fig. H18a. A few elastic fibers (stained black) are scattered here and there. 20×.

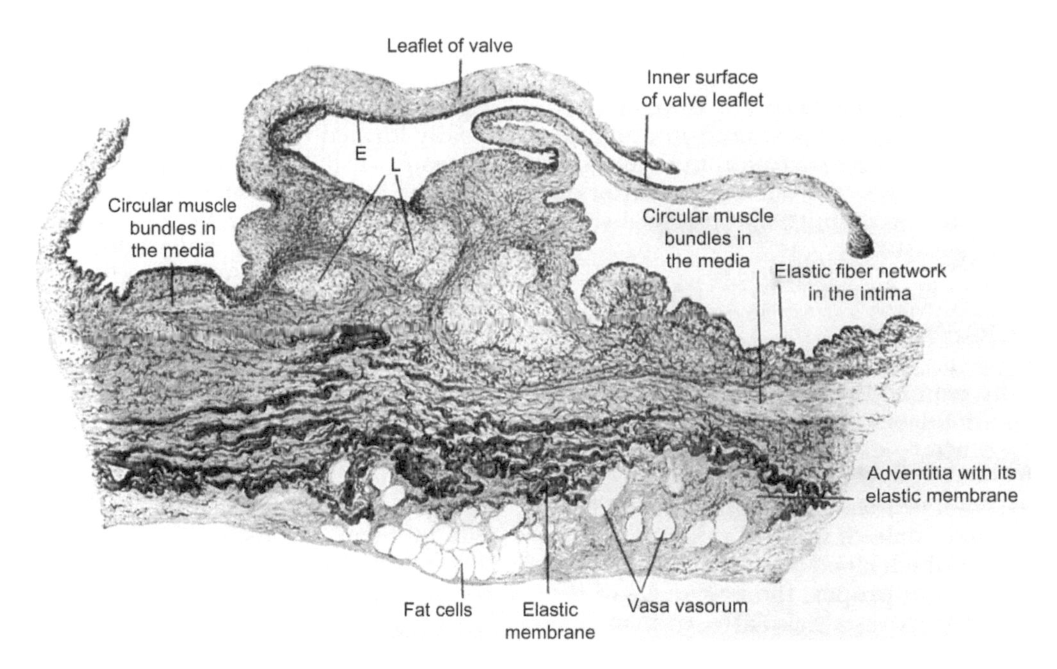

FIGURE H18c Drawing of a valve in the wall of the femoral vein of a human. Elastic fibers are abundant. *E*, endothelial cells; *L*, longitudinal muscle cells.

inconspicuous. In birds and mammals, especially those with long extremities, many of the medium-sized veins of the limbs are provided with valves on the inner walls that prevent the backflow of blood (Figs. H18a–H18c). The entire tunica intima extends into the lumen in the form of a flap of connective tissue covered with endothelium. The TUNICA MEDIA of veins is a thin layer of muscle and fibers of connective tissue; it is usually composed entirely of circular fibers but occasionally has longitudinal fibers. It has more collagenous fibers than the tunica media of an artery and includes elastic fibers only in the largest veins. The TUNICA ADVENTITIA is the thickest layer in the wall of veins. It often contains a large number of smooth muscle fibers.

COMPARISONS

Compare large and small arteries and veins noting the ratio of the thickness of wall to size of lumen; the shape of vessels in section; the relative amounts of muscle, elastic tissue, and collagenous tissue; the relative thickness of the tunics; and the size and distribution of vasa vasorum and relate any differences to the functions of the vessels.

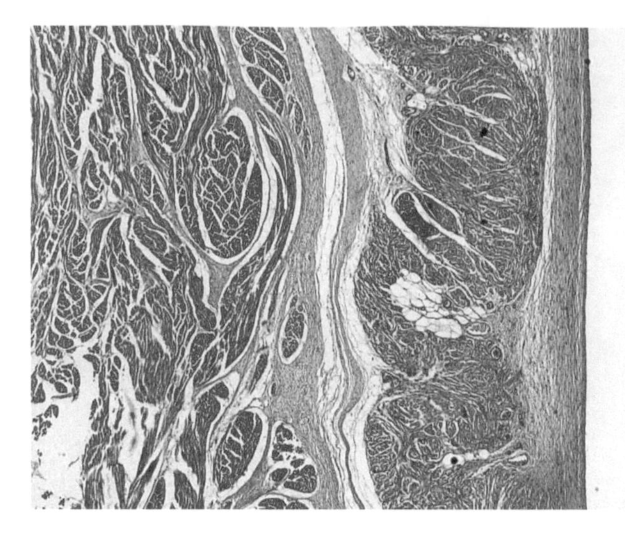

FIGURE H19a Low-powered photomicrograph of a section of mammalian heart. The endocardium at the right is subtended by a layer of connective tissue, smooth muscle, and elastic fibers. Most of the substance of the heart consists of cardiac muscle. Fibers of the conducting system penetrate the cardiac muscle. 2.5 ×.

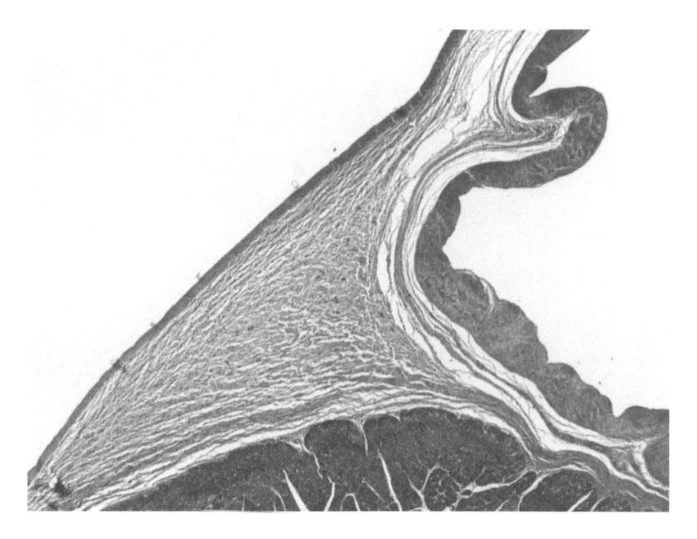

FIGURE H19b Photomicrograph of section through a mammalian semilunar valve. A leaflet of the valve extends upward. It has a tough core of connective tissue. The lumen of the aorta occupies the lower right corner. All surfaces are lined with endothelium of the endocardium. 2.5 ×.

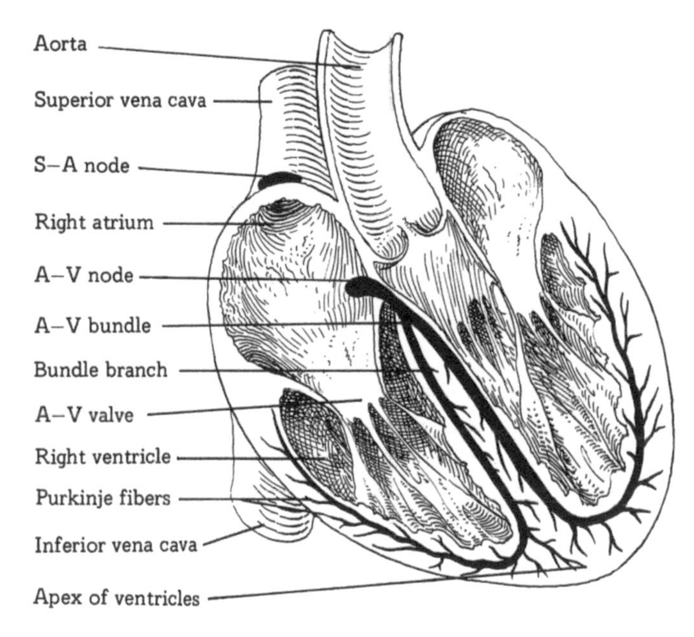

FIGURE H19c Drawing of the human heart cut open to demonstrate the interior of the heart and the main parts of the impulse-conducting system (Purkinje system). Backwash of the leaflets of the valves is prevented by stout cords of connective tissue, the fibrous chordae tendinae (not labeled).

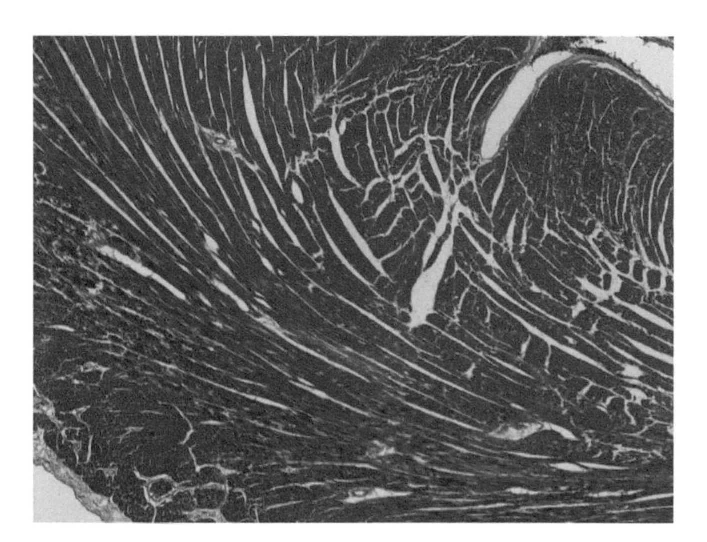

FIGURE H19d A poor section of the ventricle of an entire mammalian heart. The lumen, lined with endocardium is at the upper right and a small amount of the epicardium is seen at the lower right corner. 2.5 ×.

The Heart

The heart is the thick, muscular, contracting portion of the vascular system. It consists of three main layers homologous to the three tunics of blood vessels (Fig. H19a). ENDOCARDIUM corresponds to the tunica intima of the vessels and includes an endothelial lining and a relatively thick subendothelial layer of connective tissue, smooth muscle, and elastic fibers. The valves of the heart are folds of the endocardium in which the fibroelastic elements are prominent (Fig. H19b). The free borders of the valves of the heart are connected to the PAPILLARY MUSCLES by the fibrous CHORDAE TENDINEAE (Fig. H19c).

MYOCARDIUM constitutes the main mass of the heart and corresponds to the tunica media (Fig. H19d). It consists of sheets of cardiac muscle that wind around the atria and ventricles in a spiral fashion. Collagenous interstitial

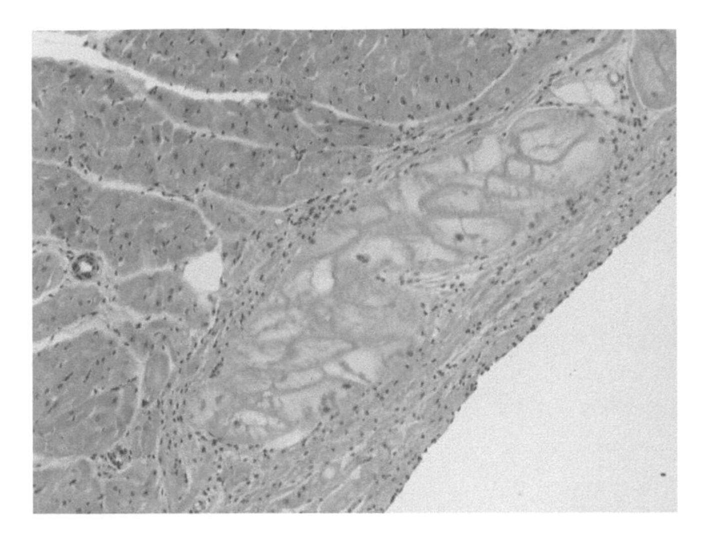

FIGURES H20a Photomicrographs of a section of beef heart showing a group of conducting fibers embedded in the loose connective tissue of the endocardium. The lumen of the heart (right) is lined with squamous endothelial cells. 10 ×.

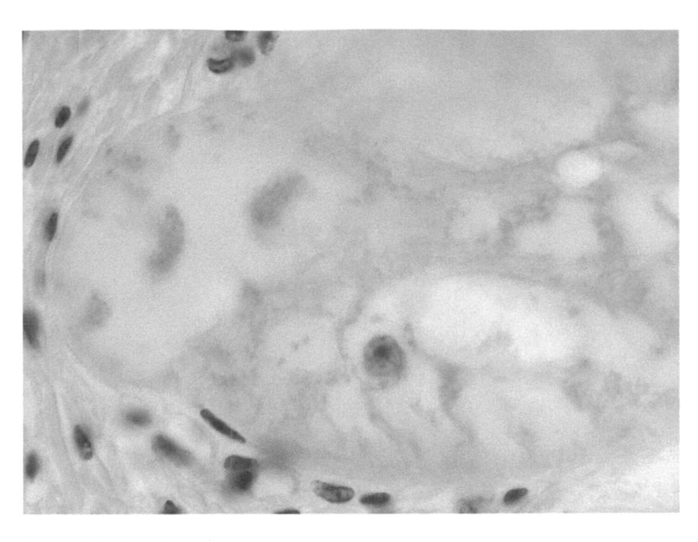

FIGURE H20b Photomicrographs at higher maginfication of the section of beef heart in H20a showing conducting fibers in the loose connective tissue of the endocardium. 63 ×.

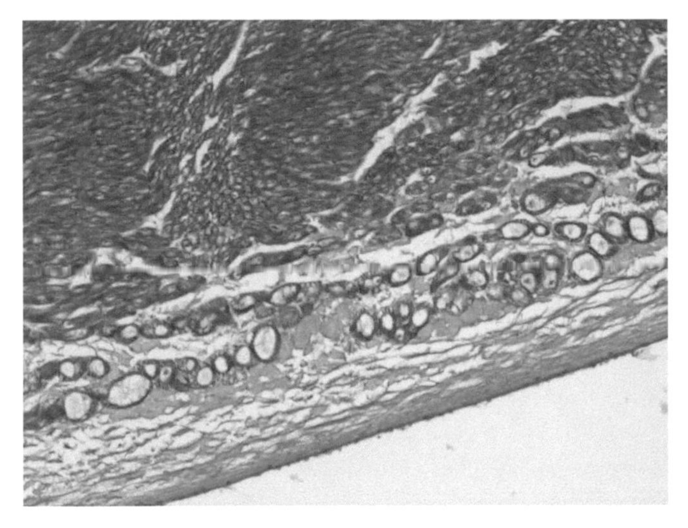

FIGURE H20c Cross sections of conducting cells from the ventricular wall of a mammal. The cells are larger than regular heart muscle cells and the myofibrils characteristically occupy a peripheral position. 10 ×.

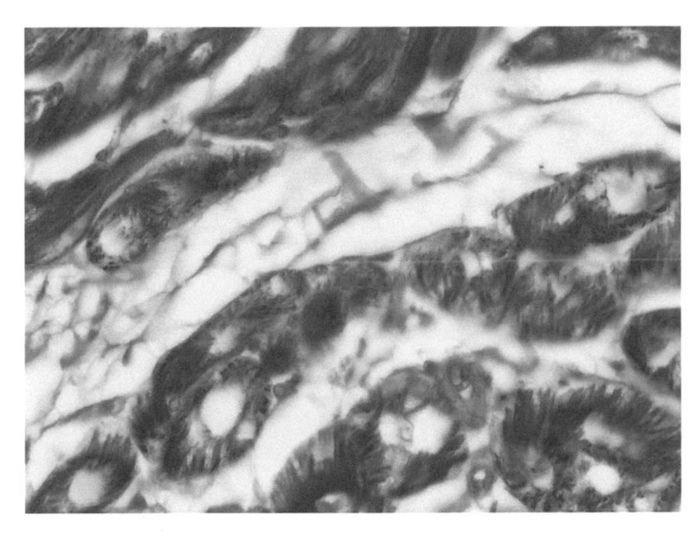

FIGURE H20d At higher magnification the conducting cells of the ventricular wall of a mammal. The myofibrils are clearly seen occupying the characteristically peripheral position. 40 ×.

fibers of the ventricle on which the muscle fibers insert constitute the CARDIAC SKELETON. A specialized portion of the myocardium is a peculiar group of fibers known as the CONDUCTING SYSTEM (Figs. H20a–H20d). This includes the SINOATRIAL NODE and the ATRIOVENTRICULAR BUNDLE (of His). These fibers are larger and paler than the usual cardiac muscle fibers. They were described by Purkinje and are often called PURKINJE FIBERS. This system of fibers has a faster rate of conduction than ordinary fibers of heart muscle and correlates the contractions of different parts of the heart. CONDUCTING MYOFIBERS are atypical fibers of cardiac muscle and, just as in ordinary cardiac muscle, they form a continuous sarcoplasmic network. They are larger than ordinary muscle fibers, are rich in glycogen, and contain fewer myofibrils. They anastomose freely with themselves and with ordinary fibers. They are relatively large in the hearts of herbivores.

The myocardium of some of the more sluggish vertebrates forms a spongy, avascular mesh whose fibers receive their nutrients from the blood passing through the interstices (Figs. H21a–H21d). The myocardium of more active vertebrates, such as birds and mammals, is compact and vascular. Intermediate forms exist where the outer layer of the myocardium is compact and vascular and the inner layer is spongy. Note in each trabecula of spongy cardiac muscle that the fibers are separated from the blood by a thin sheet of endothelium.

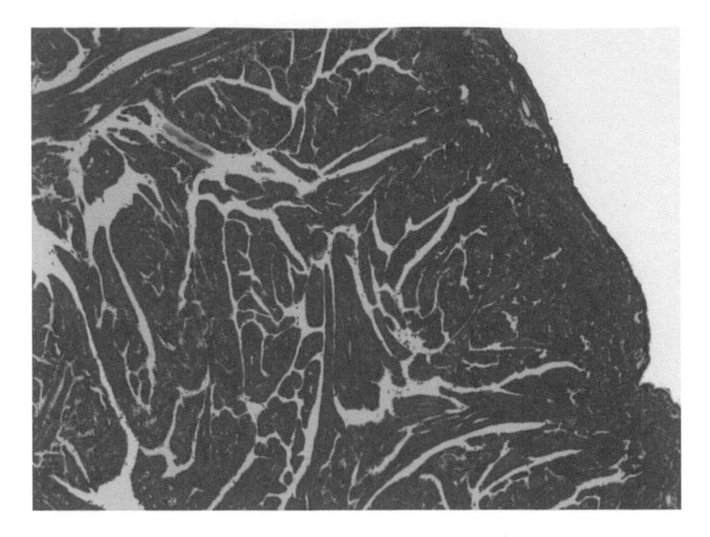

FIGURE H21a Photomicrograph of a section of the heart of a dogfish. The cardiac muscle is spongy at the center of the heart, dense at its periphery. The muscle cells pump blood through the heart and are nourished as it passes. 2.5 ×.

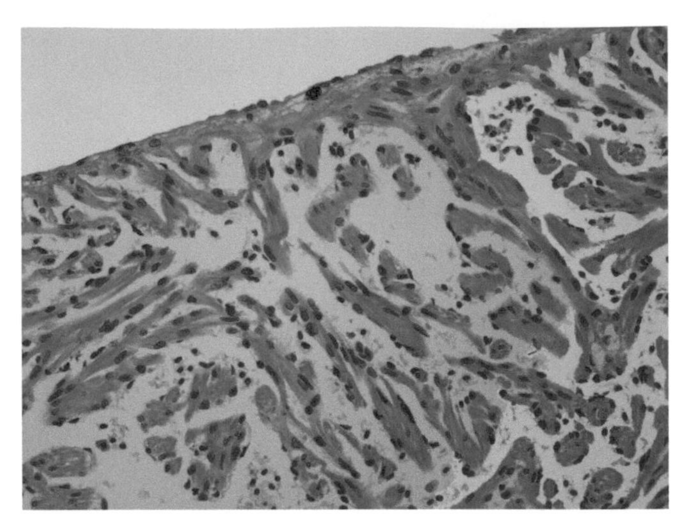

FIGURE H21b Photomicrograph through the heart of *Protoptrerus*, a tropical lungfish. Most of the heart muscle forms a loose mesh enclosed by a thin, denser arrangement at the periphery. Blood is pumped through the mesh, nourishing the cells as it passes. 10 ×.

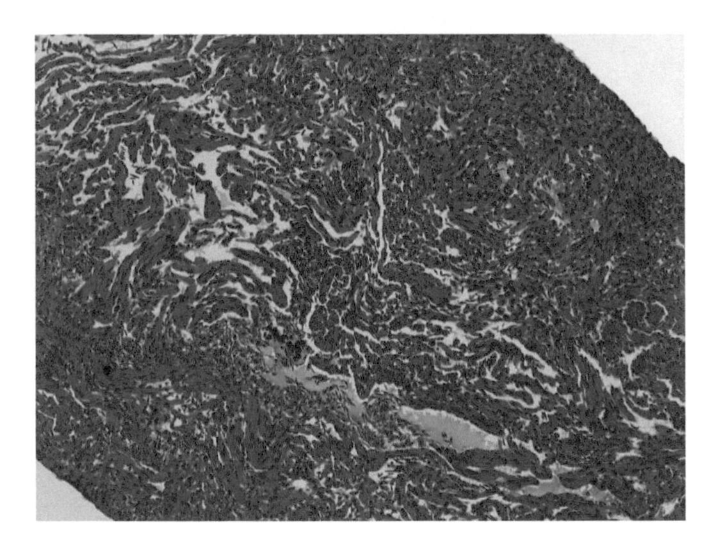

FIGURE H21c Photomicrographs of sections through the heart of a salamander, *Necturus*. The mesh-like arrangement of the muscle fibers is evident. 2.5 ×.

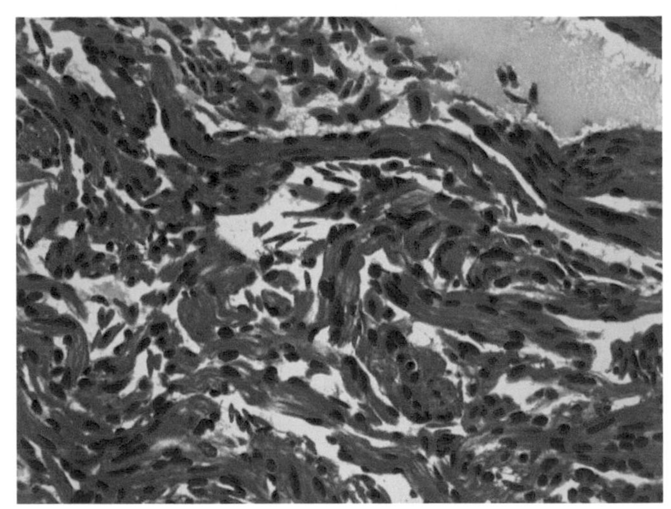

FIGURE H21d At higher power a sections through the heart of a salamander, *Necturus* the mesh-like arrangement of the muscle fibers is clearly seen 10 ×.

EPICARDIUM is the connective tissue covering of the heart and corresponds to the tunica adventitia of vessels. It is bounded on the outside by VISCERAL PERICARDIUM (mesothelium).

Lymph Vessels

Fluids escape into the connective tissue from blood capillaries and venules and are returned by lymph vessels to the large veins entering the heart (Fig. H22a). Only a few organs lack a lymph drainage (central nervous system, eyeball, bone marrow) and lymph vessels are well developed in all vertebrates (Fig. H22b). Lymph capillaries, composed only of delicate endothelium, arise as blind, swollen tubules in the tissues, often in company with blood capillaries.

Electron micrographs show that lymph capillaries resemble blood capillaries but are larger and less regular (Fig. H22c). Their ultrastructure is similar except that the borders of the endothelial cells often overlap. There are

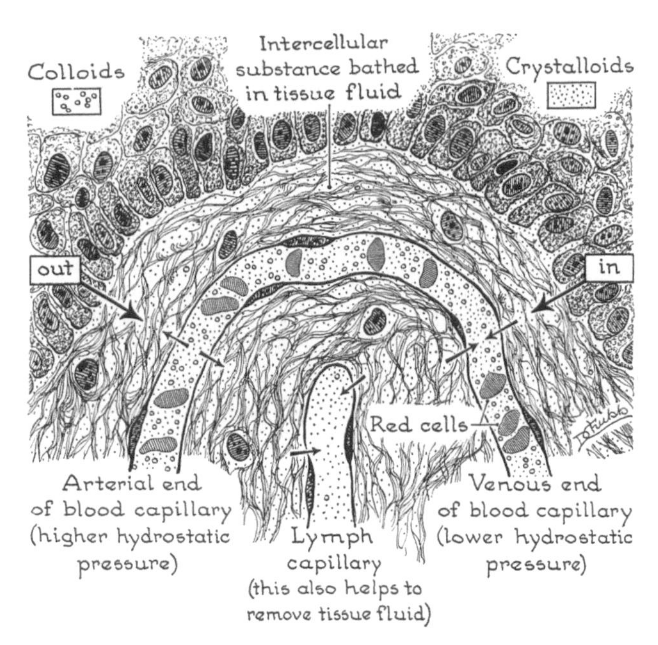

FIGURE H22a Diagram showing the leakage of tissue fluid from capillaries and its subsequent uptake by capillaries and lymphatics. Proteins of the blood (small circles) and crystalloids of the blood and tissue fluid (dots). Normally little colloid escapes from the capillaries; the small amount that does escape is returned to the circulation by the lymphatics.

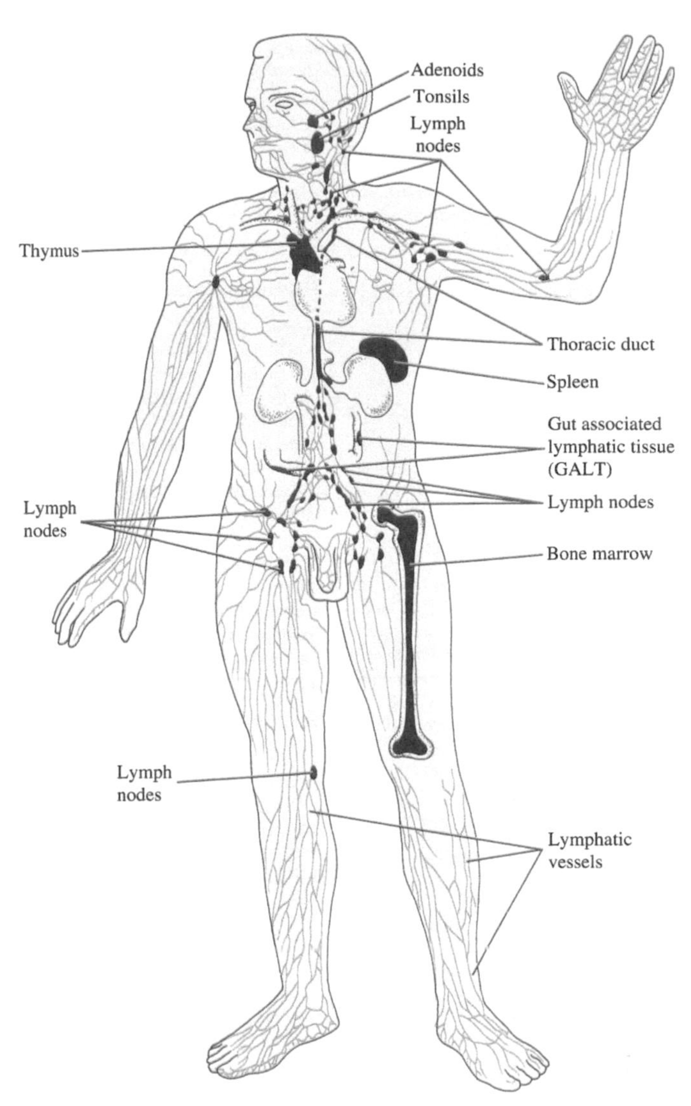

FIGURE H22b Diagram showing the human lymphatic system. The major lymph vessels are shown in green. These unite as the thoracic duct to carry the tissue fluid back to the heart. It passes through many lymph nodes on the way.

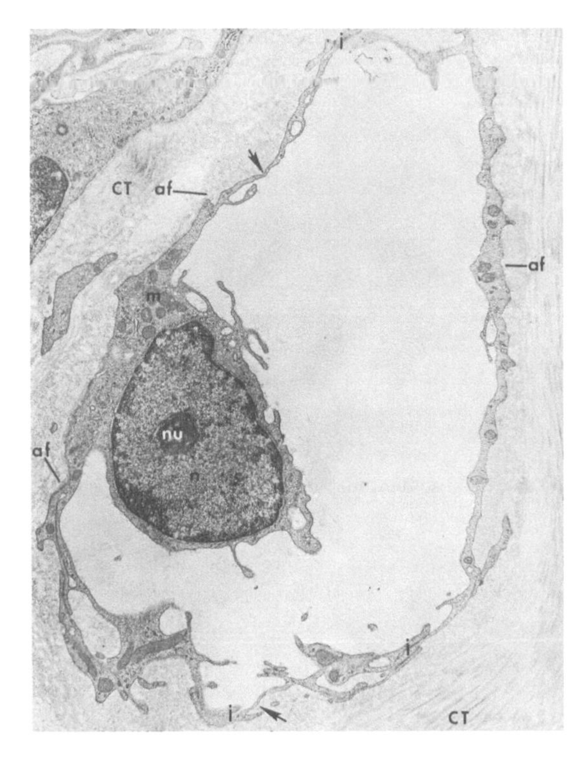

FIGURE H22c Transmission electron micrograph of a cross section of a lymphatic capillary. Numerous anchoring filaments (af) attach to connective tissue in the area (CT) providing stability. Arrows indicate attenuations of the endothelial cells. *i*, intercellular junctions; *m*, mitochondria of the endothelial cell; *n*, nucleus of the endothelial cell; *nu*, nucleolus of the endothelial cell. 11,000 ×.

zonulae adhaerentes between the endothelial cells of lymph capillaries but there are no fenestrae. The basement lamina is discontinuous and there are no pericytes. Anchoring filaments extend from the endothelial cells to the surrounding connective tissue.

Lymph capillaries unite to form larger lymph vessels that exhibit indistinctly the three layers of the blood vessels (Fig. H22d). Try to distinguish the tunics in a section of mammalian thoracic duct. Note the endothelial lining and the distribution of collagenous and elastic fibers and smooth muscle.

In all classes of vertebrates, except mammals and cartilaginous fishes, pulsating lymph hearts aid in the propulsion of lymph into the veins. These chambers are modified lymph vessels and contain striated muscle that resembles cardiac muscle. In most forms the

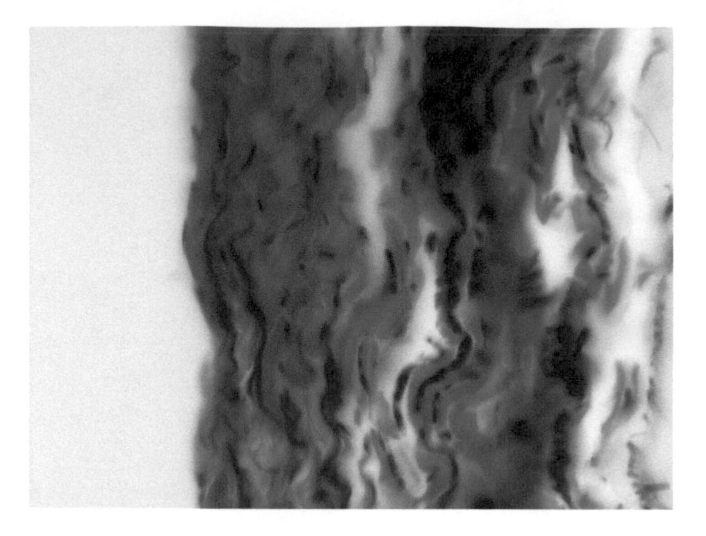

FIGURE H22d Photomicrograph of a transverse section of the thoracic duct of a horse, the biggest lymph vessel in a huge mammal. An elastic tissue stain has been used. Although the layers are not clearly delineated the wall of this vessel resembles that of a vein; endothelial cells can be distinguished covering the tunica intima and an internal elastic membrane separates the tunica intima from the ill-defined layers below.

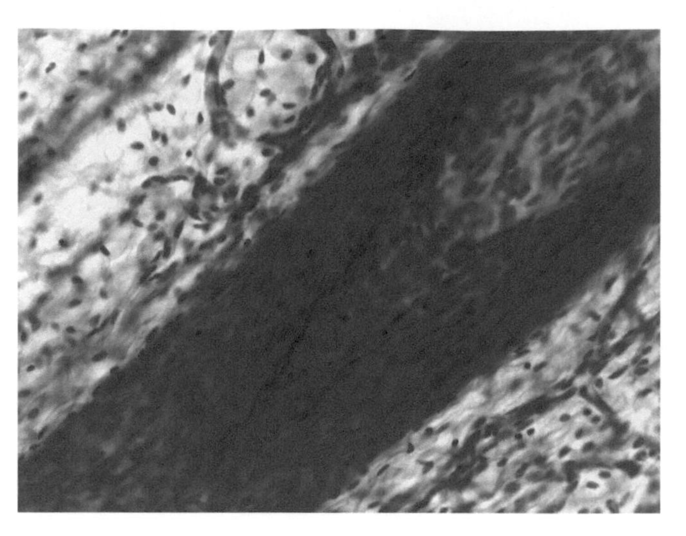

FIGURE H22e Photomicrograph of a whole mount of a valve in a mammalian lymph vessel. 20 ×.

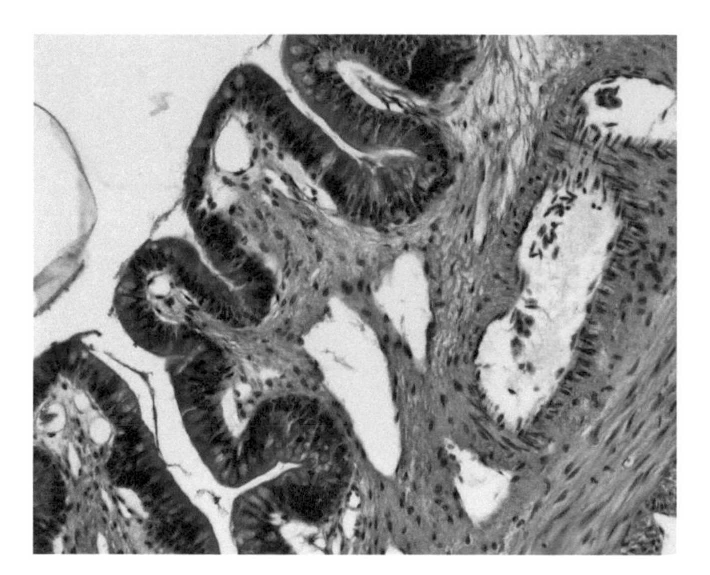

FIGURE H22f Photomicrograph of a cross section of the intestine of a toad showing large lymph spaces in the connective tissue and muscle layers. 20 ×.

lymph is also propelled sluggishly by movements of the viscera and other parts of the body; in tetrapods, valves in the lymph vessels prevent a backflow (Fig. H22e). Large lymph spaces lined with endothelial cells are often interposed between various organs of lower vertebrates, especially aquatic forms (Fig. H22f).

In mammals and, to a lesser extent, in birds the lymph vessels are interrupted at certain places by LYMPH NODES; these will be considered in Chapter I, Haemopietic Organs.

CHAPTER

I

Hemopoietic Organs

Blood-forming organs of vertebrates are distributed throughout the body in quiet backwaters where hemopoiesis occurs undisturbed. In higher vertebrates these functions are segregated where agranulocytes are produced in LYMPHOID TISSUE and granulocytes and erythrocytes produced in MYELOID ORGANS. The basic structural plan of all hemopoietic organs is the same: a framework or stroma of vascular reticular tissue in the meshes of which blood cells mature (Fig. I1a). Hemopoietic organs may also have secondary functions in filtering body fluids and in the production of antibodies.

The reticulum is well shown in sections of lymph node or spleen where the reticular fibers have been stained with silver (Fig. I1b). The reticular cells produce reticular fibers and surround them so that the fibers are essentially separated from the compartment that is occupied by the free cells (Fig. I1c). Contained within the meshes of this reticulum is a spectrum of all stages of developing blood cells (Fig. I2).

HEMOPOIETIC ORGANS IN THE LOWER VERTEBRATES

Hemopoietic tissue has a long-standing association with the gut in all vertebrates and in the most primitive forms the main activity is centered here. Areas of hemopoietic activity are seen in sections of the gut of various vertebrates. The diagnostic feature is the presence in loose connective tissue of dense aggregations of cells with prominent nuclei.

In the lowest vertebrates the "spleen" is the fundamental hemopoietic organ. The spleen of the hagfish, a primitive cyclostome, is a diffuse mass of hemopoietic tissue scattered within the connective tissue of the gut (tela submucosa) (Figs. I3a and I3b). In the lamprey, a more advanced cyclostome, the spleen is confined to the typhlosole (spiral valve) region of the gut (Figs. I4a and I4b).

The spleen of the lungfish is still within the gut wall but is no longer diffuse and has become segregated within the wall of the stomach (Figs. I5a and I5b). In the higher fishes the spleen is a well-defined organ outside the gut but attached to it by peritoneum (Figs. I6a and I6b). In the forms with a distinct spleen there is still considerable hemopoietic activity in the tela submucosa of the gut.

Secondary aggregates of blood-forming tissue can be seen in the connective tissue layers elsewhere in the gut. There is a tremendous development as LEYDIG'S ORGAN in the selachians (sharks and rays) where a mass of hemopoietic tissue extends throughout the esophagus and even into the stomach (Figs. I7a and I7b).

Hemopoietic tissue occurs elsewhere in various species. The intertubular connective tissue of the kidney of a bony fish, is its major center of hemopoiesis (Fig. I8). A considerable amount of hemopoietic activity can also be found in the loose connective tissue of the ovary of the dogfish (Fig. I9) and in the subcapsular tissue of the liver of urodeles (Fig. I10).

HEMOPOIETIC ORGANS OF HIGHER VERTEBRATES

The functions of hemopoiesis are segregated in the higher vertebrates with the advent of hollow bones: bone marrow produces erythrocytes and granulocytes (MYELOID ELEMENTS) and lymphoid tissues produce agranulocytes (LYMPHOID ELEMENTS). The transition is seen in amphibians where the spleen is still the essential erythropoietic

An Atlas of Comparative Vertebrate Histology.
DOI: https://doi.org/10.1016/B978-0-12-410424-2.00009-3

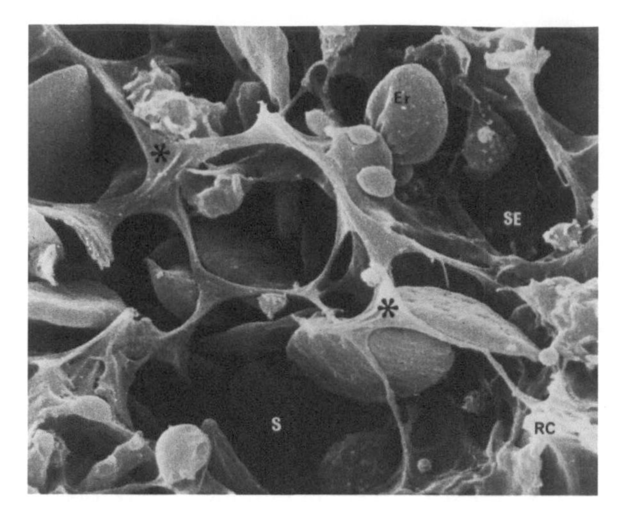

FIGURE I1a Scanning electron micrograph of the red pulp of the spleen of a sunfish (*Lepomis* sp.) showing its reticular stroma and entrapped blood cells. Long, slender cytoplasmic processes (*) of reticular cells (*RC*) extend along reticular fibers and support the sinusoids (*S*). Gaps in the luminal surface of the sinusoidal endothelium (*SE*) permit inward and outward migration of blood cells; a distorted erythrocyte (*Er*) is passing through a sinusoidal wall. 7920×.

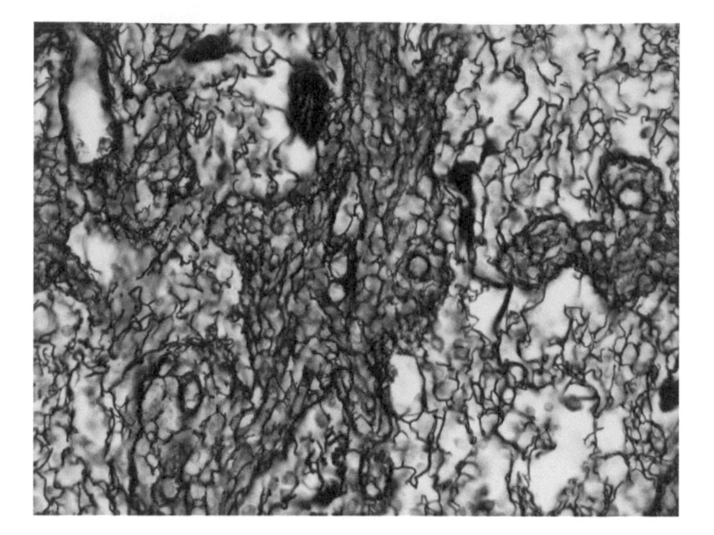

FIGURE I1b This silver preparation of mammalian lymph node shows the reticular fibers that support the loose cells, little else is seen other than a ghostly image of reticular cells extended on the fibers. 40×.

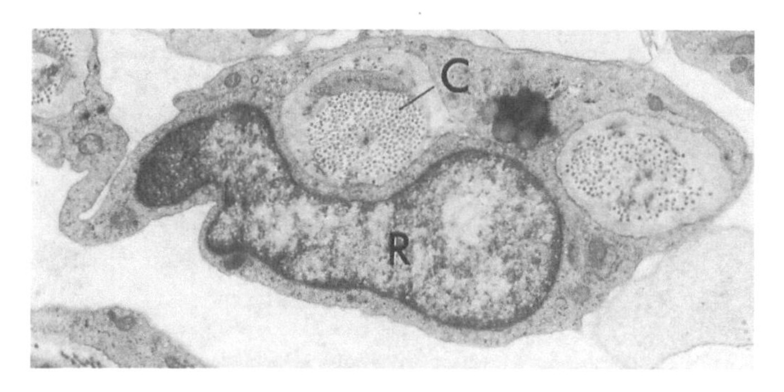

FIGURE I1c Transmission electron micrograph of a reticular cell (*R*) enclosing collagen fibrils (*C*) in the marginal sinus of human lymph node. 12,940×.

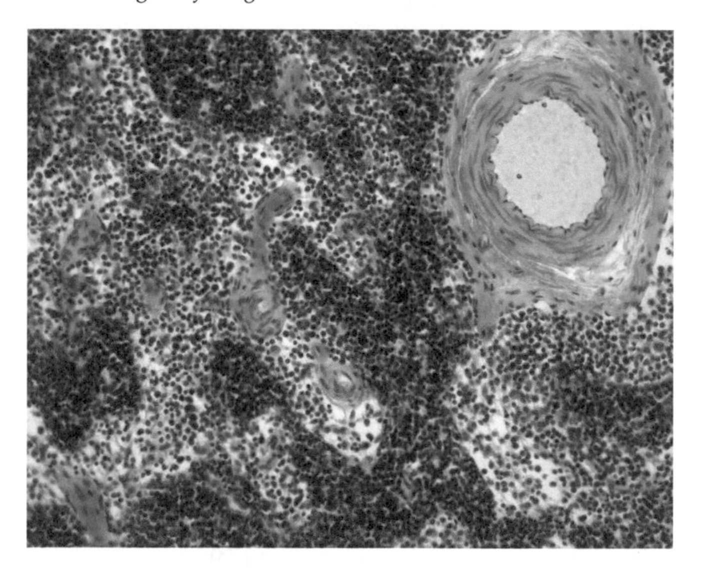

FIGURE I2 Reticular tissue forms a scaffold containing loose cells—largely small lymphocytes—in this section of mammalian lymph node. The scaffold consists of reticular fibers (which cannot be seen) and long reticular cells. A small artery is seen at the upper right. 20×.

organ of the urodeles (tailed amphibians) but, with the development of hollow bones in the anurans (tailless forms), the spleen assumes less importance in erythropoiesis. Although the spleen retains the functions of erythropoiesis and granulopoiesis in some reptiles, the bone marrow becomes the primary center for these in others. In birds and mammals the separation of function is complete: formation of granulocytes and erythrocytes is restricted to the bone marrow and agranulocytes are produced in lymphoid tissue.

LYMPHOID ORGANS OF HIGHER VERTEBRATES

Lymphoid tissues are scattered throughout the body in the form of lymph nodes, thymus, tonsils, spleen (Fig. I11), and as lymphoid nodules in the reticular tissue of the tunica mucosa of the digestive, respiratory, and urinogenital systems. In consists of a stroma of reticular cells and fibers with lymphoid cells in the interstices. Lymphoid tissue may be DIFFUSE and infiltrate the connective tissue of many regions of the body or it may form compact aggregations, the LYMPHOID NODULES.

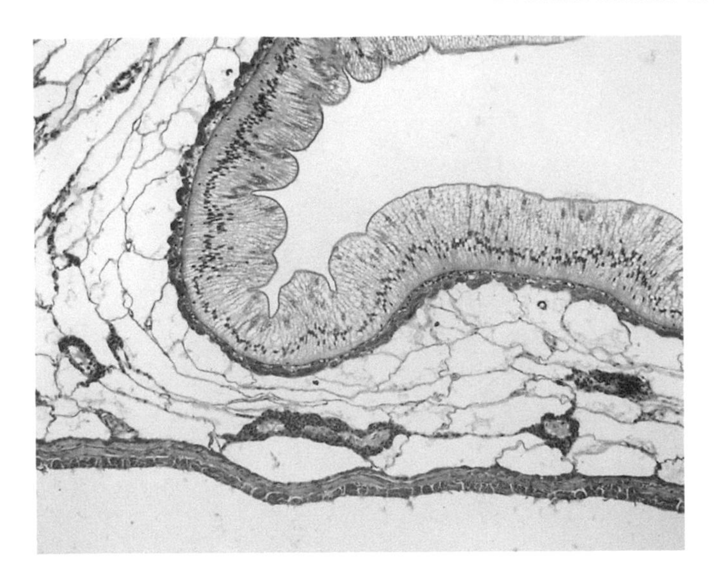

FIGURES I3a Low power of the connective tissue layer of the gut of hagfish showing islets of hemopoietic tissue scattered throughout the loose connective tissue. 10×.

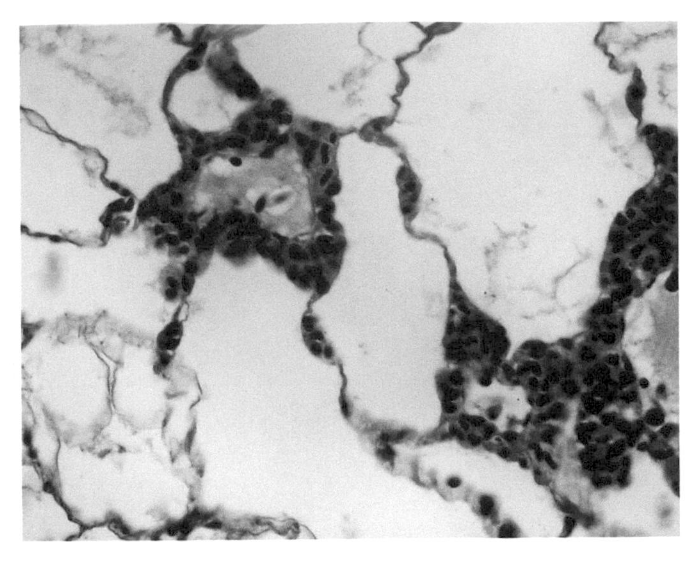

FIGURES I3b Higher power of the connective tissue layer of the gut of hagfish showing islets of hemopoietic tissue scattered throughout the loose connective tissue. 40×.

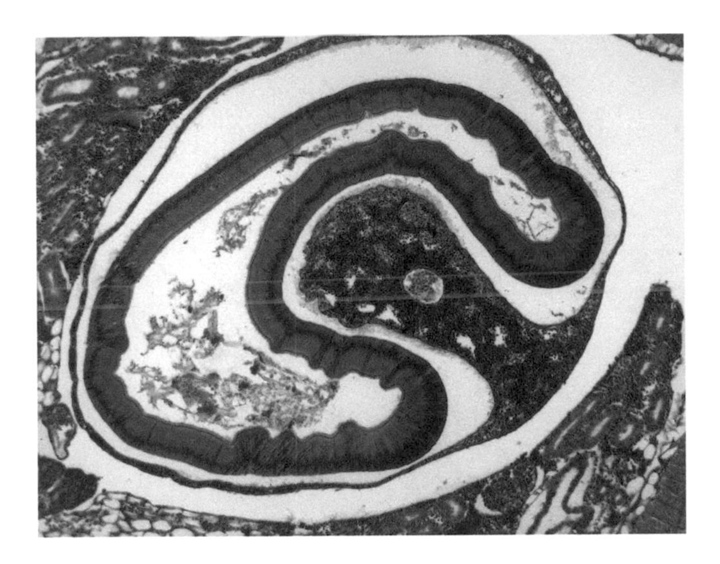

FIGURES I4a In the larval lamprey (ammocoete) the hemopoietic tissue is still contained within the wall of the gut, but it is concentrated into a single mass at one side. The hemopoietic tissue is permeated with sinusoids. Considerable shrinkage has occurred in this preparation. 10×

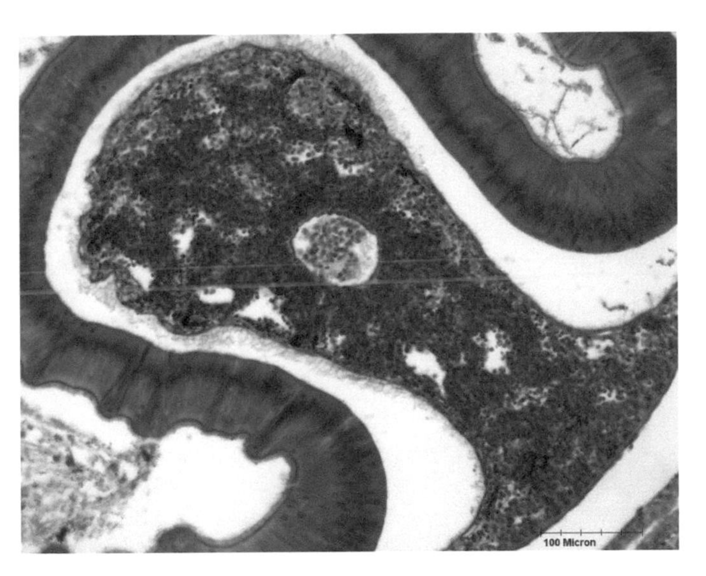

FIGURES I4b In the ammocoete hemopoietic tissue is still contained within the wall of the gut, but concentrated into a single mass. The hemopoietic tissue is permeated with sinusoids both large and small. Considerable shrinkage has occurred. 20×.

Figs. I12a and I12b show DIFFUSE LYMPHOID TISSUE in sections of the subepithelial connective tissue of the human stomach. Similarly the reticular stroma of the connective tissue of the respiratory and genital tracts harbors masses of lymphoid cells—lymphocytes, monocytes, and plasmocytes as well as their precursors.

In diffuse lymphoid tissue, especially in the gut, there is a tendency for lymphoid cells to aggregate in spherical masses, the LYMPHOID NODULES. Lymphoid nodules are compact masses of cells scattered on meshworks of reticular cells in the diffuse lymphoid tissue of the connective tissue layers of various organs, especially the gut (Figs. I13a and I13b). They are not demarcated by a sheath. PRIMARY NODULES consist chiefly of small lymphocytes. SECONDARY NODULES show a GERMINAL CENTER of lighter density than the periphery. The germinal center develops when a lymphocyte that has recognized an antigen returns to the primary nodule and the proliferation of lymphoblasts and plasmablasts ensues resulting in the production of antibodies. These cells, with their greater

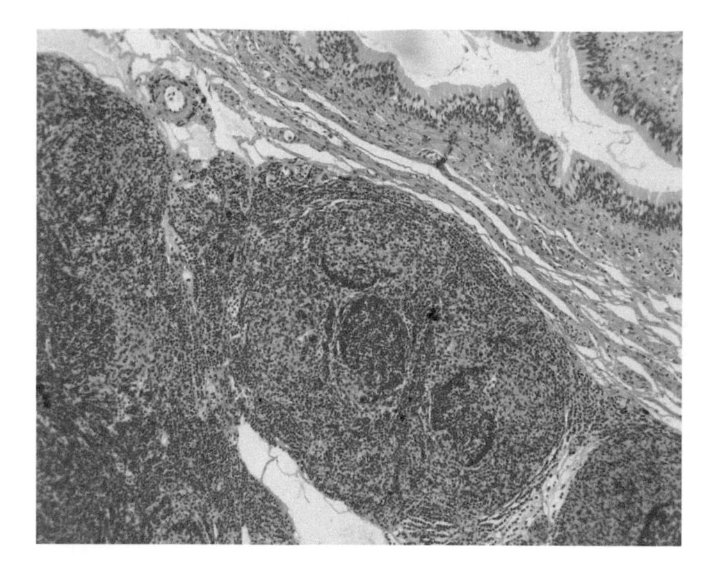

FIGURES I5a In the lungfish (*Protopterus*), hemopoietic activity is confined to the connective tissue layers in the wall of the stomach. Note the characteristic granular appearance of hemopoietic tissue. The black spots are cells that have unaccountably picked up a black pigment. 5×.

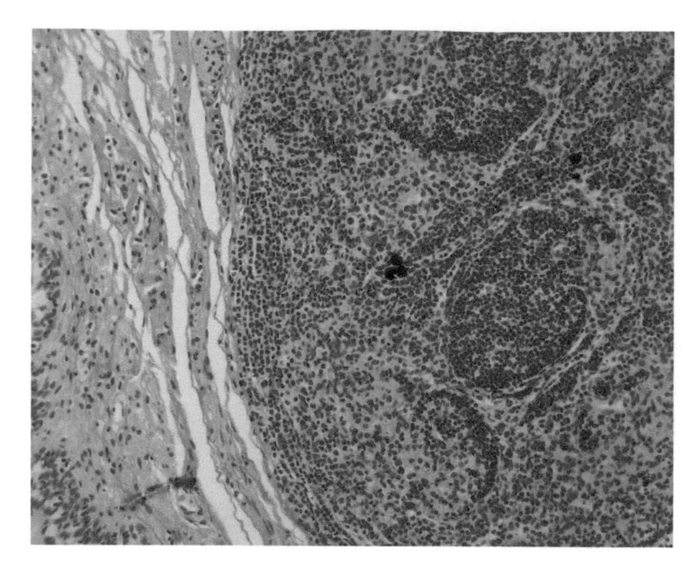

FIGURES I5b In the lungfish (*Protopterus*), hemopoietic activity is confined to the connective tissue layers in the wall of the stomach. At higher power the characteristic granular appearance of hemopoieic tissue is more apparent. 10×.

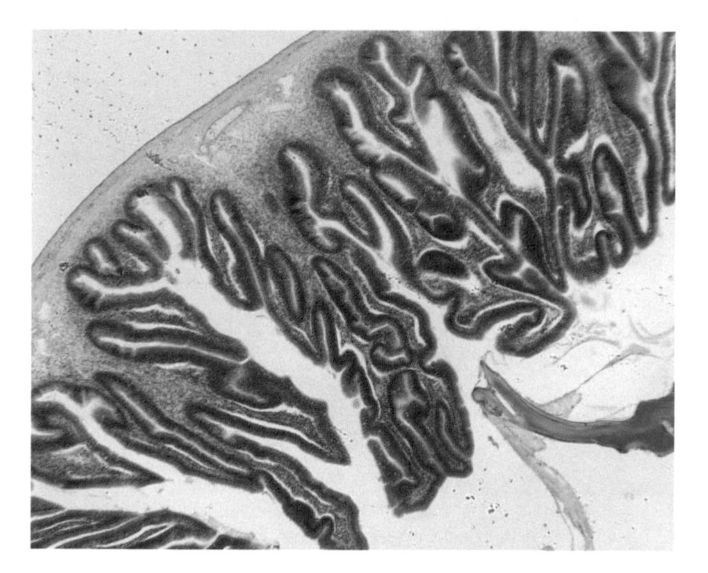

FIGURES I6a Although bony fishes have a distinct spleen where active hemopoiesis proceeds, there is also hemopoietic activity diffusely spread out in the connective tissue layers of the gut wall, even within the tissue folds of absorptive layers lining the gut. 4×.

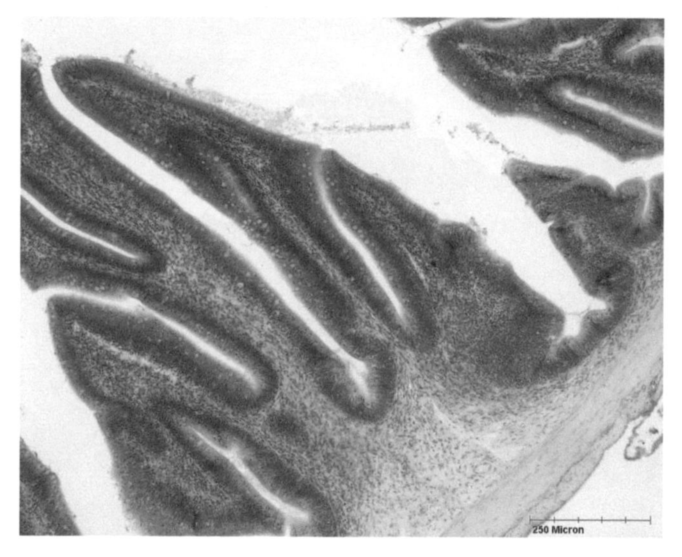

FIGURES I6b Hemopoietic activity is diffusely spread out in the connective tissue layers of the gut wall, even within the tissue folds of absorptive layers lining the gut. 10×.

content of euchromatin and more abundant cytoplasm, produce the pale staining of the germinal center. Germinal centers are found only in the nodules of birds and mammals.

In some regions (e.g., ileum and appendix) several nodules form an aggregate mass, the AGGREGATE NODULES (in the ileum these constitute Peyer's patches) (Fig. I14a–I14c). No capsule is present. The diffuse lymphoid tissues and nodules of the gut constitute the GUT ASSOCIATED LYMPHOID TISSUE (GALT).

Tonsils

Aggregations of lymphoid nodules in close association with stratified squamous epithelium of the respiratory and digestive tracts are called tonsils (Fig. I15a) and (Figs. I15b–I15d). Lingual, palatine, and pharyngeal tonsils of

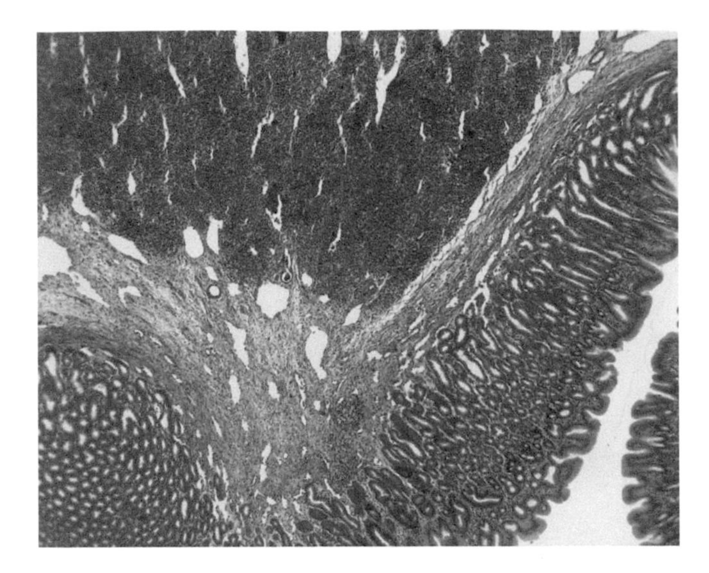

FIGURES 17a Leydig's organ is a mass of hemopoietic tissue lying in the connective tissue of the wall of the esophagus of selachians. The tubular digestive glands of the gut are seen at the bottom of Fig. 17a and Leydig's organ is at the top. Leydig's organ is permeated with sinusoids. Note the characteristic granular appearance of hemopoietic tissue. 5×

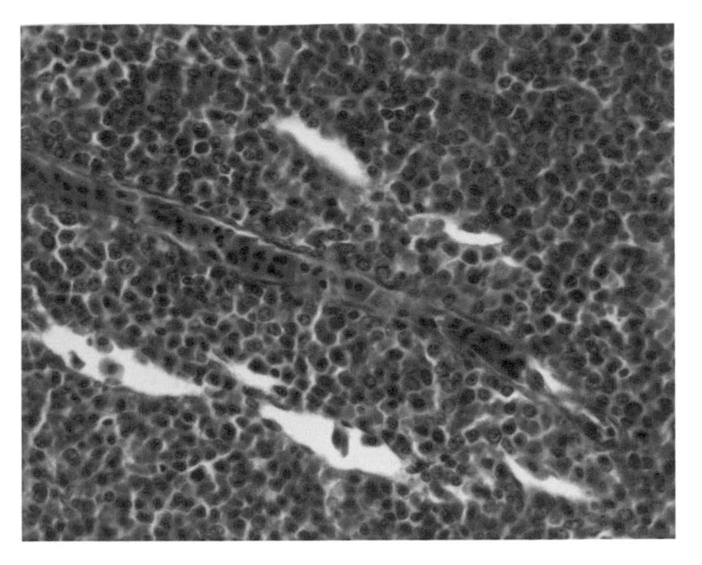

FIGURES 17b Leydig's organ is a mass of hemopoietic tissue lying in the connective tissue of the wall of the esophagus of selachians. The characteristic granular appearance of hemopoietic tissue and the simple squamous endothelial walls of the sinusoids are evident in this higher power view of Leydig's organ. 40×.

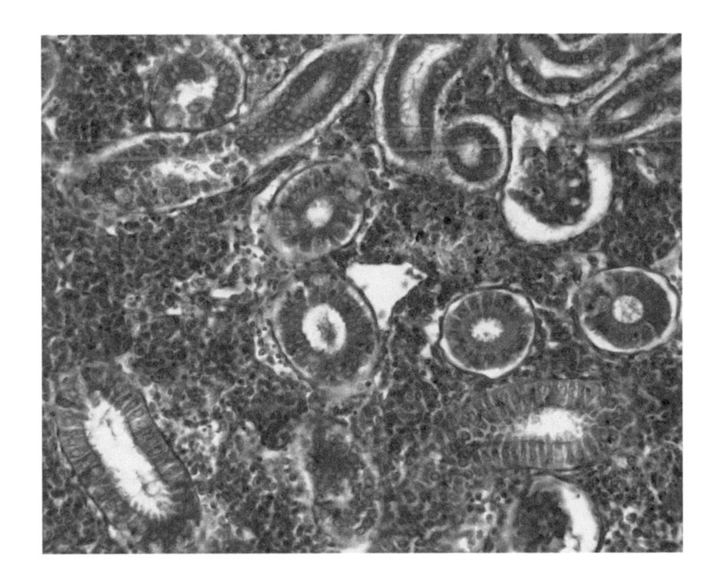

FIGURE 18 Photomicrograph of a section of the kidney of the flounder (a bony fish). The loose connective tissue between the kidney tubules is packed with hemopoietic tissue. The brown blob at the upper right is a phagocyte busily digesting erythrocytes. The brown color results from the breakdown of the pigment hemoglobin. 40×.

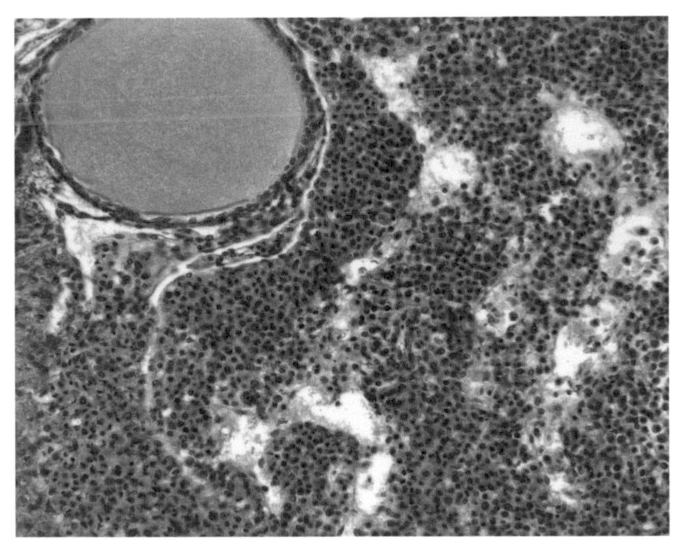

FIGURE 19 Photomicrograph of a section of the ovary of a dogfish where hemopoiesis is active in the loose connective tissue between the developing oocytes. One oocyte is seen at the upper left. 20×.

mammals are found over the root of the tongue, in the palatine arches, and in the mucous membrane lining the nasal pharynx, respectively. The epithelial surface may form pits or CRYPTS and is infiltrated with lymphocytes. Lymphoid nodules occur in the layers of connective tissue under the epithelium, the dense LAMINA PROPRIA MUCOSAE and the looser TELA SUBMUCOSA. The tela submucosa forms a capsule that sends TRABECULAE between the nodules. Blood vessels in the capsule and trabeculae supply the lymphoid tissue. The nodules may contain germinal centers.

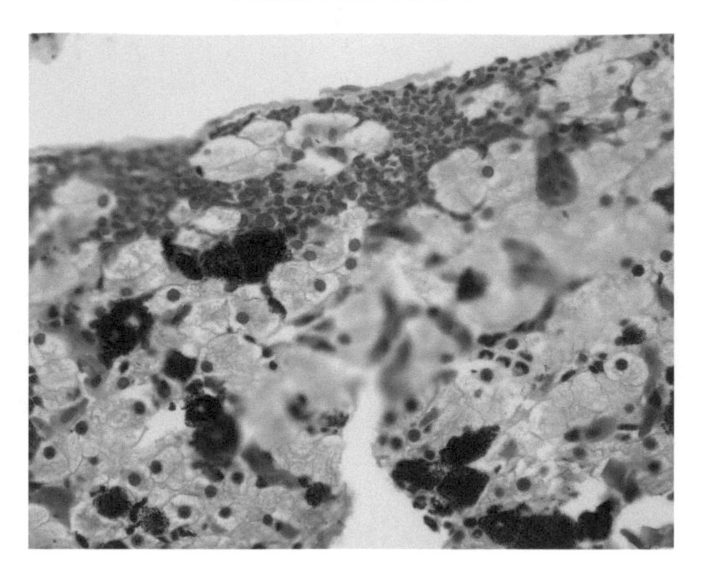

FIGURE I10 The granular appearance of hemopoietic tissue is unmistakable under the capsule in this photomicrograph of a section of the liver of a tailed amphibian (*Amphiuma*). Inexplicably the livers of amphibians often contain pigment cells packed with coal-black granules. 20×.

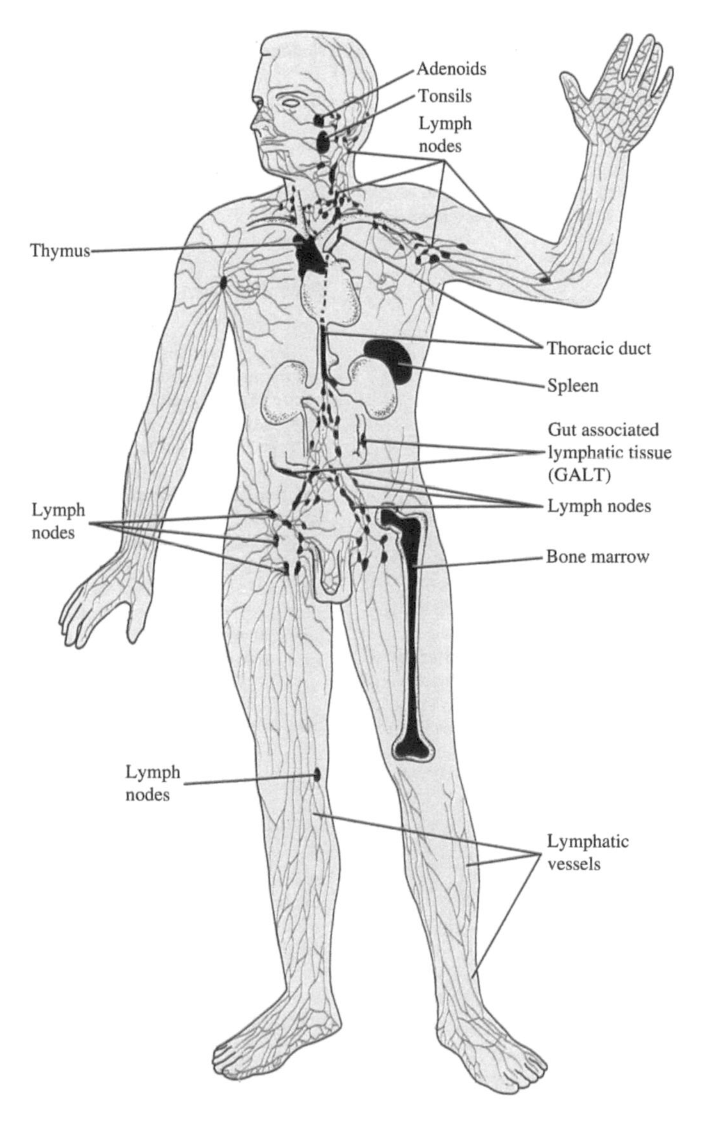

FIGURE I11 The human lymphatic system. Although lymphatic tissue is a secondary part of many organs (red marrow of bones, lymphatic nodules of the digestive tract), this drawing shows only those whose main component is lymphatic tissue.

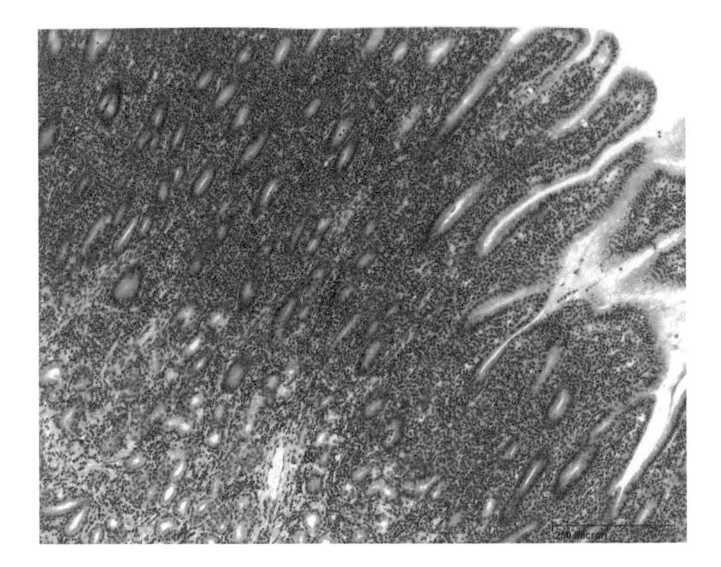

FIGURES I12a Although mammals have highly developed specialized hemopoietic organs, there is still considerable hemopoietic activity in the diffuse lymphoid tissue of the loose connective tissue in various organs. In this photomicrograph of a section of the human pyloric stomach, diffuse lymphoid tissue of unmistakable appearance is packed between the simple tubular glands. 10×.

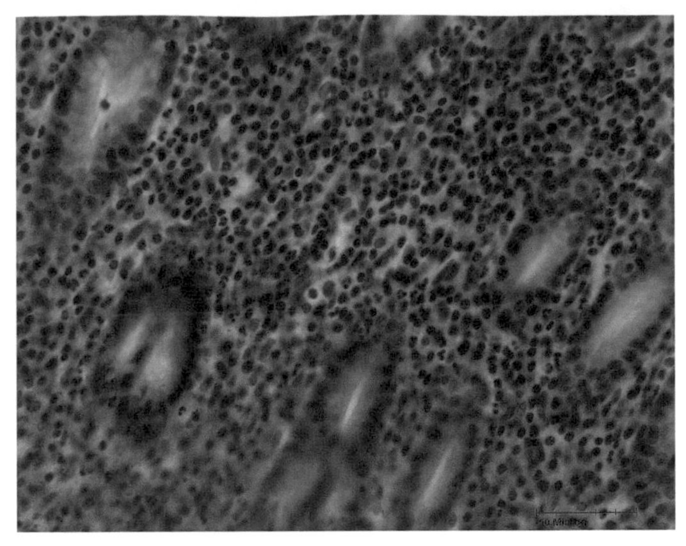

FIGURES I12b In this higher magnification image of a section of the human pyloric stomach, diffuse lymphoid tissue of unmistakable appearance is packed between the simple tubular glands. 40×.

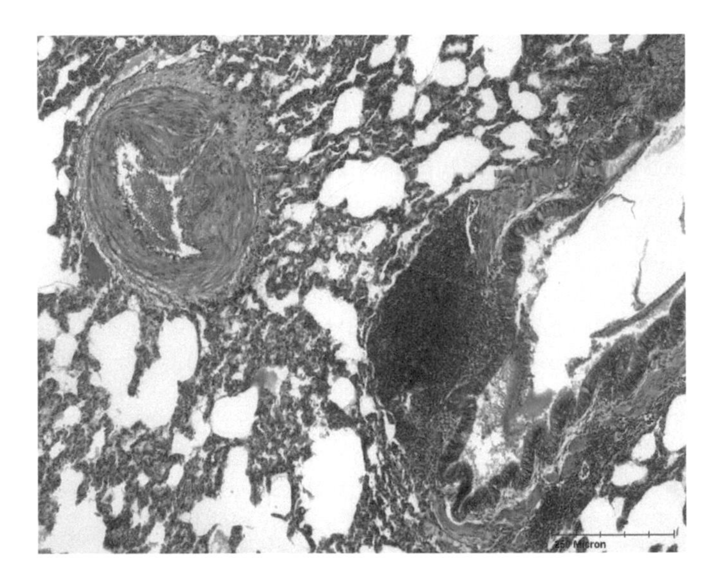

FIGURES I13a A lymphoid nodule is seen here beside an air duct in this section of the lung of a rabbit. Gas exchange takes place in the small air spaces throughput the lung. There is a small, muscular artery at the top left. 10×.

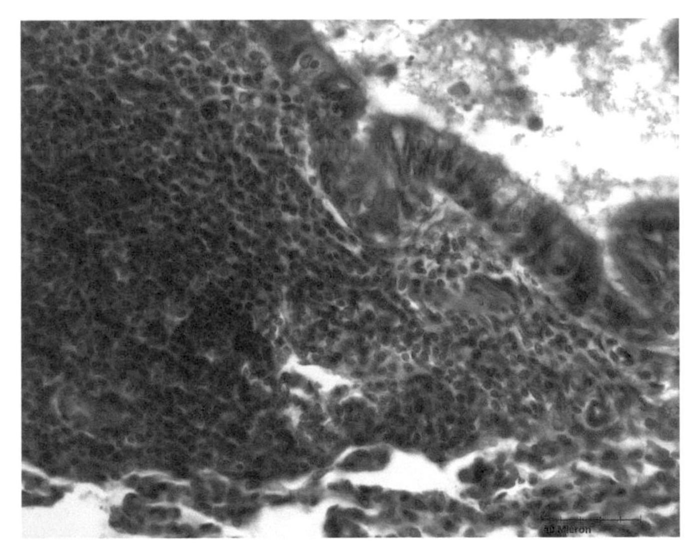

FIGURES I13b At higher magnification the lymphoid nodule demonstrates the characteristically granular appearance of lymphoid tissue. 40×.

Diffuse lymphoid tissue fills the spaces between the nodules. Lymph capillaries permeate the lymphoid tissue and drain into efferent lymph vessels.

The upper regions of the gut are richly supplied with hemopoietic tissue in many vertebrate classes. Definite tonsil-like groupings are seen in the amphibians where the epithelium of the mouth cavity is densely infiltrated with lymphocytes. True tonsils with crypts lined with long lymphoid nodules are seen in some reptiles. Birds possess considerable aggregations of lymphoid tissue in the pharynx and often a large esophageal "tonsil" as well. In lower vertebrates these hemopoietic aggregations may be responsible for the development of all types of blood cells, not just the agranulocytes, and are more properly referred to as LYMPHOMYELOID ORGANS.

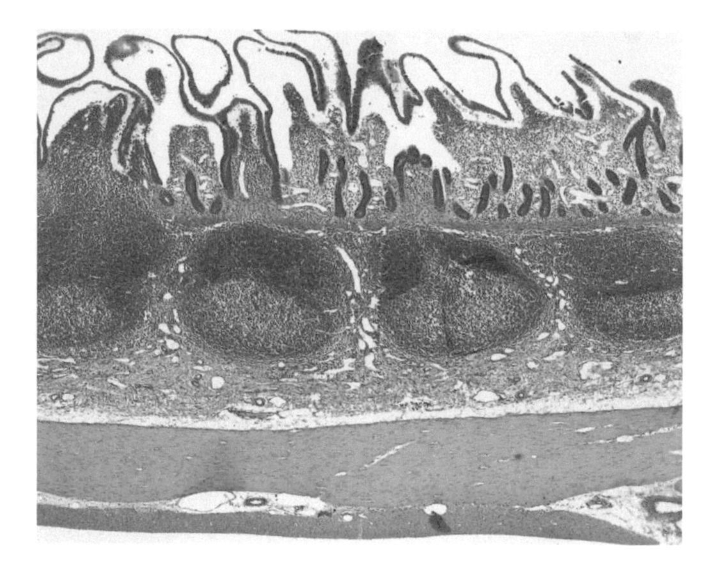

FIGURES I14a Aggregate nodules (Peyer's patch) in the wall of the ileum of a mammal. The bulging nodules distort the tunica submucosa of the small intestine. 5×.

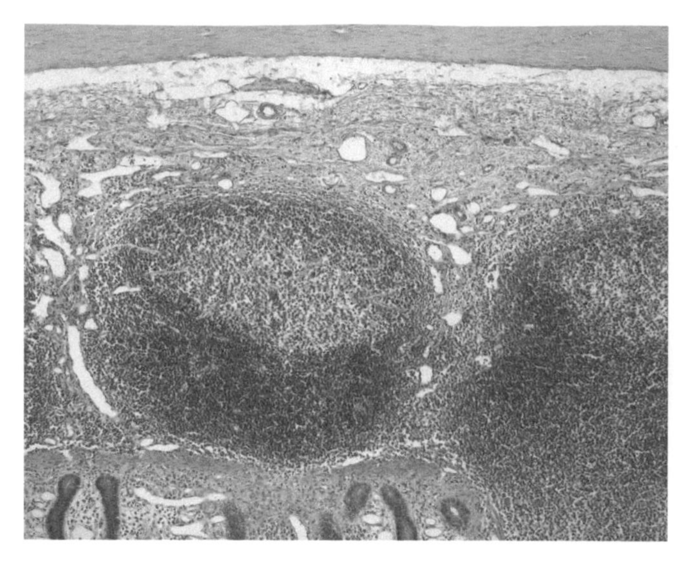

FIGURES I14b Aggregate nodules (Peyer's patch) in the wall of the ileum of a mammal. There is no capsule around the nodule and small lymphocytes are seen migrating outward into adjacent loose connective tissue. Secondary nodules consist of an outer rim of small lymphocytes and inside is the paler germinal center consisting of large lymphocytes, lymphoblasts, and plasmoblasts. 10×.

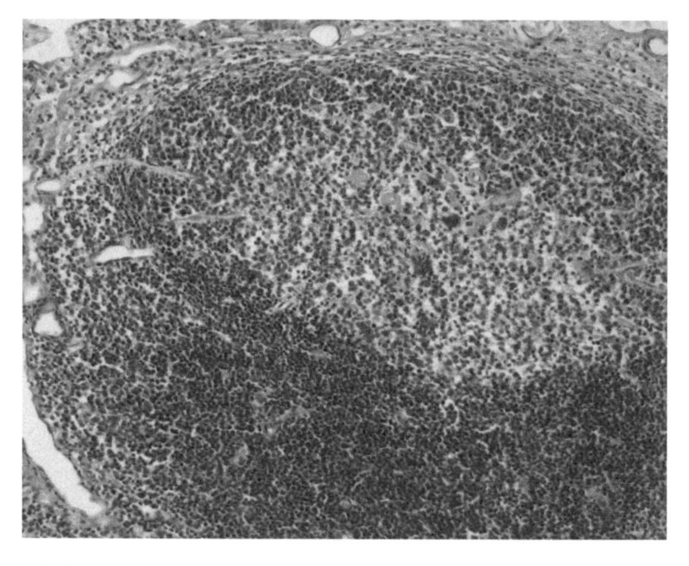

FIGURES I14c Aggregate nodules (Peyer's patch) in the wall of the ileum of a mammal. No capsule is evident around the nodule and small lymphocytes are seen migrating outward into adjacent loose connective tissue. The paler germinal center, consisting of large lymphocytes, lymphoblasts, and plasmoblasts is evident. The germinal center develops when a lymphocyte that has recognized an antigen returns to the primary nodule and brings about the proliferation of lymphocytes and plasmocytes. 20×.

Appendix

The appendix is a blind pouch at the end of the cecum of some mammals that contains an especially dense aggregation of lymphoid nodules (Figs. I16a–I16c) (it reaches its highest development in herbivores). Its thick wall is derived from the large intestine and surrounds a narrow lumen, often filled with debris. Its massive nodules are often confluent and are contained within the connective tissue layers of the wall (lamina propria mucosae and tela submucosa). The other layers are distorted and greatly reduced.

The paired CAECA occurring at the junction of the small and large intestine of birds are similar to the mammalian appendix. Here again lymphoid nodules overwhelm other poorly developed structures (Fig. I17).

Lymph Nodes

For the most part, lymph nodes are specialized mammalian structures. Although lymph vessels are present in all vertebrates from the bony fish upward, no notable aggregations of hemopoietic tissue are associated with them in classes below the birds and mammals. In waterbirds a few pairs of nodular lymphoid structures may be present but in mammals the lymph nodes are plentiful, both under the skin and in the interior of the body. It is said that there are over 500 in man.

Lymph nodes are bean-shaped structures from 1 to 20 mm in their longest diameter, lying along the course of lymph vessels (Figs. I18, I19a–I19e). AFFERENT LYMPHATICS enter at various points on the convex surface and a single EFFERENT LYMPHATIC leaves at the HILUS. Blood vessels and nerves also enter and leave at the hilus and penetrate the node by way of the connective tissue stroma.

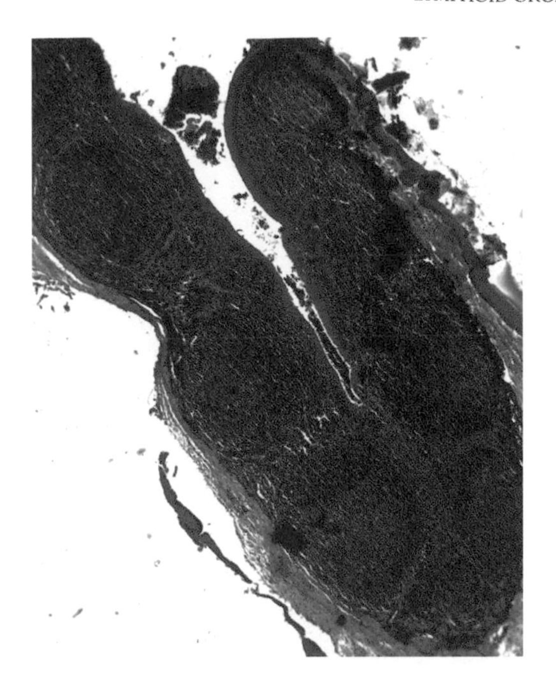

FIGURE I15a Poor section of the tonsil of a cat (it was all we could find). We used this section because it clearly shows the close relationship of lymphoid nodules beneath the stratified squamous epithelium of a crypt in the wall of the pharynx. 5×.

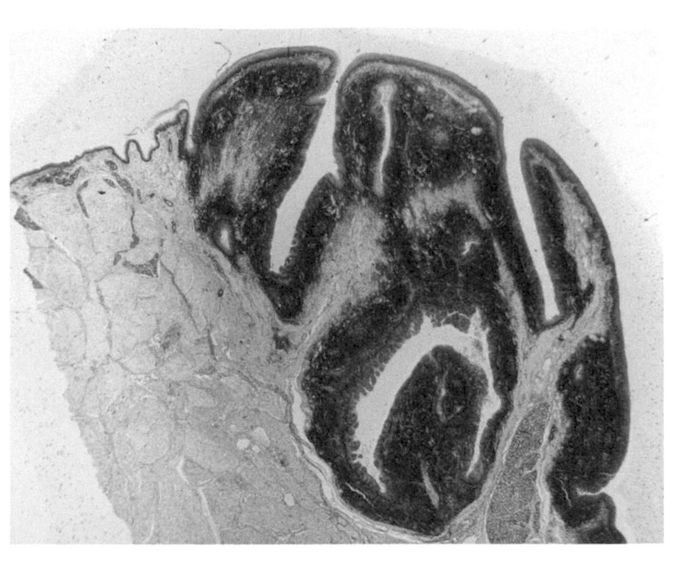

FIGURES I15b This celloidin section shows the relationship of the lymphoid nodules to the stratified squamous epithelium of the pharynx of the mammalian tonsil. 1×.

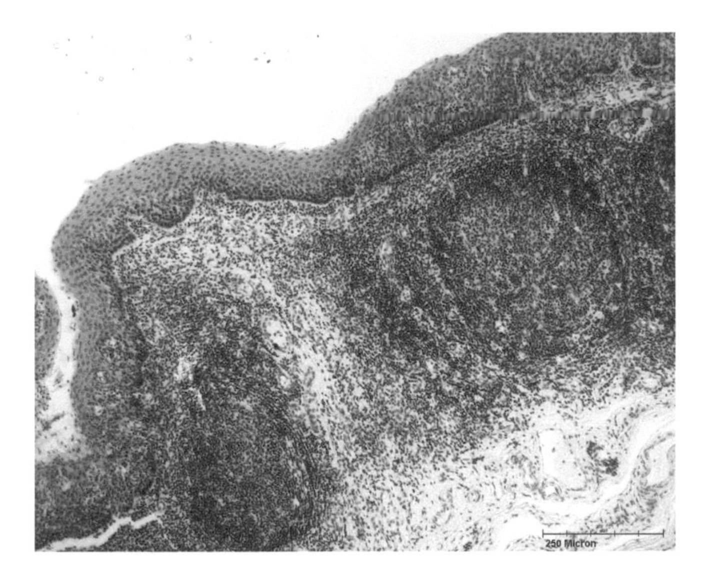

FIGURES I15c This celloidin section shows the relationship of the lymphoid nodules to the stratified squamous epithelium of the pharynx of the mammalian tonsil but it is difficult to discern the nature of the crypt. 10×.

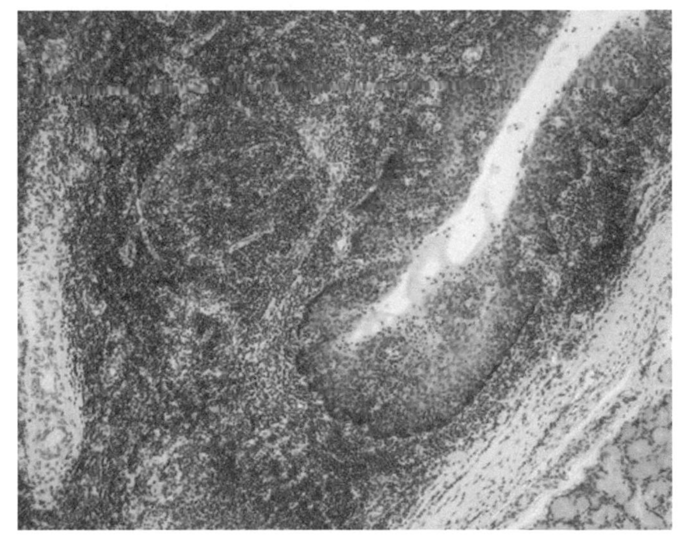

FIGURES I15d This celloidin section shows the relationship of the lymphoid nodules to the stratified squamous epithelium of the pharynx of the mammalian tonsil but even at higher power it is still difficult to discern the nature of the crypt. 40×.

Stroma (framework of connective tissue). The lymph node is enclosed in a CAPSULE of fibrous connective tissue that contains a few smooth muscle fibers around the points of entrance and exit of the lymph vessels (Fig. I19a). At various points on the inner surface the capsule gives off TRABECULAE that extend into the substance of the organ and partially divide the parenchyma into compartments. In the interior of the node there is a scaffolding of RETICULAR TISSUE consisting of reticular cells and macrophages extended on the reticular fibers upon which the remaining elements of the node are contained (Figs. I19f and I19g). The fibers are continuous with the collagenous fibers of the trabeculae and are so arranged that they form a distinct CORTEX and MEDULLA (Fig. I19b).

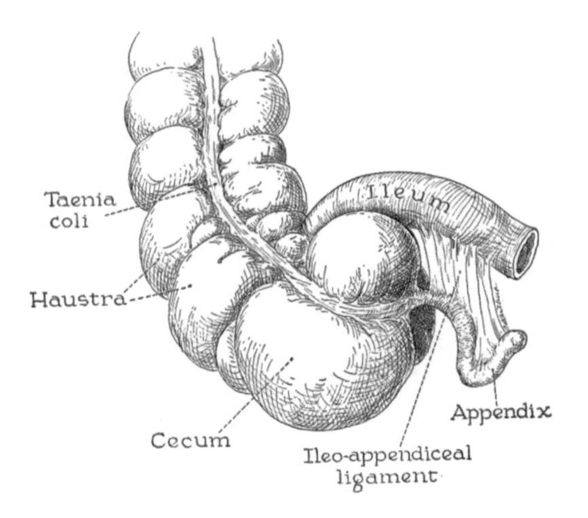

FIGURE I16a Drawing of the human cecum and appendix.

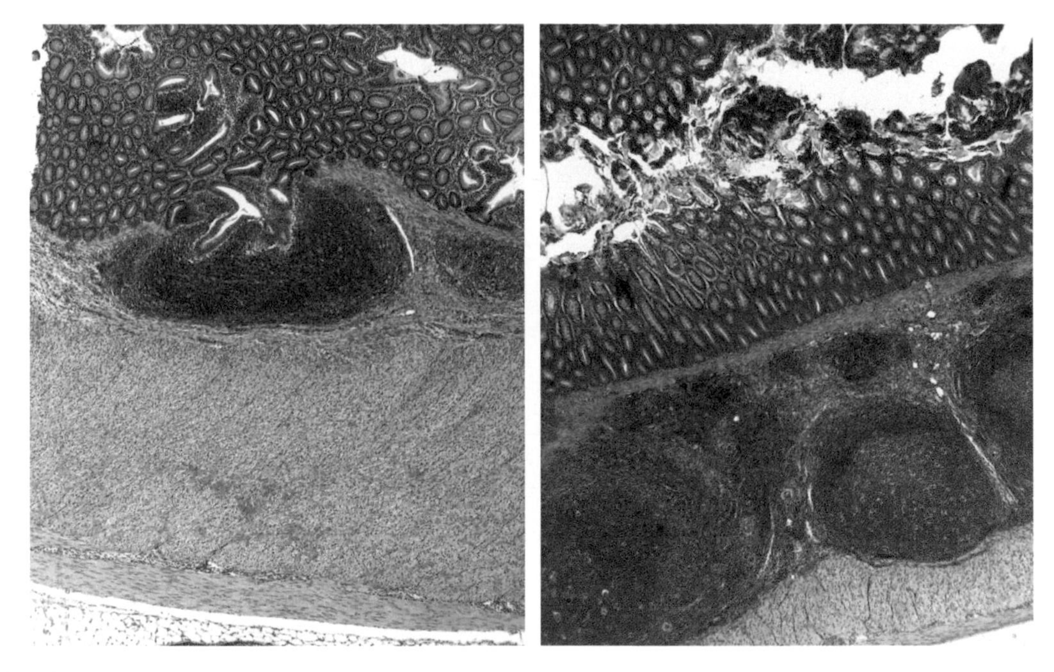

FIGURES I16b and I16c Two views of cross sections of the mammalian caecum (left) and appendix (right). The lining of the large intestine is packed with simple tubular glands that open into the lumen. Beneath this layer of simple glands is connective tissue brimming with lymphoid tissue. 5×.

Sections of lymph node stained to show the reticular fibers show well the arrangement of the parts of the stroma (Fig. I19e).

Parenchyma (the essential cellular substance of the node). In sections stained by the usual methods, only some members of the parenchyma can be identified (Fig. I20). The CORTEX is the deeply stained peripheral area containing both primary and secondary nodules. Reticular cells, lymphocytes, macrophages, and plasmocytes are present. The loosely arranged central portion of the node is the MEDULLA. The dense bands of lymphoid material are MEDULLARY CORDS.

Sinuses (loose spaces in the node through which lymph filters) (Fig. I19d). Sinuses are incompletely lined by endothelial cells supported by reticular fibers. The sinuses between the capsule and the cortex are SUBCAPSULAR SINUSES (Figs. I21a–I21c). These drain into TRABECULAR SINUSES (Fig. I21d) along the trabeculae and open in turn into MEDULLARY SINUSES (Figs. I21e and I21f) between the medullary cords. RETICULAR BRIDGES frequently span the sinuses. Blood vessels do not communicate with the sinuses.

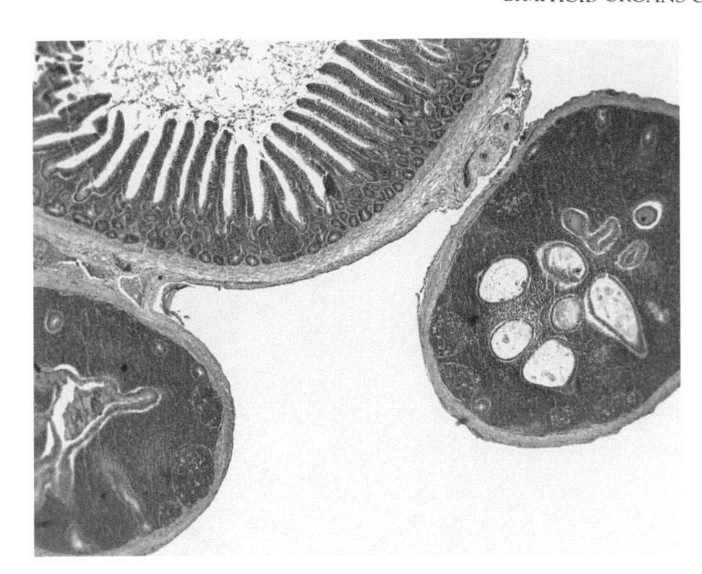

FIGURE I17 Paired caeca open into the intestine of birds at a point arbitrarily declared as the site where small and large intestine meet. Their appearance is strikingly similar to that of the mammalian appendix and they probably have a similar function. At the top of this micrograph is a section of the small intestine. 5×.

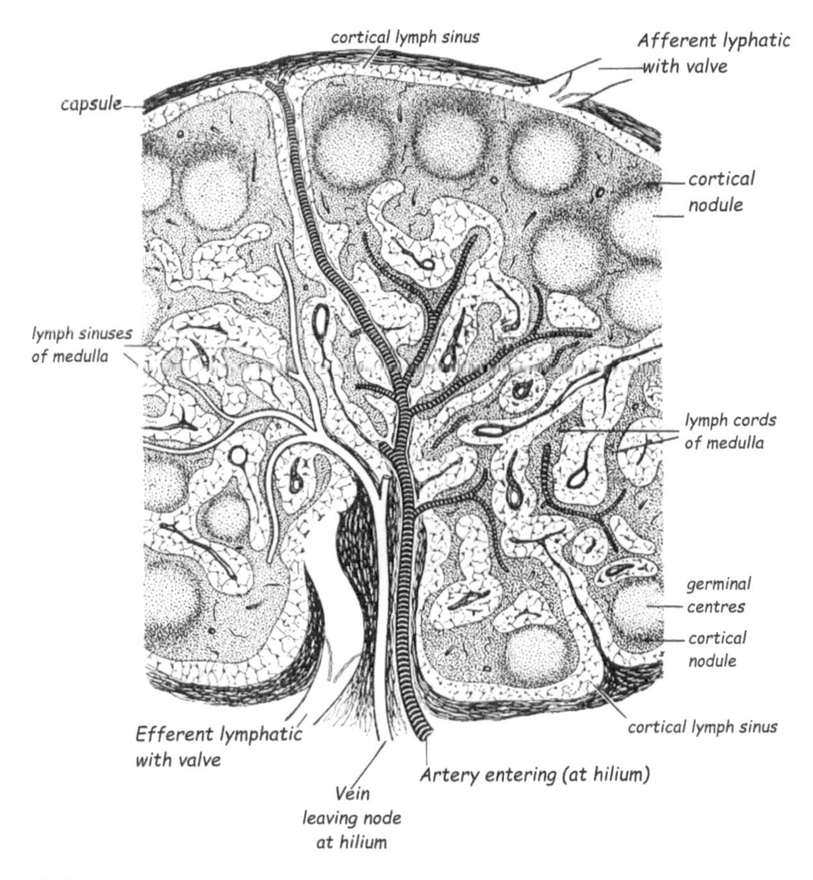

FIGURE I18 Diagram of a section of a mammalian lymph node. The node is poised between afferent lymphatics at the top and efferent lymphatics at the bottom. It filters materials from the lymph as it passes through. Note the valves in the lymphatics; these prevent the back flow of lymph.

Spleen

Beginning as an aggregation of scattered hemopoietic elements associated with the gut, the spleen assumes a high degree of development in higher vertebrates. The ganoid fish is the first to show a discrete spleen and this is located near the junction of the stomach and intestine with the mesentery. It is only in the spleen and lymph nodes that hemopoietic organs appear as discrete organs.

The vascular relations of the spleen are fundamentally alike from cyclostomes to mammals so that a description of the mammalian spleen should suffice to give a general picture of this organ in vertebrates (Figs. I22a–I22c).

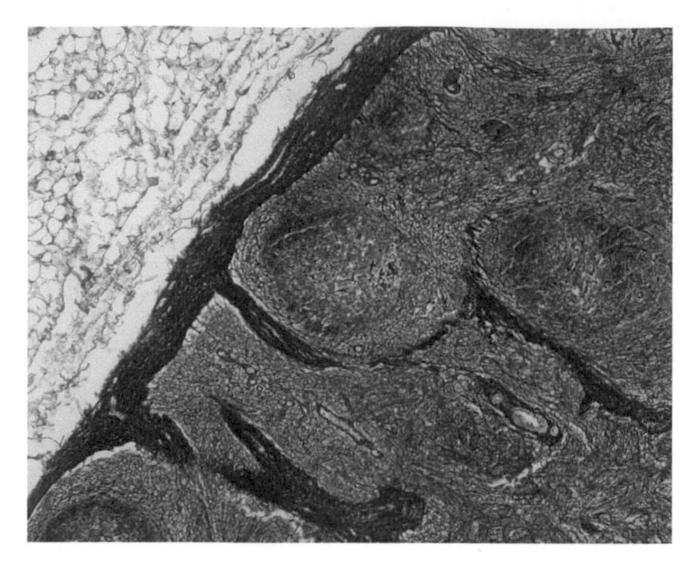

FIGURE I19a Section of a mammalian lymph node impregnated with silver to show its reticular stroma. The stout collagenous fibers of the capsule cross the screen from lower left to upper right. Trabecular fibers, continuous with the fibers of the capsule, partition the parenchyma of the gland. Lymphoid nodules, with their own reticular support, inhabit these small compartments in the cortex of the node. Clearings in the parenchyma adjacent to the capsule and trabeculae are the capsular and trabecular sinuses. 10×.

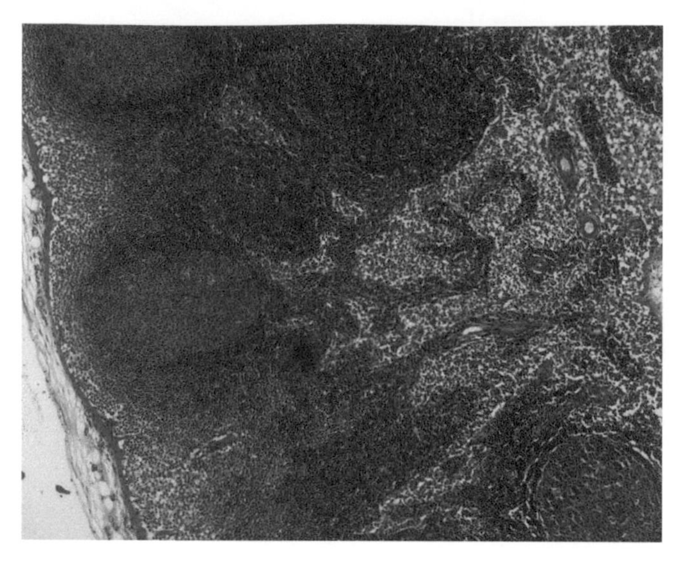

FIGURE I19b A corresponding section of mammalian lymph node stained with conventional methods. The deep blue capsule is seen at the left. A nodule is near the lower center. Medullary cords, invaded by medullary sinuses occupy the right. Connective tissue fibers of trabeculae are sparse. 10×.

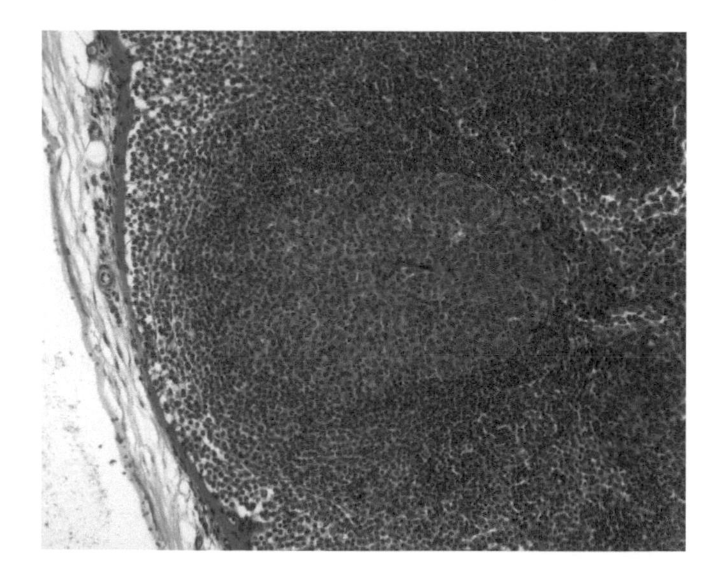

FIGURE I19c A slightly higher power view of the same slide shows the delicate endothelial wall of the subcapsular sinus under the deep blue capsule. Below the sinus is a lymphoid nodule with a germinal center. Hints of trabecular sinuses are seen on both sides of the nodule. 20×.

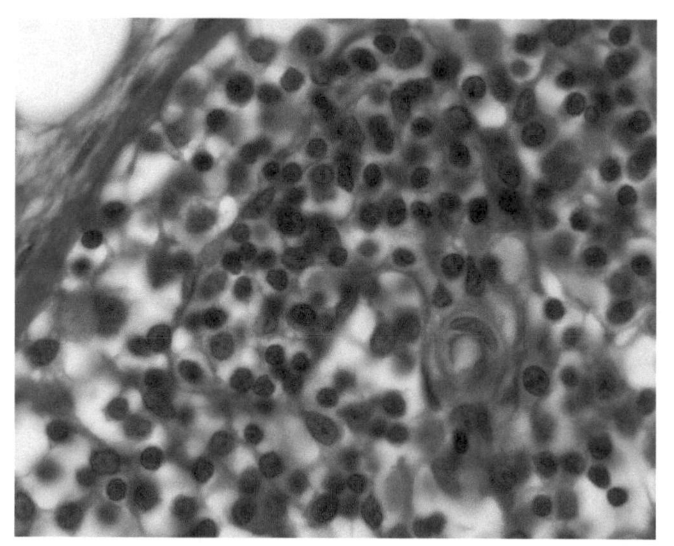

FIGURE I19d Increasing the magnification yet again, shows the subcapsular sinus beneath the capsule; its delicate endothelial wall is clearly visible. The cells flowing in the sinus are captured. Reticular cells supporting the cortical nodule are clearly visible. 100×.

All lymphoid tissue that filters blood is located in the spleen (except for a few hemal nodes in some species). It is the largest accumulation of lymphoid tissue in the body. It is an encapsulated organ of variable shape. The splenic artery and vein, nerves, and lymph vessels enter at the HILUS on the medial surface.

In many species the capsule and trabeculae of the spleen contain large amounts of smooth muscle that is capable of rapid contraction, thereby providing the animal with a quick "transfusion"—of the right blood type—in times of need.

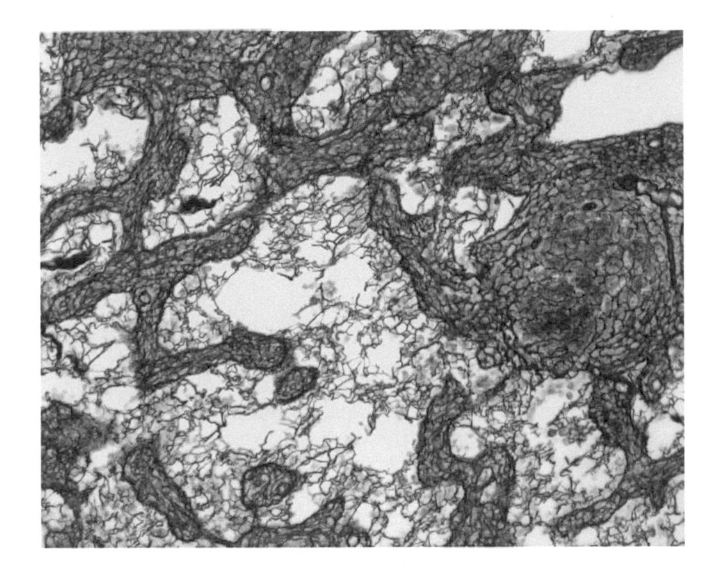

FIGURE I19e Silver impregnation showing the reticular stroma in a section of the medulla of a mammalian lymph node. The stroma is denser in the medullary cords than in the medullary sinuses. 40×.

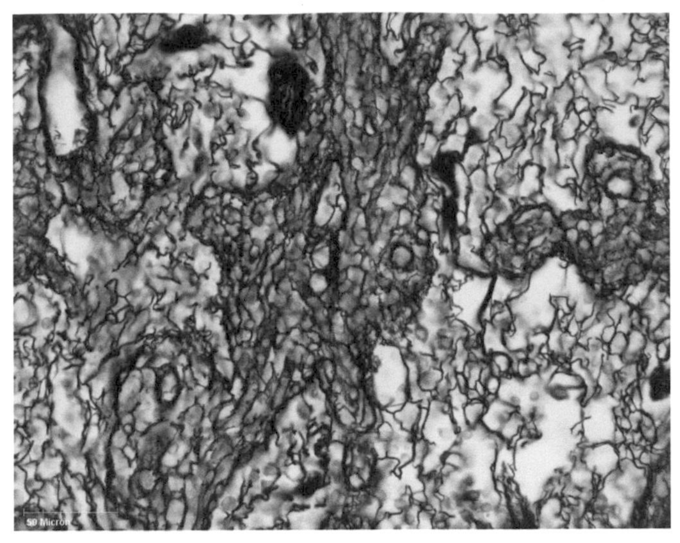

FIGURE I19f Silver impregnation of the reticular stroma of a mammalian lymph node. The ghostly images of free cells within the stroma can be seen. 40×.

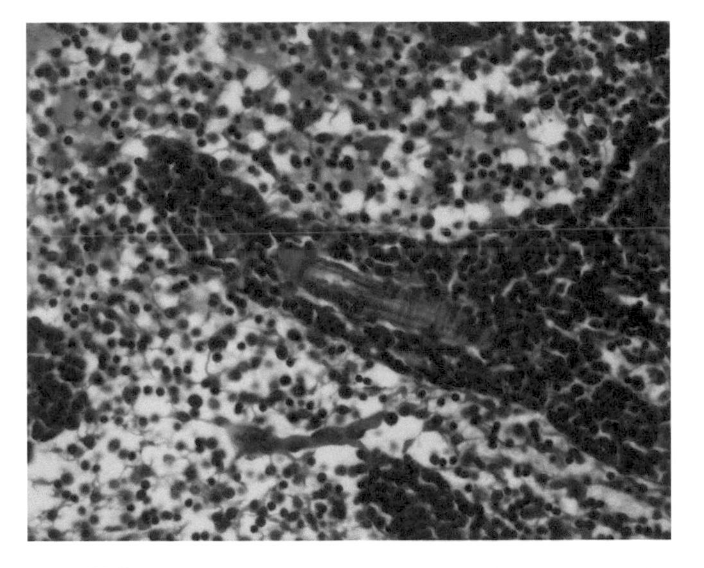

FIGURE I19g Conventional stain of the medulla of a mammalian lymph node. A glancing section of the muscular wall of a trabecular arteriole is at the center. Reticular fibers and capillaries can be seen here and there. 40×.

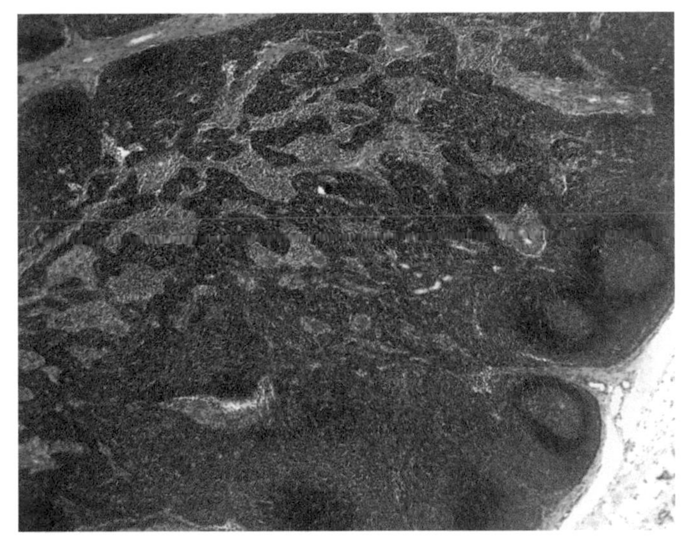

FIGURE I20 Section of mammalian lymph node stained with hematoxylin and eosin. The stout collagenous capsule is at the top left. Beneath it is the tightly packed parenchyma of the cortex. The parenchyma of the medulla consists of trabeculae of lymphoid tissue permeated with sinuses. 5×.

Stroma. The capsule is made up of layers of connective tissue containing contractile cells (Fig. I23). These cells may be referred to as myofibroblasts because, although contractile, they produce connective tissue fibers. trabeculae extend from the capsule into the spleen (Fig. I24a); some extend from one side to the other. reticular tissue forms the supporting framework for the parenchyma and its fibers are continuous with collagenous fibers of the capsule and trabeculae (Fig. I24b). Superficially the capsule is covered with reflected peritoneum so that there is a layer of mesothelium on the free surface.

Parenchyma or splenic pulp. trabecular arteries arise from the splenic artery, ramify through the trabeculae, and pass into the parenchyma where their tunica adventitia loosens and becomes infiltrated with a sheath of dense lymphoid tissue, the periarterial lymphoid sheath (PALS; Figs. I23 and I24a). This sheath constitutes the white pulp of the spleen and may be expanded to form lymphoid nodules, the splenic nodules (Malpighian corpuscles),

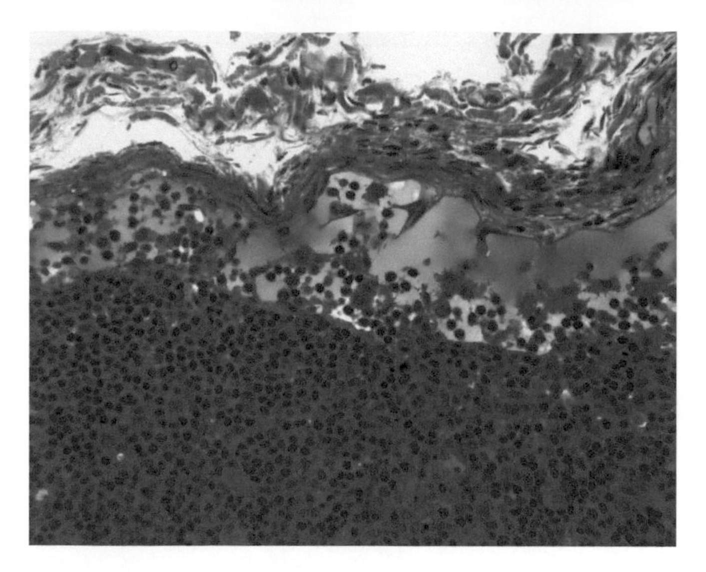

FIGURE I21a The subcapsular sinus flows beneath the capsule in this semithin section of a mammalian lymph node. A few reticular fibers and an array of loose cells are seen in the sinus. Below is the tightly packed lymphoid tissue of the cortex. 40×.

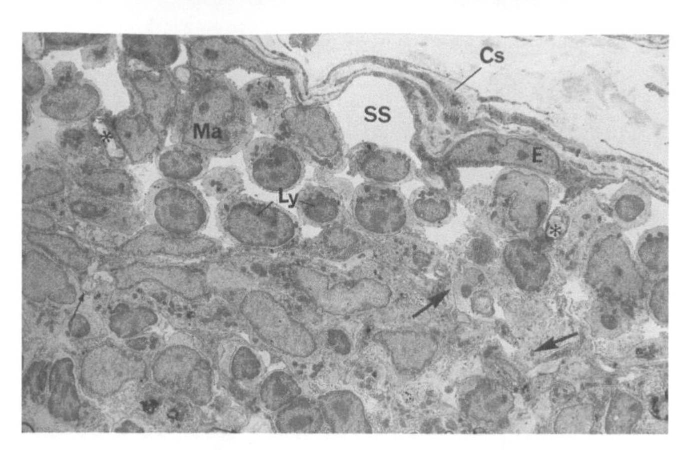

FIGURE I21b Transmission electron micrograph of a section of the lymph node of a rat. The subcapsular sinus (SS) lies between the capsule (Cs) and lymphoid parenchyma. The sinus is lined by squamous endothelial cells (E) with gaps (large arrows) between them that permit the passage of cells in and out of the adjacent lymphoid tissue. The sinus contains macrophages (M), lymphocytes (Ly), and bundles of collagen fibers surrounded by endothelial cells (*). Trabeculae of collagen fibers and fibroblasts (small arrow) support the parenchyma. 2600×.

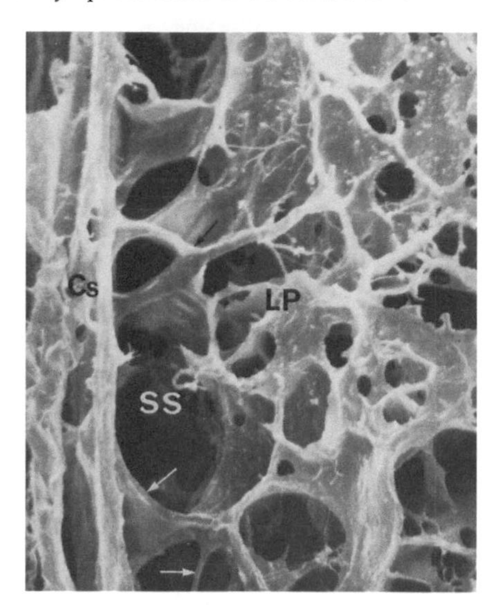

FIGURE I21c Scanning electron micrograph of the cortex of a lymph node of a rat. The meshwork of trabeculae (arrows) traverses the subcapsular sinus (*SS*) and connects the capsule (*CS*) to the lymphoid parenchyma *(LP)*. 1400×.

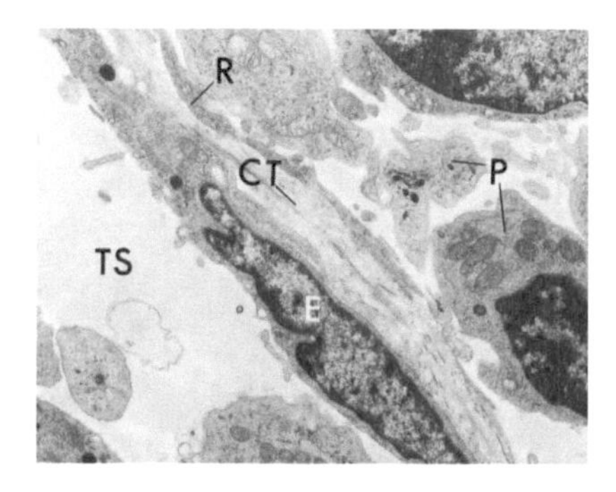

FIGURE I21d The parenchymal wall of the trabecular sinus (TS) in the human lymph node consists of sinus endothelium (*E*), connective tissue (*CT*) and reticular cells (*R*). *P*, lymphoid parenchyma. 9775×.

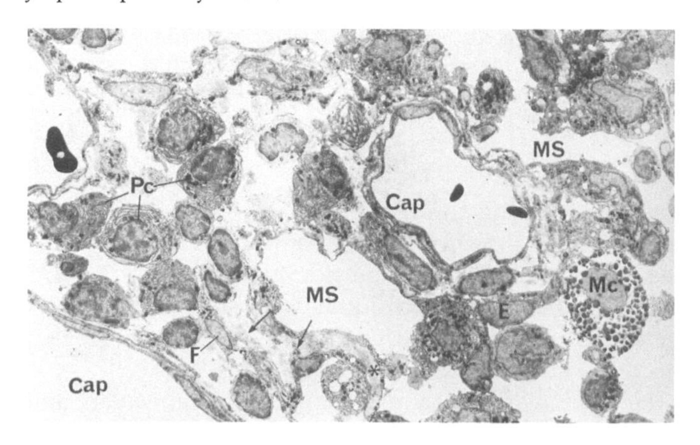

FIGURE I21e In this section of the lymph node of the rat the medullary sinus (*MS*) is lined by endothelial cells (*E*) which also cover the trabeculae of collagenous fibers (*) within the sinuses. In the medullary cords, collagenous bundles (arrows), accompanied by extensions of fibroblasts (*F*) are connected with those in the sinus trabeculae. *Ma*, macrophage; *Pc*, plasmocytes; *Cap*, capillaries; *Mc*, mast cell. 2200×.

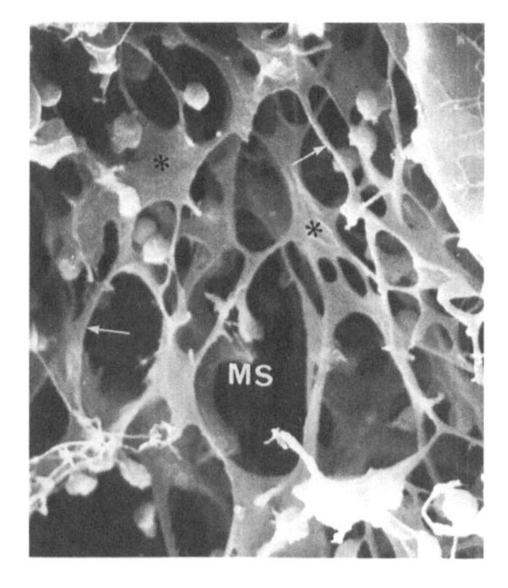

FIGURE I21f Scanning electron micrograph of the medullary sinus (*MS*) in the lymph node of a rat. Cord-like (arrows) and plate-like (*) trabeculae anastomose freely with one another forming a loose meshwork. 1200×.

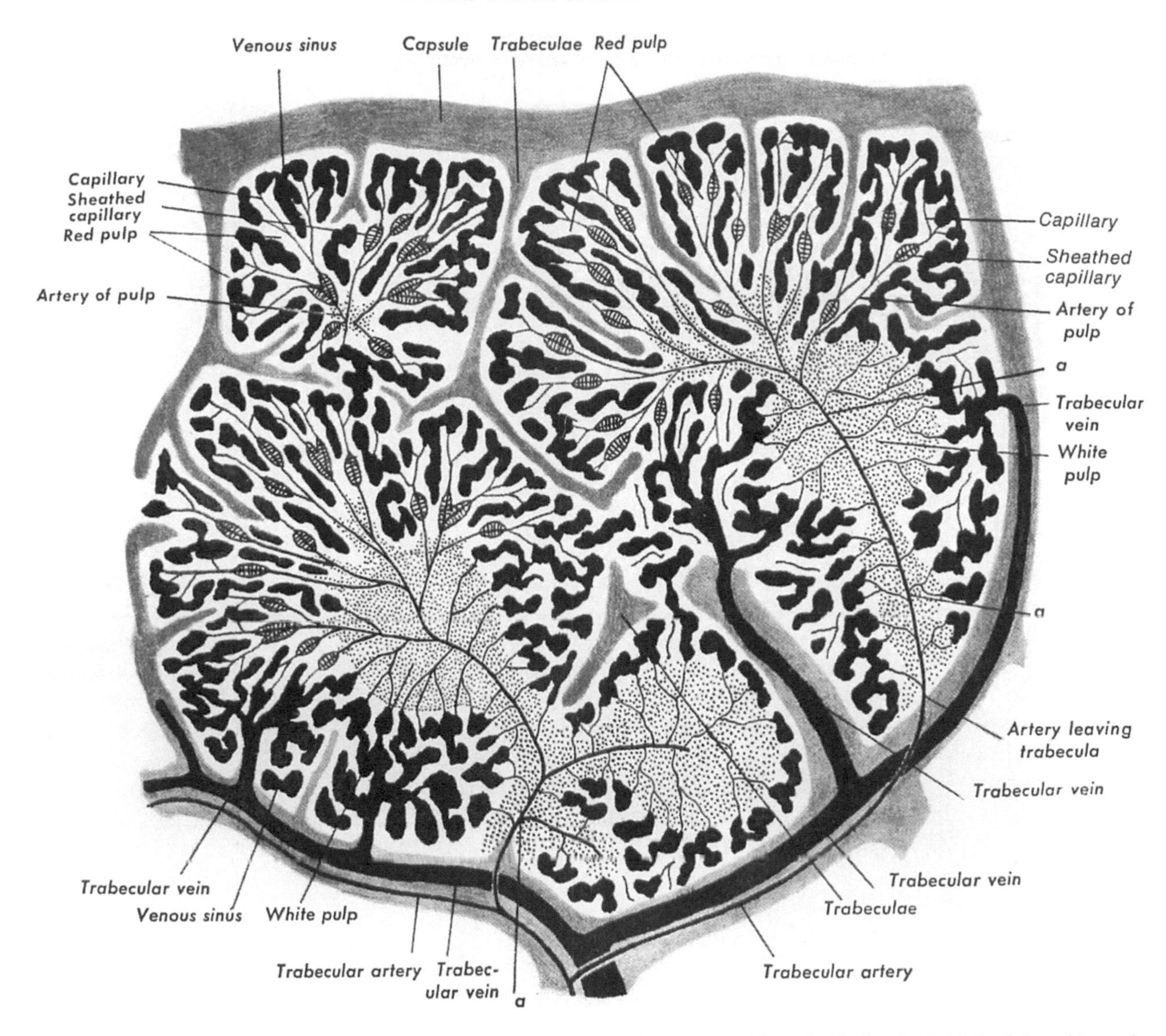

FIGURE I22a Diagram of the vascular tree of the spleen of a mammal. The periarterial lymphoid sheaths (PALS) of the white pulp are the stippled areas around the central arteries (a) leaving the trabeculae. The circulation depicted here is CLOSED, with the capillaries opening directly into the sinusoids shown in black. In an OPEN circulation the capillaries would open into the interstitial spaces in the cords of the red pulp.

often with germinal centers. Identify the eccentric CENTRAL ARTERIOLE in a cross section of the white pulp (Fig. I23). The central artery sends branches to the white pulp itself and to sinusoids at the perimeter of the white pulp, the MARGINAL SINUSOIDS. The central arteriole loses its investment of white pulp, and enters the red pulp where it branches into several straight arterioles, the PENICILLI (Figs. I24b and I24c). The penicilli may be surrounded by concentrically arranged macrophages and reticular fibers; these constitute the ELLIPSOIDS, sheathed arterioles, or PERIARTERIAL MACROPHAGE SHEATHS (PAMS) (Fig. I24d). The tissue filling the spaces between the white pulp and the capsule is the RED PULP (Figs. I24b and I24d). It consists of a reticular mesh containing all types of blood cells, reticular cells, macrophages, and monocytes. It may be permeated by the VENOUS SINUSOIDS, from 12 to 40 μm in width, that drain into veins. Note in scanning electron micrographs that the sinusoidal walls consist of relatively thick endothelial cells arranged much in the manner of barrel staves with large gaps between them through which an erythrocyte can easily pass (Figs. I25a–I25f). The transmission electron microscope shows that the cells contain abundant free ribosomes, mitochondria, and a few Golgi complexes (Fig. I26). Pinocytotic pits and vesicles, similar to those seen in the endothelial cells of capillaries, are abundant. The endothelial cells are supported by thick, ring-like formations of basement membrane containing a few scattered reticular fibers in the manner of hoops around a barrel (Fig. I27). The reticular mesh between the sinusoids is permeated with blood and constitutes the PULP CORDS (Billroth's cords).

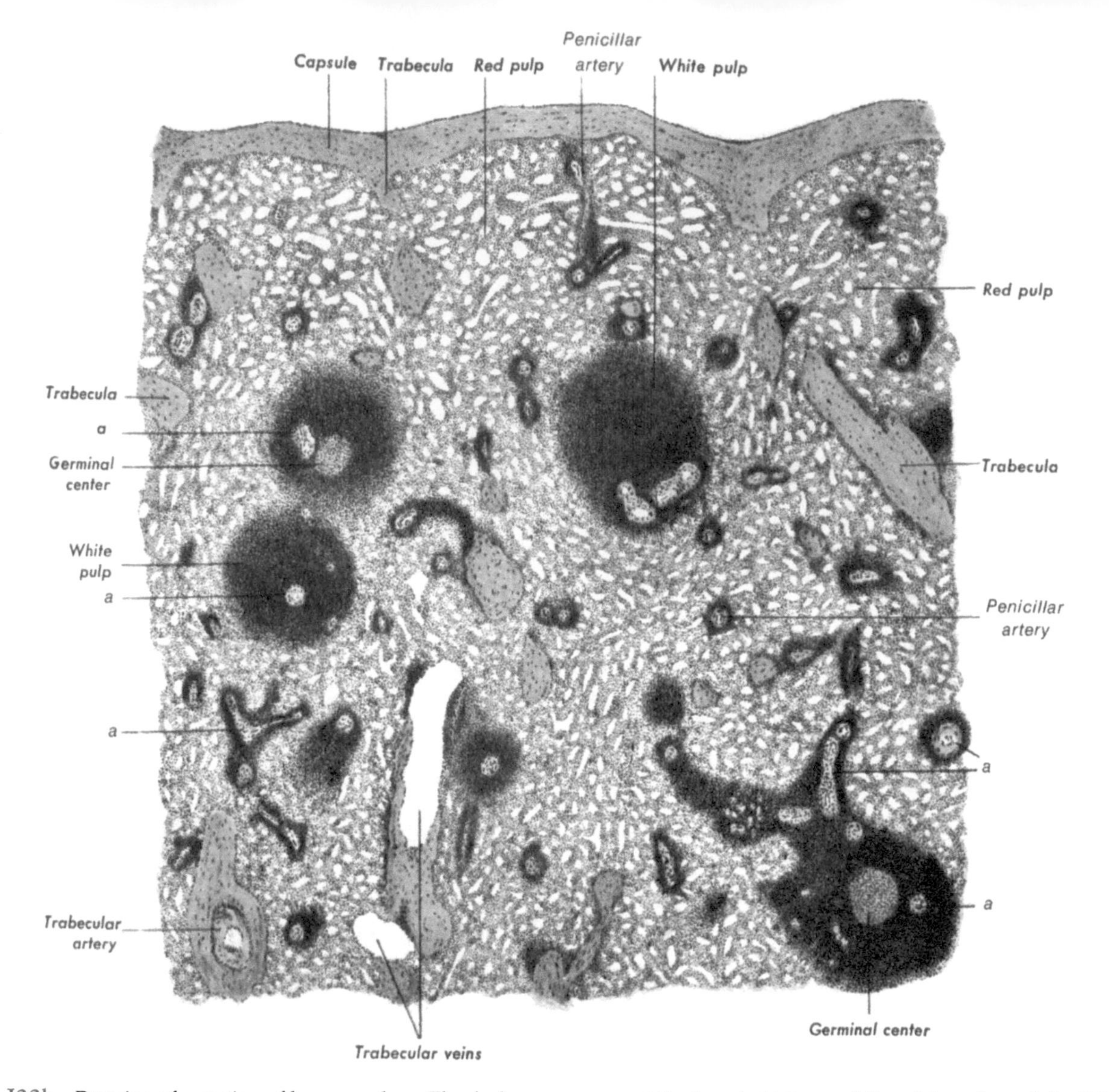

FIGURE I22b Drawing of a section of human spleen. The dark areas represent the lymphoid tissue of the white pulp and the lighter areas, the red pulp. Arterial branches (a) sheathed in white pulp.

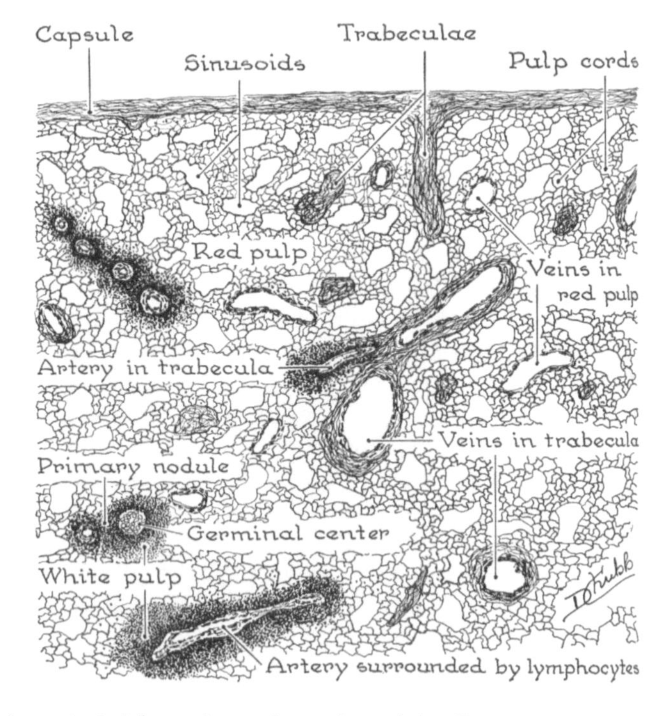

FIGURE I22c Diagram of a section cut at right angles to the surface of the distended mammalian spleen. The white pulp consists of aggregations of lymphocytes and the red pulp is an open mesh with sinusoids running through it.

FIGURE I23 Photomicrograph of a section of the spleen of a Northern pike (*Esox lucius*). Trabeculae, rich in smooth muscle, criss-cross the pulp. The capsule at the upper right is also muscular. A nodule at the lower left displays a germinal center. 10×.

FIGURE I24a Photomicrograph of a semithin section of the spleen of a monkey showing a trabeculaemerging from the muscular capsule. 20×.

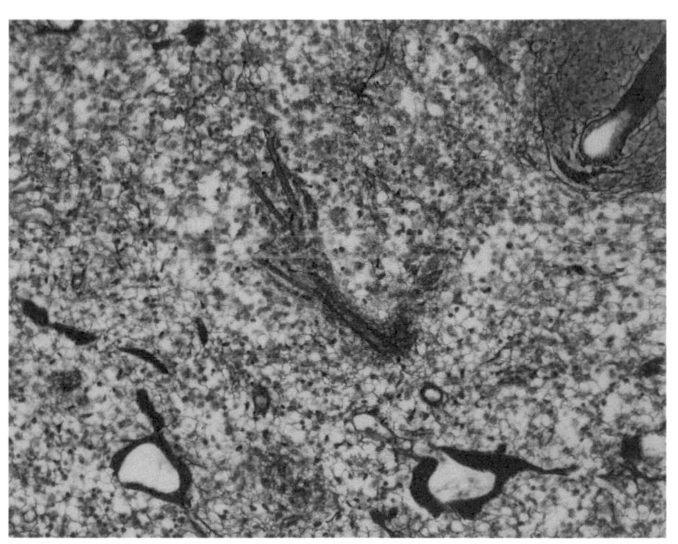

FIGURE I24b Photomicrograph of a silver impregnation of a section of mammalian spleen showing a sheathed artery at the right of center and two arteries at the bottom. An artery at the upper right is surrounded by densely packed lymphocytes. 20×.

The foregoing description applies to the SINUSAL SPLEEN found in the human, dog, rat rabbit, and others. In the NONSINUSAL SPLEEN, as seen in the cat and mouse, there is no arterial continuity between the arterial terminals and the venous vessels so that the blood oozes, extravascularly, through the reticular sponge.

Hemal Nodes

Hemal nodes resemble lymph nodes except that numerous erythrocytes are seen within the lymphatic sinuses and lymphoid tissue (Figs. I28a and I28b). They are found in birds and some mammals, being most abundant in ruminants. Their occurrence in man is doubtful. They are scattered near large blood vessels in the retropleural and retroperitoneal tissues along the vertebral column from the neck to the pelvis. At the HILUS, an artery, vein, and efferent lymphatic penetrate the capsule. Their functions are similar to those of the spleen.

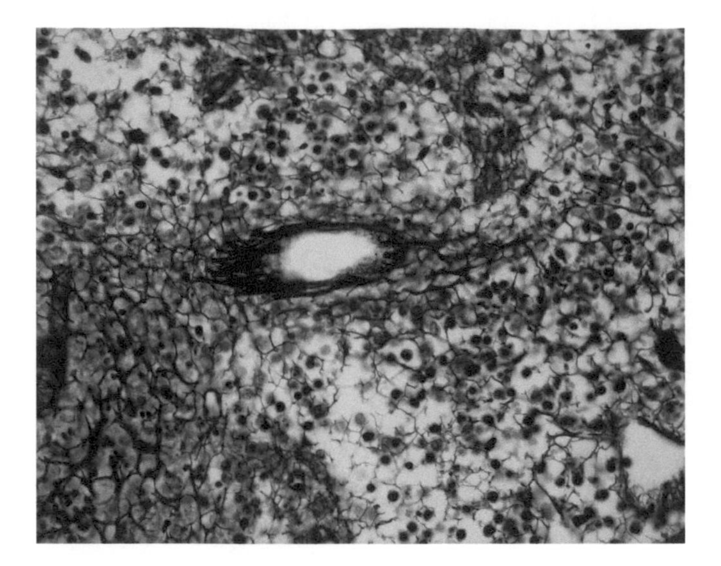

FIGURE I24c Photomicrograph of a silver impregnated section of mammalian spleen showing its reticular stroma. There is an artery sheathed with white pulp (PALS) a little above center; its sheath consists of packed lymphocytes and continues down to the left corner. The sparser areas are the labyrinth of sinusoids of the red pulp, which are penetrated by the red pulp cords. 40×.

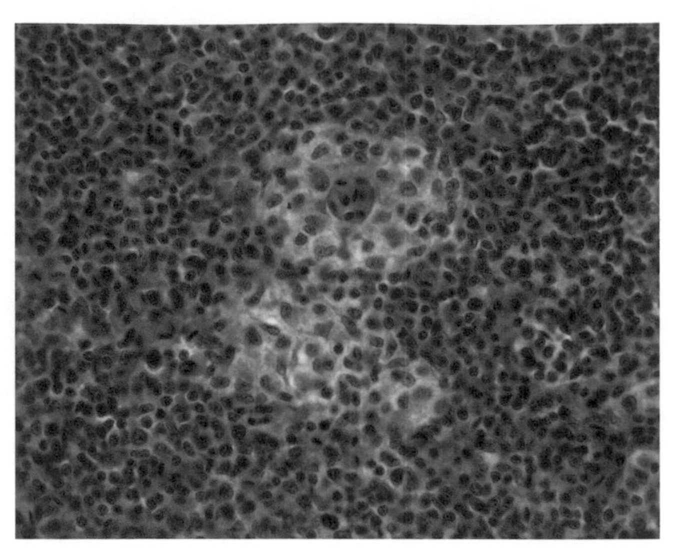

FIGURE I24d Photomicrograph of a section through a sheathed capillary in the spleen of a dogfish. The sheath is composed of reticular cells. 40×.

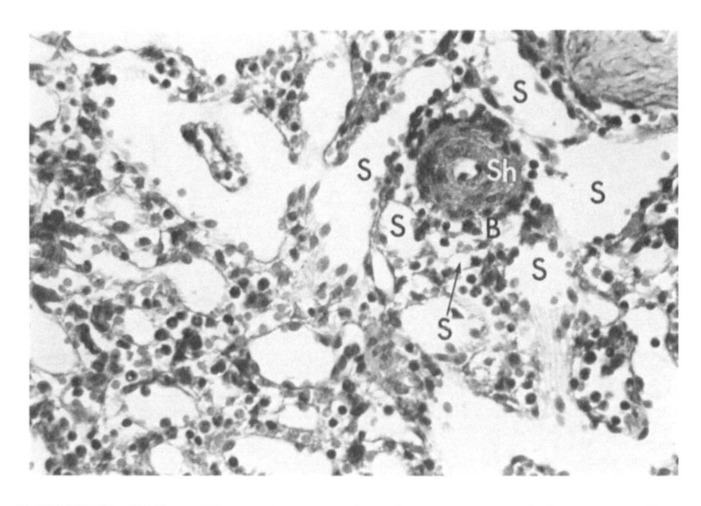

FIGURE I25a Photomicrograph of a section of human spleen showing a sheathed artery (Sh) enveloped by a thin layer of red pulp cords (B) and surrounded by sinusoids (S). Note the glancing section of the barrel-stave configuration of the sinusoidal wall at the lower right.

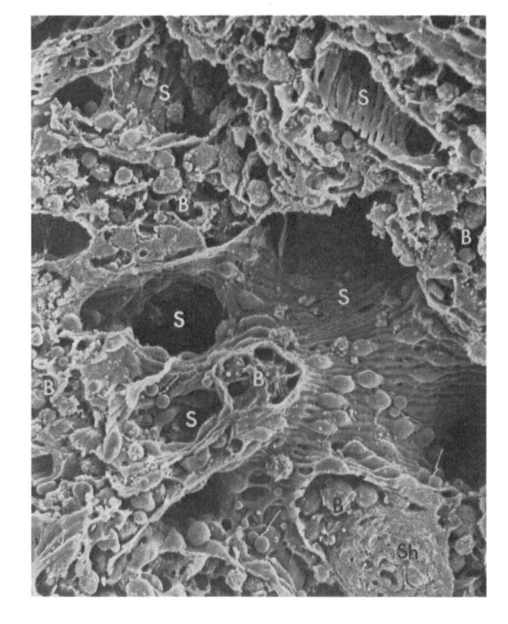

FIGURE I25b Scanning electron micrograph of the red pulp of human spleen. The red pulp cords (cords of Billroth) (B) are fractured. Various configurations of the sinusoids (S) are opened so that the cells of their walls are seen. The arrows indicate erythrocytes squeezing between the cells of the sinusoidal wall. A small sheathed artery (Sh) is seen at the lower right. 700×.

The structure of a hemal node is similar to that of a lymph node except that there are few afferent lymphatics (Fig. I29). It contains masses of lymphoid tissue that is permeated by lymphatic sinuses. The node is highly vascular and is riddled with capillaries that have thin basement membranes. Erythrocytes leak from postcapillary venules and dilated capillaries into the lymphoid tissue where they are phagocytosed by macrophages.

Scanning electron micrographs of a hemal node show the familiar picture of the reticular stroma (Fig. I30). In a section of hemal node, note the dense CAPSULE of fibrous connective tissue loosely connected with the surrounding tissue (Figs. I31a and I31b). Lymphatic sinuses permeating the lymphoid parenchyma have an incomplete

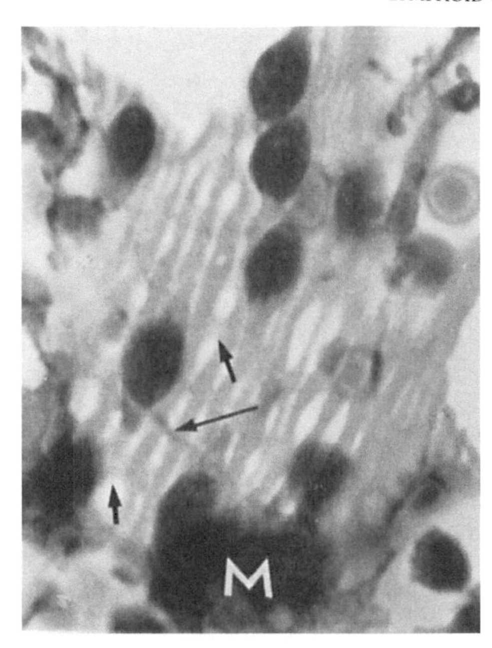

FIGURE I25c Photomicrograph of a glancing section showing the barrel-stave nature of the endothelial cells of the sinusoidal wall in the red pulp of a human spleen. The long arrow shows the reinforcing "barrel hoop" of reticular fibers. The short arrows indicate bridges between the cells of the sinusoidal wall. *M*, macrophage.

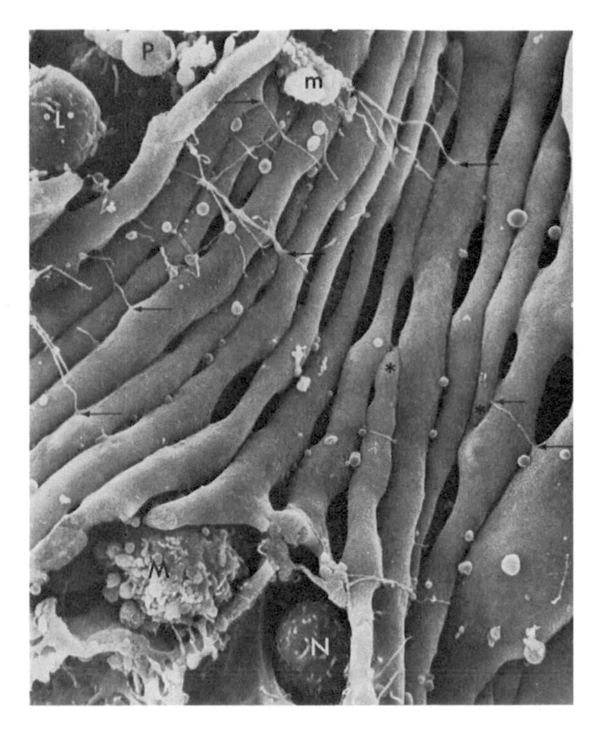

FIGURE I25d Scanning electron micrograph of the sinusoidal wall of a human spleen. *L*, lymphocyte; *M*, macrophage; *m*, macrophage process; *N*, neutrophil; *P*, platelet; the *arrows* indicate thread-like fibers of unknown function. *, an endothelial cell has apparently dived below the plane of the other cells to re-emerge a bit further along. 3700×.

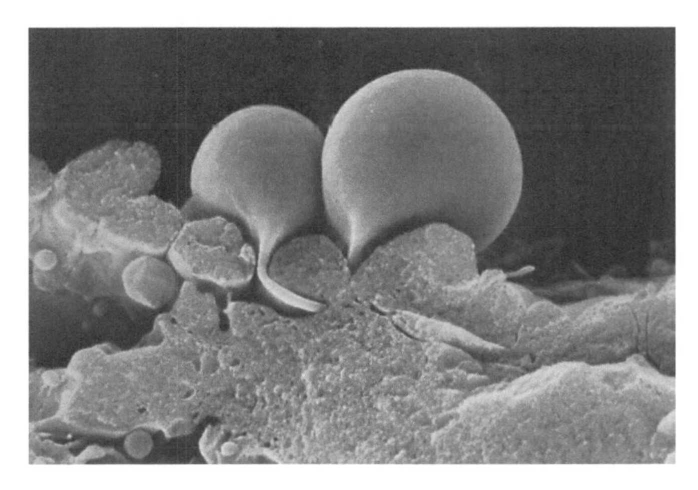

FIGURE I25e Scanning electron micrograph of erythrocytes passing between the lattice of cells forming the sinusoidal wall of the red pulp of the human spleen. 8500×.

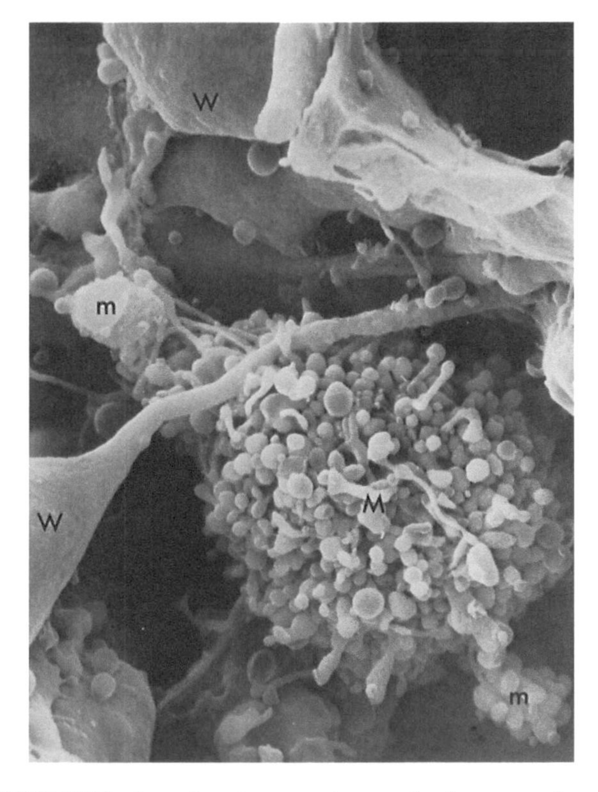

FIGURE I25f Scanning electron micrograph of a macrophage (M) extending its processes (m). It is likely wandering through a space between the wall cells (W) of the sinusoid. 7500×.

endothelial lining. There are more numerous blood vessels than in a lymph node. In places, dilated blood capillaries are separated from lymph sinuses only by the basement membrane of the capillary (Fig. I32). Transmission electron micrographs show erythrocytes crossing this thin barrier into the lymph sinuses (Fig. I33). Macrophages in the sinuses phagocytose many erythrocytes as well as particulate matter (Fig. I34).

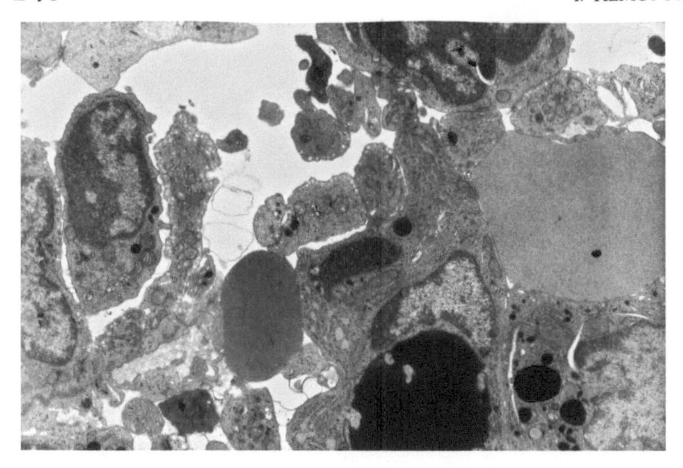

FIGURE I26 Transmission electron micrograph of a sinusoid in the spleen of a guinea pig. The lumen of a sinusoid occupies the lower right of this micrograph. It is surrounded by a leaky wall of endothelial or littoral cells. Although sinusoids contain flowing blood, their littoral cells differ from ordinary endothelial cells in that they are thicker and may even be columnar. The littoral cell at the extreme right is sectioned through its nucleus; its cytoplasm is rich in mitochondria and ribosomes. A few Golgi complexes may be seen. Pits and small vesicles, reminiscent of those seen in endothelial cells, are seen in littoral cells. The deep black bobs in these cells are lysosomes. An erythrocyte, slightly to the right of center, is seen squeezing between littoral cells. At the left border of the micrograph is an erythrocyte that has left the sinusoid entirely. Reticuloendothelial cells inspect erythrocytes and destroy worn or damaged cells (top, slightly left of center). 17,000×.

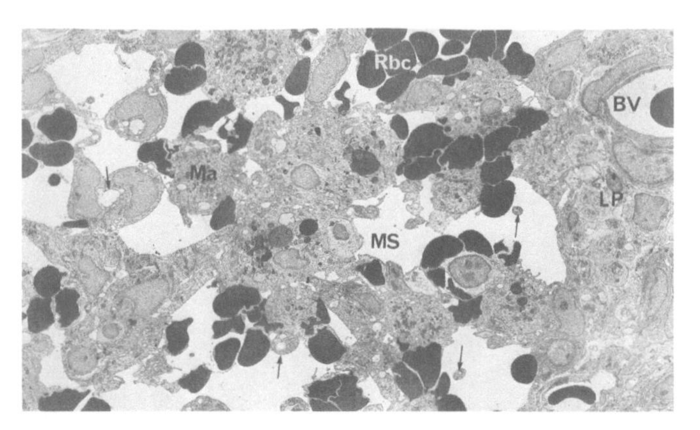

FIGURE I27 Photomicrograph of a glancing section showing the barrel-stave nature of the endothelial cells of the sinusoidal wall in the red pulp of a human spleen. The long arrows shows the reinforcing "barrel hoop" of reticular fibers. The short arrows indicate bridges between the cells of the sinusoidal wall. *Ma*, macrophage; *Rbc*, red blood cells; *BV*, blood vessel; *LP*, lymphoid cells.

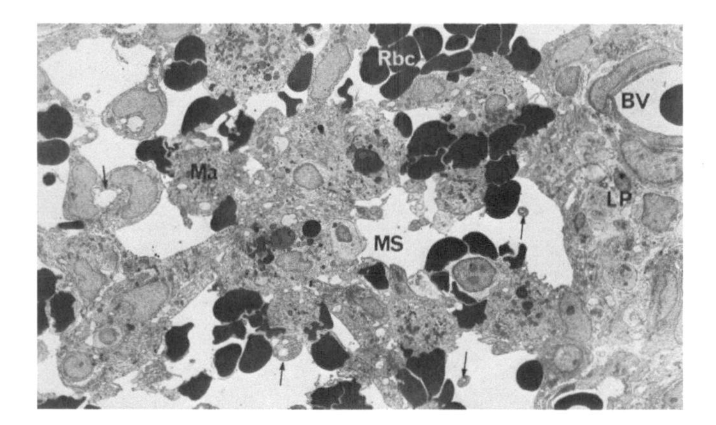

FIGURE I28a Transmission electron micrograph of a section of a hemolymph node of a rat. Macrophages (*Ma*) clinging to the surface of trabeculae (*arrows*) trap erythrocytes (*Rbc*) with their micropseudopods to form rosettes. *BV*, blood vessel; *LP*, lymphoid parenchyma. 2400×.

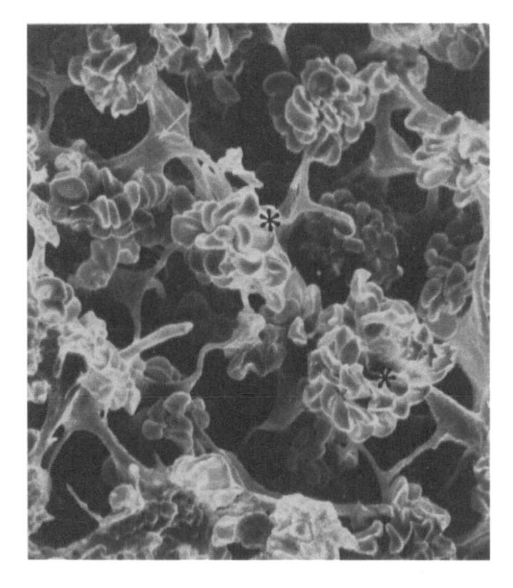

FIGURE I28b Scanning electron micrograph of rosettes of erythrocytes (*) in the medullary sinus of a hemolymph node of a rat. These grape-like formations are attached to trabeculae (*arrow*). 1300×.

On superficial examination a section of hemal node may be confused with spleen. The correct diagnosis is based on the presence of the eccentric arterioles in the spleen and their absence from the hemal node.

Thymus

The thymus is found in all vertebrates except cyclostomes and arises from dorsal evaginations of the third and fourth visceral or pharyngeal pouches (Fig. I35a). It is located in the thorax or pharyngeal region of vertebrates (Fig. I35b). The thymus receives stem cells from the bone marrow and causes them to differentiate into T cells. It involutes during youth and puberty. It retains its lumina in the lower vertebrates but the epithelial cells of higher

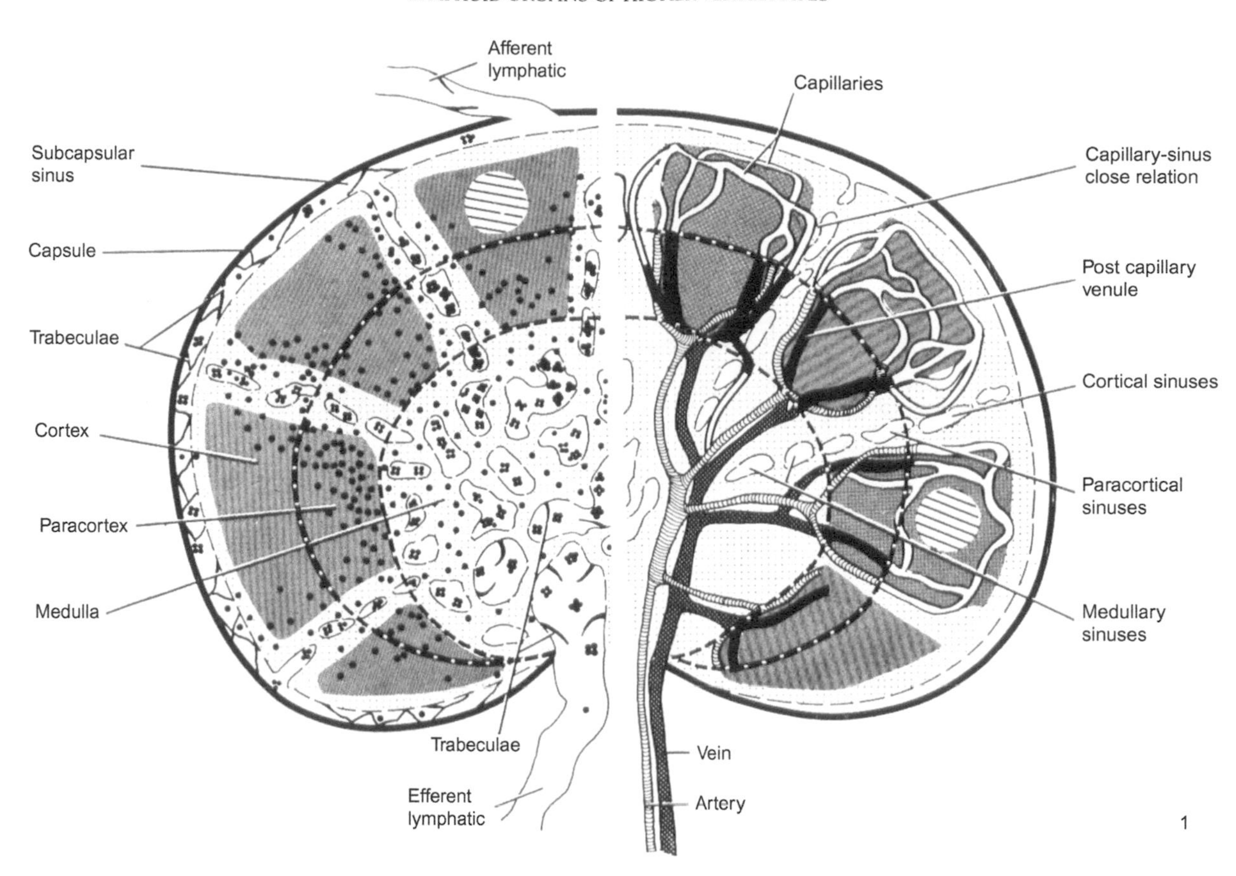

FIGURE I29 Diagram of a section of a hemolymph node of a rat. On the left side the distribution of red cells is indicated; on the right the vascularization is shown.

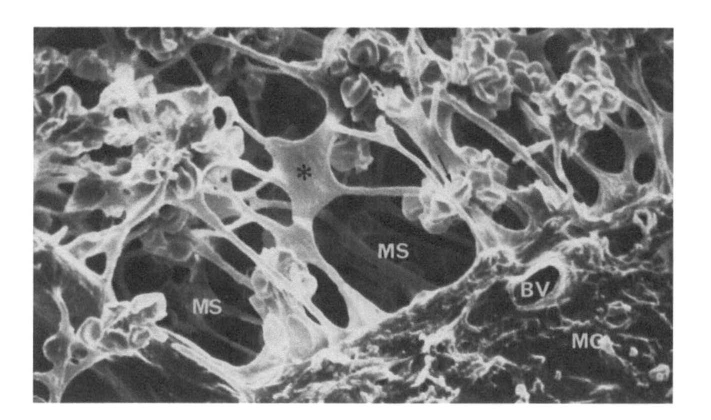

FIGURE I30 Scanning electron micrograph of a medullary sinus (*MS*) in the hemolymph node of the rat. The meshwork of trabeculae (*) is attached to the lymphoid parenchyma of the medullary cord (*MC*) *BV*, blood vessel. 1900×.

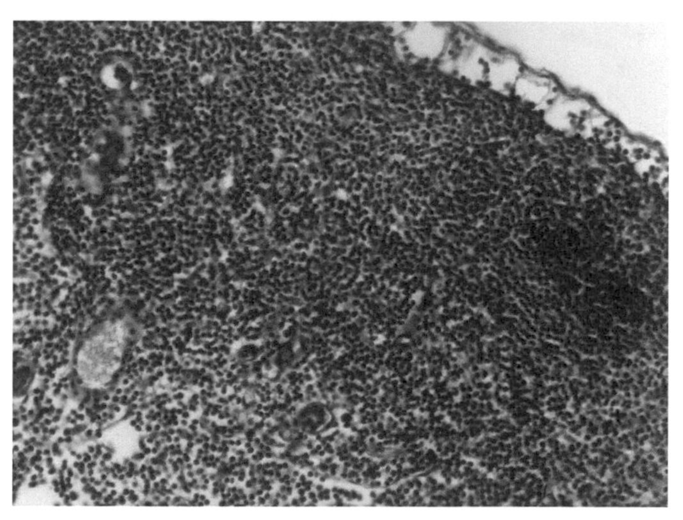

FIGURE I31a The capsule and subcapsular sinus are at the upper right. The cortex is at the upper right, the medulla at the lower left; the cortex is more densely packed with blood cells than the medulla. Sinuses (with thin walls) and blood vessels are clustered at the lower left.

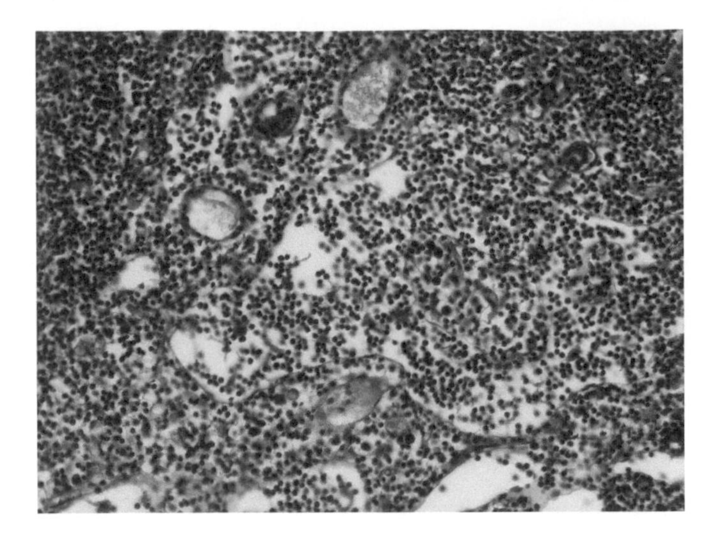

FIGURE I31b A group of sinuses and blood vessels at the junction of the cortex and medulla.

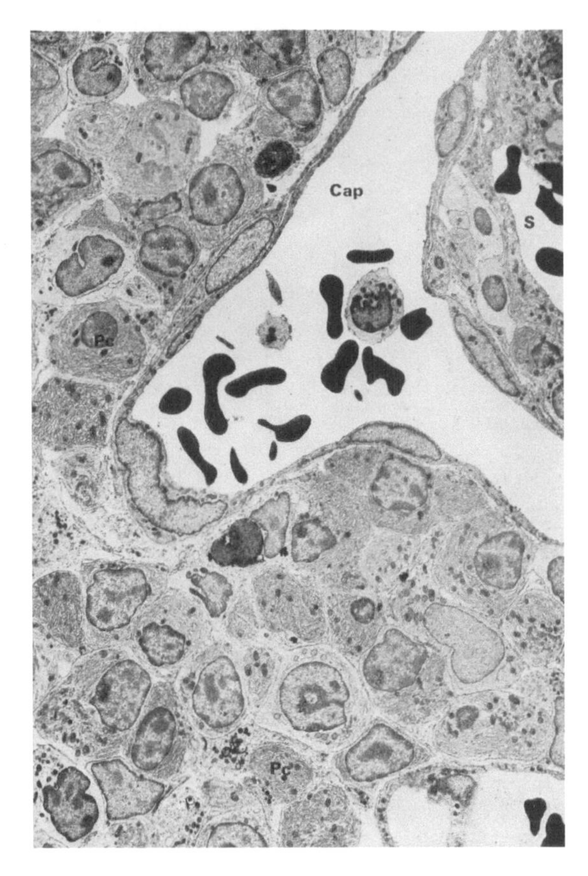

FIGURE I32 Transmission electron micrograph of a large capillary (*Cap*) with a thin wall in the paracortex of a hemolymph node of the rat. Many plasmocytes (*Pc*) are present. Note the proximity of the capillary with a sinus (*s*) at the upper right. 2600×.

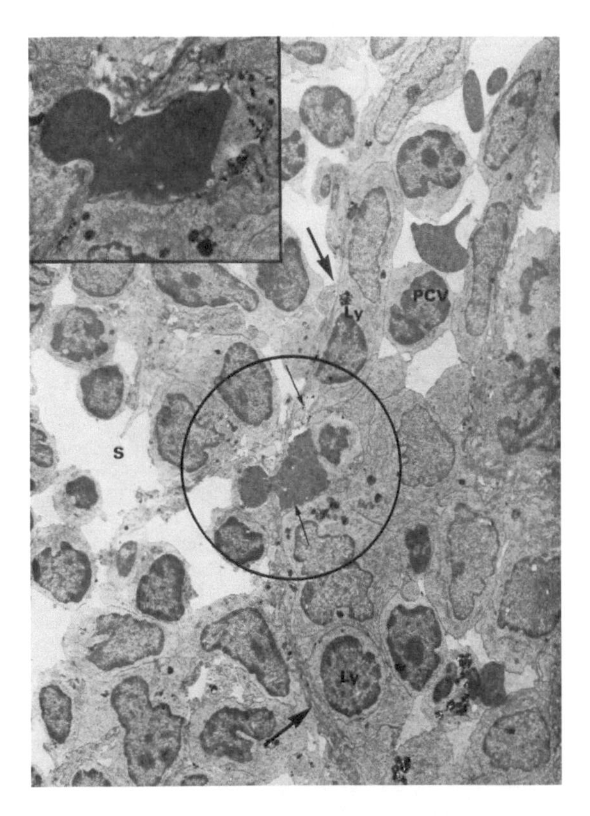

FIGURE I33 Transmission electron micrograph of the paracortex of the hemal node of a rat one minute after intravenous injection of colloidal carbon. The sinus (*S*) and postcapillary venule (*PCV*) are separated only by basement membranes (*large arrows*); the *small arrows* indicate the passage of colloidal carbon through this wall. The *large circle* encloses a red blood cell migrating from the venule to the sinus. *Ly*, migrating lymphocyte. 3500×.
The inset shows the detail of the passage of the red blood cell accompanied by colloidal carbon. 7300×.

forms form cords that invade the surrounding mesenchyme. The thymic primordia of higher vertebrates migrate to the thoracic region. Lymphocytes invade this epithelial rudiment and occupy the spaces between the epithelial cells. The reticular stroma, therefore, unlike that of all other hemopoietic organs, has an epithelial, rather than mesenchymal, origin and its cells are referred to as EPITHELIORETICULAR CELLS. Fig. I36a. Scanning (Fig. I36b) and transmission electron micrographs (Fig. I36c) of the thymic reticulum and show the differences resulting from its epithelial heritage: the absence of reticular fibers and the desmosomal attachments between the broad, leafy epithelioreticular cells.

In sections of mammalian or avian thymus the CAPSULE of fibrous connective tissue can be seen (Fig. I37a). SEPTA extend from the capsule dividing the organ into LOBES and LOBULES. The peripheral, deeply staining portion of each lobule constitutes the CORTEX and the lightly staining central portion is the MEDULLA (Fig. I37b). The cortex contains densely packed small lymphocytes, the medulla large lymphocytes and relatively more epithelioreticular cells. The medulla is a central core that sends a lateral projection into each lobule; it is covered throughout by the cortex

(Fig. I37c). Identify THYMIC CORPUSCLES (Hassall's corpuscles) in the medulla (Figs. I37d and I38). Thymic corpuscles vary considerably in structure but the center is usually acidophilic and contains whorls of a few hyaline central cells surrounded by a lamellated periphery that may be nucleated. What is the origin of thymic corpuscles?

In transmission electron micrographs of thymus note that the endothelial cells of the capillaries are relatively thick, lack fenestrae, and are supported by a thick basement lamina containing elaborate pericytes (Fig. I39). Around these layers, a perivascular sheath of connective tissue, a sheath of epithelioreticular cells, and its basement membrane, create a "blood–thymus barrier" to all but the smallest molecules. Note desmosomes between processes of the epithelioreticular cells. These cells have a lower electron density than lymphocytes and contain conspicuous tonofibrils. They contain vesicles, ribosomes, and Golgi complexes and secrete the hormone THYMOSIN that stimulates the transformation of stem lymphocytes into T lymphocytes.

The thymus reaches its maximum size in youth and begins to INVOLUTE during adolescence (Fig. I40a). Comparing thymi of young and old mammals we note the overall decrease in size with age and the replacement of masses of thymocytes by adipose tissue age (Fig. I40c). Thymic corpuscles persist, even in old age (Fig. I40b).

A BLOOD–THYMUS BARRIER is formed that seals the small blood vessels of the thymus from outside influences. It consists of the capillary endothelium, its basal lamina, the perivascular connective tissue, as well as the epithelioreticular cells and their basal lamina (Figs. I41a and I41b).

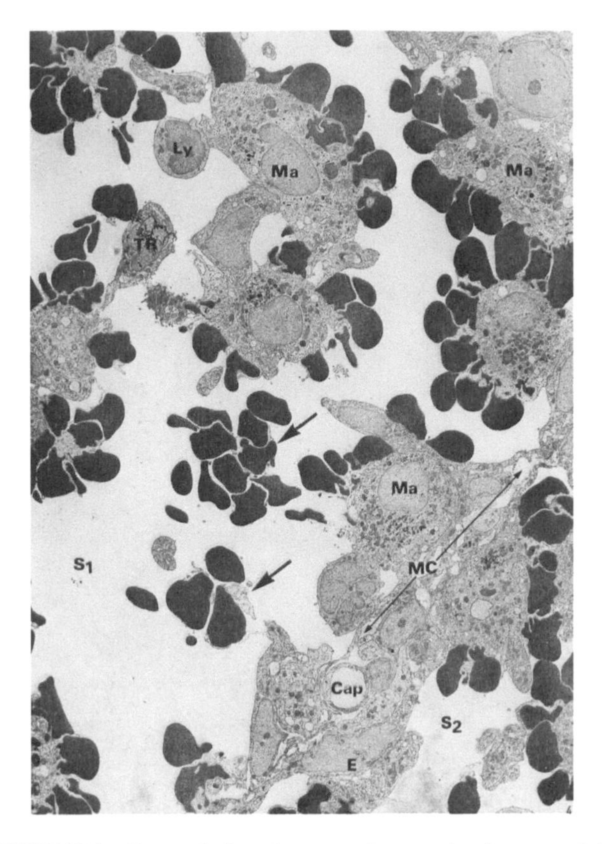

FIGURE I34 Transmission electron micrograph of two medullary sinuses (S_1 and S_2) separated by thin medullary cord (*MC*) containing large macrophages (*Ma*) to which many red blood cells are attached. Many phagosomes in the cytoplasm of the macrophages represent the degradation products of the red cells. Although groups of red cells in *S1* appear to be free, they are partly surrounded by pseudopods of macrophages (*large arrows*). *Cap*, capillary in medullary cord; *E*, lymphatic endothelial cell; *L*, lymphocyte; *T*, trabecula. 2400×.

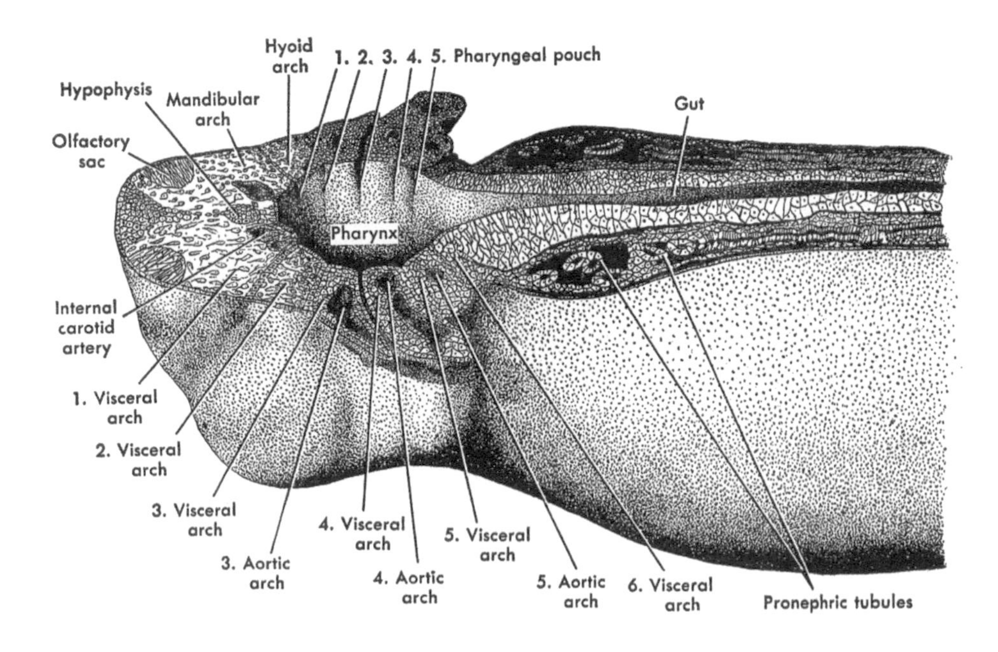

FIGURE I35a Drawing of the anterior half of a frog larva. The thymus of vertebrates develops from dorsal evaginations of the third and fourth pharyngeal pouches.

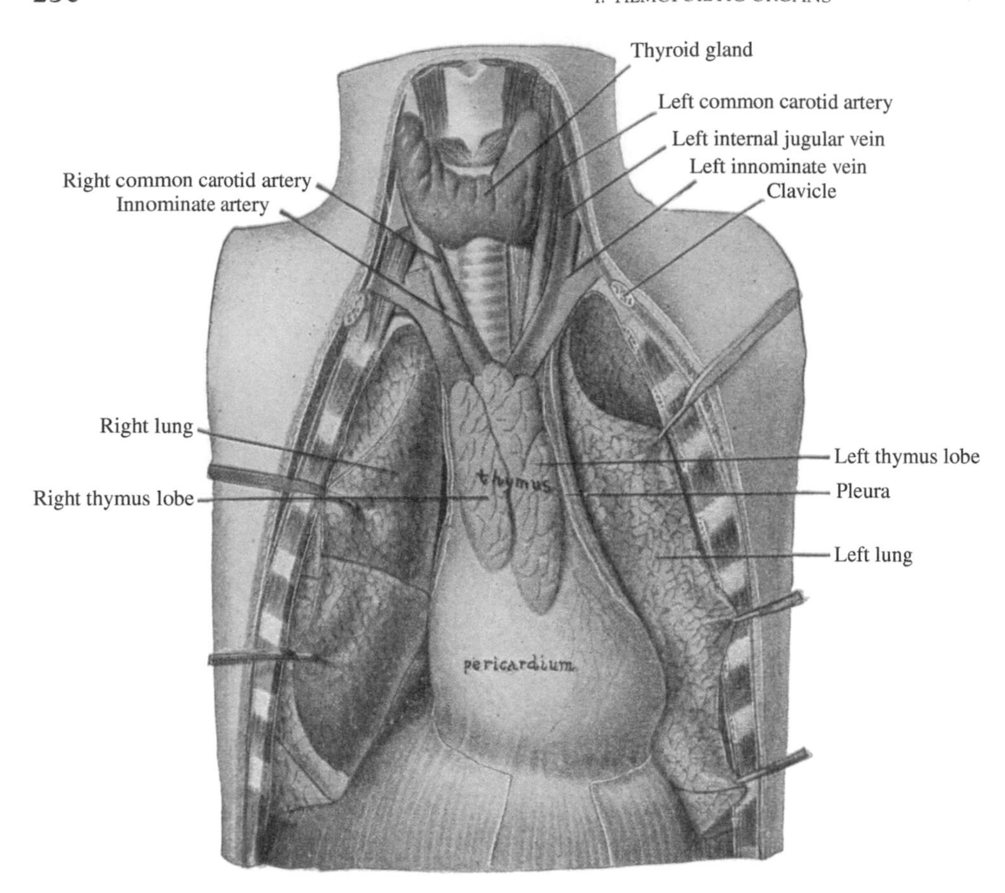

FIGURE I35b Ventral view of the thymus in a child.

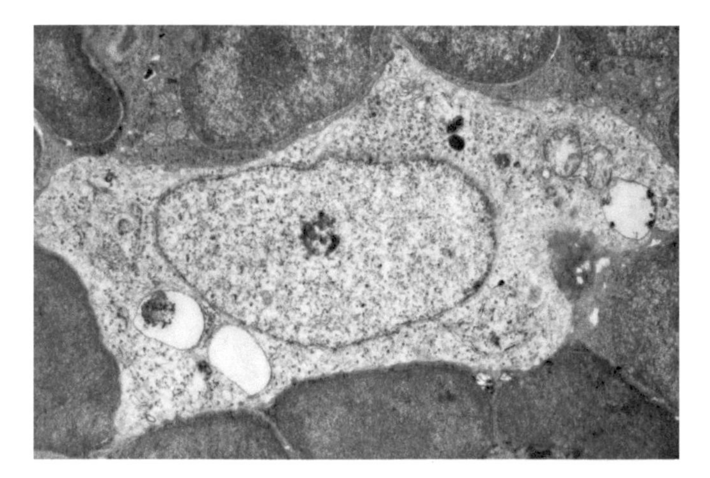

FIGURE I36a Transmission electron micrograph of a section of the thymus of a newborn rat. At the center is an epithelioreticular cell surrounded by several lymphocytes. 18,400×.

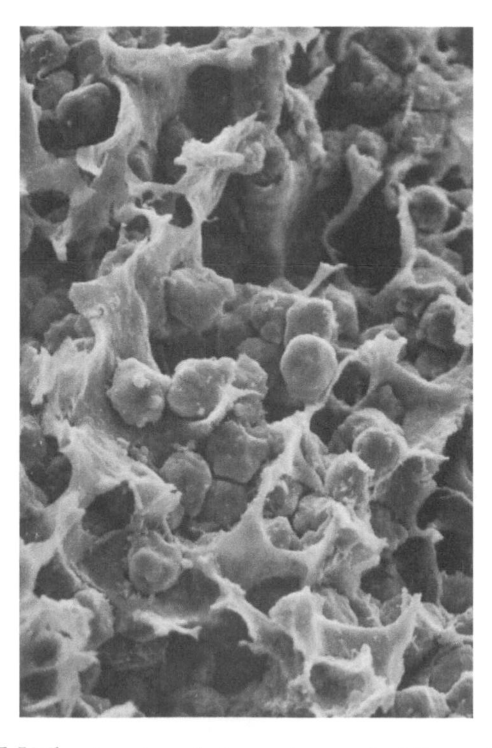

FIGURE I36b Scanning electron micrograph of the cortex of the thymus of a rat in which many cortical lymphocytes have been washed away to reveal the epithelioreticular cells: the cells resemble the individual leaves of a head of lettuce. 5700×.

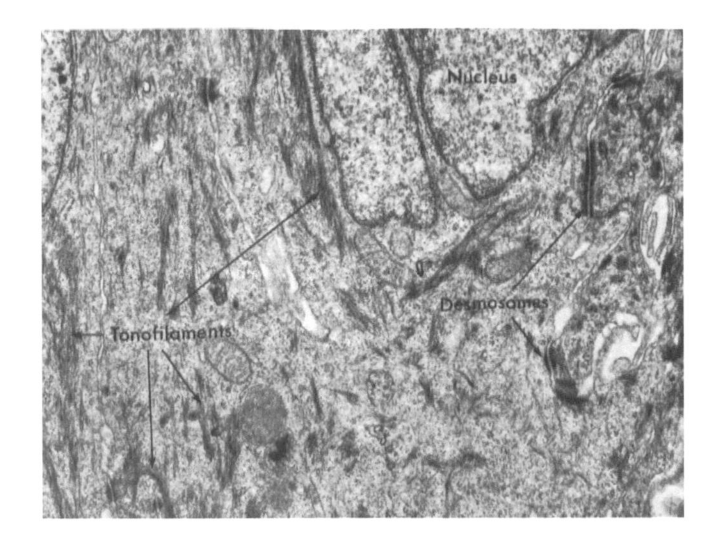

FIGURE I36c Transmission electron micrograph of s section of the thymus of a rat. Processes from several epithelioreticular cells are joined by desmosomes. These cells are strengthened by bundles of tonofilaments.

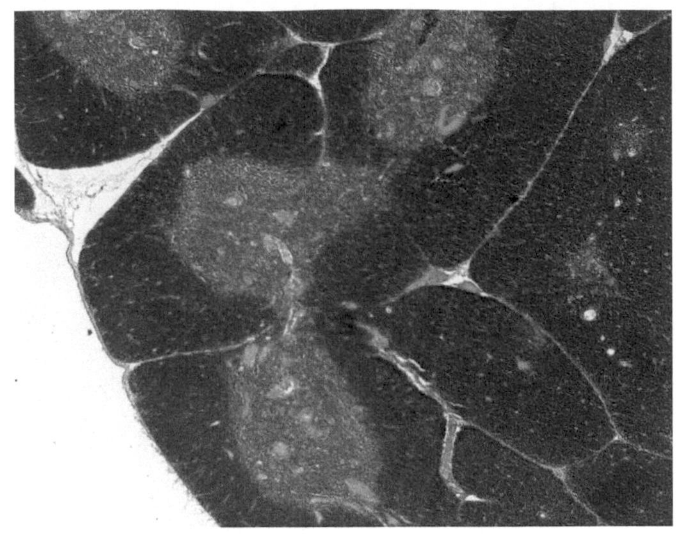

FIGURES I37a Photomicrograph of a section of the thymus of a young mammal. The thymus consists of lobes which contain several lobules. Each lobule consists of a densely staining peripheral cortex and a lighter staining central region, the medulla. Several thymic corpuscles (Hassal's corpuscles) are scattered throughout the medullas. 5×.

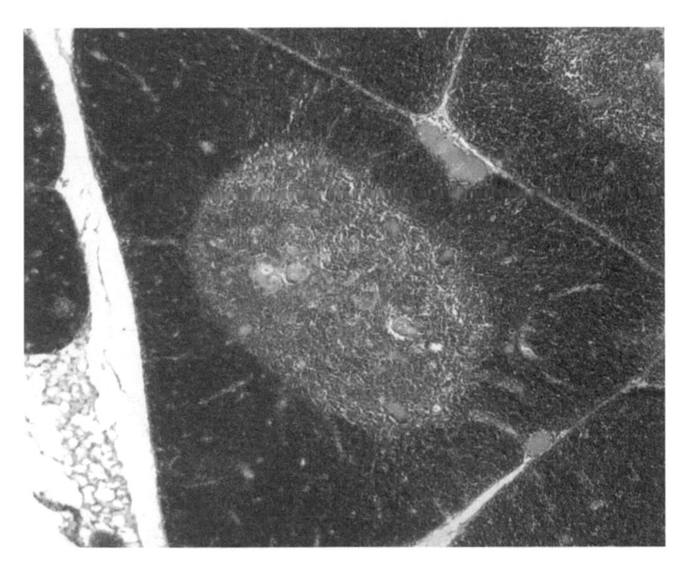

FIGURES I37b Photomicrograph at higer magnifiction of the section in I37a - thymus of a young mammal. This lobe clearly shows a densely staining peripheral cortex and a lighter staining central region, the medulla. Several thymic corpuscles (Hassal's corpuscles) are scattered throughout the medullas. 10×.

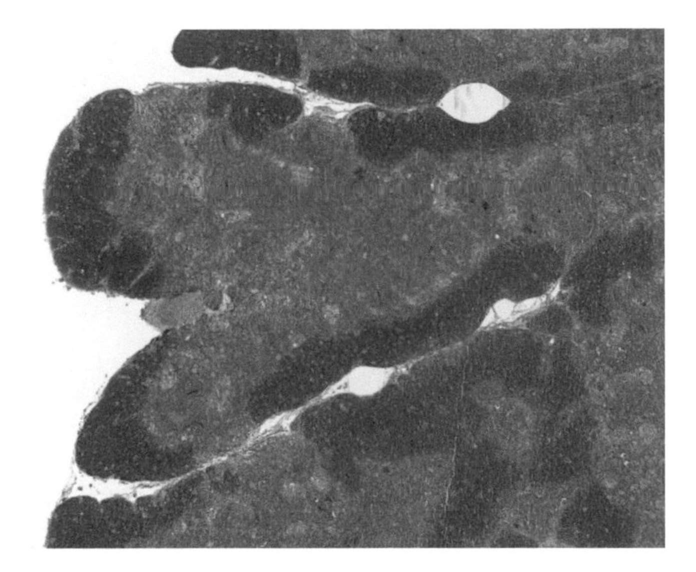

FIGURE I37c Photomicrograph of a section of the thymus of a young mammal cut to show that the medulla is continuous from lobule to lobule. 5×.

Cloacal Bursa (Bursa of Fabricius)

The bursa is a dorsal diverticulum of the cloaca of young birds that is richly infiltrated with lymphoid tissue (Fig. I42a). Its development as an outpouching of the gut wall, its functions in the development of immunity, and the fact that it undergoes involution with age have earned it the sobriquet of CLOACAL THYMUS. Unlike the thymus, however, it retains its connection to the cloaca by the CLOACAL CANAL. But, like the thymus, its reticular stroma is not derived from connective tissue but from epithelium and consists of a supporting mesh of epithelioreticular cells joined to each other by desmosomes; there are eno collagenous fibers in the lymphoid tissue (Fig. I42b).

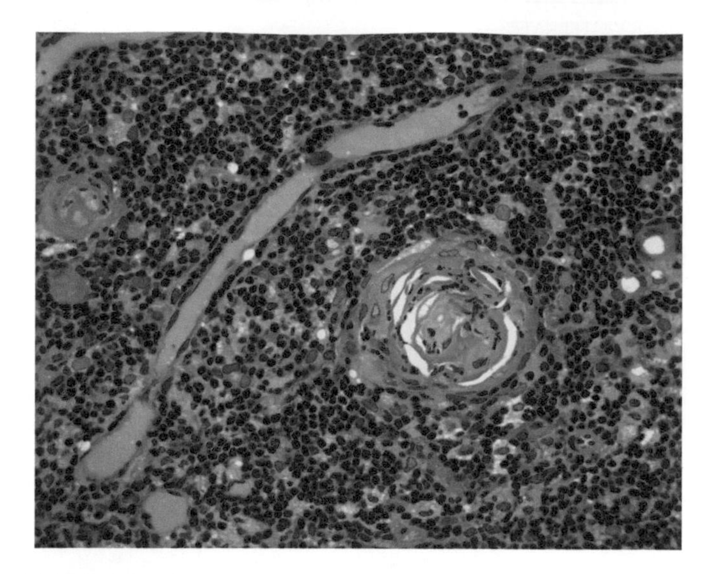

FIGURE I37d Photomicrograph of a semithin section of the thymus of a squirrel showing a thymic (Hassal's) corpuscle at the right. A capillary crosses the field from the lower left to the upper right. 40×.

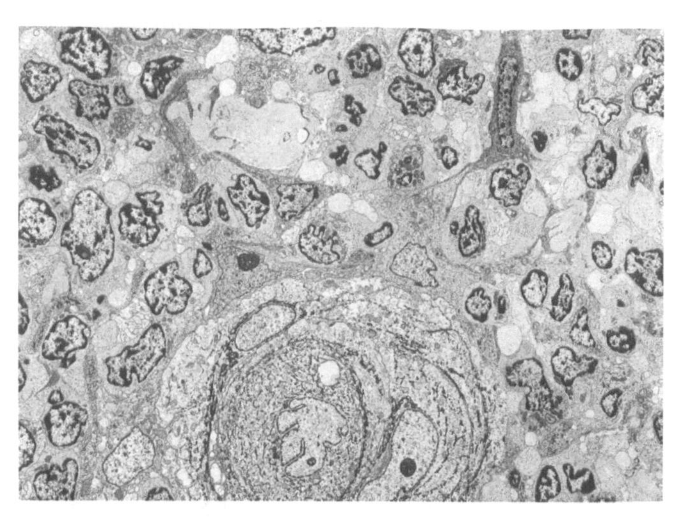

FIGURE I38 Transmission electron micrograph of a section of the human thymus. There is a thymic corpuscle at the lower edge.

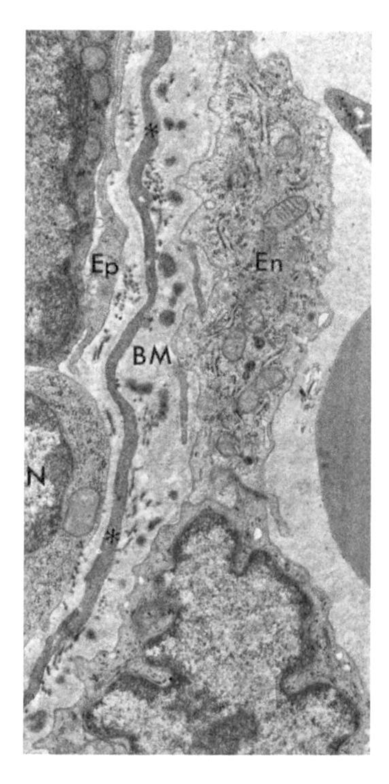

FIGURE I39 Transmission electron micrograph of a section of the thymus of a newborn rat illustrating the blood/thymus barrier surrounding a capillary at the right. The endothelium of the capillary (*En*) is thick and lacks fenestrae. A stout basement membrane (*BM*) supports the endothelial cell and encloses pericytes (*). Portions of an epitheloreticular cell (*Ep*) and a lymphocyte (*N*) are seen at the left. 9000×.

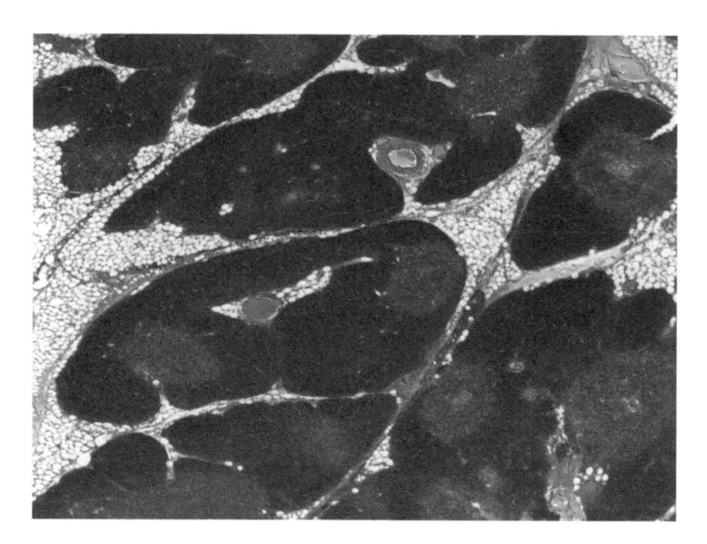

FIGURE I40a Photomicrograph of a section of involuting mammalian thymus. The lobules atrophy and the septa widen. The connective tissue between the lobules is invaded by fat. Cortical lymphocytes and epithelioreticular cells are depleted to be replaced by fibrous fatty tissue. 2.5×.

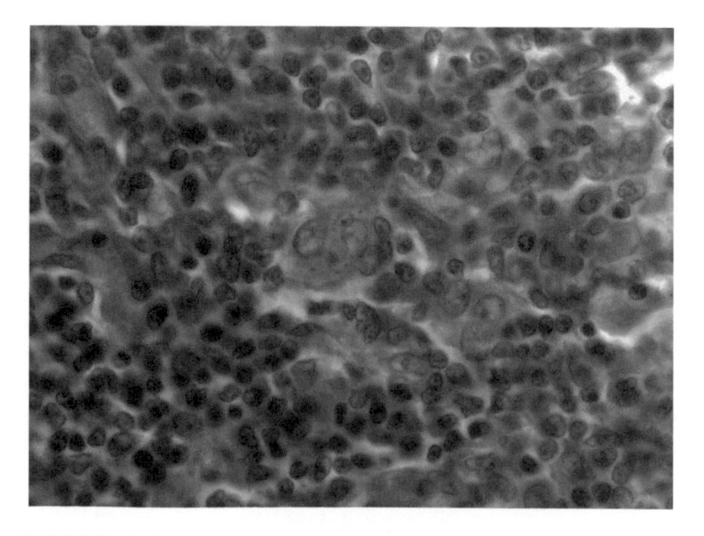

FIGURE I40b Photomicrograph of the section of the involuting mammalian thymus shown in Fig. I40a. Although there is desolation and involution taking place, the thymic corpuscles, although somewhat depleted, persist. 63×.

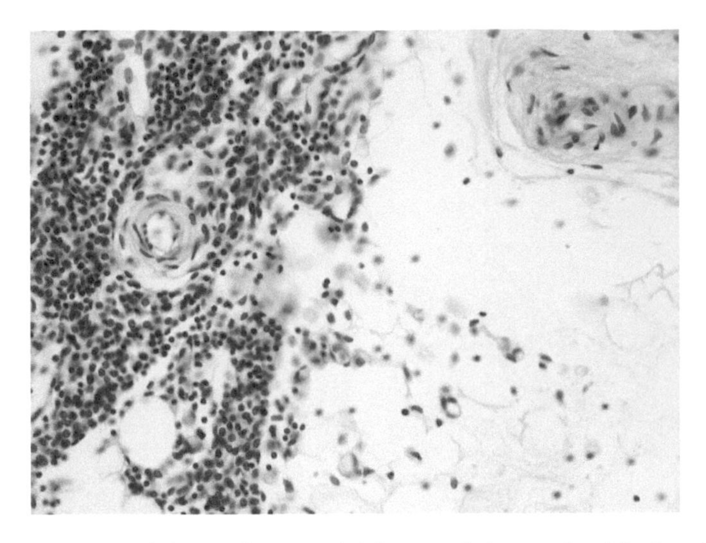

FIGURE I40c Photomicrograph of a section of thymus from an adult human. Only strands of the lymphoid tissue remain. Large amounts of fat and a thymic corpuscle (center left) persist. 20×.

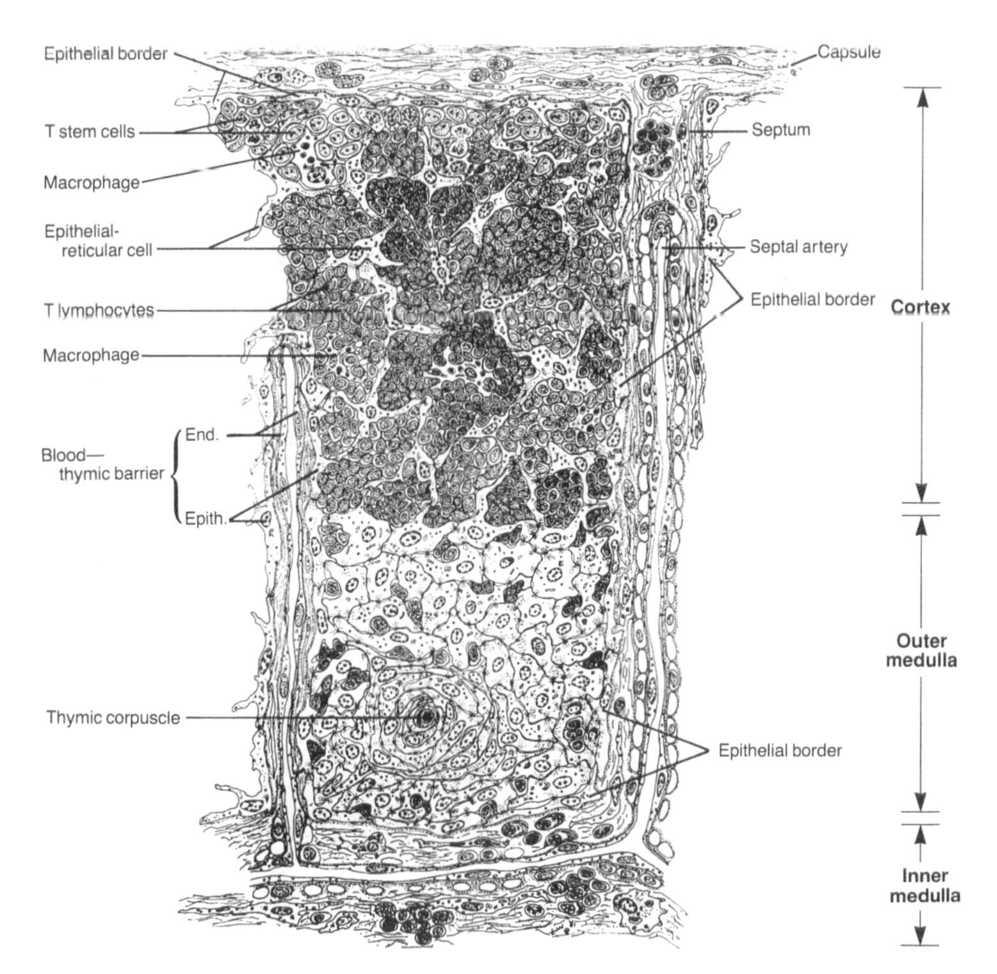

FIGURE I41a Schematic drawing of a portion of a thymic lobule. The cortex is heavily infiltrated with lymphocytes that stretch the epithelioreticullar cells and their desmosomal connections. The proportion of epithelioreticular cells is greater in the medulla than in the cortex. A thymic corpuscle is seen in the medulla. The capsule and trabeculae are rich in collagenous fibers and contain blood vessels, plasmocytes, granulocytes, and lymphocytes. A border of flattened epithelial cells surrounds the cortex and outer medulla.

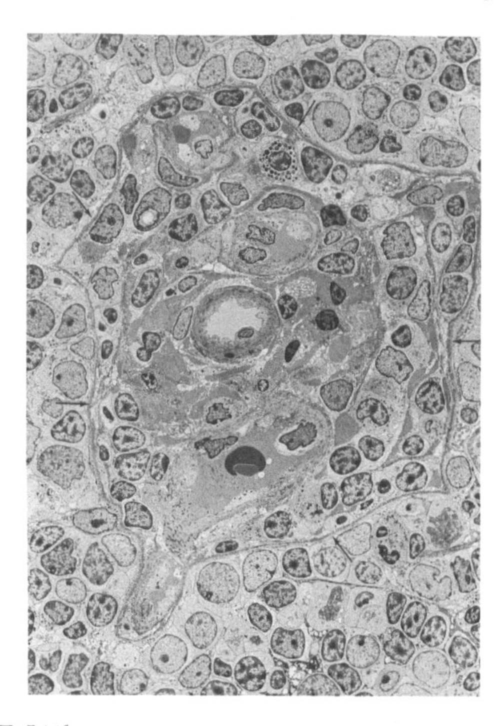

FIGURE I41b Photomicrograph of a section of human thymus near the border of the cortex and medulla showing the blood–thymus barrier. An arteriole, with several branches, is surrounded by a perivascular connective tissue space limited by epithelioreticular cells (*arrows*). The barrier consists of endothelium of the blood vessels in the area, endothelial basal lamina, perivascular connective tissue, epithelial basal lamina, and epithelium. 1850×.

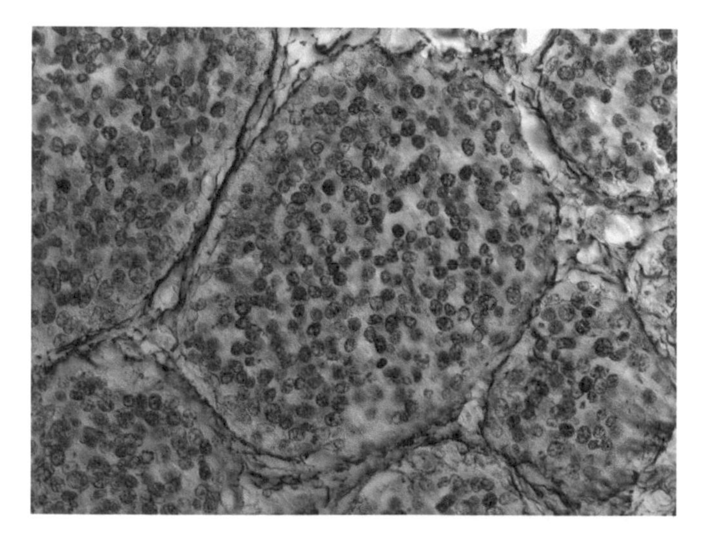

FIGURE I42b Photomicrograph of a section of the bursa of Fabricius of a chick stained with Wilder's silver reticular stain. Note that silver has been deposited on reticular fibers within the interlobular connective tissue but there is none in the individual lobules where the reticulum is formed from epithelial cells. 40×.

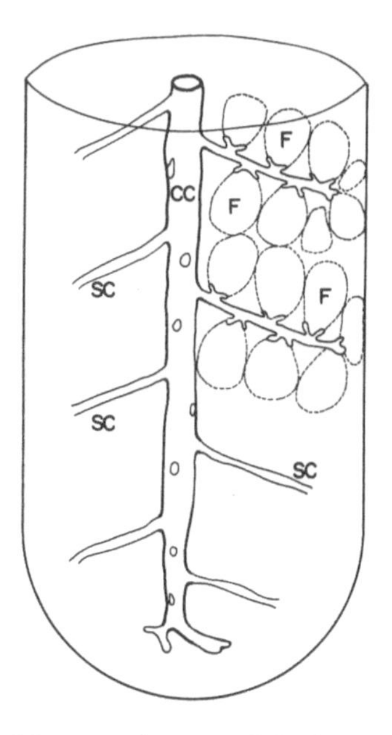

FIGURE I42a Schematic drawing of the bursa of a starling. The central canal (*CC*) originates from the cloaca and branches into secondary canals (*SC*). The bursal follicles (*F*) are associated with the secondary canals.

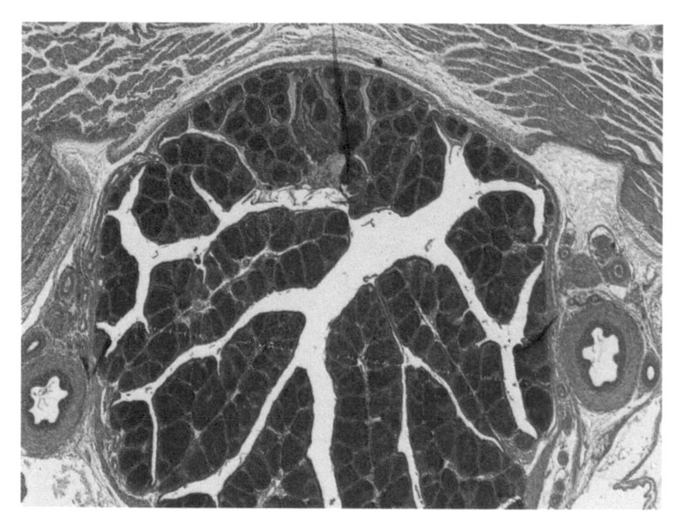

FIGURES I42c Photomicrograph of a section of the bursa of Fabricius of a chick. The bursa is a diverticulum of the cloaca of the bird and its lumen is continuous with that of the cloaca. The outer layers of the gut wall, although greatly attenuated, are visible. The tunica mucosa is thrown up into large folds or PLICAE covered with a simple columnar or pseudostratified columnar epithelium. The loose vascular connective tissue of the lamina propria mucosae and the tela submucosa is packed with spherical lymphoid follicles. 2.5×.

Lymphocytes arise from stem cells in the bone marrow and divide to form uncommitted lymphocyte precursors. In birds, some of these are carried in the blood to the bursa of Fabricius where they proliferate and differentiate into antibody producing cells: bursa-dependent lymphocytes or B lymphocytes. Mammals have no bursa and B lymphocytes probably mature in the bone marrow as well as other sites.

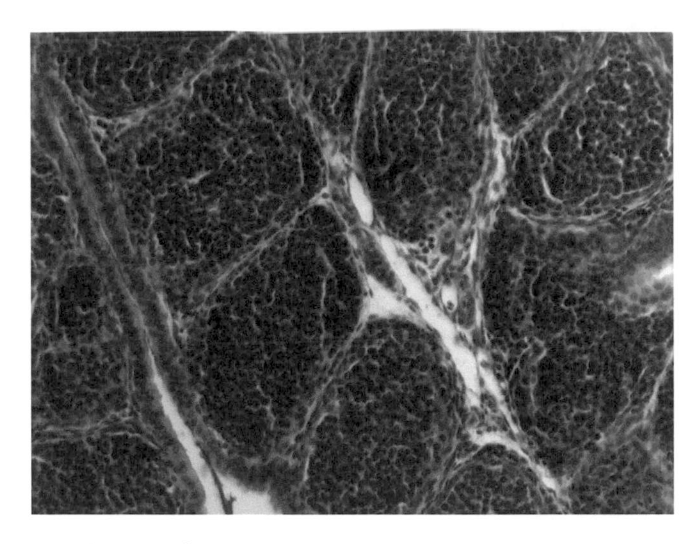

FIGURES I42d A higher power micrograph of the bursa of Fabricius of a chick. The loose vascular connective tissue of the lamina propria mucosae and the tela submucosa is seen to be packed with spherical lymphoid follicles separated into two roughly parallel rows by TRABECULAE of denser connective tissue and into individual compartments by SEPTA. 20×.

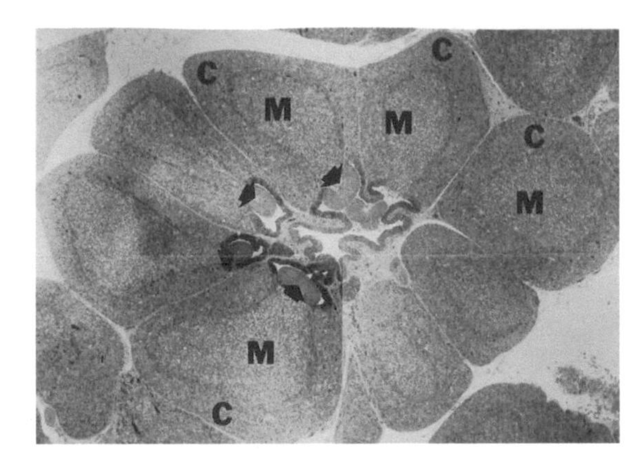

FIGURE I42e Bursal follicles line the secondary canal in this photomicrograph of a cross section of the bursa of a starling. The cortex (*C*) and medulla (*M*) of the follicles are clearly defined. 75×.

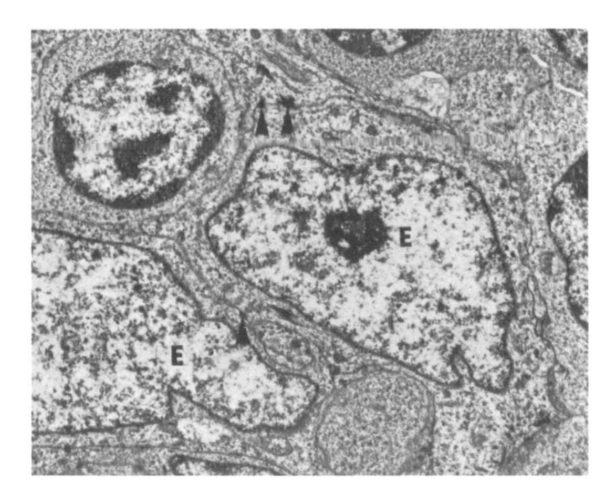

FIGURE I42f Transmission electron micrograph of a section of the bursa of a 1-day-old chick showing reticuloepithelial cells (*E*) joined to other cells by desmosomes (arrows). 7100×.

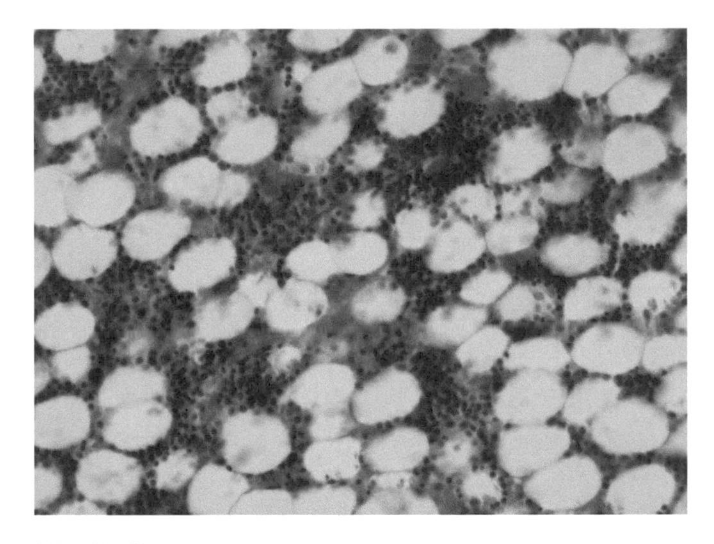

FIGURE I43 Photomicrograph of a section of mammalian red bone marrow. Large numbers of fat cells intrude on the actively proliferating myeloid cells. 20×.

In photomicrographs of sections of the bursa and the lumen persists and the layers of the gut wall can be recognized (Fig. I42c). The tunica mucosa is thrown up into large folds or PLICAE covered with a simple columnar or pseudostratified columnar epithelium. Loose vascular connective tissue of the lamina propria mucosae and the tela submucosa is packed with spherical lymphoid follicles separated into two roughly parallel rows by TRABECULAE of denser connective tissue and into individual compartments by SEPTA (Fig. I42d). The follicles consist of an epithelioreticular stroma infiltrated with lymphocytes and may show a CORTEX and MEDULLA separated by a basement lamina continuous with that of the surface epithelium (Fig. I42e). Note that the, like the epithelioreticular cells of the thymus, epithelioreticular cells of the bursa are joined by desmosomes i (Fig. I42f). The tela submucosa is surrounded by a thin layer of smooth muscle (TUNICA MUSCULARIS) in which circular fibers predominate. The TUNICA ADVENTITIA forms a capsule of loose connective tissue and is continuous with the sheaths of other organs in the region.

MYELOID ORGANS: THE BONE MARROW

Hemopoietic RED BONE MARROW occurs in all bones of young mammals but it is later replaced in some places by fatty YELLOW MARROW that is inactive in hemopoiesis but can regain this ability. Photomicrographs of red marrow demonstrate its similarity to lymphoid tissue (Fig. I43). There is a reticular STROMA with reticular cells stretched out on it. Ordinary blood vessels connect with large, tortuous SINUSOIDS lined with endothelium and surrounded by flattened reticular cells. MYELOID ELEMENTS, consisting of maturing erythrocytes and granulocytes, are found in the meshes of the stroma. Megakaryocytes and blood platelets are also developed here. Specific types of blood cells develop in nests. Erythrocyte nests near the sinus wall always contain a macrophage. Granulocytes develop in nests further from the wall. Megakaryocytes occur near the sinus wall; these enormous cells should present no difficulty in identification. Single fat cells are scattered within the stroma of the red marrow; in yellow marrow they are so concentrated that they can exclude nearly all other elements.

CHAPTER

J

Integument

The integumentary system includes the skin and the structures derived from it: hair, feathers, hooves, nails, scales, glands, etc.

SKIN

To study the basic structure of vertebrate skin without the distraction of specialized structures, examine images of the plantar skin of a mammal (Figs. J1, J2a–J2c, J3a–J3e). The EPIDERMIS, the ectodermal layer of stratified squamous epithelium, and the DERMIS, the underlying mesodermal layer of connective tissue. The skin is bound to the fascia of muscle or the periosteum of bone by the HYPODERMIS, a layer of areolar connective tissue.

Five layers can be recognized in the epidermis of this thick skin. The superficial STRATUM CORNEUM is the widest layer and consists of flattened, desiccated, cornified cells whose cytoplasm is shiny and highly refractile because of its high content of pliable SOFT KERATIN. No nuclei or organelles are seen in these cells. The spirally coiled ducts of the sweat glands are often seen penetrating this layer. In some preparations, the deepest layer of the stratum corneum, the STRATUM LUCIDUM, appears as a thin refractile layer of cells undergoing keratinization. Below is the STRATUM GRANULOSUM, a double or triple layer of cells containing basophilic keratohyaline granules; the cells are flattened and angular with indistinct nuclei. The next layer is the STRATUM SPINOSUM. Its cells are cuboidal at the base and gradually become flattened as they near the stratum granulosum. The prickle cells of this layer are firmly joined to each other by INTERCELLULAR BRIDGES. Review the electron microscopic study of intercellular bridges in Chapter C on pages C-xx and C-xx. The deepest layer of the epidermis is the STRATUM GERMINATIVUM or STRATUM BASALE and consists of a single layer of cells, many of which are undergoing mitosis. The melanin pigment of the skin tends to be concentrated in the stratum germinativum and in the basal layers of the stratum spinosum but, if abundant, is also found in other layers. These layers are shown a sketch of human plantar skin (Fig. J4). There are no blood vessels in the epidermis. Most of the cells are KERATINOCYTES (Figs. J5a–J5c). Their abundant rough endoplasmic reticulum and free ribosomes are progressively overwhelmed by increasing amounts of tonofilaments, keratohyaline granules, and lamellar bodies as the cells become keratinized. Melanin is elaborated in stellate cells, the MELANOCYTES, that extend their long processes between the other epidermal cells and distribute their pigment to the keratinocytes (Fig. J5d). Note the characteristic granules in the stellate INTRAEPIDERMAL MACROPHAGES (Langerhans cells); these cells are found mainly in the stratum spinosum (Figs. J6a and J6b). TACTILE CELLS (Merkel cells) are mechanoreceptors associated with a nerve fiber (Figs. J7a and J7b).

The dermis is divisible into the STRATUM PAPILLARE (Fig. J1), a superficial mat of fibroelastic connective tissue underlying the epidermis, and the STRATUM RETICULARE, the deeper, dense, irregularly arranged fibroelastic connective tissue that is not sharply demarcated from the stratum papillare. DERMAL PAPILLAE project into the epithelium from the dermis and contain either special nerve endings (nervous papillae) or capillary loops (vascular papillae).

Compare thick skin from the plantar surface with thin skin from another region of the human body such as the eyelid (Fig. J8). The layers outlined previously are poorly defined in thin skin.

Study photomicrographs of sections of the skin from representatives of each class of vertebrates as well as skin of amphioxus and note the stratification of the epithelium, type of epithelium, cornification of the epithelium, and modifications of the dermis (Figs. J9, J10a–J10c, J11, J12, J13a and J13b, J14a–J14c, J15, J16a–J16f). These characteristics will differ depending on the habits of the animal and the functions of the skin in the particular region from which the section was taken.

An Atlas of Comparative Vertebrate Histology.
DOI: https://doi.org/10.1016/B978-0-12-410424-2.00010-X

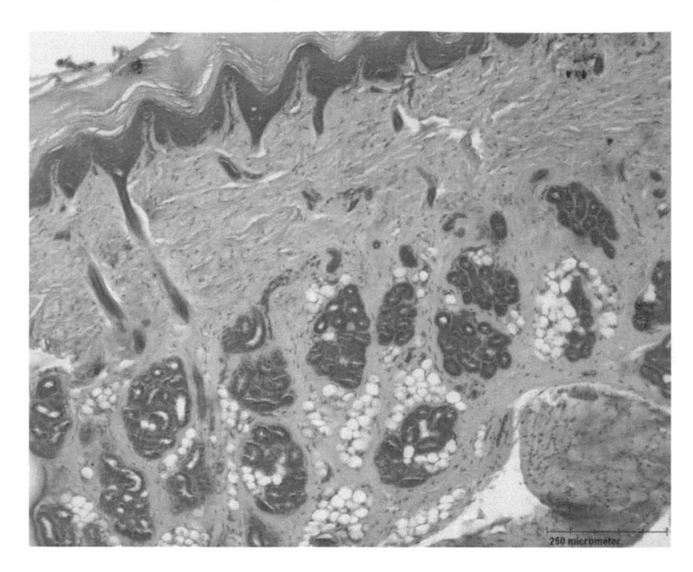

FIGURE J1 The basic structure of vertebrate skin is clearly demonstrated in this low-power micrograph of the footpad of a squirrel. The EPIDERMIS is the epithelial layer that rests on connective tissue of the DERMIS. The epidermis is deeply invaginated by the vascular connective tissue projections from the dermis beneath. This irregular border greatly increases the surface area between the two layers, optimizing the flow of nutrients and wastes between the dermis and the nonvascular epidermis. The dermis is firmly attached to bone or muscle beneath by areolar connective tissue of the hypodermis. The hypodermis contains fat and the coiled secretory regions of the simple tubular sweat glands. Sections of ducts can be seen extending from the sweat glands to the surface. 10 ×.

In whole mounts of vertebrate skin, note the presence of CHROMATOPHORES in the dermis (Figs. J17a–J17e). Large variations occur in the pigmentation of human skin (Figs. J17f and J17g). Introduced pigments, in the form of tattoos, are lodged in the dermis (Fig. J17h). The body of many vertebrates may have a protective covering, derived from the skin, scales, hair, or feathers.

Scales

1. *Epidermal scales.* The scales of reptiles, birds, or mammals are cornified derivatives of the stratum germinativum (Figs. J18, J19a–J19f and Fig. J19bc montage). These scales are usually shed and replaced. The epidermal scales of the pangolin (spiny anteater) are especially well developed. Beaks and bills of turtles, tortoises, and birds are modified epidermal scales.
2. *Dermal scales* are derived from mesenchyme and are remnants of the dermal skeleton. They are found chiefly in fish but may be present in reptiles (e.g., bony plates of the turtle shell) and a few other forms (e.g., some toads). Dermal bones forming the roof of the skull are the only dermal skeletal structures of most mammals. Elasmobranch and teleost scales are shown in whole mounts in Figs. J21a and J21b.
   a. *Elasmobranch scales.* Unlike the scales of bony fishes, portions of the PLACOID SCALES of elasmobranchs may perforate the epidermis and become exposed on the surface (Figs. J20a–J20c). So striking is their resemblance to mammalian teeth that many of the same terms are used in describing them. The bulk of the scale consists of bone-like DENTINE surrounded by a PULP CAVITY. A layer of ODONTOBLASTS lines the cavity and sends fine processes outward through the dentine. Blood vessels enter through the BASAL PLATE. The dentine is covered by harder VITRODENTINE. Both the dentine and vitrodentine are of mesodermal origin. Although the scale may be formed under the influence of an ectodermal ENAMEL ORGAN, it has been shown that it is not covered by enamel (Figs. J20d and J20e).
   b. *Scales of bony fish.* Teleost scales are hardened dermal plates of acellular bone impregnated with collagenous fibers. The fibers may be arranged in alternating strata of parallel arrays, plywood fashion. The scales are held in place by fibrous dermal structures that form the SCALE POCKET (Figs. J22a and J22b). Usually the scales lie completely in the loose outer layer of the dermis, slanting upward and backward in a shingle-like arrangement. They differ from the scales of elasmobranchs in that they do not pierce the epidermis. The stratified squamous epithelium of the epidermis completely covers the dermis but is pushed out into overlapping folds by the scales below (Fig. J21b). The general epidermal covering is slimy and relatively soft, although keratinization of the outer layers does occur and is important in the osmotic regulation of the fish. The sliminess of the skin is due to the presence of many goblet cells that move upward with the other cells in the epidermal layers, eventually to discharge upon the surface. Sometimes, epidermal and dermal scales occur concurrently in the same animal (Fig. J23).

FIGURE J2(a–c) The series of micrographs was taken from a section of the thick skin of a human fingertip.

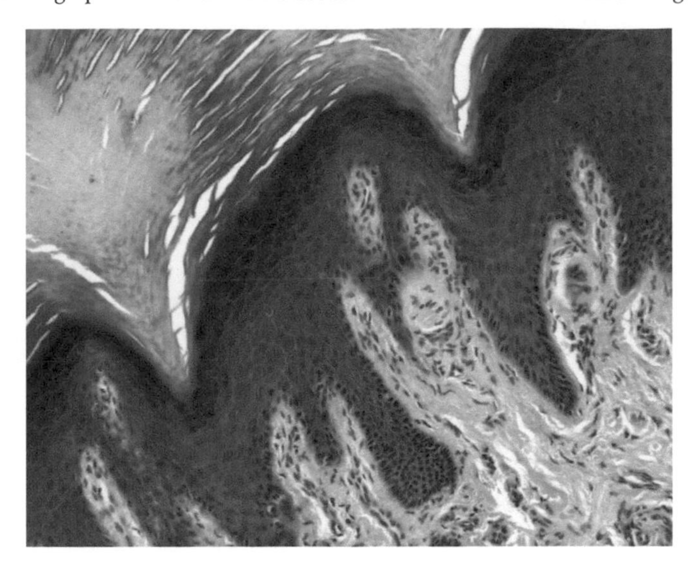

FIGURE J2A Five layers can sometimes be recognized in the epidermis of thick skin. The STRATUM CORNEUM is the translucent external layer of dead cells (upper left). These cells are constantly sloughing off and spaces can be seen between the cells. This preparation does not distinguish the refractile, deep layer of the STRATUM CORNEUM, the STRATUM LUCIDUM. Below the stratum is a pigmented layer, the STRATUM GRANULOSUM, whose cells are packed with basophilic granules of KERATOHYALINE. The next layer is the stratum spinosum which rests on a single layer of dividing cells, the STRATUM GERMINATIVUM that rests on the lightly stained connective tissue of the dermis (lower left). 20 ×.

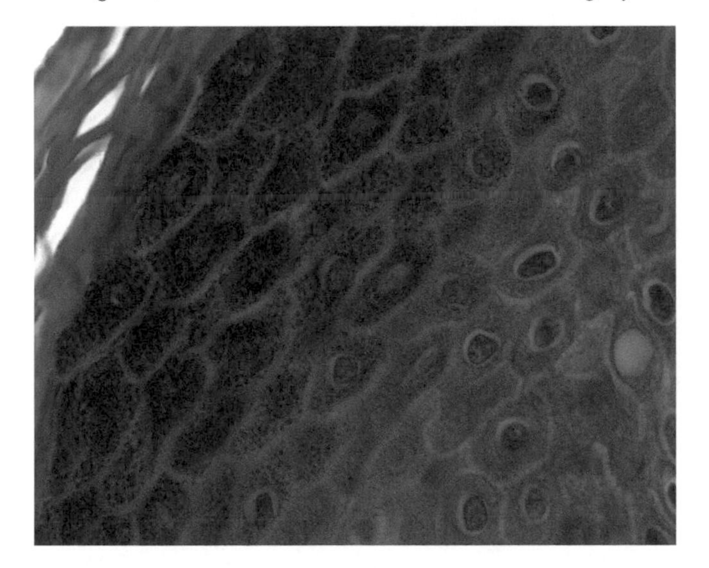

FIGURE J2b At a higher power, dead cells of the stratum corneum can be seen to be pulling apart at the upper left. The basophilic granules of the stratum granulosum are distinctly visible. At the lower right the "spiny" nature of the stratum spinosum is evident—the spines are intercellular bridges that extend between adjacent cells. Note the transition between the stratum spinosum and stratum granulosum as the cells accumulate basophilic granules. 100 ×.

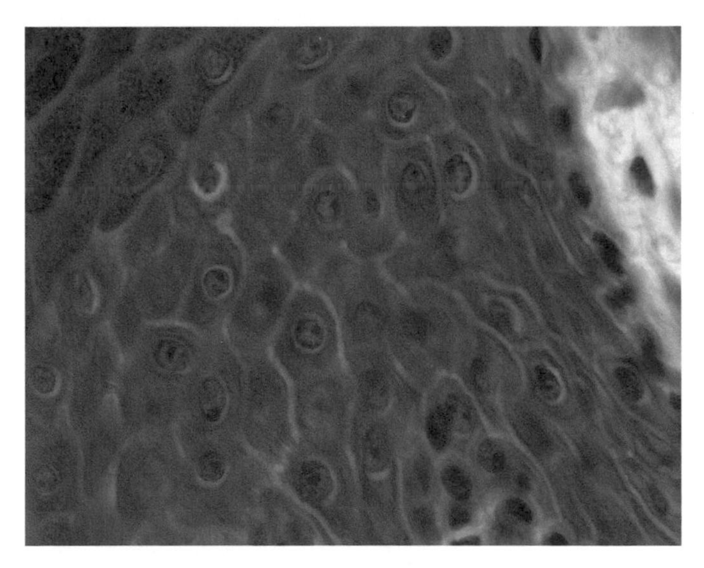

FIGURE J2c In the deeper layers of the skin, intercellular bridges are clearly shown not only between the cells of the stratum spinosum but also between the cells of the stratum granulosum (upper left). The single germinal layer of cells of the stratum germinativum rests on the connective tissue of the dermis at the upper right. Germinal cells divide slowly by mitosis; no mitotic figures are visible in this section. 100 ×.

Hair

Mammalian hairs are cylindrical shafts of adherent keratinized epithelial cells. Since some protection is afforded by the hair, the cornified layer of the epidermis is much thinner than it is on the plantar surfaces (Fig. J24a). Sometimes it is reduced to less than half the thickness of the germinative layer. The stratum lucidum is much reduced or entirely lacking and there are few granular cells. The spines of the porcupine, hedgehog, and spiny anteater are modified hairs.

FIGURE J3(a–e) The series of photomicrographs was taken from a semithin section of the plantar pad of a monkey.

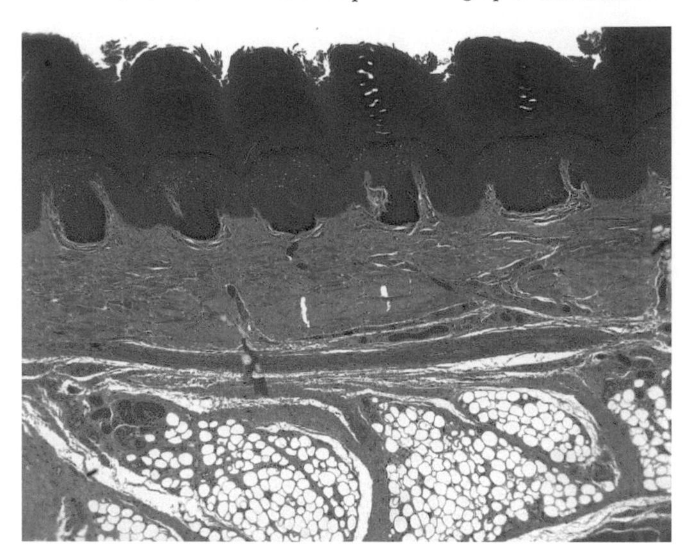

FIGURE J3a Since this area is subject to much abrasion, the stratum corneum is thick and well developed, the dermis is a tough, layer of leather, providing some cushioning with adipose tissue in the hypodermis. These features can be seen in this survey of the area. Sweat glands at the extreme left and right occur just below the dermis. Ducts from other glands penetrate the dermis and epidermis to release their secretion at the surface (thereby assisting traction)—two ducts penetrating the stratum corneum are well shown to the right of center. 5 ×.

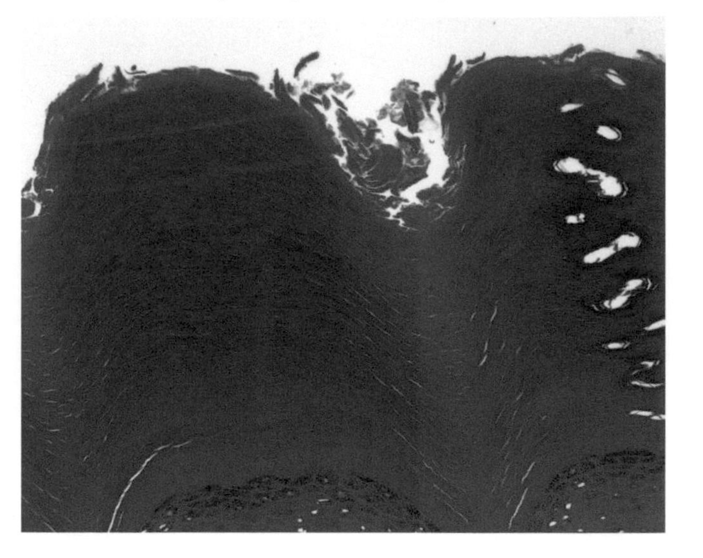

FIGURE J3b Cells of the stratum corneum are dead and provide a waterproof coating for the skin; they resist the severe abrasion that would be experienced in the day-to-day life of a monkey. Cells are constantly sloughing off, as can be seen in the groove at the upper center and are constantly being replaced from below. There is a sharp transition from the cells of the stratum granulosum, which have accumulated the basophilic granules of the precursor of keratin, KERATOHYALINE, to the cornified, translucent cells of the stratum corneum. The duct of a sweat gland penetrates the stratum corneum at the right. 20 ×.

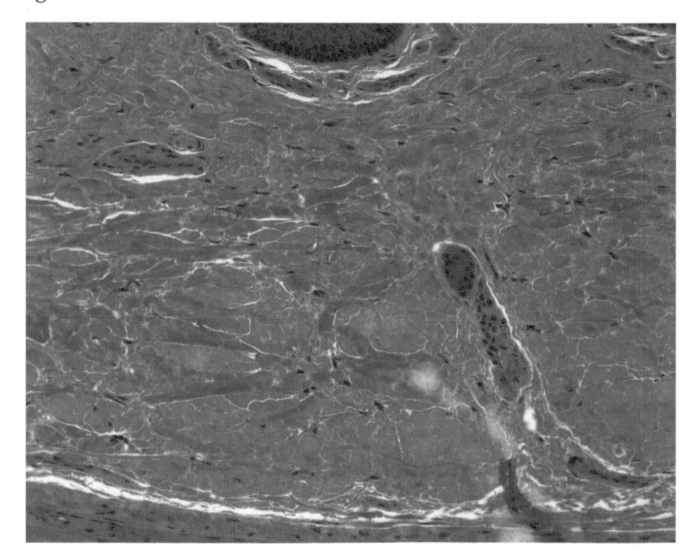

FIGURE J3c The dermis provides an excellent example of irregular dense, collagenous tissue where tough, collagenous fibers form an irregular feltwork. This is the "leather" layer of the skin. Pigmented cells can be seen in a small portion of the stratum germinativum shown at the top. There is a small segment of the duct of a sweat gland at the right. 20 ×.

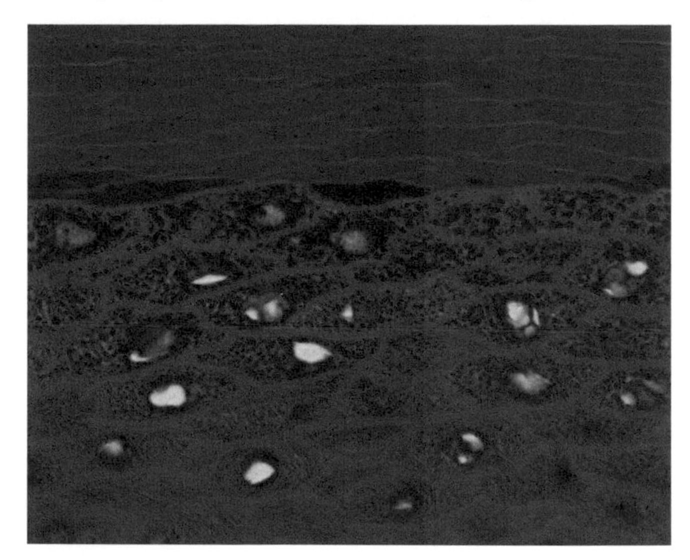

FIGURE J3d The transition from the stratum granulosum to the stratum corneum is abrupt. Cells of the stratum spinosum, at the bottom of the micrograph, are firmly attached to each other by intercellular bridges. These cells gradually accumulate basophilic droplets of keratohyaline (center). Death seems sudden as their source of nutrients disappears and the keratohyaline transforms into hyaline thereby forming the stratum corneum. The clear areas are artifacts. 100 ×.

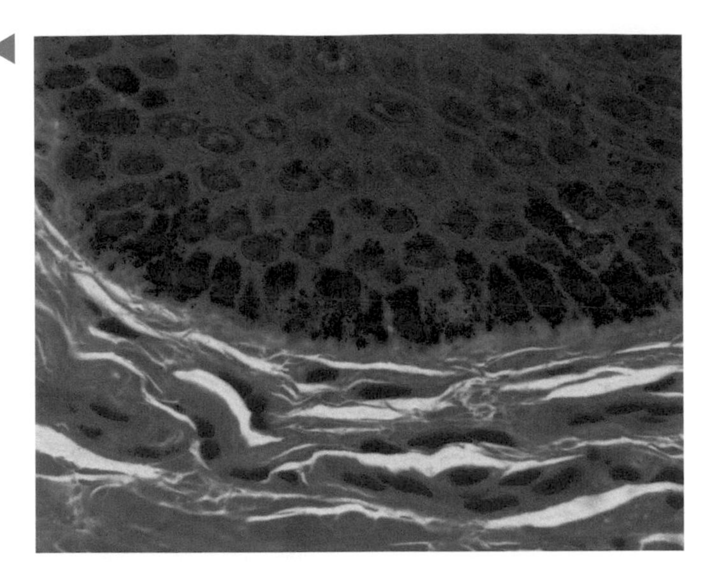

FIGURE J3e Melanin pigment of the skin tends to be concentrated in the stratum germinativum and, to some extent, in the stratum spinosum. Intercellular bridges between cells of the stratum spinosum are well demonstrated. 100 ×.

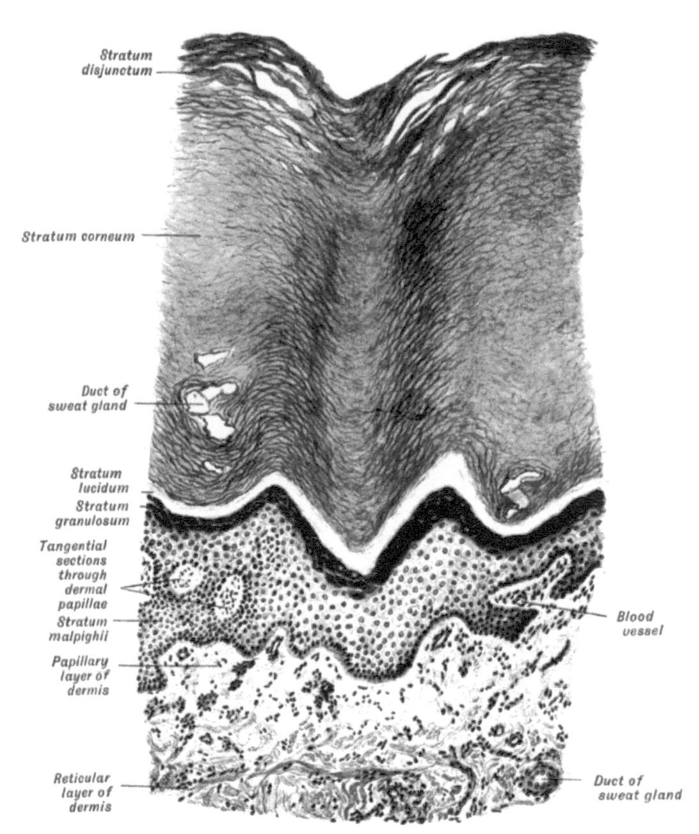

FIGURE J4 Drawing of a vertical section of the skin from the sole of the human foot.

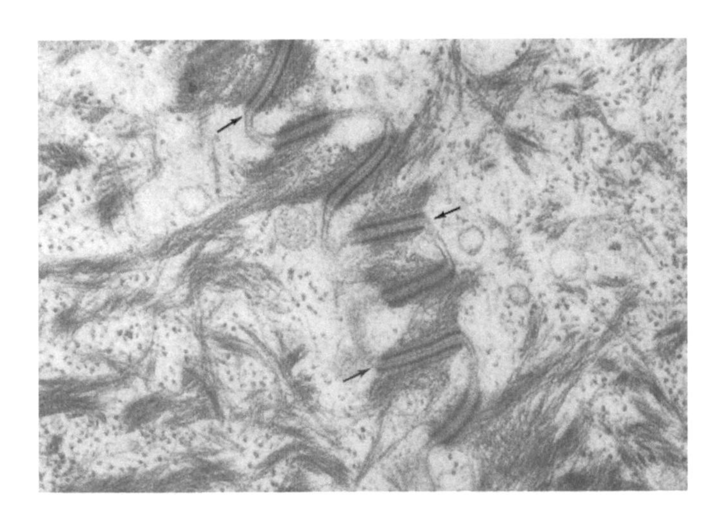

FIGURE J5a Transmission electron micrograph of a section through the interdigitating junction between two adjoining cells of the stratum granulosum of mammalian skin. Dense bundles of filaments in the cytoplasm terminate in the dense plaques of desmosomes (arrows).

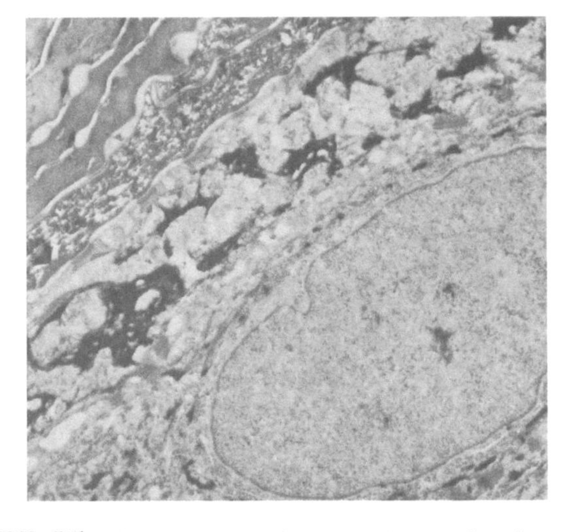

FIGURE J5b Transmission electron micrograph of a section through the stratum corneum (upper left) and the stratum granulosum (lower right) of mammalian skin. Note the irregularly shaped granules of the stratum granulosum.

In this image from a section of mammalian hairy skin, the portion of the hair projecting above the skin consists of the CUTICLE, the peripheral layer of cells arranged like shingles, and the CORTEX that forms the greater portion of the shaft (Figs. J24b and J24c, J25, J26). If a MEDULLA is present, it consists of the central two or three rows of cells. The tissue that surrounds the embedded portion of the hair is the HAIR FOLLICLE and consists of two layers. The outer layer is a poorly defined CONNECTIVE TISSUE SHEATH and the inner layer is the EXTERNAL ROOT SHEATH, a down

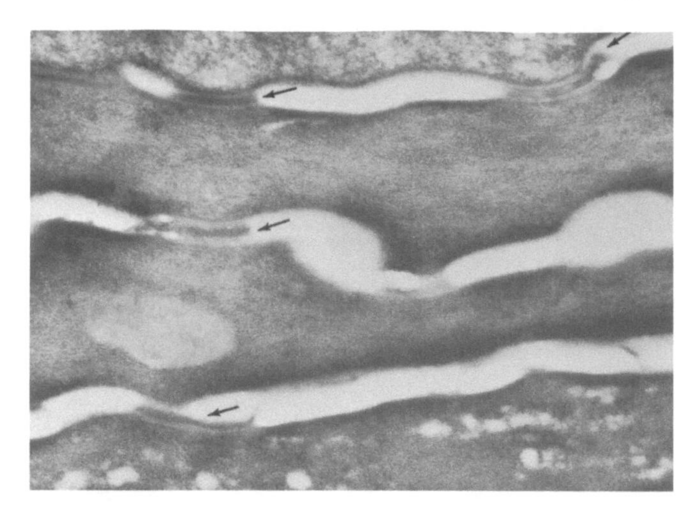

FIGURE J5c Transmission electron micrograph of a section through four cells in the human epidermis. Their cytoplasm is filled with keratin filaments embedded in a dense matrix. The arrows indicate modified desmosomes. The spaces between the cells are, in part, artifacts of specimen preparation.

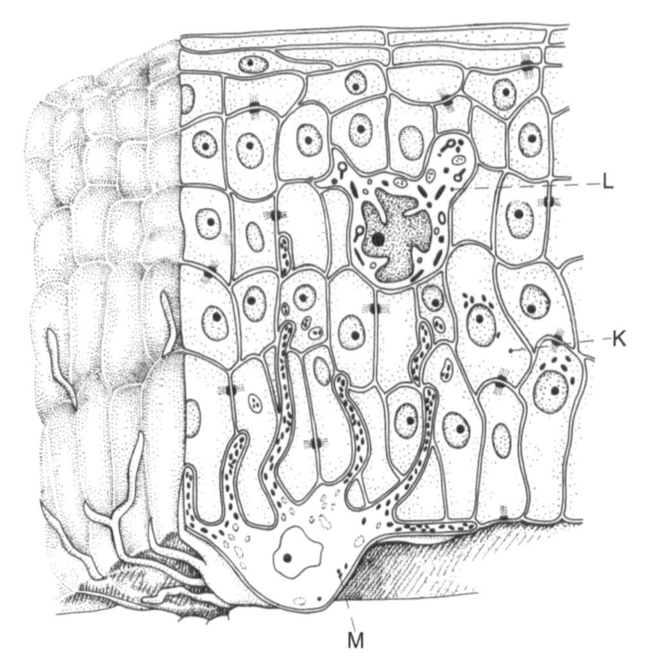

FIGURE J5d Sketch of a stellate melanocyte (M) extending its processes between the cells of mammalian epidermis. Granules of melanin are formed within melanocytes and relayed to the other epidermal cells—keratinocytes (K) and intraepidermal macrophages (L)—by way of the intercellular procscesses.

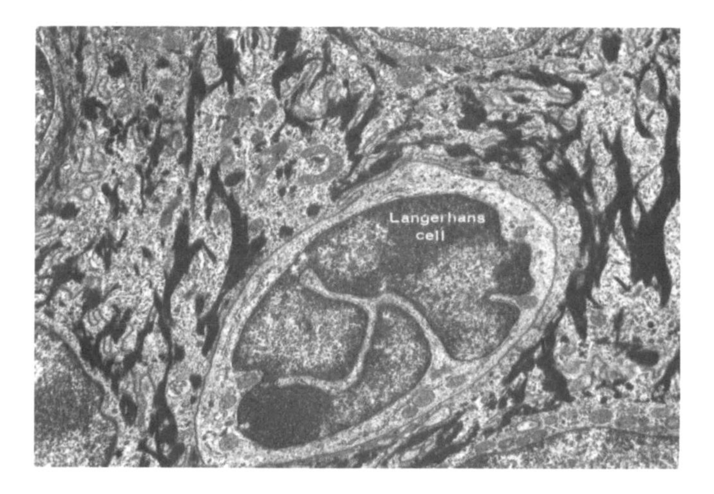

FIGURE J6a Transmission electron micrograph of an intraepidermal macrophage (Langerhans cell) surrounded by keratinocytes containing dense bundles of keratin filaments. None of the processes of this cell is included in this section.

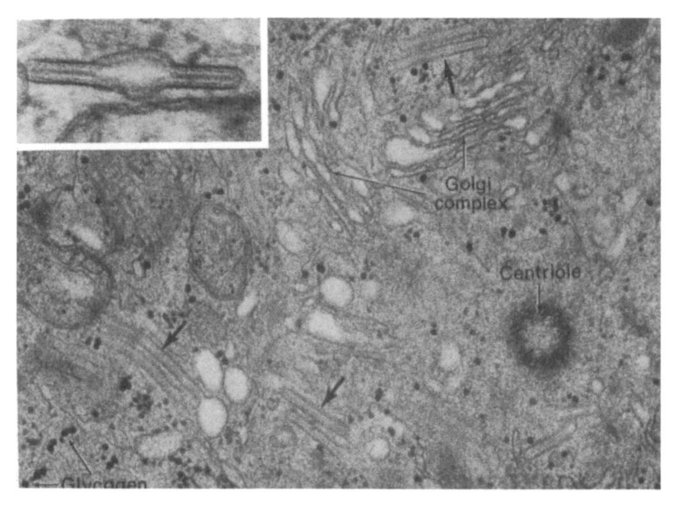

FIGURE J6b Electron micrograph of a small area of the cytoplasm of an intraepidermal macrophage. A centriole and Golgi complex are included in this section. The arrows indicate the specific granules of this cell; the inset shows one of these granules at a higher magnification.

growth of the stratum germinativum of the epidermis. The two sheaths are separated by the GLASSY MEMBRANE that corresponds to the basement membrane beneath the epidermis. At the base of the follicle the connective tissue forms a HAIR PAPILLA that is capped by epithelial cells of the HAIR MATRIX (Figs. J27a and J27b). Active mitosis of these epithelial cells produces the cells of the hair as well as the INTERNAL ROOT SHEATH, a sleeve of epithelial cells surrounding the hair. Soft keratin develops in the cells of the internal root sheath, hard keratin in those of the hair shaft. Basophilic keratohyaline granules do not develop in cells destined to form hard keratin. Both the internal root sheath and the hair are pushed up by the addition of new cells at the matrix.

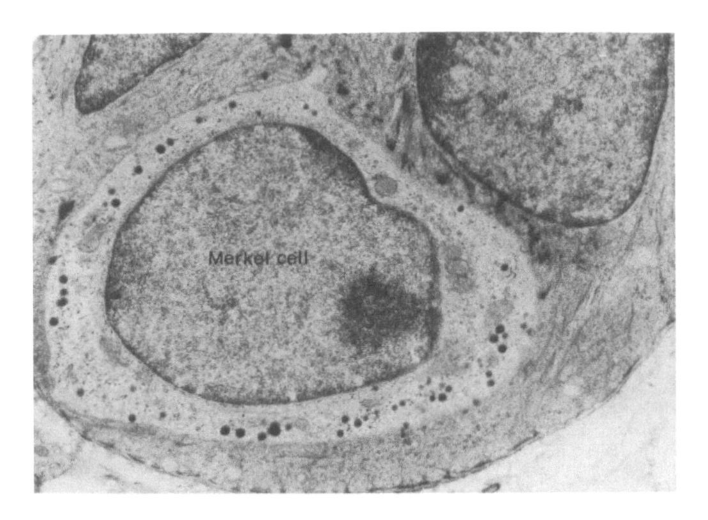

FIGURE J7a Electron micrograph of a section of the base of human epidermis showing a tactile cell (Merkel cell).

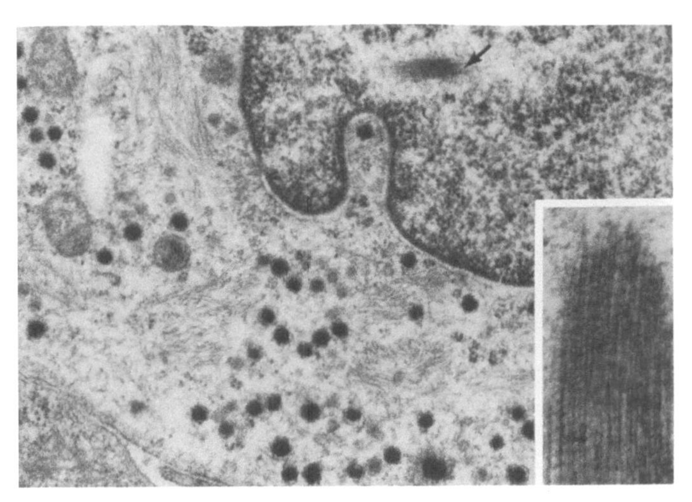

FIGURE J7b Electron micrograph of a portion of a tactile cell from the epithelium of a human gum. The cytoplasm contains numerous dense granules and intermediate filaments. The *inset* shows a higher magnification of the dense granule in the nucleoplasm (arrow).

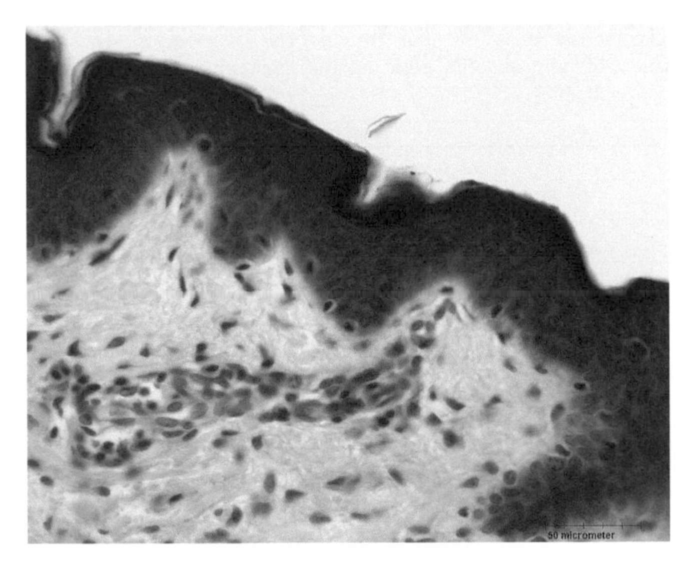

FIGURE J8 The skin over the human eyelid is extremely delicate. The cells of the dermis are clearly visible but their differentiation into layers is not defined. There are a few mitotic figures in the stratum germinativum. Basophilic keratohyaline has accumulated in the stratum granulosum and a thin stratum corneum of sloughing cornified cells forms the outer surface. The dermis consists of areolar tissue; this tissue must be loose and pliable to permit rapid and delicate movement. The swirl of cells approaching the epidermis from the lower left is a blood vessel. 40 ×.

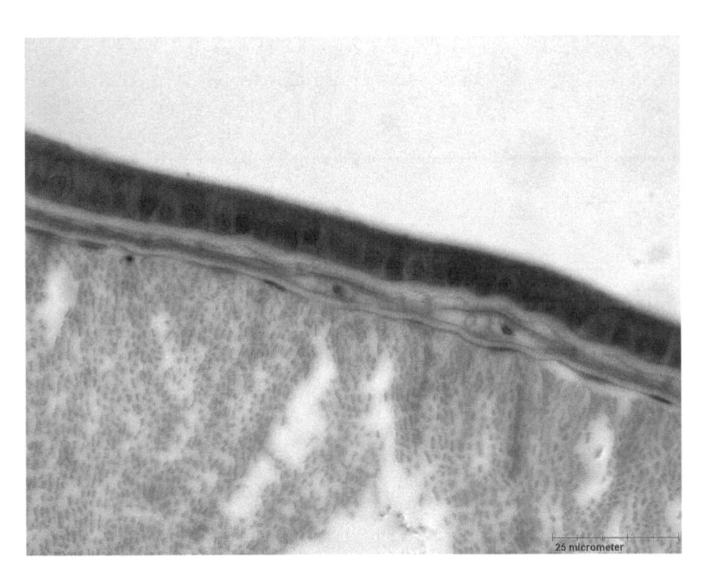

FIGURE J9 The epidermis of the cephalochordate AMPHIOXUS is a simple, columnar epithelium in this cross section. This single layer of cells rests on the loose areolar connective tissue of the DERMIS. Sections of a capillary(ies) can be seen in the dermis at the right. Below the dermis is a thick layer of striated muscle of the body wall. 100 ×.

A SEBACEOUS GLAND and a strand of smooth muscle, the ARRECTOR PILI, are often associated with a hair follicle (Figs. J28a and J28b). The axis of the follicle is never exactly perpendicular to the surface of the skin and the muscle and the gland lie in the wider angle formed between the follicle and the surface (Figs. J28c and J40c). Sebaceous glands are branched alveolar holocrine glands consisting of epithelial cells resting on a basement membrane and encapsulated by a thin layer of connective tissue. Their secreting alveoli are not composed of a single layer of cells grouped around a lumen but are rounded masses of cells. The central cells are filled with vacuoles. Secretion of the sebaceous gland is accompanied by the breakdown of these cells and their remains are poured out with the oily secretion into the hair follicle. The cells destroyed are replaced by mitotic activity of the BASAL CELLS at the periphery. Electron micrographs of the basal cells show the usual evidence of active cells:

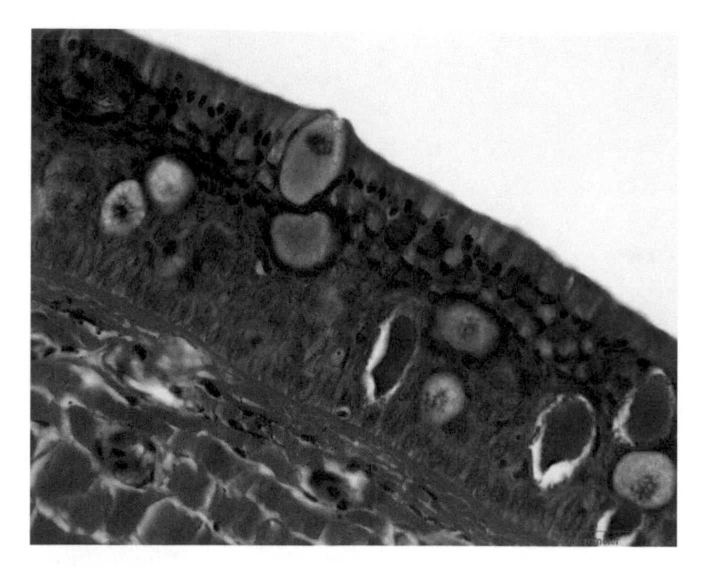

FIGURE J10a The EPIDERMIS of an agnathan is stratified and shows a complex array of different cell types. This is a section through the body wall of a HAGFISH. Two types of secretory cells can be differentiated in the epidermis: mucous cells (blue–gray), and thread cells (orange). The thread cells are packed with a tightly coiled, single thread consisting of a large number of intermediate filaments bundled in parallel (reddish). Most of the epidermal cells remain less differentiated and form a protective layer over the body. The DERMIS consists of tough parallel bands of collagenous fibers. 40 ×.

FIGURE J10b Three types of cells can be identified in the stratified epidermis of the LARVAL LAMPREY: the basal, pale club cells, round, acidophilic granular cells at a mid-level of the epithelium, and mucous cells. The epidermis rests on a thin, grayish band of collagenous tissue of the DERMIS. The HYPODERMIS, consisting mainly of adipose cells, rests on the thick muscular layer of the body wall. 40 ×.

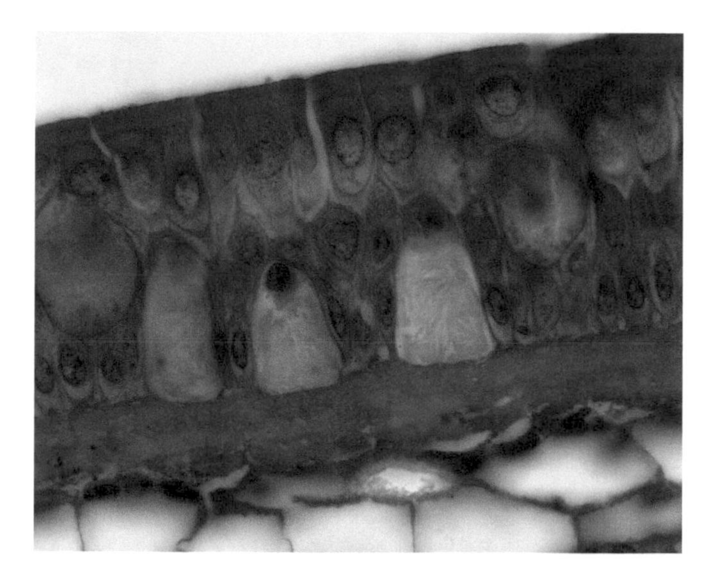

FIGURE J10c The highly differentiated cells of the epidermis of the LAMPREY skin arise from division in basal cells. The most abundant cells in this section are mucous cells. The largest cells are the enigmatic club cells, containing whorls of material. Their nuclei are apical; often two nuclei are seen in a single club cell. Two round granular cells hover at a mid-level in this section—they contain acidophilic granules. It has been suggested that these cells release a poison, which renders the ammocoete unattractive to potential predators, or that they assist in the release of mucus from mucous cells. A few other round cells at this level are probably younger granular cells. 100 ×.

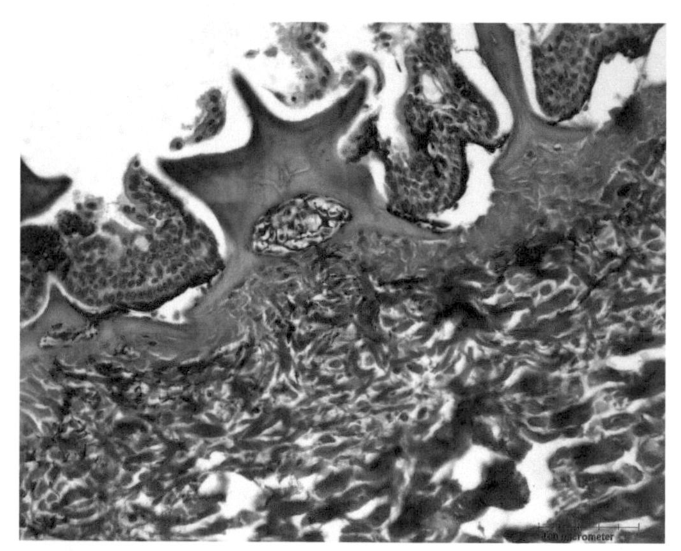

FIGURE J11 This is a section of the skin of a DOGFISH SHARK. The presence of tooth-like scales has caused difficulties in sectioning, and the epidermis is broken and has pulled away from the dermis. The epidermis is made up of undifferentiated cells but a great deal of specialization has occurred with the formation of a dermal scale that pokes through the epidermis to appear on the surface, giving the skin the feel of sandpaper. Pigment cells occur at the base of the epidermis and in the upper levels of the dermis. The dermis is a tightly woven layer of collagenous tissue. 20 ×.

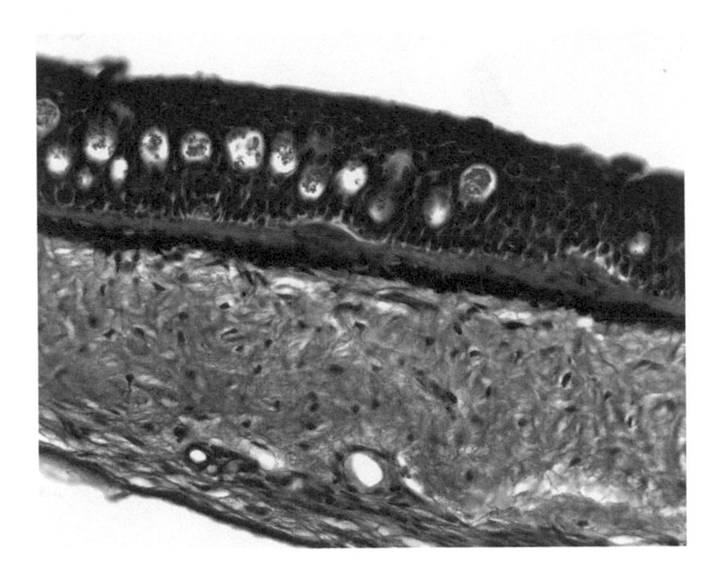

FIGURE J12 The dermis and epidermis are readily distinguished in this section of the skin of a darter (BONY FISH). Most of the cells of the epidermis are undifferentiated and form a stratified, squamous epithelium. The cells of the lowest layer of this epithelium divide to produce new cells that move toward the surface. A few mucous cells have differentiated among the epithelial cells. 40 ×.

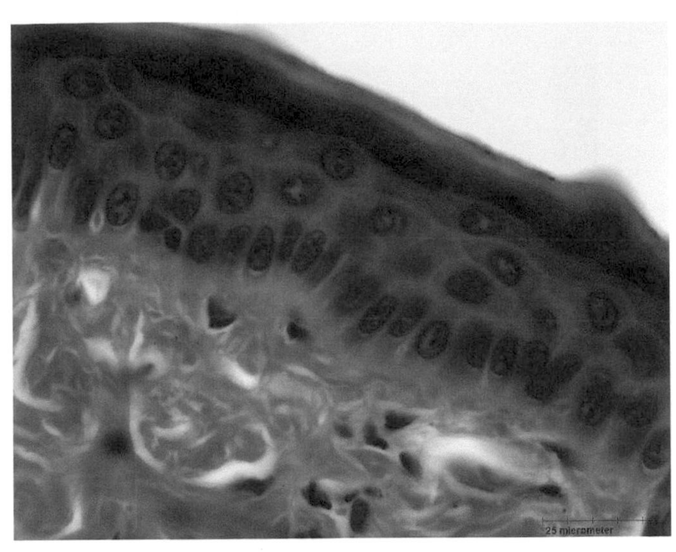

FIGURE J13a The epidermis of the skin of a marine TOAD (*Bufo marinus*) is simple: it rests on the dermis of vascular connective tissue. Immediately below the epidermis, the connective tissue forms a dense collagenous layer below containing a few pigment cells. Most of the dermis consists of areolar connective tissue. A few arterioles can be seen at the bottom of the micrograph. 40 ×.

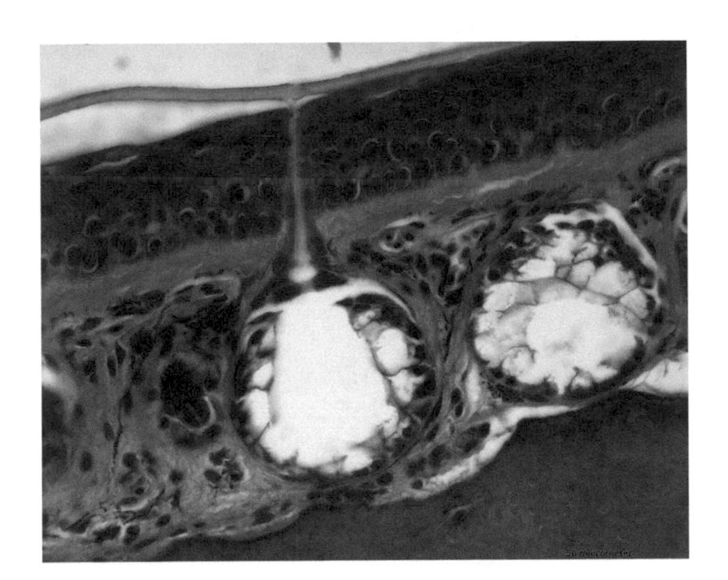

FIGURE J13b This section from a highly specialized AMPHIBIAN, a limbless caecilian from Tanzania, shows that the characteristics of amphibian skin are still recognizable. Epidermal cells can form highly specialized glands that penetrate into the dermis and pour their secretions onto the surface. Again, the lowest layer of the epidermis is a germinal layer. As the cells rise to the surface, they cornify, die, and ultimately slough off. The dermis is a thick layer of vascular connective tissue that accommodates the large alveolar epidermal glands. A few brown chromatophores extend their bodies throughout this layer. In this region, the dermis is connected to bone at the lower right by loose connective tissue of the hypodermis; the hypodermis is pulled apart on the right of the photomicrograph. Note an osteocyte within its lacuna in the acidophilic bony matrix. 40 ×.

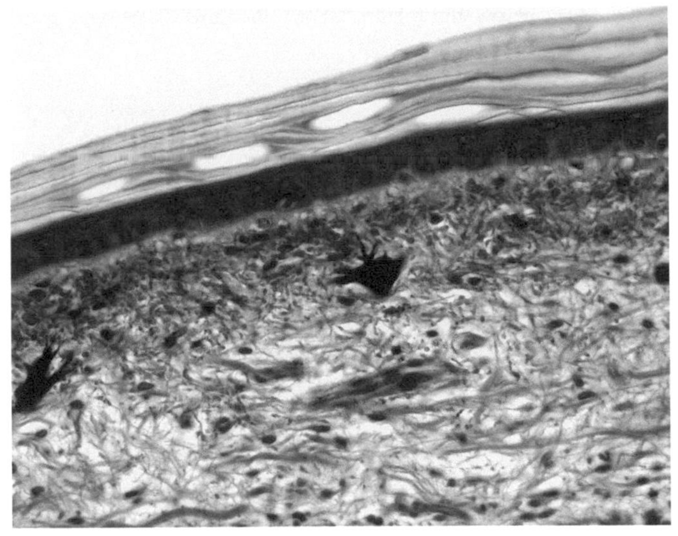

FIGURE J14a This is a section through the dry skin of a RATTLESNAKE; the skin is thrown up into a scale a distraction which we will consider later. The epidermis is thin; its cells rapidly cornify and die as they move to the surface. Sloughing of the cornified cells is not a continuous process as it is in many other vertebrates—the snake loses its cornified layer at well-defined intervals of molting. Abundant pigment granules are scattered throughout the epidermal cells. The dermis is areolar connective tissue and contains abundant chromatophores in its upper regions; the dermis is broken in few places. A longitudinal section of a capillary courses through the dermis from the lower right to a point about halfway across the micrograph; near this point, an arteriole is seen in cross section. 10 ×.

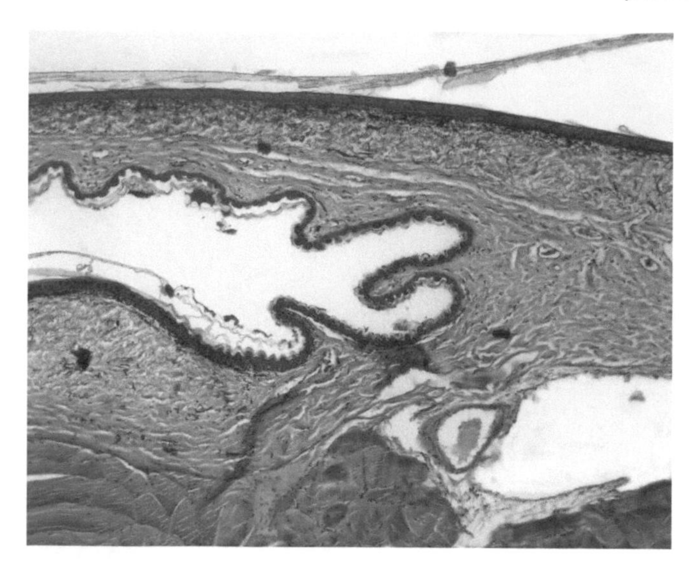

FIGURE J14b A similar picture is seen in this section of the skin of a CAIMAN, a reptile in the alligator family. The stratified, squamous, cornified epidermis contains only a few layers of living cells and a thicker layer of cornified cells but, unlike the snake, dead cells slough off continuously from the surface. The prominent, dark chromatophores in the dermis extend delicate, pigmented processes in all directions (these were cut off in sectioning) forming the brownish mesh beneath the epidermis. Longitudinal sections of two capillaries occur halfway across the picture, about one-third of the way from the bottom. 40 ×.

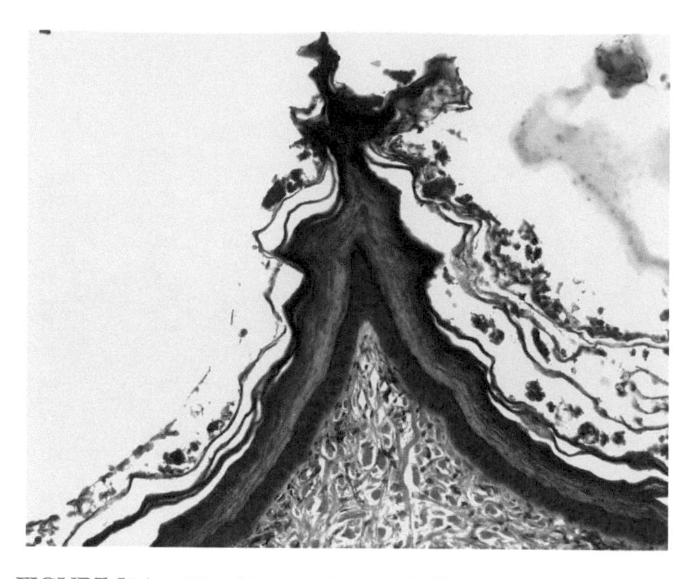

FIGURE J14c The skin covering the shell in a SNAPPING TURTLE has risen up into oddly shaped simple scales. This section demonstrates the characteristics we saw in other specimens of reptilian skin: an epidermis of stratified squamous cornified epithelium resting on a pigmented dermis of areolar connective tissue. Dead cells of the stratum corneum that have been sloughed off still cling, along with some débris, to the surface. The outer layers have adhered to particles of dirt as they go. This scale is in the form of a spike and is subtended by areolar connective tissue containing a few scattered chromatophores. 20 ×.

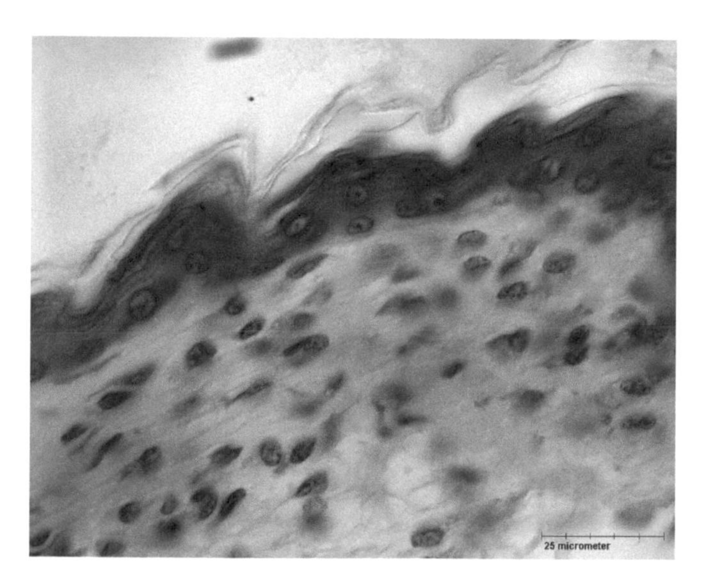

FIGURE J15 The delicate skin of a newly hatched CHICK is protected against abrasion by rigid feathers and has no need for a thickened stratum corneum. The stratified squamous epidermis is cornified, minimizing dehydration. The dermis consists of densely packed collagenous fibers and provides a firm base for the attachment of feathers. 100 ×.

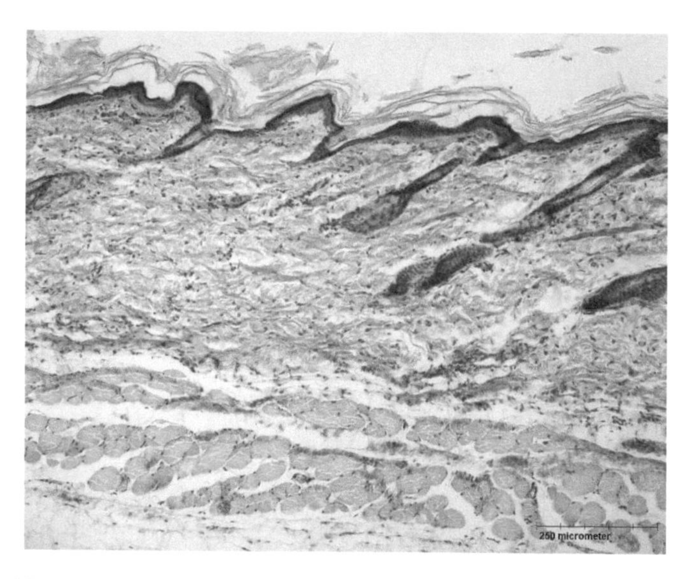

FIGURE J16a Since there is far less abrasion on the skin of other regions of the mammalian body, the epidermis is thinner and not as elaborately developed as was seen in the plantar surfaces. Unlike the situation on most regions of the human body, this skin of other mammals is well protected by hair. The epidermis forms the thin, irregular dark line at the top of this micrograph from a section of the body skin of a FURRY MAMMAL. The well-defined layers of the epidermis that were seen in the plantar skin cannot be distinguished. Cornified cells are sloughing from the surface. Hair follicles extend diagonally into the thick dermis (or leather layer) of dense collagenous tissue. This is the layer that gives the skin its strength to resist tearing. The bottom quarter of the picture contains the striated muscle of the skin, which inhabits the hypodermis and permits the animal to move its skin. 10 ×.

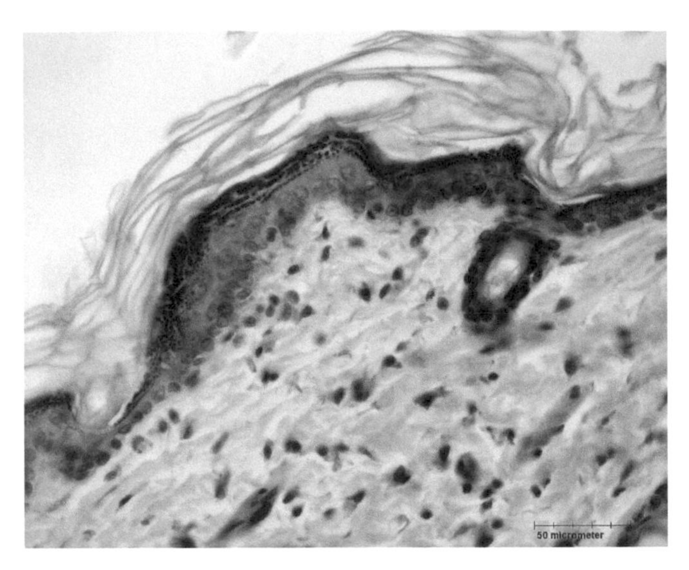

FIGURE J16b Details of the epidermis can be seen in this higher power of Fig. J16a. Although not obvious in this view, mitosis occurs in the deepest layer of the epidermis. Basophilic granules of keratohyaline are apparent in the layer below the stratum corneum. Most of the lower half of the photograph consists of dense collagenous tissue of the dermis. 40 ×.

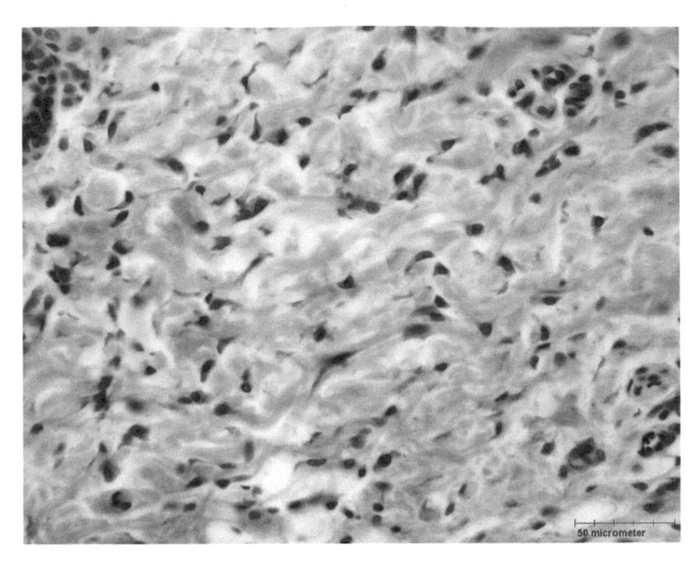

FIGURE J16c The densely packed fibers of collagen form a tough feltwork from the slide shown in Fig. J16a. 40 ×.

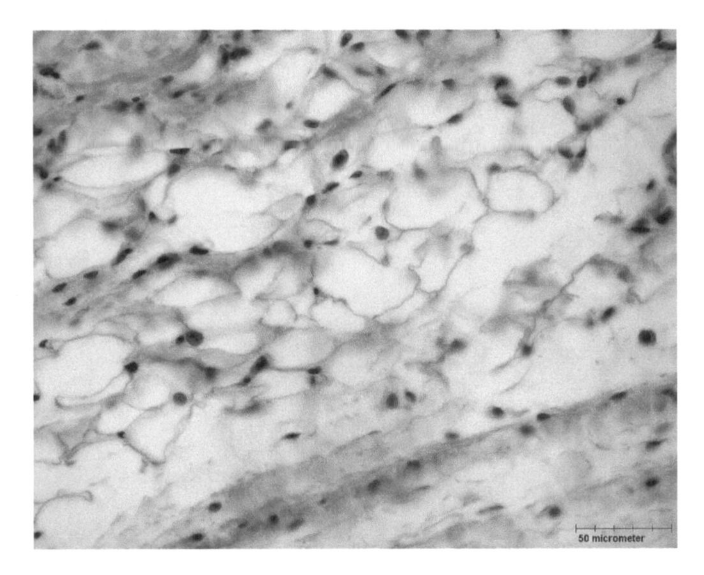

FIGURE J16d Abundant adipose tissue provides insulation and cushioning in the hypodermis—micrograph from the slide shown in Fig. J16a. 40 ×.

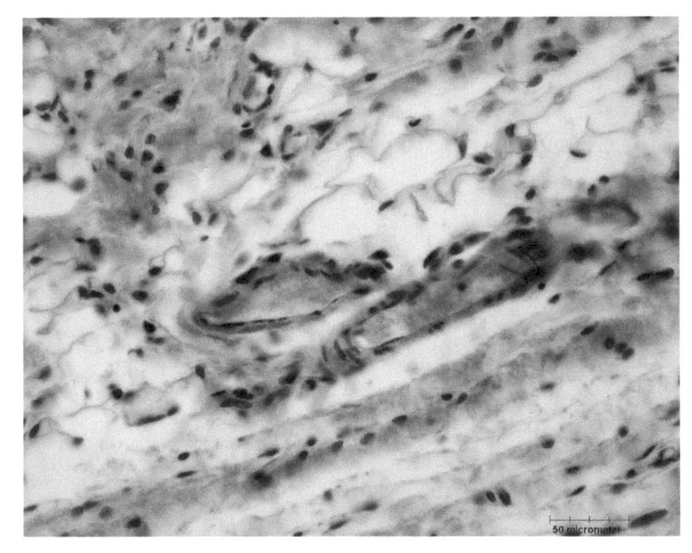

FIGURE J16e An arteriole and vein course through the hypodermis from the slide shown in Fig. J16a. The arteriole is cut tangentially showing the circular smooth muscle of its wall. 40 ×.

agranular and granular endoplasmic reticulum, free ribosomes, glycogen, mitochondria, and Golgi complexes; no oil droplets are apparent. As the lipid accumulates, the smooth endoplasmic reticulum increases in amount.

Feathers

The stratum corneum reaches its climax of elaboration and specialization in feathers. Feathers are said to be modified reptilian scales, each being formed from a dermal papilla covered with epidermis (Figs. J29a and J29b). Note the similarity of the FEATHER FOLLICLE to the hair follicle. In both there is a tubular invagination of the epidermis producing a structure of hard keratin. At the base of both the hair and feather is a vascular PAPILLA that is

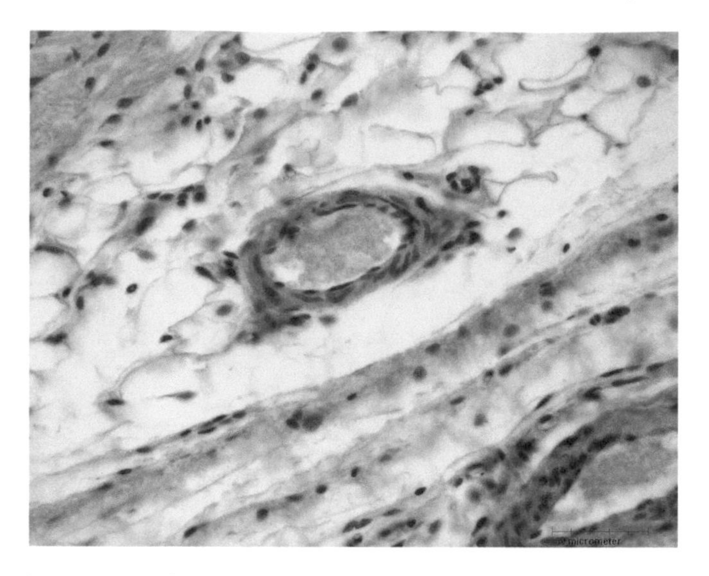

FIGURE J16f Cross section of an arteriole in the hypodermis from the slide shown in Fig. J16a. 40 ×.

FIGURE J17a Whole mount of the skin of a darter (bony fish) showing an elaborate array of chromatophores. The yellow background is from picric acid in the fixing solution. 10 ×.

FIGURE J17b Pigment cells lurking in the connective tissue of the dermis of the fish may assume elaborate shapes as seen in this unstained whole mount of the skin of a shiner. 10 ×.

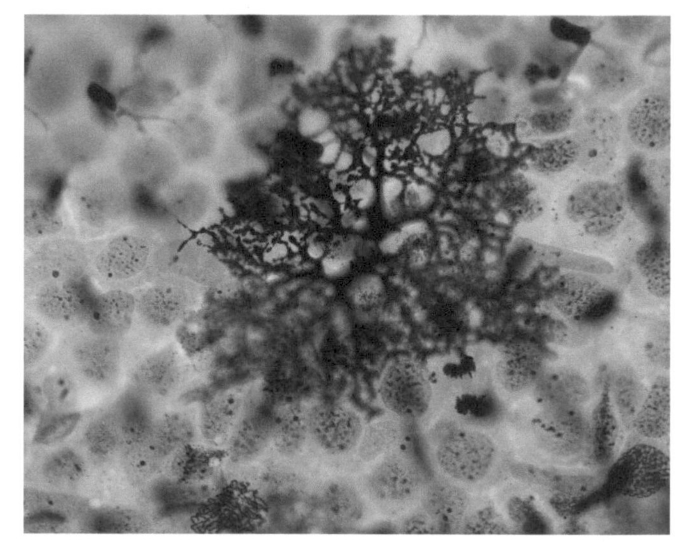

FIGURE J17c An elaborate chromatophore in a whole mount of the skin of the salamander, *Ambystoma*. Granules of the pigment melanin roll out into the processes to darken the skin; when they roll to the center of the cell, the skin becomes paler. 40 ×.

considerably more elaborate in the feather than in the hair. Vascular connective tissue fills the hollow core of the shaft of the developing feather. Blood vessels enter by the INFERIOR UMBILICUS at the base and by the SUPERIOR UMBILICUS just below the surface of the skin at one side. When the feather is mature this tissue dries to a pith. Like the hair follicle, the feather follicle often has a smooth muscle attached to it, the ARRECTOR PLUMAE, and the feather is able to move.

It is difficult to imagine how a structure as complex as a feather can be molded of cornified squamous cells in a structure as apparently simple as a feather follicle.

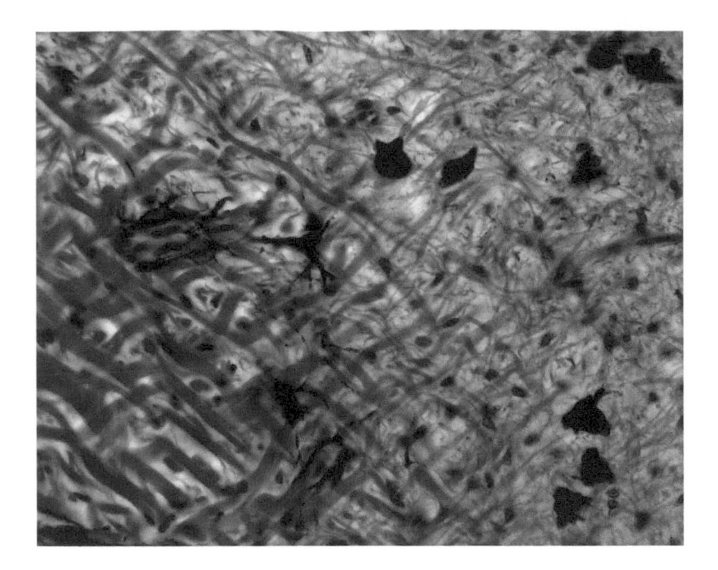

FIGURE J17d Dark brown chromatophores inhabit the meshes between collagenous fibers in this section of the dermis of a caiman (a reptile that has the appearance of a small alligator). Many processes of the chromatophores have been cut off in sectioning. 40 ×.

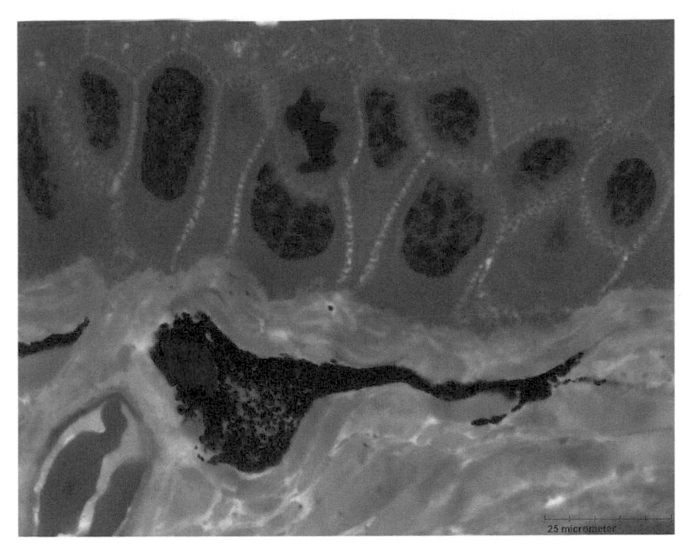

FIGURE J17e Details of a chromatophore in the dermis of *Ambystoma* show up well in this semithin section. Most of the processes have been lopped off during sectioning, but it can be seen that the cytoplasm and the single process shown are packed with granules of melanin. The nucleus, not often seen in preparations of chromatophores, is clearly shown. 100 ×.

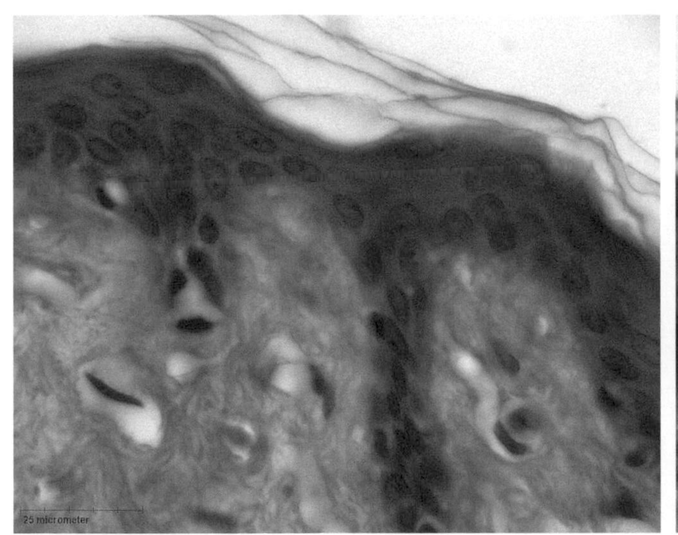

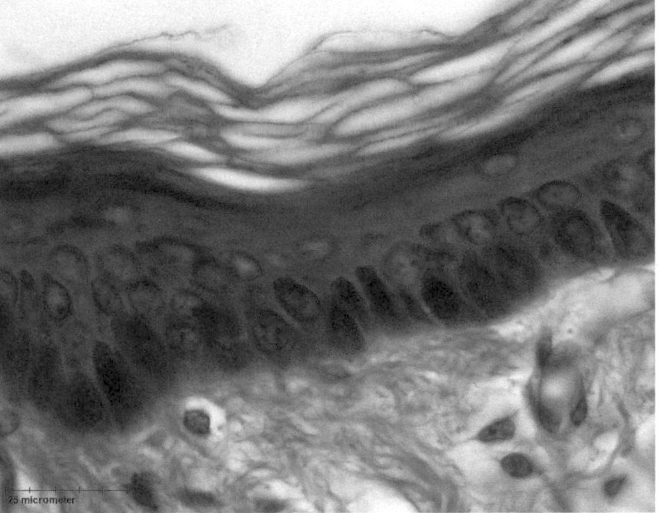

FIGURES J17f and J17g Variations in the pigmentation of HUMAN epidermis are obvious in these two sections. Large amounts of melanin have accumulated in the basal layers of NEGROID skin (g) while little is present in the skin of a CAUCASIAN (f); much pigment is lost as the cells move toward the surface. 100 ×.

OTHER INTEGUMENTARY GLANDS

Glands of Nonmammals

In sections of the skin of representative chordates it is apparent that glands are developed in the epidermis of almost all. Only merocrine GOBLET CELLS scattered among the columnar epithelial cells are present in the epidermis of *amphioxus* (Fig. J30). Elaborate unicellular glands are developed in the skin of *cyclostomes*. In the lamprey, GOBLET CELLS secrete mucus; as the epithelial cells move toward the surface because of proliferation of underlying layers, they become filled with more mucus (Figs. J32a and J32b). Acidophilic GRANULAR CELLS secrete a substance

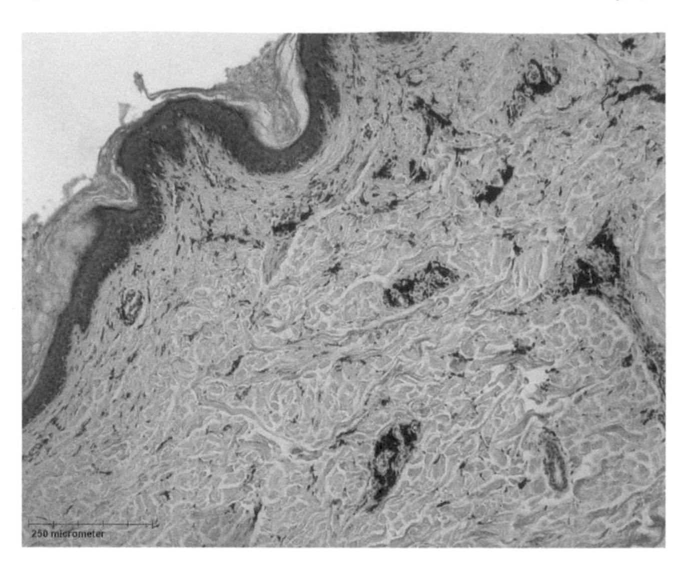

FIGURE J17h As an interesting aside, we included this slide of tattooed human skin. The section shows the dispersed accumulation of ink granules in the dermis. 10 ×.

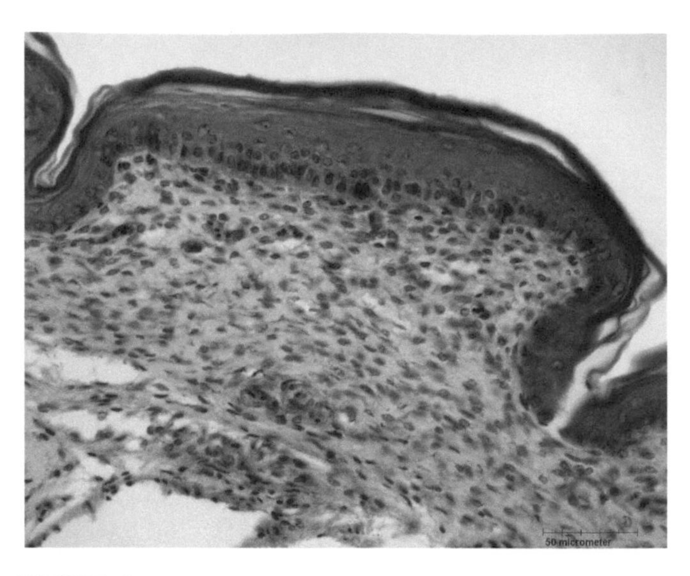

FIGURE J18 Scales in birds are simple structures. This section through the skin from the leg of a chick shows a simple plate of thickened skin surrounded by similar plates. The skin throughout is a stratified squamous epithelium with a cornified outer layer that is constantly sloughing off. Flexibility is provided by the deep grooves between the scales. A few mitotic figures can be seen in the stratum germinativum. Occasional capillaries course through the areolar connective tissue of the dermis. The dermis blends imperceptibly into the hypodermis. There is a section through a lymph vessel at the lower left. 40 ×.

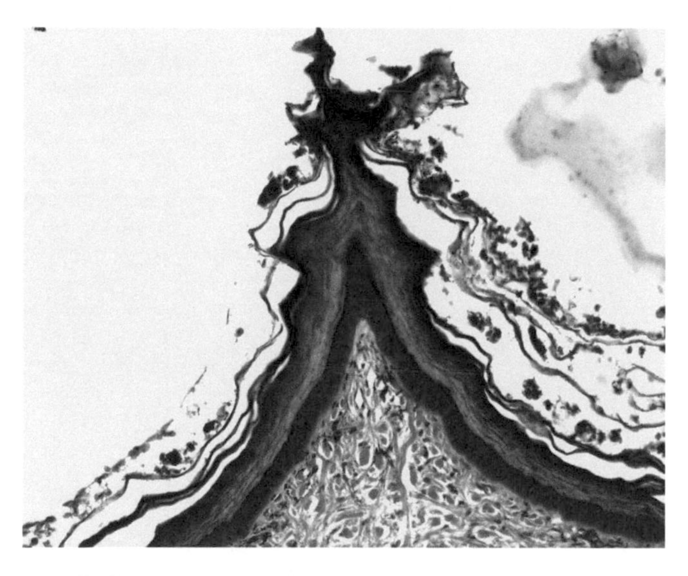

FIGURE J19a The skin covering the shell of a snapping turtle is a stratified, cornified epithelium that is thrown up into oddly shaped projections. The outer layers of the stratum corneum slough off and adhere to particles of dirt as they go. The scale is in the form of a spike and is subtended by areolar connective tissue containing a few scattered chromatophores. 10 ×.

distasteful to would-be predators. Tall, pale, acidophilic club cells, of obscure function, rest on the basement membrane. Similar thread cells in the skin of the hagfish secrete coiled strands of mucus (Figs. J31a and J31b). *Elasmobranch* and *Teleost fish* have unicellular mucous glands similar to those of cyclostomes but they also have multicellular alveolar mucous glands (Fig. J33). The luminous photophores of deep-sea fish are modified alveolar mucous glands in which light is produced either by symbiotic bacteria in the gland or by chemical reaction. These glands may even develop a reflector behind and a lens in front, much like an automobile headlamp. A few fish have developed alveolar poison glands, usually in association with spines (e.g., catfish). The abundant mucous glands of water-living *amphibians* are all alveolar (Figs. J34a–J36c). Many amphibians also possess

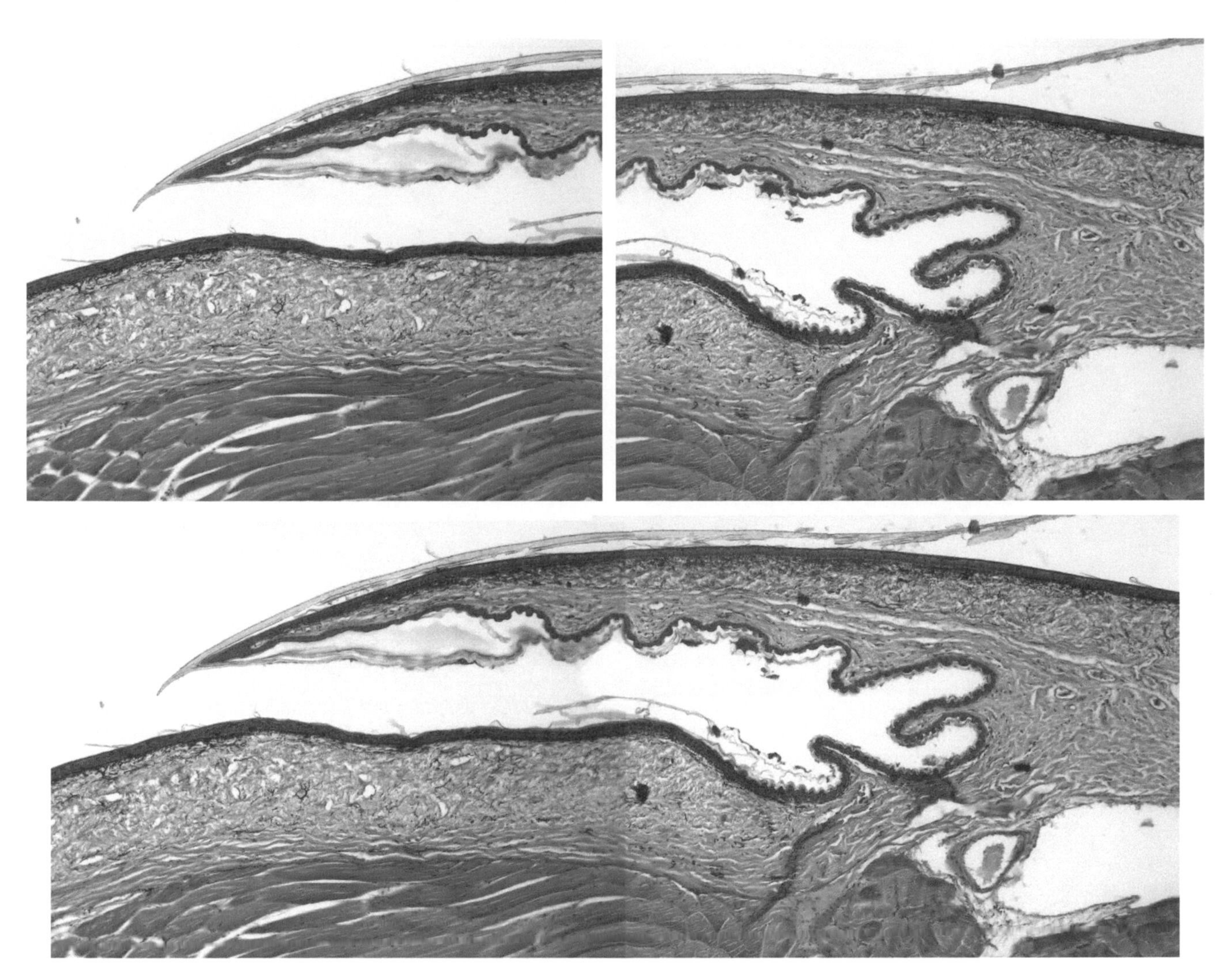

FIGURES J19b, J19c, and J19bc montage The scales on the slick skin of a RATTLESNAKE appear to overlap like shingles on a roof. The stratified squamous cornified epithelium of each scale forms an overlapping plate, which is much thicker on the outer surface than on the inner. A groove, enclosing a flexible joint, still of stratified squamous cornified epithelium, connects each scale. There is little specialization in the areolar connective tissue forming the core of the scale. Chromatophores congregate under the outer surface of the scale. A few capillaries can be seen in the deeper layers of the dermis. The trailing edge of the scale terminates in a sharp edge of cornified cells. The skin is subtended by collagenous fibers of the hypodermis and striated muscle fibers of the body wall. 10 ×.
These two micrographs can be fitted together to accommodate the entire scale as in *J19bc montage.*

FIGURE J19d This flat mount of the skin of a LIZARD (*Anolis*) illustrates the overlapping nature of the epidermal scales of a reptile. Abundant chromatophores inhabit the connective tissue of the dermal cores of the scales. 10 ×.

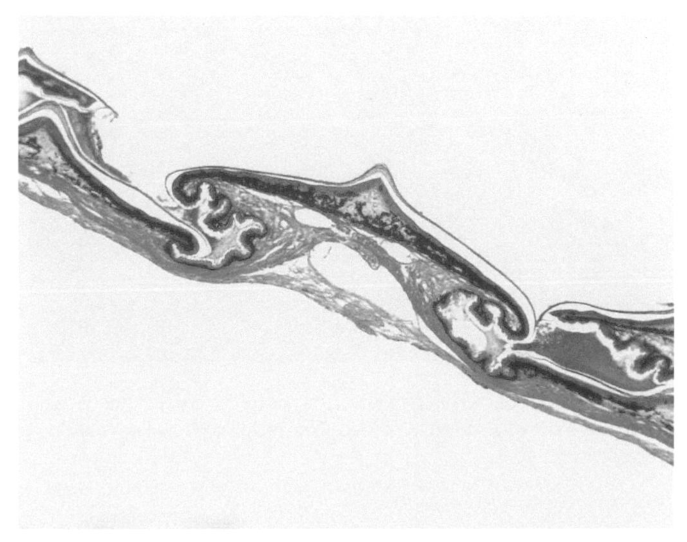

FIGURE J19e This is a section of the skin of a SNAKE showing that epidermal scales can assume elaborate shapes. 10 ×.

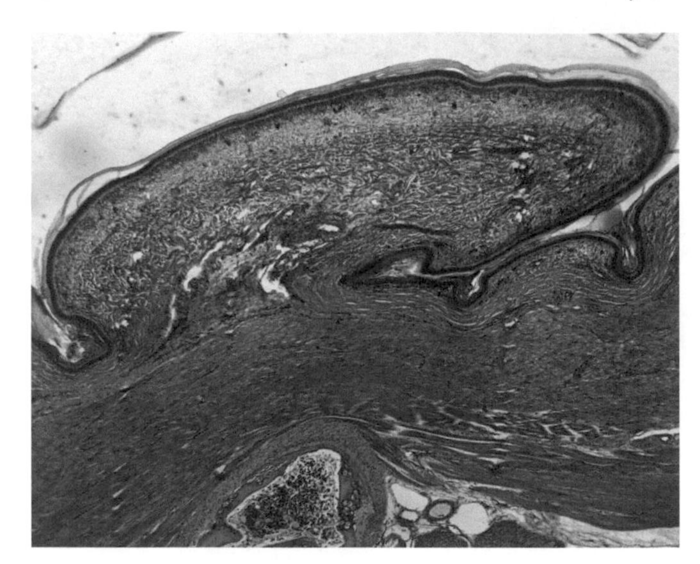

FIGURE J19f The plate-like scale of the CAIMAN has a stratified squamous cornified epithelium and a core of areolar connective tissue. Flexible grooves connect it to other scales on the surface. Large numbers of chromatophores in the dermis darken the skin; they are especially dense immediately below the stratum germinativum. A few blood vessels pass through the core of the scale. Most of the lower part of the micrograph consists of striated muscle of the body wall. 5×.

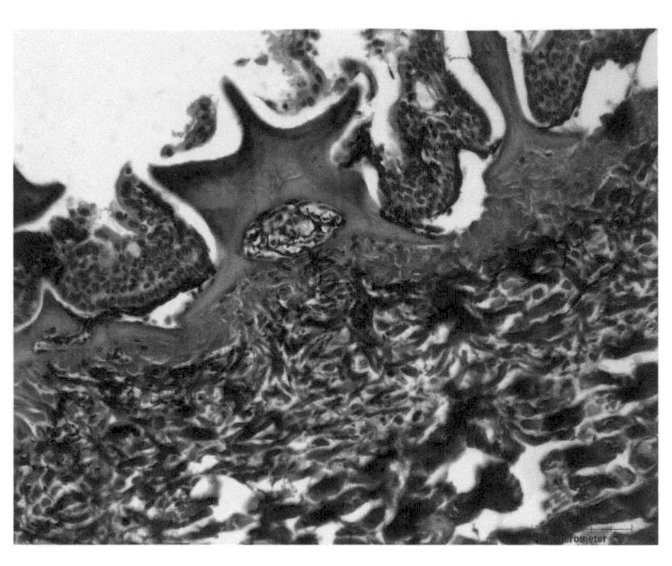

FIGURE J20a As we have seen, epidermal scales represent specializations of the epidermis with little involvement of the dermis. DERMAL SCALES, on the other hand, are specializations of the dermis—they may remain in the dermis, covered by epidermis, or they may penetrate the epidermis, like a mammalian tooth, and emerge on the surface. The placoid scales of the SHARK are dermal scales of this type, giving the skin the feeling of sandpaper.
Parts of three dermal scales are seen in this section through the skin of a dogfish SHARK. The scale is formed of hard DENTINE. The dentine is coated with harder VITRODENTINE, which has dissolved away in the preparation of the slide. These scales puncture the epidermis and appear on the surface but are rooted in the dermis. Here the epidermis has pulled away from the dermis during sectioning. Abundant chromatophores are seen in the dermis, especially at the base of the epidermis. The scale at the center is cut medially, showing its PULP CAVITY. 20×.

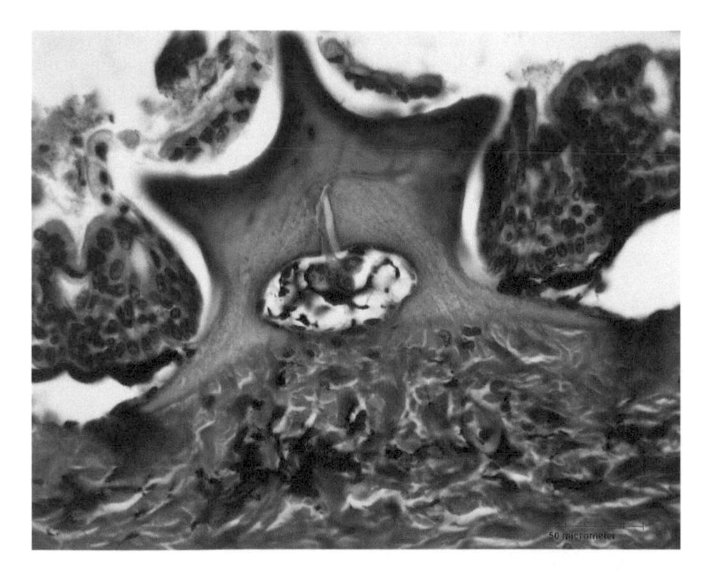

FIGURE J20b ODONTOBLASTS line the pulp cavity of this placoid scale in the skin of a dogfish SHARK. The pulp cavity also contains a chromatophore. 40×.

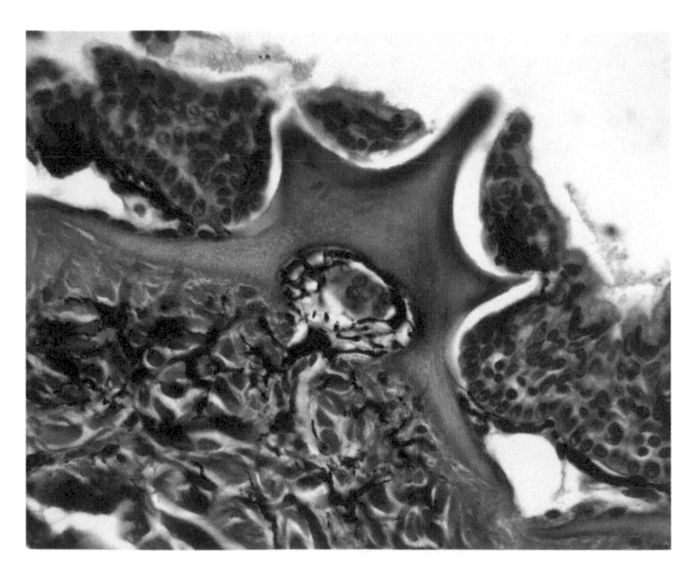

FIGURE J20c A blood vessel enters the pulp cavity through the BASAL PLATE of this placoid scale of a dogfish SHARK. 40×.

FIGURE J20d An early stage in the formation of a PLACOID scale is seen in this section of the skin of a young dogfish SHARK. Cells of the dermis are clustered beneath a group of specialized cells that form the ENAMEL ORGAN in the basal region of the epidermis. Although the enamel organ appears to guide the formation of the scale, it is not thought to produce enamel. The specialized clump of cells in the dermis is a SCALE BUD that will form the placoid scale. Note that the epidermis over the enamel organ and the dermis below the scale bud appears to be unspecialized. 40 ×.

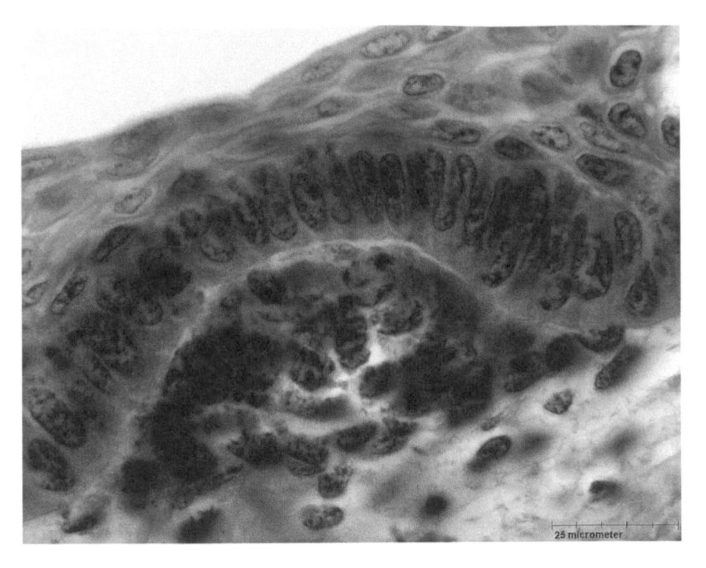

FIGURE J20e Note the features mentioned in Figure J20d, shown in this higher power view of the same scale bud in the skin of a young dogfish shark. 100 ×.

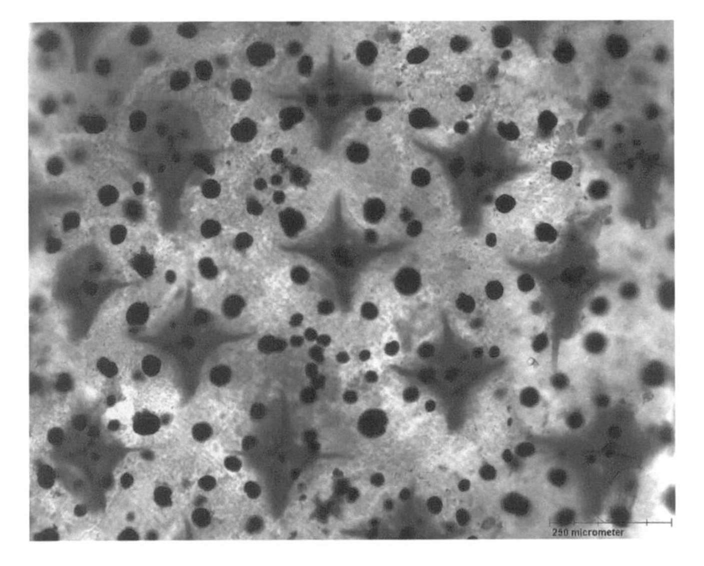

FIGURE J21a The regular spacing of placoid scales and their four-point star shape can be seen in this flat whole mount of the skin of a dogfish SHARK. Chromatophores (which have withdrawn pigment granules from their tentacles) are dispersed throughout. 10 ×.

FIGURE J21b The overlapping nature of dermal scales in a bony fish can be seen in this flat mount of the skin of a SHINER. The skin is liberally speckled with chromatophores. No stain has been used—the yellow color is from picric acid in the fixative. 5 ×.

alveolar POISON GLANDS (GRANULAR GLANDS) that are more eosinophilic than mucous glands. The "warts" and "parotid glands" of toads are aggregations of poison glands (Figs. J37a–J37c). Glands are rare in the hard, dry skin of REPTILES; those that are present are all multicellular alveolar glands, usually holocrine, and are scent glands, for the most part associated with sexual activity. The only integumentary gland present in BIRDS is the holocrine UROPYGIAL or PREEN GLAND located at the base of the tail (Figs. J38a–J38f). In this branched, alveolar gland it is easy to trace the gradual breakdown of the glandular cells in the production of the oily secretion.

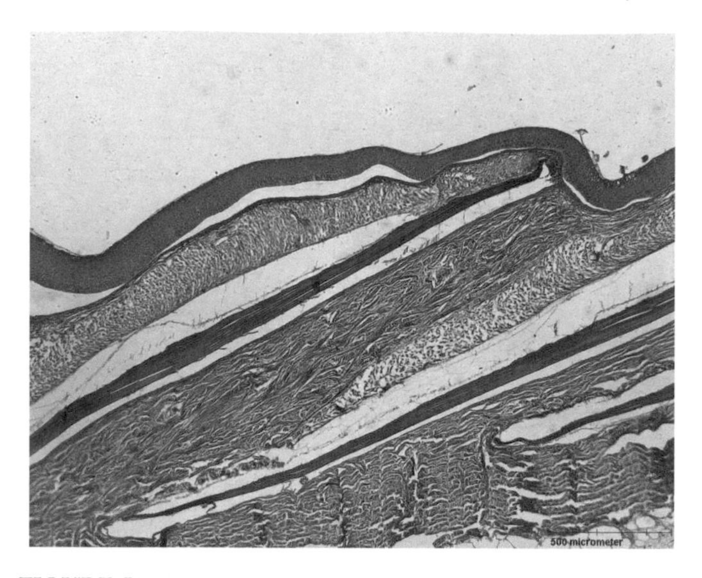

FIGURE J22a It is a misconception that the scales of a bony fish lie on the surface, much like shingles on a roof. On the contrary, these bony plates are contained within SCALE POCKETS in the dermis which in turn is covered by a blanket formed by the stratified squamous epithelium of the epidermis. The scale is a highly specialized formation produced by and contained within the dermis. The epidermis is unspecialized. Two scales, in their individual pockets, are shown in this section of the skin of a FLOUNDER. As a result of the hardness of the scale, there has been considerable distortion of the section. 5 ×.

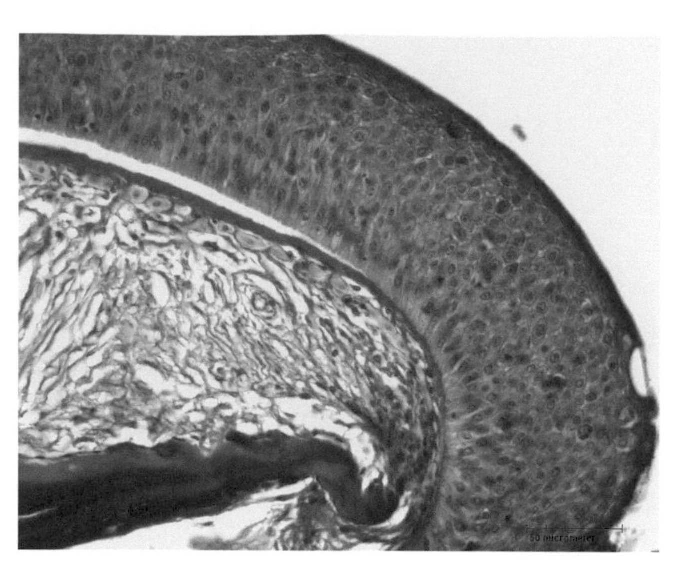

FIGURE J22b This enlargement of part of the previous micrograph shows a few mitotic figures in the stratified squamous epidermis. The epidermis rests on a dense layer of fibrous connective tissue (it has pulled away slightly). The scale is contained within a pocket of loose areolar tissue and is covered by thin squamous connective tissue cells. A few eosinophilic cells of unknown significance lurk among the epidermal cells. 40 ×.

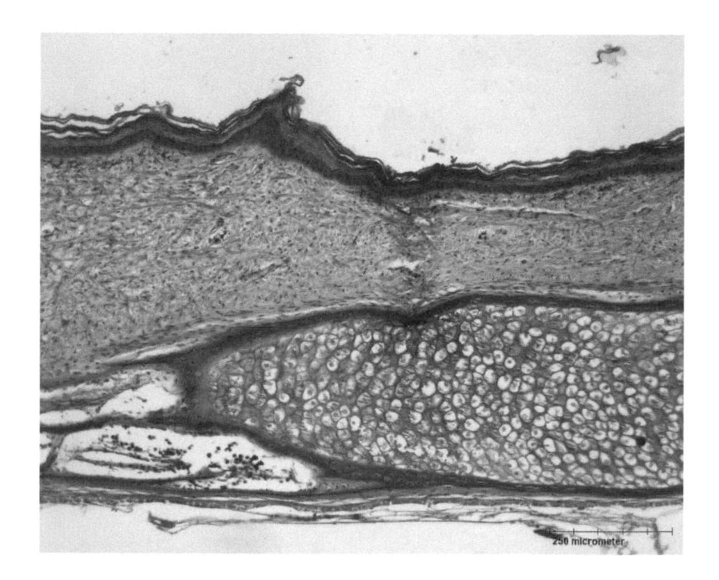

FIGURE J23 An interesting concurrence of epidermal and dermal scales is seen in the body wall of the snapping TURTLE. An epidermal scale exists in this section of the cornified, stratified squamous epidermis while, in the connective tissue of the dermis, a thick, cartilaginous dermal scale is seen. 10 ×.

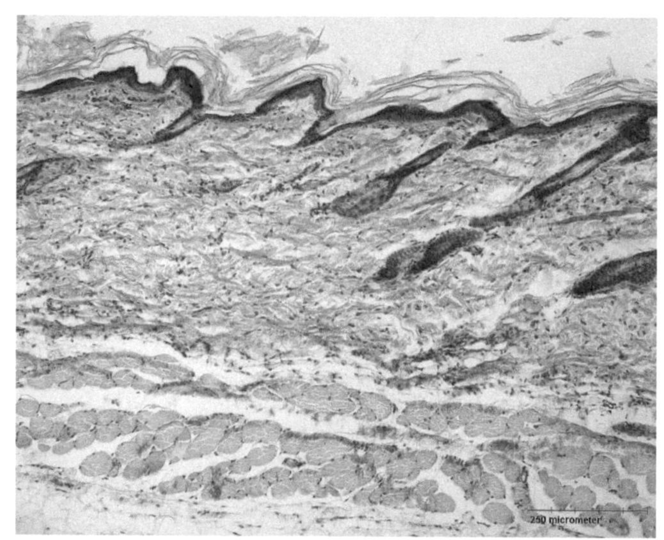

FIGURE J24a Mammalian hairs are cylindrical shafts of adherent, keratinized epithelial cells that emerge at an angle to the skin. Because of the thinness of each section, it is unlikely that a single hair will appear from its root to its emergence from the hair follicle. This is a section of the furry skin of a MAMMAL. The epidermis is a stratified, squamous, cornified epithelium: the outer layers of squamous cells are sloughing off. Beneath is a thick layer of leather, the dermis. Tubular extensions of the epidermis penetrate deeply into the dermis and constitute the hair follicles. Below the dermis is the areolar connective tissue of the hypodermis and subcutaneous striated muscle. 10 ×.

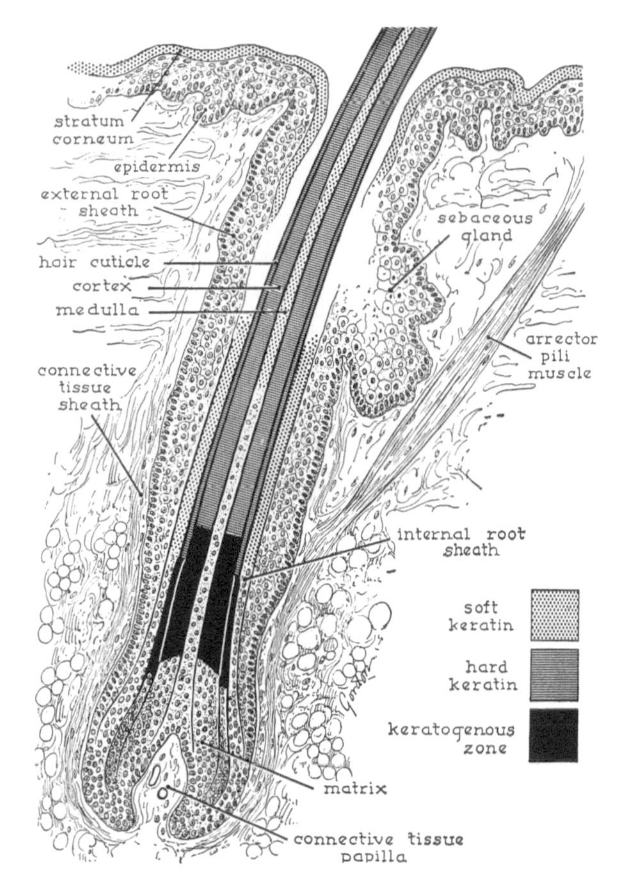

FIGURE J24b Diagram of a human hair in a hair follicle.

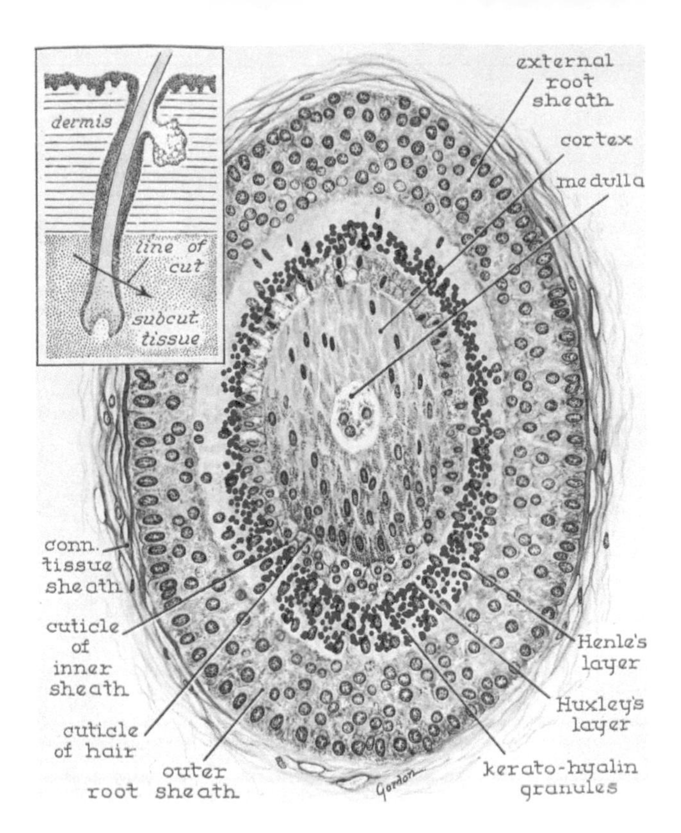

FIGURE J24c Drawing of an oblique section through the keratogenous zone of a human hair in its follicle. The colored granules of keratohyalin indicate the formation of soft keratin. The insert in the upper left shows where the hair follicle was sectioned.

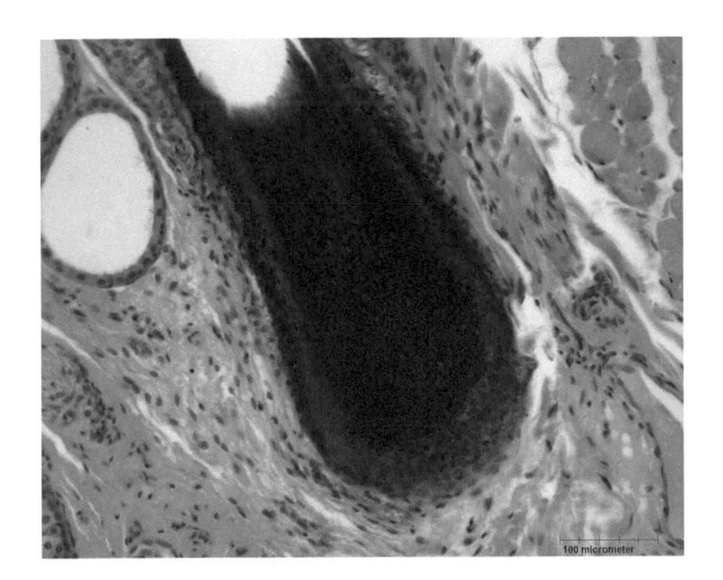

FIGURE J25 The hair follicle is seen to good advantage in this longitudinal section of a human eyelash; only the basal portion of the follicle is shown. The section does not pass through the hair papilla. The CORTEX is deeply pigmented; there is no medulla. A poorly defined layer at the periphery of the follicle forms the CONNECTIVE TISSUE SHEATH. Immediately inside this layer is a sleeve of actively dividing cells of the ROOT SHEATH that have grown down as a tube from the stratum germinativum of the epidermis, maintaining their propensity for division as they go. Two layers of this sheath are recognized, the EXTERNAL and INTERNAL ROOT SHEATHS. Like other cells that have arisen from the stratum germinativum, cells of the internal root sheath cornify and form a scaly covering of dead cells on the hair, the CUTICLE; in humans, these cells emerge on the surface of the scalp and, together with epidermal cells desquamated from the surface of the scalp, form the dreaded dandruff of television commercials. A few striated muscle cells that move the eyelid can be seen in the upper right corner. The large opening at the top left is a section through the ciliary gland (of Moll), a modified apocrine sweat gland that opens onto the free surface of the eyelid or into the hair follicle. 20 ×.

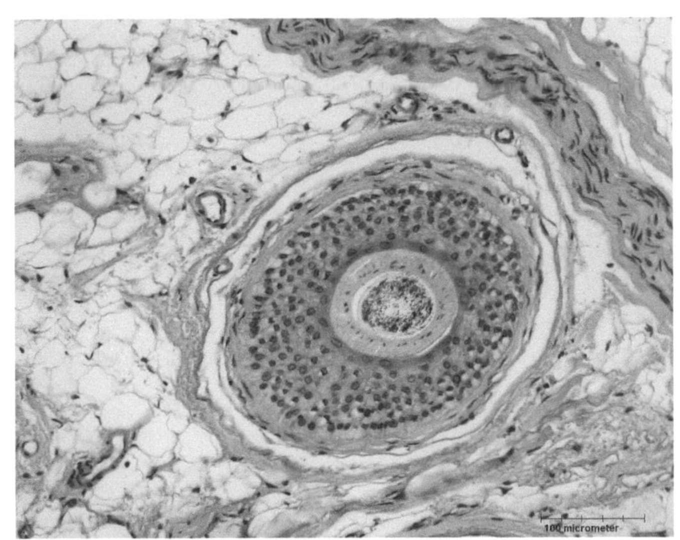

FIGURE J26 This tangential section of a HUMAN SCALP shows a cross section of a hair follicle surrounded by fatty connective tissue. The hair shaft consists of the pigmented MEDULLA at its center surrounded by pale cells of the CORTEX and a single layer of cells of the CUTICLE. Most of the cells surrounding these layers constitute the EXTERNAL ROOT SHEATH; the internal root sheath has become indistinct at this level. A CONNECTIVE TISSUE SHEATH enwraps the cylinder of the hair. The hair, surrounded by these layers, has pulled away from the connective tissue of the scalp. 20 ×.

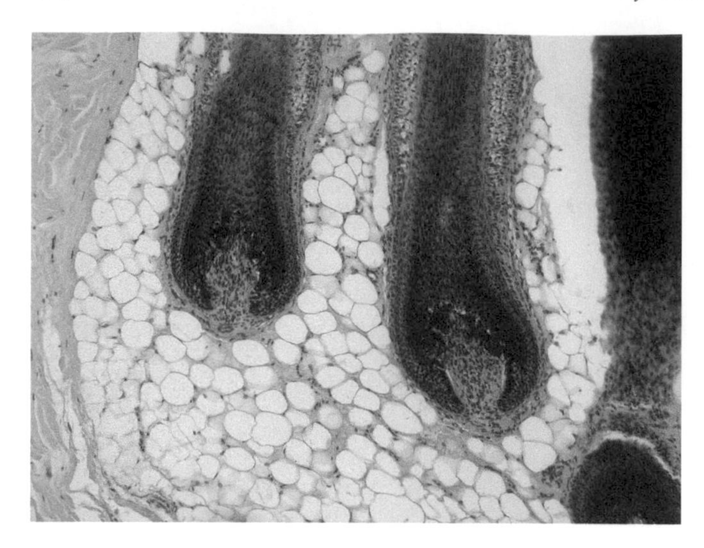

FIGURE J27a Medial longitudinal sections of three hair follicles are shown in this vertical section of human scalp. Connective tissue from the dermis forms the HAIR PAPILLA that is capped by epithelial cells of the HAIR MATRIX. Active mitosis (not seen) of these epithelial cells produces the cells of the hair as well as the internal root sheath, a sleeve of epithelial cells surrounding the hair. The dermis contains abundant adipose cells. The cells of the internal root sheath and the hair are pushed up by the addition of new cells at the matrix. MELANIN PIGMENT is present in the cells at the tip of the papilla and in the hair itself. 10 ×.

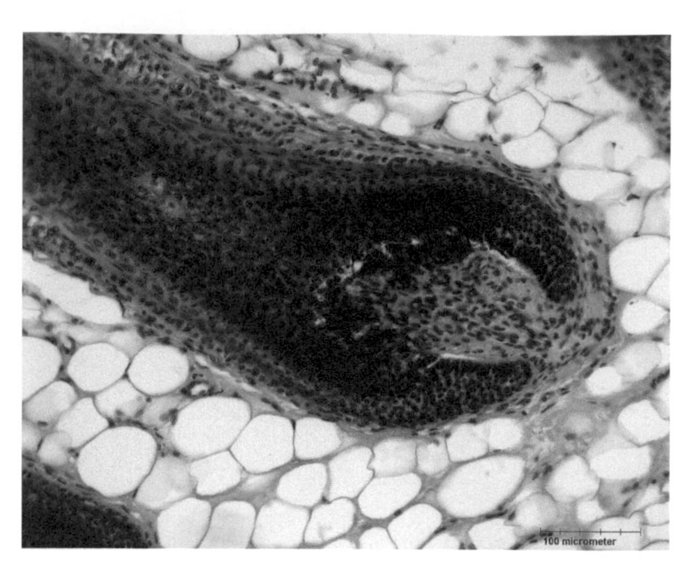

FIGURE J27b Greater detail of one of the follicles is shown Figure J27a. A sleeve of connective tissue surrounding the hair consists of OUTER AND INNER ROOT SHEATHS. The cells of the hair itself form an ill-defined CORTEX and MEDULLA and may be distinguished by their content of pigmented cells. 20 ×.

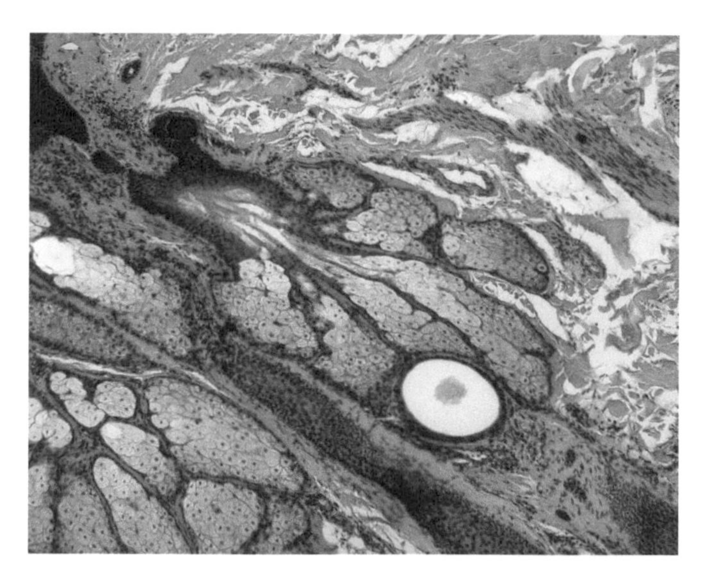

FIGURE J28a This section passes through a sebaceous gland in the HUMAN SCALP. This holocrine gland is made up of several sac-like structures that pour their oily secretion into the lumen of the gland at the upper left of center. Cells divide by mitosis at the periphery of each sac and are pushed to the center. As they go the cells accumulate an oily material and die leaving a mixture of oil and débris in the lumen. With the assistance of a nudge from the ARRECTOR PILI muscle, this secretion pours out onto the scalp at the base of the hair; the débris contributes to the dandruff mentioned in Figure J25 above. Fragments of the arrector pili are seen at the upper right. The round structure at the lower right is a section through a duct. 10 ×.

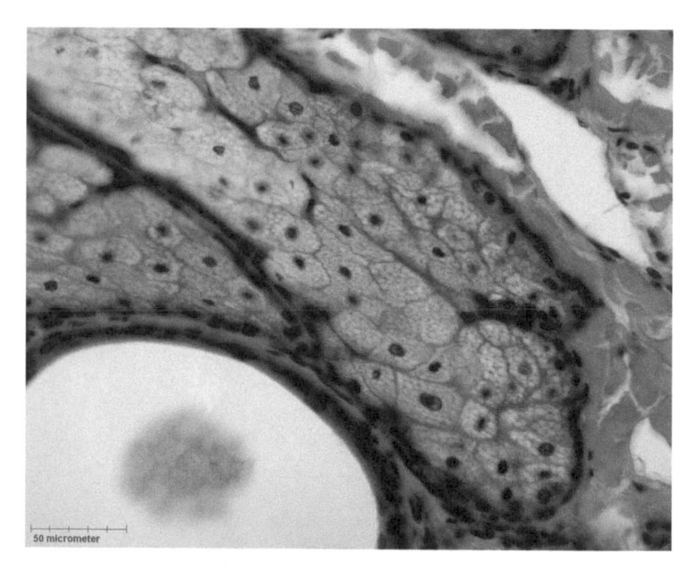

FIGURE J28b The birth and death of cells in a holocrine gland is seen in this enlargement of the previous micrograph. Cells are produced by mitosis at the periphery of the lobule; they can be seen to accumulate discrete droplets of oily secretion and, at the upper left, the cells die leaving débris and the oily secretion to be released into the hair follicle and from there to the surface of the scalp. 40 ×.

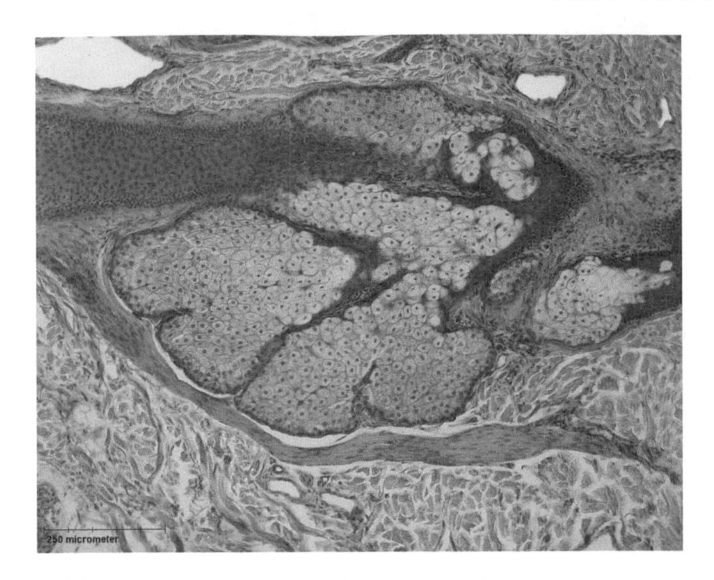

FIGURE J28c This section of human scalp shows the ARRECTOR PILI to advantage. This band of smooth muscle attaches to the hair follicle at the upper left and passes to the lower right, enclosing the sebaceous gland; involuntary contractions of this smooth muscle force the oily secretions into the hair follicle and from there to the surface. 10 ×.

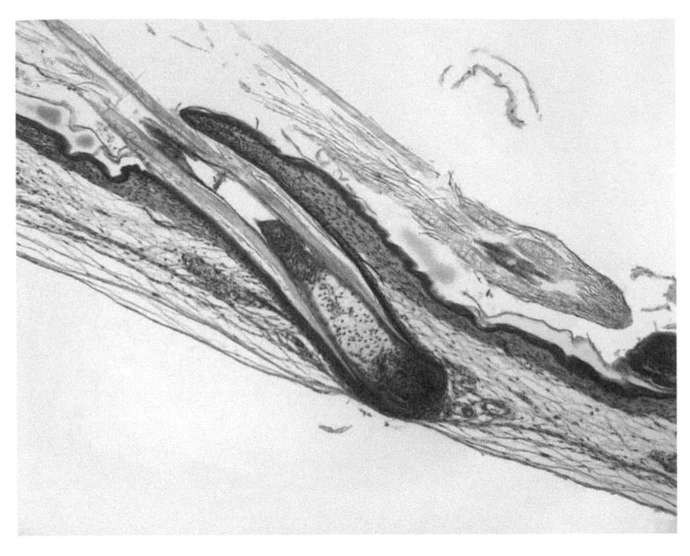

FIGURE J29a The FEATHER FOLLICLE of a newly hatched CHICK bears a superficial resemblance to a hair follicle. It is a tubular invagination of the epidermis that produces a structure of hard keratin. At the base of both the hair and the feather is a vascular PAPILLA derived from the stratum germinativum of the epidermis that becomes invaded by vascular dermal connective tissue. Division of cells in both the hair and feather follicles produces keratinized, squamous epithelial cells that adhere together. While these cells form the simple, cylindrical structure of a hair, they unite to produce the incomprehensibly complex feather, consisting of hollow CALAMUS and rachis; the vanes, barbs, and barbules. This is a section of the skin of a newly hatched chick; vascular connective tissue fills the hollow core of the developing feather. When the feather is mature, this tissue dries to a pith. The follicle is firmly rooted in the connective tissue of the dermis. A few fibers of smooth muscle of the ARRECTOR PLUMAE serve to move the feather. 10 ×.

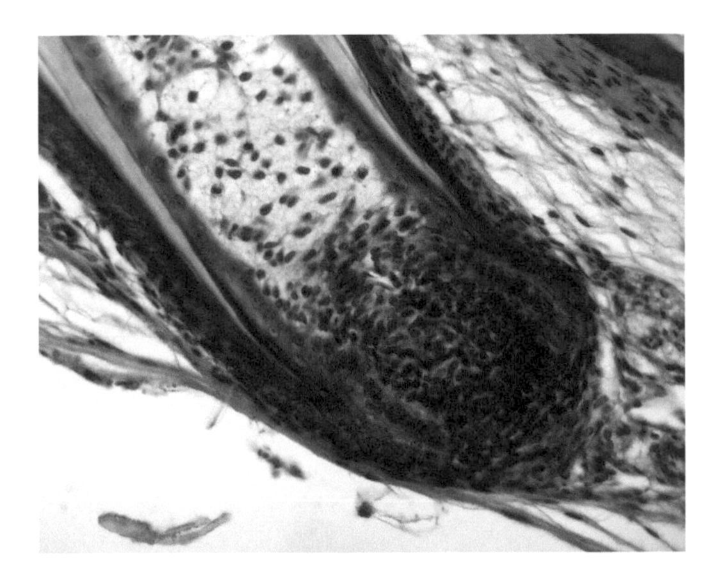

FIGURE J29b This is a higher power picture of the same feather follicle as seen in the previous micrograph. A few strands of smooth muscle of the ARRECTOR PLUMAE are attached to the base of the feather follicle. PULP of loose areolar tissue fills the hollow cavity of the feather. Keratin is formed by the activity of the cells of the invaginated stratum germinativum of the epidermis. 40 ×.

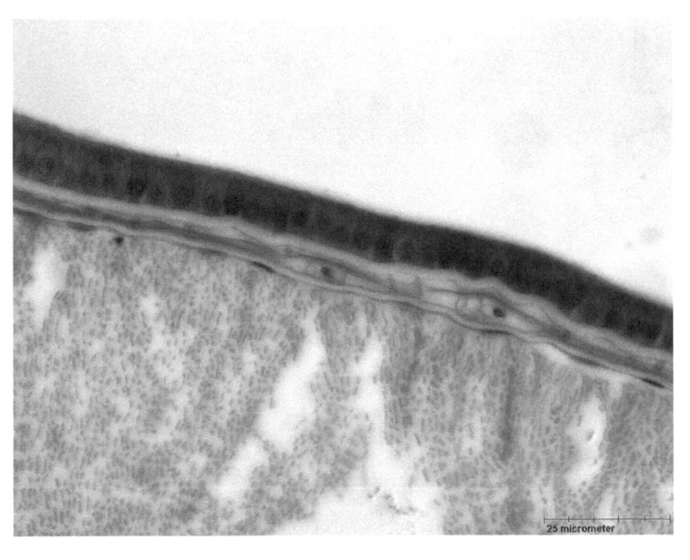

FIGURE J30 A few merocrine goblet cells are scattered among the simple columnar epithelial cells in the epidermis of AMPHIOXUS. Below the epidermis is the loose vascular connective tissue of the dermis, which rests on the striated muscle of the body wall. 100 ×.

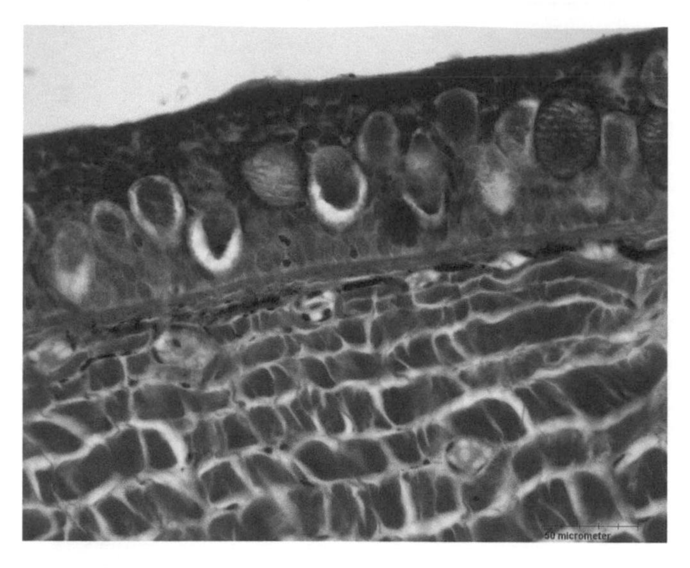

FIGURE J31a The EPIDERMIS of an agnathan is stratified and shows a complex array of different cell types. This is a section through the body wall of a HAGFISH. Two types of secretory cells can be differentiated in the epidermis: mucous cells (blue–gray) and thread cells (orange). The thread cells are packed with a tightly coiled, single thread consisting of a large number of intermediate filaments bundled in parallel (reddish). Most of the epidermal cells remain less differentiated and form a protective layer over the body. The DERMIS consists of tough parallel bands of collagenous fibers. 40 ×.

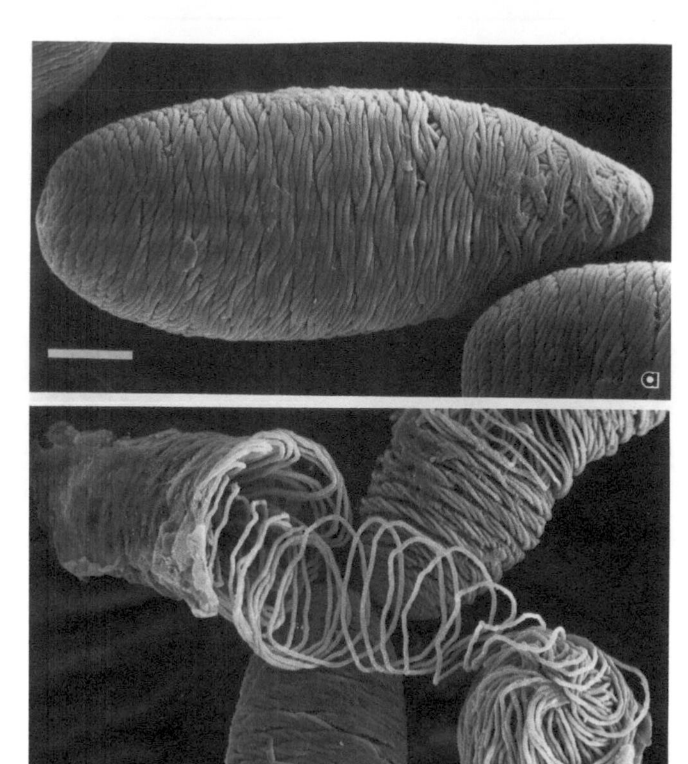

FIGURE J31b Scanning electron micrographs of two isolated thread cells from the skin of a hagfish. Bars = 20 μm. (Top) An isolated thread cell without its plasma membrane. The precise packaging of a single thread contained in each thread is evident. (Bottom) A thread cell that has been pulled apart, revealing how the thread is packaged within the cell as a series of sequential loops.

FIGURE J32a Three types of cells can be identified in the stratified epidermis of the larval lamprey: the basal, pale CLUB CELLS, round, acidophilic GRANULAR CELLS, at a mid-level of the epithelium, and MUCOUS CELLS. The epidermis rests on a thin, grayish band of collagenous tissue of the DERMIS. The HYPODERMIS, consisting mainly of adipose cells, rests on the thick muscular layer of the body wall. 40 ×.

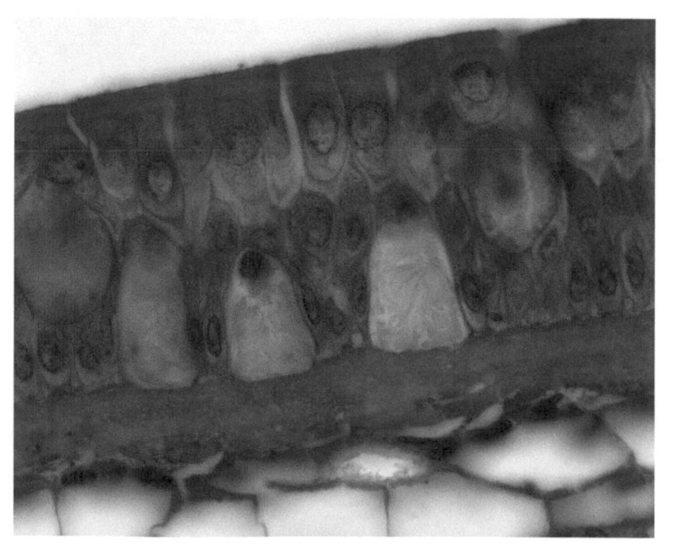

FIGURE J32b The highly differentiated cells of the epidermis of the lamprey skin arise from division of BASAL CELLS. The most abundant cells in this section are MUCOUS CELLS. The largest cells are the enigmatic CLUB CELLS, containing whorls of material. Their nuclei are apical—often two nuclei are seen in a single club cell. Two round GRANULAR CELLS hover at a mid-level in this section—they contain acidophilic granules. It has been suggested that these cells release a poison, which renders the ammocoete unattractive to potential predators, or that they assist in the release of mucus from mucous cells. A few other round cells at this level are probably younger granular cells. 100 ×.

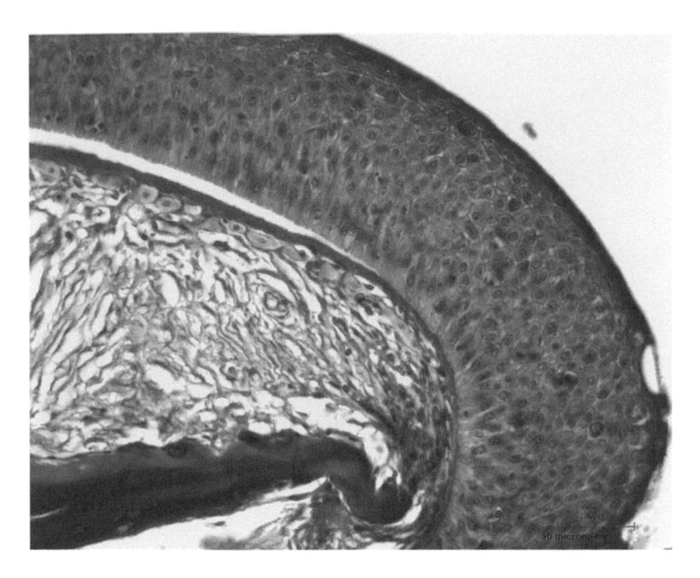

FIGURE J33 Simple goblet cells are seen at the surface this section of the skin of a FLOUNDER. They are produced by basal cells in the epidermis and accumulate mucus as they rise to the surface to discharge their contents. A scale, in its pocket, lies at the lower left. 40 ×.

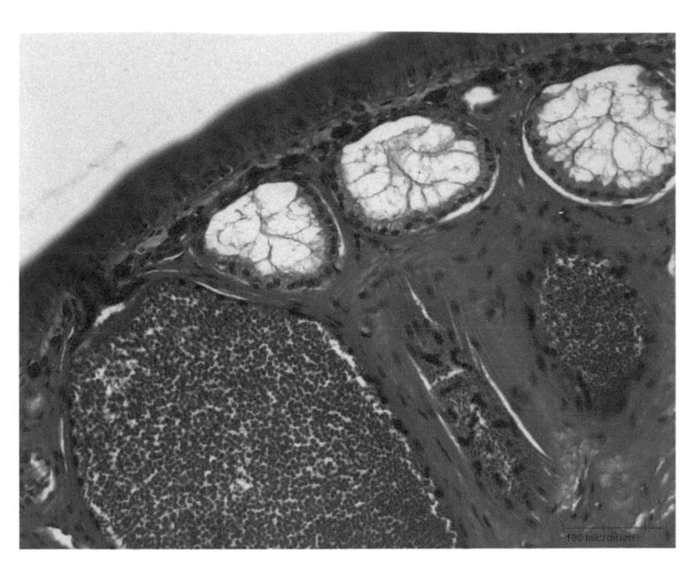

FIGURE J34a This section from a frog is as a typical example of the skin of an amphibian. The stratified squamous epithelium provides a tough outer jacket. It encloses groups of pigment cells at the junction of epidermis and dermis. Below are three alveolar mucous glands. To their right is an enormous alveolar poison gland packed with granules. 40 ×.

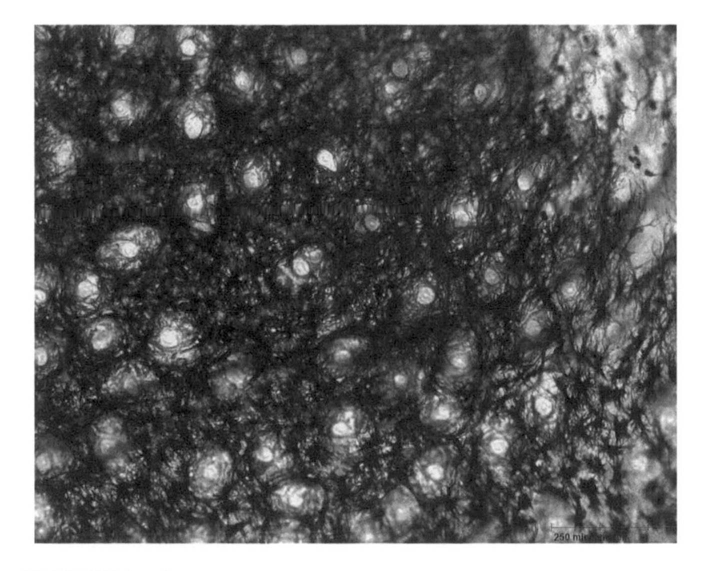

FIGURE J34b This is an unstained whole mount of the skin of a frog. The regularly spaced openings of alveolar glands are seen within the fibrous mesh formed by the delicate processes of chromatophores. 10 ×.

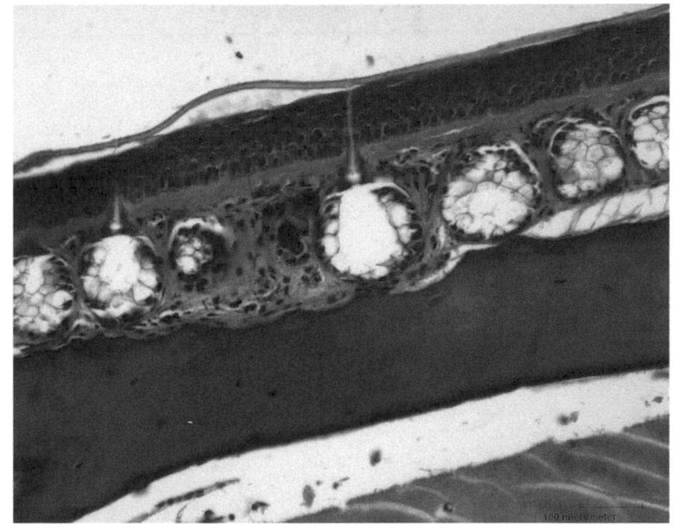

FIGURES J35a Section of a limbless amphibian from Africa, Order Gymnophiona. Although the animal looks much like an earthworm, its skin resembles that of other amphibians. This section shows only alveolar mucous glands. Elsewhere in its skin alveolar poison glands are visible. The lower half of the section consists of dense collagenous tissue. 20 ×.

Glands of Mammals

1. *Sebaceous glands.* The holocrine sebaceous glands are epidermal in origin and are scattered in the dermis of the skin. They are usually, though not always, associated with a hair follicle.
2. *Sweat or sudoriferous glands.* These simple, tubular glands are scattered in the skin of man and certain other mammals. The secretory tubules are coiled and occur in the dermis. Two types are recognized: merocrine ECCRINE SWEAT GLANDS and APOCRINE SWEAT GLANDS.
   a. *Eccrine sweat glands* function throughout life. They are lined by three types of cells (Figs. J39a–J39c, J40a, J40b). MYOEPITHELIAL CELLS are flattened and parallel to the axis of the tubule; they rest on the basement

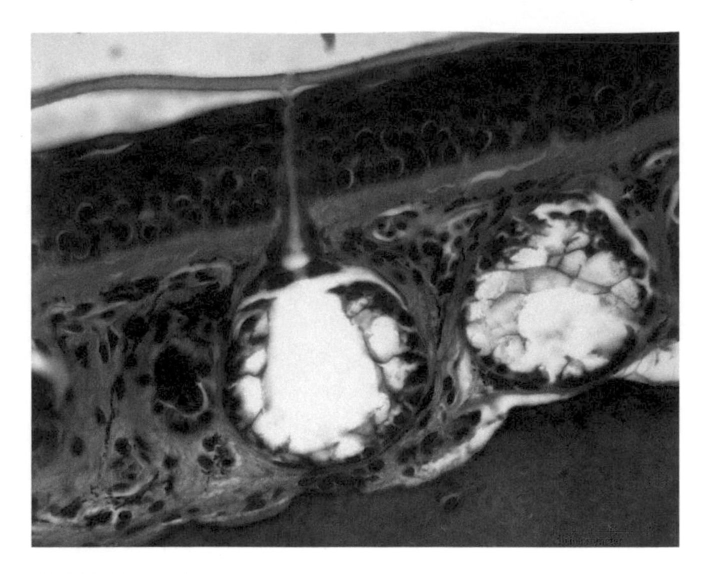

FIGURES J35b Section of a limbless amphibian from Africa at higher magnification. Only two alveolar mucous glands are shown and the duct of one gland penetrates to the surface. The lower right corner of this image show the dense collagenous tissue. 40 ×.

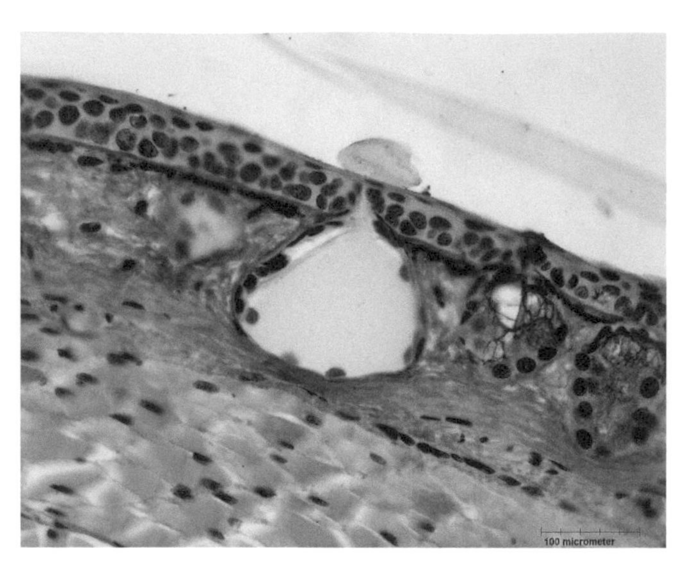

FIGURE J36a Section of the skin of *Amphiuma* showing an alveolar mucous gland discharging on the surface; there are two alveolar poison glands to its right. The epidermis consists of a stratified squamous epithelium about four cells in thickness. Below is areolar connective tissue with a few chromatophores in the upper levels. It rests on dense collagenous tissue. 20 ×.

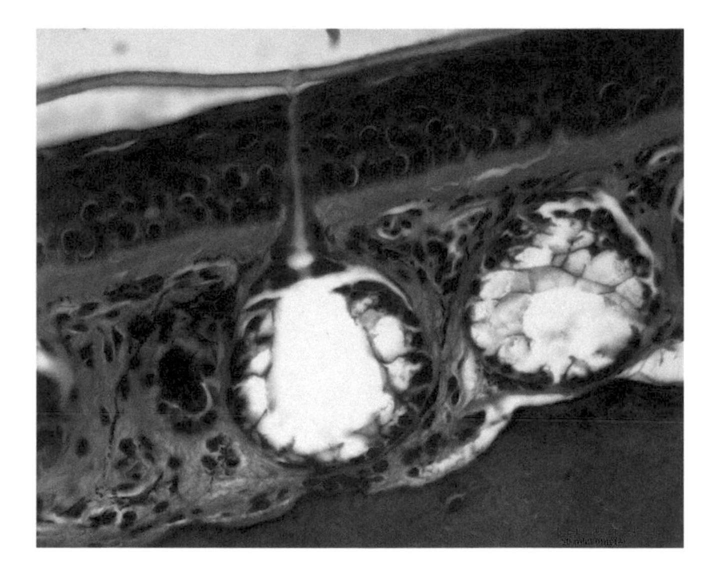

FIGURE J36b Enlarged view of the section described in Figure J36a, above. The large alveolar mucous gland is at the center; part of a poison gland is at the right. 40 ×.

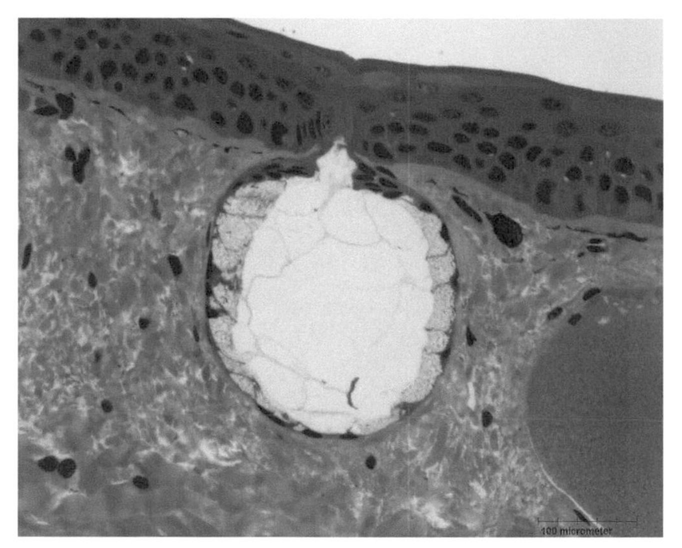

FIGURE J36c Alveolar mucous gland from a semithin section of the skin of *Amphiuma*; a portion of its duct is shown. On its right is a poison gland. Both glands have invaded the dermis. 20 ×.

membrane. The GLANDULAR EPITHELIUM consists of simple columnar CLEAR SEROUS CELLS whose cytoplasm may be clear and vesicular when the cell contains secretion. Scattered between these cells are basophilic DARK MUCOUS CELLS. The duct is usually coiled and bordered by a double row of flattened cells. In man the duct opens onto the surface by a PORE but in some mammals it opens into a hair follicle.

The functions of the cells of the eccrine sweat gland may be inferred from electron micrographs (Fig. J41). Contractile filaments are abundant in the myoepithelial cells. The mucous cells appear secretory with their rough endoplasmic reticulum and proteoglycan granules; they are thought to secrete glycosaminoglycans. The clear cells show the elaborate surface modifications of actively transporting cells: apical folds and microvilli and interfolding of the lateral plasma membranes. Agranular endoplasmic reticulum, mitochondria, and glycogen are present but the clear cells contain no granules and only small amounts of granular endoplasmic reticulum.

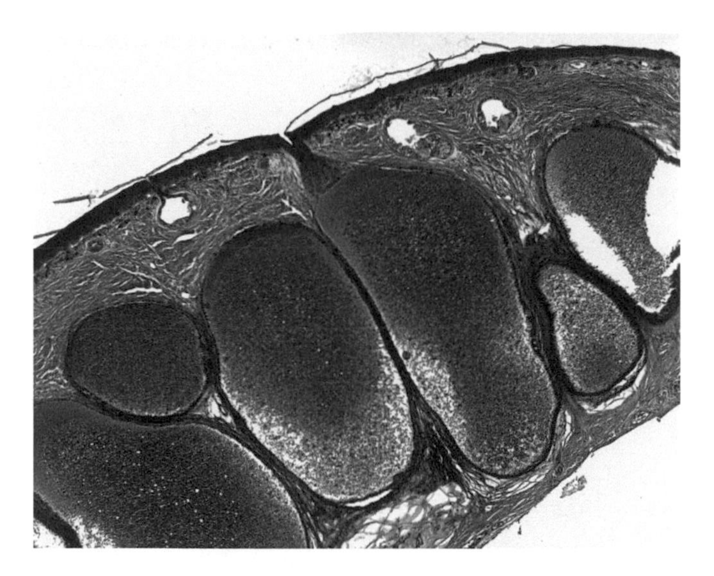

FIGURE J37a Alveolar poison glands are especially conspicuous in the skin of a toad, groups of them forming the characteristic "warts." Two of the largest warts lie between the ears and are known as PAROTID GLANDS. Several large poison glands are seen in this section through a parotid gland. Note the duct to the surface from the large poison gland to the right of center. There are a few small alveolar mucous glands immediately below the epidermis; the duct of one of them makes its way to the surface of the skin. 5×.

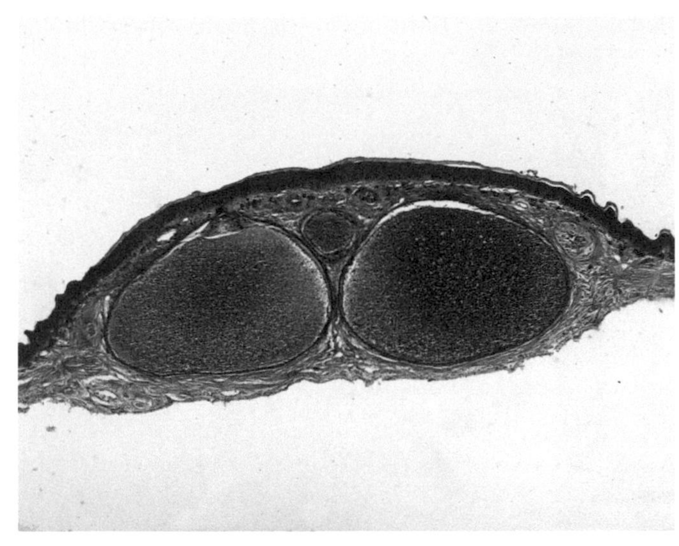

FIGURE J37b Two large alveolar poison glands of the parotid gland of the toad lie side by side in the dermis. A duct on the alveolus at the left is directed toward the surface. 5×.

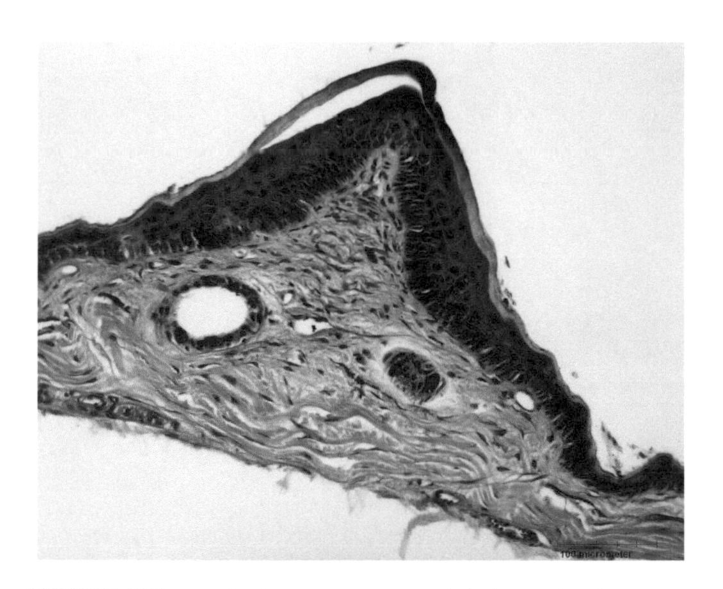

FIGURE J37c Only the upper portion of the huge alveolar poison gland from the parotid gland of a toad is shown. Note the lower portion of its duct to the surface. A small alveolar mucous gland lurks in the connective tissue immediately below the epidermis near the center of the picture. Chromatophores are abundant in the upper levels of the dermis. 20×.

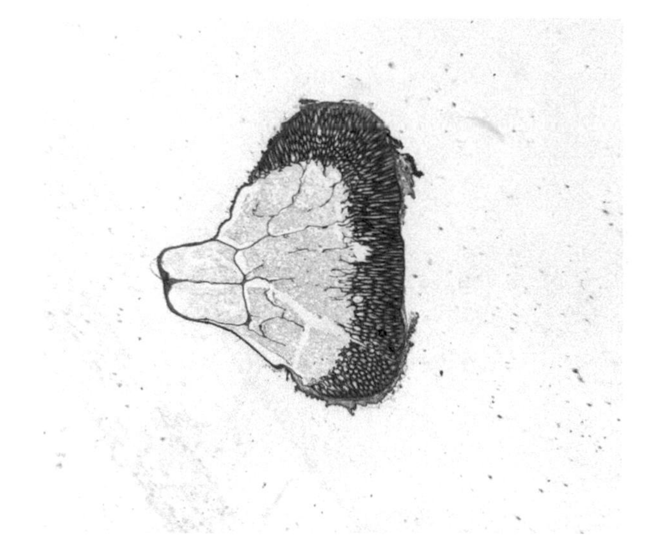

FIGURE J38a The only integumentary gland present in birds is the large holocrine UROPYGIAL or PREEN GLAND located at the base of the tail. Since it would be counterproductive to produce oil at the base of each large feather, where it would form an oily mess over the surface of the skin, the bird has consolidated all of its sebaceous glands in one spot where the secretion can be withdrawn from a nipple by means of the beak and distributed to the feathers. The gland shown here is from a cowbird and consists of a mass of simple tubular, holocrine glands that open into a central reservoir containing the oily secretion and débris from breakdown of the glandular cells. Cells at the base of the glands divide by mitosis and produce cells that develop droplets of oily secretion as they move toward the nipple. During this migration the cells breakdown so that only the oily secretion and cellular débris arrives in the reservoir to be secreted into the bird's mouth when it squeezes on the nipple before preening its feathers. The nipple is shown at the left of the micrograph. 1×.

FIGURE J38(b–e) This is a series of micrographs of the uropygial gland from the base to the uropygial nipple from a small BIRD, the junco. 40×.

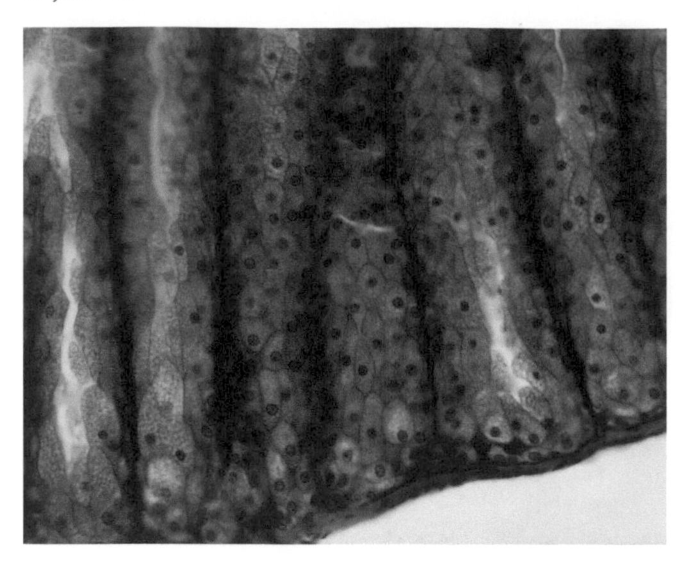

FIGURE J38b The simple tubular glands rest on a thin capsule of connective tissue and smooth muscle. Droplets of oily secretion have accumulated within the cytoplasm of the secretory cells. 40×.

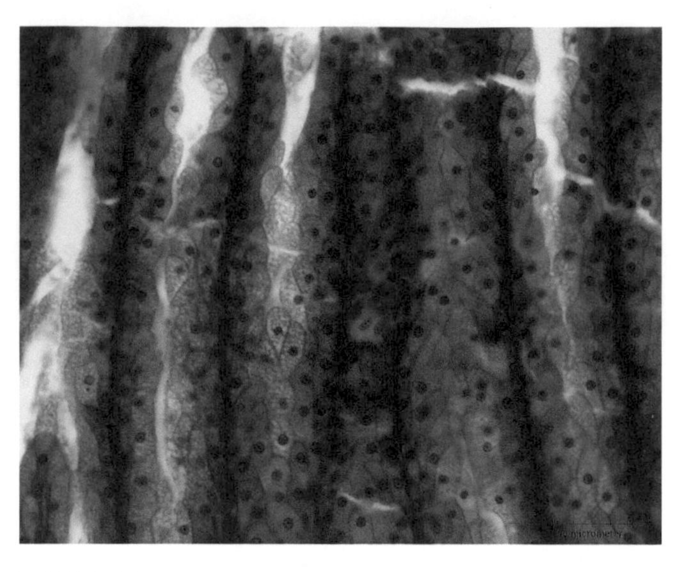

FIGURE J38c Cells containing the oily secretion migrate toward the nipple. 40×.

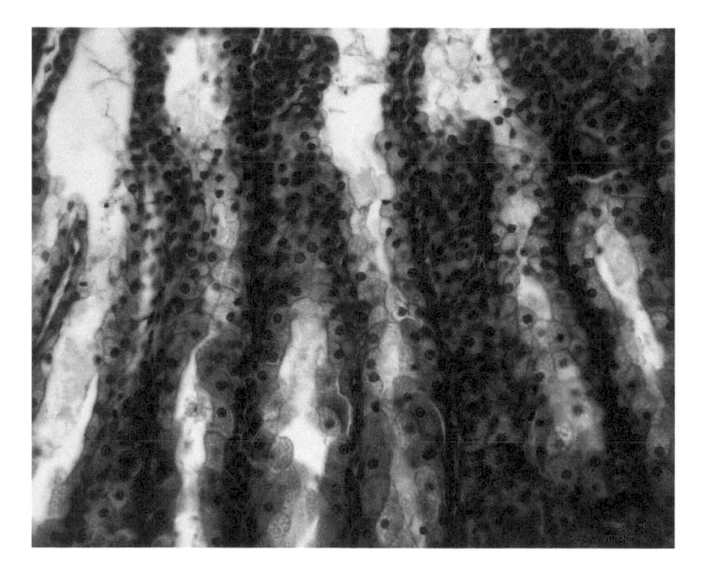

FIGURE J38d As they approach the reservoir the secretory cells breakdown and the lumina of the glands become distended with the oily secretion and cellular débris. 40×.

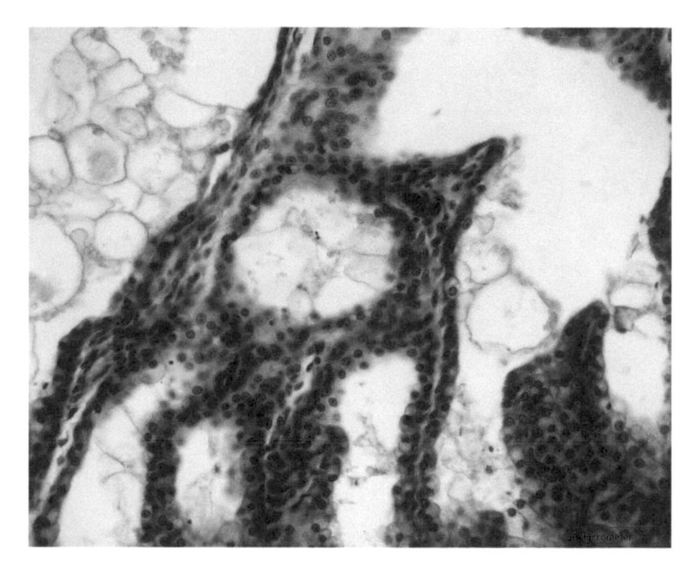

FIGURE J38e The oily secretion, cellular débris, and a few epithelial cells along with some connective tissue remain in the lumen of the reservoir. 40×.

b. *Apocrine sweat glands.* Certain specialized sweat glands produce an apocrine secretion: the CERUMINOUS or WAX GLANDS of the external auditory canal, the CILIARY GLANDS (of Moll) of the eyelids (Figs. J42a–J42d), some sweat glands of the axilla, etc. They do not function until puberty. Various mammalian scent glands are all of this type. Like eccrine sweat glands, apocrine glands are coiled and tubular; they may be distinguished, however, by their great distension with stored secretory products. The duct is similar to that of eccrine sweat glands. There is only one type of secretory cell: it may be cuboidal to squamous, depending upon the degree of distension of the gland, and rests on a thick, hyaline basement membrane. Myoepithelial cells are present. The term *apocrine* was applied to these glands at a time when it was thought that blebs from the secretory cells were lost into the lumen along with the secretion. Electron micrographs have shown that there is no such loss but the name has persisted and is still used to designate these cells.

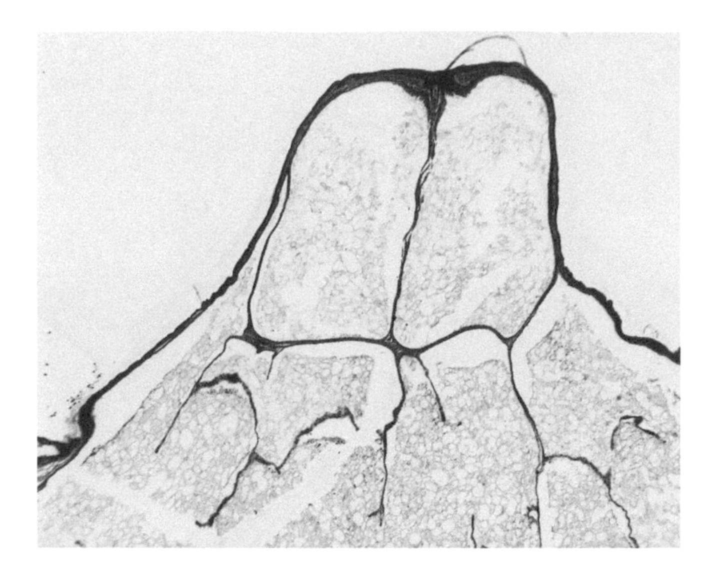

FIGURE J38f The bird squeezes the nipple with its beak and receives a mouthful of oily secretion which it applies to its feathers during preening. 4×.

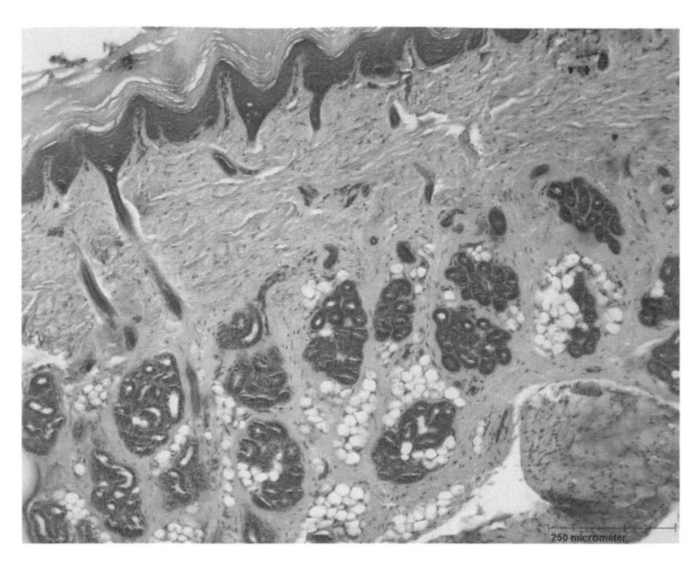

FIGURE J39a This is a section of the skin from the footpad of a squirrel. The coiled secretory portions of several sweat glands, along with a few fat deposits, are scattered throughout the dermis. Portions of the ducts from these glands, seen in longitudinal section, extend from the stratum germinativum of the epidermis and carry sweat to the surface where it keeps the sole of the foot moist and assists in traction. 10×.

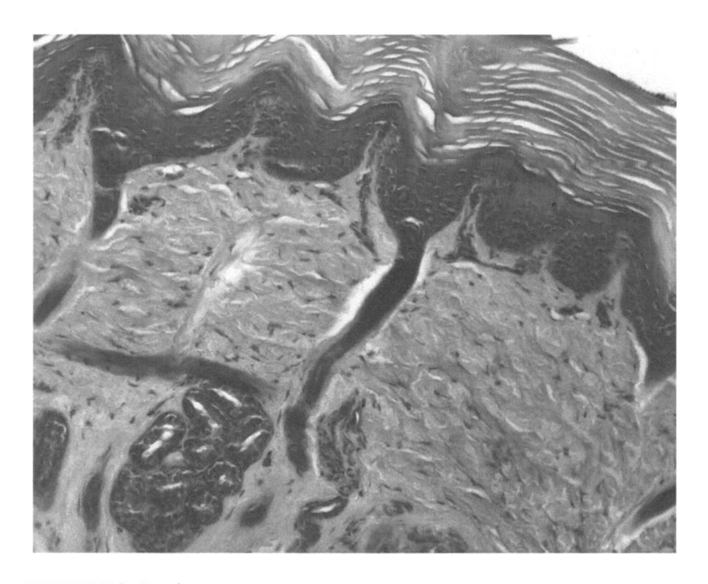

FIGURES J39b The body of a sweat gland. A duct connects to the base of the epidermis. 20×.

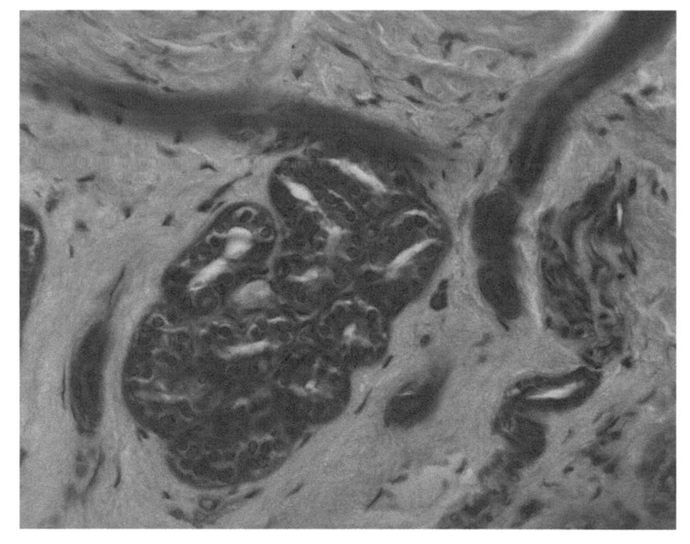

FIGURES J39c The glandular epithelium of the sweat gland is simple columnar cells with a clear and vesicular cytoplasm when the cells contain secretions. 40×.

3. *Mammary gland*. Mammary glands are present in both sexes and develop only slightly before puberty, when they enlarge rapidly in the female, largely as the result of the development of adipose and other connective tissue (Figs. J43a and J43b). They remain incompletely developed until pregnancy occurs (Fig. J43c). A section of active mammary gland is composed of compound, branched, tubuloalveolar glands (Fig. J43d). ALVEOLI are numerous in the active gland (Fig. J43e). The SECRETORY EPITHELIUM varies from simple cuboidal to columnar and the cells often contain droplets of fat in their distal portions. The spherical nuclei are usually located in the basal portion of the cell. These cells rest on a basement membrane. There are MYOEPITHELIAL CELLS resembling those of sweat glands (Fig. J44a). The INTRALOBULAR DUCTS are lined by a single or double row of cells that rest on a basement membrane surrounded by the condensed adjacent connective tissue.

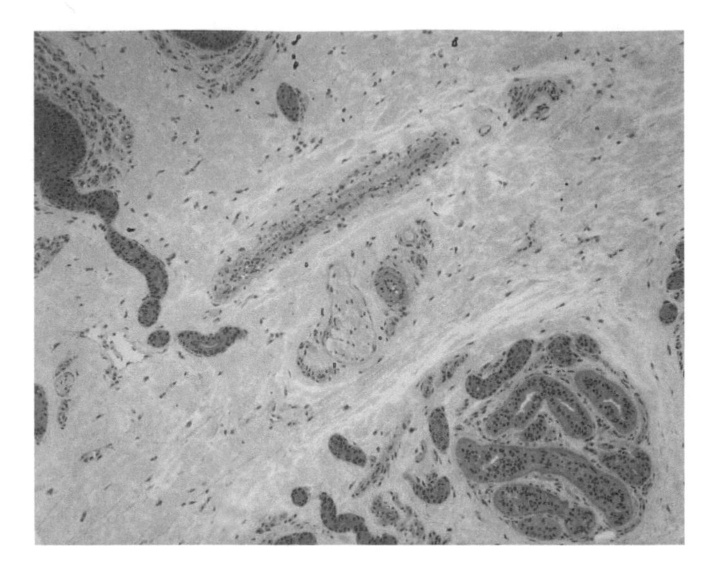

FIGURE J40a This is a semithin section through the palmar skin of a monkey showing the secretory portion of a sweat gland coiled in the dermis at the lower right. A duct, which could have arisen from this gland, ascends to the surface where it penetrates the epidermis to pour its watery secretion onto the surface. 10 ×.

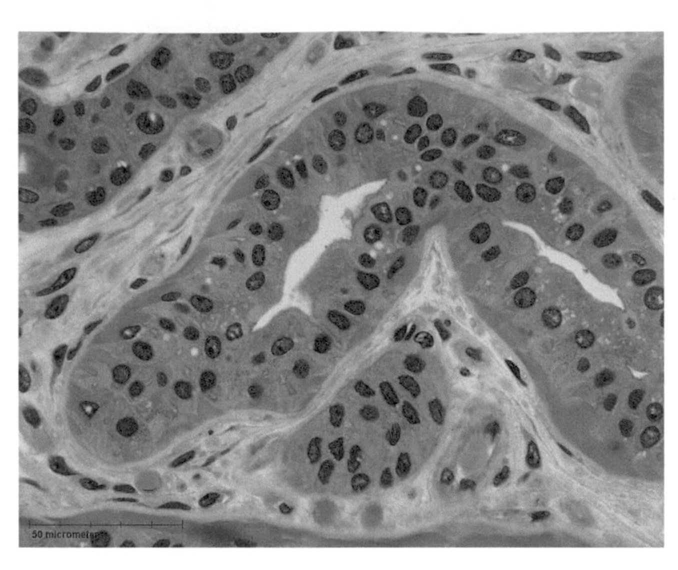

FIGURE J40b A few cells from the coiled portion of a sweat gland from the palm of a monkey are enclosed by a few spindle-shaped MYOEPITHELIAL CELLS, which assist in the expulsion of the secretion. 63 ×.

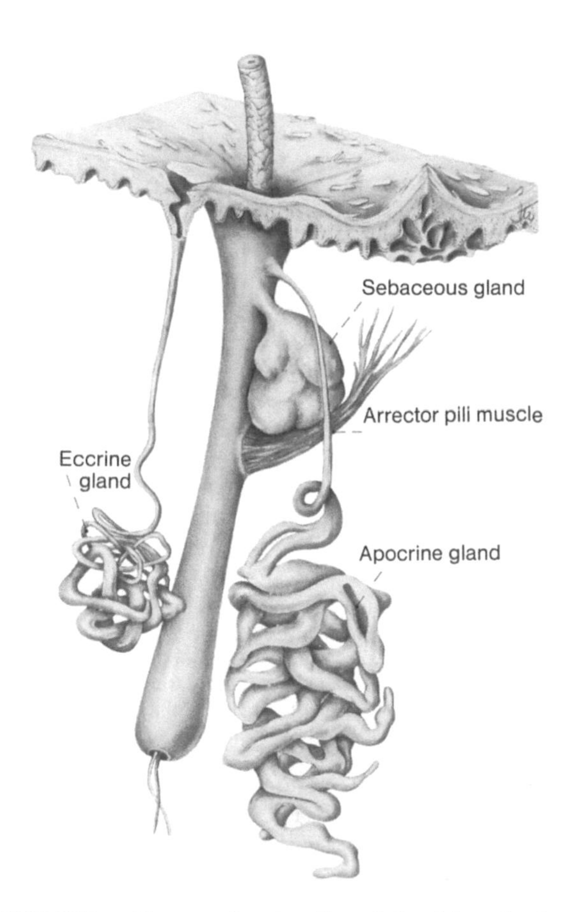

FIGURE J40c Diagram showing the relationships of sebaceous, eccrine, and apocrine glands in the mammalian skin.

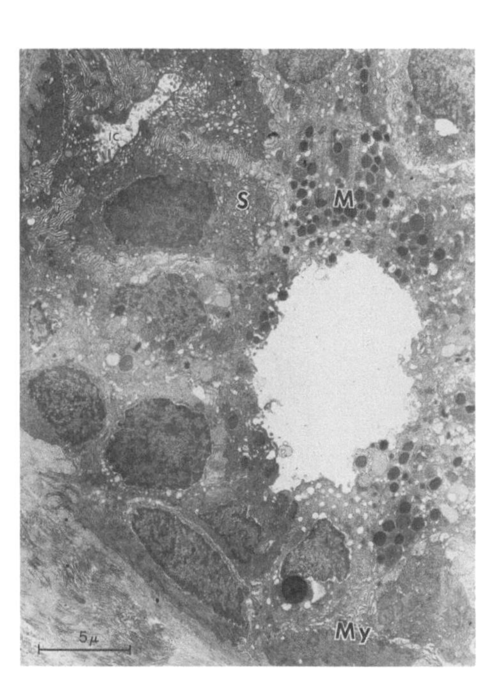

FIGURE J41 Transmission electron micrograph of the secretory portion of an eccrine gland. Two types of glandular cells surround the lumen of the gland: serous cells (S) and mucous cells (M). Myoepithelial cells (My) enwrap the glandular epithelium. Eccrine sweat glands are capable of producing large amounts of sweat as evidenced by the microvilli (IC) line the intercellular channels and the appearance of apical and basal elaborations of the cell membranes.

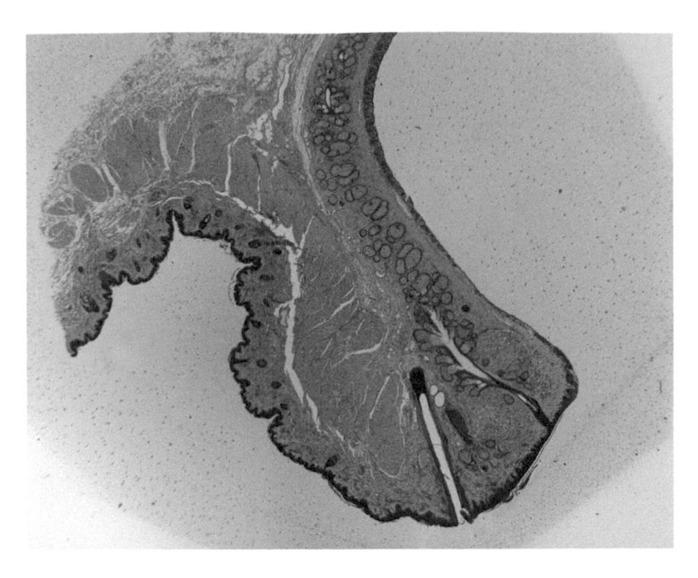

FIGURE J42a This low-powered view of the mammalian eyelid shows several glands. The largest is a modified sweat gland at the conjunctival surface of the eyelid, the ciliary gland (of Moll). Immediately adjacent to the hair follicle (of the eyelash) is a sebaceous gland, the tarsal (Meibomian) gland. 1 ×.

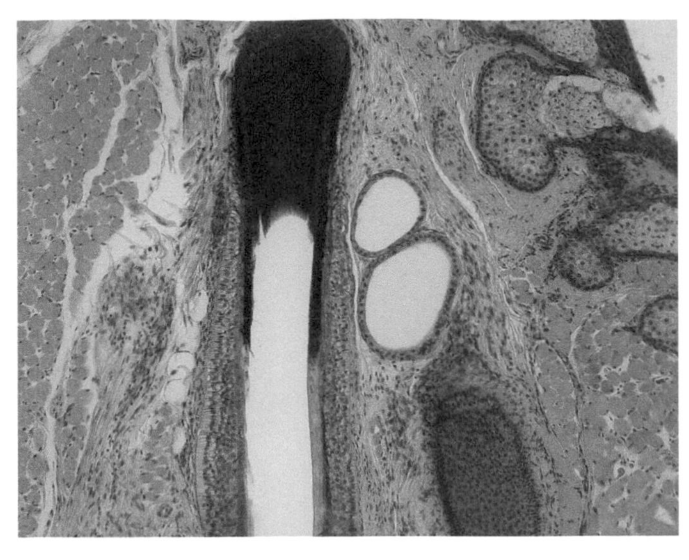

FIGURES J42b A portion of the apocrine sweat gland of the human eyelid is shown to the left of a follicle of an eyelash. Its epithelium consists of a single type of cuboidal secretory cell. 10 ×.

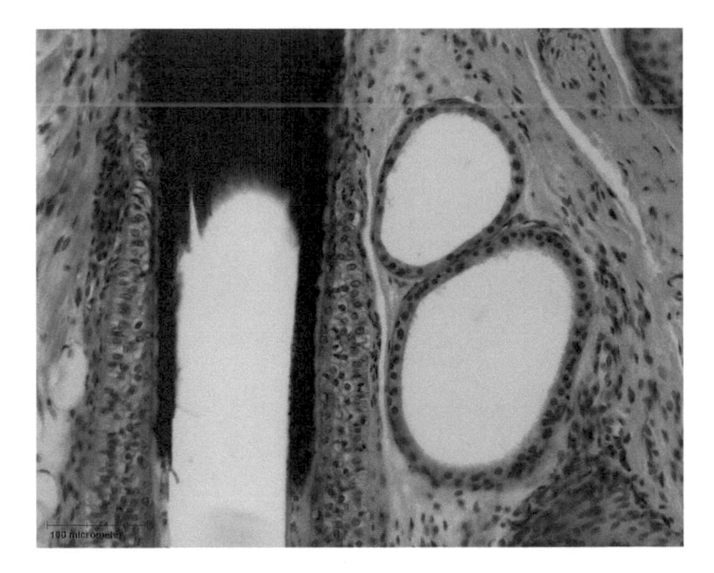

FIGURES J42c At higher magnification a portion of an apocrine sweat gland is shown to the left of a follicle of an eyelash. Its epithelium clearly consists of a single type of cuboidal secretory cell. 20 ×.

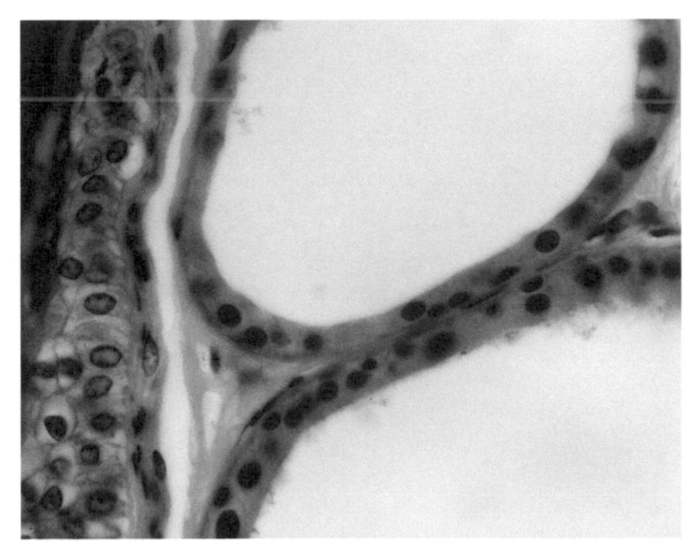

FIGURE J42d The epithelium of the apocrine sweat gland from the eyelid is simple cuboidal. It rests on a basement lamina subtended by slender myoepithelial cells. 63 ×.

Electron micrographs of columnar secretory cells show the familiar characteristics of a secretory cell: well-developed granular endoplasmic reticulum and Golgi complex, abundant mitochondria, and apical secretory droplets (Figs. J44b and J44c). Milk proteins are released by exocytosis (Fig. J45). Note that a thin layer of cytoplasm encloses each droplet of lipid as it is released from the apical portion of the cell. Although this cytoplasmic loss is minimal the lipid secretion of the mammary gland is considered to be apocrine. This is an example of both merocrine and apocrine secretion occurring simultaneously in the same cell.

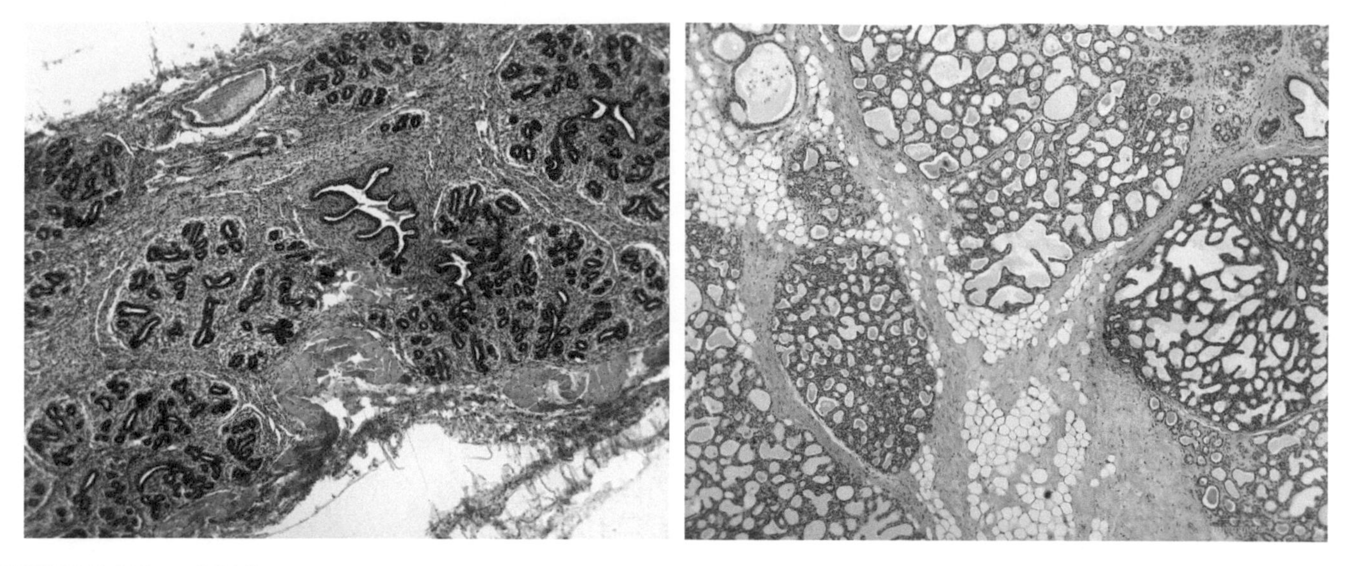

FIGURES J43a and J43b Compare these two sections of human mammary gland. Fig. J43a is inactive, while the tissue in Fig. J43b is actively lactating. The lobular nature of the mammary gland is apparent in the section of an inactive gland. The LOBULES are separated by INTERLOBULAR DENSE CONNECTIVE TISSUE. Scattered, inactive TUBULES are surrounded by abundant loose connective tissue. There is a branching LACTIFEROUS DUCT at the center and an INTRALOBULAR DUCT at the upper right.
The active gland is packed with distended SECRETORY ALVEOLI—traces of secretion can be seen in their lumina. There is some adipose tissue at the lower center and upper left. The gland is divided into LOBES by INTERLOBULAR CONNECTIVE TISSUE. 5 × . both

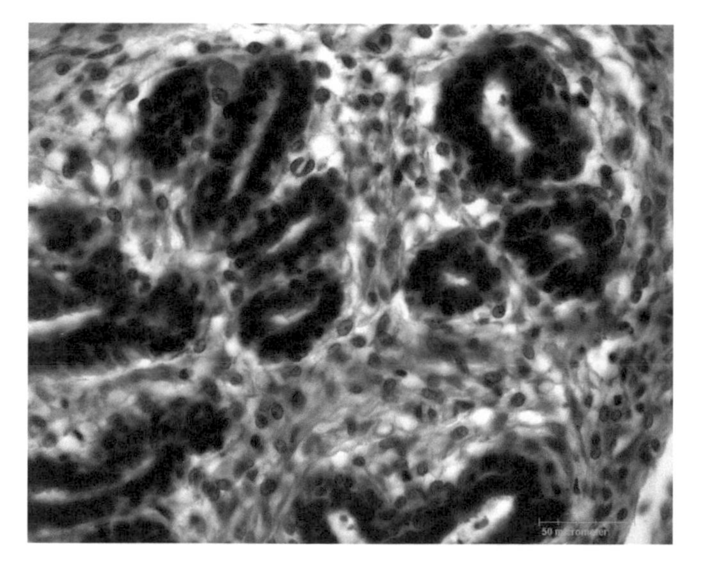

FIGURE J43c The inactive mammary gland contains scattered alveoli separated by abundant loose connective tissue. 40 × .

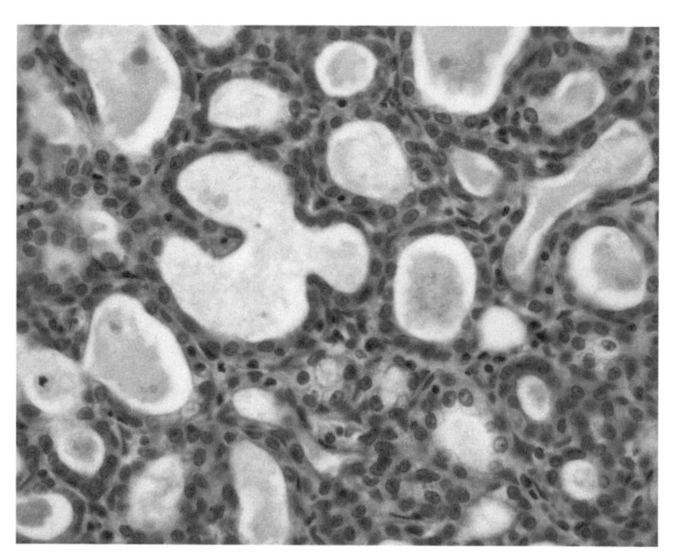

FIGURE J43d Irregular lobules in this active mammary gland are separated by intralobular connective tissue. Some secretory product is seen within the lobules. All alveoli are not in the same state of activity at the same time: some are storing secretion in their cells while others are in a resting state. Occasional slender MYOEPITHELIAL CELLS can be seen at the periphery of some alveoli. 40 × .

DIGITAL TIPS

An exceptional development of the stratum lucidum in amniotes has resulted in a variety of keratinized epidermal structures: the claws, nails, and hooves. Figs. J46a–J46c are from a longitudinal section of human finger showing the fingernail. The NAIL PLATE or UNGUIS consists of closely welded keratinized epithelial cells firmly attached to the skin below, the NAIL BED. The ROOT is protected by a fold of skin, the NAIL FOLD, and is inserted in the NAIL GROOVE. The nail fold has all the layers of skin, but as it turns inward to the nail groove it loses its outer layers and in the nail bed only the stratum germinativum remains. New nail is constantly being formed in the region of the root by the proliferation of epidermal cells in the MATRIX (Fig. J47a). Cells in the matrix divide, migrate toward the root of the nail, differentiate (without producing keratohyaline granules), and produce the hard keratin of the nail. The LUNULA or "half moon" of the nail is a reflection of partly keratinized cells (Fig. J47b). As the nail glides forward, a few cornified epidermal cells from the nail fold remain attached to its upper surface; these constitute the EPONYCHIUM, or the cuticle. The nail itself consists of two layers, which are better defined in claws and hooves. The harder, upper part of the nail plate is the UNGUIS. Under the free edge of the nail the stratum corneum of the epidermis of the nail bed reforms and constitutes the thickened HYPONYCHIUM or SUBUNGUIS.

Claws and hooves are similar in basic structure to the primate nail. In CLAWS the unguis forms a curved, V-shaped trough enclosing the subunguis. The unguis of HOOVES curves around the tip of the digit and

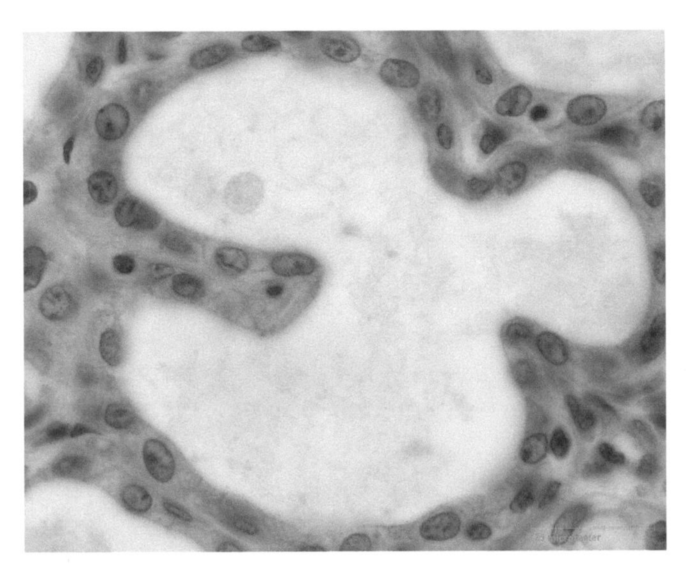

FIGURE J43e The epithelium of an alveolus from an active mammary gland is simple cuboidal. Secretory product is contained within the lumen. Only sparse connective tissue separates this alveolus from adjacent alveoli. Can you detect MYOEPITHELIAL CELLS at the periphery of the alveolus? 100 ×.

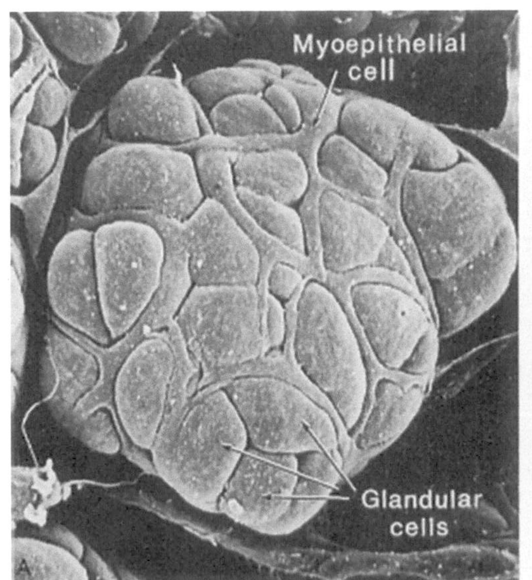

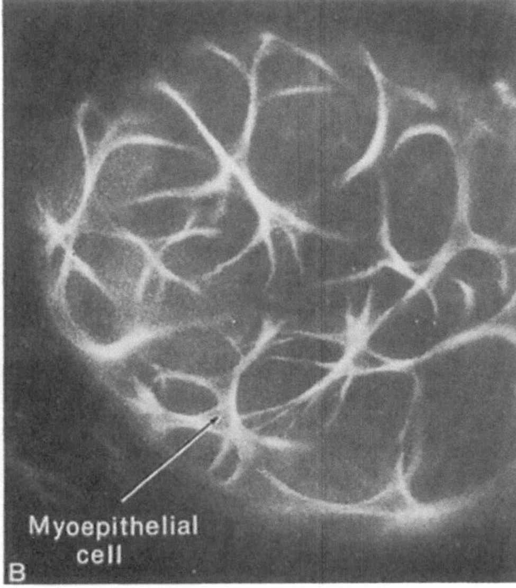

FIGURE J44a (Left) Scanning electron micrograph of an acinus of a mammary gland showing myoepithelial cells occupying the grooves between the bases of secretory cells. (Right) Whole mount of an acinus of a mammary gland stained with a fluorescent probe for actin.

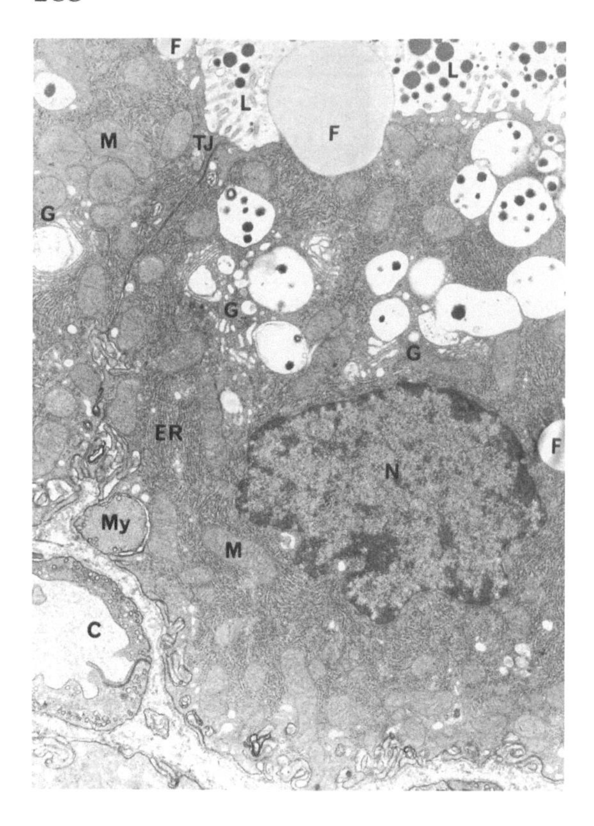

FIGURE J44b Transmission electron micrograph of an alveolar cell from the mammary gland of a mouse in full lactation. The dense, round bodies within the lumen and secretory vacuoles are milk proteins. *C*, capillary; *ER*, granular endoplasmic reticulum; *F*, fat droplet; *G*, Golgi complex; *L*, lumen of the excretory alveolus; *M*, mitochondrion; *My*, myoepithelial cell process; *N*, nucleus; *TJ*, tight junction. 13,000 ×.

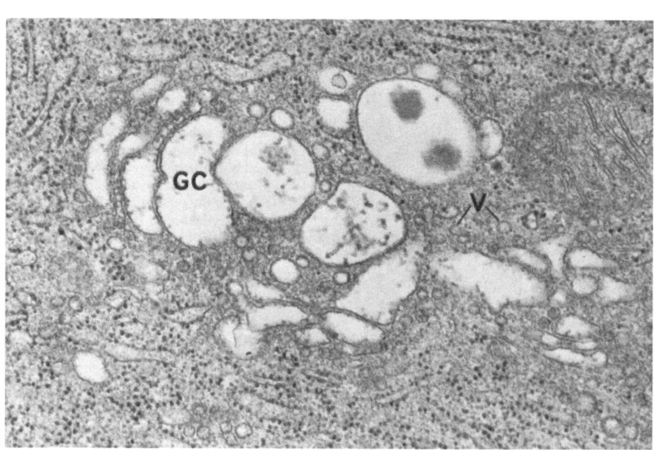

FIGURE J44c The Golgi region of a lactating alveolar cell showing abundant microvesicles (V), inflated Golgi cisternae (GC), and abundant granular endoplasmic reticulum. 40,600 ×.

encloses the subunguis. Since the softer subunguis wears away faster than the unguis, it forms a pad within the curve of the hoof tip and a sharp edge is maintained on the hoof.

HORNS AND ANTLERS

The HOLLOW HORN of cattle, sheep, goats, buffalo, etc., is a projection of the frontal bone covered by thick, keratinized epidermis. This horny layer is not shed. ANTLERS are solely bone, but they are formed under the influence of soft skin, the VELVET, that covers them until they are fully grown. Antlers of deer are shed each year and new extensions of the frontal bone are formed in the next year under the influence of a covering layer of velvet. The small antlers of the giraffe are not shed and remain permanently "in velvet."

The unique KERATIN–FIBER HORN of the rhinoceros is a conical mass of hardened, keratinized epidermal cells covering a cluster of long dermal papillae. Fibers grow from the papillae and cells growing between the papillae cement the whole together.

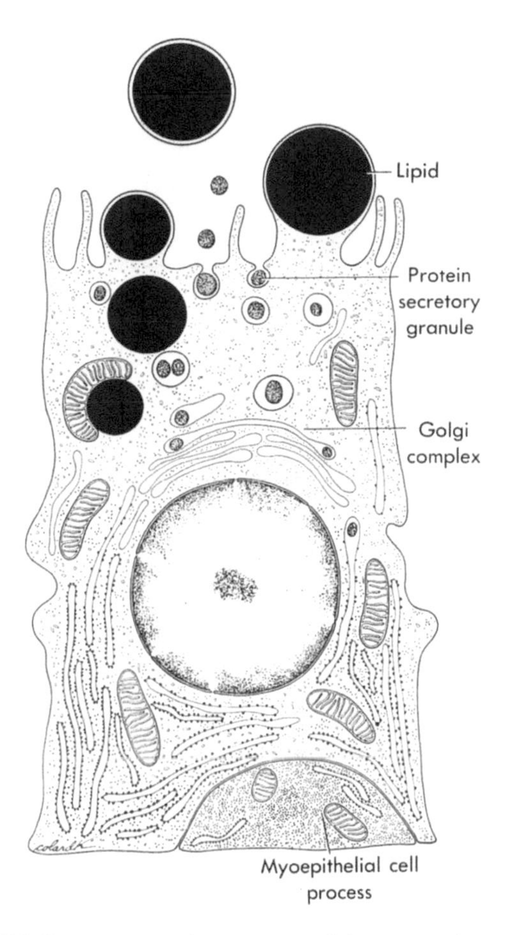

FIGURE J45 Diagram of a secretory cell from an active mammary gland. The protein components of the milk are released by fusion of their limiting membrane with the plasmalemma: *merocrine secretion*. A lipid droplet, however, cannot be constrained by a lipid bilayer so that its release from the cell would damage the cell membrane. As a solution, a thin layer of cytoplasm, surrounded by a lipid bilayer, envelops each droplet, and fusion of the membranes takes place during release. The cytoplasmic loss is minimal but the secretion of lipids is considered to be *apocrine*.

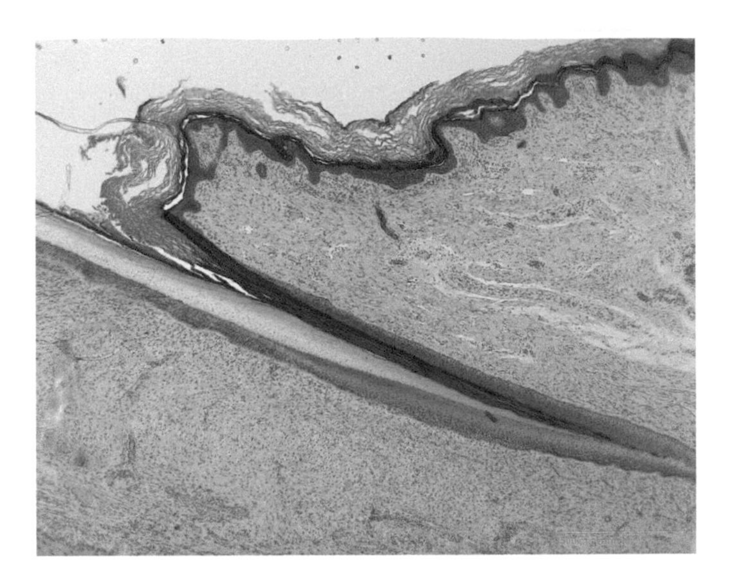

FIGURES J46a The NAIL PLATE or UNGUIS consists of closely welded epithelial cells that form HARD KERATIN. It is formed by the proliferation and keratinization of epidermal cells in the NAIL MATRIX. (This is parallel to the formation of hard keratin in a hair follicle.) The NAIL ROOT is located within the NAIL FOLD (lower right), a pocket formed by infolding of the epidermis. The UNGUIS is firmly attached to the skin of the NAIL BED below. New keratinized cells are constantly being formed at the root by the proliferation of cells of the stratum germinativum, which synthesize large amounts of KERATOHYALIN GRANULES. The epithelial cells are transformed into flat anuclear elements consisting of keratin and a hard interfibrillar material. As the nail grows it slides over the thin epidermis of the nail bed, which contributes nothing to its formation. The nail appears pink because blood in capillaries of the dermis below show through the semitransparent nail plate. The LUNULA is the white area at the base of the nail where the capillaries do not show through. At the outer edge of the nail fold the stratum corneum of the epidermis extends for a short distance onto the upper surface of the nail; this is the EPONYCHIUM. These cornified cells contain SOFT KERATIN. 5×.

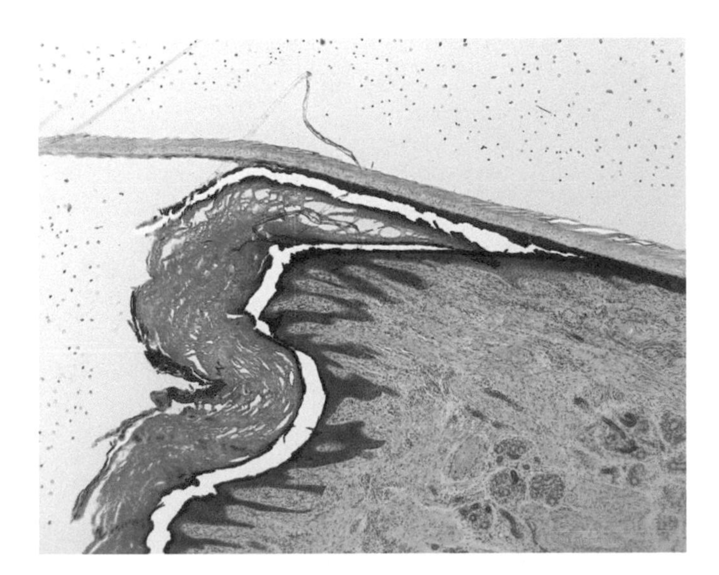

FIGURE J46b At the free edge of the nail the epithelium of the nail bed becomes continuous with the epidermis of the skin. The stratum corneum of the skin beneath the free edge of the nail is thickened and is known as the HYPONYCHIUM or SUBUNGUIS. Its cells contain SOFT KERATIN. 5×.

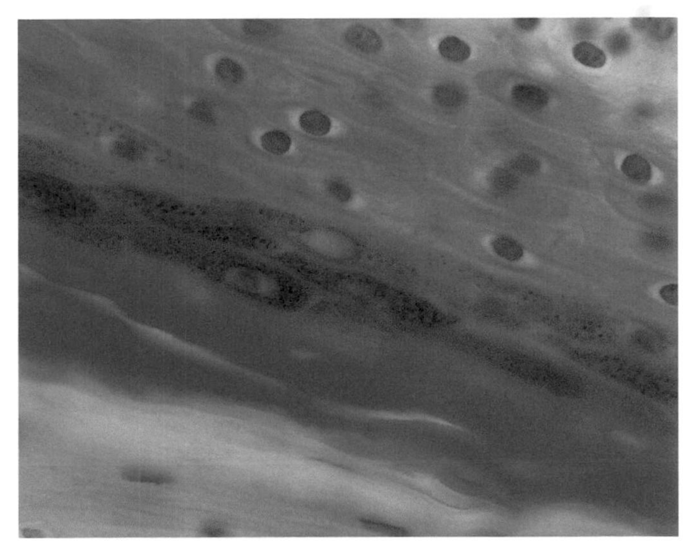

FIGURE J46c This view shows the formation of keratinized epithelial cells within the nail matrix. A few cells of the STRATUM GERMINATIVUM appear at the upper right corner. Cells of the STRATUM SPINOSUM are connected by intercellular bridges. cells of the STRATUM GRANULOSUM below this layer have accumulated basophilic droplets of keratohyalin. These cells transform into the hard, flat cells of the UNGUIS, which replaces the stratum corneum (bottom). Cells that form hard keratin do not accumulate keratohyalin granules—it is concluded, therefore, that these basophilic cells are engaged in forming soft keratin of the hyponychium. 100×.

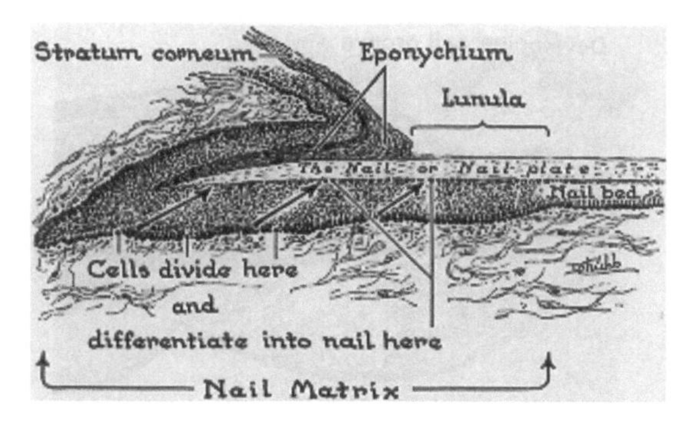

FIGURE J47a Diagram of a longitudinal section of a human fingernail showing the root of the growing nail.

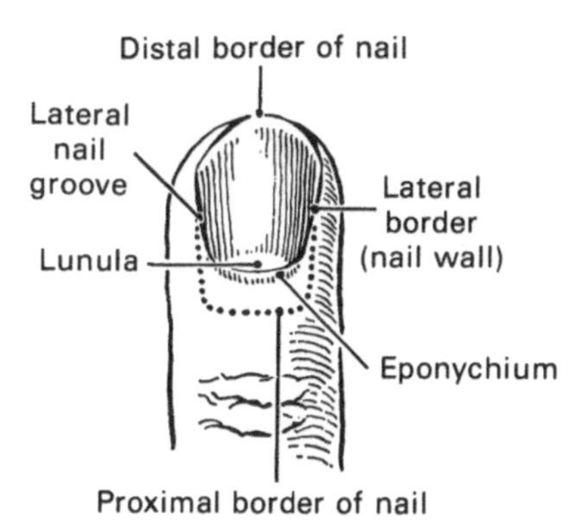

FIGURE J47b Diagram of the end of a human finger showing its fingernail.

C H A P T E R

K

The Digestive System

GENERAL ARRANGEMENT OF THE DIGESTIVE TUBE

Much of the digestive system takes the form of a tube that twists and turns through the body, largely within the abdomen (Fig. K1c). In addition, there are several glands pouring their secretions into the tube. A fundamental arrangement of the layers of the wall of the vertebrate gut is seen, with modification, in all regions. This arrangement is shown diagrammatically in Fig. K1a. In this study the digestive system try to relate the modifications with the functions of the particular region. Identify the following layers in a section of gut (Fig. K1b).

1. *Tunica mucosa.* The internal mucous membrane consists of three layers (Fig. K2):
   a. *Epithelium mucosae,* often simple columnar, resting on a basement membrane.
   b. *Lamina propria mucosae,* the subepithelial layer consisting of loose reticular tissue and areolar connective tissue with lymphoid nodules, capillaries, and lymphatics.
   c. *Lamina muscularis mucosae,* the narrow band of circularly and longitudinally arranged smooth muscle. It is the outer limit of the tunica mucosa, separating the connective tissue of the lamina propria mucosae from the connective tissue of the next layer, tela submucosa.
2. *Tela submucosa.* This layer of areolar connective tissue contains the larger blood vessels and lymphatics of the gut. The SUBMUCOUS PLEXUS (Meissner's plexus) (Figs. K2, K65d) of autonomic nerve fibers and ganglia is located here. Some glands may penetrate into the tela submucosa. Masses of hemopoietic tissue often appear in the connective tissue of the lamina propria mucosae and the tela submucosa (Fig. K71c). While these are restricted to the production of agranulocytes in the higher vertebrates, all types of blood cells may be produced here in lower vertebrates lacking hollow bones and have, therefore, no bone marrow.
3. *Tunica muscularis* (Fig. K3). There are usually two layers of muscle (usually smooth) surrounding the tela submucosa, an inner CIRCULAR LAYER and an outer LONGITUDINAL LAYER. This relationship is quite constant the plane of any section may be determined by observing these layers. Between the layers is the MYENTERIC PLEXUS (of Auerbach) (Figs. K4a, K65c, K65e) a plexus of the autonomic nervous system.
4. *Tunica serosa or tunica adventitia* (Fig. K3). The peripheral layer may be either a tunica serosa or tunica adventitia. A tunica serosa is visceral peritoneum surrounding the portion of the tract contained within the coelom; it is continuous with the mesentery and consists of an inner layer of areolar connective tissue covered by a layer of flattened mesothelial cells with microvilli on their free surface (*Mesentery* is a thin sheet of areolar connective tissue, often containing many fat cells, sandwiched between two layers of squamous mesothelial cells. Nerves, blood vessels, and lymphatic vessels pass through this connective tissue on their way to and from the gut). Tunica adventitia surrounds portions of the gut not located in the coelom (e.g., pharynx, anterior esophagus) and consists of areolar connective tissue continuous with the surrounding connective tissue.

Glands

Glands arise as invaginations of the epithelium mucosae but vary in their position being located in the lamina propria mucosae, the tela submucosa, or external to the tract (salivary glands, liver, pancreas).

An Atlas of Comparative Vertebrate Histology.
DOI: https://doi.org/10.1016/B978-0-12-410424-2.00011-1

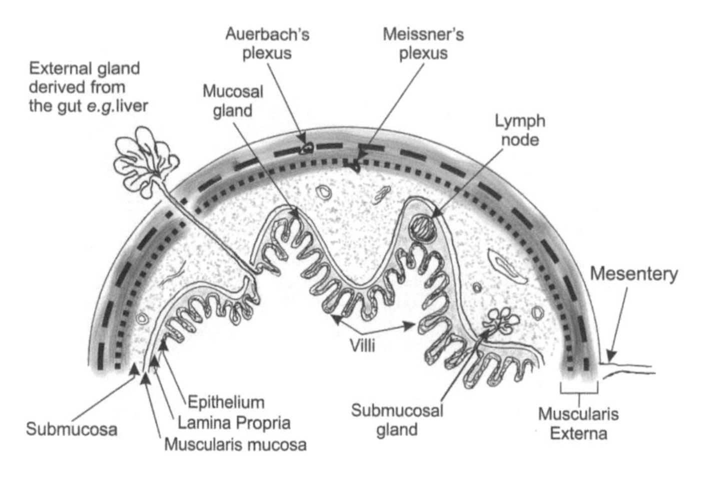

FIGURE K1a Generalized view of a cross section of the vertebrate gut. *Authors.*

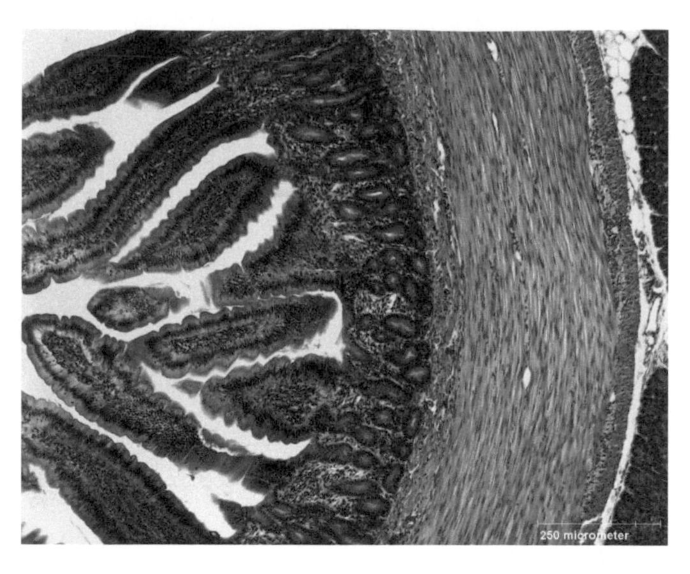

FIGURE K1b This cross section of the duodenum of a cat shows the fundamental arrangement of the layers of the wall that is seen throughout the gut. The gut is lined by the TUNICA MUCOSA, which is the bluish region on the left. In the duodenum the tunica mucosa is thrown up into finger-like projections, the VILLI, which greatly increase its surface area for absorption. Villi are seen only in the intestine. In other regions of the gut, the tunica mucosa demonstrates other appearances appropriate to their functions. The tunica mucosa consists of three sublayers, the EPITHELIUM MUCOSAE,[1] a simple columnar epithelium, which rests on a subepithelial layer of delicate connective tissue, the LAMINA PROPRIA MUCOSAE, which contains the delicate capillaries and lymph vessels, filling the cores of the villi and extending between the tubular glands beneath. Vessels of this layer carry absorbed substances away from the epithelium. The outside of the tunica mucosa is wrapped by a thin layer of smooth muscle (LAMINA MUSCULARIS MUCOSAE), which can hardly be seen in this section. 10 ×
The next layer of the gut is the TELA SUBMUCOSA; it has an orange hue in this section. It consists of coarser connective tissue than the lamina propria mucosae and contains the larger blood vessels and lymphatics of the gut. Also present in the tunica submucosa (but not visible in this section) is the SUBMUCOSA PLEXUS of autonomic nerve fibers and ganglia.
The TUNICA MUSCULARIS of smooth muscle is well developed in this section. It consists of an inner thick CIRCULAR LAYER and an outer LONGITUDINAL LAYER of fibers cut in cross section. A few lymph vessels can be seen between the fibers of the circular layer. The grayish masses between the layers of muscle are parts of the MYENTERIC PLEXUS of nerves and ganglia.
The loose strands of connective tissue surrounding the longitudinal layer of muscle are made up of peritoneum and constitute the TUNICA SEROSA. Some fat, blood vessels, and lymphatics are seen in this layer. The pancreas is stained a deeper purple and lies on the extreme right of the micrograph.

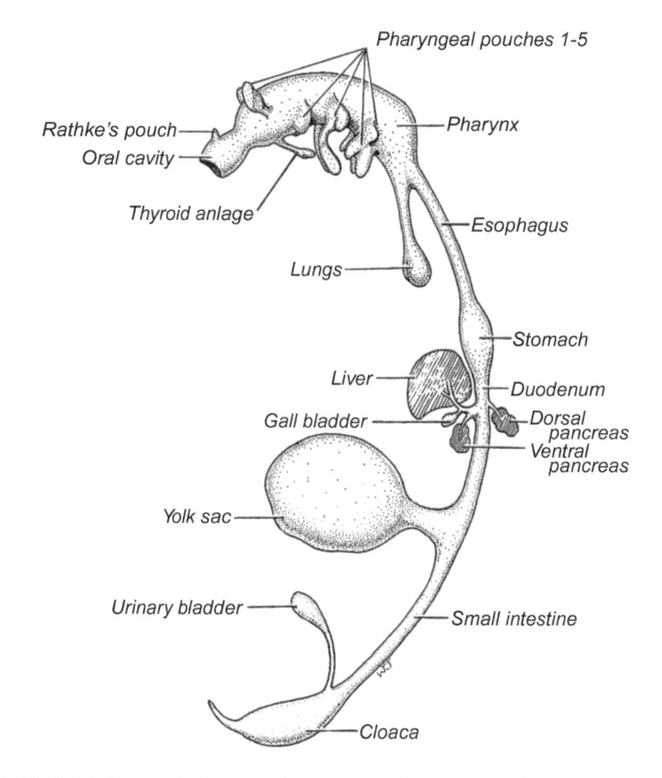

FIGURE K1c A diagram illustrating the parts and outgrowths of the embryonic digestive tube.

Nerve Supply

Neurons and nonmedullated fibers of the autonomic nervous system are grouped in the tela submucosa as the SUBMUCOUS PLEXUS and between the two layers of the tunica muscularis as the MYENTERIC PLEXUS (Figs. K4a and K4b). These networks consist of afferent fibers from various receptors in the gut and both parasympathetic and sympathetic fibers of the autonomic nervous system; the ganglion cells are part of the parasympathetic system.

[1]Note the spelling of *mucosae*. This is the Latin genitive case and indicates the meaning *of* the mucosa. Spelled without the terminal *e*, mucosa is simply an adjective describing the tunic.

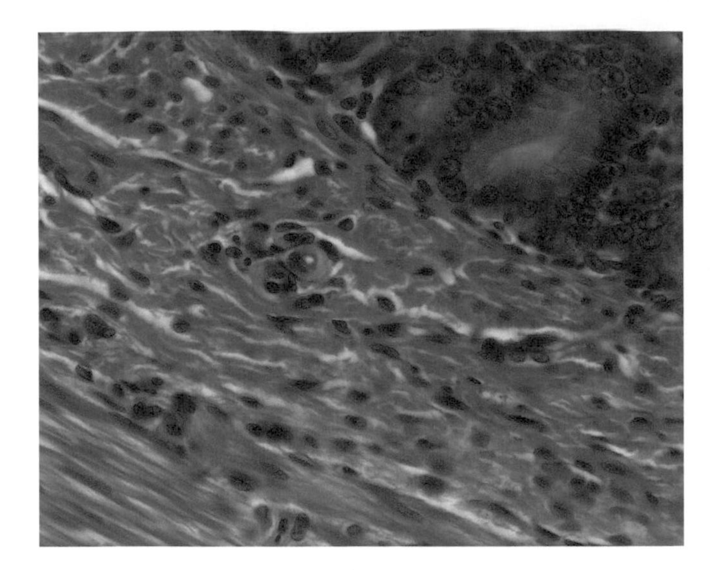

FIGURE K2 The deep portions of the simple tubular glands of the duodenum occupy the upper right of this photomicrograph. An anaphase mitotic figure can be seen at the base of one of the glands. Delicate vascular connective tissue of the lamina propria mucosae separates the glands from each other and from the lamina muscularis mucosae, a thin layer of smooth muscle only one or two cells thick. Coarser vascular connective tissue of the tela submucosa occupies a large portion of the field and passes from the upper left to the lower right. Note the submucous nerve plexus near the center. Smooth muscle of the inner circular layer of the tunica muscularis occupies the lower left corner. 63 ×.

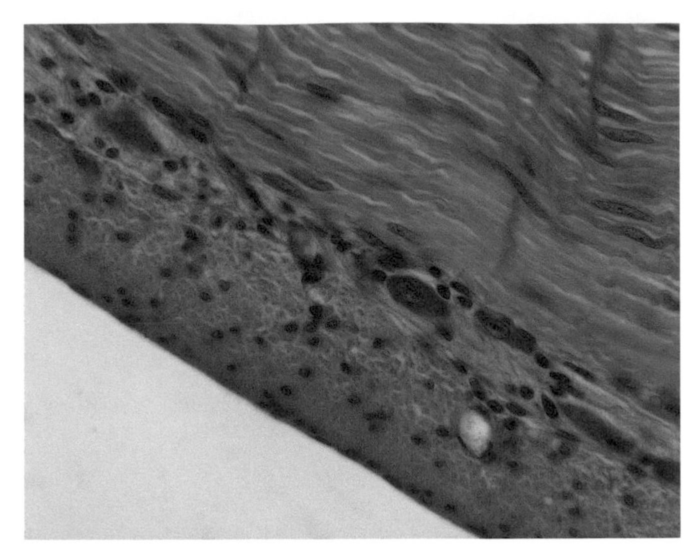

FIGURE K3 Two ganglia of the myenteric plexus of the autonomic nervous system lie between the layer of circular muscle at the upper right and the longitudinal layer at the lower left. The ganglia consist of the bodies of nerve cells and nonmyelinated nerve fibers. An extremely thin layer of peritoneum, the TUNICA SEROSA, encloses the longitudinal muscle. 63 ×.

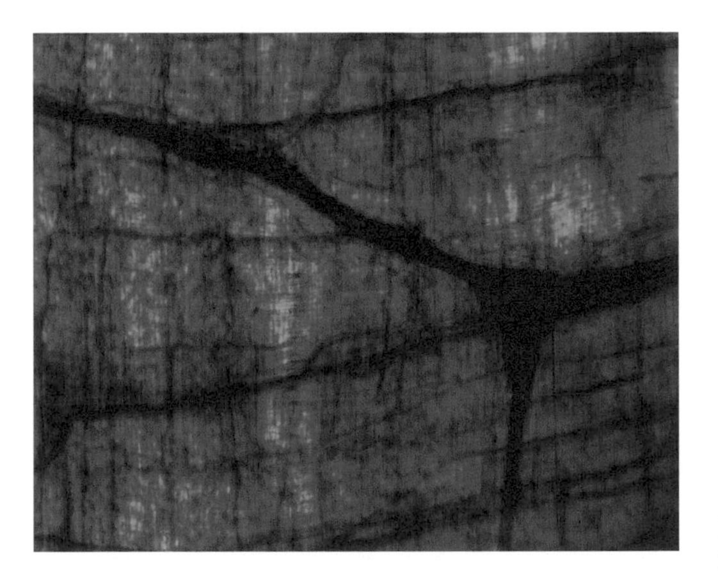

FIGURE K4a This is a whole mount of the muscular wall of the stomach of a rabbit. It has been stained with silver to show the elaborate MYENTERIC PLEXUS between the muscle layers; a ganglion is at the center right and nonmyelinated nerves radiate from it. Delicate nerve fibers form a mesh between the muscle layers. 20 ×.

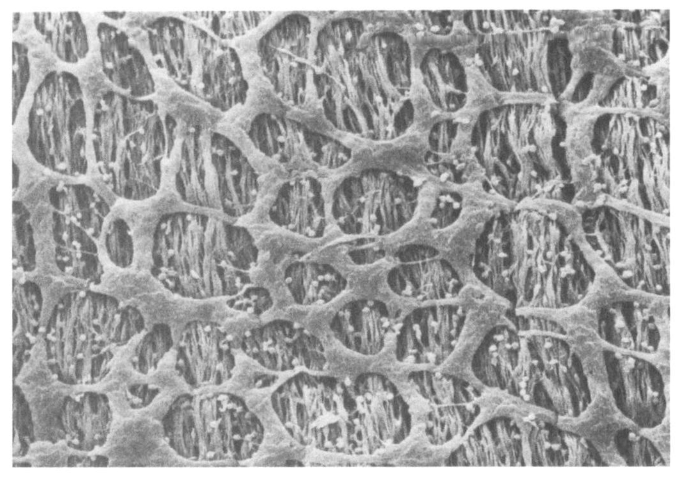

FIGURE K4b Scanning electron micrograph of the myenteric plexus of the intestine of a rat. The overlying longitudinal muscle and connective tissue have been removed by dissection and enzymatic digestion.

Vascularization

Arteries of the gut usually penetrate the tunica muscularis to form an elaborate plexus in the tela submucosa; this plexus sends branches to the capillaries of the tunica mucosa and tunica muscularis (Figs. K5a and K5b). Veins draining the tunica mucosa form plexuses in both the lamina propria mucosae and tela submucosa and

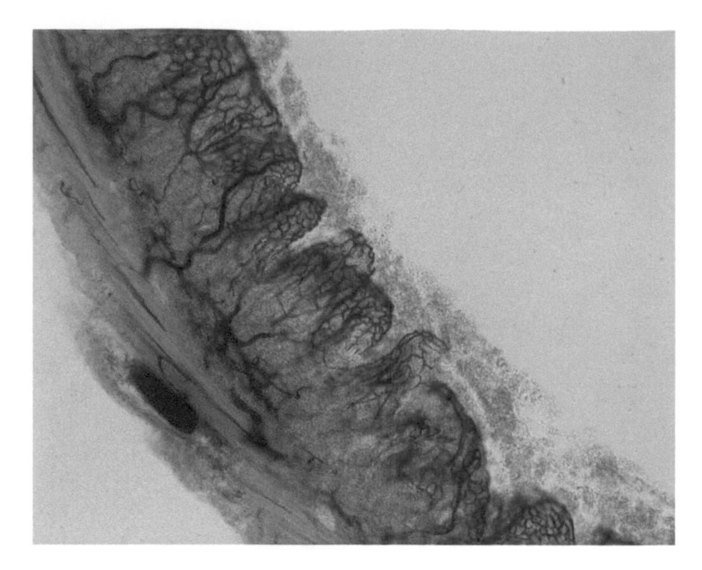

FIGURE K5a This is a section of the intestine of a cat. The tissue was injected with colored gelatin to delineate the blood vessels. The background tissue is stained yellow. The large arteries in the tela submucosa form a plexus that distributes blood to both the tunica mucosa and to the tunica muscularis. One large vessel can be seen entering the gut at the lower left. The elaborate capillaries in the lamina propria mucosae bring oxygen to the tissues of the villi and carry away absorbed food. 10 ×.

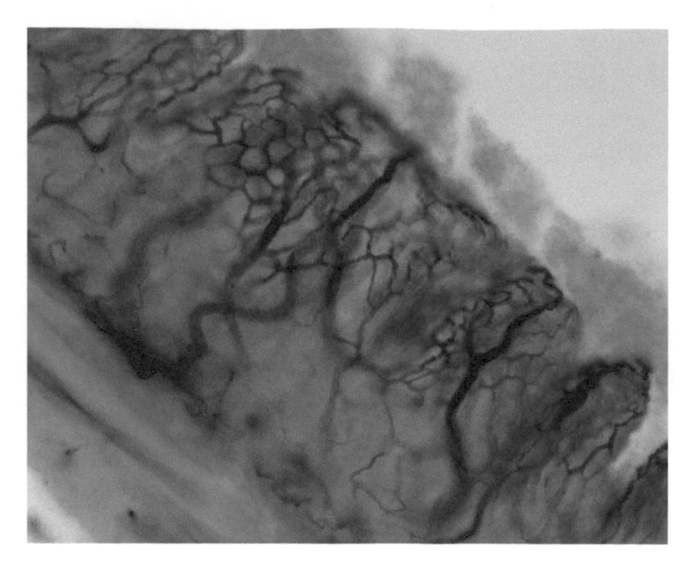

FIGURE K5b The elaborate vascularization of the tunica mucosa is seen at this higher powered micrograph of the same section. 20 ×.

these are drained by large veins running through the tunica muscularis parallel to the arteries. Lymph from the tunica mucosa drains into lymphatic plexuses in the lamina propria mucosae and tela submucosa and is carried through the tunica muscularis in lymph vessels that parallel the blood vessels, receiving lymph from the muscle layers as they pass.

ORAL CAVITY

The oral cavity of many vertebrates is lined with a tunica mucosa of stratified squamous epithelium, which does not keratinize completely as in the skin. It rests on the tela submucosa of loose vascular areolar connective tissue that is often permeated by lymphoid nodules and salivary glands (Figs. K6a–K6c). Compare an image of the tunica mucosa of the oral cavity of a mammal with the images of mammalian skin studied previously. Superficial cells of the partially keratinized stratified squamous epithelium are constantly being sloughed off and are found in the saliva. The lamina propria mucosae contains a network of fine collagenous and elastic fibers, terminal branches of the trigeminal nerve, blood vessels, and lymphoid tissue, the latter especially at the entrance to the pharynx. The tela submucosa is loose areolar tissue and may contain adipose tissue. Numerous PAPILLAE of connective tissue extend into the epithelium and a rich vascular supply accompanies these, accounting for the red color of the lips and lining of the mouth.

Tongue

Mammalian tongue is composed of interlacing bundles of striated muscle covered by a tunica mucosa and tela submucosa (Figs. K6d and K6e). The lower surface is covered by a tunica mucosa similar to that in the rest of the oral cavity. On the upper surface four types of small elevations, the PAPILLAE, may be seen.

Filiform papillae, the most numerous, with pointed tips, arranged in rows parallel to the sulcus terminalis (the V-shaped boundary between the body and the root of the tongue (Figs. K7a and K7b). The connective tissue core is covered by keratinized stratified squamous epithelium whose superficial cells slough off.

Fungiform papillae, mushroom-shaped papillae scattered singly among filiform papillae; the connective tissue core has slight secondary papillae; the epithelium has a smooth surface and may have occasional TASTE BUDS on the dorsal surface (Figs. K8a and K8b).

FIGURES K6(a–c) These three photomicrographs are of the mammalian soft palate just below the nasal cavity.

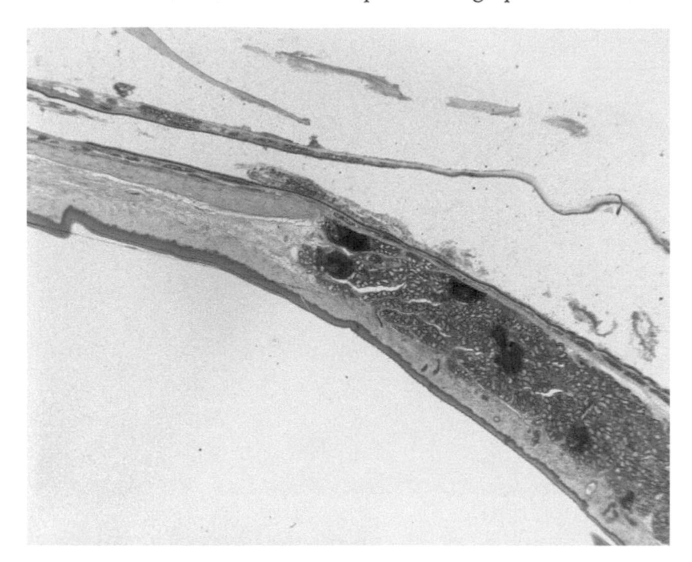

FIGURE K6a The stratified squamous epithelium of the oral cavity is at the bottom, subtended by the areolar connective tissue of the tela submucosa. Permeating this connective tissue are masses of salivary gland and a few lymphoid nodules. The epithelium of the nasal cavity is torn and is at the top. 1 ×.

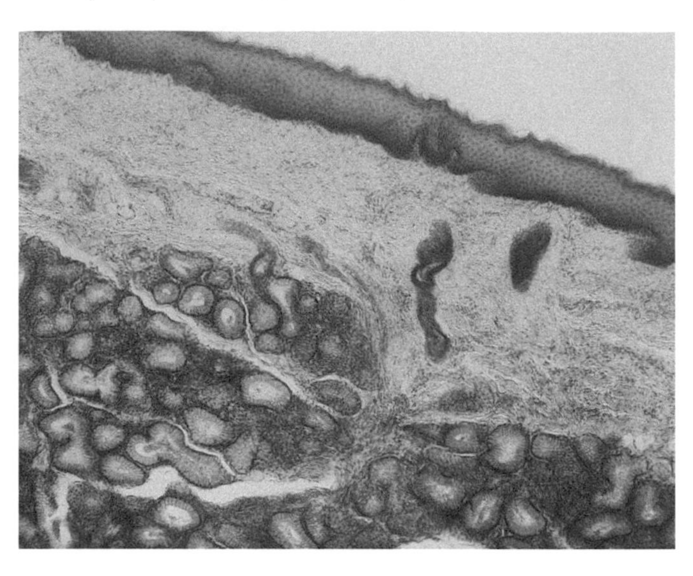

FIGURE K6b The stratified squamous epithelium of the mouth cavity is at the top; it rests on areolar connective tissue that contains several salivary glands and scattered lymphoid cells. 10 ×.

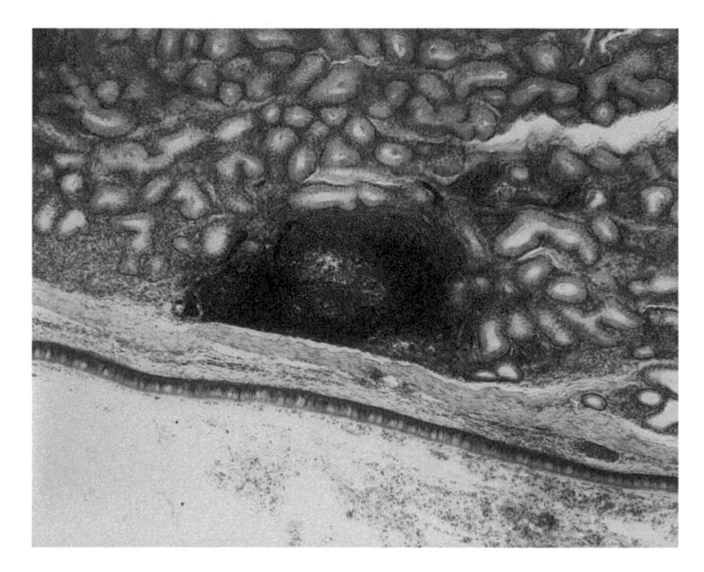

FIGURE K6c Most of the connective tissue of the soft palate is permeated with salivary glands. A single lymphoid nodule is shown at the bottom. 10 ×.

Foliate papillae are well developed in the rabbit but only poorly so in many other species, including man; these broad, leaf-like papillae are closely arranged and separated by deep, narrow, oblique grooves—think of a stack of books—(Fig. K9). Each possesses two or three connective tissue papillae. The lateral epithelium has numerous taste buds.

Vallate or circumvallate papillae, located along the sulcus terminalis (Figs. K10a and K10b). Each contains a core of connective tissue and is surrounded by a deep circular furrow into which open the ducts of serous salivary glands embedded in the mass of the tongue (glands of von Ebner).

TASTE BUDS are discrete sense organs embedded in the substance of the tongue, usually in furrows away from the intense physical activity in the mouth (Figs. K10a, K10b, K11a–K11c). A TASTE PORE is the small aperture of the bud that pierces the stratified squamous epithelium. It is thought that the deeply stained, rod-shaped cells are gustatory or taste cells. They have a central nucleus and small hair-like processes that penetrate the taste pore. Between the taste cells are the SUSTENTACULAR CELLS: lightly stained, banana-shaped cells with small vesicular

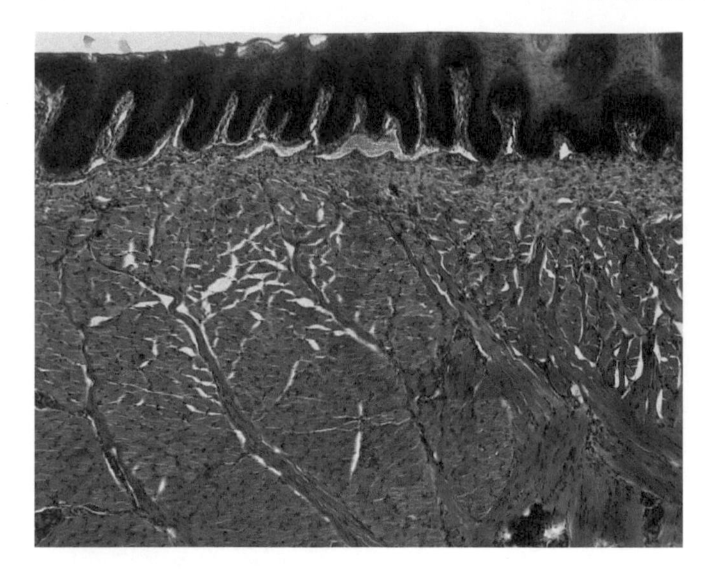

FIGURE K6d The tongue is a highly motile mass of striated muscle (which we can consciously control) covered with the epithelium of the tunica mucosa and a thin layer of connective tissue. The striated muscle of the tongue is unique in that its fibers run in three axes thereby providing amazing agility. Tongues of many vertebrates show the same woven appearance of striated muscle.
The TUNICA MUCOSA in this section of the tongue of a rabbit is at the top of the micrograph. It is a tough layer of stratified squamous epithelium that readily withstands the abrasions of coarse food. It is invaded by projections of the vascular connective tissue of the TELA SUBMUCOSA, which brings nutrients and carries away wastes. Most of the micrograph consists of striated muscle whose fibers can be seen to run in various directions. 10×.

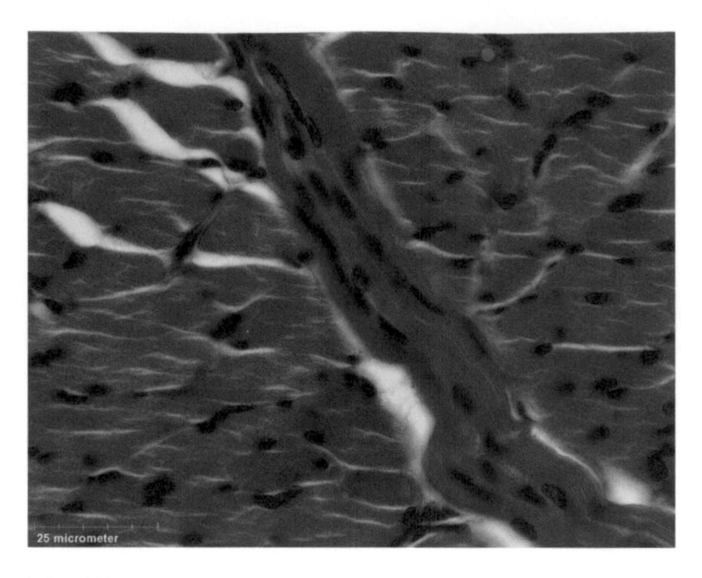

FIGURE K6e Striations are visible in this higher powered view of the same section. The fibers of the tunica muscularis run in various directions. There are longitudinal sections of capillaries at the upper left. 63×.

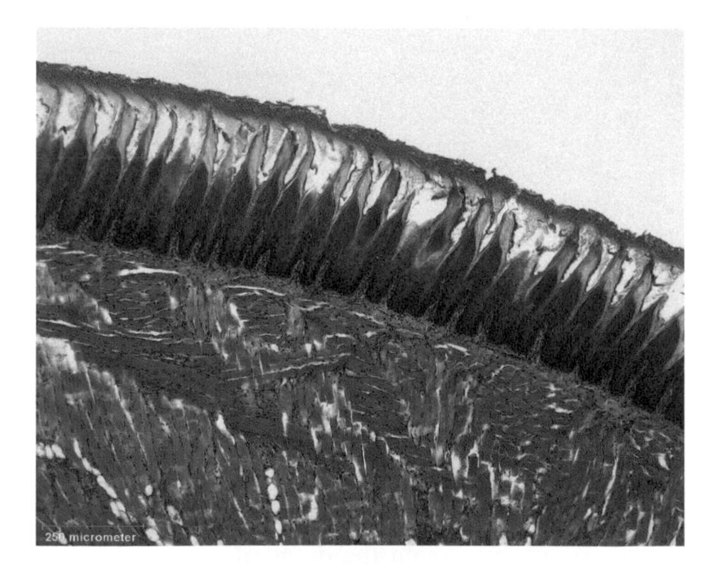

FIGURE K7a This section of the tongue of a rat shows the woven appearance of the striated muscle; this lattice of fibers permits the amazing motility of the tongue. The epithelium covering the surface is cornified, reflecting the abrasive food eaten by the rat. In this region the surface is thrown up into many hair-like projections, the FILIFORM PAPILLAE. Filiform papillae in some animals are large and stiff resulting in a sandpaper texture of the tongue. 10×.

FIGURE K7b Under a higher power the cornification of the epithelium of the tongue is obvious. The abrasive cells of this layer are constantly sloughing off, to be replaced by mitosis from a germinal layer below. At the bottom of the micrograph is the connective tissue of the TELA SUBMUCOSA and a small amount of striated muscle of the TUNICA MUSCULARIS at the lower left. 20×.

nuclei. BASAL CELLS occur peripherally, near the basement membrane; they are thought to give rise to the other cells whose life span is about 10 days. The electron microscope shows three types of cells, which are probably different phases of one type since both are innervated and both have apical microvilli. It has been suggested that the pale sustentacular cells differentiate from the darker taste cells that in turn arise from the division of surrounding basal cells (Figs. K12a–K12c).

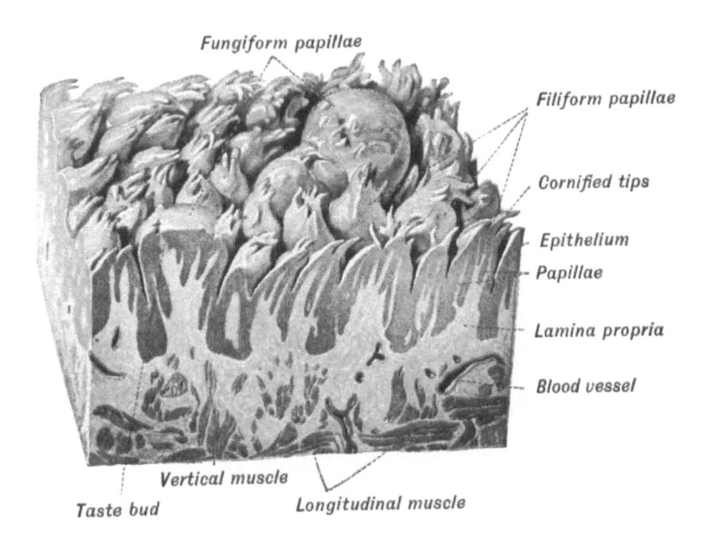

FIGURE K8a Photograph of the dorsal surface of a human tongue showing the distribution of filiform, fungiform, foliate, and circumvallate papillae.

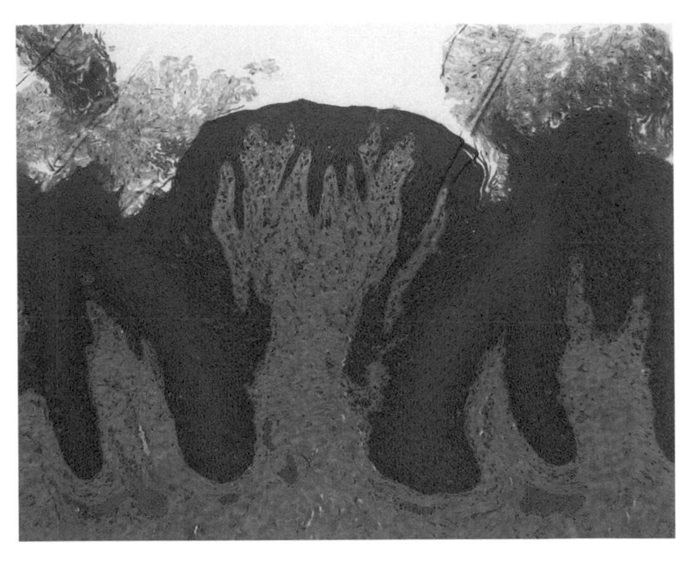

FIGURE K8b Photomicrograph of a vertical section through a FUNGIFORM PAPILLAE whose core of fibrous connective tissue extends secondary papillae that touch the basal cells of the stratified squamous epithelium. 10 ×.

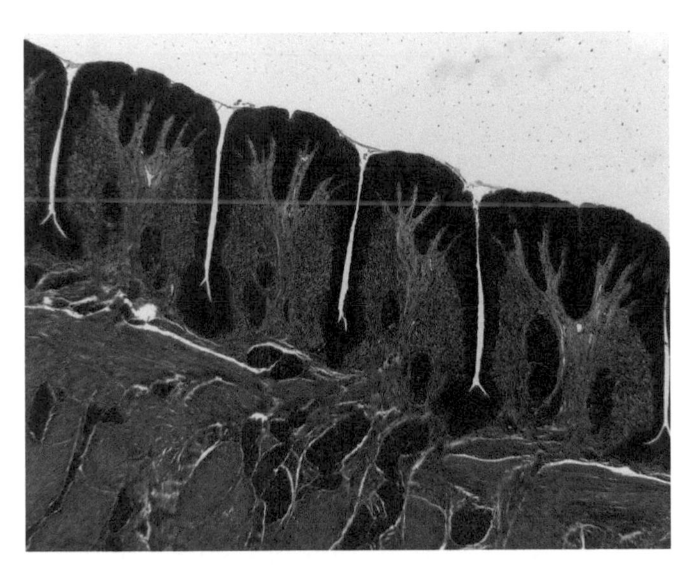

FIGURE K9 FOLIATE PAPILLAE are well developed in some mammals and have the appearance of a ploughed field with long ridges separated by furrows. To some fanciful observer, this must have resembled the pages in a book, hence the name. The papillae have a core of vascular connective tissue and are covered with a stratified squamous epithelium in which taste buds are embedded. Below the surface are compound alveolar lingual salivary glands and striated muscle. A few ducts from the glands cross the micrograph. 2.5 ×.

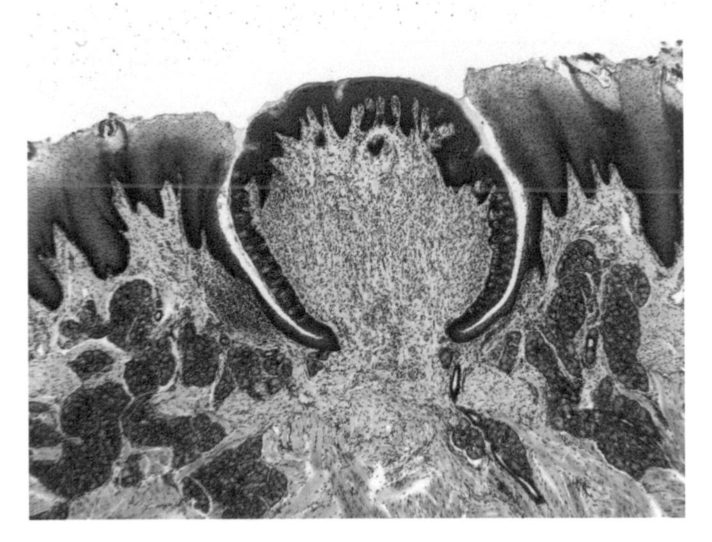

FIGURE K10a This is a section from the tongue of monkey. From 6–14 large CIRCUMVALLATE or VALLATE PAPILLAE are situated in the form of an obtuse *V* at the back of the primate tongue. They are cylindrical and roughly 1–2 mm in diameter and have a core of vascular connective tissue. Each occupies a recess in the tunica mucosa and is surrounded by a "moat" or SULCUS. The lateral surfaces are dotted with taste buds, which are distanced from exposure to the abrasive food being eaten. 4 ×.

Tonsils

Embedded in the posterior portion of the tongue are the LINGUAL TONSILS, lymphoid nodules densely massed around irregular clefts in the stratified squamous surface epithelium, the CRYPTS (Fig. K12d). These aggregations of lymphoid nodules in close association with stratified squamous epithelium of the respiratory and digestive tracts are the TONSILS, already considered in Chapter I, Hemopoietic Organs. Lingual, palatine, and pharyngeal tonsils of mammals are found over the root of the tongue, in the palatine arches, and in the mucous membrane

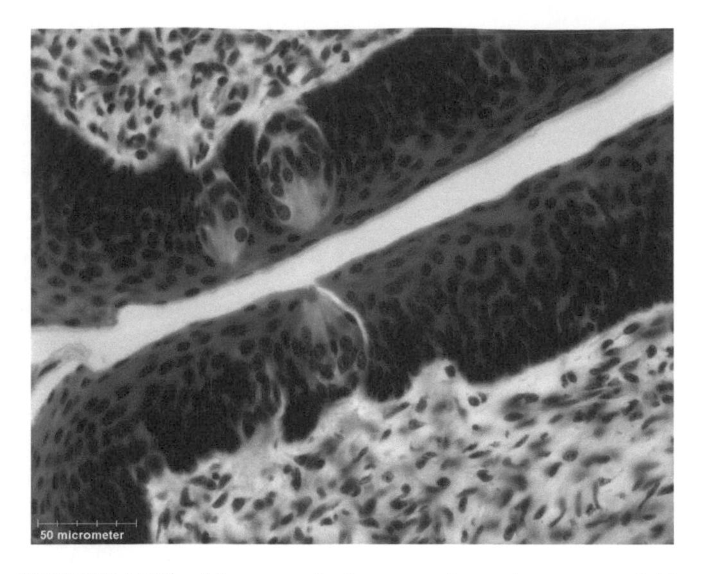

FIGURE K10b Many taste buds occupy the tunica mucosa of this fungiform papillae from the tongue of a monkey. 10 ×.

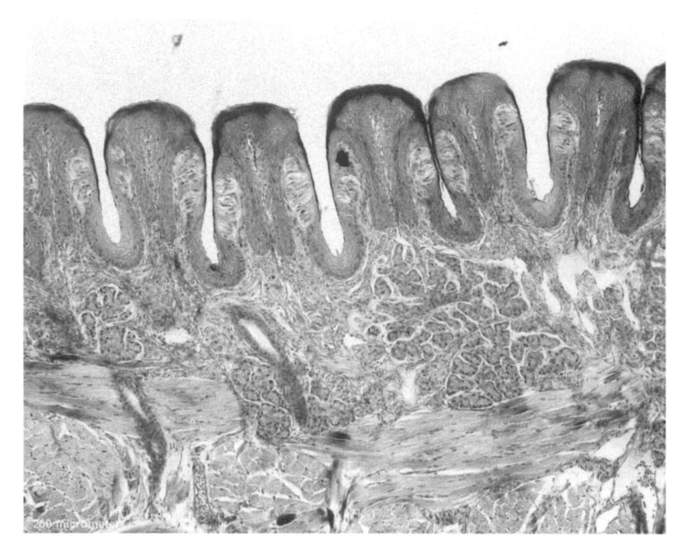

FIGURE K11a Taste buds abound on the sides of these FOLIATE PAPILLAE from the tongue of a rabbit. (The blue-black stain is iron hematoxylin.) Masses of lingual salivary gland are distributed in the connective tissue of the tela submucosa and between bundles of striated muscle. 10 ×.

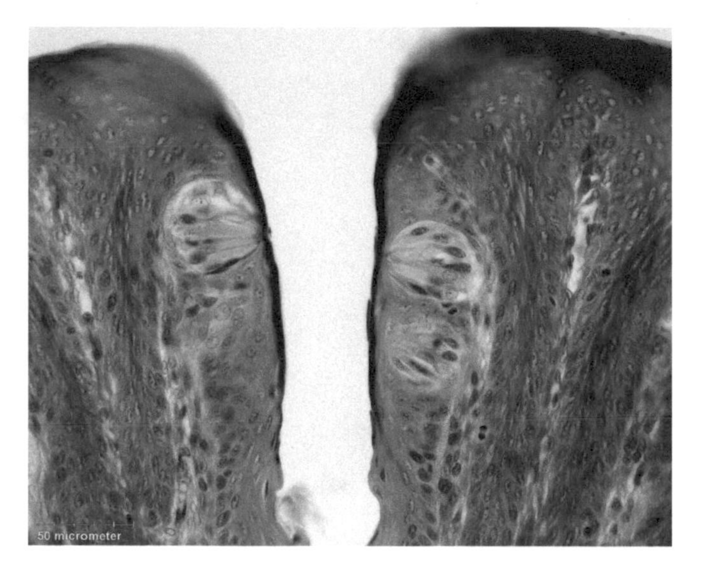

FIGURE K11b Two taste buds occur on opposite sides of two foliate papillae. Each taste bud shows the dark, banana-shaped GUSTATORY or TASTE CELLS. These cells have a central nucleus and a small hair-like process that reaches the surface through the TASTE PORE. Only the taste bud on the left shows the taste pore; its content of taste hairs can be imagined. The lightly stained cells between the taste cells are SUSTENTACULAR CELLS. Rounded cells at the base of the taste bud may be BASAL CELLS. 40 ×.

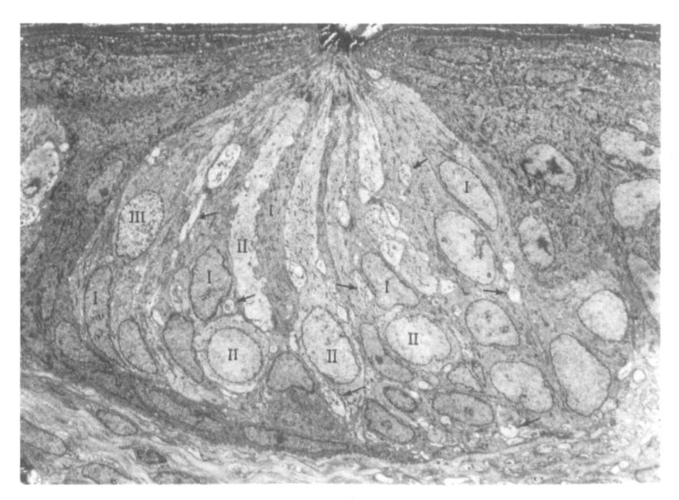

FIGURE K11c Electron micrograph of a section of the taste bud of a rabbit. Dark cells (I) are presumed to be sensory (gustatory) and light cells (II and III) are presumed to be supportive. Nerve axons (arrows) penetrate the taste bud. Microvilli extend from the cells of the taste bud into the TASTE PORE (TP).

lining the nasal pharynx, respectively. The epithelial surface may form pits or CRYPTS and is infiltrated with lymphocytes. Lymphoid nodules occur in the layers of connective tissue under the epithelium, the dense LAMINA PROPRIA MUCOSAE and the looser TELA SUBMUCOSA. The tela submucosa forms a capsule that sends TRABECULAE between the nodules. Blood vessels in the capsule and trabeculae supply the lymphoid tissue. The nodules may contain germinal centers. Diffuse lymphoid tissue fills the spaces between the nodules. Lymph capillaries permeate the lymphoid tissue and drain into efferent lymph vessels.

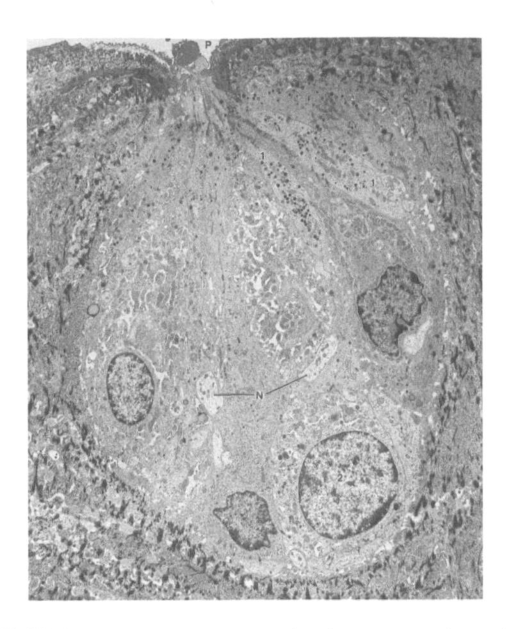

FIGURE K12a Electron micrograph of a section through a taste bud in a vallate papillae from the tongue of a monkey. Two types of cells inhabit the structure (marked 1 and 2). Microvilli from the banana-shaped cells extend through the taste pore. Note that the cells are entwined by nerve fibers (N). 5000 ×.

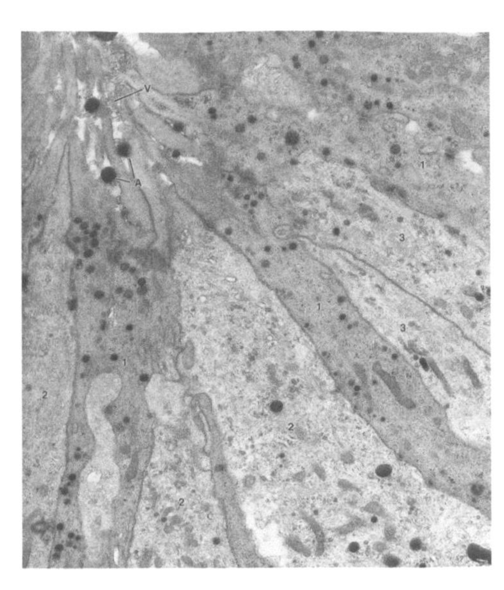

FIGURE K12b Electron micrograph of a section through the apical portion of a taste bud from the vallate papillae on the tongue of a monkey. Microvilli extend into the taste pore (with electron dense bodies - A and free vesicles- v) from the three types of banana-shaped cells (1,2,3) of the taste bud. 9700 ×.

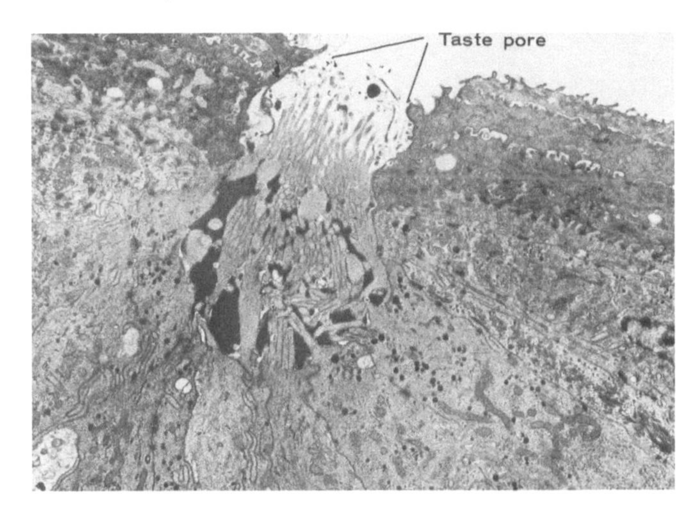

FIGURE K12c Electron micrograph of a section through the taste pore in the taste bud of a rabbit. Long microvilli on the sensory cells are surrounded by an amorphous secretory product.

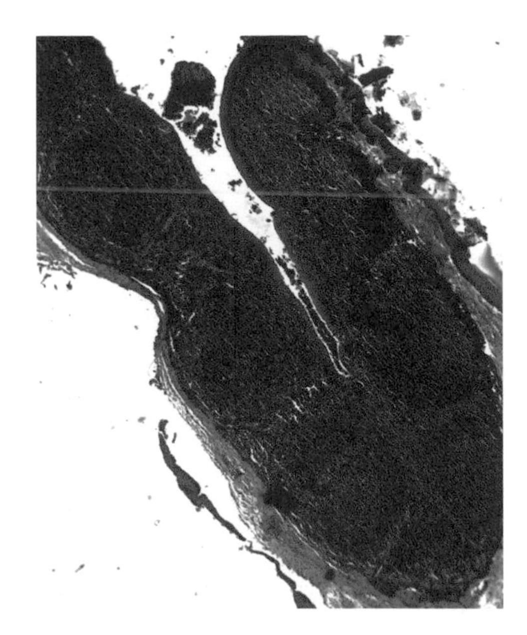

FIGURE K12d Poor section of a tonsil of a cat. It was used here because it shows a simple association between the crypt and the surface. Note the lymphoid nodules within the connective tissue, the stratified squamous epithelium of the lining of the mouth, and the debris within the crypt. 5 ×.

Adaptations of Tongues

Compare images of the tongue and mouth lining of various vertebrates, note the type of epithelium, the presence of glands, and whether or not the epithelium is ciliated. The ciliated epithelium lining the mouth of most amphibians is thin and richly vascularized, providing for respiratory exchange (Figs. K13a and K13b). Taste buds and papillae are not necessarily confined to the tongue and may be found throughout the lining of the mouth and even in the skin. Many taste buds similar to those of the mammalian tongue are located on the external barbels (whiskers) of fish (e.g., carp or catfish) (Figs. K14a and K14b).

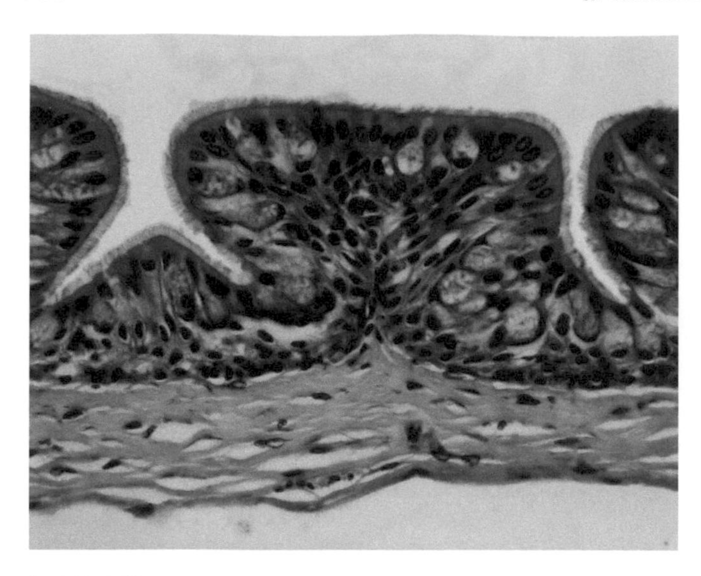

FIGURE K13a The roof of the mouth of the toad is provided with several longitudinal ridges covered with stratified ciliated columnar epithelium. These cilia carry materials to the stomach in a stream of mucus provided by many goblet cells. A delicate lamina propria mucosae carries capillaries that nourish the epithelial cells. The tunica mucosa is subtended by the tela submucosa formed by a dense layer of collagenous tissue. Some of the spaces between the fibers are probably lymphatics. 40 ×.

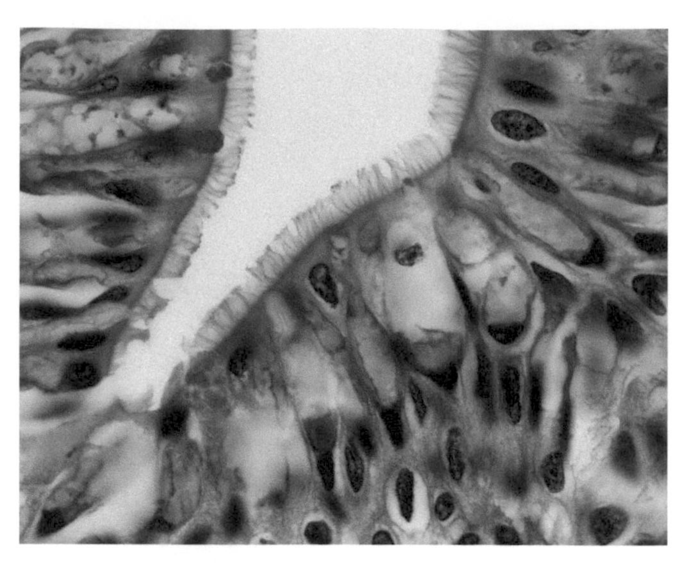

FIGURE K13b The columnar epithelial cells lining the roof of the mouth of the toad bristle with cilia; these are inserted on BASAL BODIES, which form the deep pink line just under the apical membrane of the epithelial cells. Abundant goblet cells are seen in various stages of accumulation of mucous granules. 100 ×.

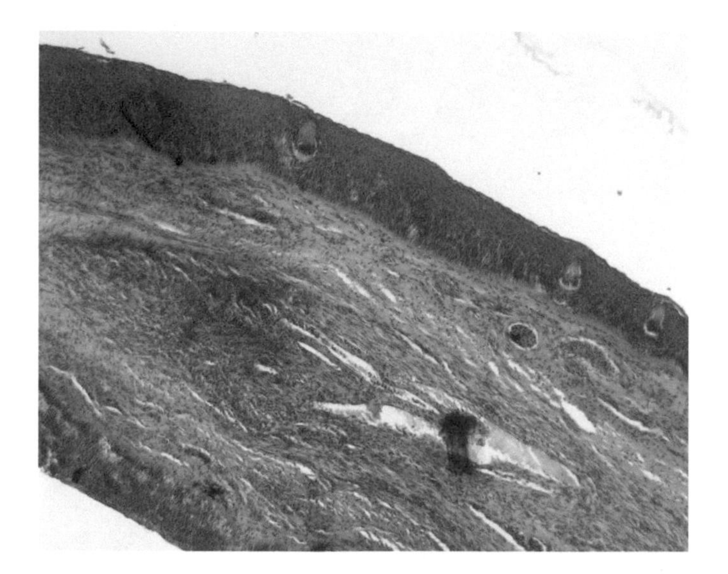

FIGURE K14a Taste buds may be found in locations other than the tongue. This is a portion of a longitudinal section of the barbel (whisker) of a carp. Note the taste buds interposed in the stratified squamous epithelium of the external skin of the fish. These taste buds keep the fish informed about dangers lurking in the surrounding water. Taste cells may be found over the entire surface of the naked skin of catfishes. 10 ×.

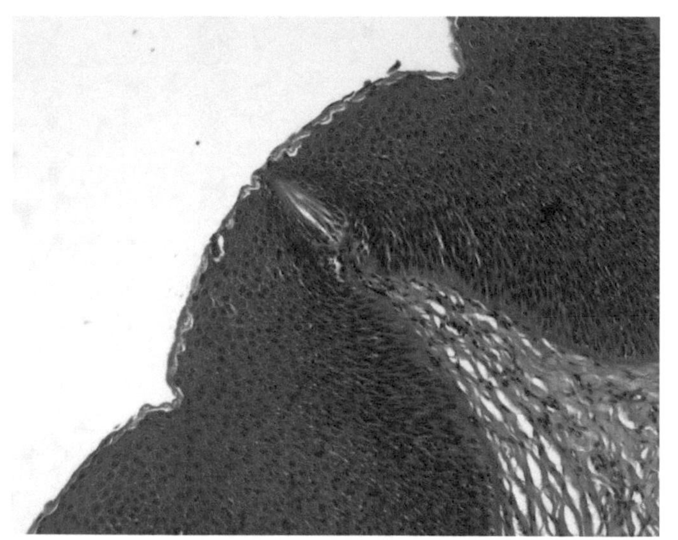

FIGURE K14b This taste bud located on the barbel of a carp resembles those seen on the tongue of mammals: banana-shaped cells poking their ends through a taste pore. 20 ×.

The tongues of other vertebrates are diverse structures and not all are homologous with the mammalian tongue (Figs. K15a–K15g). The tongue of most BIRDS has no intrinsic muscles but is moved by the action of muscles inserted on the hyoid apparatus. In contrast the tongue of many REPTILES possesses amazing motility. Note the complex arrangement of muscles in sections of the forked tongue of a snake (Figs. K16a–K16c). The smooth,

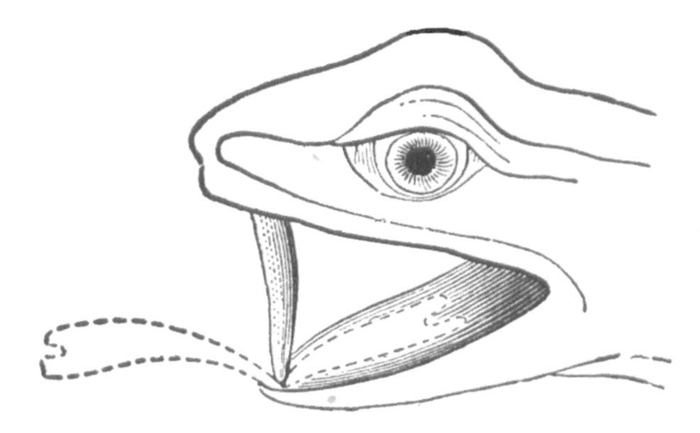

FIGURE K15a Food is delivered to the stomach of many amphibians by ciliary action. The surface of the tongue as well as the lining of the mouth and esophagus are richly provided with cilia which transport the food, unassisted by any peristaltic action. The tongue of a frog is shown in three different positions.

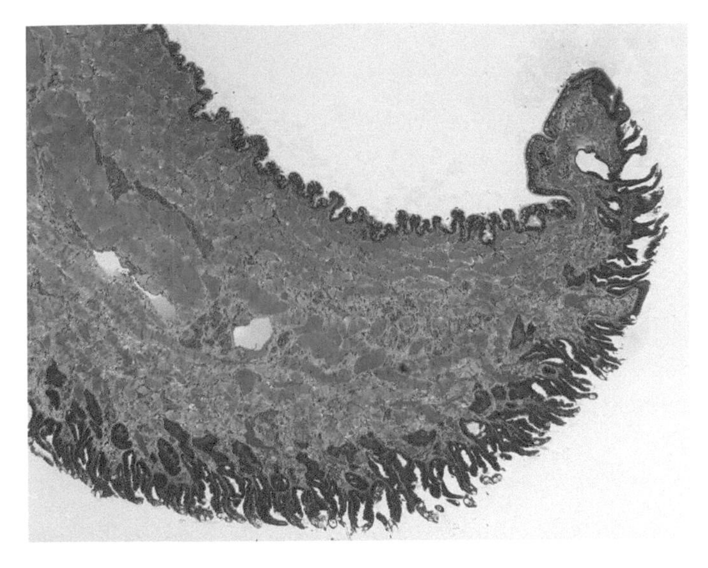

FIGURE K15b The tongue of the frog flips outward and catches its prey in adhesive secretions of the simple tubular glands at the bottom of the micrograph. The woven nature of the striated muscle apparent in these micrographs, together with rich innervation, provide the precise control required by this activity. A nerve trunk is visible at the upper left among the cells of striated muscle. The upper surface is ridged, and richly endowed with cilia. It is provided with goblet cells that produce mucus that assists food in sliding down the esophagus. 4 ×.

FIGURE K15c The upper surface of the tongue is a ridged stratified columnar epithelium, only a few cells thick, consisting of goblet cells and heavily ciliated cells. When the mouth is closed and the tongue is at rest, these cilia carry objects held in the mouth down to the esophagus where they are carried to the stomach. The goblet cells provide mucus, which lubricates this function. 10 ×.

cornified, stratified squamous epithelium that covers most of the surface possesses no glands or sense organs (Figs. K16d, K16e, K16f). The tongue, when retracted, is inserted into the blind pockets of the VOMERONASAL ORGAN (Jacobson's organ) located in the roof of the mouth (Figs. K17a–K17d). This sensory organ detects the presence of volatile chemical substances clinging to the surface of the tongue. Dendritic receptor cells in the sensory epithelium bristle with microvilli, reminiscent of the olfactory epithelium of mammals.

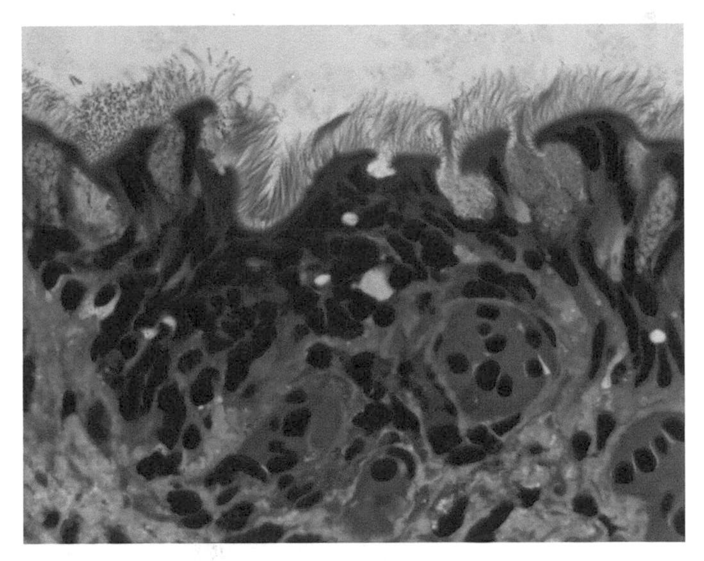

FIGURE K15d This epithelium of the upper surface is covered with ciliated cells, liberally interspersed with goblet cells. Two goblet cells at the right are releasing their content of mucus. The connective tissue below contains a few blood vessels. 100 ×.

FIGURE K15e Woven fibers of striated muscle constitute most of the mass of the tongue. These fibers are richly innervated—nerve trunks traverse the field. 10 ×.

FIGURE K15f When the tongue is everted, prey is affixed to its sticky surface. The simple columnar epithelium forms many simple tubular glands that secrete the sticky substance. These glands penetrate fairly deeply into the muscle mass. The deepest cells are packed with eosinophilic granules while the cells nearer the surface appear to be a mixture of goblet cells and other cells. 10 ×.

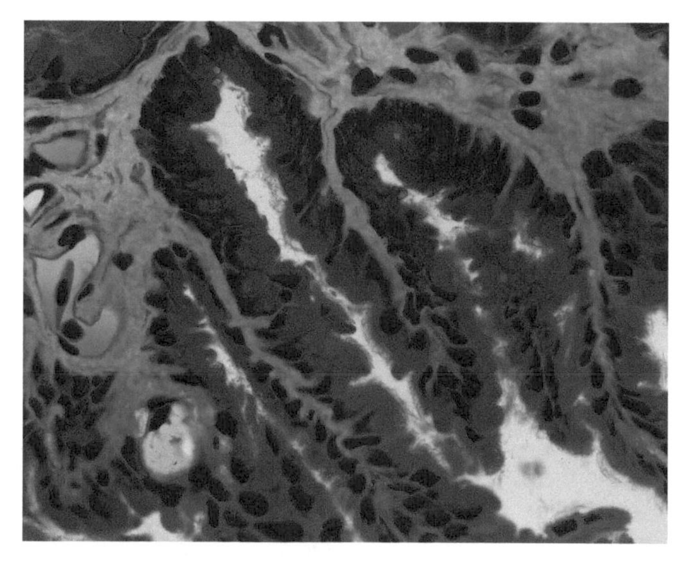

FIGURE K15g The basal regions of the secretory cells of this epithelium are basophilic; a few of the epithelial cells bear cilia. 63 ×.

The roof of the mouth, or PALATE, of a mammal separates the mouth from the nasal cavity (Fig. K18a). Its forward portion is stiffened by bone and is the HARD PALATE (Fig. K18b); there is no bone in the SOFT PALATE (Figures K18c–K18e) which lies behind and terminates in a fleshy flap, the UVULA.

Teeth

Two types of teeth are found in vertebrates: EPIDERMAL TEETH and TRUE TEETH.

FIGURES K16(a–c) The tongue of snakes is probably the most agile of any vertebrate tongue. It darts out of the mouth, capturing scent molecules which it brings into be sampled by a sensory organ in the roof of the mouth: Jacobson's VOMERONASAL ORGAN. The forked tongue of many reptiles (a snake in this example) is made up of woven fibers of striated muscle, which are characteristic of vertebrate tongues. Large nerve trunks provide amazing control. Unlike most tongues the surface is a smooth stratified squamous epithelium with no trace of a sensory organ even at the tip of the forked portion. Pigmented cells lurk between the basal cells of the epithelium and among the connective tissue fibers. 4 ×.

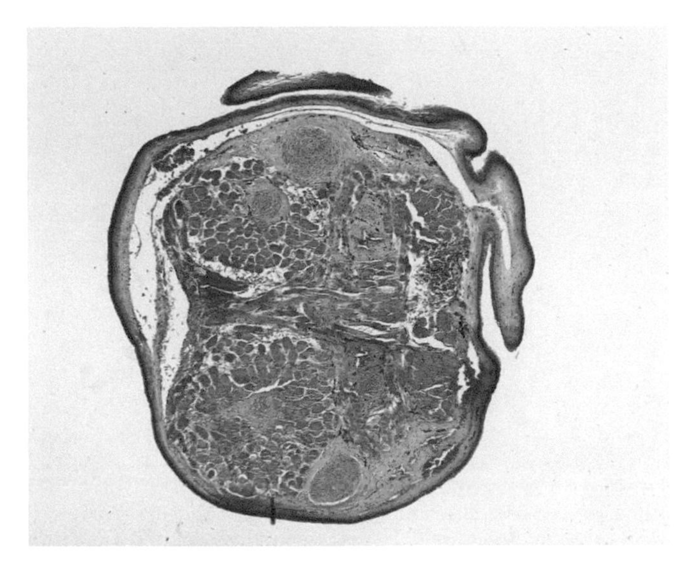

FIGURE K16a Base of the tongue. It consists mostly of a mass of woven striated muscle. The epithelium is smooth, pigmented, stratified squamous. 4 ×.

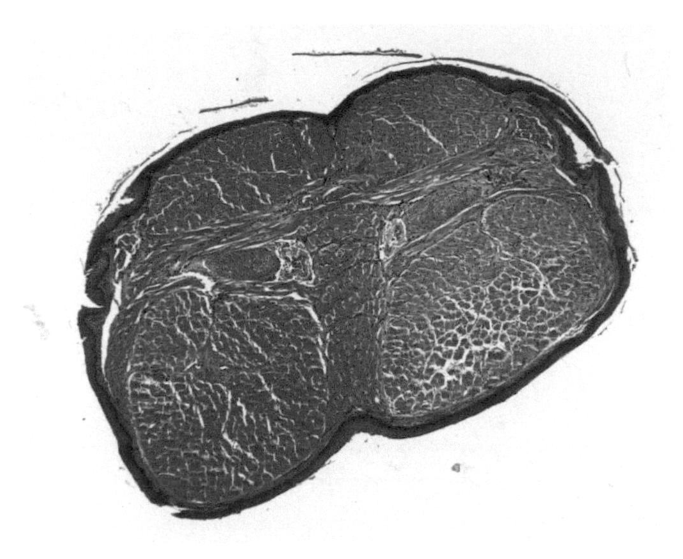

FIGURE K16b Midpoint of the tongue. Woven striated muscle, large nerve trunks, and smooth, pigmented stratified squamous epithelium. 4 ×.

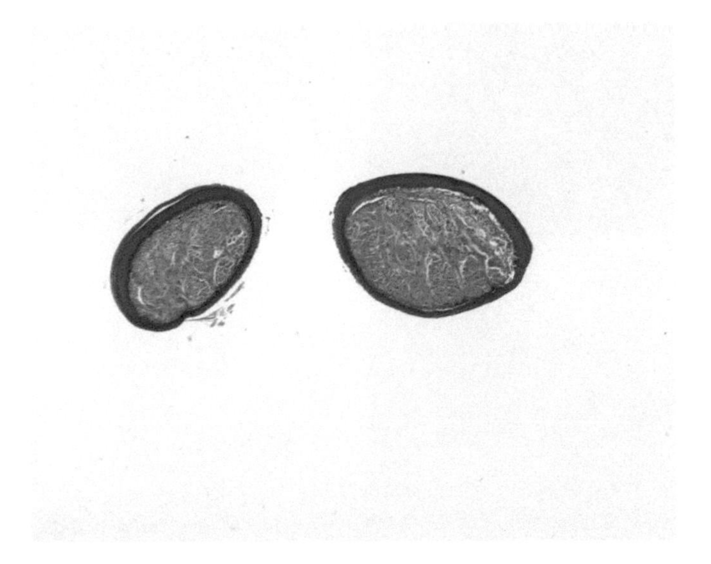

FIGURE K16c The forked tip of the tongue. It consists of the woven striated muscle, nerve trunks, and the pigmented stratified squamous epithelium. There is no evidence of sensory organs. 4 ×.

Epidermal Teeth

These consist of cornified projections of the epithelium of the mouth lining and are not homologous with true teeth. Perhaps the best developed epidermal teeth are those lining the mouth cavity of the lamprey. The BALEEN or "whalebone" of the whale consists of stacks or laminae of cornified epithelial cells forming the long papillae making up the enormous strainers the animal uses to feed on microorganisms it filters from the water.

FIGURE K16(d and e) It can be seen under higher magnification that the epithelium of the tongue is devoid of chemosensory structures. The tongue is covered with a smooth, pigmented stratified squamous epithelium and has the typical woven striated muscle seen in many tongues.

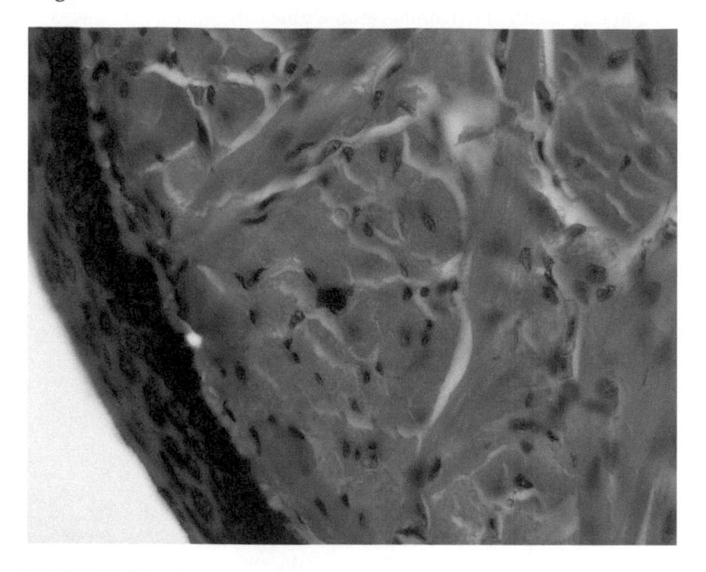

FIGURE K16d This section is taken from the forked tip of the tongue where one would expect to find some sort of structure resembling taste buds. 63 ×.

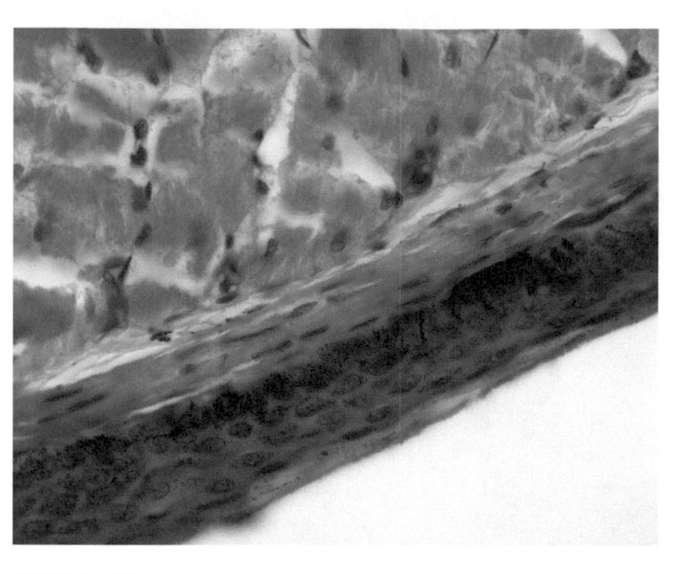

FIGURE K16e The stratified squamous epithelium covers the shaft of the mid portion of the tongue but there is less pigmentation. Several capillaries can be seen between the muscle cells. 63 ×.

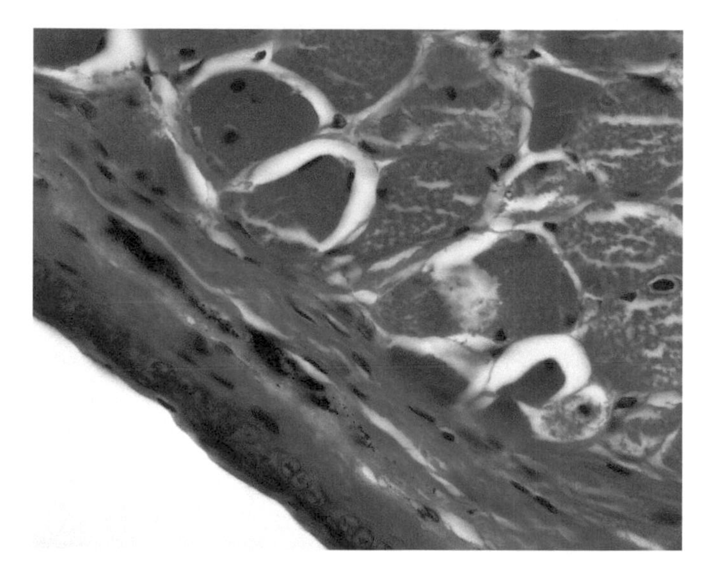

FIGURE K16f There is less pigment in the stratified squamous epithelium at the base of the tongue although some is seen in the connective tissue of the tela submucosa. 63 ×.

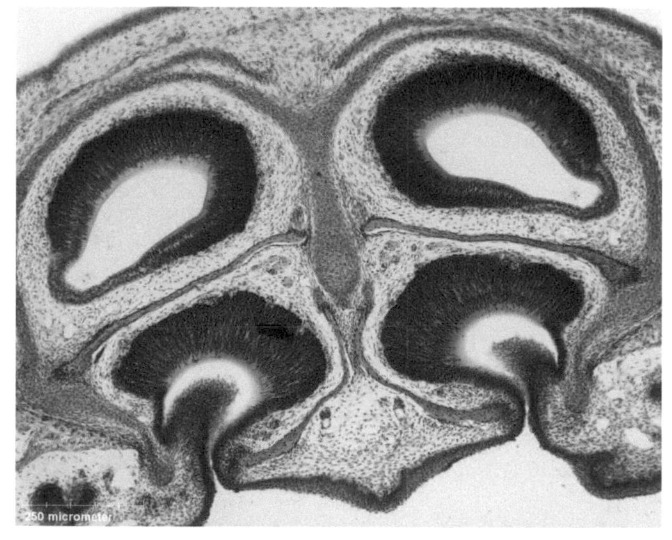

FIGURE K17a This cross section through the head of a lizard should be considered in light of the previous slide. The two large cavities at the top are nasal passages; below are the cavities of the vomeronasal organ (Jacobson's organ) showing their communicating ducts to the mouth cavity. When the tongue is withdrawn into the mouth, the two forked parts are poked into these ducts and scent molecules are detected by the sensory epithelium. 10 ×.

True Teeth

All other teeth found in vertebrates have the same fundamental plan (Fig. K19) and are similar to the placoid scales of the elasmobranchs (Fig. K20 and Figs. K21a–K21c). Placoid scales are covered with a layer of hard VITRODENTINE of mesodermal origin but the exposed surface of true teeth is composed of ENAMEL derived from

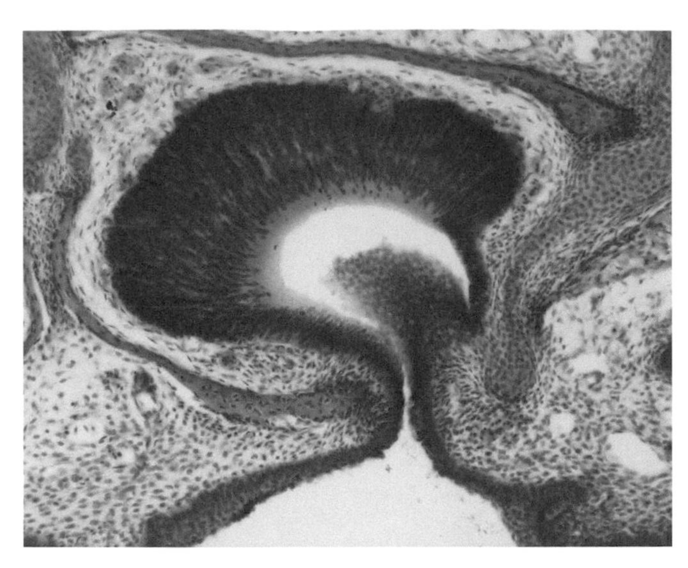

FIGURE K17b The sensory portion of the one side of the vomeronasal organ from the lizard is surrounded by delicate plates of bone (pale pink) and enclosed by loose connective tissue. The cavity of the organ and the duct accommodate the forked portion of the tongue. The sensory epithelium is thick. 20×.

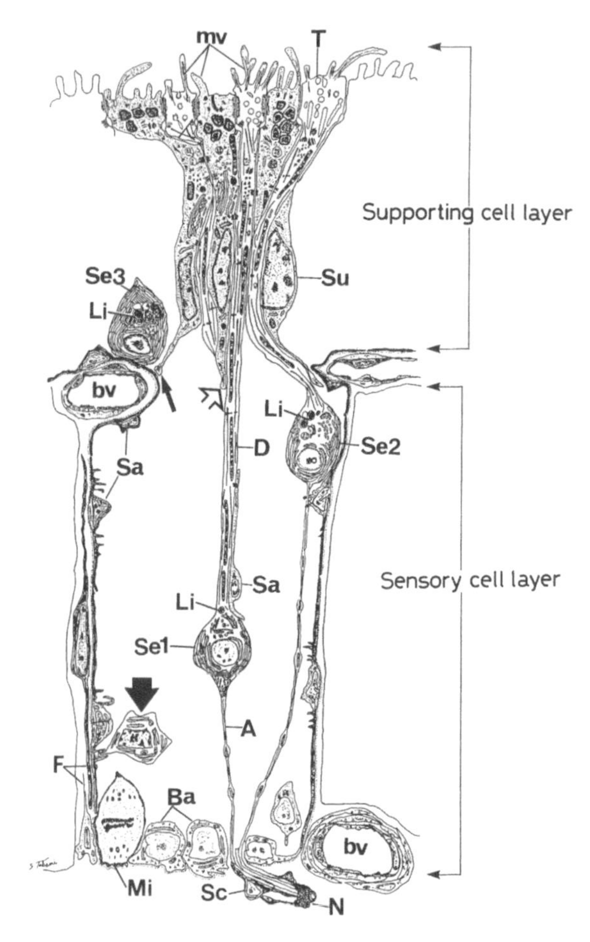

FIGURE K17c Schematic diagram of the vomeronasal epithelium of a crotaline snake. The sensory epithelium consists of a superficial SUPPORTING LAYER and an underlying layer of SENSORY CELLS. Dendrites (D) from the sensory cells penetrate the supporting layer to terminate as knobs (T) from which microvilli (mv) project into the vomeronasal lumen. Satellite cells (Sa) in the sensory cell layer isolate individual groups of sensory cells into columns (Se), one of which is shown. The apical regions of the sensory dendrites contain numerous microtubules.

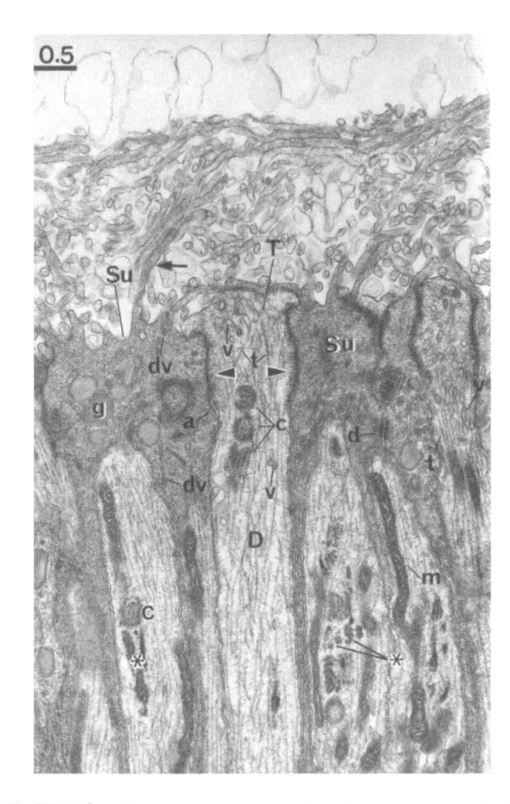

FIGURE K17d Electron micrograph of a section of the apical region of the supporting cell layer (Su) of the vomeronasal epithelium of a crotaline snake. Dendritic knobs from the receptor neurons (T) bristle with microvilli. *D*, dendrites: *m*, mitochondira; *arrow*, microvilli; *a*, zonula adherens.

ectoderm (Fig. K22). True teeth are considered to be homologous with placoid scales, both structures being modified remnants of bony dermal plates that formed the hard armor of ancestral forms. True teeth are found in fish, amphibians, reptiles, and mammals. The following description is based on studies of mammalian teeth.

It is difficult to section teeth because they consist of extremely hard materials. ENAMEL is the hardest substance found in the body of vertebrates and is 99% mineral (hydroxyapatite crystals). DENTINE, the foundation of teeth, is less hard, being 20% organic and 80% inorganic. We study ground sections of tooth, similar to the type used by geologists to study the composition of rocks but these methods destroy the soft tissues (Fig. K23). Alternatively, decalcified teeth, which have been treated with acids, retain the soft tissues but lose the hard materials. With the advent of modern plastics,

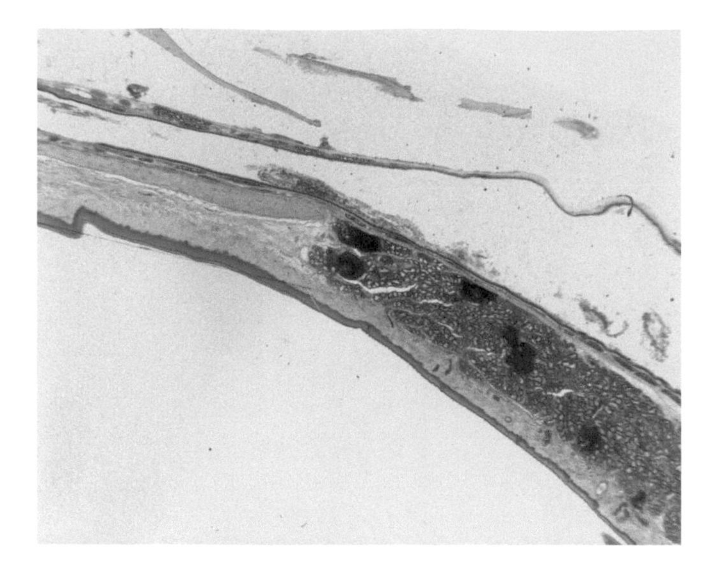

FIGURE K18a An overview of the mammalian palate that separates the mouth from the nasal cavity; the forward part is stiffened by bone and is the HARD PALATE shown at the left; there is no bone in the posterior SOFT PALATE (right), which terminates in a fleshy flap, the UVULA (not shown). The floor of the nasal cavity is at the top of the figure, and the roof of the mouth is below. Note the ridge in the roof of the mouth at the left. 1 ×.

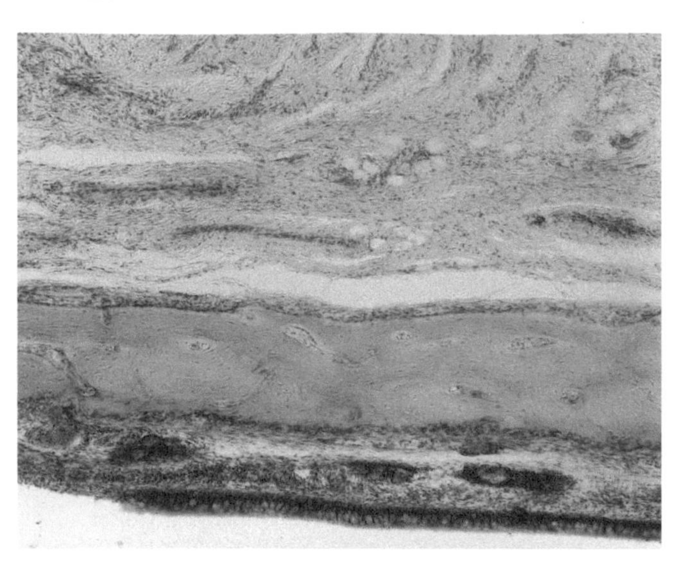

FIGURE K18b Mammalian hard palate. The floor of the nasal cavity consists of a stratified columnar epithelium that is richly supplied with goblet cells. It is separated from the bone of the hard palate by the LAMINA PROPRIA MUCOSAE, which contains a few mucous glands. The bone is surrounded by dense collagenous tissue of the TELA SUBMUCOSA that constitutes the remainder of the micrograph. 10 ×.

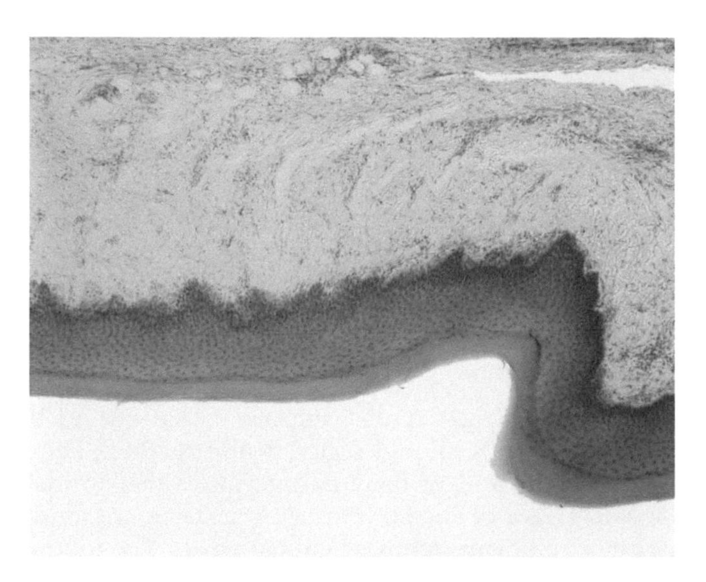

FIGURE K18c The dense collagenous tissue of the tela submucosa subtends the tough stratified squamous cornified epithelium, which is thrown up into several transverse ridges, one of which is shown here. Bone lies beneath the epithelium. 10 ×.

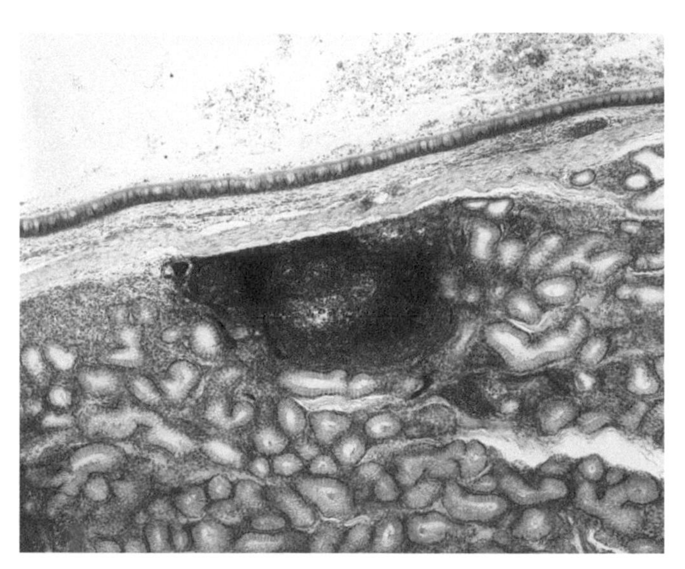

FIGURE K18d The soft palate forms the posterior roof of the mouth and contains no bone. The floor of the nasal cavity is similar to that seen in the region of the hard palate: a stratified columnar epithelium that is rich with goblet cells and subtended by a tela submucosa of collagenous tissue. Most of the tela submucosa is packed with compound alveolar mucous glands. A solitary LYMPH NODULE is seen at the center. 10 ×.

sections have been produced that retain their entire structure (Figs. K24a–K24c). Compare images of dry ground sections and decalcified stained sections of mammalian teeth noting features depicted in Fig. K19. Soft tissues are lost in ground sections but mineralized portions are preserved (Fig. K23). In decalcified sections, soft tissues are preserved while mineralized portions, notably the enamel, are lost (Figs. K24a–K24c). The general form of the

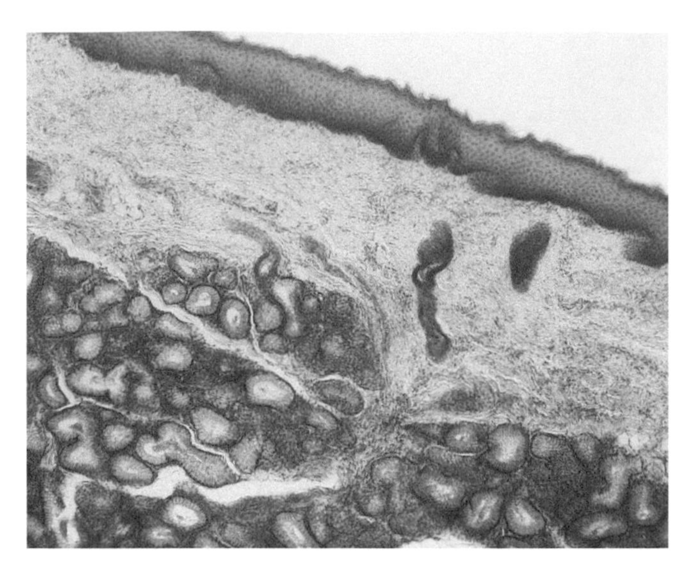

FIGURE K18e The oral surface of the soft palate and is continuous with the previous picture. Diffuse LYMPHOID TISSUE is liberally dispersed throughout the compound alveolar mucous glands. A duct from these glands is making its way through the collagenous tissue of the tela submucosa toward the cornified stratified squamous ORAL EPITHELIUM. 10 ×.

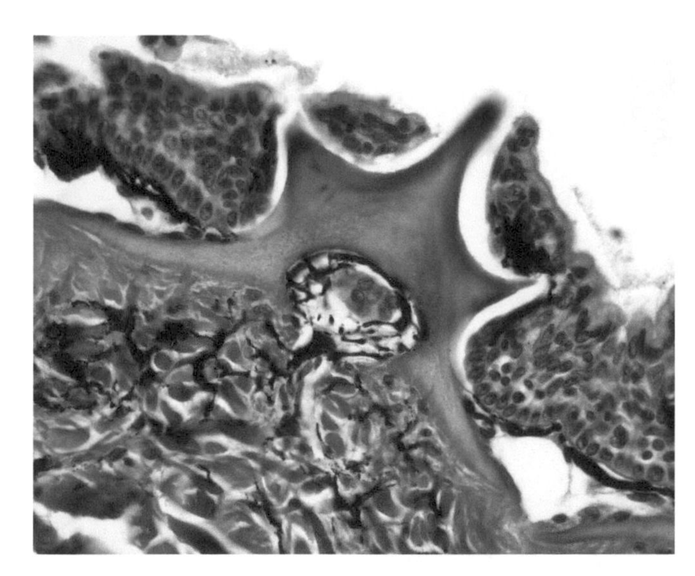

FIGURE K20 Recall the placoid scale seen in the skin of the dogfish. Placoid scales have much in common with the true teeth of vertebrates. These specializations of the dermis penetrate the epidermis, like a tooth, to emerge on the surface. The placoid scales of the shark are dermal scales of this type, giving the skin the feeling of sandpaper. The scale is formed of hard DENTINE coated with harder VITRODENTINE, which has dissolved away in the preparation of the slide. These scales puncture the epidermis and appear on the surface but are rooted in the dermis. In this slide the epidermis has pulled away from the dermis during preparation. Abundant CHROMATOPHORES are seen in the dermis, especially at the base of the epidermis. The scale at the center is cut medially, showing its pulp cavity lined with ODONTOBLASTS. It also contains a chromatophore. A blood vessel enters the pulp cavity through the BASAL PLATE. 40 ×.

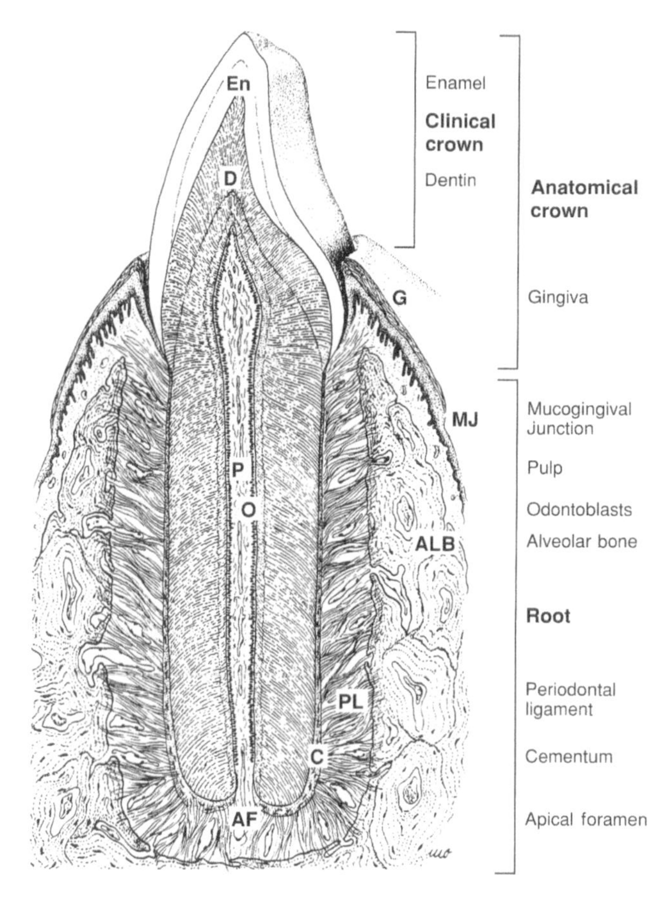

FIGURE K19 Drawing of a longitudinal section through an erupted lower human incisor. Source: *Drawing, M. Oeltzschner.*

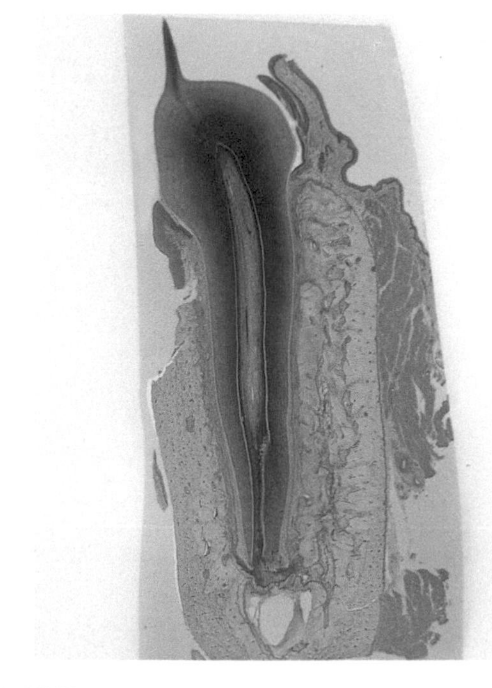

FIGURE K21a This mammalian tooth is still in position in its socket in the alveolar bone of the jaw. The specimen has been decalcified so that the enamel coating and other calcified portions of the tooth and bony matrix have been lost. The PULP CAVITY occupies the core of the tooth and is filled with material resembling vascular mesenchyme. 1 ×.

FIGURE K21b The mesenchymatous nature of the pulp is apparent at the left of this section of decalcified bone. Straight blood vessels course through the pulp. A layer of ODONTOBLASTS at the outer edge of the pulp is plastered against the thick layer of DENTINE. Dentine constitutes most of the mass of the tooth and, although much harder than bone, is similar in its chemical composition: 80% of dentine consists of inorganic material, largely crystals of HYDROXYAPATITE; most of the organic portion is collagen whose fibers are arranged in a parallel fashion. The dentine has a striated appearance due to the presence of parallel DENTINAL TUBULES, fine canaliculi that radiate from the pulp cavity and contain long processes from the odontoblasts.
Comparing ODONTOBLASTS and OSTEOBLASTS (Chapter D: Connective Tissues), we see both cell types produce long processes that inhabit CANALICULI contained within a hard matrix made up of hydroxyapatite and collagen fibers. The cell body of an osteoblast lives within a lacuna in the bony matrix; the cell body of an odontoblast, however, resides in the pulp cavity alongside the hard matrix. The absence of lacunae within the matrix itself allows dentine to be denser and therefore harder than bone. 10 ×.

FIGURE K21c This micrograph was taken at a higher power of the same tooth as described in Fig. K21b above. The pulp cavity, at the left, contains the vascular mesenchymatous PULP. Outlines of the odontoblasts are indistinct; they lie apposed to the dentine which takes up most of the right half of the picture. The parallel nature of the dentinal tubules is clear. 40 ×.

tooth is maintained by the DENTINE upon which are laid the ENAMEL of the CROWN and CEMENTUM of the ROOT. The PULP resembles vascular mesenchyme and fills the central pulp cavity (Fig. K24b). Its basophilic ground substance contains fine collagenous fibers and fibroblasts, macrophages, mast cells, and cells outwandered from the blood.

Outlining the pulp cavity is a single row of ODONTOBLASTS each sending one or more processes into fine DENTINAL TUBULES in the dentine (Fig. K25a). Dentinal tubules are fine, dark S-shaped lines that pass through the dentine from the pulp cavity to the dento–enamel junction. These tubules contain DENTINAL FIBERS (Tomes' fibers),

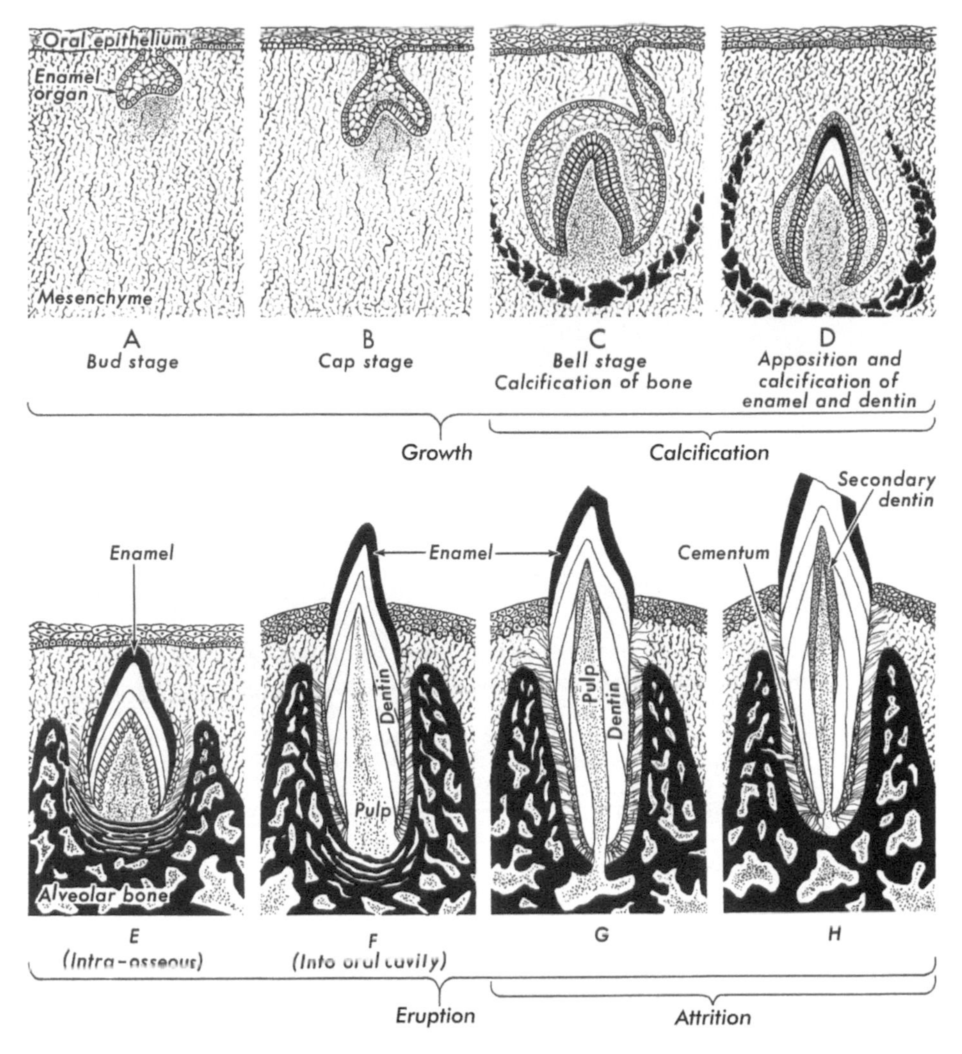

FIGURE K22 Diagrams of the early development of the human incisor tooth: the formation of its root, its eruption, and beginning attrition. Enamel and bone are drawn in black.

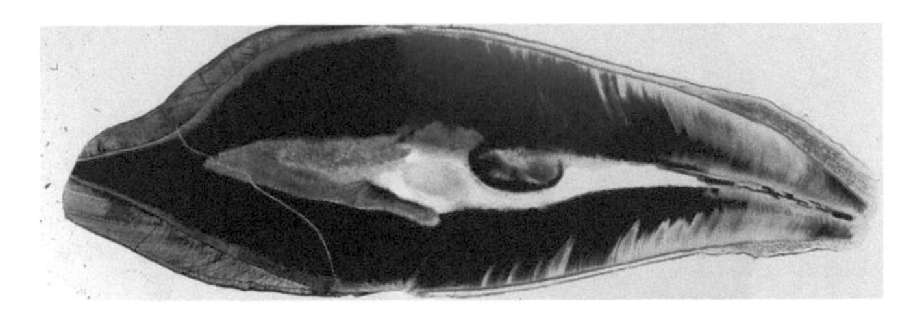

FIGURE K23 A montage of a nondecalcified ground dried section of human tooth. The cells and all soft tissue have been removed in processing—only the hard, calcified parts of the tooth remain. Grinding compound is retained within some of the cavities of the specimen and appears black.
Note the outer enamel, the central pulp cavity, and the dentine in between. DENTINAL TUBULES form a pattern in the dentine reminiscent of wavy hair. The tip of the tooth has worn away exposing the dentine. The outer coating of enamel is replaced by CEMENTUM below the gum line. The pulp cavity contains only the débris left from grinding and no remnant of the pulp. There is a crack crossing the section.
The tooth is firmly held within a socket or ALVEOLUS in the bone of the jaw and is separated from the bone by the cementum, similar to the matrix of bone. It is firmly attached to the bone by collagenous fibers of the PERIODONTAL LIGAMENT (the periosteum of the bone socket). The upper one-third to one-half of the cementum is acellular but the remainder has cells in its matrix and is similar to bone with lacunae and canaliculi present. The cells resemble osteocytes and are called CEMENTOCYTES, 1 ×.

FIGURES K24(a–c) The next series of micrographs was taken from a decalcified section of the jaw of a dog with the tooth cut in cross section.

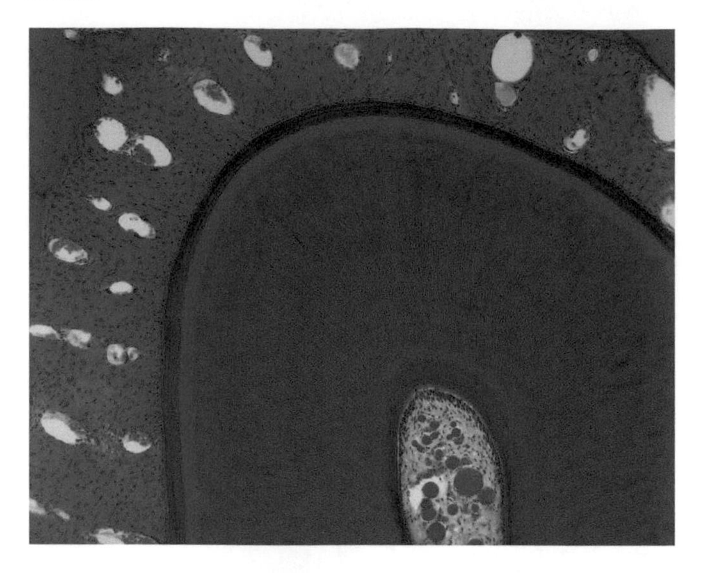

FIGURE K24a The tooth is surrounded by bone of the jaw—the irregular nature of OSTEONS is apparent in this ALVEOLAR BONE; the bone is penetrated by many OSTEON CANALS. The PULP CAVITY at the center of the tooth contains conspicuous fine blood vessels. ODONTOBLASTS form a membrane apposed to the dentine. Parallel DENTINAL FIBERS radiate from the pulp cavity to the dark CEMENTUM, which affixes the tooth to the bony socket in the jaw. 20×.

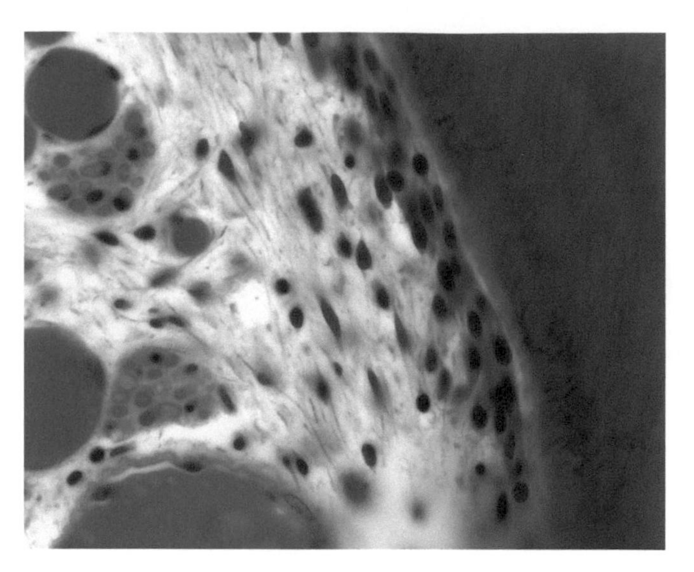

FIGURE K24b At a higher magnification the layer of odontoblasts apposed to the dentine and the dentine itself are seen. DENTINAL TUBULES penetrate the dentine. Pulp is a loose aggregation of spindle-shaped fibroblast-like cells, collagenous fibers, delicate blood vessels with thin endothelial walls, and bundles of nerve fibers. The small BLOOD VESSELS are packed with erythrocytes and appear bright red. Two NERVE BUNDLES are shown; they contain several fibers cut in cross section. 100×.

FIGURE K24c The parallel nature of the DENTINAL TUBULES is evident in this section of dentine. Each tubule contains a process from an odontoblast. 100×.

the protoplasmic processes of odontoblasts (Figs. K25b and K25c). Between the dentinal fibers is a meshwork of collagenous fibers embedded in calcified ground substance upon which the calcified portion of the dentine is laid (Fig. K25d).

Meanwhile a stalk of epithelial cells from the lining of the mouth grows down to envelop the developing dentine. These cells become the ENAMEL ORGAN (Fig. K26), which secretes the extremely hard ENAMEL RODS that stand upright on the dentine; 99% of the enamel is hydroxyapatite crystals.

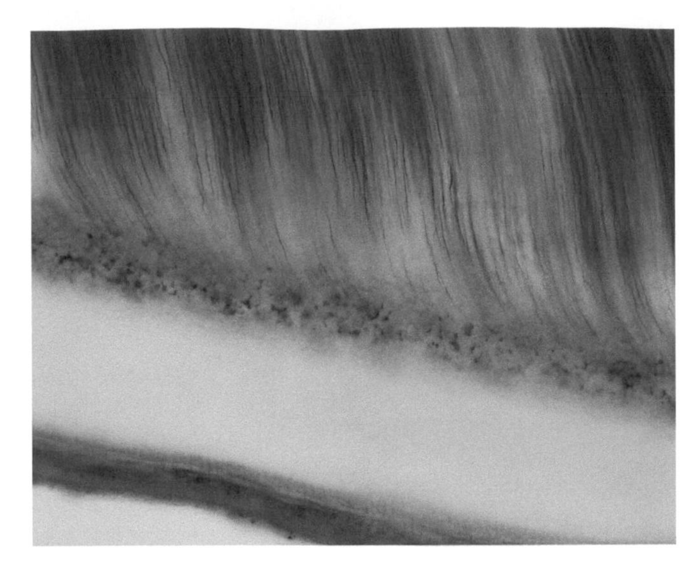

FIGURE K25a Odontoblasts line the pulp cavity and send one or more processes, the dentinal tubules, into the dentine. These are fine lines that resemble wavy human hair and pass through the dentine from the pulp cavity to the DENTO–ENAMEL junction.

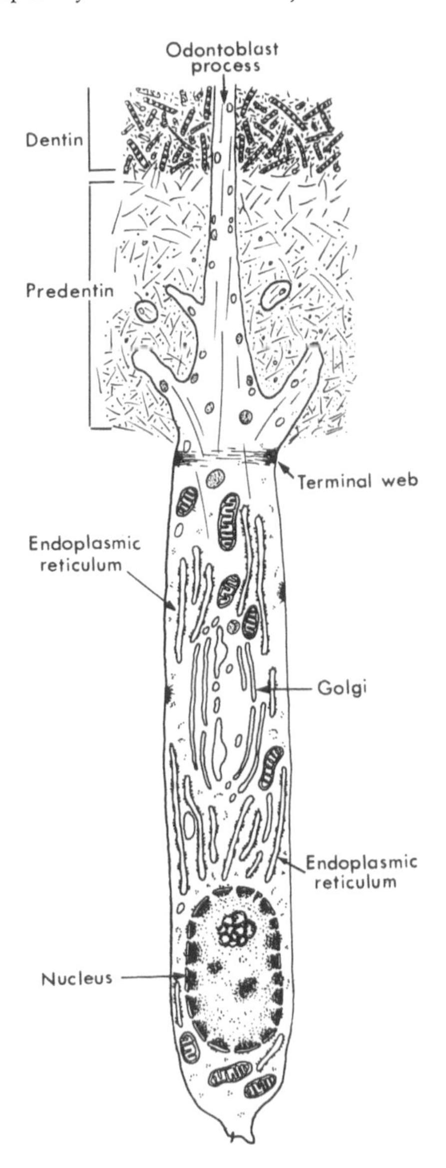

FIGURE K25c Drawing showing the ultrastructure of an osteoblast from the developing incisor of a rat.

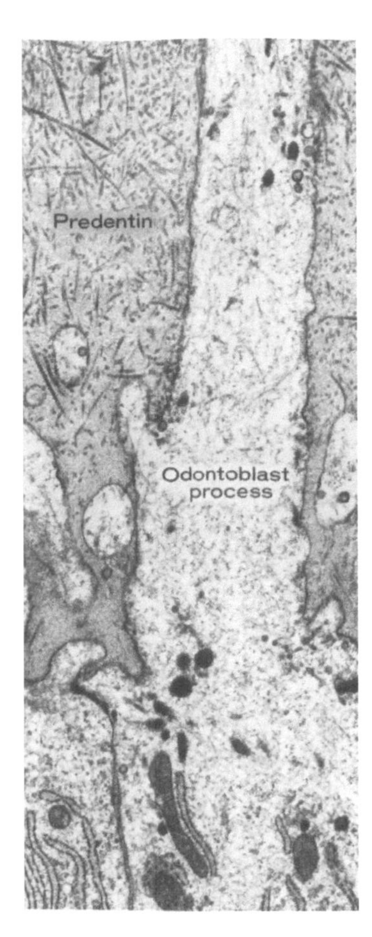

FIGURE K25b Electron micrograph of the apical region of an odontoblast showing the odontoblast process and the surrounding predentine.

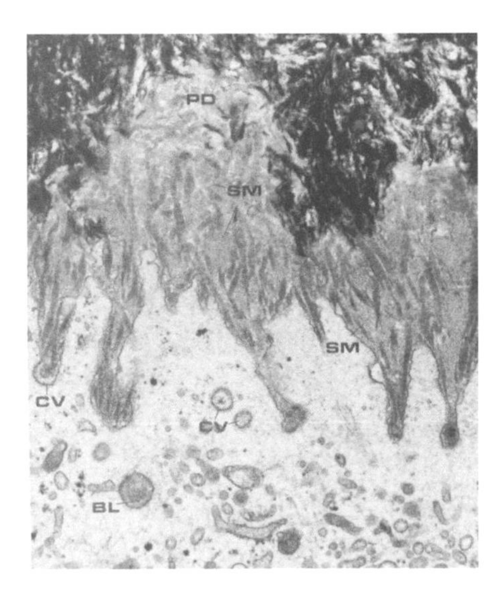

FIGURE K25d Electron micrograph of a developing tooth of a cat. The dentine at the top is heavily calcified; below is the predentine (PD) with conspicuous collagen fibers. At the bottom is the cytoplasm of an odontoblast with coated vesicles (CV), a remnant of basal lamin (BL) and some stippled material (SM). 32,000 ×.

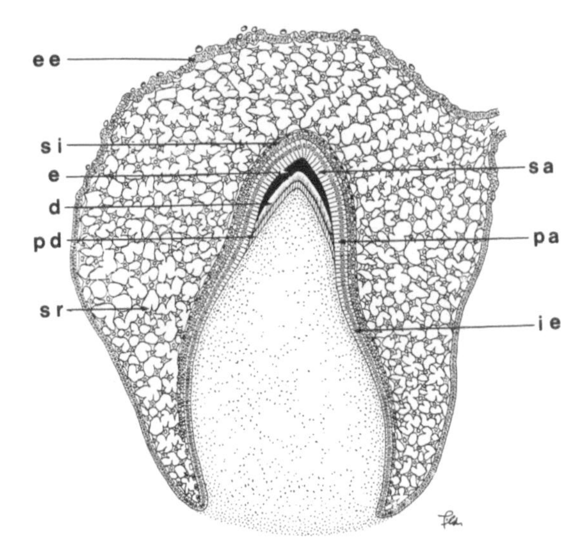

FIGURE K26 Diagram of a section through a human tooth germ in the bell stage. *d*, dentine; *e*, enamel; *ee*, external enamel epithelium; *ie*, internal enamel epithelium; *pa*, preameloblasts; *pd*, predentine; *sa*, secretory ameloblasts; *si*, stratum intermedium; *sr*, stellate reticulum.

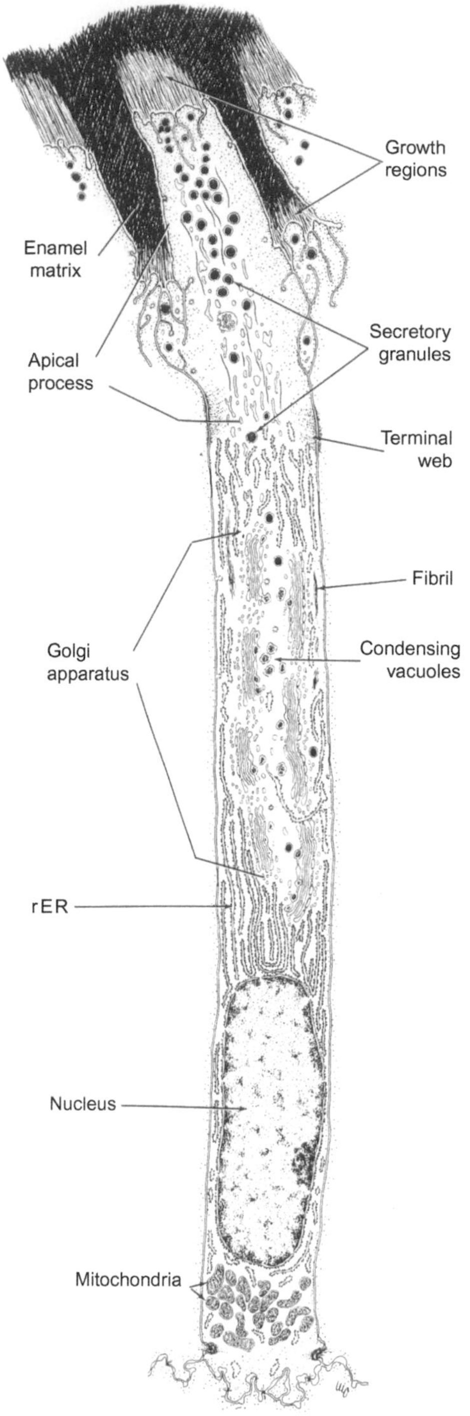

Odontoblasts and ameloblasts are columnar cells with basal nuclei located along the inner and outer margins, respectively, of the dentine. Apically the cells bear processes, which are instrumental in forming the dentine and enamel.

The processes of ODONTOBLASTS are branching A terminal web around the "waist" of odontoblasts—at the point of origin of the processes—anchors intercellular junctions that unite the odontoblasts into an epithelial sheet isolating the dentine from the pulp cavity (Fig. K25c). The basal cytoplasm is packed with granular endoplasmic reticulum surmounted by a large Golgi complex. Odontoblasts secrete procollagen, which becomes the collagen of PREDENTINE which later becomes calcified as DENTINE (Fig. K25d).

Columnar AMELOBLASTS, applied to the outer surface of the dentine, are cells directly adjacent to the developing enamel. Each bears a stout process (Tomes' process) that nestles in a cavity of the forming enamel (Figs. K27a and K27b). Each ameloblast process produces one crystal of hydroxyapatite, an ENAMEL ROD, that stands upright on the dentine and extends through the enamel layer. After the tooth erupts, ameloblasts degenerate and enamel becomes incapable of repair.

Growth regions
Enamel matrix
Secretory granules
Apical process
Terminal web
Fibril
Golgi apparatus
Condensing vacuoles
rER
Nucleus
Mitochondria

FIGURE K27a Diagrammatic representation of the ultrastructure of a secretory ameloblast from the maxillary incisor of a rat.

The root of the tooth is affixed into an alveolus in the jaw by CEMENTUM, which resembles bone (Fig. K28). The upper cementum is ACELLULAR but CEMENTOCYTES, which resemble osteocytes of bone, are scattered throughout the deeper, thicker layer of CELLULAR CEMENTUM (Figs. K25a and K28) Collagenous fibers in the cementum extend from the surrounding bone, affixing the tooth firmly in the jaw. The ROOT CANAL, containing nerves and small blood vessels, passes from the pulp cavity to the tip of the root. In ground sections, ENAMEL RODS are seen to be arranged in bundles lying in different planes. The upper one-third to one-half of the CEMENTUM is noncellular but

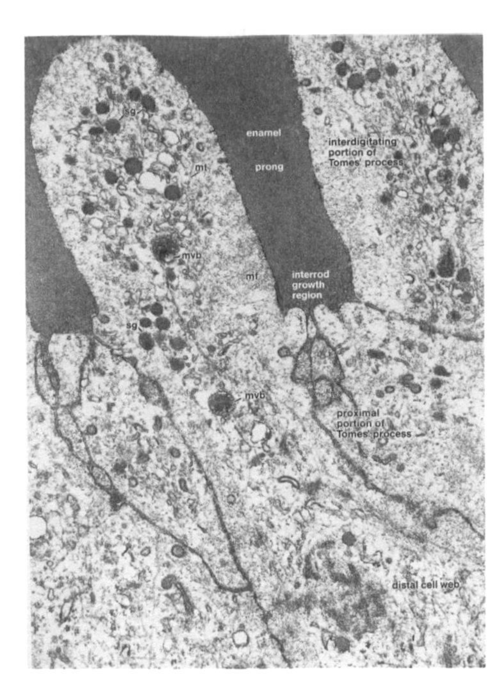

FIGURE K27b Electron micrograph of processes (Tomes' processes) from the apical cytoplasm of secretory ameloblasts extending between prongs of enamel matrix. Secretory granules (sg) and microtubles (mt) are evident in the processes *mvb*, electron dense multivesicular body. 25,000 ×.

FIGURE K28 Photomicrograph of cellular cementum in a ground section of human tooth. The cells (CEMENTOCYTES) have disappeared but their lacunae, full of air and grinding compound, are clearly visible. 40 ×.

the remainder has cells in the matrix and is similar to bone. Note its LAMELLAE and the presence of LACUNAE and CANALICULI. The tooth is fastened into the ALVEOLUS of the jaw bone by collagenous fibers of the PERIODONTAL LIGAMENT, i.e., the periosteum of the alveolus. These fibers extend from the bone of the alveolus to the cementum.

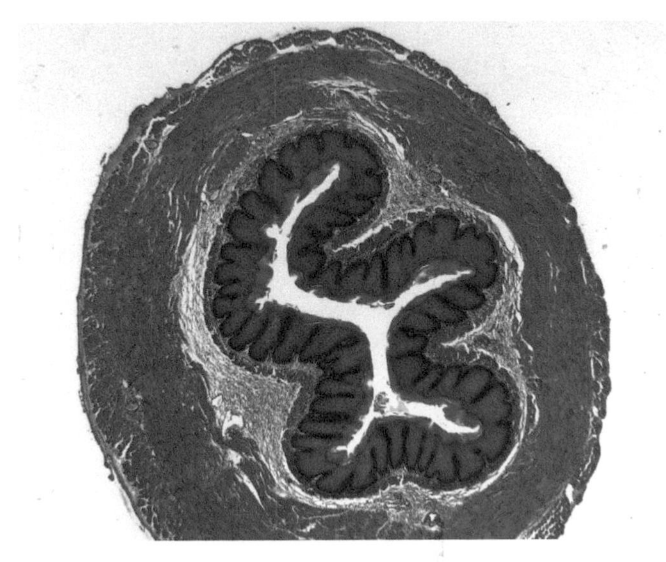

FIGURE K29a The four layers seen in all regions of the gut are apparent in this cross section of the esophagus of a *mammal*: TUNICA MUCOSA, TELA SUBMUCOSA, TUNICA MUSCULARIS, and TUNICA ADVENTITIA. This section was probably taken above the diaphragm where the esophagus is not enclosed by peritoneum so that use of the term tunica serosa would be inappropriate. 2.5 ×.

ESOPHAGUS

The esophagus of *mammals* displays the four layers typical of the digestive tube (Figs. K29a–K29c and K30). The EPITHELIUM is stratified squamous and in some mammals it becomes keratinized. Below the epithelium the LAMINA PROPRIA MUCOSAE forms the subepithelial papillae. The TUNICA MUSCULARIS is variable at different levels and there is a gradual transition from striated muscle in the upper regions to smooth muscle in the lower (Fig. K31a). In order to initiate swallowing, *striated* muscle, not smooth, must be used. Most of the esophagus is outside the coelom and is covered by a tunica adventitia of loose connective tissue that blends with the surrounding tissues. After entering the abdominal cavity, part of the esophagus is covered by a tunica serosa.

Two types of GLANDS may be present (Figs. K31b and K31c). DEEP GLANDS or ESOPHAGEAL GLANDS are mucus secreting, compound tubuloalveolar glands

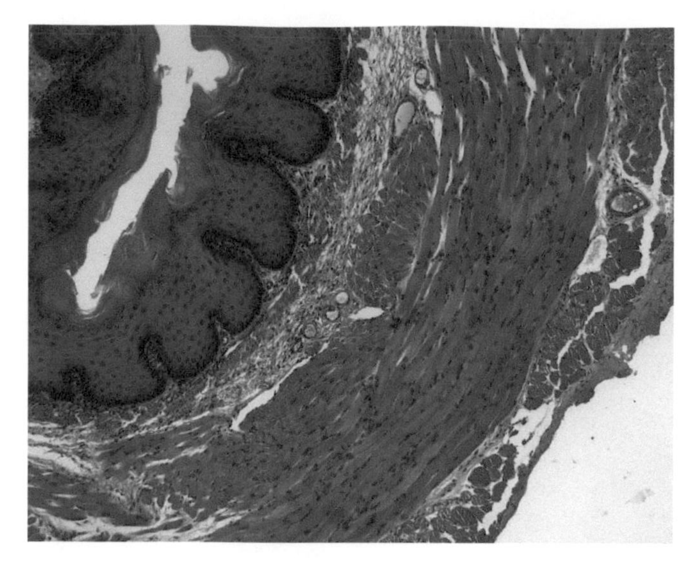

FIGURE K29b The three layers of the TUNICA MUCOSA can be distinguished in this section of a mammalian esophagus. The lining EPITHELIUM MUCOSAE is cornified stratified squamous. It can be seen that the superficial cells are sloughing off with the ingestion of coarse food. Below is a thin layer of delicate connective tissue of the LAMINA PROPRIA MUCOSAE. This connective tissue is separated from the TELA SUBMUCOSA by a few smooth muscle cells of the LAMINA MUSCULARIS MUCOSAE. The inner layer of the TUNICA MUSCULARIS is circular and the outer layer is longitudinal. Note that these muscle fibers are striated. The upper portions of the esophagus are under the control of the will thereby enabling swallowing at will and are made up of striated fibers. The TUNICA ADVENTITIA of connective tissue encloses the layer of muscle. 10 ×.

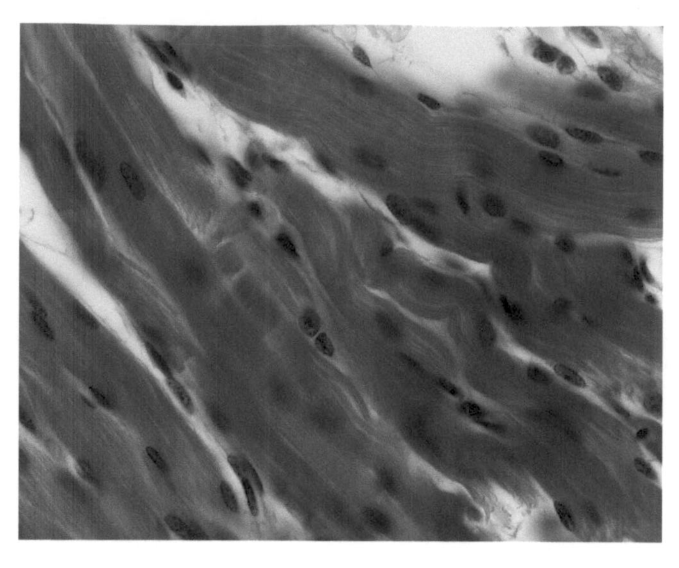

FIGURE K29c This gray-scale micrograph, the section has been stained with iron hematoxylin, demonstrates the striated muscle fibers in the tunica muscularis of the mammalian esophagus. 63 ×.

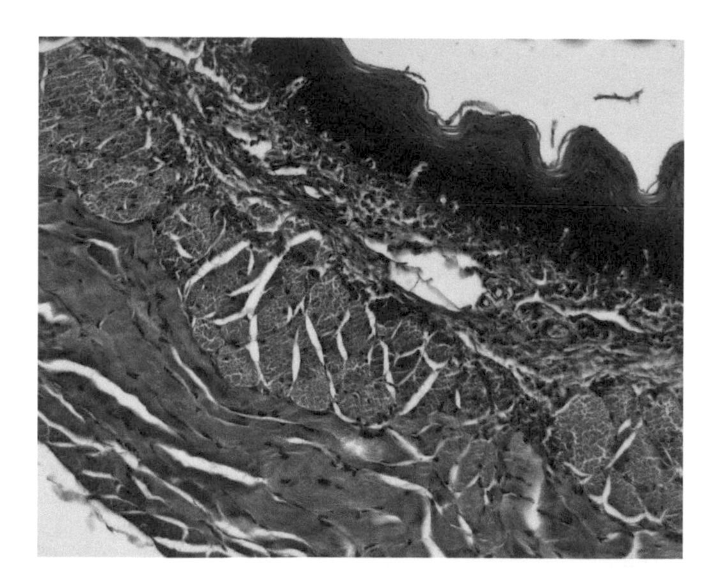

FIGURE K30 Cross section of rabbit esophagus. Sometimes, the lamina muscularis mucosae cannot be distinguished in the esophagus and the epithelium mucosae appears to rest on the tela submucosa. This section shows an unusual condition: the inner layer of muscle is longitudinal and the outer is circular. In parts of the section (not shown here) there is an outer layer of longitudinal muscle. All of this muscle is striated. The tunica adventitia has largely been torn away in sectioning. 20 ×.

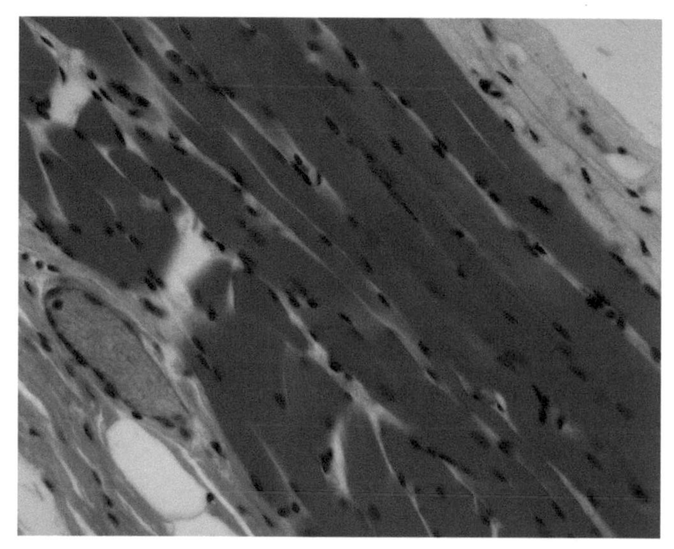

FIGURE K31a This section of the upper portion of the esophagus of a dog has been cut obliquely so that it is difficult to distinguish the layers of muscle. The muscle of the tunica muscularis is striated. The dog's well-known ability to gulp large amounts of food and then reject any undesirable bits is related to the large amounts of voluntary striated muscle throughout the length of its esophagus. 40 ×.

located in the tela submucosa. They appear throughout the length of the esophagus but are most numerous in the upper two-thirds. SUPERFICIAL or CARDIAC GLANDS are mucus secreting, tubular glands located in the lamina propria mucosae. They are always located at the lower end of the esophagus. Esophageal glands assist in smooth sliding of food into the stomach (Figs. K32a and K32b). In many birds the lower portion of the esophagus is dilated to form a CROP lined with

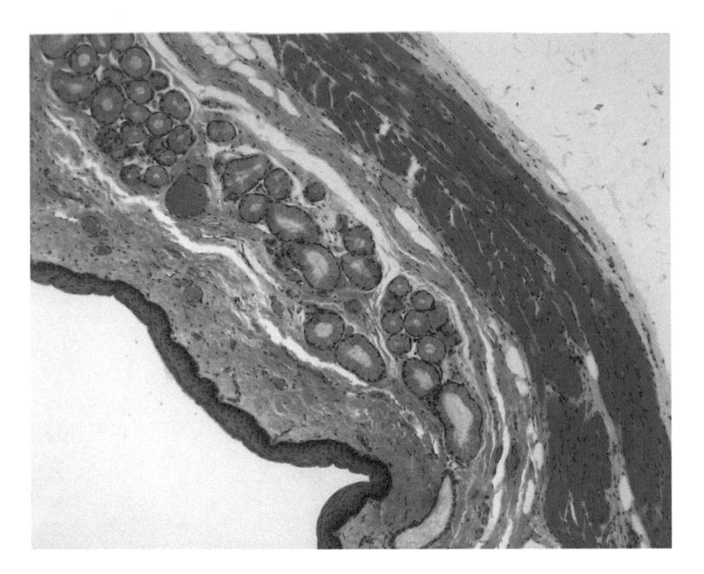

FIGURE K31b There are DEEP ESOPHAGEAL GLANDS in the tela submucosa in this section of the upper part of the esophagus of a dog. 10 ×.

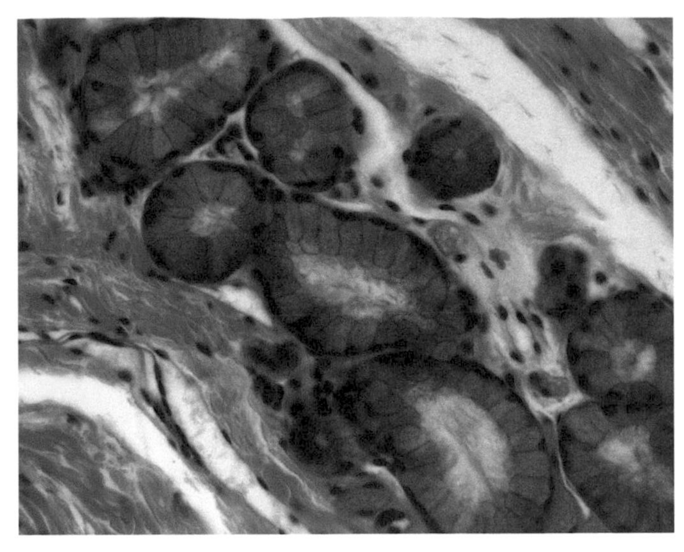

FIGURE K31c The compound tubuloalveolar deep esophageal glands in the tela submucosa secrete mucus. 40 ×.

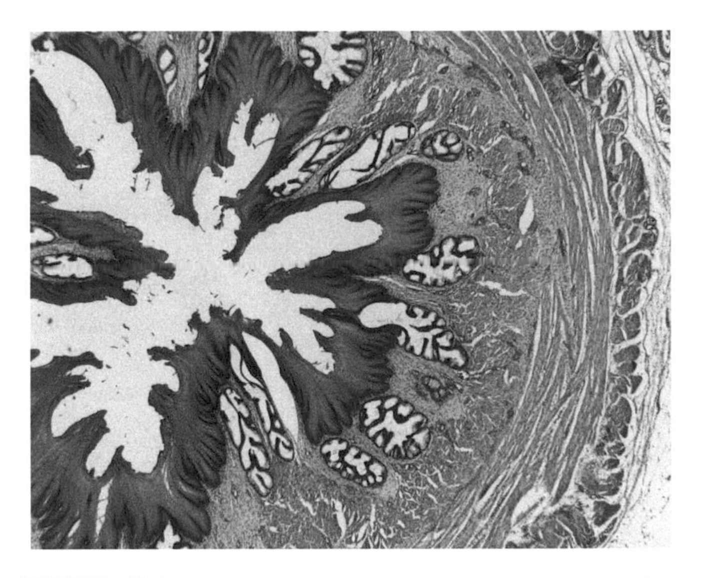

FIGURE K32a The esophagus of the chicken is lined with a tough, stratified squamous epithelium, resistant to the coarse food and gravel ingested. Materials are assisted in their passage by mucus secreted by the large glands in the tela submucosa; the duct from one of these glands, making its way to the lumen, can be seen to the right of center of the micrograph. 2.5 ×.

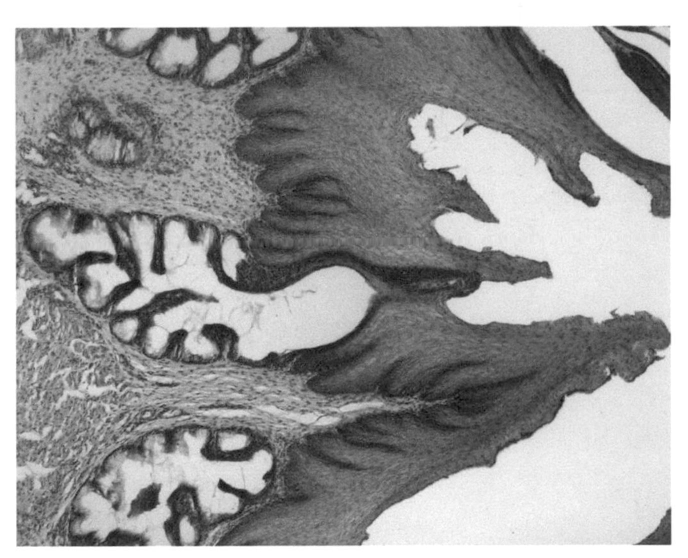

FIGURE K32b A duct from a deep esophageal gland in the chicken penetrates the tough stratified squamous epithelium of the esophagus to carry mucus to the lumen. 10 ×.

stratified squamous epithelium whose wall is similar in structure to that of the remainder of the esophagus (Figs. K33d). In the doves the epithelial cells accumulate fat droplets in their cytoplasm and masses of these are cast off to constitute "crop milk" that is fed to the young by both male and female parents (Figs. K33a–K33c, K34a–K34c). The activity of these glands is controlled by the "lactogenic" hormone secreted by the anterior lobe of the pituitary gland. The esophagus of *reptiles* shows little specialization in the preparations available to us (Figs. K35a–K35c, K36a and K36b). The lining of the esophagus of *amphibians* is made up of ciliated epithelium liberally interspersed with goblet cells (Figs. K37a–K37c). Note the condensation of hemopoietic tissue (*Leydig's organ*) in the tela submucosa extending the length of the esophagus of *elasmobranchs* (Fig. K38).

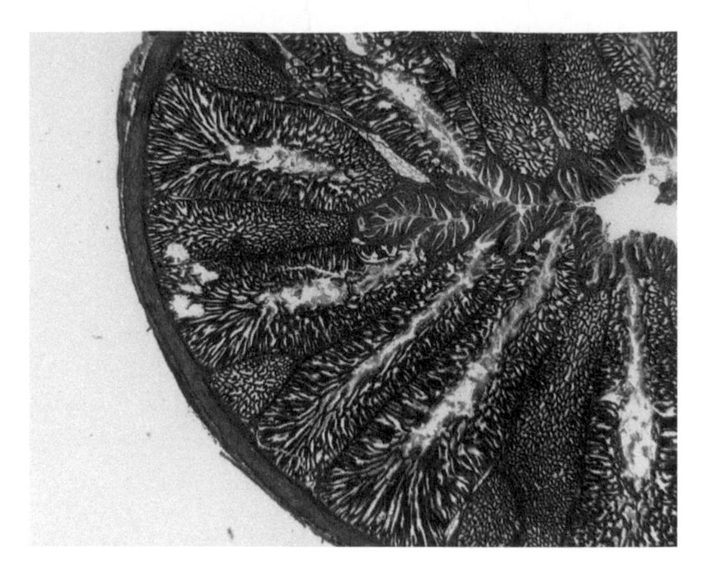

FIGURE K33a The crop of birds is an enlargement of the esophagus and is used to store ingested food pending pulverization in the gizzard. This strategy enables the bird to spend a minimal time exposed to potential predators while it forages for food. Secretory glands in the wall of the crop are capable of producing "crop milk" that is fed to the young of some species. This is a section of the crop of a pigeon, which was not feeding young. Note that the submucosal glands have an "arid" appearance. 2.5 ×.

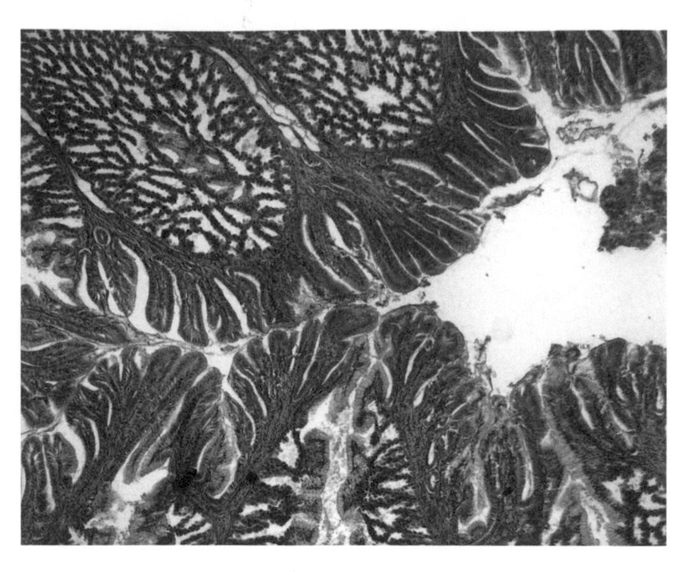

FIGURE K33b This section is from the same nonlactating crop. 10 ×.

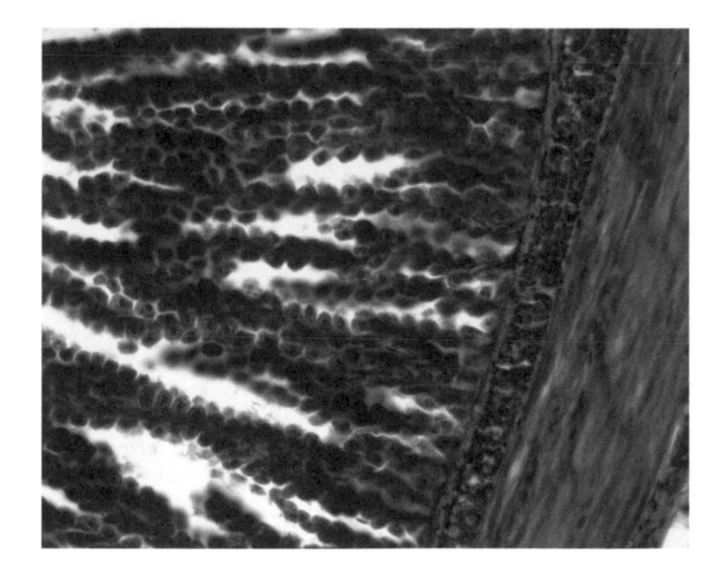

FIGURE K33c Detail of the wall and nonsecreting glands of the crop of a pigeon. Note the tela submucosa and the tunica muscularis. A fragment of the tunica adventitia may be seen in the lower right corner. 40 ×.

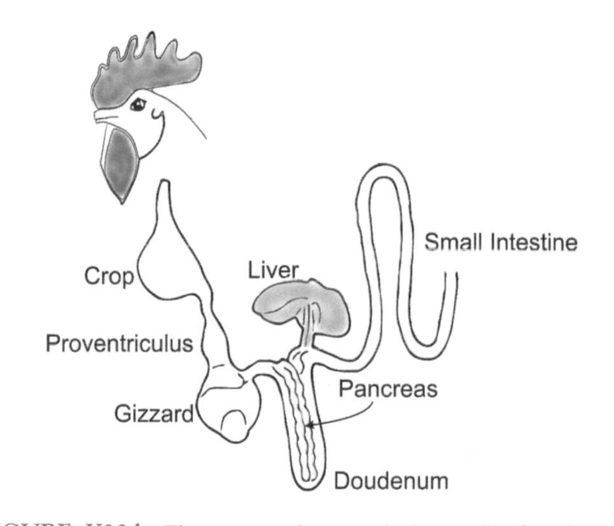

FIGURE K33d The crop and stomach (gizzard) of a domestic chicken. *Authors.*

JUNCTION OF STOMACH AND ESOPHAGUS

At the junction of the stomach and esophagus of a mammal there is an abrupt change of the epithelium from stratified squamous to the simple columnar of the surface epithelium of the stomach (Figs. K39a and K39b). The tela submucosa is unchanged at the level of transition and the tunica muscularis retains its two layers.

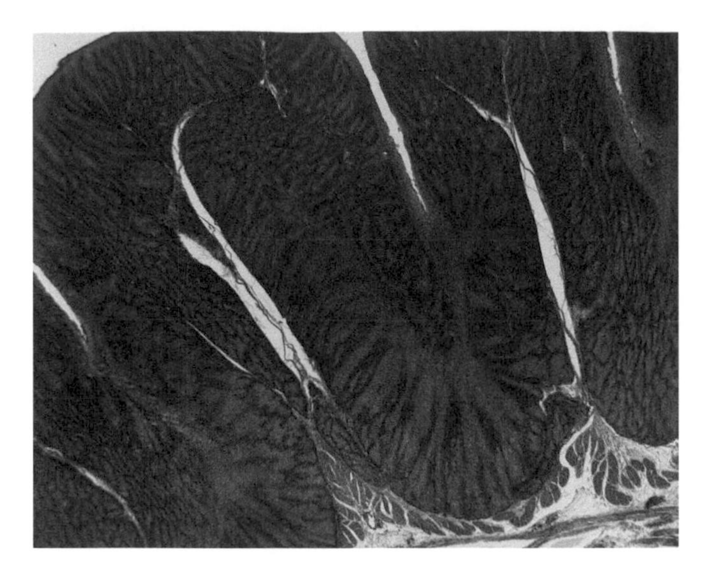

FIGURE K34a The alveolar glands of the crop of a lactating pigeon are distended with "crop milk." Both male and female birds produce this secretion. This section is from a male. Overview. 2.5 ×.

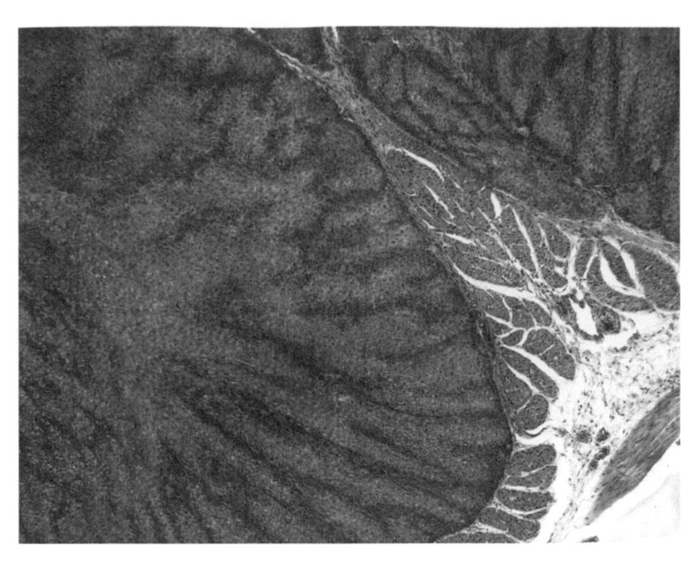

FIGURE K34b The alveolar glands of the crop of a lactating pigeon are distended with "crop milk." This is the same section as Fig. K34a at higher magnification. 10 ×.

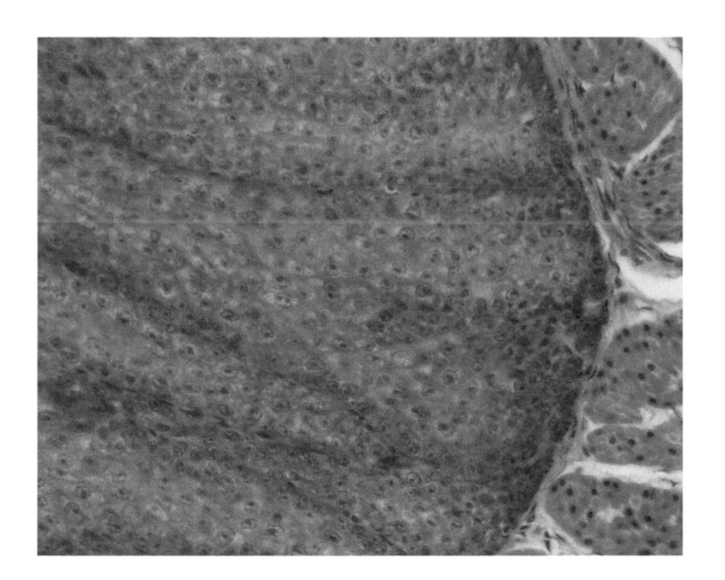

FIGURE K34c The alveolar glands of the crop of a lactating pigeon are distended with "crop milk." This is a high power view of the section in Fig. K34a. 40 ×.

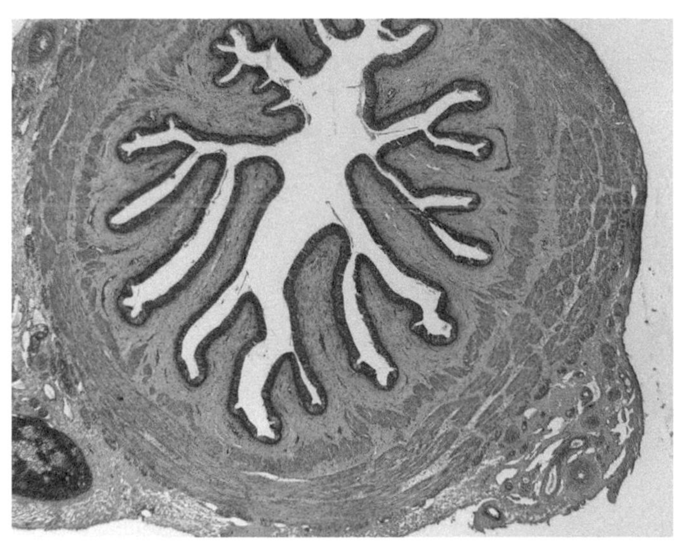

FIGURE K35a The esophagus of a caiman (a reptile that was often sold in pet shops as a "baby alligator") shows little specialization. Its lining is extensively folded and accommodates large chunks of food passing down. While difficult to see at this low magnification goblet cells in the epithelium provide mucus to assist this passage. 2.5 ×.

STOMACH

The stomach of primitive vertebrates is basically a storage organ. Secretory and digestive functions appear to have developed later. There are three regions in the stomach of a mammal: CARDIAC, FUNDIC, and PYLORIC (Fig. K40). The three regions may be distinguished in the monkey on the basis of the type of glands located in the tunica mucosa (Figs. K41a–K41c). Variations exist in the structure of mammalian stomachs but many fit the following description for the human, especially those of the carnivores and insectivores.

The TUNICA MUCOSA is thrown up into branching folds, the RUGAE (Figs. K42a–K42c), whose core is the tela submucosa. In the full stomach the rugae are almost completely "ironed out." Shallow grooves subdivide the

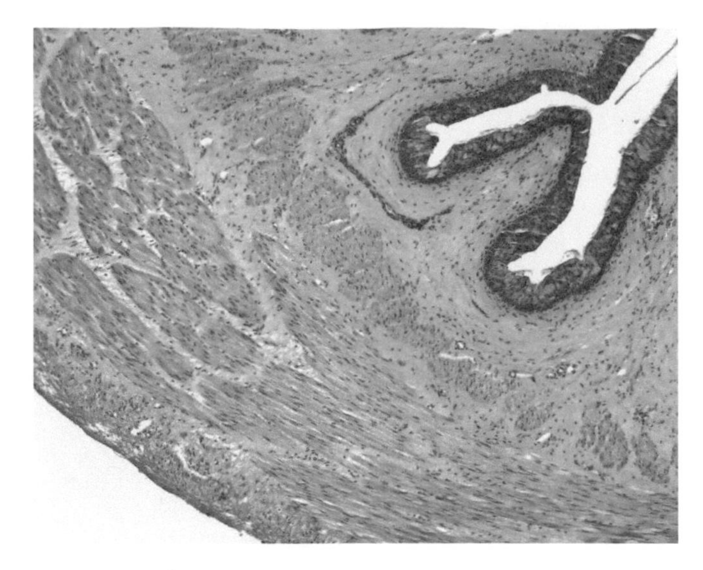

FIGURE K35b The esophagus of a caiman shows little specialization. Its lining is extensively folded and accommodates large chunks of food passing down. Goblet cells in the epithelium provide mucus and bands of muscle work together to pass the food down to the stomach. 10 ×.

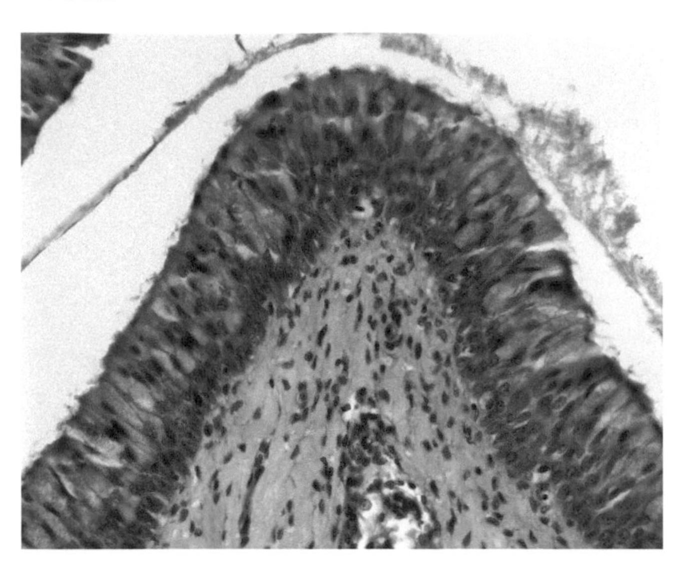

FIGURE K35c The esophagus of a caiman. Its lining is extensively folded and accommodates large chunks of food passing down. Goblet cells in the epithelium are readily visible and numerous. 40 ×.

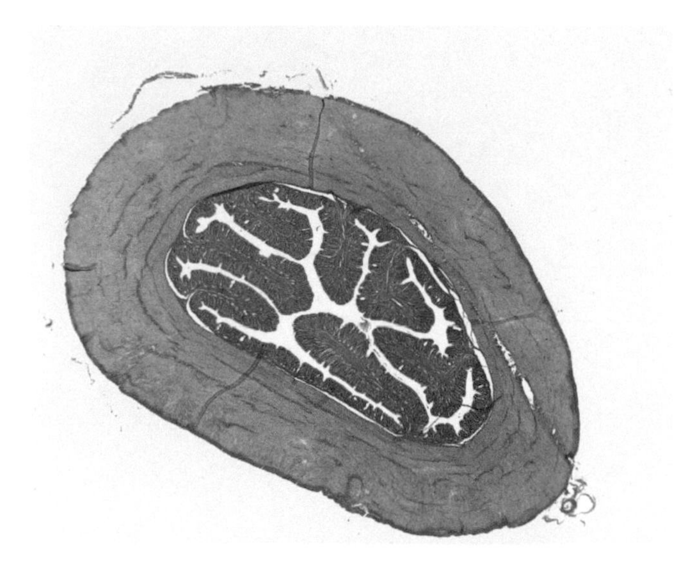

FIGURE K36a The esophagus of a tortoise is similar to that of the caiman. Note the well-developed lamina muscularis mucosae separating the lamina propria mucosae and the tela submucosa. 1 ×.

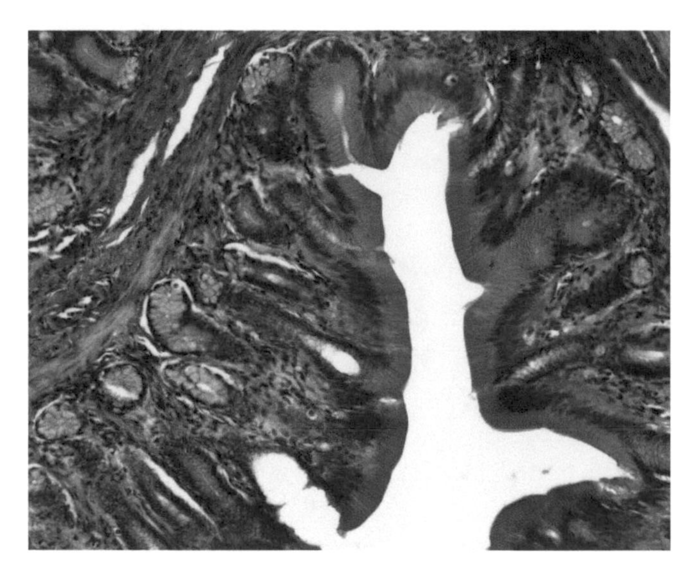

FIGURE K36b The esophagus of a tortoise is similar to that of the caiman, but there are simple tubular glands which produce lubricating mucus obvious here at higher magnification. 20 ×.

stomach surface into smaller regions, the MAMILLATED AREAS (Fig. K42c). The surface of the tunica mucosa is marked by closely set GASTRIC PITS that are lined with simple columnar epithelium (Fig. K43). Beneath the epithelium there is a LAMINA PROPRIA MUCOSAE of reticular or fine areolar tissue and this layer contains the GASTRIC GLANDS, which open into the pits. The shape and proportionate depth of the pits and the characteristics of the glands vary in different parts of the stomach. The LAMINA MUSCULARIS MUCOSAE is continuous throughout the stomach and includes both longitudinal and circular fibers.

The SURFACE EPITHELIUM lining the stomach consists of a simple layer of tall columnar cells. The apical portion of the cells is packed with granules of mucin that force the nucleus into a basal position (Figs. K43 and K44). The apical borders are united by JUNCTIONAL COMPLEXES, the TERMINAL BARS of the light microscope. The granules, endoplasmic reticulum, microvilli, Golgi complex, and mitochondria appear in electron micrographs.

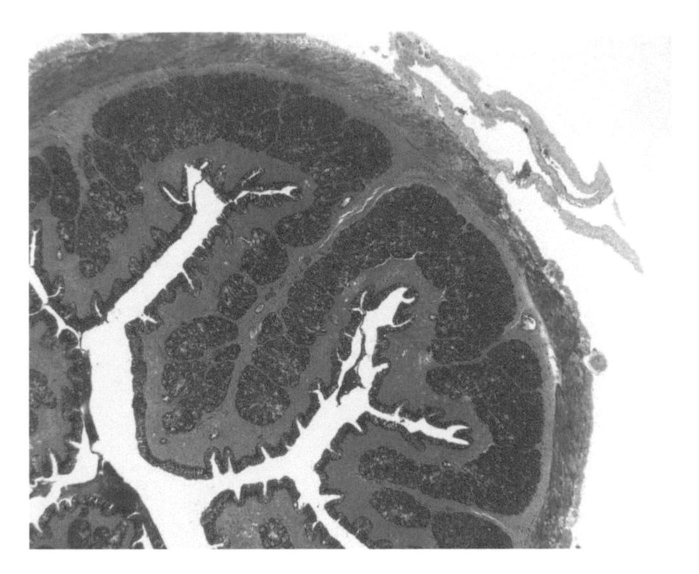

FIGURE K37a Recall that the mouth cavity and the surface of the tongue of many amphibians are lined with a ciliated epithelium that carries ingested food to the esophagus where its journey to the stomach is continued under the aegis of another ciliated epithelium. There are large, compound alveolar glands in the tela submucosa. The muscle of the two layers of the tunica muscularis is smooth; there is no need for voluntary control of swallowing since food simply slides down into the stomach by ciliary action. 2.5 ×.

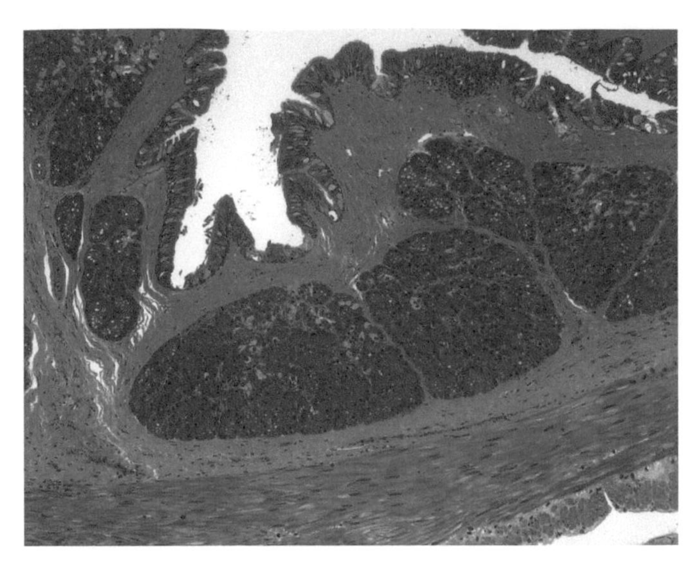

FIGURE K37b At a higher power the simple columnar epithelium lining the esophagus of a frog is obvious. Goblet cells are scattered between the ciliated cells. There are large, compound alveolar glands in the tela submucosa. The muscle of the two layers of the tunica muscularis is smooth; there is no need for voluntary control of swallowing since food simply slides down into the stomach by ciliary action. 10 ×.

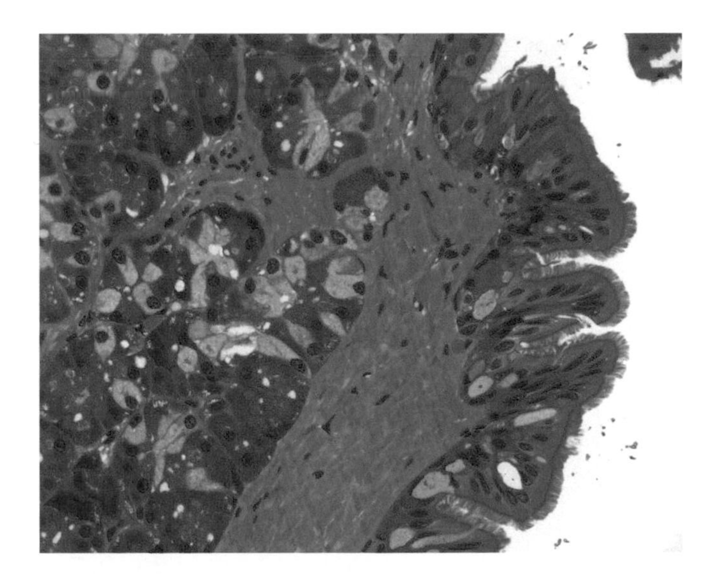

FIGURE K37c Basal bodies form a deep red line underlining the cilia of the epithelium of the esophagus of a frog. A few goblet cells, in various stages of formation of mucus, are seen between the ciliated cells. The mixed compound alveolar gland in the tela submucosa contains both mucous (pale) and serous (basophilic) cells. 40 ×.

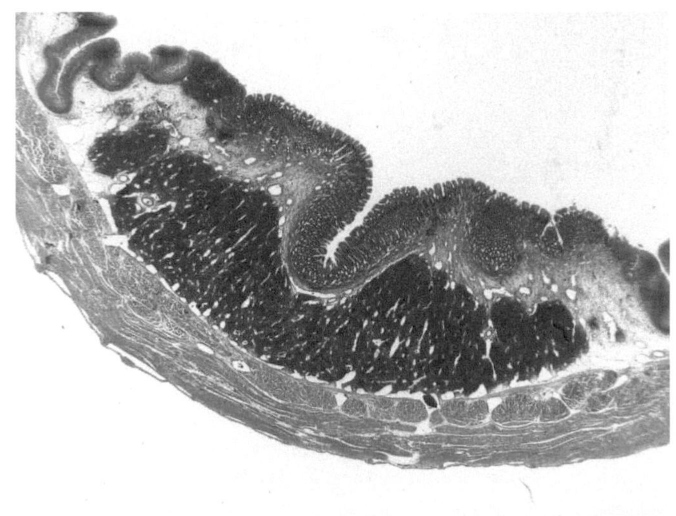

FIGURE K38 A mass of hemopoietic tissue develops in the tela submucosa and extends the full length of the esophagus of elasmobranchs. This is LEYDIG'S ORGAN and was considered in the section on hemopoietic organs. 1 ×.

In the CARDIAC REGION, near the entrance from the esophagus, the pits are shallow and the glands secrete mucus. The glands are lined with simple cuboidal epithelium and have wide lumina.

In the FUNDIC REGION the tunica mucosa is much thicker than it is in the cardiac and it contains a greater number of glands (*compare* Figs. K41a and K41b). The lamina propria mucosae is reduced to a fine interglandular stroma in its thickest portion and the pits extend only about one-fourth of the distance from the surface to the lamina

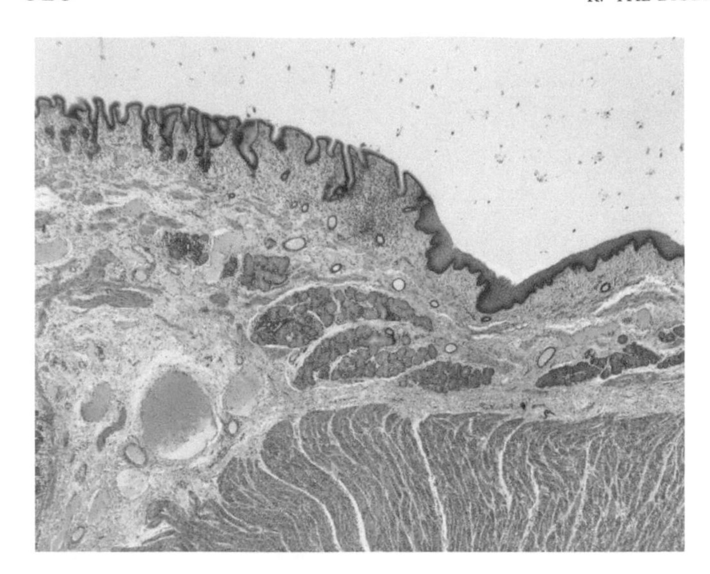

FIGURE K39a Junction of stomach and esophagus of a dog. The transition is abrupt from the stratified squamous epithelium of the esophagus (at the right) to the simple columnar epithelium of the upper regions of the stomach. This transition of the epithelium does not occur at the same level in the other layers and deep esophageal glands are seen in the tela submucosa of the stomach. A few simple tubular glands are typical of the stomach. 2.5 ×.

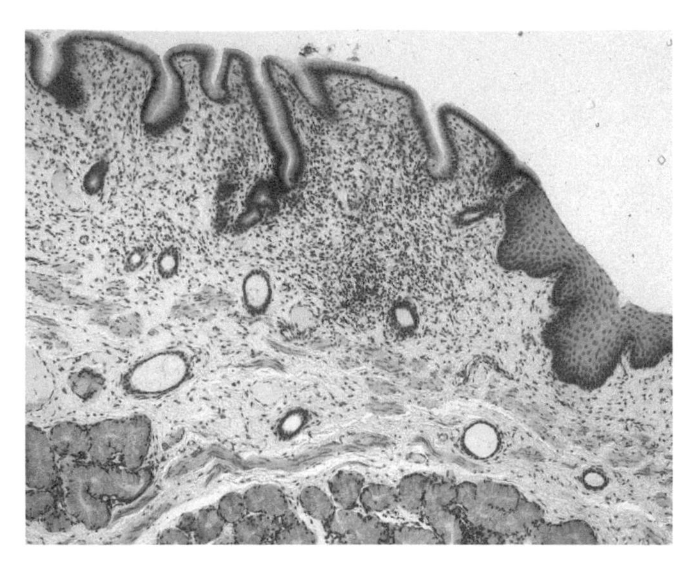

FIGURE K39b The abrupt transition from esophagus to stomach is more apparent at higher power. There are deep esophageal glands and a few strands of lamina muscularis mucosae at the bottom of the micrograph. Scattered lymphocytes have invaded the lamina propria mucosae. 10 ×.

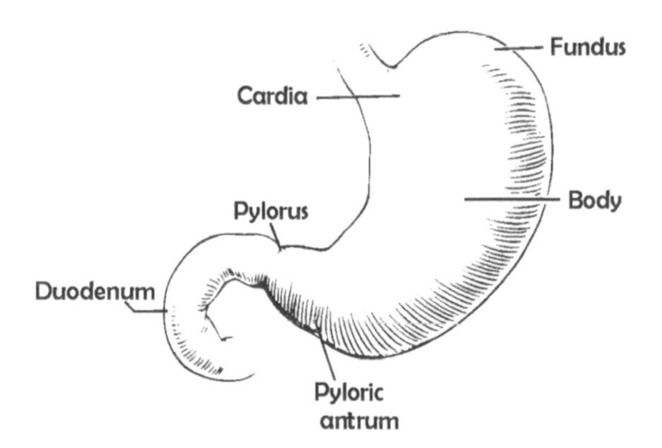

FIGURE K40 Sketch of the human stomach showing the various regions. Histologists often lump the body and the fundus together as the fundus.

muscularis mucosae. Note the following structures in a FUNDIC GLAND: the ISTHMUS is the upper constricted portion of the gland, the NECK is the central portion, and the BASE—the blind terminal portion (Figs. K45a–K45f). Four types of cells border the lumen of the gland:

Mucous neck cells form a prominent single layer in the neck. The cytoplasm contains some mucin and the nuclei are basal. The free surfaces of these cells bear microvilli and have a glycocalyx. The lateral surfaces interdigitate with neighboring cells. Mucous neck cells divide by mitosis and slide up to replace worn-out cells of the surface epithelium.
Chief or zymogenic cells are the wedge-shaped predominant cells in the deepest parts of the gland.

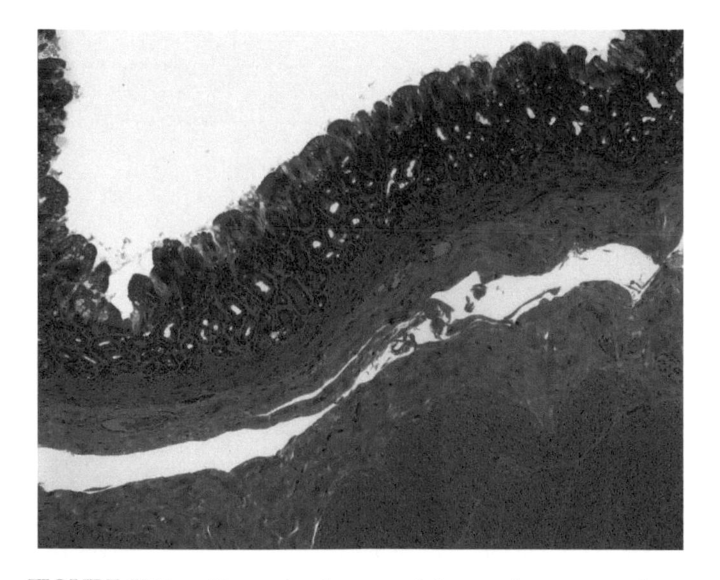

FIGURE K41a The major features of the vertebrate stomach can be seen in this section of the CARDIAC REGION of a primate. The simple tubular glands secrete mucus into gastric pits and are less elaborate than in other regions. They are enmeshed by delicate vascular connective tissue of the lamina propria mucosae and are subtended by a thick lamina muscularis mucosae, which separates them from the tela submucosa. The thick connective tissue of the tela submucosa is split in this section. A portion of the tunica muscularis is shown at the lower right. 10 ×.

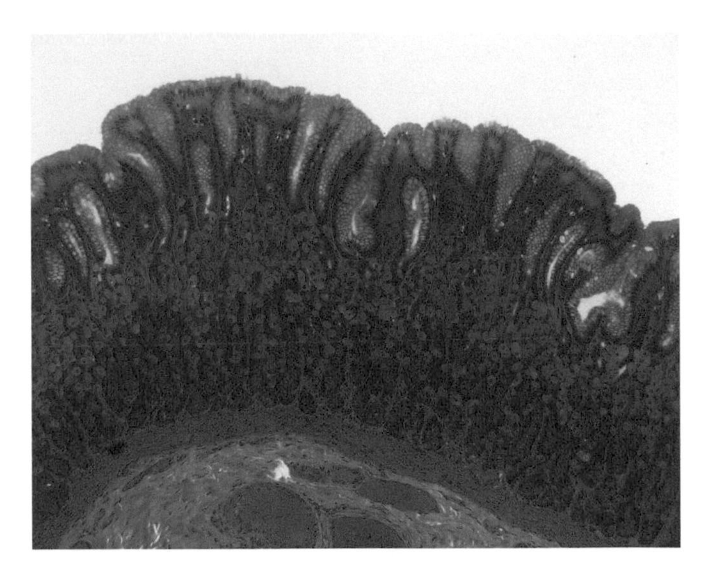

FIGURE K41b The tunica mucosa in this section of the FUNDUS of the stomach forms three islands or MAMMILLATED AREAS at the top of the micrograph. Simple tubular glands of the epithelium mucosae extend into the lamina propria mucosae. They open to the surface via GASTRIC PITS that are abundantly lined with goblet cells. A thick lamina muscularis mucosae separates the tunica mucosa from the vascular connective tissue of the tela submucosa. 10 ×.

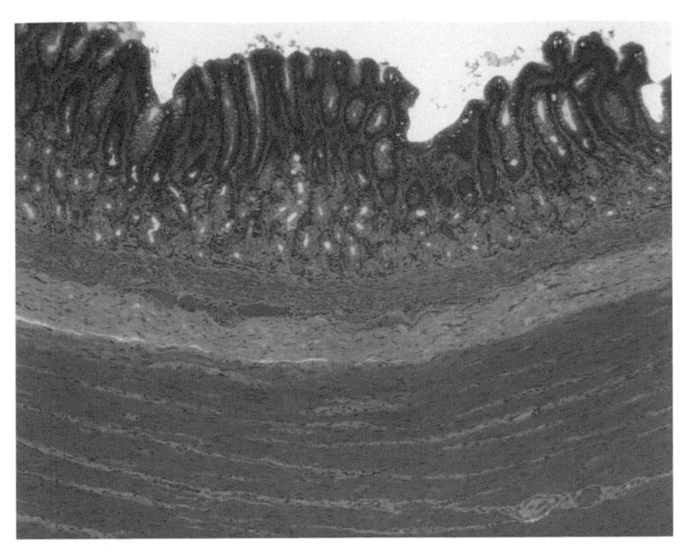

FIGURE K41c The tunica mucosa is thrust up into three mammillated areas in this section of the PYLORIC REGION. The secretory portions of the simple tubular glands in the pyloric region are less elaborate than those seen in the fundus. In this section, it is difficult to see that the glands are branching. The glands are subtended by a thick lamina muscularis mucosae and pour their secretions into the gastric pits, which are amply lined with goblet cells. The tela submucosa consists of loose vascular connective tissue. The thick tunica muscularis of smooth muscle is well developed. Note that nerve fibers (pale pink) and blood vessels course between the muscle fibers. 10 ×.

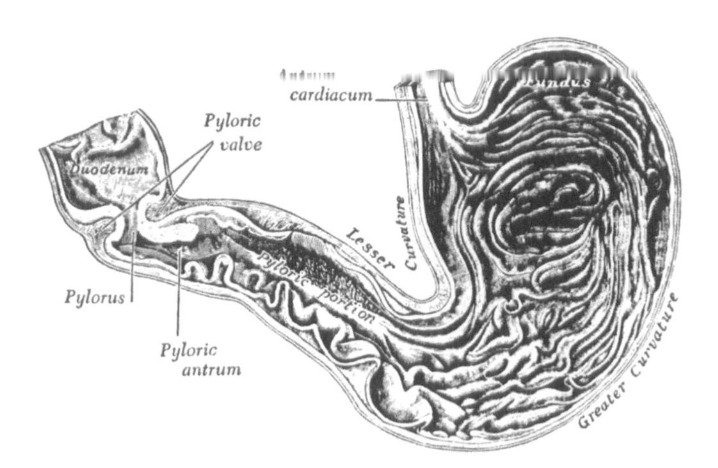

FIGURE K42a Drawing of the interior of the human stomach. The tunica mucosa of the mammalian stomach is thrown up into folds, the RUGAE; these flatten out when the stomach fllls.

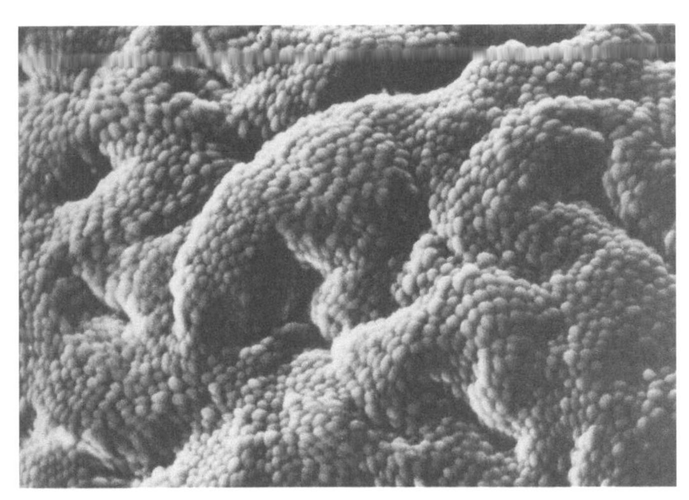

FIGURE K42b Scanning electron micrograph of the surface of the gastric mucosa showing the convoluted pattern of ridges and pits.

They contain pale zymogen granules apically and basophilic material basally. In electron micrographs note the typical appearance of cells secreting proteins: short apical microvilli, the parallel arrays of granular endoplasmic reticulum sandwiching mitochondria between them, and the supranuclear Golgi complex. The apical granules are membrane bound and are of low electron density (Fig. K45g).

Parietal or oxyntic cells are scattered between the chief cells and are most numerous near the neck. They have an acidophilic, finely granular cytoplasm, one or more large spherical nuclei, and are usually crowded away from the lumen of the gland. Electron micrographs show that the luminal surface of these cells is deeply invaginated to form many complex channels. The free surfaces, including those of the channels, lack a glycocalyx and are arrayed with long, slender microvilli (Fig. K45h). The cytoplasm contains an extensive

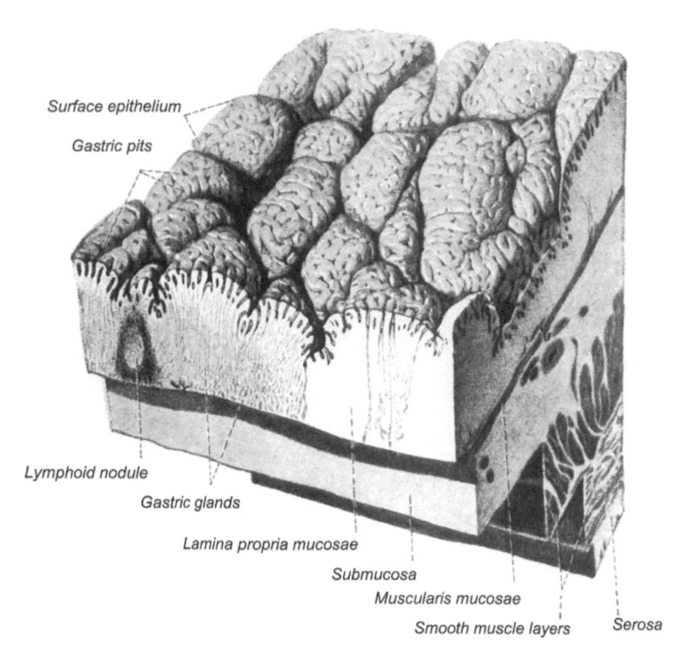

FIGURE K42c Drawing of the surface of the human gastric mucosa as seen with the dissecting microscope. Note the grouping of gastric glands in mammillated areas.

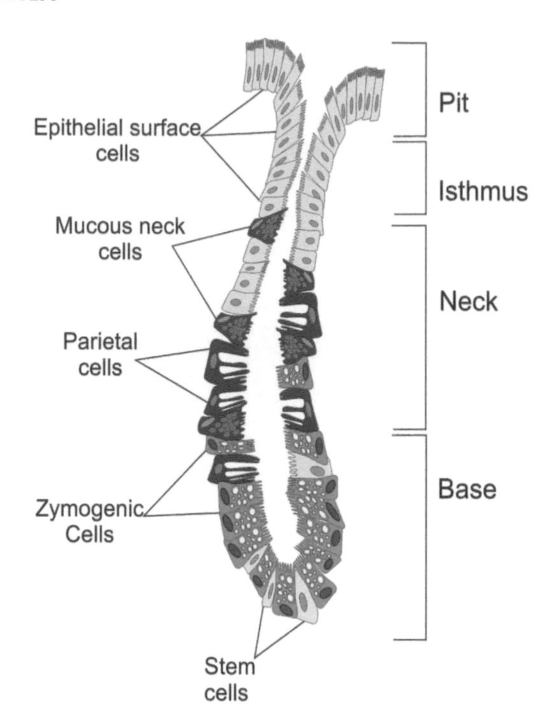

FIGURE K43 Schematic diagram of a fundic pit and gland in the stomach. *Authors.*

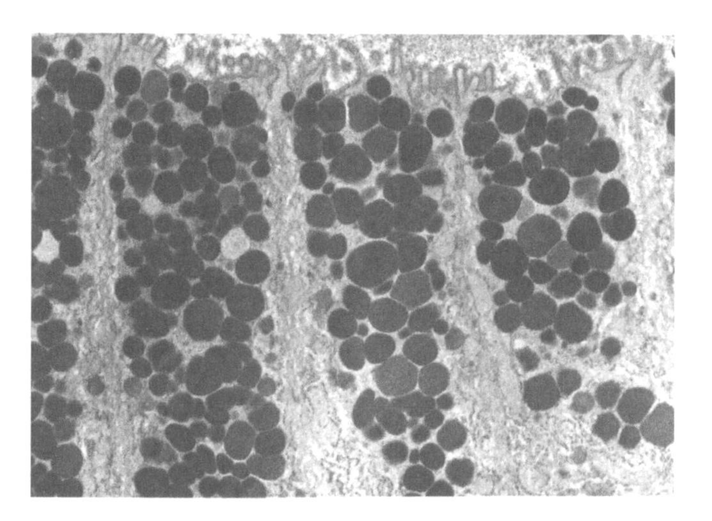

FIGURE K44 Electron micrograph of the apical portions of several mucous cells from the inner lining of the stomach of a mammal. The short microvilli have a conspicuous glycocalyx and the surface is covered with a layer of mucus.

tubular agranular endoplasmic reticulum, numerous free ribosomes, and large mitochondria; the Golgi complex is small. These cells produce the hydrochloric acid of the stomach juice.
Enteroendocrine or APUD (amine precursor uptake decarboxylation) cells are scarce. They are present in the base of the gland and are difficult to identify with the light microscope except when the sections are stained with chromium or silver salts. They are flask-shaped cells with large spherical nuclei and finely granular cytoplasm. Several types of enteroendocrine cells may be distinguished by the appearance of the granules in electron micrographs (Fig. K45i). The basal position of the granules indicates that they are secreted into the surrounding connective tissue rather than into the lumen of the stomach.

PYLORIC GLANDS are also the simple, branched, tubular type (Figs. K46a–K46c). They may be differentiated from the fundic glands by the following characteristics: branchings are more numerous, lumina are larger,

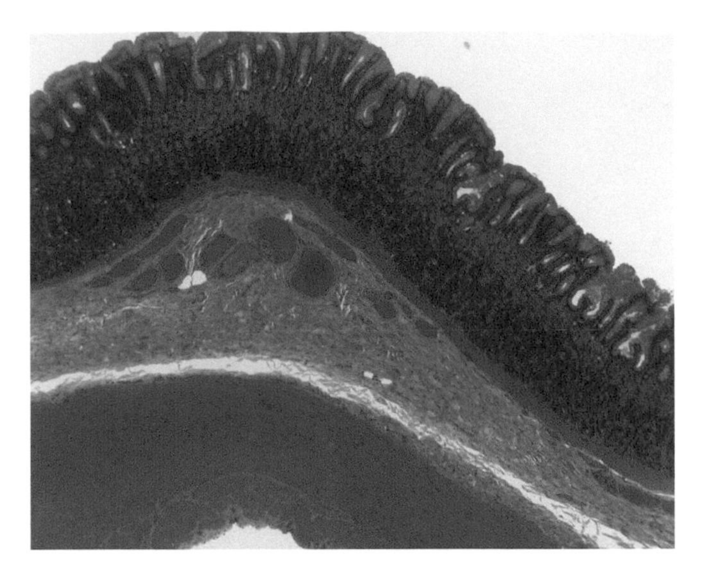

FIGURE K45a The tunica mucosa is thick in the fundus of the stomach of a monkey and contains many glands. The distinct lamina muscularis mucosae separates the depths of these glands from the richly vascular tela submucosa below. The split between the tela submucosa and the tela muscularis is an artifact. The inner circular layer of smooth muscle is much thicker than the outer longitudinal layer. 4 ×.

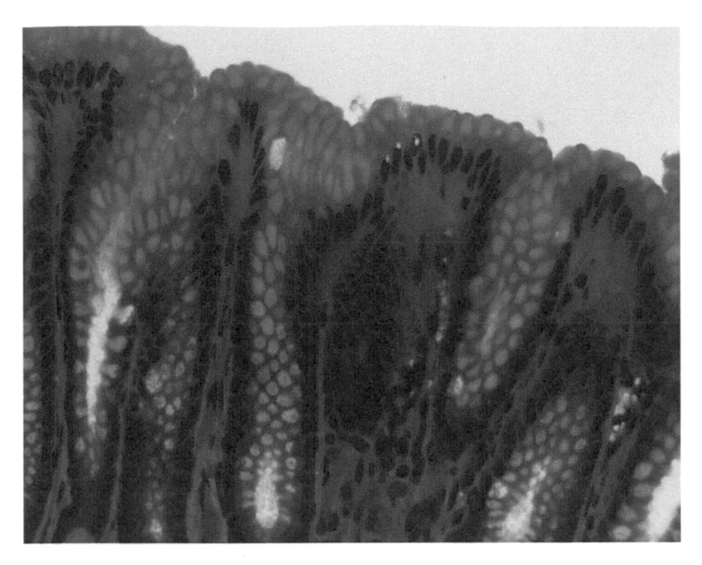

FIGURE K45b Several gastric pits in the fundic stomach of a monkey, lined with mucous neck cells, constitute the inner portion of the wall of the fundic region of the stomach. Existence is difficult for the surface cells in withstanding the rigors of the stomach contents, both chemical and mechanical and cells are constantly sloughing off to be replaced by cells migrating from below. A few capillaries of the lamina propria mucosae can be seen between the pits. A few elongate fibroblasts also inhabit this space. The pits become constricted as they open into the isthmus of the fundic gland. 40 ×.

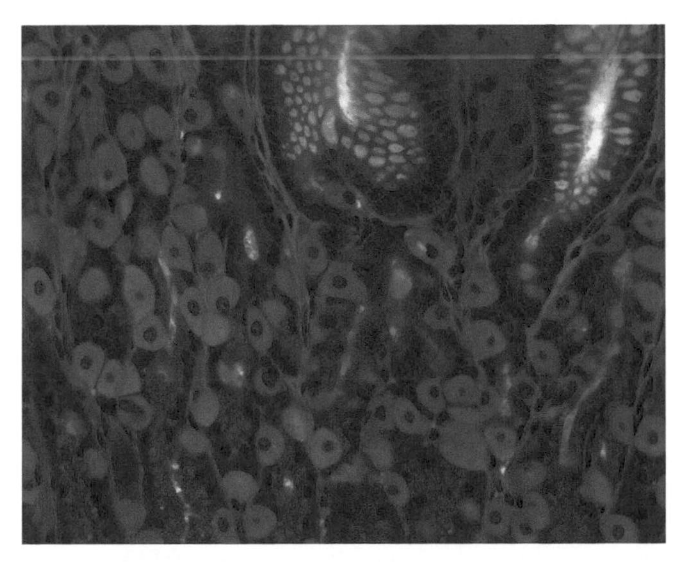

FIGURE K45c Two gastric pits at the top are from the fundic stomach of a monkey; they are lined with GOBLET CELLS. The basophilic cells surrounding the deeper portions of the gland are MUCOUS NECK CELLS. The rounded acidophilic cells seen throughout the micrograph are PARIETAL OR OXYNTIC CELLS. A few elongate FIBROBLASTS are contained within the delicate connective tissue of the lamina propria mucosae. Most of the granular cells at the bottom of the micrograph contain basophilic cytoplasm at their bases; these are the chief or ZYMOGENIC CELLS. A few deeply acidophilic ENTEROENDOCRINE or APUD cells lurk near the periphery of the tubular glands; these cells secrete various peptides into the connective tissue around the gland and do not secrete into the lumen. 40 ×.

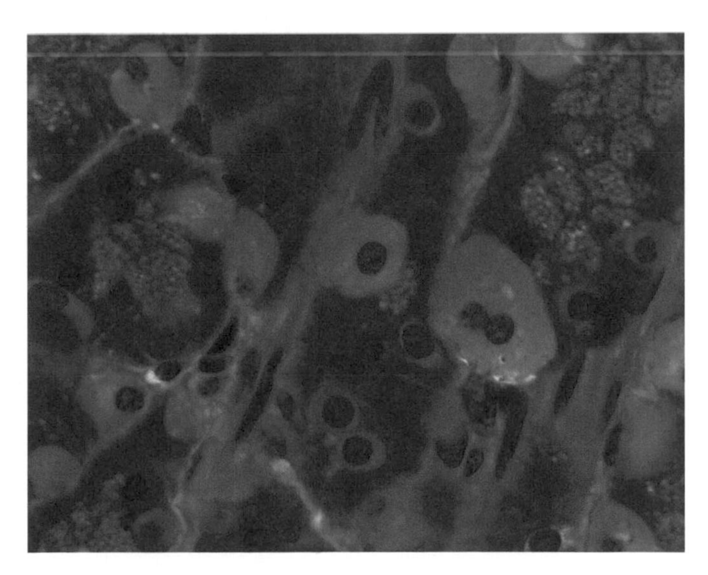

FIGURE K45d This higher power micrograph is from the same region as Fig. K45c above, the fundic stomach of a monkey. Most of the cells are either chief cells or parietal cells. The chief cells contain abundant basophilic cytoplasm and are packed apically with acidophilic granules. Note the binucleate parietal cell at the center right. A few smooth muscle cells have strayed from the lamina muscularis mucosae and are interposed in the lamina propria mucosae between the tubular glands. There are also few delicate fibroblasts in this space. 100 ×.

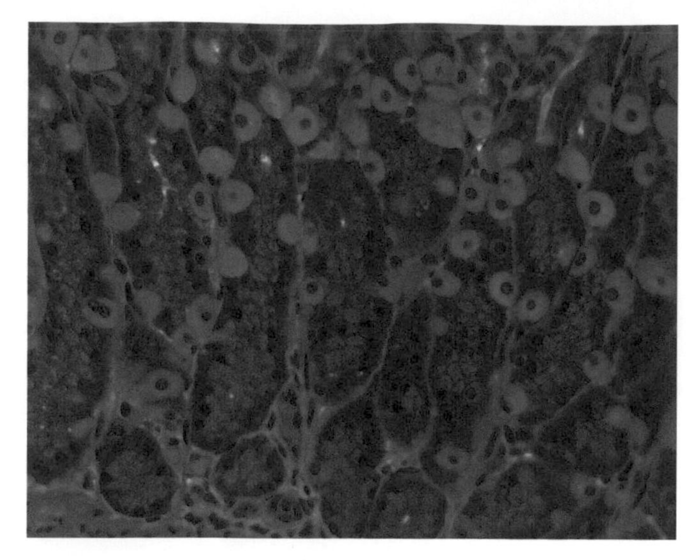

FIGURE K45e Deeper portion of the tunica mucosa from the same section in Fig. K45c - the fundic stomach of a monkey. Chief cells with their basophilic cytoplasm and apical granules dominate this picture. Deeply acidophilic enteroendocrine cells are present near the periphery of the tubular glands. A few capillaries pass in and out of the picture. The lamina muscularis mucosae is at the extreme bottom left. 40 ×.

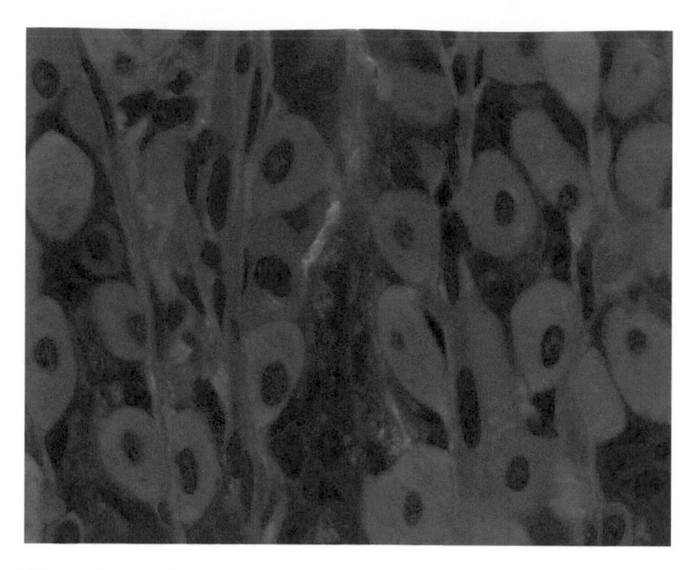

FIGURE K45f Deeper portion of the tunica mucosa from the same section in Fig. K45c - the fundic stomach of a monkey. Pale pink balloon-shaped parietal cells dominate this picture. Between them is the occasional chief cell with its conspicuous basophilic cytoplasm. Note the enteroendocrine cell, loaded with acidophilic granules, at the lower left. Smooth muscle cells and fibroblasts inhabit the connective tissue of the lamina propria mucosae. 100 ×.

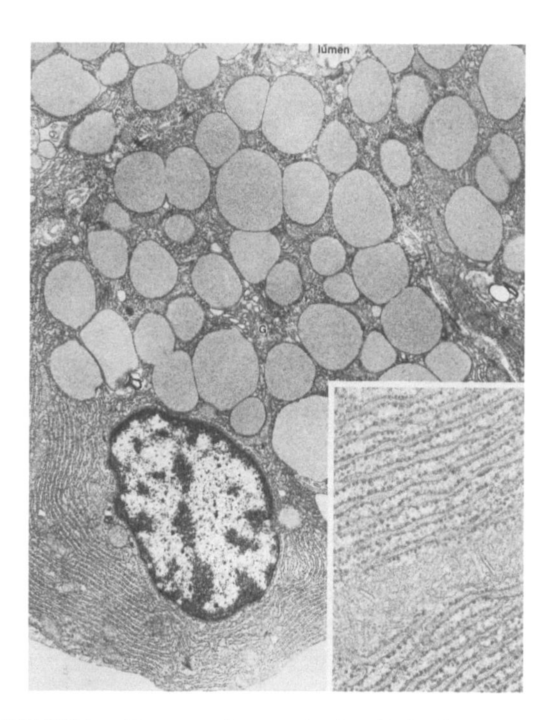

FIGURE K45g Electron micrograph of a chief cell within a gastric gland of a monkey. The basal cytoplasm is filled with parallel cisternae of granular endoplasmic reticulum (inset at higher power). The apex of the cell is packed with secretion granules and elements of the Golgi complex (G).

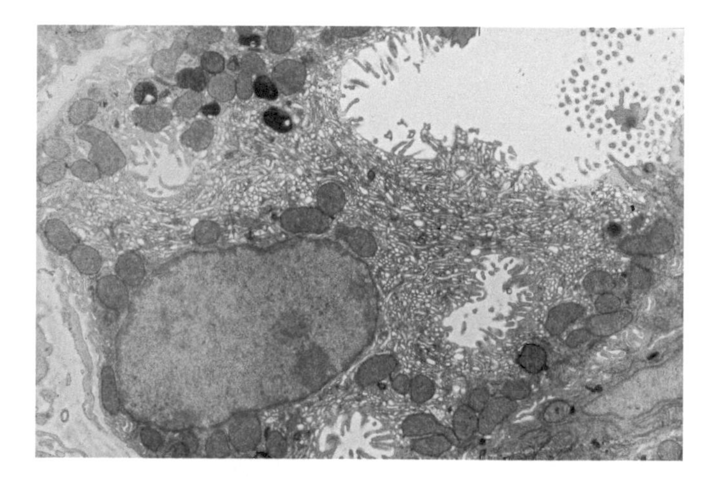

FIGURE K45h Electron micrograph of a parietal cell from the stomach of a mouse. Parietal cells, with their characteristic cytoplasm laden with tubules, secrete hydrochloric acid solution of approximately 0.15 N. Secretory canaliculi penetrate the cytoplasm of these cells (e.g., at mid-bottom) and connect to a lumen of the gastric gland (upper right). The canaliculi are covered with microvilli (upper right). The dark blobs are lysosomes. Mitochondria are clustered nearby. 17,000 ×.

tubules are more coiled, and the lining cells are of the mucous type. The glands do not extend into the tela submucosa. The predominant cell type is a mucus-secreting cell that resembles the neck cells of the oxyntic glands. The nuclei are frequently flattened basally. Occasional parietal or enteroendocrine cells may be found in pyloric glands.

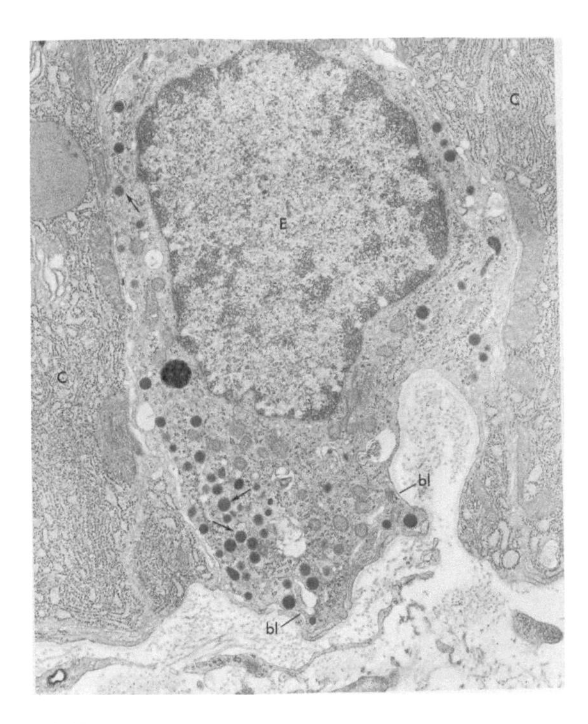

FIGURE K45i An enteroendocrine cell (E) from the gastric gland of a bat is situated between two chief cells (C). Its basal cytoplasm rests against the basal lamina (bl) and the cytoplasm contains small, dense, membrane-bound granules (arrows). 18,000 ×.

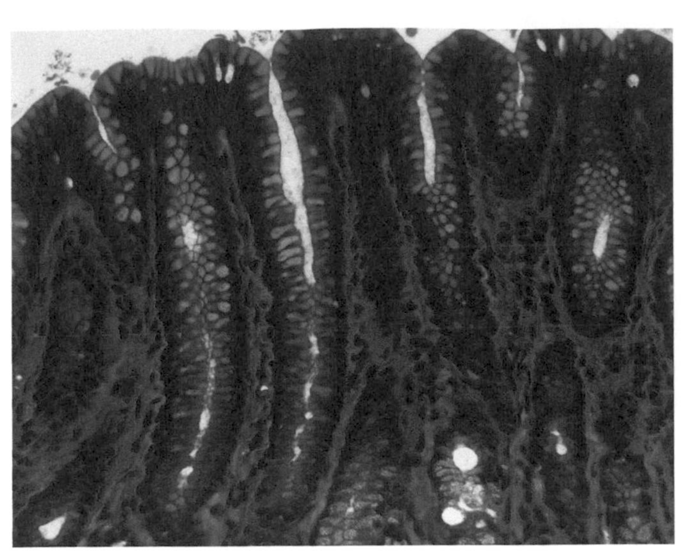

FIGURE K46a The gastric pits in the pylorus of a monkey are deep and are lined with mucous cells. The delicate vascular connective tissue of the lamina propria mucosae is well developed between the pits. 40 ×.

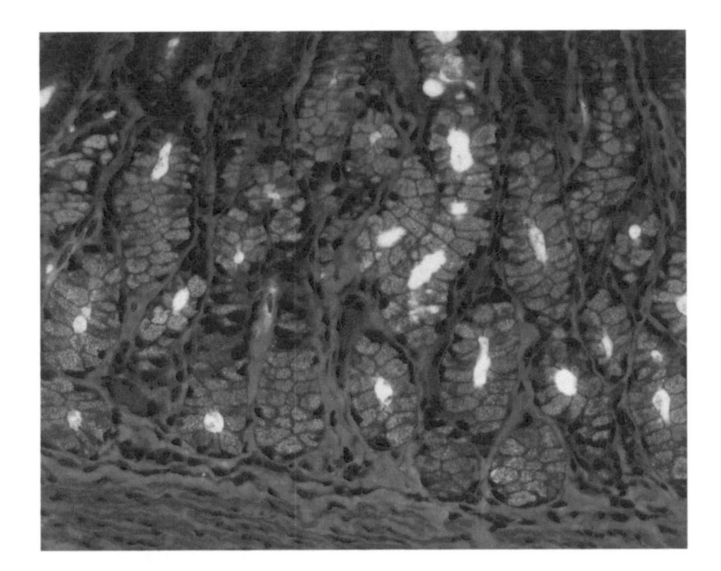

FIGURE K46b The deeper portions of the simple tubular glands of the tunica mucosa from the pyloric region of the stomach of a monkey are packed with pale granules of mucin. An occasional capillary meanders between the glands. The smooth muscle of the lamina muscularis mucosae is at the bottom of the micrograph. 40 ×.

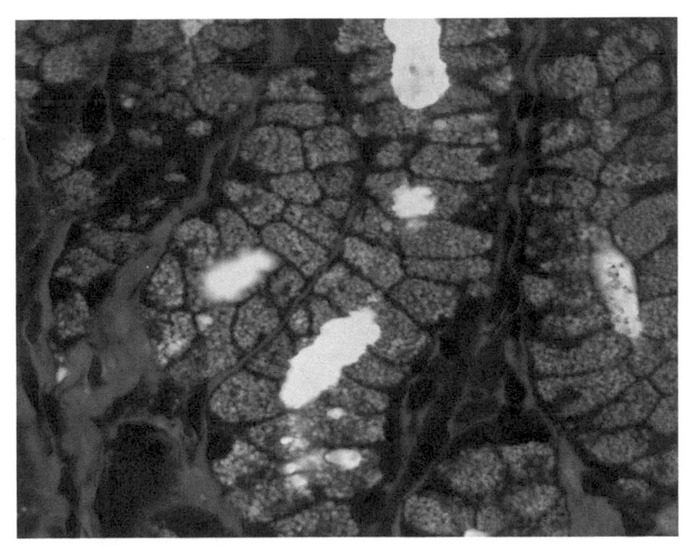

FIGURE K46c Secretory cells of the pyloric glands are packed with mucous droplets. Enterochromaffin cells can be distinguished adjacent to the basal lamina of the glands. 100 ×.

The LAMINA PROPRIA MUCOSAE of the stomach may be heavily infiltrated with diffuse lymphoid tissue and lymphoid nodules. The LAMINA MUSCULARIS MUCOSAE consists of two thin layers of smooth muscle.

The TELA SUBMUCOSA is composed of areolar connective tissue containing small blood vessels, lymphatics, and, the SUBMUCOUS PLEXUS (Figs. K47a–K47c, K62d). It does not contain glands but at the junction of the esophagus and the stomach some of the end pieces of deep mucous glands may extend into the tela submucosa of the stomach. Since their ducts open into the esophagus they should be considered as part of the esophagus.

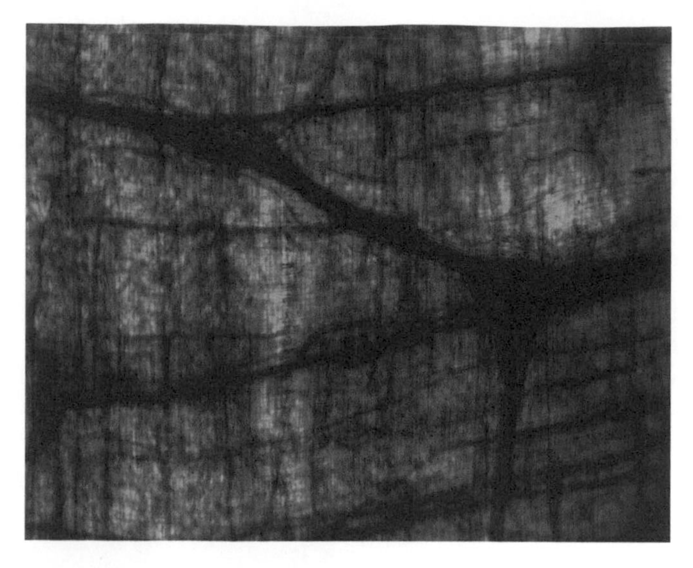

FIGURE K47a This is a whole mount of the muscular wall of the stomach of a rabbit. It has been stained with silver to show the elaborate MYENTERIC PLEXUS between the muscle layers; a ganglion is at the center right and nonmyelinated nerves radiate from it. Delicate nerve fibers form a mesh between the muscle layers. 20 ×.

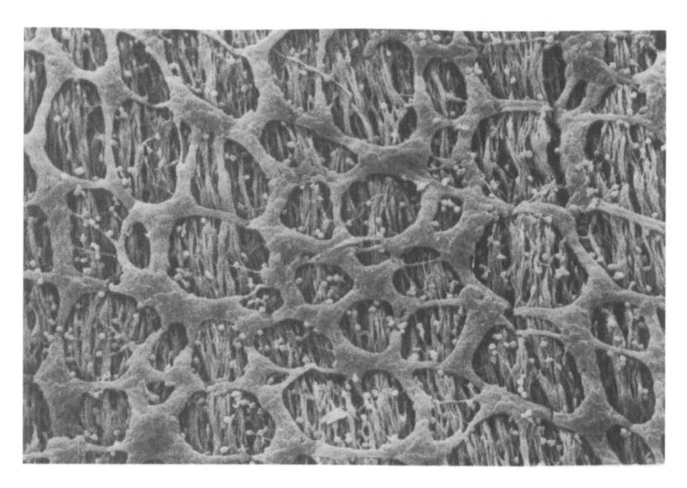

FIGURE K47b Scanning electron micrograph of the myenteric plexus in the intestine of a rat. The overlying longitudinal muscle and connective tissue have been removed.

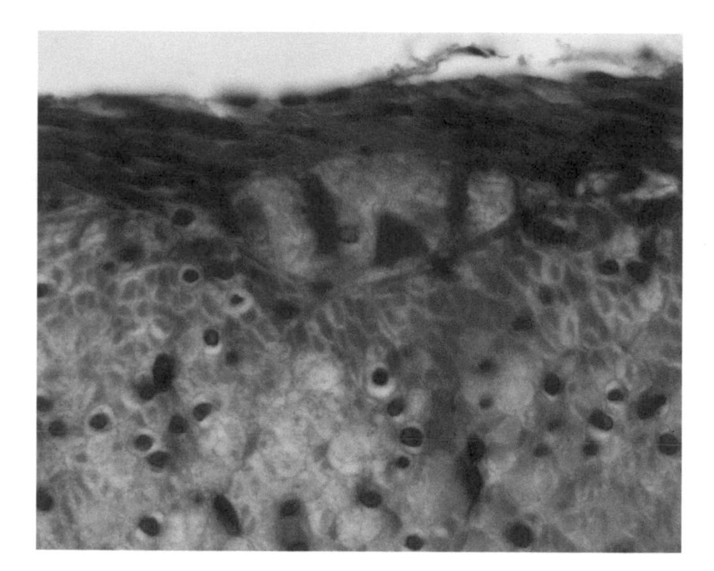

FIGURE K47c At the upper center is a grayish ganglion of the myenteric plexus in the wall of the stomach of a rat. It is wedged between the longitudinal layer of the tunica muscularis (above) and the circular layer. Fibers from the ganglion extend between these layers of muscle but cannot be distinguished. 100 ×.

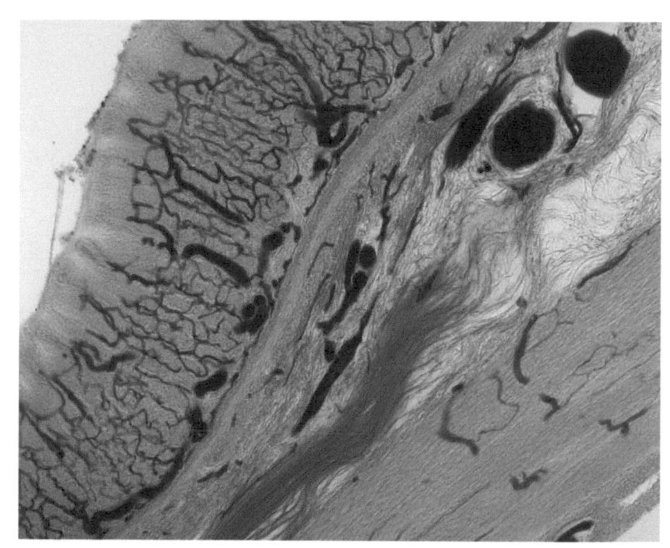

FIGURE K48 This is a section of the stomach of a cat. The lumen is at the left. The stomach was injected with red and green gelatin solutions to show the elaborate capillary circulation around the glands of the tunica mucosa. The outer layers have much sparser vascularization than the tunica mucosa. Note the large vessels in the tela submucosa. 10 ×.

The TUNICA MUSCULARIS consists of an inner CIRCULAR and an outer LONGITUDINAL LAYER (Fig. K61d) with an incomplete layer of obliquely arranged fibers between the circular layer and the tela submucosa. The circular layer is the thickest of the three coats. The arrangement of the fibers is irregular and there may be some difficulty in distinguishing the three coats in some sections. The MYENTERIC PLEXUS is present between the circular and longitudinal fibers (Fig. K47c).

Most of the stomach is covered with a TUNICA SEROSA of visceral peritoneum. This layer is often partially torn away in the preparation of the section.

Observe the *vascularization* around the glands and in the various coats in injected sections of the stomach (Fig. K48).

The stomachs of herbivores are generally larger in relation to body size than those of carnivores and are often divided into two or more compartments (Figs. K49a–K49d). Only the posterior chamber (ABOMASUM) is provided

FIGURES K49(a–d) The herbivore stomach. Herbivores must eat large quantities of food that is low in nutrients. They reduce their exposure to predators by ingesting large quantities of food quickly and processing it at their leisure in more protected areas. The storage portion of the stomach is lined with stratified squamous epithelium where the food may be processed mechanically; it then passes to the digestive portion of the stomach where it is broken down enzymatically.

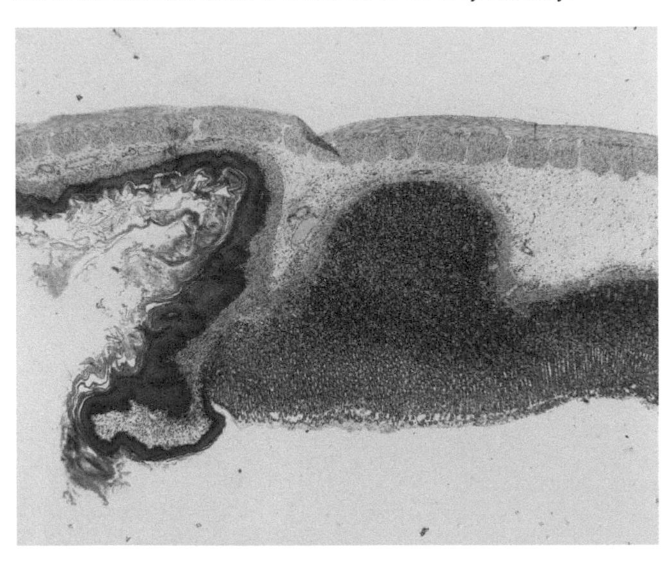

FIGURE K49a The rat has a simple storage stomach where about half is lined with stratified squamous epithelium (left) and the other half is glandular (right). A fold in the wall loosely separates these parts. 2.5 ×.

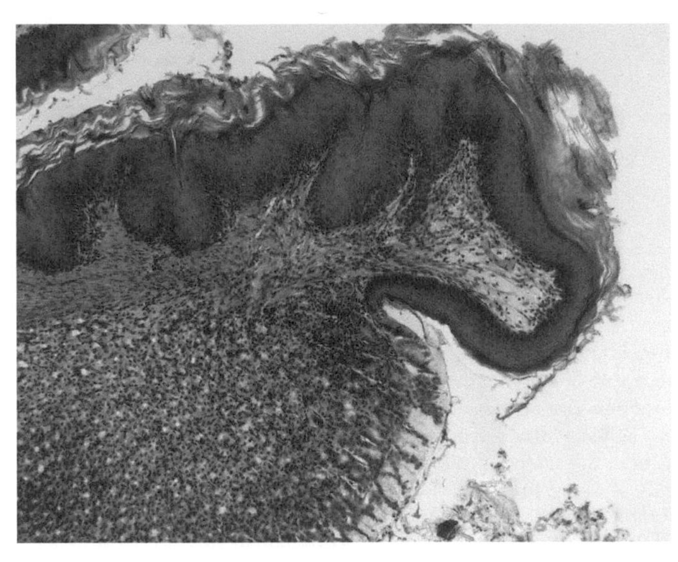

FIGURE K49b At this fold the stratified squamous epithelium of the storage portion gives way abruptly to the glandular portion, which consists of closely packed simple tubular glands. 10 ×.

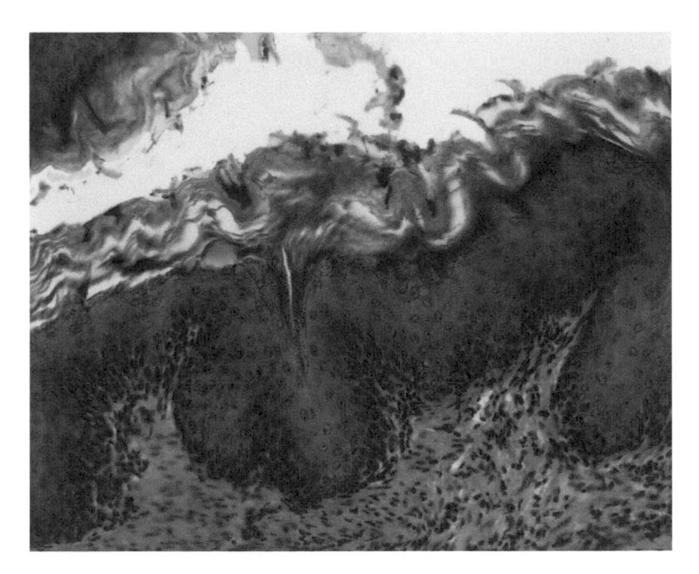

FIGURE K49c Sloughing cells from the surface of cornified stratified squamous epithelial cells of the storage portion of the stomach reflect the rigors of their life in this hostile environment. Active mitosis in the basal portions of the epithelium replaces lost cells. Note the presence of keratohyaline granules in the stratum granulosum of this epithelium, indicating the formation of soft keratin. 20 ×.

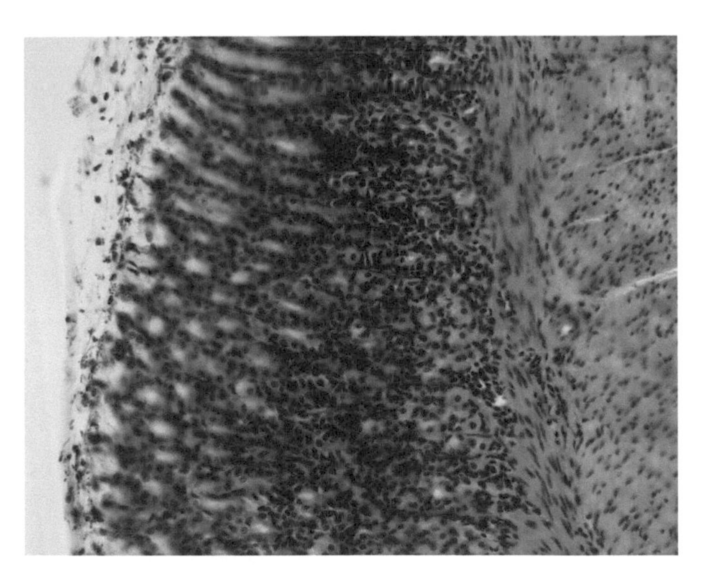

FIGURE K49d Simple tubular glands of the secretory portion of the stomach rest upon the lamina muscularis mucosae. There is a thin tela submucosa and a thick tunica muscularis at the right. 20 ×.

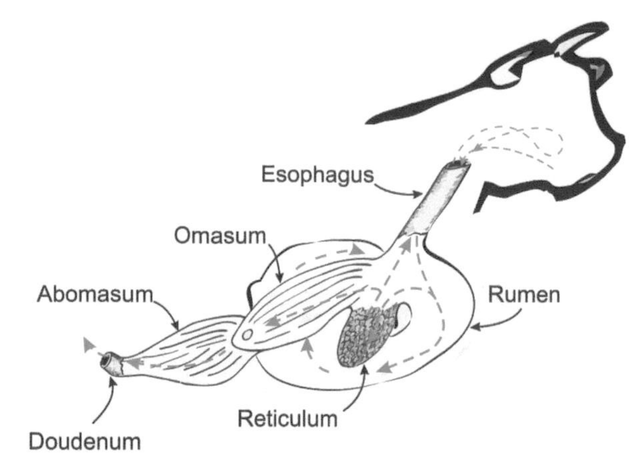

FIGURE K50a The ruminant stomach of cattle is a highly developed example of a herbivore stomach. Grasses and other foods high in cellulose pass from the esophagus into the rumen. They may be regurgitated to the mouth for further chewing or pass along to the reticulum. Here, further digestion takes place. While the food is being further digested, it passes along to the omasum and ABOMASUM. Digestive juices from simple tubular glands in the walls of these chambers complete digestion and pass the digested material to the small intestine. The following series of micrographs is taken from the ruminant stomach of a cow.

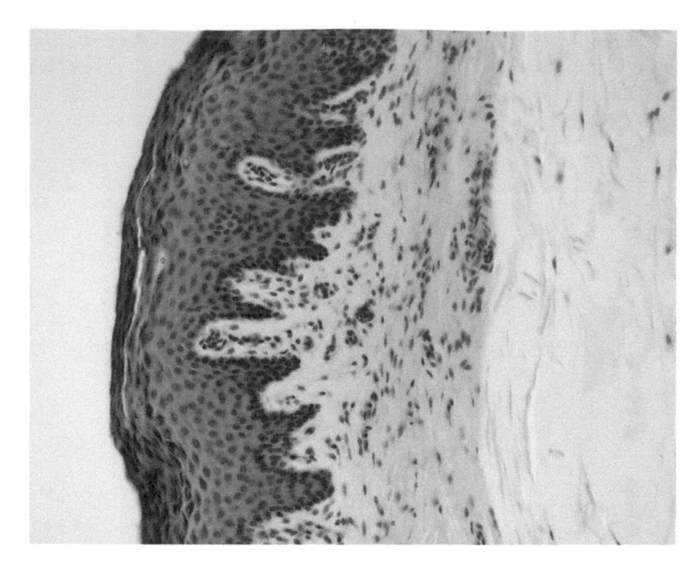

FIGURE K50b The RUMEN is lined with smooth, stratified squamous epithelium. 20 ×.

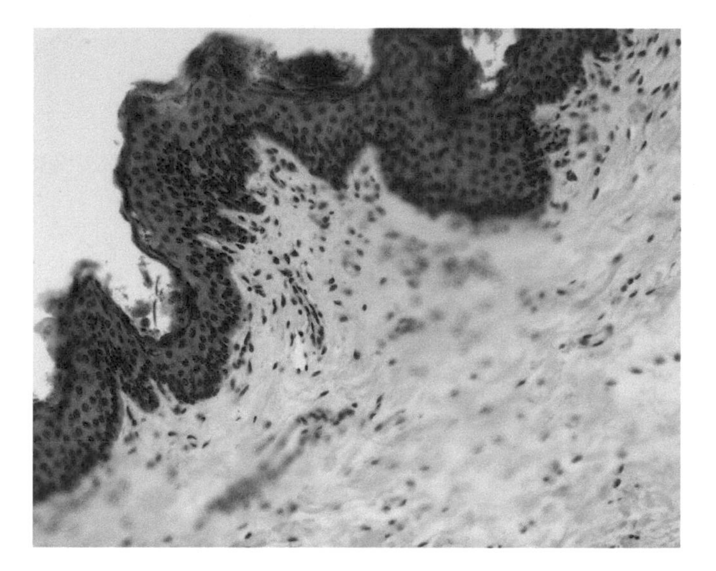

FIGURE K50c The stratified squamous lining of the RETICULUM is thrown up into folds that form a honeycomb of irregular chambers in the wall. 20 ×. *Authors.*

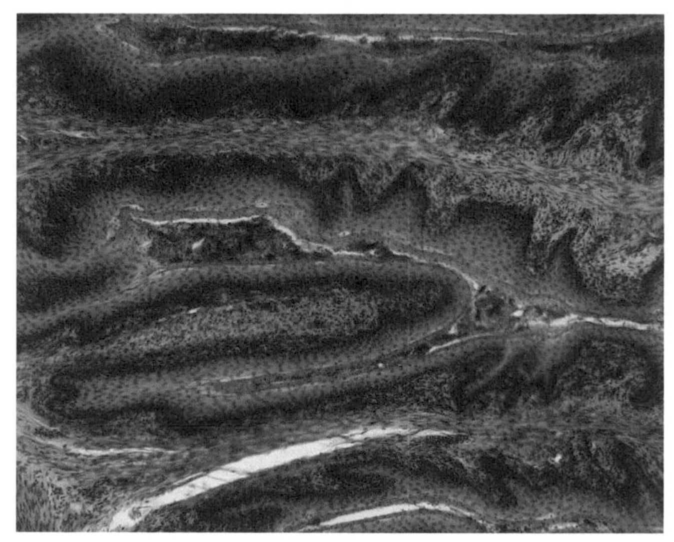

FIGURE K50d There are no digestive glands in the OMASUM but the stratified squamous epithelium is thrown up into elaborate folds. These folds permit distension as the stomach fills with food. 10 ×.

with gastric glands. The storage chambers are lined with stratified squamous epithelium that is often cornified and may be thrown up into elaborate folds. The ruminant stomach reaches its highest development in the artiodactyls (even-toed hoofed mammals) where four chambers, RUMEN, RETICULUM, OMASUM, and ABOMASUM may be recognized (Figs. K50a–K50f). Stratified squamous epithelium is found in at least part of the stomach of many mammals, especially those on diets of low food value. In a section of the single-chambered stomach of the rat note that only about one-third of the lining is glandular (Fig. K49a). One immediately suspects that these nonglandular areas are actually esophagus; embryological studies show, however, that they are true stomach.

Make the attempt to relate any differences in the stomach of vertebrates with the habits of the particular animal. Note especially the following points. The stomach of the higher *birds* is separated into two portions, the glandular thin-walled PROVENTRICULUS or true stomach and the thick-walled muscular GIZZARD or VENTRICULUS representing the pyloric region (Figs. K51a and K51b). The gizzard has a thick horny lining produced by the surface

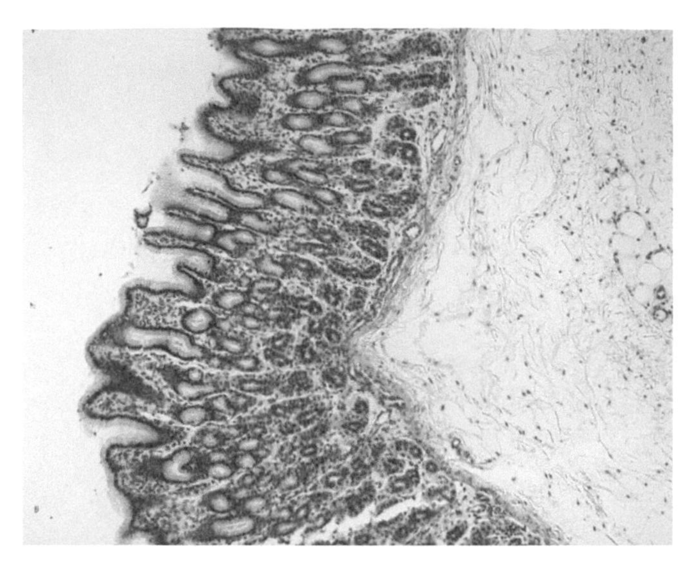

FIGURE K50e Only in the fourth chamber, the ABOMASUM, is there the presence of simple tubular gastric glands. This is a section of the upper portion of the abomasum. 10 ×.

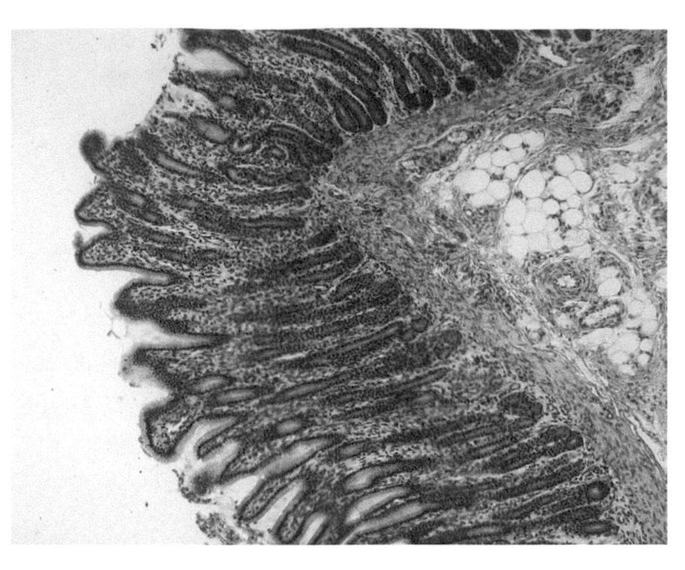

FIGURE K50f The lower region of the ABOMASUM demonstrates deep simple tubular glands. 10 ×.

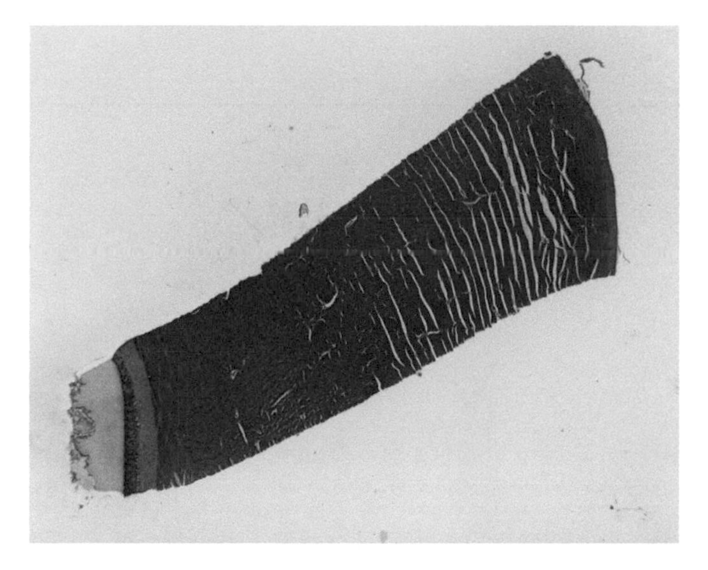

FIGURE K51a This is a section taken from the gizzard (VENTRICULUS) of a pigeon. Most of the wall consists of a thick layer of smooth muscle. The lumen (left) is lined by a tough, horny, abrasive substance that is formed by secretions of the simple tubular glands of the tunica mucosa. 1 ×.

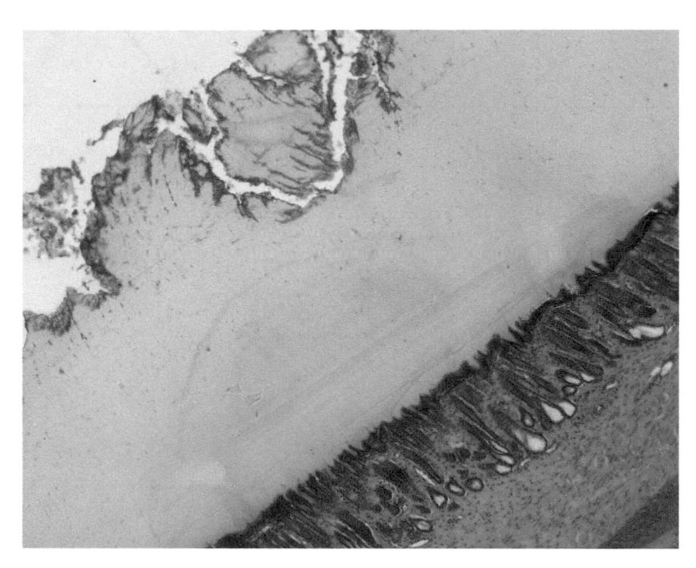

FIGURE K51b The mucosal lining of the gizzard resembles the lining of many stomachs until one notices that a thick translucent layer is imposed above it. Instead of secreting digestive juices, these simple tubular glands secrete a gelatinous material that hardens to form the thick, resistant lining. The surface wears away unevenly so that a rough, abrasive surface is formed. 10 ×.

epithelium and the glands. The tubular glands produce tightly packed "pencils" of hard material. Each gland produces a pencil and each pencil is arranged perpendicular to the surface. The outer longitudinal coat of muscle is almost entirely lacking but the circular coat is greatly developed. A gizzard is also seen in some fish (e.g., anchovy, gizzard shad) and in some reptiles (e.g., crocodiles).

In general the stomachs of *fish, amphibians*, and *reptiles* are similar, carrying out several essential functions: their tubular glands secrete digestive enzymes, they support hemopoietic activity in the tela submucosa; their longitudinal and circular layers of smooth muscle aid digestion; and by active mitosis they replace epithelial cells lost from the epithelium (Figs. K52a, K52b, K53a–K53c, K54a–K54f, K55a, K55b, K56a–K56e). *Elasmobranchs* are the lowest group exhibiting a secretory stomach; there is no distinction between esophagus and stomach in *cyclostomes.*

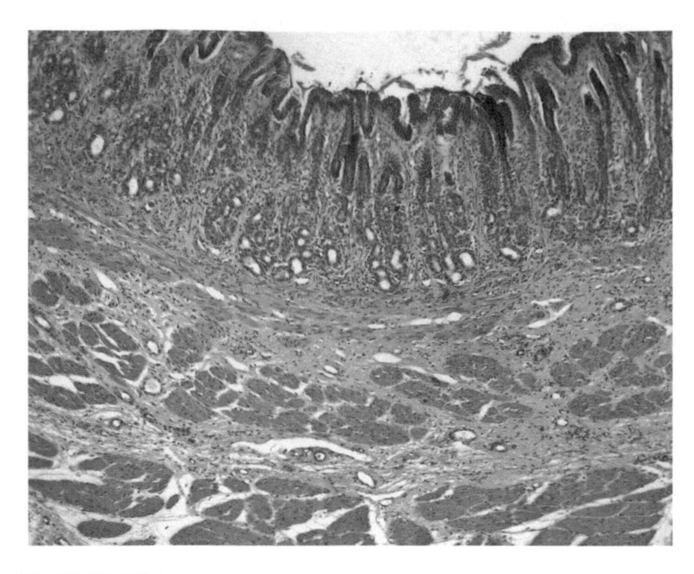

FIGURE K52a The stomach of the caiman is unexceptional with thick muscle layers and tubular glands that pour their secretion onto the surface. 10 ×.

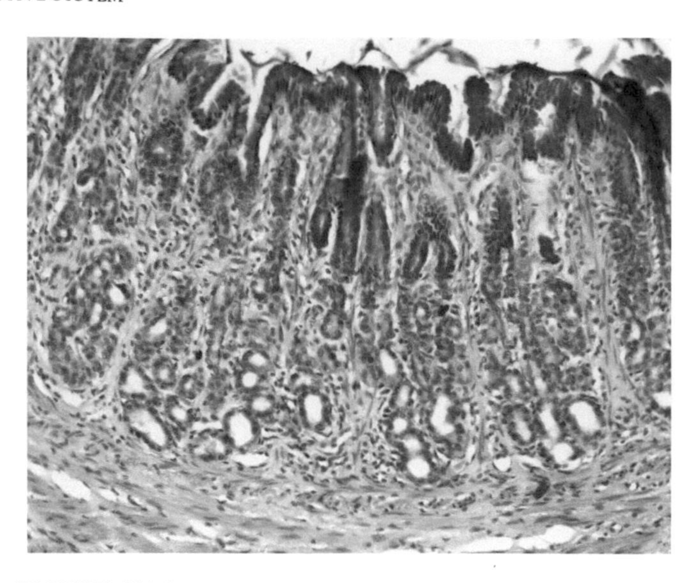

FIGURE K52b The gastric glands of the caiman are branching tubular and pour their secretions into common gastric pits. 20 ×.

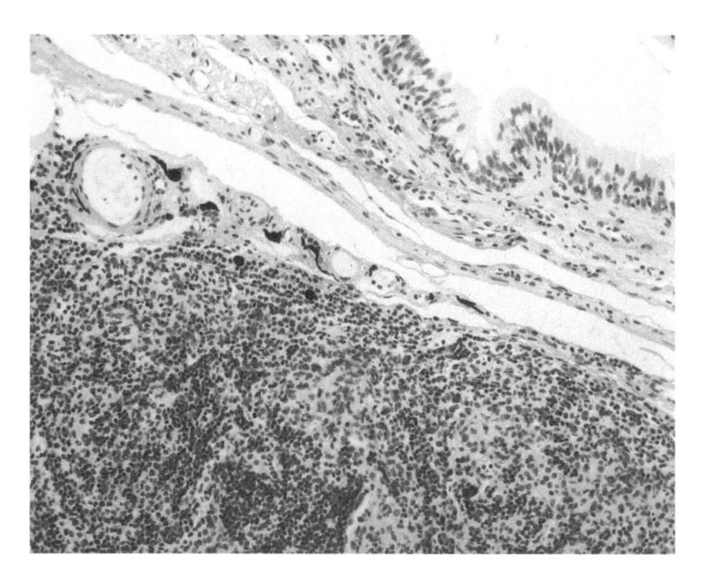

FIGURE K53a The stomach of the lungfish, *Protopterus*, is simple and shows no tubular glands. The pseudostratified columnar epithelium of the tunica mucosa rests on the vascular connective tissue of the lamina propria mucosae and is separated from the tela submucosa by the lamina muscularis mucosae. These layers are hard to differentiate in this pale section. The tela submucosa contains larger blood vessels and is packed with hemopoietic tissue. In animals lacking hollow bones, hemopoiesis often takes place in the congenial atmosphere of the richly vascular connective tissue of the gut. There are large lymphoid spaces, lined with simple squamous epithelial cells, in the tela submucosa. 10 ×.

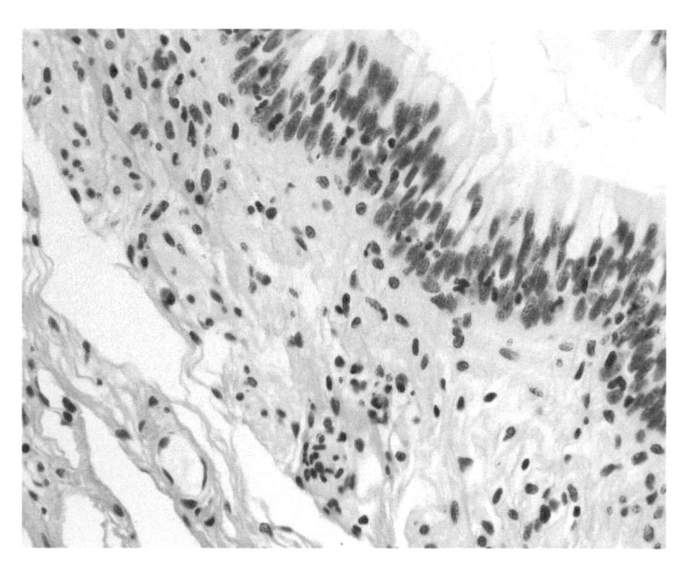

FIGURE K53b Some mitosis can be seen in the cells of the pseudostratified columnar epithelium lining the stomach of *Protopterus*. A few lymphatics pass through the tela submucosa at the lower left. 20 ×.

GASTRODUODENAL JUNCTION

In a longitudinal section through the junction of the pylorus and the duodenum of a mammal the tunica mucosa of the pylorus is relatively thick and contains branched tubular glands while that of the duodenum is somewhat thinner and contains simple tubular glands, the INTESTINAL CRYPTS (crypts of Lieberkühn) (Fig. K57a).

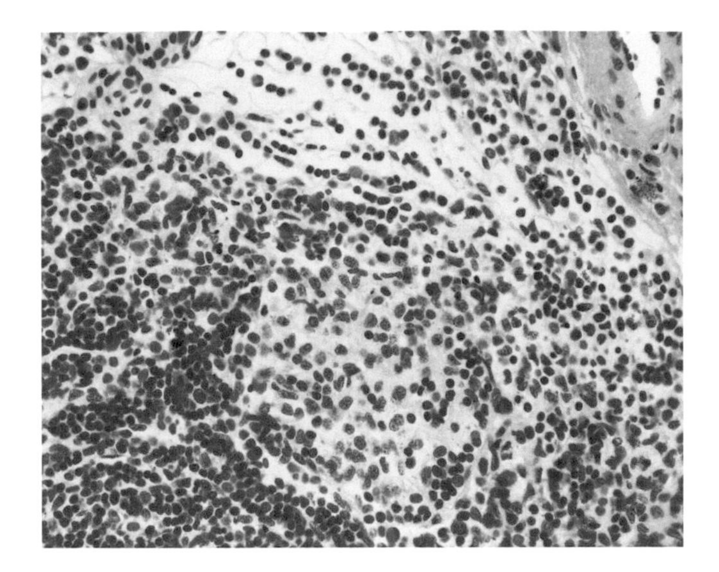

FIGURE K53c The characteristic granular nature of hemopoietic tissue is seen in this section through the tela submucosa of the stomach of *Protopterus*. Cells produced here migrate into the ample blood vessels of the region and make their way throughout the animal. 20 ×.

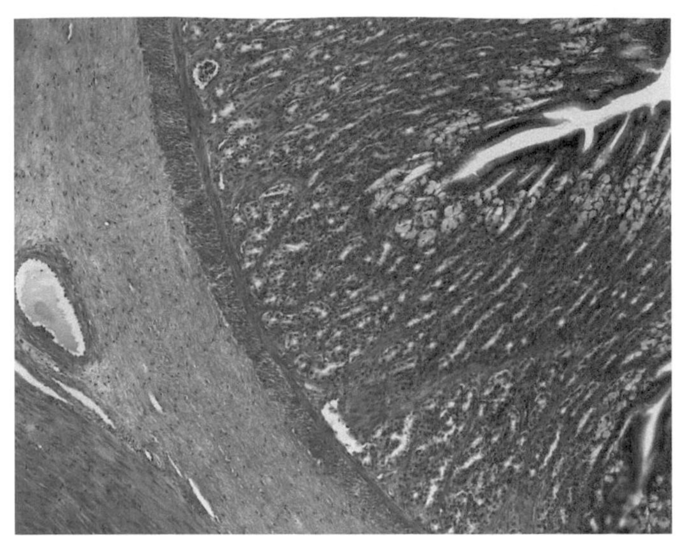

FIGURE K54a The stomach of the marine toad, *Bufo marinus*, is provided with densely packed, deep tubular glands in the tunica mucosa. 10 ×.

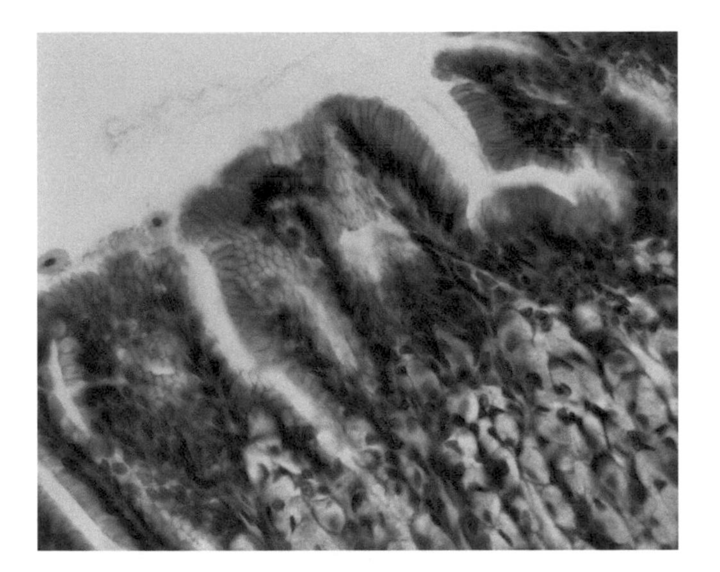

FIGURE K54b Columnar cells form the lining of the stomach and also line the gastric pits. Note the terminal bars in oblique section of the epithelium of the gastric pit at the left of center. Large vacuolated mucous cells occur at the lower left corner, just below the pits. 40 ×.

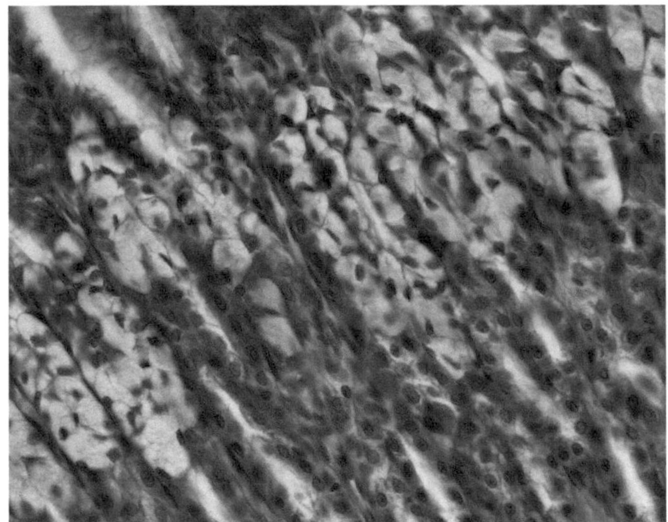

FIGURE K54c Still deeper in the tunica mucosa, the clear mucous epithelial cells give way to lightly basophilic cells that secrete digestive enzymes. Note the strands of smooth muscle cells between the tubular glands. 40 ×.

At the point of transition the tubules of the pyloric glands break through the lamina muscularis mucosae and lie to some extent in the tela submucosa. Glands similar to those of the pylorus but with a more complex system of branching occur in the tela submucosa of the duodenum, the SUBMUCOSAL GLANDS (Brunner's glands). The inner layer of the tunica muscularis increases in thickness at the zone of transition to form the PYLORIC SPHINCTER.

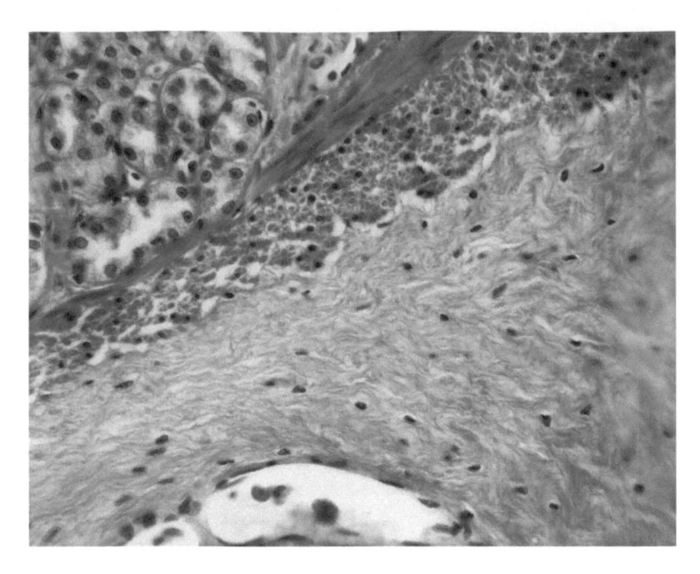

FIGURE K54d The strands of smooth muscle between the tubular glands are more obvious. These have migrated from the lamina muscularis mucosae to lie between the tubular glands. Paler cells form the fundi of the tubular glands. The lamina muscularis mucosae is thick and consists of an inner circular and out longitudinal layer. Connective tissue of the tela submucosa constitutes the lower right of the micrograph. A large vein of the tela submucosa occurs at the center bottom. 40 ×.

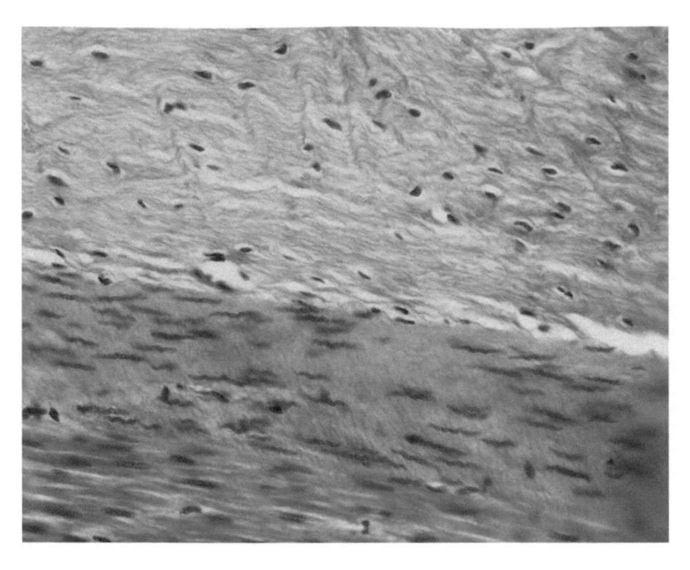

FIGURE K54e The tela submucosa is subtended by a thick layer of circular smooth muscle. 40 ×.

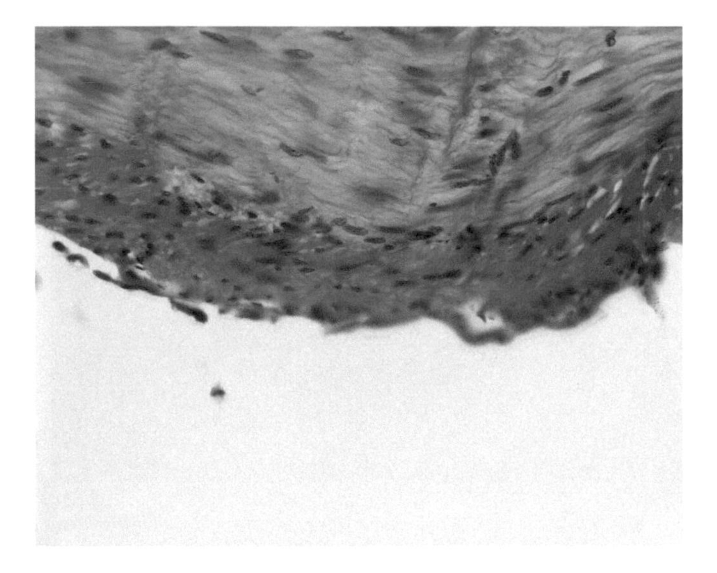

FIGURE K54f A myenteric plexus is seen at the left in the tunica muscularis between the upper circular layer of smooth muscle and the lower longitudinal layer. 40 ×.

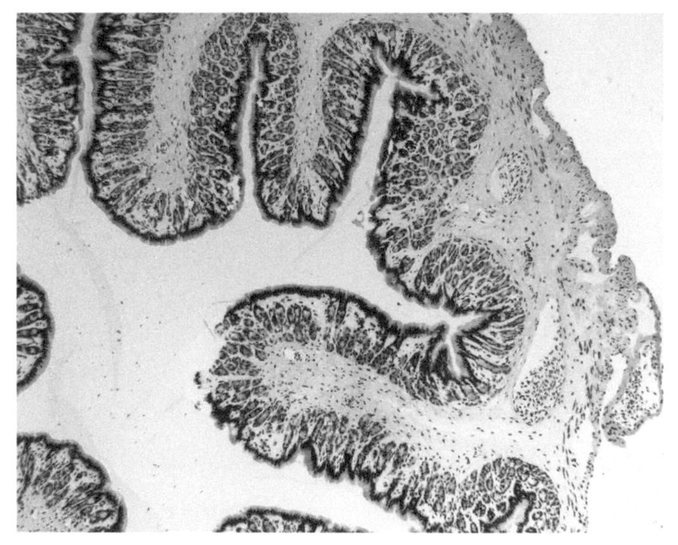

FIGURE K55a The stomach of this mudpuppy, *Necturus*, is folded permitting distension. 2.5 ×.

SMALL INTESTINE

Digestion continues in the small intestine and absorption of the digested products begins. The absorptive surface is enhanced in four ways: its length is many times that of the body; the inner surface is raised up in circular ridges, the PLICAE CIRCULARES, that project into the lumen (Fig. K57b); each ridge is covered by the tunica mucosa (Fig. K57c); in higher vertebrates the presence of minute motile finger-like VILLI cover the surface of each plica giving the lining a velvety appearance (Fig. K57d); each villus has a core of connective tissue of the lamina propria mucosae and is covered by epithelium that bristles with a striated border of microvilli (Figs. K58a and

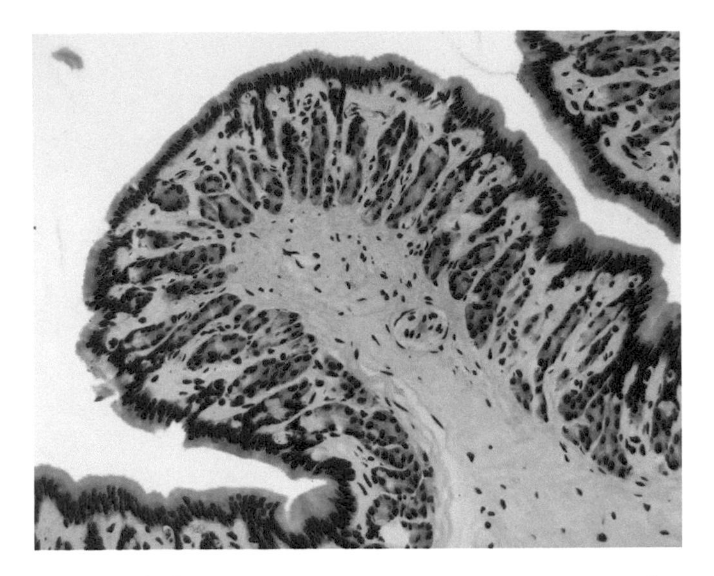

FIGURE K55b The simple columnar epithelium accommodates shallow, simple tubular glands. A cross section of a capillary can be seen in the tela submucosa of a RUGA near the center of the micrograph. This is not a picture of the stomach of an animal that lives in the fast lane. Compare it with the stomach of another amphibian, the marine toad (above), whose stomach appears to be highly developed. 10 ×.

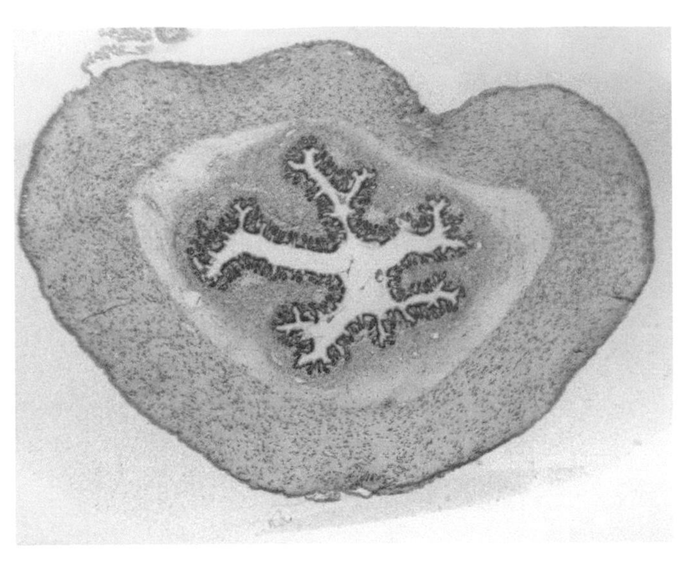

FIGURE K56a Nor does *Amphiuma* live in the fast lane – its organs reflect its relaxed life style. The stomach is thrown up into several low folds. 1 ×.

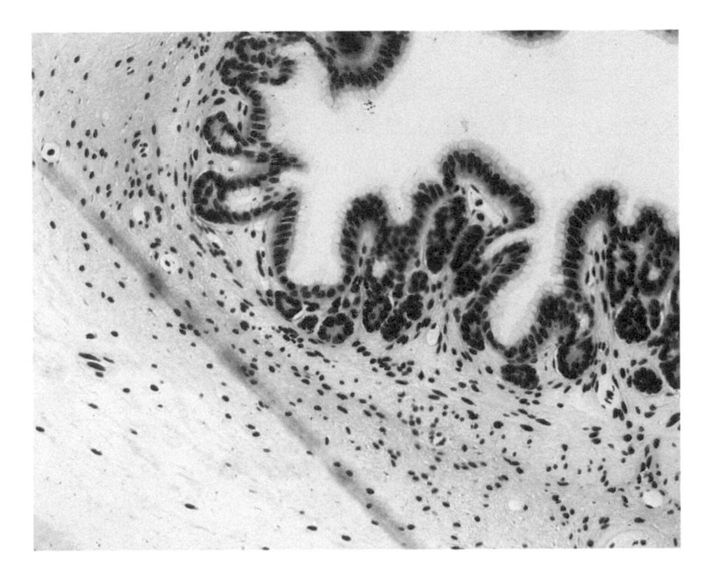

FIGURE K56b Simple tubular glands probably secrete mucus and may also produce digestive enzymes. Balls of epithelial cells seem to have penetrated into the tela submucosa. 10 ×.

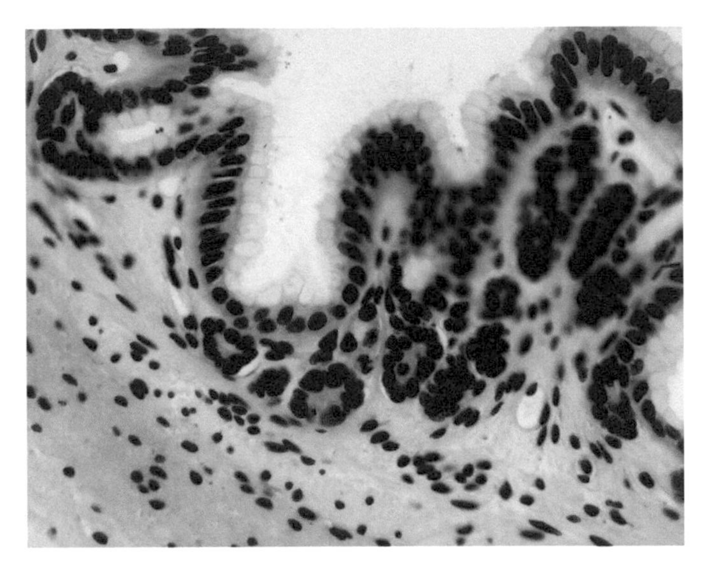

FIGURE K56c Masses of epithelial cells occur just below the epithelium of the intestine of *Amphiuma*. 20 ×.

K58c). Villi may be variously shaped but minor differences have little significance for the zoologist. Surrounding the tela submucosa are two layers, longitudinal and circular, of smooth muscle of the TUNICA MUSCULARIS (Fig. K58b). Ganglia of the autonomic nervous system are often seen between the muscle layers. Small ganglia are also found in the tela submucosa (Fig. K58d).

Villi are found in the mammalian small intestine; each is a projection of the lamina propria mucosae covered by simple columnar epithelium (Figs. K60a–K60c and K61a) with a STRIATED BORDER of microvilli (Fig. K60d). The lamina propria mucosae consists of reticular tissue containing blood vessels, lymphatics, and a few scattered smooth muscle fibers (Figs. K58a and K58c) derived from the inner layer of the lamina muscularis mucosae.

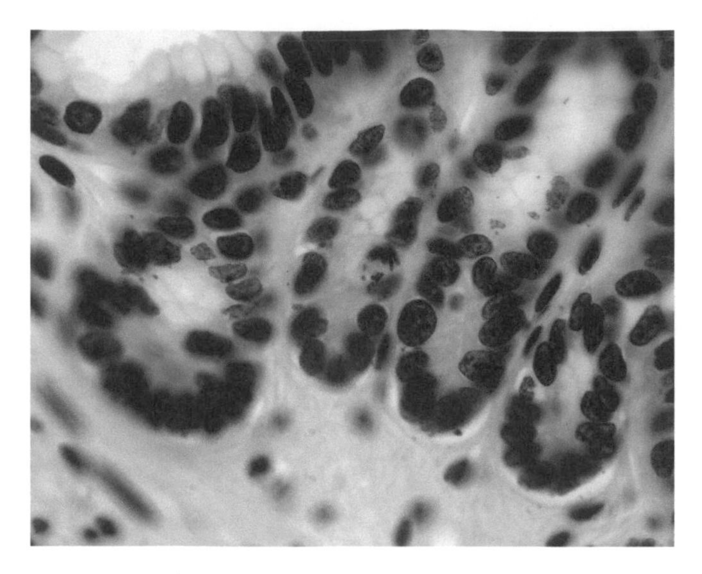

FIGURE K56d A higher power view of the section in Fig. K56c. Mitotic activity can be seen in these cell masses. It is assumed that they are CELL NESTS and provide replacements for cells lost from the surface of the stomach. 40 ×.

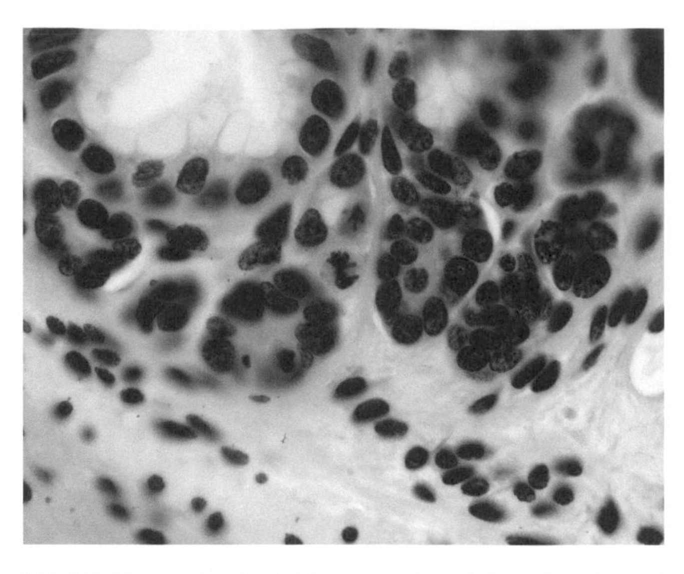

FIGURE K56e Another high power view of the region shown in Fig. K56c. 40 ×.

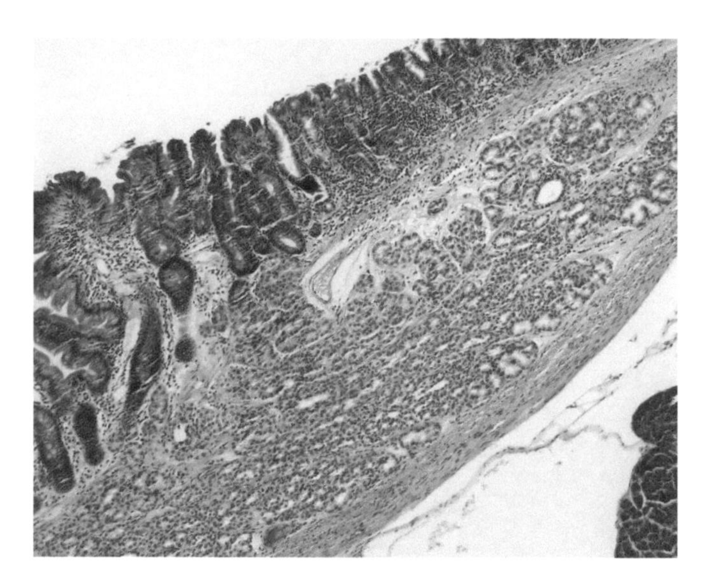

FIGURE K57a This is a poor section of the gastroduodenal junction of a rat. It is used to show the transition between the stomach and the duodenum. The epithelium forms simple tubular glands in the wall of the stomach at the right. At the left it is projected into the finger-like VILLI, characteristic of the small intestine. The large SUBMUCOSAL GLAND is a duodenal gland that has encroached on the stomach. A fragment of pancreas appears at the lower right. 10 ×.

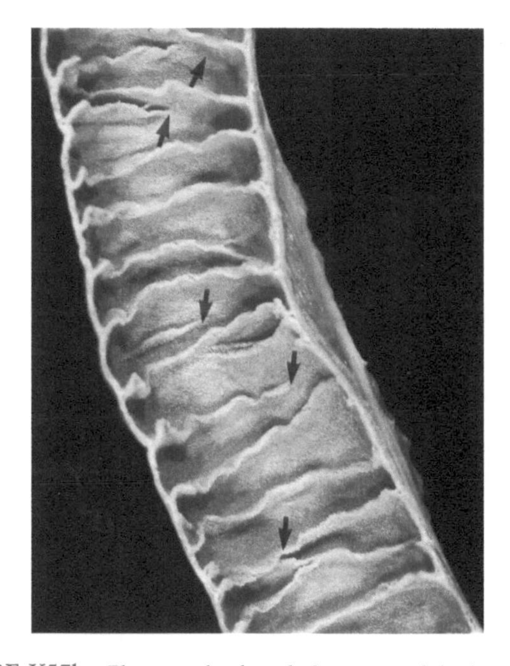

FIGURE K57b Photograph of a whole mount of the human duodenum with one side removed to reveal the PLICAE CIRCULARES that project, like shelves, into the lumen. These ridges greatly increase the surface area for absorption.

Note the distribution of blood vessels in an injected specimen where the extensive capillary network is seen (Figs. K59a and K66). A central lymphatic, the LACTEAL, is located in each villus but can be seen only in special preparations (Fig. K59c).

Between the bases of the villi are the openings of the tubular INTESTINAL GLANDS or CRYPTS (crypts of Lieberkühn) (Figs. K59b and K61b). Several types of cells are found in the crypts. It has been suggested that four lines of epithelial cells of the small intestine are derived from stem cells in the crypt base.

In the hostile environment of the lumen, epithelial cells are constantly being lost and replaced by maturing cells that migrate upward from undifferentiated stem cells at the base of the crypts (Fig. K60f). It has been shown

FIGURE K57c Photomicrograph of a cross section of the small intestine of the amphibian *Amphiuma*. A plica rises up from the surrounding tissue providing additional absorptive surface. Goblet cells are conspicuous in the epithelium. 10 ×.

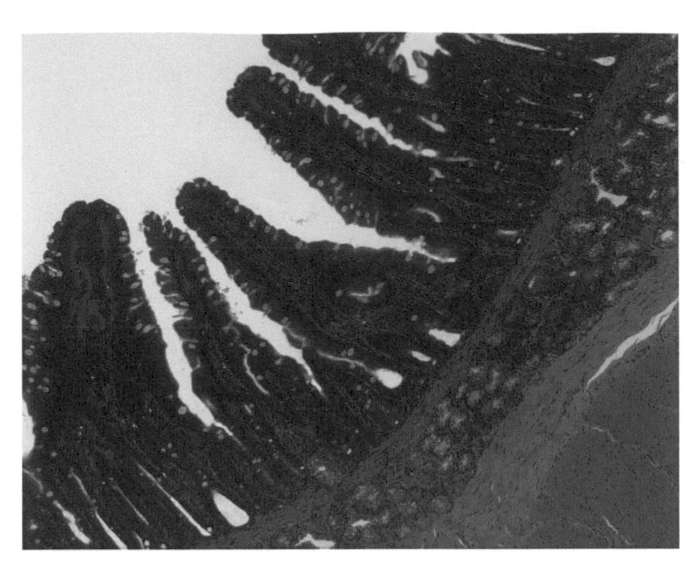

FIGURE K57d Several VILLI extend into the lumen of the intestine of a monkey. Simple tubular CRYPTS open between the bases of the villi and produce digestive juices. Branched, tubular submucosal glands, the DUODENAL GLANDS, pour their sections of mucus through ducts into the intestinal crypts between the villi. The inner circular and outer longitudinal layers of smooth muscle are evident. 10 ×.

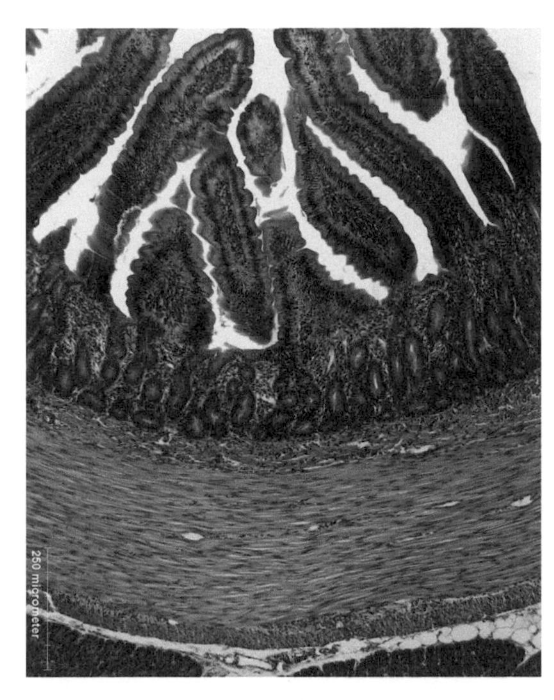

FIGURE K58a The surface area for absorption is vastly increased by the VILLI projecting into the lumen and the MICROVILLI on the surface of each absorptive cell. The core of each villus consists of vascular connective tissue of the lamina propria mucosae and a few muscle cells from the lamina muscularis mucosae. Between the bases of the villi, and extending down toward the lamina muscularis mucosae, are the simple tubular glands or CRYPTS (of Lieberkühn). They are surrounded by the delicate vascular connective tissue of the lamina propria mucosae. 10 ×.

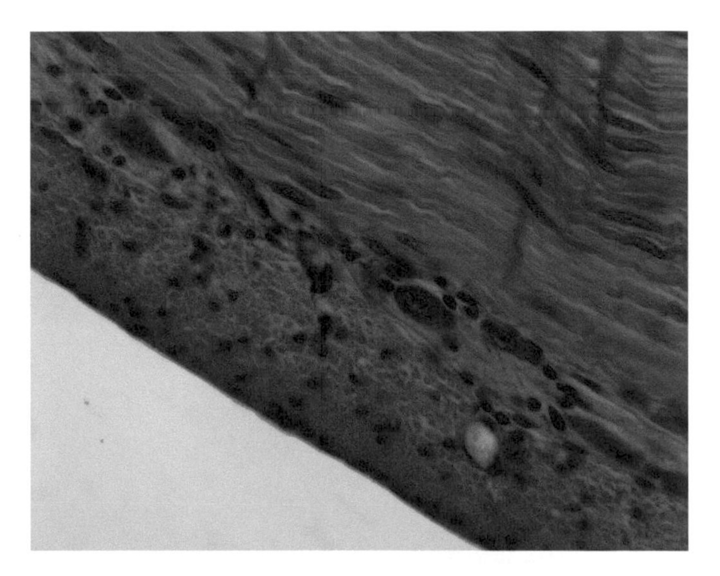

FIGURE K58b Surrounding the tela submucosa is a thick layer of smooth muscle, the TUNICA MUSCULARIS. The inner layer consists of CIRCULAR MUSCLE and the outer is the thinner layer of LONGITUDINAL FIBERS. Between the two layers of muscle note the grayish neurons of ganglia of the MYENTERIC PLEXUS (Auerbach's). A thin peritoneal layer adheres to the outer surface of the longitudinal muscle. Since this sheath is peritoneal, it is referred to as the TUNICA SEROSA. 63 ×.

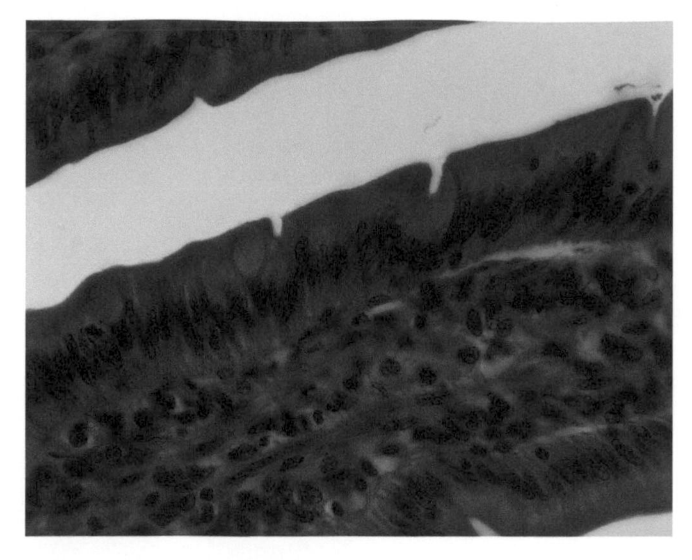

FIGURE K58c This figure shows portions of two villi from the same section as the section in Fig. K58b above. The villi are covered with a simple columnar epithelium of ENTEROCYTES, which have a STRIATED BORDER consisting of microvilli. The apical regions of adjacent enterocytes are sealed by JUNCTIONAL COMPLEXES, which are often seen with the light microscope as TERMINAL BARS but are not visible in this section. A few GOBLET CELLS are scattered among the enterocytes. Goblet cells contain an apical mass of mucus that they release into the lumen of the intestine. The core of the villus contains delicate vascular connective tissue. The simple squamous epithelium of capillaries and lymphatics are visible in this area. The vessel at the center may well be a LACTEAL, which drains LYMPH from the area. 63 ×.

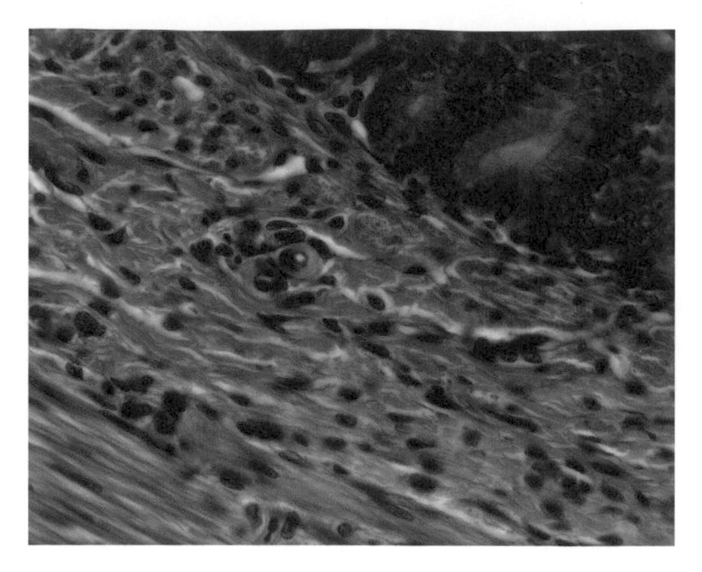

FIGURE K58d Mitosis is active at the bases of the crypts and one mitotic figure can be seen at the upper center of the micrograph. The crypts are separated from the tela submucosa by a few strands of smooth muscle from the lamina muscularis mucosae. A few neurons form a ganglion of the SUBMUCOUS PLEXUS (Meissner's) in the tela submucosa, near the center of the micrograph. A few blood vessels and/or lymphatics can be seen. At the lower left is a portion of the circular layer of smooth muscle of the tunica muscularis. 63 ×.

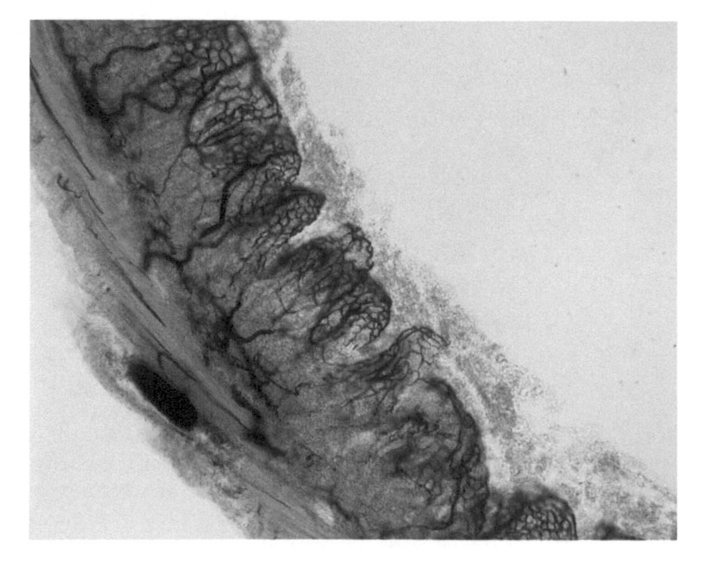

FIGURE K59a The elaborate vascularization of the mammalian intestine, especially the villi, is seen in this section where the blood vessels were injected with a blue/green material before the sections were made. The green material has penetrated most of the vessels while the red appears only in the larger vessel seen only faintly in the deeper tissue. 10 ×.

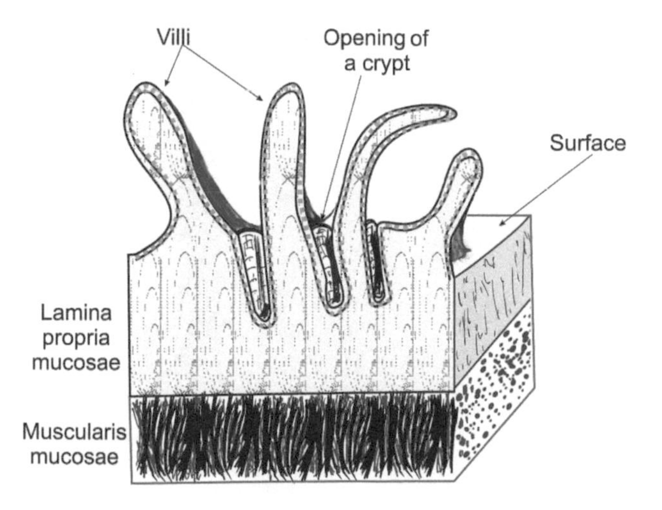

FIGURE K59b Three-dimensional schematic drawing of the lining of the mammalian small intestine. Finger-like villi have cores of connective tissue from the lamina propria mucosae. Between the villi are the glandular crypts. *Authors.*

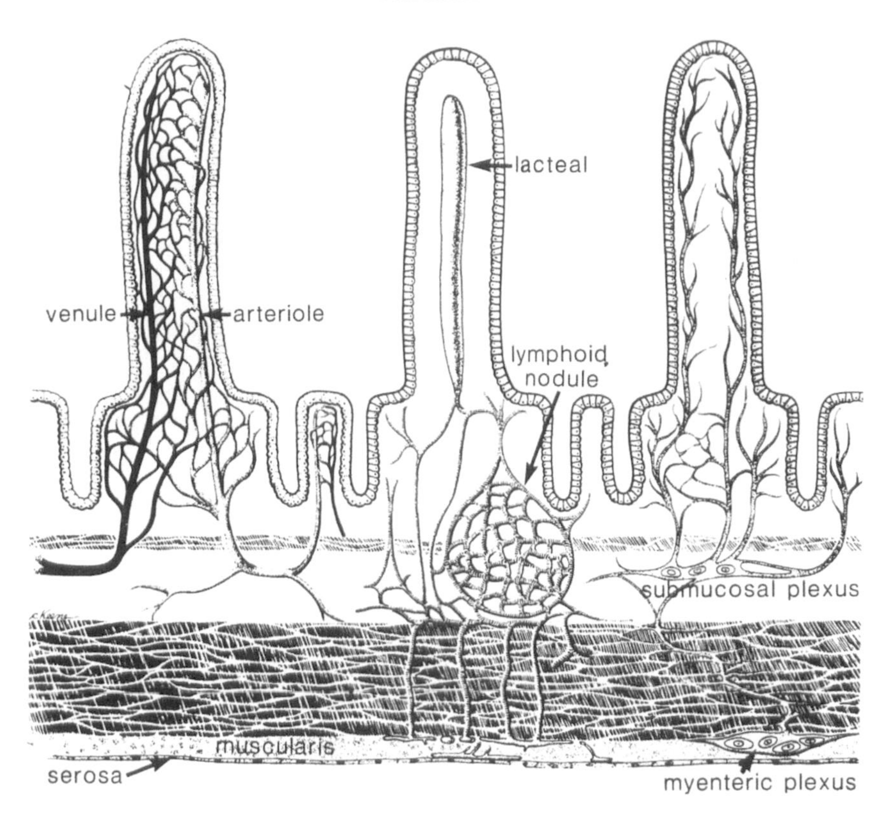

FIGURE K59c Diagram illustrating the blood circulation (left), lymphatic drainage (center), and innervation (right) of the villi in the mammalian small intestine.

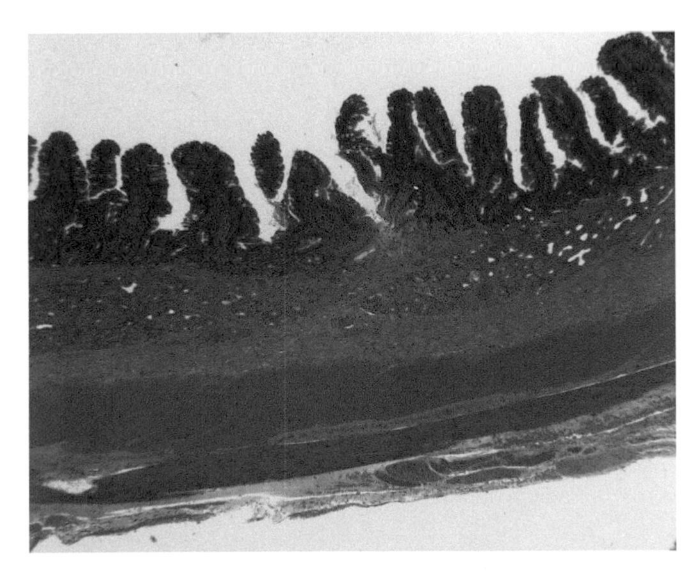

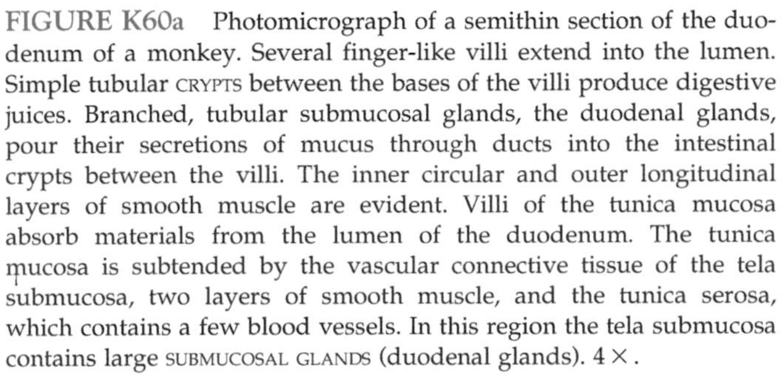

FIGURE K60a Photomicrograph of a semithin section of the duodenum of a monkey. Several finger-like villi extend into the lumen. Simple tubular CRYPTS between the bases of the villi produce digestive juices. Branched, tubular submucosal glands, the duodenal glands, pour their secretions of mucus through ducts into the intestinal crypts between the villi. The inner circular and outer longitudinal layers of smooth muscle are evident. Villi of the tunica mucosa absorb materials from the lumen of the duodenum. The tunica mucosa is subtended by the vascular connective tissue of the tela submucosa, two layers of smooth muscle, and the tunica serosa, which contains a few blood vessels. In this region the tela submucosa contains large SUBMUCOSAL GLANDS (duodenal glands). 4×.

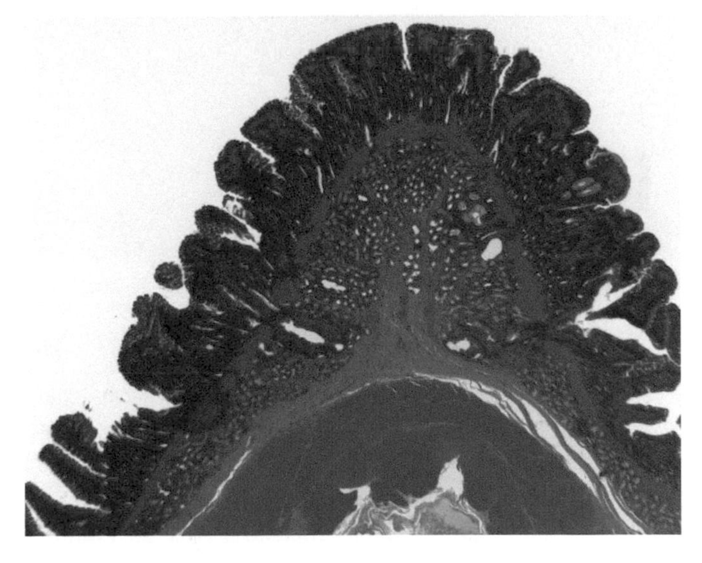

FIGURE K60b In this section the villi are firmly planted on a ridge, the PLICA CIRCULARIS (plural: plicae circulares), formed by a thickening of the tela submucosa. This fold provides a further increase of absorptive area in the duodenum and contains a well-developed submucosal gland. The major blood vessels of the intestine are distributed from the tela submucosa. 2.5×.

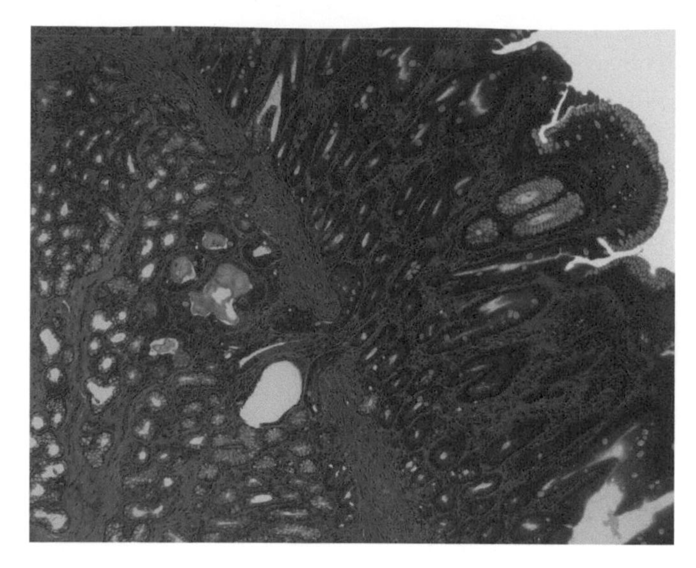

FIGURE K60c The submucosal gland occupies most of the left portion of this micrograph. Its duct can be seen passing through the lamina muscularis mucosae to pour its secretions into the base of one of the crypts. 10 ×.

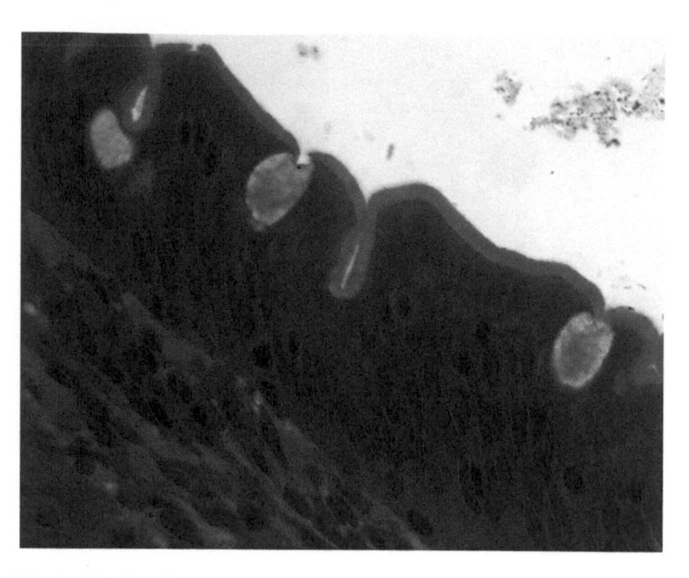

FIGURE K60d The simple columnar ENTEROCYTES of a villus are covered with microvilli that greatly increase the surface area for absorption. Among these cells are a few GOBLET CELLS, packed with mucus. The apical surfaces of the enterocytes are sealed by junctional complexes, which cannot be seen in this micrograph. 100 ×.

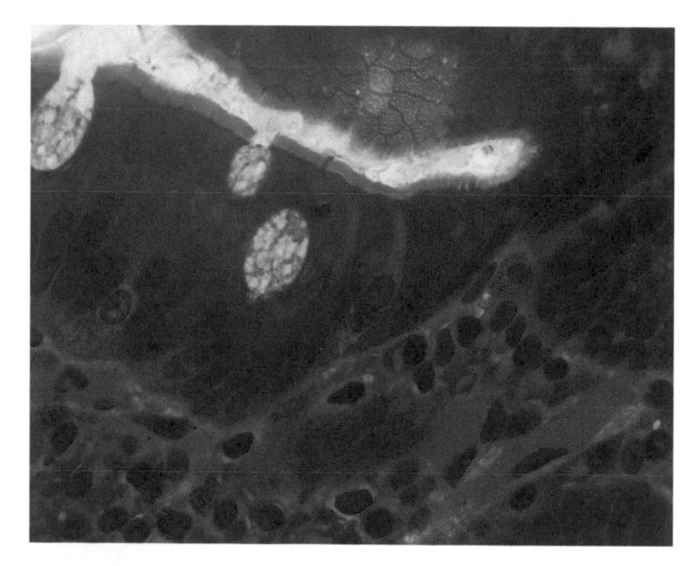

FIGURE K60e This section from the same slide as Fig. K60d above was taken through the upper region of a crypt. At the top, an oblique slice of the apical portions of a few enterocytes, demonstrates the "chicken wire" appearance of terminal bars that seal the space between these cells. A few goblet cells and occasional ACIDOPHIL CELLS (Paneth cells) filled with acidophilic granules may be seen between the enterocytes. 100 ×.

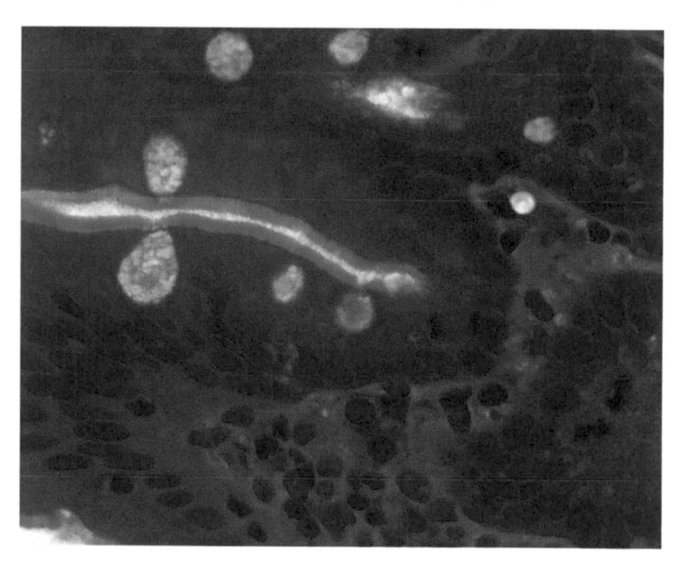

FIGURE K60f Mitosis of the cells at the bases of the crypts provides new cells that slide up the epithelium of the crypts and up the epithelium of the villi, to replace cells lost at their tips. Note the anaphase at the bottom of this crypt. 100 ×.

that new cells are constantly gliding upward to replace cells that have worn away from the surface of the villi (Fig. K67c). There are four lines of epithelial cells lining the crypts of mammals—the cells assume their mature characteristics as they rise to the tips of the villi. The cell lines include:

Enterocytes or absorptive cells. The striated border of the columnar enterocytes consists of many closely packed MICROVILLI and provides a huge surface area for absorption. The microvilli are coated with a glycocalyx. Adjacent cells are sealed by JUNCTIONAL COMPLEXES, the terminal bars of light microscopy (Figs. K60e, K61e–K61h). Note the terminal web, prominent mitochondria, granular and agranular endoplasmic reticulum, and free ribosomes. Apical pits and vesicles indicate the pinocytotic uptake of nutrients.

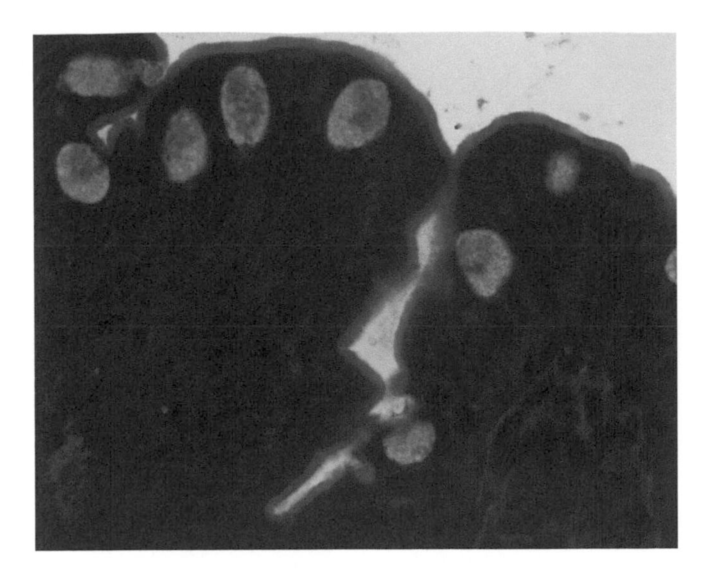

FIGURE K61a The epithelium of the jejunum of the monkey is simple columnar. The cells have a striated border and scattered goblet cells are seen. 100 ×.

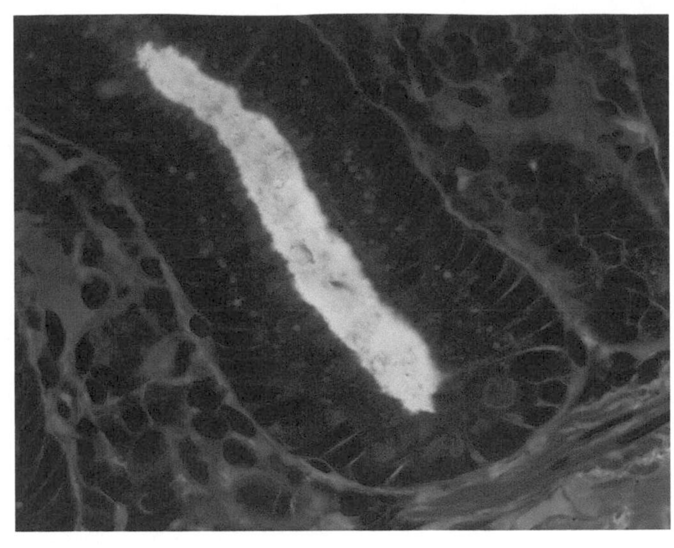

FIGURE K61b Most of the cells in this crypt are enterocytes that contain pale droplets of digestive enzymes. The conspicuous cells at the base of the crypt are acidophil cells and there is an ENTEROENDOCRINE CELL, containing fine acidophilic granules in the cytoplasm, wedged between enterocytes near the bottom left of the gland. The gland is surrounded by vascular connective tissue of the lamina propria mucosae, and a few strands of the lamina muscularis mucosae are seen at the lower right. 100 ×.

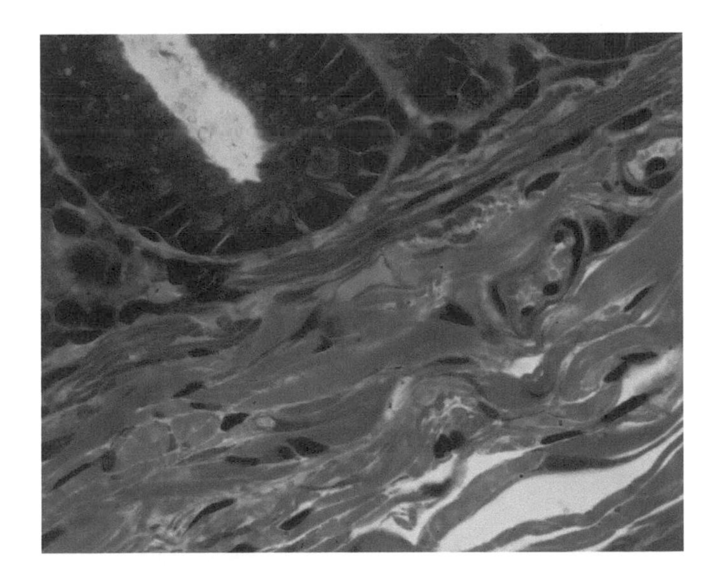

FIGURE K61c The acidophil cells and enteroendocrine cells appear once again. The lamina muscularis mucosae clearly delineates the vascular connective tissue of the lamina propria mucosae from the vascular connective tissue of the tela submucosa. Collagenous fibers of the tela submucosa (bottom) appear coarse. 100 ×.

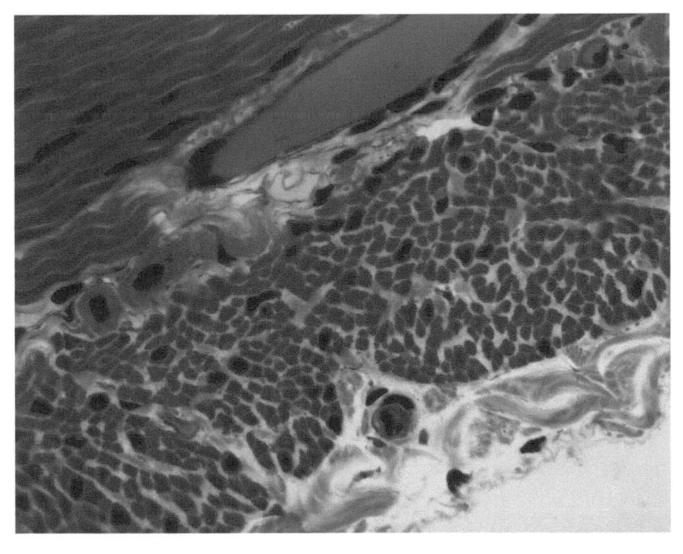

FIGURE K61d The inner circular layer of smooth muscle of the tunica muscularis and the outer layer of longitudinal muscle are separated by loose connective tissue containing a blood vessel. 100 ×.

Goblet cells are scattered between the absorptive cells (Fig. K61f) *and* (Figs. K62a–K62c, K62e, K62f). The basal cytoplasm contains extensive granular endoplasmic reticulum. Between the nucleus and the apical mass of mucous droplets is the Golgi complex.

Acidophil cells (cells of Paneth) do not migrate and are restricted to the base of the gland for their lives of about 3 weeks in humans (Fig. K61b). Their apical cytoplasm contains numerous coarse acidophilic granules. Their nuclei are large and oval. In electron micrographs these cells show the typical characteristics of protein secretion: well-developed granular endoplasmic reticulum, Golgi complex, and apical membrane-bound droplets (Fig. K61h).

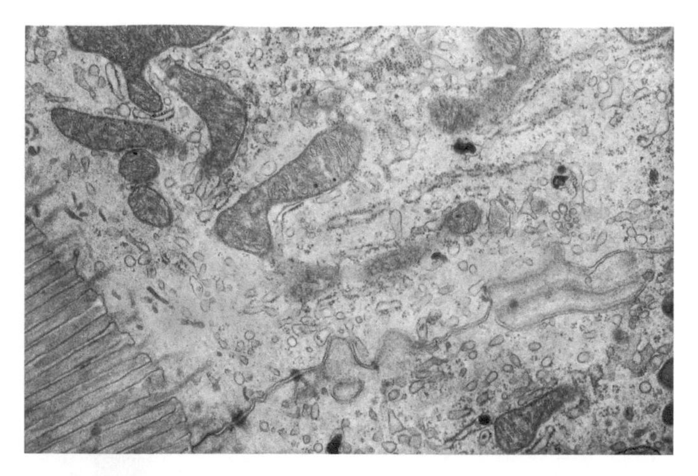

FIGURE K61e Electron micrograph of the absorptive epithelium of the jejunum of a rat. Microvilli on the surface are aligned with military precision at the surface of the columnar cells (upper right). Bundles of fine filaments at the cores of the microvilli are rooted in the terminal web just below the surface of the cell. Several junctions couple the adjacent cells (top center): a tight junction (zonula occludens) at the top of the micrograph, a terminal bar (zonula adhaerens) below. A few desmosomes (maculae adhaerentes) strengthen the union between the two cells. The cytoplasm contains several mitochondria and endoplasmic reticulum, both granular and agranular. Note the precise separation of the plasma membranes of the two adjacent cells. 42,000 ×.

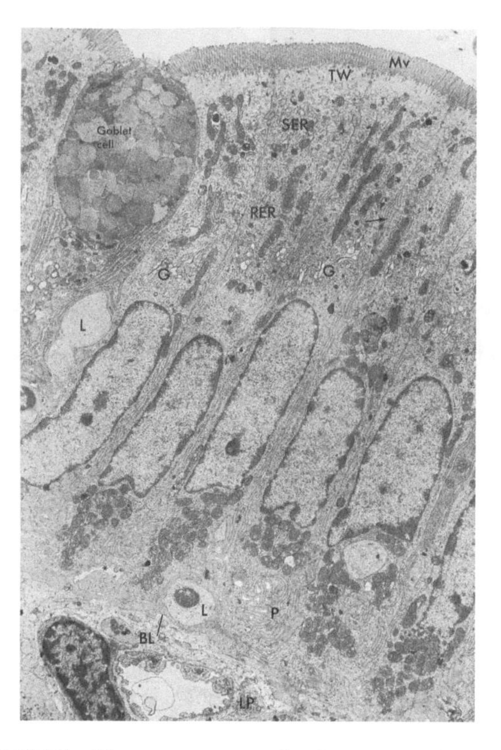

FIGURE K61f Electron micrograph of a section of the intestinal villus of a starved rat. Several absorptive cells are fringed with microvilli (Mv) rooted in a terminal web (TW) at the luminal surface. A portion of a goblet cell is shown. The cytoplasm below the terminal web contains agranular and granular endoplasmic reticulum (SER and RER). *BL*, basal lamina; *G*, Golgi complex; *LP*, lamina propria mucosae; *P*, interdigitating lateral cell projections; *L*, lateral intercellular spaces. 6000 ×.

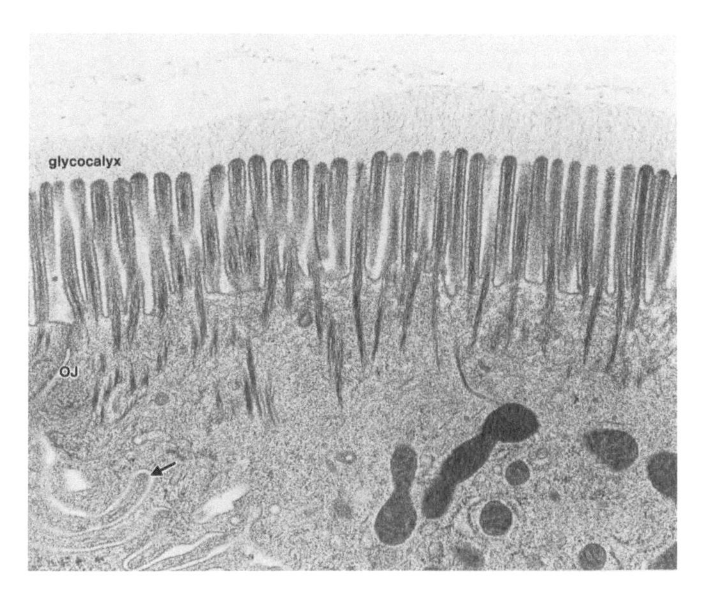

FIGURE K61g Electron micrograph of the apical border of an absorptive cell on the villus in the intestine of a cat. The microvilli, standing with military precision, contain bundles of actin microfilaments that are rooted in the cytoplasm of the terminal web. Cells are bound tightly together with occluding junctional complexes (*OJ*) and interdigitating lateral cell processes (*arrow*).

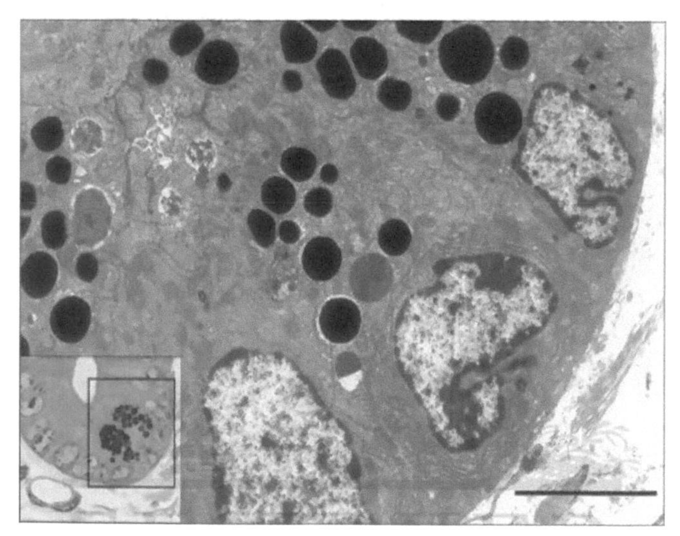

FIGURE K61h Electron micrograph of cells in the region of the crypt base of the small intestine of a mouse. A columnar cell (*crypt base columnar cell*) sits surrounded by paneth cells packed with acidophilic zymogen granules. The columnar cell may be a stem cell. Stem cells in the crypt give rise to other cells in the epithelium of the small intestine. The inset shows a toluidine blue stained semithin section with the area of the electron micrograph noted. (Netter Images # 14406 paneth cells with toluidine blue inset). *Netter Images # 14406 © 2005–2018 Elsevier.*

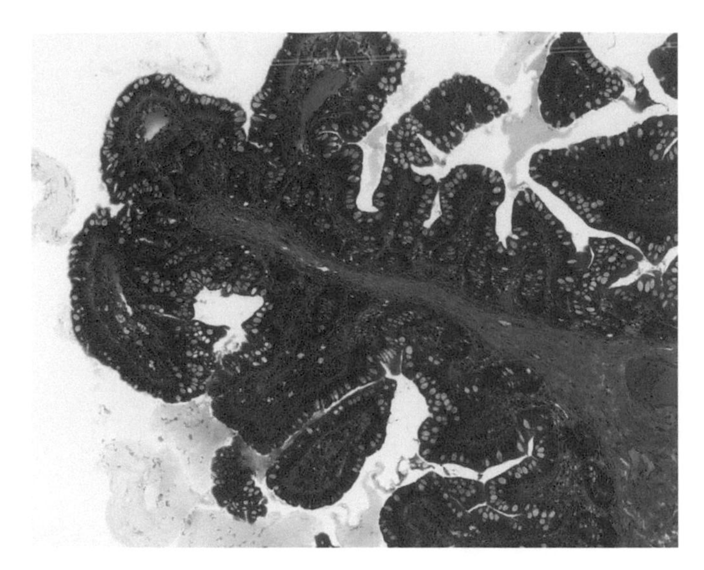

FIGURE K62a This photomicrograph shows an elaborate plica circularis in the wall of the ileum of a monkey. The surface area of the intestine is increased in several ways: by the fold, by the villi growing on the fold, and by the microvilli on the surface of the absorptive cells (not seen at this power). The fold has a core of vascular connective tissue. Note the crypts between the bases of the villi and the large arteriole at the base of the fold. 10 ×.

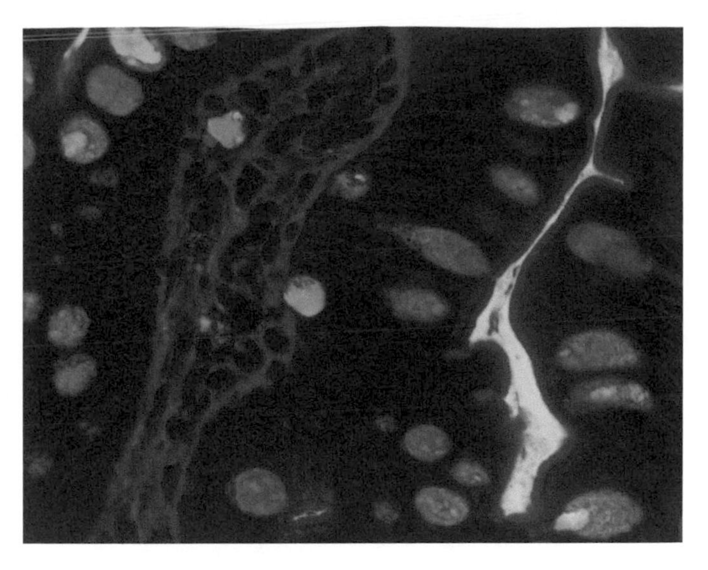

FIGURE K62b The epithelium of a villus of the ileum has the usual composition of simple columnar enterocytes interspersed with a few goblet cells. The enterocytes have a striated border and their intercellular spaces are sealed apically by junctional complexes, seen here as a dark line at the base of the striated border. Vascular connective tissue of the lamina propria mucosae forms the core of the villus. 100 ×.

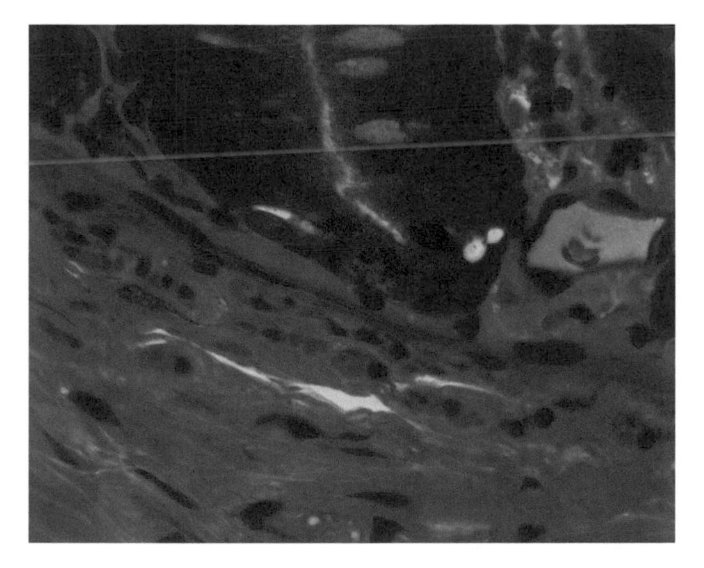

FIGURE K62c There are a few acidophil cells at the base of this crypt. 100 ×.

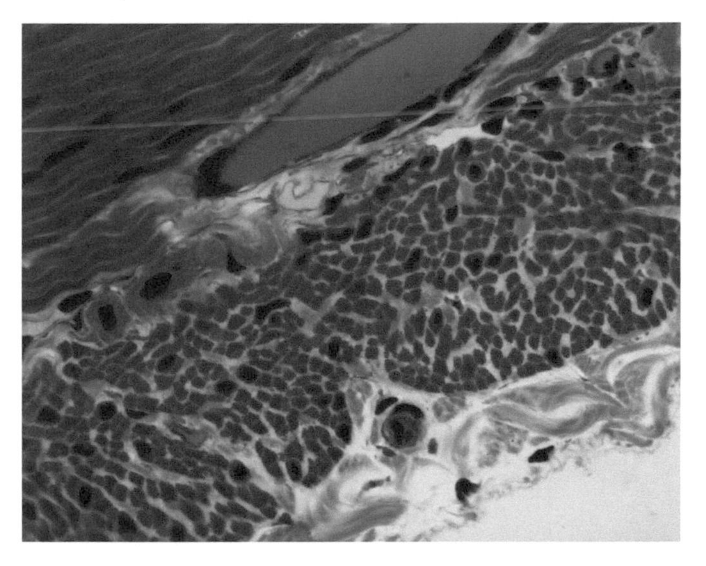

FIGURE K62d A small section through a ganglion of the submucous plexus appears between the connective tissue fibers of the tela submucosa at the top center. There are a few smooth muscle fibers of the circular layer of the tunica muscularis at the lower left. 100 ×.

Enteroendocrine or *APUD cells* are the rarest of the cells of the epithelium of the small intestine and constitute a heterogeneous group of small endocrine cells (Fig. K61b, K61c, K65b) resemble those previously described in the gastric glands and occur at the bases of the crypts.

There is intense mitotic activity in undifferentiated cells of the base of the crypts and new cells are constantly gliding upward to replace cells that have been worn away from the surface of the villi.

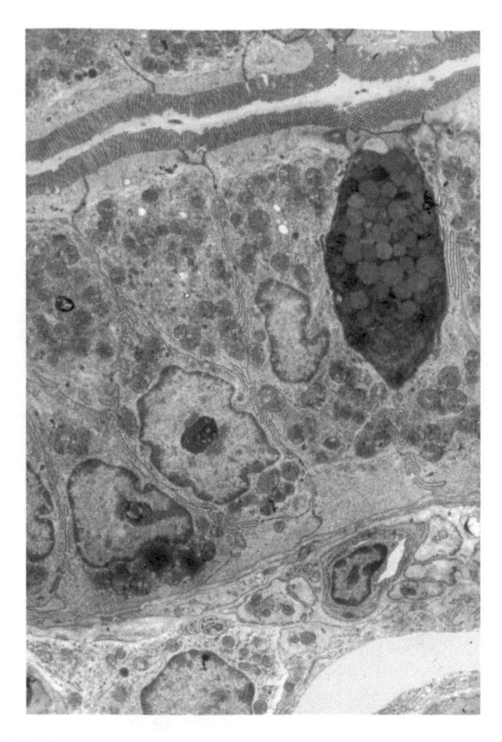

FIGURE K62e Electron micrograph of a section of the duodenal epithelium of a bat. Forests of microvilli, rooted in the terminal web, cover the free surfaces of the two adjacent cells at the right. Junctional complexes seal these borders. A goblet cell is ready to disgorge its mucus into the lumen. Mitochondria abound in the cytoplasm of these absorptive cells. Identify the basal lamina of the epithelium as well as capillaries, nerve fibers, collagen fibers, and smooth muscle in the connective tissue of the lamina propria mucosae. 7500 ×.

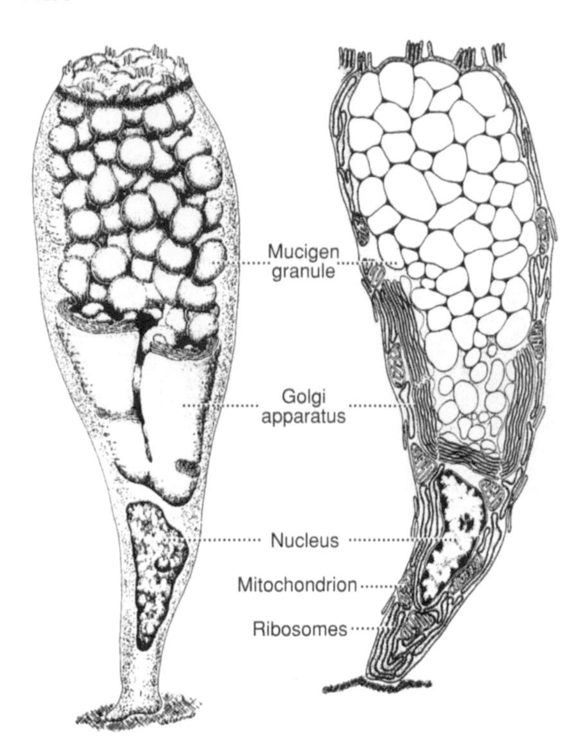

FIGURE K62f Schematic drawings of a goblet cell as imagined in three dimensions (left) and in electron microscopic section (right).

LYMPHOID TISSUE is widely distributed throughout the lamina propria mucosae (Figs. K63a–K63c) becoming more abundant along the small intestine(Figs. K64a and K64b). In the ileum it forms the aggregate nodules (Peyer's patches)(Fig. K65a) and fills not only the tunica mucosa but also the tela submucosa and may cause the epithelium to bulge into the lumen of the intestine, obliterating the villi.

The LAMINA MUSCULARIS MUCOSAE consists of two thin layers of smooth muscle, an inner circular and an outer longitudinal layer. Some fibers of the inner layer extend up into the villi (Figs. K58d, K61b, K62e).

Other Layers

The TELA SUBMUCOSA of the small intestine is a loose layer of areolar connective tissue containing blood vessels and nerves (Figs. K65d, K58a, K58d, K71). In parts of the duodenum of mammals it contains SUBMUCOSAL GLANDS (Brunner's glands), branched tubular mucous glands that pour their secretions through ducts into the intestinal crypts (Fig. K57d). Duodenal glands are found only in the duodenum but there are parts of the duodenum that do not have them. Sections of the ileum may be identified by the presence of lymphoid patches, although not all parts of the ileum have them. There are a few lymphoid nodules in the jejunum. A feature diagnostic of the jejunum of some animals is the branching nature of the plicae. There are no unusual features of the TUNICA MUSCULARIS and the TUNICA SEROSA. Note the myenteric plexus between the two layers of muscle (Figs. K65c, K67a, K67b).

FIGURE K63a Diffuse lymphoid tissue has invaded most of the lamina propria mucosae and the tela submucosa in this section of the duodenum of a dog. At the center, some has aggregated to form a simple NODULE without a germinal center. Note that lymphoid cells are present between the acini of the duodenal gland. 10×.

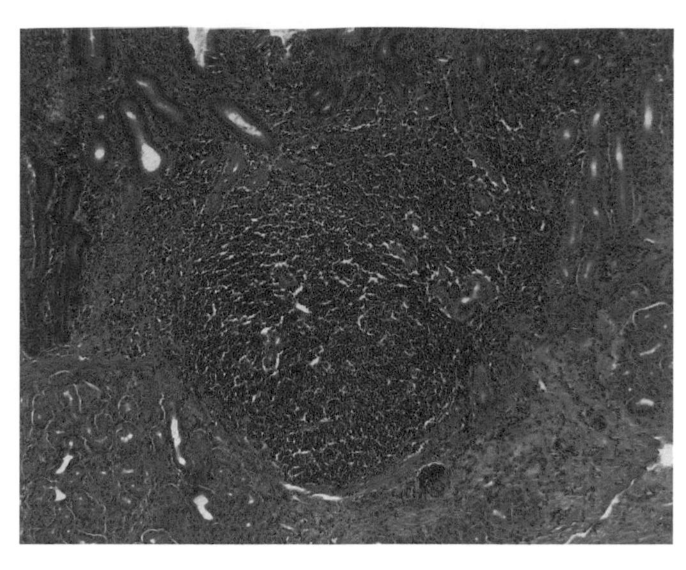

FIGURE K63b A lymphoid nodule has formed in the lamina propria mucosae of the duodenum of a dog. Note that it has stretched the lamina muscularis mucosae downward and distorted the villi of the tunica mucosa. A few crypts were enmeshed in the nodule as it expanded. 10×.

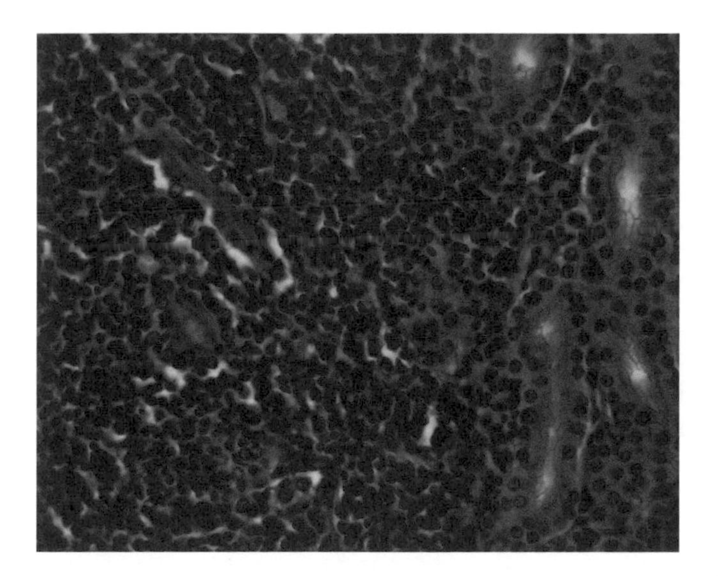

FIGURE K63c Packed lymphoid tissue at the periphery of a simple nodule comprises most of the micrograph. There are a few crypts at the right. Support for the loose cells is provided by a reticular STROMA of RETICULAR FIBERS (which cannot be seen) and long, acidophilic RETICULAR CELLS (which can be seen). 40×.

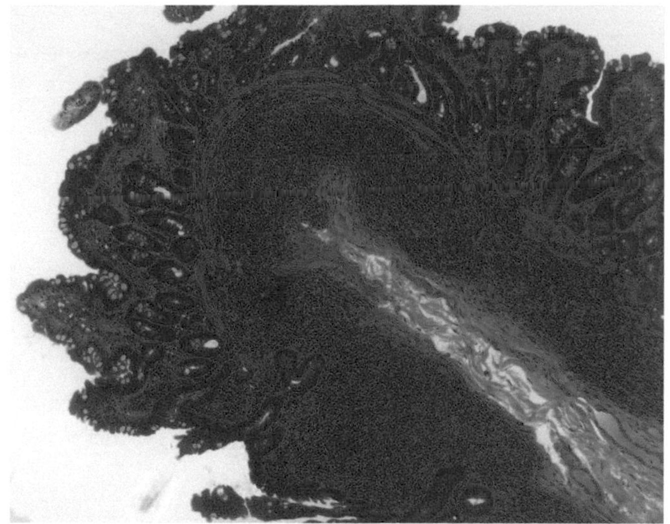

FIGURE K64a Diffuse lymphoid tissue is densely abundant in the tela submucosa of this section of the ileum. 10×.

COMPARATIVE HISTOLOGY OF THE SMALL INTESTINE

Note that villi are not present in the small intestines of all vertebrates. Villi (as in the mammal Fig. K57d) are solely extensions of the tunica mucosa with a core of connective tissue derived from the lamina propria mucosae while folds have a core derived from the tunica submucosa (Fig. K57c). Most folds (as in the frog) are seen in sections as continuous from top to bottom; few villi are seen in their entirety. Villi are said to be present in some bony fish. Various ways have evolved that increase the surface area for secretion of digestive juices and absorption of food. In addition, many structures delay the passage of materials through the gut, thereby allowing an adequate period for digestion and absorption to take place.

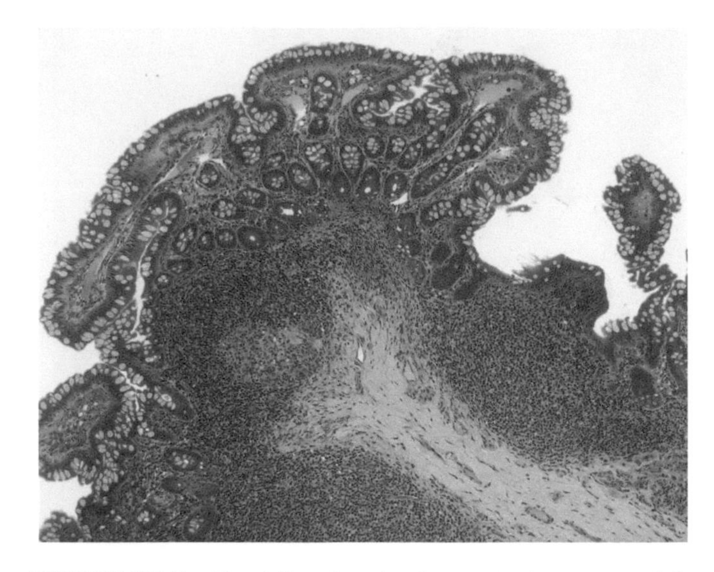

FIGURE K64b The diffuse lymphoid tissue in this section of the ileum has distorted the semicircular fold. 10 ×.

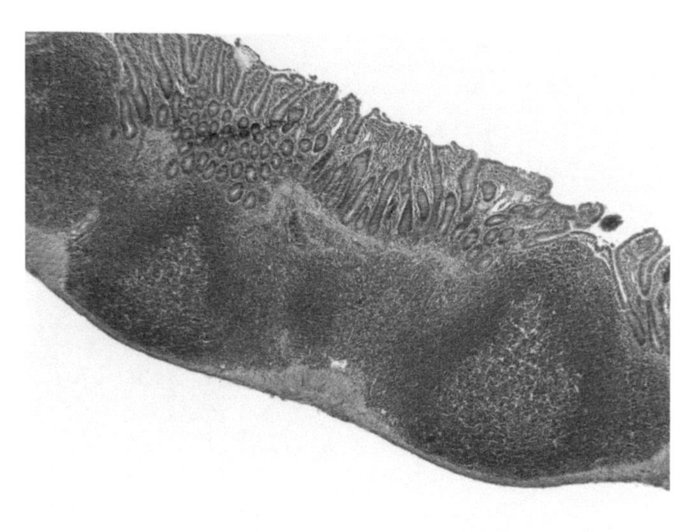

FIGURE K65a The ileum of this rat has been grossly distorted by the presence of AGGREGATE LYMPHOID NODULES (Peyer's patch) in the tela submucosa. Some lymphoid cells have migrated across the lamina muscularis mucosae and have settled in the connective tissue of the lamina propria mucosae between the tubular glands and the villi. 4 ×.

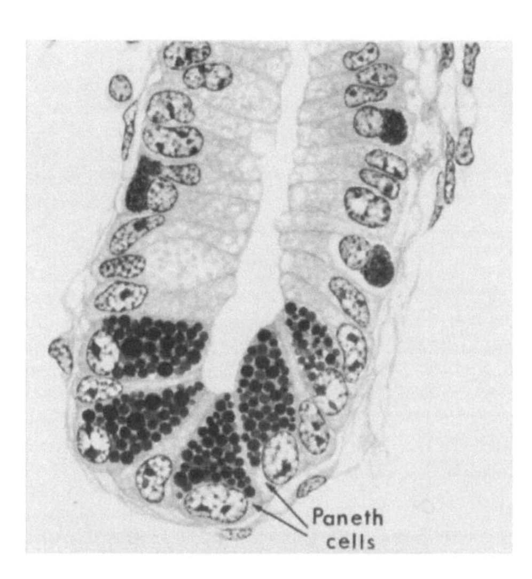

FIGURE K65b Drawing of a section of a mammalian intestinal crypt (of Lieberkühn) showing acidophil cells (of Paneth).

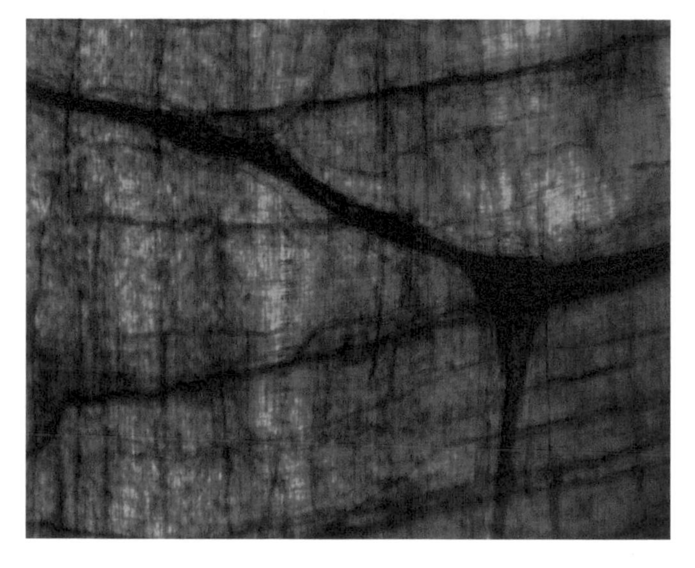

FIGURE K65c This is a whole mount of the MYENTERIC PLEXUS (Auerbach's plexus) sandwiched between the two layers of muscle in the intestine of a rabbit. It has been stained with silver so that all of the branching fibers that radiate from the ganglion may be seen. 20 ×.

The intestine of the hagfish has the four layers typical of the vertebrate intestine but the tunica muscularis is greatly reduced and the tunica serosa is thick; it is packed with oil droplets and hemopoietic islets. The lamina muscularis mucosae and lamina propria mucosae are thin but well developed (Figs. K68a–K68e). Ammocoetes have a simple type of spiral valve associated with the TYPHLOSOLE (Figs. K69a–K69c).

The intestine of elasmobranchs contains an elaborate SPIRAL VALVE that increases the surface area of the lining (Figs. K70a–K70d).

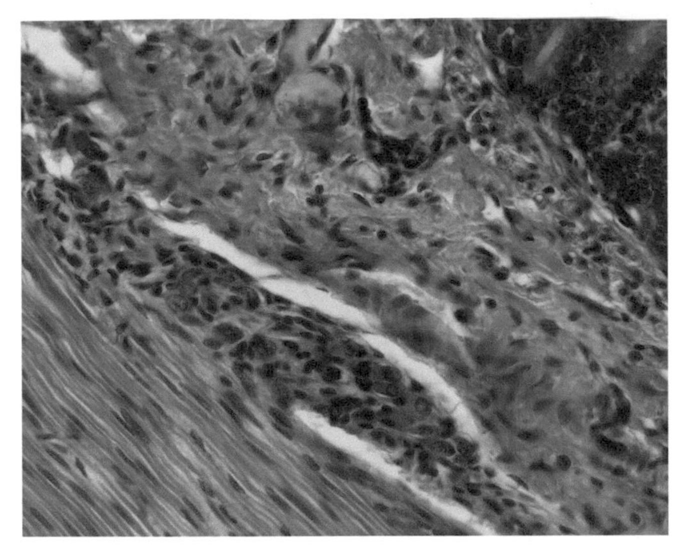

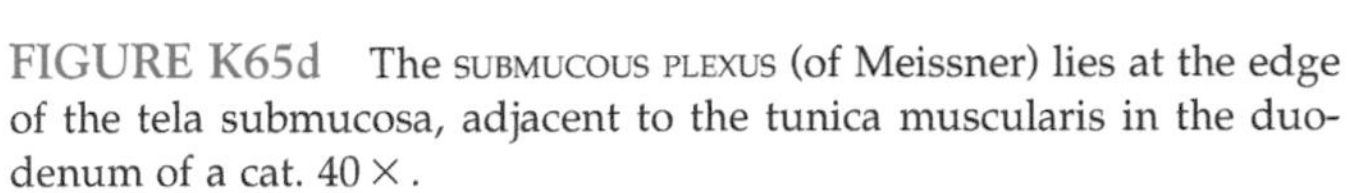

FIGURE K65d The SUBMUCOUS PLEXUS (of Meissner) lies at the edge of the tela submucosa, adjacent to the tunica muscularis in the duodenum of a cat. 40×.

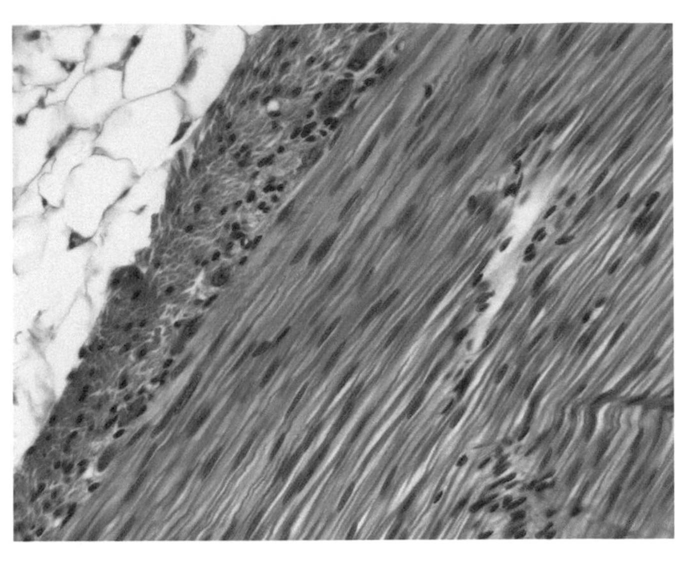

FIGURE K65e The MYENTERIC PLEXUS (of Auerbach) lies between the two layers of muscle of the tunica muscularis. 40×.

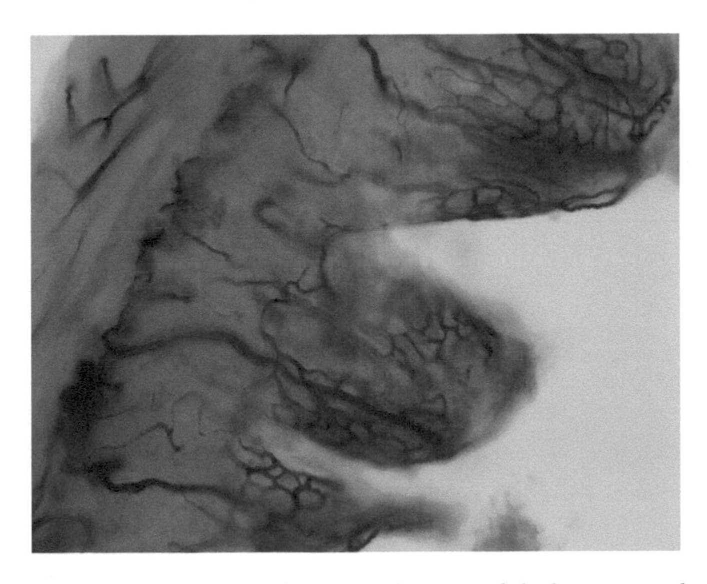

FIGURE K66 Red and green gelatin has been injected into the intestinal artery and the hepatic portal vein before fixation of this intestine. The green material has penetrated most of the vessels while the red appears only in the larger vessel seen faintly at the left. Note the remarkable network of capillaries in the villi. Larger vessels, probably venules, drain these capillary beds. Vascularization of the tela submucosa and tunica muscularis is less extensive. 20×.

Intestinal folds are seen in the small intestine of the bowfin and the sunfish (bony fish) (Figs. K71a, K71b, K72a, K72b). To the noncritical eye, these structures could be interpreted as villi. Many teleosts show a number of diverticula extending from the intestine near the pylorus: the PYLORIC CAECA (Figs. K73a and K73b). They are generally similar in histological structure to the corresponding portion of the intestine and seem to be concerned with increasing the surface area for the absorptive processes.

Cell replacement in the small intestine of birds and mammals comes from mitotic activity in the crypts. In reptiles, cell division occurs more generally over the epithelial surface. There seems to be little or no mitotic activity in the surface epithelium of amphibians; instead solid masses of the epithelium pass into the underlying lamina

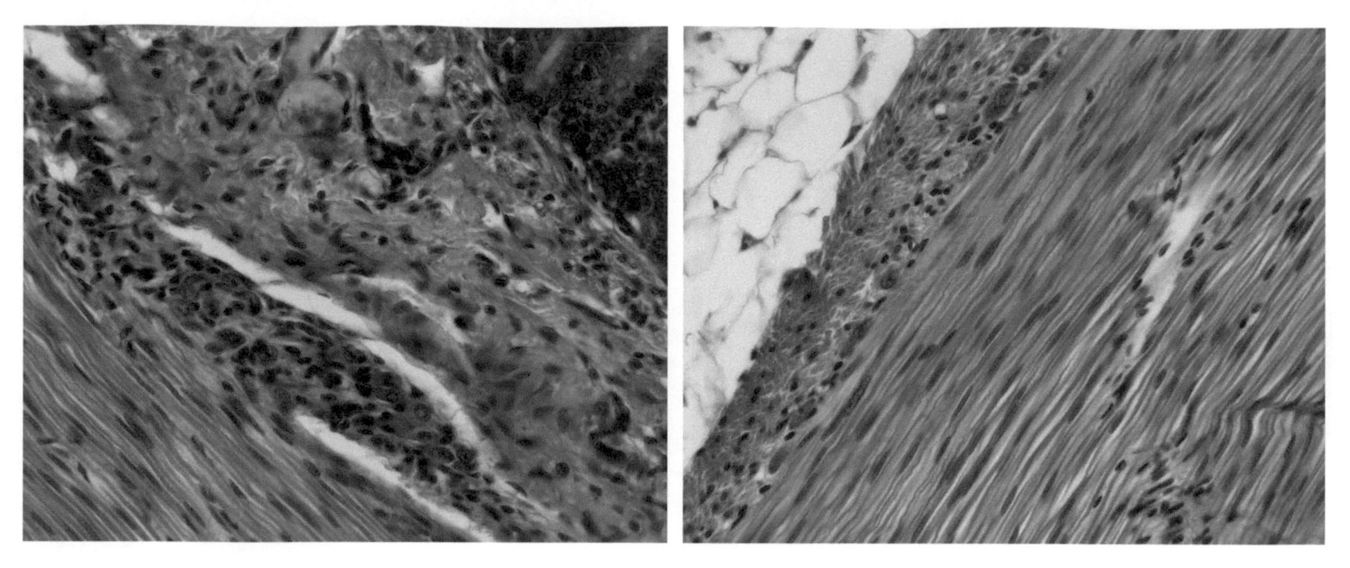

FIGURES K67a and K67b The SUBMUCOUS PLEXUS (of Meissner) lies at the edge of the tela submucosa, adjacent to the tunica muscularis in the duodenum of a cat (01), and the MYENTERIC PLEXUS (of Auerbach) between the two layers of muscle of the tunica muscularis (02).

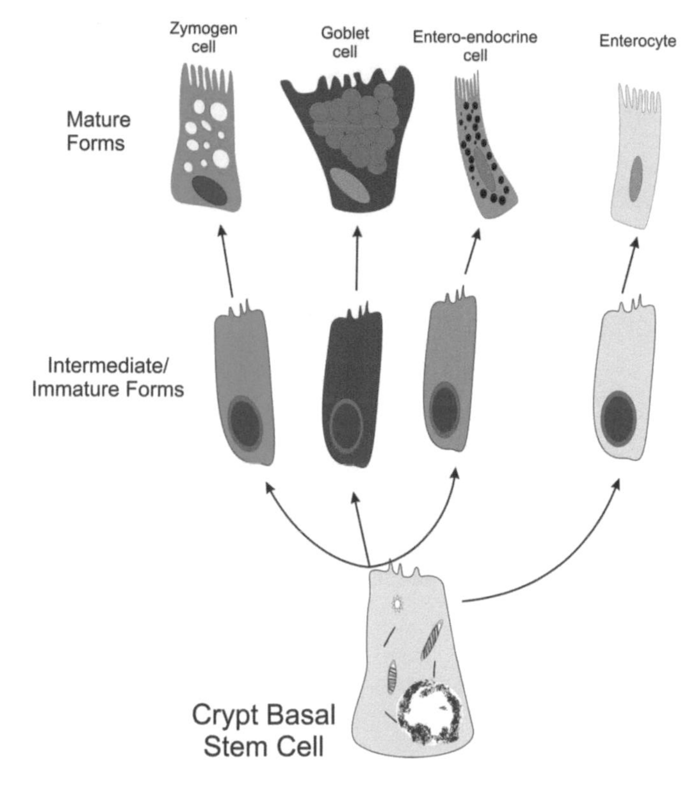

FIGURE K67c The new cells that replace cells that have worn away from the villi surface are derived from stem cells at the base of the crypts. Basal crypt stem cells give rise to four lines of epithelial cells lining the crypts of mammals, which assume their mature form as they move up the villi. *Authors.*

propria mucosae. Mitotic division occurs in these CELL NESTS and from them new cells are furnished to the epithelium lining the gut (Figs. K56b–K56d).

All vertebrates have muscle layers that churn the intestinal contents and move them along. In addition, some of the lower forms have cilia assisting that movement.

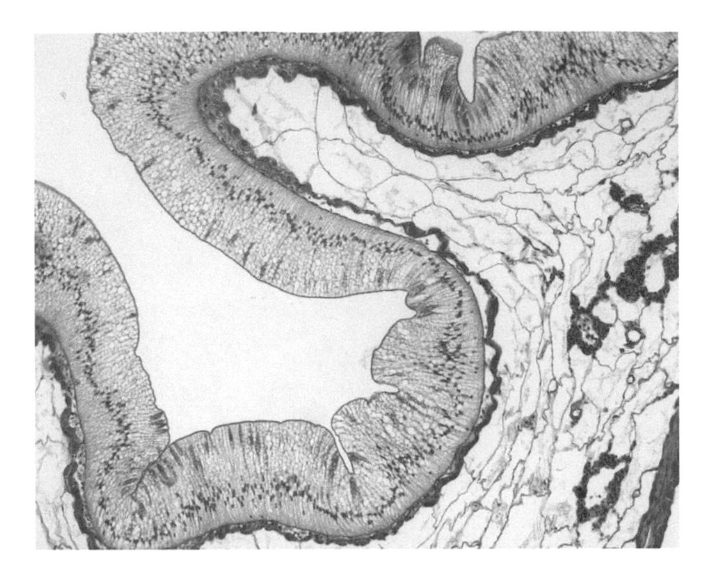

FIGURE K68a The TUNICA MUCOSA of the intestine of the hagfish (*Myxine glutinosa*) consists of a simple epithelium of extremely tall cells that are packed with inclusions, probably lipid. Loose, delicate, vascular connective tissue of the LAMINA PROPRIA MUCOSAE, separates these cells from the thin layer of smooth muscle of the LAMINA MUSCULARIS MUCOSAE. The tunica mucosa is surrounded by a layer of cells that appear empty but are probably filled with lipid. These cells constitute the TELA SUBMUCOSA. Note the blood vessels that permeate the tela submucosa. They are surrounded by masses of hemopoietic tissue. The large folds in the intestinal wall increase the absorptive area and flatten to accommodate large masses of food that pass through. A small portion of the tunica muscularis can be seen at the lower right corner. Although extremely thin, this layer displays the typical arrangement of an inner layer of circular and outer layer of longitudinal smooth muscle. It may even be possible to see a MYENTERIC PLEXUS. 10 ×.

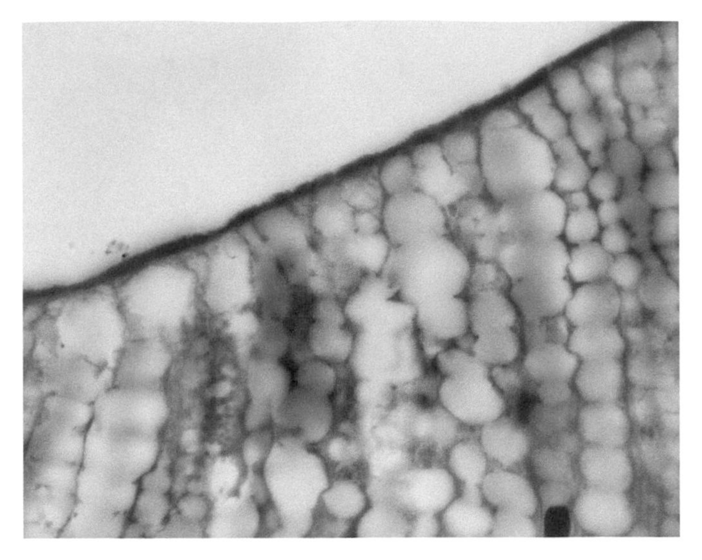

FIGURE K68b The epithelial cells of the intestine of the hagfish are packed with droplets of oily material and appear to be provided with a border of microvilli increasing their absorptive area. 100 ×.

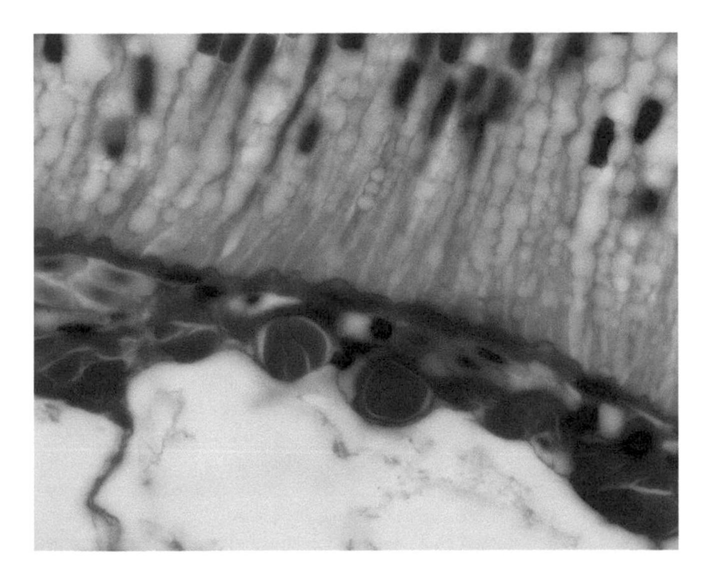

FIGURE K68c The epithelial cells rest on a thin layer of vascular connective tissue of the lamina propria mucosae which, in turn, rests on the lamina muscularis mucosae consisting of an inner circular and outer longitudinal layer of smooth muscle. There are several blood vessels in this layer. Below is the loose connective tissue of the tela submucosa. 100 ×.

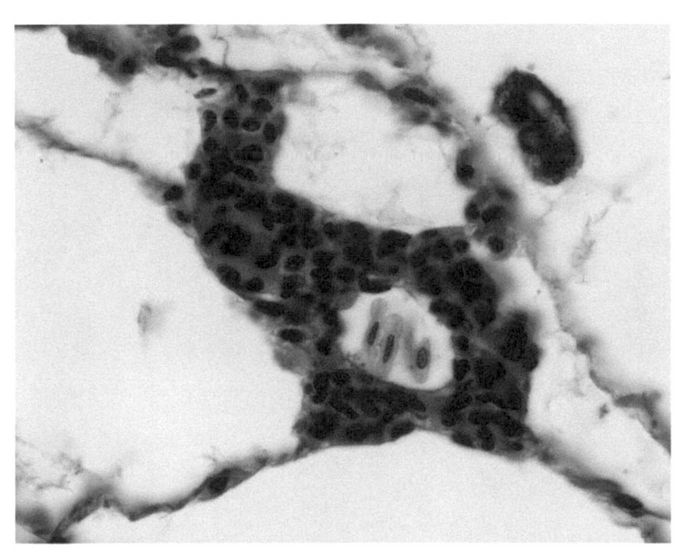

FIGURE K68d Huge vacuolated cells presumably filled with an oily material comprise most of the tela submucosa. Here and there are scattered hemopoietic islands penetrated with sinusoids lined with squamous endothelial cells and containing a few erythrocytes. 63 ×.

ILEOCECAL VALVE

The opening from the ileum into the cecum is guarded by folds of the intestinal wall that consist of the tunica mucosa, tela submucosa, and a central plate of smooth muscle (Fig. K74a).

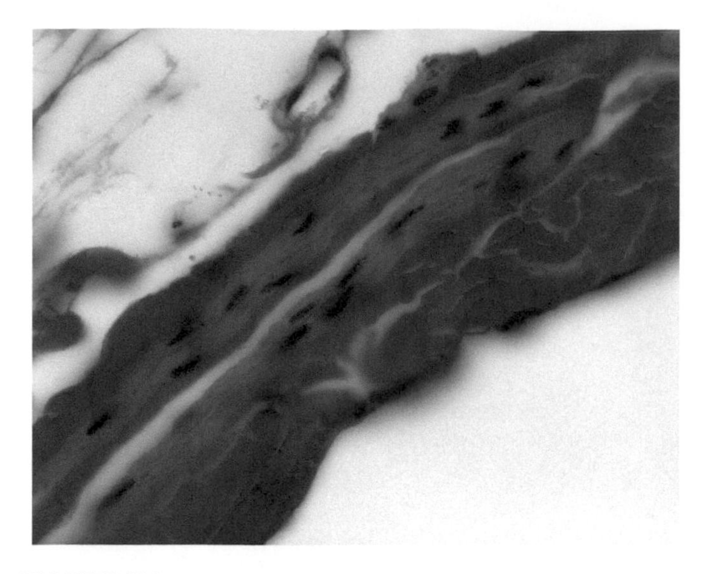

FIGURE K68e The large vacuolated cells of the tela submucosa are in the upper left corner. The tunica muscularis is not much thicker than the lamina muscularis mucosae. It consists of an inner circular and outer longitudinal layer of smooth muscle. A nerve ganglion appears to be nestled between the two layers at the left. 100 ×.

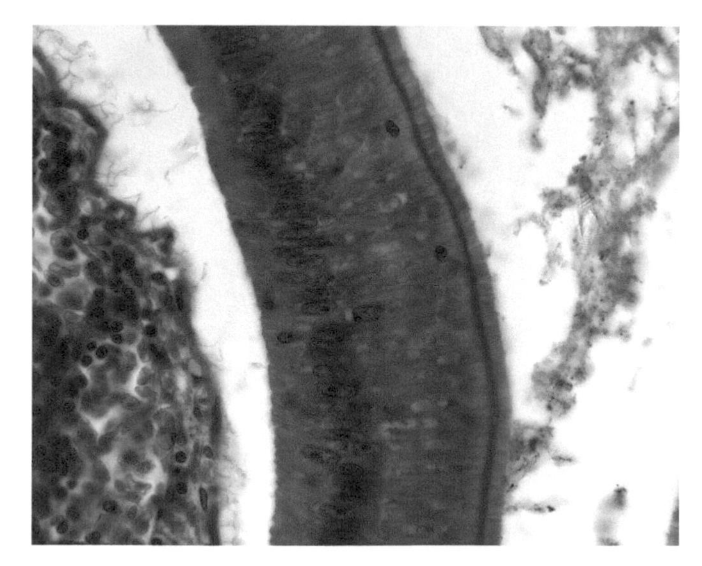

FIGURE K69b The EPITHELIUM MUCOSAE is made up of columnar cells and appears to be pseudostratified. Some of the cells are absorptive, others are secretory. The occasional nucleus, seen near the apical surface of the epithelium, may be cells that undergo mitosis and help to regenerate the epithelium. The surface of the columnar epithelial cells is greatly increased by prominent microvilli. The dark band just below the apical surface represents JUNCTIONAL COMPLEXES. An artefactual space separates the epithelium from the vascular connective tissue of the tela submucosa at the left. Careful examination reveals a thin band of smooth muscle cells of the lamina muscularis mucosae between the tela submucosa and the epithelial cells. The loose strands in this space are all that remains of the LAMINA PROPRIA MUCOSAE. 63 ×.

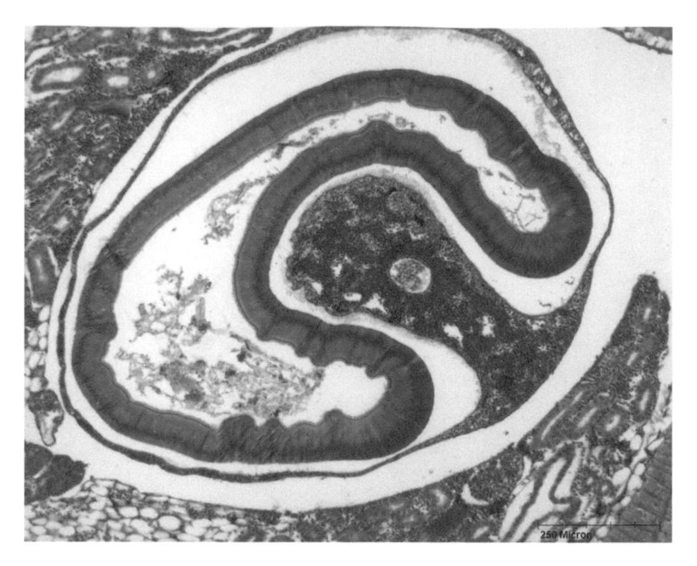

FIGURE K69a The intestine of the ammoceoete or larval lamprey (*Petromyzon marinus*) is characterized by a deep fold on one side, the SPIRAL VALVE or TYPHLOSOLE, which increases the surface area for the absorption of nutrients and accommodates a large mass of hemopoietic tissue in the tela submucosa. The intestine appears between masses of kidney tubules at the upper left and upper right. Unfortunately in this preparation there has been considerable shrinkage, producing large spaces between the various layers. The TELA SUBMUCOSA consists of vascular connective tissue containing the mass of hemopoietic tissue on one side. This distorts the EPITHELIUM MUCOSAE, producing a fold that increases the surface area for absorption. Because it vaguely resembles a structure found in the earthworm, this fold on one side of the intestine is sometimes called the typhlosole. Since this is not a term used elsewhere in histology, its use should be discouraged. 10 ×.

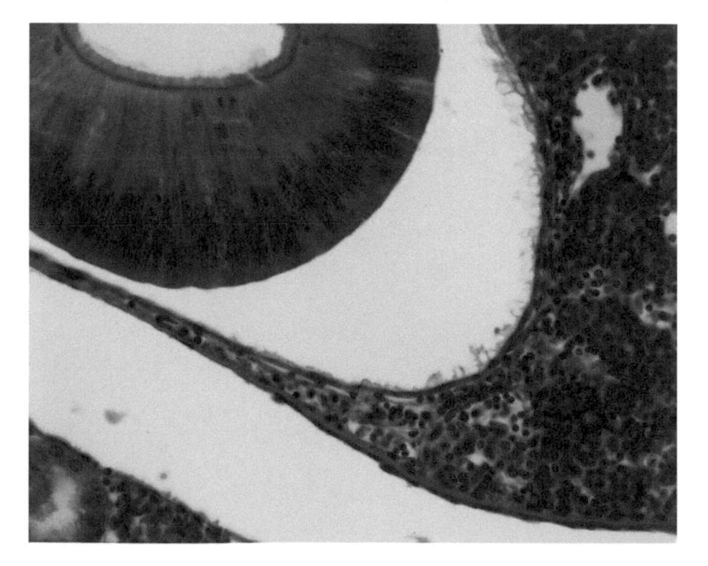

FIGURE K69c The epithelium at the top left has pulled away from the extremely thin layers below. These layers can be discerned at the point where they widen to accommodate the hemopoietic mass within the spiral valve. The loose "hairs" at the inside of the hemopoietic mass are connective tissue fibers of the lamina propria mucosae. Surrounding this are the usual layers of the gut in drastically reduced form: a layer of smooth muscle of the lamina muscularis mucosae; vascular connective tissue of the tela submucosa that widens to form the hemopoietic mass; a thin tunica muscularis of inner circular and outer longitudinal fibers of smooth muscle. Squamous cells of the tunica serosa surround the gut. Kidney tissue occupies the lower left corner of the micrograph. 40 ×.

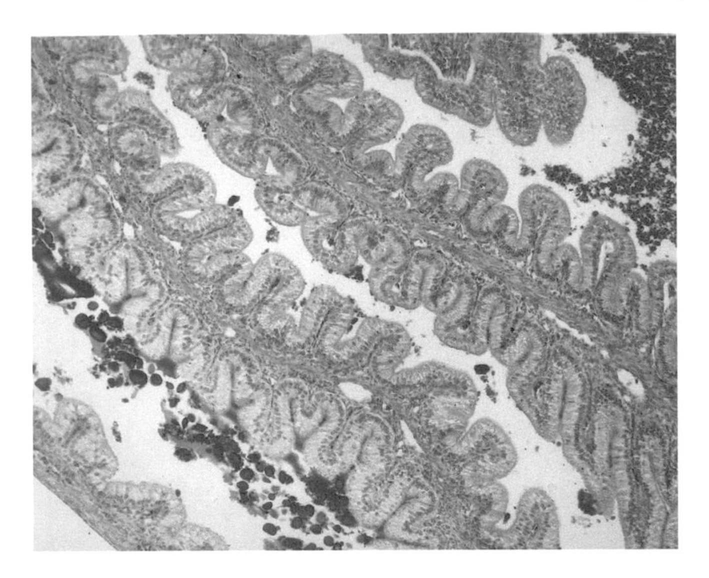

FIGURE K70a Partitions (folds) grow from the walls of the intestine of elasmobranchs to form the SPIRAL VALVE in this section of an immature shark. These folds slow down the passage of food, thereby providing greater time for digestion and absorption and also increase the surface area. The shark in this preparation was still feeding on yolk left in the gut from its early days; this yolk stains bright red and is seen throughout the lumen. The spiral valve consists of highly developed folds of the intestinal wall containing a core of vascular connective tissue of the TELA SUBMUCOSA. Growing on these folds are lesser folds, which also have a core of connective tissue from the tela submucosa. 10 ×.

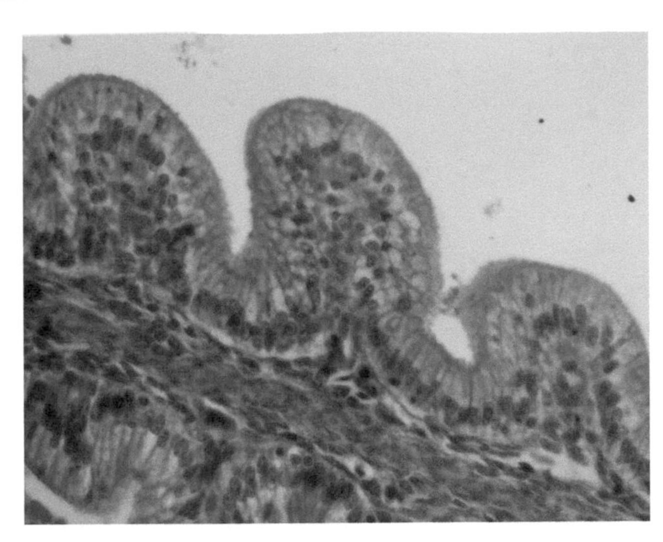

FIGURE K70b Three of these lesser folds rest on the vascular connective tissue of the tela submucosa in the immature shark. It appears that terminal bars seal the luminal borders of the epithelial cells; a tangential section of the epithelium of the middle fold reveals the "chicken-wire" appearance of these junctions. A portion of the other free surface of this fold is seen at the bottom left. Unfortunately this tissue was not well fixed. 40 ×.

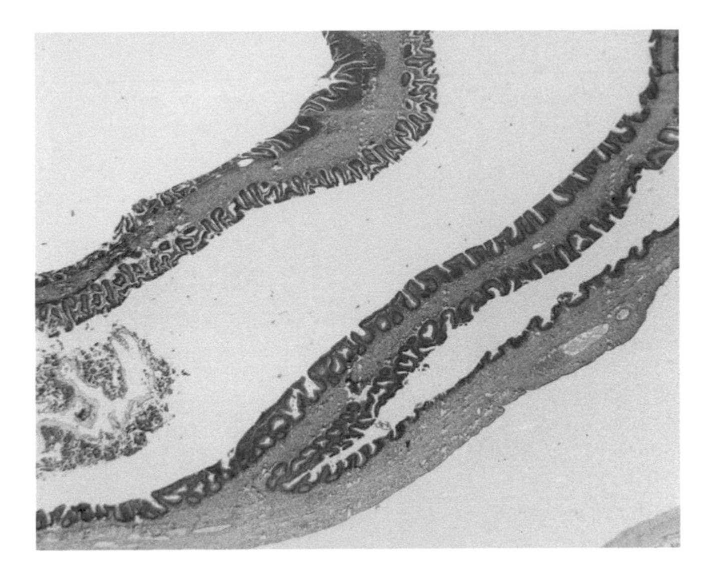

FIGURE K70c This low-power picture indicates the attachment of the intestinal fold that constitutes the spiral valve. It contains a core of vascular connective tissue from the tela submucosa and is covered with the tunica mucosa. The flap winds around for several turns inside the lumen. 1 ×.

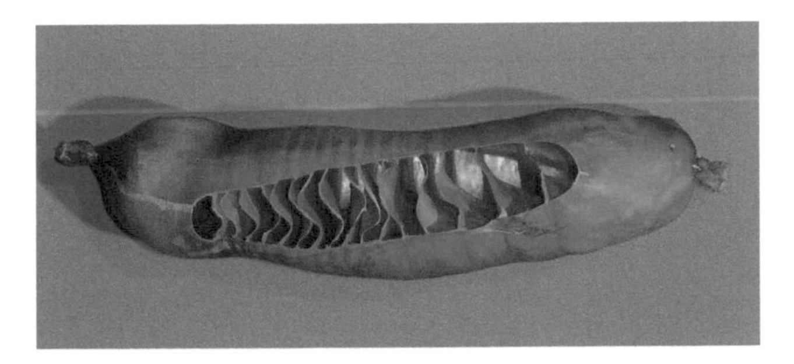

FIGURE K70d The spiral valve in the small intestine of a nurse shark (*Ginglymostoma cirratum*). A part of the wall has been removed to show the inner partitions. *Haplochromis—Own work, CC BY-SA 3.0, https://commons.wikimedia.orgcurid=5825837 2009.*

LARGE INTESTINE

The large intestine of mammals may include the colon, cecum, appendix, and rectum. The cecum and appendix reach their highest development in herbivores but are poorly developed or even absent from carnivores. The large intestine resembles the small intestine with the villi shaved off; although villi are present late in fetal life these disappear. No further digestion takes place in the large intestine and the crypts of this region secrete mucus instead of digestive juices.

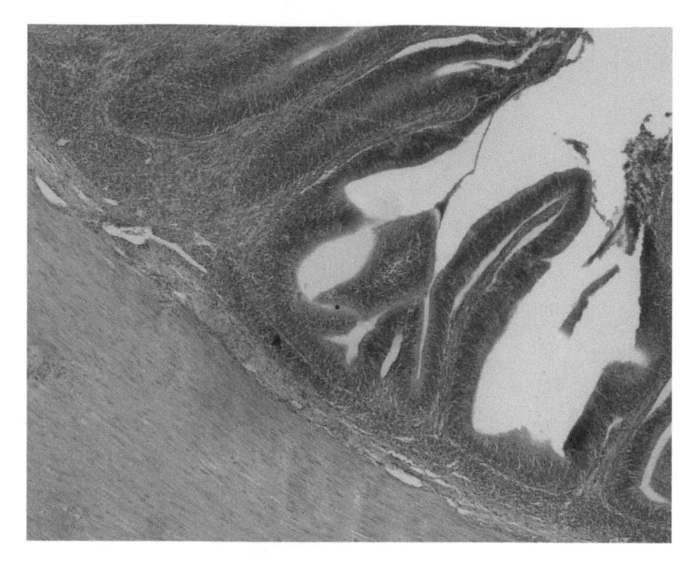

FIGURE K71a The lining of the small intestine of the bowfin (*Amia calva*) is thrown up into complex folds that have a core of connective tissue of the tela submucosa. The folds not only flatten when the gut is full but also increase the surface area for absorption. These folds are covered with an epithelium of absorptive cells of the tunica mucosa. There is a large vein at the left in the tela submucosa. The large spaces in the tela submucosa are lymphatics—note their endothelial lining. No endothelium surrounds the spaces in the folds of tissue between the epithelium and the underlying connective tissue and it may be concluded that these are artifacts of shrinkage. A thick layer of circular smooth muscle is at the lower left. 10 ×.

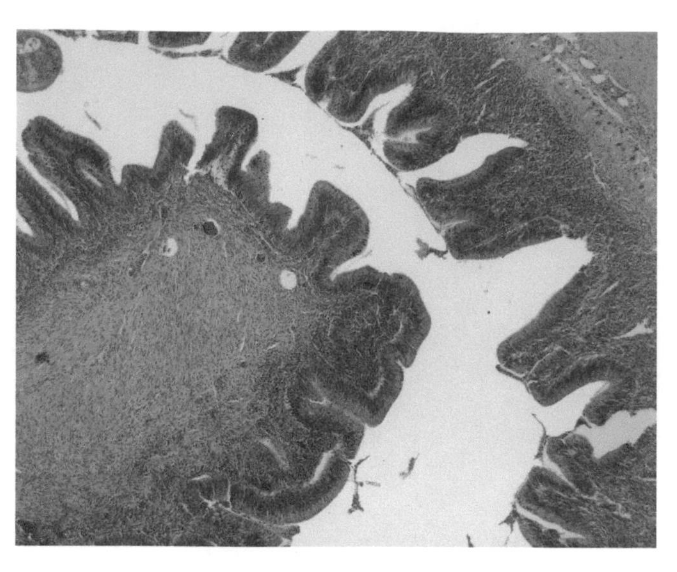

FIGURE K71b In another region of the small intestine, a flap of tissue at the left contains a core of connective tissue of the tela submucosa and constitutes a simple "spiral valve." Upon this are several smaller folds that also have a core of connective tissue of the tela submucosa, and are FOLDS, not villi. 10 ×.

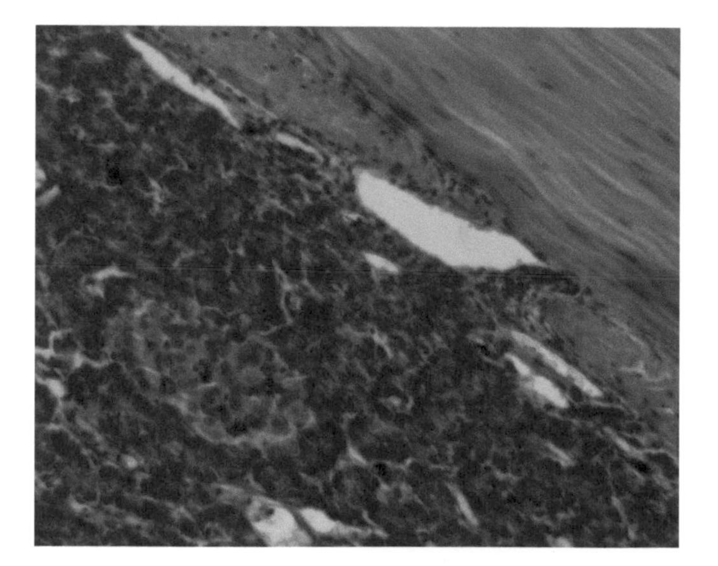

FIGURE K71c Section of the small intestine of the bowfin showing the double function of the tela submucosa (hemopoietic and connective tissue) as well as the tunica muscularis. Blood vessels and lymphatics pass through the connective tissue of the tela submucosa. 40 ×.

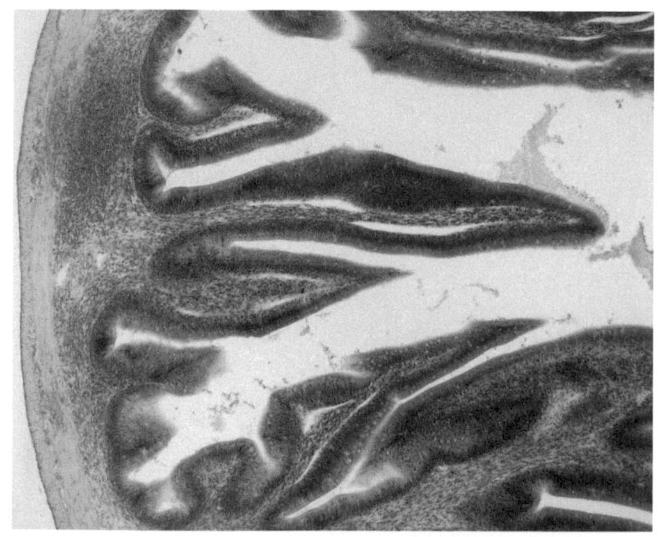

FIGURE K72a The lining of the small intestine of this bony fish (sunfish—*Lepomis*), is thrown up into complex folds that not only flatten to accommodate masses of food passing through but also increase the surface area for absorption. These are not villi: they have a core of connective tissue derived from the tela submucosa and are therefore folds. 10 ×.

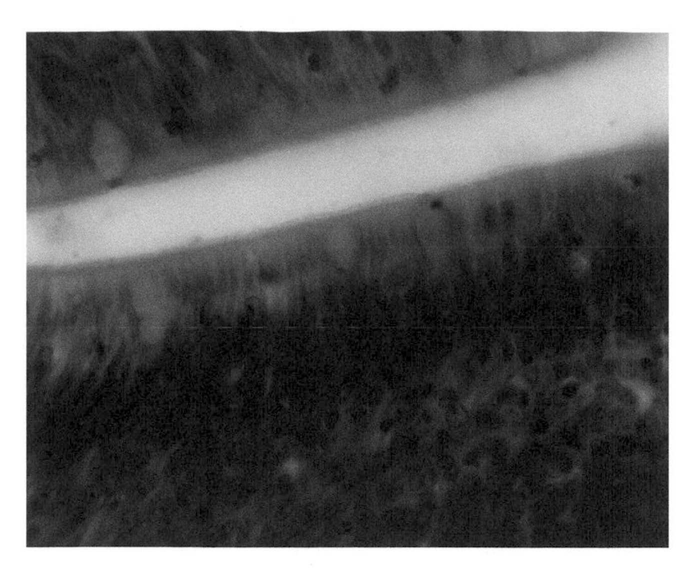

FIGURE K72b The epithelium of the small intestine of the sunfish is simple columnar. The epithelial cells are tall and narrow; they are cut obliquely in this section, giving a confused appearance. Goblet cells are plentiful and a striated border is evident on most of the epithelial cells. 100 ×.

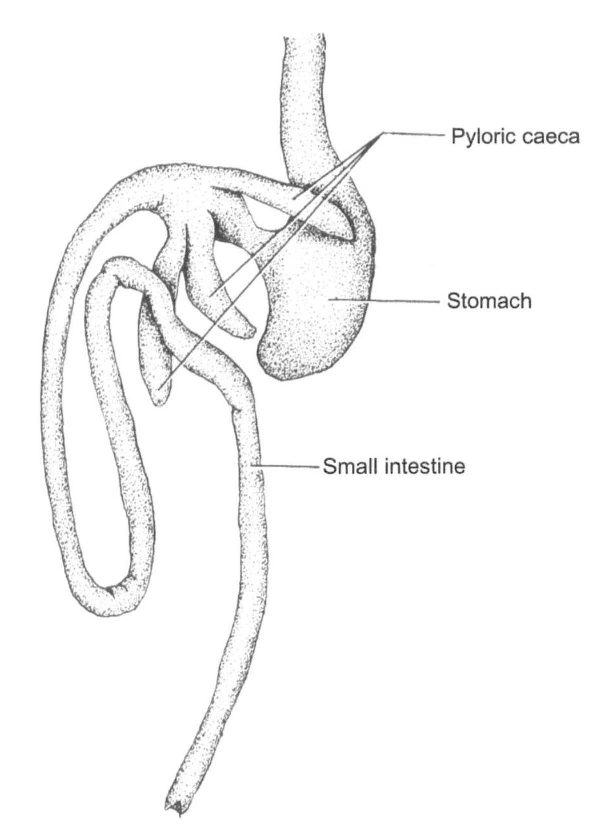

FIGURE K73a Many fishes have PYLORIC CAECA coming off that part of the intestine immediately following the pylorus. There may be several of these blind sacs, which provide further area for absorption of nutrients.

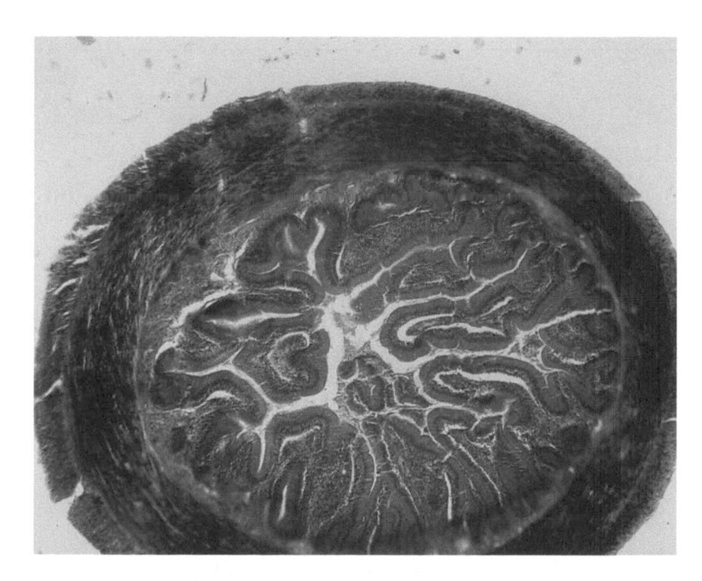

FIGURE K73b This section of a pyloric cecum of a sunfish indicates that all of the layers of the intestine are present. The tunica mucosa is greatly folded, thereby increasing the area for absorption even more. 10 ×.

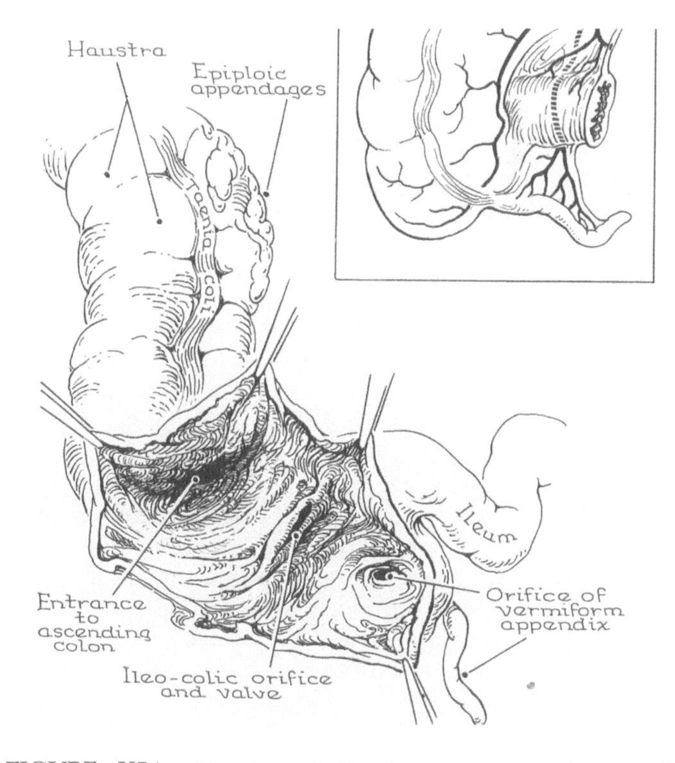

FIGURE K74a Drawing of the human cecum showing its entrance to the ascending colon. *Source: From Taylor, N.B., McPhedran, M.G., 1965. Basic Physiology and Anatomy. Putnam and Sons, NewYork, and MacMillan Co Ltd., Toronto, Hodder & Stoughton.*

COLON

The most striking feature of the mammalian colon is the abundance of large, simple tubular mucus-secreting glands packed into the tunica mucosa (Figs. K74b, K74d, K75a, K75b, K76c, K76d). There are no plicae circulares in the large intestine but there are three bands of longitudinal smooth muscle, the TAENIAE COLI, which have the effect of a drawstring run through

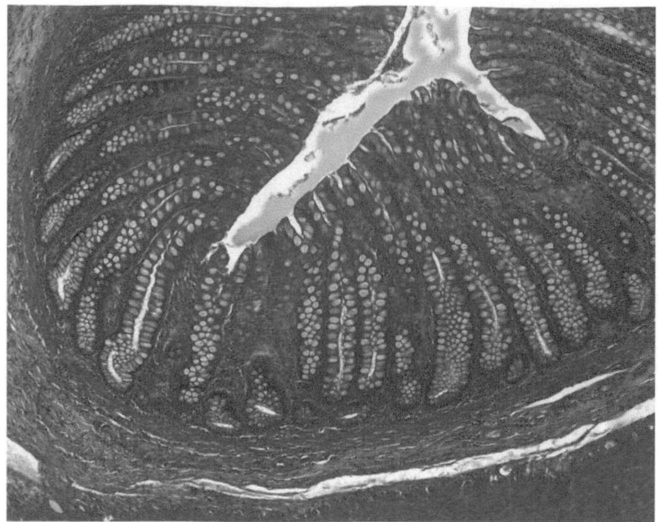

FIGURE K74b The colon is a portion of the gut that passes along the waste material from digestion and removes water as it goes. It secretes mucus and is well provided with smooth muscle assist this passage. The simple tubular glands of the tunica mucosa in this section of the colon of a dog are amply provided with goblet cells that produce mucus. The apical regions of the glands as well as the surface cells of the lining absorb water from the waste while the deeper portions of the glands produce mucus. Although there are no villi, as seen in the small intestine, these tubular glands are still referred to as CRYPTS—and, as in the small intestine, cell renewal is provided by mitosis at their depths. The tela submucosa brings the major blood vessels to the colon. Note the large lymphatic at the lower right. Below the lymph vessel is an artifact, a split that separates the tela submucosa from the tunica muscularis. 10 ×.

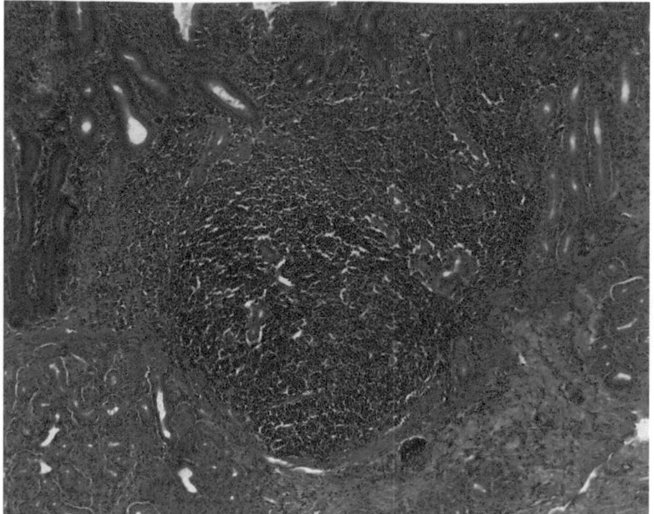

FIGURE K74c Lymphoid nodules appear in the tela submucosa of the colon of a dog. 10 ×.

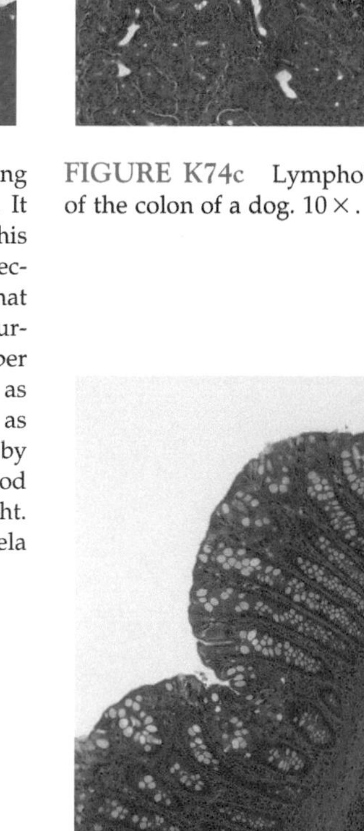

FIGURE K75a A large mass of lymphoid tissue is contained within the tela submucosa in the colon of a monkey. 10 ×.

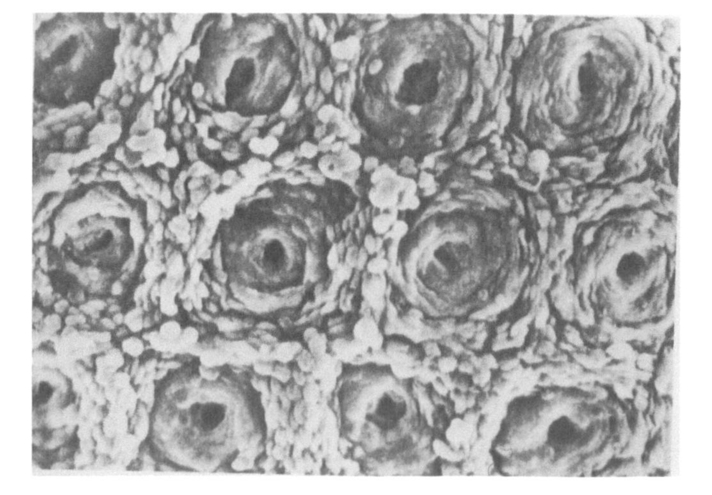

FIGURE K74d Scanning electron micrograph of the inner surface of the descending colon of a monkey (*Macaca mulatta*) showing the regular array of the opening of the crypts.

a piece of cloth, causing sacculations, the HAUSTRA, to form between the crescentic SEMICIRCULAR FOLDS (Figs. K74a, K76a, K76b). These folds extend into the lumen over about one-third of the circumference and involve the entire wall, being visible from the outside of the colon.

The surface epithelium of the TUNICA MUCOSA is simple columnar and consists of mature absorptive cells with interspersed goblet cells. The absorptive cells have a striated border. Four cell types occur within the

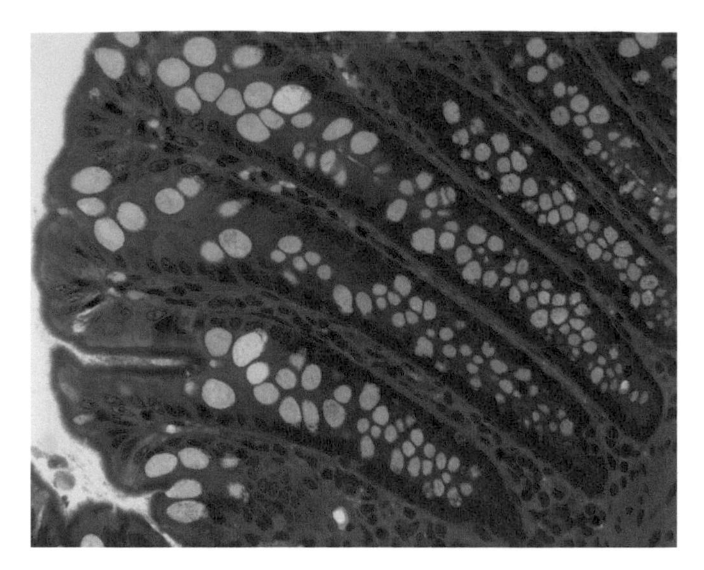

FIGURE K75b The epithelium mucosae is simple columnar and consists of absorptive cells interspersed with goblet cells in the colon of a monkey. A striated border can be seen on the surface cells. Careful inspection reveals a few enteroendocrine cells in the epithelium; these are packed with acidophilic granules. 40 ×.

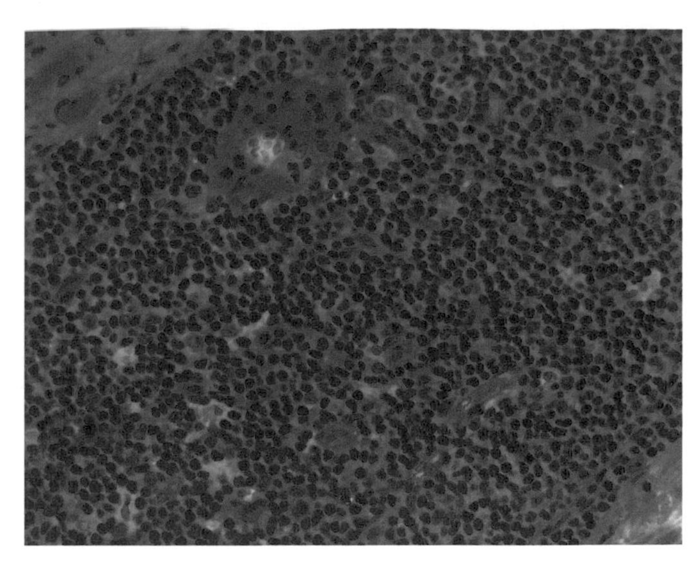

FIGURE K75c Capillaries and larger blood vessels penetrate the mass of lymphoid tissue in the colon of a monkey. 40 ×.

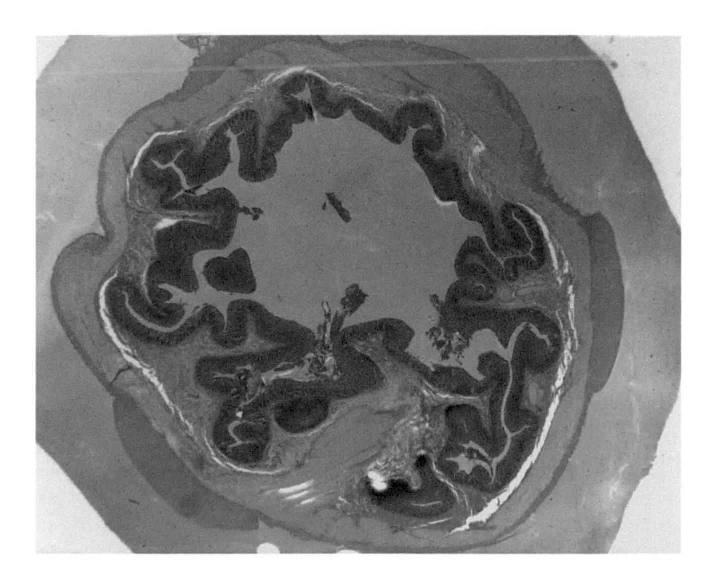

FIGURE K76a This portion of mammalian colon has been embedded in celloidin, which remains in the section to produce the distracting pink border. The stain is mucicarmine, which has a strong affinity for the mucin of goblet cells. Most of the features cited previously can be identified. The longitudinal smooth muscle forms a fairly thin layer outside the circular muscle but at three points it thickens to form three bands, the TAENIAE COLI, which tighten and have the effect of a drapery cord forming pleats or chambers, the HAUSTRA, in the wall of the colon, These pleats have the effect of slowing the passage of materials through the gut and providing a great surface area for absorption. 1 ×.

FIGURE K76b A MYENTERIC PLEXUS is lodged between the longitudinal muscle of a TAENIA COLI and the circular muscle of the tunica muscularis. 10 ×.

simple tubular crypts: undifferentiated cells, immature absorptive cells, goblet cells, and enteroendocrine (APUD) cells. The entire epithelium is a continuous sheet that is constantly renewed; cell replacement is provided by division of cells in the lower half of the crypts. The other layers of the tunica mucosa are not unusual except that lymphoid nodules are numerous in the lamina propria mucosae and are often so large that they break through the lamina muscularis mucosae into the tela submucosa (Figs. K74c, K75a, K75c).

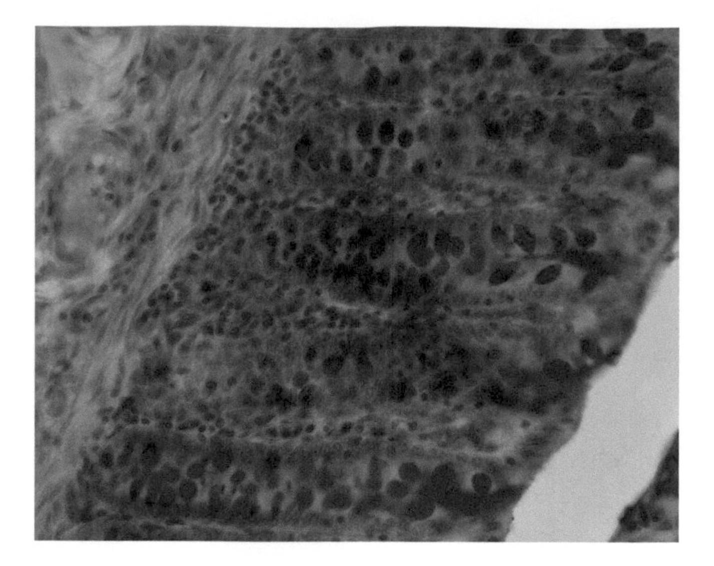

FIGURE K76c Connective tissue of the lamina propria mucosae extends between the tubular glands of the epithelium mucosae. A band of smooth muscle of the lamina muscularis mucosae delimits the tunica mucosa from the tela submucosa below. 40 ×.

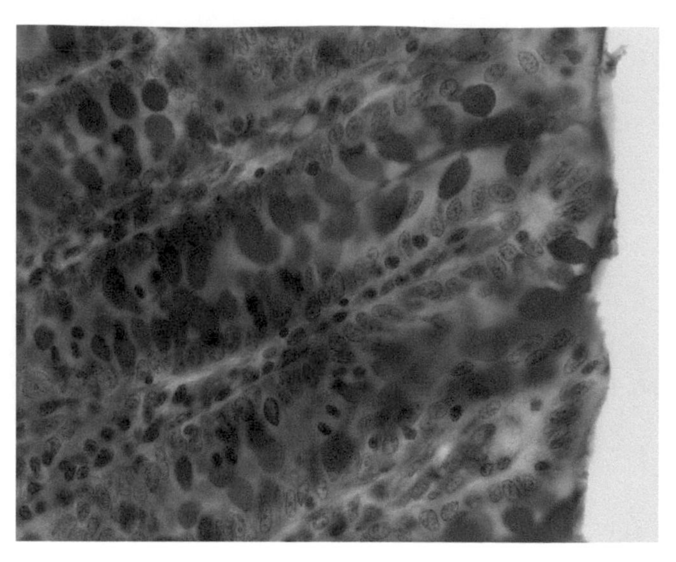

FIGURE K76d Goblet cells abound in the simple tubular glands of the colon; their content of MUCIN is stained bright red. Several mitotic figures in these glands provide replacement for cells lost. Vascular connective tissue of the lamina propria mucosae extends between the tubular glands. 63 ×.

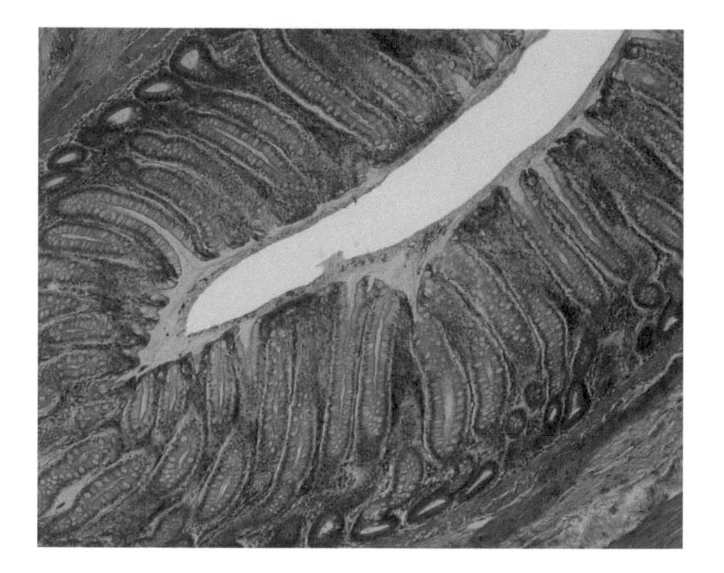

FIGURE K77a The structure of the mammalian rectum is similar to that of the colon with conspicuous simple tubular mucous glands—crypts—dominating the scene. 10 ×.

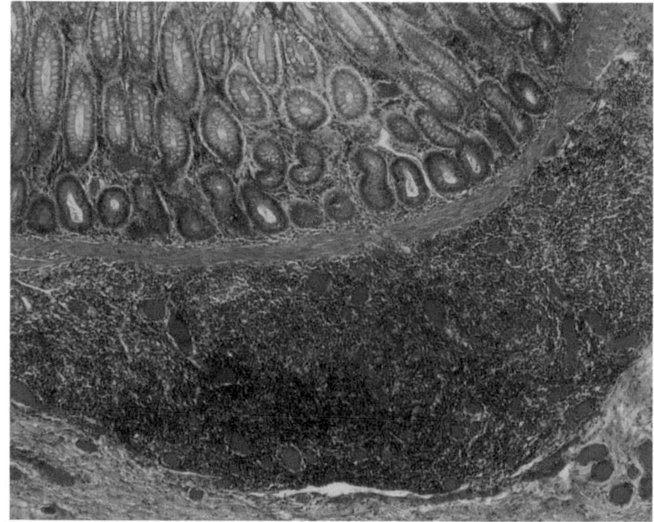

FIGURE K77b A band of smooth muscle of the lamina muscularis mucosae separates the vascular connective tissue of the tela submucosa from that of the lamina propria mucosae. The tela submucosa is packed with lymphoid tissue. Loose lymphoid tissue also invades the lamina propria mucosae. Large blood vessels, stained deep red, penetrate the lamina propria mucosae, the tela submucosa, and the tunica muscularis. 10 ×.

RECTUM

The upper parts of the rectum are similar to the colon although its tubular glands are deeper and there is less lymphoid tissue present (Figs. K77a–K77d). There are no taeniae coli and the tunica muscularis is composed of the typical two layers. The lining forms several large SEMILUNAR FOLDS. Near the anus the glandular epithelium is replaced by stratified squamous epithelium that is continuous with the epidermis. The circular muscle forms two sphincters: an inner ring of smooth muscle and an outer ring of striated muscle.

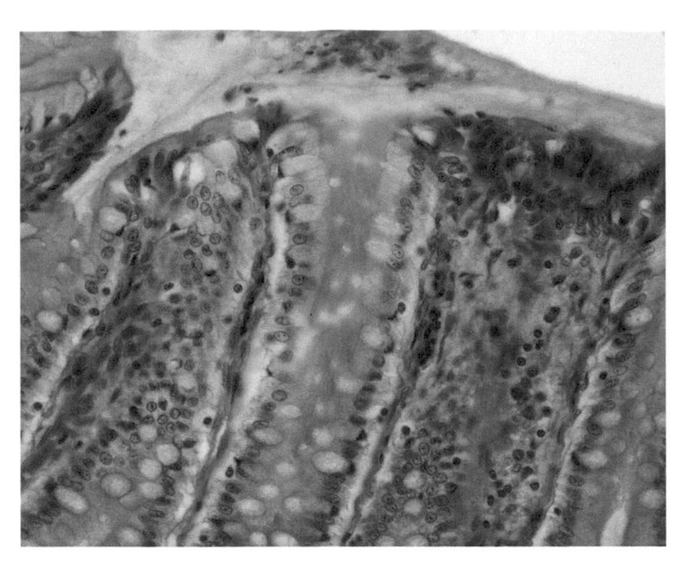

FIGURE K77c Vascular connective tissue of the lamina propria mucosae penetrates the spaces between the simple tubular glands of the rectum. Goblet cells are abundant in these glands. 40 ×.

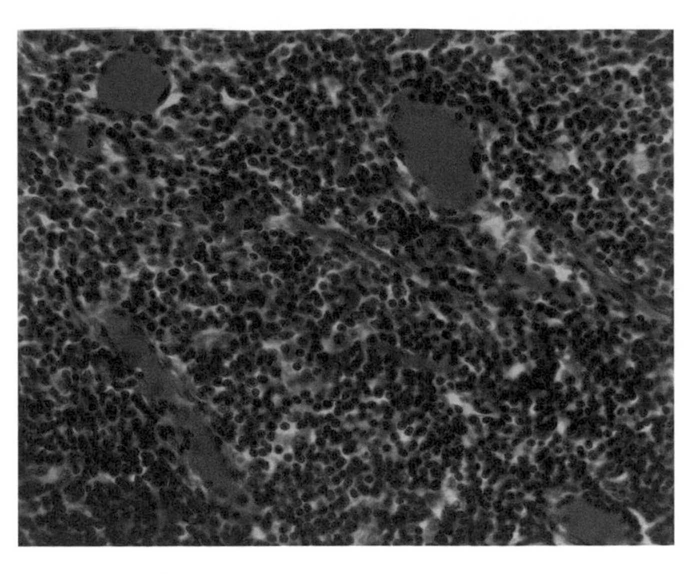

FIGURE K77d Lymphoid tissue is present in the tela submucosa of the rectum. Capillaries and larger blood vessels are abundant. 40 ×.

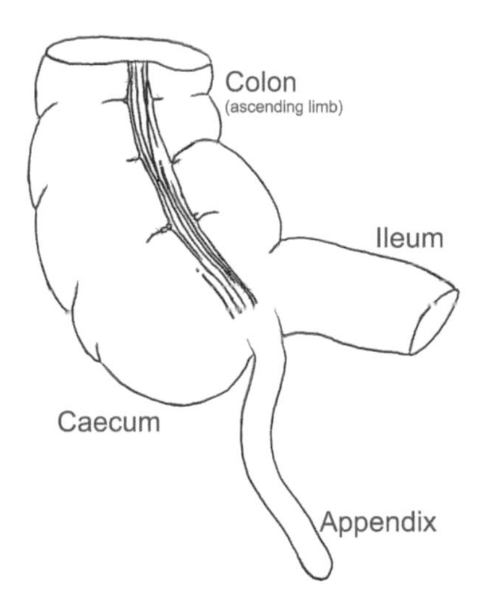

FIGURE K78 Drawing of the junction of the ileum and colon of man showing the cecum and vermiform appendix. *Authors.*

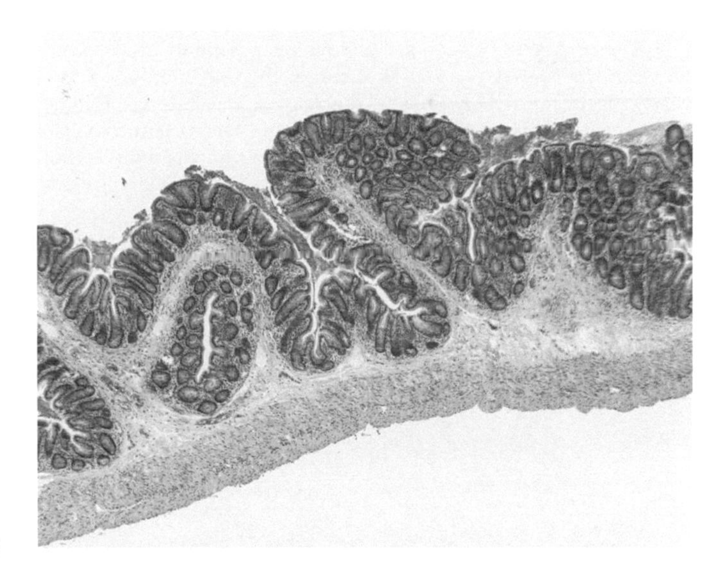

FIGURE K79a This is a low-powered photomicrograph of a section of the cecum of a rat. The cecum is a blind pouch attached to the gut at the proximal end of the colon. Its distal tip is the vermiform appendix. The cecum is well developed in herbivorous animals, such as the rabbit, and almost nonexistent in carnivores, such as the cat. Its structure is similar to that of the colon. It produces mucus that helps its content of waste material to slide along and it absorbs water to solidify the feces. 1 ×.

CECUM

The cecum is a blind pouch at the proximal end of the colon; its tip is the vermiform appendix (Fig. K78). Compare the structural characteristics of the mammalian cecum with other parts of the small and large intestine (Figs. K79a–K79c). In the rabbit (a herbivore), the cecum is by far the largest part of the gut having about 10 times the capacity of the stomach. It is almost absent from the cat (a carnivore).

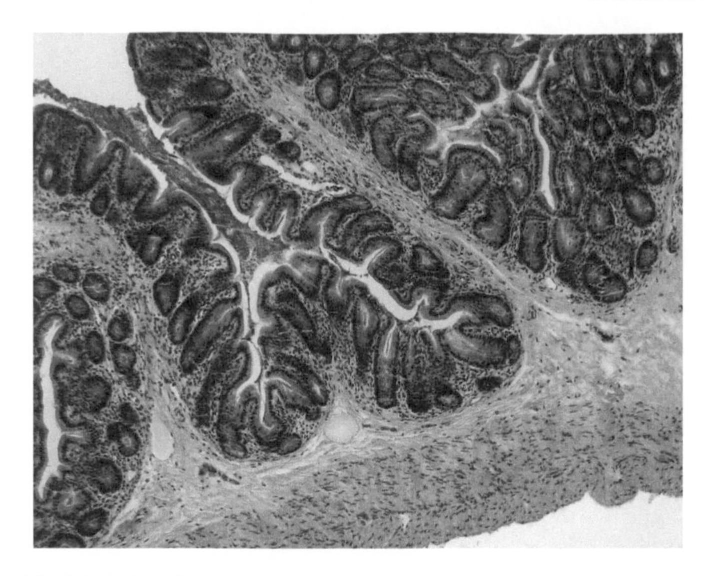

FIGURE K79b In this section of the cecum from a rat the wall exhibits many folds. Some of the waste material left after digestion may be seen between the folds. The epithelium invaginates to form many simple tubular mucous glands that are surrounded by the delicate vascular connective tissue of the lamina propria mucosae. Water absorption occurs at the surface and in the upper portions of the crypts. The connective tissue of the lamina propria mucosae is separated from the coarser connective tissue of the tela submucosa by a few strands of smooth muscle of the lamina muscularis mucosae. Blood vessels and lymphatics are seen in the tela submucosa. The two layers of the tunica muscularis can be seen but, since this section has been cut obliquely, the usual appearance of inner circular and outer longitudinal smooth muscle is not apparent. 10 ×.

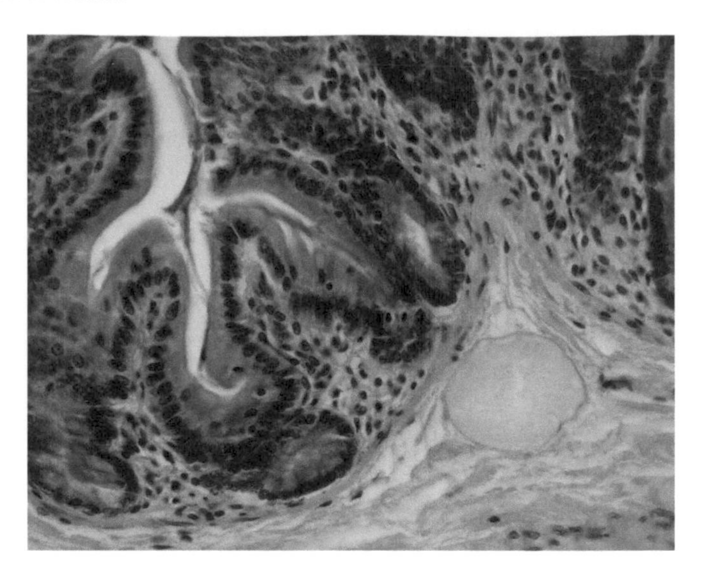

FIGURE K79c The tela submucosa is seen at the right lower corner. It contains a large lymph space. There is a mitotic figure (telophase) in a crypt above and to the left of the lymphatic. Enterochromaffin cells are difficult to see in paraffin sections but one may be seen at the right, halfway down the section. There are a few mitotic figures in the epithelium. 40 ×.

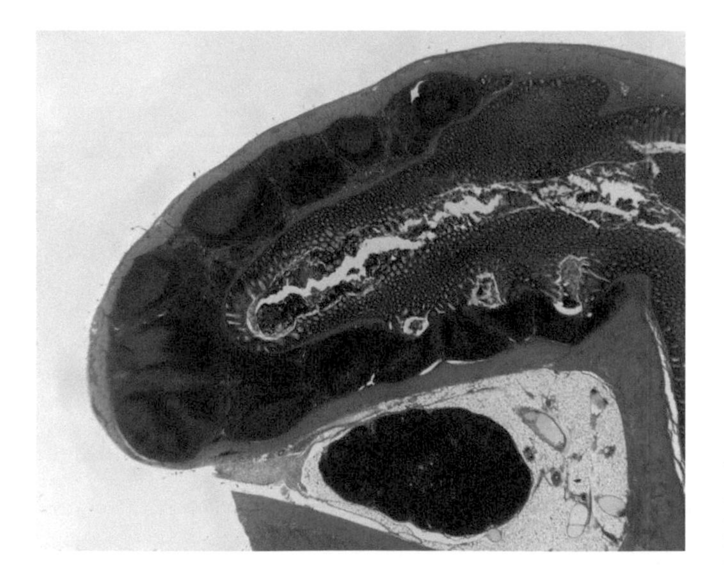

FIGURE K80a The mammalian appendix is a blind sac at the end of the cecum. Like the cecum, it is well developed in herbivorous mammals and reduced (or nonexistent) in carnivores. In omnivores, such as ourselves, its development is intermediate. The wall of the appendix shows all of the layers seen in the intestine, but the tela submucosa is greatly distorted by the presence of lymphoid nodules. Detritus is often seen in the lumen. The tunica mucosa contains abundant simple tubular mucous glands—it is separated from the tela submucosa by a well-developed lamina muscularis mucosae. There is a small lymph node at the bottom of the micrograph. 1 ×.

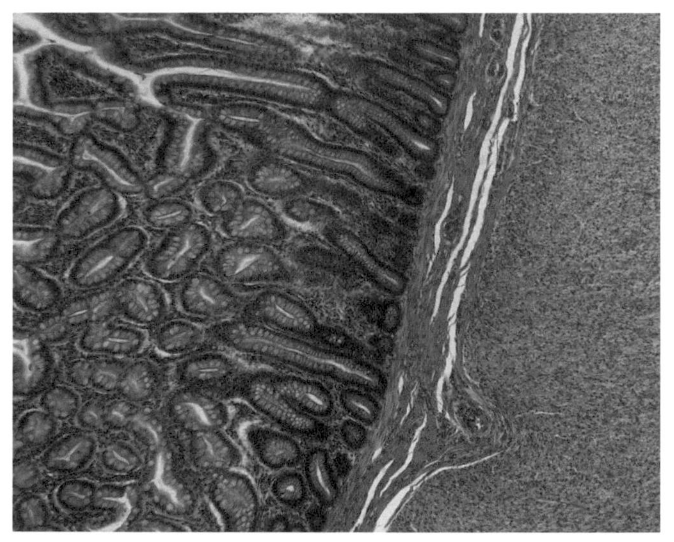

FIGURE K80b In this section of mammalian appendix, simple tubular mucous glands of the tunica mucosa are separated from the lymphoid tissue of the tela submucosa by the smooth muscle of the lamina muscularis mucosae. 1 ×.

VERMIFORM APPENDIX

Review the structure of the appendix in Chapter I, Hemopoietic Organs. It is similar to the colon but this relationship is obscured by the enormous development of lymphoid tissue in the lamina propria mucosae and tela submucosa (Figs. K80a and K80b). The lumen is

FIGURE K81a There is no sharp distinction between the small and large intestines in birds but, passing posteriorly, there is a gradual decrease in the length of the villi and depth of the crypts and an increase in the number of mucous glands. Two richly lymphoid ceca, similar to the mammalian appendix, are usually found at the junction of the parts sometimes called small and large intestine. 2.5 ×.

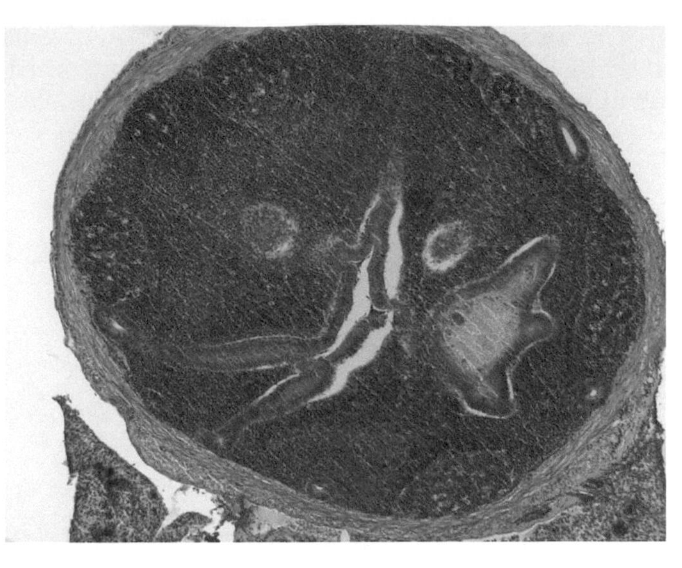

FIGURE K81b The walls of the ceca have the same basic structure as the rest of the intestine, but the connective tissue of the tela submucosa is greatly distended by lymphoid tissue. 10 ×.

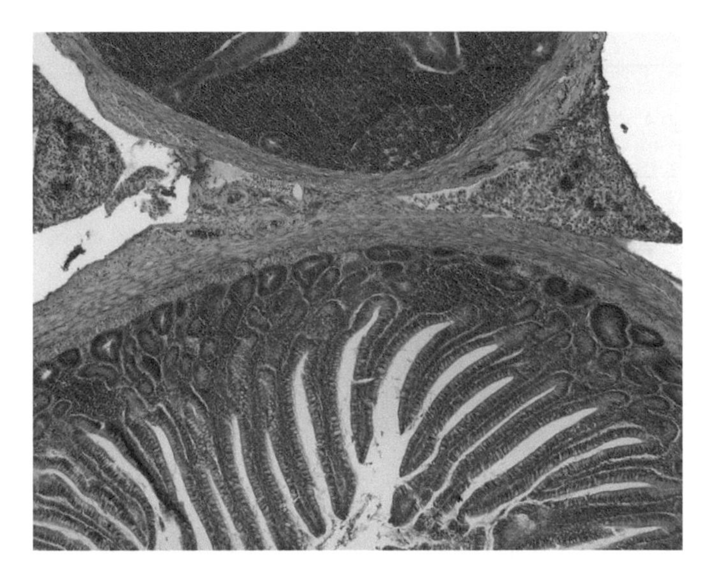

FIGURE K81c Simple tubular crypts are interspersed between the villi of the small intestine of a bird. 10 ×.

reduced and contains debris. The lamina muscularis mucosae is so riddled with lymphoid nodules that pass between the two layers of connective tissue that only a few strands of smooth muscle remain. The two layers of the tunica muscularis are complete but reduced.

COMPARATIVE HISTOLOGY OF THE LARGE INTESTINE

There is no sharp distinction between the small and large intestine in birds, but passing posteriorly there is a gradual decrease in the length of the villi and depth of the crypts, and an increase in the number of mucous glands. Two ceca are usually found at the junction of the parts arbitrarily called small and large intestine (Figs. K81a–K81c). Compare these structures with the mammalian appendix. In the lower classes there is usually no clear demarcation of a large intestine. An increase in the number of goblet cells posteriorly is the only generalization that usually holds.

The rectal gland of elasmobranchs is a diverticulum of the intestine just posterior to the spiral valve (Fig. K82a). It is a branched tubular gland organized around a central lumen and connects to the intestine by a short duct. It is involved in the excretion of salts. We will consider the rectal gland more fully in our study of the excretory system (Chapter M: Excretory Systems).

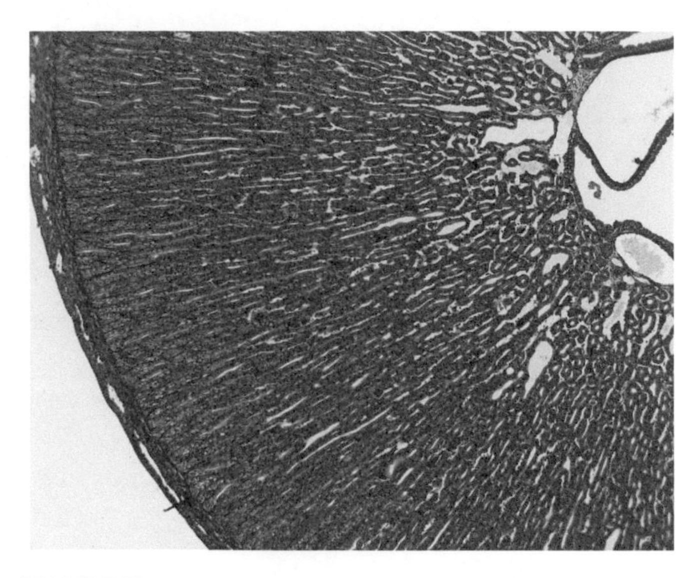

FIGURE K82a The RECTAL GLAND of elasmobranchs is an excretory organ that removes salt from the blood. It is a finer-shaped diverticulum of the intestine just posterior to the spiral valve. It is suspended in the dorsal mesentery and receives an abundant blood supply by way of the posterior mesenteric artery. The rectal gland will be considered in greater detail in Chapter M, Excretory System. 5 ×.

GLANDS ASSOCIATED WITH THE DIGESTIVE TRACT

In addition to glands in the lining of the mouth, in the tongue, and in the wall of the digestive tract, there are large masses of glandular tissue lying outside the limits of the tube that pour their secretions into it through ducts. These are the salivary glands, pancreas, and liver.

Salivary Glands

Mammalian salivary glands as well as the exocrine portion of the pancreas are merocrine glands of the tubuloalveolar type. Their alveoli secrete directly into an INTERCALATED DUCT, which carries the secretion to a larger STRIATED DUCT and ultimately into the oral cavity (Fig. K82b). They consist of varying proportions of serous and mucous alveolar cells. Spindle-shaped MYOEPITHELIAL CELLS surround secretory acini (Fig. K84b)—presumably they assist in expelling the secretion.

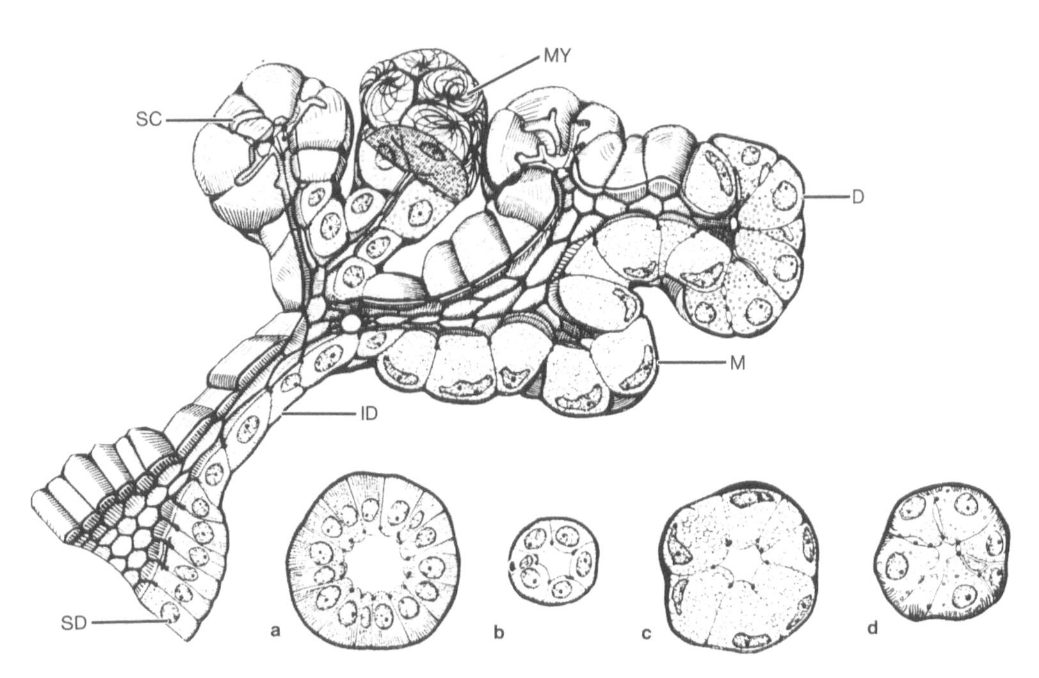

FIGURE K82b Mammalian salivary glands and the exocrine portion of the pancreas are tubuloacinar merocrine glands. They consist of varying proportions of serous and mucous acinar cells. This is a reconstruction of the acinus and intralobular ducts of the mammalian submaxillary gland. *D*, demilune composed of serous cells; *ID*, intercalated duct; *M*, mucous cells; *My*, myoepithelial cells; *SC*, secretory capillaries in a serous acinus; *SD*, striated duct. Cross sections below: A, striated duct; B, intercalated duct; C, mucous acinus; D, serous acinus.

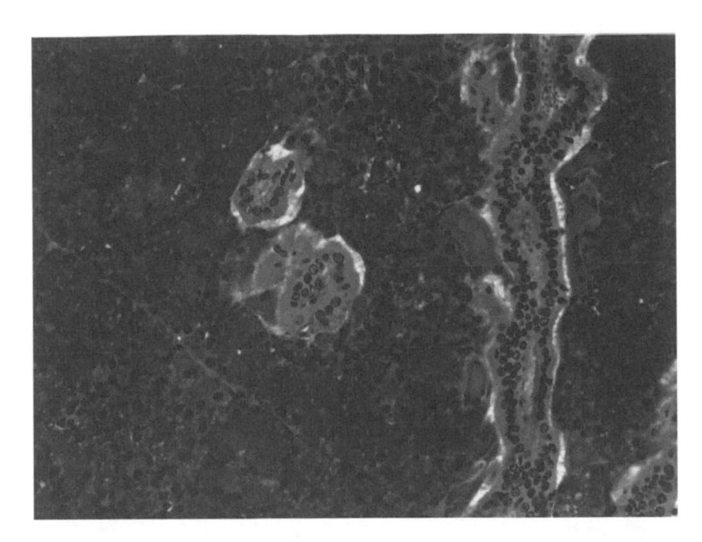

FIGURE K82c Photomicrograph of a semithin section of the parotid gland of a monkey showing two intercalated ducts near the center and a longitudinal section of a striated duct at the right. 20 ×.

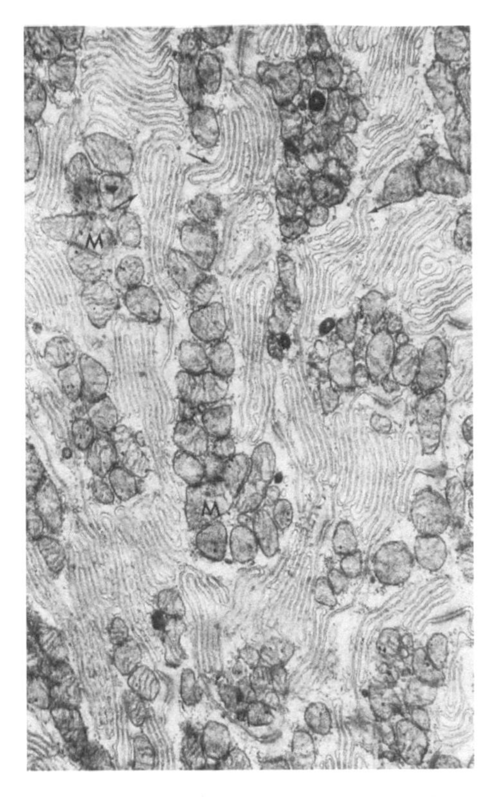

FIGURE K82d Electron micrograph of a cell from the striated duct of a human submaxillary gland showing the close association between mitochondria (M) and the infoldings of the basal plasma membrane (arrows) an indication of a transporting epithelium. 17,000 ×.

Parotid Gland

The largest salivary gland is purely serous in the human but this is not true of all mammals (Figs. K83a–K83e). Note the connective tissue CAPSULE and SEPTA dividing the gland into LOBES and LOBULES. Nerves, blood vessels, and lymph vessels extend throughout the septa. A fine STROMA of connective tissue supports the ACINI and ducts. The acini are elongate and may be branched. They are composed of pyramidal SEROUS CELLS grouped about a small LUMEN. Cytoplasm of the serous cells contains apical droplets of secretion and basophilic material below a spherical nucleus. Electron micrographs of the serous cells show that they are typical secretory cells and are sealed with junctional complexes between their apical borders. Immediately outside the acini is a basement membrane containing stellate MYOEPITHELIAL or BASKET CELLS—difficult to see in ordinary slides (Figs. K82c, K83f–K83h). The form of the myoepithelial or basket cells is well displayed in scanning electron micrographs; in transmission electron micrographs we note their processes are seen to be packed with contractile filaments.

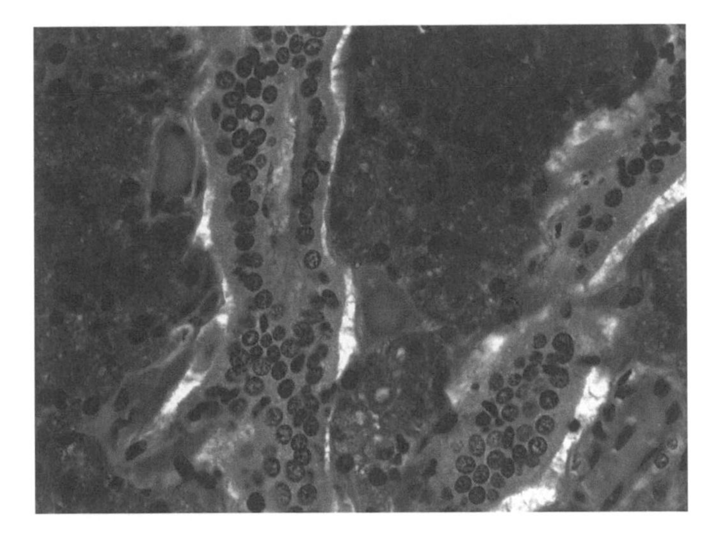

FIGURE K82e Photomicrograph of branches of an excretory duct in the parotid gland of a monkey. 40 ×.

Several acini open together into a fine INTERCALATED DUCT composed of flattened cells surrounding a narrow lumen (Fig. K82b). The electron microscope shows droplets of secretion in some of these cells; the duct is surrounded by myoepithelial cells. Several intercalated ducts open into the SECRETORY or STRIATED DUCT, which is intralobular in position (Fig. K82c). Try to identify vertical striations in the base of the epithelial cells. The electron microscope shows that these striations are infolded pockets of the basal plasma membrane that enclose erect mitochondria (Fig. K82d). In addition, folds of the basal plasma membrane interdigitate with similar folds on adjacent cells. Apical junctional complexes occur between these cells. These are the characteristics of transporting epithelia. Their apical cytoplasm contains vesicles.

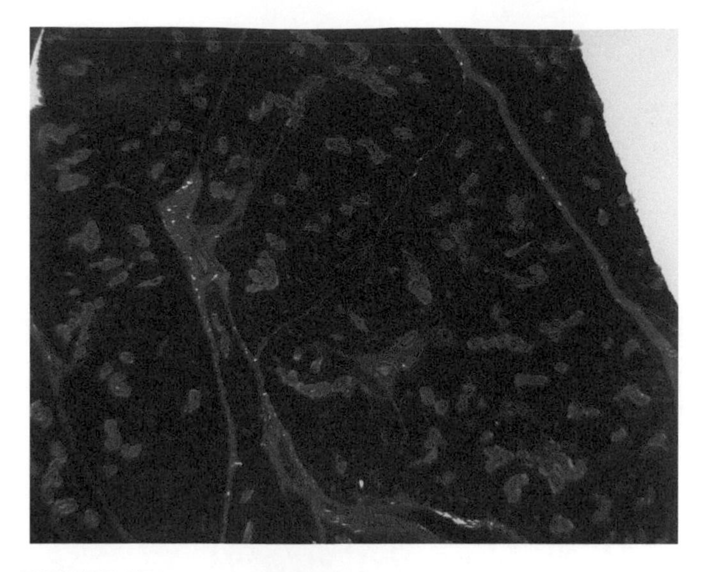

FIGURE K83a The acini of this section of the parotid gland of a monkey are purely serous. The secretory cells are deeply basophilic while the ducts are acidophilic. The gland is divided into LOBES and LOBULES by SEPTA of connective tissue that contain nerves and blood vessels. 2.5 ×.

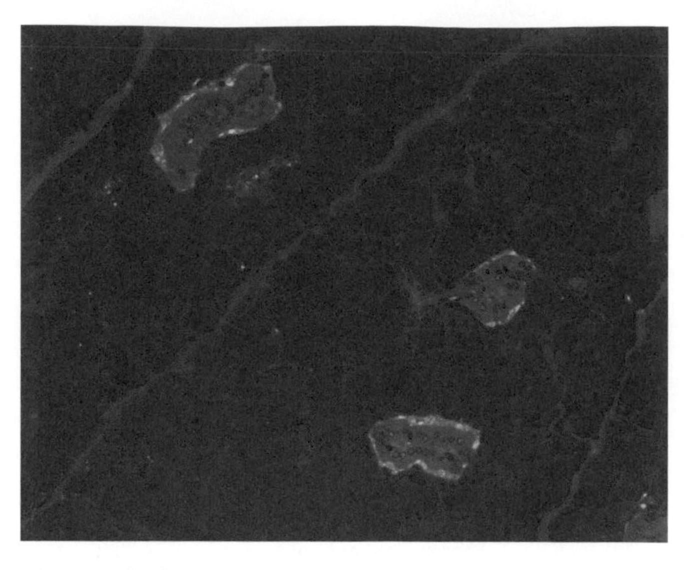

FIGURE K83b In this semithin section of the parotid gland of a monkey, the acini are supported by a fine STROMA of connective tissue; blood vessels penetrate throughout. Sections of acidophilic DUCTS can be seen. 20 ×.

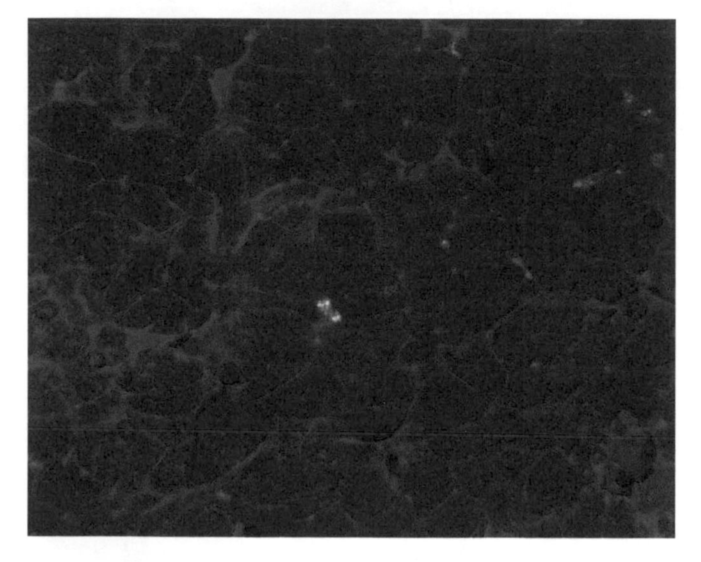

FIGURE K83c Pyramidal serous cells of the alveoli are so tightly packed with basophilic cytoplasm that their small lumina are often not seen. The vascular connective tissue of the stroma provides support for the acini. 63 ×.

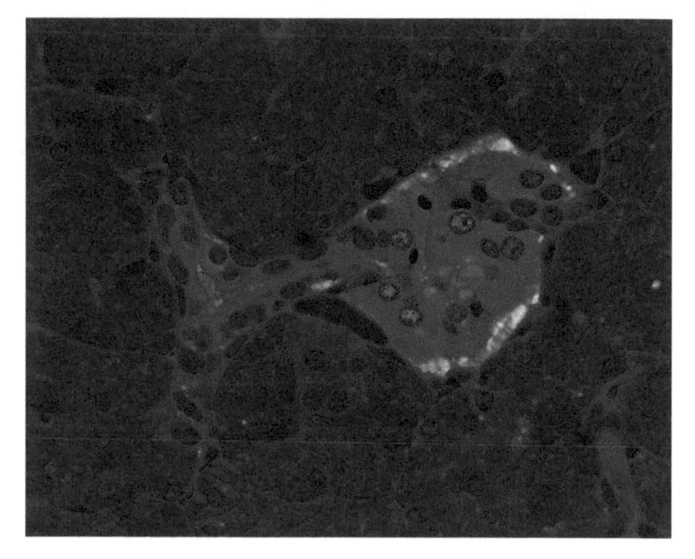

FIGURE K83d Two small ducts enter a larger duct within the septum of this gland. 63 ×.

The secretory ducts open into the EXCRETORY DUCTS including the INTERLOBULAR DUCTS, the INTERLOBAR DUCTS, and the MAIN DUCT (Fig. K82e). These may be seen in the broader septa between the lobes and lobules. Excretory ducts are lined with columnar epithelium. As one traces the ducts toward the opening into the oral cavity, the epithelium is seen to change first to pseudostratified and then to stratified columnar.

Submaxillary Gland

The submaxillary gland is similar to the parotid but contains MIXED ACINI in addition to the serous acini (Figs. K84a–K84c). In a mixed acinus the central cells surrounding the lumen are pale, have basal nuclei, and

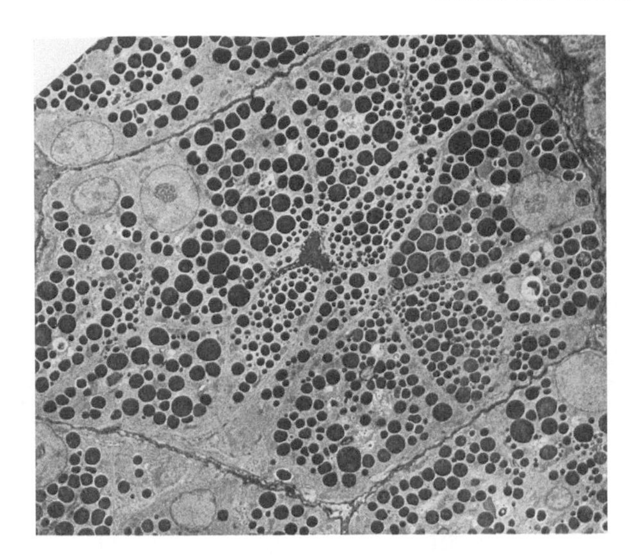

FIGURE K83e Electron micrograph of an entire acinus packed with dense serous granules in the parotid gland of a baboon. The inconspicuous lumen of the acinus is near the center. 2100 ×.

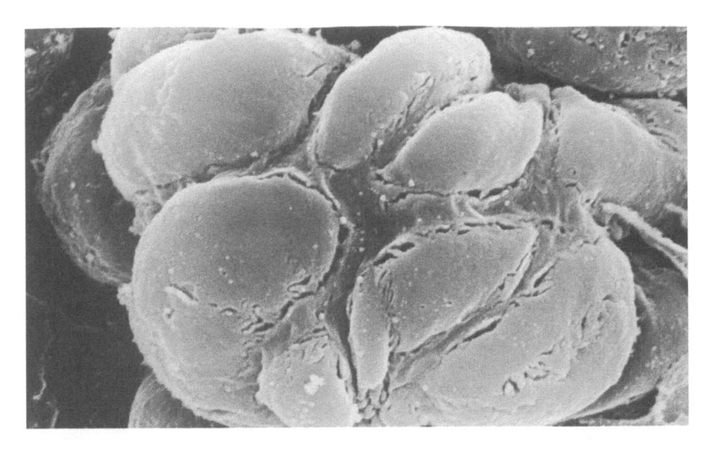

FIGURE K83f Scanning electron micrograph of a maceration preparation of a human parotid gland. A star-shaped myoepithelial cell encircles an entire acinus. 6800 ×.

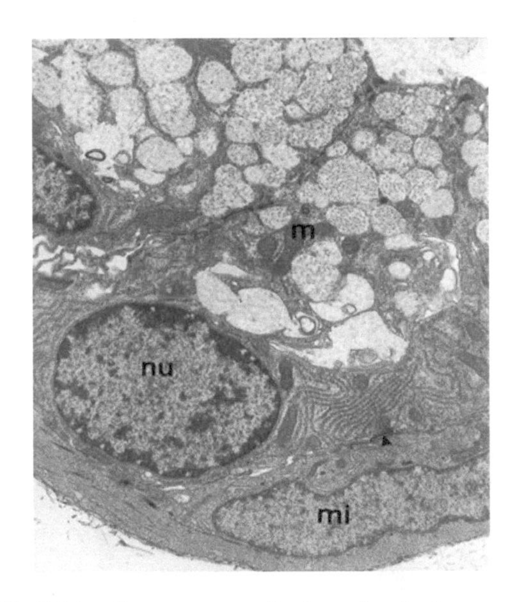

FIGURE K83g Seromucous cell show in its base, part of a myoepithelial cell (mi) joined by desmosomes (*arrow*). Mitochondrion (m); nucleus (nu). 10,000 ×. *Tirapelli, L.F., Tirapelli, D.C.P., Tamega, O.J., 2002. Ultrastructural alterations in the submandibular glands of rats (Rattus norvegicus) submitted to experimental chronic alcoholism. Rev Chil Anat. 20(1). Temuco https://scielo.conicyt.cl/scielo.php?script=sci_arttext&pid=S0716-98682002000100001#**CC by 4.0.*

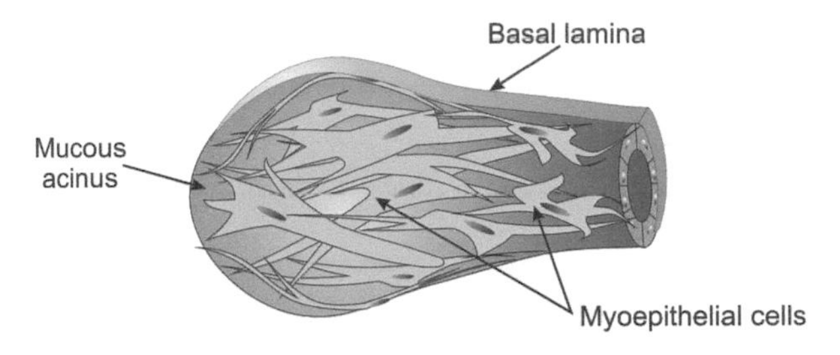

FIGURE K83h Reconstruction of myoepithelial cells embracing a mucous acinus in the salivary gland. *Authors.*

secrete mucus. Capping this acinus is a SEROUS DEMILUNE of acidophilic cells that pour their secretions into the lumen by passing it through minute channels between the mucous cells. Slanting sections may show some mixed acini as purely mucous, but all mucous acini are capped with serous demilunes. In human there are about five times as many serous acini as mixed, in carnivores most acini are mixed, and in insectivores and rodents most are serous. Intercalated ducts are short and difficult to find.

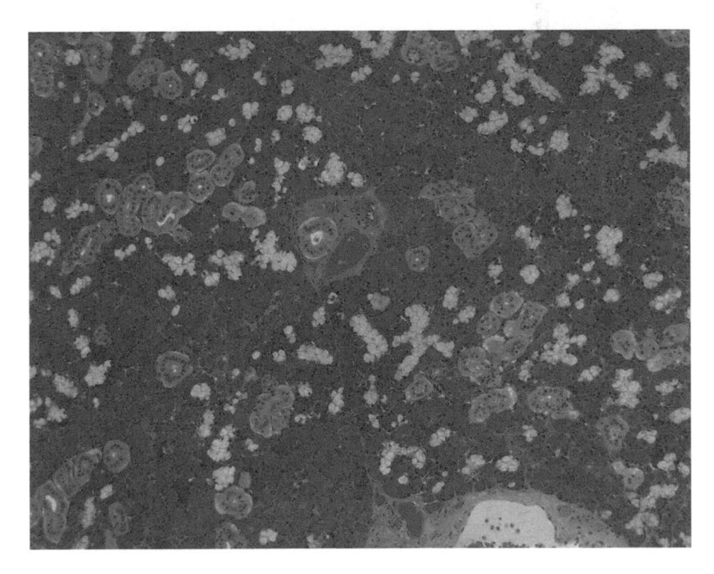

FIGURE K84a The submaxillary gland of a monkey is partitioned into lobules and lobes. SEROUS ACINI are dispersed among MIXED ACINI, which are serous acini sporting caps of serous cells. Basophilic serous acini are abundant; there are fewer pale mixed acini. Pink sections of ducts are seen throughout. 10 ×.

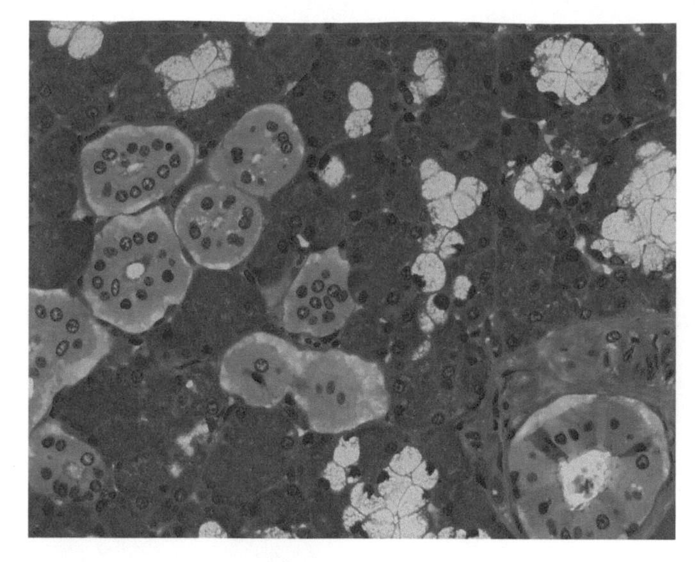

FIGURE K84b The ducts of salivary glands are contained within the connective tissue partitions of the lobes and lobules. The large duct at the lower right is an EXCRETORY DUCT. It receives the secretions of the INTERLOBAR and INTERLOBULAR DUCTS. The small ducts near the center are contained within a lobule and are STRIATED DUCTS. An INTERCALATED DUCT peeps out just above and to the left of these. 40 ×.

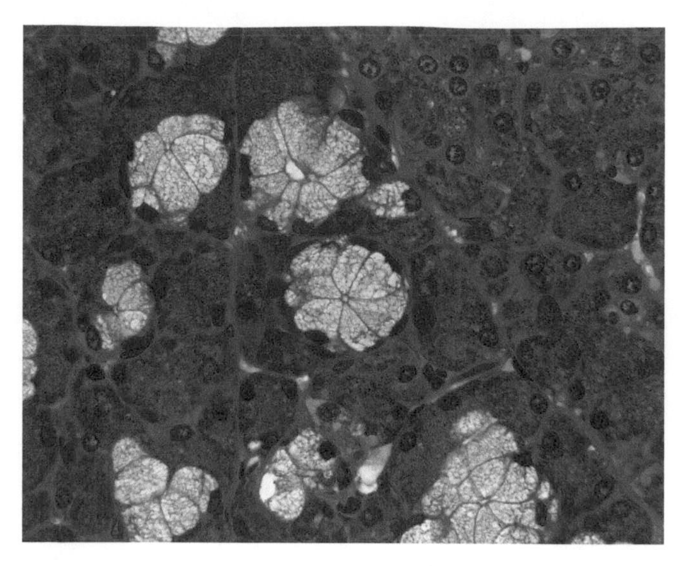

FIGURE K84c The distinction between serous and mixed acini is clear in this section of the submaxillary gland of a monkey. All mucous acini bear a cap of serous cells, but these may not be visible in all sections. These "hats" are called SEROUS DEMILUNES. These serous cells pour their secretions in to the lumen of the mucous acini by passing it through minute channels between the mucous cells. Due to the vagaries of sectioning, many of the serous cells are parts of demilunes and not serous acini. 63 ×.

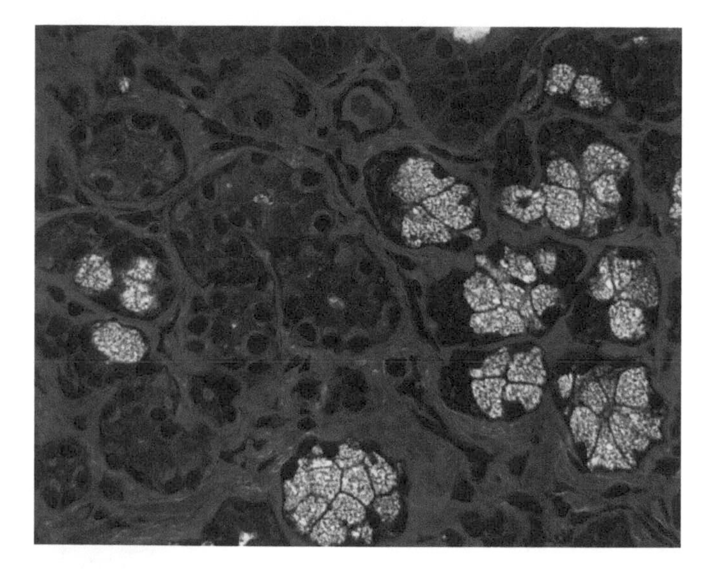

FIGURE K85a Several mixed acini are present in this section of the sublingual gland of a monkey. SEROUS DEMILUNES form caps on most of the serous acini. Striations in the cells of the STRIATED DUCT at the top to the right of center are evident. 63 ×.

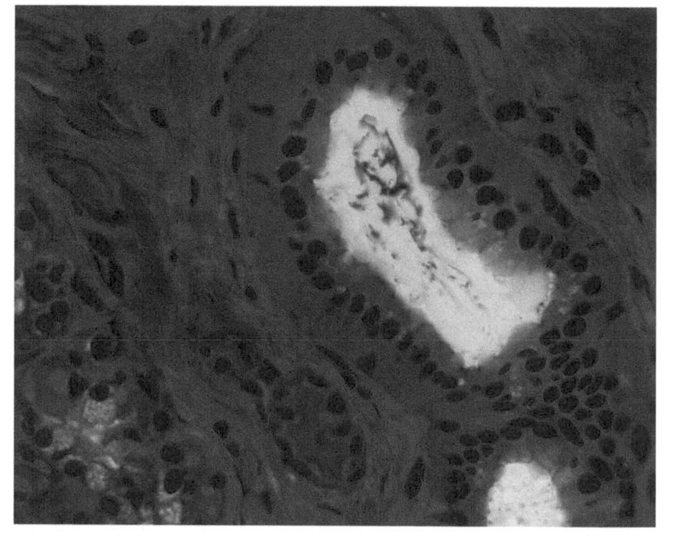

FIGURE K85b Two excretory ducts are present in the connective tissue between lobes of the sublingual gland of a monkey. 63 ×.

Sublingual Gland

There are about equal numbers of mucous and serous acini in the sublingual glands of human (Figs. K85a and K85b) but the mucous elements predominate in sublingual glands of the cat, dog, and pig, and in rodents the gland is completely mucous. Intercalated ducts are short and inconspicuous.

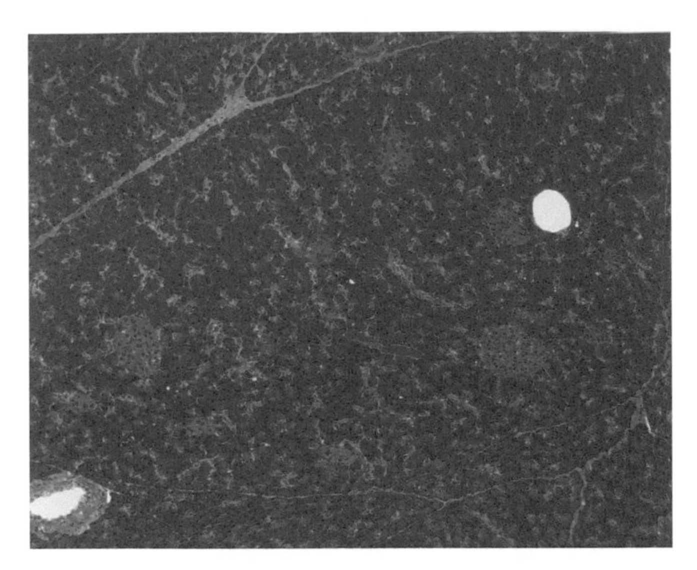

FIGURE K86a The pancreas bears strong resemblance to the tubuloacinar salivary glands. In this semithin section from a monkey, a few pale masses, the ISLETS (of Langerhans) are scattered among the acini. The gland is divided into LOBES and LOBULES. 10 ×.

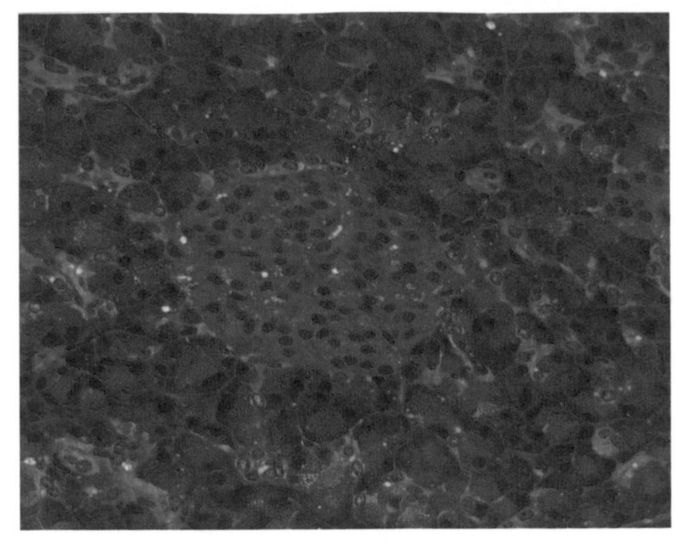

FIGURE K86b An endocrine PANCREATIC ISLET is surrounded by exocrine ACINAR TISSUE. The bases of the wedge-shaped acinar cells are basophilic and the apical portions are packed with acidophilic droplets of secretion. The islet consists of cords of polygonal cells separated by thin-walled capillaries. 40 ×.

COMPARATIVE HISTOLOGY OF THE SALIVARY GLANDS

The lower chordates do not possess salivary glands. The *fishes* generally lack them although unicellular mucous glands occur in the stratified squamous epithelium lining the mouth. In *amphibians*, salivary glands are generally mucous in type consisting of cylindrical cells organized into acini or tubules lying below the epithelium. Ducts, which are often ciliated, empty into the mouth cavity from these small groups of acini. The salivary glands of *reptiles* are similar to those of amphibians but show an advance in that there is more than one type of secretory cell present; serous cells are also found and their secretion may be venomous. Salivary glands are poorly developed in *birds* and their function is primarily in lubricating food.

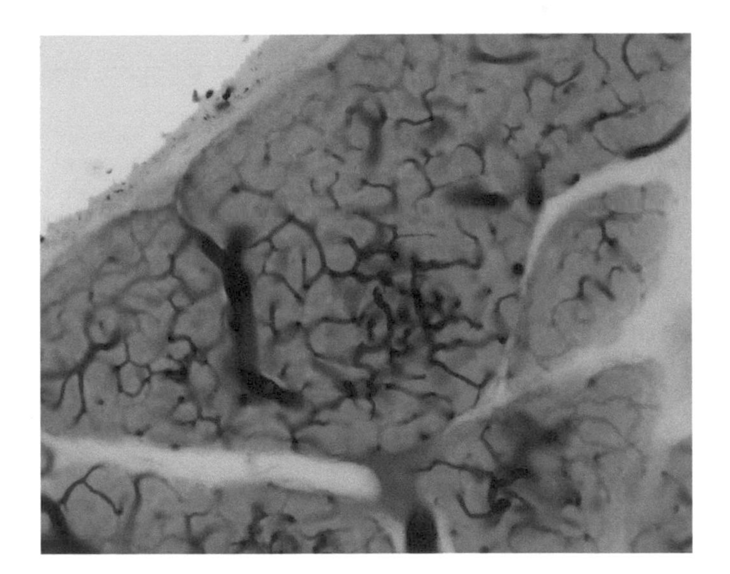

FIGURE K86c This mammalian pancreas was injected with colored gelatin before fixation, red into the arteries, green into the veins, to show the elaborate distribution of blood vessels. The denser aggregations of blood vessels are probably islets. 20 ×.

PANCREAS

In the pancreas, an exocrine gland and an endocrine gland function together within a common capsule (Fig. K86a). The exocrine portion secretes a digestive juice and is a tubuloacinar gland that closely resembles the salivary glands. The endocrine portions are scattered among the acini in the form of small, pale-staining PANCREATIC ISLETS (islets of Langerhans) that secrete insulin (Fig. K86b). We will consider the islets in the Chapter N, Endocrine Organs.

The STROMA of connective tissue includes the thin CAPSULE and SEPTA that divide the gland into LOBES and LOBULES (Fig. K86a). Nerves and blood vessels as well as lymph vessels and ducts are found in the septa. The PARENCHYMA consists of acinar cells and cells of the islets. Note the distribution of blood vessels in an injected preparation (Fig. K86c).

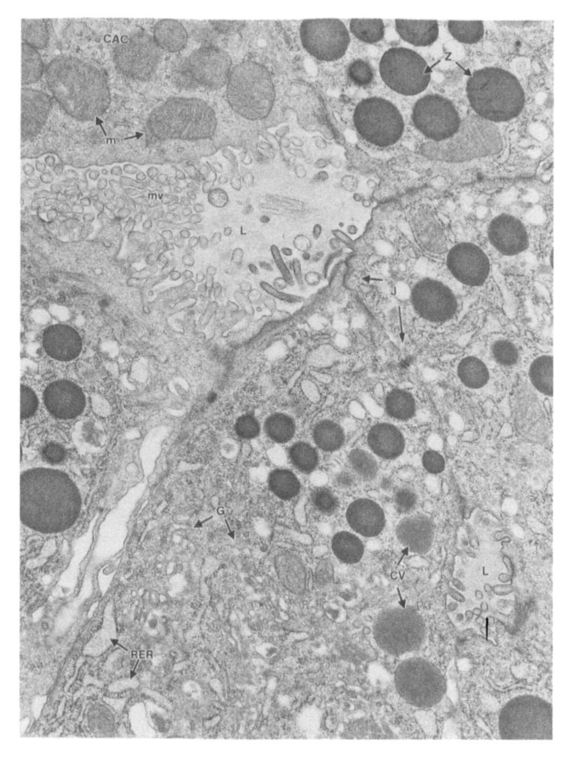

FIGURE K87a Electron micrograph of a thin section through the apical region of a mammalian pancreatic acinus. A portion of a centroacinar cell (CAC) is at the top left. Junctional complexes (J) join adjacent acinar and centroacinar cells, sealing the lumen from the intercellular space. Stubby microvilli (mv) extend into the centroacinar lumen (L) from the apical surfaces of both the acinar and centroacinar cells. *CV*, condensing vacuoles; *G*, elements of the Golgi complex; *Z*, zymogen droplets; *m*, mitochondria; *RER*, rough endoplasmic reticulum.

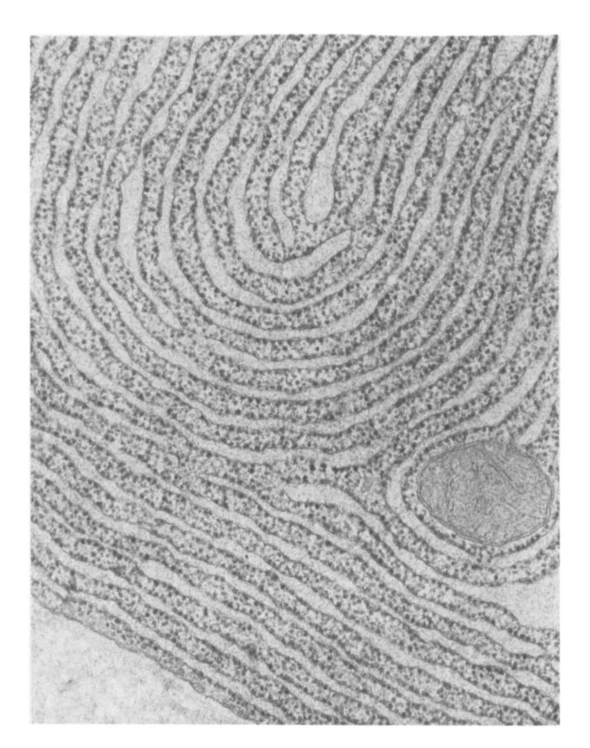

FIGURE K87b Electron micrograph of a section of the basal cytoplasm of a human pancreatic acinar cell. Extensive lamellar arrays of granular endoplasmic reticulum are characteristic of these cells and are indicative of protein synthesis. They engulf a mitochondrion at the lower right. 65,000 ×.

In the EXOCRINE portion of the pancreas the ACINI consist of indistinct wedge-shaped cells surrounding a central LUMEN and resting on a reticular membrane (Fig. K86b). The intercalated duct begins within the acinus so that the lumen is hard to distinguish. The squamous duct cells within the lumen are CENTROACINAR CELLS; they stain lightly with eosin. The apical portions of the acinar cells contain highly refractile, acidophilic ZYMOGEN or SECRETION GRANULES. The nucleus is spherical and lies near the base. The basal cytoplasm is basophilic and may show striations. Note the droplets of secretion in electron micrographs of pancreatic acinar cells, supranuclear Golgi complex, extensive granular endoplasmic reticulum, and mitochondria (which may be oriented perpendicular to the base of the cell) in pockets of inturned plasma membrane (Fig. K87a). The acini are enclosed by a thin basement membrane supported by reticular fibers. There are no myoepithelial cells.

The lumen of the INTERCALATED DUCTS is long and narrow with either a simple squamous or simple cuboidal epithelium supported by reticular tissue. EXCRETORY DUCTS in the septa have either a simple cuboidal or simple columnar epithelium. The PANCREATIC DUCT, which leads to the duodenum, has a simple columnar epithelium surrounded by fibrous connective tissue containing fibers of smooth muscle. The cytoplasm in all regions is pale staining and contains few organelles. Electron micrographs of epithelial cells show apical microvilli, interdigitations of the lateral plasma membranes with desmosomes and junctional complexes, and a thin basement membrane (Figs. K87a–K87c). The large ducts also contain goblet cells and APUD cells.

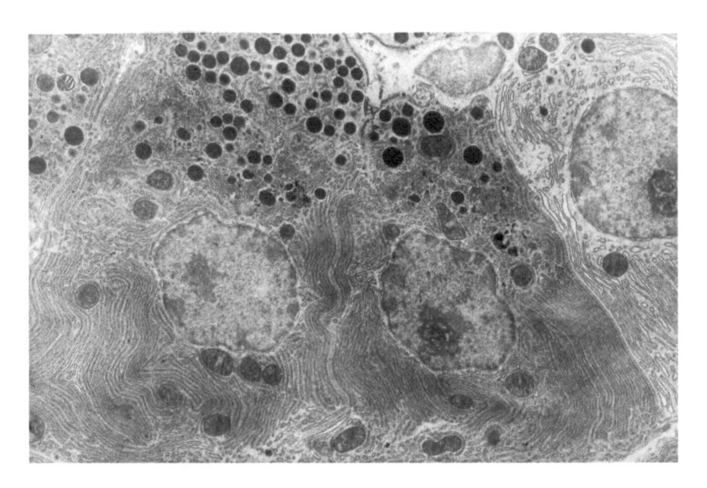

FIGURE K87c Electron micrograph of a section of acinar cells from the pancreas of a bat. These cells contain dark zymogen droplets but are fairly inactive, containing few mitochondria. They are packed with granular endoplasmic reticulum.

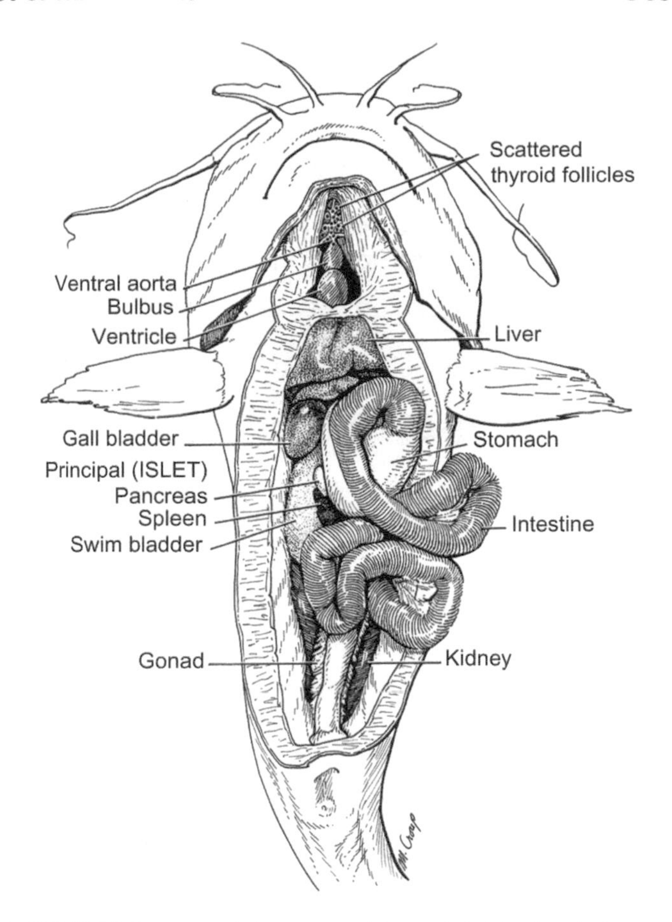

FIGURE K88 Ventral dissection of the catfish (*Ameiurus*) showing the location of the principal (islet) pancreas with reference to other organs. Scattered thyroid follicles are shown diagrammatically around the ventral aorta.

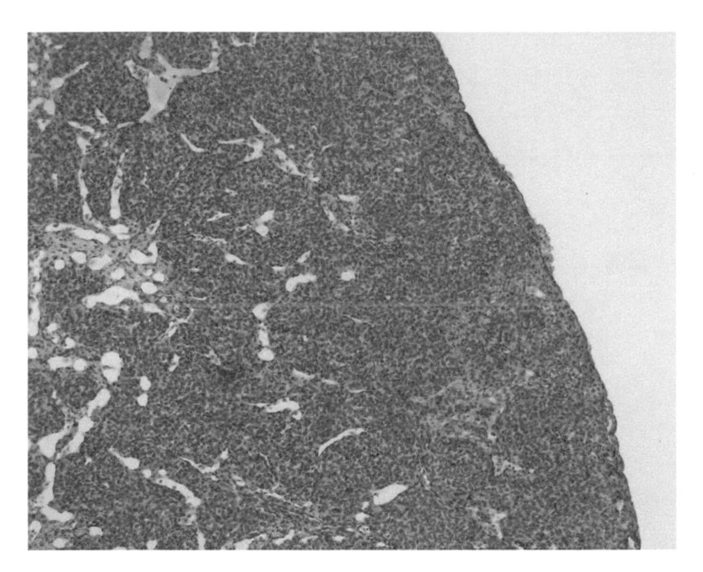

FIGURE K89a The endocrine and exocrine portions of the pancreas of the hagfish (*Myxine glutinosa*) are separate. An islet organ occurs alongside the midgut and shares its peritoneal connections. 10×.

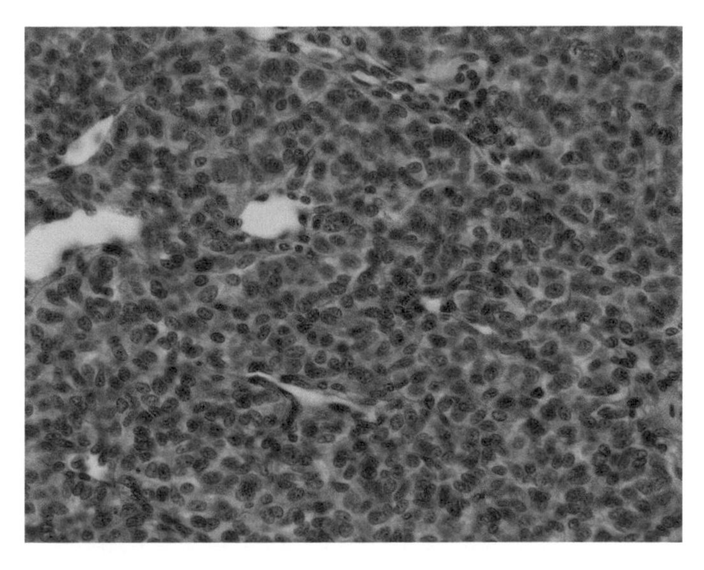

FIGURE K89b The islet organ of the hagfish is a mass of endocrine cells permeated by vascular sinusoids. 40×.

COMPARATIVE HISTOLOGY OF THE PANCREAS

Pancreas-like tissue of *cyclostomes* consists of masses of glandular cells embedded within the liver and in the tunica serosa of the upper part of the small intestine (Figs. K89a and K89b). It probably represents only the endocrine portion and no traces of the exocrine portion have been found. The pancreas is present in all *fishes* and shows the transition from a diffuse organ to a compact one (Fig. K88). Parts of the DISSEMINATED pancreas are so widely distributed that they are visible

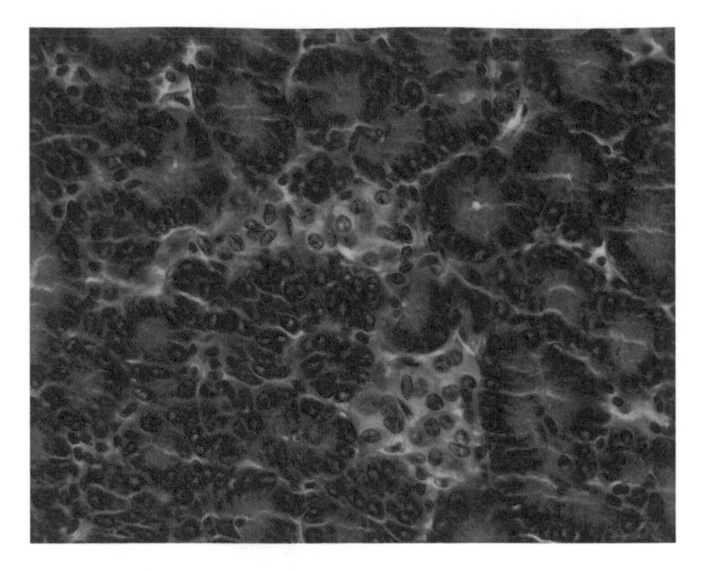

FIGURE K90 Pancreatic islets are distributed among exocrine acini in the pancreas of the dogfish (*Squalus acanthias*). 40 ×.

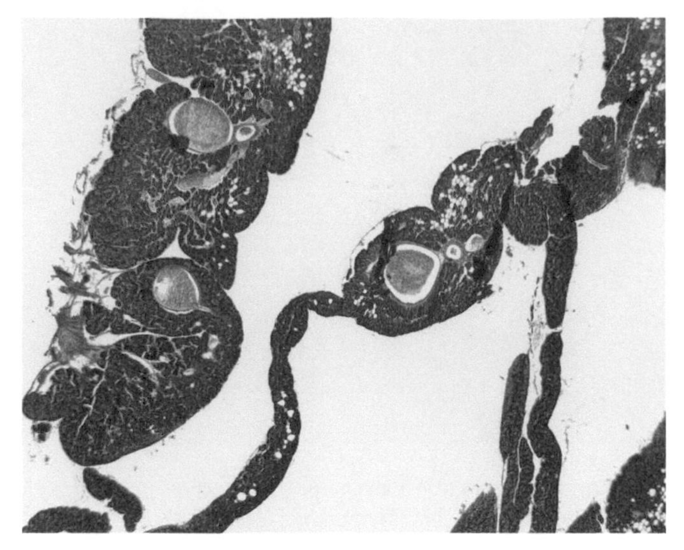

FIGURE K91a The pancreas of some bony fish is spread throughout the mesentery in small masses: a DIFFUSE PANCREAS. This example was taken from a blenny. 2.5 ×.

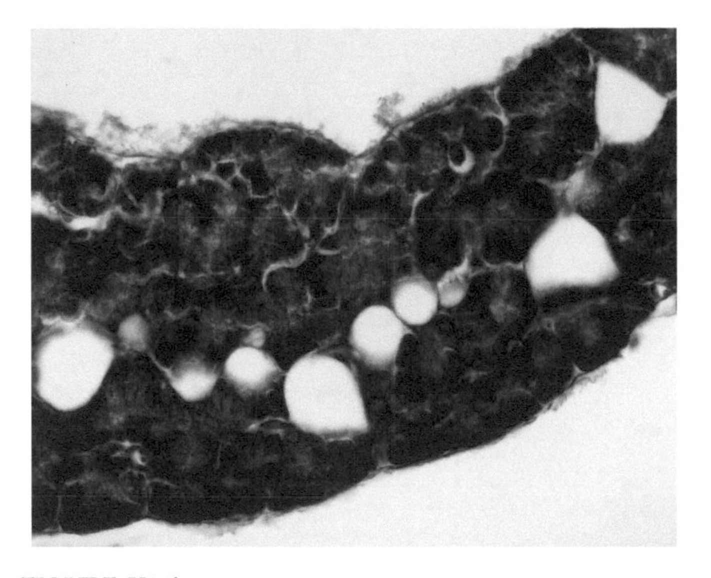

FIGURE K91b An occasional endocrine islet is scattered among the pancreatic exocrine acini of the blenny. A few large adipose cells distend the pancreas. 40 ×.

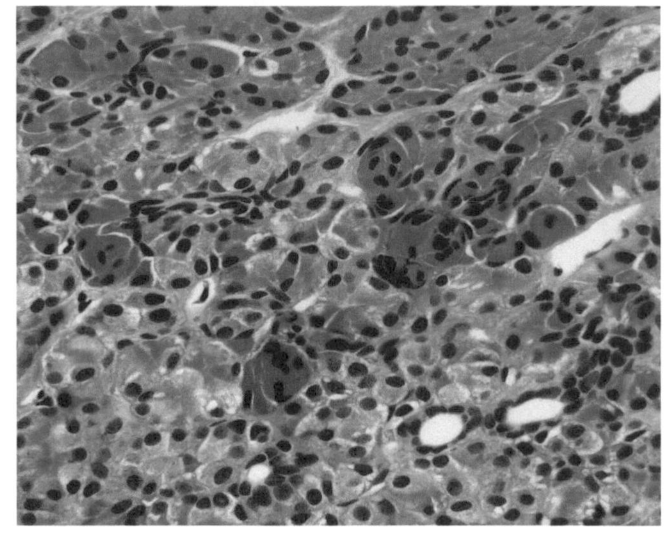

FIGURE K92a In this preparation from the amphibian *Amphiuma*, masses of endocrine cells occupy most of the bottom of the screen. At the top are exocrine acini. 20 ×.

only with the microscope although their total volume may be considerable (e.g., perch, *Perca;* carp, *Cyprinus*). The DIFFUSE pancreas is the commonest: the parts are grossly visible but spread out (e.g., cod, *Gadus;* some flounders, *Pleuronectes;* marine gar, *Belone*) (Figs. K90, K91a, K91b). The compact form is the rarest and is found in the pike (*Esox*), some catfish (*Silurus*), and others. The islet tissue of fish is frequently separated from the acinar tissue as a PRINCIPAL ISLAND inside a separate capsule (Fig. K88). The pancreas of *amphibians* is divided into a dorsal portion lying in the dorsal mesentery and a ventral portion lying between the intestine and the liver (Figs. K92a, K92b, K93a–K93c). Usually there is some connection between the parts. This double condition is also seen in the embryos of reptiles, birds, and mammals (Figs. K94a, K94b, K95a, K95b, K96a, K96b). The acinar tissue resembles that of fishes; the islet tissue occurs as cords of cells in close association with the intralobular ducts. *Birds* have been shown to have two types of islets; the first is pale and consists of B cells while the second is dark and consists of A and D cells.

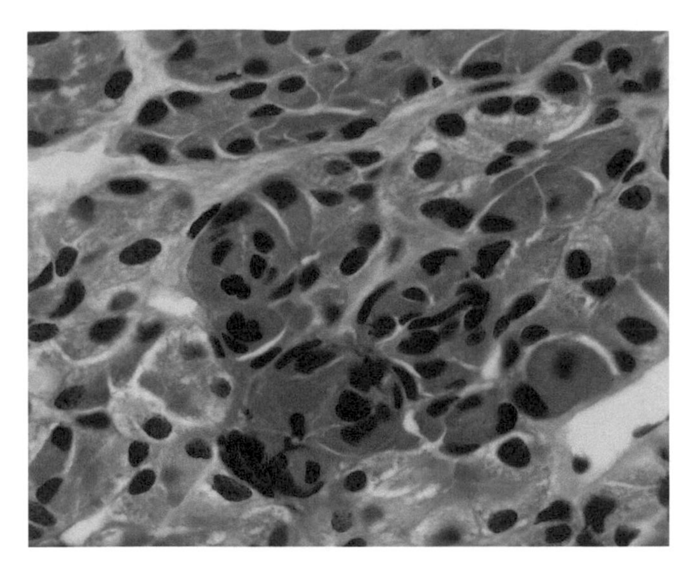

FIGURE K92b Exocrine acini occupy most of the upper right in this preparation from the pancreas of *Amphiuma*. The basal portions of the cells are basophilic; acidophilic secretion droplets are apical. At the lower right are pale islet cells. 40 ×.

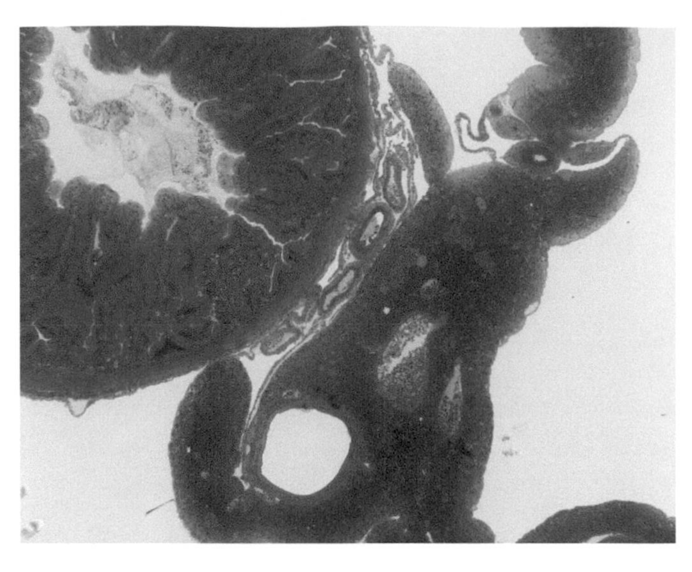

FIGURE K93a The pancreas of the frog forms an irregular mass surrounding large blood vessels in the mesentery. At this low power, pale islets can be discerned scattered throughout the exocrine tissue. 2.5 ×.

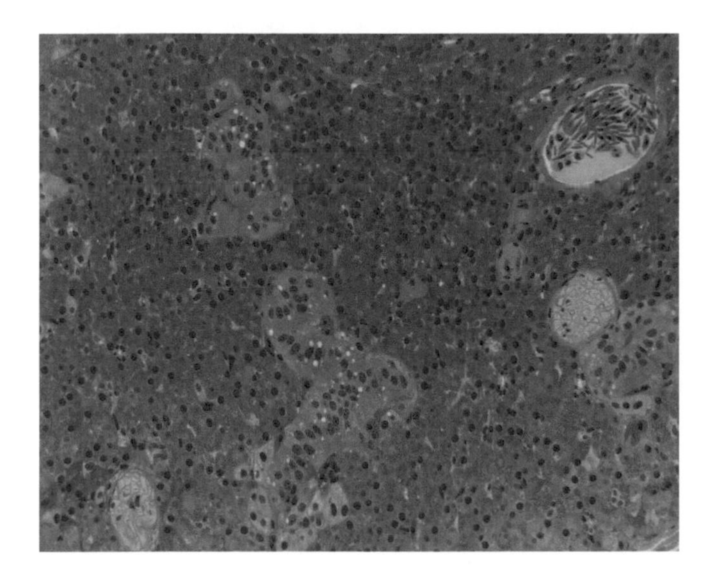

FIGURE K93b The exocrine acini in the pancreas of the frog exhibit the classic appearance of basal basophilia and apical acidophilia. There are cross sections of nerve trunks at the lower left and right center, and a large artery at the upper right. 20 ×.

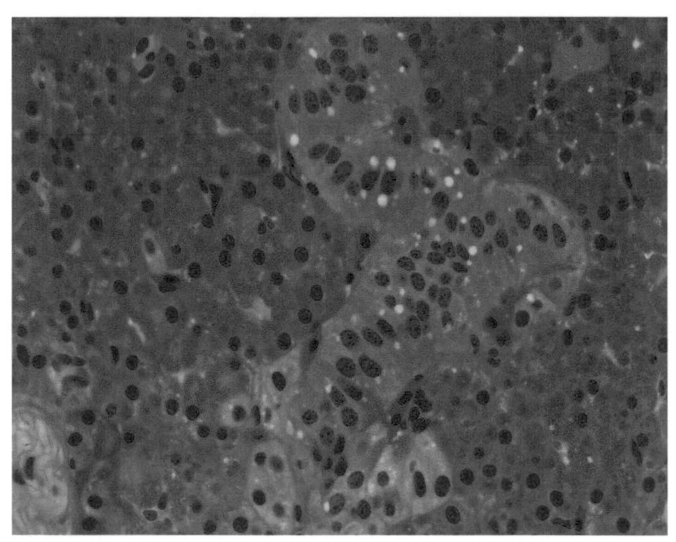

FIGURE K93c An islet in the pancreas of a frog is surrounded by exocrine acini.

LIVER

The liver is unusual in that it has two blood supplies (Fig. K97a) It receives fresh blood directly from the AORTA through the HEPATIC ARTERY. It also receives deoxygenated blood from the capillaries of the small intestine by way of the HEPATIC PORTAL VEIN. This venous blood has lost much of its oxygen and has picked up nutrients and other substances in its passage through the intestine. Blood from the liver makes its way back to the heart *via* THE HEPATIC VEIN to the VENA CAVA.

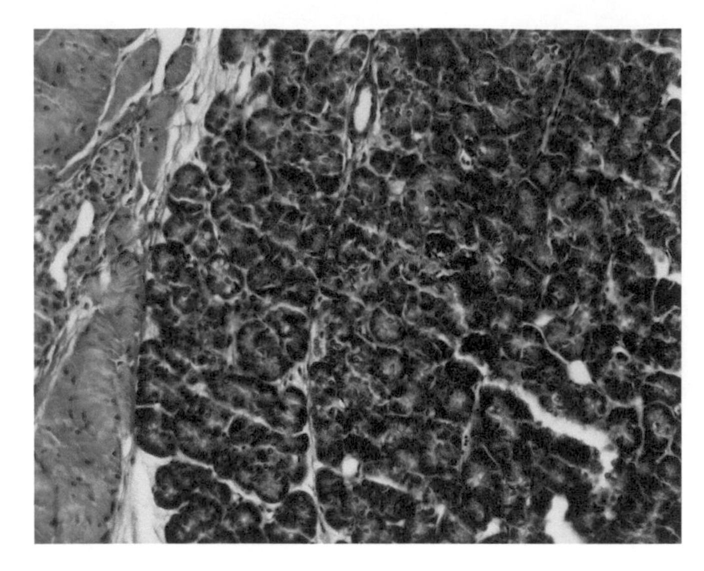

FIGURE K94a The pancreas of a caiman (reptile) is squeezed into the mesentery within a loop of the intestine. 20 ×.

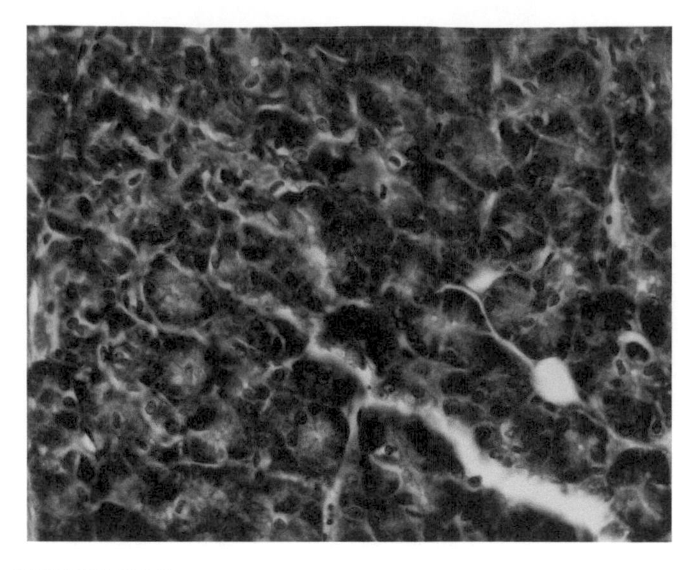

FIGURE K94b The exocrine pancreas of the caiman is a mass of exocrine acini. This layout shows little variation throughout the higher vertebrates. 40 ×.

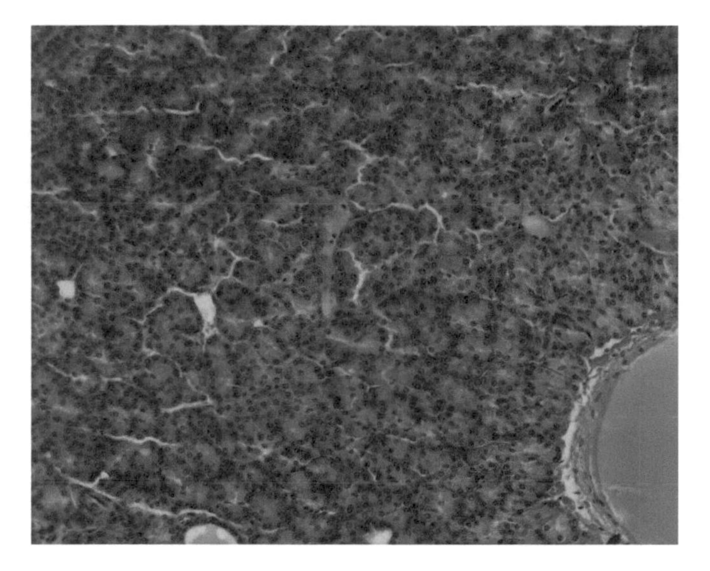

FIGURE K95a The pancreas of the tuatara (*Sphenodon punctatum*: reptile) consists of occasional endocrine islets surrounded by masses of acinar tissue. There are ducts at the lower left and lower right. 20 ×.

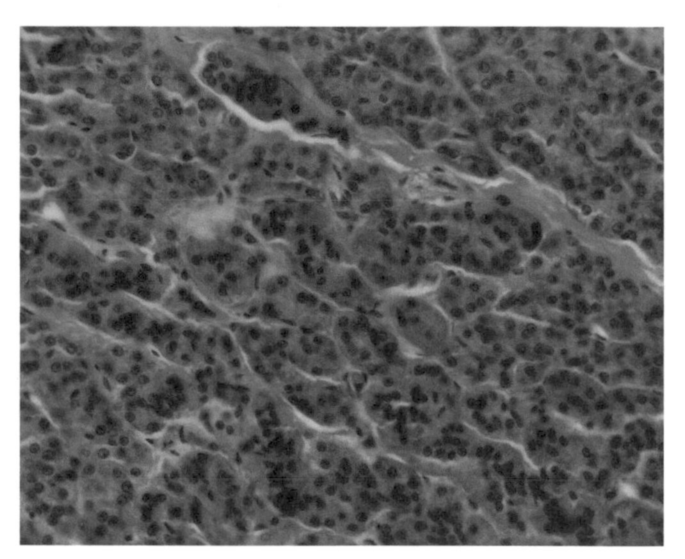

FIGURE K95b The oblong mass of cells in the pancreas of a tuatara at the top center may be islet tissue. 40 ×.

Fig. K97b is a section of pig liver seen under low power. (The connective tissue stroma is especially well developed in the pig.) The PARENCHYMA is divided into roughly hexagonal LOBULES by SEPTA of connective tissue that are continuous with the superficial covering of the whole liver. The septa contain blood vessels and ducts and are especially prominent at certain angles of the lobules where they form PORTAL AREAS. These vessels are branches of the hepatic artery, hepatic portal vein, and bile duct and constitute the PORTAL TRIAD (Fig. K98a); the smaller blood vessels are INTERLOBULAR ARTERIES and the larger are INTERLOBULAR VEINS (Figs. K97c and K97d). At the center of each lobule is the CENTRAL VEIN into which open the SINUSOIDS. Blood from the central veins is drained into SUBLOBULAR VEINS, which fuse to form the hepatic veins. The sublobular veins arise at right angles to the central veins and are never associated with the bile ducts or arteries. Between the sinusoids LAMINAE or CORDS of PARENCHYMAL CELLS radiate toward the periphery of the lobule.

FIGURE K96a Acinar tissue of the usual form constitutes most of the pancreas of the turtle (reptile).

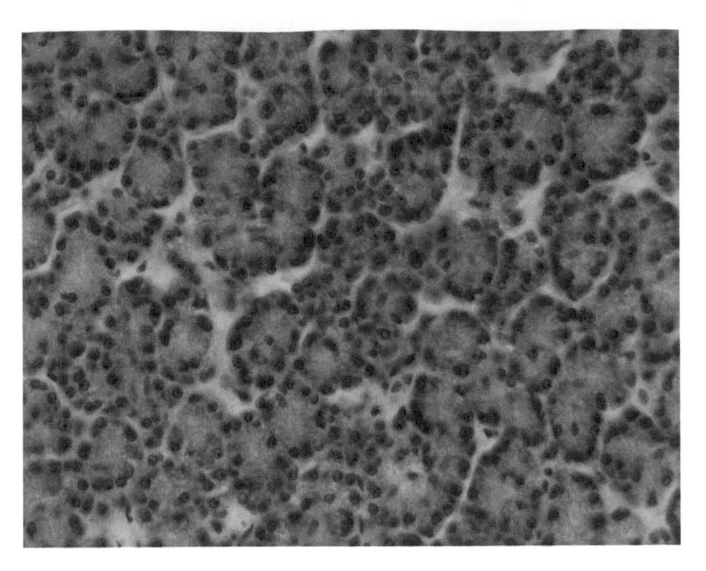

FIGURE K96b An islet is surrounded by acinar tissue in this section of the pancreas of the turtle. 40 ×.

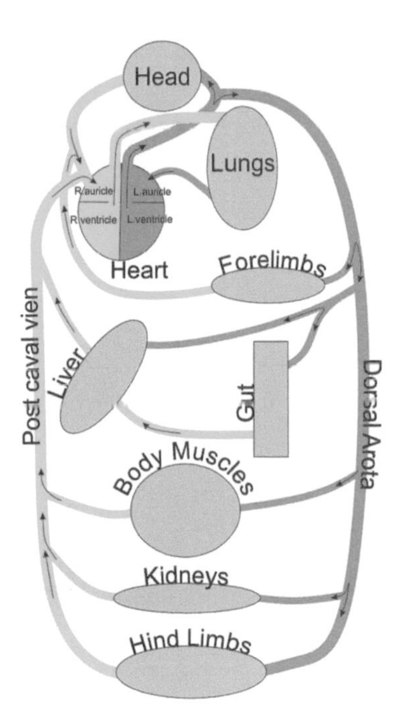

FIGURE K97a There is a double blood supply to the liver of birds and mammals: blood coming from the intestine in the hepatic portal vein (blue) is laden with food but low in oxygen; blood rich in oxygen (red) arrives directly from the aorta through the hepatic artery. *Authors.*

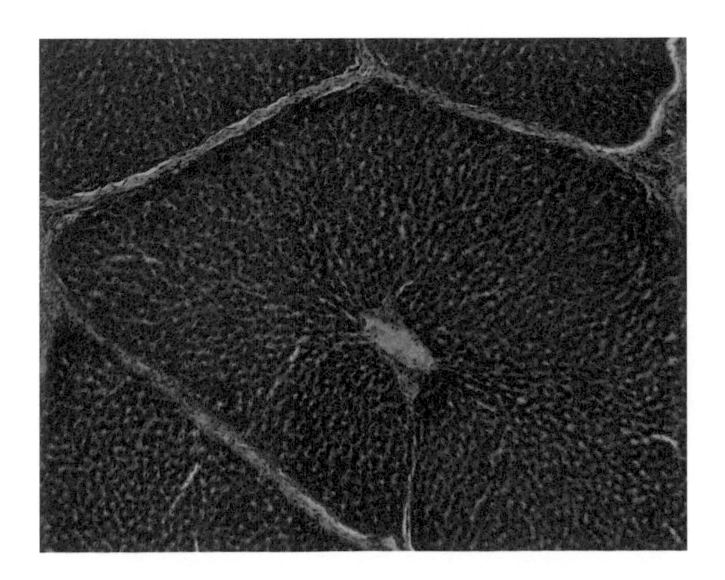

FIGURE K97b The connective tissue STROMA of the liver of the pig is especially well developed. The PARENCHYMA is divided into roughly hexagonal LOBULES by SEPTA of connective tissue that are continuous with the superficial covering of the entire liver. In this micrograph, there is some evidence of the blood vessels and ducts in the septa, but at the corners of the lobule, these are especially prominent. These are the PORTAL AREAS and contain the PORTAL TRIAD consisting of branches of the HEPATIC ARTERY, branches of the HEPATIC PORTAL VEIN and branches of the BILE DUCT. The smaller vessels are INTERLOBULAR ARTERIES and the larger are INTERLOBULAR VEINS.

The parenchyma consists of irregular LAMINAE or cords of HEPATOCYTES that radiate toward the periphery of the lobule and contain sinusoids between them.

Blood comes to the liver from two sources: the hepatic artery and the hepatic portal vein. These vessels break up into interlobular arteries and veins, respectively and run through the liver in the portal areas. From them, blood enters the sinusoids, mixes, and flows toward the CENTRAL VEIN at the center of the lobule. Blood from the central veins is collected by the sublobular veins that unite to form the hepatic vein opening into the inferior vena cava. 10 ×.

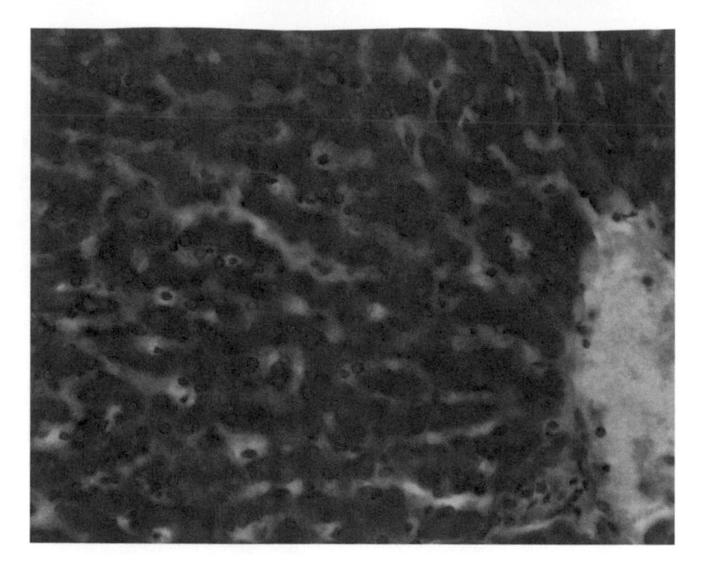

FIGURE K97c Liver parenchymal cells or hepatocytes are polyhedral and contain finely granular basophilic cytoplasm. The sinusoids are lined with pale blue squamous ENDOTHELIAL CELLS continuous with the endothelial lining of the central vein. 40 ×.

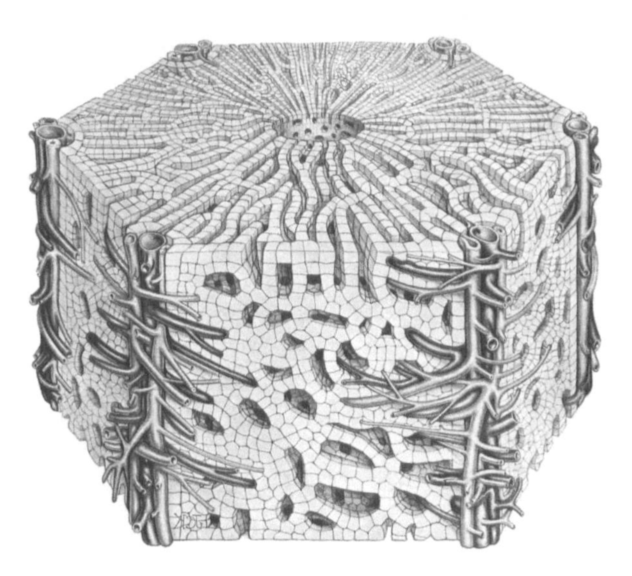

FIGURE K97d Schematic view of a mammalian liver lobule. The central vein at the center of the lobule is surrounded by laminae (erroneously called "cords") of hepatocytes. At the periphery of the lobule are six portal triads consisting of branches of the hepatic portal vein, hepatic artery, and the bile duct.

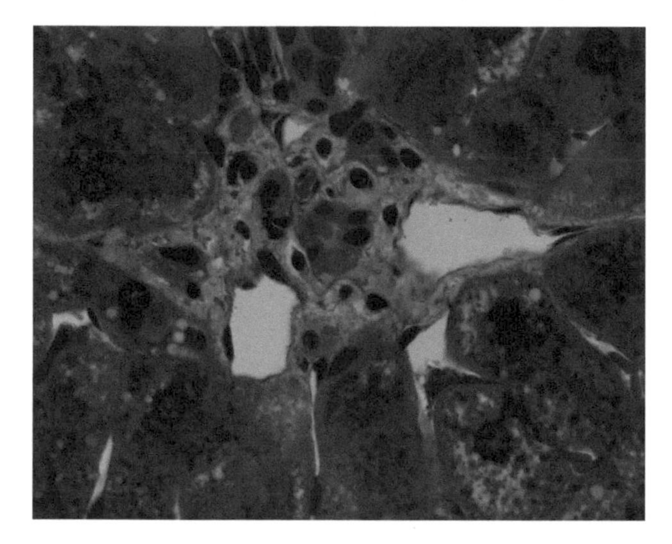

FIGURE K98a A PORTAL TRIAD, consisting of PORTAL VEINS, HEPATIC ARTERY, and BILE DUCTS lies in the connective tissue of the PORTAL AREA of the liver of a rabbit where three lobules come together. The large, empty spaces are branches of the hepatic portal vein; the small vessel containing an erythrocyte is a branch of the hepatic artery; and the lightly basophilic duct near the center is a branch of the bile duct. A sinusoid, lined with endothelial cells, opens into a branch of the portal vein at the right. Microvilli on the surface of the hepatocytes hold aloft the endothelial cells—like a blanket over blades of grass; one imagines that one can see the pale PERISINUSOIDAL SPACE thus formed. 100 ×.

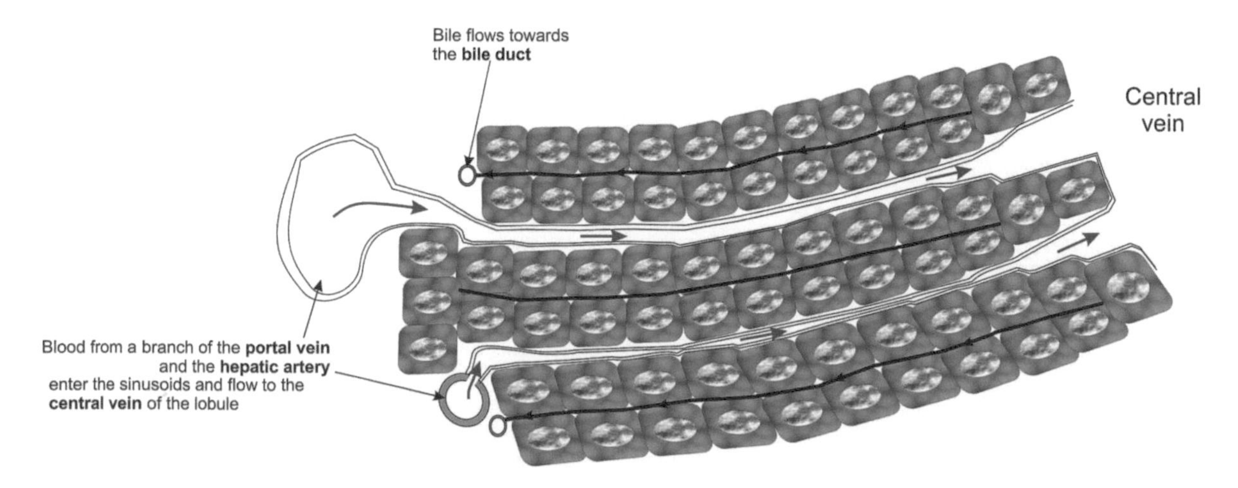

FIGURE K98b Diagram illustrates the flow of blood and bile through the portal triad. Blood in branches from the portal vein and hepatic artery flows into the sinusoids, which in turn empty into the central vein. Bile travels in the opposite direction in canaliculi emptying into bile duct in the portal area. *Authors.*

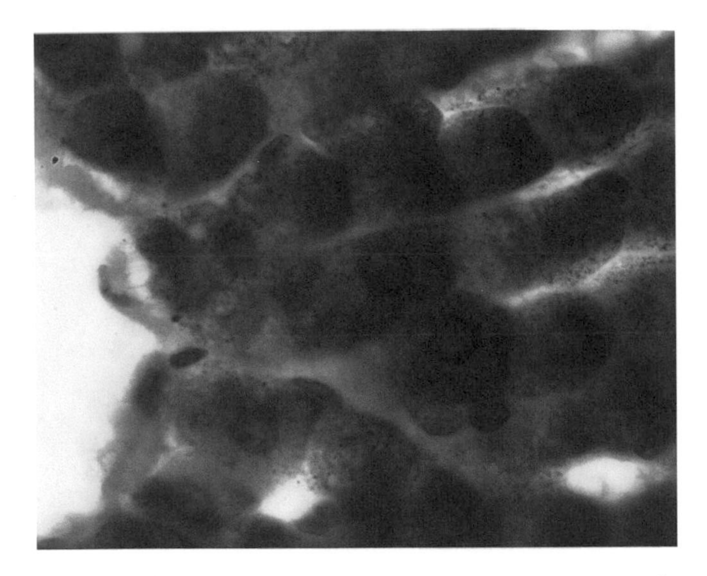

FIGURE K99a Hepatocytes are packed with GLYCOGEN that washes out in ordinary preparations when water is used. This slide of mammalian liver has been prepared to preserve the glycogen that appears as red granules in the cytoplasm. Note the pale blue endothelial cells lining the sinusoids. 100 ×.

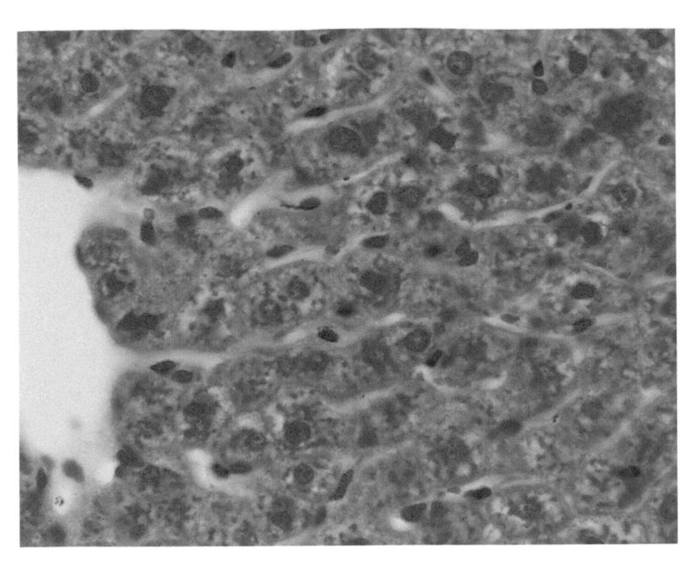

FIGURE K99b Lurking within the sinusoids are STELLATE MACROPHAGES that remove particulate material from the blood. The mammal from which this section was taken was injected with a particulate dye, Trypan blue, a few hours before the specimen was taken and the macrophages have picked up the blue dye. Note that a hepatocyte on the left, halfway up from the bottom, has two nuclei, a common feature of liver cells. Also note the round nuclei with prominent nucleoli in the hepatocytes. 63 ×.

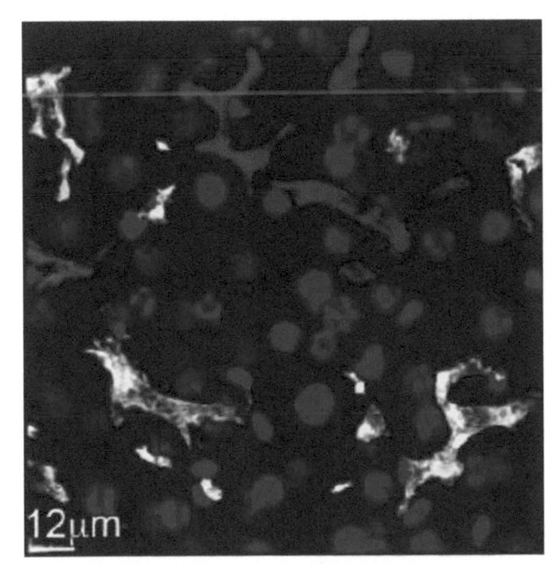

FIGURE K99c A confocal image illustrates stellate macrophages in the sinusoids of liver from a post-irradiated mouse. Two macrophage populations are present, resident and bone marrow derived. Nonchimeric resident macrophage cells appear red (expressing F4/80—red), while nonchimeric macrophage cells from bone marrow appear green (expressing GFP—green). Chimeric macrophage cells express both fluorescent proteins and appear yellow. *Beattie, L., Sawtell A., et al., 2016. Bone marrow-derived and resident liver macrophages display unique transcriptomic signatures but similar biological functions. J. Hepatol. 65(4), 758–768. http://doi.org/10.1016/j.jhep.2016.05.037 http://dx.doi.org/10.1016/j.jhep.2016.05.037.*

Blood from the hepatic artery and hepatic portal vein enters the interlobular arteries and veins, respectively, and permeates the liver in the portal areas (Fig. K98b). It enters the sinusoids (Fig. K99b), mixes, and flows toward the central vein. From the sinusoids the central veins collect blood and carry it to the sublobular veins, which unite to form the hepatic vein that opens into the inferior vena cava. In the injected preparation seen in Figs. K99e and K99f try to identify these vessels.

Hepatic cells are polyhedral and are loaded with glycogen (Fig. K99a); they possess finely granular, basophilic cytoplasm (Fig. K101c–K101e). Cell membranes are typically indistinct. The round, vesicular nuclei have prominent nucleoli and often two nuclei are seen in one cell. Fat droplets and glycogen granules may be demonstrated in hepatic cells by special techniques (Fig. K99a). The irregular hepatic sinusoids have an incomplete lining of endothelial cells and STELLATE MACROPHAGES (cells of Kupffer) (Figs. K99b–K99d, K99h).

The laminae or cords are irregular plates of cells one cell thick; there is always a tiny lumen between adjacent surfaces of two cells, the BILE CANALICULUS, into which the liver cells secrete bile (Fig. K98b). Bile

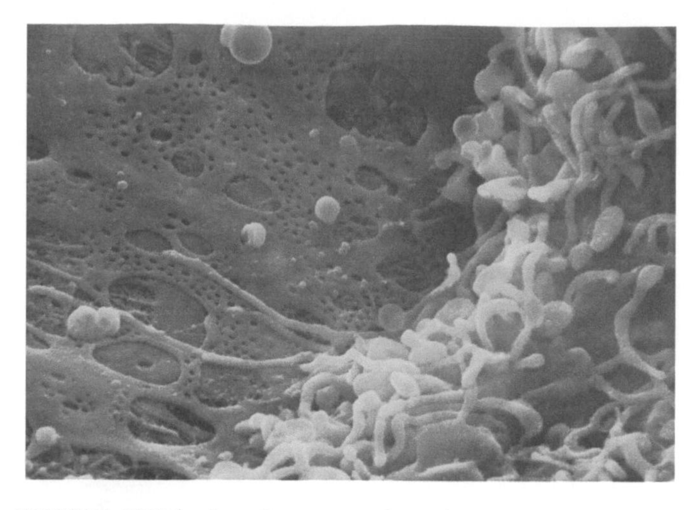

FIGURE K99d A stellate macrophage (with an active surface) extends its processes to the endothelial cell lining a sinusoid of the mammalian liver. The endothelial cell is pierced by large fenestrations and groups of small pores. 8400 ×.

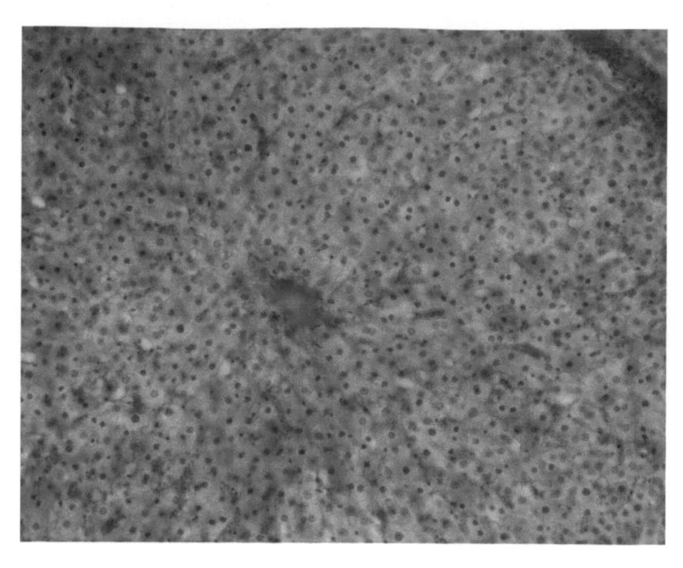

FIGURE K99e Colored gelatin has been injected into the veins supplying the liver of this mammal and shows up the sinusoids, the central vein, and the vessels at the periphery of the lobule. 20 ×.

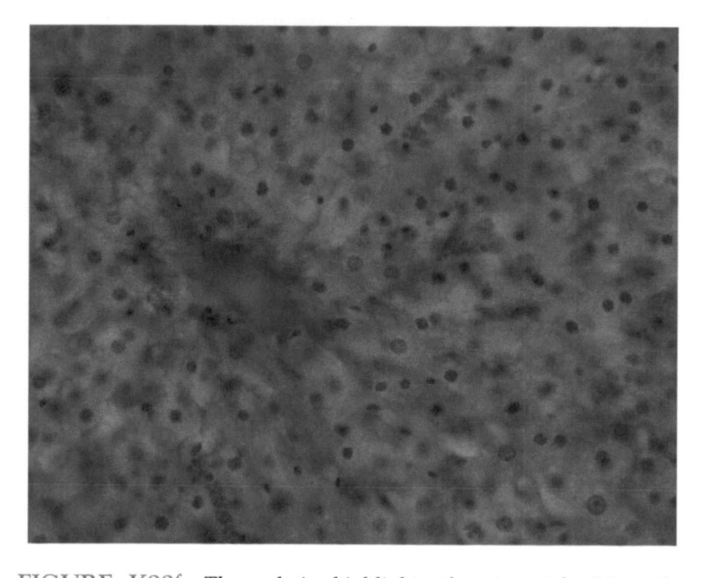

FIGURE K99f The gelatin highlights the sinusoids. Note the hepatocyte with two nuclei near the left border and about half way from the bottom. 40 ×.

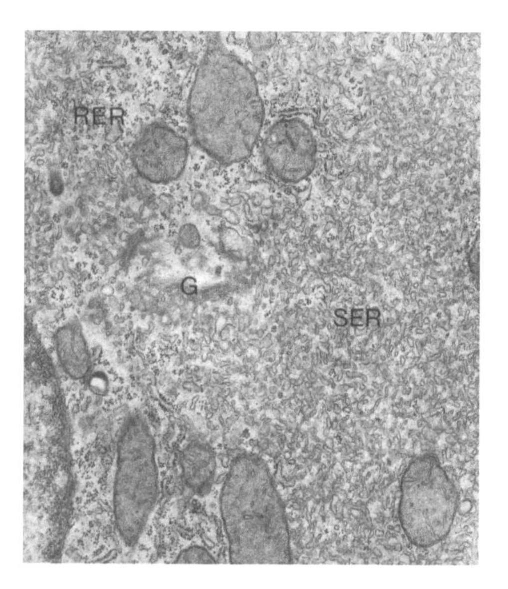

FIGURE K99g Section of a liver from a mammal treated with phenobarbital showing hypertrophy of the agranular endoplasmic reticulum (SER). This "territorial aggression" seems to crowd the granular endoplasmic reticulum (RER) into localized areas. The Golgi complex (G) shows several areas of communication with the *SER*. 40,000 ×.

canaliculi are best demonstrated by silver staining techniques (Fig. K100a). The branching, plate-like laminae, with their tiny bile canaliculi, extend from the center of the lobule and the canaliculi empty at or near the periphery into small branches of the bile ducts, the BILE DUCTULES (canals of Hering). The complex three-dimensional arrangement of bile canaliculi is clearly evident when we reconstruct a focal series through a thick section of liver as an animated 3-D image (Fig. K100b). The bile then passes into the INTERLOBULAR BILE DUCTS in the portal areas. These ducts are lined with simple cuboidal epithelium continuous with that of the bile ductules and surrounded by a sheath of connective tissue.

Hepatic parenchymal cells have three surfaces: the SINUSOIDAL surface, the INTERCELLULAR surfaces, and the BILIARY surface (Fig. K101d). Irregular microvilli on the sinusoidal surface hold aloft the endothelial cells,

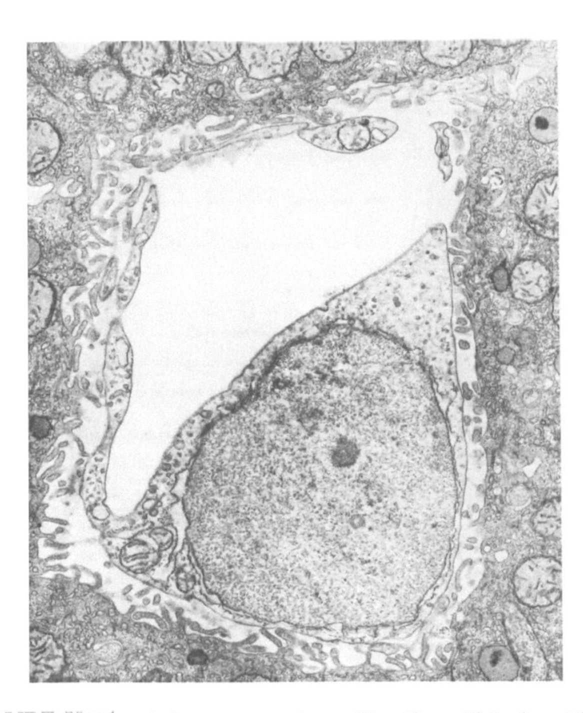

FIGURE K99h A large macrophage (Kupffer cell) lurks within a sinusoid of the liver of a rat. Microvilli from hepatocytes hold aloft the endothelial cells of the sinusoid, creating the perisinusoidal space (of Disse) beneath them. The microvilli greatly increase the surface area for the exchange of materials between hepatocytes and plasma. Since the bile canaliculi constitute the luminal part of the hepatocyte, the surface lining the perisinusoidal space is basal. 10,000 ×.

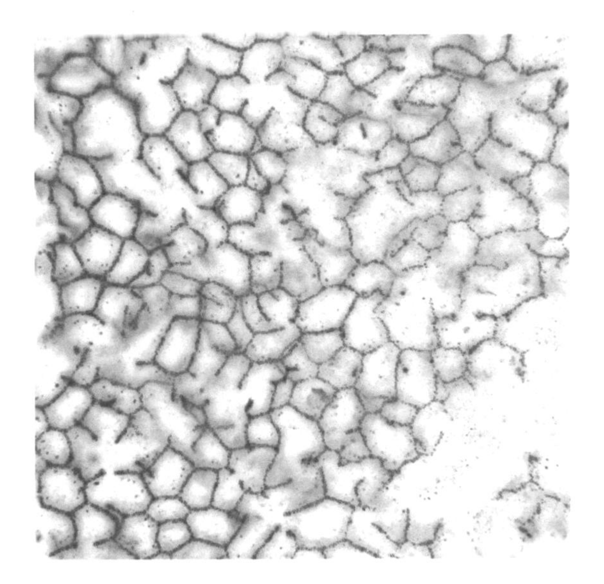

FIGURE K100b The three-dimensional "chicken-wire" configuration of bile canaliculi is clearly demonstrated in this animated image of a thick section of liver treated with a silver stain. A series of micrographs were taken sequentially through the section and then reconstructed as a three-dimensional image using image analysis software. The focal series is sometimes referred to as a Z-series. Using an image analysis program (Northern Eclipse), these images were converted to gray scale, deconvolved (to reduce out of focus information) and processed to emphasize the silver stained canaliculi prior to creating the three-dimensional rendering.

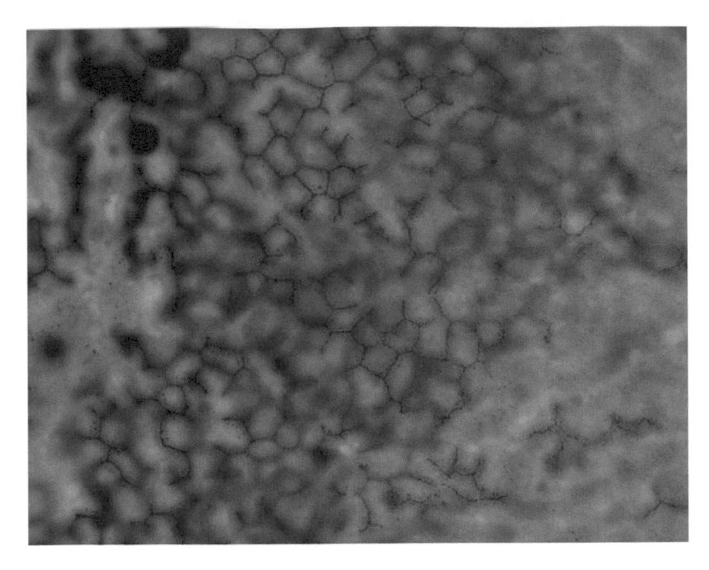

FIGURE K100a The liver arises as an outpouching of the wall of the embryonic gut lying in the pathway of the vitelline veins and umbilical veins, which are broken up into the multitude of small sinusoids between the parenchymal cells. The outpouching is an epithelial sac. During the invasion by blood vessels, it retains its epithelial nature by maintaining a surface that opens directly into the gut lumen. This connection is small and difficult to see in ordinary preparations, but it consists of the BILE CANALICULI, small vessels between adjacent hepatocytes of the parenchyma. Since these cells are roughly polyhedral, the canaliculi form a structure that resembles a three-dimensional chicken wire. (Imagine a hexagonal nut with a groove around its periphery. Then construct a lamina from several of these nuts and imagine the spaces that are formed as the grooves come together.)

producing the PERISINUSOIDAL SPACE (space of Disse) between the two cells (Fig. K99h). Delicate reticular fibers permeate this space. The endothelium of the sinusoids is incomplete: there are gaps between the cells and large fenestrae without diaphragms. The basement membrane surrounding the sinusoids is also incomplete. Pinocytotic vesicles are abundant in the endothelial cells. Large stellate macrophages within the lumen of a sinusoid can be identified (Figs. K99c and K99d). Their elaborate shape may be best appreciated in scanning electron micrographs. Bile canaliculi contain microvilli and are sealed by zonulae occludentes. Both granular and agranular endoplasmic reticulum is present in abundance in parenchymal cells (Figs. K101a–K101c). Agranular endoplasmic reticulum is especially well developed in animals exposed to drugs and other noxious agents (Fig. K99g). Ribosomes often aggregate to form polyribosomes. Glycogen appears as dense particles, 30–40 nm in diameter, often aggregated into clumps. Membrane bound lysosomes and microbodies (peroxisomes) are seen in the cytoplasm. Lysosomes are dense and contain fragments of organelles in various stages of disintegration. Microbodies have a less dense, homogeneous matrix enclosing a characteristic crystalline nucleoid.

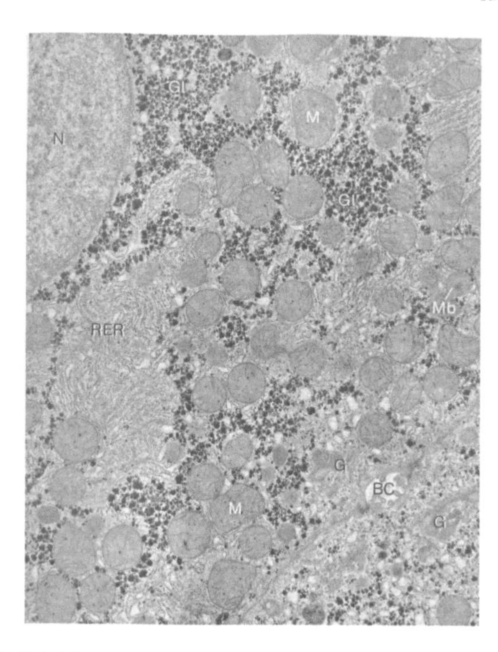

FIGURE K101a Transmission electron micrograph of a section of rat liver. *BC*, bile canaliculus; *G*, Golgi complexes; *Gl*, glycogen; *M*, mitochondria; *Mb*, a microbody; RER, granular endoplasmic reticulum. 20,000 ×.

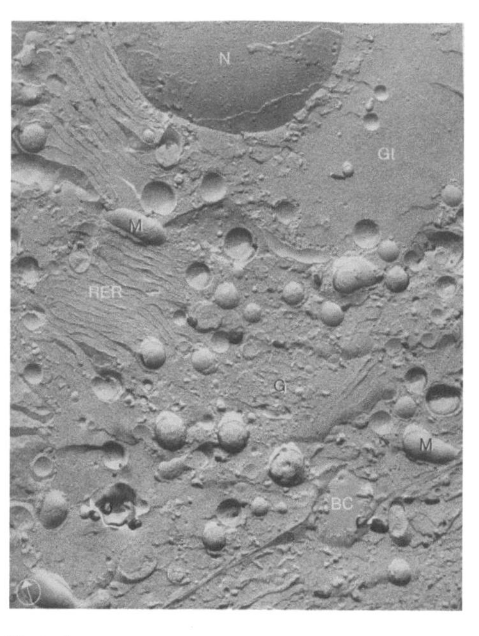

FIGURE K101b Corresponding freeze–etching preparation of the liver of the rat. *BC*, bile canaliculus; *G*, Golgi complex; *Gl*, glycogen; *M*, mitochondria; *N*, nucleus; *RER*, granular endoplasmic reticulum. 21,000 ×.

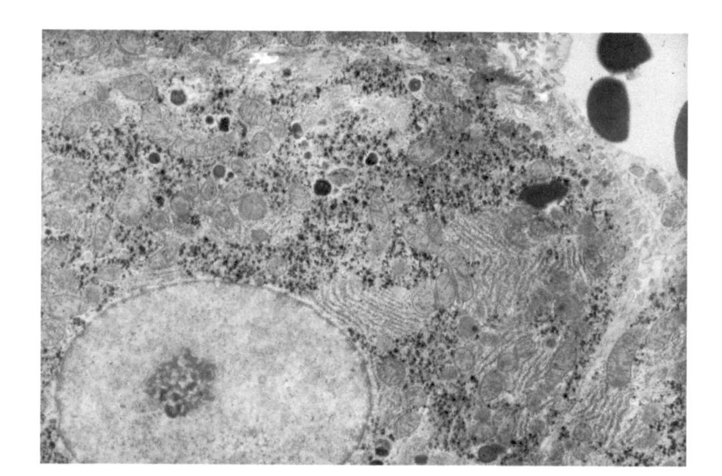

FIGURE K101c Electron micrograph of a section of a hepatocyte of a rat. The large oval nucleus looms at the lower left. Dark glycogen granules are abundant throughout. There are a few dark lysosomes here and there. The endothelial lining of the sinusoid in the upper right corner is indistinct. It is supported by microvilli from the hepatocytes thereby creating the perisinusoidal space (of Disse). 13,000 ×.

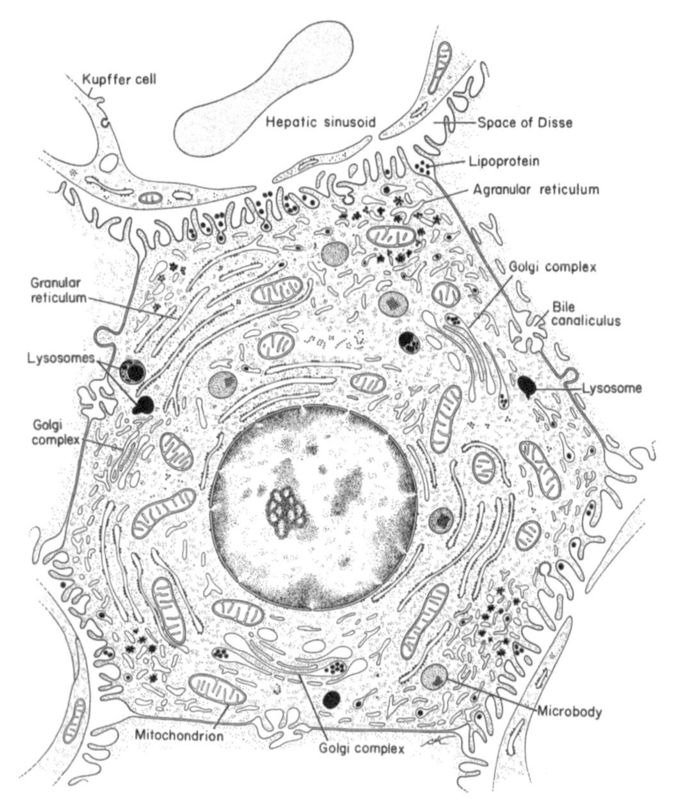

FIGURE K101d Drawing of a mammalian hepatocyte showing its relationship to bile canaliculi (apical) and sinusoids (basal) and its array of organelles.

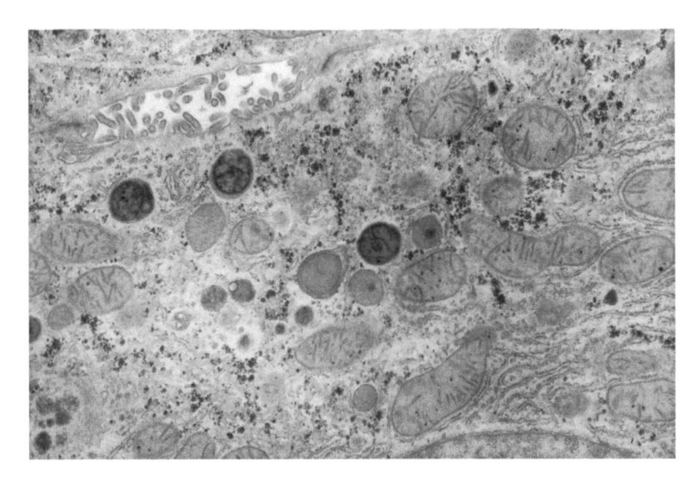

FIGURE K101e Electron micrograph of a thin section of a hepatocyte from a rat. A small portion of its nucleus appears at the top left. Large mitochondria with their well-developed cristae are seen throughout as well as a few large, round lysosomes. Both granular and agranular endoplasmic reticulum are seen. Small dark granules of glycogen are scattered throughout. There is a bile canaliculus at the lower right corner; microvilli extend into its lumen from the plasma membrane of the hepatocyte. 30,000 ×.

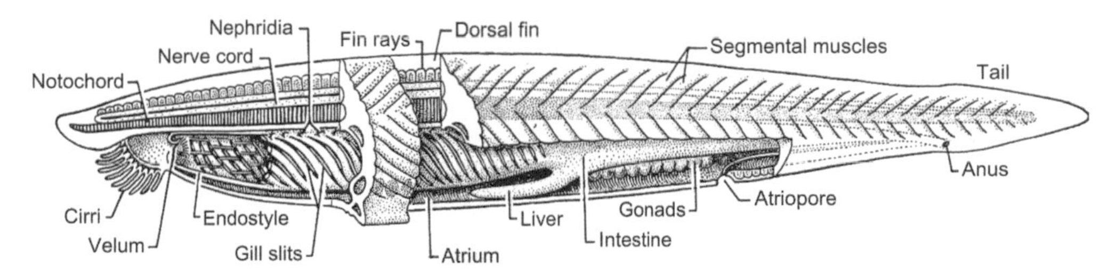

FIGURE K102a Drawing of amphioxus showing the liver diverticulum, which may recall the ancestry of the vertebrate liver that originates as a diverticulum of the gut and imposes itself in the path of aggressive vascular mesenchyme that vastly alters its appearance. The liver cords are remnants of the epithelial lining of the diverticulum although they do not appear epithelial. (Remember: an epithelium is a layer of cells that has a free surface bordering a lumen.) The remnant of the lumen of the diverticulum in this case is greatly reduced but exists as the bile canaliculi.

COMPARATIVE HISTOLOGY OF THE LIVER

The hepatic diverticulum of amphioxus may represent the first stage in the development of the gland as we know it in the vertebrates (Figs. K102a–K102d). It arises as a ventral outpouching of the intestine and grows forward beneath the pharynx but remains a hollow sac with a ciliated lining throughout life. A system of veins coming from the intestine breaks up into capillaries on its surface, resembling the hepatic portal system of higher forms. The liver of hagfish, *Myxine*, is a mass of branching tubules embedded within vascular mesenchyme and receives blood from both the hepatic artery and hepatic portal veins (Figs. K103a and K103b). In higher vertebrates the tubular arrangement is lost and the laminar form appears. Note, however, that the lumina of the tubules in the hagfish liver, and the lumina of the bile canaliculi of the higher forms are all derived from the lumen of the gut and retain their continuity with it.

The laminae of mammalian liver cords are one cell thick. In the lower vertebrates these may be two cells thick so that, while each hepatocyte still has a frontage on a blood sinusoid, its blood supply is not as rich as that of mammals. Compare the thickness of the laminae in images of liver sections from various vertebrates. Note also that true lobules occur only in the livers of birds and mammals (Figs. K104, K105a, K105b, K106a, K106b, K107a, K107b, K108a, K108b, K109a, K109b, K110a–K110j, K111a–K111d, K112a and K112b, K113a, K113b.)[2]

[2]Hagfish Fig. K104; shark Figs. K105a and K105b; dogfish Figs. K106a and K106b; Lungfish Figs. K107a and K107b; Figs. K108a and K108b; Protopterus Figs. K109a and K109b; Necturus Figs. K110a and K110b; AmphiumaFigs. K110c–K110f; toad Figs. K110g; marine toad Figs. K110i and K110j; tuatara Figs. K111a and K111b; rattlesnake Figs. K111c and K111d; chicken Figs. K112a and K112b; horse Fig. K113a; fetal mammal Fig. K113b.

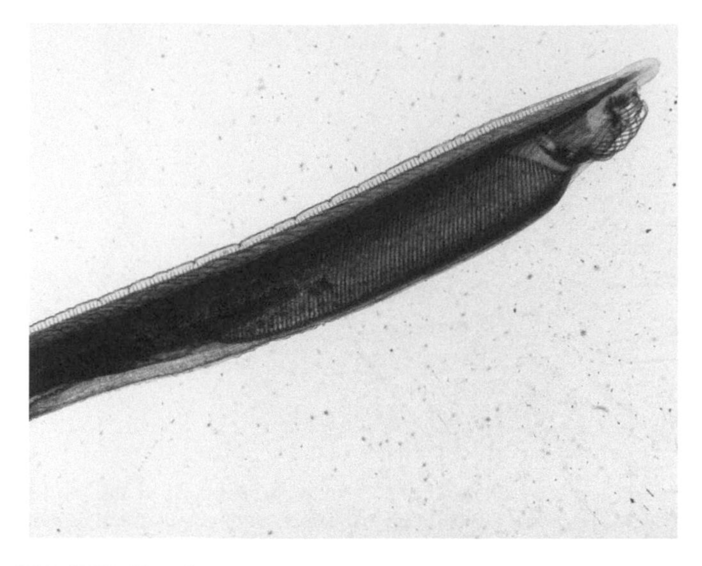

FIGURE K102b This is a whole mount of amphioxus or the lancelet, a cephalochordate. The mouth, surrounded by the oral hood, lies at the upper right corner. The dark band extending from the upper right to the lower left is the NOTOCHORD, the chief support of the body. The PHARYNX is enclosed by a basket of GILL BARS with GILL SLITS between. The LIVER DIVERTICULUM extends forward from the anterior portion of the INTESTINE and lies alongside the pharynx. 1 ×.

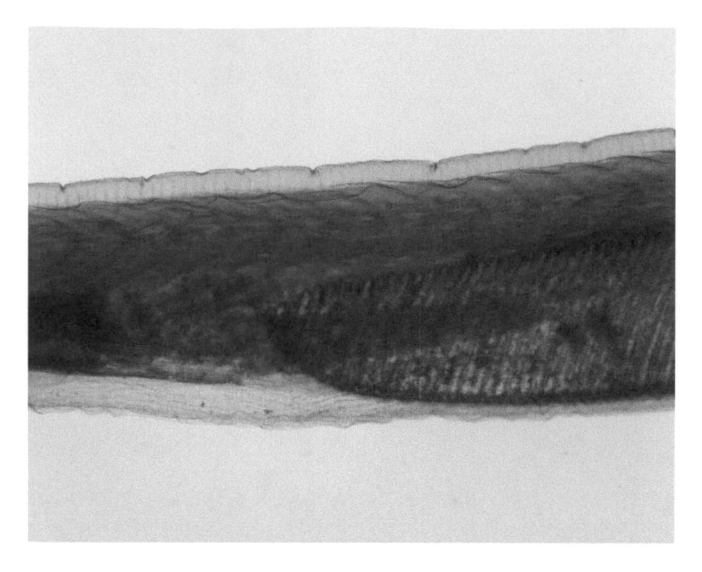

FIGURE K102c The LIVER DIVERTICULUM is the shadowy structure just below the center of the micrograph lying beneath the gill bars of the pharynx. 3.2 ×.

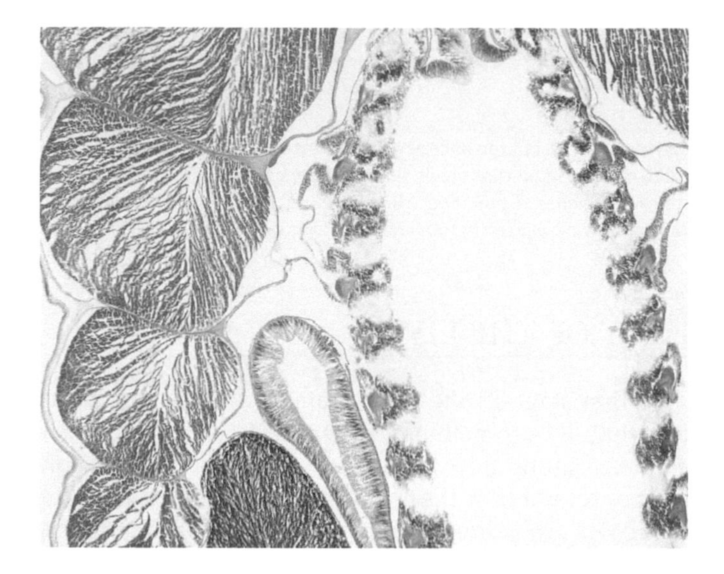

FIGURE K102d This cross section through the pharynx of amphioxus shows the LIVER DIVERTICULUM immediately to the left of the gill bars. Masses of muscle of the body wall surround the pharynx; the TESTIS is at the lower left. 10 ×.

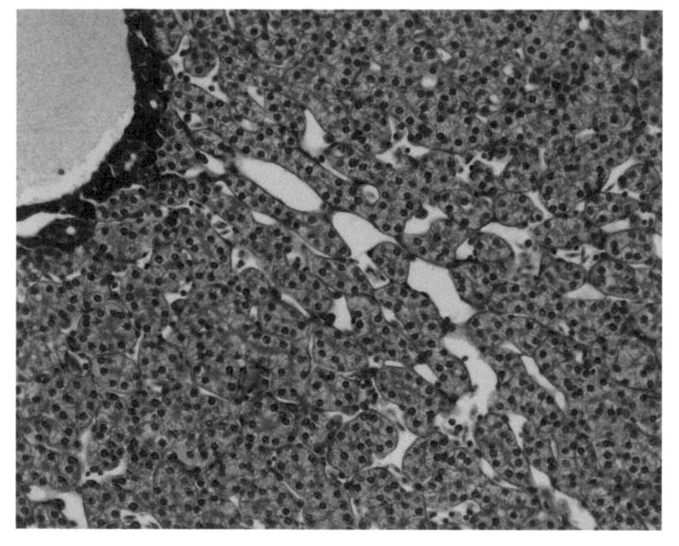

FIGURE K103a The double nature of the liver cords is obvious in the liver of a hagfish (*Myxine glutinosa*). The sinusoids at the center are distended and demonstrate their thin lining of squamous endothelial cells. Erythrocytes may be seen in a few sinusoids. Since lobulation is not apparent in this liver, it is probably misleading to refer to the vein at the upper left as a "central vein." 20 ×.

Pigmented cells of unknown significance may be scattered between the parenchymal cells in livers of some vertebrates, notably the amphibians (Fig. K116c). Recall the presence of hemopoietic activity in the subcapsular tissue of the liver of some amphibians (Fig. K110c). The storage product of the liver of higher forms is glycogen. Droplets of oil are also stored in many vertebrates and give sections of liver prepared by routine methods the appearance of Swiss cheese.

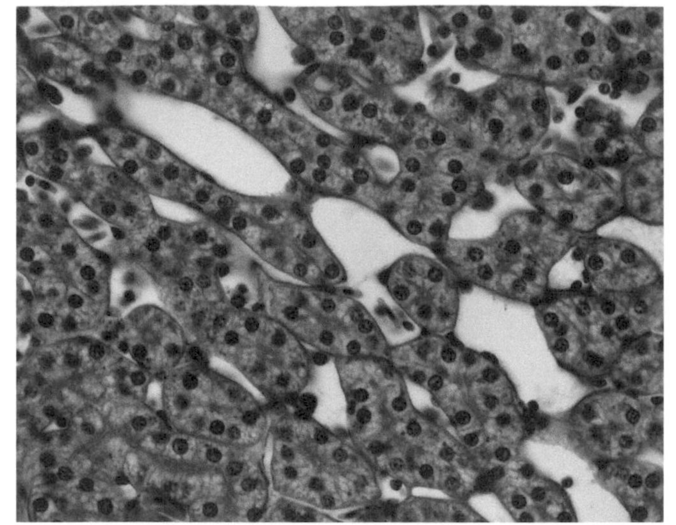

FIGURE K103b In the course of the development of the hepatic diverticulum in the hagfish, epithelial cells form alveoli that are invaded by vascular mesenchyme. This alveolar nature is retained and can be seen in this micrograph. Note the tiny bile canaliculi at the center of some of these "alveoli." The blood spaces or sinusoids are surrounded by thin endothelial cells and contain a few blood cells. 40×.

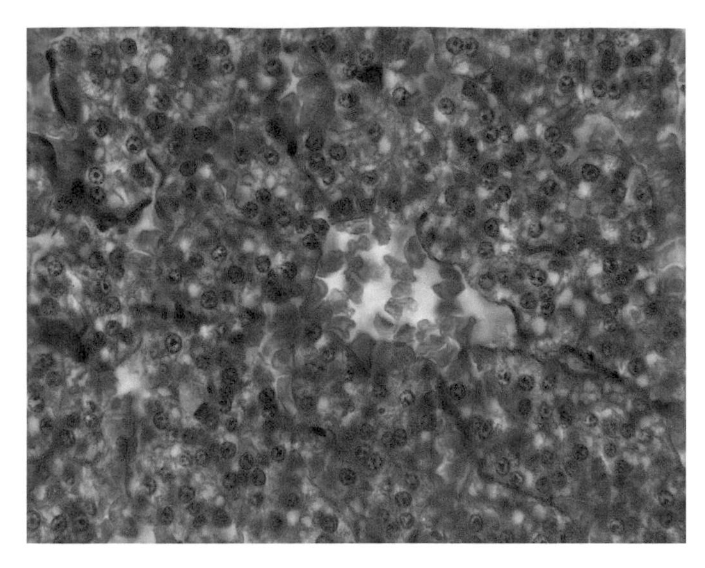

FIGURE K104 The bright orange staining of the erythrocytes in this section well delineates the sinusoids and "central "vein of the liver of the lamprey *(Petromyzon marinus)*. A few endothelial cells may be seen on some of the blood spaces. The hepatocytes have assumed an alveolar appearance. 63×.

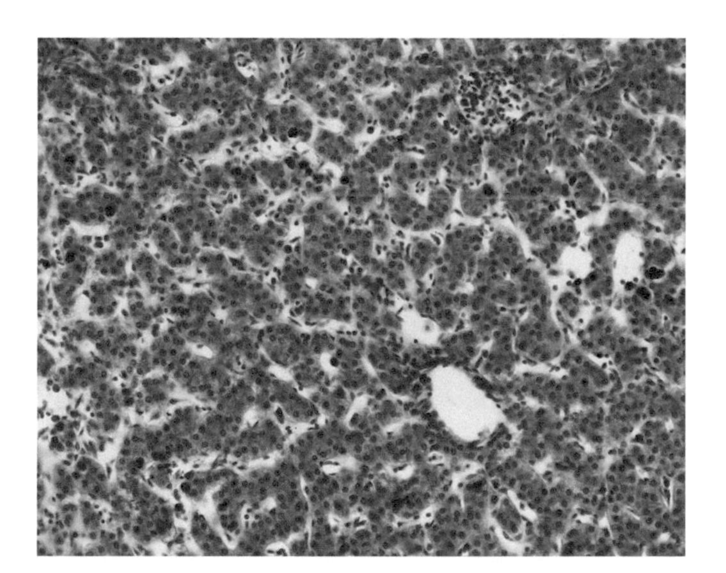

FIGURE K105a The double nature of the liver cords with sinusoids between is well shown in this micrograph from a shark. A "central" vein is near the center. A few melanocytes are scattered throughout. 40×.

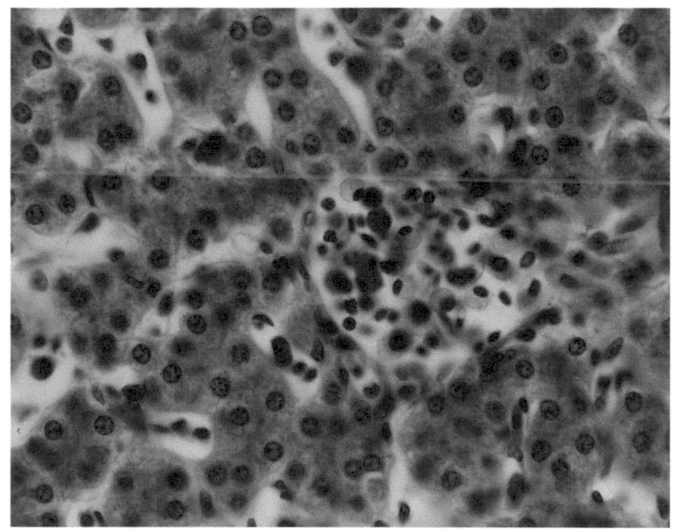

FIGURE K105b Several sinusoids pour into the "central" vein, which is full of blood cells. The endothelial lining of blood spaces can be seen. 100×.

GALL BLADDER

The gall bladder is also derived from an outpouching of the gut and as such bears some resemblance to it (Fig. K114). It is present in most vertebrates and is remarkably similar throughout. Although the four typical layers of the gut are not well defined in the hagfish, they are recognizable (Fig. K115). Histological structure of the gall bladder from an assortment of different vertebrates can be compared in Figs. K116a–116c, K117, 118a–K118c, K119, K120a–K120c.[3]

[3]Amphiuma Figs. K116a–116c; Frog Fig. K117; Toad Figs. 118a and 118b; Marine toad Fig. K118c; Rattlesnake Fig. K119; Mammal Figs. K120a–120c.

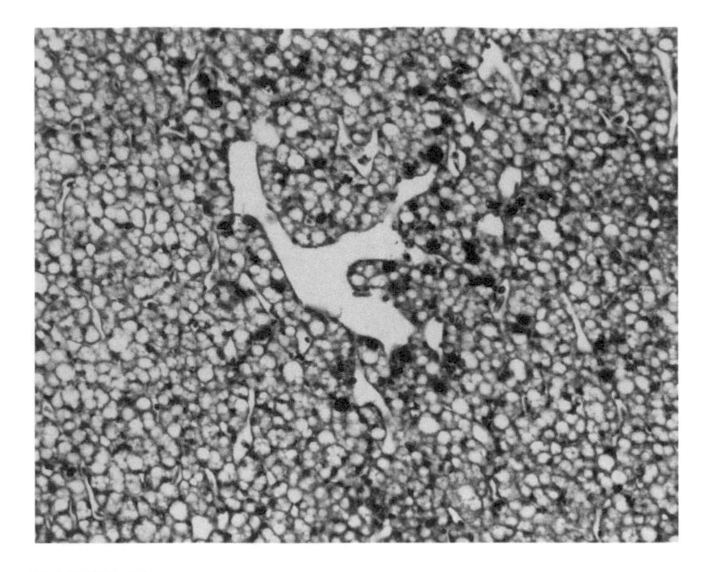

FIGURE K106a The storage product of the liver of many animals is glycogen. This fairly insoluble material is able to store large amounts of fuel without upsetting the osmotic balance of the liver. However, in some animals, such as shark, the storage product is oil. The cells of the liver of this dogfish are greatly distorted by their content of droplets of oil and the picture of liver cords permeated by sinusoids is obscured. Several sinusoids can be discerned in this section, some entering the collecting vein at the center. Pigment cells are often seen in the liver of vertebrates. Their presence is an enigma. 20 ×.

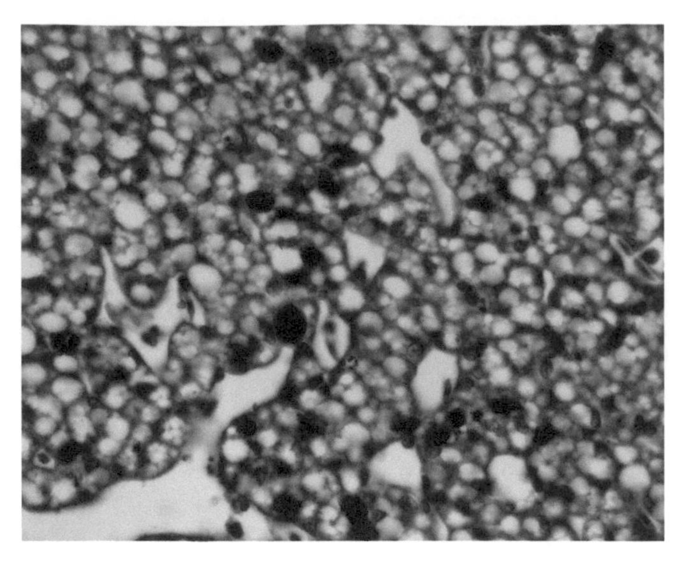

FIGURE K106b A sinusoid at the lower left pours into the collecting vein at the bottom left. The endothelial lining of these vessels is so thin that it can hardly be distinguished. The hepatocytes are bulging with their content of oil. Pigment cells are distributed throughout. 40 ×.

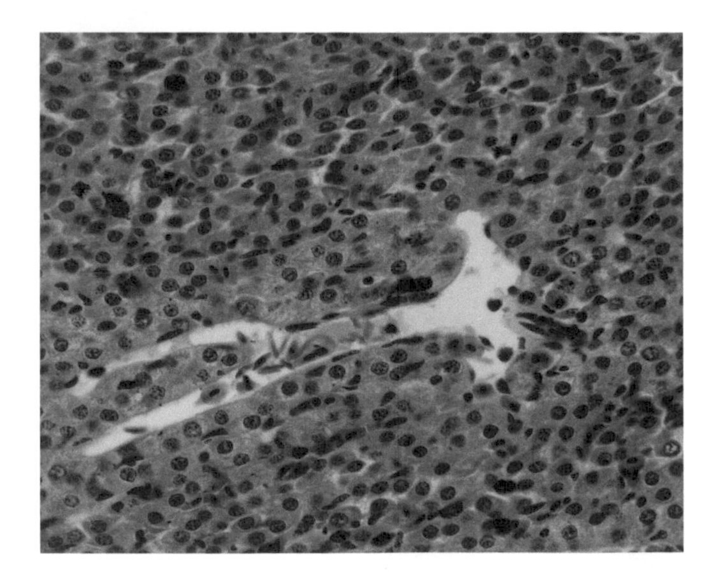

FIGURE K107a The liver of the lungfish has double cords of hepatocytes with sinusoids in between. Two sinusoids at the bottom left come together to pour into a larger venous space. The presence of large numbers of eosinophilic cells is enigmatic. 20 ×.

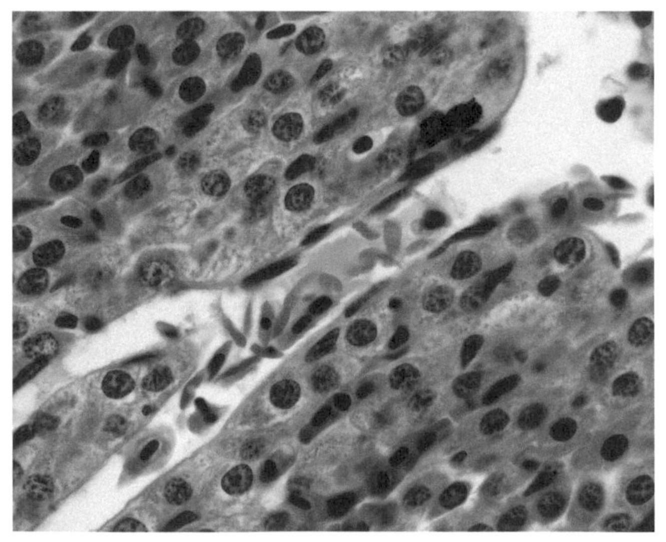

FIGURE K107b Sinusoids in the liver of the lungfish are lined by endothelial cells. They come together in a larger venous space at the upper right. 40 ×.

The simple columnar epithelium of the tunica mucosa of the gall bladder in mammals rests on fibroelastic connective tissue that represents an intermingling of the lamina propria mucosae and the tela submucosa (Figs. 121a and 121b). The only secretion in the gall bladder is that of a small group of tubuloacinar mucous glands near its neck. Outside the connective tissue is a thin tunica muscularis of smooth muscle in

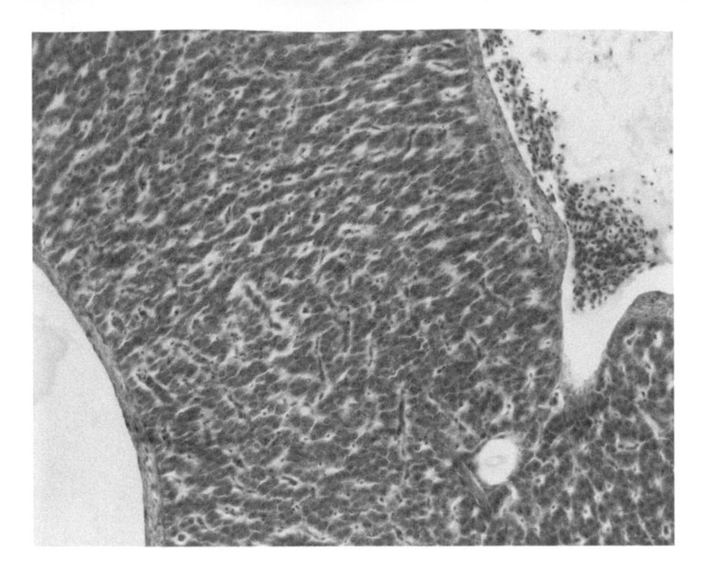

FIGURE K108a In the liver of an eel (*Anguilla anguilla*) several sinusoids, packed with blood cells, make their way to the central vein at the upper right. 20 ×.

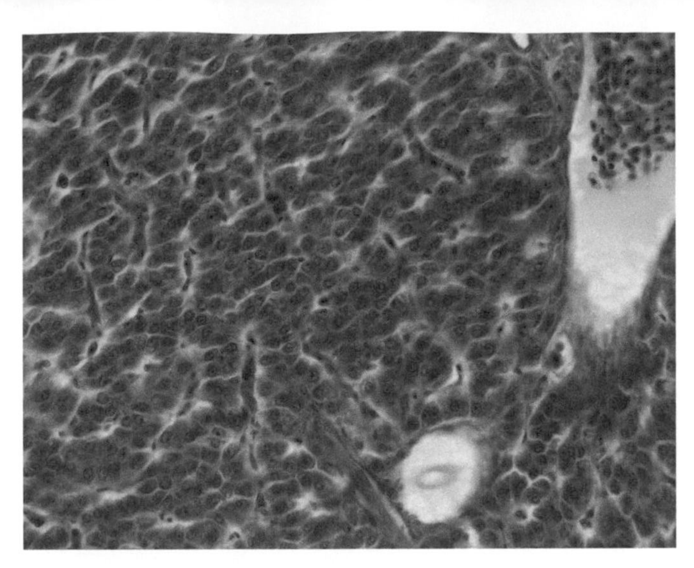

FIGURE K108b The double nature of the cords of hepatocytes is apparent in the eel. 40 ×.

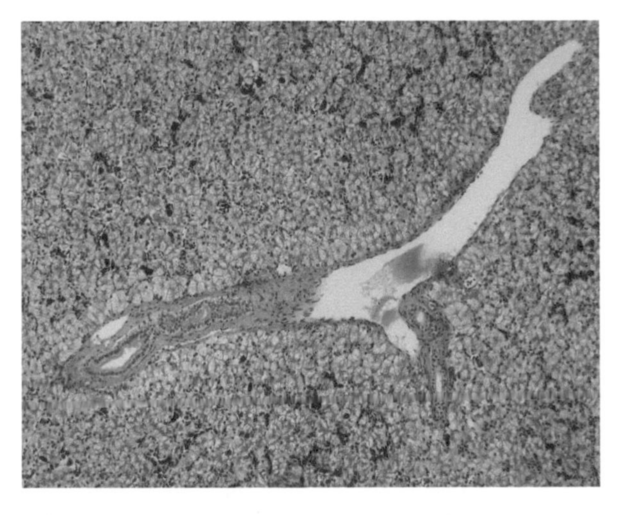

FIGURE K109a The liver of the toad shows double cords, several melanophores, branches of the biliary system (note the simple columnar epithelium), and a branch of the portal vein traversing the screen diagonally. 10 ×.

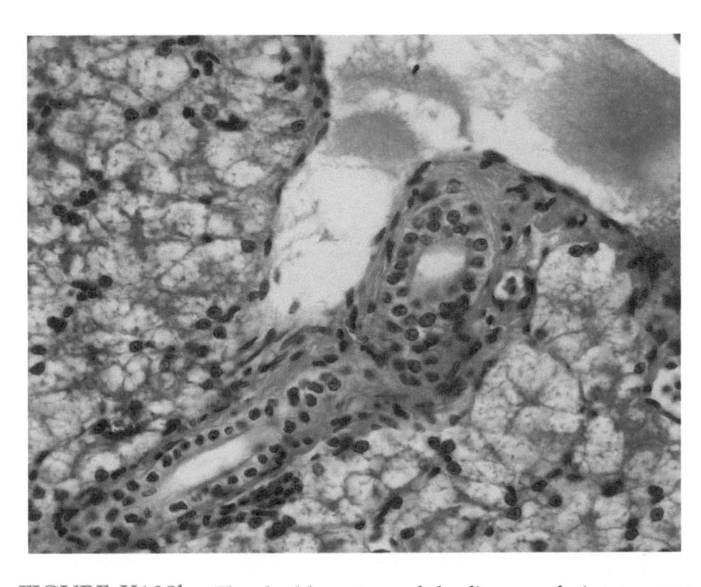

FIGURE K109b The double nature of the liver cords is apparent in this micrograph of the liver of a toad. Branches of the biliary system are recognized by their simple cuboidal/columnar epithelium. 40 ×.

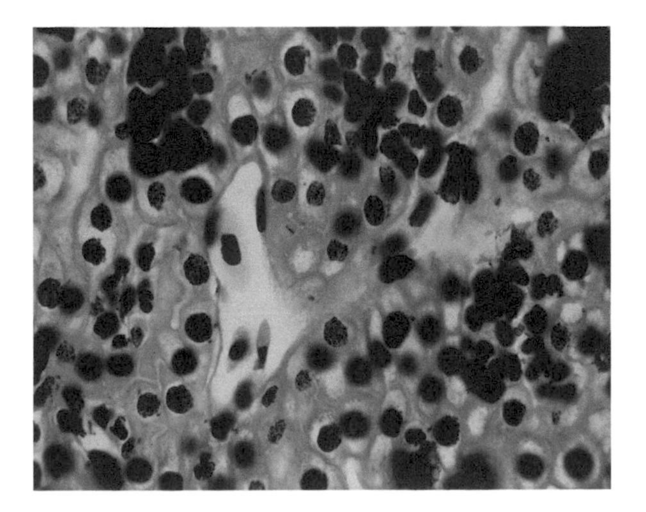

FIGURE K110a A branching sinusoid, lined with endothelial cells, occupies the center of this picture of the liver of a mudpuppy (amphibian, *Necturus*). The abundance of nuclei here and there probably indicates some diffuse hemopoietic activity. Melanocytes are scattered throughout. 40 ×.

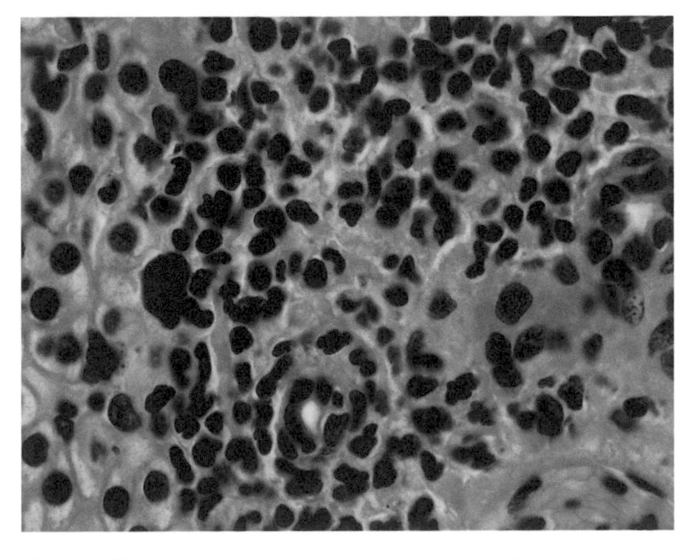

FIGURE K110b Hemopoiesis is abundant in this region of the section of the liver of a mudpuppy. (A small branch of the biliary system is at the lower center.) 40 ×.

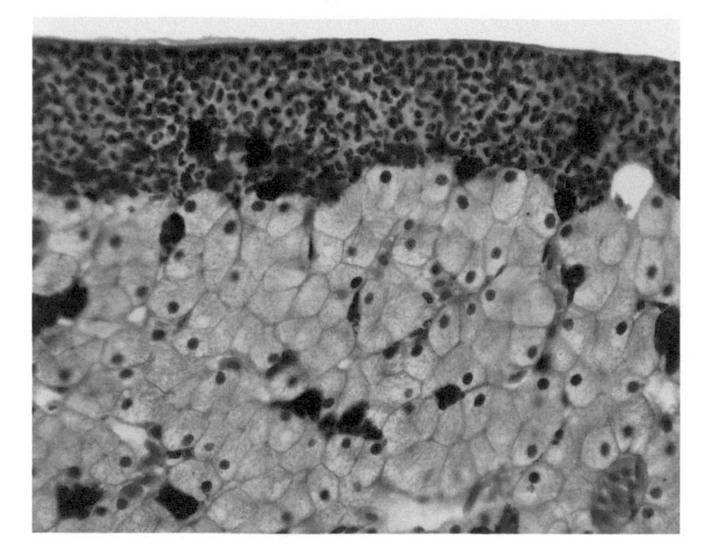

FIGURE K110c There is active hemopoiesis in the livers of some urodeles. In this section of the liver of *Amphiuma*, it forms a thick band under the capsule, obscuring the characteristic structure of the liver. Below, it is easy to discern the double nature of the liver cords with sinusoids between. A few melanocytes are scattered here and there. Some bile capillaries are evident. 20×.

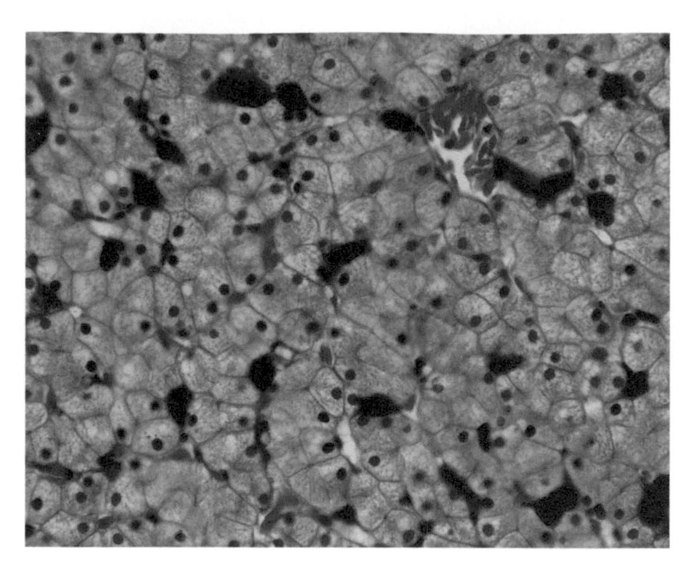

FIGURE K110d A "central" vein is full of erythrocytes at the upper right of this section of the liver of *Amphiuma*. The endothelial lining of sinusoids can be seen occasionally. Pigment cells abound. 20×.

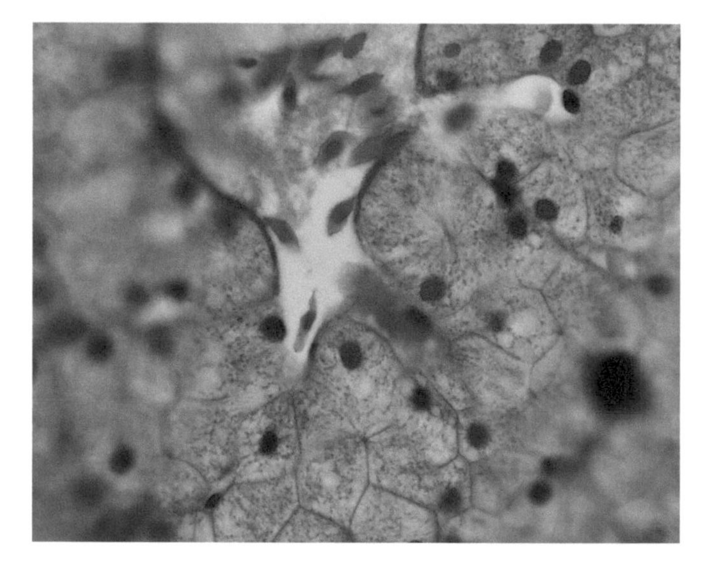

FIGURE K110e While this section is the liver of *Amphiuma* may not be as flat as might be desired; a sinusoid opening into a "central" vein at the center is plainly visible and another smaller sinusoid opens into the vein at its right side. 40×.

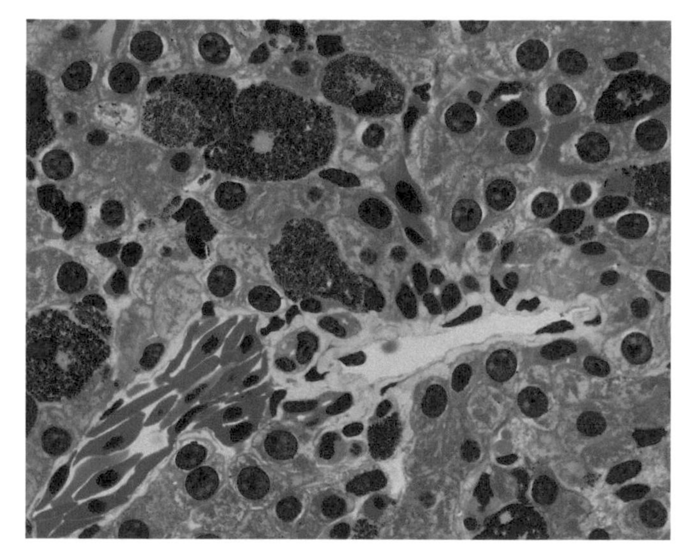

FIGURE K110f A great deal of detail can be seen in this semithin section of the liver of *Amphiuma*. A sinusoid, with its thin wall of endothelial cells, runs across the center. Melanocytes, packed with brown melanin granules, are scattered throughout. 40×.

intermingled groups of circular, longitudinal, and oblique fibers. There is a fairly thick tunica serosa of loose connective tissue covered by mesothelium. Electron micrographs of the epithelial cells display the features of transporting epithelia: apical microvilli, apical junctional complexes, and elaborate interdigitating lateral cell borders (Fig. K122).

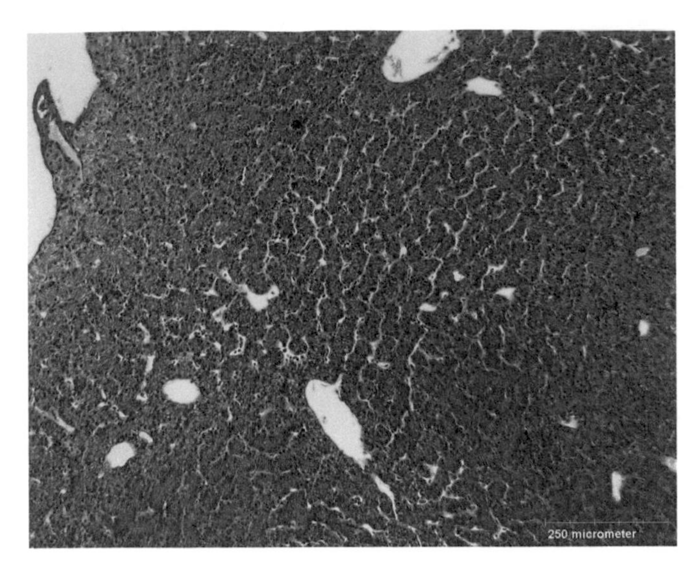

FIGURE K110g The double nature of the liver cords is apparent in this section of this toad liver (amphibian). A sinusoid enters a central vein at the lower center. A few pigment cells are scattered here and there. 10 ×.

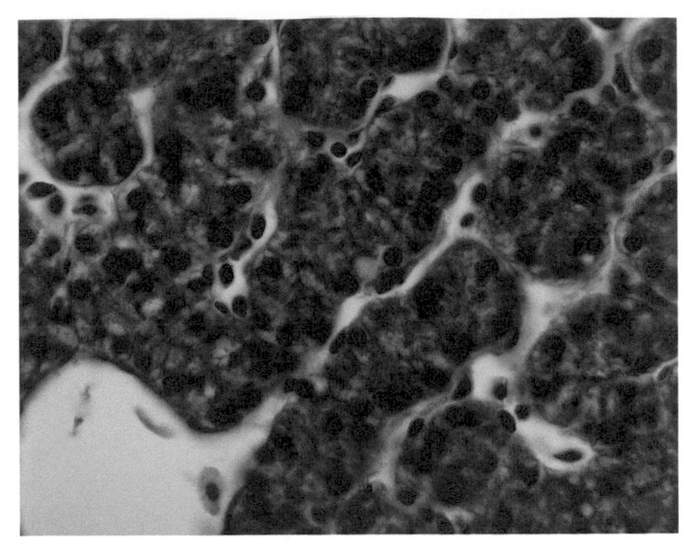

FIGURE K110h The delicate central vein is lined with squamous endothelial cells in the liver of a toad. The sinusoid seen above enters at the right. The endothelial lining of these vessels is continuous. 63 ×.

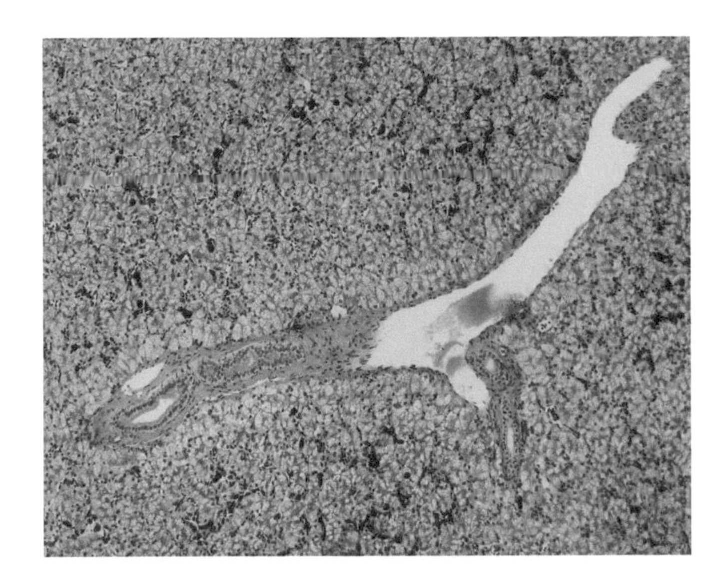

FIGURE K110i Portions of the portal triad can be seen crossing the field diagonally in this section of the liver of a marine toad (*Bufo marinus*). A central vein opens into a branch of the portal vein from the bottom. There are sections of the biliary system nearby. At the lower left is a larger branch of the biliary system. Erythrocytes have stained a cherry red and indicate the presence of sinusoids. A few melanocytes are scattered throughout. 10 ×.

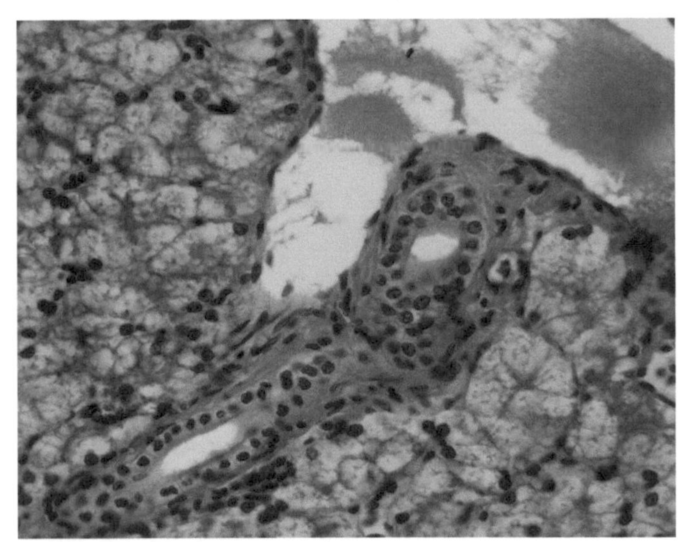

FIGURE K110j A central vein enters a branch of the portal vein at the upper right of this section of the liver of a marine toad. Note its lining of squamous endothelial cells. Branches of the biliary system, with a simple cuboidal epithelium, are seen below the veins. Glycogen has been washed out of the hepatocytes, leaving them with a bubbly appearance. The liver cords are double. 40 ×.

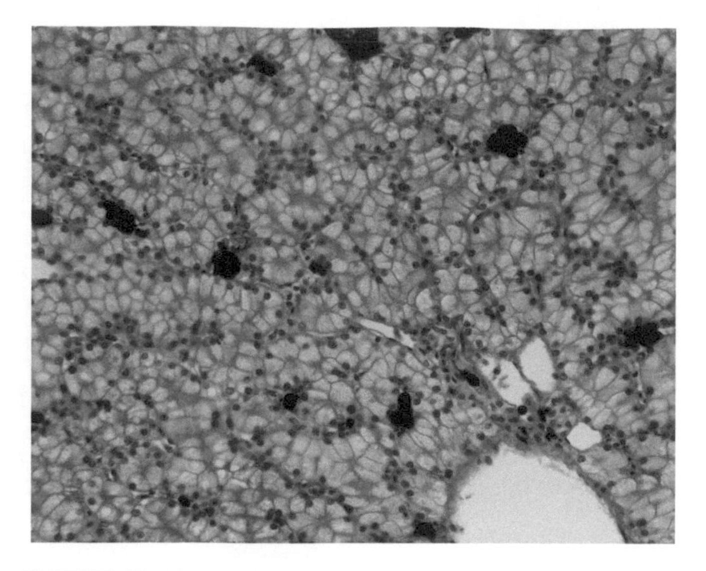

FIGURE K111a The liver of this reptile, the tuatara *Sphenodon punctatum* from New Zealand, resembles many of those seen previously. Its double cords radiate from the central vein at the lower right. Several pigment cells are seen in the parenchyma. The endothelial lining of several sinusoids can be seen. 20 ×.

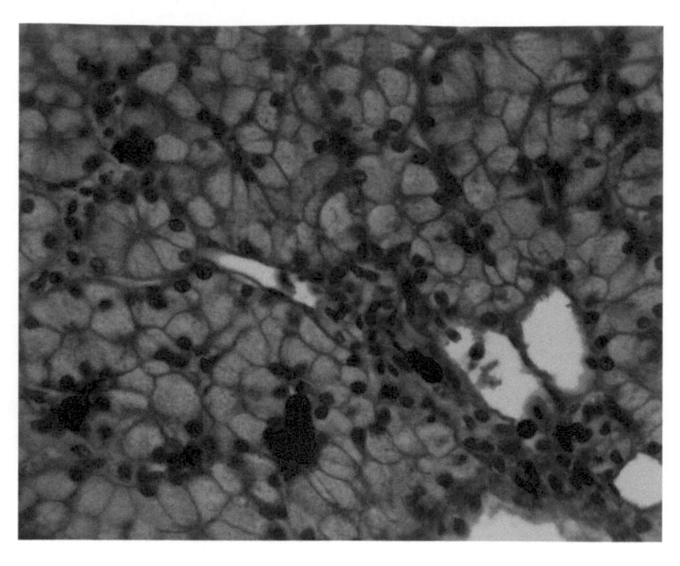

FIGURE K111b A bile canaliculus appears at the left of this micrograph of the liver of tuatara. 40 ×.

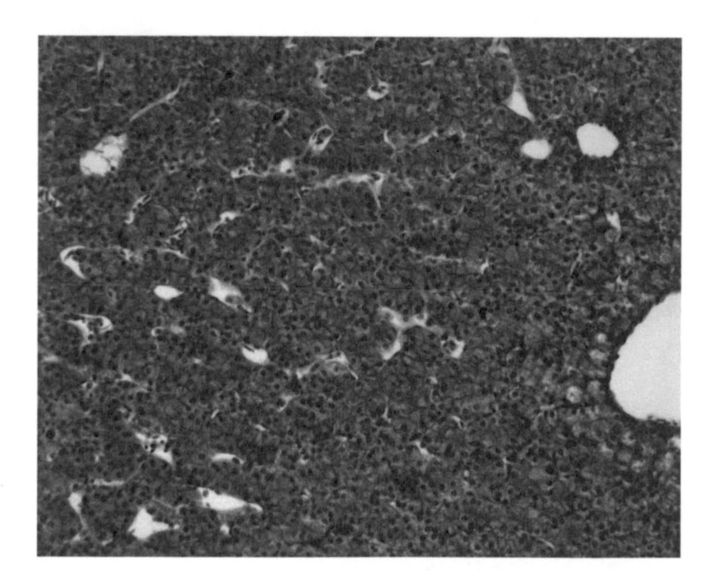

FIGURE K111c Several sinusoids with thin endothelial linings penetrate between double cords of hepatocytes in the parenchyma of the liver of a rattlesnake. A few pigment cells are scattered throughout. There is a central vein at the right. 20 ×.

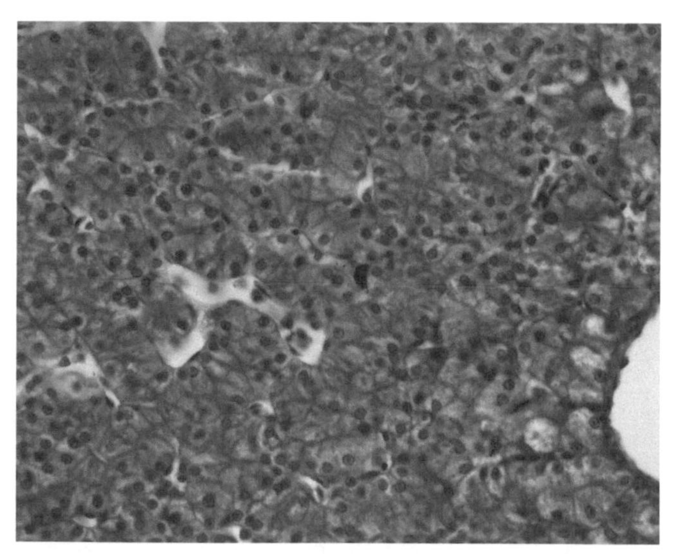

FIGURE K111d The sinusoids with thin endothelial linings may be seen between double cords of hepatocytes in the liver of this rattlesnake. Endothelial cells are well demonstrated lining the sinusoids and central vein. 40 ×.

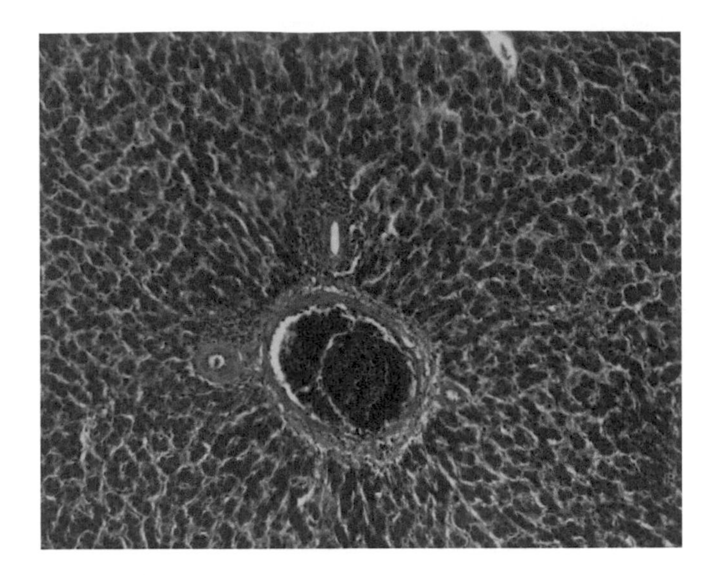

FIGURE K112a A lobular pattern is not distinct in the liver of the chicken. Double cords of hepatocytes abut on this portal triad of vein, artery, and bile passage. Note that a small amount of hemopoietic tissue is clustered near the artery and bile passage. 20 ×.

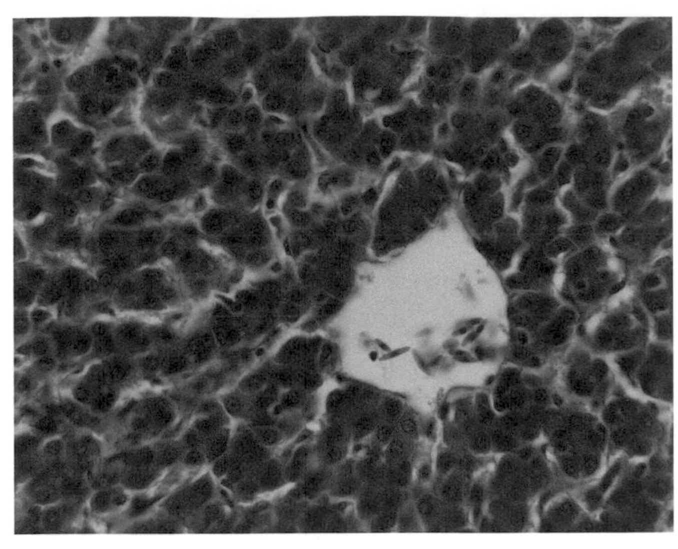

FIGURE K112b Several sinusoids open into the central vein of the chicken liver. Their double nature is clear. There is a simple squamous endothelial lining within the sinusoids and lining the central vein. 63 ×.

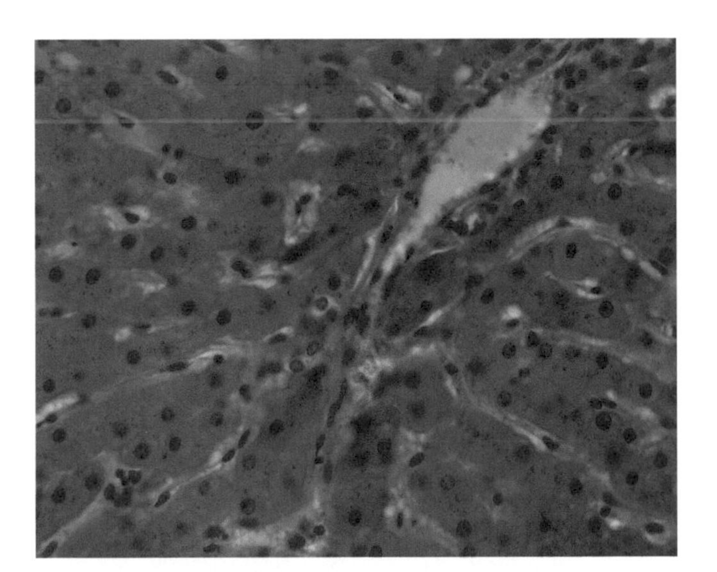

FIGURE K113a Even in a huge mammal such as the horse, the characteristic mammalian pattern of single-walled hepatic cords with sinusoids between is maintained. 40 ×.

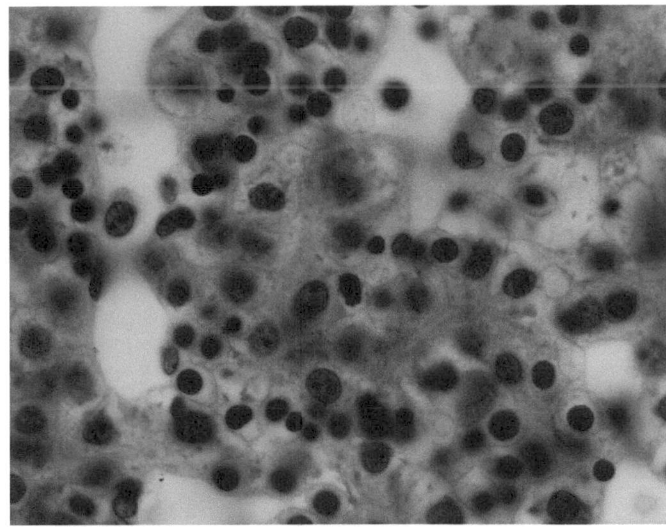

FIGURE K113b As was mentioned in the section on hemopoiesis the liver of fetal mammals functions as a hemopoietic organ for some time. Although a sinusoid can be distinguished in this section, the lobular nature of the liver is obscured by the active formation of blood cells. Note the abundant mitotic figures. 100 ×.

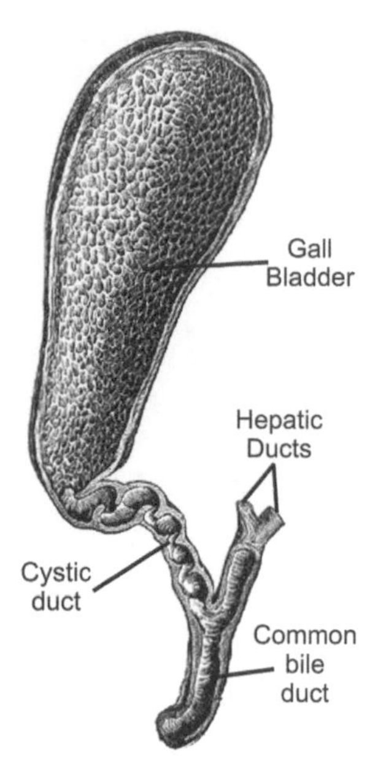

FIGURE K114 Drawing of a human gallbladder opened to show its lining of tunica mucosa and the bile ducts. ***Public Domain****, Henry G. (1918) Anatomy of the Human. Body Plate 1095, https://commons.wikimedia.org/w/index.php?curid=297359.*

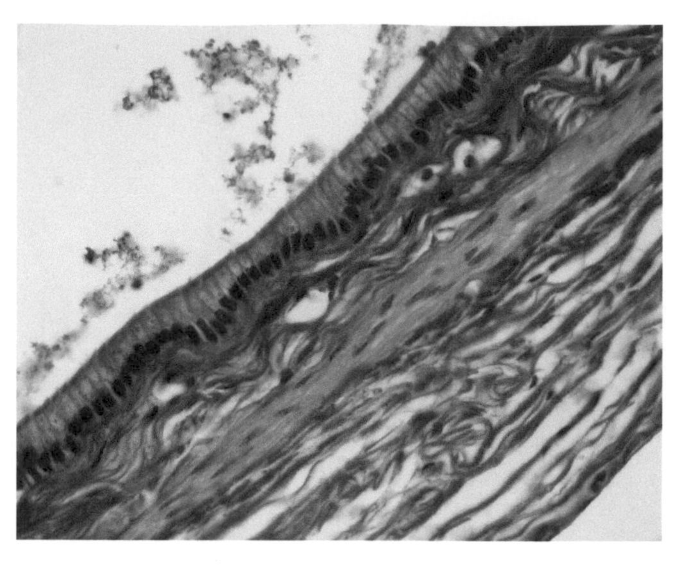

FIGURE K115 The wall of the gall bladder of the hagfish shows a simple columnar epithelium resting on a vascular connective tissue of the tela submucosa. 40 ×.

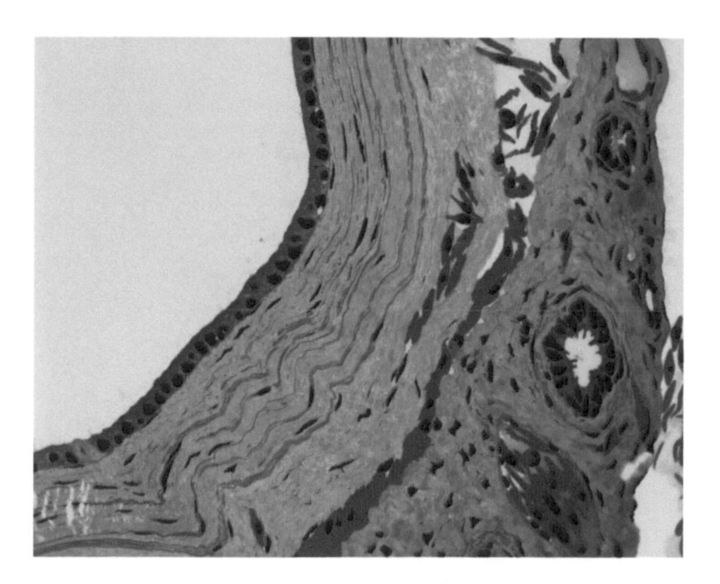

FIGURE K116a The gall bladder of *Amphiuma* demonstrates the layers of the gut—a simple columnar or cuboidal epithelium resting on vascular connective tissue of the tela submucosa. Small branches of the biliary system can be seen at the right; their epithelium is similar to that of the gall bladder. 20 ×.

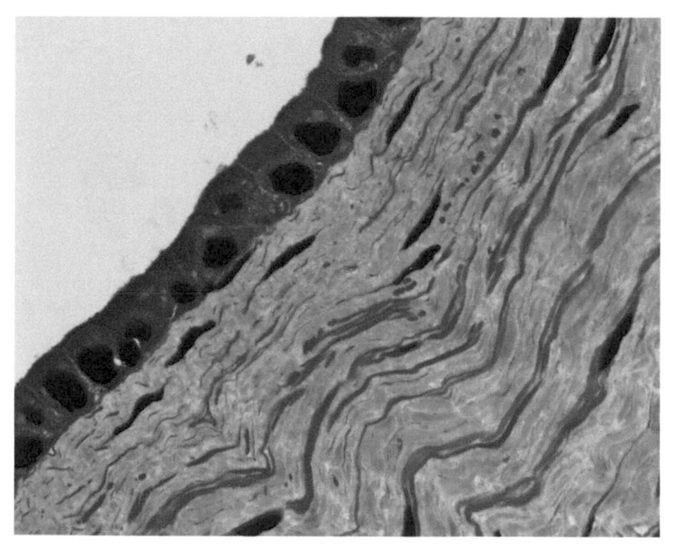

FIGURE K116b The low columnar cells of the epithelium of the gall bladder of *Amphiuma* have a ragged border indicating the presence of microvilli. 63 ×.

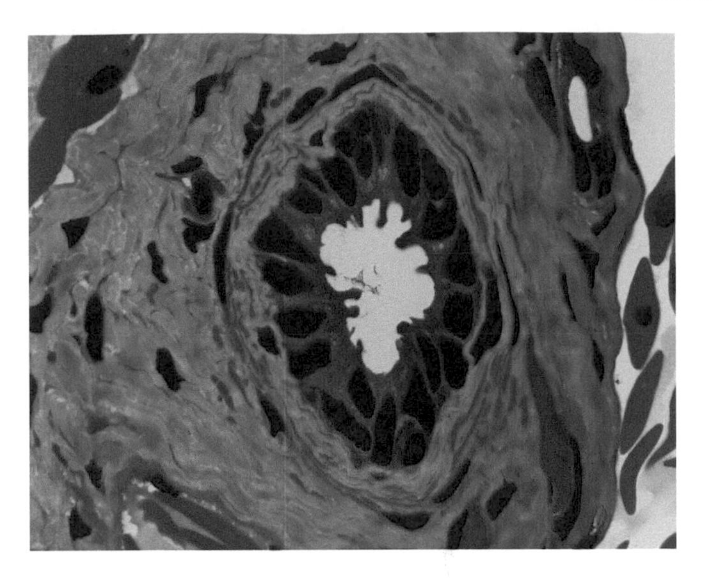

FIGURE K116c The epithelium of a small branch of the biliary system in *Amphiuma* resembles that of the gall bladder. 63 ×.

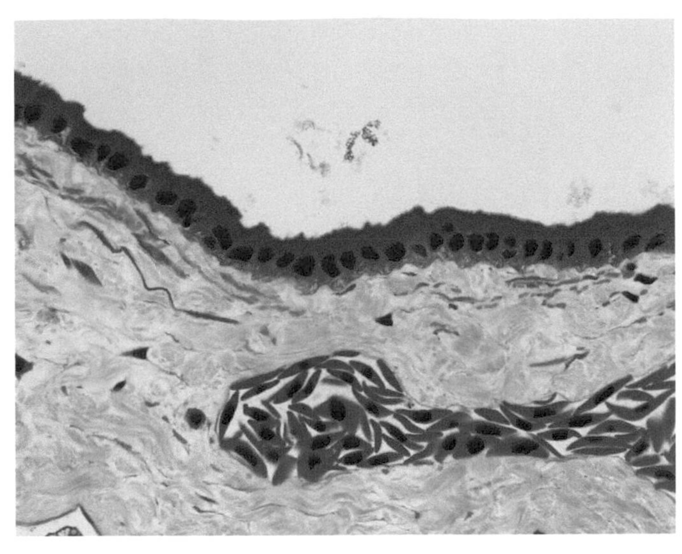

FIGURE K117 A low, simple columnar epithelium lines the gall-bladder of a frog. The ragged apical border of the epithelial cells indicates the presence of microvilli. A large vein inhabits the tela submucosa. 63 ×.

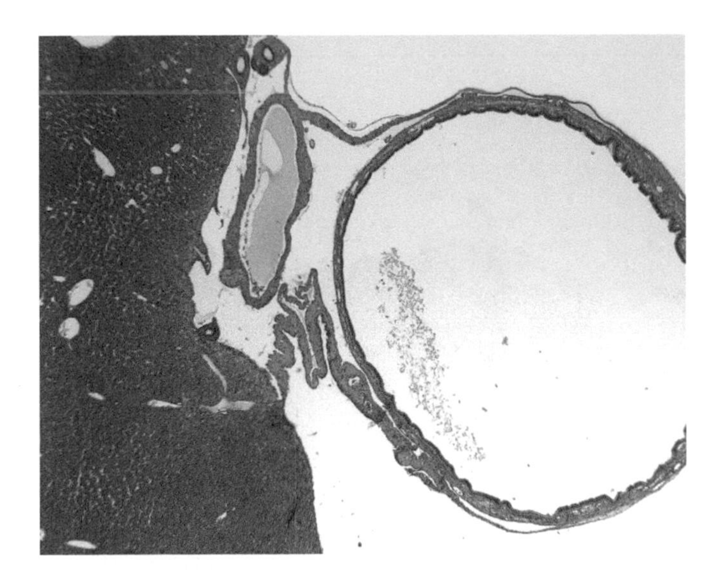

FIGURE K118a The gall bladder of a toad lies alongside a large vein within the peritoneum covering the liver. 2.5 ×.

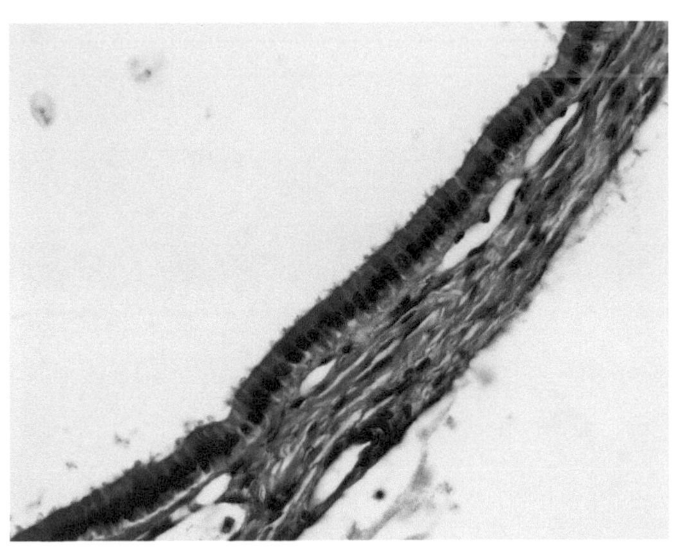

FIGURE K118b A simple columnar epithelium lines the gallbladder of a toad. Note the large lymph spaces, lined with endothelium in the tela submucosa just below the epithelium. 40 ×.

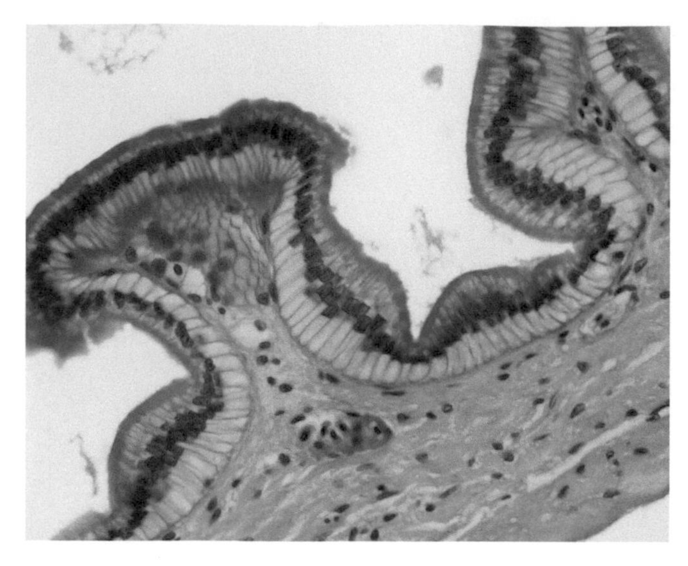

FIGURE K118c A simple columnar epithelium lines the gallbladder of the marine toad. The lining is thrown up into folds with a core of vascular connective tissue from the tela submucosa. These folds accommodate stretching as the organ fills. 40 ×.

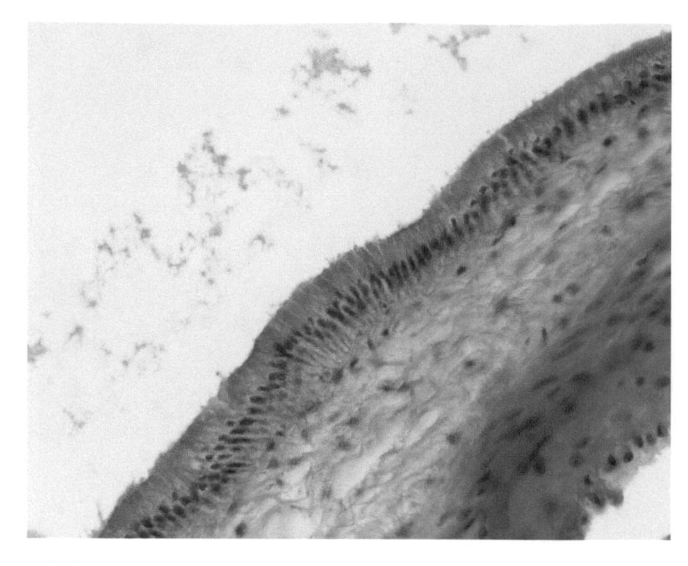

FIGURE K119 The gall bladder of the rattlesnake shows the typical simple columnar epithelium resting on the loose connective tissue of the tela submucosa. 40 ×.

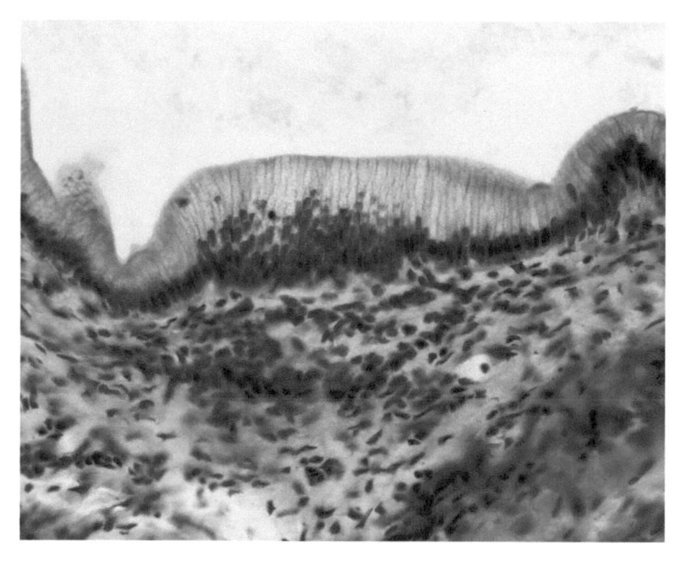

FIGURE K120a A simple tall columnar epithelium lines the mammalian gall bladder. It rests on a lamina propria mucosae whose connective tissue merges imperceptibly with that of the tela submucosa. A layer of smooth muscle is seen at the lower right. 40 ×.

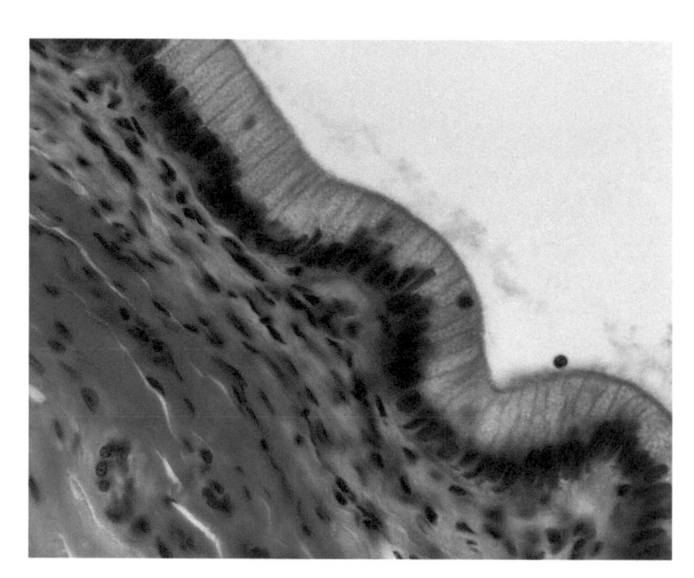

FIGURE K120b The apical portion of the columnar epithelial cells appears to possess a striated border. The lamina propria mucosae provides a blood vessel that approaches the epithelium. There is a mucous gland at the lower left. 63 ×.

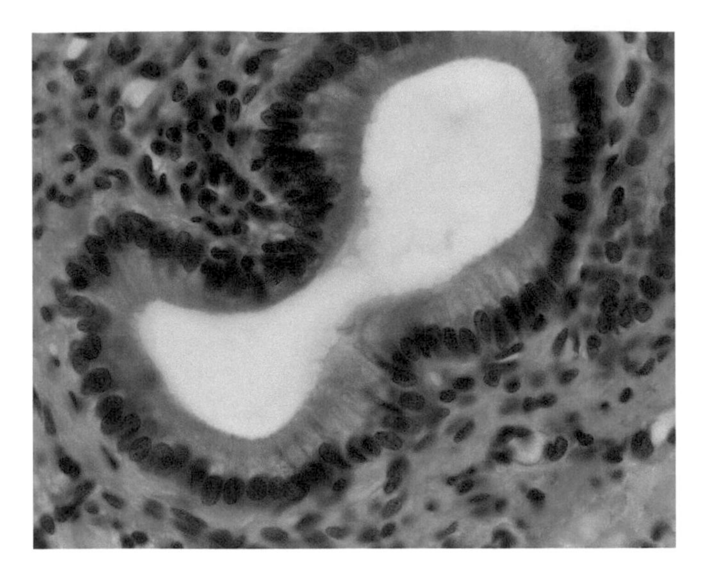

FIGURE K120c Branches of the bile duct show the characteristic simple columnar epithelium seen in the gall bladder itself. 63 ×.

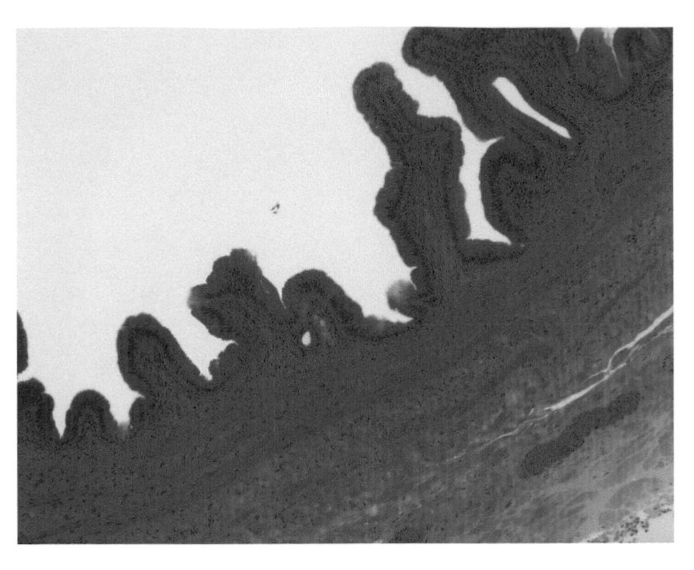

FIGURE K121a The epithelium lining the gall bladder is usually seen to be folded because the organ contracts as it empties during processing. In life the folds accommodate stretching as the organ fills. Below the epithelium, smooth muscle maintains a degree of tension on the organ and assists in the expulsion of bile at the appropriate moment. The washboard effect is "chatter," an artifact that sometimes occurs during sectioning. 10 ×.

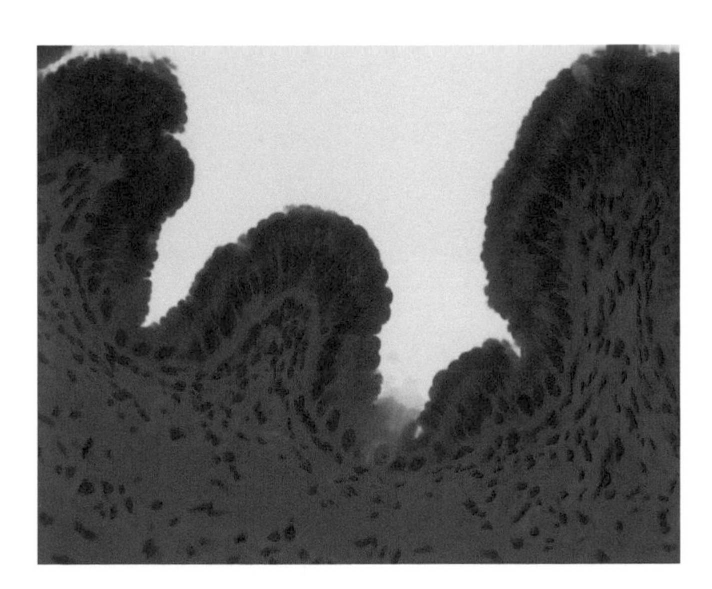

FIGURE K121b The simple columnar epithelium lining the gall bladder of a monkey is absorptive with a ragged border, indicative of the presence of microvilli. 40 ×.

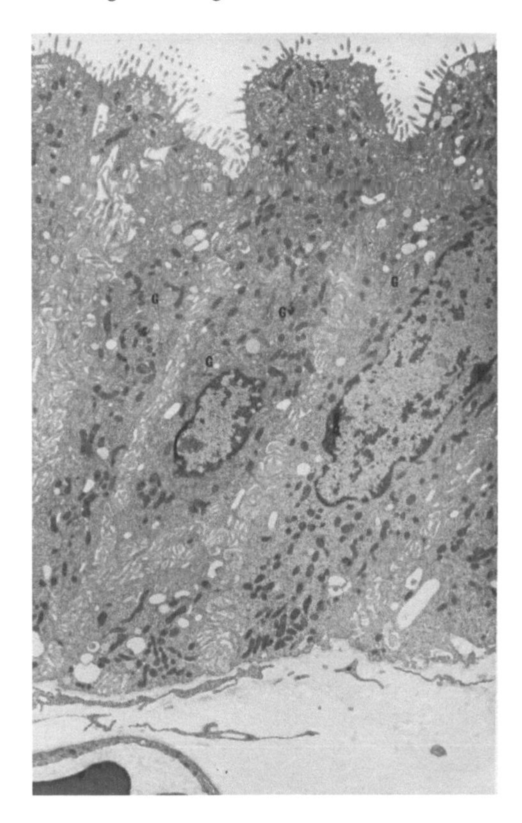

FIGURE K122 Electron micrograph of a section of typical simple columnar epithelial cells from the gall bladder of a guinea pig. Note the long microvilli and characteristic bulging of the apical region into the lumen. The Golgi complex (G) is supranuclear and mitochondria are distributed randomly in the cytoplasm. Numerous pinocytotic vesicles and larger PAS-positive granules are seen in the apical cytoplasm. The lateral cell borders are interdigitated the cells rest on a continuous basal lamina. 7000 ×.

CHAPTER

L

Respiratory Systems

Vertebrates require a constant supply of oxygen for the production of energy from the breakdown of food. In this process, water and carbon dioxide are produced. The oxygen is transferred from the ambient air or water to the blood in gills, lungs, and sometimes the skin. Several characteristics are common to the respiratory organs of all vertebrates and these include

- *Extensive vascularization* closely applied under:
- a *thin epithelium* that offers little resistance to the passage of gases,
- a *moist epithelium* of *large surface area*, and
- a means of *moving the water or air*.
- Accessory *chemoreceptors* usually accompany respiratory systems.

WATER BREATHING

Ancestors of all chordates were aquatic and respired by gills. There are no extant nor fossil forms of terrestrial chodates of the Chondrichthyes, or Agnathans class. In addition, all vertebrate embryos show traces of the primitive branchial apparatus at some stage in their development, although in higher forms, it never becomes functional. All the characteristics of respiratory tissue of higher forms may be seen in the basket-like gills of lower chordates such as amphioxus and a tunicate (Figs. L1a–L1c, L2a, and L2b). Water is passed through the GILL SLITS of the pharynx or branchial sac by the action of the cilia.

Vertebrates have two types of gills: external and internal. EXTERNAL GILLS are richly vascularized extensions of the visceral arches covered with the epithelium of the integument. They are usually branched and filamentous (Figs. L3a–L3d, L4a, and L4b). The gill is a wholly ectodermal structure with a mesodermal support. External gills occur in all larval amphibians but are lost in terrestrial adults. They are also found in embryonic lungfishes. A second set of gills develop within the opercular cavity of amphibians and are sometimes referred to as internal gills (Figs. L3e and L3f). The entire opercular cavity is lined with ectoderm and the gills are covered with ectoderm. These gills should not be homologized with the internal gills of fishes. External gills remain in some adult amphibians (Figs. L4a and L4b).

The INTERNAL GILLS of fish are more complex than external gills (Figs. L5a–L5g, L6a–L6d, L7a, L8a, and L8b). It is generally assumed that the internal gills are covered with endoderm but there is some controversy on this point. In the developing vertebrate, embryo endodermal pouches grow outward from the lining of the pharynx and contact corresponding pouches growing inward from the skin. In the lower vertebrates, these become pierced with GILL SLITS. The number of gill slits varies, with the greatest number in the lower forms.

Amphioxus	About 140
Hagfish	7–14
Lamprey	7
Shark	6–7
Teleosts	4
Amphibians	4 or less

An Atlas of Comparative Vertebrate Histology.
DOI: https://doi.org/10.1016/B978-0-12-410424-2.00012-3

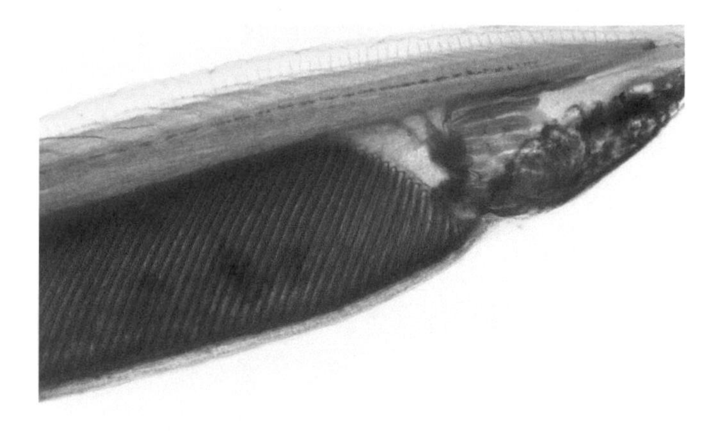

FIGURE L1a There is a large number of gill bars and gill slits in amphioxus. Water, driven by cilia on processes surrounding the mouth, enters at the right and passes into the large PHARYNX. Food is trapped by the sieve-like gill basket and passes posteriorly into the intestine. Water oxygenates the blood as it passes through the GILL SLITS—it enters the ATRIUM and is expelled through the ATRIOPORE, 3.2×

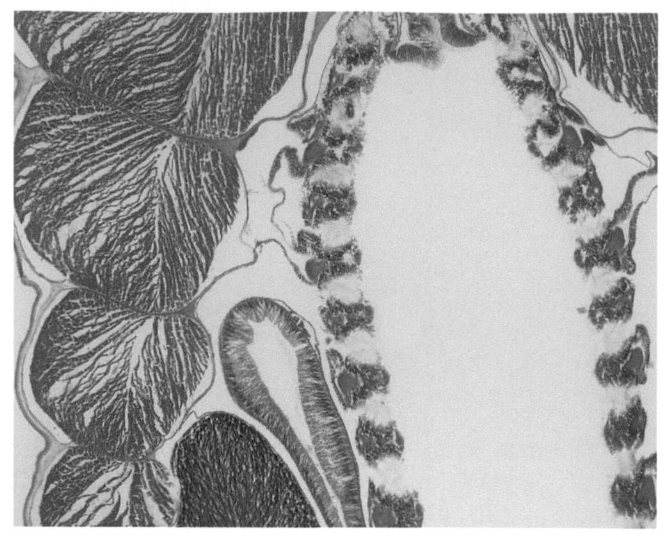

FIGURE L1b This cross-section of amphioxus shows the gill bars and gill slits surrounding the pharynx. The liver diverticulum is wedged between the pharynx and the testis and muscle of the body wall. Fixation of this specimen was not good, and little detail of the gill bars can be discerned, 10 ×

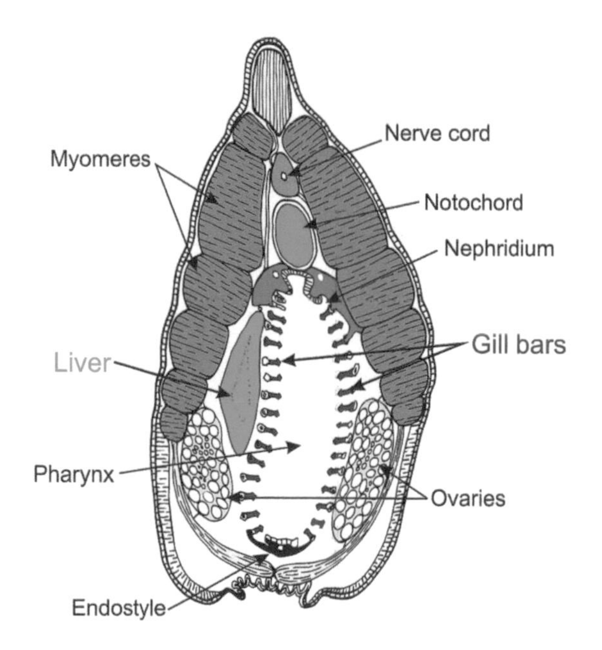

FIGURE L1c Tracing of a section through the pharynx of amphioxus illustrating the positions of the gill bars and liver. *Authors.*

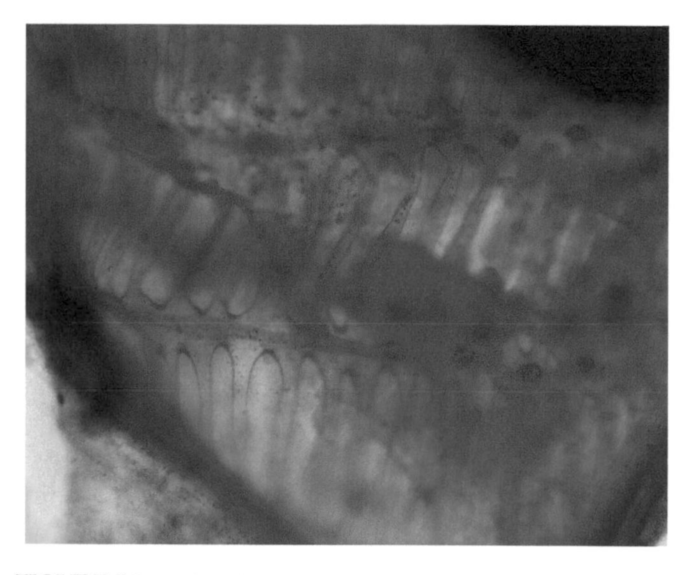

FIGURE L2a The gills of the tunicate are in the form of a basket, the BRANCHIAL SAC. Water enters at the INCURRENT SIPHON at the top, propelled by ciliary action passes through the GILL SLITS, and leaves by way of the EXCURRENT SIPHON at the right. The gill slits are surrounded by cilia which are not shown in this micrograph of a whole mount. This simple respiratory structure exemplifies the features of a respiratory organ: it is extensively vascularized, and is of large surface area, 10 ×.

An individual gill (HOLOBRANCH) consists of a cartilaginous or bony GILL ARCH with an accordion-pleated DEMIBRANCH or FILAMENT hanging down on each side of a median connective tissue SEPTUM (Fig. L7a). There may be cartilagenous GILL RAKERS embedded in the septum (Fig. L7b). Striated muscle is present in the gill making limited movement possible.

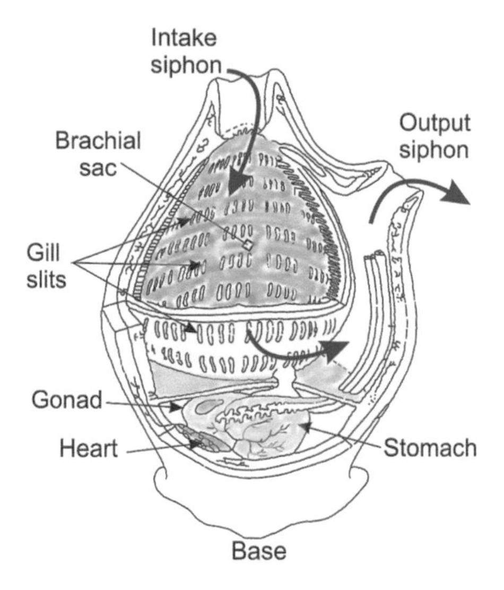

FIGURE L2b Diagram shows the structure of a tunicate. Parts have been removed to expose the elaborate gill basket. The arrows indicate the path of water currents coursing through the animal. *Jon Houseman, CC BY-SA 3.0 commons.wikimedia.org/w/index.php?curid = 25855290.*

FIGURE L3a A salamander larva shows external gills. *Brian Gratwicke—IMG_3547, CC BY 2.0, commons.wikimedia.org/w/index.php?curid=8436840.*

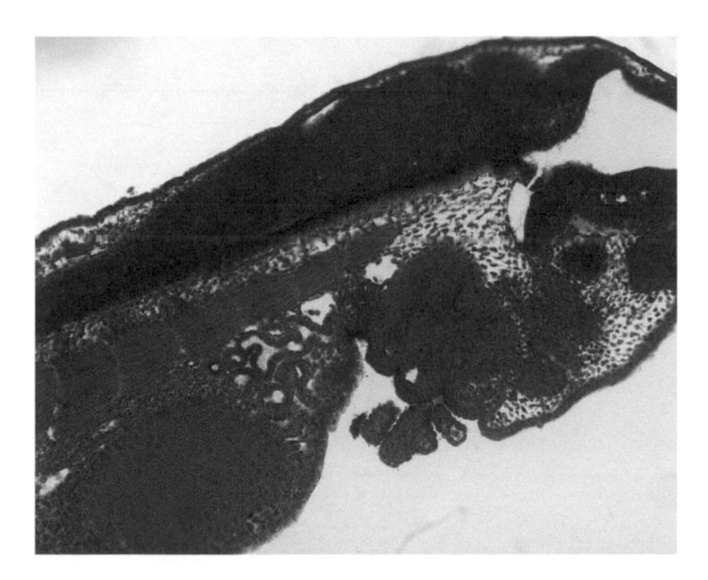

FIGURE L3b This tissue was stained *into*, before the tadpole was embedded and sectioned, with the result that details of the gills are difficult to discern. External gills are branching extensions of the visceral arches. They are covered with the epithelium of the integument. Only a few branches are visible in this section, 10 ×.

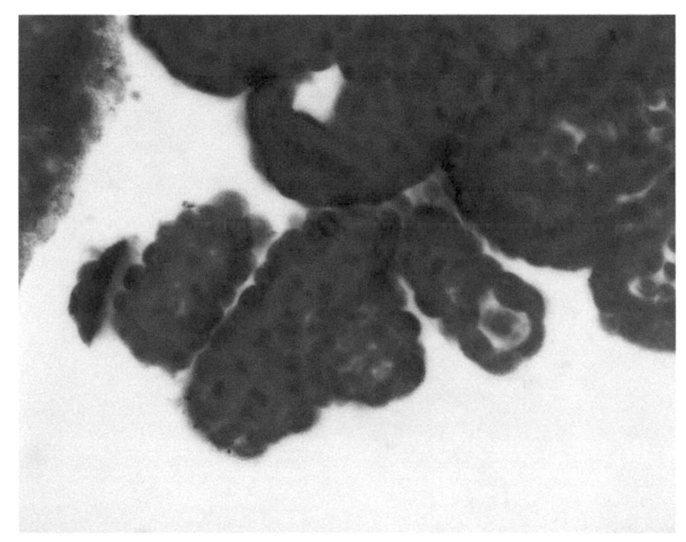

FIGURE L3c For a small animal such as a larval salamander that breathes through its skin, there is no pressing need for efficient gills. Nevertheless, these external gills increase the surface area for gaseous exchange and are highly vascularized, as seen in the branch at the lower right, 40 ×.

Each fold in a demibranch is a LAMELLA and consists of highly vascular loose connective tissue covered by a thin epithelium (Figs. L8a–L8c). The blood channels within lamellae are not tubular; they are formed by spool-shaped PILLAR CELLS whose thin flanges are arranged parallel to the epithelium and enclose a low flat space (Fig. L8d). Pillar cells create the channels that achieve directional blood flow from the afferent to the efferent side of the lamellae, approximating a counter-current exchange of oxygen and carbon dioxide (Fig. L8e). The flanges of adjacent pillar cells interconnect and thus form a continuous border to the blood channels. Since their flanges constitute the entire wall of the blood channels in the lamellae, the pillar cells may be considered endothelial. The nucleus, organelles, and bundles of fine cytoplasmic filaments are contained within the shaft of the spool, the PERIKARYON (Fig. L8c). Gas exchange occurs from both lower and upper surfaces with minimal impediment to the flow of blood. The pressure of the blood is not excessively reduced as it passes from the gills into the dorsal

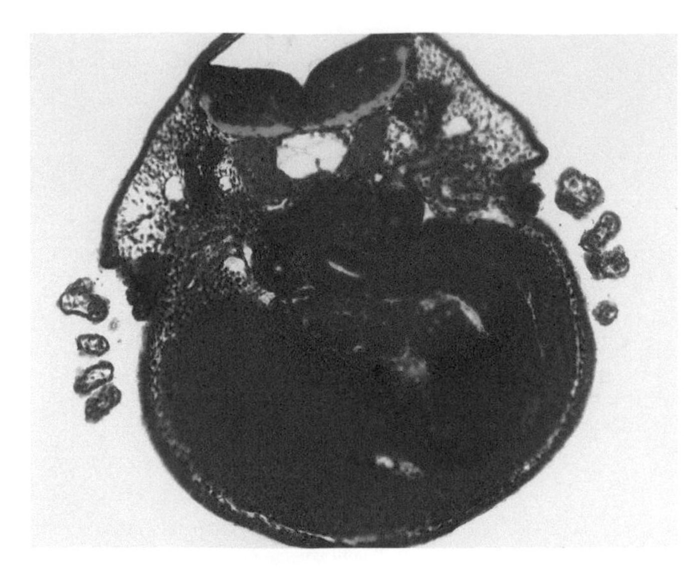

FIGURE L3d The vascularization of the external gills is seen to good advantage in this cross-section of a tadpole, 10 ×.

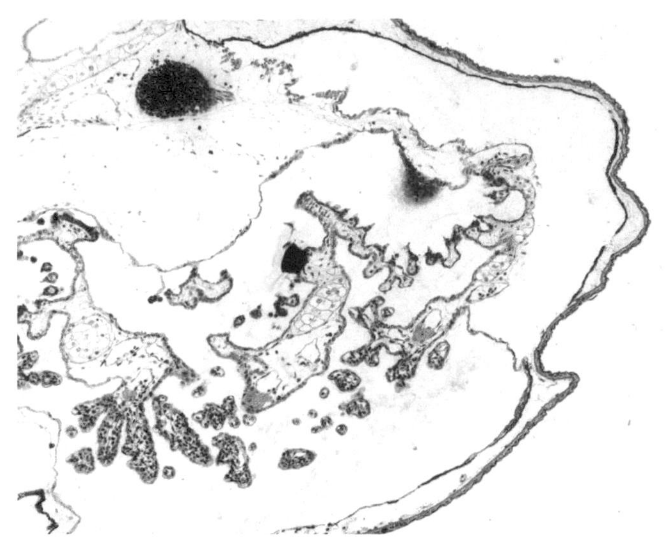

FIGURE L3e This section of the gill region of a tadpole was stained in the conventional way following sectioning. This sagittal section passes well to one side of the midline so that the external gills are visible; the stain, iron hematoxylin, was used. Although the gills are covered with a flap of skin (OPERCULUM) and appear to be internal, they are considered to be external gills, 10 ×.

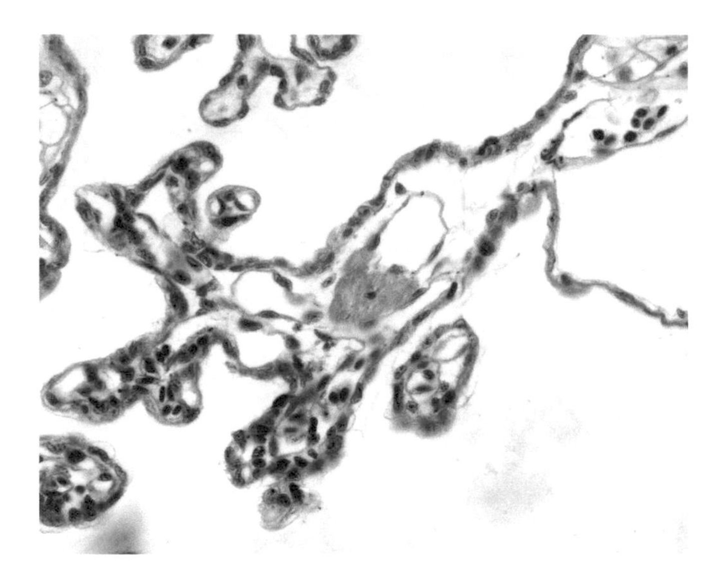

FIGURE L3f The thin epithelium covering the gills of a tadpole and their rich vascularization are apparent, 40 ×.

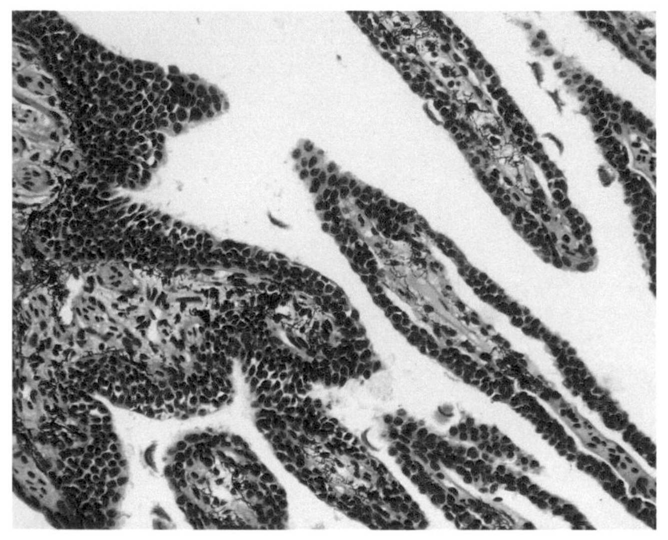

FIGURE L4a In the mudpuppy, *Necturus*, gaseous exchange occurs in three locations: in the lungs, across the external gills, and through the skin. There is a modest increase in surface area in these gills but the relatively thick respiratory epithelium probably permits little exchange, 10 ×.

aorta (Fig. L9a). The pillar cells are invaginated by reinforcing columns of collagenous fibers that intrude as far as the central shaft of the cell, extending perpendicular to the epithelia (Figs. L9b–L9d). One cell may be invaginated by as many as six stout columns of collagenous fibers. These collagenous columns remain external to the plasmalemma (Fig. L9e). Because of the presence of the bundles of fine filaments within these cells, it has been proposed that the pillar cells are contractile and they may be able to regulate blood pressure by controlling the flow of blood through the gills (Figs. L10a and L10b). Since the blood vessels of the gills join arteriole to arteriole, they constitute a rete mirabile.

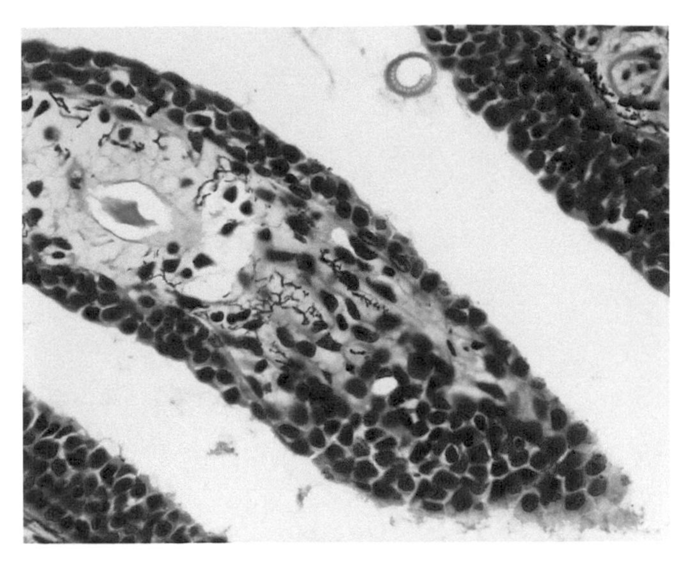

FIGURE L4b The thick, stratified respiratory epithelium in the mudpuppy is separated from the capillaries by a thick layer of loose connective tissue. Elaborate MELANOPHORES are abundant, 20 ×.

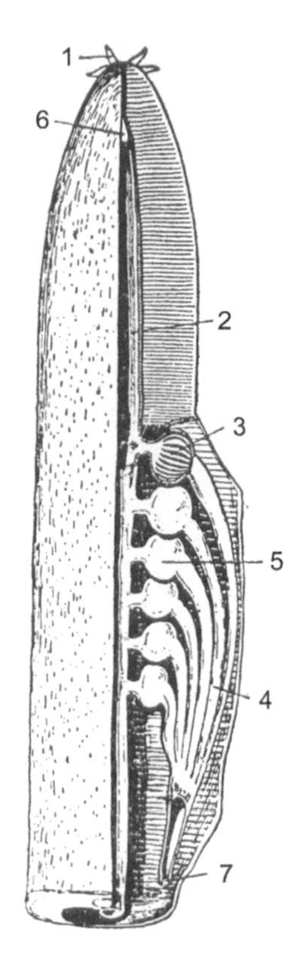

FIGURE L5a Water enters the hagfish, by a NOSTRIL at the anterior end of the body (1), passes posteriorly to enter the PHARYNX (2) from which several GILL SACS open on either side (3, 5). Water flows through the gill sacs and is expelled through the branchial duct (4) to the common branchial aperture (7). The gill sacs are packed with richly vascular gills. (6) Mouth.

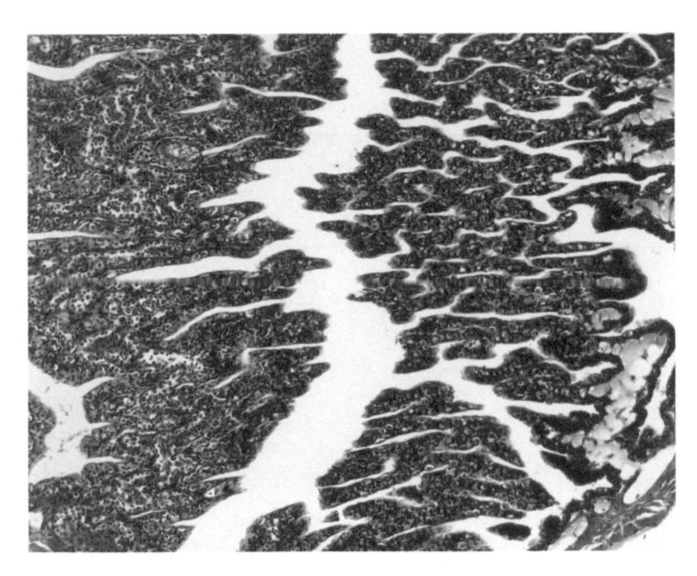

FIGURE L5b In this photomicrograph of the gill sac of a hagfish (*Myxine glutinosa*) the great surface area of the gills and the thin barrier between blood and water that contribute to the efficiency of respiration are exemplified, 10 ×.

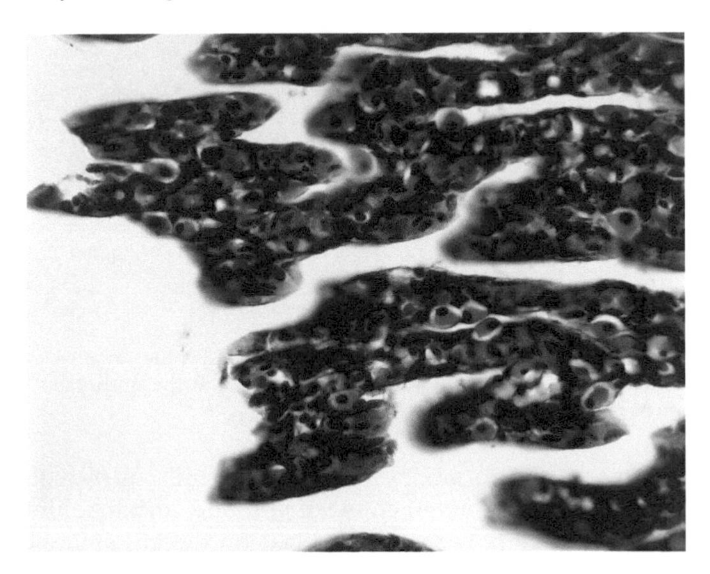

FIGURE L5c The blood vessels in the internal gills of the hagfish are covered with a thin layer of endothelial cells and packed with blood cells—an ideal situation for the uptake of oxygen and the elimination of carbon dioxide, 40 ×.

The epithelium covering each lamella consists largely of flattened PAVEMENT CELLS (Figs. L11a–L11e) which may have a ridged surface, giving it the appearance of a fingerprint. Scattered among the pavement cells are mitochondria-rich CHLORIDE CELLS (Figs. L12a–L12e). These cells are present in both freshwater and saltwater species where they function in transporting ions for the maintenance of blood levels in the body fluids. Evidence for their transporting function is seen in the elaborate tight junctions between adjacent cells. The apical surface of the chloride cell may form an apical pit, partially overlain by the surface epithelial cells (Figs. L12f and L12g). MUCOUS CELLS are also present in the epithelium of the lamellae (Fig. L13).

FIGURE L5d A gill arch spans the pharynx of this ammocoete. Water passes through gill clefts on either side. Gill filaments bearing gill lamellae extend outward from the gill arch, 10 ×.

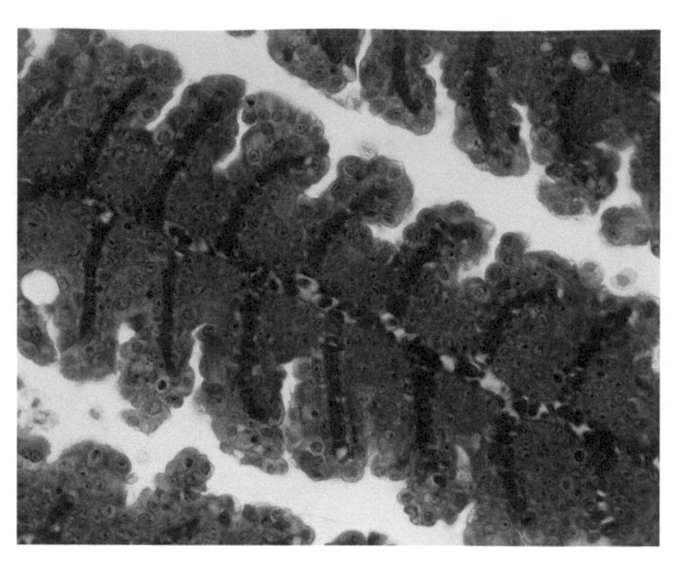

FIGURE L5e Lamellae extend in regular fashion from the gill filament of an ammocoete. Each lamella contains what appears to be an ordinary tubular capillary. Since each lamella bears a complete capillary—none is cut obliquely—these "capillaries" must not be tubular, 40 ×.

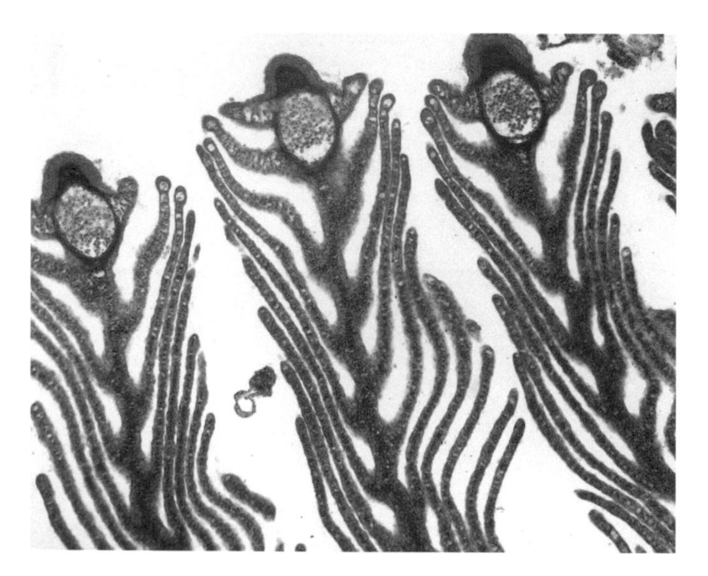

FIGURE L5f The gill of the ammocoete shows the same general structure as seen in bony fish. The large space at the tip of the filament is the EFFERENT FILAMENT ARTERY, 10 ×.

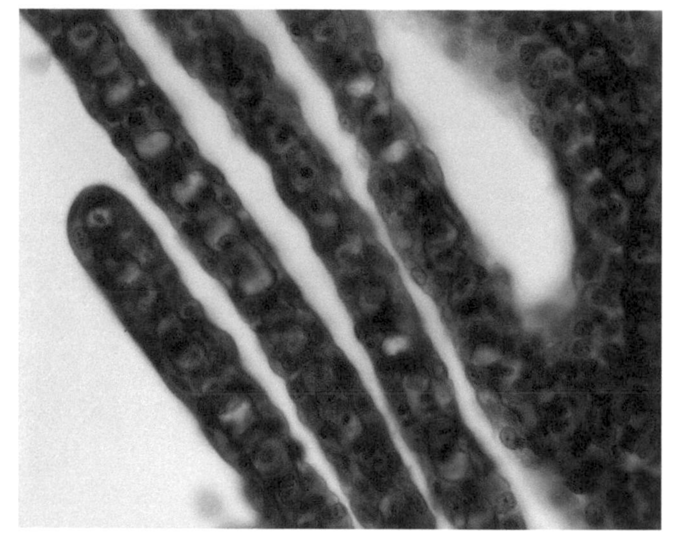

FIGURE L5g The rich vascularization of the gill lamellae is obvious. Blood spaces are formed by spool-shaped pillar cells, 63 ×.

The BARRIER to gas exchange in the gill consists of the thin epithelium of the lamella, a basement membrane with small amounts of connective tissue, and the thin flanges of the pillar cells (Fig. L14).

Taste buds have been described on the gill arches of the mullet (*Mugil cephalus*) and the killifish (*Fundulus heteroclitus*) (Figs. L15a–L15c). It is presumed that these organs sense the presence of inhospitable surroundings.

AIR BREATHING

The respiratory tract of tetrapods consists of the lungs and a system of tubes by which air passes to them from the pharynx (Figs. L16a and L16b). The tract develops as a ventral outpouching of the anterior portion of the gut

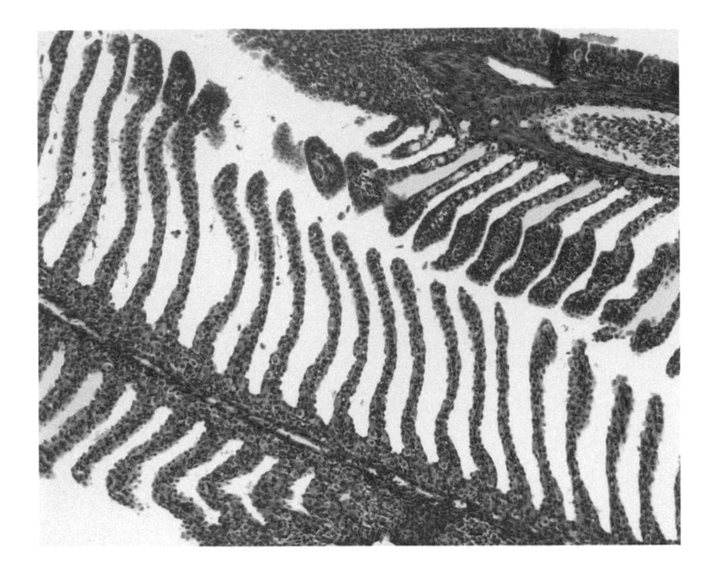

FIGURE L6a The gill of the dogfish is similar to that of the catfish seen above. Richly vascular lamellae extend from the GILL FILAMENTS into the rushing water of the GILL SLITS. The gill filaments are supported by the connective tissue of the GILL RAY, 10 ×.

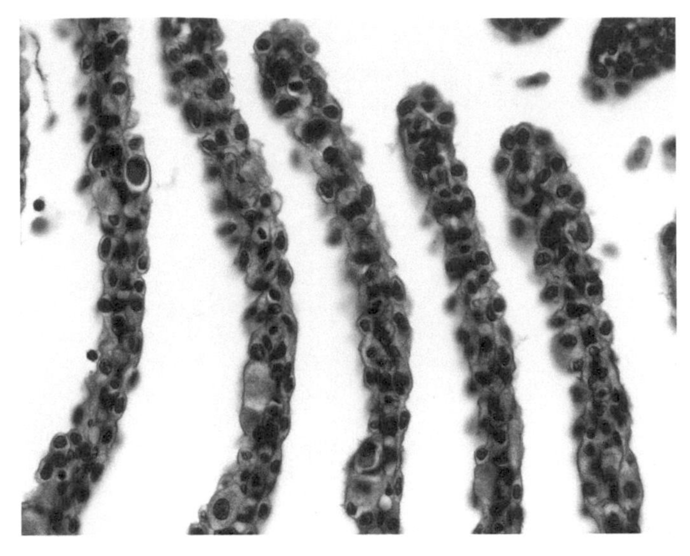

FIGURE L6b The blood spaces of the lamellae are formed by PILLAR CELLS which are not as clear as those of the previous example (Fig. L5g), 40 ×.

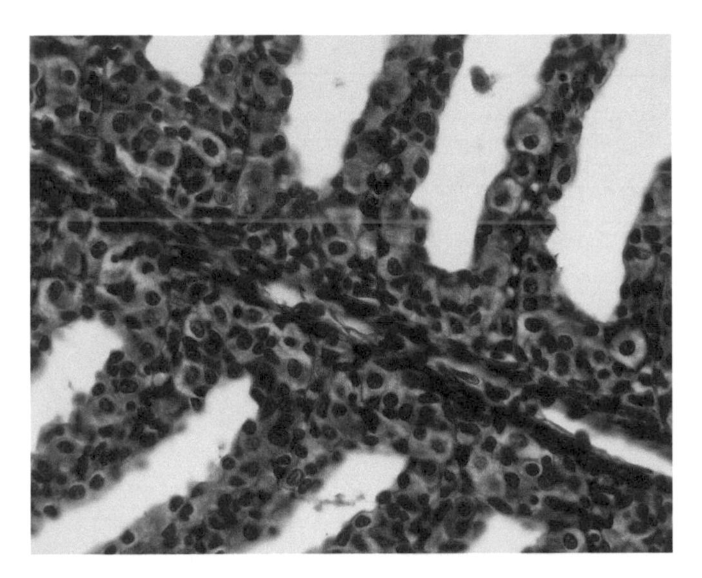

FIGURE L6c Several mucous cells can be seen in the epithelium of the lamellae. The connective tissue of the gill ray contains blood vessels, 40 ×.

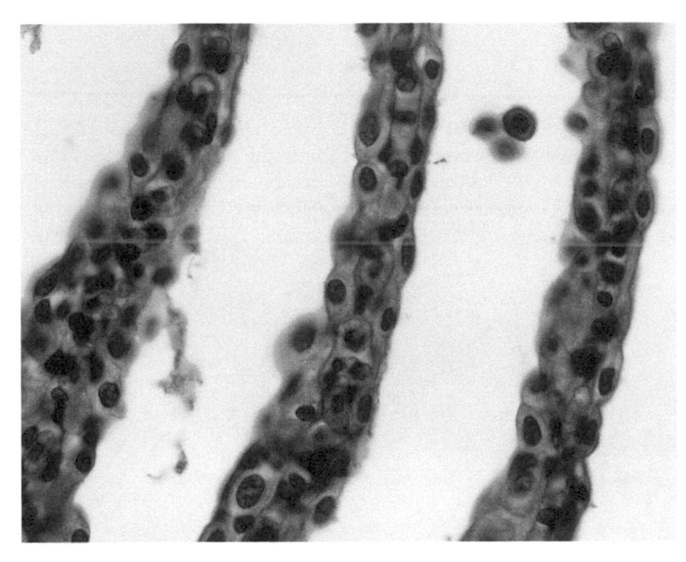

FIGURE L6d Several greyish balloon-shaped CHLORIDE CELLS are seen in the epithelium of these lamellae. The outer epithelium of the gill lamella, in contact with the water, consists of pillar cells, chloride cells, and squamous pavement cells, 63 ×.

and as such bears structural similarities to the gut. In structure and method of development the tetrapod lung so closely resembles the swim bladder of *Protopterus* and certain other fishes that it is hard to deny that these organs are homologous. The air bladder of bony fish appears to have evolved from the lung-like bladder of the lungfishes.

Olfactory organs of air-breathing vertebrates are closely associated with the respiratory passages (Figs. L17a–L17c). In lower aquatic forms, however, these structures are usually separate. There is no connection between the nostrils and mouth cavity of most *fishes*, and the nasal pit is purely olfactory. In the *lungfishes*, a connection between the nasal pit and the pharynx is seen, and the olfactory epithelium is located in the dorsal part of each nasal passage. It is thought that the nasal passages serve primarily to increase the effectiveness of olfaction (Figs. L16a and L16b). Ascending the vertebrate classes, the nasal passages become increasingly complex, containing the

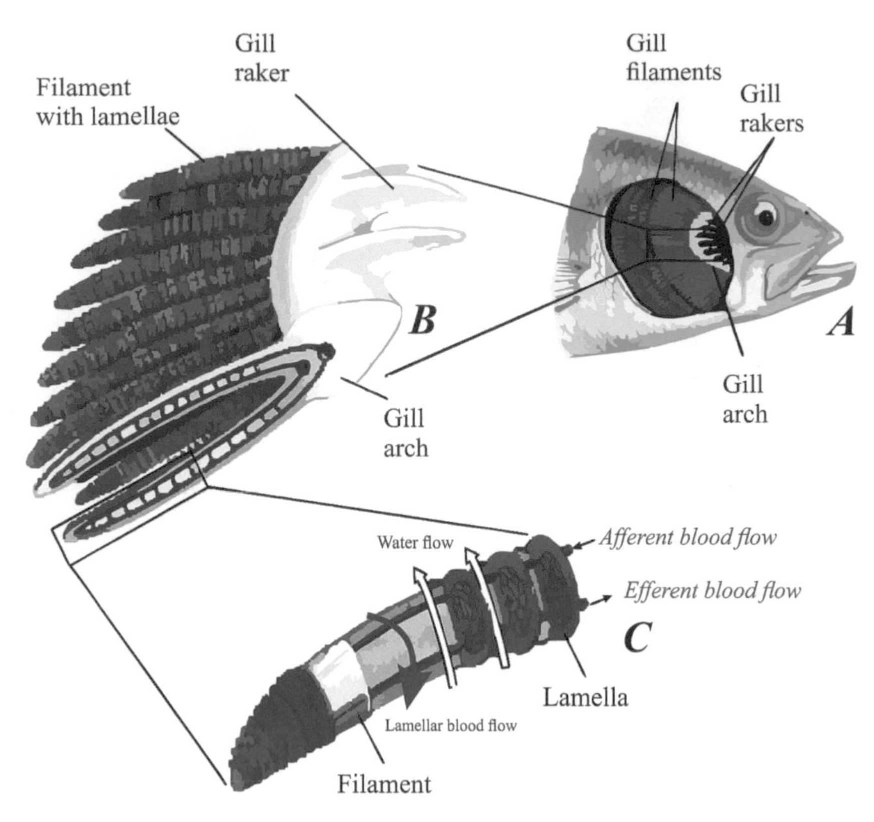

FIGURE L7a Internal gills of a bony fish. **(A)** Shows the gills in the gill chamber with the operculum cut away. **(B)** The segment of the gill consisting of five holobranchs each made up of gill rakers and filaments. The arrows indicate the path of blood in the filaments. **(C)** i Part of one filament much enlarged. Each lamella contains blood vessels where oxygen is taken on and carbon dioxide is expelled. It should be noted that water flows posteriorly and the blood in the gill flows in the opposite direction enabling a counter-current exchange system. *Modified from AreeshA09—Own work, CC BY-SA 3.0, commons.wikimedia.org/curid=1184299.*

FIGURE L7b Scanning electron micrograph of a gill arch and respiratory lamellae (*arrows*) of the gill arch of a mullet *Mugil cephalus*. *F*, filaments, *R*, gill rakers, 26 ×.

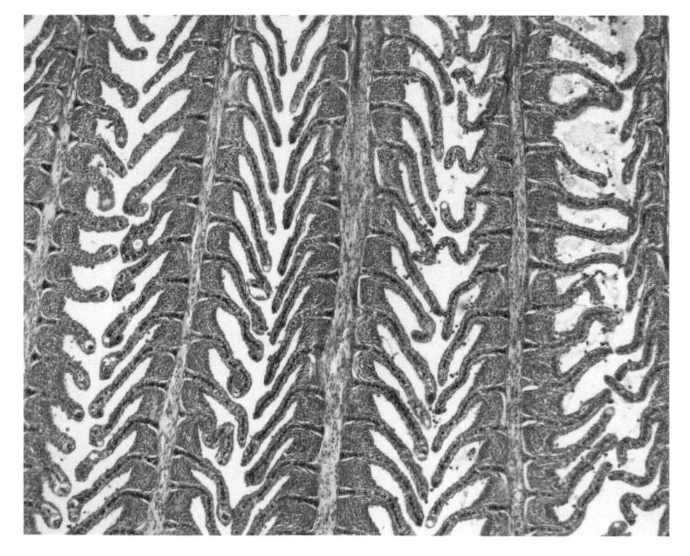

FIGURE L8a An individual gill or HOLOBRANCH consists of a cartilaginous or bony GILL ARCH with an accordion-pleated DEMIBRANCH or FILAMENT hanging down on each side of a median connective tissue SEPTUM—each pleat is a LAMELLA. This picture is from a longitudinal section of a catfish. It was taken at right angles to the conventional view of the gill arch with its gill filaments. This is a lateral view of several GILL RAYS with their LAMELLAE extending out on either side. The ERYTHROCYTES STAIN, a bright cherry red, indicating the extensive vascularization of the lamellae, 10 ×.

elaborate, scroll-like nasal conchae, thereby increasing the surface area available for olfaction. The mucous membrane lining the nasal passages is often provided with goblet cells and cilia (Figs. L16c and L16d).

Electron micrographs of the olfactory mucosa of a mammal show that the NEUROSENSORY CELLS are bipolar

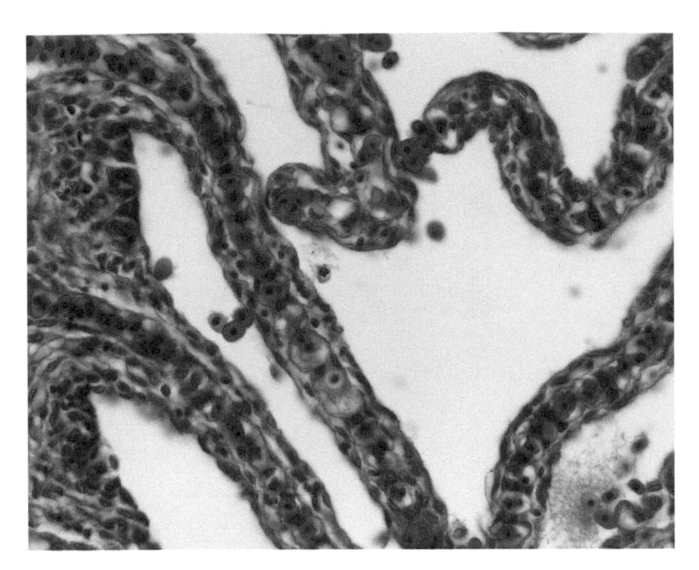

FIGURE L8b The question arises about the vascularization of the lamellae that traverse the screen in this micrograph. Are they vascularized by capillaries? If these blood vessels are capillaries—i.e., small tubes—all sections would not appear as if they were cut through the center of each tube. These are blood spaces unique to gills: flat open spaces akin to a level of a parking garage. The floors and roofs of these spaces are formed by the flanges of PILLAR CELLS which can be seen in areas devoid of red cells—e.g., to the left of center. These pillar cells take the form of spools, with thin flanges extending from the central column to form the intact wall of these blood spaces. Since they form the only barrier between the lumen of the blood spaces and the surrounding tissue, it would seem reasonable to conclude pillar cells are *endothelium*. The epithelium covering the lamella is thin. Together with the flanges of the pillar cells, it presents little barrier to the exchange of gases, 63 ×.

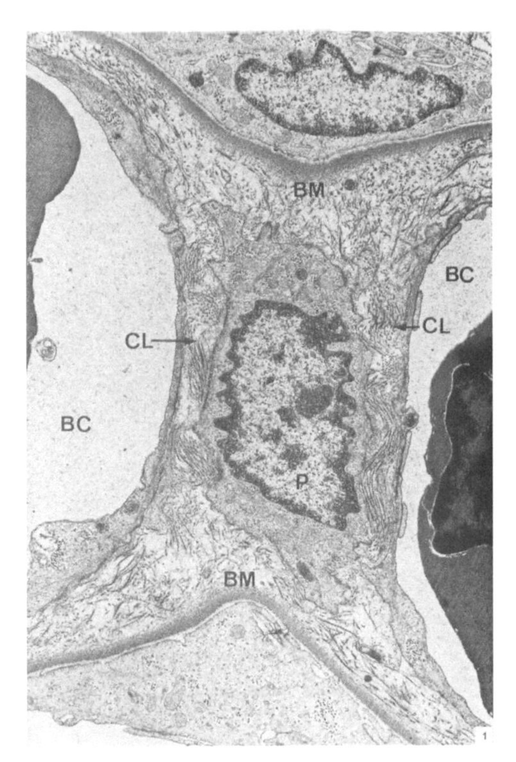

FIGURE L8c Transmission electron micrograph of a transverse section cut through a secondary lamella from the gill of a toadfish (*Opsanus tau*). A pillar cell (P) separates two blood channels (BC). Two collagenous columns (CL) invaginate the pillar cell but are external to it. They connect the collagenous layer of the basement lamina (BM). Contraction of the pillar cell has caused folding of the nuclear envelope, 13,500 ×.

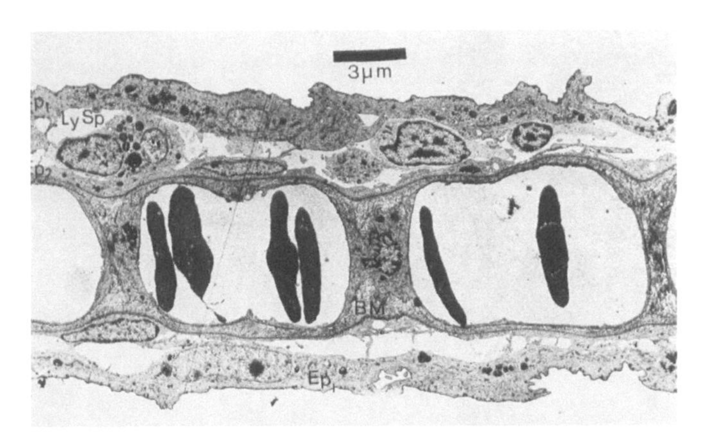

FIGURE L8d Transmission electron micrograph of a section, cut transversely to the direction of blood flow through the gill lamella of the toadfish, *Opsanus tau*. Three pillar cells (PCP) are visible forming and lining the blood channels. The barrier to gaseous exchange is clear: it consists of an epithelium (Ep), which includes pavement cells, and a fine layer of areolar tissue and a lymphoid space (LySp). The epithelium can often include chloride cells, mucous cells, and nerve endings.

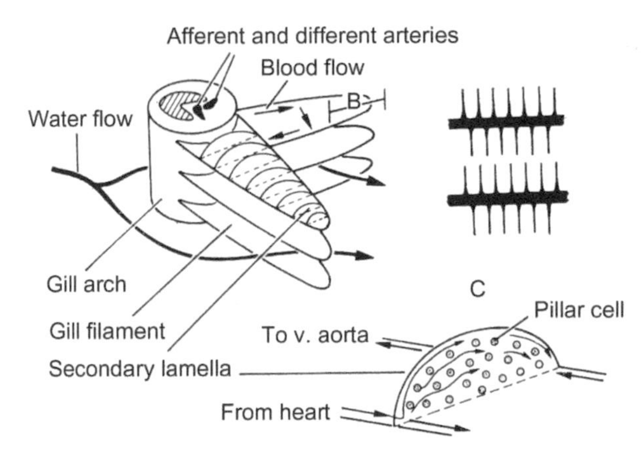

FIGURE L8e These diagrams show the structure of fish gills and the flow of water and blood in opposite directions within them. (a) A single arch with two rows of filaments. Lamellae are shown on the upper surface of one filament. The arrows indicate the flow of blood and water. (b) A longitudinal section of two filaments showing lamellqe projecting above and below each filament. (c) A single lamella showing blood flow between pillar cells.

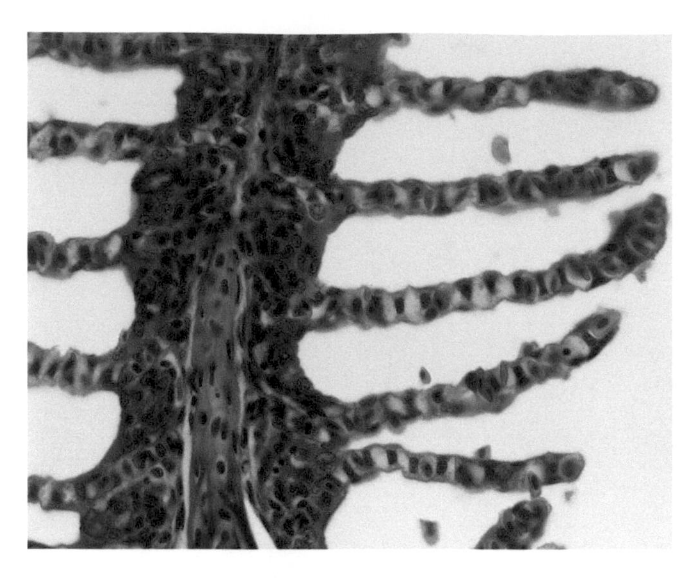

FIGURE L9a Pillar cells are especially clear in the lamella at the lower right in this section of the gill of a bony fish, 63 ×.

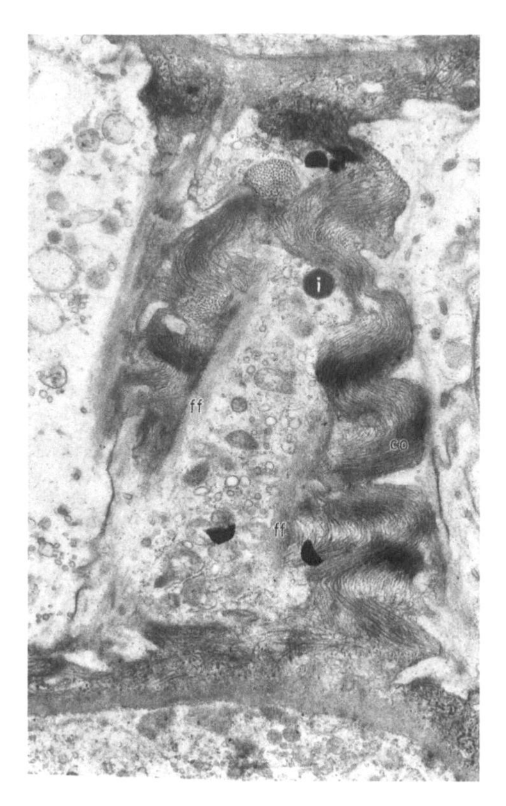

FIGURE L9b Transmission electron micrograph of a contracted pillar cell. The sinuous nature of the collagenous fibers in columns (Co) suggests that the fine fibrils (ff) are involved in contraction. *I*, inclusion, 13,500 ×.

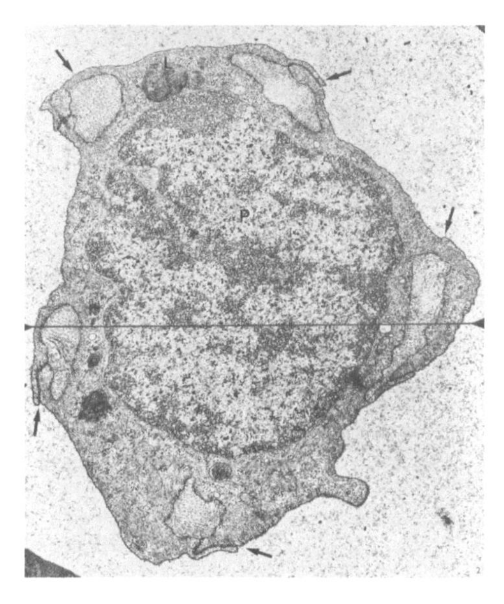

FIGURE L9c Transmission electron micrograph of a section taken at right angles to Fig. L8c. A pillar cell (P) is cut transversely exposing cross-sections of five collagenous columns (*arrows*) enfolded in the pillar cell but outside of it. Small desmosomes (D) maintain the gap between the enveloping flaps of protoplasm from the pillar cell, 24,600 ×.

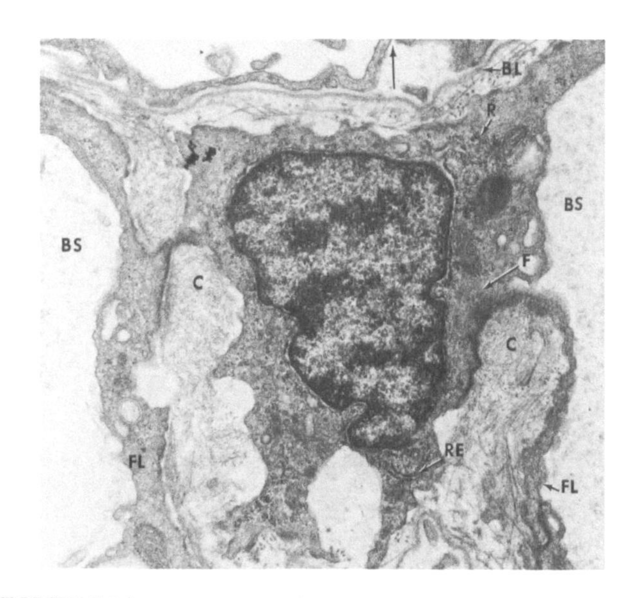

FIGURE L9d Transmission electron micrograph of a pillar cell cut transversely revealing several columns of collagen (C), each enfolded by the plasma membrane of the pillar cell. The cytoplasm of the pillar cell contains fragments of granular endoplasmic reticulum (RE), free ribosomes (R), and microfilaments (F). *BS*, blood space; *D*, desmosome; *FL*, flanges of the pillar cell, 20,800 ×.

neurons and form an integral part of the columnar epithelium (Figs. L17d and L17e). The slender distal process is a dendrite, the OLFACTORY ROD, that extends to the surface and terminates in a small bulb, the OLFACTORY VESICLE. The vesicle bears several modified cilia, the OLFACTORY HAIRS. The shape of this vesicle and the number of cilia present are highly variable among the vertebrates. The proximal part of the neurosensory cell narrows into an unmyelinated axon and passes into the underlying connective tissue to join with axons

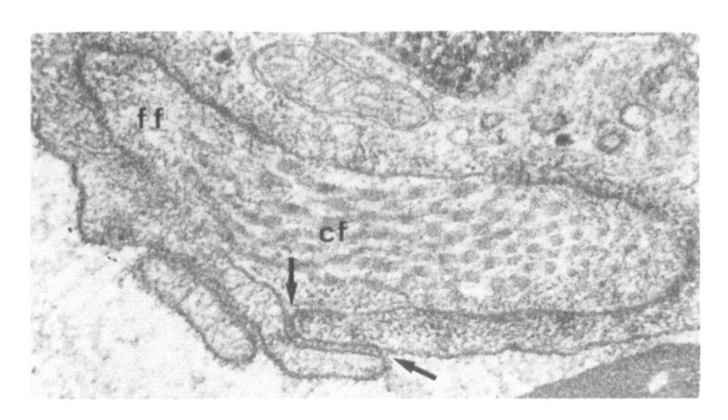

FIGURE L9e Transmission electron micrograph of a part of a pillar cell enfolding a collagenous column showing the narrow connection with the blood space (*arrows*). *cf*, collagen fibrils; *ff*, fine fibrils, 76,000 ×.

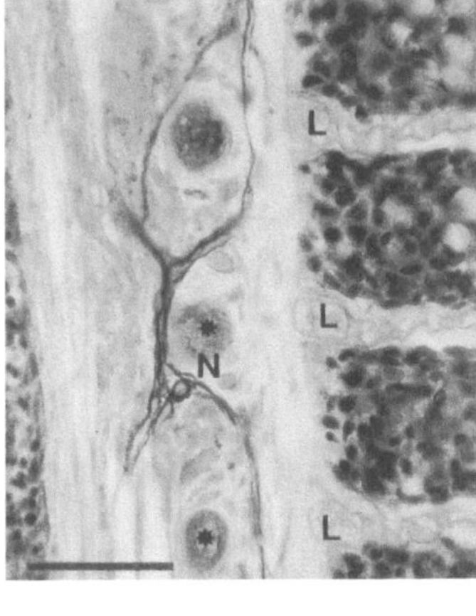

FIGURE L10a Photomicrograph of a silver preparation (Cajal) of the gill filament of the pike-perch, *Sander lucioperca*. Processes of a neuron (N) encompass arterioles. *L*, lamellae. Bar = 50 μm.

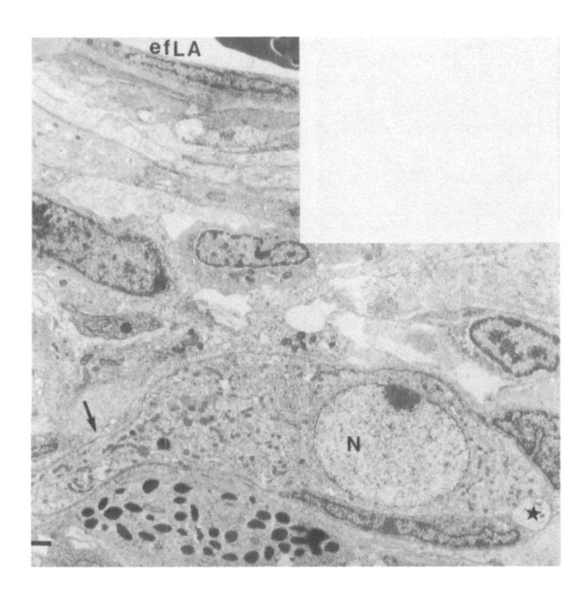

FIGURE L10b Transmission electron micrograph of a section of a neuron (N) located between two efferent lamellar arterioles (*efLA*; only one shown) in the gill of a black bass, *Micropterus dolomieui*. This neuron sends a process (*arrow*) toward another arteriole (not shown). An axonal profile (*star*) in close contact with the neuron. Bar = 1 μm, 5200 ×.

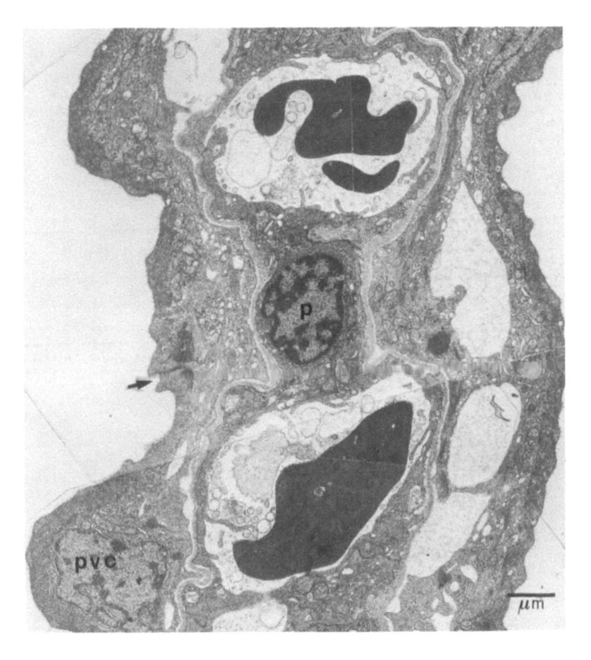

FIGURE L11a A composite transmission electron micrograph of a section of a gill lamella from the marine sole *Solea solea* showing a pillar cell (p) at the center enclosing two blood channels. In the epithelium the perikaryons of two large pavement cells (pvc) are joined by junctional complexes at the arrow. Note the double ridge on both sides of the junction, 12,000 ×.

from other olfactory cells and become the olfactory nerve. Neurotubules throughout the cytoplasm of the neurosensory cells can be seen in electron micrographs.

Neurosensory cells are interspersed among tall columnar SUSTENTACTULAR CELLS that reach to the surface, and small conical BASAL CELLS between the bases of the sustentacular cells. Sustentacular cells secrete mucus and bear long slender microvilli that penetrate the layer of mucus coating the surface. Basal cells are undifferentiated and are able to transform into either of the mature types. Dispersed sparsely throughout the olfactory epithelium are BRUSH CELLS with large, blunt, apical microvilli. These cells form synaptic contact with nerve fibers in the epithelium, and it has been suggested that they provide sensory input for the sneeze reflex. Sustentacular and olfactory

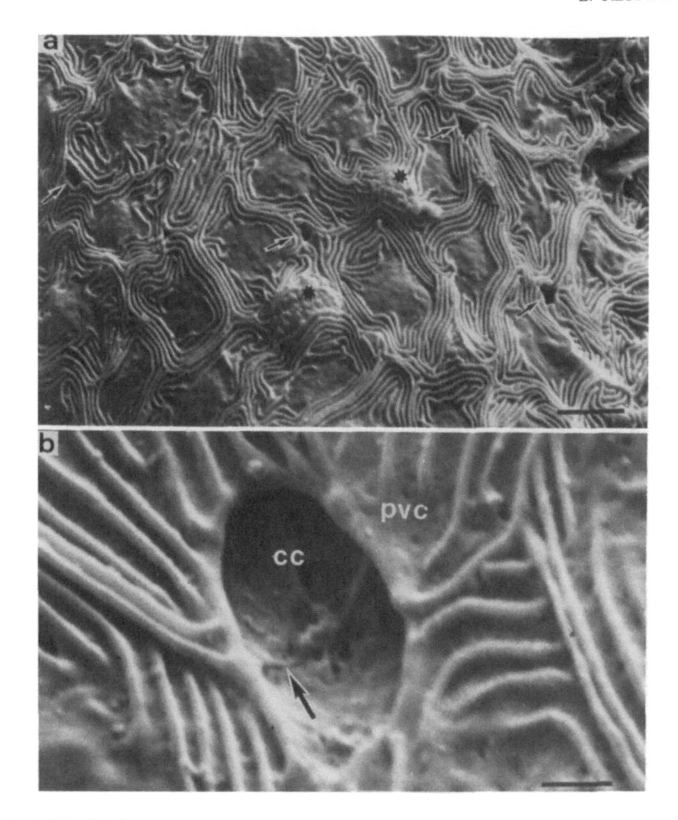

FIGURE L11b Scanning electron micrographs of the lamellar epithelium from the gill of a cichlid teleost, *Oreochromis alcalicus grahami*. (a) The arrows indicate chloride cells. Scale bar = 5 μm. (b) The apical crypt of a chloride cell (cc). *pvc*, pavement cell. The arrow indicates an accessory cell. Scale bar = 1 μm.

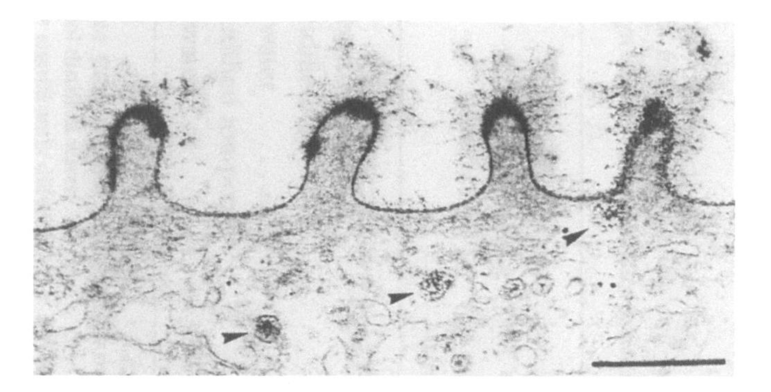

FIGURE L11c Transmission electron micrograph showing the glycocalyx on a section of gill lamellar pavement cell from a rainbow trout (*Salmo gairdneri*). A glycocalyx coats the apical membrane. The arrows indicate vesicles apparently transporting glycoproteins (black filamentous material) to the apical membrane. Bar = 0.5 μm.

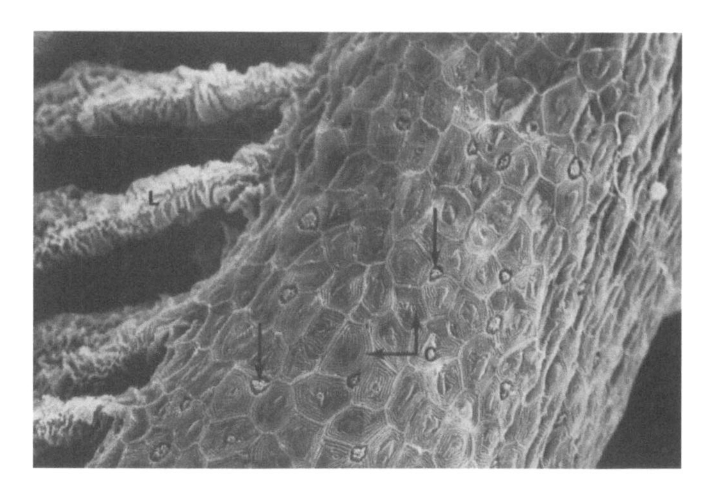

FIGURE L11d Scanning electron micrograph of the surface of the gill arch of a mullet (*Mugil cephalus*). Lamellae (*L*) extend from the right side, 1400×.

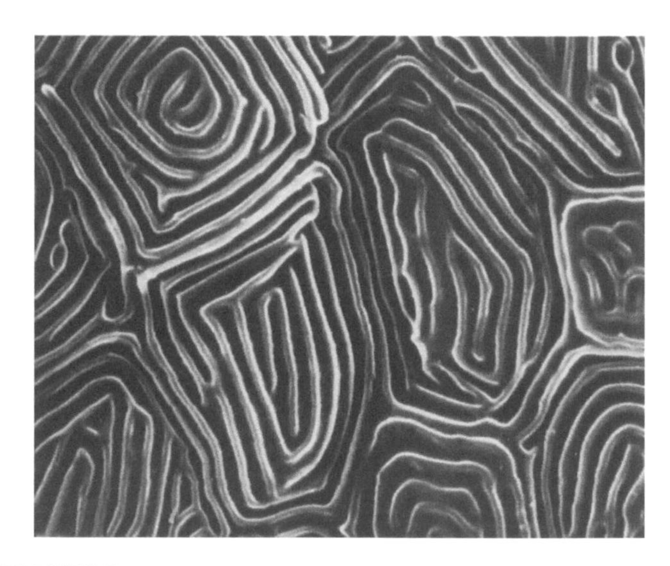

FIGURE L11e The epithelial cells of both the arch and the lamellae display a fingerprint pattern of ridges. The arrows indicate pores in the epithelium, 9000×.

cells have oval and round nuclei respectively and are apical to the nuclei of the basal cells. Special methods are needed to distinguish these cells clearly with the light microscope. Note the cilia of the olfactory cells. Branched tubuloacinar serous OLFACTORY GLANDS (Bowman's glands) in the lamina propria mucosae moisten the surface. This type of olfactory epithelium, with modification, is characteristic of all vertebrates.

VOMERONASAL ORGAN

The VOMERONASAL organ (Jacobson's organ) is associated with the nasal cavity of most tetrapods and consists of a pair of blind pouches that usually open into the nasal cavity (Figs. L18a and L18b). It receives branches from cranial nerves O, I, and V and is thought to aid recognition of food. In snakes and lizards, where it is especially

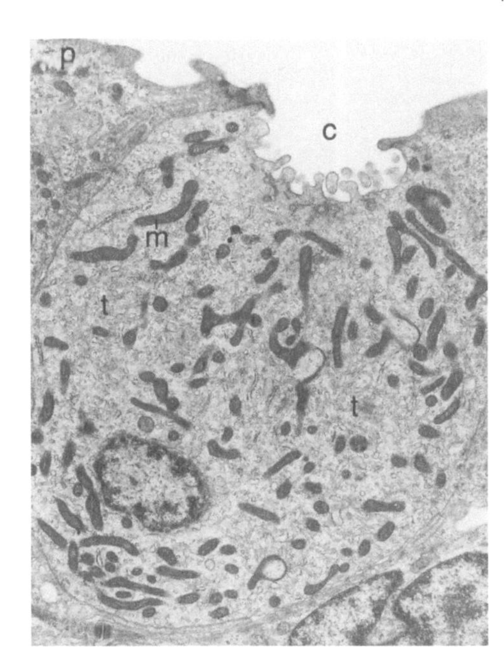

FIGURE L12a Transmission electron micrograph of a section of a mature chloride cell in the epithelium of the gill filament of the teleost, *Oreochromis mossambicus*. Note the tubular system (t) in the cytoplasm and the abundance of mitochondria (m). *c*, crypt, 8500 ×.

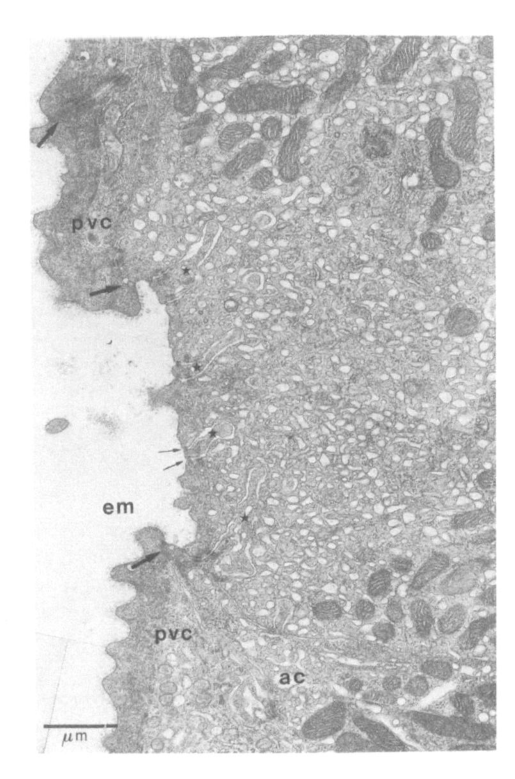

FIGURE L12b Transmission electron micrograph of the apical crypt of a chloride cell in a gill lamella of the sole, *Solea solea*. The crypt is bordered by pavement cells (pvc). An accessory cell (ac) beside the chloride cell sends out four processes (stars) which burrow into the protoplasm of the chloride cell. Junctions have formed between the pavement cell and the chloride cell (large arrows), and the accessory cells and the chloride cell (small arrows). External milieu, *em*. 20,000 ×.

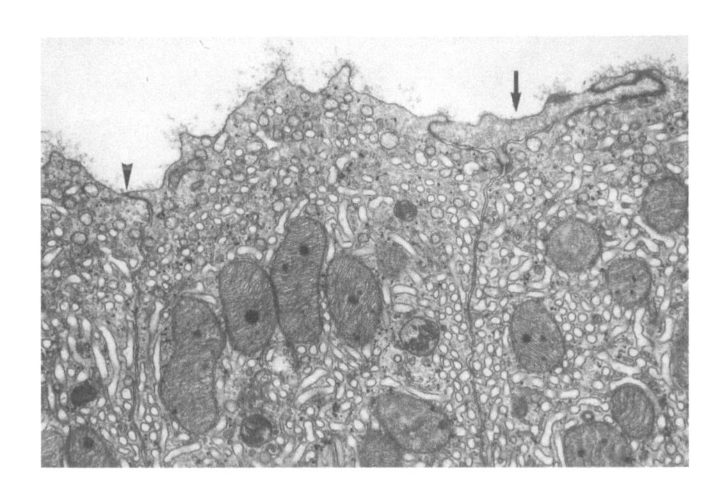

FIGURE L12c Transmission electron micrograph showing the apical regions of three chloride cells in a cross-section of a gill filament from the cyclostome *Geotria australis* held in freshwater. The central chloride cell is separated from the one on the right by a process of a pavement cell (*arrow*), whereas it shares a zonula occludens (*arrowhead*) with the cell on the left, 24,000 ×.

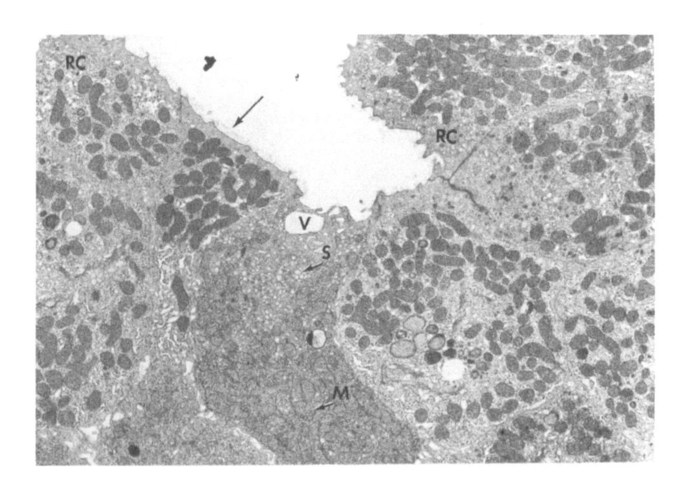

FIGURE L12d Transmission electron micrograph of a section of the interlamellar region of the gill of an ammocoete of the sea lamprey, *Petromyzon marinus*. Mitochondria-rich cells (RC) with a fibrillar apical coat (*arrow*) surround a cell containing a large apical vacuole (V), agranular endoplasmic reticulum (S), and numerous basal mitochondria (M), 5000 ×.

well developed, it opens into the oral cavity. Chemical substances brought into the mouth on the forked tongue of these animals are detected when the tips are thrust into the pits (Figs. L19a–L19c). A vomeronasal organ is absent from adult turtles, crocodilians, bats, primates, and aquatic mammals, but usually appears in the embryo. The tall pseudostratified columnar epithelium lining the pouch is similar to the olfactory

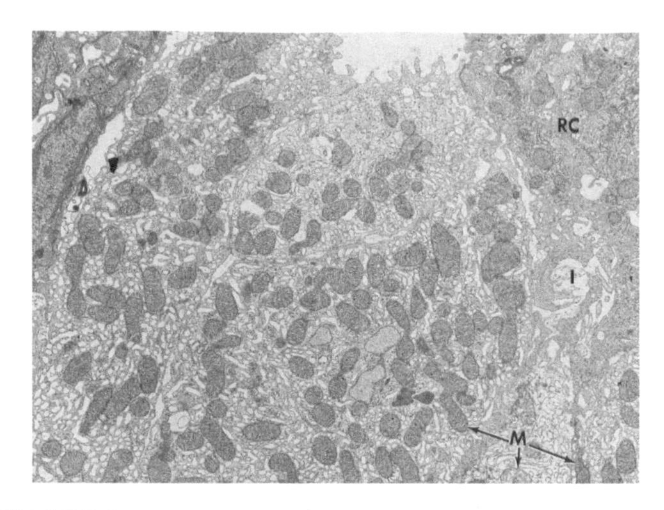

FIGURE L12e Transmission electron micrograph of a section of the interlamellar region of the gill of an adult lamprey, *Petromyzon marinus*. Chloride cells are separated from basal (B) and mitochondria-rich cells (RC) by wide intercellular spaces (I) and are distinguished by their abundant agranular endoplasmic reticulum, 6000 ×.

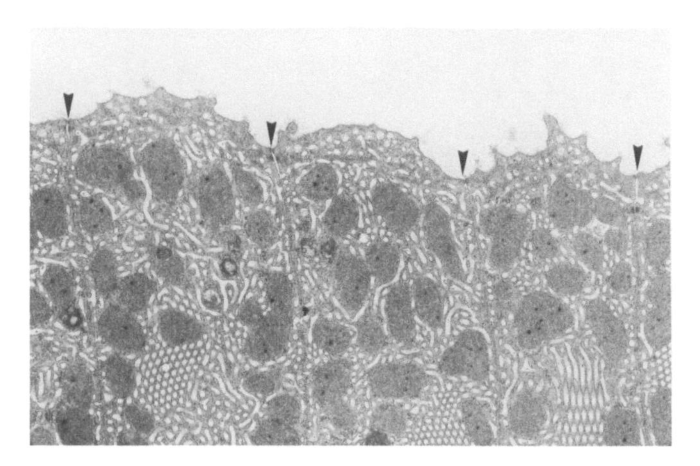

FIGURE L12f Transmission electron micrograph of a cross-section of a gill filament from the lamprey *Geotria australis* acclimated to seawater. Chloride cells are not separated by pavement cell processes at the level of the zonula occludens (*arrowheads*). Cytoplasmic tubules form bundles, pavement cells, 18,000 ×.

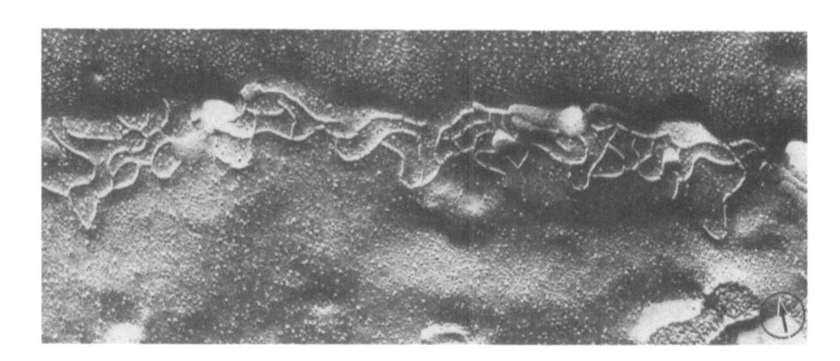

FIGURE L12g Transmission electron micrograph of a freeze-fractured replica of the zonula occludens between adjacent chloride cells in the young adult lamprey *Geotria australis* following acclimation to seawater (arrow indicates the direction of shadowing.), 60,000 ×.

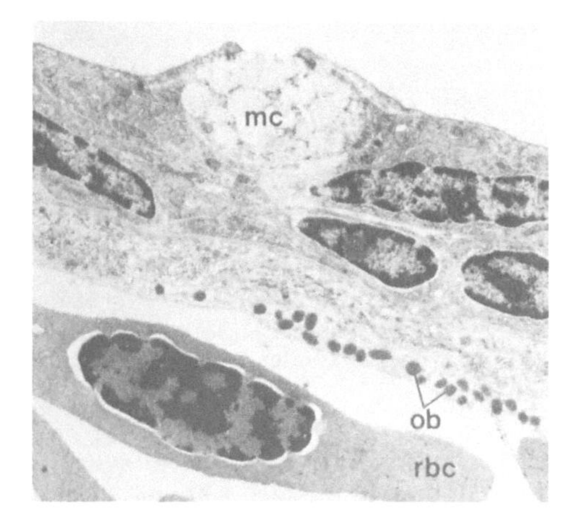

FIGURE L13 Transmission electron micrograph showing a mucous cell (mc) in a section of the epithelium of a gill lamella of the zebrafish, *Brachydanio rerio*. An erythrocyte (rbc) is seen in the central blood vessel, 15,300 ×.

epithelium and contains neurosensory, sustentacular, and basal cells. The neurosensory cells differ from those in the olfactory epithelium in that their surfaces do not bear cilia but are covered with microvilli. The sustentacular cells may be ciliated.

LUNGS AND AIR DUCTS

The respiratory tract of tetrapods consists of the lungs and a system of tubes by which the air passes to them from the pharynx (Fig. L16a). The tract develops as a ventral outpouching of the anterior portion of the gut and as such bears structural similarities to the gut. In structure and method of development the tetrapod lung so closely resembles the swim bladder of *Protopterus* and certain other fishes that it is hard to deny that these organs are homologous. The air bladder of the bony fish appears to have evolved from the lung-like bladder of the lungfishes.

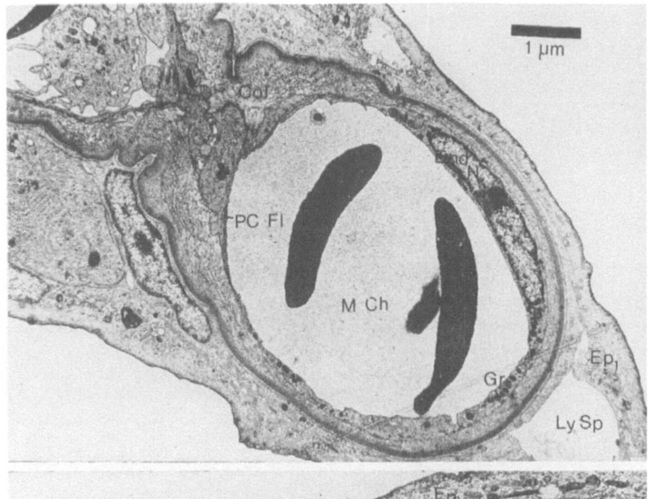

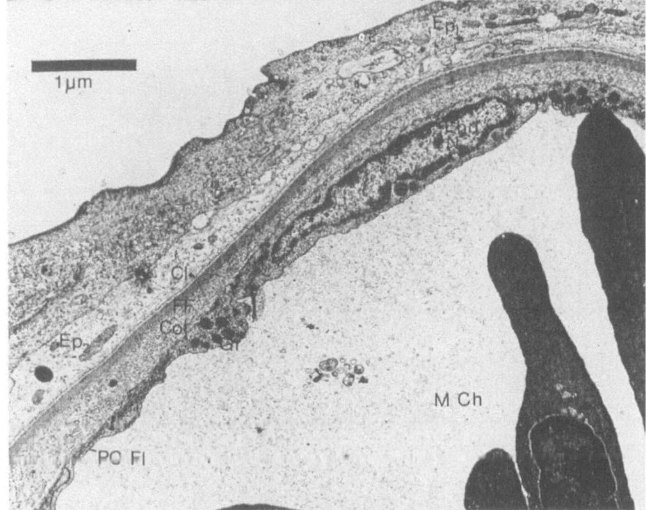

FIGURE L14 Transmission electron micrographs of the barrier to diffusion in the marginal channel (MCh) of the lamellae in the gill of the toadfish, *Opsanus tau*. *Top*: The nucleus of an endothelial cell is visible but the section only passes the edge of a pillar cell (PC Fl) of the outer row. Bordering the pillar cell is a thickened collagen layer (Col). *Ep*, epithelium of the lamella; *Gr*, granules; *LySp*, lymph space. *Bottom*: Higher magnification of the marginal channel showing the components of the water-blood barrier in the gill of the fish. The section passes through a flattened endothelial cell nucleus (EndN) and typical granules.

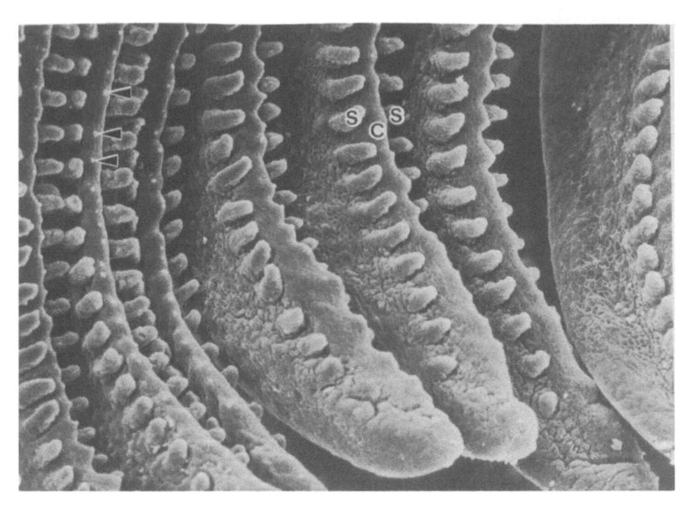

FIGURE L15a Low-power scanning electron microscope image of gill of the Mullet, *Mugil cephalas*, showing the distribution of taste buds (arrows) along the gill arches. *C*, central ridge; *S*, secondary projection, 350 ×.

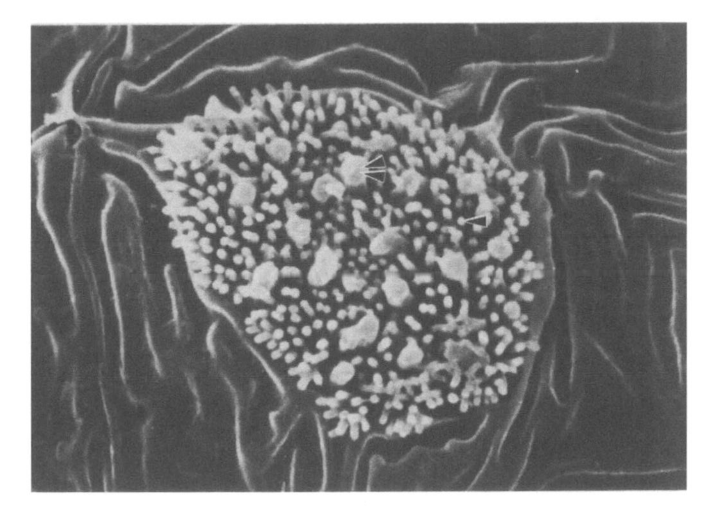

FIGURE L15b Scanning electron micrograph of a tuft of microvilli in the taste bud of a mullet, *Mugil cephalus*. The double arrowhead indicates a thick microvillus, the single indicates a thin microvillus, 27,000 ×.

Trachea

In a cross-section of mammalian trachea, note the relationship of the layers with those of the gut (Figs. L20a, L20b, and L21). The trachea is lined with a TUNICA MUCOSA consisting of a ciliated pseudostratified columnar epithelium in which CILIATED CELLS and GOBLET CELLS predominate (Figs. L22a–L22c). Also present in the epithelium are the BASAL CELLS that provide a reserve population for cell replacement. You may be able to recognize the rarer BRUSH CELLS by their large, blunt microvilli. Because sensory nerve endings make contact with these cells, they are believed to be sensory. The SMALL GRANULE CELLS are difficult to see in ordinary preparations with the light microscope. They are a heterogeneous population of tracheal endocrine cells similar to the enteroendocrine cells of the digestive tract. These are amine precursor uptake and decarboxylation (APUD) cells filled with small, dense granules. (Recall the description of APUD cells in the epithelium of the stomach, p. K-15 (Fig. K45i) and small intestine, p. K-22 (Figs. K61b, K61c, and K65b)). Many are argyrophilic. Some are innervated, some not. Identify the various types of epithelial cells and try to relate the appearance of their organelles to their differing functions.

Beneath the epithelium is a lamina propria mucosae of reticular or fine areolar tissue, and an elastic membrane. Lymphocytic cells and occasional lymphoid nodules are found in the lamina propria mucosae. The TELA

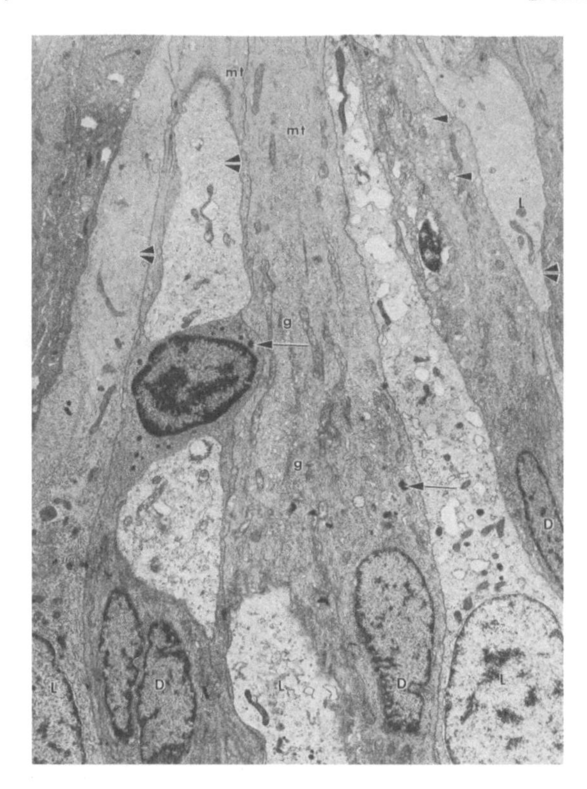

FIGURE L15c Transmission electron micrograph of a longitudinal section near the center of the taste bud of a killifish (*Fundulus heteroclitus*). The *arrows* indicate perinuclear dense vesicles of dark cells; *double arrowheads*, tubular membrane systems of light cells; *single arrowheads*, apical vesicles of dark cells; 17,800 ×.

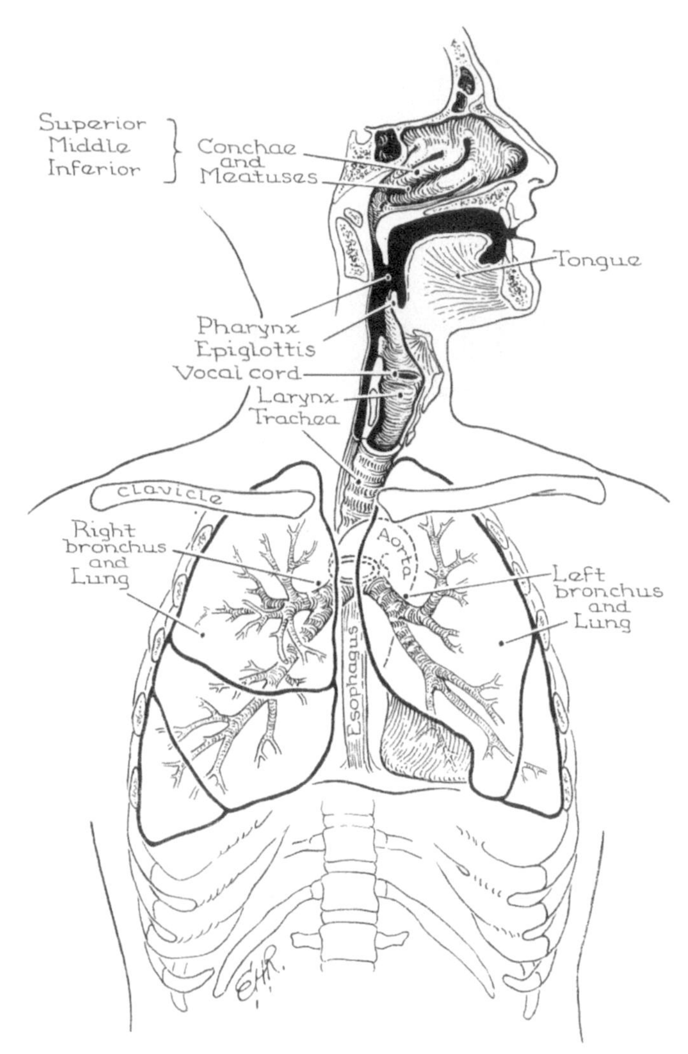

FIGURE L16a Diagram of the human respiratory system. Source: *From Taylor, N.B., McPhedran, M.G., 1965. Basic Physiology and Anatomy. Putnam and Sons, NewYork, and MacMillan Co Ltd., Toronto, Hodder & Stoughton.*

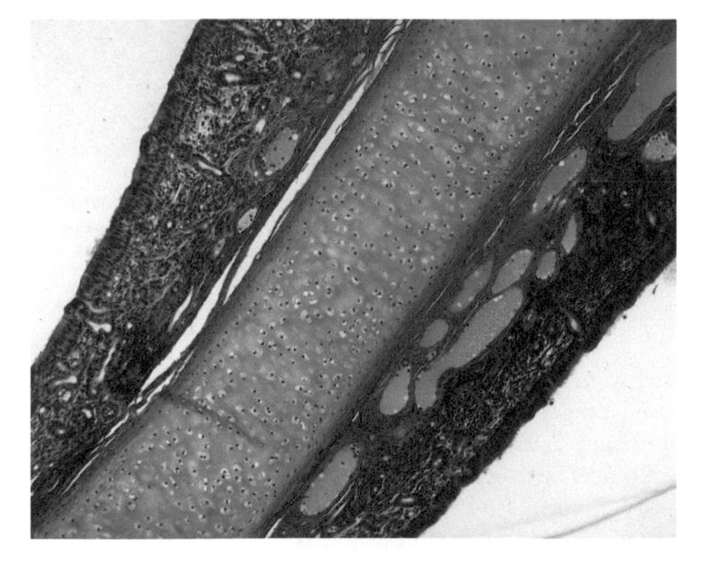

FIGURE L16b The nasal septum runs down the center of the nose from the external nares to the nasopharynx, separating the right and left NASAL CAVITIES. Flexible support is provided by a cartilaginous plate. The lamina propria mucosae at the left is richly provided with mucous glands. Distended veins occupy much of the lamina propria mucosae on the right. These vessels form a heat exchanger, warming inhaled air, and cooling exhaled air, thereby minimizing heat loss. A few distended capillaries are seen on the left, 10 ×.

SUBMUCOSA lies outside the elastic membrane and consists of areolar connective tissue and the secreting portions of scattered mixed serous and mucous glands. Ducts from these glands empty on the inner surface of the trachea. The trachea is held permanently open by C-shaped rings of cartilage corresponding in position to the tunica muscularis of the gut; in the space between the ends of the cartilage are interlacing fibers of smooth muscle knitted together by connective tissue. Because they are incomplete dorsally, the rings accommodate the distension of the esophagus during swallowing. In a longitudinal section of the trachea the rings are cut transversely and are seen to be oval in cross-section. The space between adjacent rings is filled with dense connective tissue continuous with that of the perichondrium of each ring. The outer layer of the trachea is continuous with the loose

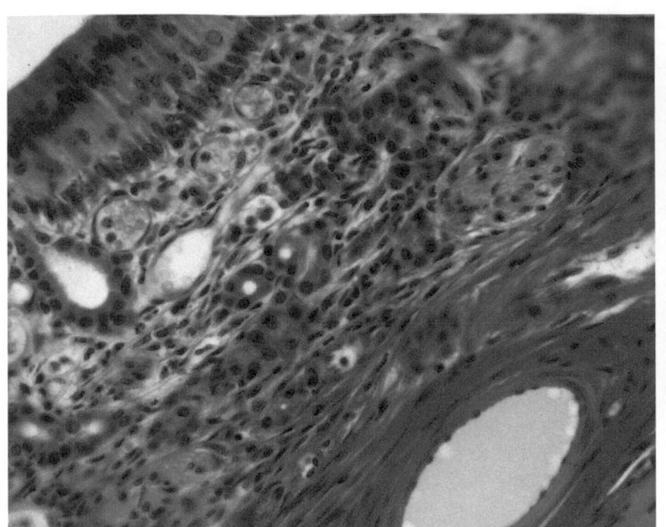

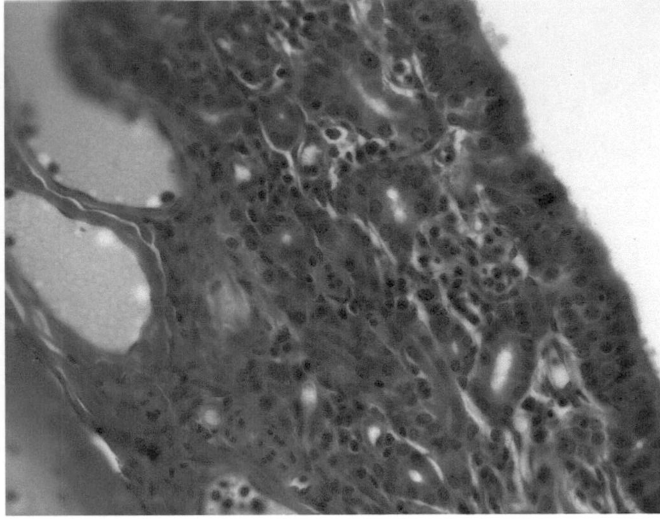

FIGURES L16c and L16d The distended veins and mucous glands of the lamina propria mucosae are shown in these two micrographs of the nasal septum. The ciliated, pseudostratified columnar epithelium consists of several cell types including ciliated cells, goblet cells, and basal cells that give rise to the other cells, 40 ×.

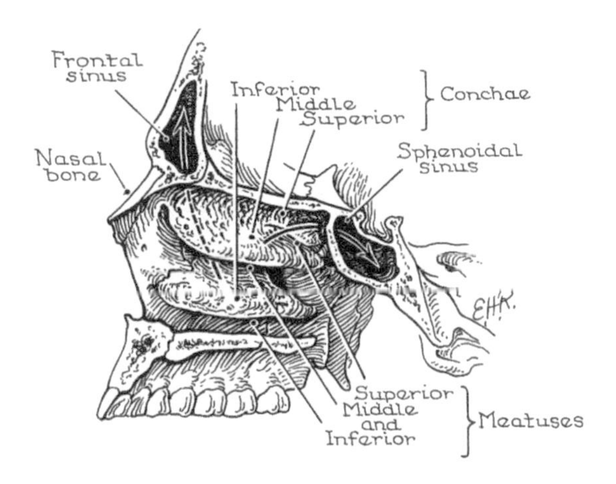

FIGURE L17a Drawing of the lateral wall of the right nasal cavity of a human showing nasal conchae. Source: *From Taylor, N.B., McPhedran, M.G., 1965. Basic Physiology and Anatomy. Putnam and Sons, NewYork, and MacMillan Co Ltd., Toronto, Hodder & Stoughton.*

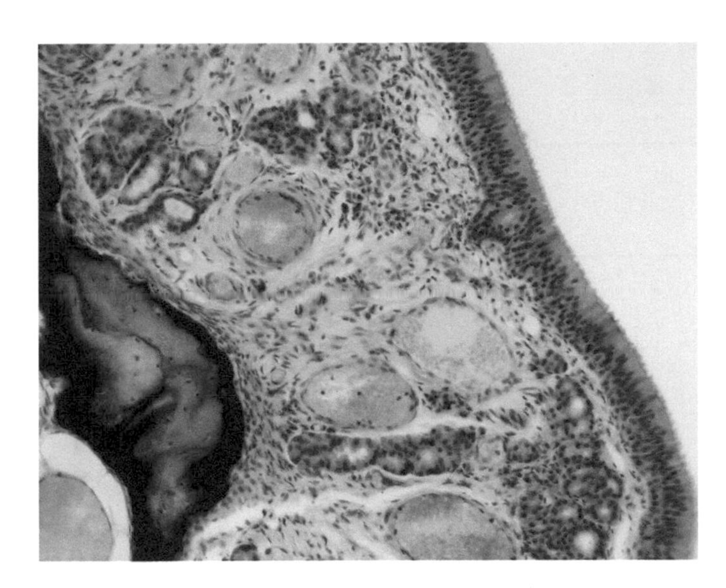

FIGURE L17b The surface area of the nasal cavities is increased by scroll-like projections from the lateral walls, the NASAL CONCHAE. A portion of one of these bones is at the left of this micrograph. The lamina propria mucosae is similar to that seen previously and harbors several distended veins (which function as heat exchangers) and several mucous glands. This is covered by the OLFACTORY EPITHELIUM which is the organ of the sense of smell. It is a pseudostratified columnar epithelium, 20 ×.

connective tissue of the mediastinum and contains abundant blood and lymphatic vessels and nerves. Internally the connective tissue is continuous with the perichondrium.

The histological structure of the trachea of other tetrapods is similar to that of mammals although in some amphibians the cartilaginous rings are less regular. In some amphibians the trachea is so short that it can hardly be said to exist.

The upper end of the mammalian trachea is modified as the LARYNX (Figs. L23 and L24a–L24g). The larynx is supported by separate cartilages of both the hyaline and elastic fibrous type. The lining is largely pseudostratified columnar ciliated epithelium except over the vocal cords and the epiglottis where it is stratified squamous. Below the epithelium is a layer of connective tissue containing many small striated muscles associated with movements of the larynx and production of sound.

The larynx of mammals is highly developed, being able to produce a wide range of sounds. Birds produce sounds by a SYRINX, a structure at the lower end of the trachea; the larynx is poorly developed and silent.

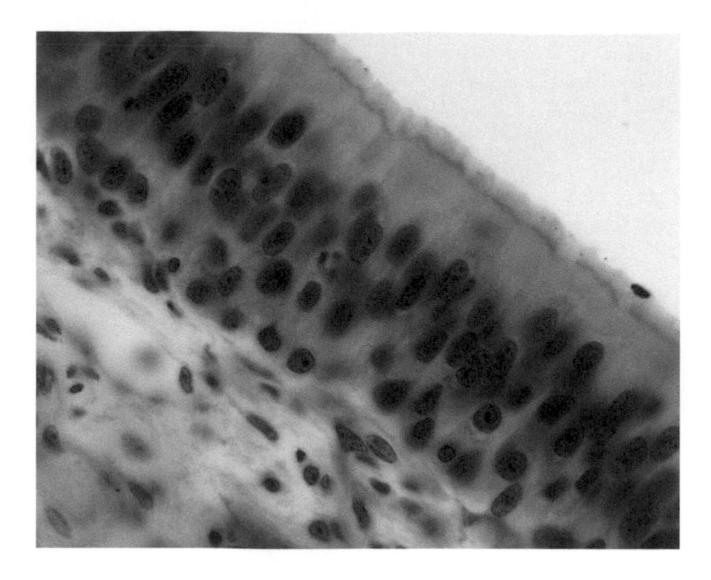

FIGURE L17c The olfactory epithelium consists of three types of cells: OLFACTORY CELLS, SUSTENTACULAR CELLS, and BASAL CELLS. The neurosensory olfactory cells are bipolar neurons evenly distributed among the sustentacular cells. The most apical nuclei in this micrograph are probably those of sustentacular cells, those nearer the center are probably olfactory cells, and those at the base of the epithelium are probably basal cells. Sustentacular cells secrete mucus which coats the surface. They bear long, slender microvilli which penetrate the mucus. A slender process, the OLFACTORY ROD, is a dendrite and extends to the surface from each olfactory cell—this is difficult to see in this preparation and should be studied in electron micrographs. The olfactory rod terminates at the surface in a small bulb, the OLFACTORY VESICLE. The vesicle bears several modified cilia, the OLFACTORY HAIRS. The proximal part of each olfactory cell narrows into an unmyelinated axon and passes into the underlying connective tissue to join with axons from other olfactory cells to become the olfactory nerve. Basal cells are undifferentiated and transform into either of the mature types, 100 ×.

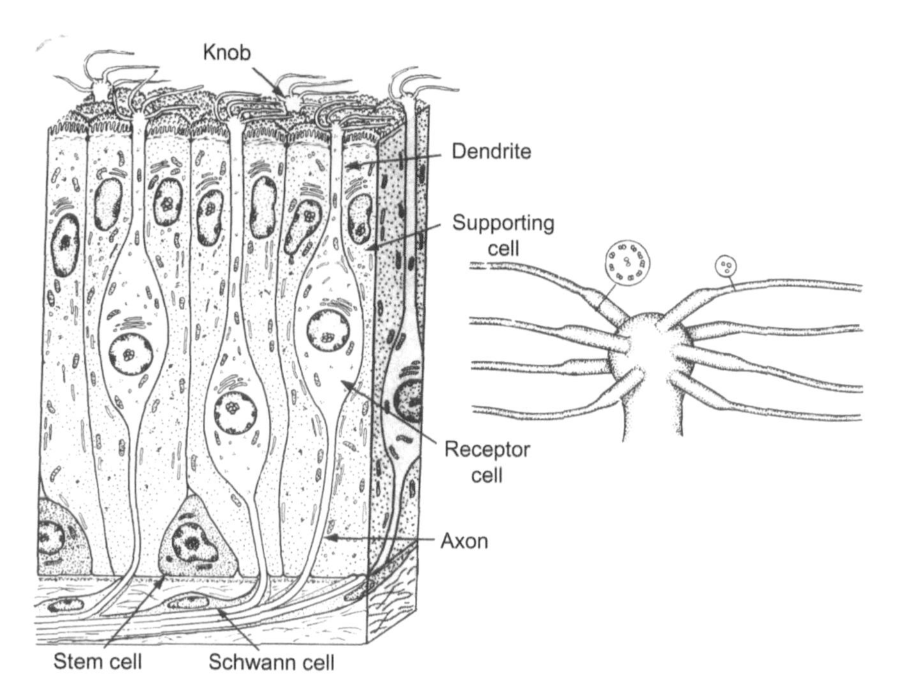

FIGURE L17d *Left:* Sketch of mammalian olfactory epithelium showing receptor cells is interspersed among tall, columnar SUSTENTACULAR CELLS that reach to the surface as well as small conical BASAL CELLS between their bases. The sustentacular cells secrete mucus and bear long slender microvilli that penetrate the layer of mucus coating the surface. Basal cells are undifferentiated and are able to transform into either of the mature types. BRUSH CELLS, dispersed throughout the olfactory epithelium have large, blunt, apical microvilli, and form synaptic contact with nerve fibers in the epithelium, and it has been suggested that they provide sensory input for the sneeze reflex. *Right:* Drawing of the knob at the exposed end of the dendrite of the neurosensory neuron. Its modified cilia bear the receptors for various odors.

The larynx in reptiles is also poorly developed and thus they are able to produce little more than hisses. The wide range of sounds produced by the amphibian larynx is modified by saccular extensions of the mouth cavity that act as resonators, the VOCAL SACS.

Bronchi

Bronchi are similar in structure to the trachea but differences may be seen: the elastic membrane of the tracheal tunica mucosa is gradually replaced by a layer of smooth muscle; mucous glands are more numerous in the bronchi and often extend from the tela submucosa into the cartilaginous layer; and the C-shaped cartilages are replaced by overlapping plates of irregular shape. In the lower parts of the bronchi the cartilages and glands diminish in number, the latter extending further down the tubes than the former.

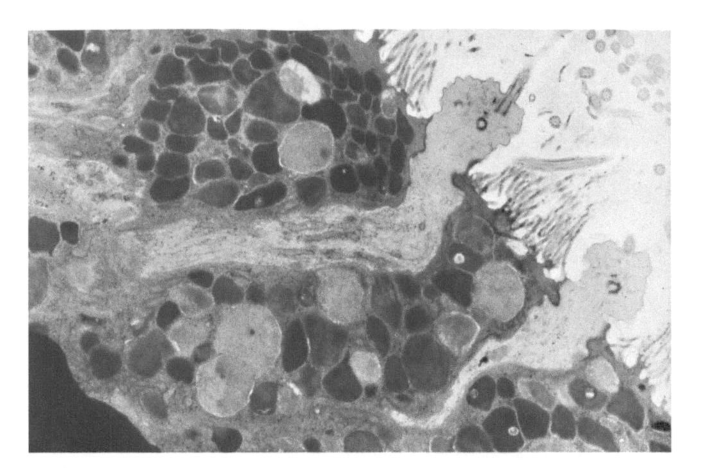

FIGURE L17e Transmission electron micrograph of a section of the olfactory epithelium of a leopard frog (*Rana pipiens*). The pseudostratified columnar epithelium bristles with microvilli and its cells are packed with mucous droplets and microtubules. An elbow-shaped pale cell at the center of the micrograph is the RECEPTOR CELL; this neurosensory cell is a bipolar neuron. The slender distal process is a dendrite, the OLFACTORY ROD, that extends to the surface and terminates in a small bulb, the OLFACTORY VESICLE, which bears several modified cilia, the OLFACTORY HAIRS. The shape of the vesicle and number of cilia present are highly variable among the vertebrates. The proximal part of the neurosensory cell narrows into an unmyelinated axon and passes into the underlying cells to become the olfactory nerve, 20,000 ×.

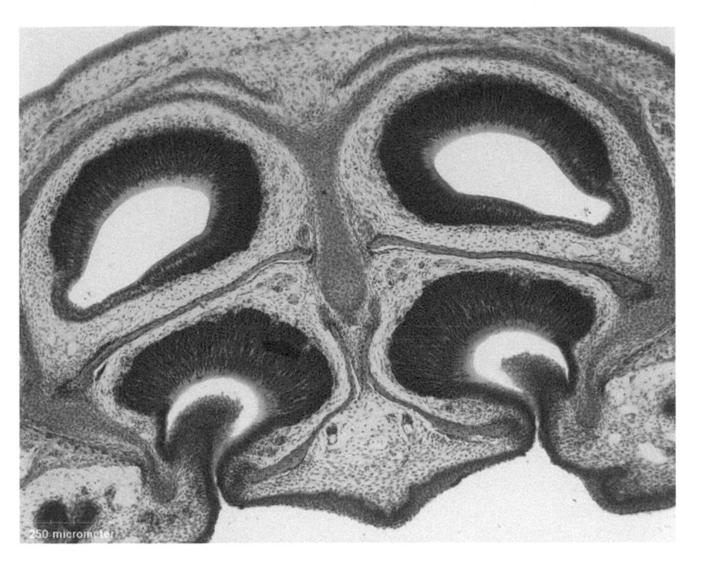

FIGURE L18a Recall the structure of the VOMERONASAL ORGAN seen in the roof of the mouth of many vertebrates. In this section of the head of a small lizard, two blind pouches extend dorsally from the roof of the mouth—their lumina are continuous with the oral cavity. Chemical substances brought into the mouth on the forked tongue are detected when the tips are thrust into the pits, 10 ×.

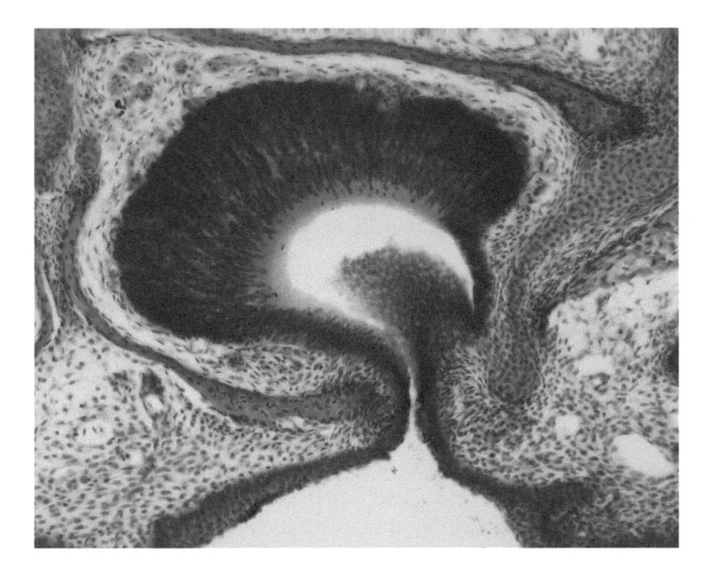

FIGURE L18b The tall pseudostratified epithelium lining the pouch is similar to an olfactory epithelium, 20 ×.

Saccular Lungs

We have seen the appearance of a respiratory function in the swim bladder (or more properly the lung) of some fishes where the wall has become highly vascularized and its surface area increased by the presence of sacculations, the ALVEOLI. A similar organ is seen in the simple lung of amphibians and reptiles (Fig. L25a).

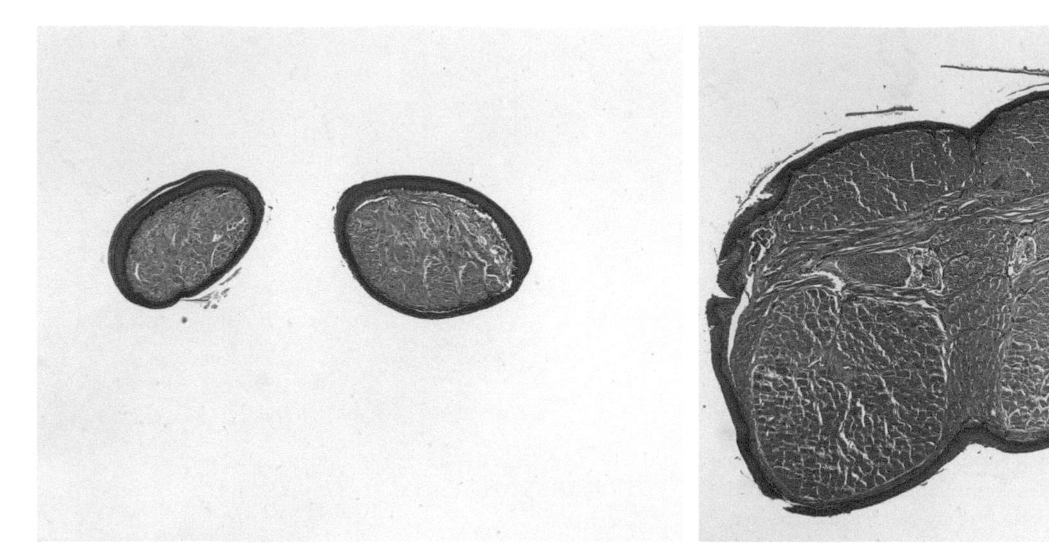

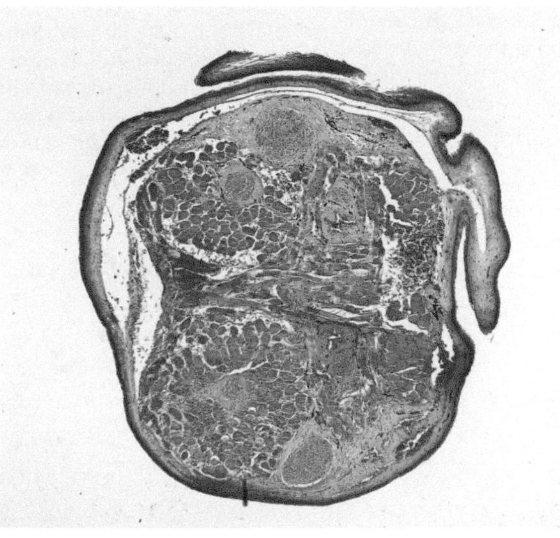

FIGURES L19a, L19b, and L19c The tongue of snakes is probably the most agile of any vertebrate tongues. It darts out of the mouth, capturing scent molecules which it brings into be sampled by a sensory organ in the roof of the mouth: Jacobson's VOMERONASAL ORGAN. In this series of micrographs from tip (L19a) to base (L19c), we see the forked tongue of many reptiles (a snake in this example) is made up of a core woven fibers of striated muscle covered with a stratified squamous epithelium; this is characteristic of most vertebrate tongues. Large nerve trunks provide amazing control. Unlike most tongues, however, the surface is a smooth stratified squamous epithelium with no trace of a sensory organ even at the tip of the forked portion. Pigmented cells lurk between the basal cells of the epithelium and among the connective tissue fibers, 4 ×.

Lungs of *amphibians* may be wholly smooth on the inside (e.g., *Necturus*), partly smooth and partly alveolar, or completely alveolar (e.g., frogs and toads). In all amphibians the lung is a thin-walled, highly vascular sac, lined with a simple epithelium whose cells stand beside the capillaries and cover them with a thin protoplasmic flap or flange. The wall contains a network of elastic fibers that assist in exhalation. The lining may be raised into many delicate vascular ridges, the SEPTA, to form a honeycombed surface; the ALVEOLI are the chambers thus formed (Figs. L25b–L25e). The rims of the septa are provided with stout bands of smooth muscle (Figs. L25f–L25j). Note the structure of the alveoli in other species (Figs. L26a–L26c). Generally only one side of the capillary of an alveolus is in contact with the alveolar lining. On this side the endothelial cells are flatter than those on the connective tissue side. In addition the part of the cell body containing the nucleus often lies away from the alveolus. These adaptations appear to lessen resistance to respiratory exchange.

In a number of terrestrial *urodeles* (tailed amphibians), no traces of larynx, trachea, or lungs occur, even after the gills are resorbed. Respiration takes place through the skin and walls of the mouth and pharynx where there is a great development of capillaries. The skin of all amphibians is respiratory and many amphibians have three functional respiratory organs in their lives: gills, lungs, and skin.

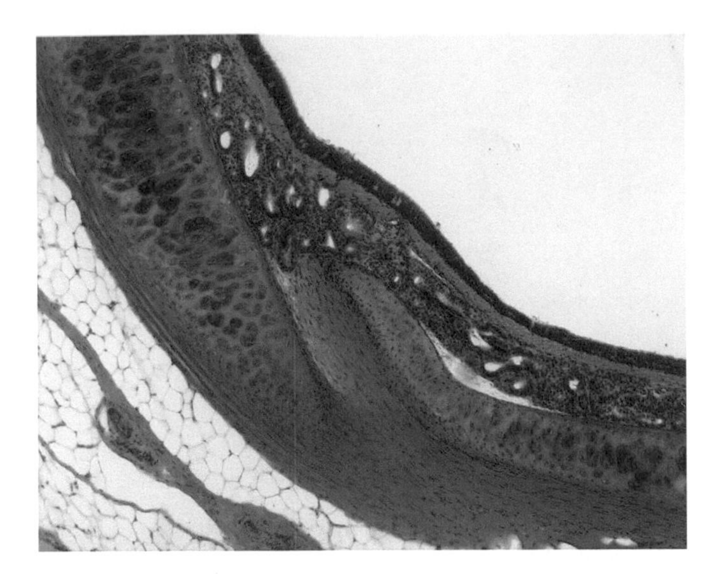

FIGURE L20a The layers of the wall of the trachea of most vertebrates are homologous to those of the wall of the gut. The trachea is lined by a TUNICA MUCOSA where ciliated cells and goblet cells predominate in the epithelium. A few compound mucous glands lie in the TELA SUBMUCOSA. The trachea is held open by C-shaped rings of cartilage which occupy the position of the tunica muscularis of the gut. The ends of the C's overlap in this view. The outer layer of the trachea is continuous with the loose connective tissue of the mediastinum, 10 ×. Source: *From Taylor, N.B., McPhedran, M.G., 1965. Basic Physiology and Anatomy. Putnam and Sons, NewYork, and MacMillan Co Ltd., Toronto, Hodder & Stoughton.*

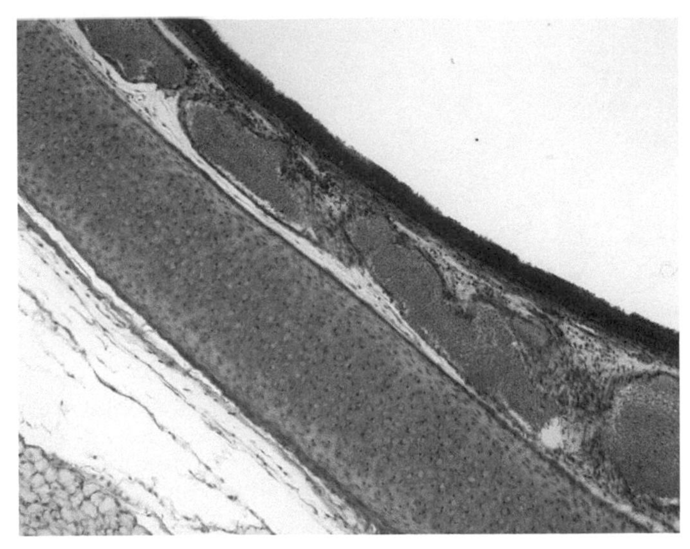

FIGURE L20b Large veins in the tela submucosa are presumed to assist in the control of body temperature by radiating heat to the air passing through the trachea, 10 ×. Source: *From Taylor, N.B., McPhedran, M.G., 1965. Basic Physiology and Anatomy. Putnam and Sons, NewYork, and MacMillan Co Ltd., Toronto, Hodder & Stoughton.*

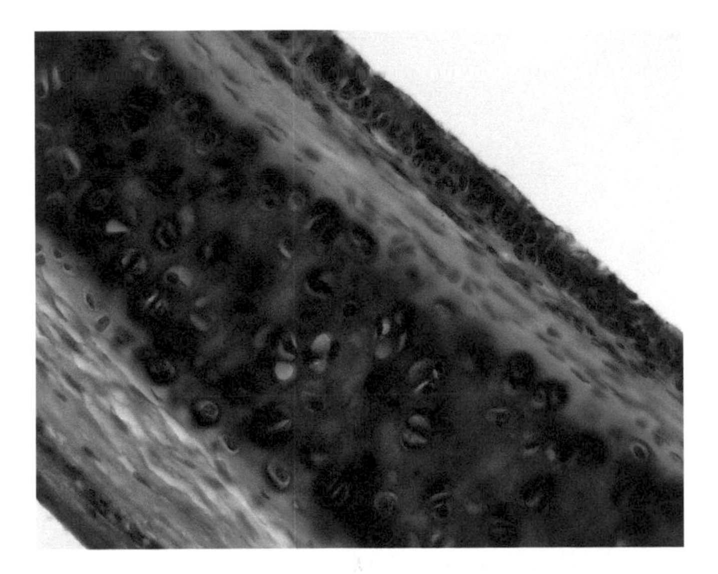

FIGURE L21 The trachea of a small reptile, the caiman, bears a striking resemblance to the trachea of mammals. It is lined with a tunica mucosa in which ciliated cells and goblet cells predominate. C-shaped rings of cartilage prevent its collapse, 40 ×.

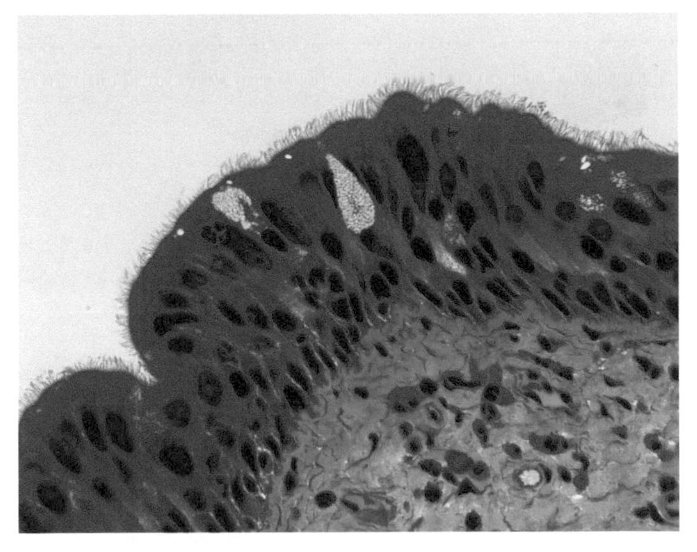

FIGURE L22a GOBLET CELLS and CILIATED COLUMNAR CELLS abound in the epithelium; individual cilia are easily recognized in this semithin section. Small cells at the base of the epithelium are BASAL CELLS that divide to provide replacements for surface cells that are lost over time—the epithelium is constantly regenerating, 63 ×.

In some, such as *Necturus*, which retains its gills in the adult, all the three function simultaneously although it is doubtful that the lungs have much effectiveness (Fig. L26d). Terrestrial amphibians generally have a larger lung respiratory surface than aquatic forms or those remaining close to water.

The lung of a garpike is similar to that of the frog and toad with septa forming alveoli and the presence of smooth muscle around the rims of the septal mouths (Figs. L27a–L27c).

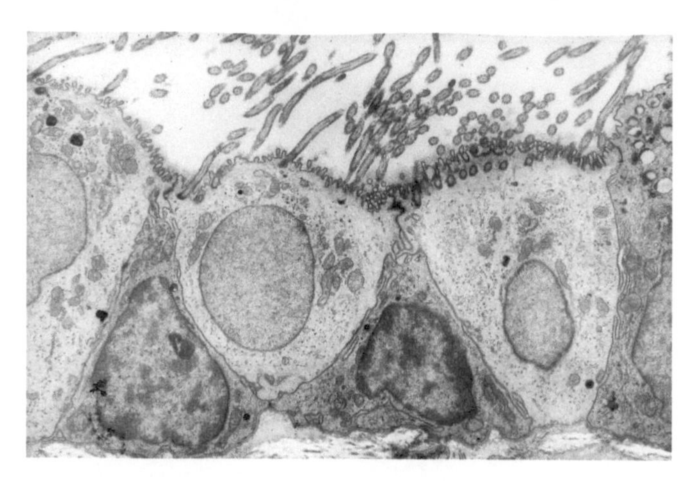

FIGURE L22b Transition electron micrograph of a section of the ciliated epithelium lining the trachea of a bat. The ciliated cells alternate with mucus-secreting cells. Inhaled foreign particles are trapped in the sticky mucus and the beating cilia drive them up the trachea where they are expelled, 97,000 ×.

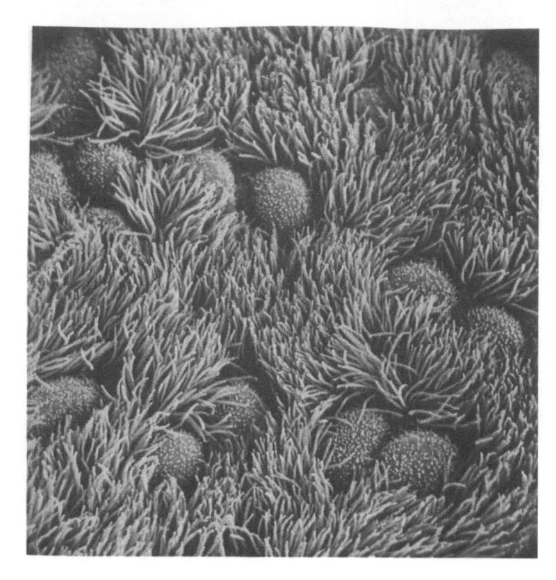

FIGURE L22c Scanning electron micrograph of the lining of the trachea of rat. The epithelium is made up of ciliated columnar cells interspersed with groups of mucous cells, 4200 ×.

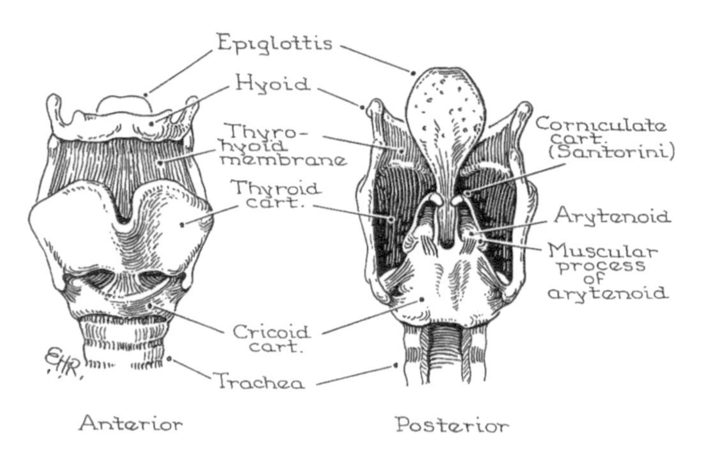

FIGURE L23 Drawings of the human larynx, anterior and posterior views.

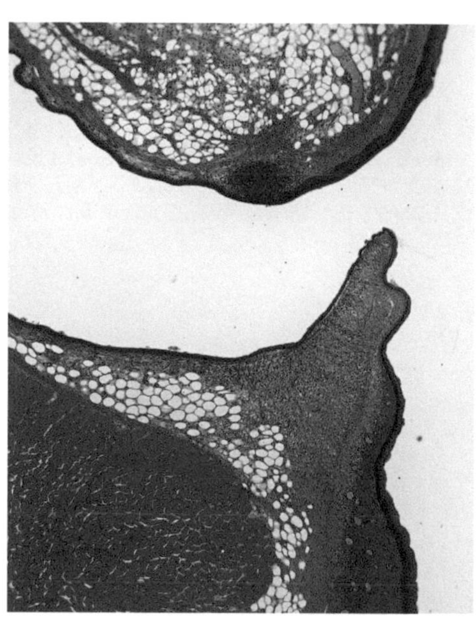

FIGURE L24a Frontal section of the mammalian larynx. Cartilages at the upper end of the trachea are modified to form the framework of the "voice box" or LARYNX. The epithelium lining the larynx is pseudostratified ciliated columnar except over the vocal cords and the epiglottis where it endures considerable abuse and is stratified squamous. Beating of the cilia causes entrapped particles to move toward the mouth, 5 ×. The cartilaginous flap at the top is the EPIGLOTTIS which closes the air passage, or GLOTTIS, during swallowing. The epiglottis is supported by elastic cartilage and is covered by stratified squamous epithelium on the upper side, pseudostratified columnar on the lower, more protected side. A few alveolar mucous glands are scattered throughout. Projecting into the lumen of the larynx are the FALSE VOCAL CORDS (*upper*) and TRUE VOCAL CORDS (*lower*). Between them is a narrow recess, the SINUS OF THE LARYNX. Sound is produced by the vibration of the free edges of the cords as a result of air passing between them. The false vocal cords have a core of loose connective tissue liberally provided with fat cells. Note the small lymphoid nodule at the edge of one of the folds. Their epithelium is stratified squamous and pseudostratified ciliated columnar. The true vocal cords, however, contain a large mass of striated muscle, the THYROARYTAENOID MUSCLE, which is under the control of the will; changes in the tension of this muscle modify the sounds produced by the larynx.

Among *reptiles*, the lungs of snakes and many lizards are little advanced over the amphibian condition. They are still simple sacs with a modest development of alveoli, the distal portions of the wall often remaining smooth. In some reptiles (e.g., chameleons) the smooth wall is prolonged into small air sacs that extend among the viscera, somewhat like those of birds. The most advanced reptilian lungs are those of some lizards, turtles, and crocodiles (Figs. L28a–L28c). The septa, originally confined to the outer walls have grown inward to subdivide the lung into chambers lined with alveoli. A main central tube runs the length of the lung as a continuation of the bronchus.

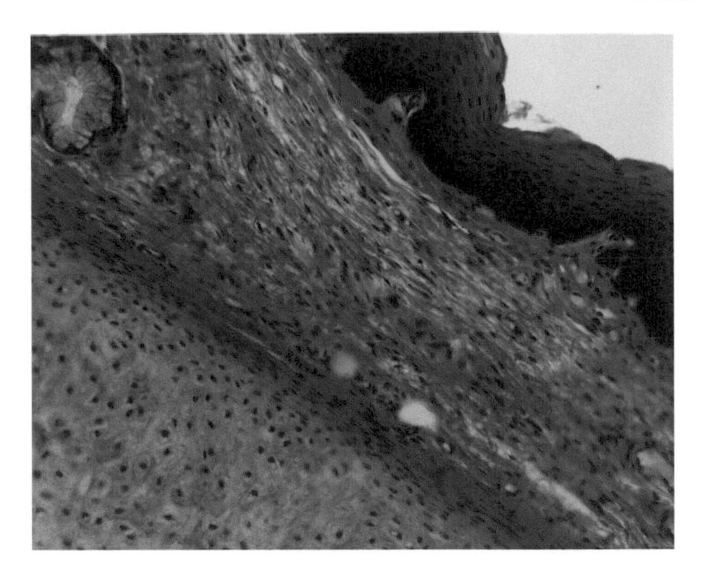

FIGURE L24b The epiglottis is supported by a plate of elastic cartilage. The MUCOUS MEMBRANE on the anterior and posterior surfaces is nonkeratinizing stratified squamous epithelium. It rests on the loose connective tissue of the LAMINA PROPRIA MUCOSAE which is continuous with the PERICHONDRIUM of the cartilage. Mucous glands are present in the loose connective tissue. The posterior surface of some regions of the epiglottis (*not shown*) is covered with pseudostratified ciliated columnar epithelium; the cilia move mucus and particular matter to the pharynx, 5×.

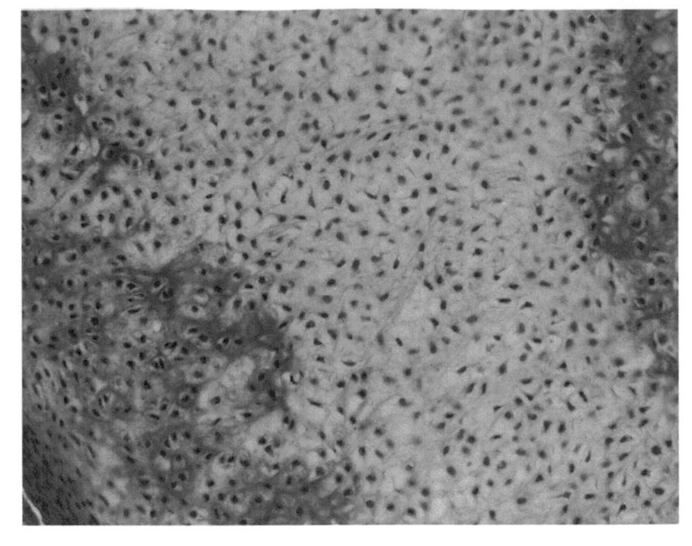

FIGURE L24c Elastic fibers abound in the matrix of the cartilaginous plate of the epiglottis, 20×.

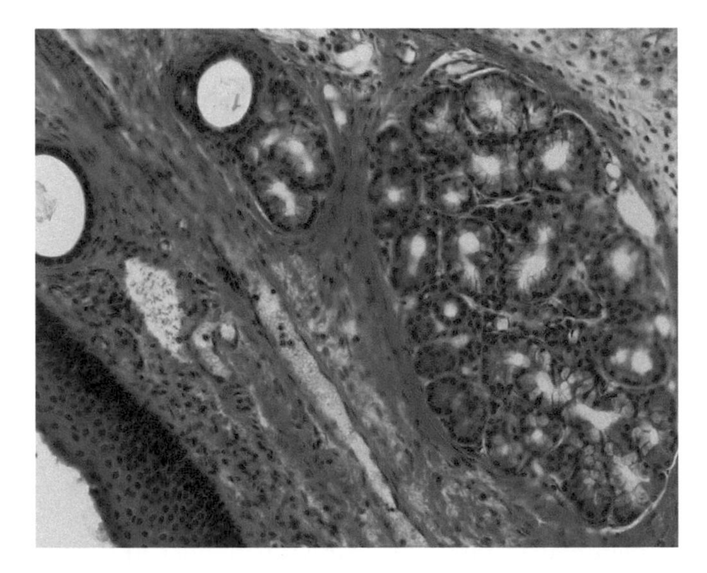

FIGURE L24d Mucous alveoli are more abundant in the lamina propria mucosae of the posterior portion of the epiglottis. The stratified squamous epithelium of the posterior surface is at the lower left. A few ducts from the mucous glands are seen at the upper left, 20×.

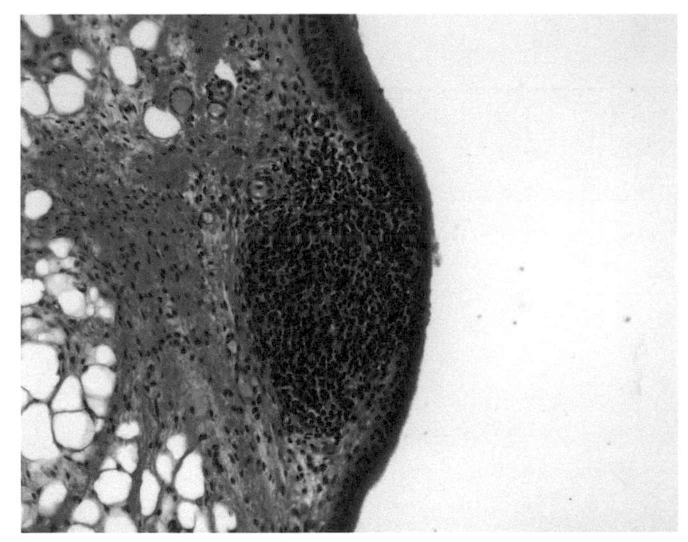

FIGURE L24e The loose connective tissue in the false vocal cord is liberally provided with fat. It is enclosed by stratified squamous epithelium. A lymphoid nodule invades the connective tissue, 20×.

Histologically the reptilian lung is similar to that of amphibians except that the lining is a simple squamous epithelium. Often thin flat processes of the epithelial cells overlie the capillaries while the portions containing the nuclei lie beside the capillary. In this way the nucleus does not impede respiration. In many apodous amphibians and reptiles, only one lung, usually the right, attains any degree of development in the long slender body.

The respiratory system of mammals develops as a median ventral diverticulum of the foregut that forms a tree which divides into two main branches and extends into the surrounding mesenchyme. The epithelium lining the larynx, trachea, bronchioles, and alveoli develops from this outpouching of the gut. The surrounding connective tissue, muscle, and blood vessels arise from the mesenchyme (Fig. L30a).

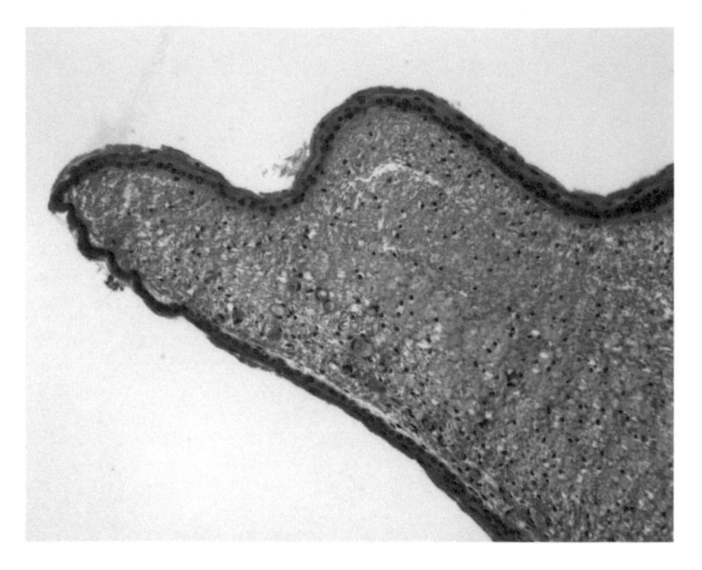

FIGURE L24f The tip of the true vocal cord has a core of loose connective tissue and is covered with stratified squamous epithelium, 20 × .

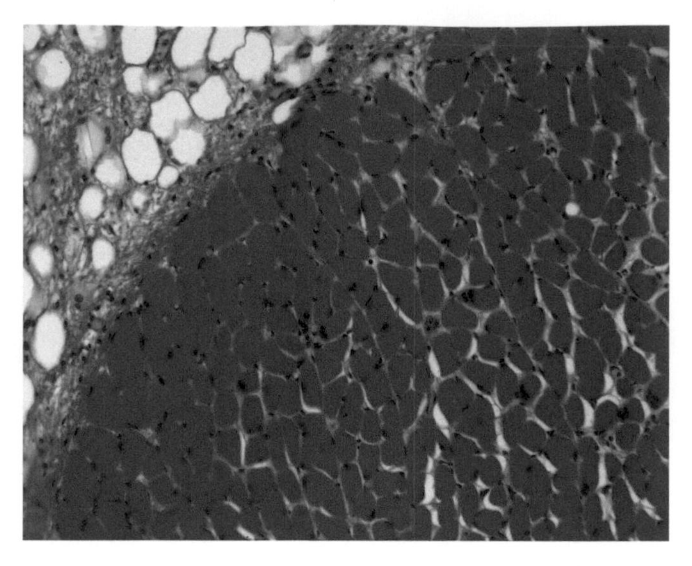

FIGURE L24g Embedded within the connective tissue core of the true vocal cord, however, is the thyroarytaenoid muscle, which modifies the sounds produced, 20 × .

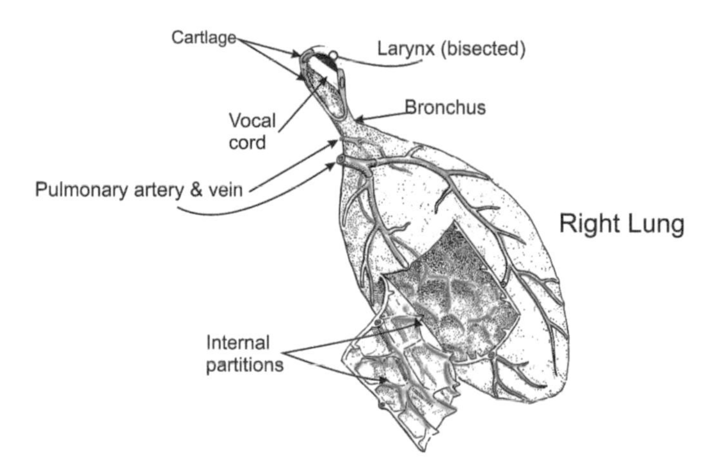

FIGURE L25a Laterodorsal view of a portion of the respiratory organs from bullfrog (*Bufo* sp.), the left lung has been removed and the right cut open to show partitions on the inner surface. The larynx is shown bisected along the midline with a vocal cord in place. *Authors.*

In many ways the *mammalian lung* is like a number of frog lungs bound together with connective tissue and supplied with air by the BRONCHIOLES branching off the bronchi (Figs. L29a, L29b, L29h, and L29i). Each "frog lung" is called an ALVEOLAR SAC. A TERMINAL BRONCHIOLE in a mammalian lung has a structure similar to that of a bronchus except that all goblet cells, submucosal glands, and cartilage have been lost. In the layers of the wall it may be possible to identify lymphoid tissue (Figs. L29c–L29g). Each "frog lung" is called an ALVEOLAR SAC. In sections taken through a mammalian lung we can find TERMINAL BRONCHIOLES. Its structure is similar to that of a bronchus except that all goblet cells, submucosal glands, and cartilage have been lost. Within the layers of the wall, we may see lymphoid tissue. In electron micrographs of the epithelium, NONCILIATED BRONCHIOLAR CELLS (Clara cells) can be distinguished. These are abundantly distributed among the ciliated cells and display all the features of secretory cells: basal rough endoplasmic reticulum, well-developed Golgi complex, and apical secretory granules. They secrete surface active agents.

Air from the terminal bronchiole passes into the RESPIRATORY BRONCHIOLE, a short tube with epithelium varying from ciliated columnar to nonciliated low cuboidal. Interlacing fibers of collagenous connective tissue, smooth muscle, and elastic connective tissue form the remainder of the wall. A few alveoli branch off from the side of

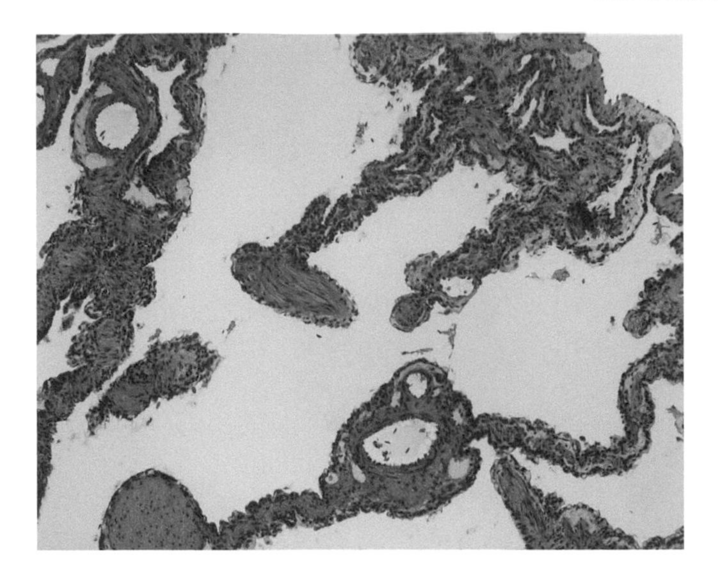

FIGURE L25b Surface area in the lining of the frog lung is increased by the presence of SEPTA that form a honeycomb of ALVEOLI that open from an ALVEOLAR DUCT. Bands of smooth muscle run along the free surfaces of the septa. Pumping action of the floor of the mouth cavity fills the lungs; air is expelled by the contraction of these smooth muscles. This section was taken from a contracted lung, 10 ×.

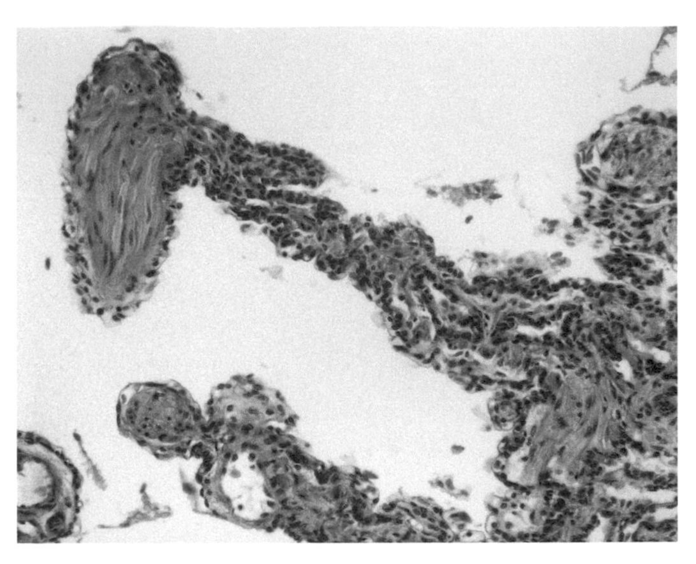

FIGURE L25c Bands of smooth muscle within the septa of the frog lung bring about the exhalation of air, 20 ×.

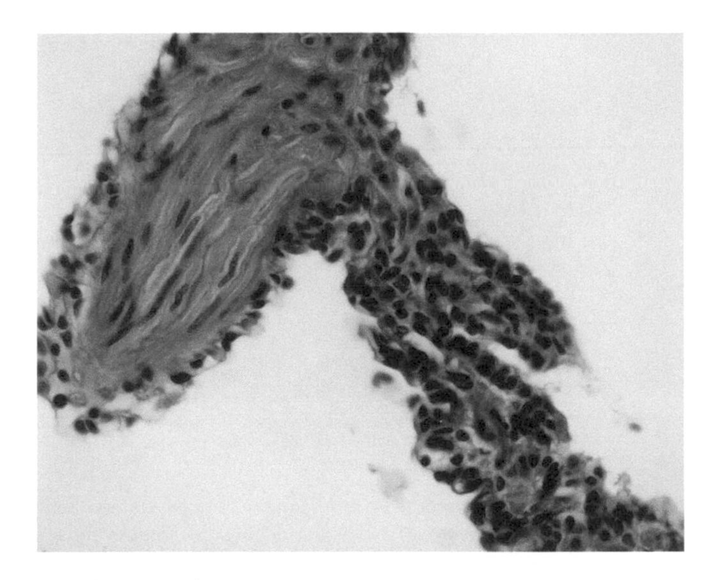

FIGURES L25d The lung is lined with a simple epithelium that forms a thin covering over the capillaries in the connective tissue below. Smooth muscle is present in the tissue at the rim of the septa, 40 ×.

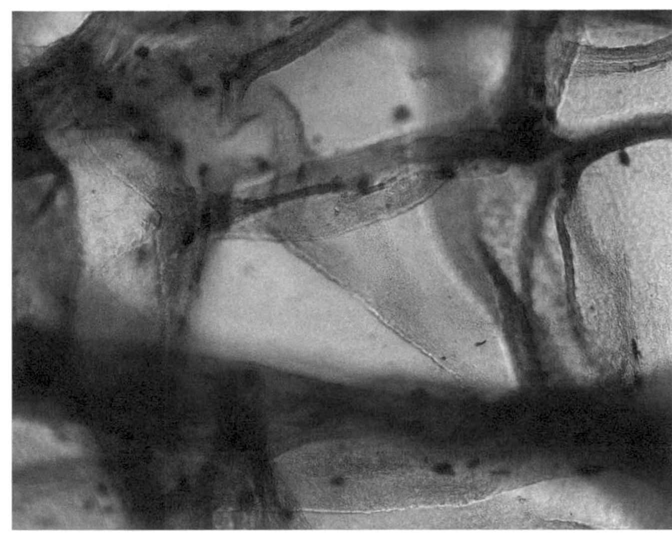

FIGURE L25e This is a whole mount of a portion of the lung of a frog. Bands of smooth muscle outline the tips of the septa that form the rims of the alveoli, 10 ×.

the respiratory bronchiole; these are the first respiratory structures of the lung and are responsible for the term "respiratory bronchiole" (Figs. L29c and L29d).

Coming off the respiratory bronchioles are several thin-walled ALVEOLAR DUCTS. From these arise single ALVEOLI and ALVEOLAR SACS containing two or more alveoli. Alveoli are thin-walled polyhedral chambers opening from the alveolar ducts. Dense single networks of capillaries anastomose freely in the walls of the alveoli (Fig. L30d). The alveolar walls also contain a network of branching reticular fibers and numerous elastic fibers (Figs. L29e, L30b, and L30c). The capillaries are so situated that the greater portion of their surface is toward the alveolar air while the larger reticular and elastic fibers occupy a central position in the septa. Smooth muscle fibers occur in

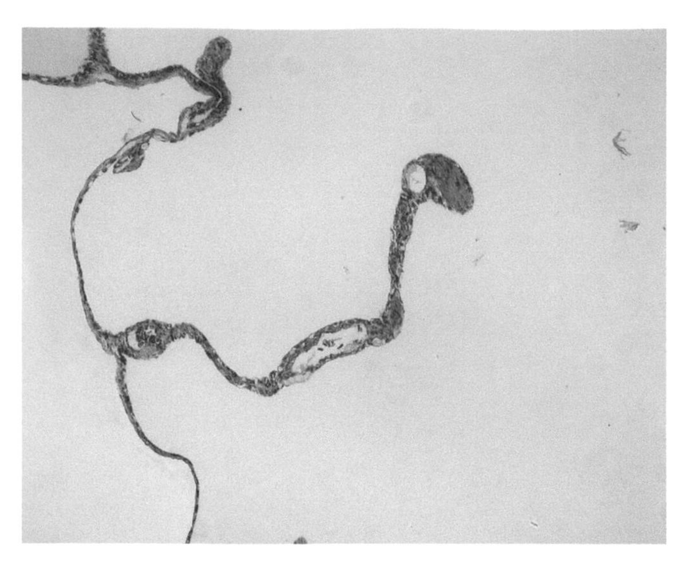

FIGURE L25f The alveolar walls are thin in the expanded lung. The rich supply of blood can be discerned as well as a band of smooth muscle at the septal rim, 10 ×.

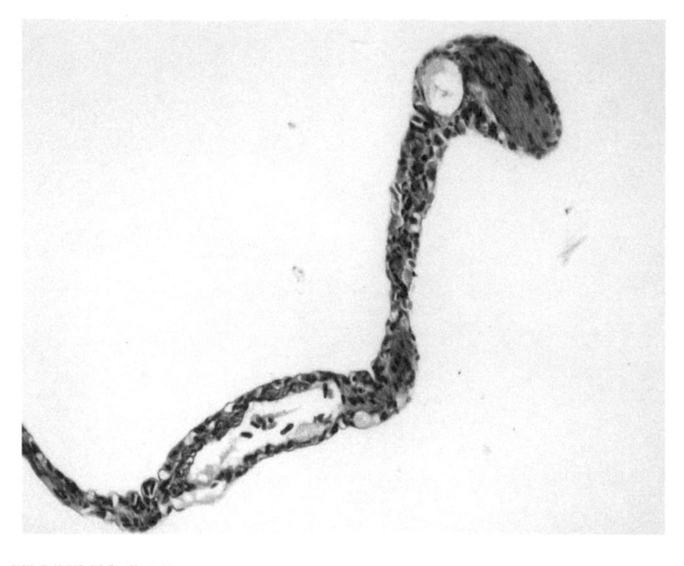

FIGURES L25g Large blood vessels are accommodated within the septa, 20 ×.

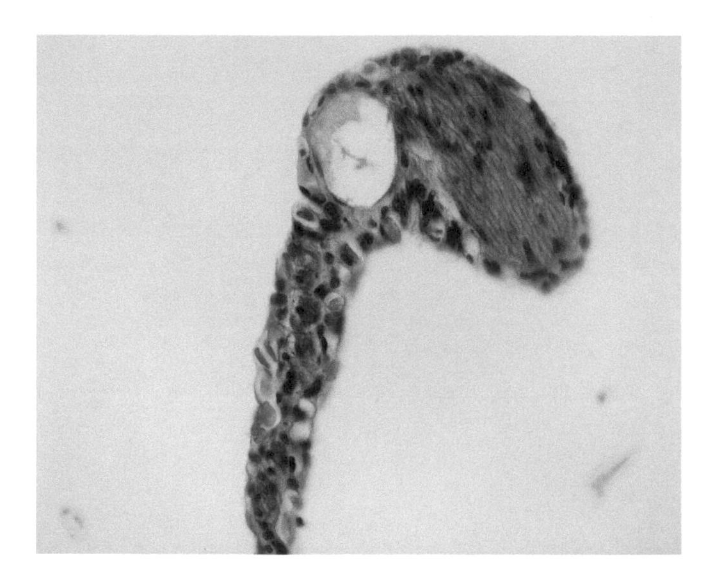

FIGURES L25h Large blood vessels are accommodated within the septa, 40 ×.

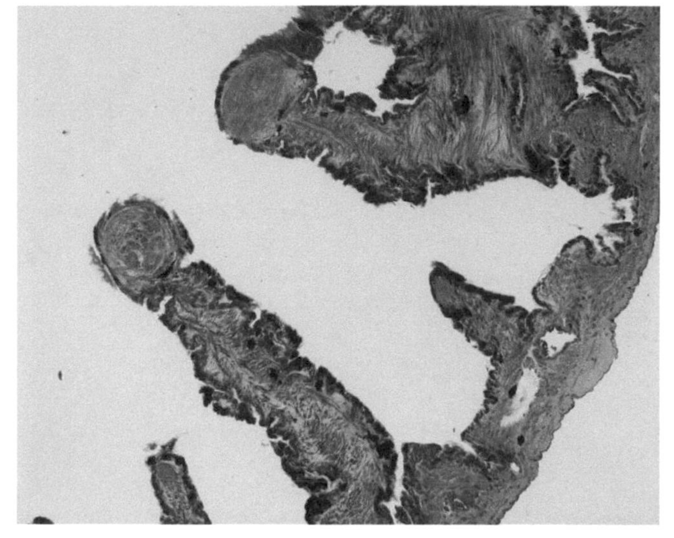

FIGURE L25i This section of frog lung has been stained with methylene blue and eosin clearly distinguishing the smooth muscle cells. Not only is there smooth muscle in the bands at the tips of the alveolar septa, smooth muscle fibers are liberally dispersed within the septa themselves, 10 ×.

the wall of the alveolus around the opening into the alveolar duct. Small lymphatics permeate the alveolar septa and drain into larger lymph vessels parallel to the blood vessels. Capillary networks in a section of lung, in which the blood vessels have been injected with colored material, are obvious (Figs. L30d and L30e).

The nature of the covering of the alveolar wall has long been a controversial problem in histology; its completeness and even its presence in the mammal were a matter of dispute. Electron microscope studies have shown a thin, apparently continuous covering of SQUAMOUS EPITHELIAL CELLS (pneumocytes type I) that is often just below the range of visibility of the light microscope (Figs. L29f and L29h). Separating the epithelium and the endothelium in most places is a thin basement lamina containing bundles of collagenous fibrils and occasional elastic fibers. Lying free on the epithelium of the alveoli and also within the connective tissue of the

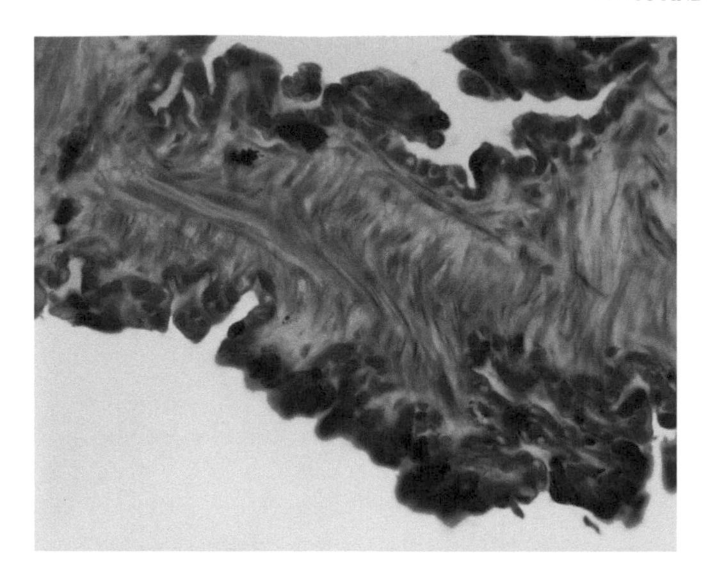

FIGURE L25j Contraction of these smooth muscle fibers assists the contraction of the frog's lung. Extremely delicate capillaries near the surfaces of the septum can be identified by their content of cherry red erythrocytes. A few melanocytes occur in the connective tissue of the septum, 40 ×.

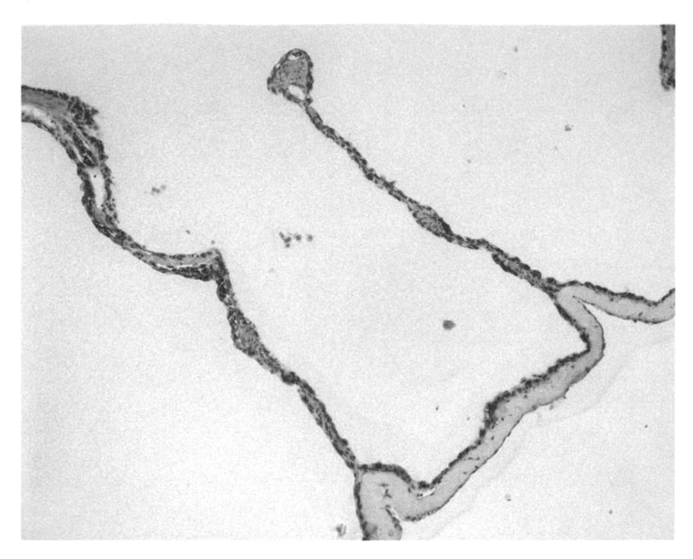

FIGURE L26a The lung of the toad appears similar to that of the frog. It may be that, in this section, the epithelium is thinner, and the capillaries more abundant—i.e., this lung is more efficient than that of the frog, 10 ×.

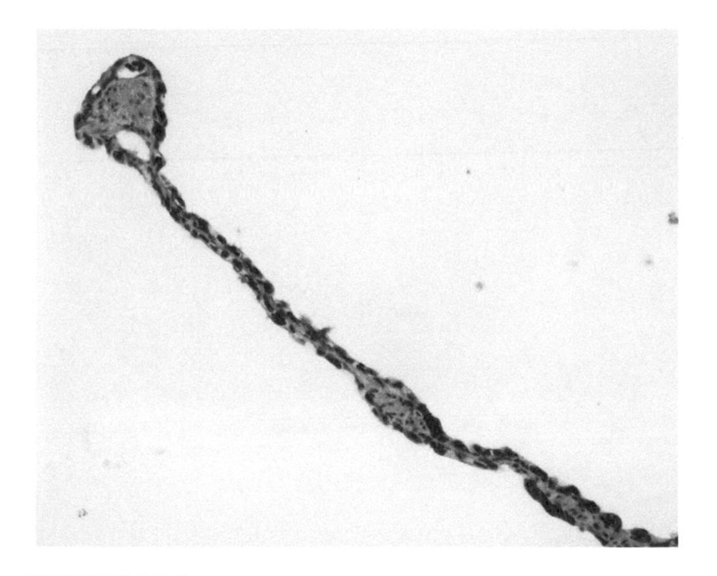

FIGURES L26b Capillaries are especially evident in the septa of the toad lung. The erythrocytes stain a brilliant cherry red in this section, 20 ×.

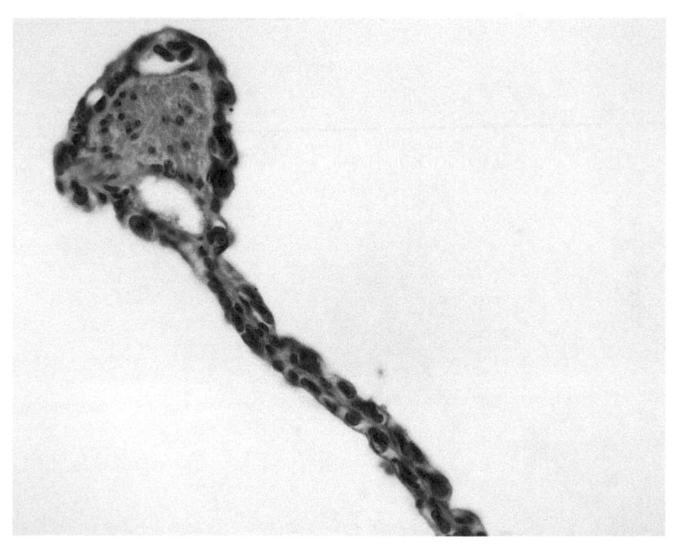

FIGURES L26c Capillaries are especially evident in the septa of the toad lung. The erythrocytes stain a brilliant cherry red in this section, 40 ×.

septa are the ALVEOLAR MACROPHAGES. When they contain particles of dust they are called DUST CELLS (Figs. L31a and L31b). A third cell type, the GREAT ALVEOLAR CELLS (pneumocytes type II), may be identified between the surface epithelial cells (Fig. L29g). These cuboidal cells synthesize and secrete pulmonary surfactant and occur singly or in small groups. They have small microvilli projecting into the alveoli and contain multivesicular bodies and lamellar bodies in their apical cytoplasm. What is their function? Continuous tight junctions connect the squamous epithelial cells to each other and to the great alveolar cells.

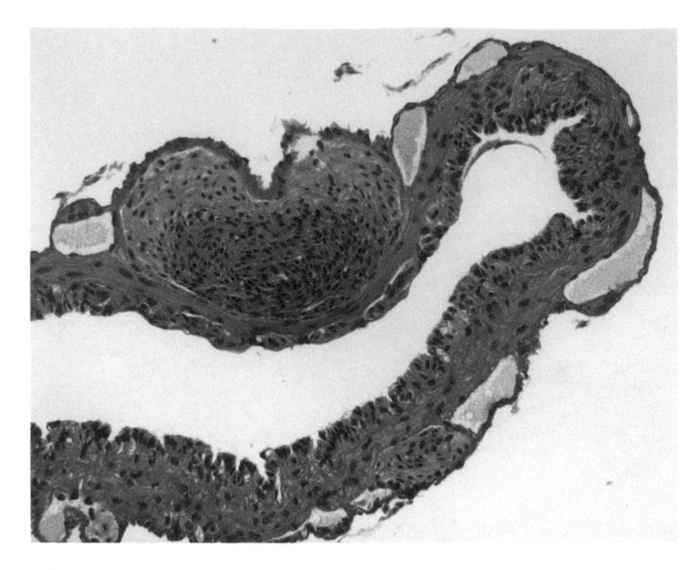

FIGURE L26d The mudpuppy has three functioning means of gaseous exchange: gills, the skin, and lungs. Its lung is a simple sac, without the presence of alveoli. The epithelial lining is thick, but a few capillaries can be seen, especially beneath the large blood vessel at the center of the screen. There is a smaller blood vessel at the bottom. Note the abundance of lymphatics (lined with simple squamous endothelium) around the outside of the lung, 10 ×.

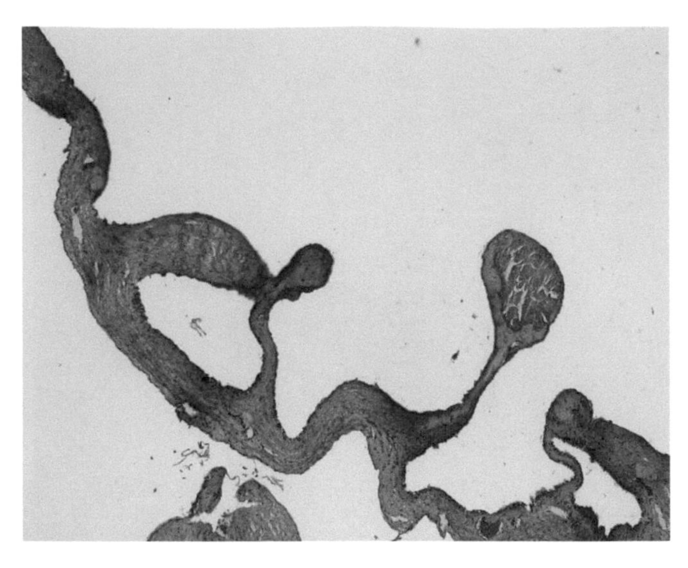

FIGURE L27a The lung of the garpike is similar to that of the frog and toad with septa forming alveoli and the presence of smooth muscle around the rims of the septal mouths, 10 ×.

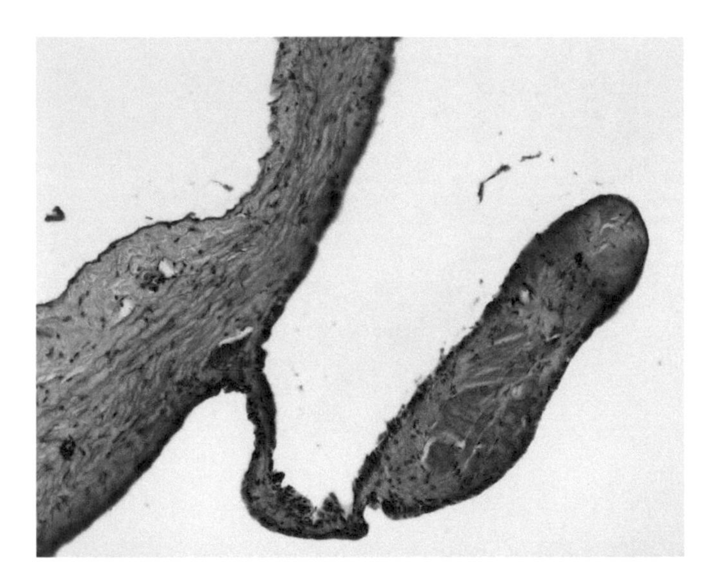

FIGURES L27b A simple epithelium lines the alveoli. Capillaries are not as apparent as in the lungs of frogs, but a large vessel can be seen within the smooth muscle band, 20 ×.

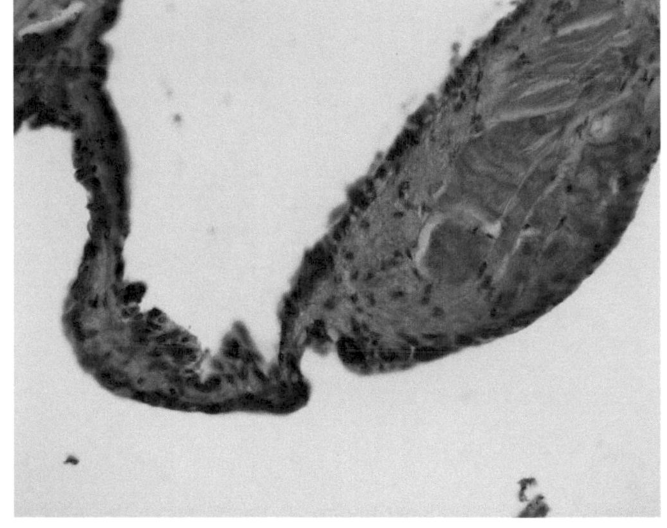

FIGURES L27c A simple epithelium lines the alveoli. Capillaries are not as apparent as in the lungs of frogs, but a large vessel can be seen within the smooth muscle band, 40 ×.

Bird Lungs

Birds have the most complex and highly evolved lungs of the vertebrates. This is probably associated with the fact that their high metabolic rate requires a constant supply of large amounts of oxygen. The lungs are small, spongy, and inelastic and from them the paired AIR SACS extend into every major part of the body, even the hollows of several bones (Fig. L33a). The lining of the sacs is smooth and probably of little value as a respiratory surface.

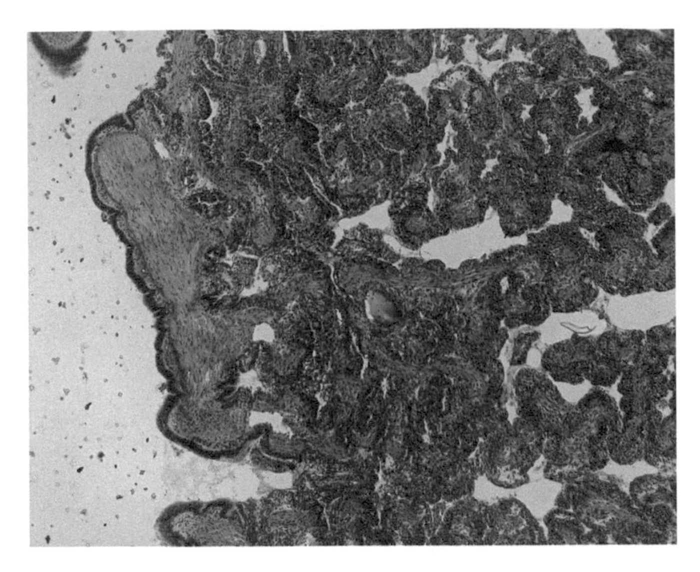

FIGURE L28a The rims of the alveolar septa in the lung of a turtle (Chelonia) are liberally supplied with bands of smooth muscle. The lung is inflated by the pumping action of the mouth cavity, and air is expelled by contraction of these bands of muscle, 5×.

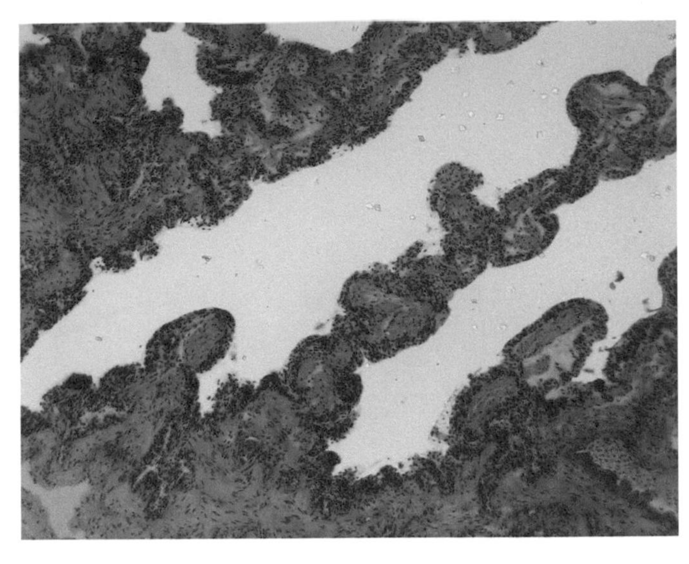

FIGURE L28b The surface area of the lung is greatly enhanced by the partitions separating the alveolar spaces, 10×.

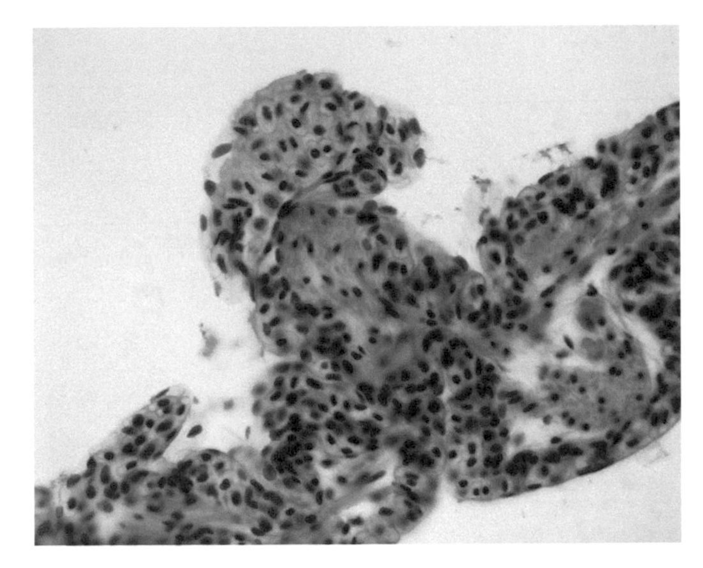

FIGURE L28c The rims of the alveolar septa are liberally supplied with bands of smooth muscle. The lung is inflated by the pumping action of the mouth cavity, and air is expelled by contraction of these bands of muscle. Although not readily apparent in this section, the alveolar SEPTA are rich in blood vessels. The efficiency of this lung is fairly low, reflecting the relatively sluggish life style of the turtle, Note the presence of smooth muscle within the septa—the lung is inflated by the pumping action of the mouth cavity and is deflated when the lung is made to contract by tension of this smooth muscle, 40×.

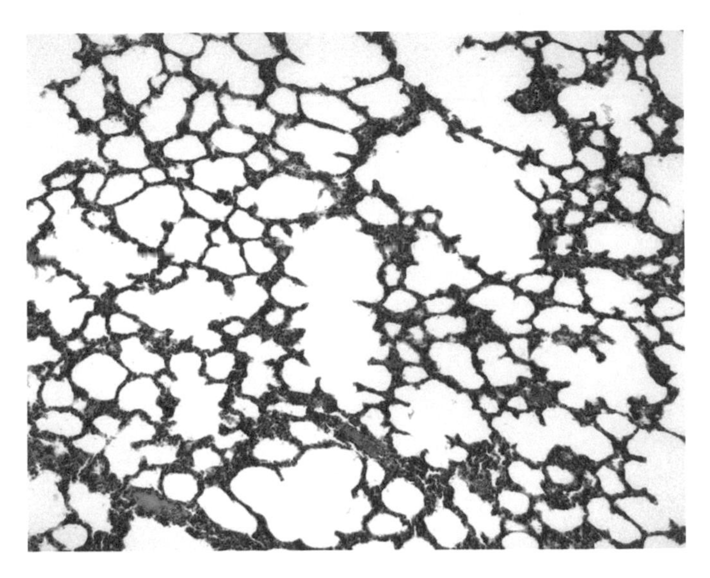

FIGURE L29a In many ways the mammalian lung is like a number of frog lungs (called ALVEOLAR SACS) bound together with connective tissue. Each alveolar sac has several rooms, or ALVEOLI, opening off a corridor or ALVEOLAR DUCT. Alveolar ducts are supplied with air by the BRONCHIOLES branching off the bronchi. No bronchioles are visible in this section, 10×.

Air is drawn through the lungs and passes into the air sacs; these function much as the bulb of an eyedropper drawing air through an inflexible tube (Fig. L33c). Quiet respiratory movements are produced by the intercostal muscles (inspiratory) and abdominal muscles (expiratory) acting to enlarge and contract the body cavity, drawing air in and out of the air sacs through the lungs. During flight, movements of the pectoral muscles provide ventilation, the sternum moving toward and away from the vertebral column.

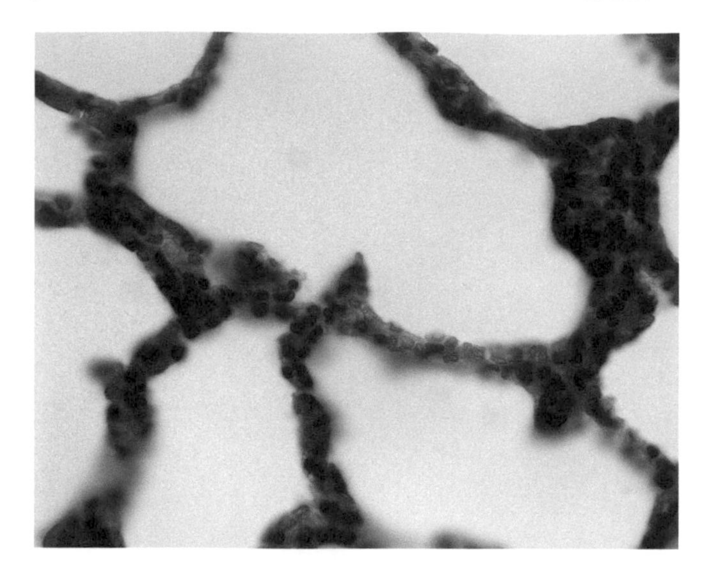

FIGURE L29b The alveolar walls of mammalian lungs are extremely thin and present little barrier to the exchange of gases. Indeed, before the advent of electron microscopy, there was doubt about existence of an epithelium lining the alveoli. The walls of the capillaries in the alveoli are extremely thin and cannot be seen in this section. Abundant erythrocytes crowd through capillaries near the surface, 63 ×.

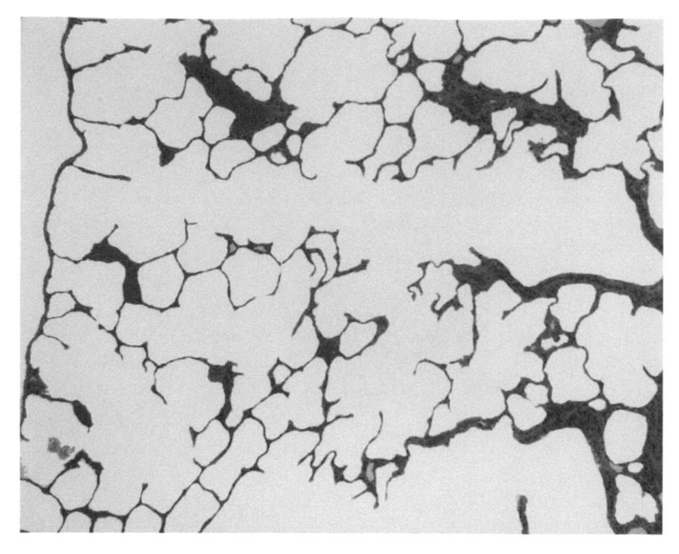

FIGURE L29c In this semithin section of the lung of a rabbit, there are sections of a few BRONCHIOLES at the right. These ducts are lined with a cuboidal epithelium that does not engage in gaseous exchange. Bronchioles open into RESPIRATORY BRONCHIOLES and ALVEOLAR DUCTS—as seen crossing the field from left to right, just above the center—which are capable of gaseous exchange. Alveoli open off the alveolar ducts, 10 ×.

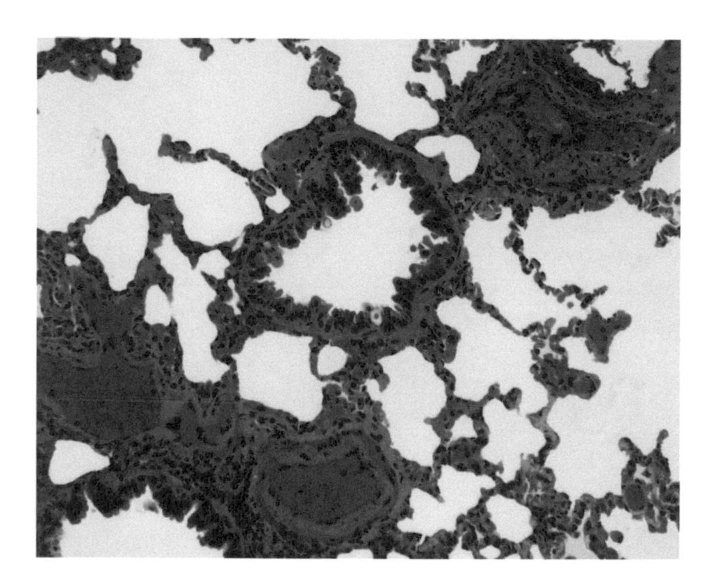

FIGURE L29d At higher magnification, we see a BROCHIOLE near the center, 20 ×.

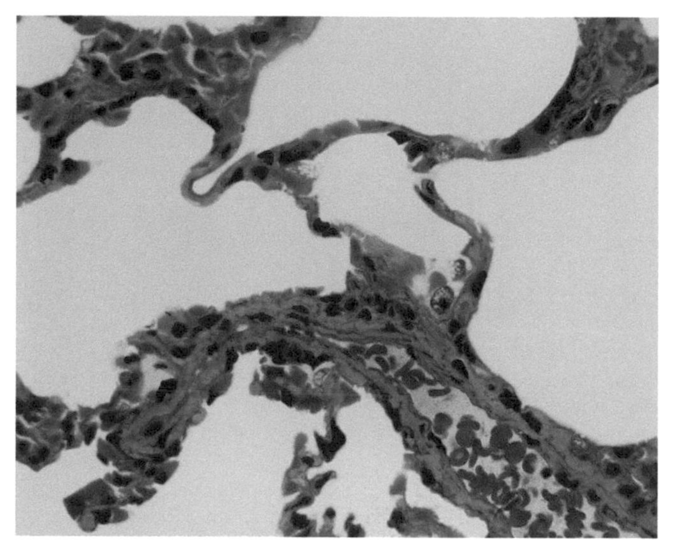

FIGURE L29e Large blood vessels lie deep in the alveolar septa. Semithin section, 63 ×.

A BRONCHUS on each side passes, as the MESOBRONCHUS, through the lung to the posterior end. On the way it loses its cartilaginous rings and gives rise to several SECONDARY BRONCHI. These branch into numerous small tubes, the PARABRONCHI, that form loops connecting with other secondary bronchi; therefore there is no "bronchial tree" as in mammals but a series of anastomosing tubules forming circuits within the lungs (Figs. L33a–L33c). No tubes end blindly. Parabronchi are surrounded by respiratory tissue in the form of a richly vascular sponge (Figs. L34a–L34d). The air-blood barrier is contained within the spongy outpocketings of the parabronchial wall and is similar to that of mammals but about three times thinner: a continuous squamous epithelium, basement membrane, and the endothelium of capillaries. The mesobronchus and several secondary bronchi continue on through the walls of the lung and open into the AIR SACS. In addition, RECURRENT BRONCHI connect the sacs with adjacent portions of the lung.

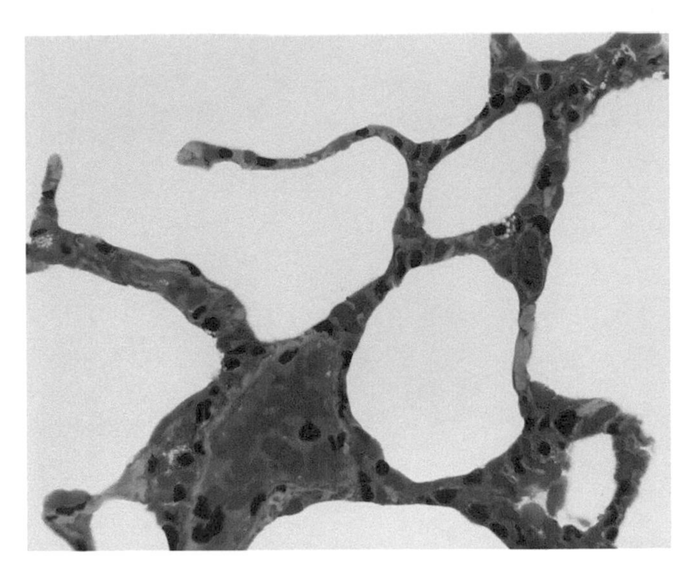

FIGURE L29f Alveolar walls are extremely thin and richly supplied with blood. Semithin section, 63 ×.

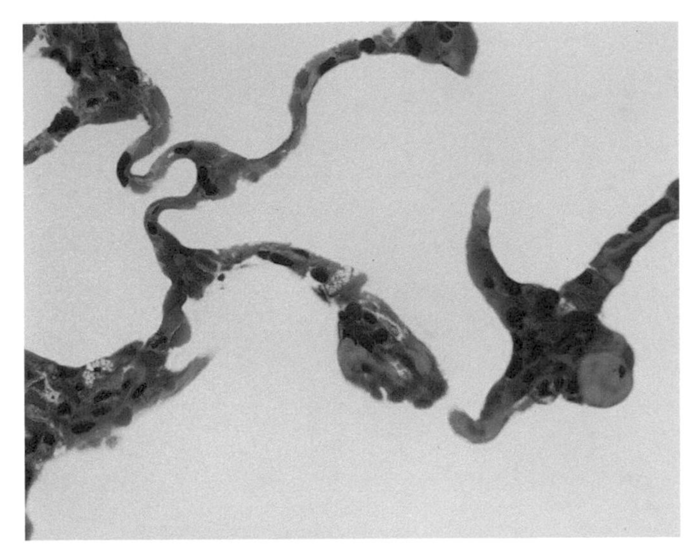

FIGURE L29g A few bubbly GREAT ALVEOLAR CELLS are embedded in the epithelium of the alveolar walls. They secrete the SURFACTANT that lowers surface tension and prevents collapse of the lung, 63 ×.

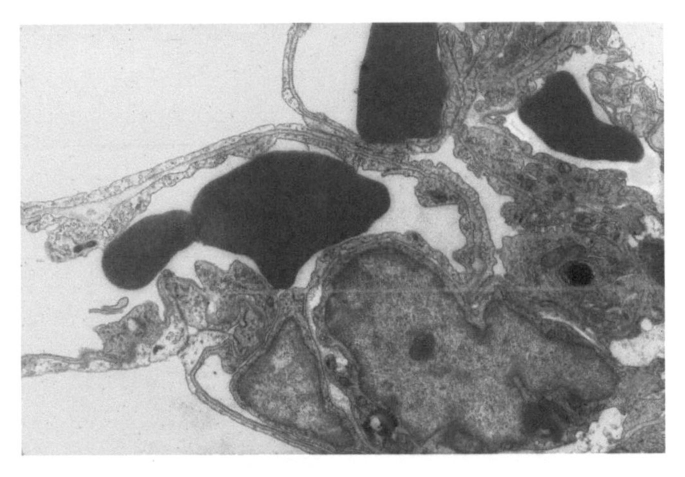

FIGURE L29h Transmission electron micrograph of a section of the lung of a mouse. Note the barrier between the inspired air within alveolar sacs and the erythrocyte in the blood. The epithelium lining the alveolar sacs and the endothelium of the capillaries are extremely thin, offering little resistance to the passage of gases, 18,000 ×.

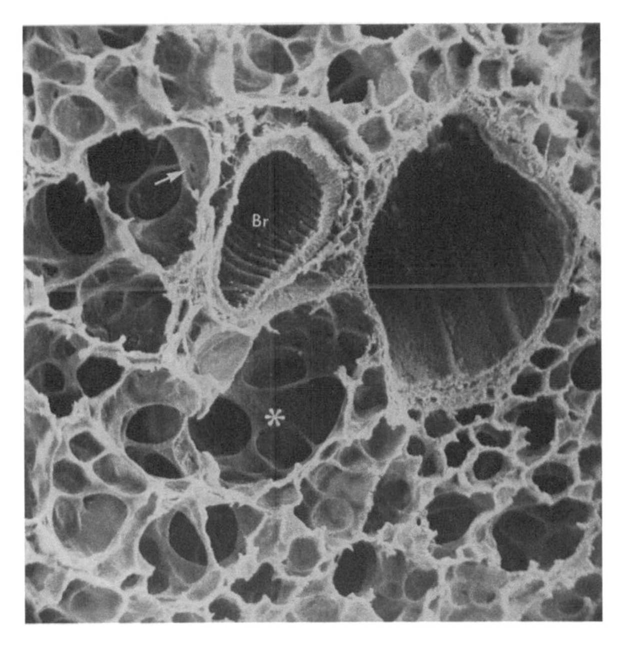

FIGURE L29i Scanning electron micrograph of the dried surface of a broken lung of a rat revealing some of its air passages. The lung was fixed by perfusion and there are no erythrocytes remaining. A bronchiole is slightly to the left of center showing its folded inner lining. Identification of the larger vessel to its right is uncertain—it may be a venule. Between these two and just below is an alveolar duct with alveolar sacs opening into it. Occasionally pores are seen between adjacent alveolar sacs; they are thought to equilibrate air within the sacs as inspiration and expiration occur, 200 ×.

Fig. L32 is a thick section of bird lung. Cylindrical LUNG LOBULES: parabronchi surrounded by spongy respiratory tissue may be seen here. Thin stained sections reveal further details of the tissue structure (Figs. L34a–L34d).

A "two cycle" system has been proposed for the passage of air through the bird lung (Fig. L33c). At the first inspiration, air is thought to pass back through the trachea, bronchi, mesobronchi, and secondary bronchi into the expanding posterior air sacs with little respiratory exchange on the way. At expiration, it passes from the constricting air sacs into the recurrent bronchi and through the intercommunicating system of parabronchi where respiratory exchange takes place. This air is then forced into the expanding anterior air sacs on the second inspiration (while a fresh mass of air is entering the posterior air sacs), and it is expelled through the mesobronchus on the second expiration. Since there are no valves in the system, various aerodynamic theories have been

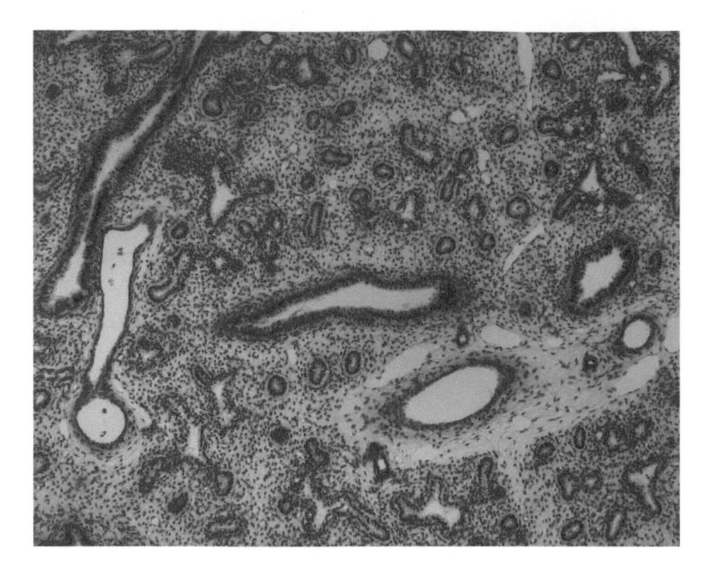

FIGURE L30a In this section the various levels of the pulmonary tree penetrate the mesenchyme of the fetal human lung. Since the fetus is within the uterus, the future air spaces are filled with fluid. As development proceeds, the future air spaces expand and the mesenchymatous tissue becomes compressed. Surfactant must be present when the first breaths are drawn in order to prevent collapse of the lung, 10 ×.

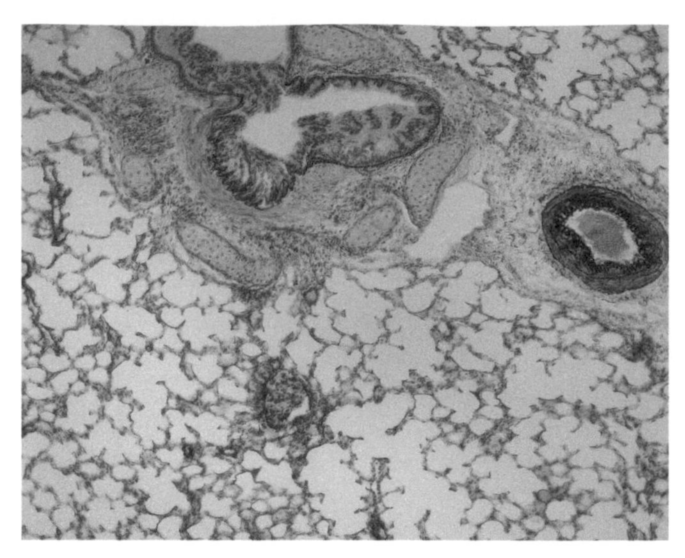

FIGURE L30b Alveolar walls are strengthened by a network of reticular and elastic fibers. Reticular fibers are not seen in this section but the ELASTIC FIBERS are demonstrated by the use of an elastic tissue stain. Note the rich presence of elastic fibers in the artery at the right, 10 ×.

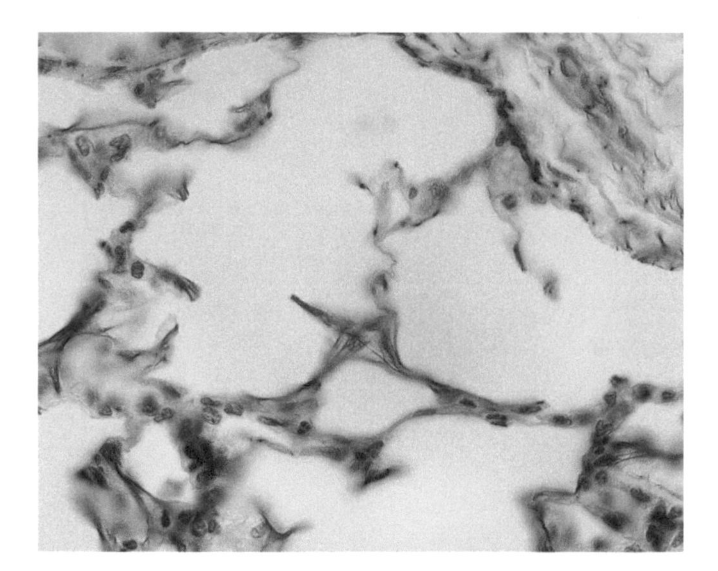

FIGURE L30c The delicate supporting framework of elastic fibers is apparent in this section of mammalian lung stained with an elastic tissue stain, 63 ×.

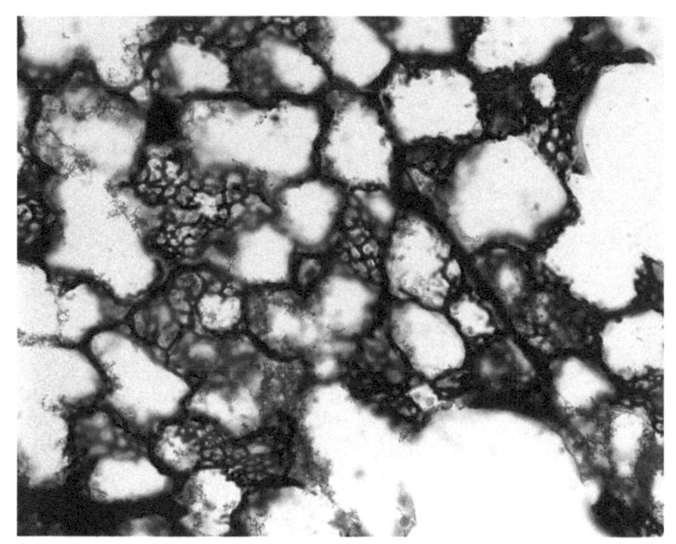

FIGURE L30d The elaborate network of capillaries in alveolar walls is demonstrated in this injected preparation of the lung of a cat. In this preparation, the pulmonary arteries were injected with red gelatin, the veins with green, 20 ×.

proposed to explain why the air does not back up but passes in one direction over the respiratory surfaces. A constantly renewed supply of fresh air passes over the respiratory surfaces since there are no blind alveoli where stagnant air can collect. Blood and air flow in opposite directions in the parabronchi thereby enabling the efficient exchange of gases made possible by a counter-current exchange mechanism. In diving birds, air may be passed back and forth between the anterior and posterior air sacs, through the lung several times allowing for sustained dive times.

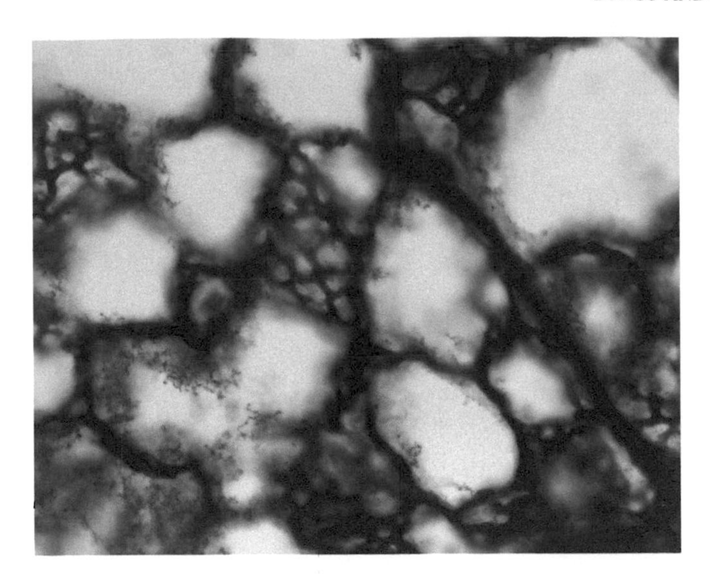

FIGURE L30e The elaborate network of alveolar capillaries is evident at higher power—resembling a fishing net embedded in the alveolar walls, 40×.

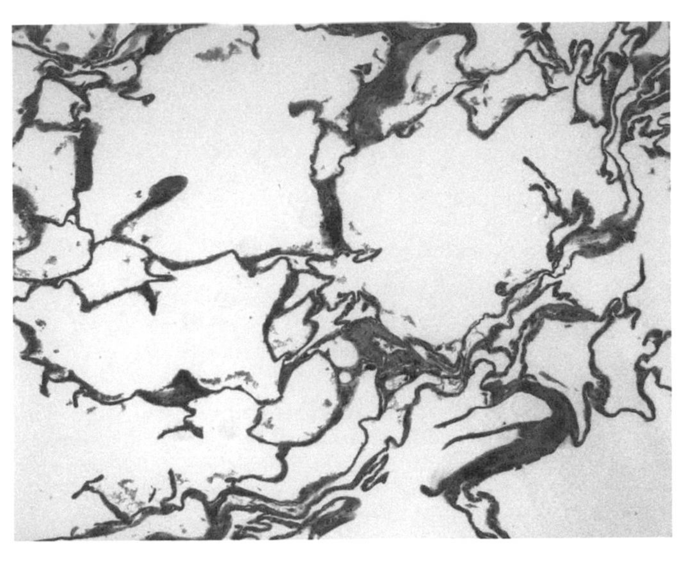

FIGURE L31a ALVEOLAR MACROPHAGES lurk in the connective tissue and on the surface epithelium of the lungs. These cells become distended with carbon in the lungs of smokers, miners, and those working in dusty surroundings; they are also called DUST CELLS and here appear filled with black granules, 10×.

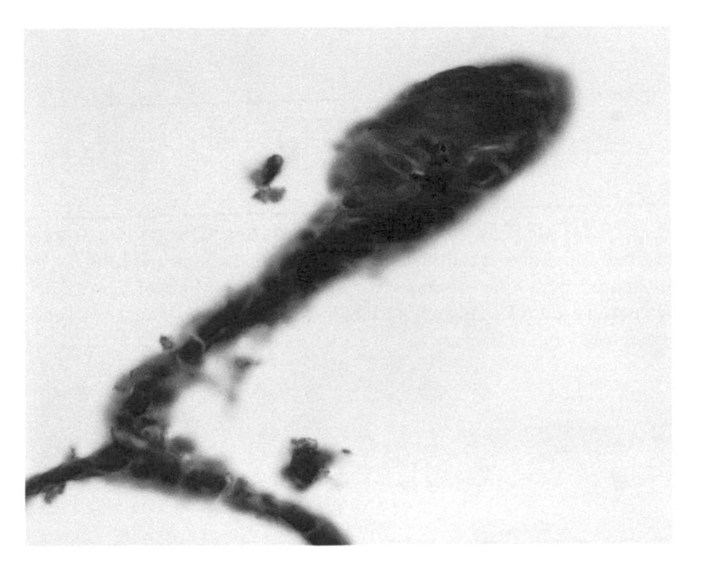

FIGURE L31b It is presumed that these macrophages become engorged with particles and eventually pass through the epithelium of the alveolus, migrate along the air passages to the bronchioles and bronchi where they are carried upward by ciliary action to be swallowed or spat out, 63×.

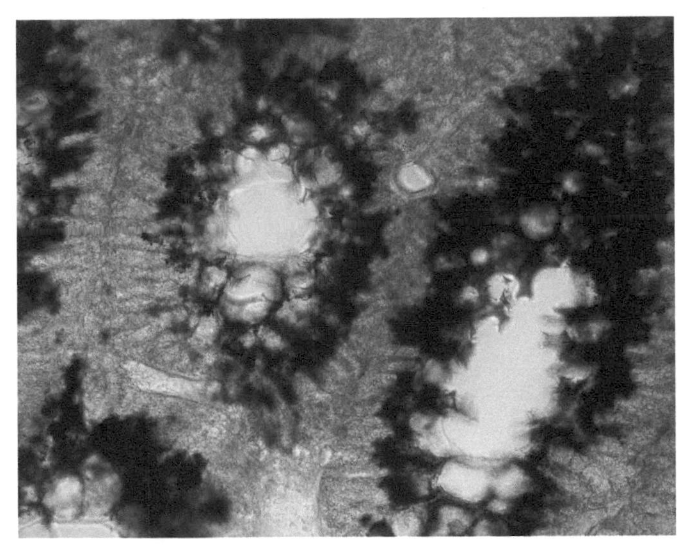

FIGURE L32 This thick section of cowbird lung was cut by hand with a sharp blade. Unlike the lungs of other vertebrates, the lung of the bird is inelastic and air passes through in one direction only, propelled by the bellows action of the air sacs which respond to changes in the size of the body cavity. Air from the trachea, passes directly through the lungs in a large passage, the MESOBRONCHUS. Little or no gaseous exchange takes place and the air enters the posterior air sacs. On exhalation, the air follows a different route and passes through channels in the lung tissue itself, the PARABRONCHI. Two parabronchi are seen in this thick section. The black material is fixed, coagulated blood, 10×.

The bird lung can be more efficient than mammalian lungs in that air passes through the air capillaries in one direction, opposite to the direction of the flow of blood. Because the flow of air and blood is more or less in opposite directions, a counter-current exchange of gases is approximated; and the pattern is designated a CROSS-CURRENT TYPE of flow (Figs. L33d and L33e). In addition, since exhalation is never complete in mammalian lungs, there is always a quantity of stale air that mixes with fresh air from the outside.

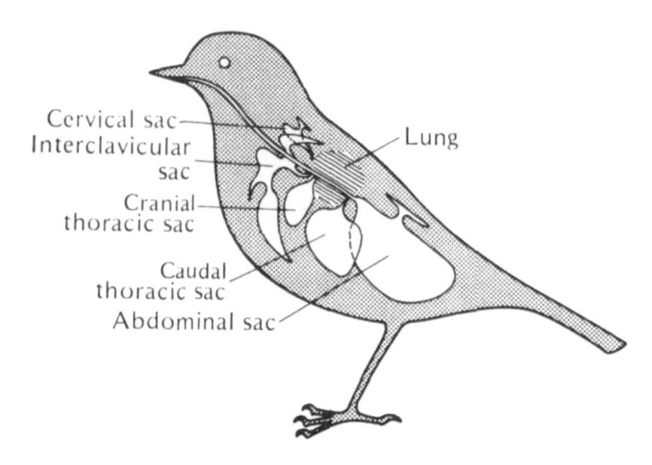

FIGURE L33a The body of a bird contains several large, thin-walled air sacs. The paired lungs are small and located along the vertebral column. The parabronchus, which runs through the lungs, has connections to the air sacs as well as to the lung. Source: *From Salt, G.W., 1964. Respiratory evaporation in birds. Biol. Rev. 39, 113–136.*

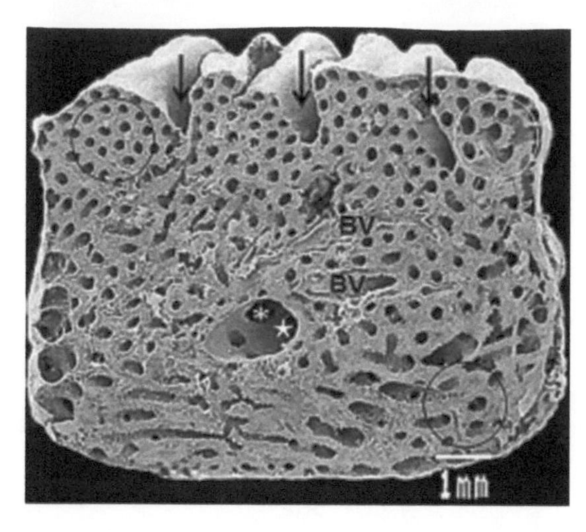

FIGURE L33b A low power scanning electron micrograph of the lung of a 16-day-old chick showing well-developed airways (circles) throughout the lung. Star, primary bronchus; asterisk, secondary bronchus; arrows, costal sulci; BV, blood vessel. *Maina, J.N., 2015. The design of the avian respiratory system: development, morphology and function. J Ornithol 156 (Suppl 1):S41–S63.*

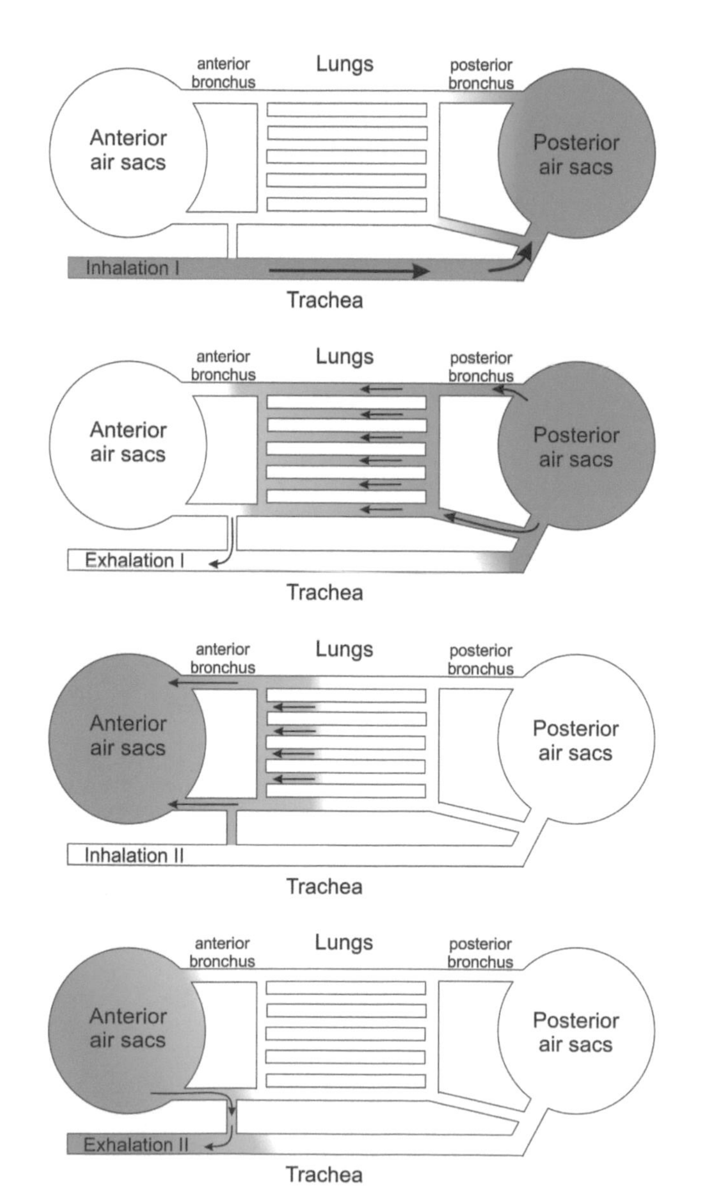

FIGURE L33c Diagram illustrates the movement of a single "slug" of air through the avian respiratory system. It takes two full respiratory cycles to move the air through its complete path. *Authors.*

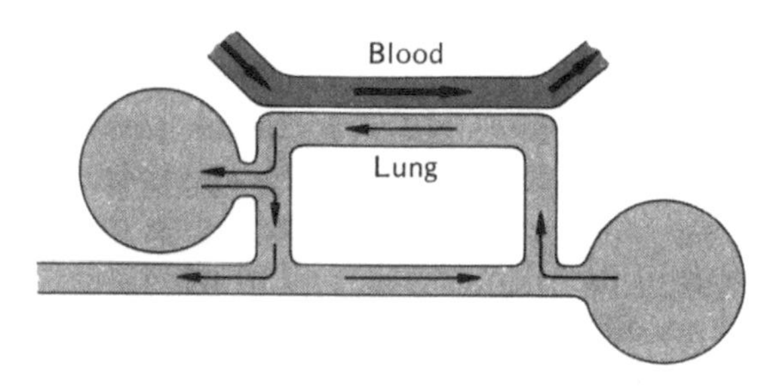

FIGURE L33d Gas exchange in the bird lung. This highly simplified diagram shows the flow of blood and air through the lung, each represented by single stream with opposite directions of flow. This flow pattern permits oxygenated blood to leave the lung with high oxygen tension.

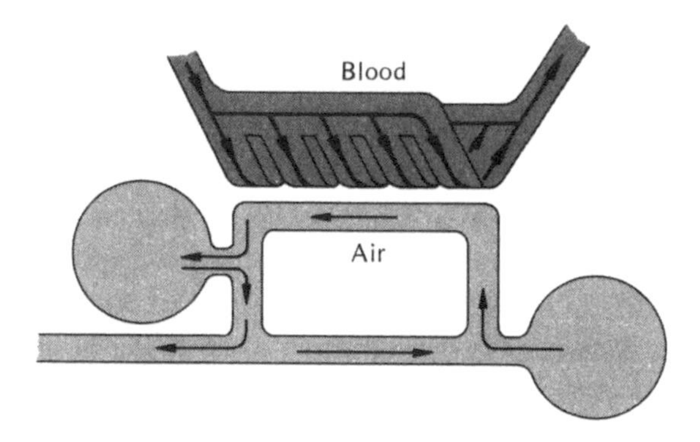

FIGURE L33e Flow of blood in the avian lung does not flow in parallel capillaries but rather in an irregular complex network. This simplified diagram shows that blood leaving the lung is a mixture of blood flowing through different parts of the lung having different degrees of oxygenation. This is a cross-current flow.

SWIM BLADDER

In some fishes, there is a swim bladder that may have a respiratory function (Figs. L35a and L35b). Because it is formed as an embryonic evagination of the esophageal region its epithelial lining is related to the lining of the vertebrate lung. A swim bladder is

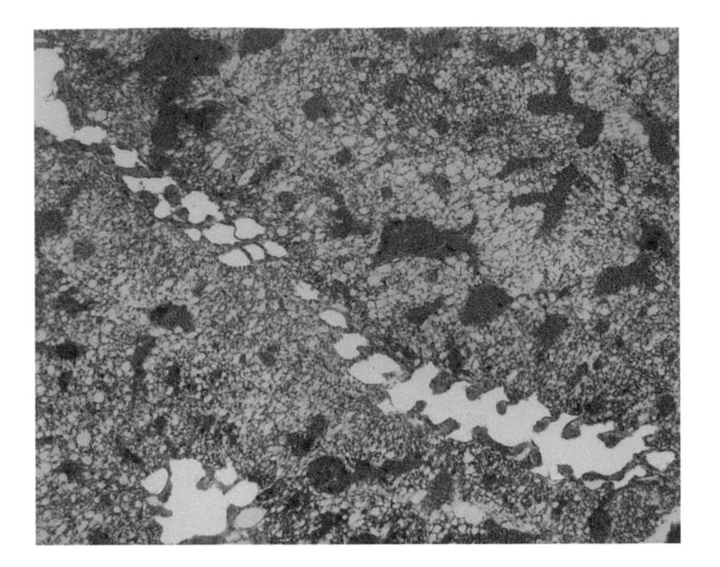

FIGURE L34a A PARABRONCHUS crosses the screen from left to right in this section of the lung of a duck; another parabronchus is seen in cross-section at the lower left. Parabronchi are surrounded by a loose tissue consisting of small tubes, the AIR CAPILLARIES, which pass through richly vascular tissue. Air from the posterior air sacs is pumped forward and, instead of returning the way it came (through the mesobronchus) it is whistles through these air capillaries, where gaseous exchange takes place, and makes its way to the anterior air sacs. On a second exhalation, this air passes to the trachea and to the outside, 10 ×.

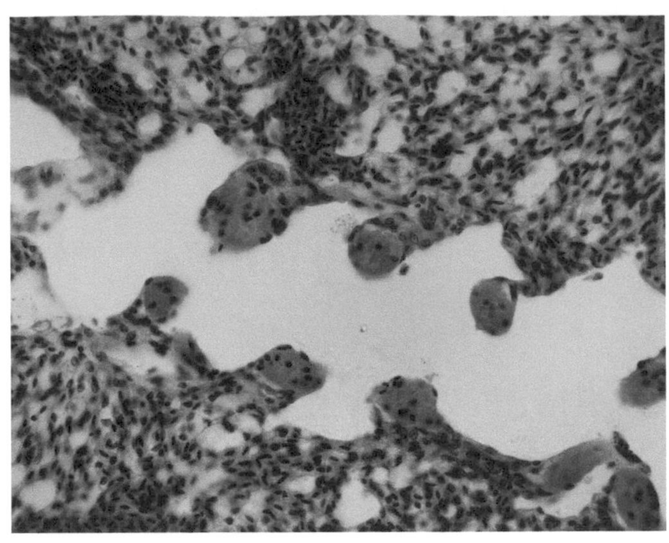

FIGURE L34b A parabronchus is at the center. It is surrounded by loose tissue of richly vascularized air capillaries. Note that the rims of the respiratory tissue contain masses of SMOOTH MUSCLE; this muscle maintains tension on the respiratory tissue, presenting a maximal surface to the flow of air, 40 ×.

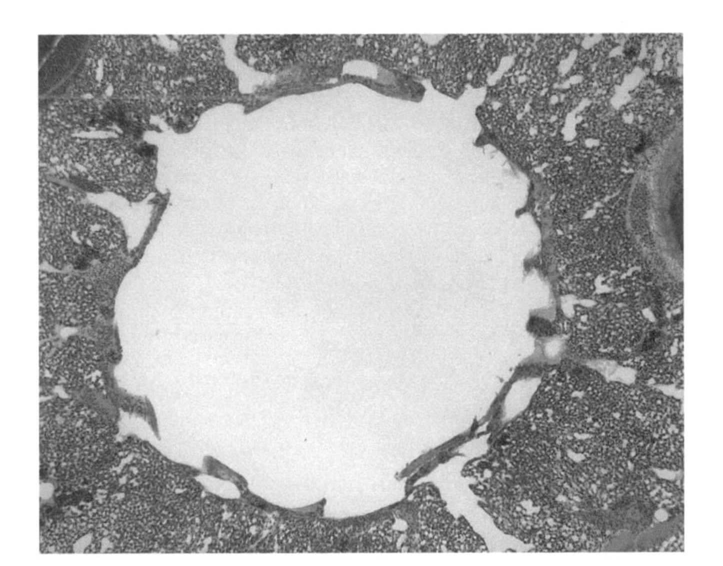

FIGURE L34c The PARABRONCHUS at the center of this chicken lung is surrounded by air capillaries intertwined with blood capillaries, 10 ×.

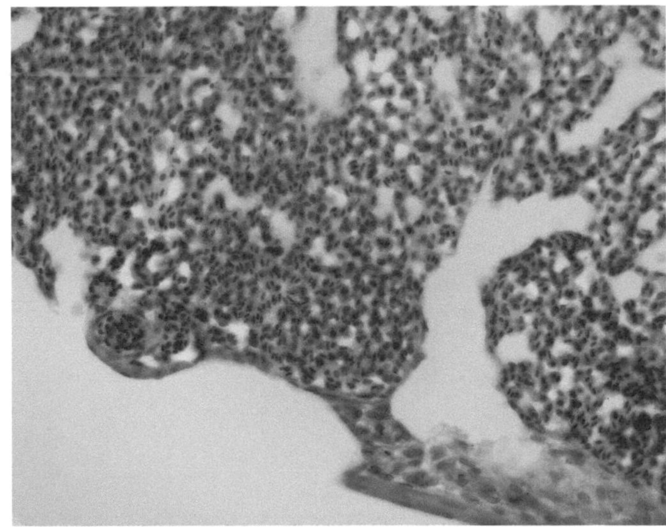

FIGURE L34d The lumen of a parabronchus is at the bottom left with respiratory tissue above. A band of smooth muscle at the lower right assists in stretching the tissue thereby exposing maximal a surface for air exchange, 40 ×.

lacking in the cyclostomes and elasmobranchs and also in bottom-living teleosts. The PNEUMATIC DUCT connecting the bladder to the gut may persist in PHYSOSTOMATOUS fish, as *Protopterus* (lungfish), *Polypterus*, *Amia* (bowfin), and *Lepisosteus* (garpike); or it may become closed so that the bladder exists as a blind sac in PHYSOCLISTOUS fish, as the carp and many other teleosts.

Since the swim bladder is formed as an evagination of the embryonic esophagus, layers of the wall of the swim bladder are equivalent to the gut: tunica mucosa, tela submucosa, tunica muscularis, and tunica adventitia. It is possible in some preparations to distinguish a lamina propria mucosae and lamina muscularis mucosae.

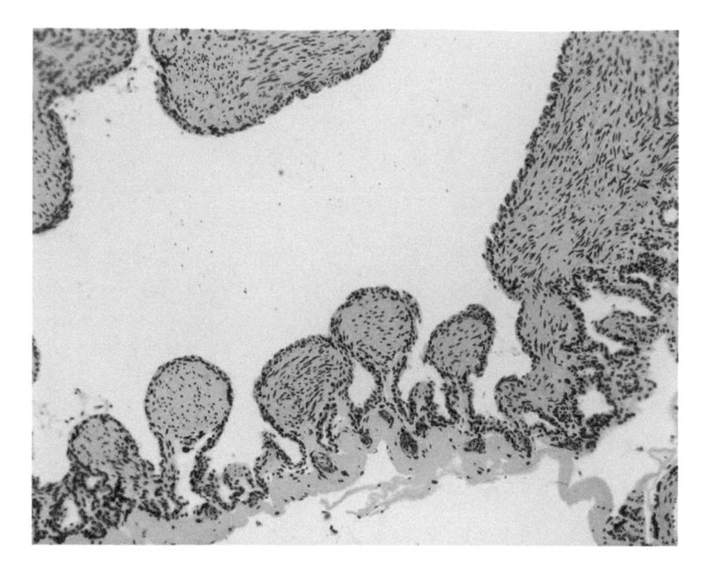

FIGURE L35a *Protopterus* has paired swim bladders which connect to the esophagus by the way of the PNEUMATIC DUCT. The respiratory surface of the lungs is increased by interior SEPTA containing well-developed bands of smooth muscle. Air is pumped into the lungs by the rising and falling of the floor of the mouth cavity and expelled by the contraction of the smooth muscle in the septa, 10 ×.

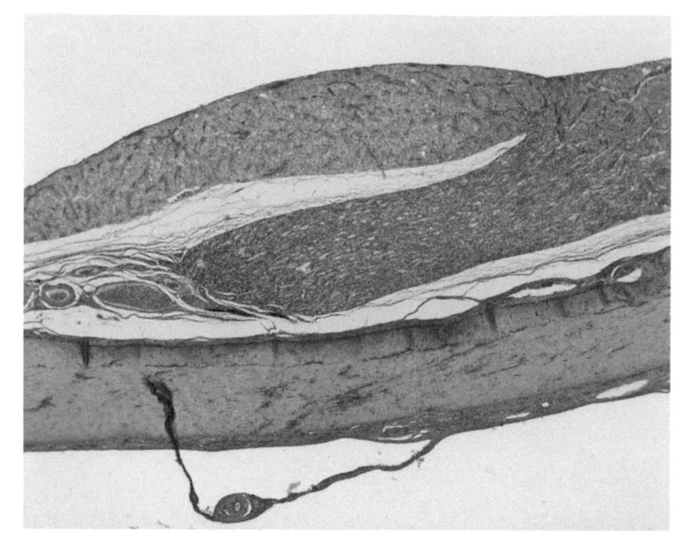

FIGURE L35b The lumen of the physosclistous swim bladder of the sheepshead (*Arcosargus probatocephalus*) is not connected to the exterior by a pneumatic duct. The wall of the bladder (bottom) is extremely tough and is rich in dense collagenous tissue in the TUNICA EXTERNA. As a hydrostatic organ that helps orient the fish vertically in the water, it must withstand drastic changes in pressure as the fish rises and falls. Within the lumen is the RED BODY or GAS GLAND (*top*) which releases gas when the fish rises. Circulation to and from the gas gland is effected by a double RETE MIRABILA of parallel vessels that unite to form larger vessels that supply and drain the capillaries of the gas gland. A rete mirabile is a capillary bed that is supplied by an artery or vein that reunites to form an artery or venule that then supplies a second capillary bed. In this case, arterioles at the left supply a bed of parallel capillaries at the center of the screen. These reunite to form arterioles at the upper right which subsequently divide into the capillary bed of the red gland. Blood from the red gland is then collected into venules at the upper right which redivide to form a third capillary interspersed with and parallel to the first. In this example, blood in these straight arterial capillaries runs in opposite direction to the blood in the venous capillaries, thereby providing ideal circumstances for a COUNTER-CURRENT EXCHANGE mechanism, 5 ×.

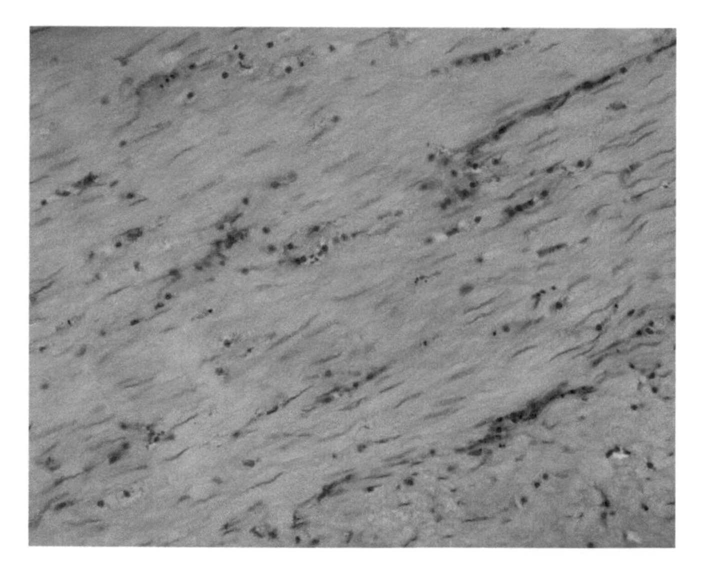

FIGURE L35c The tough WALL of the swim bladder of the sheepshead consists largely of densely packed, parallel collagenous fibers, 40 ×.

Swim bladders with ducts may be highly vascular and lung-like and may have the lining thrown into a series of highly vascular folds and sacculations of an alveolar nature, thus providing an excellent respiratory surface. These are perhaps more properly referred to as lungs.

Note that the walls of swim bladders that function as hydrostatic organs are greatly reinforced with collagenous fibers in the position of the tunica muscularis (Figs. L35b and L35c); since this layer contains little muscle, it is more appropriately designated as the TUNICA EXTERNA. The wall of the anterior parts of physoclistous swim bladders contains the RED BODY or GAS GLAND which secretes gases (Figs. L135e and L35f). It consists of two parallel RETIA MIRABILIA (double rete mirabile), arterial and venous, that supply and drain a capillary bed in the epithelium of the bladder (Fig. L35d). Thin sheets of crystalline guanine line the swim bladder and inhibit the outward diffusion of gases through the lining. A posterior vascular pouch on the dorsal wall, the OVAL, serves to resorb the gases from the bladder. It is surrounded by a ring of muscle that can be relaxed to expose its vascular surface to the gases in the main bladder or constricted to separate the two areas.

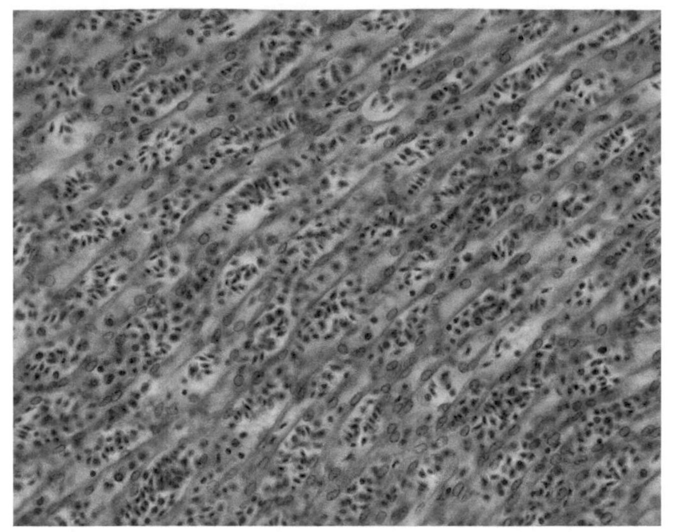

FIGURE L35d The double RETE MIRABILE carries blood to and from the gas gland and consists of parallel arterial and venous vessels that form a counter-current exchange system, 40×.

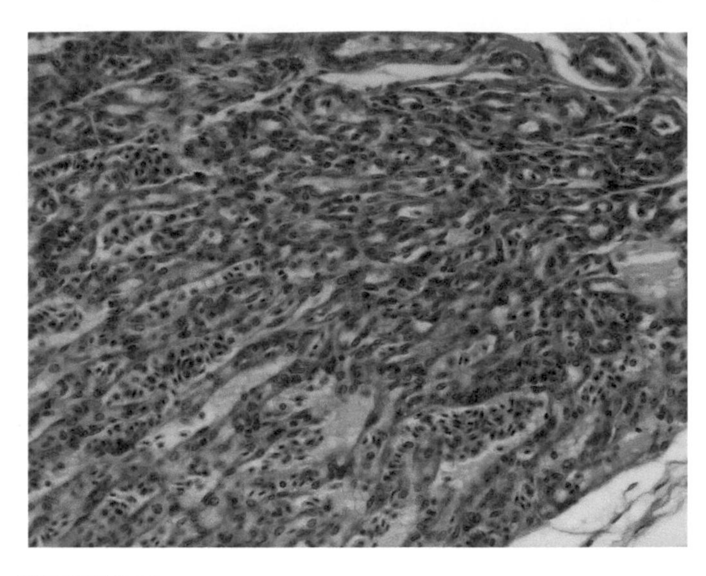

FIGURE L35e As the vessels of the double rete mirabile approach the gas gland, they unite to form larger vessels that supply its capillaries, 40×.

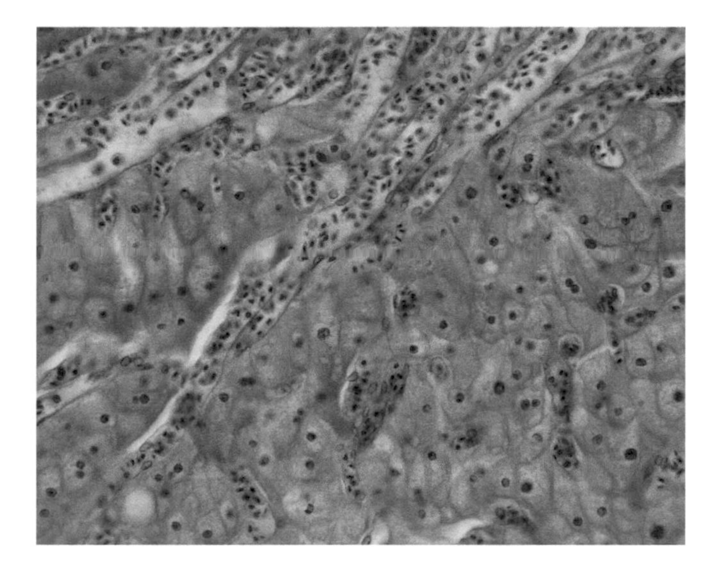

FIGURE L35f The GAS GLAND secretes gases into the lumen and consists of parenchymal cells richly infiltrated with capillaries, 40×.

CHAPTER

M

Excretory Systems

Excretion involves the separation and elimination of metabolic waste products from the body. Various organs are involved in this process: lungs, gills, skin, etc. These have already been considered. The kidneys and their ducts are the major full-time excretory organs and comprise the "excretory system."

In addition to the elimination of metabolic wastes, the excretory system functions in the maintenance of a proper water balance in the body: a balance of water, inorganic salts, and other substances in the internal environment of the organism. There is a great difference in the problems of water balance encountered by marine, fresh water, and terrestrial vertebrates, and it is remarkable that their kidneys are as much alike as they are.

The kidney of all vertebrates consists of knots of blood vessels, either GLOMERULI or GLOMERA, closely associated with masses of kidney TUBULES. A single tubule, with its associated blood vessels, is a NEPHRON.

EVOLUTION OF THE KIDNEY

It is believed that the kidney of the vertebrate ancestor was an ARCHINEPHROS or (HOLONEPHROS), a type found in larval hagfish and in larvae of some caecilians (Fig. M1). It consisted of a series of several KIDNEY TUBULES, one per segment, on each side, that connected to a pair of longitudinal ducts, the ARCHINEPHRIC DUCTS, running along the dorsal wall of the coelom. Each tubule opened into the coelom by a ciliated, funnel-shaped NEPROSTOME and was associated with a knot of blood vessels, the GLOMERULUS (*plural:* glomeruli) (Fig. M2). The glomerulus presses against the wall of the tubule to form a double-walled egg-cup structure around the glomerulus: the RENAL CORPUSCLE (Figs. M3 and M4).

Fluids removed from the blood in the renal corpuscle and passed down tubules which are associated with blood vessels. A selective reabsorption by the blood of much of the water and other constituents occurs so that a concentrated urine enters the excretory duct (Figs. M5a–M5f). In marine and terrestrial vertebrates, where the conservation of water is essential, either the glomerulus is lost or reduced (marine fish, reptiles) or the tubules are well developed (birds, mammals). Terrestrial reptiles effect a further saving by the absorption of water from the urine held in a cloacal bladder with absorptive walls. Fresh-water vertebrates, on the other hand, are constantly threatened with "water logging" and have well-developed renal corpuscles that expel large amounts of water. Although sharks live in salt water, they appear to contradict these generalizations; they have large glomeruli and kidneys similar to those of fresh-water fish. Excessive water loss is prevented by the retention in the blood of large amounts of urea thereby raising the internal osmotic pressure to that of the surroundings. Observation of the relative development of the parts of the nephron can be associated with a vertebrate's mode of life and its environment.

The ARCHINEPHRIC DUCT persists in anamniotes (cyclostomes, fishes, amphibians) (e.g., Fig. M4) but not in amniotes (reptiles, birds, mammals) (e.g., Fig. M5f) where the URETER assumes the function of transporting urine.

An Atlas of Comparative Vertebrate Histology.
DOI: https://doi.org/10.1016/B978-0-12-410424-2.00013-5

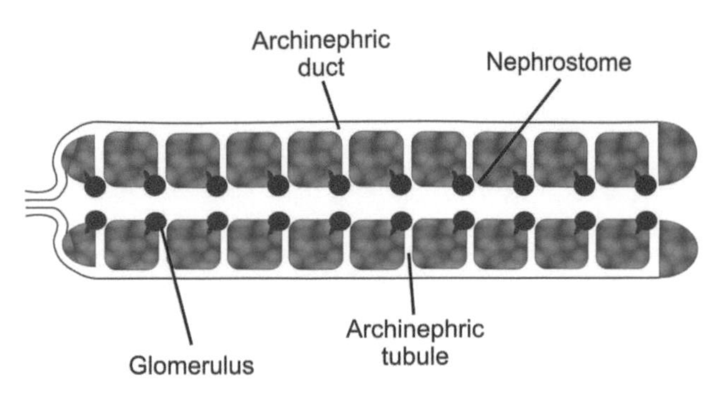

FIGURE M1 Diagram shows the hypothetical structure of the archinephros of a developing vertebrate. *Authors.*

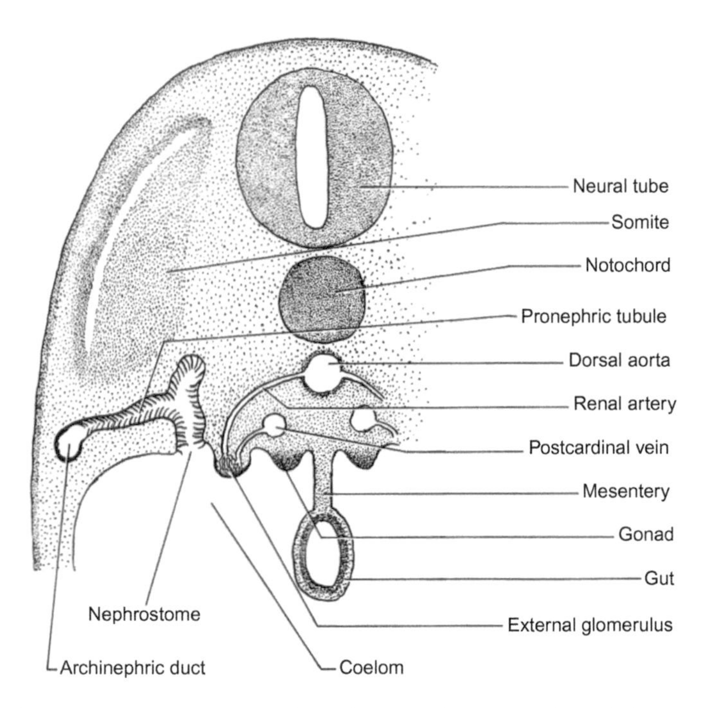

FIGURE M2 Diagrammatic section through a vertebrate embryo showing the relationship of the external glomerulus to the pronephric tubule and the nephrostome.

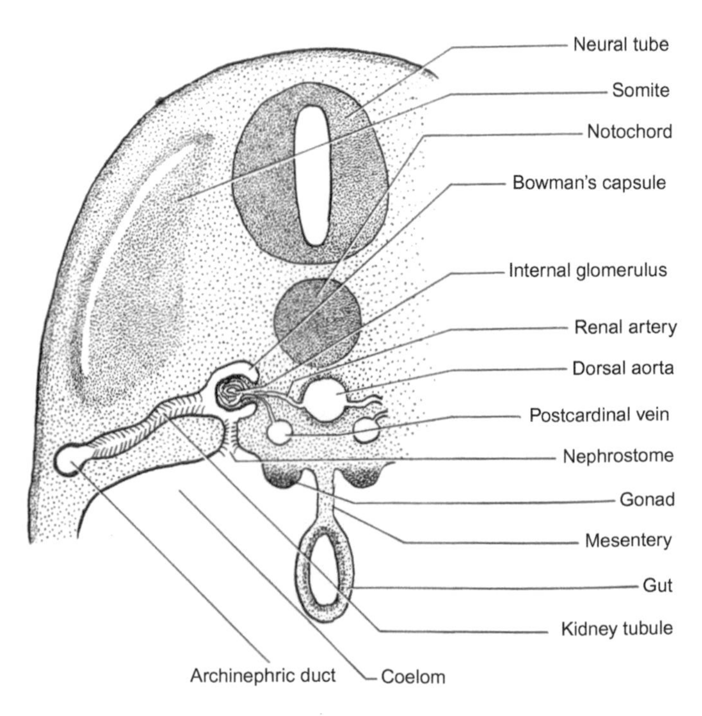

FIGURE M3 Diagrammatic section through part of a vertebrate embryo showing the relationship of the glomerulus to the kidney tubule and archinephric duct.

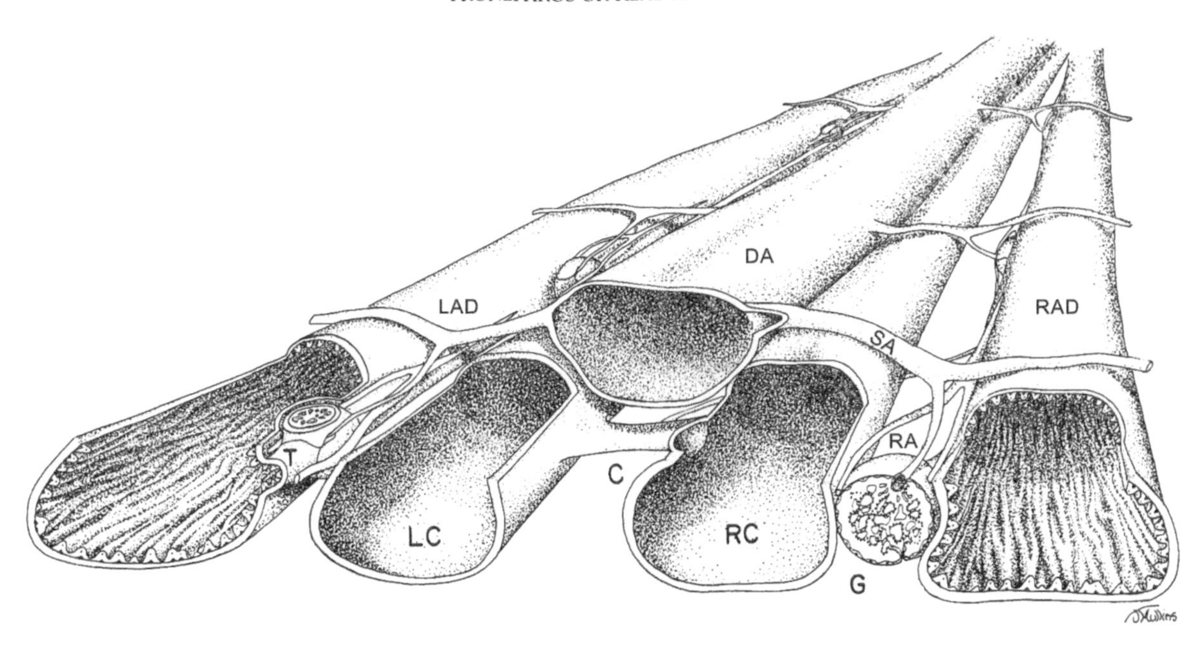

FIGURE M4 Schematic drawing of a portion of the corpuscular region of the hagfish kidney. The dorsal aorta (DA) gives off segmernal arteries (SA) which give rise to renal arteries (RA). Glomeruli (G) located within each segment receive blood from the renal arteries. Note the close relationship of a glomerulus to the archicnephric duct. *C*, cardinal anastomosis; *LAD* and *RAD*, left and right archinephric ducts; *LC*, *RC*, left and right posterior cardinal veins; *T*, tubule.

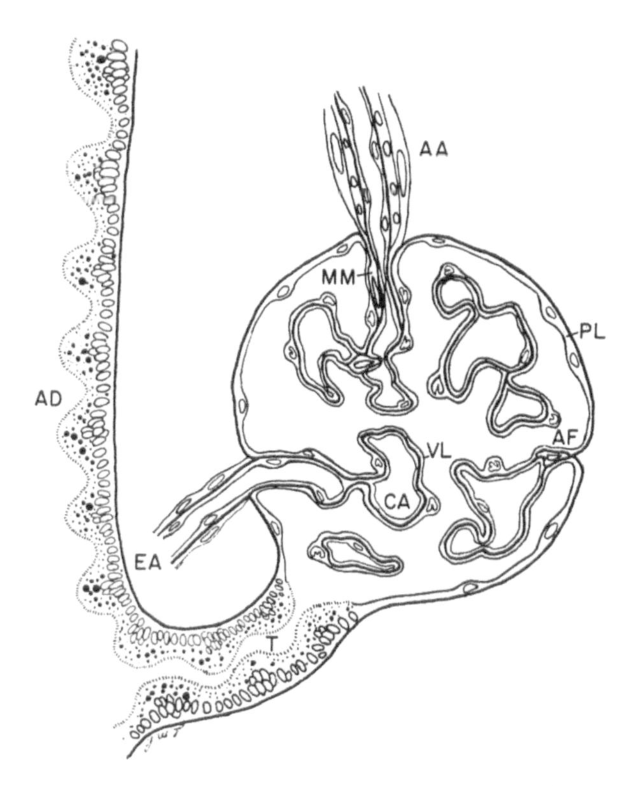

FIGURE M5a Schematic drawing of a frontal section (anterior at top) to demonstrate the relationship of the renal corpuscle, tubule (T), and archnephric duct (AD) in the hagfish kidney. The afferent artreriole (AA) enters the corpuscle and breaks up into a capillary network (CA) (reduced here for simplicity). Note the arterial fold (AF). One of the efferent arterioles is shown leaving the glomerulus. *MM*, mesangial cell; *PL*, parietal layer of Bowman's capsule; *VL*, visceral layer.

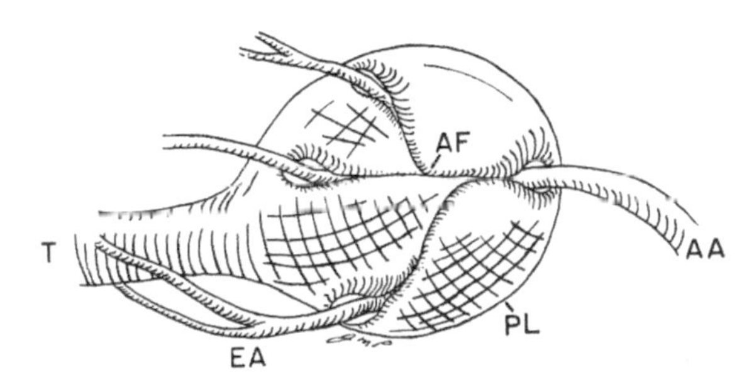

FIGURE M5b Schematic sketch of a renal corpuscle from the kidney of a hagfish. The connective tissue capsule has been omitted in order to demonstrate the infolding of the parietal layer (PL) of the renal (Bowman's) capsule to form an arterial fold (AF). Three efferent arteriorles (EA) drain the glomerulus. *AA*, afferent arteriole; *T*, tubule.

PRONEPHROS OR HEAD KIDNEY

The most anterior portion of the archinephros, the PRONEPHROS or HEAD KIDNEY, first appears in developing vertebrates embryos and in most it is a transitory structure. Frog tadpoles clearly show pronephric tubules (Figs. M6a and M6b) and (Figs. M7a–M7c). The remainder of the kidney posterior to the pronephros persists as the OPISTHONEPHROS (=back kidney) and is the functional kidney of most adult anamniotes. Only the posterior, unsegmented portions of the kidney tissue are retained in amniotes and these constitute the

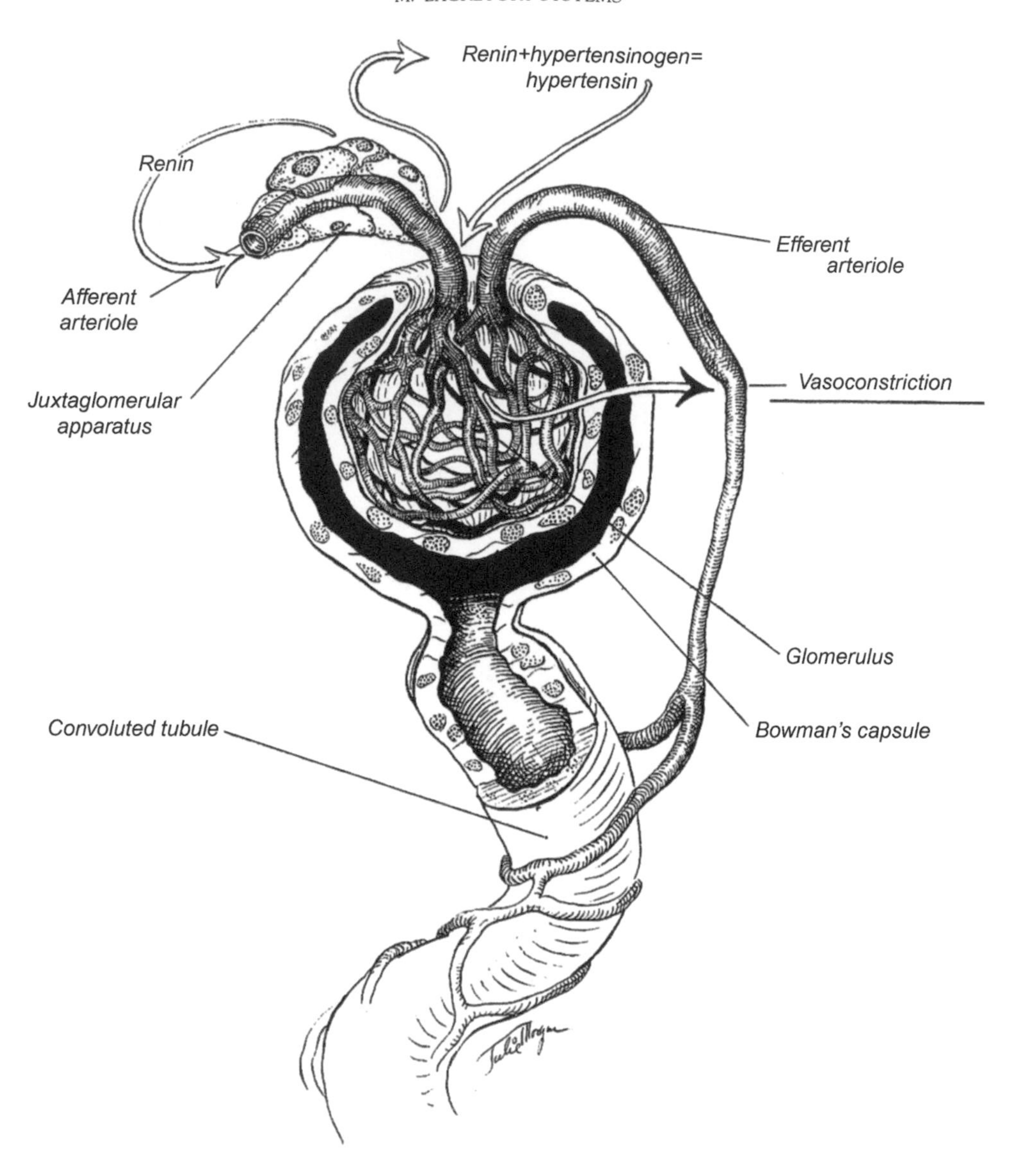

FIGURE M5c Schematic diagram of a typical vertebrate renal corpuscle. Fluids lost from the glomerulus and enter the renal capsule (Bowman's capsule), and pass through a convoluted tubule to be released. During this passage, useful materials are reabsorbed and wastes are lost to the outside world.

METANEPHROS (=hind kidney). Note that a close relationship is often seen between excretory tubules and hemopoietic tissue (Figs. M10a and M10b). Early in the development of many vertebrates, the excretory function of the pronephros is lost and the hemopoietic function takes over.

Pronephric tubules may be seen in sections of vertebrate embryos such as ammocoetes or frog tadpoles (Figs. M8a–M8c). As development proceeds, the number of tubules involved in the pronephros is usually reduced so that only one to three are found in older embryos. Each coiled pronephric tubule begins as a CILIATED FUNNEL or NEPHROSTOME whose flattened epithelium is continuous with that of the peritoneum (Fig. M7c). From this a CONVOLUTED TUBULE, lined with simple cuboidal epithelium, leads to the lateral ARCHINEPHRIC DUCT. Near the mouth of each nephrostome is an EXTERNAL GLOMERULUS, a capillary net branching off the dorsal aorta.

The pronephros of an adult hagfish shows segmental renal corpuscles closely associated with the archinephric duct (Fig. M9a). The coils of the glomerular capillaries bristle with podocytes, the cells involved in the initial filtering of body fluids (Figs. M9b–M9e). The cells of the pronephros of the hagfish show the podocytes, characteristic cells of filtering epithelia (Fig. M9f). The signs of active transport are seen in the junctional complexes between cells of the epithelium lining the archinephric duct.

Unlike most vertebrates, there are no renal corpuscles in the kidney of the larval lamprey (ammocoete) nor are there glomeruli. Instead masses of blood vessels, GLOMERA (*sing.* glomus) on each side of the body in the region of

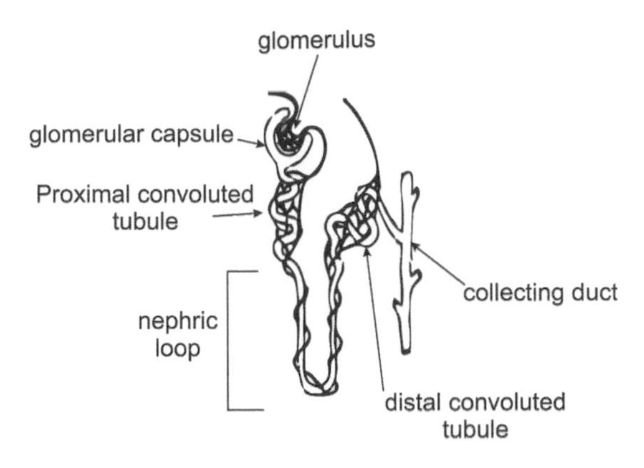

FIGURE M5d Sketch of a vertebrate nephron. The nephron is made up of two parts: the glomerulus, a knot of blood vessels surrounded by the double-walled glomerular capsule; and renal tubules. *Authors.*

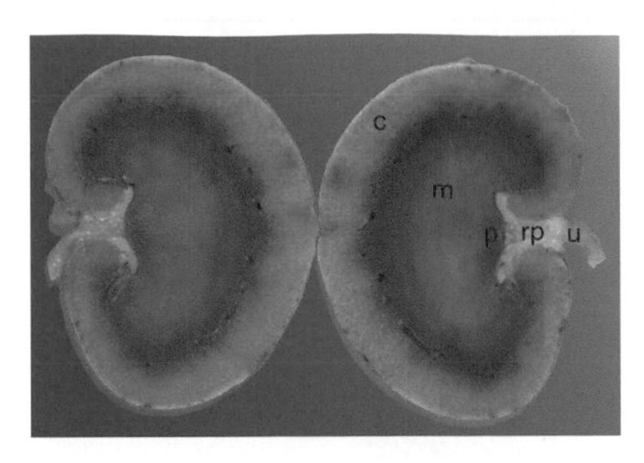

FIGURE M5e The nephrons of the higher vertebrates are compactly arranged in the KIDNEYS. In this midsagittal section of a cat kidney the blood vessels and tubules are arranged in a single renal pyramid. The convoluted portions of the tubule plunge down into the MEDULLA of the kidney losing fluids as they go and rise again to the CORTEX. *C*, cortex; *m*, medulla; *p*, papilla of pyramid; *rp*, renal pelvis; *u*, ureter.

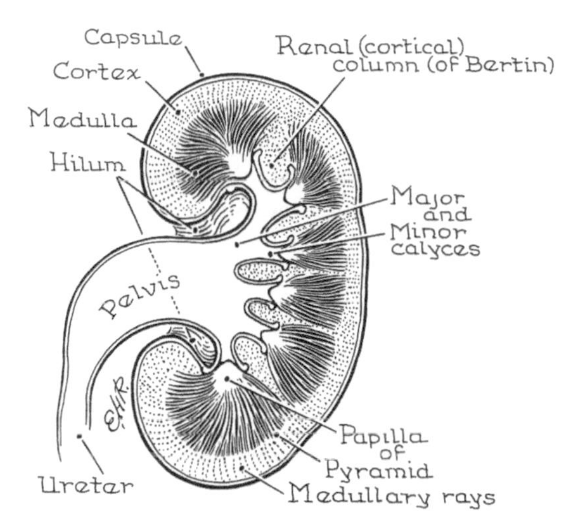

FIGURE M5f Drawing of a midsagittal section of a human kidney. The medulla contains a number of renal pyramids. *From Taylor NB, McPhedran MG. 1965 Basic Physiology and Anatomy. Putnam and Sons NewYork, and MacMillan Co Ltd Toronto, Hodder & Stoughton.*

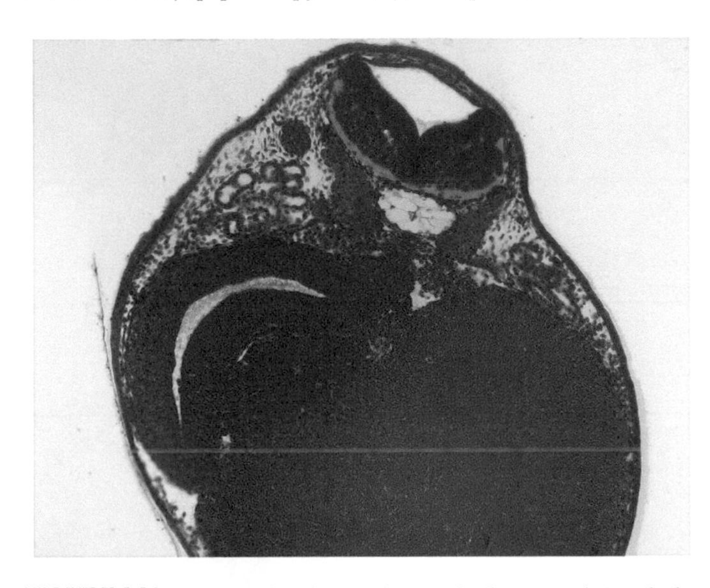

FIGURE M6a In the developing frog, tadpole pronephric tubules course through the vascular mesenchyme on either side of the notochord. There is a large mass of yolk at the lower right and a portion of the gut at the left, 10 ×.

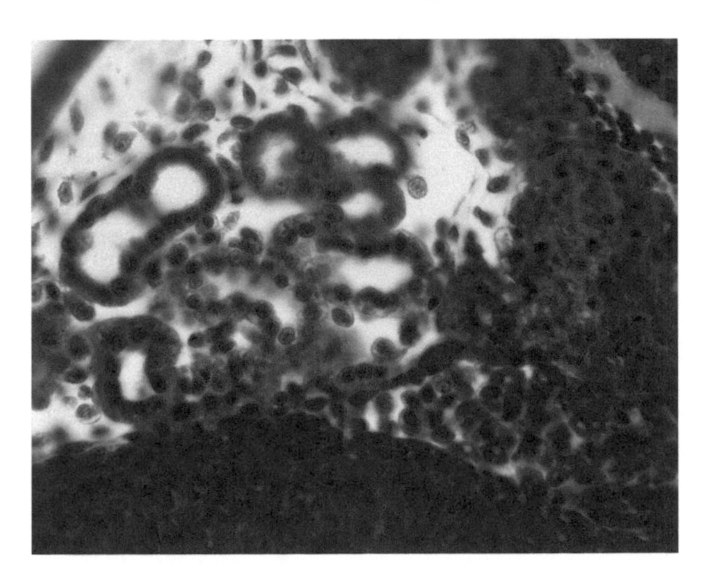

FIGURE M6b Pronephric tubules in the frog tadpole are surrounded by masses of hemopoietic mesenchyme. A nephrostome is visible at the center right, 40 ×.

FIGURE M7(a–c) Cross-sections of a 10-mm frog tadpole stained *in toto*.

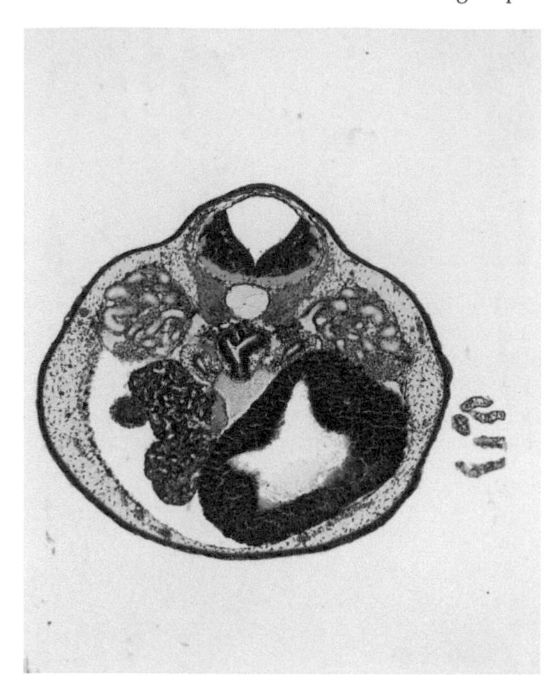

FIGURE M7a Pronephric tubules in section of frog tadpole stained *in toto*, 5 ×.

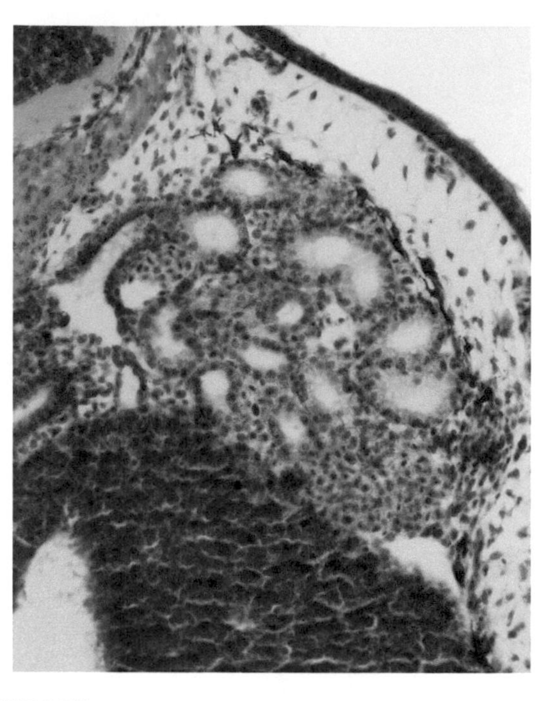

FIGURE M7b A nephrostome lies at the left of this pronephros. Note the active hemopoiesis between the tubules, 20 ×.

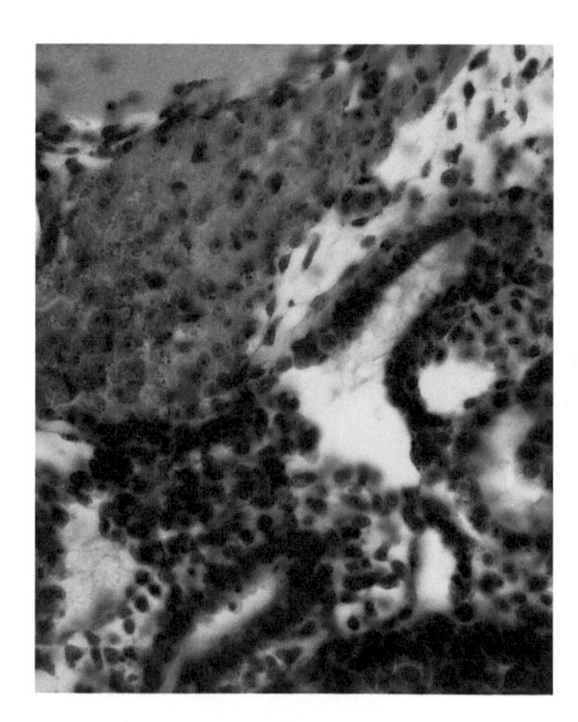

FIGURE M7c PRONEPHRIC TUBULES course through the vascular mesenchyme on either side of the notochord. There is a large mass of yolk at the lower right, and a portion of the gut at the left. Cilia on the cells lining the nephrostomes draw in fluids from the body cavity. There is an indistinct glomerulus near the entrance of the nephrostome, 40 ×.

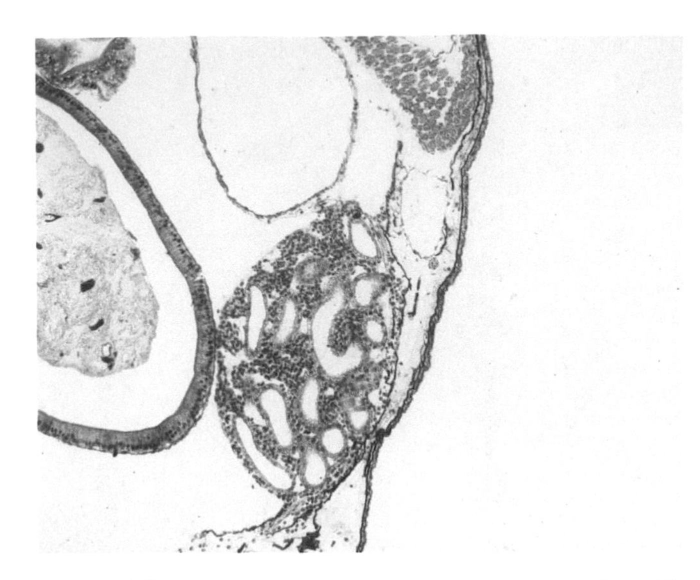

FIGURE M8a Pronephric tubules are seen in sections of an older tadpole stained with iron hematoxylin. The pronephros has become a more discrete organ; there is active hemopoiesis between the tubules. A section of the gut is seen at the left and the lung above, 10 ×.

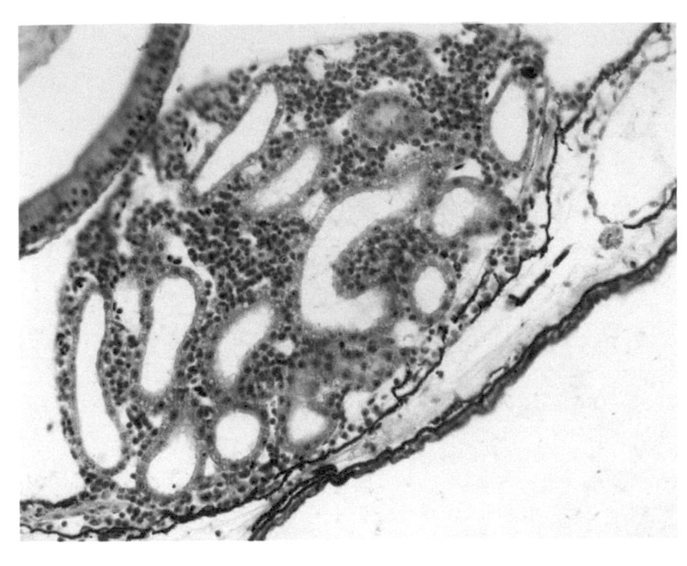

FIGURE M8b Abundant hemopoietic tissue fills the spaces between the pronephric tubules, 20 ×.

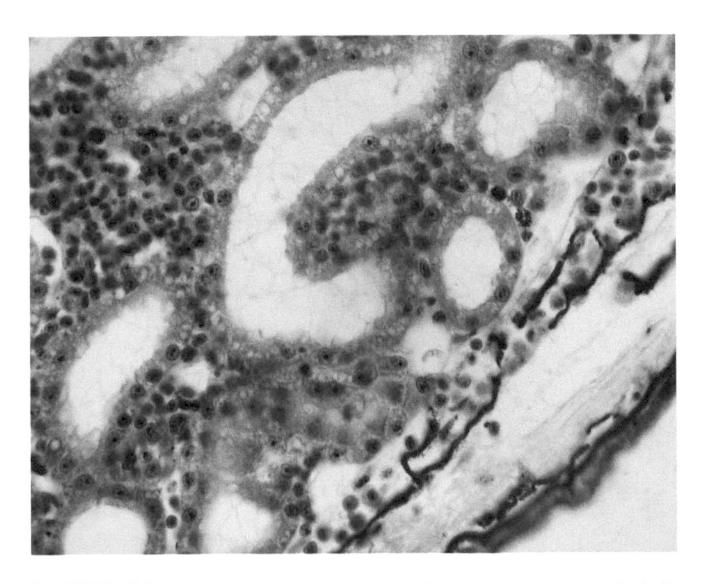

FIGURE M8c A few pronephric tubules are surrounded by masses of hemopoietic tissue, 40 ×.

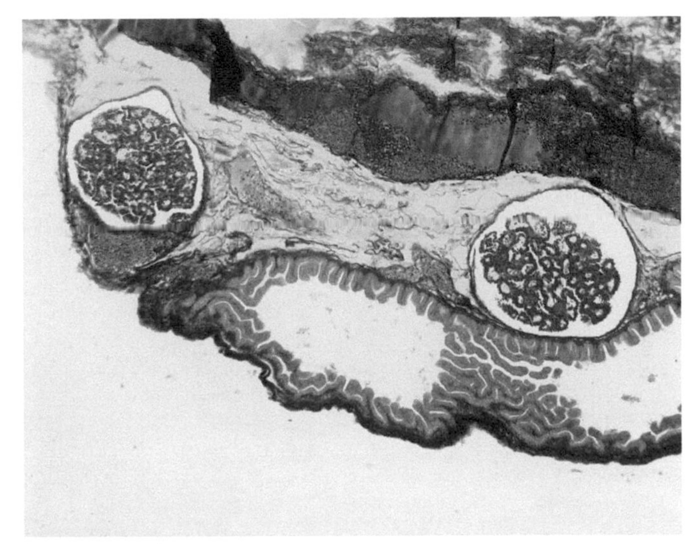

FIGURE M9a Frontal section through the pronephros of an Atlantic hagfish (*Myxine glutinosa*) showing two segmental RENAL CORPUSCLES containing GLOLMERULI. The ARCHINEPHRIC duct lies lateral to them. A section of the RENAL ARTERY lies to the left of the right corpuscle and a portion of the portal VEIN LIES to the right of the left corpuscle, 5 ×.

the heart are invaded by flattened balloons at the ends of the kidney tubules (Figs. M9g–M9i). These absorb a filtrate from the blood as it passes through the capillaries of the glomus. Three types of cells make up the walls of the tangled knot of capillaries in the glomus: PODOCYTES, MESANGIAL CELLS, and ENDOTHELIAL CELLS of the capillaries (Fig. M9j). The filter between the blood in the glomus and the lumina of the kidney tubules is formed from the combined basement laminae of the capillary endothelium and the podocytes (Fig. M9k). This filter shows an amazing similarity to the filter found in most vertebrate kidneys. Podocytes are elaborate cells whose foot processes (pedicels) gird the capillaries and provide support for the delicate filter within. Filtrate from the glomus passes into the coelom and is collected by the ciliated nephrostomes and carried to the proximal tubule, the intermediate segment, and the distal tubule, to be emptied in the cloaca (Fig. M9l).

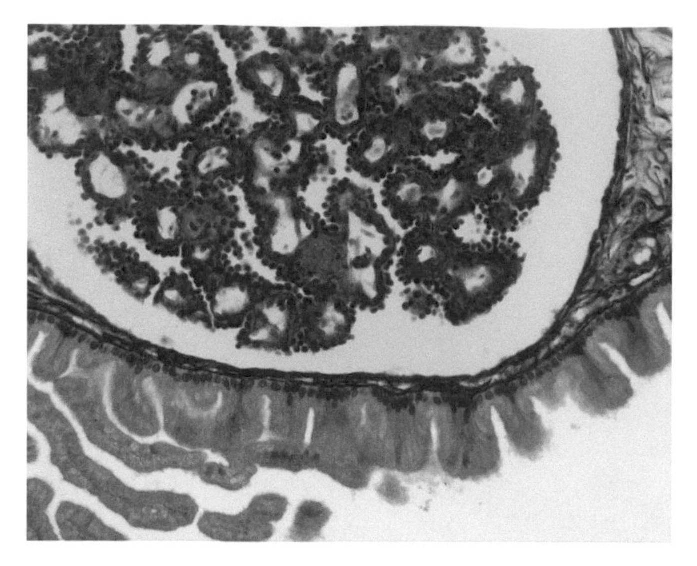

FIGURE M9b The PARIETAL LAYER of the GLOMERULAR CAPSULE is closely applied to the archinephric duct. The epithelium of the ARCHINEPHRIC DUCT varies from simple cuboidal to simple columnar producing a characteristic fan-like appearance in sections. Nuclei are basal. The apical cytoplasm contains brownish droplets of unknown significance. The glomerular capillaries bristle with podocytes, 20 ×.

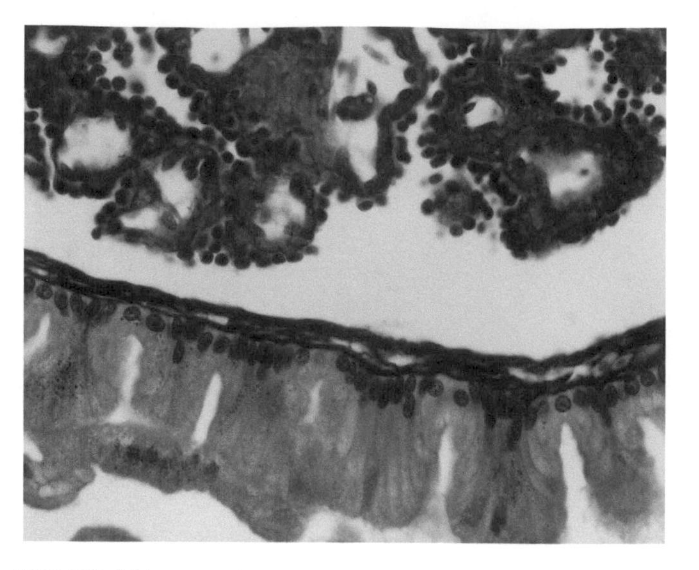

FIGURE M9c Capillaries of the GLOMERULUS (top) are formed of simple squamous endothelial cells. They are separated by a BASEMENT MEMBRANE from the bulbous PODOCYTES that invest the capillaries and form the VISCERAL LAYER of the GLOMERULAR CAPSULE. The PARIETAL LAYER of the glomerular capsule consists of a simple squamous epithelium. It is separated by a basal lamina from dense fibrous connective tissue that merges imperceptibly with the connective tissue coat of the archinephric duct. A few capillaries course through this connective tissue, 40 ×.

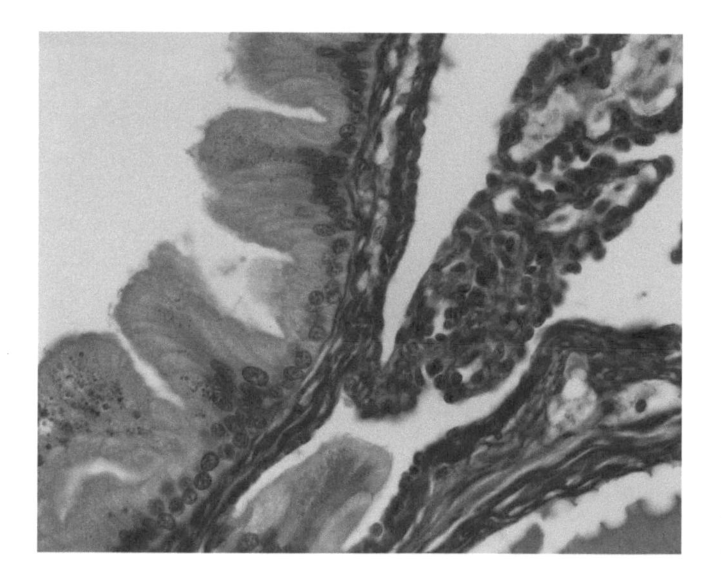

FIGURE M9d An ARTERIOLE approaches the renal capsule; it would penetrate the capsule in another section. A portion of the TUBULE that drains the glomerular capsule is seen at the bottom center, 40 ×.

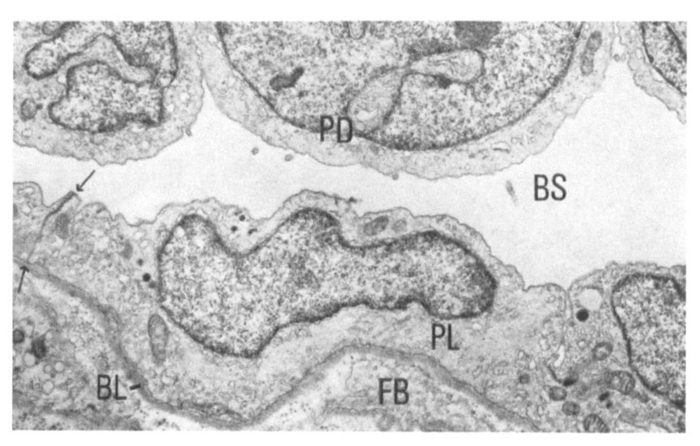

FIGURE M9e Podocytes are at the top of this electron micrograph of a renal capsule (Bowman's capsule) from the kidney of a hagfish. The capsular space (Bowman's space) separates the visceral and parietal layers of the glomerular capsule. Squamous cells of the parietal layer (PL) rest on a basal lamina (BL) underlain by dense collagenous fibers (CO) and the cytoplasmic processes of a fibroblast (FB). The *arrows* indicate an abutment between cells, 7700 ×.

During their first year the glomus begins to degenerate. Capillaries become occluded with dense amorphous material—cellular debris from endothelial and mesangial cells (Fig. M9m). The tubules begin to atrophy and by the last year of larval life all traces of the tubules, except the nephrostomes, have disappeared (Fig. M9n).

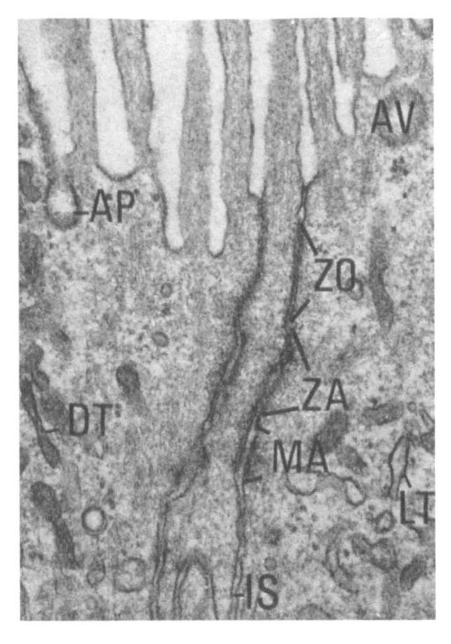

FIGURE M9f Electron micrograph of a cell from the archinephric duct of a hagfish. Two junctional complexes, a zonula occludens (ZO) and a zonula adherence (ZA), can be seen as well as a macula adherence (MA), dense tubules (DT), and light tubules (LT) below apical pits (AP), 21,500 ×.

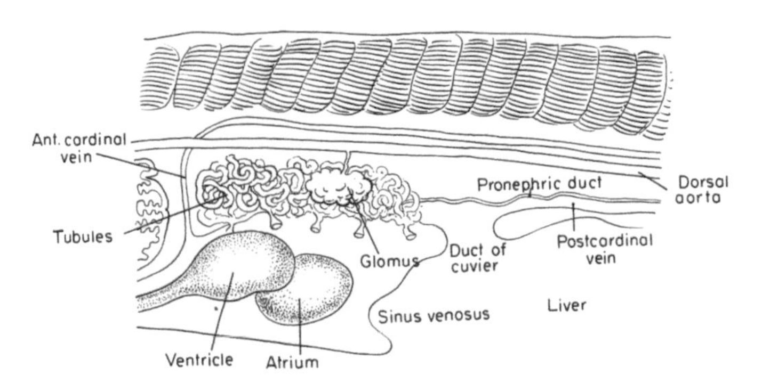

FIGURE M9g Diagrammatic sagittal section through the pronephric region of a first-year ammooete.

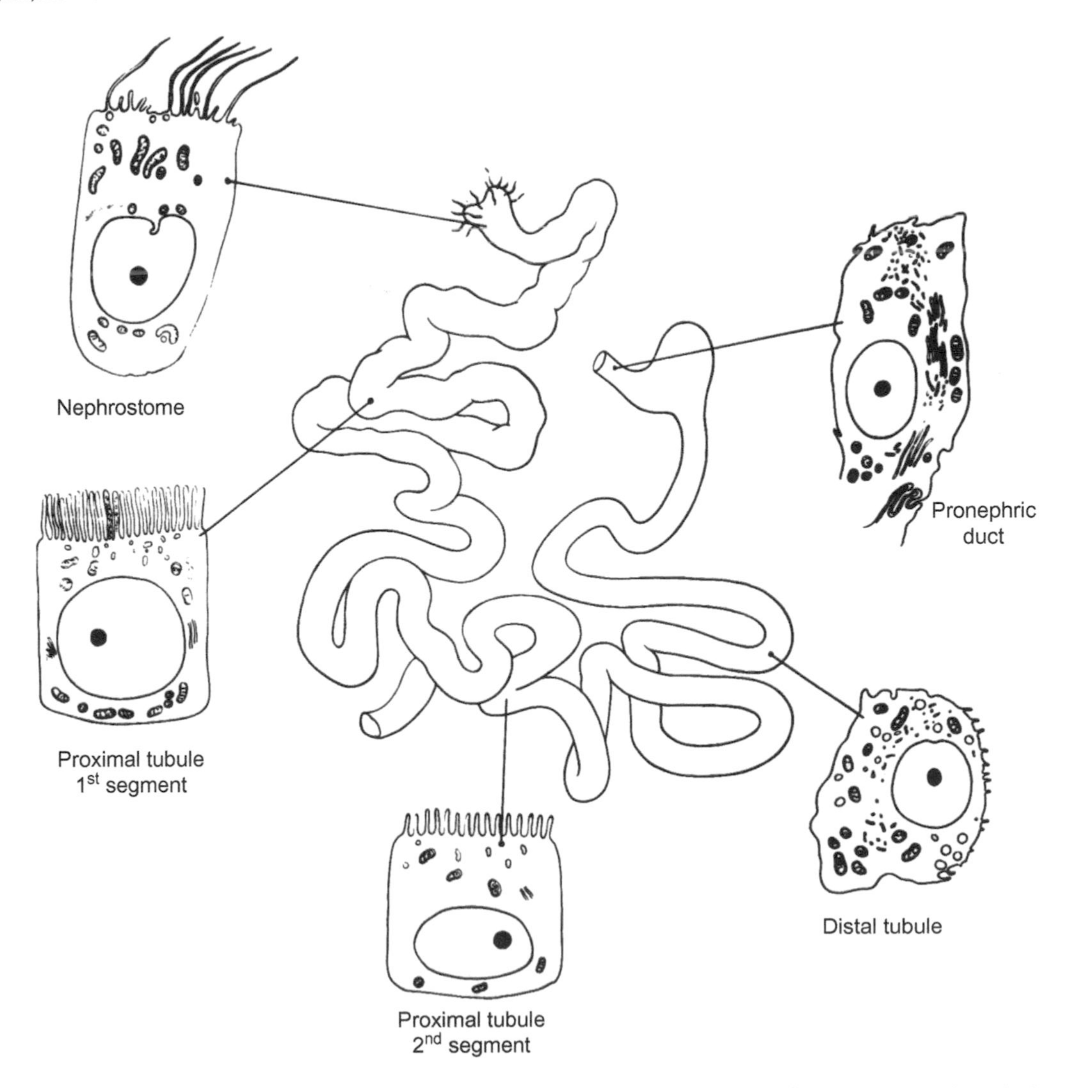

FIGURE M9h Diagram of the cell types in a single pronephric tubule of a first-year ammocoete and their positions along the length of the tubule.

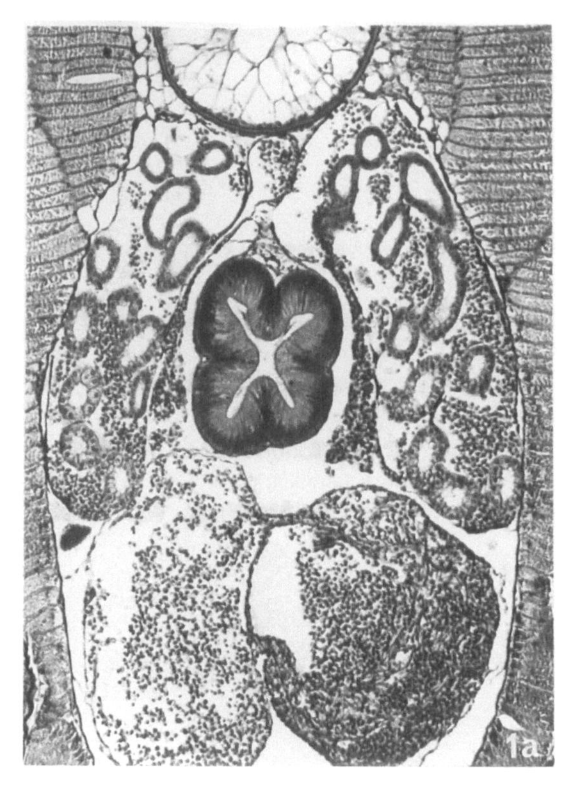

FIGURE M9i Photomicrograph of a cross-section through the pronephric region of a 15-mm ammocoete. Glomera are on both sides of the central esophagus. The glomerular arteriole, which supplies the left glomus, arises from the dorsal aorta, immediately ventral to the notochord (*top*). The tubules are surrounded by blood corpuscles in the lumen of the anterior cardinal vein which covers the tubular portion of the pronephros. The section passes through the heart (*bottom*).

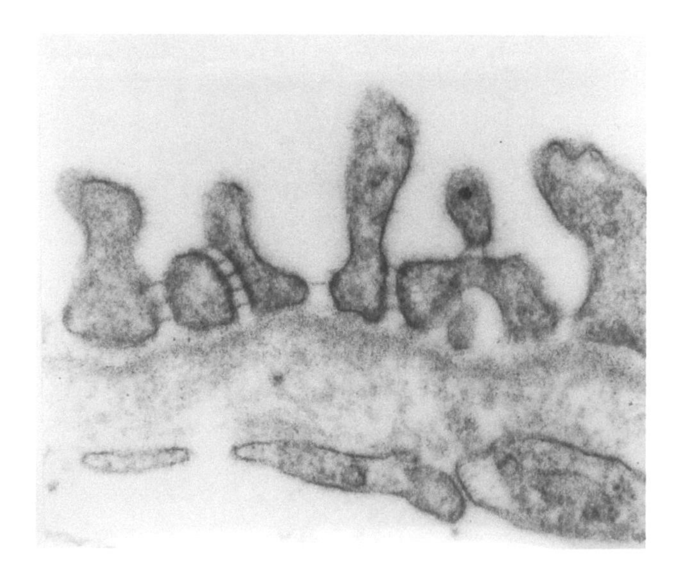

FIGURE M9k High-power electron micrograph of the glomerular capillary wall. The epithelial foot processes (podocytes) are at the top of the micrograph, below is the dense component of the basement membrane, the *lamina densa*. On either side are the less dense regions, the *lamina rara externa* and the *lamina rara interna*. Note: multiple filtration slit membranes between some of the foot processes. Lead citrate stain, 70,000 ×.

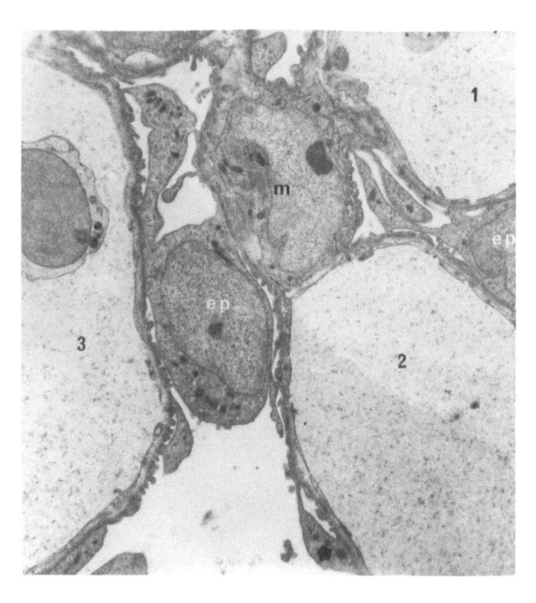

FIGURE M9j Low-power electron micrograph of a portion of one lobe of the glomus. The three capillaries (1, 2, 3) surround a mesangial cell (m), each of which is lined by a thin cytoplasmic rim. Two epithelial cells are seen (ep), 9000 ×.

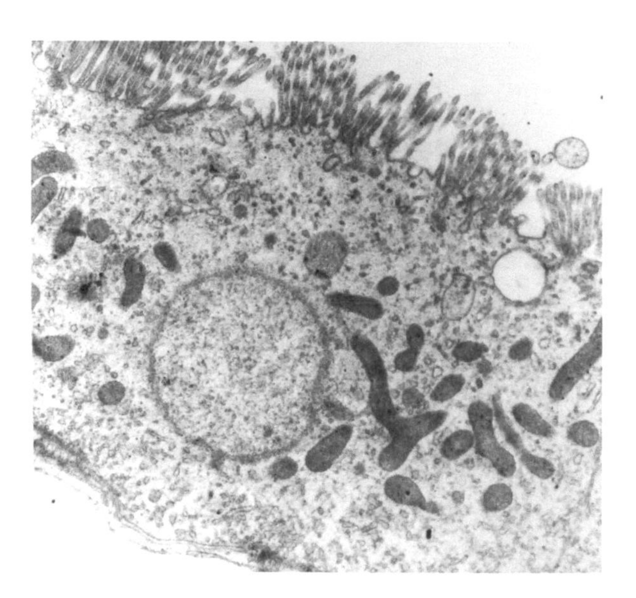

FIGURE M9l Electron micrograph of a cell of the second segment of the proximal tubule. The cell rests on a smooth basement membrane beneath which is a thin cytoplasmic layer, 13,500 ×.

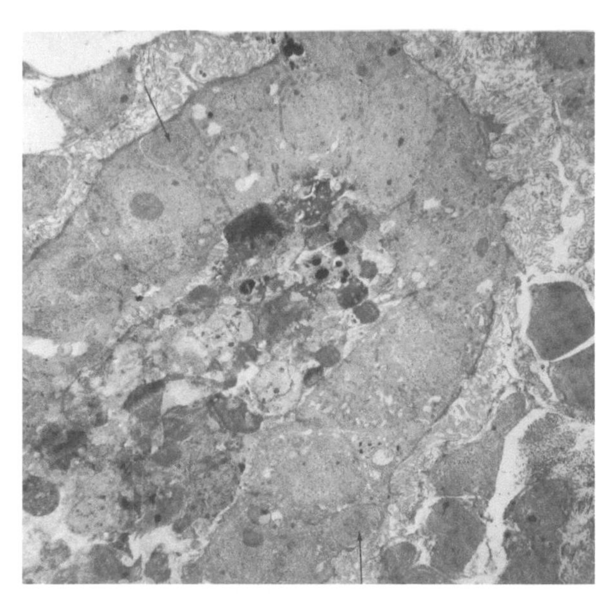

FIGURE M9m Low-power electron micrograph of a blocked proximal tubule of *Lampetra planeri* during the first year. Although some of the cells appear to be intact, microvilli are reduced or absent and the nuclei are no longer spherical. Portions of invading phagocytes (arrows) are seen between some cells. The tubule is surrounded by cells and by connective tissue which appears to be proliferated basement membrane, 6000 ×.

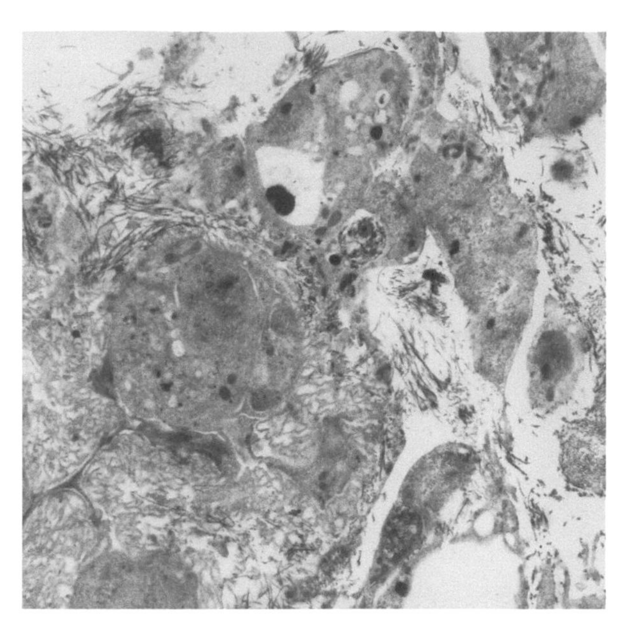

FIGURE M9n Electron micrograph of the pronephros of a 5th year specimen of *Petromyzon marinus*. The only traces of tubules are a small amount of unidentifiable cytoplasm and folds of basement membrane. Collagenous fibrils are abundant in the intertubular regions which also contain cytoplasmic processes and phagocytes, 12,300 ×.

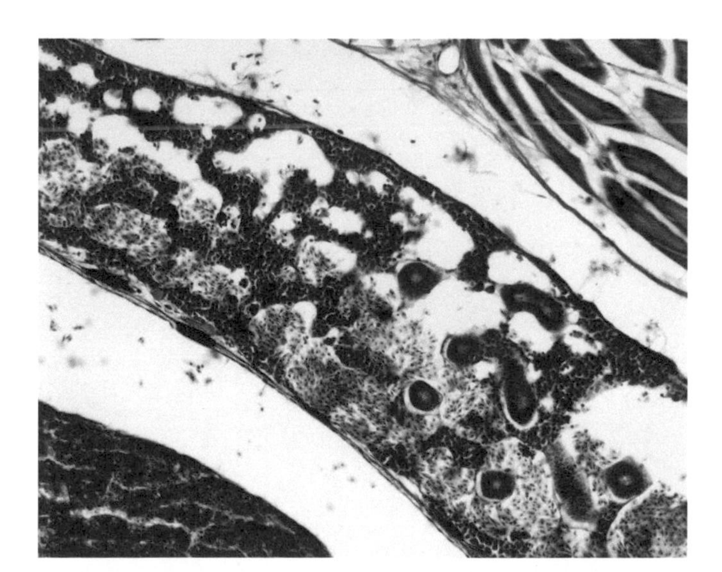

FIGURE M10a Active hemopoiesis takes place in the pronephros of a small, fresh-water teleost, a mature darter. A few tubules remain and the pronephros consists of richly vascular, irregular masses of hemopoietic tissue permeated by vascular sinusoids, 20 ×.

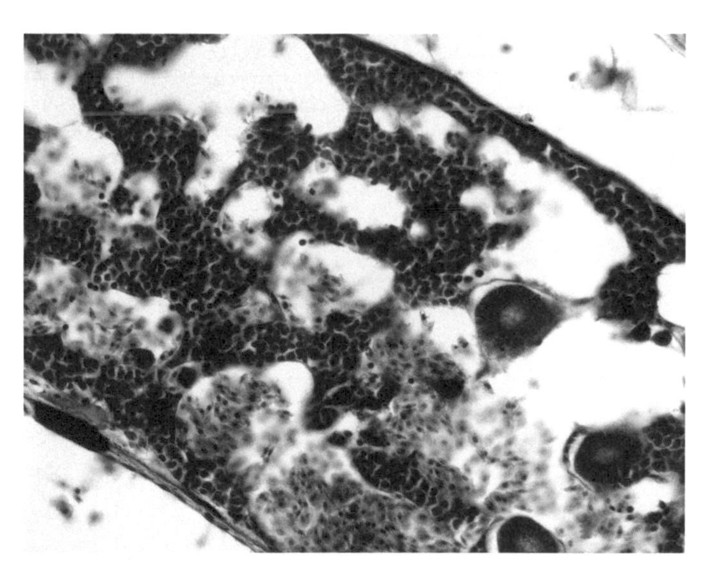

FIGURE M10b Sinusoids with thin walls of squamous endothelial cells permeate the actively hemopoietic loose connective tissue of the pronephros of the darter, 40 ×.

OPISTHONEPHROS

In adult anamniotes the pronephros has disappeared and only loose masses of hemopoietic tissue remain (Figs. M10a and M10b). The new kidney is more compact than its predecessor. It is a mass of tubules and blood vessels supported on a framework of loose fibrous connective tissue and contained within a capsule of connective tissue. Often hemopoietic activity can be seen in the connective tissue between the tubules. The segmental arrangement of tubules is lost in the OPISTHONEPHROS and several tubules may lie within a single segment.

Figs. M11a–M11d ventral strands of ADRENAL TISSUE are visible in various views of a section of adult bullfrog kidney under different magnifications. A nephrostome appears at the top of Fig. M11f on the ventral surface;

FIGURES M11(a–i) The opisthonephros of the bullfrog is a mass of tubules and blood vessels supported on a framework of loose fibrous connective tissue; there seems to be no hemopoietic activity in the interstitial tissue. (*BC*, nephric capsule; *CD*, collecting duct; *D*, distal segment; *I*, intermediate segment; *N*, neck segment; *P*, proximal segment; *PT*, postterminal segment; *S*, sexual segment). (*Note*: A distinction should be drawn between two similar structures, the NEPHROSTOME and the CILIATED NECK SEGMENT. Nephrostomes are ciliated peritoneal funnels that drain the coelom and are found in the opisthonephros of a few vertebrates. In amphibians, they are not usually connected with kidney tubules but drain into the renal veins. The ciliated neck segment draws fluid from the nephric capsules and passes it along to the rest of the nephron.)

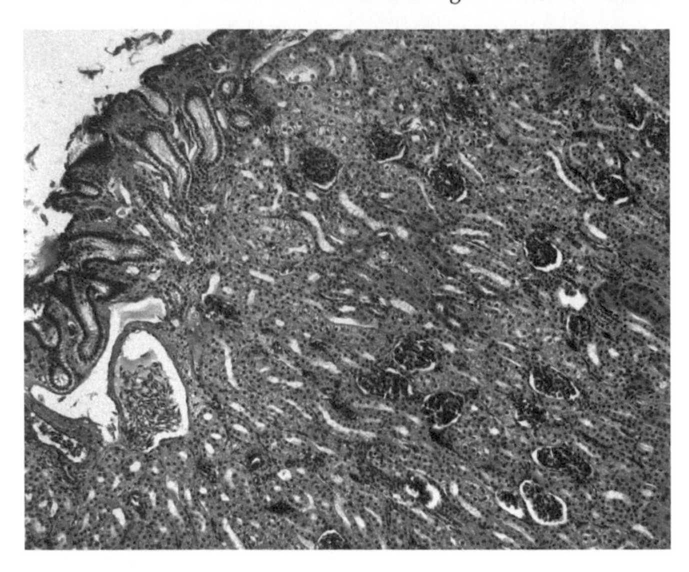

FIGURE M11a Nephric tubules comprise most of the substance of the kidney; several renal corpuscles occur between the tubules in the ventral portion. Some ciliated funnels, the NEPHROSTOMES, occur at the ventral HILUS of this figure. 10 ×.

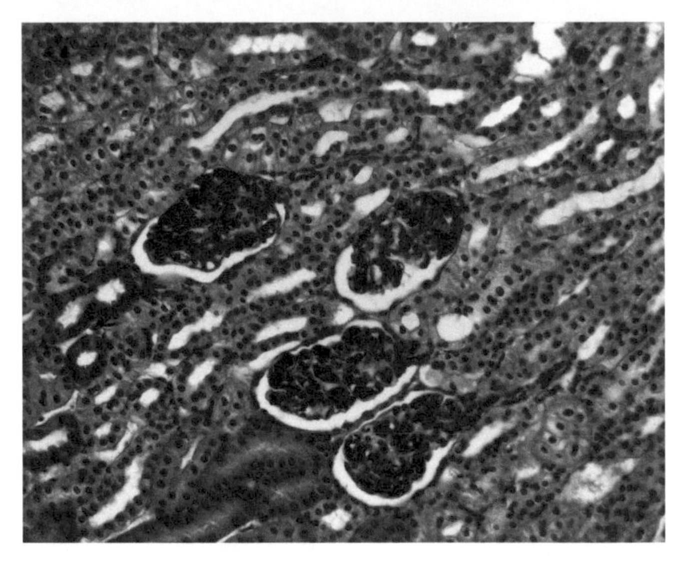

FIGURE M11b Nephric tubules comprise most of the substance of the kidney; several renal corpuscles occur between the tubules in the ventral portion. Some ciliated funnels, the NEPHROSTOMES, occur at the ventral HILUS of Fig. M11a, 20 ×.

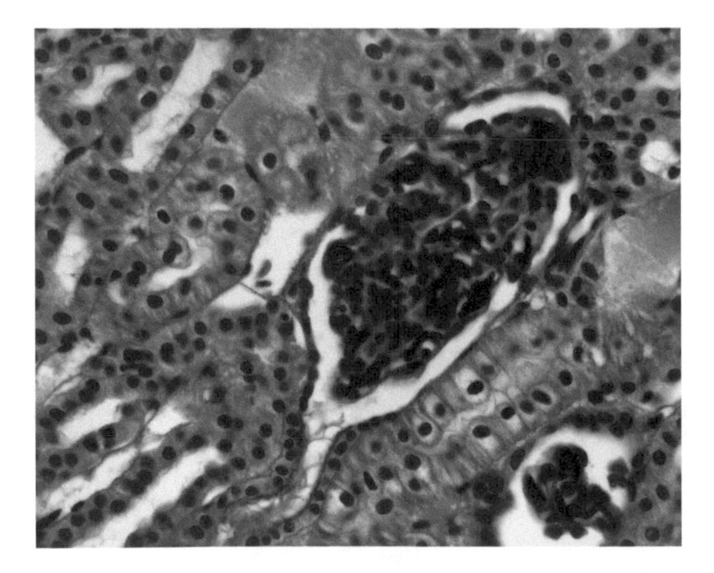

FIGURE M11c This RENAL CORPUSCLE consists of a knot of capillaries, the GLOMERULUS, that is enveloped by the GLOMERULAR CAPSULE. The inner VISCERAL LAYER of the capsule is closely applied to the capillaries and is separated from the outer PARIETAL LAYER by the CAPSULAR SPACE. This renal corpuscle is drained by a CILIATED NECK SEGMENT of the nephron, 40 ×.

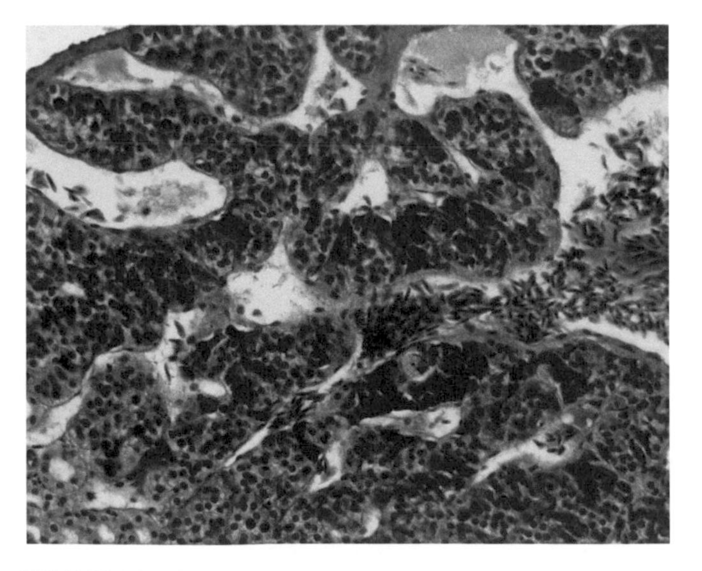

FIGURE M11d Strands of basophilic ADRENAL TISSUE occur near a large blood vessel near the ventral surface of the kidney. Large sinusoids course between the strands, 20 ×.

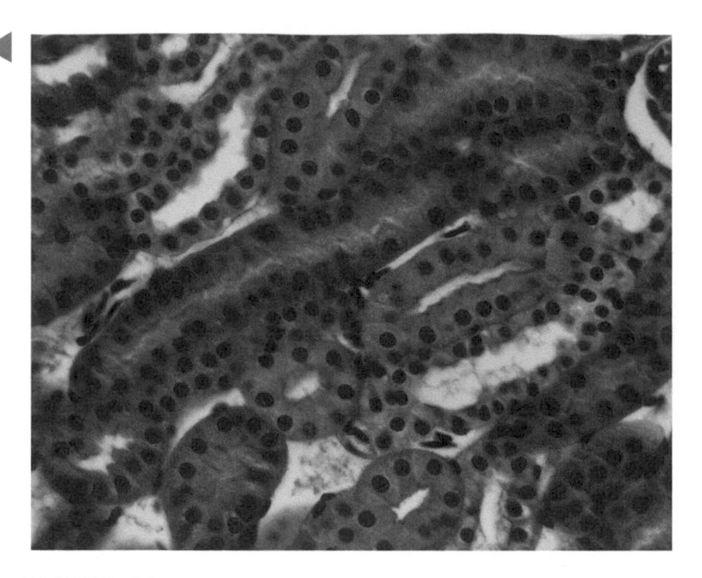

FIGURE M11e Connective tissue between the tubules of the nephron is permeated by capillaries. The deeper eosinophilia of the apical portions of the tubular cells indicates the presence of a brush border; deeper eosinophilia in the basal regions belies the presence of the mitochondria which provide the energy for active transport of fluids, 40 ×.

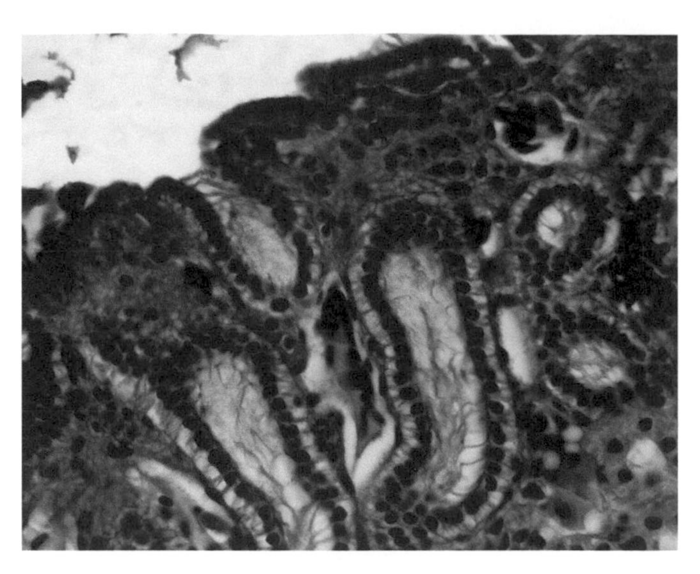

FIGURE M11f Here a ciliated NEPHROSTOME can be seen opening to the ceolomic cavity. These nephrostomes are said to drain fluids from the coelomic cavity into the renal veins, 40 ×.

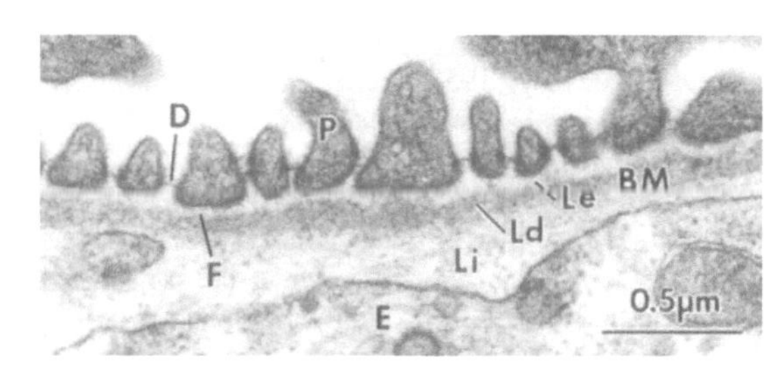

FIGURE M11g Transmission electron micrograph of the filtration membrane in the glomerulus of the frog *Rana esculenta*. The filter consists of the visceral podocyte epithelium, the basement membrane (BM) and the endothelium (E). The basement membrane is composed of a small lamina rara externa (Le), a compact lamina densa (Ld), and a lamina densa interna (Li). Pedicels (P) rest on the outer layer of the basement membrane. Slit pores are covered by a single slit diaphragm (D) interposed between adjacent foot processes. Wispy filaments (F) extend from the pedicels across the lamina rara externa to the lamina densa.

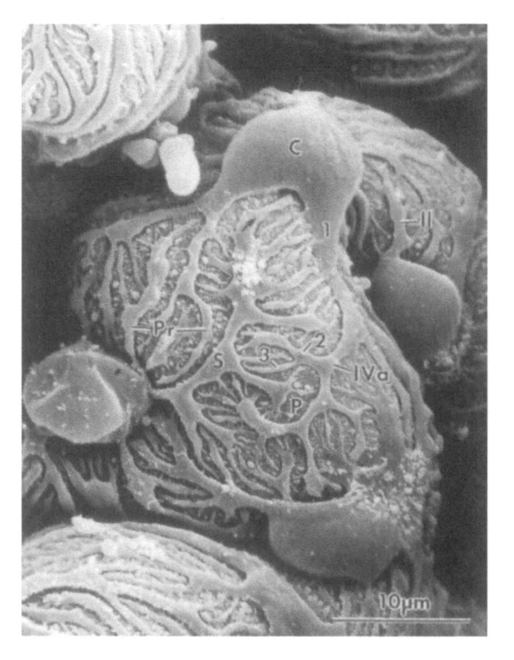

FIGURE M11h Scanning electron micrograph from the glomerulus of the frog (*Rana temporaria*). Part of the nephric capsule has been removed exposing the capillary tuft with its podocytes exposed. Anuran podocytes conform to the general vertebrate pattern with large spherical cell bodies (C), well-defined processes (Pr), and small pedicels (P).

these open into veins and drain lymph from the body cavity into the blood stream. RENAL CORPUSCLES surrounded by tubules are concentrated toward the ventral third of the kidney (Fig. M11b). The GLOMERULUS is enclosed by a double capsule of tubular tissue, the GLOMERULAR CAPSULE (Bowman's capsule). The inner VISCERAL LAYER of the capsule is closely applied to the blood vessels of the glomerulus and is separated from the outer PARIETAL LAYER by the lumen or CAPSULAR SPACE; at the ARTERIAL POLE, or point of entrance of the blood vessels, the two layers are

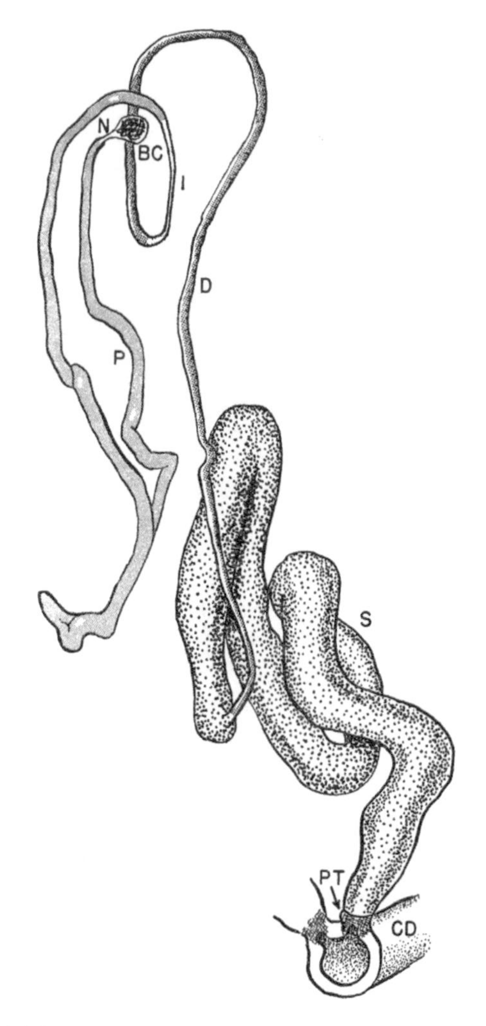

FIGURE M11i Diagram representing the complete renal tubule of a male garter snake (*Thamnophis sirtalis*).

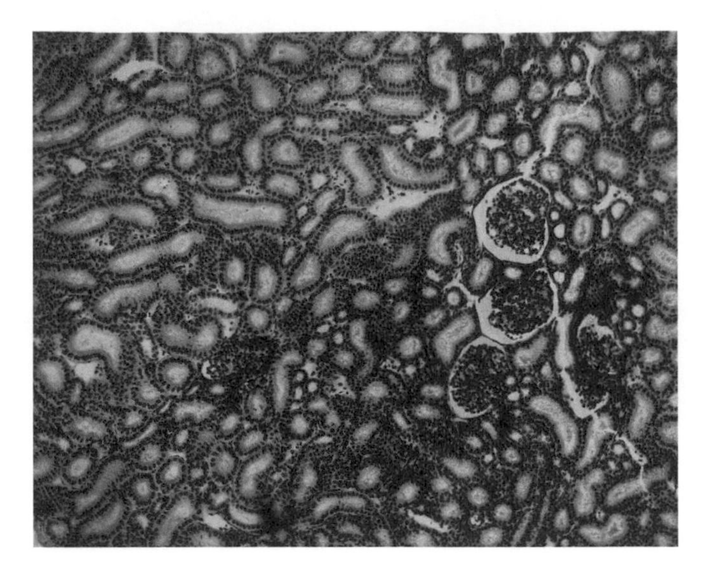

FIGURE M12a Superficially, the opisthonephros of the shark resembles that of the other vertebrates seen in this study. It is a mass of tubules interspersed with renal corpuscles. Capillaries permeate the connective tissue between the tubules. Some hemopoiesis occurs within this vascular connective tissue, 10 ×.

continuous. Fluid from the blood is filtered directly into the capsular space without first passing into the coelomic fluid. The filter consists of the endothelium of the capillary, the visceral layer of the glomerular capsule, and their combined basement membrane (Fig. M11g). Support for this delicate filter is provided by the elaborate pedicels of the podocytes (Fig. M11h).

The blood vessels of the glomerulus are larger than capillaries and are interposed in the course of an arteriole; to form a RETE MIRABILE (Fig. M5c) that is, blood enters the glomerulus by the way of an AFFERENT ARTERIOLE and leaves by the way of an EFFERENT ARTERIOLE. From the efferent arteriole, it breaks up into true capillaries that twine around the tubules (Fig. M11e). In those animals with a renal portal system the glomeruli are supplied with blood from the renal artery while blood from the renal portal vein (which drains the posterior regions of the body) forms a capillary network around the convoluted tubules.

The glomerular capsule is lined with a simple squamous epithelium and empties into a long, tortuous tubule that is divisible into four parts (Fig. M11i):

- a short ciliated NECK SEGMENT;
- the PROXIMAL CONVOLUTED TUBULE of acidophilic cells with a brush border;
- the INTERMEDIATE SEGMENT, often ciliated;
- the DISTAL CONVOLUTED TUBULE, smaller and less acidophilic than the proximal tubules and with a larger lumen.

As the tubule continues from the glomerular capsule, the cells become more columnar and more distinctly outlined. Each distal convoluted tubule empties directly into the ARCHINEPHRIC DUCT (Wolffian duct). This duct has a simple or pseudostratified ciliated columnar epithelium; a LAMINA PROPRIA MUCOSAE of loose connective tissue, a

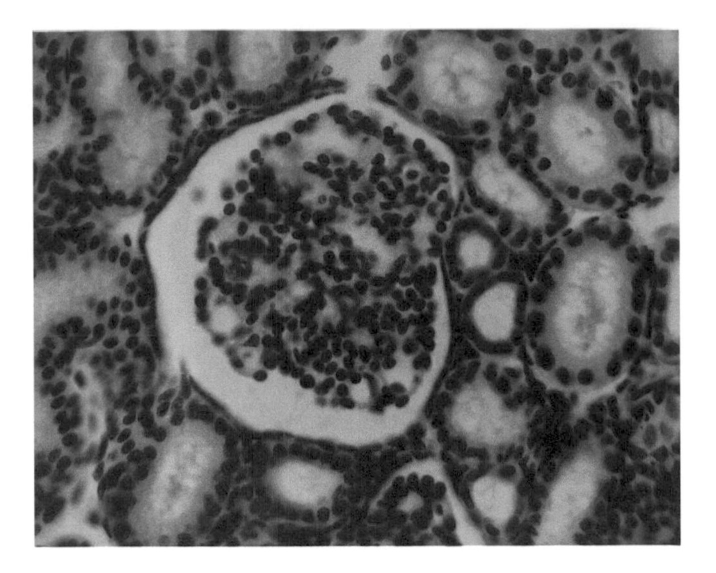

FIGURE M12b A renal corpuscle in the kidney of a shark is surrounded by tubules of the nephron, 40 ×.

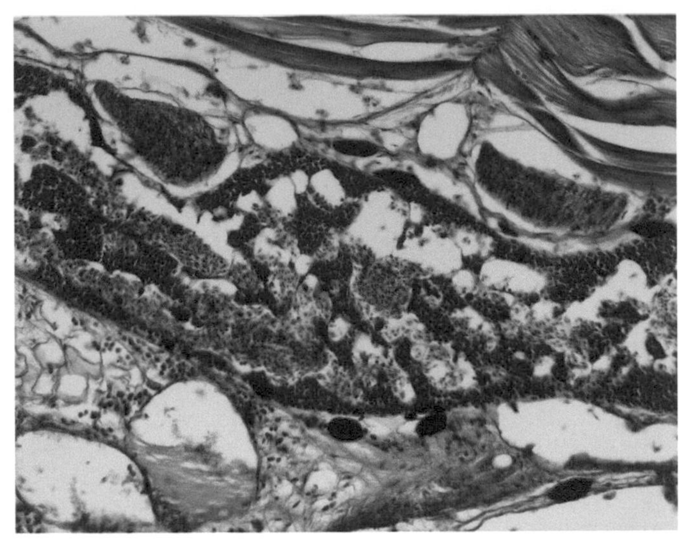

FIGURE M13a Posterior to the pronephros of the darter is the OPISTHONEPHROS. Tubules have been lost in the anterior regions, and only vascular hemopoietic strands can be seen. In the posterior regions, however, hemopoietic tissue mingles with the excretory tubules, 20 ×.

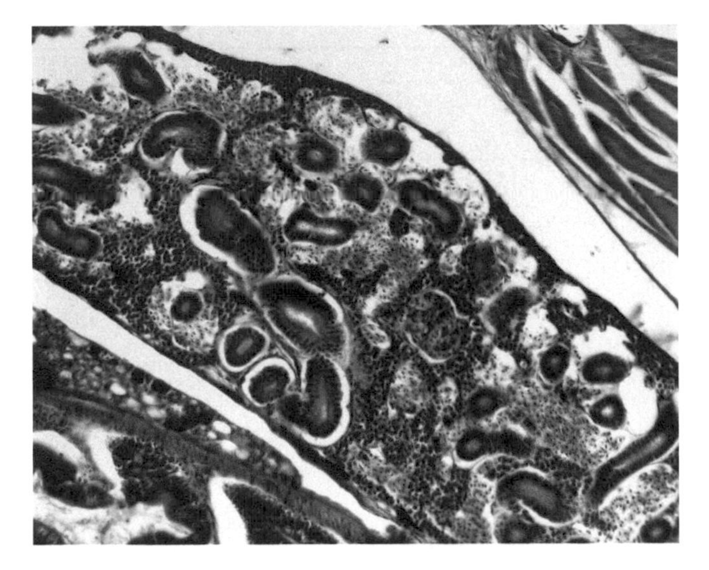

FIGURE M13b Although tubules have been lost in the anterior regions of the darter, some remain posteriorly, 40 ×.

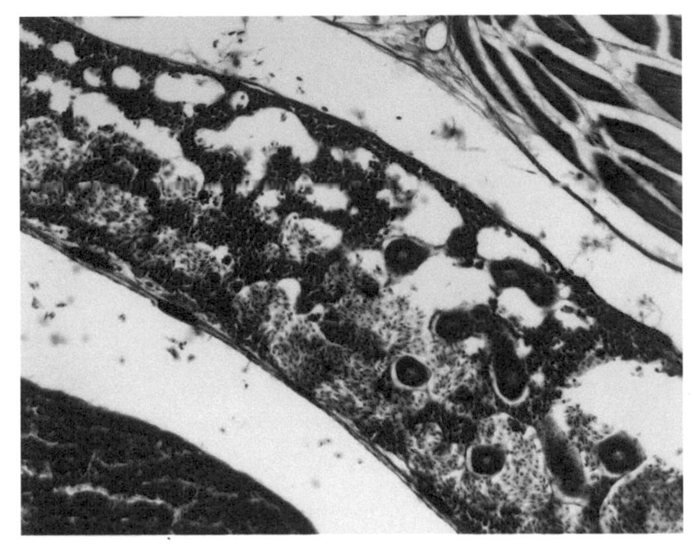

FIGURE M13c Sinusoids permeate the hemopoietic issue in the anterior portion of the kidney of the darter, 20 ×.

TUNICA MUSCULARIS of circular and longitudinal smooth muscle fibers, and a TUNICA SEROSA complete its wall. The nephrostomes have lost their attachment to the tubules and open by short narrow ducts into veins.

Compare the micrographs in Figs. M11a and M11b, and M12a and M12b, which were taken from the kidneys of a frog and the spiny dogfish shark (*Squalus acanthias*), respectively. In spite of their differing environments the structure of their kidneys is remarkably similar. Part of the kidney of the males of fish and amphibians is used for the passage of sperm which may be found filling many of the tubules during the breeding season. Glomeruli are absent or few in marine fish, thereby reducing water loss.

Secretions of the kidney are important in the control of blood pressure in many vertebrates.

The kidney of many vertebrates is an active hemopoietic organ. In many teleosts fish the kidney provides an ideal site for both hemopoietic activity and excretory functions. This dual role of the kidney is evident in both fresh water (Figs. M13a–M13f)—a darter and salt water fishes (Figs. M14a–M14d)—a winter flounder

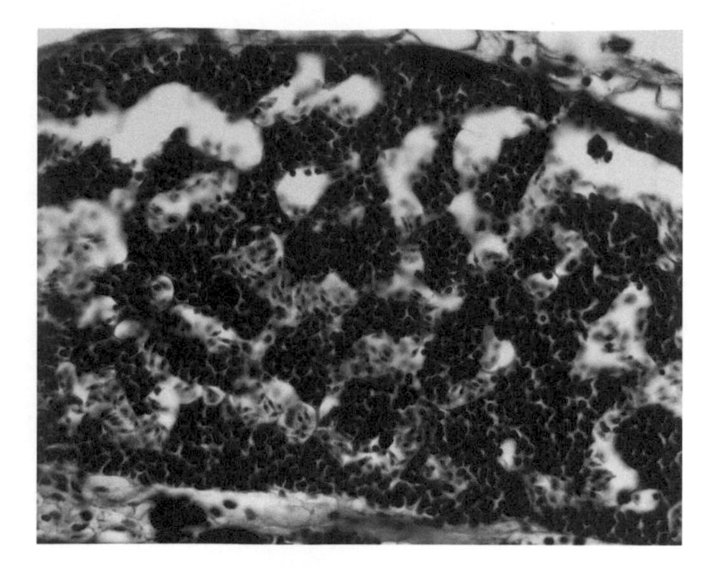

FIGURE M13d More posteriorly, a few tubules are present within the hemopoietic tissue of the opisthonephros, 40 ×.

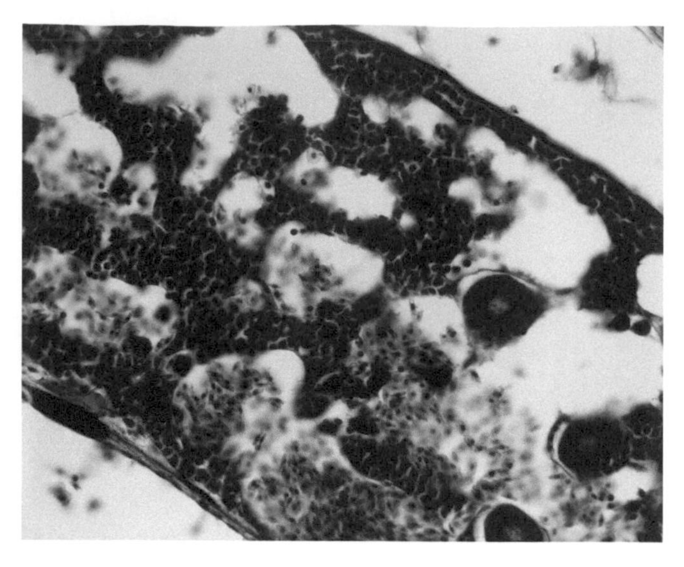

FIGURE M13e This is a view of the same region shown above, 40 ×.

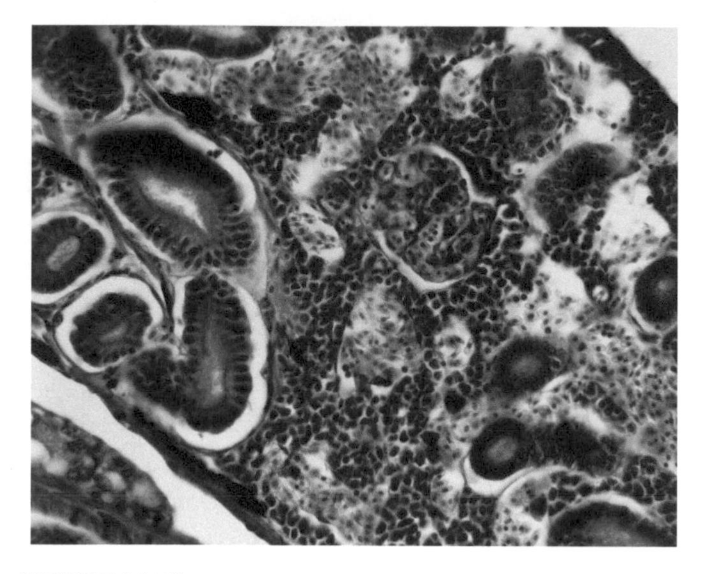

FIGURE M13f Tubules permeating hemopoietic tissue are prominent within the posterior opisthonephros. A renal corpuscle is obvious near the right of center, 40 ×.

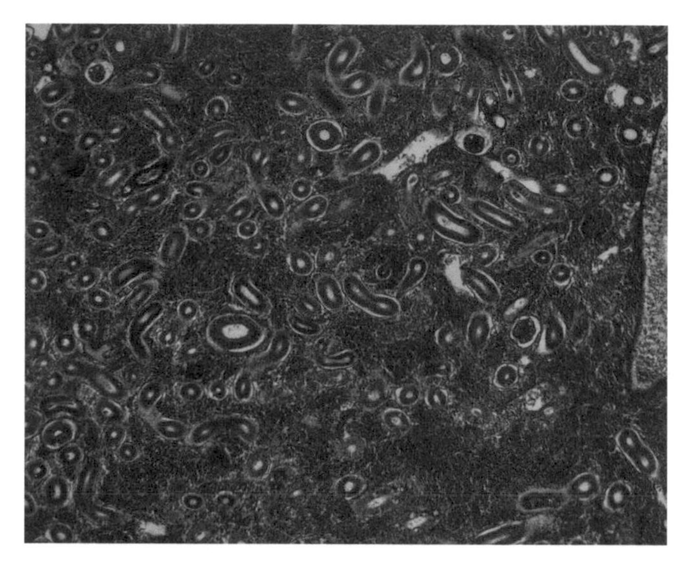

FIGURE M14a The close relationship between the kidney and hemopoietic tissue is seen in this section of the kidney of a winter flounder (*Pseudopleuronectes americanus*). Many PIGMENT NODULES are also present. These nodules are common within hemopoietic tissue of fish and are the site of destruction of effete blood cells, especially erythrocytes, 10 ×.

(*Pseudopleuronectes ameriicanus*). The interstitial tissue in both kidneys is packed with developing blood cells (Figs. M13a–M13f, and M14a–M14d). In addition to being a source of developing blood cells, the interstitial tissue of the kidney is often the site of destruction of effete blood cells (Fig. M14d). Given the substantive difference between fresh water and salt water environments, obvious differences in size and shape of both glomeruli and tubules can be seen (Figs. M13f and M14b). These differences reflect the need to retain salts and lose water in the darter and for the flounder to conserve water and lose salt.

Hemopoietic activity in the vascularized interstitial tissues of the kidney is common in both fresh water and marine amphibians, such as the mud puppy (Figs. M15a–M15c, and M14a–M14d) and the marine toad (Figs. M16a and M16b). In both animals the renal corpuscle is drained by a ciliated neck segment (Figs. M15d and M16b).

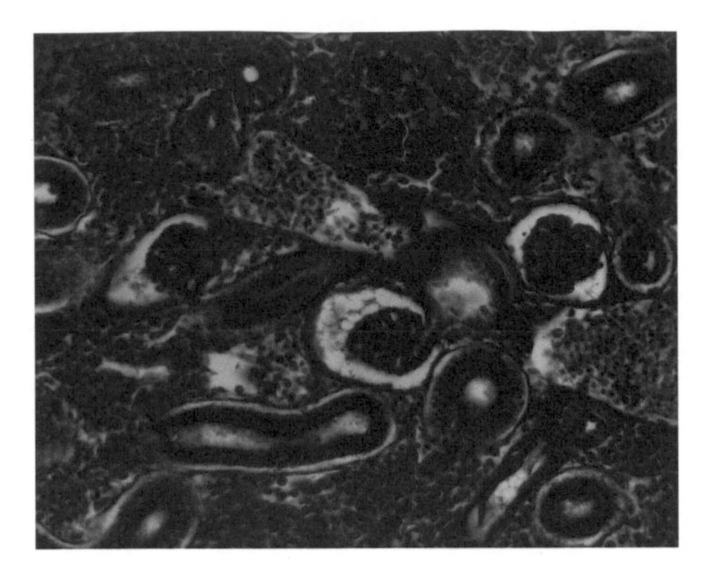

FIGURE M14b The flounder is a salt water fish. It does not produce large amounts of urine as would a fish living in fresh water. Reflecting the conservation of water, the three renal capsules (where water is lost) in this micrograph are small and the tubules (where water is reabsorbed) are well developed. Hemopoietic tissue abounds between the corpuscles and tubules. A PIGMENT NODULE appears at the upper center, 40×.

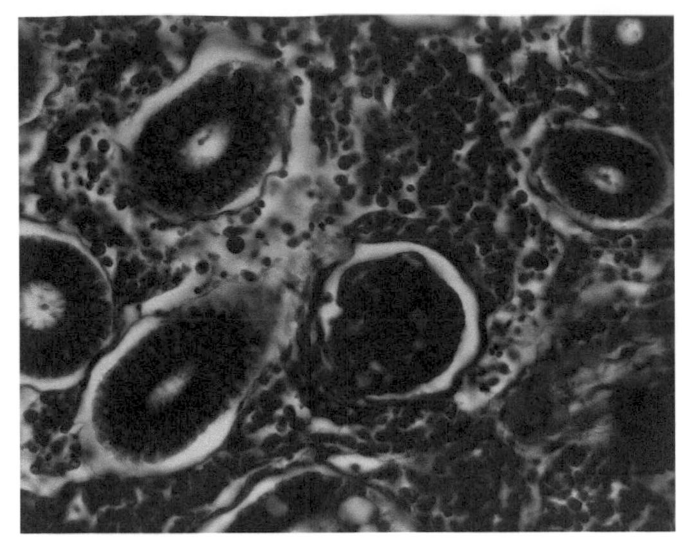

FIGURE M14c There has been considerable shrinkage in this section, especially around the tubules. Active hemopoiesis is taking place in the connective tissue between the nephrons, 63×.

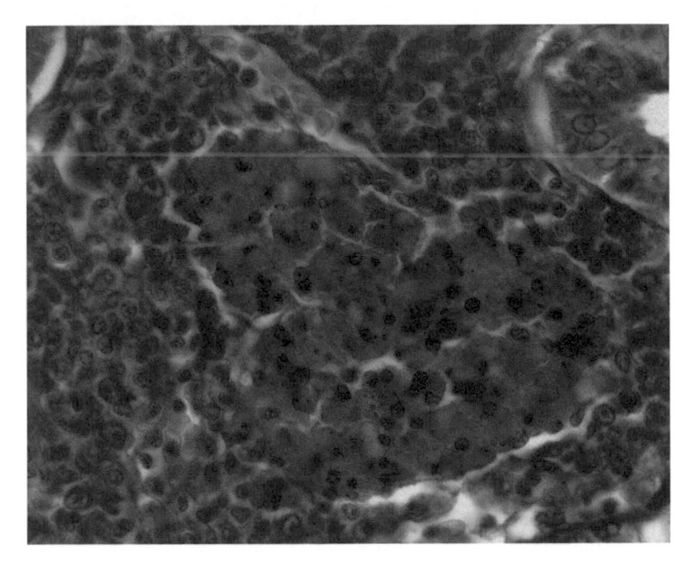

FIGURE M14d This pigment nodule is a cluster of macrophages engorged with effete blood cells. The nodule is surrounded by a thin capsule of squamous reticular cells. It is closely associated with blood vessels and is surrounded by a sinusoid. One of the products of the breakdown of hemoglobin is the golden brown pigment—hemosiderin, 100×.

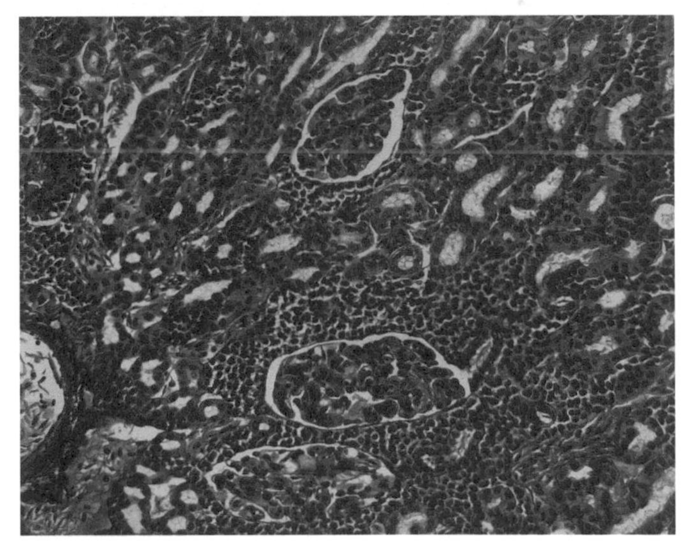

FIGURE M15a Hemopoiesis is abundant in the connective tissue between the renal corpuscles and tubules of the kidney of a mud puppy, 10×.

In larval lamprey (ammocoetes) the opistonephros exhibits hemopoietic activity packed with adipose tissue and clusters of hemopoietic cells interspersed in the vascular connective tissue between the tubules (Figs. M17a–M17c). At this stage the renal corpuscles have not yet organized into a glomus. The opisthonephros of an adult lamprey is roughly the shape of a loaf of French bread with a longitudinal trench hollowed out of the medial side (Figs. M18a–M18c). The trench contains a single complex renal corpuscle that consists of a number of lobed glomeruli lying one behind the other to form an elongate GLOMUS (Figs. M18a and M18b). Interposed

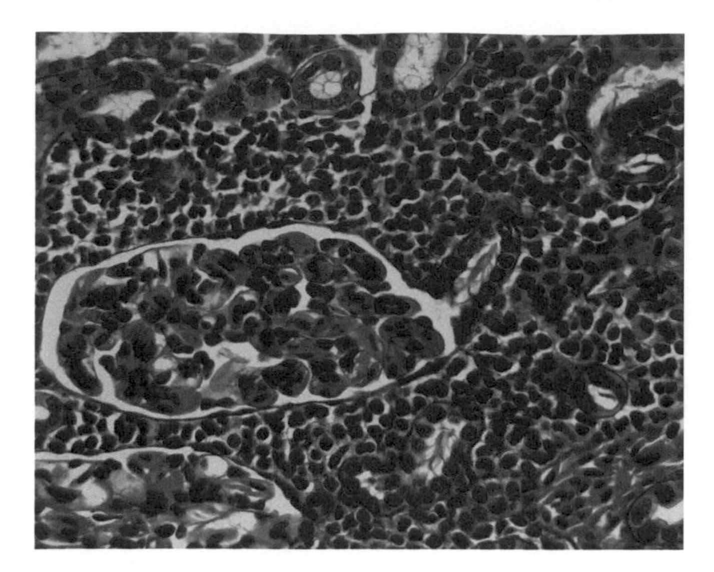

FIGURE M15b This glomerular capsule is drained by a ciliated neck segment of the tubular nephron, 20 ×.

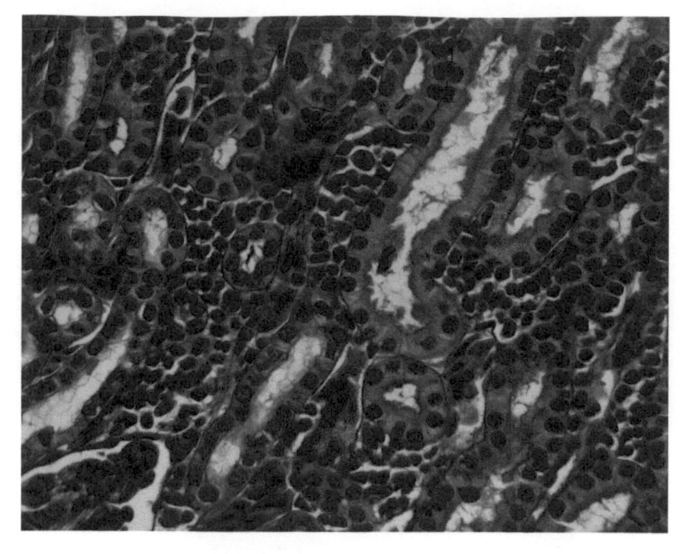

FIGURE M15c Some of the epithelial cells of these tubules display evidence of brush borders shown by the heightened eosinophilia in their apical regions. Intertubular capillaries are well shown. There is active hemopoiesis in the connective tissue between the tubules, 20 ×.

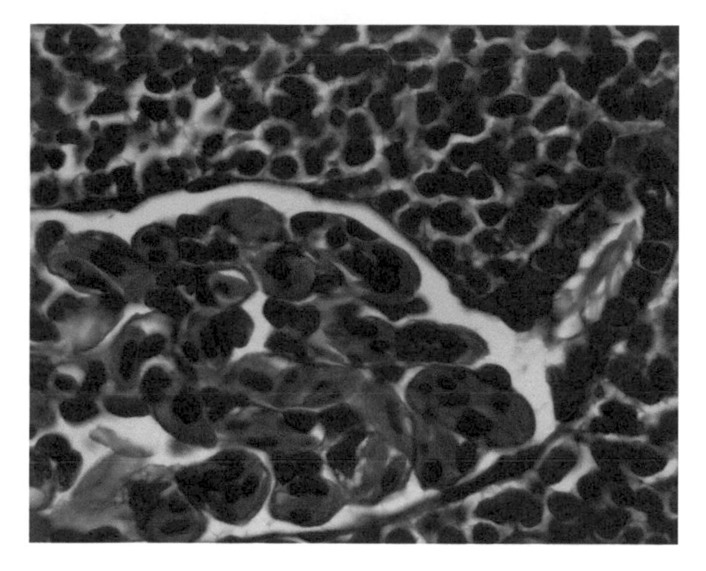

FIGURE M15d This renal corpuscle is drained by a ciliated neck segment of the tubular nephron, 40 ×.

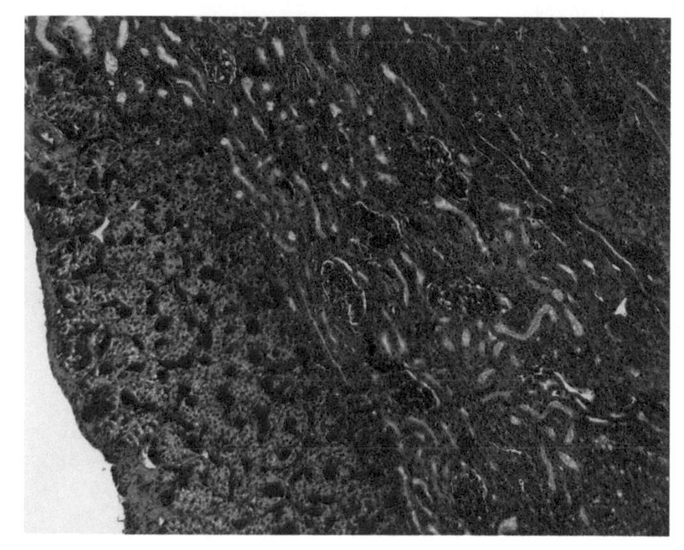

FIGURE M16a Clusters of adrenal tissue (left) are associated with the opisthonephros (right) of the marine toad, *Bufo marinus*. Several renal corpuscles are visible among the kidney tubules, 10 ×.

between each pair of lobes is the dilated end of a nephron, and this is drained by a conspicuous CILIATED NECK SEGMENT (Figs. M19a, M19b, and M20). Note: this dilation is not invaginated to form the usual egg cup-shaped glomerular capsule. The glomus is surrounded by tubules of the nephron that is basically similar to those seen in the kidney of the frog.

The kidney of the hagfish is an opisthonephros of almost diagrammatic simplicity. Each of the segmentally arranged nephrons consists of a renal corpuscle, comprising a GLOMERULUS and GLOMERULAR CAPSULE, and a short TUBULE that drains into the ARCHINEPHRIC DUCT. Each renal corpuscle is supplied with blood from one AFFERENT ARTERIOLE but is drained by several EFFERENT ARTERIOLES that in turn send branches over the tubules and the archinephric duct. The tubules are not differentiated into regions and, like the archinephric duct, are lined by a simple columnar epithelium with a brush border.

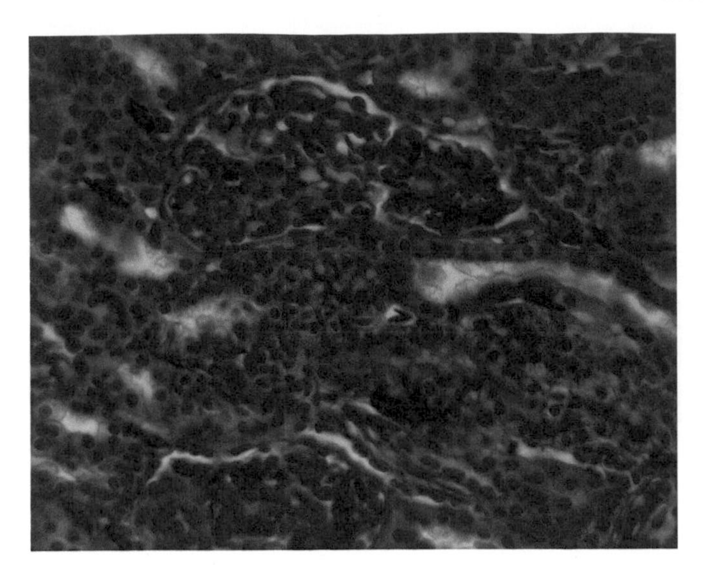

FIGURE M16b Two renal corpuscles are lodged among the renal tubules of the marine toad, 40 ×.

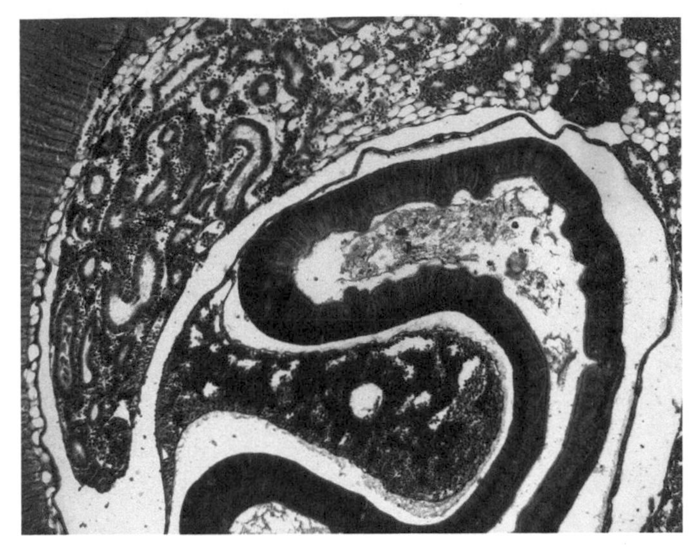

FIGURE M17a In these cross-sections of an ammocoete, muscle of the body wall appears at the left. Next is a mass of adipose tissue that is packed with tubules of the opisthonephros. Blood cells are developing in the hemopoietic tissue between the tubules. A renal corpuscle is seen on the right of the lower tip of this mass. The gut with its invading typhlosole (also containing hemopoietic tissue) occupies the lower right of the micrograph, 10 ×.

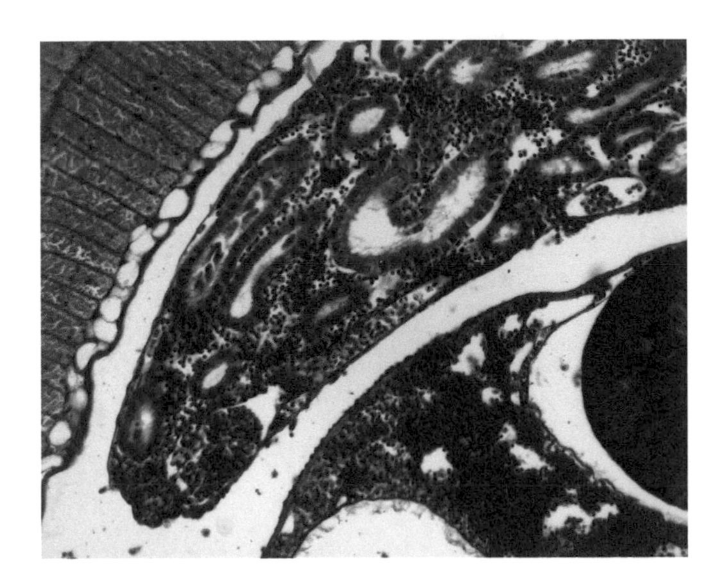

FIGURE M17b A higher magnification view of the section in Fig. M17a, centered on the tubules of the opisthonephros. The typhlosole appears on the lower right, 20 ×.

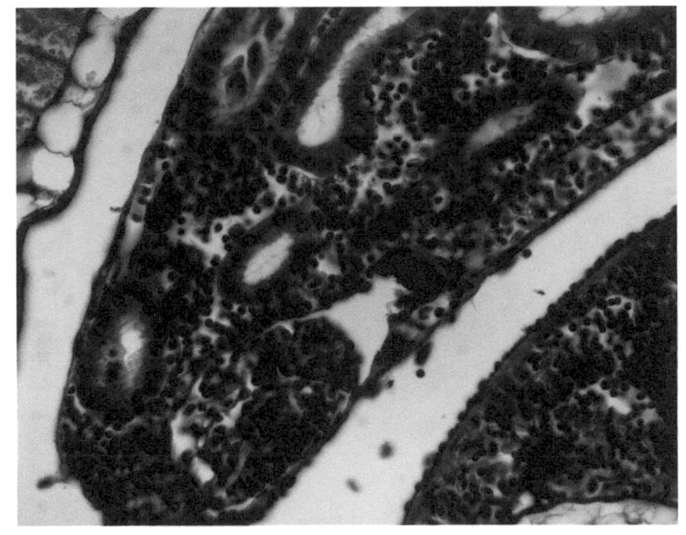

FIGURE M17c The renal capsule at the lower left is enclosed by a simple squamous epithelium. Cells of the nephric tubules have brush borders. Hemopoietic tissue is abundant between the tubules. Blood vessels run beneath the capsule at the left and right. Masses of pigment cells are distributed throughout, 40 ×.

In frogs the opistonephric kidney is not organized into cortical and medullary regions as in mammals. The long-thin adrenal gland sits atop the kidney. Adrenal tissues are often seen invading the kidney. The kidney is well-vascularized and hosts clusters of hemopoietic cells as well as a net of proximal and distal tubules typically separated by a zone of renal corpuscles (Figs. M21a and M21b). Semithin plastic sections reveal much greater cellular detail even at low magnifications (Fig. M22a). Invading adrenal tissue forms well-vascularized cords

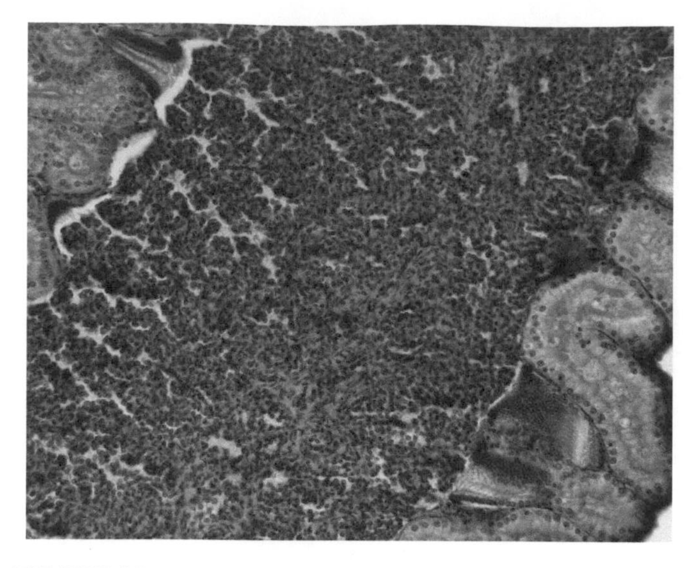

FIGURE M18a The opisthonephros of the lamprey is unusual in that the vascular tissue is consolidated into a single GLOMUS that runs longitudinally within a trench surrounded by the tubular tissue. There are no Bowman's capsules: flattened balloon-like ends of the nephrons, the NEPHRIC CAPSULES, insinuate themselves between the vessels of the glomus and carry away fluid that passes through their simple squamous epithelium, 20 ×.

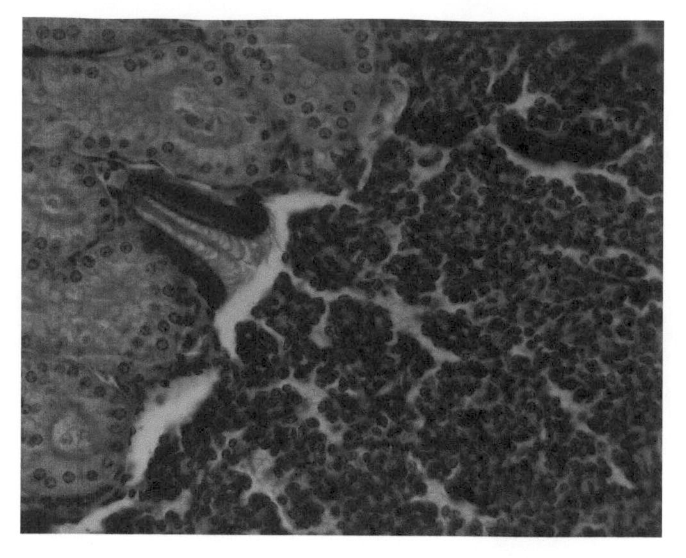

FIGURE M18b Vascular tissue of the glomus occupies the right side of this micrograph. What appear to be cracks in the tissue of the glomus are the flattened balloons at the ends of the kidney tubules that make up the NEPHRONS. CILIATED NECK SEGMENTS, at the upper left and lower right; these draw off the fluid that leaks from the glomus and pass it down the tubules, 40 ×.

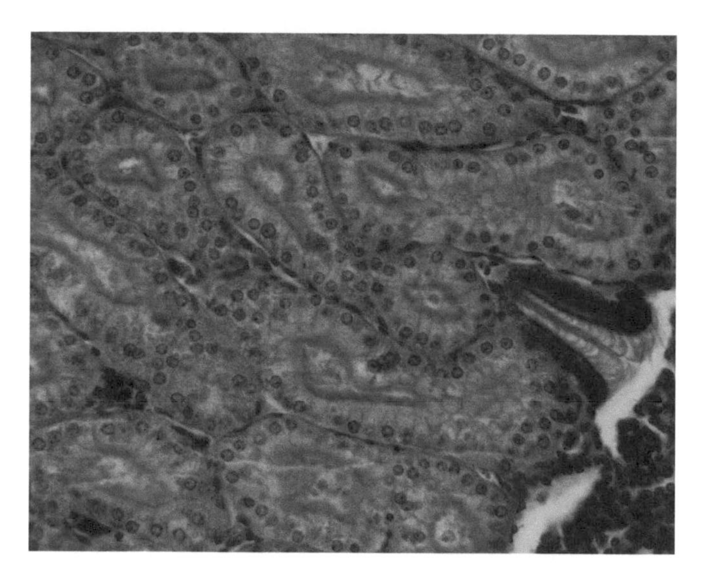

FIGURE M18c A ciliated funnel at the right draws fluid from the spaces that penetrate the glomus. A simple squamous epithelium encloses these spaces and separates them from the capillaries of the glomus. There is evidence of a BRUSH BORDER on the cells of the tubules. A few capillaries course between the tubules of the nephrons, 40 ×.

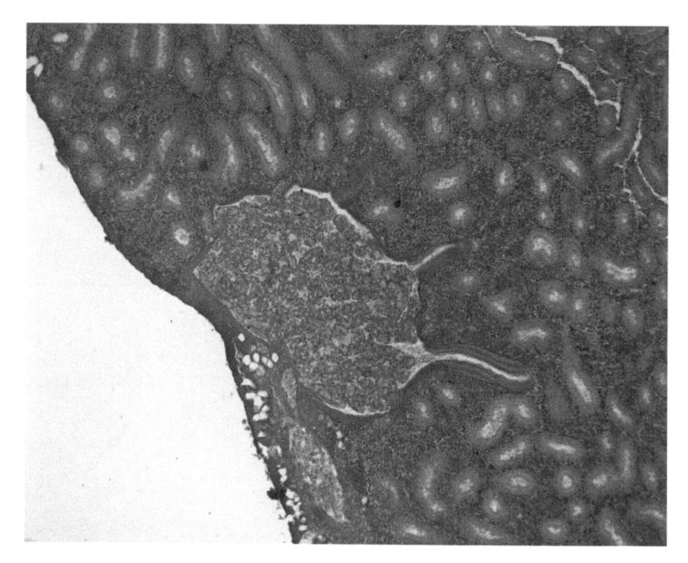

FIGURE M19a In this cross-section of a lamprey, a single glomus in the opisthonephric kidney lies near the ventral surface. Its exudate pours into nephric capsules that are drained by two ciliated neck segments of the nephrons surrounding it, 10 ×.

(Fig. M21c). Renal corpuscles are well developed (Figs. M21d, M22c, and M22d) and tend to cluster in the middle zone between the ventral zone proximal convoluted tubules and collecting ducts (Fig. M22b) and dorsal zone of paler distal convoluted tubules (Fig. M22e). Cells lining the proximal convoluted tubules typically demonstrate a brush boarder at the luminal surface (Fig. M21e). Ciliated nephrostomes are part of the frog's functional kidney, opening to the coelom and draining excretory wastes from its fluid (Fig. M22f).

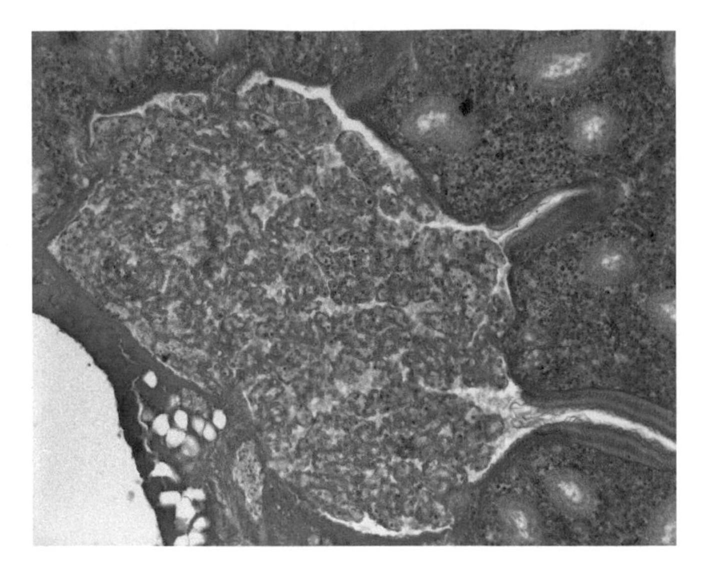

FIGURE M19b A higher power view of Fig. M19a centered on the glomus. Its exudate pours into nephric capsules that are drained by two ciliated neck segments of the nephrons surrounding it, 20 ×.

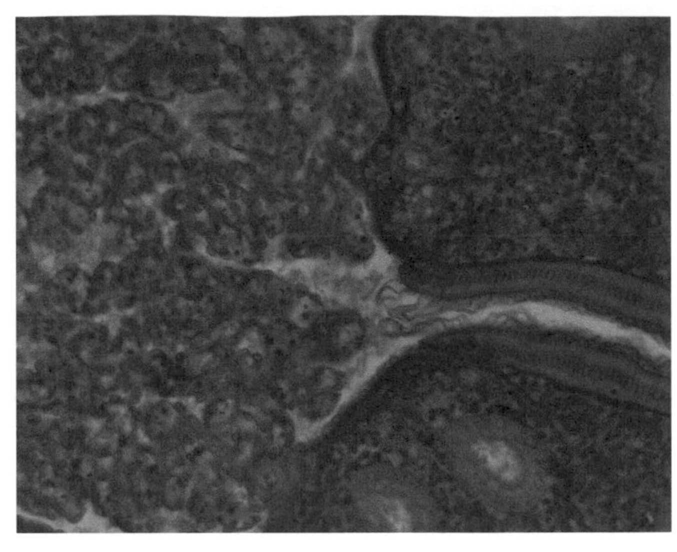

FIGURE M20 This is a longitudinal section of the opisthonephric kidney of a lamprey. Ciliated neck segments drain the spaces between the glomerular loops. Here the ciliated neck segment is plainly seen at the right center of the micrograph, 40 ×.

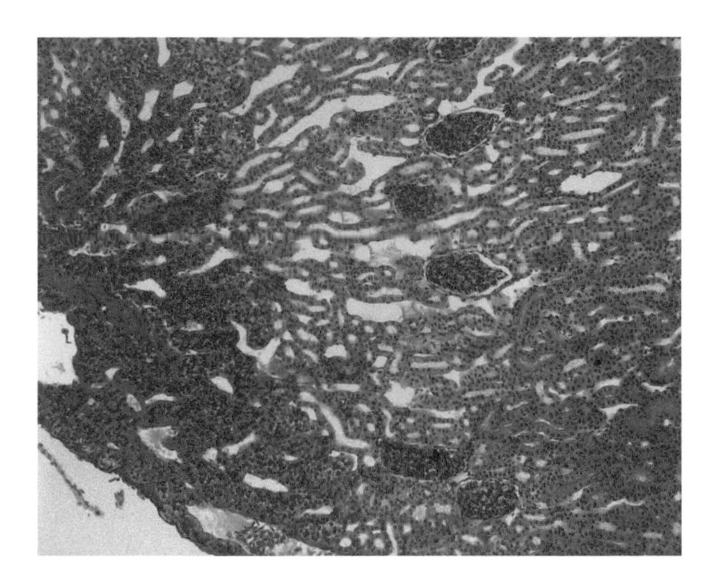

FIGURE M21a The ventral region of this opisthonephros of a frog is at the left; there is abundant basophilic adrenal tissue gathered here. The midregion of the kidney contains renal corpuscles and pale distal convoluted tubules. The dorsal region at the right consists of proximal convoluted tubules and collecting ducts, 10 ×.

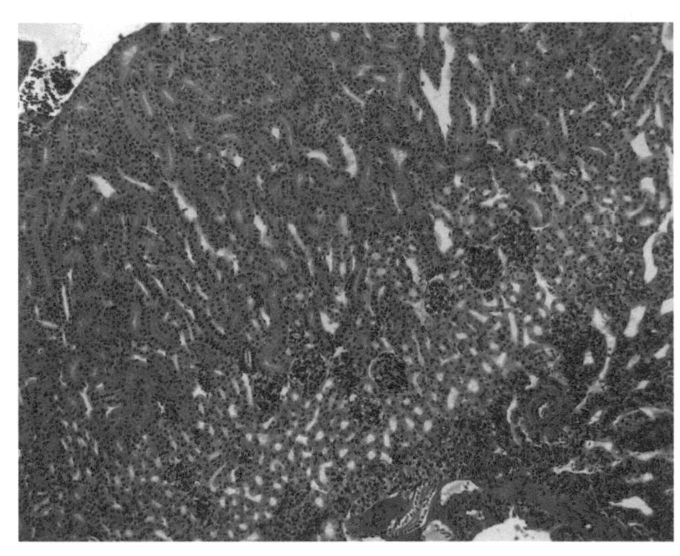

FIGURE M21b The dorsal region of this kidney is at the left. The middle region, containing several renal corpuscles and sections of distal convoluted tubules, separates it from the ventral region (lower right) which is rich in basophilic adrenal tissue, 10 ×.

METANEPHROS

A feature characteristic of the amniotes is the occurrence of a METANEPHROS, a new kidney with its separately formed duct. This leaves the male archinephric duct (Wolffian duct) free to function as the sperm carrying tube, the vas deferens. (It degenerates in the female.)

Fig. M5e is a good example of the compact metanephric kidney typically seen in mammals. The metanephric kidneys are organized into regions—a CORTEX and a MEDULLA. In the medulla the tissue appears to radiate from the renal PAPILLA. These different regions of the kidney are associated with the different components of the

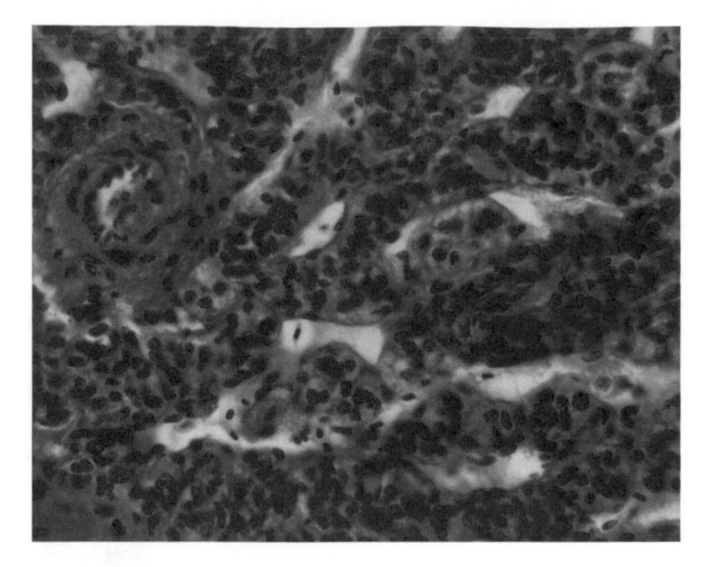

FIGURE M21c Adrenal tissue in the ventral region occurs in cords with abundant capillaries between. There is a muscular artery at the left, 40 ×.

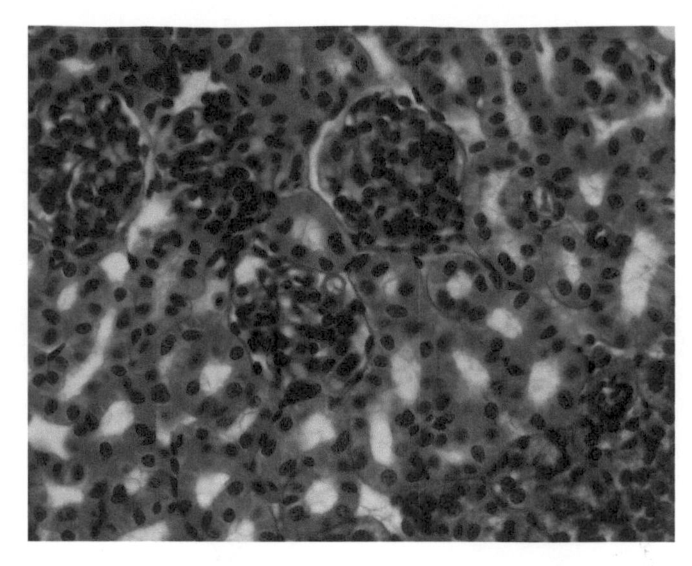

FIGURE M21d Several renal corpuscles, surrounded by sections of the distal convoluted tubule, are seen in the middle region, 40 ×.

mammalian nephron as seen in Fig. M23. There is a radial arrangement which continues to the capsule in some part of the cortex (cortical rays) but in other parts the cortical substance appears tangled (cortical labyrinth). Just below the capsule the cortex contains the renal corpuscles and both proximal and distal convoluted tubules. The medulla consists of the straight portions of the nephron and the collecting tubules. These features are readily seen in the stitched image (Figs. M24ai–M24aiii). The proximal convoluted tubule become the proximal limb of the nephric loop and descends from the cortex into the renal papilla where it takes a hairpin turn and heads back toward the cortex as the distal limb (Figs. M24bi–M24biv). In the cortex, it becomes the distal convoluted tubule (Fig. M24c). In the cortex the distal convoluted tubules press against the renal corpuscle with a thickened epithelium, known as the MACULA DENSA part of the juxtaglomerular apparatus (Figs. M29 and M24d). In the cortex the nephron empties its contents into the collecting tubules which parallel the tubules of the nephric loops as the head to the papilla to empty their contents into the renal pelvis (Figs. M24e and M24f). The ureter enters at the hilus (indentation) and its expanded end forms a cavity, the renal pelvis, into which the renal papilla projects. The renal artery and vein also enter at the hilus.

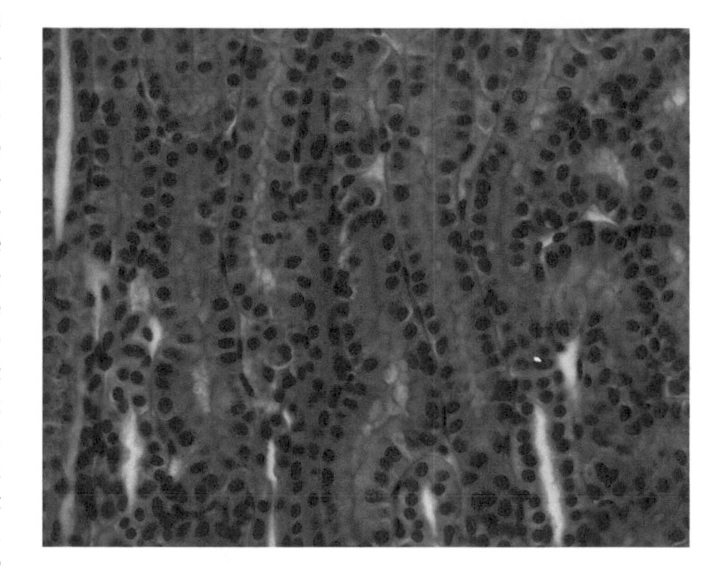

FIGURE M21e Brush borders are apparent on the cells of the proximal convoluted tubules in the dorsal region. A few paler distal convoluted tubules appear in the lower portions of the picture, 40 ×.

The nephron begins in the cortical labyrinth as the glomerular capsule (Bowman's capsule) surrounding the glomerulus. These relationships are shown especially well in the series of semithin sections of the kidney of a monkey (Figs. M25a–M25f). The outer parietal layer of the capsule is simple squamous epithelium. The inner visceral layer adheres closely to the blood vessels of the glomerulus and is difficult to see (Figs. M25a and M25b). Arterioles enter and leave the glomerulus at the arterial pole. Most glomeruli sections are cut so that neither pole can be seen.

FIGURES M22(a–f) The micrographs in this group are of semithin sections from the opisthonephros of a frog. The tissue used in his preparation was embedded in epoxy resin and sectioned at 1.5 μm. There is little of the usual shrinkage seen in paraffin sections and greater cellular detail is afforded.

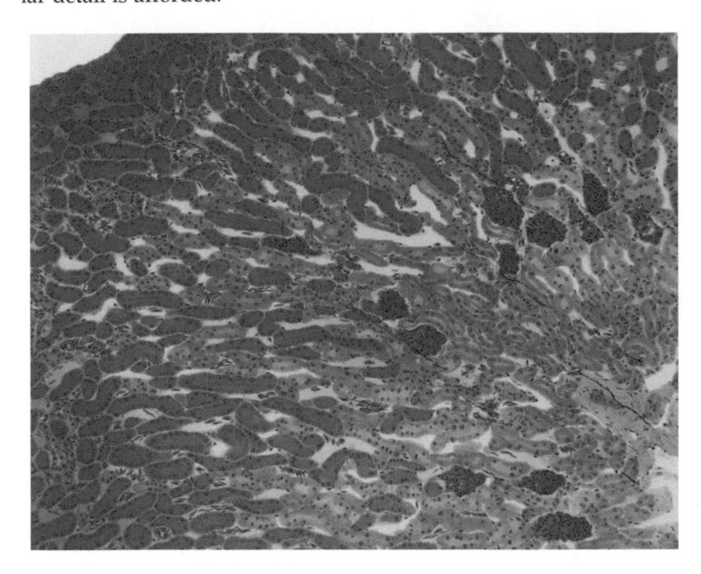

FIGURE M22a The dorsal portion of this section (left top) stains more deeply than this midzone or the ventral region. The dorsal region consists of sections of the proximal convoluted tubules and collecting ducts. Renal corpuscles are arrayed among distal tubules in the midzone. Distal tubules are abundant in the ventral zone (right bottom), 10 ×.

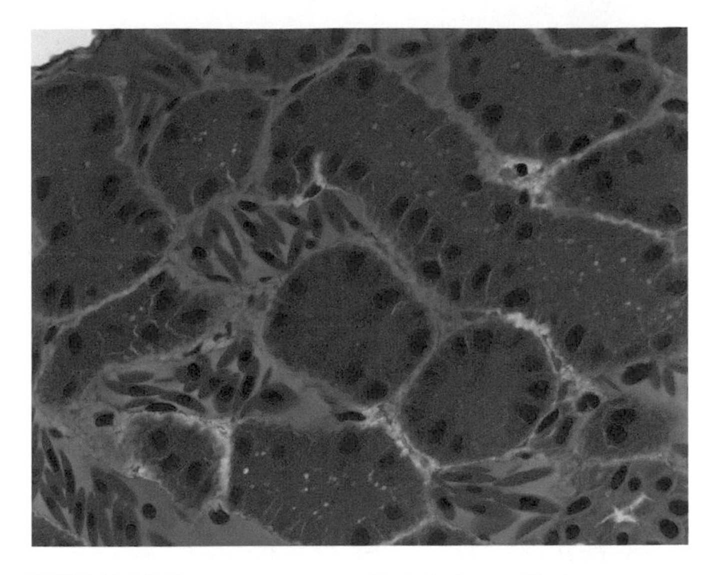

FIGURE M22b DORSAL REGION. Peritubular capillaries permeate the tissue between PROXIMAL CONVOLUTED TUBULES and COLLECTING DUCTS. Epithelial cells of the collecting ducts are columnar, while those of the proximal convoluted tubules are more cuboidal. Visible in the apical cytoplasm are endocytotic vesicles and vacuoles, 63 ×.

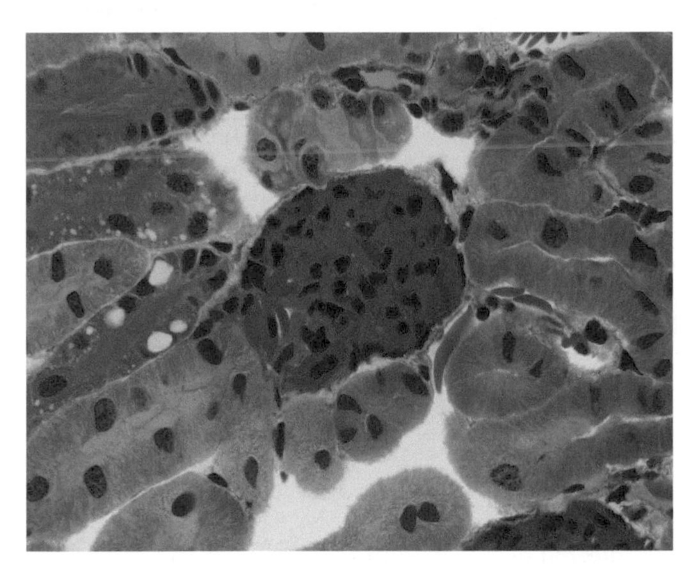

FIGURE M22c MIDDLE ZONE. A RENAL CORPUSCLE occupies the center of this micrograph. It is surrounded by pale-staining distal convoluted tubules and a few darker proximal convoluted tubules. There are a few capillaries, some containing erythrocytes. A CILIATED NECK SEGMENT appears at the top left; its lumen appears packed with cilia. The vertical striations in these dark staining cells are the evidence of the compartmentalization of mitochondria associated with active transport here in these cells, 63 ×.

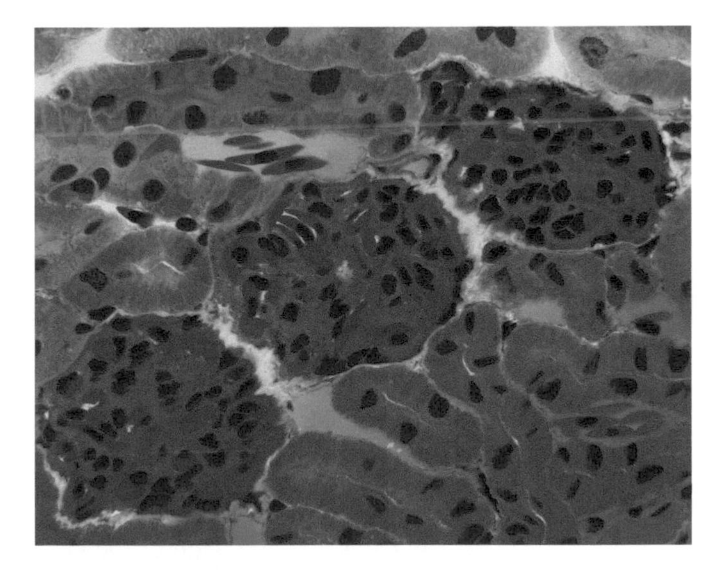

FIGURE M22d MIDDLE ZONE. Three renal corpuscles are arrayed across the center of this micrograph. They are surrounded by sections of distal convoluted tubules. Peritubular capillaries abound between the tubules, 63 ×.

Fluids leave the glomerular capsule at the urinary pole and pass into the PROXIMAL CONVOLUTED TUBULE which coils and twists in the cortical labyrinth (Figs. M25c–M25e). Its simple columnar epithelium of large acidophilic cells displays a ragged brush boarder. The tubules nest straightens out as it passes into the cortical ray and extends toward the medulla.

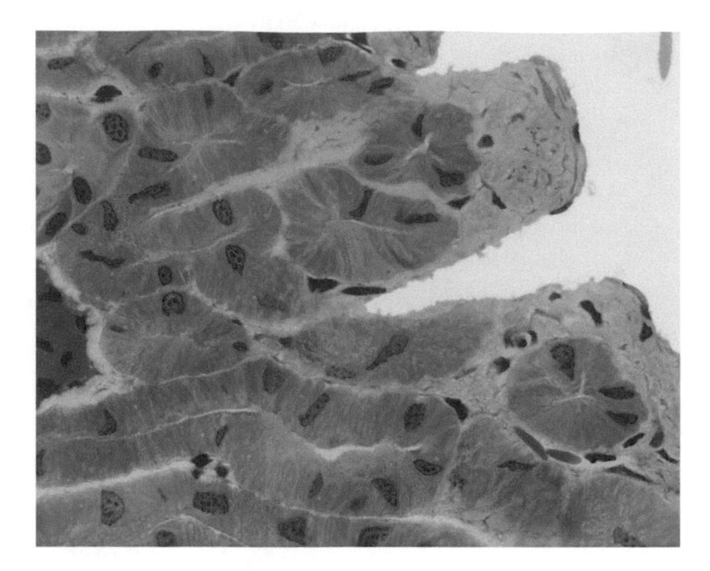

FIGURE M22e VENTRAL ZONE. Various sections of the DISTAL CONVOLUTED TUBULE are seen. The open space at the upper right is the efferent renal vein. Vertical striations in these cells are especially evident, 63 ×.

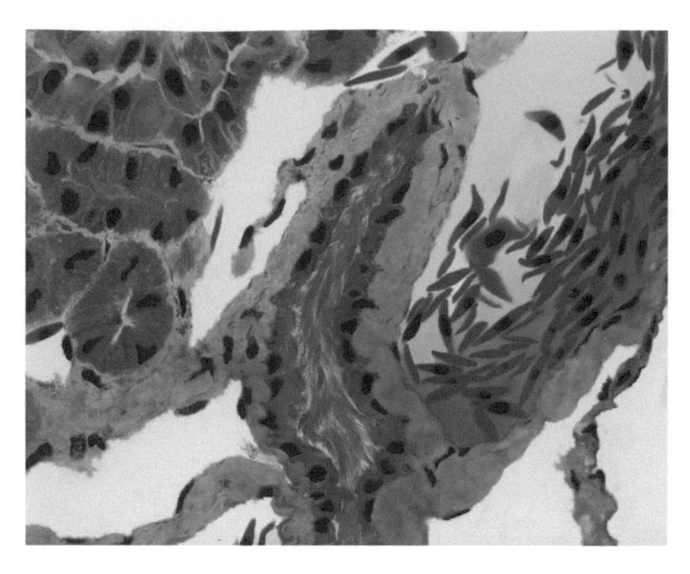

FIGURE M22f A NEPHROSTOME showing abundant cilia can be seen at the center of this micrograph—its opening to the coelom is not shown in this section. Peritubular capillaries and a larger blood vessel contain erythrocytes, 63 ×.

The tubule enters the medulla as the descending limb of the nephric loop (Henle's loop) (Fig. M25f). This straight tubule makes its way to the renal papilla; at some point it reverses its course and heads back toward the cortex as the ascending limb of the nephric loop. The hairpin portions of the loop consist of squamous cells whereas the cortical regions are low cuboidal and one may speak of the thin and thick segments of the nephric loop; the points of transition vary from nephron to nephron.

The tubule rises to the cortical ray and enters the labyrinth as the distal convoluted tubule. Its cells are smaller, less ragged, and less acidophilic than those of the proximal convoluted tubule.

The distal convoluted tubule reenter the cortical ray and pours its contents into a straight collecting tubule. The collecting tubule receives the output of many nephrons as it runs through the cortex. It passes through the medulla and opens into the renal pelvis by the way of a pore on the surface of the renal papilla. The epithelium of the collecting tubule can be distinguished by its clear cytoplasm and distinct cell membranes.

The kidney is wrapped a tough, connective tissue capsule composed of collagenous tissue and occasional elastic fibers (Fig. M25g).

A few slides from the collection of amniote metanephroi were photographed to illustrate several points (Fig. M26a). For example, the kidney of a huge mammal, the horse, illustrates little differences from the kidneys of small mammals (Fig. M27). The cortex of the kidney of a dog was stained to show mitochondria (Fig. M28a); the proximal convoluted tubules stained deeply showing an abundance of mitochondria, while the distal tubules are paler.

As the afferent arteriole enters a glomerulus, some smooth muscle cells in the tunica media appear larger and paler staining than in other parts of the arteriole and that they contain granules (Figs. M28a and M28b). These are the JUXTAGLOMERULAR CELLS. On one side of the arteriole the internal elastic membrane is absent and the juxtaglomerular cells are in direct contact with the tunica intima. On the opposite side, they are close to the MACULA DENSA (Fig. M29), an oval group of slender columnar cells in the wall of the distal convoluted tubule of the same nephron. Together the juxtaglomerular cells and the macula densa constitute the JUXTAGLOMERULAR APPARATUS.

To demonstrate the blood circulation in the mammalian kidney, a colored injection mass can be injected via the renal artery. This was done for the specimen shown in Figs. M30a and M30b followed by fixation, dehyrdration, and sectioning to produce these slides. Blood from the renal artery fans out in the medulla in the INTERLOBAR ARTERIES and reaches the ARCIFORM ARTERIES along the border between the cortex and medulla and give off branches to both. In the cortex the INTERLOBULAR ARTERIES pass to the afferent arterioles of the glomeruli, and in the medulla the arteries break up into capillaries that pass among the tubules.

Blood leaving the glomeruli by the efferent arterioles passes into capillaries among the proximal and distal convoluted tubules of the cortex and is collected at the periphery of the kidney in the STELLATE VEINS, which unite to form the INTERLOBULAR VEINS through which it returns to the ARCIFORM VEINS. Blood from the medulla also enters the arciform veins. Blood from the arciform veins passes to the renal vein through INTERLOBULAR VEINS in the medulla.

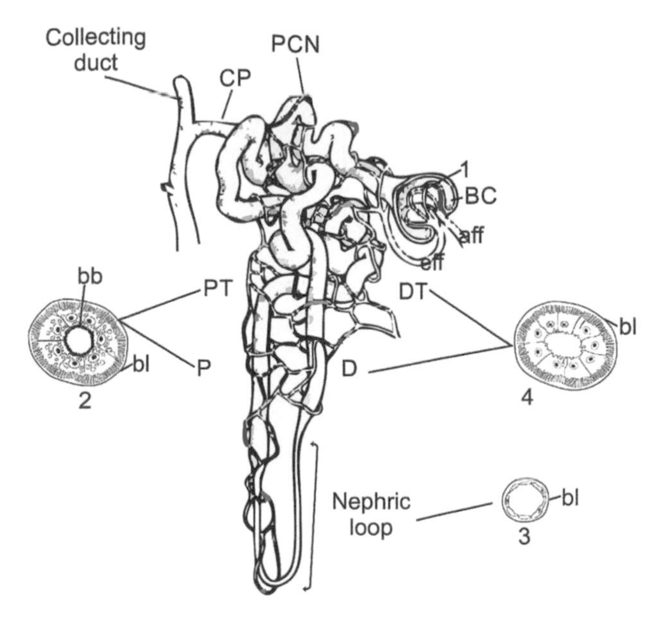

FIGURE M23 Diagram of a nephron from a mammalian kidney. Abundant fluids are lost into the glomerular capsule from the knot of blood vessels constituting the glomerulus (*1*). These fluids are filtered through the glomerular wall into the glomerular capsule (*BC*) and are passed down the proximal tubule (*2*), both straight (*PT*) and convoluted (*P*) to the hairpin-shaped nephric loop and the distal tubule, again both straight (*D*) and convoluted (*DT*). During their passage down the various tubules, valuable materials in this fluid are pumped back into the blood while waste materials continue on and are excreted by the collecting duct. Diagrams of the various segments of the tubular nephron are shown: the proximal convoluted tubule (*2*) with its brush border (*BB*) and basal lamina (*BL*); the proximal straight tubule (*P*), the thin segment (*3*), the distal straight tubule, (*D*) and the distal convoluted tubule (*DT*) (*4*). All of the tubules are lined by a basal lamina (*BL*). A short connecting piece (CP) links the distal tubule with the straight collecting duct. Aff, afferent arteriole; Eff, efferent arteriole; PCN, peritubular capillary network. *Authors.*

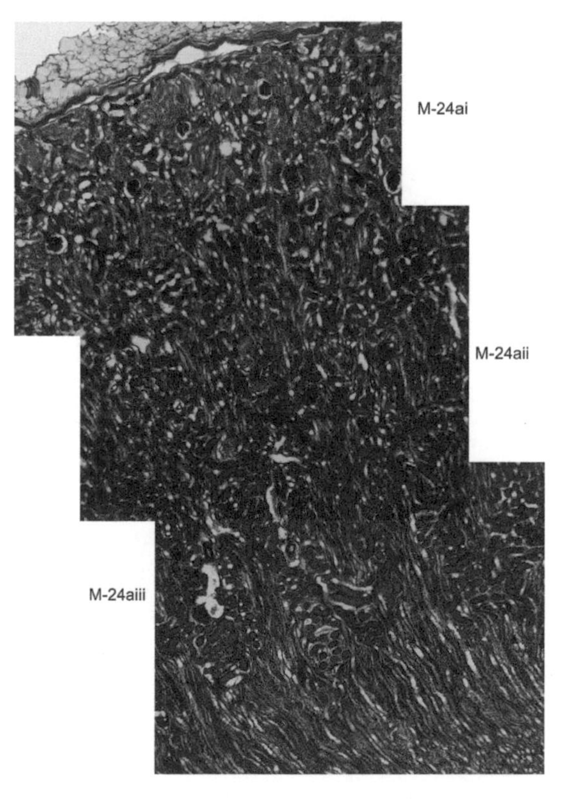

FIGURES M24ai, M24aii, and M24aiii This micrograph is stitched together from a series of images through a longitudinal section from the metanephros of a cat. It comprises the CORTEX of the kidney from the capsule (M24ai) to the border with the medulla (M24aiii). The NEPHRON begins in the cortex where the GLOMERULAR CAPSULE is surrounded by a labyrinth of tubules which includes the PROXIMAL and DISTAL CONVOLUTED TUBULES. The proximal tubules are plumper and more eosinophilic than the distal tubules. The tubules straighten as they approach the medulla and enter the medulla as the DESCENDING LIMB of the NEPHRIC LOOP. This straight tubule makes its way toward the RENAL PAPILLA. In the medulla the tissue appears to radiate from the renal papilla. This radial arrangement continues in some parts of the cortex as CORTICAL RAYS; in other parts the tubules appear tangled and form the CORTICAL LABYRINTH, 10 ×.

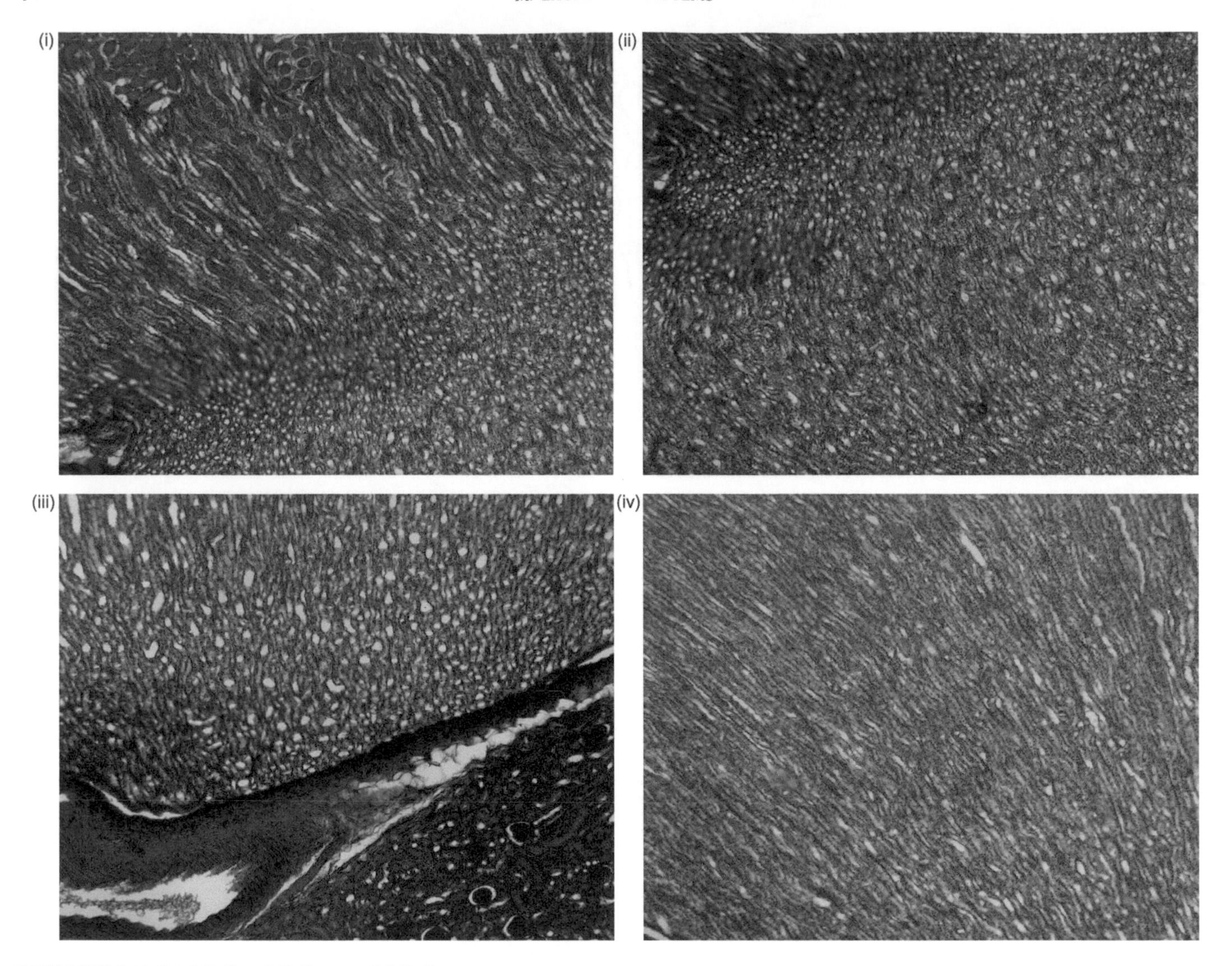

FIGURES M24bi, M24bii, M24biii, and M24biv In these sections the tubules of the nephric loop have assumed a hairpin shape, making their way straight to the renal papilla as the PROXIMAL LIMB then turning back toward the cortex as the DISTAL LIMB where they become the DISTAL CONVOLUTED TUBULE in the cortex. In the cortex, they join with other nephrons to empty their contents into a COLLECTING TUBULE that parallels the nephric loop and heads for the papilla where it opens into the renal; PELVIS by way of a pore (M24biii).

ULTRASTRUCTURAL ASPECTS OF URINE PRODUCTION

Renal Corpuscle

Electron micrographs of the visceral epithelium of the glomerular capsule reveal that it is made up of elaborate cells, the PODOCYTES, whose processes cover the capillary loops of the renal corpuscle (Figs. M26b, M26c and M26d). The extreme complexity of these podocyte cells can best be appreciated in scanning electron micrographs where they are seen to resemble a basket starfish. The cell body or PERIKARYON has several branching arms fringed with foot like processes, the PEDICELS; these interdigitate with pedicels from other podocytes to cover the glomerular capillaries (Fig. M26e). A single podocyte may extend processes to embrace several adjacent capillaries.

The relationship of podocytes to the glomerular endothelium can be seen in transmission electron micrographs (Fig. M26g). The endothelium is thin and pierced by many large fenestrations that are probably not enclosed by a diaphragm. This is particularly well demonstrated using the freeze fracture technique in electron microscopy (Fig. M26f). It is sheathed by a well defined, continuous basement lamina on which are implanted the pedicels of the podocytes. FILTRATION SLITS between adjacent pedicels open into the capsular space. A thin SLIT MEMBRANE joins adjacent pedicels and separates the basement lamina from the capsular space. Occasional stellate MESANGIAL CELLS, reminiscent of pericytes, may be seen between the endothelial cells and the basement lamina, especially where

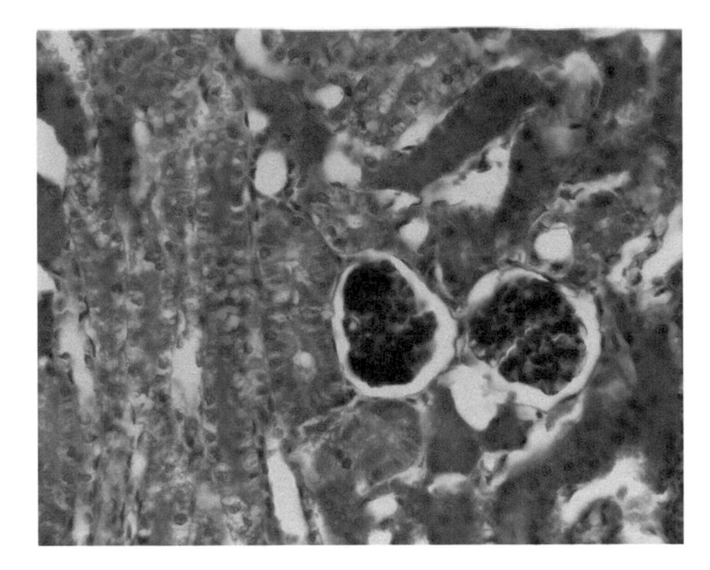

FIGURE M24c The pale DISTAL CONVOLUTED TUBULES can be distinguished from the more eosinophilic PROXIMAL CONVOLUTED TUBULES in this section of the CORTEX. Two RENAL CORPUSCLES are located at the center, 40 ×.

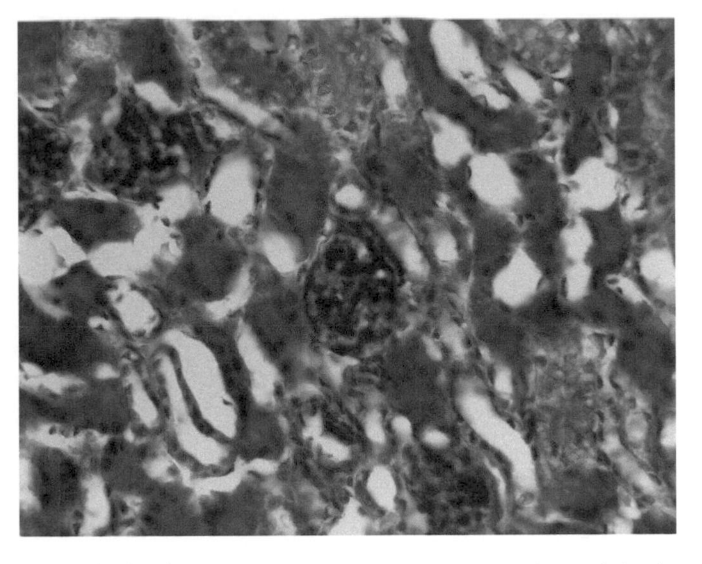

FIGURE M24d There is a thickening in the epithelium of the distal convoluted tubules as it presses against the renal corpuscle at the center. This is the MACULA DENSA, a part of the JUXTAGLOMERULAR APPARATUS, 40 ×.

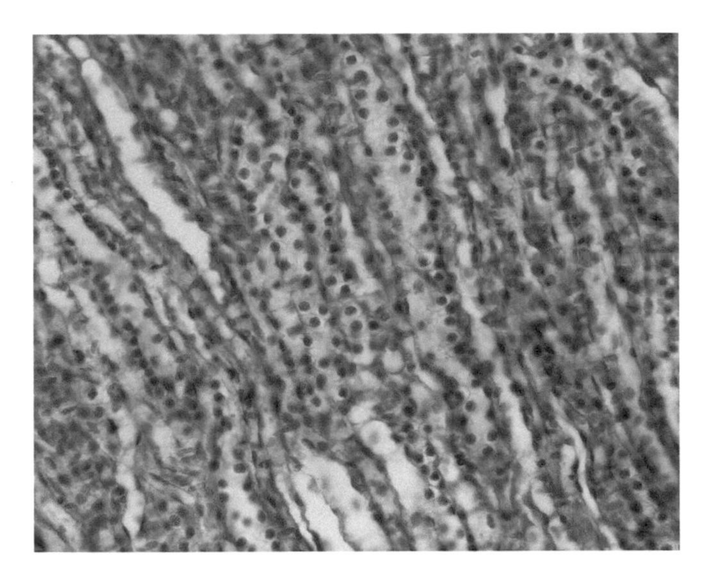

FIGURE M24e The largest tubules in this section of the MEDULLA are COLLECTING TUBULES. CAPILLARIES course between the straight section of the nephric loop and the collecting tubules, 40 ×.

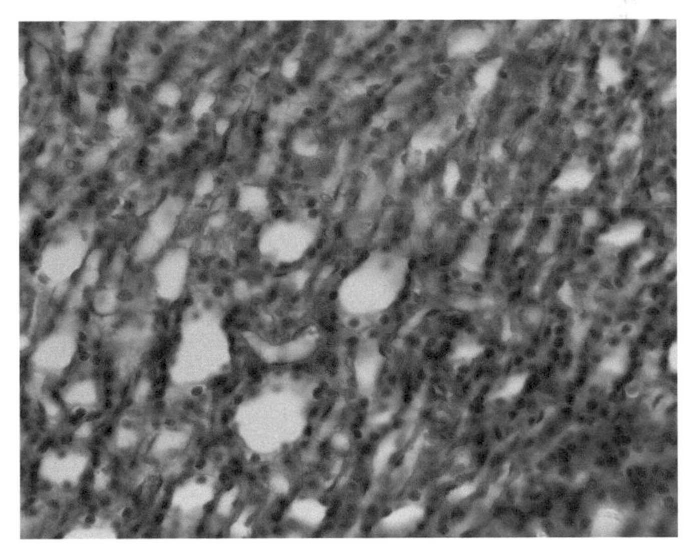

FIGURE M24f PORES open from several collecting tubules in this section of the RENAL PAPILLA, 40 ×.

the capillaries branch. An almost identical appearance of the ultrafilter in the kidney is seen in lamprey ammocoete (Fig. M9k).

The Tubular Nephron

Light and electron micrographs of sections of various regions of the nephric tubules of several species typically show a set of characteristics associated with the active transport of fluids and salts:

- Apical microvilli with pits between their bases and vesicles just below the surface (in light micrographs—a brush boarder);
- Tight junctions between the apical borders of adjacent cells;
- Complex interdigitating processes between adjacent cells;

FIGURES M25(a–g) These micrographs are of a semithin section of the cortex of the kidney of a monkey.

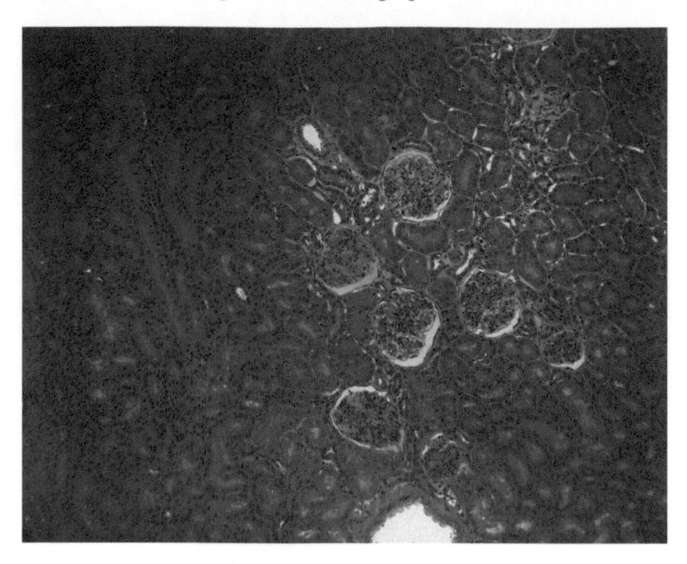

FIGURE M25a Proximal and distal convoluted tubules and a cluster of renal corpuscles are at the center. There is a section of an artery at the center of the bottom edge, and a smaller arteriole at the upper middle.

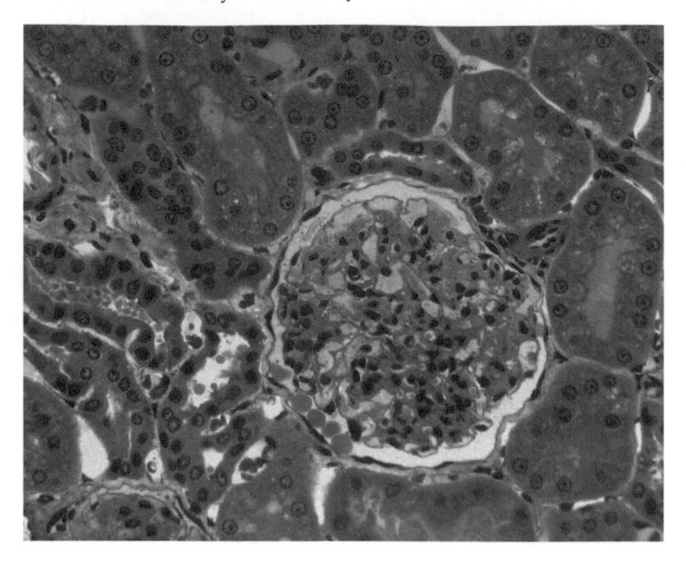

FIGURE M25b This RENAL CORPUSCLE at the center consists of the GLOMERULUS, a knot of capillary loops surrounded by the GLOMERULAR CAPSULE. The outer PARIETAL LAYER of the capsule is a simple squamous epithelium; the inner VISCERAL LAYER adheres closely to the capillaries of the glomerulus and is difficult to see. The renal corpuscle is surrounded by sections of PROXIMAL and DISTAL CONVOLUTED TUBULES. Cells of the proximal convoluted tubule are more eosinophilic than those of the distal convoluted tubule, they are plumper, and their ragged brush border is more distinct. CAPILLARIES may be detected coursing between the tubules. These capillaries receive blood from the efferent arteriole of the glomerulus and, as such, constitute a RETE MIRABILE, 40 ×.

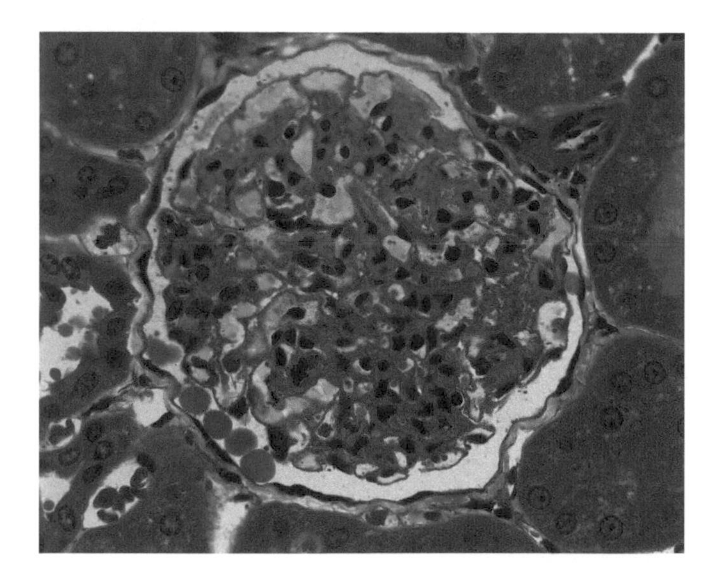

FIGURE M25c Some of the features noted above may be seen more clearly at this higher power. Note the appearance of vertical basal striations in the cells of the proximal tubules. These represent infoldings of the basal plasma membranes which enclose mitochondria, much as eggs in an egg carton. This striated appearance is characteristic of a TRANSPORTING EPITHELIUM, 63 ×.

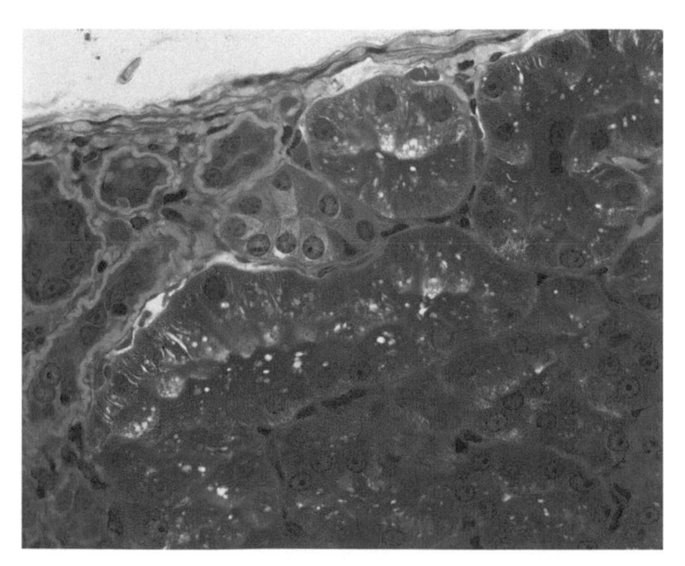

FIGURE M25d The basal striations show up especially well in this oblique section of a proximal convoluted tubule just under the capsule of the kidney, 63 ×.

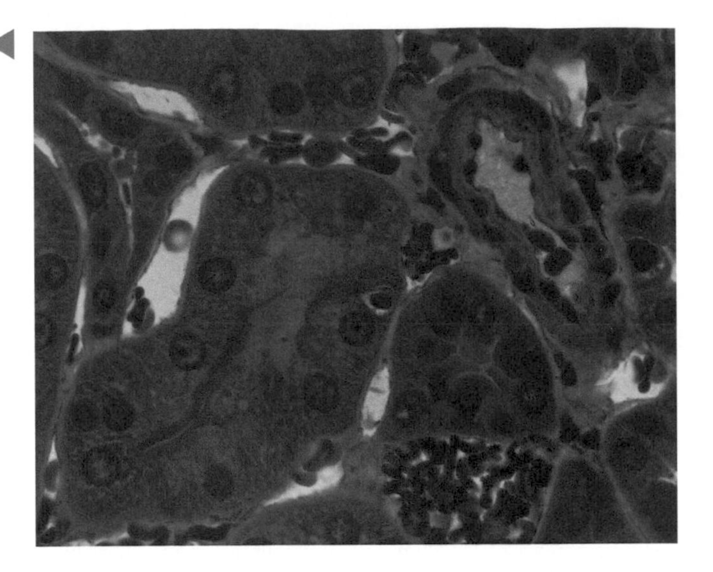

FIGURE M25e Blood vessels permeate the connective tissue between the tubules of the cortex—an arteriole lies at the upper right, a venule at the bottom to the right of center, and capillaries are abundant. The BRUSH BORDER and BASAL STRIATIONS are clear in the epithelial cells of a proximal convoluted tubule at the lower left. These features in the smaller distal convoluted tubules are less distinct, 100×.

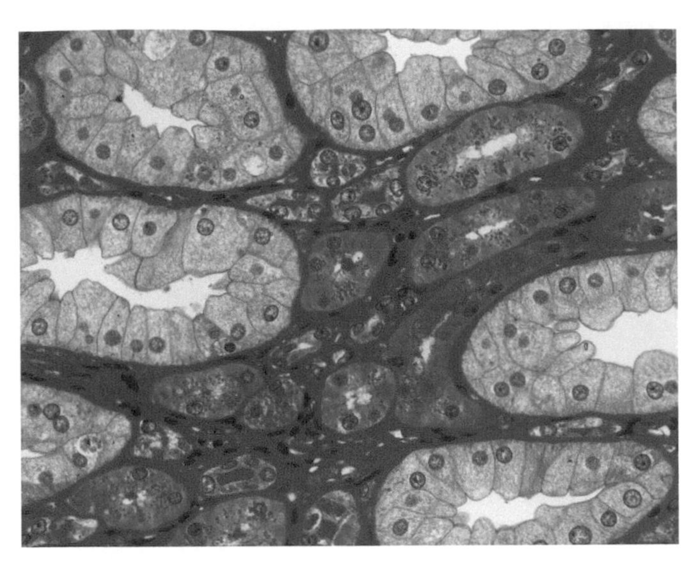

FIGURE M25f The MEDULLA consists largely of the NEPHRIC LOOP, its associated BLOOD VESSELS, and COLLECTING TUBULES. The large, pale tubules are COLLECTING DUCTS. The nephric loop is more eosinophilic and its tubules contain eosinophilic droplets; the medium-sized tubules are THICK LIMBS and the smallest tubules with a simple squamous epithelium are the THIN LIMBS. Capillaries permeate the connective tissue between the tubules, 40×.

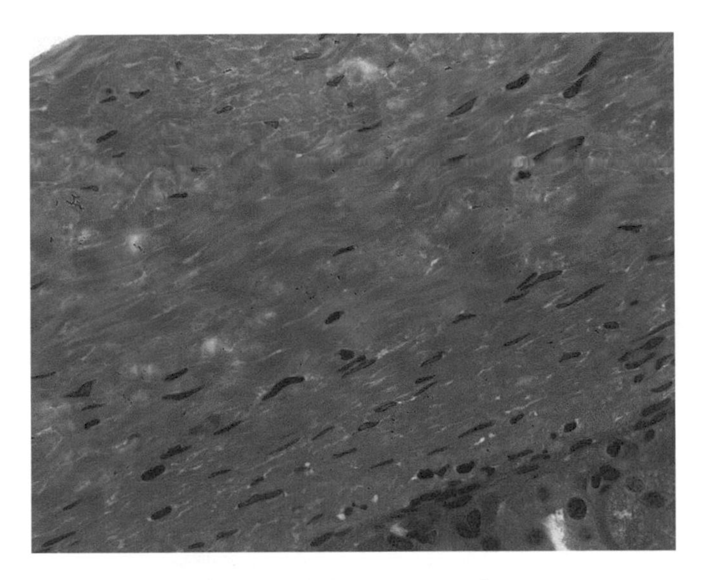

FIGURE M25g The tough connective tissue capsule of the kidney consists of DENSE COLLAGENOUS TISSUE containing occasional ELASTIC FIBERS, 63×.

- Infoldings of the basal plasmalemma (seen as striations in light micrographs) to form vertical compartments enclosing individual elongate mitochondria, much as eggs are contained within their sections of an egg carton.

  For examples see the following figures: Figs. M9f, M22e, M22f, M26h, M45d, and M46g.
  The distribution of organelles within these cells is directly related to the function of these cells.

COMPARATIVE HISTOLOGY OF THE METANEPHROS

In the rabbit, cat (Fig. M5e), rat, and many other mammals, there is a single renal papilla but in some species there may be several papillae. The human kidney (Fig. M5f), for example, has 8–18 papillae, each associated

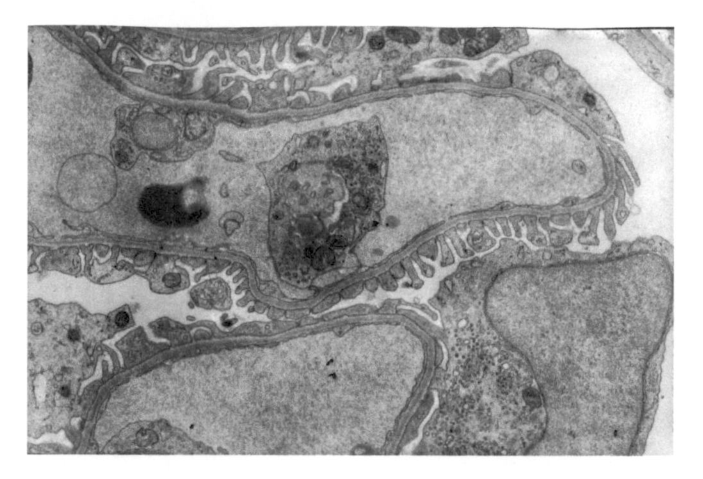

FIGURE M26a Transmission electron micrograph of a section from the kidney of a mouse showing glomerular capillaries. Podocyte feet press on the outside of the capillary, resisting the pressure of the blood within. An ultrafilter is formed by the fenestrated endothelial wall of the capillary, the double basement membrane derived from the fusion of basement membranes of the glomerular epithelium and capsular epithelium, and the slit diaphragms between the feet of supportive pedicels from podocytes. Molecules with a molecular weight below 45,000 are permitted entry into the urinary space for passage down the nephron. Part of a cell body of a podocyte is shown in the upper left corner, 29,500 ×.

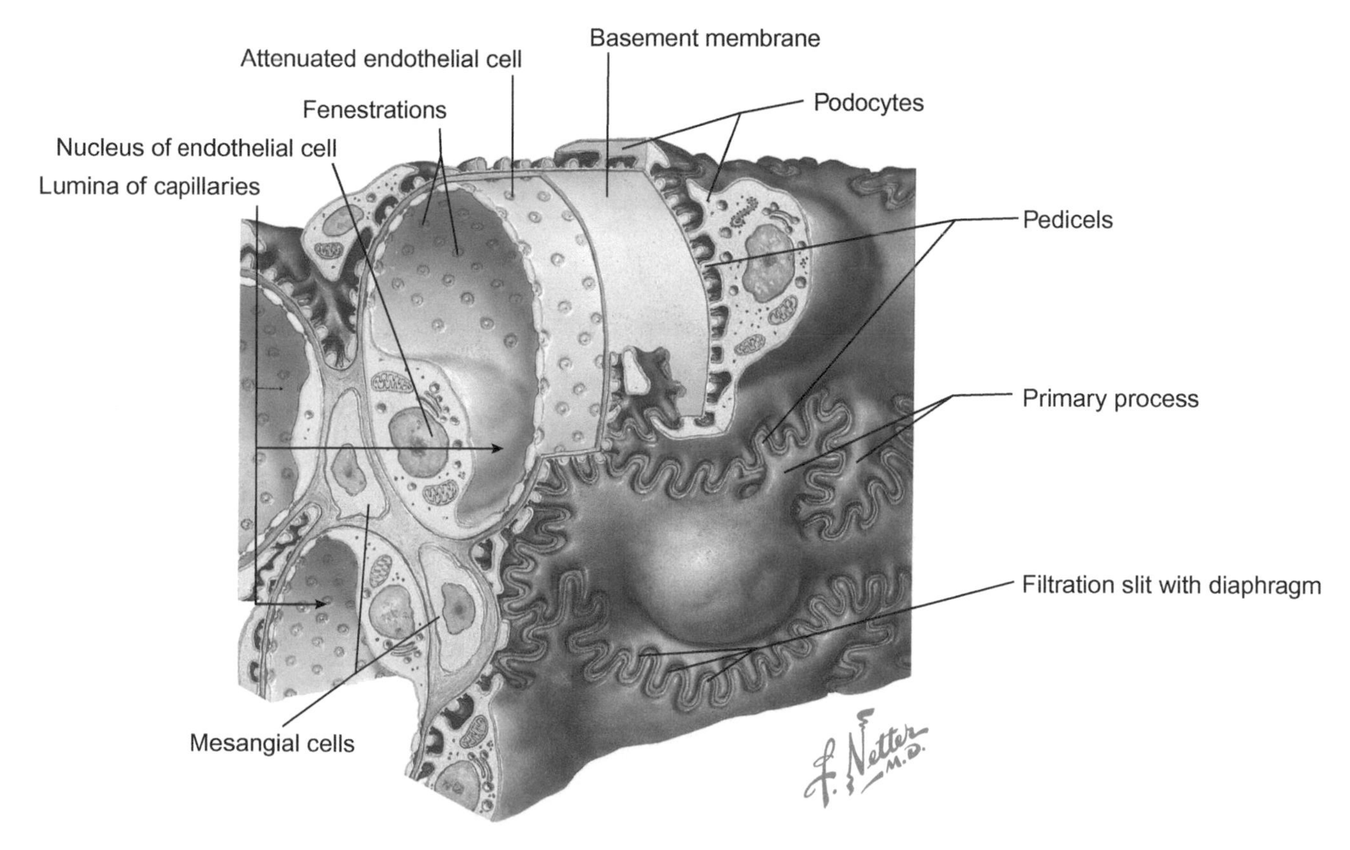

FIGURE M26b Diagram illustrates part of a capillary loop in a mammalian glomerulus with its covering of podocytes and fenestrated endothelium. A portion of a podocyte and part of the capillary have been cut away to show the components of the filtration barrier. (Netter images #15054 fine structure of renal corpuscle). *Netter Images #15054 © 2005–2018 Elsevier.*

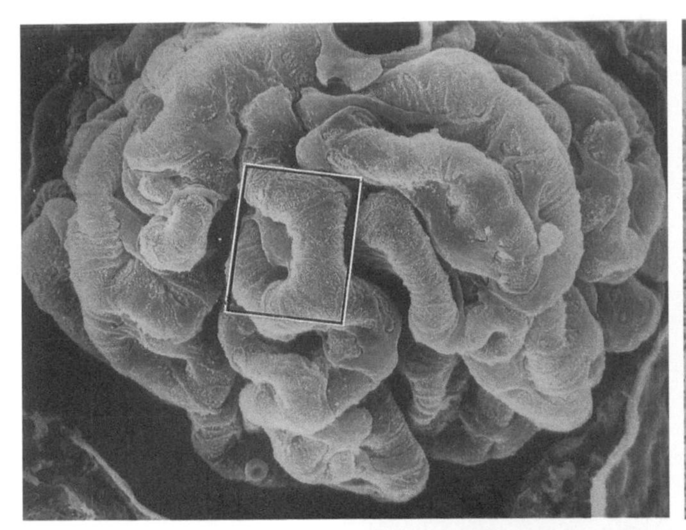

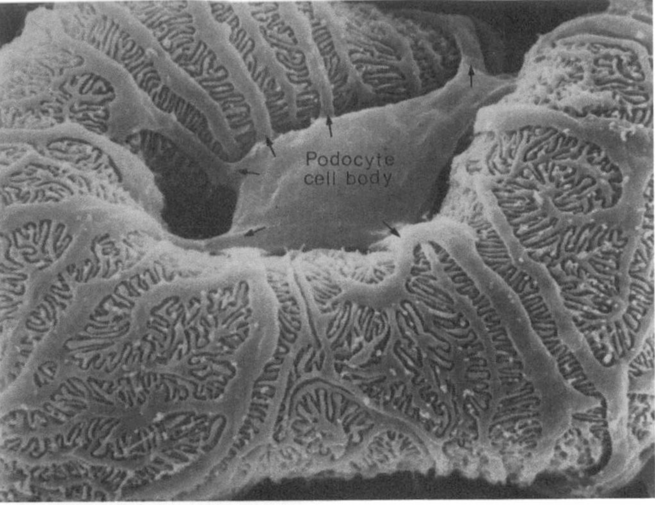

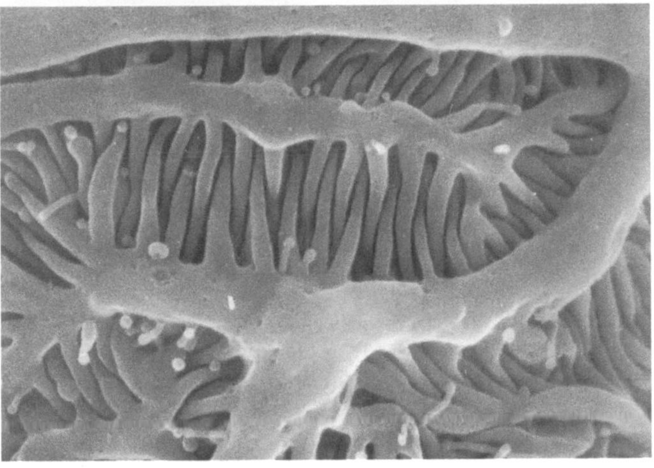

FIGURES M26c, M26d, and M26e Scanning electron micrographs of a capillary loop in a renal glomerulus. The area in the rectangle in Fig. M26c is shown at a higher power in Fig. M26d. The higher power in Fig. M26e shows the primary [processes (pedicels) of a podocyte alternating in their support of the capillary].

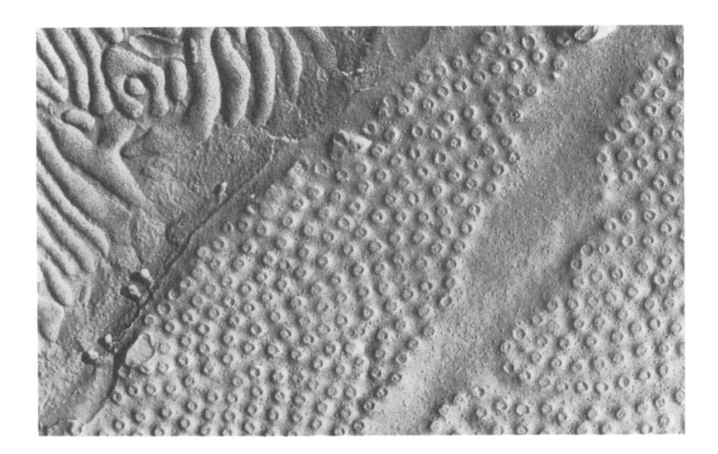

FIGURE M26f Transmission electron micrograph of a freeze-fractured preparation of the rat kidney glomerulus. At the lower right the membrane of the capillary endothelium has been cleaved showing the uniform size and distribution of the endothelial pores. At the upper left the membranes of a number foot processes have been cleaved.

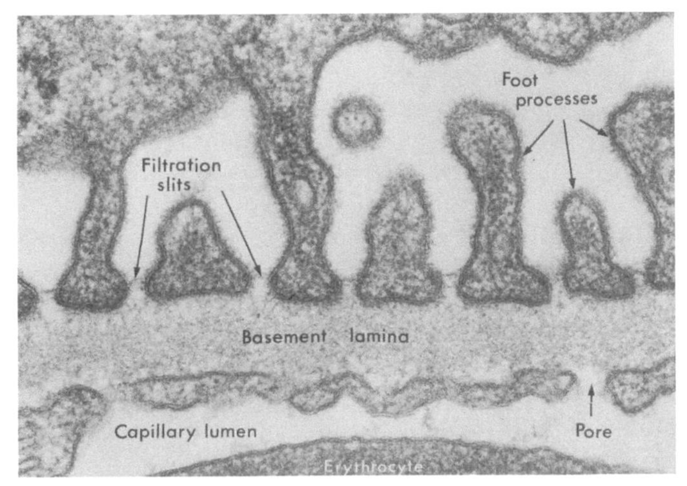

FIGURE M26g Transmission electron micrograph of a section through the wall of a glomerular capillary showing the ultrafilter. Foot processes of a podocyte on the outer surface of the basal lamina and the fenestrated endothelium form its inner aspect.

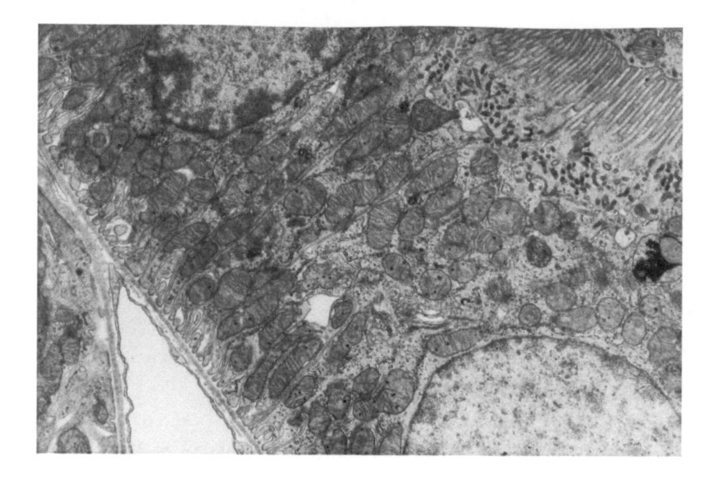

FIGURE M26h No matter where they are located or from what species cells engaged in active transport typically display a set of similar characteristics—apical microvilli; apical tight junctions; complex interdigitations between neighboring cells; basal infoldings enclosing individual mitochondria. All of these characteristics are seen here in this electron micrograph of mouse proximal convoluted tubules cells, 21,000 ×.

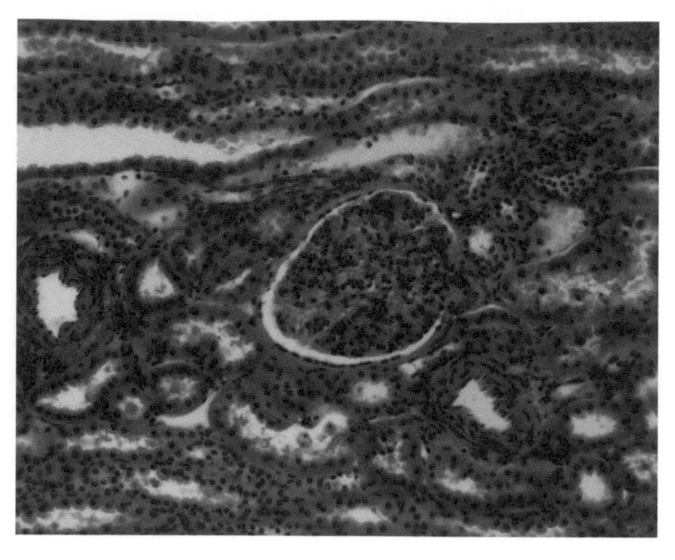

FIGURE M27 Comparing this section of the cortex of the kidney of a huge mammal, the horse, with those of the smaller mammals—there is little fundamental difference evident. Note the accumulation of JUXTAGLOMERULAR CELLS at the arterial pole of the renal capsule, 20 ×.

with its own mass of medullary tissue, the RENAL PYRAMID, separated from other like masses by cortical tissue in the form of the RENAL COLUMNS (of Bertin). At this point it should be noted that the size of an animal does not seem to have an impact on the structure, organization, or histology of the kidney. Even in animals as large as a horse there is little fundamental difference (Fig. M27).

Although the metanephros and opisthonephros are similar in many ways, note that the products of the nephrons in the opisthonephros are poured directly into the archinephric duct while those of the metanephros are gathered up into the collecting tubules and poured into the expanded end of the ureter, the renal pelvis.

The histology of the kidney of reptiles is not remarkably different than that of birds or mammals (Fig. M31a). Reptilian renal corpuscles are similar to mammalian and are seen nestled amongst the proximal and distal tubules with an abundant supply of capillaries coursing through the interstitial tissue (Fig. M31b). The major difference between reptilian and mammalian kidneys is the lack of a nephric loop. Instead a short INTERMEDIATE SEGMENT is interposed between the proximal and distal convoluted tubules.

Even in the "living fossil" tuatara (*Sphendon punctatus*) the typical reptilian kidney organization and histology is seen. Unremarkable renal corpuscles are found surrounded by proximal and distal tubules embedded in vascularized interstitial tissue's tissue (Figs. M32a–M32c).

The most conspicuous feature of the kidney of male snakes and lizards is the presence of large, acidophilic tubules of the SEXUAL SEGMENT of the nephron (Figs. M33a–M33c and M34a–M34c). These tubules have a simple columnar epithelium and a small lumen and occur at the end of the nephron before it enters the collecting tubule. They are highly developed immediately before the breeding season and are thought to provide nutriments for the spermatozoa during their long periods of storage. In the female the sexual segment secretes mucus and is little larger than the proximal convoluted tubule.

In the kidney of a bird, we see the familiar pattern of renal corpuscles surrounded by closely packed proximal and distal convoluted tubules with numerous capillaries coursing between. Occasional sites of hemopoietic activity occur in the loose vascular connective tissue between the tubules (Figs. M35a and M35b).

An INTERMEDIATE SEGMENT between the proximal convoluted tubule and the distal convoluted tubule, similar to that of amphibians, is found in the kidney of reptiles and some birds. The NEPHRIC LOOP (Henle's loop) appears in other birds and all mammals and is homologous with the intermediate segment. Relatively less fluid leaves the distal convoluted tubule in the forms with a nephric loop. The reptiles compensate by fluid absorption in the cloacal bladder.

JUXTAGLOMERULAR COMPLEX AND THE CONTROL OF BLOOD PRESSURE

In its wanderings throughout the cortex of the mammalian kidney, a part of the distal convoluted tubule, the MACULA DENSA, comes to lie at the vascular pole of the glomerulus between the afferent and efferent arterioles

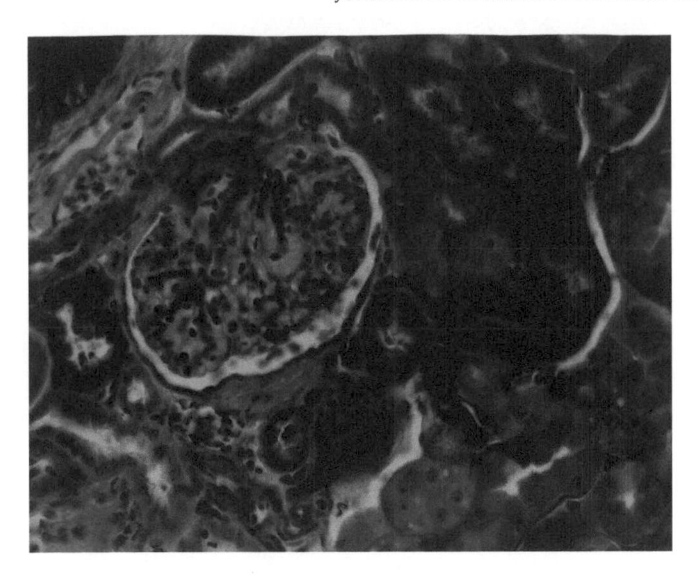

FIGURE M28a This section of the cortex of the kidney of a dog has been stained to show MITOCHONDRIA. The PROXIMAL CONVOLUTED TUBULES are stained deeply, showing an abundance of mitochondria whereas the DISTAL TUBULES are paler, 40×.

(Figs. M28a–M28c). Secretory JUXTAGLOMERULAR CELLS develop in the wall of the afferent glomerular capillaries (and sometimes the efferent arteriole) become secretory, reflecting their position by reversing their polarity (Fig. M27). Also in the region are the MESANGIAL CELLS phagocytosing débris within the walls of the glomerular capillaries. These three components make up the JUXTAGLOMERULAR COMPLEX; their secretions have a profound effect on blood pressure. Similar juxtaglomerular complexes have been noted in the toad, *Bufo bufo* (Figs. M28d–M28g).

It has been suggested that changes in the volume of fluids passing through the distal convoluted tubule distend or relaxe it thereby changing its relationships with other components of the juxtaglomerular complex, acting as a feedback mechanism that controls blood pressure (Fig. M28h). The presence of an apparent juxtaglomerular complex in the bullfrog indicates that this mechanism is widespread in vertebrates.

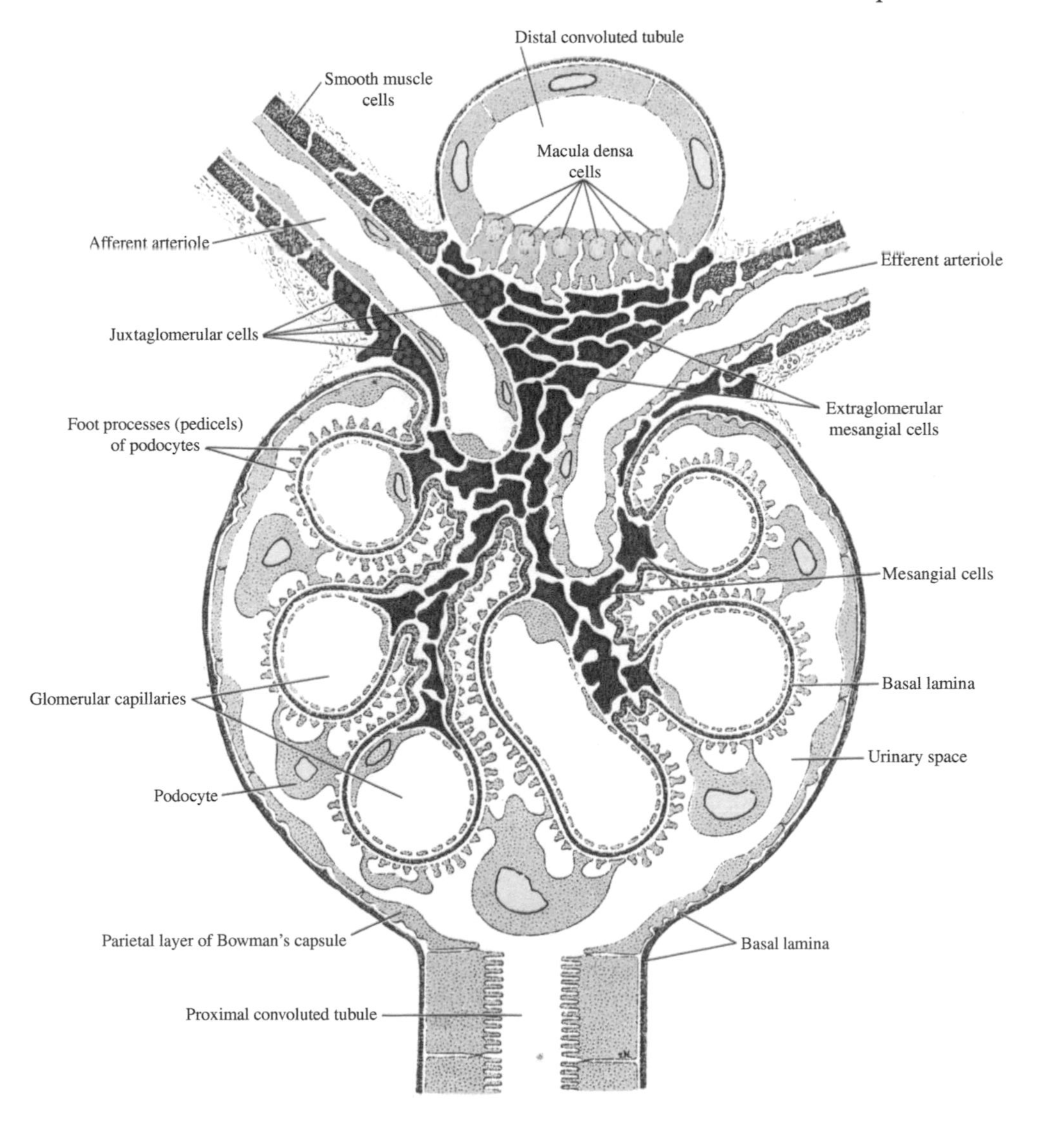

FIGURE M28b Diagram of a mammalian renal corpuscle showing the JUXTAGLOMERULAR COMPLEX. A loop of the distal convoluted tubule nestles between the afferent and efferent arterioles. Nearby a mass of mesangial cells, enclosed within the basal laminas of the glomerular capillaries and the afferent arteriole. A mass of cells forms in the wall of the distal convoluted tubule and is the MACULA DENSA. Invading the spaces between the glomerular loops are cells resembling pericytes, the MESANGIAL CELLS. In this region the smooth muscle cells in the wall of the efferent arteriole (and sometimes the afferent arteriole), become modified and contain secretory granules.

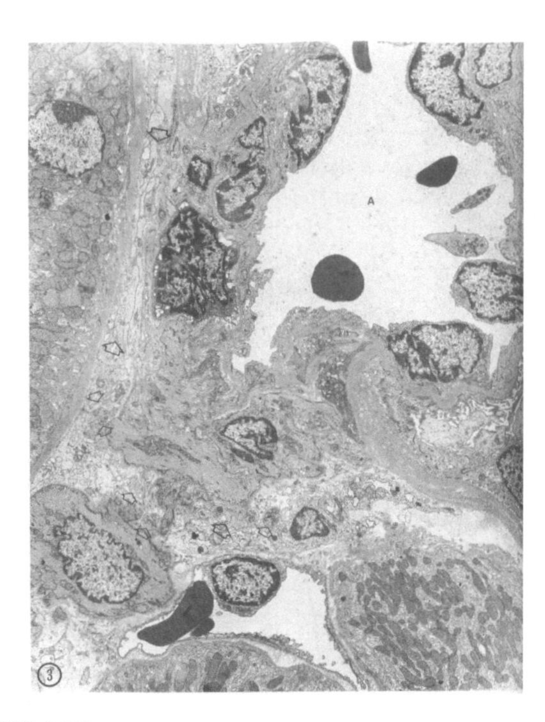

FIGURE M28c Transmission electron micrograph of a section at the hilus of the glomerulus of a kidney of a rat. Two juxtaglomerular cells are seen directly below an erythrocyte in the afferent arteriole (A). Large numbers of nerves (arrows) are associated with the afferent arteriole, 5000 ×. *From Barajas L, Wang P et al. 1976 The renal sympathetic system and juxtaglomerular cells ... Lab Investig 35(6):574, with permission from springer.*

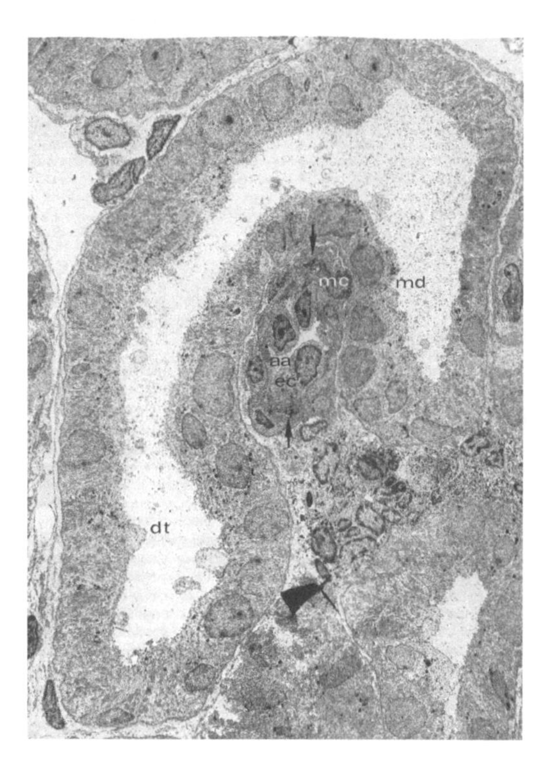

FIGURE M28d Survey electron micrograph of the contact region between an afferent arteriole (aa) and a distal tubule (dt) in the kidney of a bullfrog (*Bufo bufo*). The distal tubule partly encloses the afferent arteriole. In the wall of the tubule an accumulation of the nuclei, the macula densa (md) is present. Endothelial cells (ec) protrude into the lumen and media cells containing granules are present (arrows), 1850 ×.

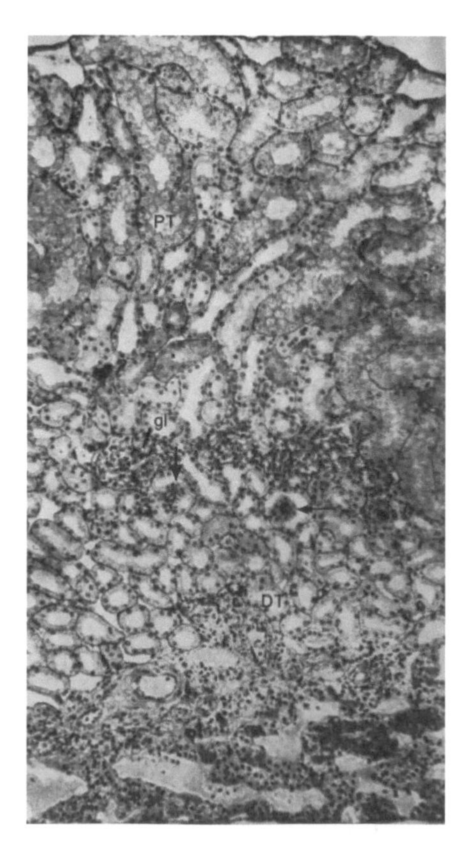

FIGURE M28e Sagittal section through the kidney of the bullfrog (*Bufo bufo*). The adrenal gland is at the bottom on the ventral surface of the kidney. Parts of the distal tubules (dt) lying in close contact with the afferent arteriole are indicated by arrows. *gl*, glomerulus; *pt*, proximal tubule, 160 ×.

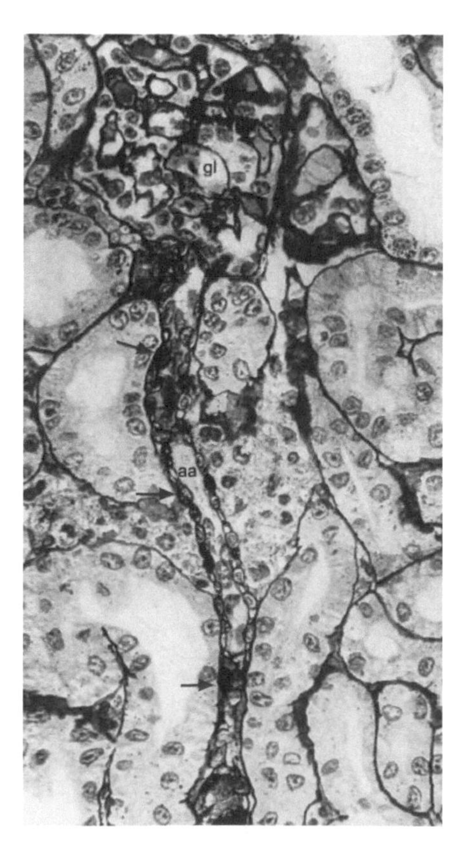

FIGURE M28f A glomerulus (gl) in the kidney of a bullfrog (*Bufo bufo*). Juxtaglomerular granules (arrows) are present in the wall of the afferent arteriole (aa). Silver impregnation, 630 ×.

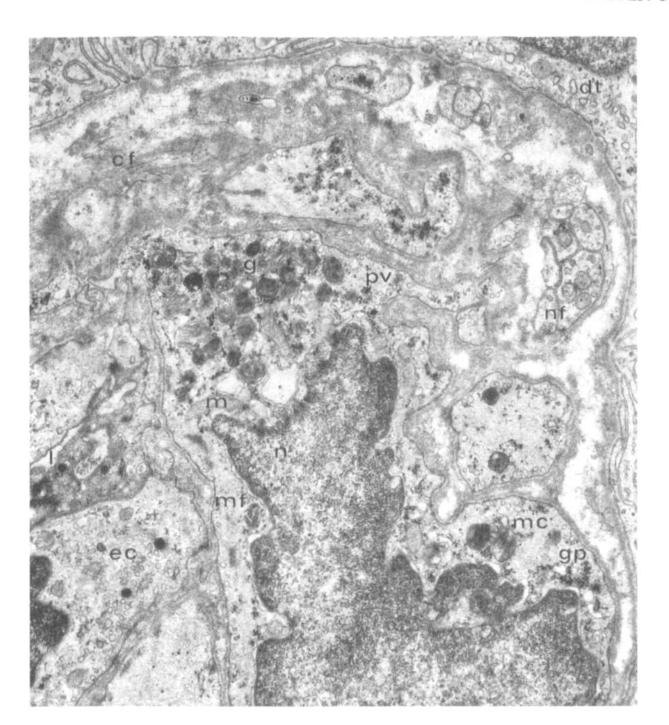

FIGURE M28g Detail from Fig. M28d. A media cell (mc) contains granules (g) with lamellar structures. The nucleus is lobed and, on the adventitial side of the cell, are many pinocytotic vesicles (pv). The media cell contains mitochondria (m), glycogen particles (gp), and myofilaments (mf). The adventitia contains nonmyelinated nerve fibers (nf), 15,000 ×. *aa*, afferent arteriole; *cf*, collagen fibers; *dt*, distal tubule; *ec*, endothelial cell; *l*, vascular lumen, 15,000 ×.

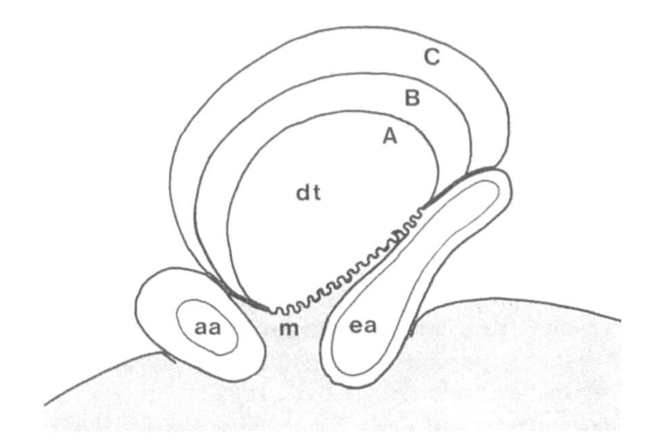

FIGURE M28h A simplified schematic representation of a proposed functional model of the juxtaglomerular complex. It suggests a possible way in which changes in volume of the distal tubule (dt) bring about changes in the extent of contact with the afferent arteriole (aa), the mesangial region (m), and the efferent arteriole in the hilar region (ea). Permanent contacts are represented by wavy lines, reversible contacts by heavy lines. As the distal tubule expands to lines B and C, the area of reversible contact with the vascular component increases.

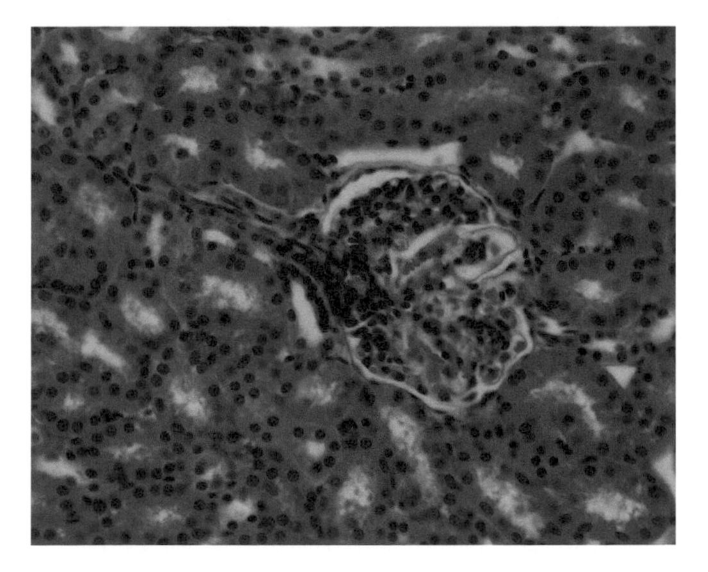

FIGURE M29 A group of specialized cells—the MACULA DENSA—in the epithelium of the distal convoluted tubules lies close to the afferent and efferent arterioles of the glomerulus. These cells are seen here immediately to the left of the glomerulus in this section of human kidney. Together with a clump of mesangial cells—which appear as a darker mass to the right of the macula densa—they constitute the JUXTAGLOMERULAR COMPLEX. Some mesangial cells occur between the basal lamina of glomerular capillaries. Others are situated outside the corpuscle, along the vascular pole where they are designated as LACIS CELLS, 40 ×.

The basal laminae of the tubules and glomerular blood vessels are rich in carbohydrates anchoring the cells to the extensive reticular framework of the kidney. A polysaccharide-rich glycocalyx is often seen on the brush boarders of epithelial cells of the tubules. These features are readily observable using a periodic acid-Schiff method (Fig. M36a). The kidney's extensive, yet delicate, framework of reticular fibers can be seen in great detail using silver impregnation methods. The reticular framework maintains a precise relationship between various components in the kidney (Figs. M36b–M36d). Components of the reticular framework are not well visualized using a more standard staining techniques such as hemotoxylin and eosin (Figs M36e–M36g).

URETER OF AMNIOTES

In amniotes (reptiles, birds, and mammals) the ureter functions to drain the waste products from the kidneys, assuming the function of the archinephric duct in anamniotes. In amniotes ureters arise from an outgrowth of the distal portion of the mesonephric ducts. The ureter is lined with a TUNICA MUCOSA of transitional epithelium resting on a LAMINA PROPRIA MUCOSAE of loose connective tissue (Figs. M37 and M38). There is no lamina muscularis mucosae. The TELA SUBMUCOSA of loose areolar tissue blends with the lamina propria mucosae on one side and with the intermuscular connective tissue of the TUNICA

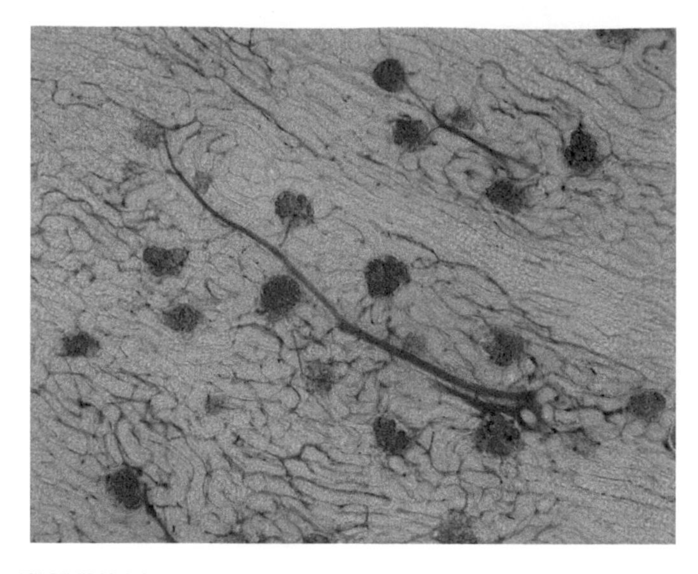

FIGURE M30a The paths of interlobular arteries in the cortex of this mammalian kidney may be seen in this injected specimen. They pass to the afferent arterioles of the glomeruli. Efferent arterioles pass from the glomeruli to capillaries that penetrate the connective tissue between the tubules, thereby forming a RETE MIRABILE, 10 ×.

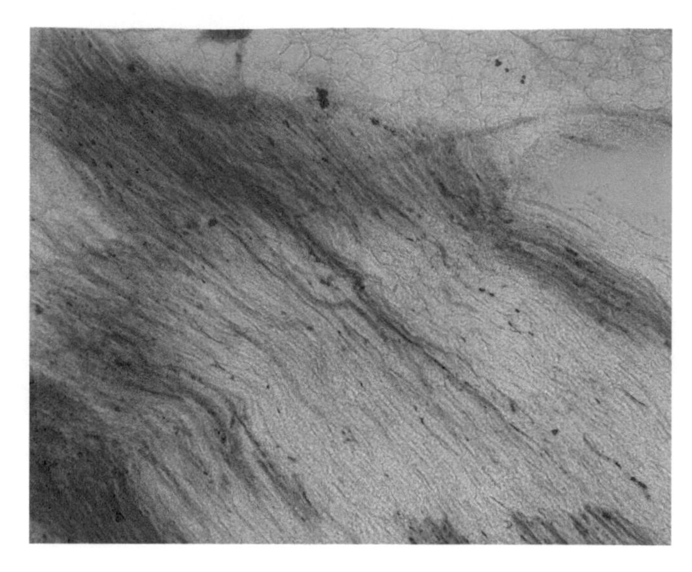

FIGURE M30b Capillaries in the medulla of a mammalian kidney penetrate between the tubules of the nephric loop and the collecting tubules, 10 ×.

FIGURE M31 The caiman is a small reptile of the family Alligoratidae.

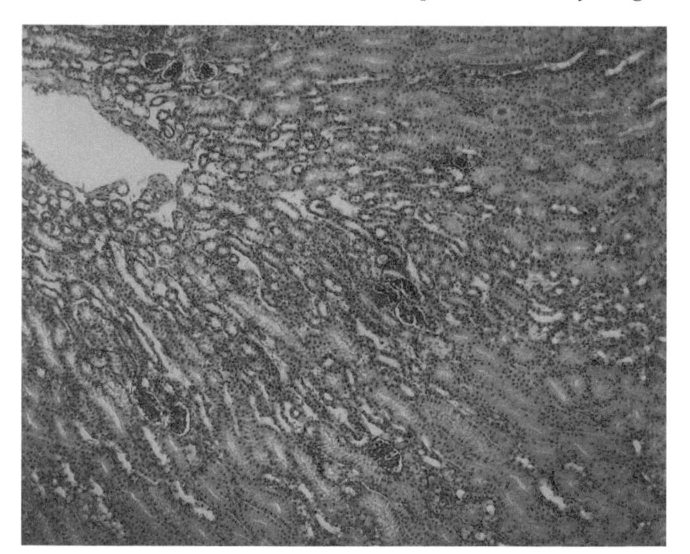

FIGURE M31a The kidney of reptiles is similar to that of mammals except that there is no NEPHRIC LOOP. Instead the proximal and distal convoluted tubules are connected through a short INTERMEDIATE SEGMENT which cannot be distinguished here, 10 ×.

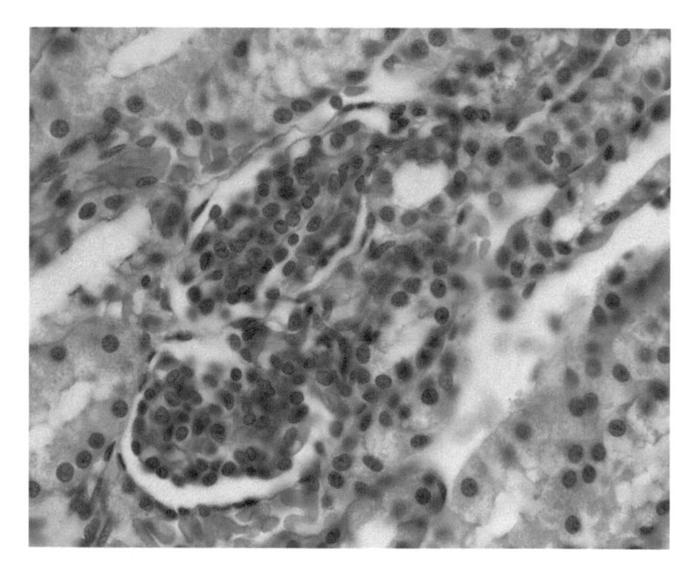

FIGURE M31b Renal corpuscles similar to those of mammals are seen. These are surrounded by segments of the proximal and distal convoluted tubules with abundant capillaries coursing between, 63 ×.

MUSCULARIS on the other (Figs. M39a and M39b). The muscle layer reverses the usual arrangement of coats, having an inner longitudinal layer and an outer circular layer. The TUNICA ADVENTITIA is formed of loose connective tissue.

FIGURES M32(a–c) These images were captured from sections of kidney from a tuatara, a primitive lizard-like reptile from New Zealand, *Spehnodon punctatum*, of the order Rhynchocephalia.

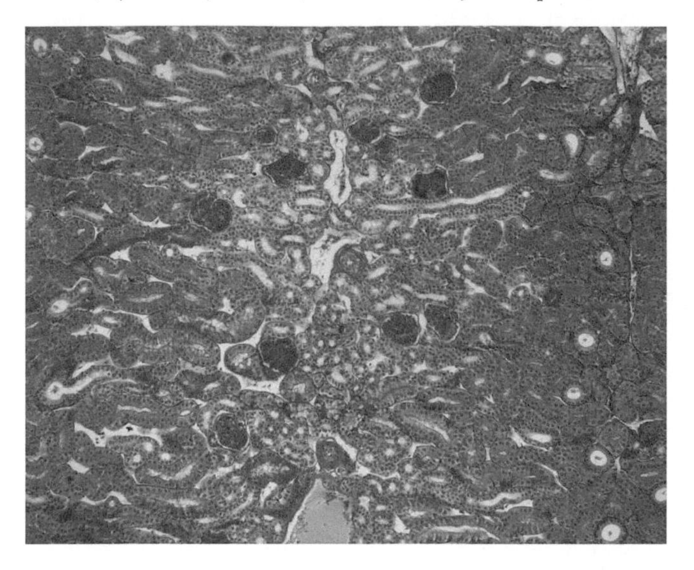

FIGURE M32a The kidney resembles that of other reptiles. Renal corpuscles are surrounded by closely packed proximal and distal convoluted tubules, 10 ×.

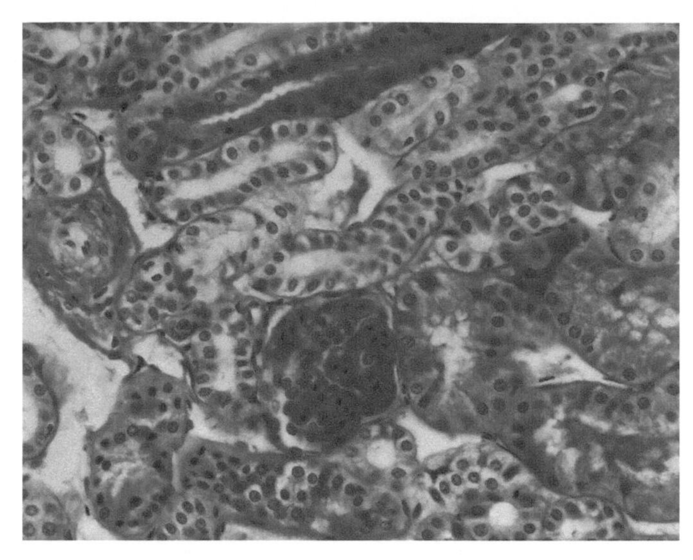

FIGURE M32b Two types of tubules can be discerned surrounding a renal corpuscle: pale and deeply acidophilic tubules, 40 ×.

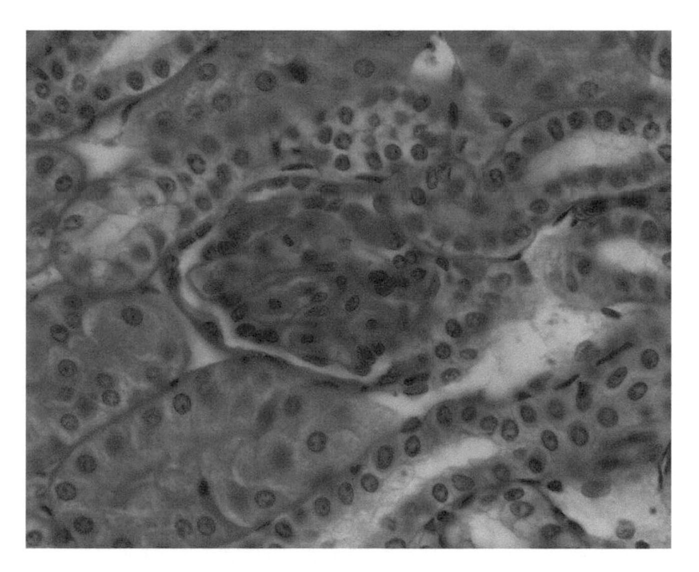

FIGURE M32c The renal corpuscle at the center is unremarkable. The outer PARIETAL LAYER of the capsule is a simple squamous epithelium. The inner VISCERAL LAYER, which adheres to the glomerular loops, is not visible. Surrounding the renal corpuscle are the sections of eosinophilic PROXIMAL CONVOLUTED TUBULES and paler, smaller, DISTAL CONVOLUTED TUBULES, 63 ×.

URINARY BLADDER

The wall of the urinary bladder of tetrapods is composed of the same elements as the lower part of the ureter: a TUNICA MUCOSA of transitional epithelium and lamina propria mucosae, a TELA SUBMUCOSA, a TUNICA MUSCULARIS of an ill-defined layer of circular muscle sandwiched between two layers of longitudinal muscle (as is the case in the lower parts of the ureter), and a TUNICA ADVENTITIA (Figs. M40a and M40b) and (Figs. M41a and M41b). The transitional epithelium varies in thickness according to the degree of distension of the organ. There is no bladder in birds other than the ostriches.

FIGURES M33(a–c) Photomicrographs of sections of the kidney of a rattlesnake.

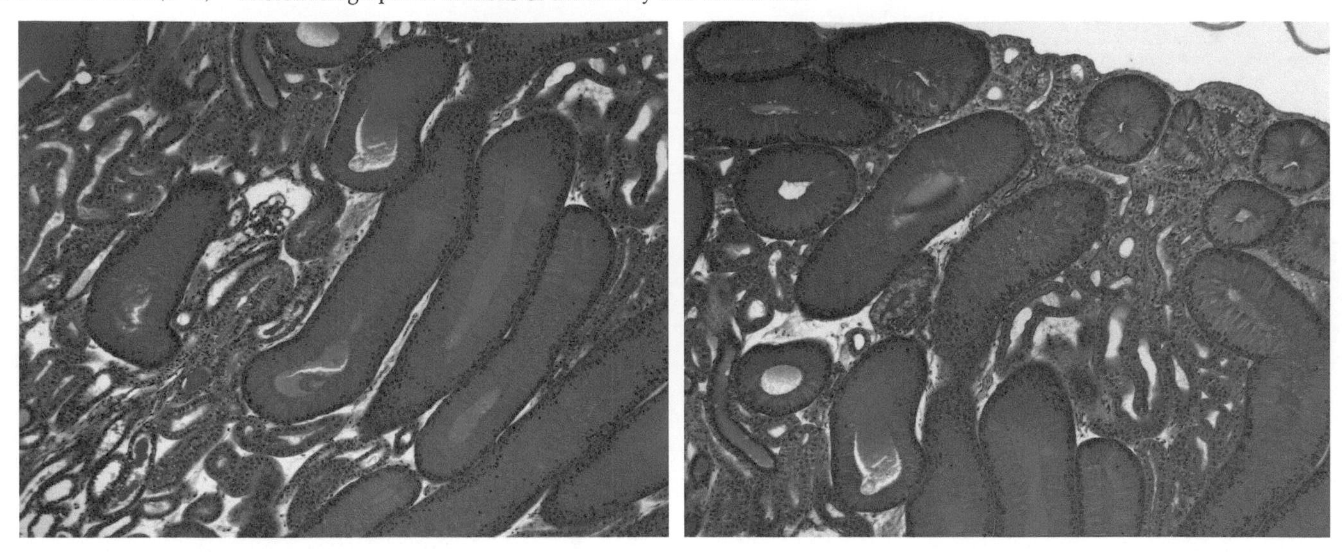

FIGURES M33a and b Two fields are shown here at 10×. The regular opisthonephric tubules of this rattlesnake kidney are overshadowed by the massive tubules of the SEXUAL SEGMENT—this snake was obviously in breeding condition. The diameter of these tubules is three to five times that of the distal tubule, depending on the time of year. Sexual segments are found only in males in reproductive condition. These tubules are enlargements of the distal portion of the nephrons and secrete a fluid vehicle for sperm. It contains various enzymes that presumably are important for the transport or capacitation of sperm. Cf. seminal secretions of mammals. Note the renal corpuscle above and to the left of center in Fig. M33a, 10×.

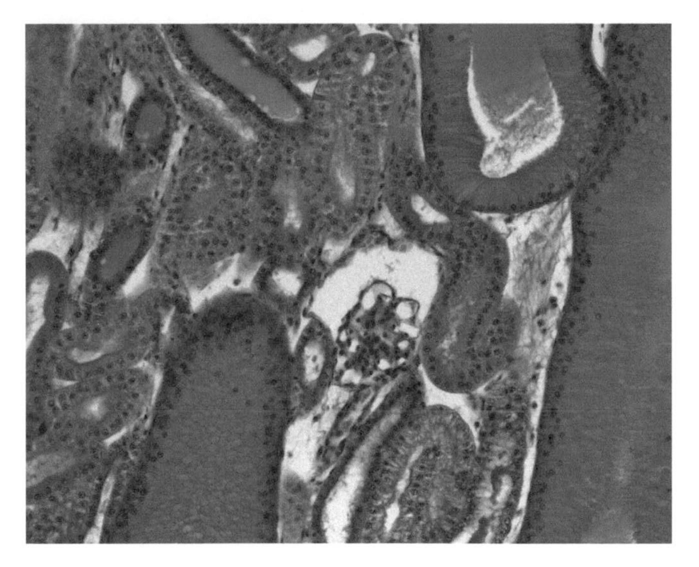

FIGURE M33c The renal corpuscle is near the center, 20×.

Electron micrographs show that the cells of the transitional epithelium rest on a basement lamina (Fig. M41c). Their stretching, sliding movement is impeded by a few small desmosomes between adjacent cells. The large, dome-shaped superficial cells are sealed by extensive tight junctions. Conspicuous discoidal FUSIFORM VESICLES abound within the cytoplasm. About 70% of the luminal plasma membrane of each superficial cell is thickened to form PLAQUES with "hinges" of ordinary plasma membrane between. Cytoplasmic filaments attach to the plaques. When the bladder contracts, the surface membrane invaginates, carrying some of the plaques below the surface. Some plaques become sealed off within the fusiform vesicles and may be retained for future reappearance at the surface or be digested by lysosomes. Numerous lysosomes contain the remnants of cell membranes constantly being replaced in the struggle to maintain a water-tight barrier.

FIGURES M34(a–c) From the kidney of a painted turtle.

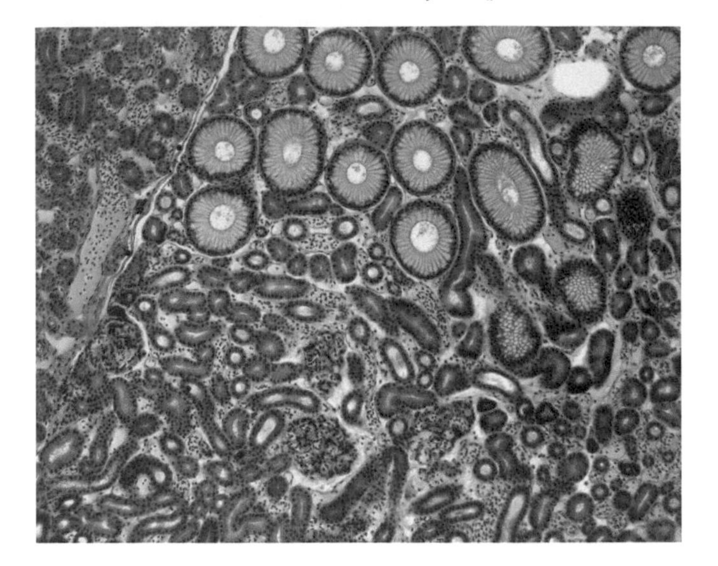

FIGURE M34a Portions of the sexual segment can be seen in upper center of this image from a section of the kidney of a turtle, 10×.

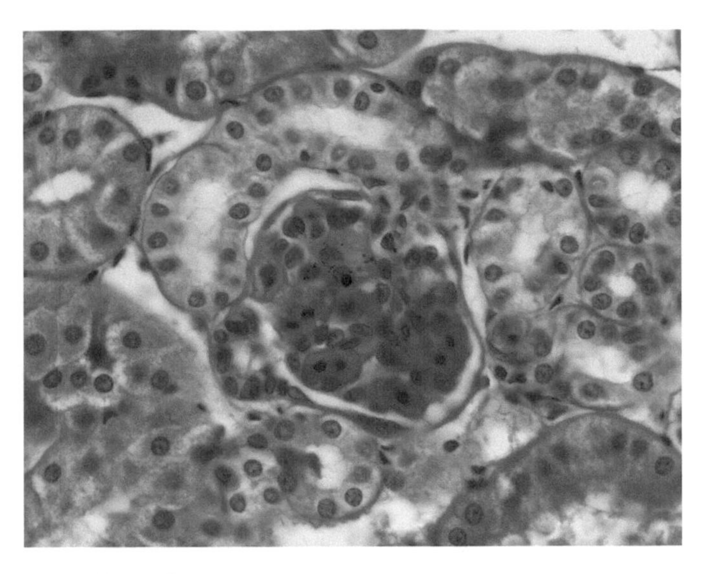

FIGURE M34b Some regions of the kidney show no sexual segments. There is a renal corpuscle at the center. The bulge at the lower left is most probably the entrance to the proximal convoluted tubule. The capsule is surrounded by the larger, more eosinophilic proximal convoluted tubules and smaller distal convoluted tubules. Brush borders are distinct on apical surfaces of the epithelial cells in the proximal convoluted tubules. Capillaries course between the tubules, 63×.

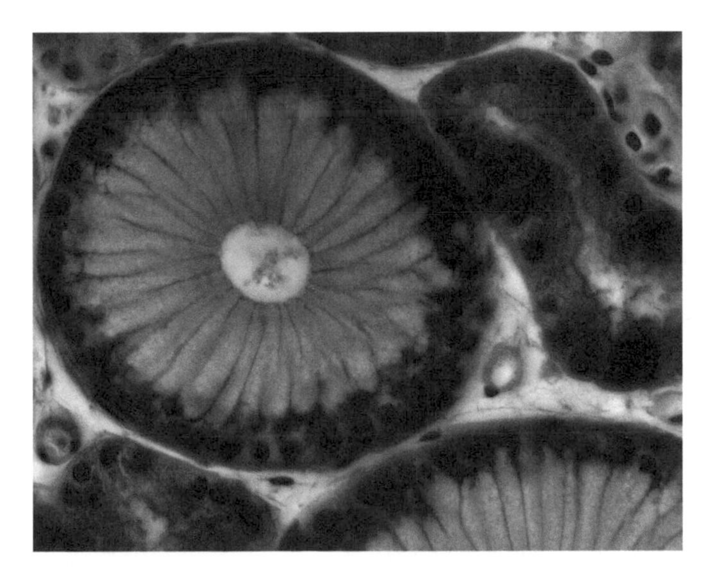

FIGURE M34c The sexual segment is lined by high columnar cells loaded with droplets that will form part of the seminal fluid. The tubule is surrounded by myoepithelial cells, which contract and assist in the expulsion of this fluid. Cross-sections of proximal convoluted tubules are seen at the upper right and lower left. Capillaries course between the tubules, 63×.

MAMMALIAN URETHRA

In mammals the bladder and urethra arise from division of the embryonic cloaca, the end of the hindgut, into the rectum and anal canal on one side and the bladder and lower urogenital tract on the other. In male mammals the urethra consists of three parts: the PROSTATIC PORTION that is surrounded by the prostate gland as it passes from the bladder, a short MEMBRANOUS PORTION, and the SPONGY PORTION that passes through the penis. The prostatic portion is lined with transitional epithelium like the bladder (Figs. M42a and M42b). The two openings of the

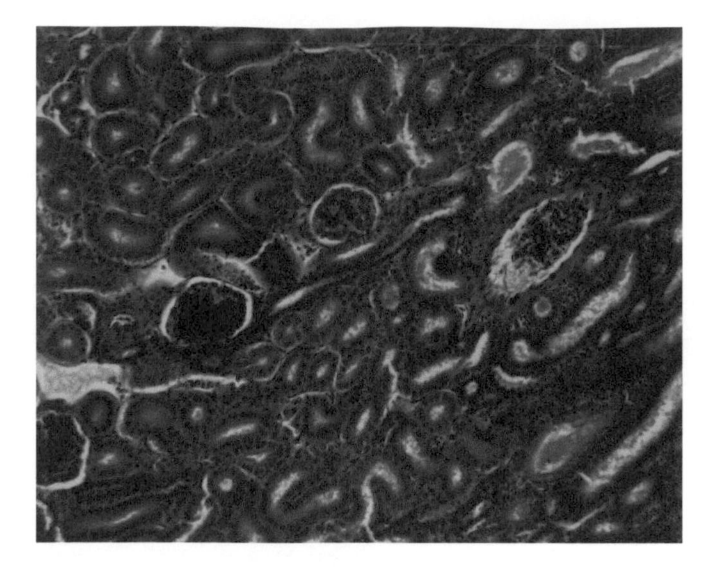

FIGURE M35a In the kidney of a bird, we see the pattern of renal corpuscles surrounded by closely packed proximal and distal convoluted tubules. There is modest hemopoietic activity in the loose vascular connective tissue between the tubules, especially in the upper right corner, 20 ×.

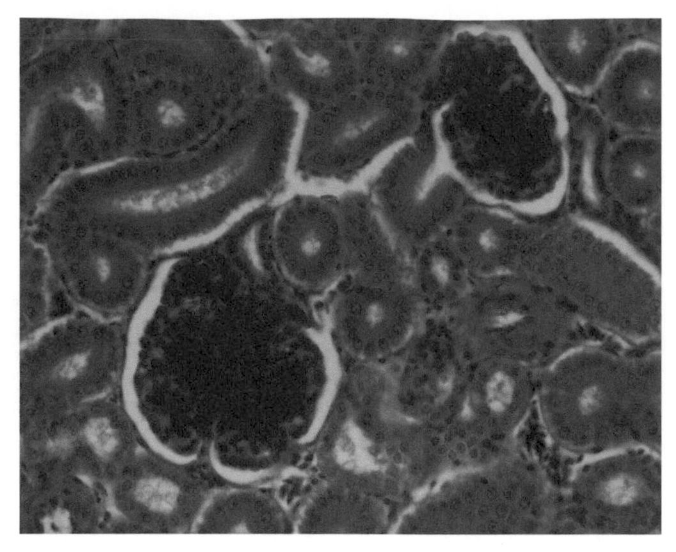

FIGURE M35b Two renal corpuscles are shown in this section of the kidney of a bird. A small portion of a distal convoluted tubule appears to be nestling against the glomerulus of the lower corpuscle, forming a MACULA DENSA. A proximal convoluted tubule drains the upper renal corpuscle. Sections of proximal convoluted tubules are larger and more eosinophilic than sections of the distal tubules, 40 ×.

ejaculatory ducts of the testes are found here as well as numerous pores of the prostate gland. The membranous portion begins as the urethra exits the prostate and passes between the bulbourethral glands (Figs. M42c and M42d). In this portion the luminal lining changes to a cuboidal or stratified epithelium. As the urethra enters the basal bulb of the penis, it undergoes a further transition to become the spongy portion. In this section epithelial lining is stratified columnar to stratified cuboidal (Figs. M42e and M42f). It is embedded in the sponge-like mass of vascular erectile tissue of the penis.

The female mammalian urethra is shorter than its male counterpart and corresponds to that portion of the urethra in the male passing through the prostate gland. The female urethra extends from the bladder to its opening in the vaginal vestibule between the labia. The stratified squamous or pseudostratified columnar epithelium lining the LUMEN is thrown up into folds (Fig. M43a). Numerous invaginations are formed by the epithelium and in many places these outpocketings are lined with clear mucous cells, similar to the glands found in the male urethra (Fig. M43b). It is surrounded by a LAMINA PROPRIA MUCOSAE of loose fibrous connective tissue and venous spaces that resembles the CORPORA CAVERNOSA of the male penis (Fig. M43c). The connective tissue merges into a layer of smooth muscle which is in turn surrounded by striated muscle.

EXTRARENAL EXCRETION OF SALT

There are other full-time excretory organs that eliminate significant amounts of salt from the blood of certain vertebrates.

Rectal Gland

The rectal gland of elasmobranchs is a finger-shaped diverticulum of the intestine just posterior to the spiral valve (Fig. M44a). It is suspended in the dorsal mesentery and receives an abundant blood supply by way of the posterior mesenteric artery. In cross-sections of the rectal gland of the dogfish note three general layers: an outer fibromuscular CAPSULE covered with visceral peritoneum, a thick intermediate zone of SECRETORY TUBULES, and an inner CENTRAL LUMEN enclosed within a transitional epithelium (Figs. M44b and M44c). The secretory tubules are lined by a simple, roughly cuboidal epithelium and are enclosed in a delicate mesh of vascular fibrous connective tissue. Blood flows centripetally from arterioles in the capsule to capillaries between the tubules and thence to

FIGURES M36(a–g) The images in this series are of sections from kidney of a gerbil stained by various common methods.

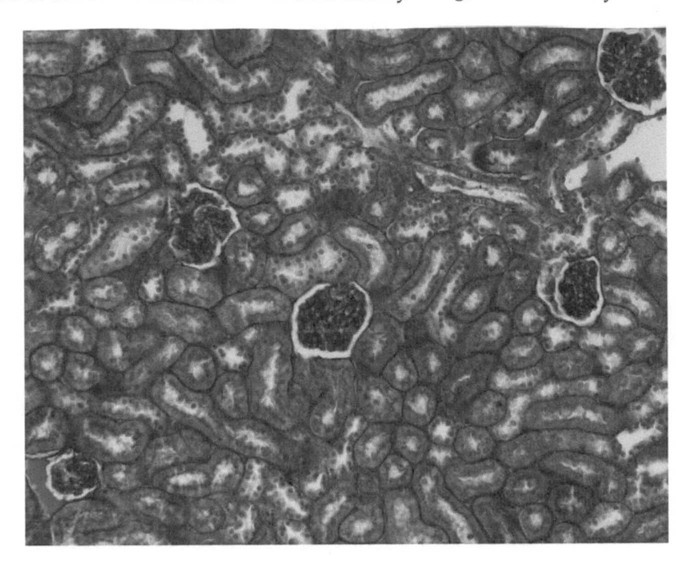

FIGURE M36a Periodic acid-Schiff method with orange G and malachite green. This method stains carbohydrates in the basal laminae of the tubules and glomerular blood vessels. Note that the brush borders of some of the epithelial cells of the tubules have stained deeply, 20 ×.

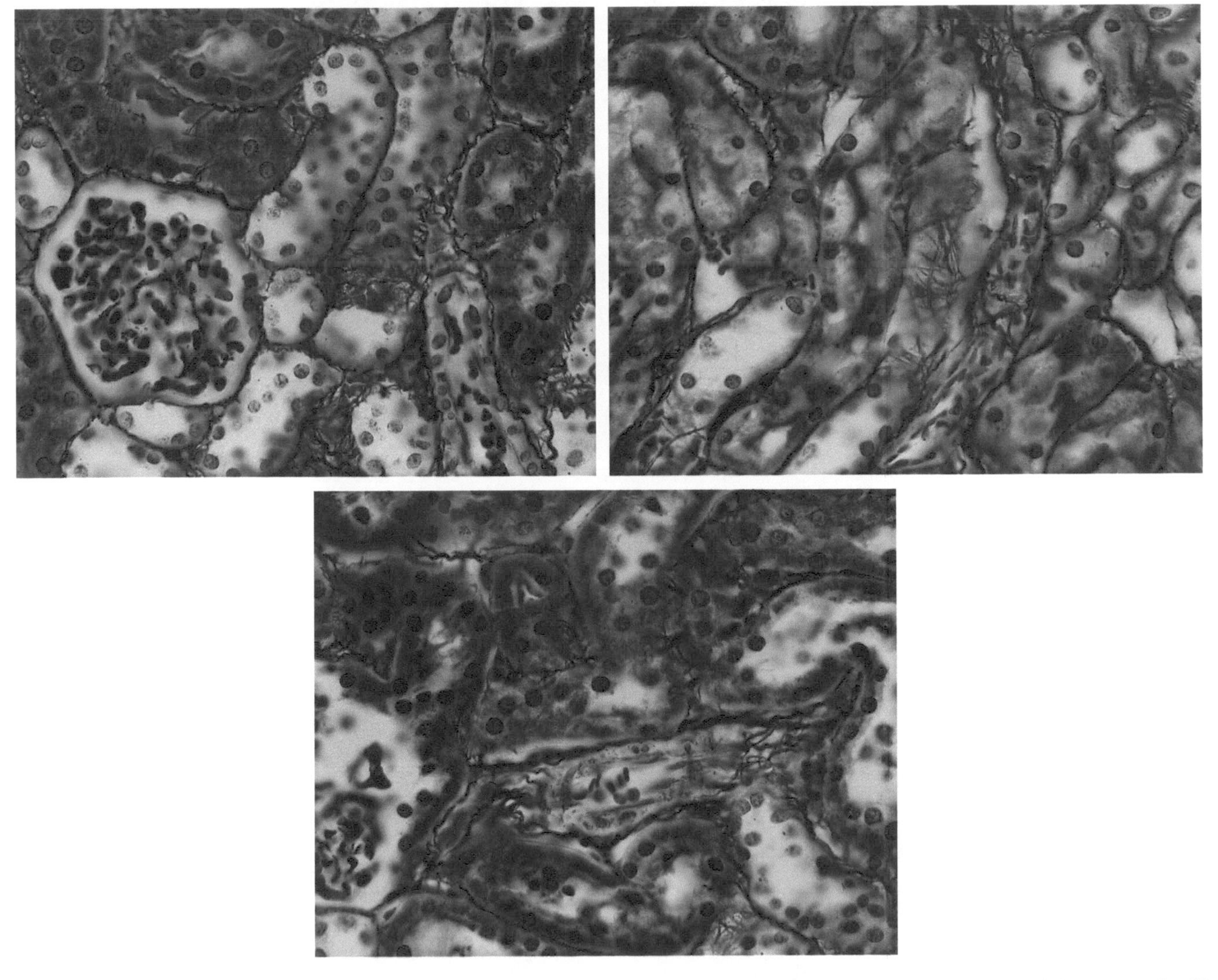

FIGURES M36(b–d) M36b (cortex), M36c (medulla), and M36d (cortex). Wilder's silver stain here demonstrates the delicate reticular fiber network typically not seen in standard hematoxylin and eosin staining. The delicate reticular fibers here are impregnated with silver and can be seen supporting tubules and glomeruli in these images, 63 ×.

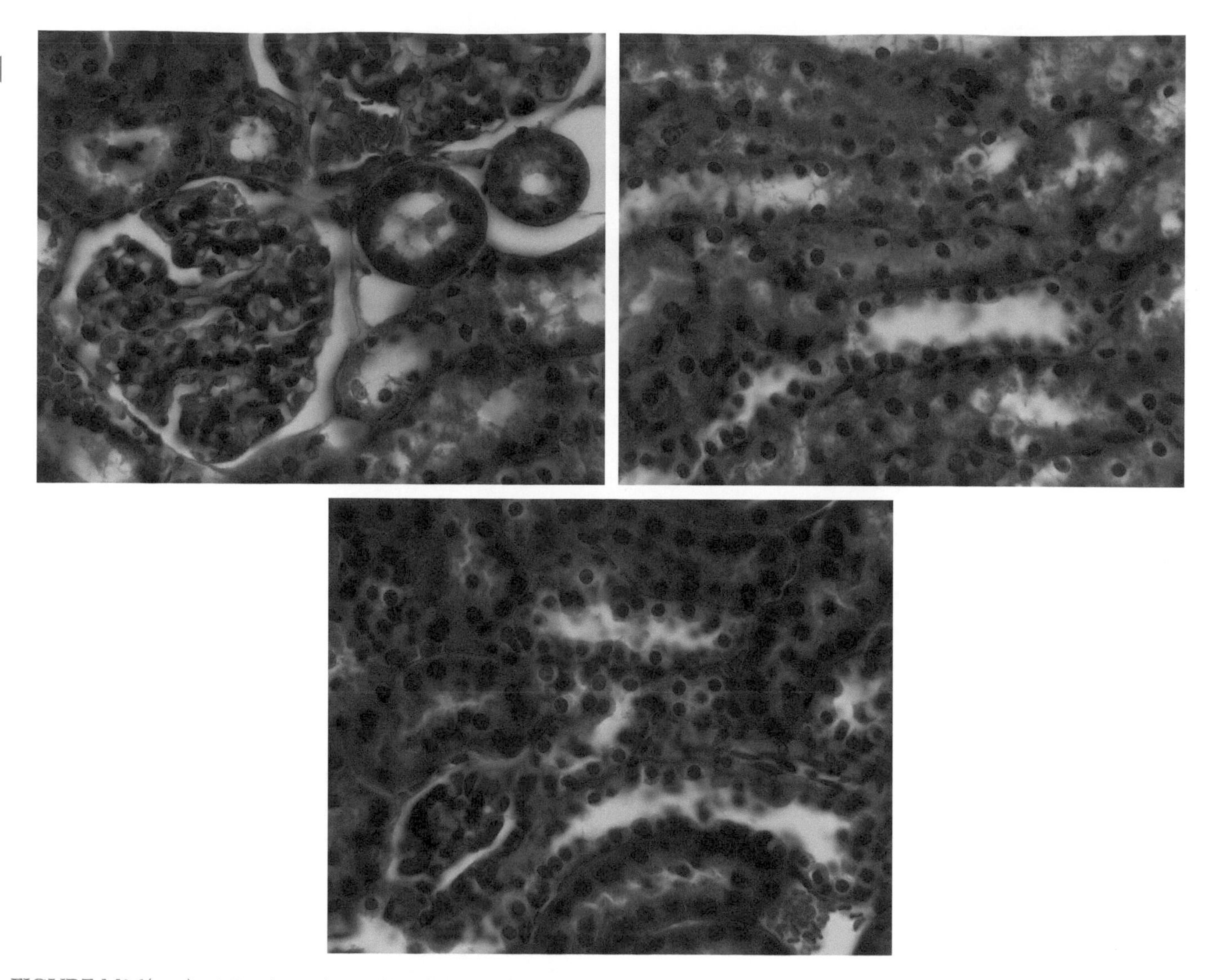

FIGURE M36(e–g) M36e (cortex), M36f (medulla), and M36g (cortex). The same kidney has been stained with hematoxylin and eosin for comparison with the stains above, 63 ×.

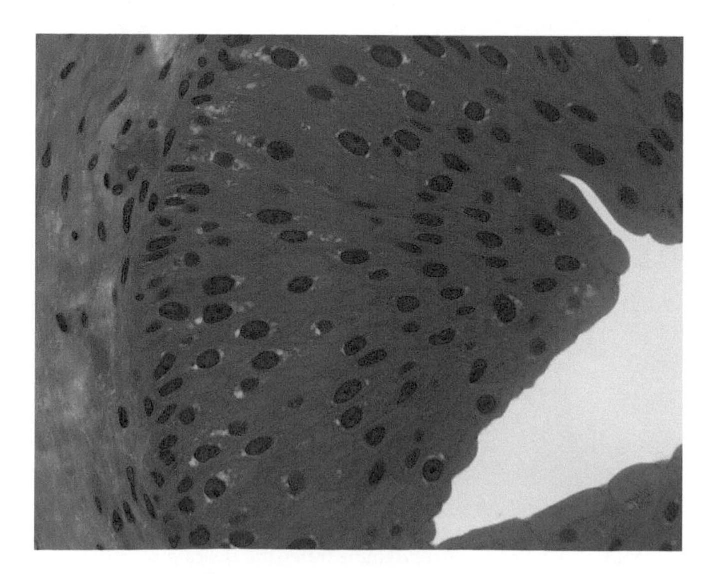

FIGURE M37 The TRANSITIONAL EPITHELIUM in the ureter of a monkey stretches to accommodate an increasing volume of urine. Its bulbous cells flatten as the organ distends, 63 ×.

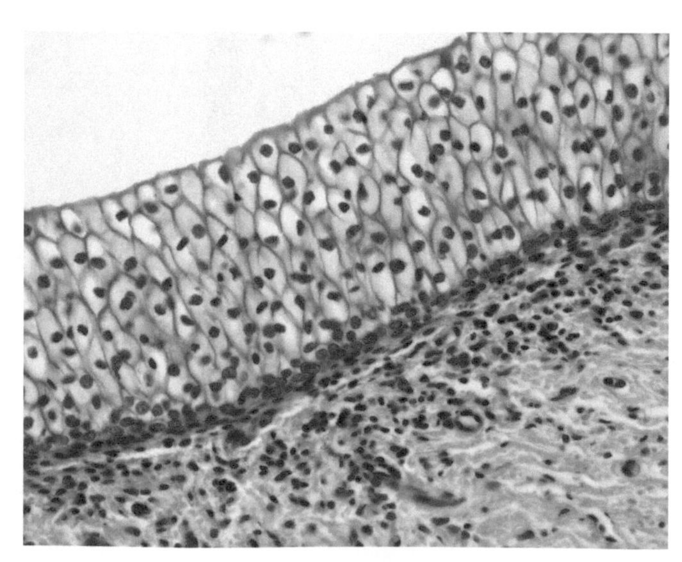

FIGURE M38 The transitional epithelium in the ureter of a large mammal (here a horse) is similar to that of all mammals, even one as small as a guinea pig, 40 ×.

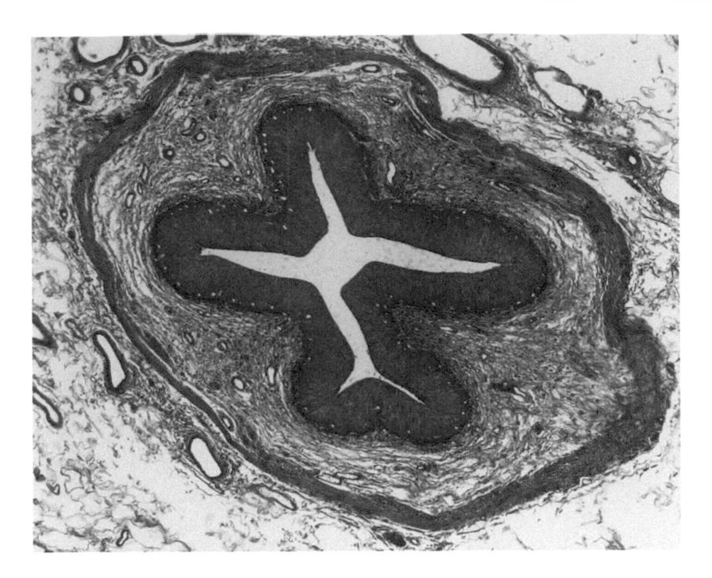

FIGURE M39a The ureter of a guinea pig is lined with a TUNICA MUCOSA of transitional epithelium resting on a LAMINA PROPRIA MUCOSAE of loose connective tissue, which blends into the TUNICA MUSCULARIS whose inner layer is longitudinal and outer layer is circular. The tunica mucosa forms longitudinal folds nearly obliterating the lumen, 10 ×.

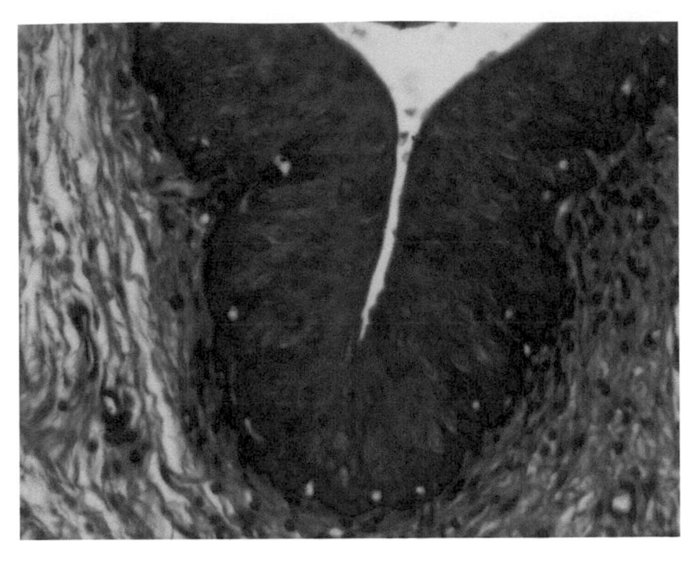

FIGURE M39b A higher magnification view of the section in Fig. M39a. Details of the transitional epithelium of the tunica mucosa and surrounding layer are more obvious here. The tunica mucosa forms longitudinal folds nearly obliterating the lumen. 40 ×.

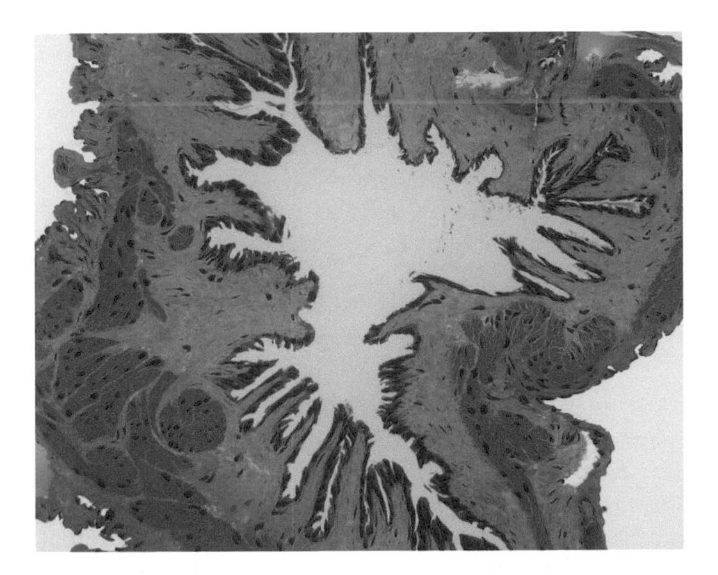

FIGURE M40a The urinary bladder of *Amphiuma*, an amphibian, is a distensible sac, which accommodates increasing amounts of fluids by flattening of the folds in the lining and deformation of the cells of the transitional epithelium. The epithelium rests on a layer of connective tissue of the TELA SUBMUCOSA. This is surrounded by the TUNICA MUSCULARIS and a TUNICA ADVENTITIA, 10 ×.

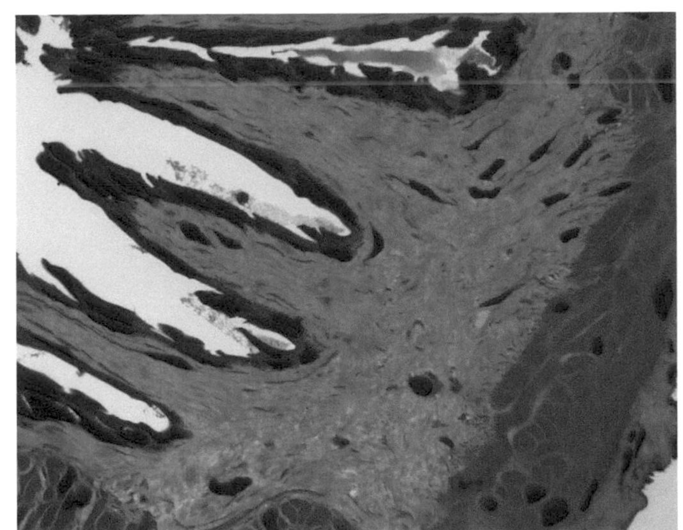

FIGURE M40b The urinary bladder of *Amphiuma*, an amphibian at a higher magnification. Dark pigment cells are scattered throughout. The epithelium rests on a layer of connective tissue of the TELA SUBMUCOSA. This is surrounded by the TUNICA MUSCULARIS and a TUNICA ADVENTITIA, 40 ×.

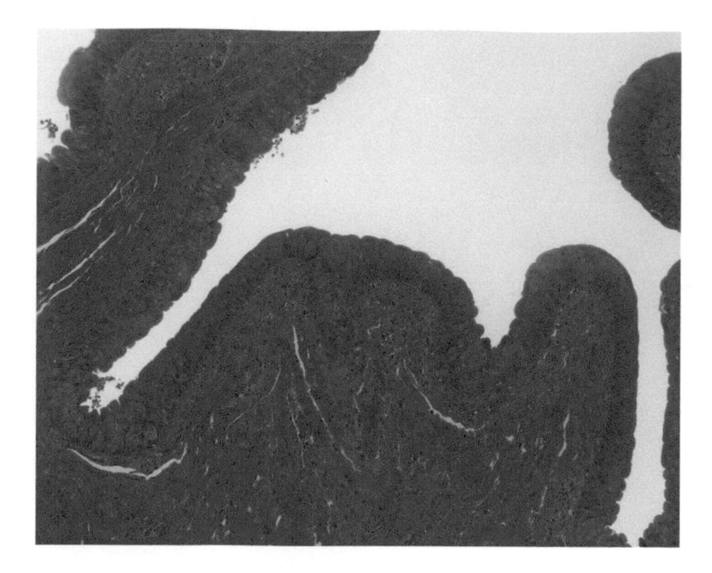

FIGURE M41a The urinary bladder of a monkey accommodates increasing accumulation of urine in the same way as that of an *Amphiuma*. The transitional epithelium of the TUNICA MUCOSA rests on a TELA SUBMUCOSA of fibrous connective tissue. This is surrounded by the smooth muscle of the TUNICA MUSCULARIS, 10 ×.

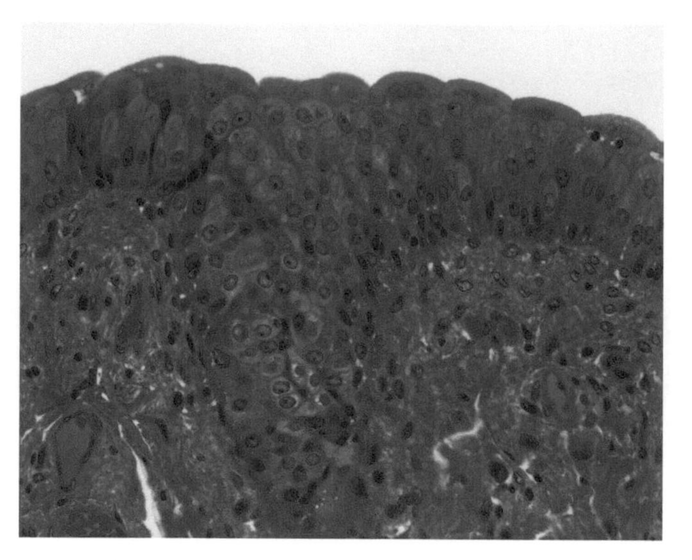

FIGURE M41b The urinary bladder of a monkey accommodates increasing accumulation of urine in the same way as that of an *Amphiuma*. The transitional epithelium of the TUNICA MUCOSA almost looking like a stratified epithelium in this higher power view, rests on a TELA SUBMUCOSA of fibrous connective tissue. This is surrounded by the smooth muscle of the TUNICA MUSCULARIS, 40 ×.

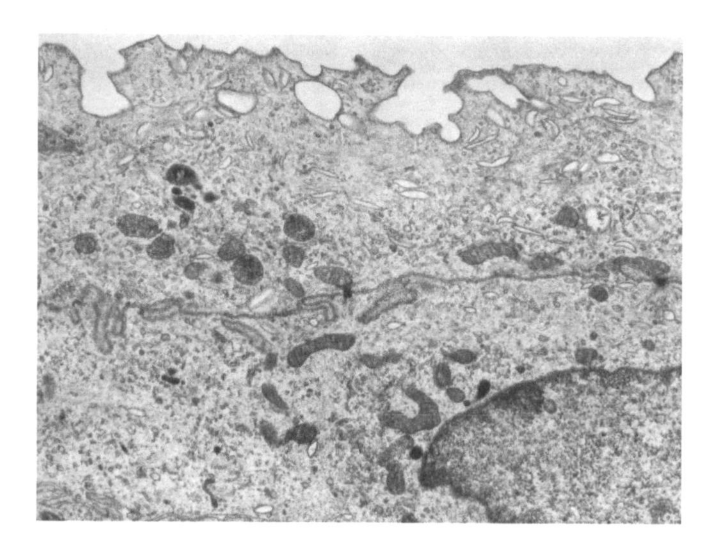

FIGURE M41c Transmission electron micrograph of the transitional epithelium in the wall of a mammalian bladder. The surface has a characteristic angular appearance which is thought to result from the presence of stiff cigar-shaped plaques of membrane.

venous sinuses in the central region (Figs. M44d and M44e). As it passes, salts are extracted by the tubules and a hypertonic salt solution flows (also centripetally) into the central lumen (Figs. M44f–M44h).

Electron micrographs of cells in the tubules of the rectal gland show the typical the characteristics associated with active ion transport: apical microvilli, pits, vesicles, tight junctions, cellular interdigitations, basal infoldings containing mitochondria, etc. (Figs. M45a–M45d).

Salt Glands

Salt glands of some birds and reptiles are also engaged in excretion of salts; indeed, in some forms they are more important than the kidneys in the elimination of salts. They are modified nasal glands in lizards and marine birds (Fig. M46a) and lachrymal glands in marine turtles but in all cases they are

FIGURES M42(a–f) The male urethra of mammals is the tube that drains the bladder.

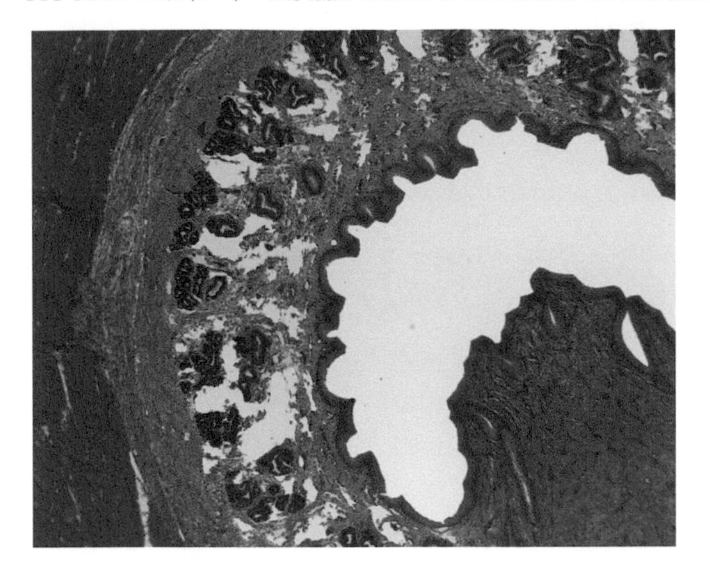

FIGURE M42a The lumen of the PROSTATIC PORTION of the urethra is at the center right. It is surrounded by the prostate gland, 5×.

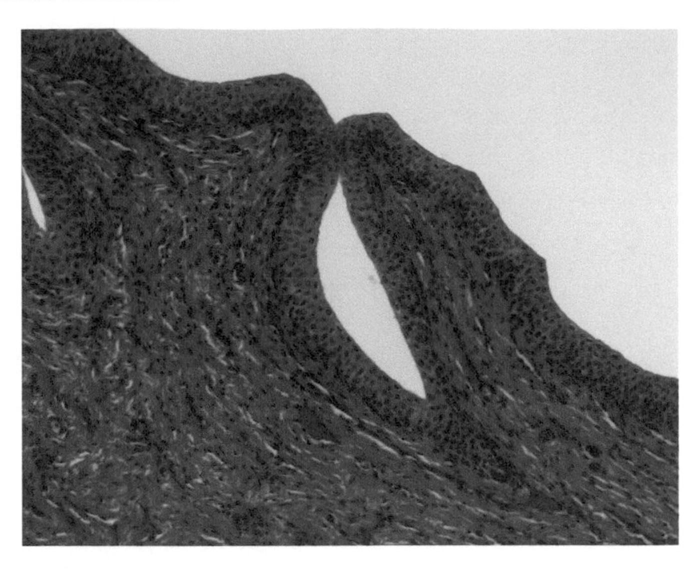

FIGURE M42b The epithelium lining the urethra is transitional, 20×.

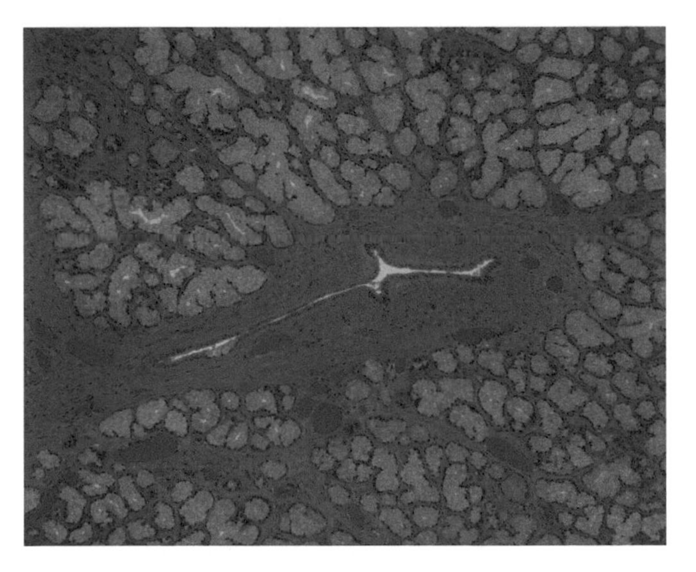

FIGURE M42c The MEMBRANOUS PORTION of the urethra passes between the bulbourethral glands. The lining is simple cuboidal, stratified in places, 10×.

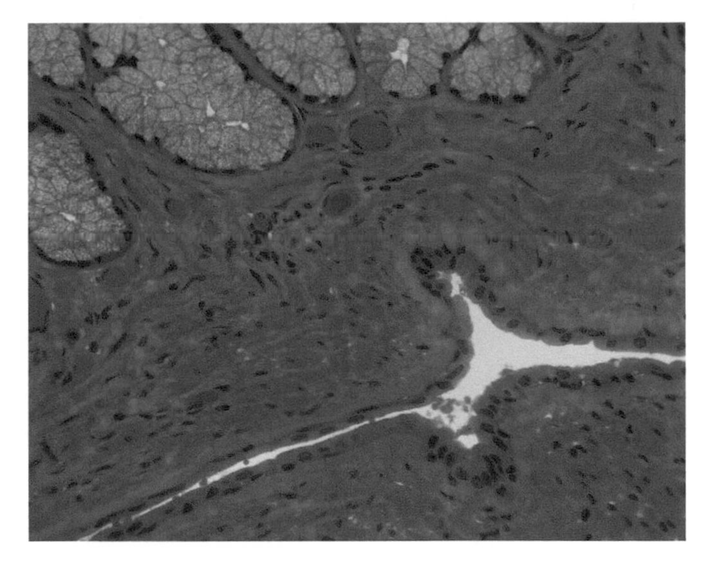

FIGURE M42d A higher power view of the section in Fig. M42c better resolves the nature of the epithelium lining the membranous portion of the urethra. 40×.

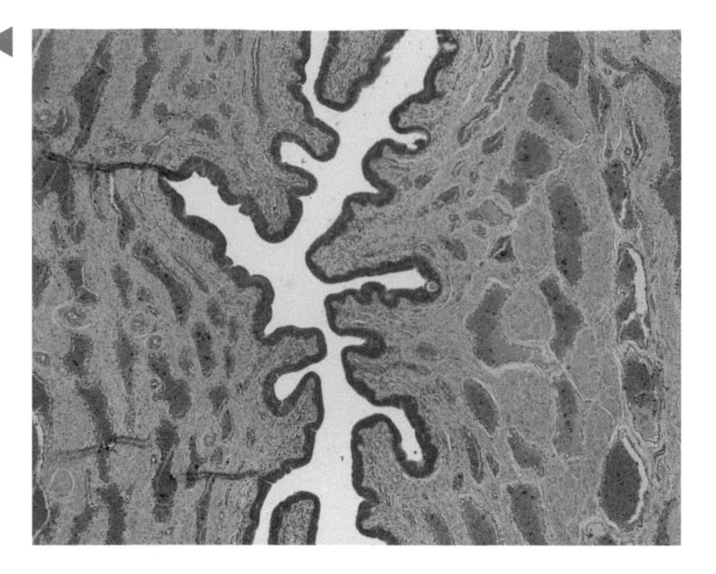

FIGURE M42e As the urethra passes down the penis, it is surrounded by erectile tissue of the CORPUS SPONGIOSUM. The epithelium lining the spongy portion of the urethra varies from stratified columnar to stratified cuboidal. It is subtended by loose vascular connective tissue, 5 ×.

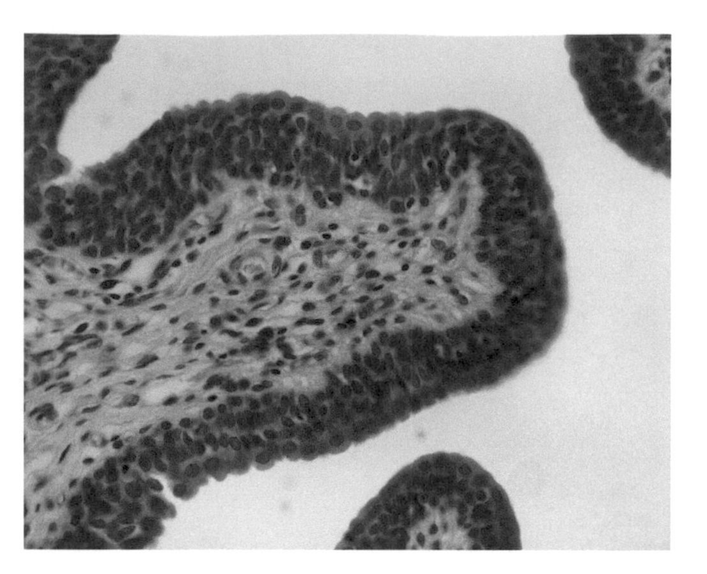

FIGURE M42f As the magnification used is increased. Here we are able to see details in both; the epithelium lining the urethra and the spongy vascularized connective tissue. The epithelium lining the spongy portion of the urethra varies from stratified columnar to stratified cuboidal, 40 ×.

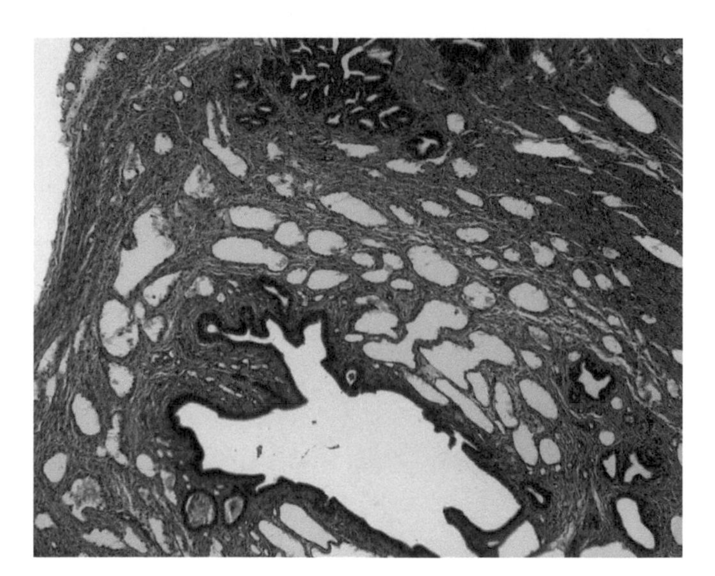

FIGURE M43a The urethra of the female mammal corresponds to the portion of the urethra in the male that passes through the prostate gland. The LUMEN is lined with a stratified squamous or pseudostratified columnar epithelium thrown up into folds. Glandular structures are formed as outpouchings of the epithelium. The loose fibrous connective tissue and venous spaces of the LAMINA PROPRIA MUCOSAE surrounding the urethra resemble the CORPORA CAVERNOSA of the male penis. The loose connective tissue merges into a layer of smooth muscle which in turn is surrounded by striated muscle, 10 ×.

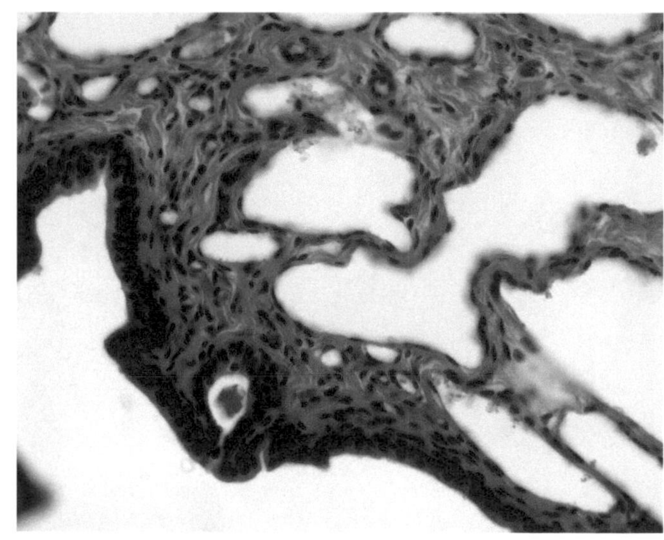

FIGURE M43b The epithelial outpouchings may contain colloidal material or concretions as at the lower left, 40 ×.

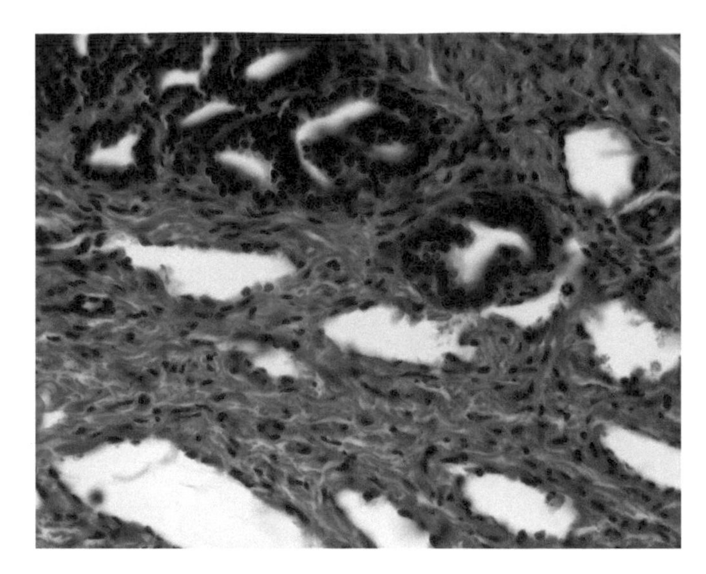

FIGURE M43c Venous spaces penetrate the loose connective tissue of the lamina propria mucosae. Evaginations of the luminal epithelium form gland-like lacunae, 40 ×.

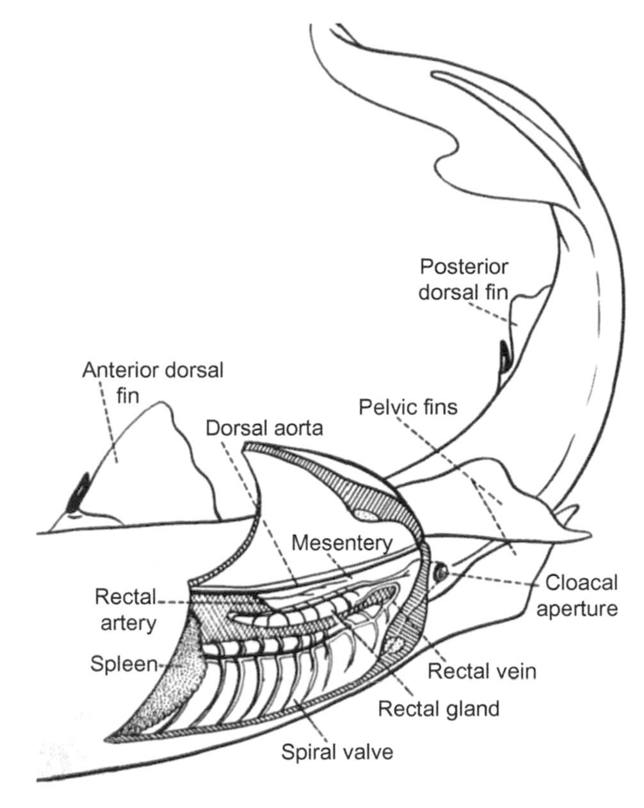

FIGURE M44a Diagram showing the location of the rectal gland in the spiny dogfish, *Squalus acanthias*. The RECTAL GLAND of elasmobranches is an excretory organ that removes salt from the blood. It is a finger-shaped diverticulum of the intestine just posterior to the spiral valve. It is suspended in the dorsal mesentery and receives an abundant blood supply by way of the posterior mesenteric artery.

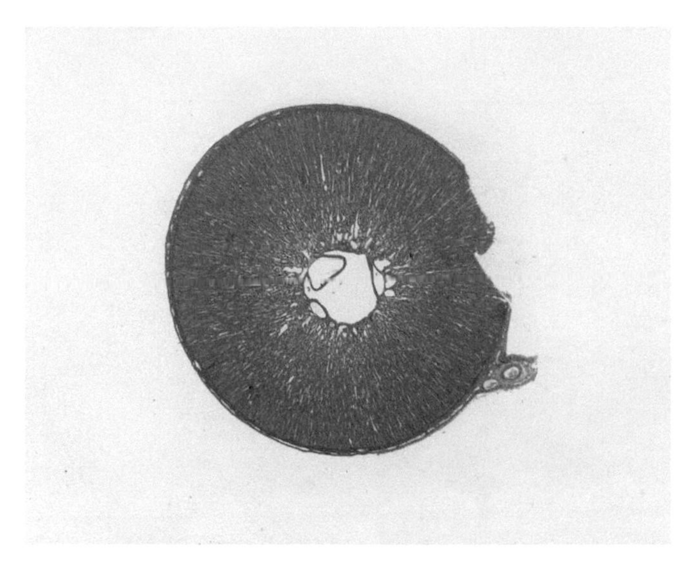

FIGURE M44b This cross-section of the rectal gland of a dogfish shows three regions of the gland: the tough, fibromuscular CAPSULE, the intermediate zone of SECRETORY TUBULES, and the central LUMEN, which is surrounded by transitional epithelium. Blood is delivered to the gland by the RECTAL ARTERY contained within the mesentery at the lower right; it is drained into RECTAL VEINS that bulge into the lumen beneath the transitional epithelium, 1 ×.

roughly similar. Their CAPSULE contains several banana-shaped LOBES, each enclosed in a stout sheath of fibrous connective tissue (Figs. M46b and M46c). The lobes seen in sections of a salt gland resemble tiny rectal glands: containing SECRETORY TUBULES of simple cuboidal epithelium radiating from a CENTRAL LUMEN of stratified cuboidal epithelium (Figs. M46d–M46f). Electron micrographs of salt glands (Figs. M46g and M46h) clearly demonstrate morphological signs of ion transport.

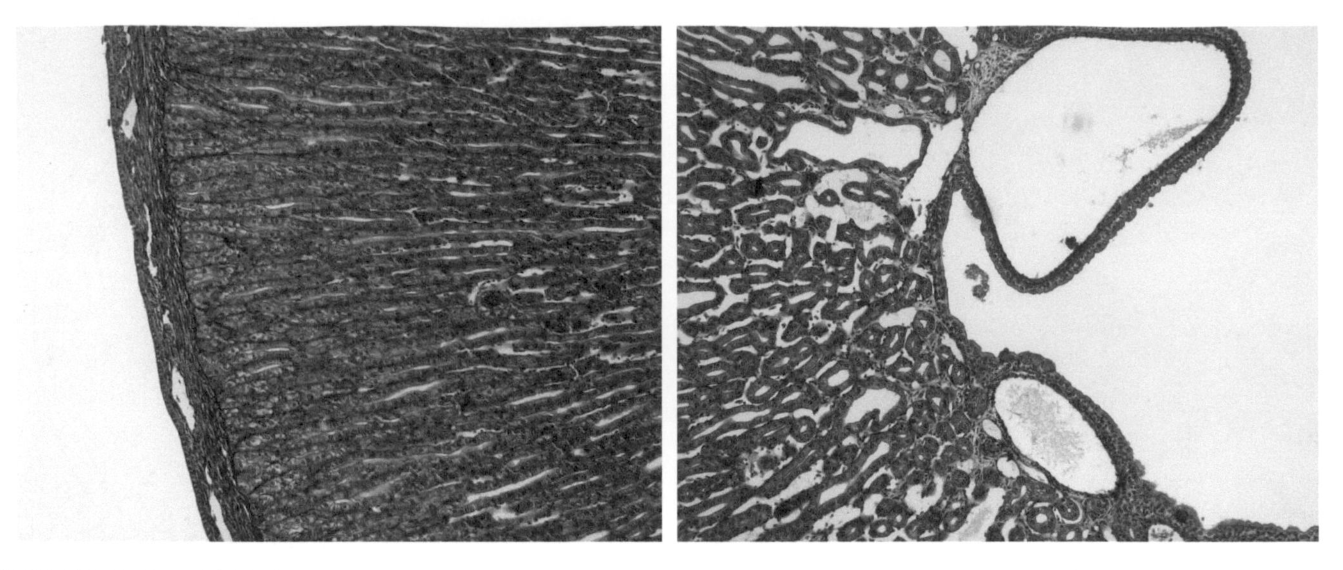

FIGURES M44c and M44d Blood flows centripetally in capillaries between the tubules; fluids containing a hypertonic salt solution flow in the same direction to be released into the lumen. Since blood and the fluids flow in the same direction, a counter-current exchange mechanism cannot function to increase the efficiency of the extraction of salt. In M44c the large rectal vein (ventral venous sinus) protrudes into the central canal at the top right and lies beneath the transitional epithelium. Smaller veins are seen below, 10 ×, 10 ×.

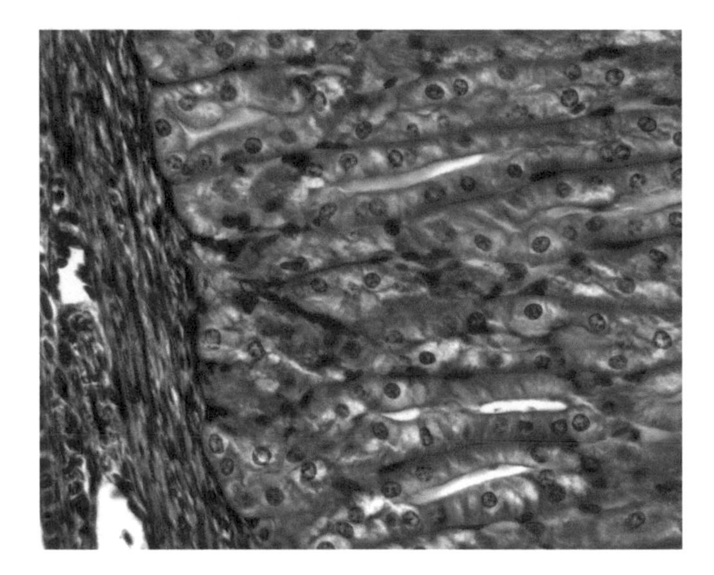

FIGURE M44e The thick capsule surrounding the gland consists of fibrous connective tissue interspersed with smooth muscle. It contains branches of the renal artery, 40 ×.

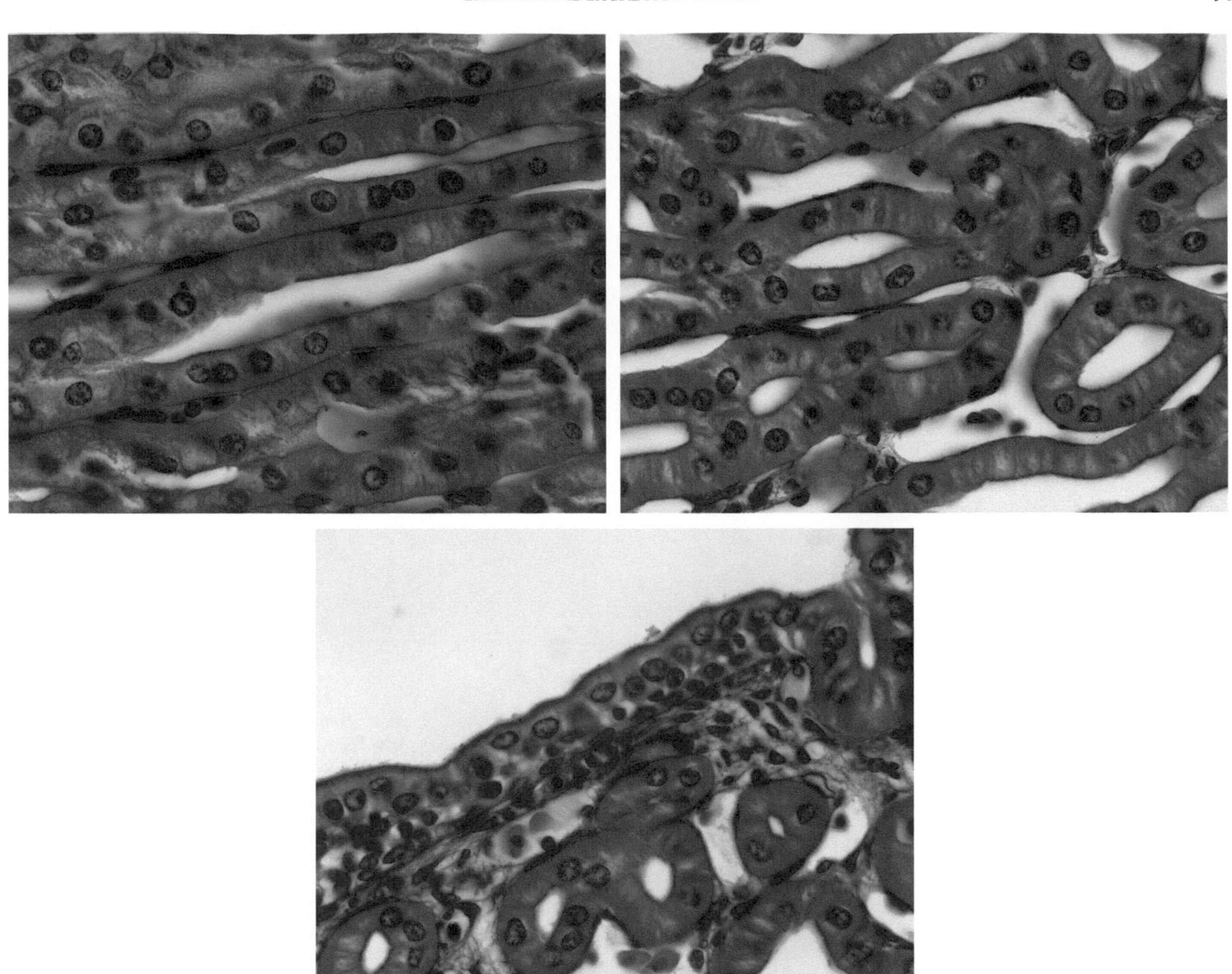

FIGURES M44f, M44g, and M44h At the periphery of the gland the tubules follow a fairly straight course toward the lumen (M44f); large blood spaces lined with a simple squamous epithelium are interposed between them. Nearer the lumen (M44g) the tubules curve somewhat and the blood spaces are more dilated. The transitional epithelium lining the central canal (M44h) lies on the wall of one of the branches of the central vein. All figures are 63×. Blood in the RECTAL ARTERY is distributed in smaller arteries within the capsule to bring it to the capillaries which course between the tubules. This blood is collected by RECTAL VEINS at the center of the gland. Active transport in the cells of the tubules cause the withdrawal of salt from the blood, so that the fluid collected in the lumen of the gland is hypertonic to both blood plasma and to seawater.

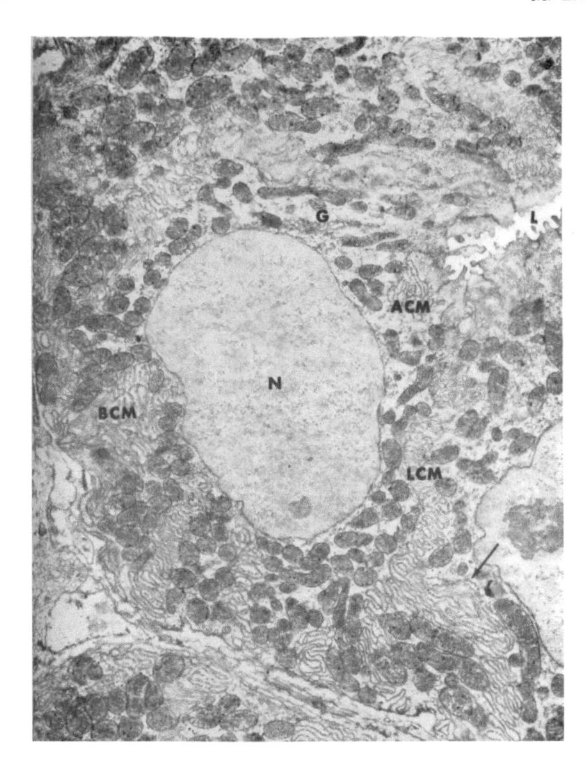

FIGURE M45a Transmission electron micrograph of cells from the outer part of the tubular layer in the rectal gland of the spiny dogfish (*Squalus acanthias*). Lateral cell membranes show elaborate infolding (ACM, BCM, and LCM). Numerous large mitochondria and a Golgi complex (G) are seen, 7600 ×.

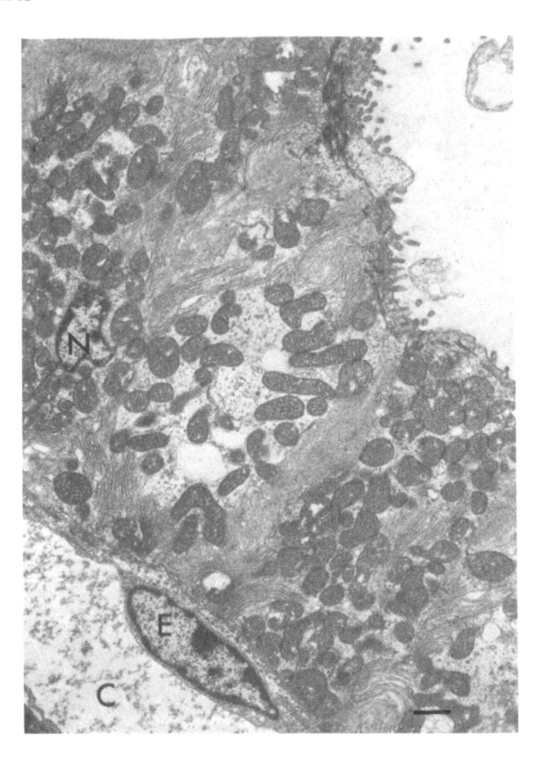

FIGURE M45b Transmission electron micrograph of the rectal gland in an elasmobranch, *Hydrolagus colliei*. Note the abundant mitochondria in the simple columnar epithelial cells. *C*, capillary; *E*, endothelial nucleus; *N*, nucleus of the rectal gland cell. *Scale*, 1 μm.

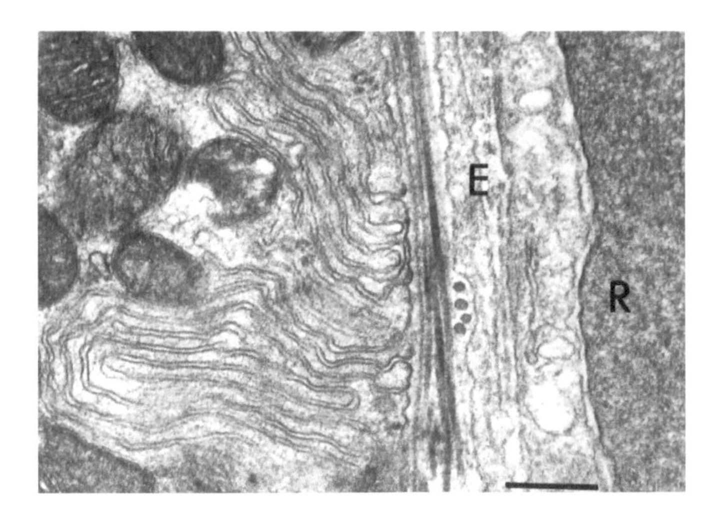

FIGURE M45c Transmission electron micrograph of the lateral infoldings of the tubular cells of the rectal gland in *Hydrolagus colliei*. *Scale*, 0.5 μm.

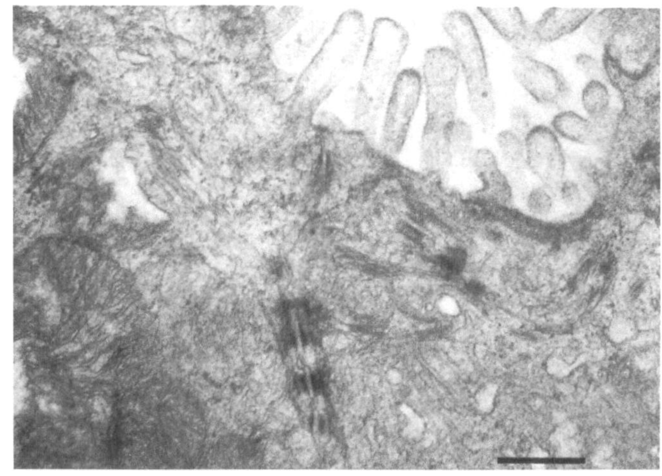

FIGURE M45d Electron micrograph of a junctional complex sealing the luminal borders of the epithelial cells in the rectal gland of *Hydrolagus colliei*. *Scale*, 0.5 μm.

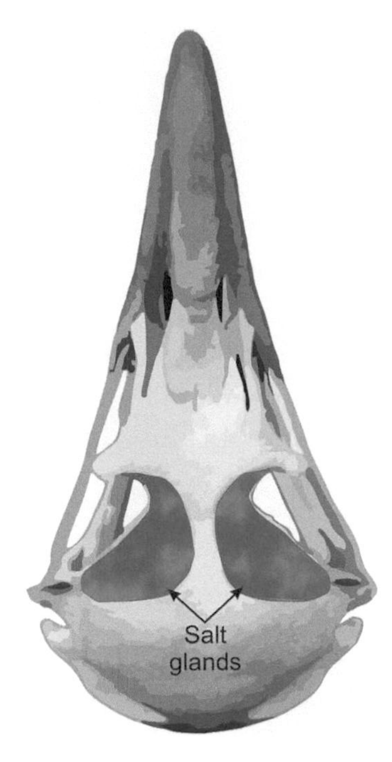

FIGURE M46a Salt glands of some birds and reptiles are also engaged in excretion of salts. In some forms, they are more important than the kidneys in the elimination of salts. They are modified nasal glands in lizards and marine birds and lachrymal glands in marine turtles but in all cases, they are roughly similar. Here, the skull of a herring gull (*Larus smithsonianus*) is shown from above; the salt glands position is just above the eyes. *Authors.*

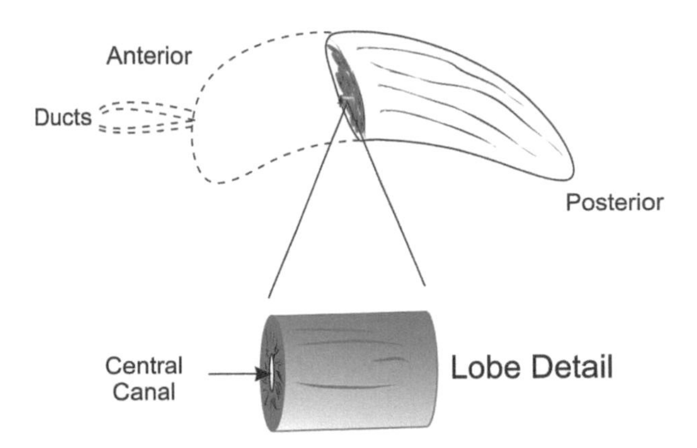

FIGURE M46b Diagram shows the gross structure of the salt gland in the herring gull. The capsule contains several banana-shaped lobes, each enclosed in a stout sheath of fibrous connective tissue. Each lobe resembles a tiny rectal gland containing secretory tubules of simple cuboidal epithelium radiating from a central lumen of stratified cuboidal epithelium. The direction of flow of blood in the capillaries surrounding the tubules is centrifugal, that is, opposite to the flow within the tubules so a counter-current exchange mechanism is set up concentrating the salts being excreted. An artery near the center of the lobe carries blood to capillaries in the inner portions of the gland; it flows centrifugally to veins at the periphery. *Authors.*

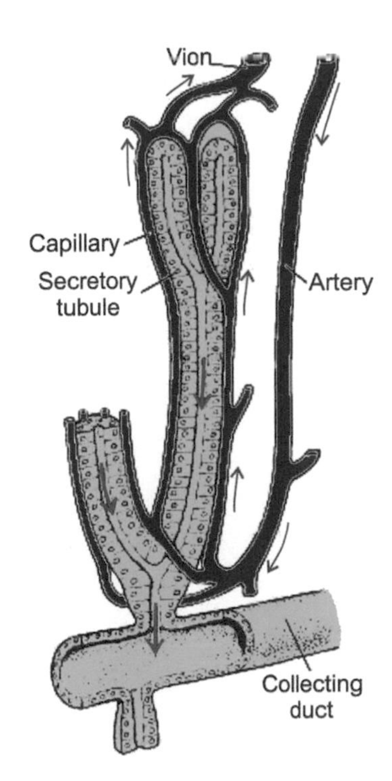

FIGURE M46c Tubules radiate from the central lumen. In the outer portion of the lobe, the tubules branch and become more closely packed, running parallel to each other. They are separated by delicate connective tissue and blood capillaries. *Modified from—By Internet Archive Book Images—https://www.flickr.com/photos/internetarchivebookimages/20381612561/Source book page: https://archive.org/stream/biologicalbulletl115mari/#page/n180/mode/1up, No restrictions.*

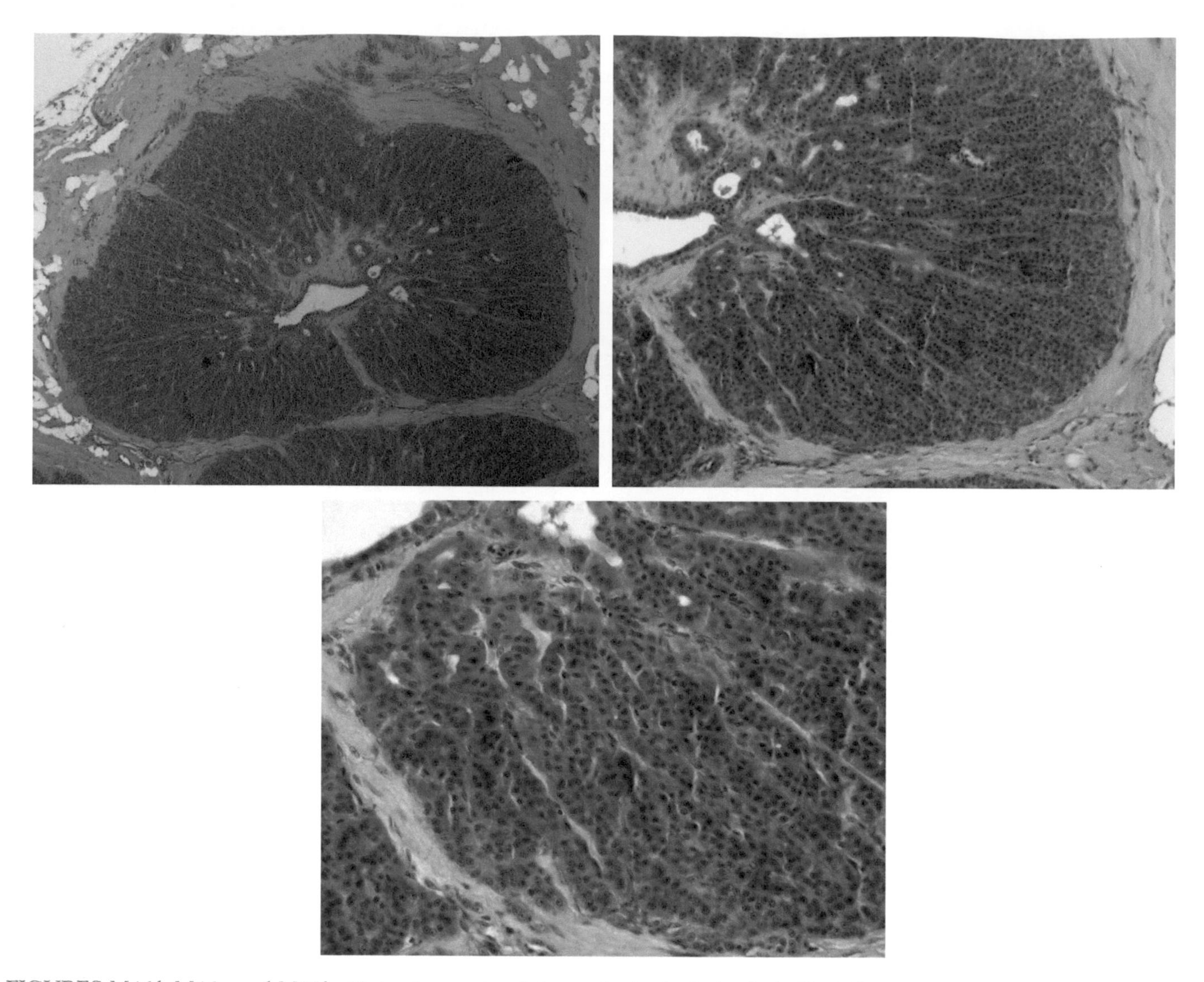

FIGURES M46d, M46e, and M46f Photomicrographs of the nasal salt gland of a duck. Blood flows centripetally in capillaries between the tubules; fluids containing a hypertonic salt solution flow in the same direction to be released into the lumen. Since blood and the fluids flow in the same direction, a counter-current exchange mechanism cannot function to increase the efficiency of the extraction of salt. The large rectal vein (ventral venous sinus) protrudes into the central canal at the top right and lies beneath the transitional epithelium. Smaller veins are seen below, 10 ×.

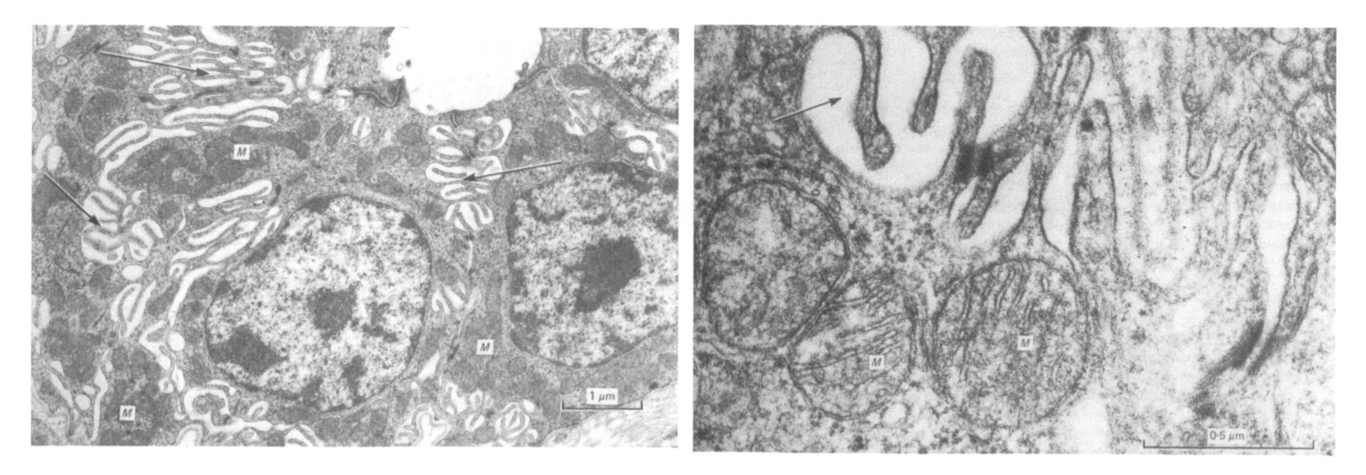

FIGURES M46g and M46h Transmission electron micrographs of a section of salt gland of the domestic duck. Mitochondria (M) are compact. Intercellular spaces are widened because of shrinkage (arrows). These spaces are packed with elaborate interlocking folds from the cell bodies. *Courtesy of Anatomical Society of GB & Ireland, with permission from Wiley.*

CHAPTER

N

Endocrine Organs

Many endocrine glands arise as invaginations of an epithelial sheet that loses its connection to the surface and must pour its secretions into the blood stream (Fig. N1a). The original epithelial arrangement of cells is lost in most cases. These glands produce polypeptide or proteinaceous hormones. Other endocrine glands are derived from masses of mesodermal cells and produce steroid hormones. In general, endocrine glands have a simple structure consisting of cords or clumps of cells supported by delicate connective tissue and richly permeated with blood capillaries or sinusoids that are often fenestrated. Most endocrine glands are discrete bodies (Fig. N1b) but some are embedded within other organs and are considered elsewhere: pancreatic islets, juxtaglomerular cells of the kidneys, corpora lutea of the ovary, and interstitial cells of the testis. In addition, there are scattered cells with endocrine function, such as the enterochromaffin or amine precursor uptake and decarboxylation (APUD) cells of the gut, that are part of the DIFFUSE NEUROENDOCRINE SYSTEM.

The diffuse endocrine system is composed of hormone-secreting cells that synthesize structurally related peptides and active amines. These products act as hormones or neurotransmitters. Neuroendocrine cells have common enzyme systems for amine handling and the production of their common secretory peptides. We have noted these cells as the APUD cells. A common origin of APUD cells from the neural crests has been suggested but not confirmed. APUD cells may act on contiguous or nearby cells (paracrine function) or on distant cells (endocrine function). Note their presence in several of the endocrine glands to be considered in this section.

PITUITARY GLAND OR HYPOPHYSIS

The pituitary gland is the most complex endocrine gland, functionally and structurally. It is a compound organ of dual embryonic origin lying at the base of the brain. The NEUROHYPOPHYSIS of tetrapods originates from a ventral evagination of the embryonic diencephalon, the INFUNDIBULUM, and the ADENOPHYPOPHYSIS arises from the HYPOPHYSEAL POUCH (Rathke's pouch), a dorsal evagination of the primitive mouth cavity or stomodaeum (Fig. N2). The pituitary is entirely ectodermal. The neurohypophysis remains attached to the brain by the INFUNDIBULAR STALK and gives rise to the POSTERIOR LOBE or PARS NERVOSA (Figs. N3 and N4). The hypophyseal pouch usually constricts off, forming a closed vesicle that comes into contact with the neurohypophysis. The posterior wall of the vesicle adjacent to the neurohypophysis becomes the PARS INTERMEDIA. The anterior wall thickens and becomes the ANTERIOR LOBE or PARS DISTALIS. This sends a trough-shaped stalk, the PARS TUBERALIS, dorsally toward the brain partially enclosing the infundibular stalk. The RESIDUAL LUMEN of the vesicle persists in some animals, separating the pars distalis and the pars intermedia; in others it is obliterated (Figs. N5a and N5b). In some sections, the pars intermedia may surround the pars nervosa. A pituitary gland is found in all vertebrates, and although the details of its gross structure vary considerably, the general description of the mammalian gland provides a good basis to begin our study. The detailed structure of the regions is shown in Figs. N5c–N5e.

The gland is ensheathed in a CAPSULE of fibrous connective tissue derived from the dura mater, one of the coverings (meninges) of the brain.

Pars Distalis (Anterior Lobe)

The pars distalis is composed of cords of cells permeated by large, thin-walled, fenestrated capillaries (Figs. N6a and N6b). This parenchyma is supported by a delicate STROMA of TRABECULAE and reticular tissue.

An Atlas of Comparative Vertebrate Histology.
DOI: https://doi.org/10.1016/B978-0-12-410424-2.00014-7

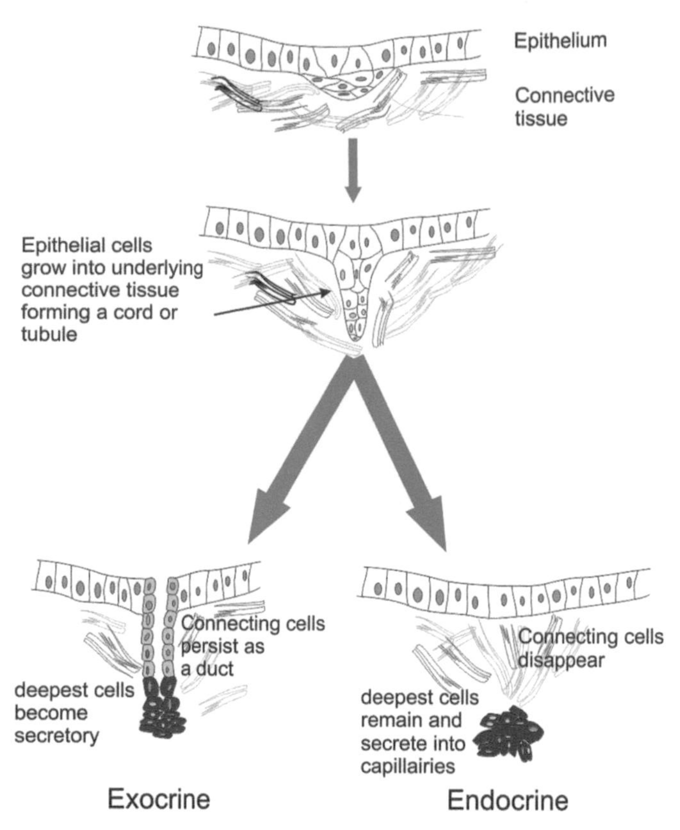

FIGURE N1a Glands may originate from stalks of epithelial cells that grow down into the underlying vascular connective tissue. These stalks may become hollow. If they retain their connections and pour their secretions onto a surface, they are EXOCRINE GLANDS. If the connection is lost and the cells secrete directly into the blood stream, they are ENDOCRINE GLANDS. *Authors.*

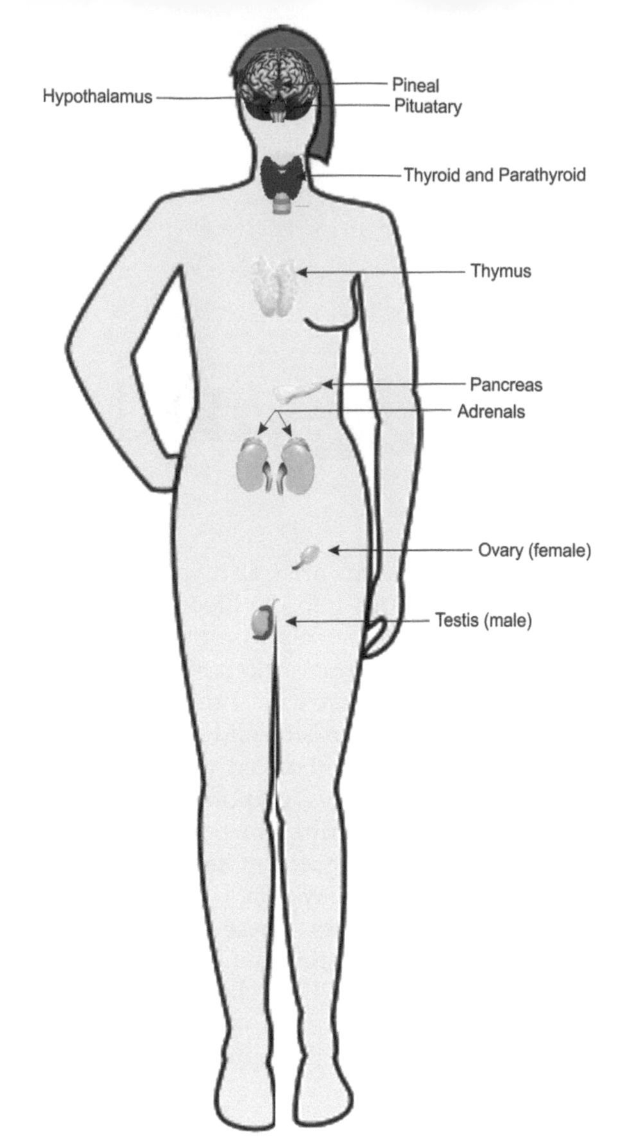

FIGURE N1b Human endocrine glands. *Authors.*

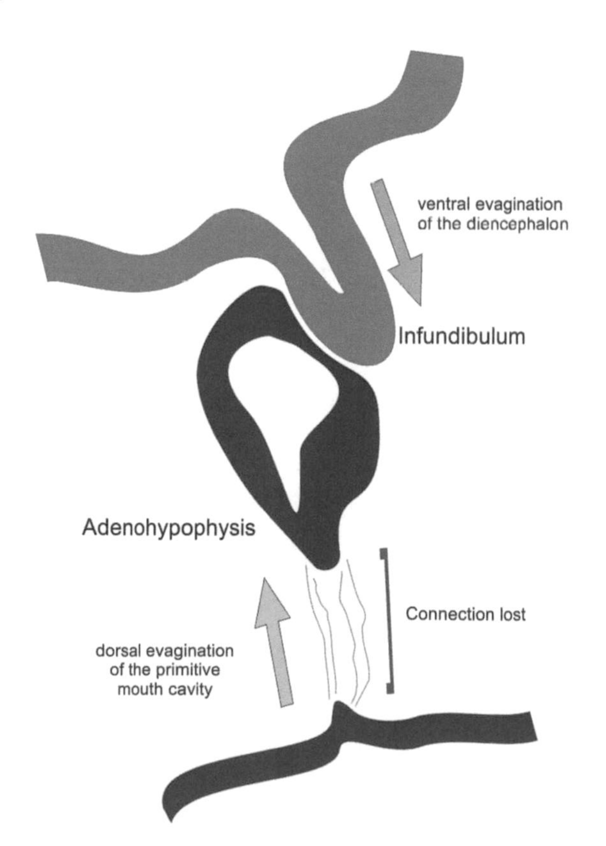

FIGURE N2 The pituitary gland of tetrapods has two embryonic origins: the NEUROHYPOPHYSIS originates from the I NFUNDIBULUM, a ventral evagination of the embryonic diencephalon, and the ADENOHYPOPHYSIS, that arises from the HYPOPHYSEAL POUCH (Rathke's pouch), a dorsal evagination of the primitive mouth cavity or stomodaeum. The pituitary is entirely ectodermal. *Authors.*

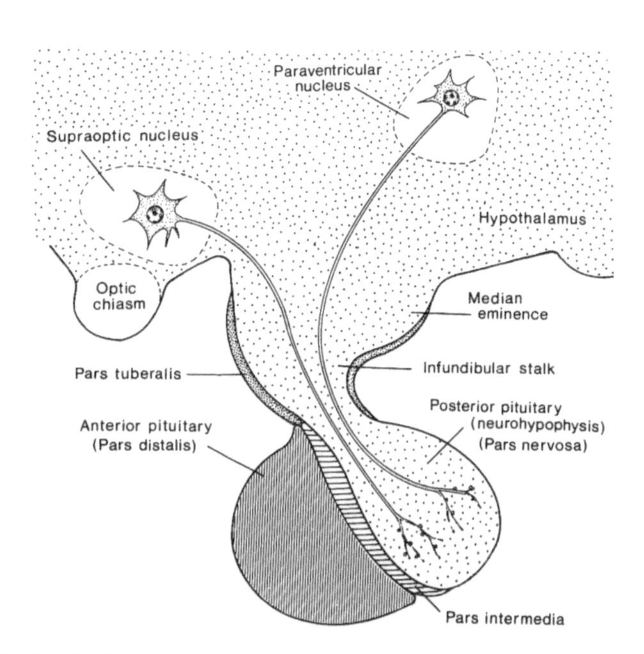

FIGURE N3 Diagram of a parasagittal section of the hypothalamus and pituitary of a mammal. Two neurosecretory neurons are shown in the hypothalamus; they pour their secretions into the pas nervosa.

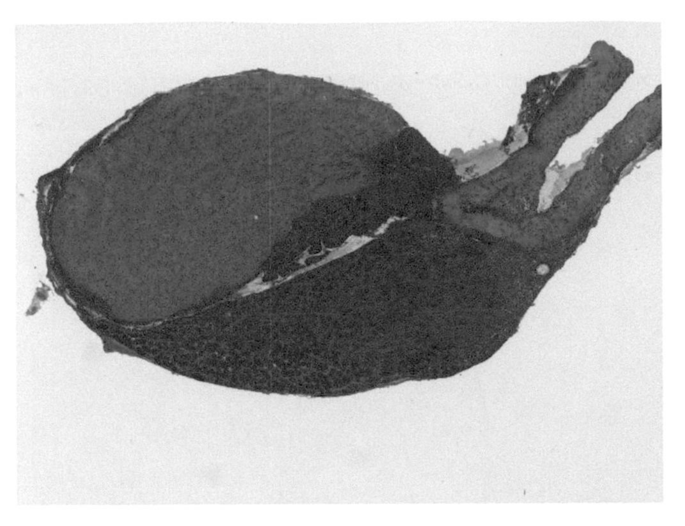

FIGURE N4 Midsagittal section of the pituitary gland of a small mammal. The PARS NERVOSA at the top is the largest area. Partially surrounding the pars nervosa is the PARS INTERMEDIA which is separated from the PARS DISTALIS at the bottom by the RESIDUAL LUMEN. Extending to the top right is the INFUNDIBULAR STALK which is partially encapsulated by the PARS TUBERALIS and connects to the hypothalamus, 2.5 ×.

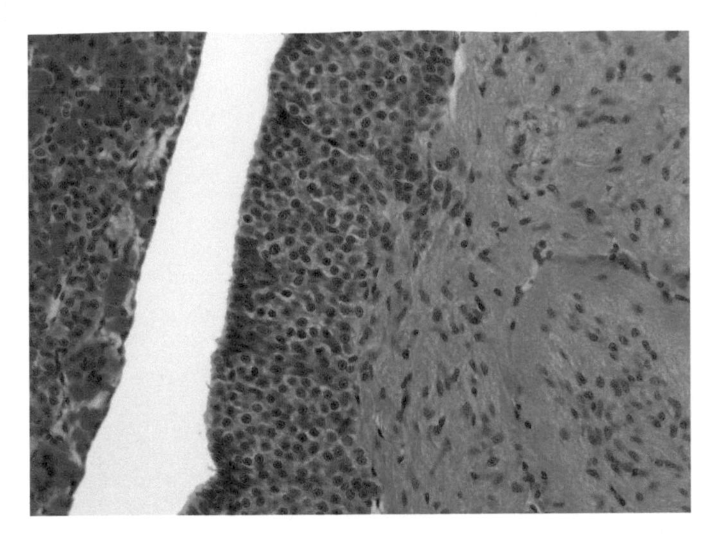

FIGURE N5a Photomicrograph of a sagittal section of the pituitary gland of a cat. The pars distalis (with conspicuous acidophils) is at the left. The pars nervosa is at the right and the pars intermedia is in the center, 20 ×.

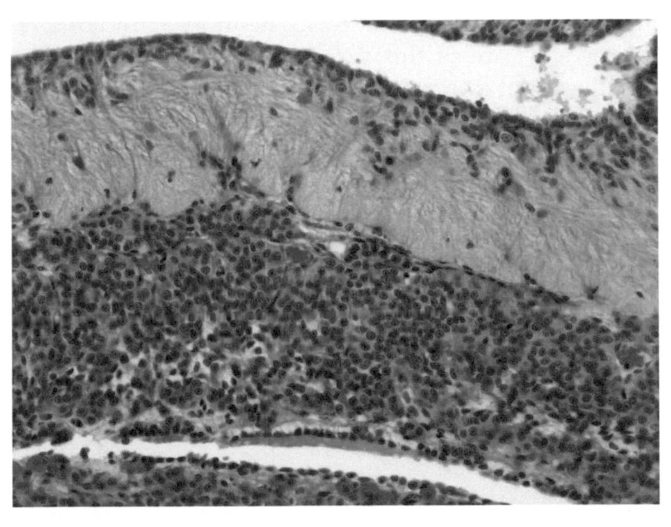

FIGURE N5b Photomicrograph of a sagittal section of the pituitary gland of a cat. The pars nervosa at the top is underlain by the pars intermedia, the residual lumen, and a small portion of the pars distalis, 20 ×.

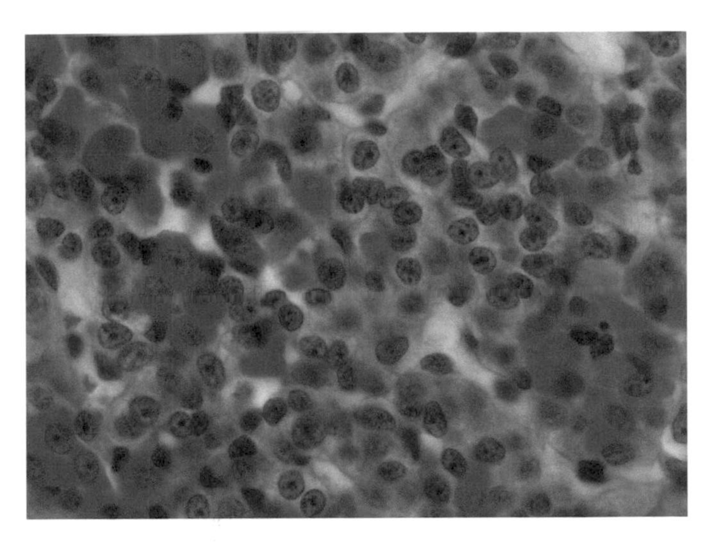

FIGURE N5c Photomicrograph of a section of the pars distalis of the pituitary gland of a cat. The ACIDOPHILS are obvious, 63 ×.

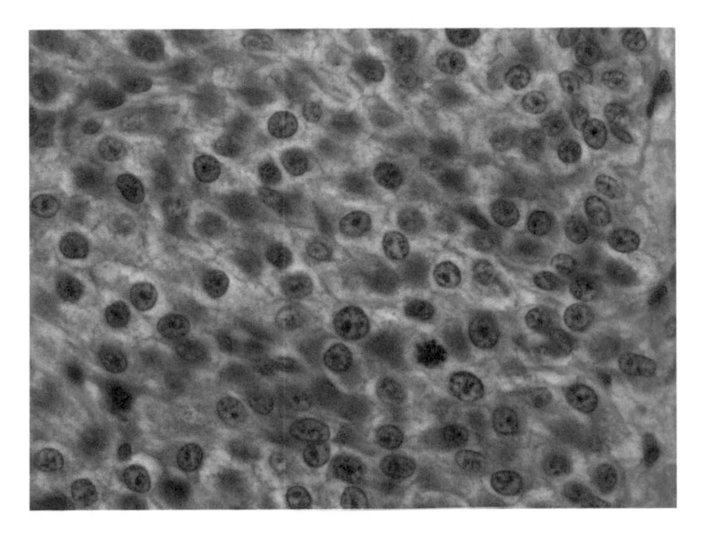

FIGURE N5d Photomicrograph of a section of the pars intermedia of the pituitary gland of a cat, 63 ×.

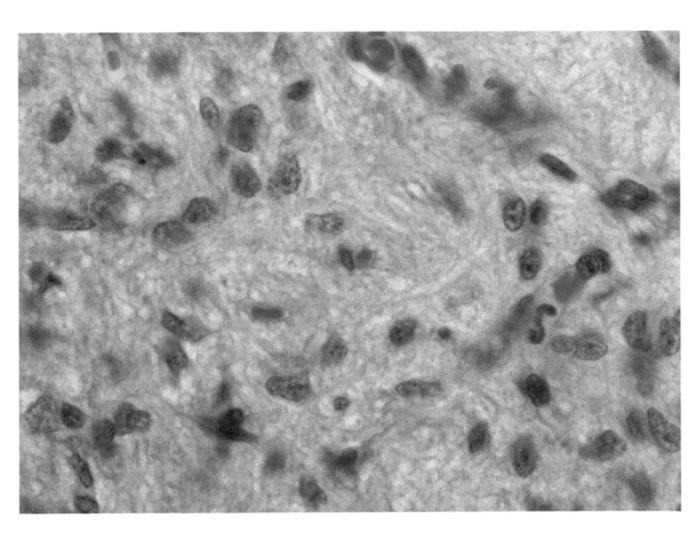

FIGURE N5e Photomicrograph of a section of the pars nervosa of the pituitary gland of a cat. Its origins from embryonic nervous tissue are obvious, 63 ×.

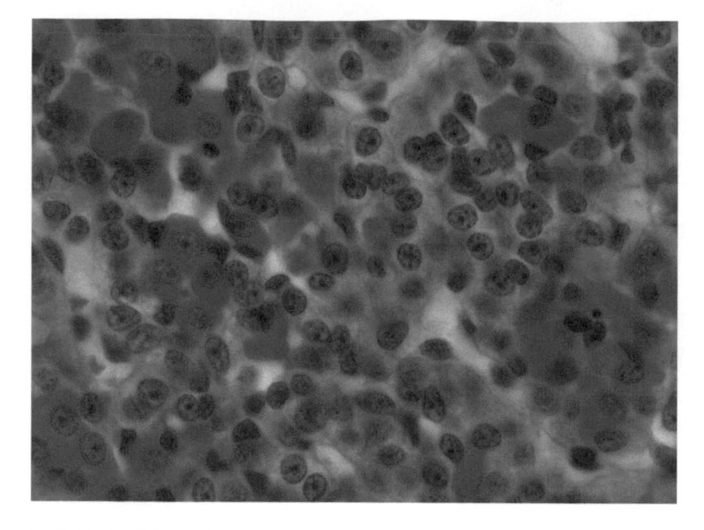

FIGURE N6a Photomicrograph of a section of the pars distalis of a cat showing conspicuous acidophils. Clusters of cells are permeated by sinusoids, 63 ×.

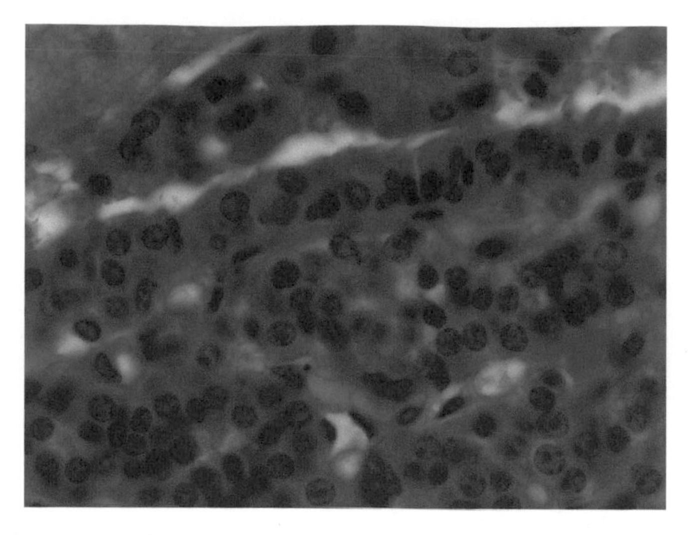

FIGURE N6b Photomicrograph of a section of the pars distalis of a small mammal. Clusters of cells are permeated by sinusoids, 63 ×.

Traditionally the cells in the cords have been named according to their staining reactions. The most abundant are small, pale, indistinctly outlined CHROMOPHOBES. About half of the cells are of this type in human. They usually clump together. About 40% of the cells are ACIDOPHILS and have abundant acidophilic granules in their cytoplasm. In special preparations, two types of acidophils may be distinguished: those with an affinity for orange G and those with an affinity for azocarmine. BASOPHILS comprise about 10% of the total and tend to be located at the periphery of the lobe. Three types of basophils may be distinguished with special stains.

The various cells may be identified immunocytochemically because they secrete specific proteinaceous hormones. Acidophils include the SOMATOTROPHS (growth hormones) and LACTOTROPHS (hormones of pregnancy). Basophils produce trophic hormones that stimulate other glands to function and include: ADRENOCORTICOLIPOTROPHS, THYROTROPHS, and GONADOTROPHS. Chromophobes are thought to be the precursors of the other types.

The anterior lobe secretes several hormones with a variety of functions: GROWTH HORMONE, PROLACTIN, FOLLICLE-STIMULATING HORMONE, LUTEINIZING HORMONE, THYROID-STIMULATING HORMONE, and ADRENOCORTICOTROPIN (ACTH).

Pars Tuberalis

Although the pars tuberalis is an outgrowth of the pars distalis, it has a different structure. It consists of small acidophilic and basophilic cells that may be arranged to form small follicles containing colloid. Its function is unknown (Figs. N7a and N7b).

Pars Intermedia

This lobe consists of a few basophilic granular cells and pale cells formed into irregular follicles containing a pale colloidal material (Fig. N7c). In elasmobranchs, amphibians, and many reptiles, it produces the polypeptide hormone INTERMEDIN (MSH: MELANOCYTE-STIMULATING HORMONE) that controls the dispersion of pigment granules in the chromatophores. This hormone stimulates the synthesis of melanin in birds and mammals but is poorly developed in primates and is absent from birds and many mammals; in these the pars distalis adheres to the pars nervosa by the pars tuberalis. (The intermedin of birds is produced by the pars distalis.)

Pars Nervosa (Posterior Lobe)

The neurohypophysis consists of three parts: the MEDIAN EMINENCE, INFUNDIBULAR STALK, and PARS NERVOSA (Fig. N3). The median eminence is the ventral portion of the hypothalamus and forms the base of the infundibular stalk. These parts bear the stamp of their origins and appear more like nervous tissue than an endocrine gland. Hormones are produced in the cell bodies of neurosecretory cells located in the supraoptic and paraventricular

FIGURE N7a Fragments of the pars tuberalis form a sleeve around the infundibular stalk (Fig. N7a). The largest portion is shown at the lower left. 5 ×.

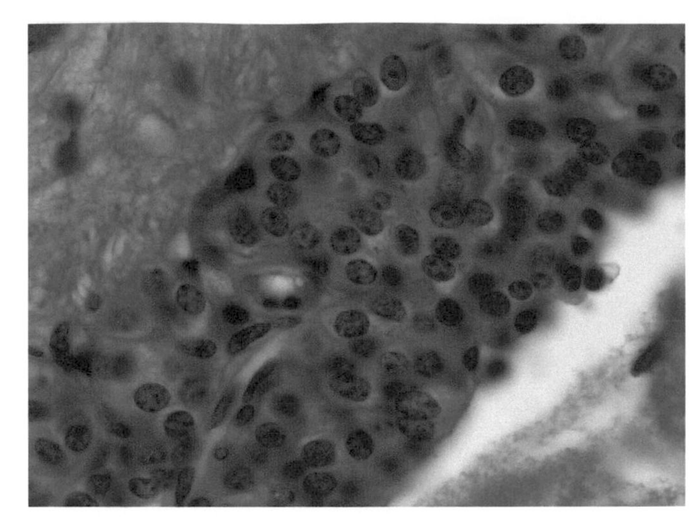

FIGURE N7b At this higher magnification several capillaries are seen coursing through the region shown in the lower right. At the top are the axonal processes of the infundibular stalk, 63 ×.

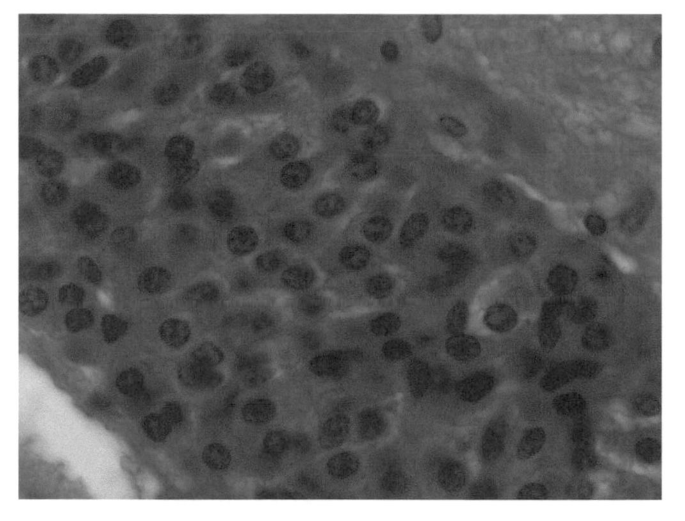

FIGURE N7c Photomicrograph of a section of the pars intermedia of a small mammal. Cords of cells are permeated by sinusoids, 63 ×.

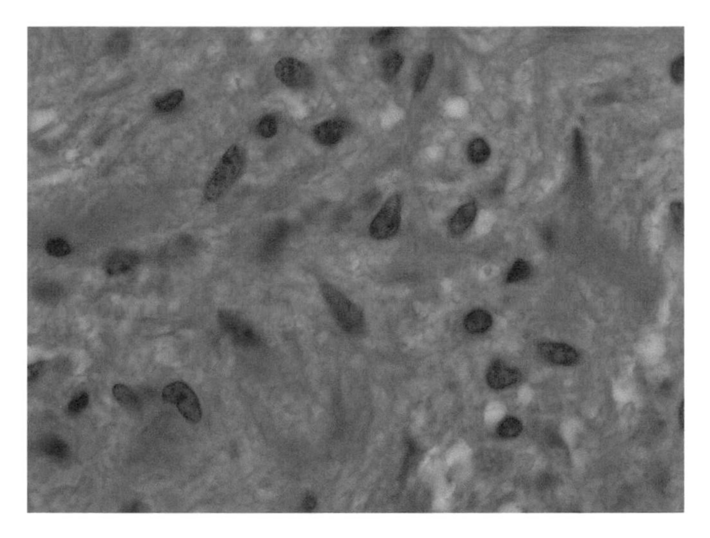

FIGURE N8 Sections of the pars nervosa of the pituitary glands of small mammals. Tracts of neurons from neurosecretory cells course in all directions. The nuclei of PITUICYTES are scattered throughout and are probably modified neuroglial cells. A few capillaries are seen in Fig. N9a, 63 ×.

nuclei of the hypothalamus and pass down the infundibular stalk to the pars nervosa in the axons of these cells (Fig. N3). Neurosecretory cells are nerve cells with all the usual characteristics expected of nerve cells, including Nissl bodies, although they usually do not synapse with other nerve cells. They produce secretory contained within a neurohaemal organ. The posterior lobe consists largely of tracts of these fibers (Figs. N5e and N8). Scattered materials that they pass along their axons and discharge into the blood from nerve terminals among the fibers are the pigmented PITUICYTES which are probably modified neuroglial cells. They vary in size and shape and in the extent of their processes which may be as much as 100 μm in length. In the tract between the hypothalamus and the pituitary, they ensheath the nerve fibers much as do the neurilemma cells of the peripheral nerves but in the pars nervosa itself they are simply scattered among the fibers, running parallel to them. Deeply staining granular aggregations of stored NEUROSECRETORY SUBSTANCE (Herring bodies) may be seen throughout the pars nervosa in the expanded ends of the axons of the neurosecretory cells (Figs. N9a and N9b). The hormones are released into the perivascular space and enter the fenestrated capillaries of this neurohaemal organ (Fig. N10a).

The neurosecretory substance contains two peptide hormones. VASOPRESSIN (ANTIDIURETIC HORMONE: ADH) stimulates cells of the mammary glands and has an antidiuretic function, promoting the reabsorption of water in the

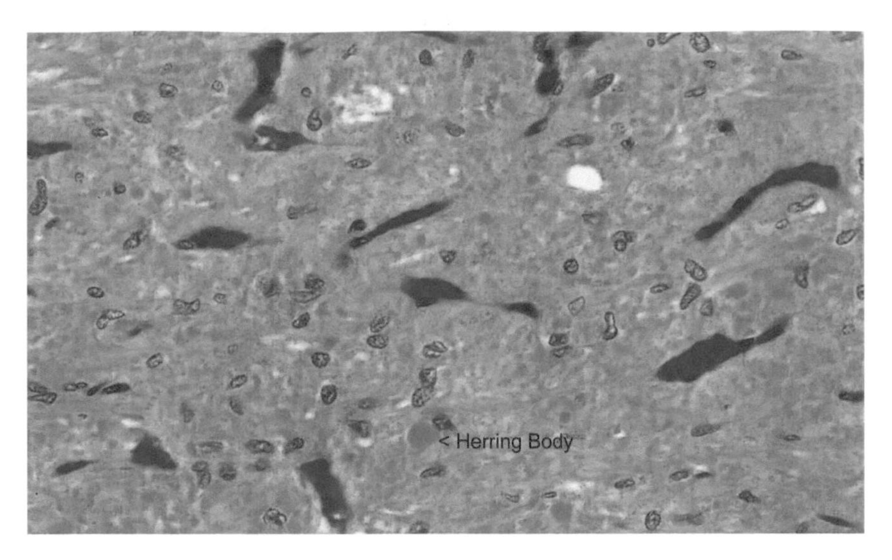

FIGURE N9a Photomicrograph of a section of the pars nervosa of a mammalian pituitary gland. A conspicuous neurosecretory granule (Herring body) at the arrow head. (63 ×) *Authors*.

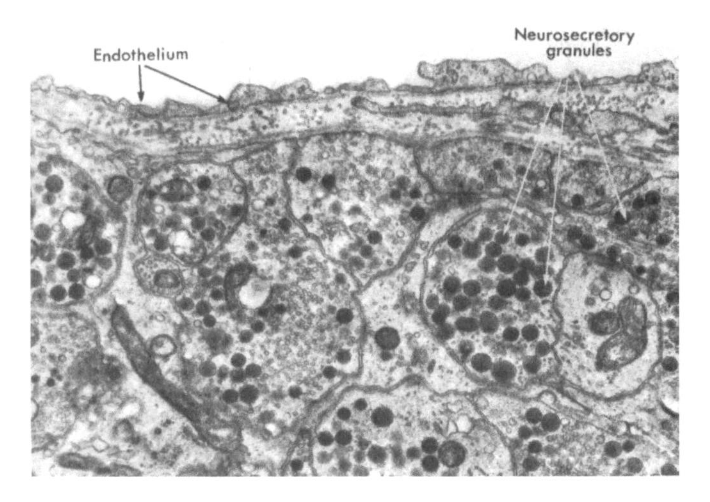

FIGURE N9b Transmission electron micrograph of a section of the neurohypophysis of a rat. The axoplasm of the fibers is packed with neurosecretory granules (Herring bodies).

kidney, and it raises blood pressure by stimulating the contraction of smooth muscle in the walls of small blood vessels. OXYTOCIN stimulates the contraction of smooth muscle in the wall of the uterus (Fig. N10b).

Blood Supply

Two pairs of arteries supply the pituitary gland (Figs. N10b and N11). Branches of the SUPERIOR HYPOPHYSEAL ARTERY pass to the median eminence and upper parts of the infundibulum. Another branch joins the INFERIOR HYPOPHYSEAL ARTERY to bring blood to the pars nervosa. The adenohypophysis has no direct arterial blood supply. Blood from the capillary beds of the neurohypophysis is drained into the HYPOPHYSEAL PORTAL VEINS and carried to the capillaries of the adenohypophysis. All parts of the pituitary are drained by the HYPOPHYSEAL VEINS.

Ultrastructure

On the basis of the appearance of their cytoplasmic granules in electron micrographs, several types of cells may be distinguished, including SOMATOTROPHS, LACTOTROPHS, ADRENOCORTICOLIPOTROPHS, THYROTROPHS, and GONADOTROPHS (Figs. N12a and N12b). CHROMOPHOBES show few or no granules in their cytoplasm. Stellate PITUICYTES can be recognized in micrographs of the pars nervosa with AXON TERMINALS of neurosecretory cells closely pressed against them

(Fig. N13). The axons contain neurotubules (microtubules), mitochondria, vesicles, and membrane-bound neurosecretory granules.

Pituitary Gland of Other Vertebrates

Although the pituitary structure of tetrapods varies considerably, it shows a basic similarity in embryonic development and mature structure in all (Fig. N14a). The pars distalis of reptiles and birds may be subdivided into the anterior ROSTRAL PARS DISTALIS and posterior PROXIMAL PARS DISTALIS. The pars intermedia is variable in amphibians and reptiles, it is absent in birds, and may be poorly defined in some mammals.

The pituitary gland of lower vertebrates lacks a true pars nervosa and the adenohypophysis lies beneath a thin-walled part of the infundibulum. The adenohypophysis consists of three regions (Fig. N14b) whose homologies with the regions of the pituitaries of higher vertebrates are not completely understood: the PROADENOHYPOPHYSIS (rostral pars distalis), the MESOADENOHYPOPHYSIS (proximal pars distalis), the MESOADENOHYPOPHYSIS (proximal pars distalis), and the METAADENOHYPOPHYSIS (pars intermedia). Nervous elements of the hypothalamus penetrate deeply into the metaadenohypophysis, less so into the mesoadenohypophysis, and least into the proadenohypophysis. Posterior to the pituitary, there may be a nervous downgrowth, the SACCUS VASCULOSUS, that contains neurosecretory fibers and may be homologous to the pars nervosa of higher vertebrates.

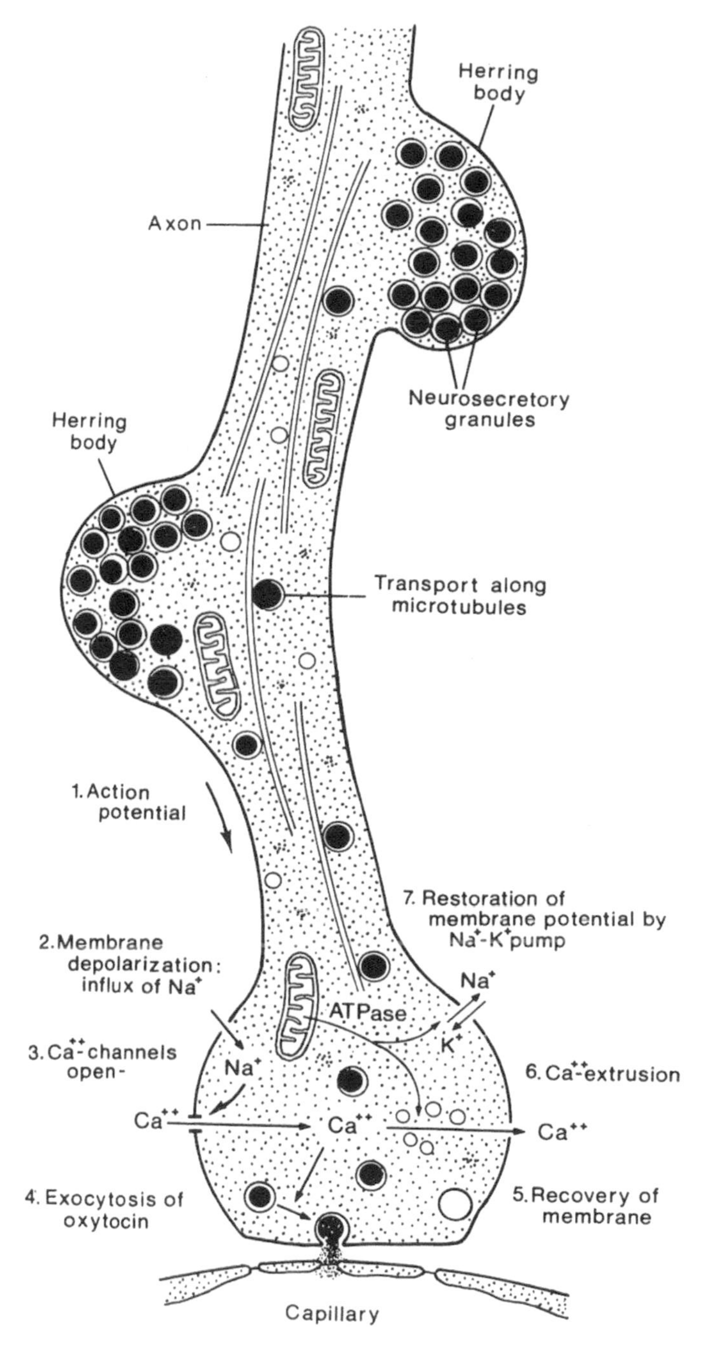

FIGURE N10a Diagram to show the travels of the NEUROSECRETORY substance from its formation in cell bodies in the hypothalamus to its release alongside capillaries in the pars nervosa, a NEUROHAEMAL ORGAN.

THE CAUDAL NEUROSECRETORY SYSTEM AND UROPHYSIS OF FISHES

Figs. N15a–N15c show the UROPHYSIS in micrographs of sections through the spinal cord of teleost fishes. These include the CAUDAL NEUROSECRETORY SYSTEM. The nervous elements in this region are gradually replaced by large basophilic NEUROSECRETORY CELLS (Dahlgren cells) lying lateral and ventral to the central canal. Axons from these cells pass, much in the manner of the neurosecretory axons of the neurohypophysis, to a NEUROHAEMAL ORGAN where they come into intimate contact with capillaries. In teleosts, this is an enlarged cord in a fossa of the last vertebral segment. Elasmobranchs lack a urophysis and the neurosecretory cells secrete directly into vascular channels in an inconspicuous neurohaemal region immediately below the spinal cord (Fig. N15d). Because of its similarity to the neurohypophysis, the caudal neurosecretory system is presumed to have an endocrine function; it is probably associated with osmoregulation.

Conspicuous aggregations of stored NEUROSECRETORY SUBSTANCE (Herring bodies) are present in localized axonal expansions of the neurosecretory cells (Figs. N16b and N16c). With the electron microscope, these cells resemble those of the hypophyseal neurosecretory system, showing the characteristics of protein secretion (Fig. 16a). The AXON

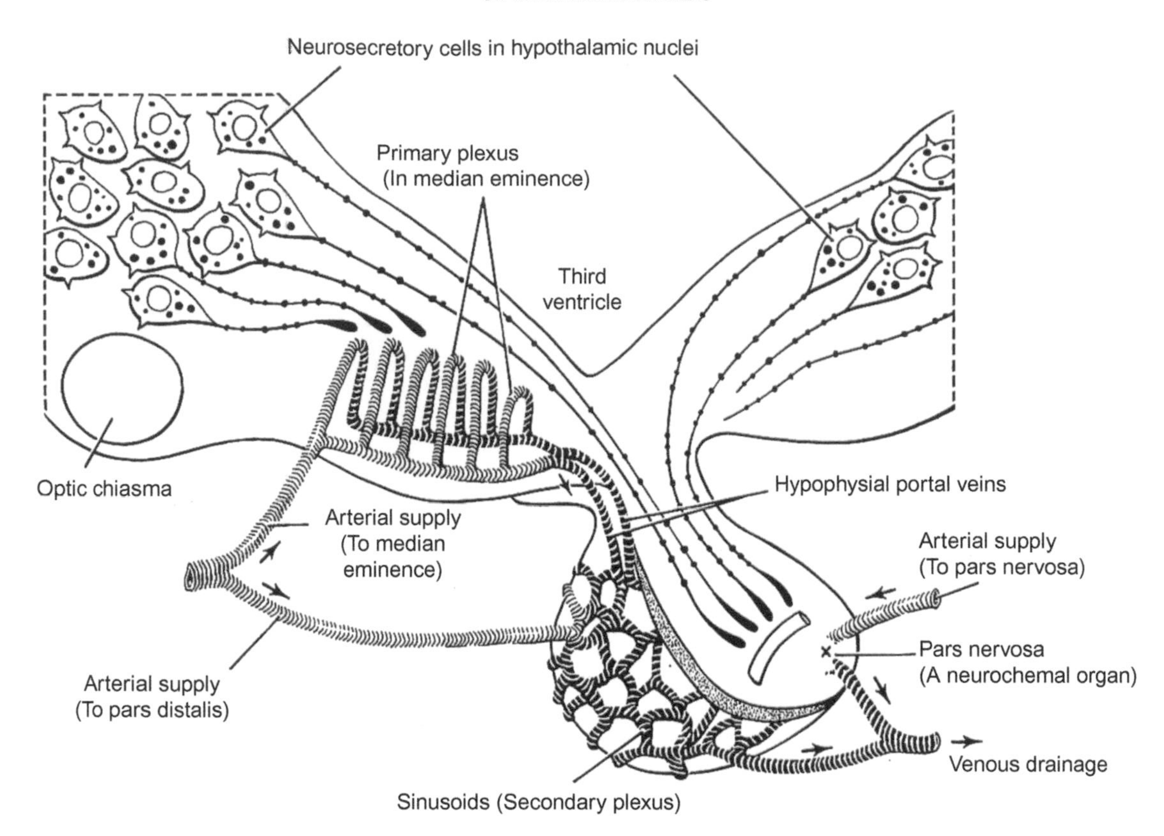

FIGURE N10b Diagram of the anatomical connections between the hypothalamus and the pituitary gland. Neurosecretory cells are present in certain hypothalamic nuclei. Some of the secretory axons pass down the infundibular stalk and terminate near blood vessels in the pars nervosa; others terminate near the capillary loops of the median eminence. The hormones of the neurohypophysis (vasopressin and oxytocin) are the products of hypothalamic neurosecretory cells and are stored and released from the pars nervosa (a neurohaemal organ). The hypophyseal portal venules begin as the primary plexus of the median eminence and convey blood downward to the sinusoids of the anterior lobe. There are strong indications that the hypothalamic axons of the median eminence liberate multiple releasing factors (probably peptide in nature) into the portal vessels and that these neural factors are concerned with the regulation of the anterior pituitary functions. It is apparent that the whole pituitary gland is predominantly subservient to and has partly evolved from the hypothalamic portion of the brain.

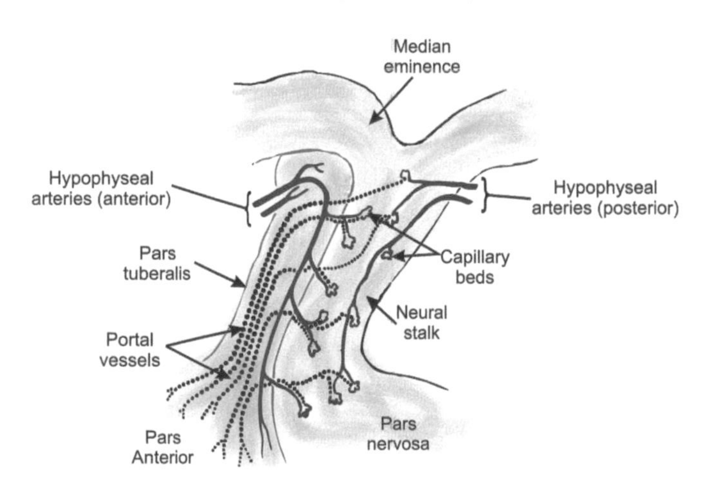

FIGURE N11 Diagram shows the blood supply to the infundibular stalk. Arteries are indicated by the red lines, and the veins by the beaded lines. *Authors.*

TERMINALS in the neurohaemal regions contain secretion granules and are interspersed with ependymal and glial elements; they abut directly on the fibrous basement membrane of a capillary.

PINEAL BODY OR EPIPHYSIS CEREBRI

The pineal body arises as a dorsal evagination of the diencephalon in all classes of vertebrates (Figs. N17a and N17b). In fishes (Fig. N17d), amphibians, some reptiles, and some birds, it remains hollow; it is solid in other reptiles and birds and all mammals. It is considered to be a vestige of a median eye that was probably a functional organ of certain extinct amphibians and reptiles. Only in cyclostomes of modern vertebrates does it form an eyelike structure. In higher forms, it responds to external stimuli, primarily visual, by changes in the synthesis and release of the indoleamine MELATONIN and other substances, thereby modifying the functions of the hypophysis and gonads in daily and seasonal rhythms.

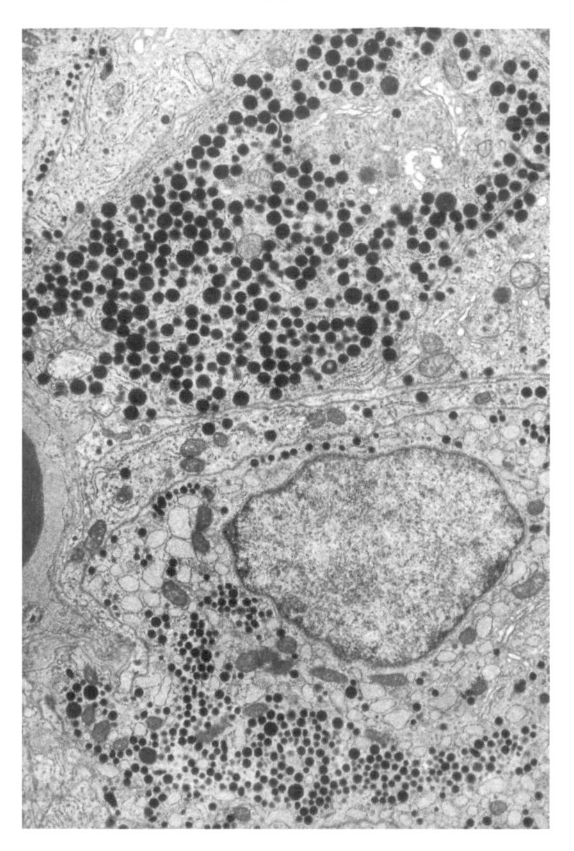

FIGURE N12a Electron micrograph of a section of the pars distalis from the pituitary of a rat. The SOMATOROPH at the top is filled with dark granules and corresponds to the acidophils of the light microscope; it produces growth hormone. The cell at the bottom with sparser granules is a GONADOTROPH; it stains as a basophil with the light microscope. Golgi complexes, and granular endoplasmic reticulum are visible in these cells, 16,000 ×.

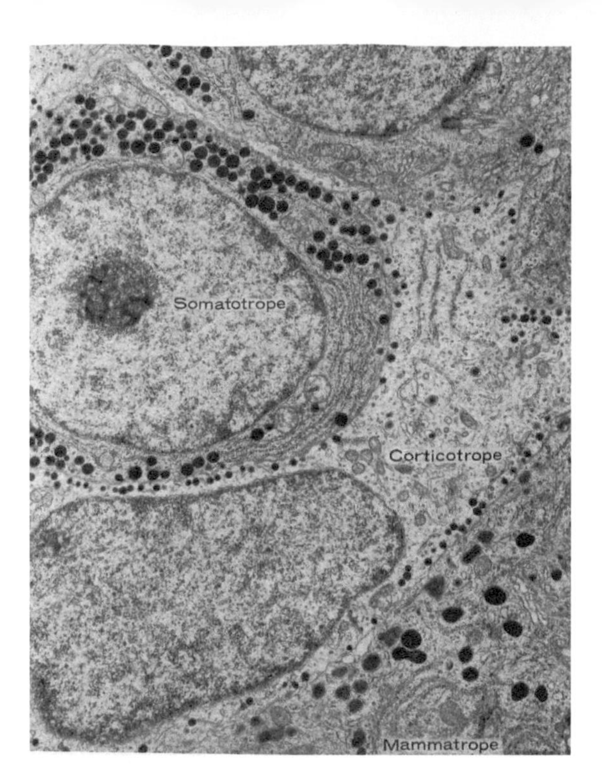

FIGURE N12b Electron micrograph of a section of the pars distalis of a rat. A SOMATOTROPE, CORTICOTROPE, and MAMMATROPE are shown.

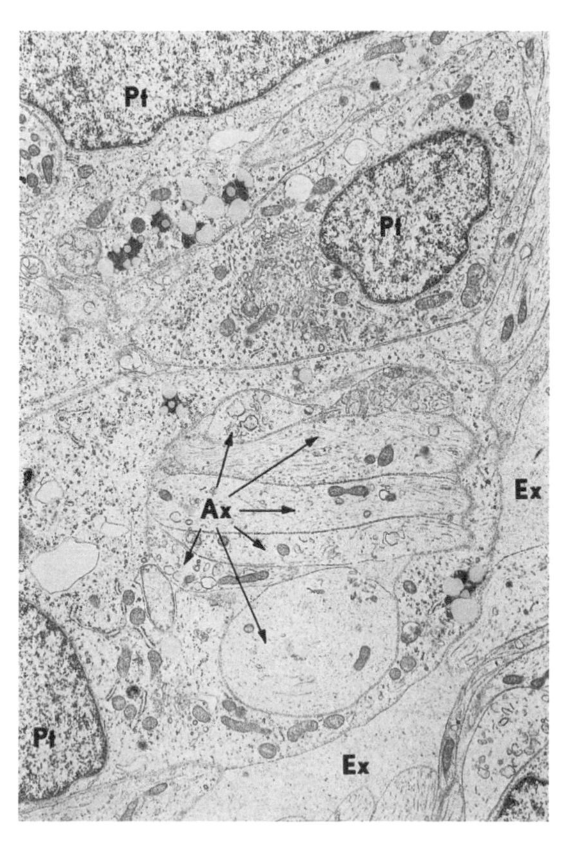

FIGURE N13 Electron micrograph of a section of the pars nervosa of the human pituitary gland. Several axon terminals (Ax) are enclosed by a single pituicyte (Pt), *Ex*, extracellular space, 5000 ×.

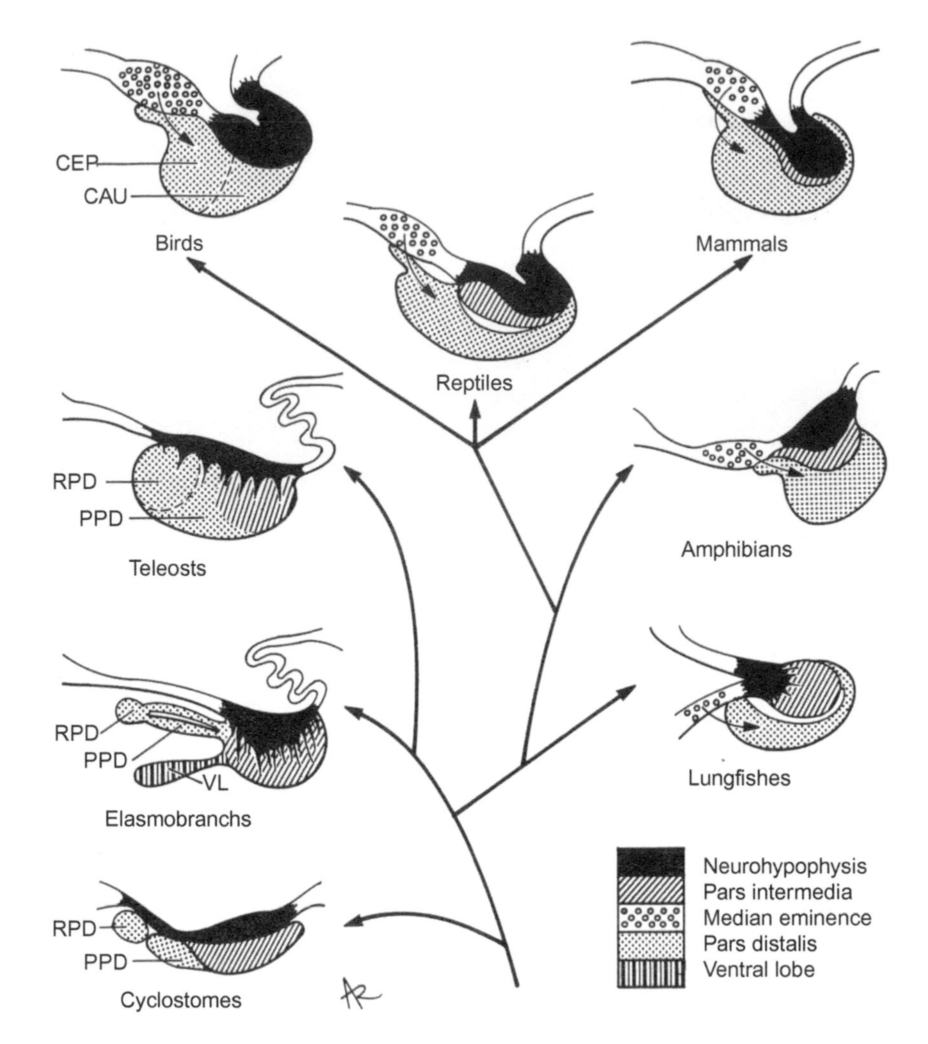

FIGURE N14a The pituitary gland in vertebrate classes. Small arrows from the median eminence to the pars distalis represent the hypophyseal portal system. The avian pars distalis is divided into two parts, the caudal (CAU) and cephalic (CEP). *PPD* is the proximal pars distalis, *RPD* the rostral pars distalis, *VL* the ventral lobe of the elasmobranch pituitary, and *SV* is the saccus vasculosus.

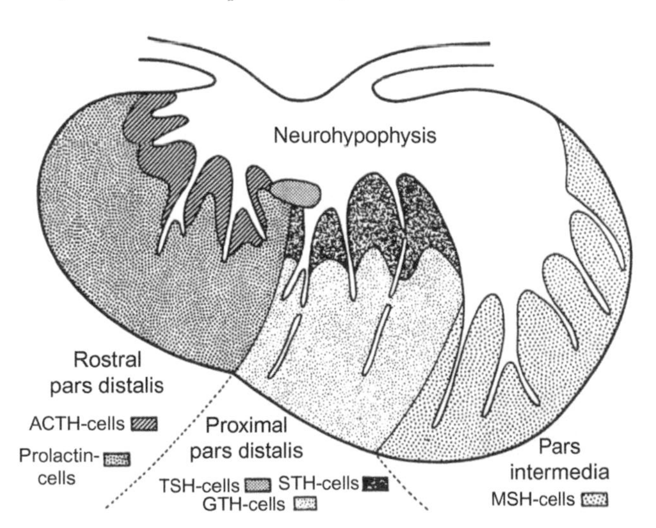

FIGURE N14b Diagram of a midsagittal section of the pituitary of a bony fish, *Tilapia mossambica*. The locations of hormone-secreting cells in the adenohypophysis are shown. *TSH*, thyrotropes; *STH*, somatotropes; *GTH*, gonadotropes; *MSH*, melanocyte-stimulating hormone.

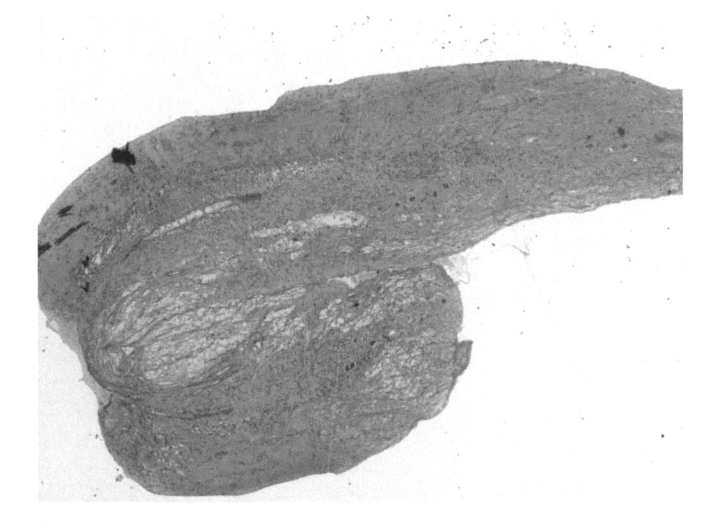

FIGURE N15a Photomicrograph of a mid-longitudinal section through the spinal cord and urophysis of a sucker (*Catostomus* sp.). Large, basophilic NEUROSECRETORY CELLS (Dalhgren cells) in the spinal cord (*top*) give rise to tracts of axons that terminate in the NEUROHAEMAL ORGAN, the UROPHYSIS (bottom). Masses of NEUROSECRETORY SUBSTANCE (Herring bodies) accumulate in the axon terminals within the urophysis. The caudal neurosecretory system synthesizes and releases two peptide hormones, UROTENSINS I and II which appear to have a role in the regulation of salt and water balance and may well be a regulator of the release of hormones of the anterior pituitary, 2.5×.

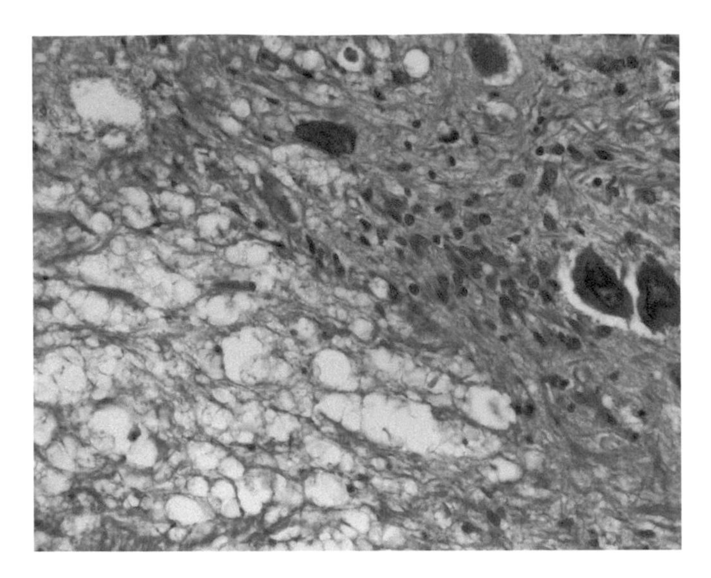

FIGURE N15b Four basophilic neurosecretory neurons are shown in the spinal cord, 40 ×.

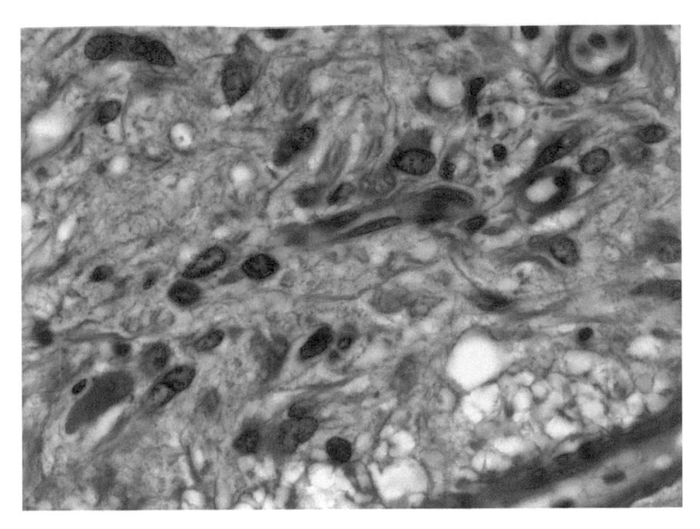

FIGURE N15c Capillaries in the urophysis are surrounded by masses of neurosecretory substance in the spinal cord off the sucker, 63 ×.

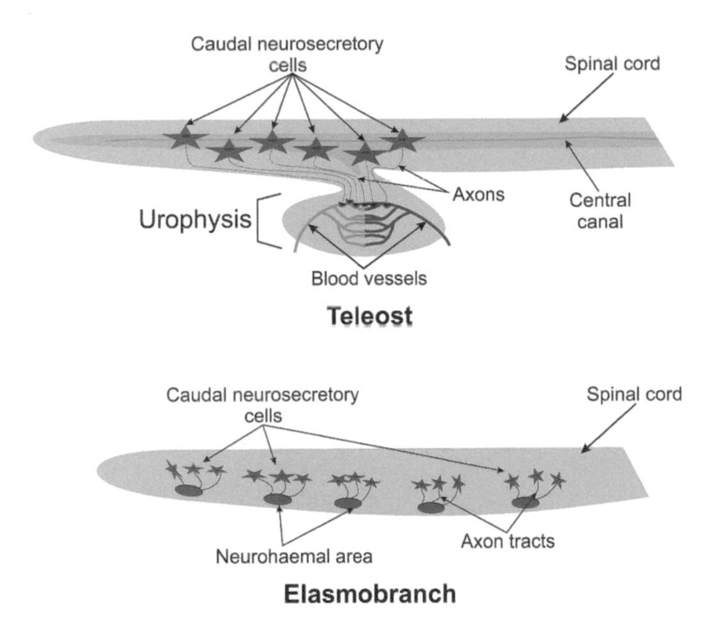

FIGURE N15d The urophysis is a neurohaemal organ in teleosts and elasmobranchs, which is associated with the posterior spinal cord and shows many parallels to the pars nervosa of the pituitary gland. This caudal neurosecretory system consists of groups of large peptide-producing neurosecretory cells in the caudal spinal cord, which give rise to axons terminating on blood vessels in the urophysis of teleosts or neurohaemal area of elasmobranchs. *Authors.*

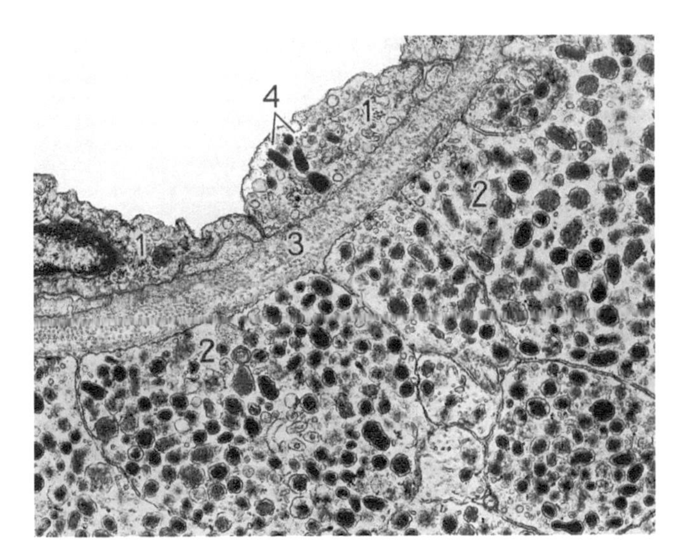

FIGURE N16a Electron micrograph of a section through the neurohaemal region of the urophysis of a pike (*Esox lucius*). Neurosecretory terminals (2) containing granules abut the basement membrane of a capillary (4) in the perivascular space (3). Granules similar to the neurosecretory granules are seen within the endothelial cell. 17,000 ×.

The mammalian pineal body is connected to the roof of the diencephalon by the PINEAL STALK (Fig. N17b). The thin connective tissue CAPSULE is a continuation of the meninges (pia mater and arachnoid) covering the brain. The hollow of the pineal stalk, the PINEAL RECESS, is an extension of the third ventricle and is lined by ependymal cells. Bundles of nerve fibers run through the stalk. The parenchyma is incompletely divided into lobules by vascular extensions of the capsule, the TRABECULAE (Fig. N17c). It consists of richly vascular and innervated aggregations of PINEALOCYTES and INTERSTITIAL CELLS. Loose connective tissue invades the parenchyma and surrounds these irregular masses. Secretions are drained by channels between the parenchymal cells, the PINEAL CANALICULI, into the PERICAPILLARY SPACES surrounding the fenestrated capillaries (Fig. N17d).

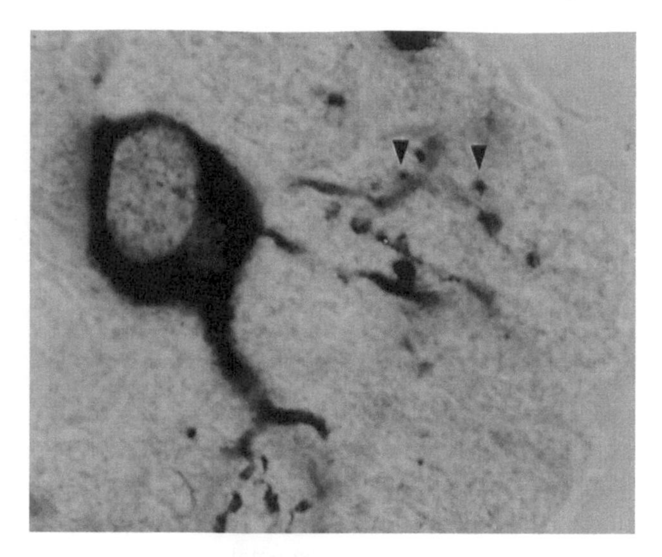

FIGURE N16b Photomicrograph of a neurosecretory cell in a section of the dogfish spinal cord treated previously with an immunostaining technique. The cell stains positive for urotensins. 1000 ×.

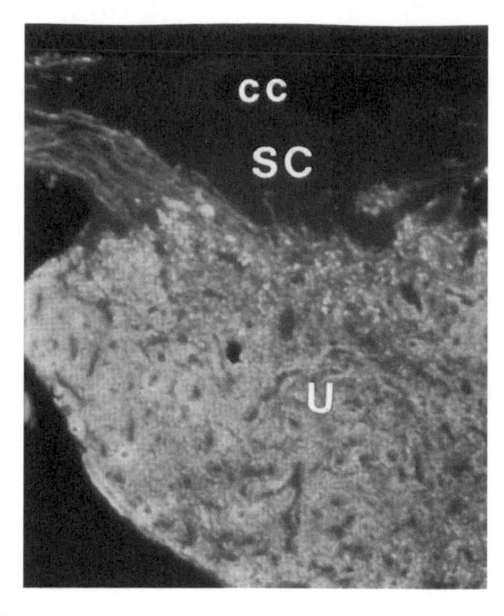

FIGURE N16c Photomicrograph of a section of the spinal cord of the teleost *Gillichthys mirabilis*. An immunofluorescence method was used to localize urotensins within cells of the caudal neurosecretory system. *cc*, central canal; *SC*, spinal cord; *U*, urophysis. 105 ×.

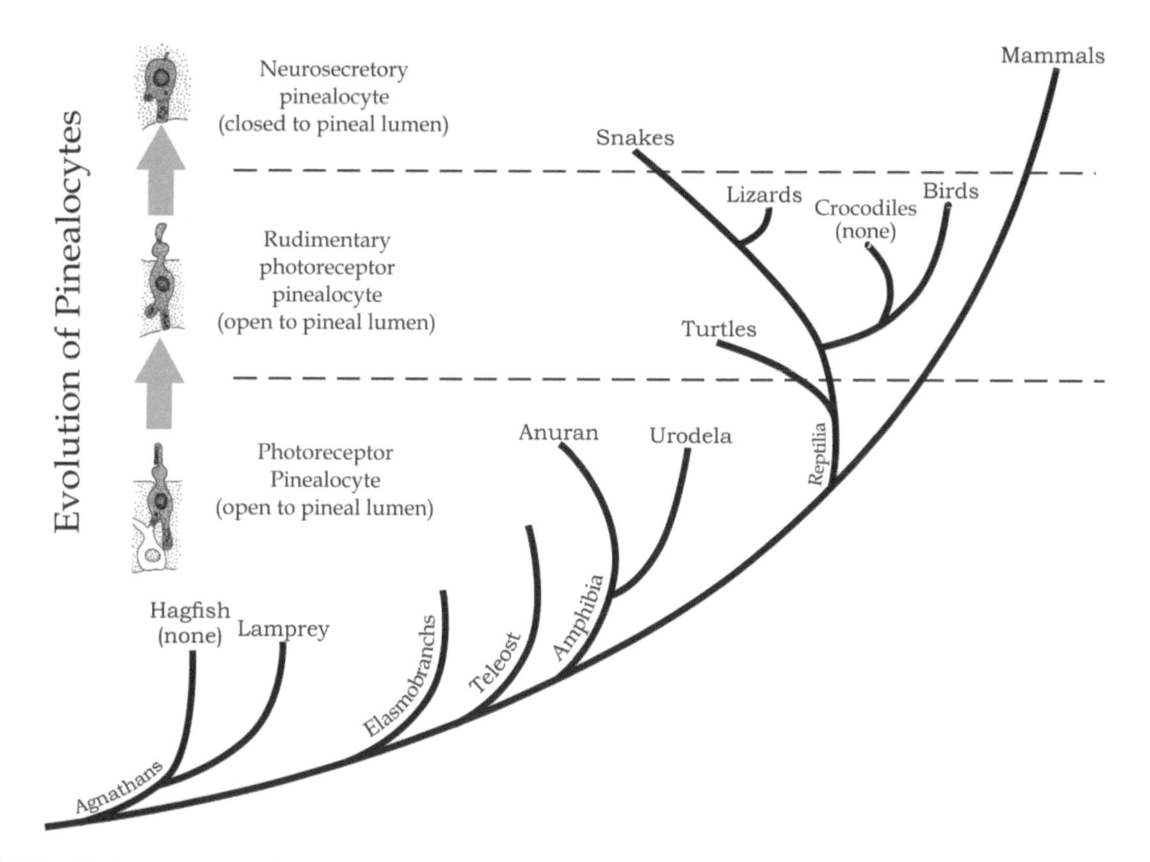

FIGURE N17a Phylogenetic tree to illustrate the development of the pineal organ in vertebrates. PINEALOCYTES may be photoreceptive or neurosecretory. Secretory pinealocytes have evolved from open photoreceptors by way of open secretory cells to closed secretory cells. *Authors.*

INTERSTITIAL CELLS may be modified astrocytes; their function is obscure although they appear to form a supporting framework for the paler, stellate PINEALOCYTES (chief cells). Long processes of the pinealocytes may end in bulbous swellings that penetrate into the pineal canaliculi or pericapillary spaces. Pinealocytes secrete melatonin, serotonin, and various peptides. With advancing age, calcified concretions, the ACERVULI or BRAIN SAND, become plentiful among the parenchymal cells (Figs. N17e and N18d).

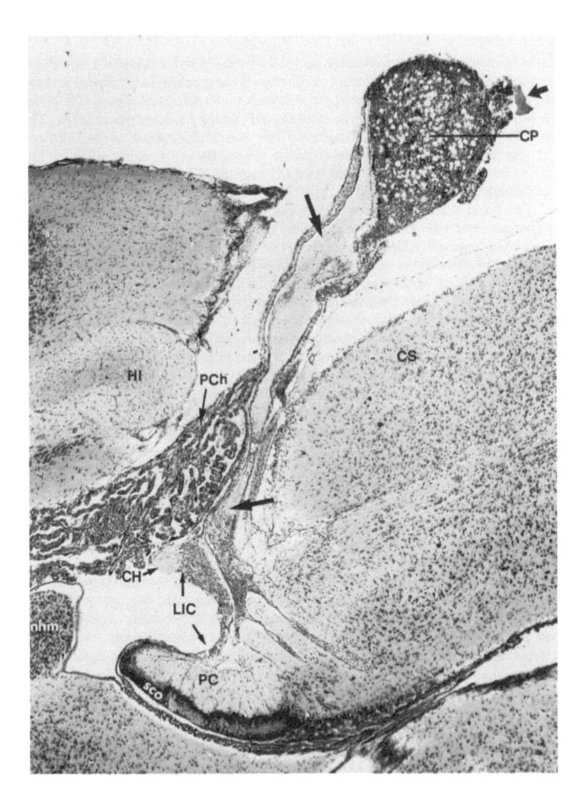

FIGURE N17b Photomicrograph of a sagittal section through the brain of a Mongolian gerbil. The superficial part of the pineal gland (CP) is connected to the brain (LIC) by a slender stalk. *CS*, superior colliculus; *HI*, hippocampus; medial habenular nucleus *nhm*; *PCh*, choroid plexus; *SCO*, subcommissural organ; nerve, *short arrow*; vein, *large arrows*, 52 ×.

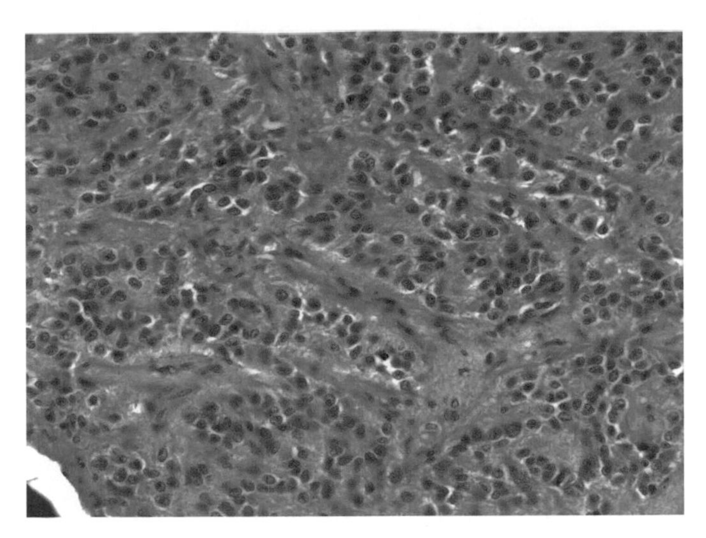

FIGURE N17c The parenchyma of a mammalian pineal body is divided into lobes by trabeculae, vascular extensions of the capsule, 20 ×.

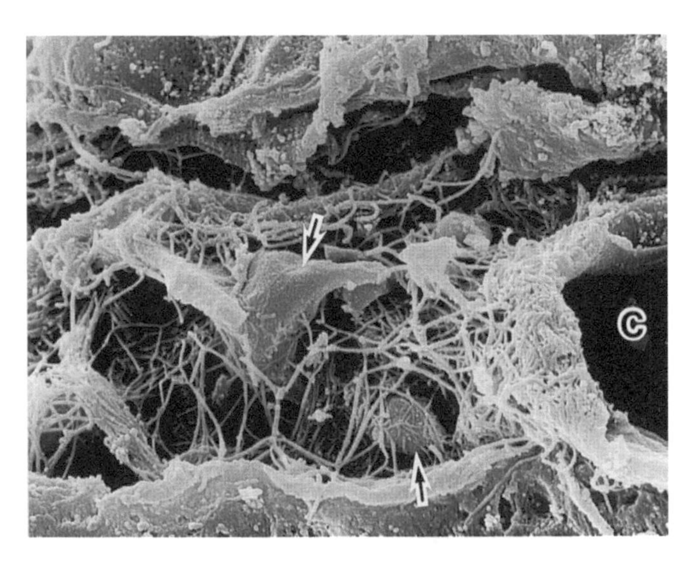

FIGURE N17d Scanning electron micrograph of a pineal organ of a Japanese fish, the ayu (*Plecoglossus altivelis*). Its capillaries are highly fenestrated and are devoid of a blood–brain barrier. The capillary (*right*) is surrounded by a fibrous perivascular space containing roaming cellular elements (*arrows*) 18,500 ×.

In electron micrographs of the pineal gland of various vertebrates, there is ultrastructural evidence of light-sensitive cells: stacks of membranous discs or LAMELLAR PLATES, and electron-dense SYNAPTIC RIBBONS aligned perpendicular to the presynaptic membrane and surrounded by synaptic vesicles (Figs. N18a–N18c)

THYROID GLAND

The thyroid gland concentrates iodine from the blood and produces the hormones TRIIODOTHYRONINE (T_3) and TETRAIODOTHYRONINE (T_4, thyroxin) that stimulate cellular metabolism throughout the body. By controlling metabolic rate, thyroid hormones have many diverse effects throughout the body. The thyroid appears in vertebrate embryos as a median ventral diverticulum of the pharynx at the level of or posterior to the first pharyngeal pouches (Fig. N19c). For this reason, attempts have been made to homologize it

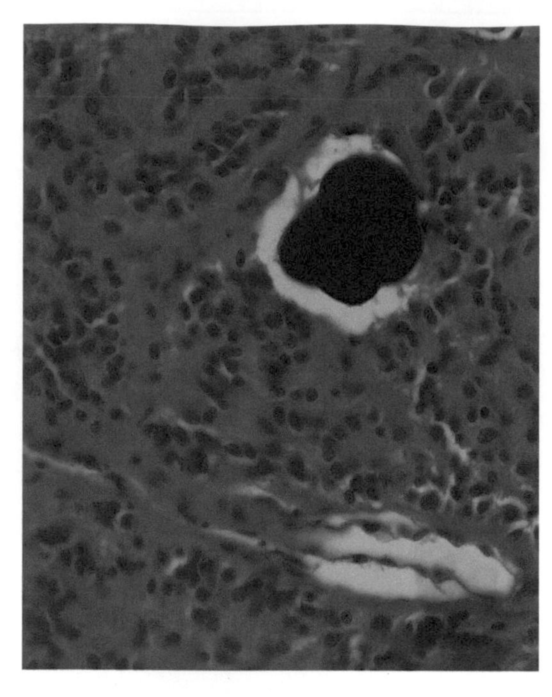

FIGURE N17e ACERVULI (calcified granules of unknown function—also called brain sand) are often seen in sections of the pineal gland of aging humans. In this section of the pineal gland of an aged human an acervulus can be seen at the upper right, 40 ×.

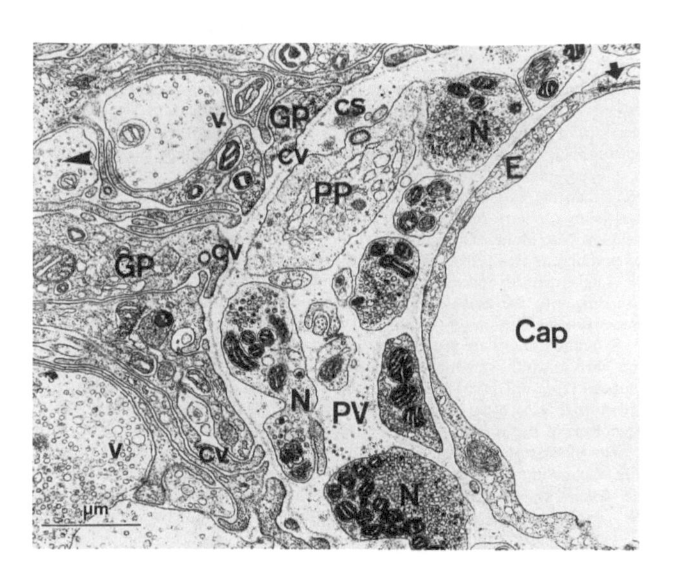

FIGURE N18a Transmission electron micrograph of the pineal of a gerbil (*Meriones unguiculatus*). A capillary at the right (Cap) is surrounded by the perivascular space (PV) which contains processes of pinealocytes (PP), and nerve endings (N). Processes of glial cells (GP) surround the perivascular area. Apposing endothelial cells are joined by tight junctions (small arrows), 19,000 ×.

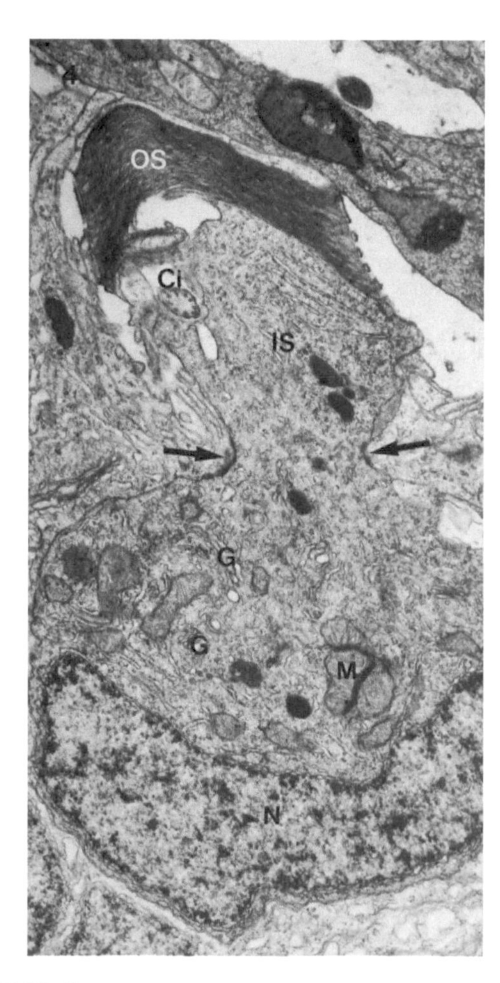

FIGURE N18b Electron micrograph of a photoreceptor cell in the pineal of a goldfish (*Carassius auratus*). Stacks of membranous discs (lamellar plates) probably indicate photosensitivity, 19,000 ×.

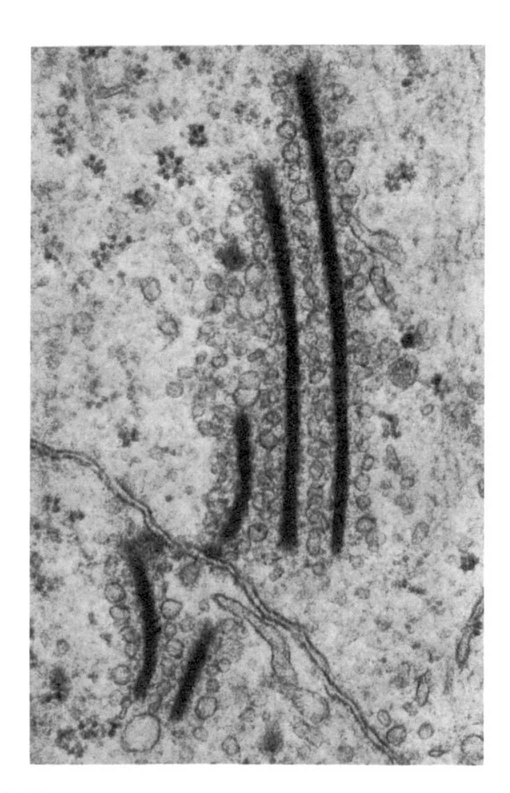

FIGURE N18c Transmission electron micrograph of synaptic ribbons in pinealocytes of the pineal of theguinea pig. Dense rods perpendicular to the surface of the cell are surrounded by small vesicles. Synaptic ribbons are similar to structures found in the sensory cells of the retina and inner ear. They do not appear to have a synaptic relationship with glial cells, nerves, or pinealocytes, 61,000 ×.

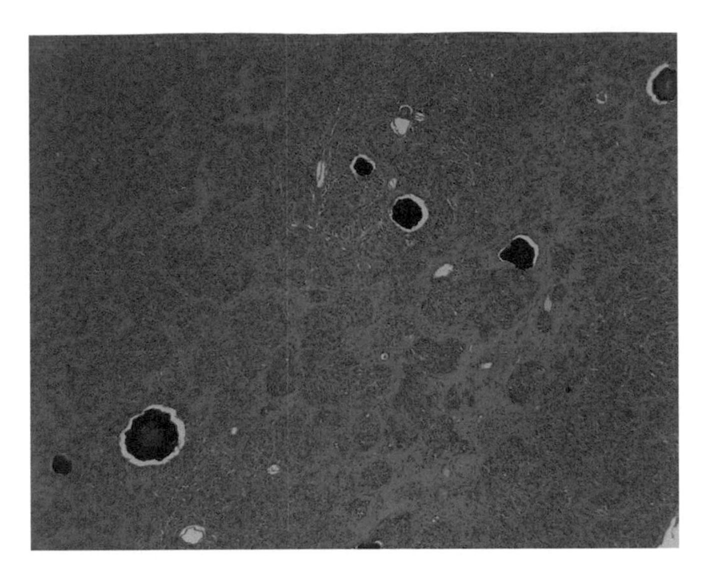

FIGURE N18d This low power photomicrograph of the pineal gland of an aged human shows several acervuli (calcified granules of unknown function—also called brain sand), 5 ×.

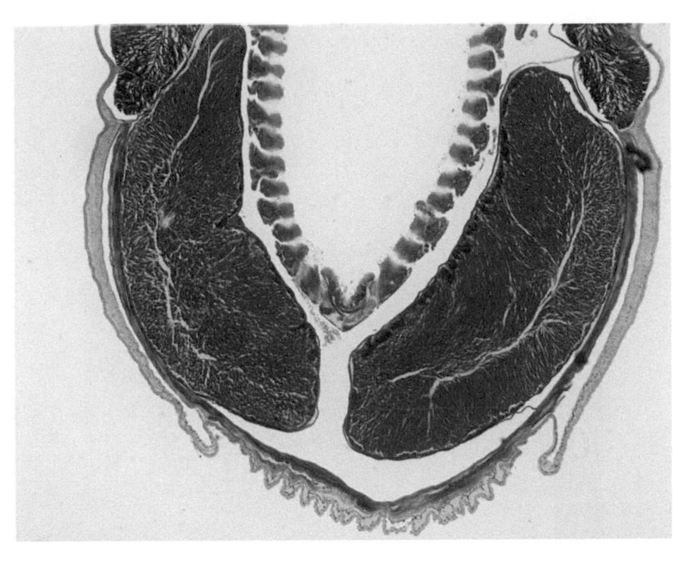

FIGURE N19a Photomicrographs of a cross-section through the pharyngeal region of amphioxus. Heavy body muscles at the lower left and lower right embrace the gill basket. The ciliated ventral groove is the ENDOSTYLE, 5 ×.

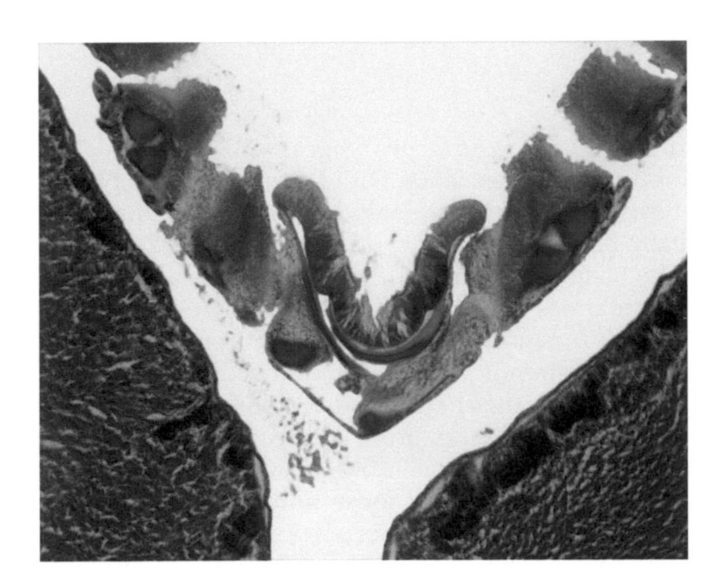

FIGURE N19b A higher magnification of the above section is the ciliated groove, the ENDOSTYLE, which has been thought by some to be the evolutionary precursor of the thyroid gland, 20 ×.

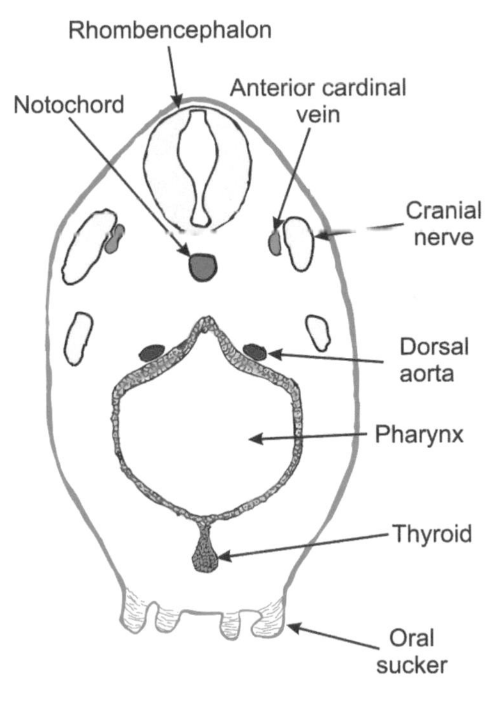

FIGURE N19c Drawing of a horizontal section through a 6.5 mm frog larva posterior to the first pharyngeal pouch shows the precursor of the thyroid gland. *Authors.*

with the ENDOSTYLE of the protochordates. This glandular, ciliated groove (which may be seen in cross sections through the pharynx of amphioxus) (Figs. N19a and N19b) not only aids in trapping food particles that are carried down the gut but has also been shown to concentrate iodine. A similar groove, the SUBPHARYNGEAL GLAND, is found in ammocoetes (Figs. N20a and N20b). It retains its connection to the pharynx and, although it is an exocrine gland, it produces thyroxin. At metamorphosis the duct is lost and some of the cells lose their cilia and assume the follicular structure of the thyroid gland. In higher vertebrates the thyroid may be paired or unpaired

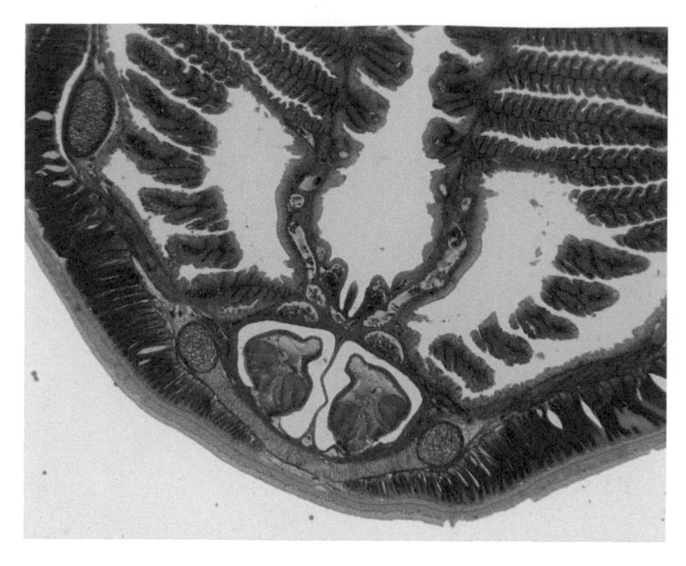

FIGURE N20a Photomicrograph of a cross-section of a larval lamprey or ammocoete. The pharynx and gills are shown in the upper part of this micrograph. Within a separate chamber, but connected to the pharynx by a duct (not shown) is the SUBPHARYNGEAL GLAND. Although this is an exocrine gland, it produces thyroxin. At metamorphosis, the duct is lost and some of the cells lose their cilia to assume the follicular structure of the thyroid gland, 5 ×.

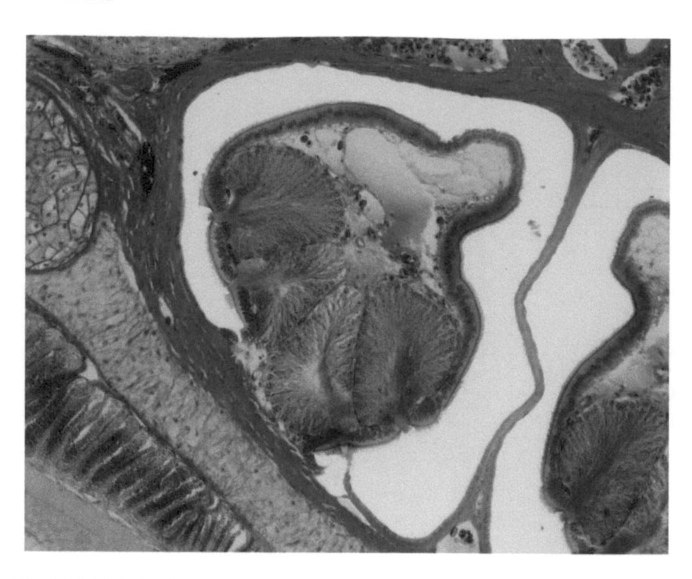

FIGURE N20b A high-powered view of the section shown in Fig. N20a. The masses of ciliated cells in the subpharyngeal gland maintain their connection to the exterior by way of a pore, 20 ×.

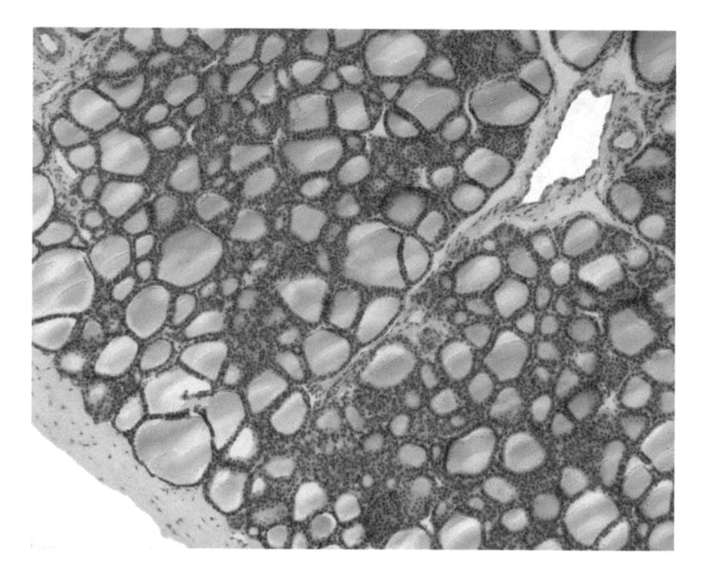

FIGURE N20c The thyroid of higher vertebrates may be paired or unpaired and lies in the ventral pharyngeal wall or, in higher forms, in close association with the trachea. Its histological structure is similar in all vertebrates: a mass of spherical THYROID FOLLICLES, separated from each other by a STROMA of delicate strands of vascular connective tissue, and contained within a CAPSULE of dense fibrous connective tissue. In this section of the thyroid of a dog, the tough CAPSULE is shown at the lower left. Two LOBULES contained within SEPTA of fibrous connective tissue. The follicles are hollow balls of simple squamous to simple columnar epithelium containing a mass of gelatinous COLLOID, 10 ×.

and lies in the ventral pharyngeal wall or, in higher forms, in close association with the trachea (Fig. N20c). In birds and mammals the thyroid is bilobed, the lobes usually being connected by an ISTHMUS (Fig. N20d).

The histological structure of the thyroid is similar in all vertebrates (Figs. N21a and N21b). It consists of spherical THYROID FOLLICLES separated from each other by delicate strands of vascular connective tissue. It is enclosed in a CAPSULE of dense, fibrous connective tissue (Fig. N21c). SEPTA from the capsule divide the gland into LOBES and LOBULES and the delicate INTERFOLLICULAR STROMA of loose fibrous connective tissue supports the follicles. FOLLICLES are hollow balls of simple squamous to columnar epithelium containing a mass of gelatinous COLLOID (Figs. N21a–N21e). In general the taller the cell, the more active it is presumed to be; cell height in any one follicle is uniform. The cells rest on a delicate basement membrane supported by a mesh of fine reticular fibers. The epithelial cells are richly supplied with capillaries from blood vessels coursing through the stroma; lymphatics are abundant (Fig. N21f).

There is a second type of cell in the follicular epithelium: the sparser, larger, rounder, and paler PARAFOLLICULAR, LIGHT, or C CELL (Fig. N21b). Although these cells are contained within the basement membrane of the follicular epithelium, they seldom

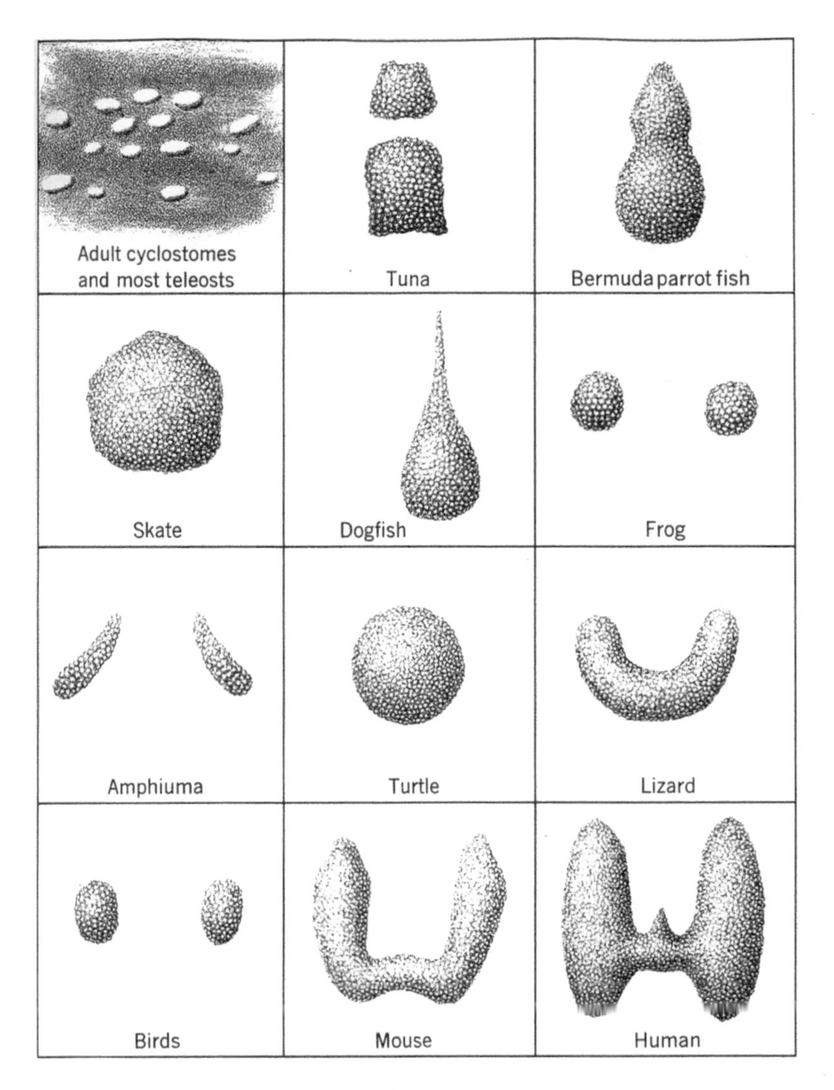

FIGURE N20d The shapes of thyroid glands in various groups of vertebrates.

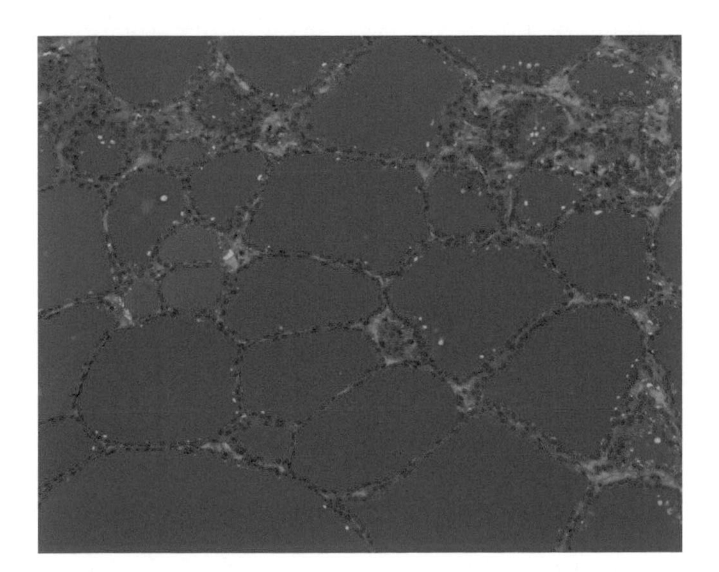

FIGURE N21a Section of the thyroid gland of a mammal. The gland consists of spherical follicles enclosed by delicate vascular connective tissue. The hollow lobules contain gelatinous COLLOID, 20×.

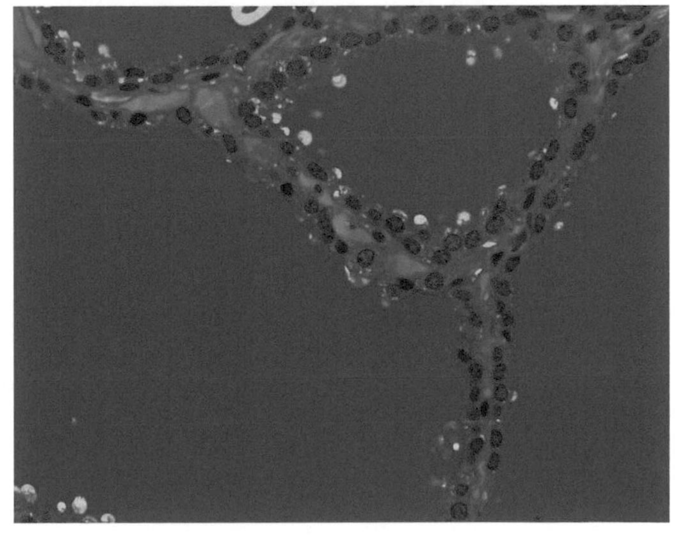

FIGURE N21b A high-powered view of the same slide shows a simple cuboidal epithelium surrounding the brightly colored colloidal material. The epithelium rests upon a delicate vascular network of collagenous fibers. A few of the epithelial cells are paler and do not reach the surface—these are CALCITONIN CELLS or C CELLS, 63×.

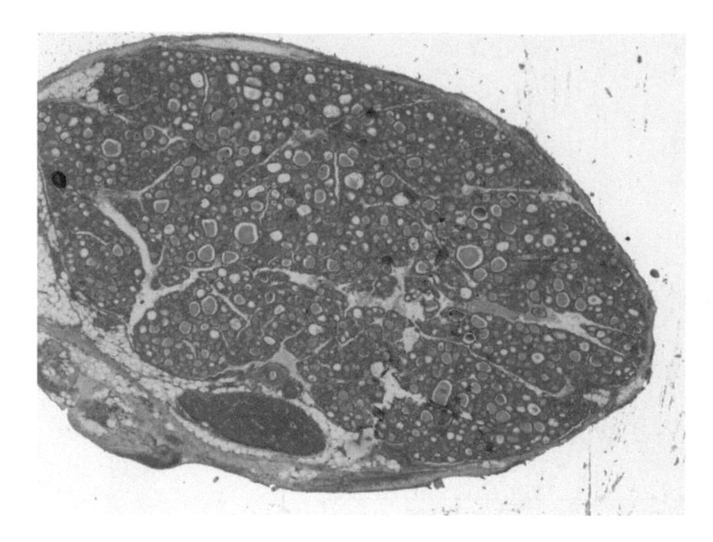

FIGURE N21c Section of the thyroid gland of a cat. SEPTA from the CAPSULE of dense fibrous connective tissue divide the gland into LOBES and LOBULES. A STROMA of loose fibrous vascular connective tissue supports the follicles. The darker structure at the bottom center of the micrograph is the PARATHYROID GLAND, 2.5 ×.

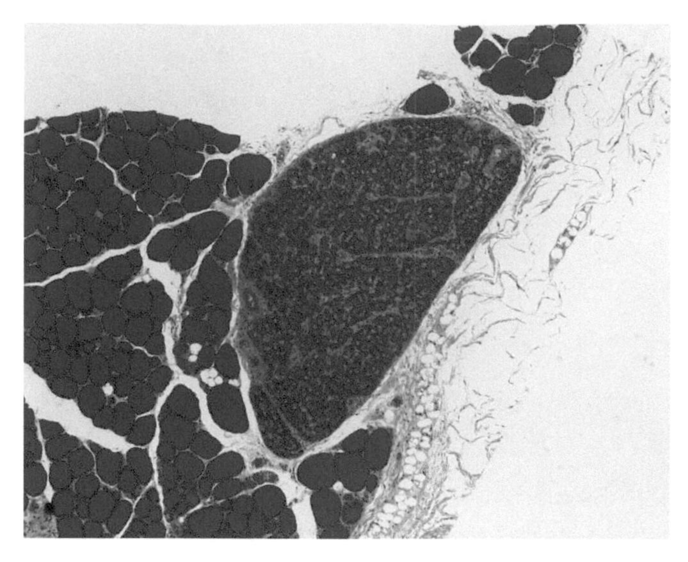

FIGURE N21d Higher power photomicrograph of the thyroid and parathyroid showing the lobulation of the glands. Semithin section, 5 ×.

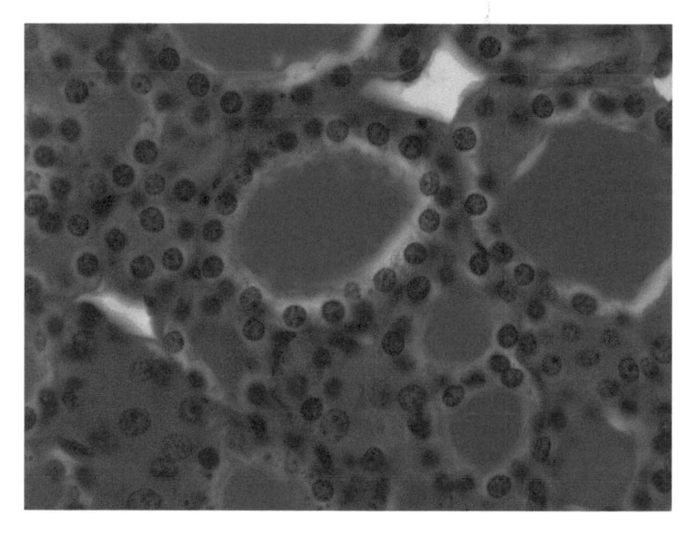

FIGURE N21e A few larger, rounder, paler cells in the epithelium do not reach the apical surface. These are CALCITONIN or C CELLS, 63 ×.

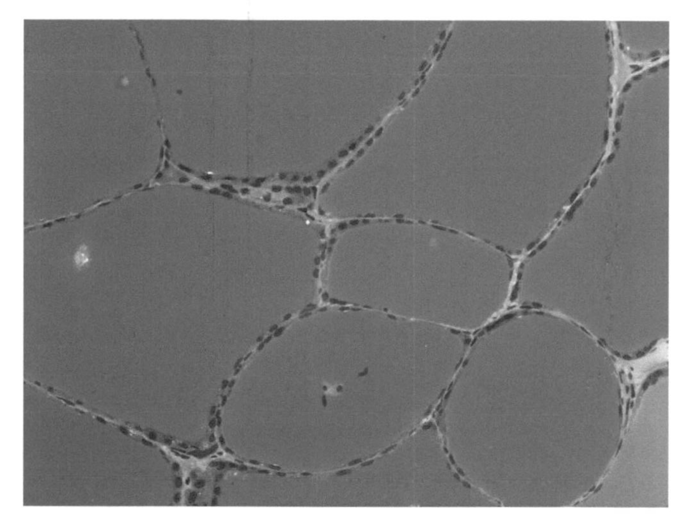

FIGURE N21f Semithin section of the thyroid of a monkey. Blood cells are squeezing through the capillaries, 20 ×.

reach the lumen. They secrete CALCITONIN (hence the name "C cells"), a polypeptide hormone that lowers the calcium levels in the plasma (hypocalcemic effect). They are derived from neural crest cells and are considered to be part of the APUD system (Fig. N22, N23a, and N23b).

Electron micrographs reveal two phases of secretion of FOLLICULAR CELLS (Fig. N24). The usual array of organelles associated with secretory and absorptive cells is seen. There is an apical Golgi complex and extensive granular endoplasmic reticulum scattered throughout. Microvilli appear on the apical borders, sometimes with bristle-coated pits between. Adjacent cells are sealed apically by tight junctions. Desmosomes, gap junctions, and interdigitations of the lateral plasma membranes are also present. The basal plasma membrane has many infoldings. Protein synthesized in the endoplasmic reticulum has a polysaccharide moiety added in the Golgi complex

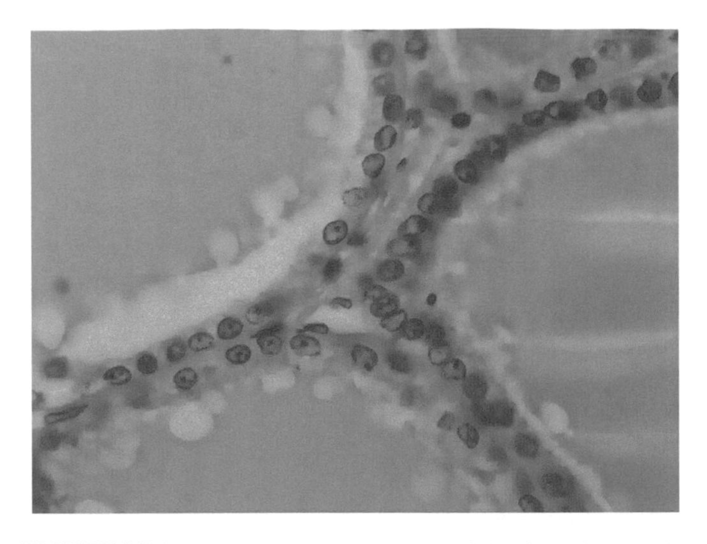

FIGURE N22 The histological structure of the thyroid is similar in all vertebrates. The follicular nature of the gland is apparent in this section of a turtle, 63 ×.

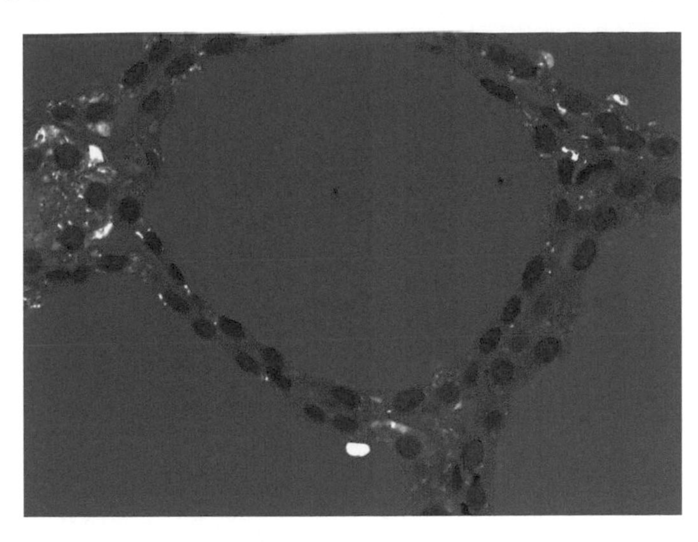

FIGURE N23a Mammalian thyroid showing detail of the follicular epithelium. Occasional C cells may be seen. Semithin section, 63 ×.

to produce the glycoprotein, THYROGLOBULIN, which is released into the lumen. This is the EXOCRINE PHASE of secretion. While the thyroglobulin is stored in the lumen it is partially iodinated. For release of the hormones, bits of colloid are sequestered by the follicular cells to form membrane-bound PHAGOSOMES. These move basally in the cytoplasm and fuse with lysosomes to form PHAGOLYSOSOMES whose enzymes degrade the thyroglobulin to active thyroid hormones. These find their way into the adjacent capillaries and lymphatics in the ENDOCRINE PHASE of secretion.

PARAFOLLICULAR CELLS may be distinguished by the fact that their granular endoplasmic reticulum is sparser and less dilated than that of the follicular cells. A well-developed Golgi complex appears close to the nucleus and many small secretory granules are scattered throughout the cytoplasm. The parafollicular cells may extend processes to neighboring capillaries.

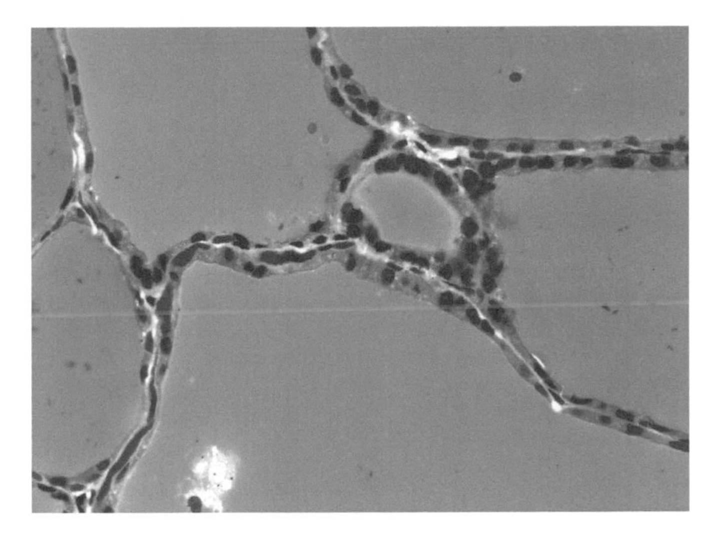

FIGURE N23b Semithin section of the thyroid gland of a monkey. Capillaries can be distinguished in the delicate epithelium of the follicles, 40 ×.

PARATHYROID GLANDS

Parathyroid glands are so named because they are closely associated anatomically with the thyroid glands in mammals (Fig. N21d). They develop from masses of endodermal epithelial cells that proliferate from the visceral pouches and are absent in animals that retain their gills. They lie on each side of the pharyngeal region, often wholly or partially embedded in the thyroid or thymus gland. They are variable in position and number: there are usually two pairs in mammals, derived from the ventral portions of the third and fourth pairs of visceral pouches (Figs. N25b and N25c); in other tetrapods, there are often three pairs, derived from pouches two, three, and four. Their cells are remarkably similar in all classes (Figs. N26a and N26b). The parathyroid glands secrete PARATHORMONE that raises the level of calcium in the blood (hypercalcaemic effect) and has a function antagonistic to that of the calcitonin secreted by the parafollicular cells of the thyroid.

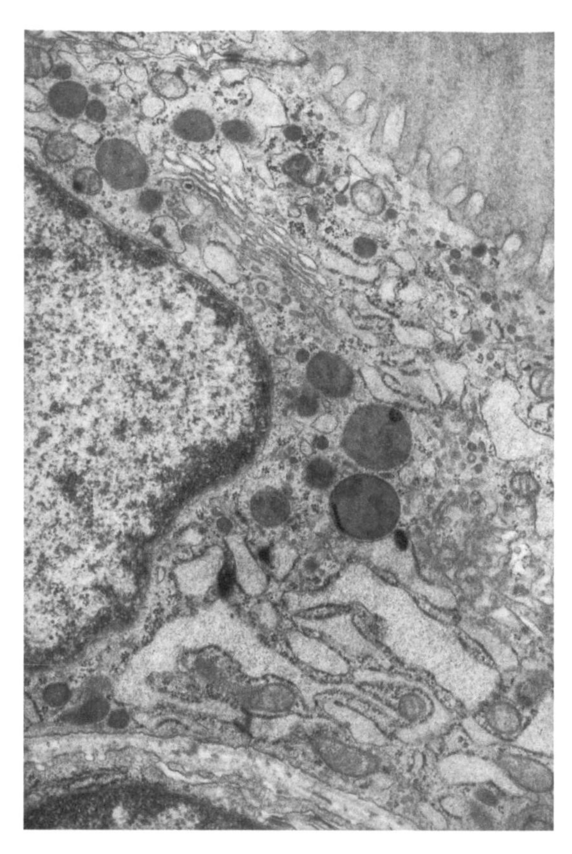

FIGURE N24 Epithelial cell from the thyroid follicle of a rat. The cell reflects two phases of secretion showing the usual array of organelles associated both with secretory and absorptive cells. There is a Golgi complex near the upper center of the micrograph and extensive granular endoplasmic reticulum scattered throughout. Microvilli appear on the apical border of the epithelial cell, projecting into the colloid in the lumen of the follicle. Protein synthesized in the endoplasmic reticulum has a polysaccharide moiety added in the Golgi complex to produce the glycoprotein THYROGLOBULIN, which is released into the lumen (the EXOCRINE phase of secretion). While thyroglobulin is stored in the lumen, it is partially iodinated. For release of the hormone, bits of colloid are sequestered in the follicular cells to form membrane-bound PHAGOSOMES. These move basally in the cytoplasm and fuse with lysosomes to form PHAGOLYSOSOMES whose enzymes degrade the thyroglobulin to active thyroid hormones. These find their way into adjacent capillaries and lymphatics (the ENDOCRINE phase of secretion, 37,000 ×.

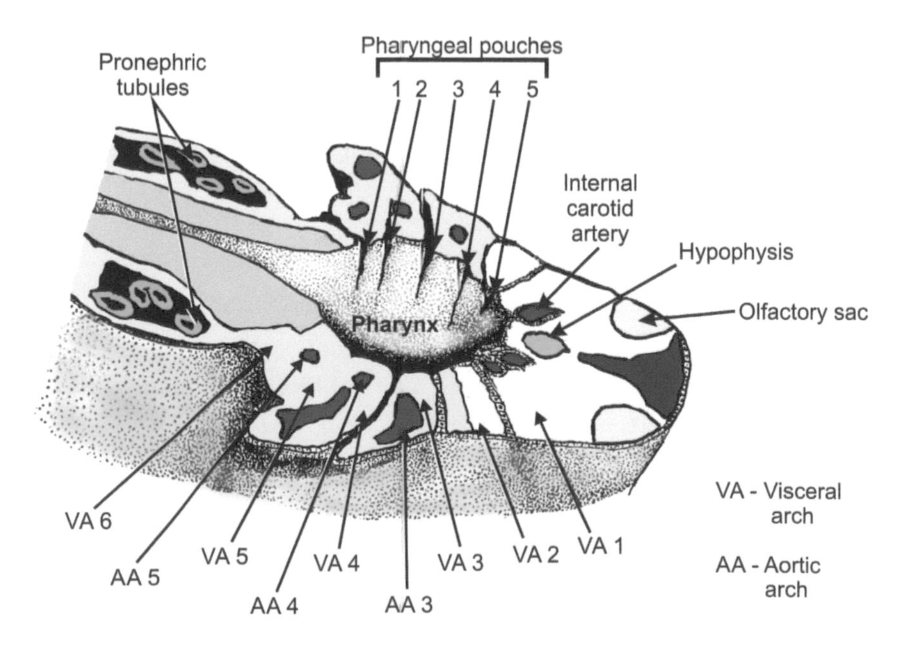

FIGURE N25a Pharyngeal pouches are numbered in this horizontal section of a frog larva. The visceral pouches are not labeled, but are on the outside of the foregut and correspond in number to the pharyngeal pouches. *Authors.*

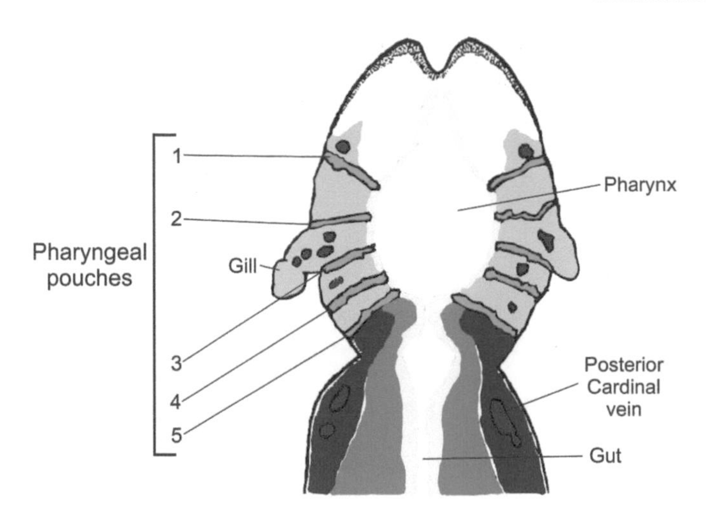

FIGURE N25b Horizontal section of a young frog larva shows the numbered pharyngeal pouches. The parenchyma of the parathyroid consists of two cell types: chief cells and oxyphil cells. The chief cells are more numerous and the oxyphil cells are larger. *Authors.*

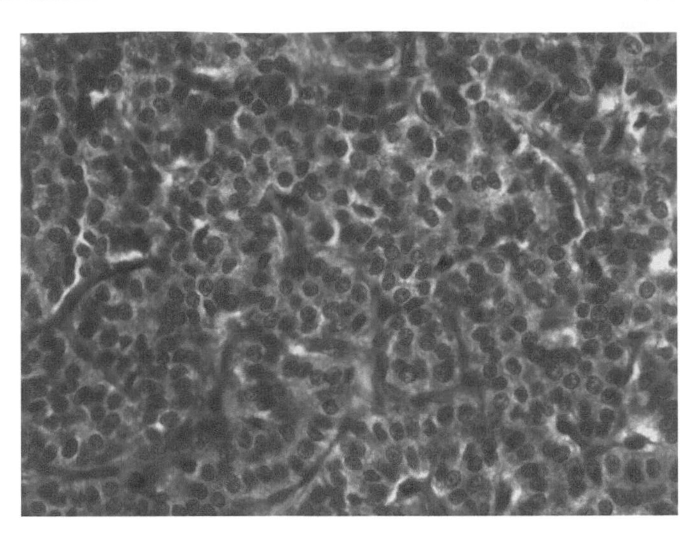

FIGURE N25d The parathyroid gland of a dog is richly invaded by capillaries. Two types of CHIEF CELLS can be distinguished: pale resting *light cells* and secretory *dark cells*. Larger OXYPHIL CELLS are found in some animals, 40 ×.

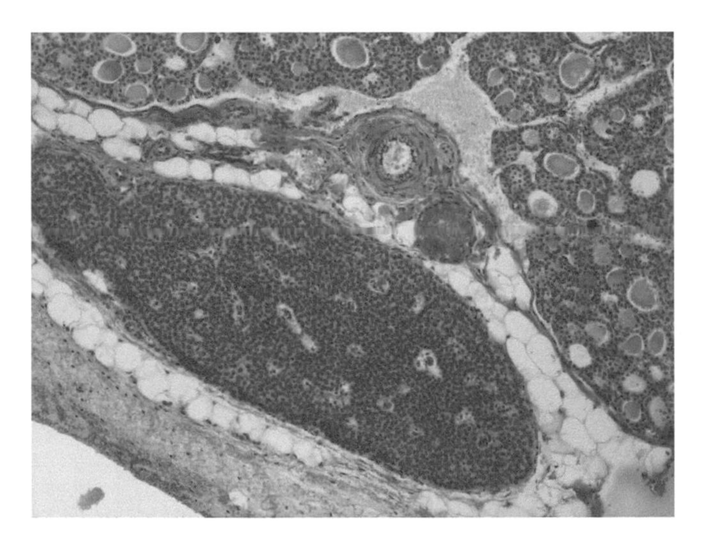

FIGURE N25c The parathyroid is closely associated with the thyroid gland—hence its name. In this micrograph of a section from a cat, the parathyroid is at the center and the thyroid is at the upper right. Wisps of fat and a muscular artery fill the space between the two glands. The parathyroid gland secretes a hormone (PTH) that raises plasma calcium levels by stimulating calcium release from bone, 10 ×.

A section of parathyroid reveals the CAPSULE, SEPTA, and STROMA of moderately dense fibrous connective tissue (Fig. N25c). The secretory cells may be arranged as irregular anastomosing cords, as a continuous mass or, less commonly, as follicles. The parenchyma is richly invaded by capillaries supported by the loose stroma of reticular tissue (Fig. N25d). It contains occasional large fat cells scattered throughout; the number of fat cells increases markedly at puberty and may come to occupy 60%–70% of the gland in old age. The polygonal parenchymal cells are of two major types. The small PRINCIPAL or CHIEF CELLS are most abundant and often appear vacuolated; they are rich in glycogen and seem to be associated with the production of the hormone. Two functional phases of the chief cells may be distinguished: LIGHT CELLS (resting phase), with large vesicular nuclei and clear agranular cytoplasm, and DARK CELLS (secretory phase), with smaller nuclei and finely granular cytoplasm. Large OXYPHIL CELLS are absent from most animals and are not present in humans until 5–7 years of age when they increase with age, especially after puberty. They occur in groups and have small dark nuclei, acidophilic finely granular cytoplasm, and numerous mitochondria. Occasional small follicles containing colloid are seen in the parathyroid, especially in old age; they are not related to thyroid follicles.

The chief cells present the ultrastructural features expected of secretory cells: well-developed granular endoplasmic reticulum, Golgi complex, mitochondria, and abundant secretory granules (Figs. N27a and N27b). Oxyphil cells have a central nucleus surrounded by large numbers of mitochondria packing the entire cytoplasm; Golgi complexes and rough endoplasmic reticulum are sparse or absent. The function of oxyphils is unknown. It may be that there is only one type of parenchymal cell, seen usually as a chief cell, but occasionally in a different guise that leads to its classification as an oxyphil cell.

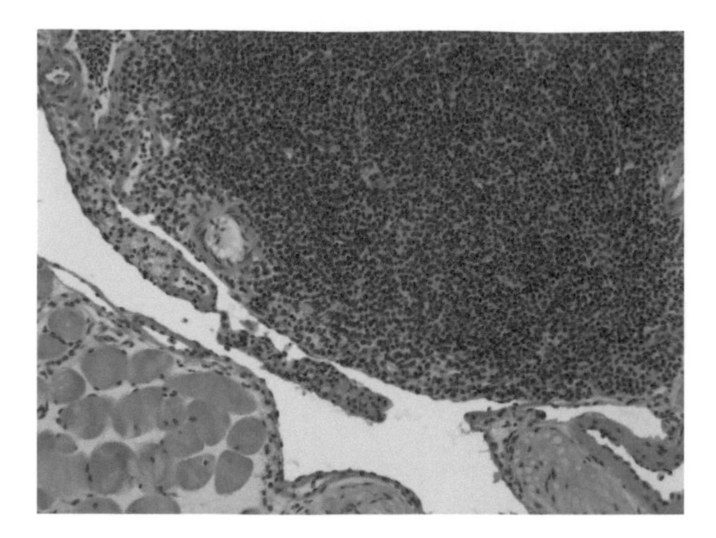

FIGURE N26a The parathyroid glands are similar in all classes of vertebrates. In the frog, masses of secretory cells permeated by vascular loose connective tissue, 10 ×.

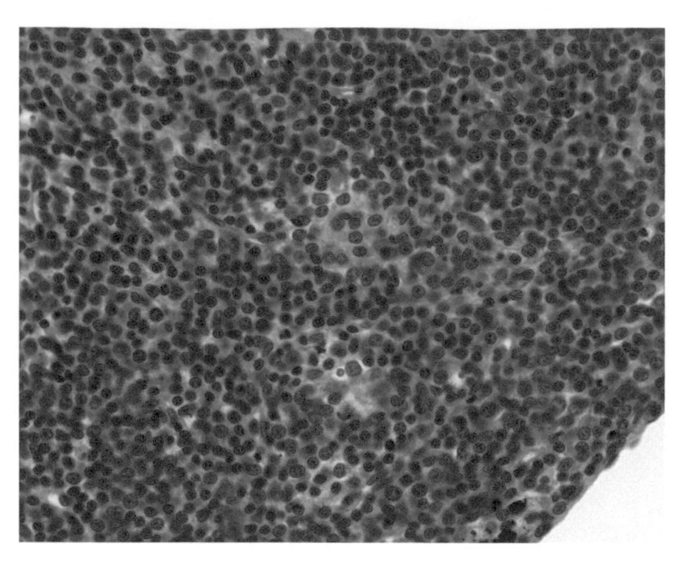

FIGURE N26b The parathyroid gland of a cat is surrounded by the thyroid gland. The parathyroid is surrounded by a connective tissue capsule. The parenchyma is penetrated by several large sinusoids. A conspicuous artery is seen at the upper center, 40 ×.

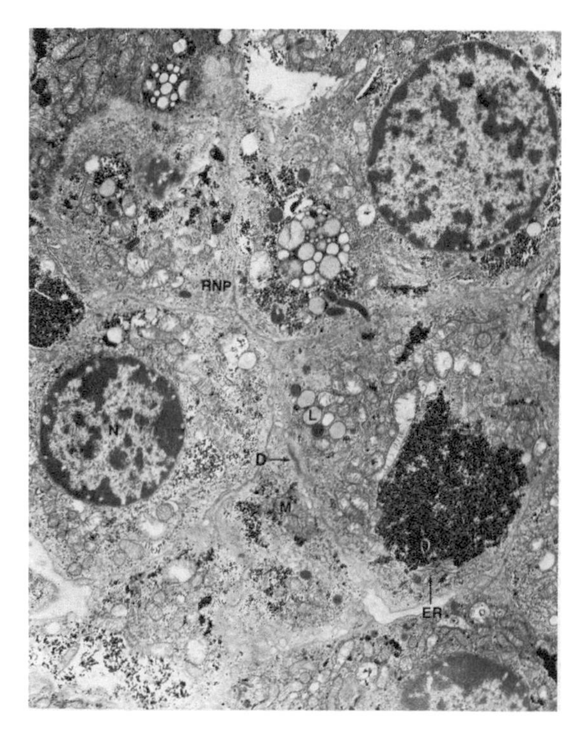

FIGURE N27a Electron micrograph of a group of chief cells from a human parathyroid gland. These cells present the appearance of secretory cells with well-developed granular endoplasmic reticulum, Golgi complex, mitochondria, and abundant secretory granules. *D*, desmosome; *ER*, endoplasmic reticulum; *G*, glycogen; *L*, lipid; *M*, mitochondria; *N*, nucleus; *RNP*, ribonucleoprotein. (Magnification not given.)

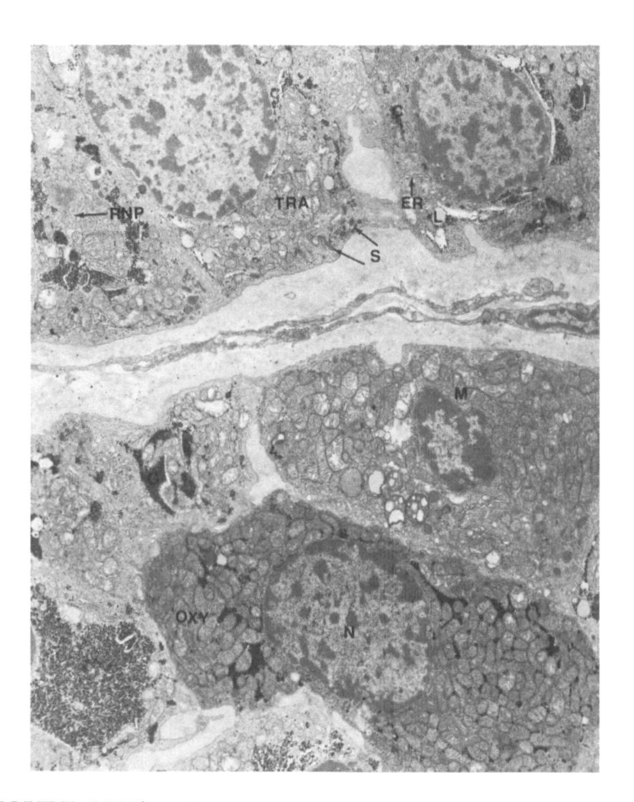

FIGURE N27b Electron micrograph of a human parathyroid gland. An oxyphil cell (OXY) has a central nucleus and is packed with mitochondria; endoplasmic reticulum and a Golgi complex are scant. A chief cell is shown above. Also shown is a transitional cell (TRA). (Magnification not given.)

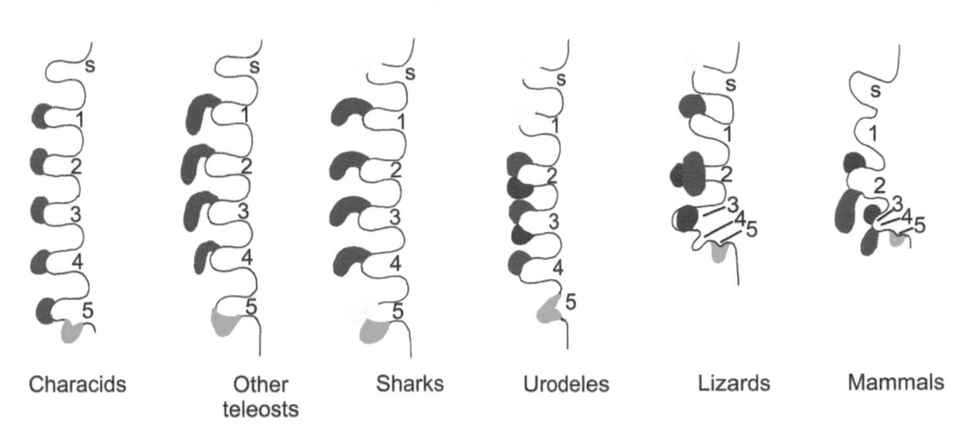

FIGURE N28a Diagrams of the gill pouches on the left side of the pharynx of several vertebrates to show the derivation of thymus, parathyroid, and ultimobranchial bodies. The dorsal part of each gill pouch is, for the purposes of this diagram, at the upper side. Yellow, variable thymus derivatives; red, thymus; red, parathyroid; and green, ultimobranchial body. s, Spiracular pouch *Authors.*

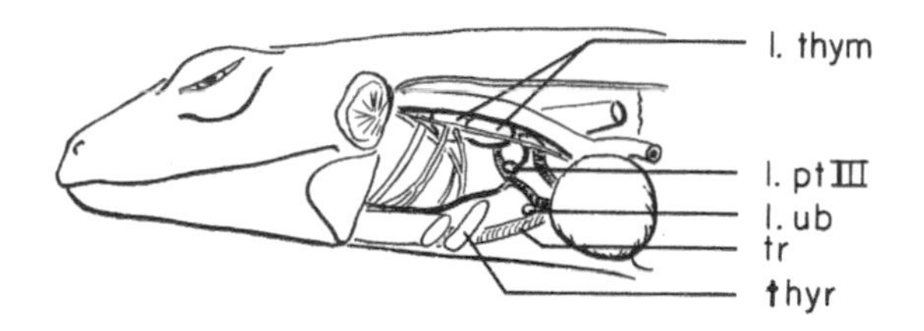

FIGURE N28b Location of ultimobranchial bodies (ub) in the turtle *Emys europea. e,* oesohagus; *thym,* thymus; *thyr,* thyroid; *tr,* trachea.

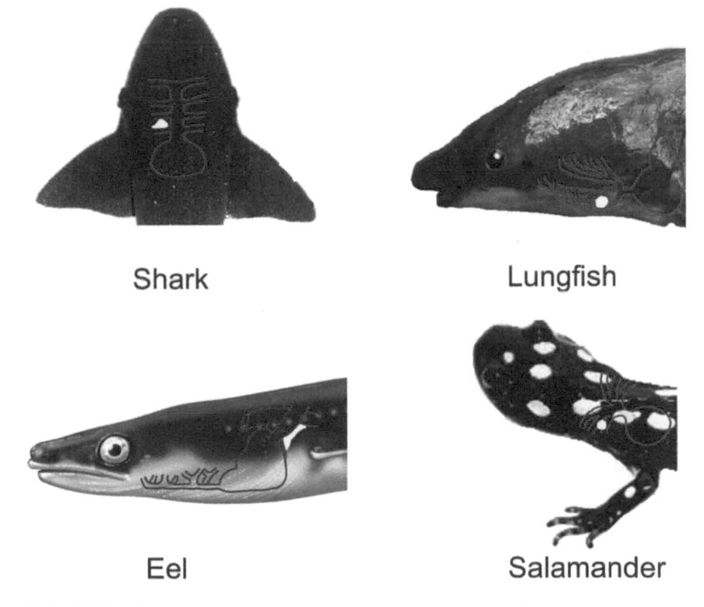

FIGURE N28c Positions (not size or shape) of the ultimobranchial bodies in four species of vertebrates. Top left, shark; top right, lungfish; lower left, eel; and lower right, salamander. *Authors.*

ULTIMOBRANCHIAL BODIES

Ultimobranchial bodies are derived from the last visceral pouches in much the same way that thymus and parathyroid tissues are derived from more anterior visceral pouches (Figs. N25a and N28a). This epithelial tissue is colonized by cells from the neural crests, giving rise to groups of calcitonin-secreting cells. The cells are part of the APUD system. Ultimobranchial bodies first appear phylogenetically in the elasmobranchs where they are asymmetric paired glands that retain their attachment to the pharyngeal epithelium. In the higher vertebrates, they are paired or single, the attachment to the pharynx is lost, and they usually lie near the thyroid gland (Figs. N28b and N28c). Often they are present only on the left side or, when paired, the left gland is the larger. In mammals ultimobranchial cells invade the developing thyroid and become the parafollicular cells or C cells (Fig. N21e); it is said that occasional follicles in the mammalian thyroid are constituted solely of ultimobranchial cells.

Ultimobranchial bodies consist of cords or follicles of secretory cells supported by a stroma of loose vascular fibrous connective tissue and enclosed within a capsule of fibrous connective tissue. In aquatic vertebrates, they are composed of a single cell type, characterized by the presence of intracytoplasmic, small, dense-cored secretory vesicles (Fig. N28d). In terrestrial species, such as toads, reptiles, and birds, a second type of cell displaying larger, homogeneous granules can be observed. Secretory cells may exist in two phases: LIGHT CELLS

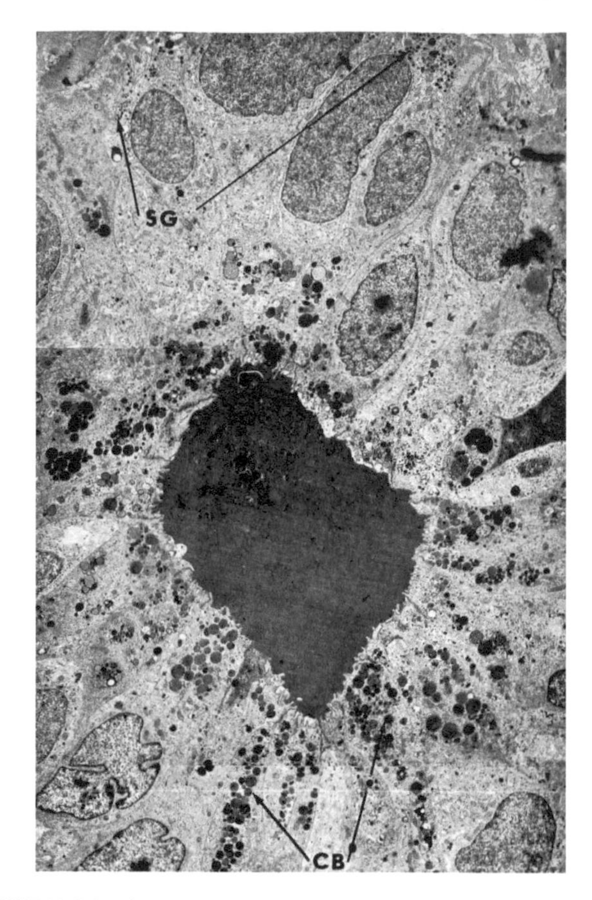

FIGURE N28d Low-powered electron micrograph of a ultimobranchial follicle of a freshwater turtle. Pseudostratified epithelial cells surround a ductless lumen filled with electron-dense material. *CB*, cytoplasmic bodies; *SG*, secretory granules, 2715 ×.

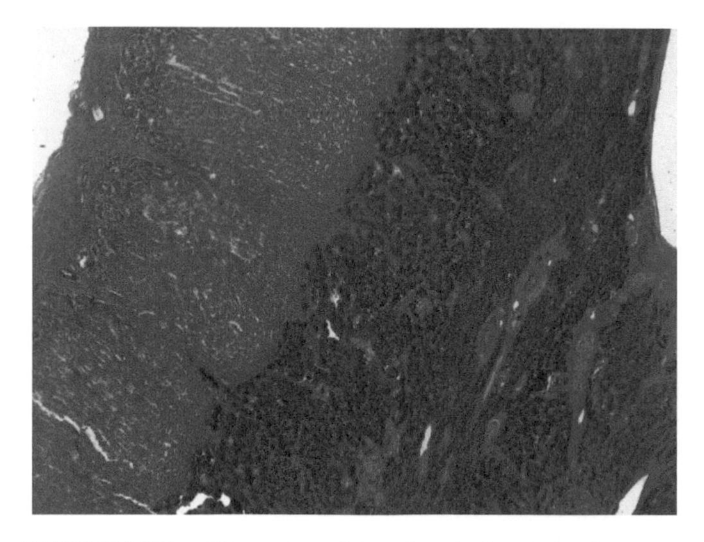

FIGURE N30a This section of the adrenal of cattle has been treated with the salts of chromic acid. Brown chromaffin cells show clearly in the medulla which is surrounded on two sides by the cortex, 2.5 ×.

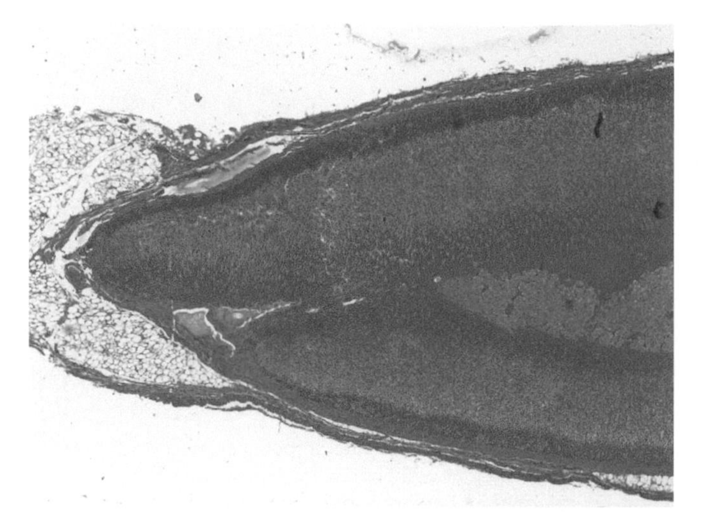

FIGURE N29 Section of mammalian adrenal gland. The *medulla* at the entre is surrounded by the *cortex*; a *capsule* encloses the entire gland. Note the brownish tint of the *chromaffin cells* in the medulla, 2.5 ×.

contain the organelles typical of secretory cells while DARK CELLS are almost devoid of organelles. It is thought that light cells are the secretory phase while dark cells are the storage phase of the same cell type.

ADRENAL GLANDS

The adrenal or suprarenal gland of mammals consists of two distinct portions, the CORTEX and MEDULLA (Fig. N29). The cortex arises from mesothelial cells and the medulla from ectodermal neural crest cells. Because they stain with chromic acid and its salts, the cells of the medulla are called CHROMAFFIN CELLS (Figs. N30a, N30b, and N36d). Adrenal medullary cells form part of the widespread APUD system. In cyclostomes and fish, these two components occur separately. The structures corresponding to the mammalian cortex are the INTERRENAL BODIES while the CHROMAFFIN or SUPRARENAL BODIES are homologous with the medulla (Fig. N31). In cyclostomes the INTERRENALS cluster along some of the large arteries and veins in the region of the kidney and may even extend into the lumina of these vessels. Chromaffin bodies occur along the paired parietal arteries that arise along the whole length of the aorta and supply the body wall. The golden interrenals of elasmobranchs are median and lie between the posterior ends of the kidneys while the chromaffin bodies are paired and segmentally arranged along the aorta on either side of the spinal cord in close association with the sympathetic ganglia. The interrenals of teleosts are arranged in a manner

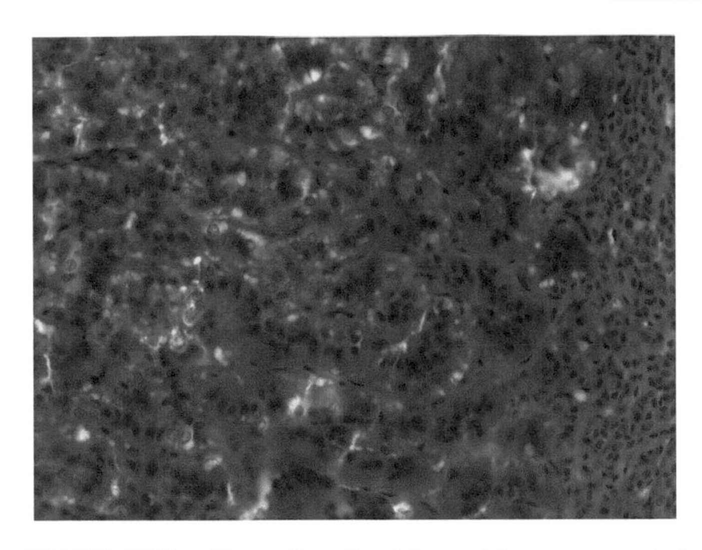

FIGURE N30b Chromaffin cells of the medulla occupy most of this micrograph; a small portion of the cortex appears at the lower right, 20 ×.

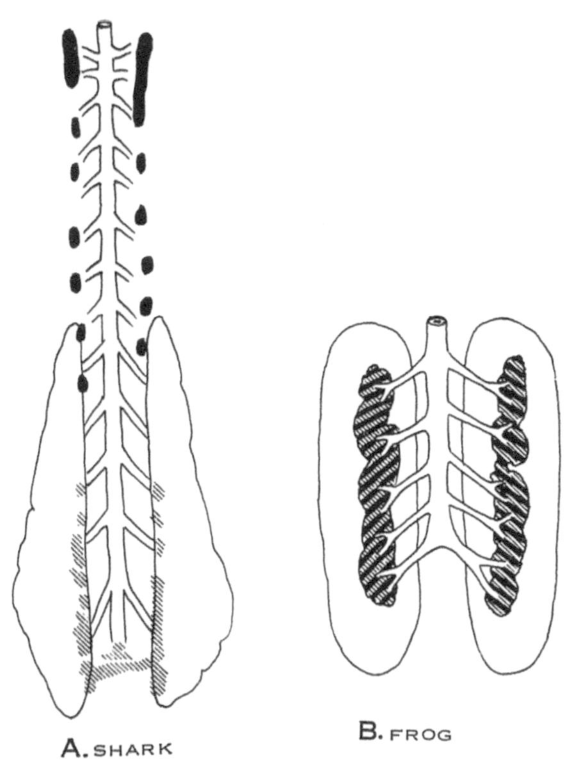

FIGURE N31 Paired chromaffin bodies occur alongside the aorta in the shark (A). In the frog (B), chromaffin cells are intermingled with corical cells and lie adjacent to the kidneys. Source: *From Ballard, W.W., 1964. Comparative Anatomy and Embryology, Fig#305. New York: Ronald Press Co, with permission from Wiley.*

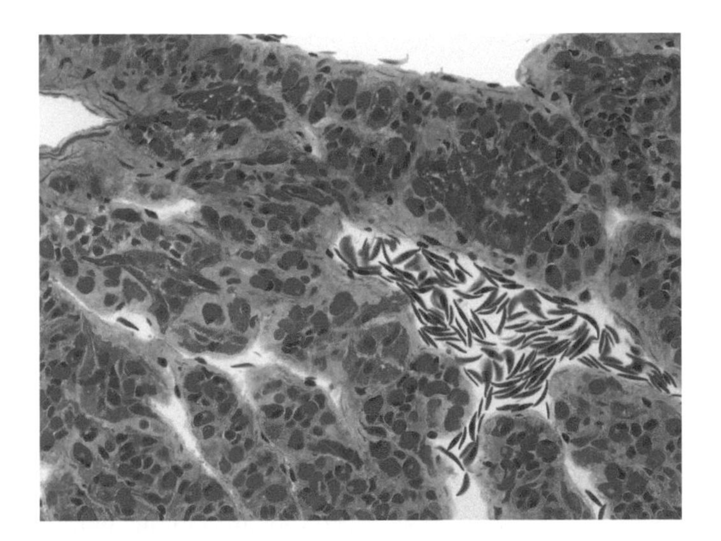

FIGURE N32a In this semithin section of the adrenal gland of a frog, basophilic chromaffin cells are intermingled with interrenal cells. (Chromic acid salts have not been used.) 20 ×.

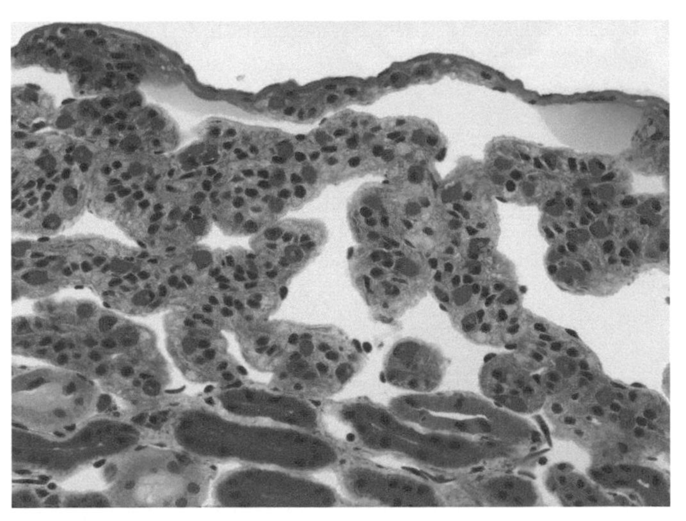

FIGURE N32b In other areas of this semithin section of the adrenal gland of a frog, the tissue is more loosely organized and the basophilic chromaffin cells intermingled with interrenal cells may be more obvious, 20 ×.

like those of elasmobranches and may lie on the surface of the kidney or partially embedded in it. The chromaffin bodies lie in the walls of the postcardinal veins.

Beginning with amphibians, there is a phylogenetic trend toward more intimate association between the two glandular tissues (Figs. N32a and N32b). The golden bands seen running along the frog kidney represent an

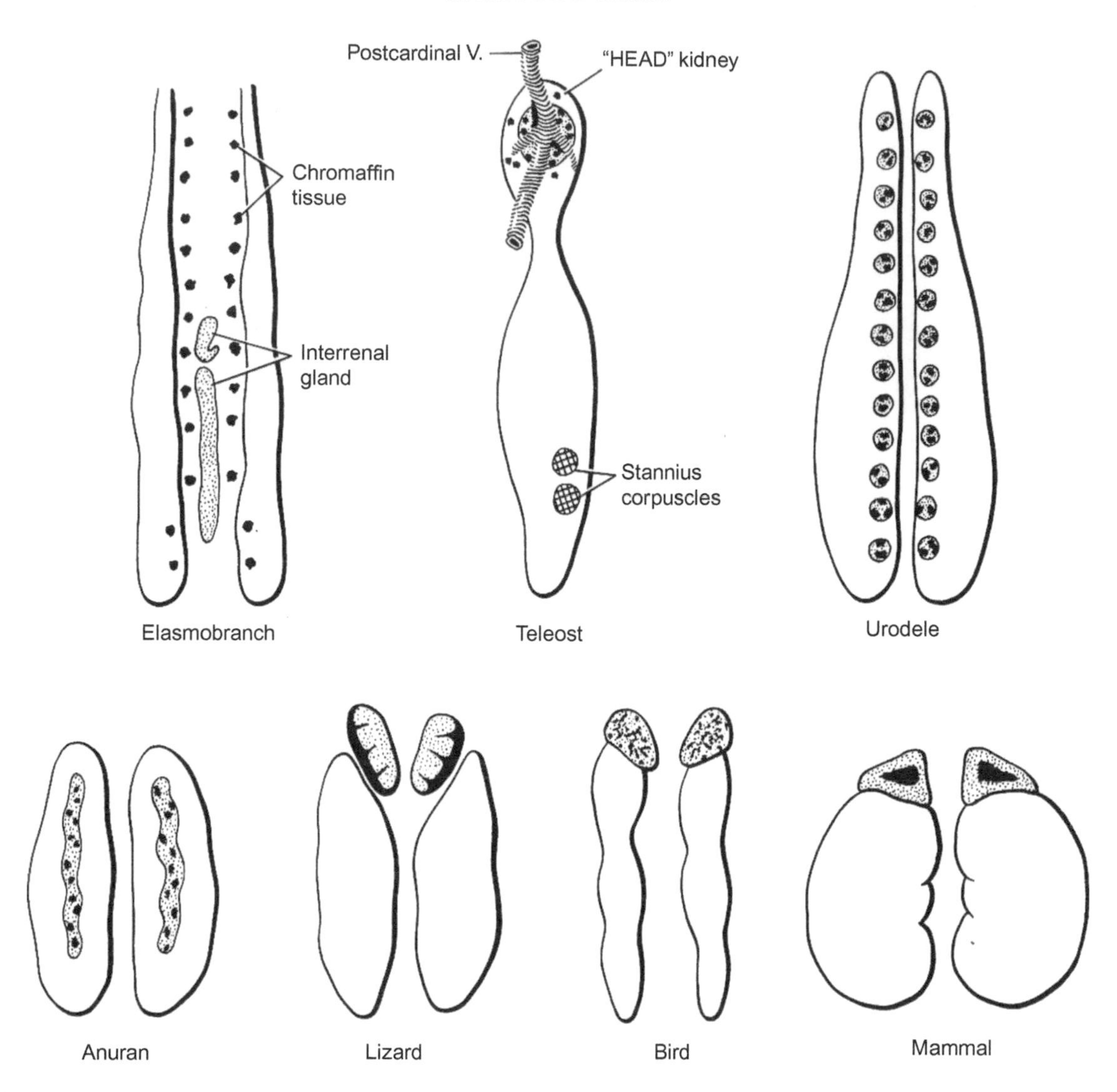

FIGURE N33a Comparative morphology of the adrenal tissues of vertebrates. Solid black indicates chromaffin tissue, stippling is steroidogenic tissue. Note that chromaffin tissue and steroidogenic tissue (interrenal gland) are spatially separate in the elasmobranch. The adrenal tissues of bony fishes are typically found in the "head" kidneys, around branches of the postcardinal veins; there may be some spatial separation of the two kinds of tissues, but they are usually intermingled. Note that the cortex (steroidogenic tissue) and medulla (chromaffin tissue) are present only in mammalian adrenals.

intermingling of the two components. They are also intermixed in reptiles and birds, but the ADRENAL GLAND has become a separate organ lying in the region of the kidney. Adrenal glands of mammals are unique in having the interrenal and chromaffin tissue separated into a distinct cortex and medulla.

In all vertebrates, but most notably in the classes other than mammals, nests of chromaffin cells, the PARAGANGLIA, occur outside the adrenal gland proper, especially in association with the autonomic ganglia (Fig. N33a). Together with the adrenal medulla they comprise the CHROMAFFIN SYSTEM. It is not clear whether or not the paraganglia function in a way similar to the medulla.

The CORTEX (Fig. N34a) exists in three somewhat indefinite regions and is made up of fairly regular cords of cells that are permeated by fenestrated SINUSOIDS (Figs. N34b–N34d). The narrow ZONA GLOMERULOSA (Fig. N34b) just inside the capsule consists of cells that are often arranged in the form of inverted Us. These merge into straight cords of cells in the ZONA FASCICULATA (Fig. N34c), the middle, widest portion of the cortex. Its cells are so packed with lipid droplets that they appear spongy in paraffin sections and are called SPONGIOCYTES. The cells of

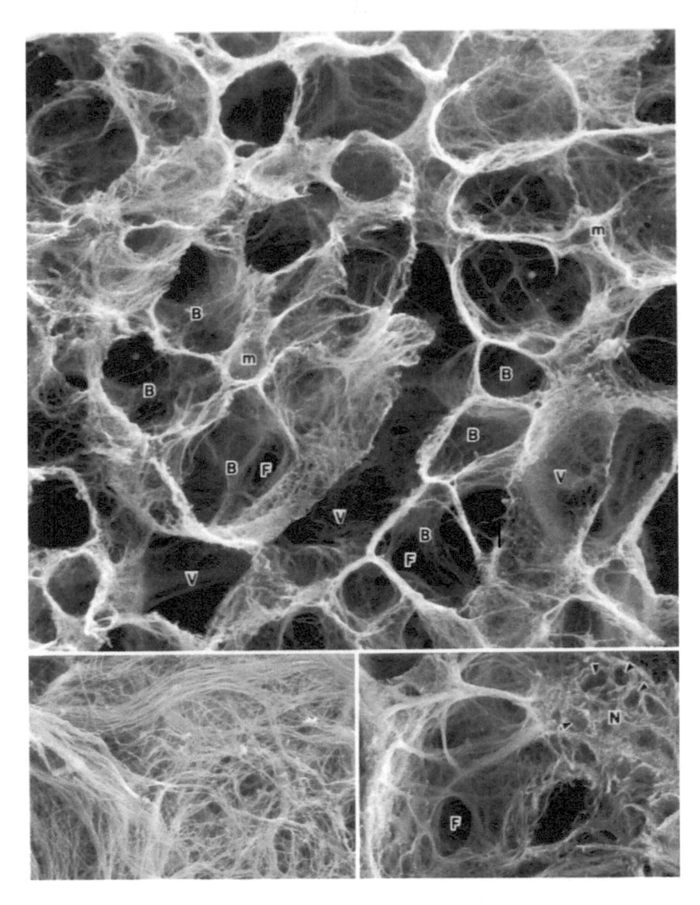

FIGURE N33b The reticular framework of the medulla of the adrenal gland of the rat has been beautifully displayed by the scanning electron micrographs—a maceration preparation where all tissue elements were removed leaving only the network of collagenous fibers. The upper micrograph shows collagenous fibrils in the medulla forming fine-meshed sheaths that embrace peripheral branches of the central vein (V). The collagen forms basket-like compartments (B) that contained clusters of chromaffin cells. Fenestrae (F) in the wall provided communication between adjacent basket-like spacer. Thin tube-like spaces in the interstices between the sheaths housed medullary capillaries (m), 900×. A closer view of the wall of the basket-like compartment interposed between chromaffin cells and the peripheral branches of the central vein is provided at the lower left, 5000×. Collagen fibrils form a tight fibrillar plexus (N) around nervous elements in the medulla. The arrowheads show tubes for nerve fibers, 1200×.

the ZONA RETICULARIS (Fig. N34d), the innermost layer of the cortex, are less regularly arranged and their cytoplasm may contain fewer lipid droplets. DARK and LIGHT CELLS may be distinguished in the zona reticularis but the significance of their differences is not understood. These cells, particularly the dark cells, contain large accumulations of LIPOFUSCIN PIGMENT. It is likely that new cells produced in the zona glomerulosa migrate toward the medulla and complete their life cycle in the zona reticularis (Fig. N34e).

The steroid hormones of the adrenal cortex have a multitude of functions. MINERALOCORTICOID HORMONES are secreted primarily by the cells of the zona glomerulosa and are associated with electrolyte and water balance; GLUCOCORTICOSTEROIDS secreted by the zona fasciculata are associated with carbohydrate balance and maintenance of connective tissues throughout the body; SEX HORMONES are secreted in the zona reticularis.

The more loosely arranged MEDULLA (Fig. N33b) and (Fig. N34c) surrounds the CENTRAL VEIN that drains both cortex and medulla. Its cells, chiefly CHROMAFFIN CELLS, are loosely arranged in short cords surrounding fenestrated capillaries and venules. Small cells, probably small lymphocytes, also occur in the medulla. The medulla secretes the catecholamines ADRENALIN (epinephrin) and NORADRENALIN (norepinephrin). The "emergency hormone" adrenalin increases the rate of metabolic functions; noradrenalin has little metabolic action but increases blood pressure by causing constriction of peripheral blood vessels throughout the body. The cells that secrete noradrenalin are

FIGURES N34a–e Semithin sections of the adrenal gland of a monkey.

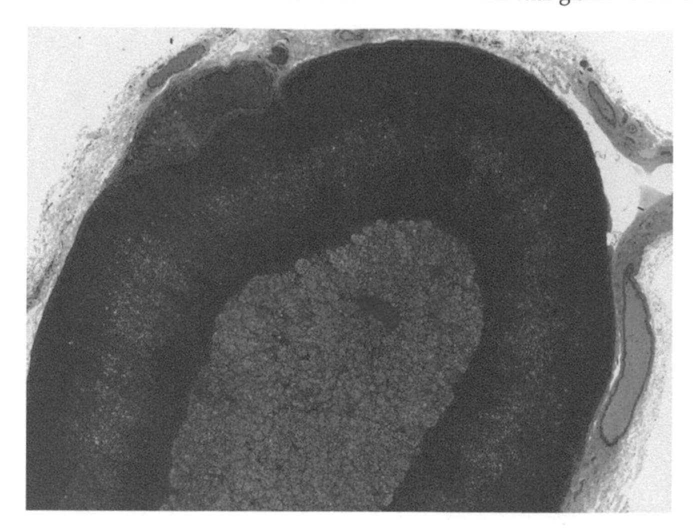

FIGURES N34a Overall view of the adrenal gland. A thin CAPSULE of fibrous connective tissue encloses the gland. The CORTEX surrounds the paler MEDULLA. Note blood vessels in the capsule. The CENTRAL VEIN passes through the cortex, 2.5 ×.

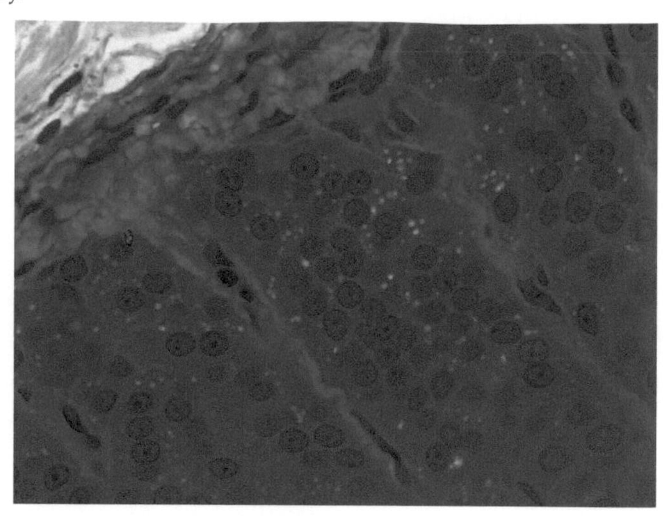

FIGURE N34b Cells of the zona glomerulosa are the outermost layer of the cortex and occupy most of the right side of this micrograph. At the upper left is the fibrous capsule. The columns of cells are arranged in the form of inverted Us and are supported by trabeculae from the capsule.

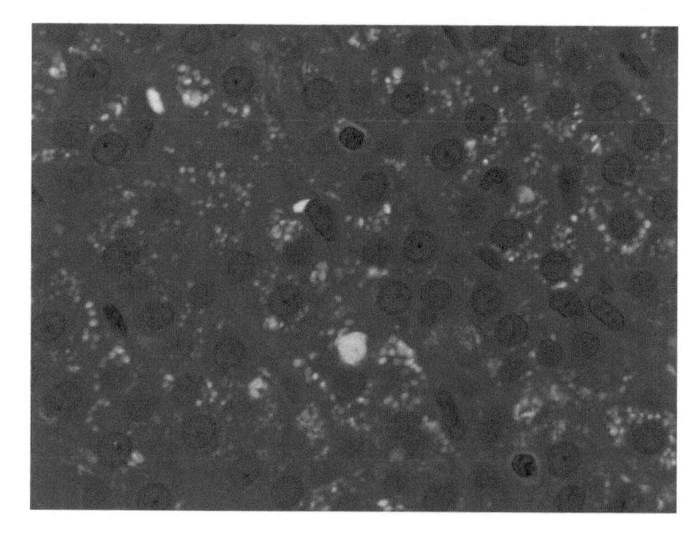

FIGURE N34c The cells of the ZONA FASCICULATA accumulate droplets of lipid that give them a spongy appearance and are called SPONGIOCYTES, 63 ×.

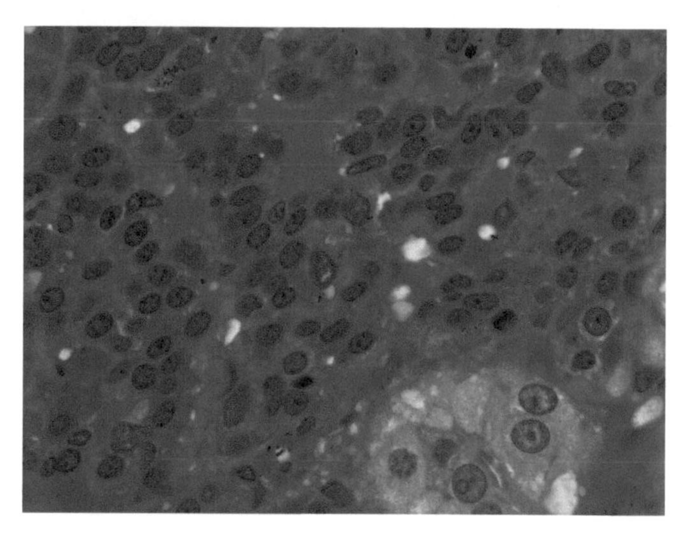

FIGURE N34d In the zona reticularis, the innermost later of the cortex, the cells are less regularly arranged and their cytoplasm contains fewer lipid droplets, 63 ×.

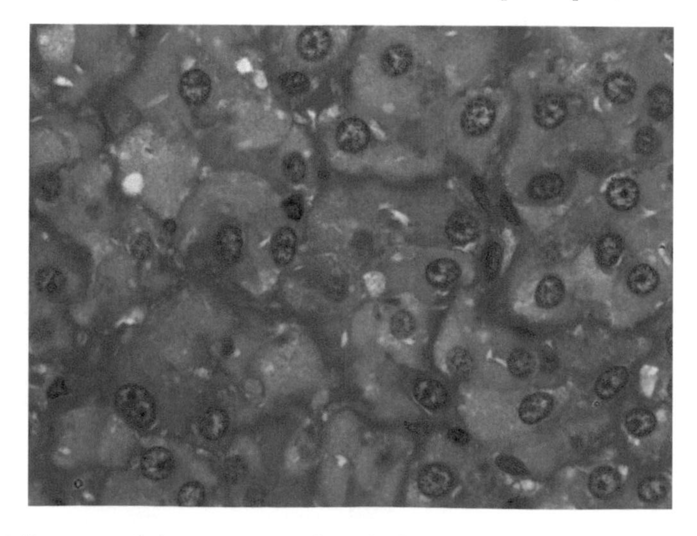

FIGURE N34e Cells of the medulla surround the CENTRAL VEIN (seen in Fig. N34a). If chromium salts had been used in the preparation of these slides, these cells would have assumed a brownish color—they are CHROMAFFIN CELLS, 63 ×.

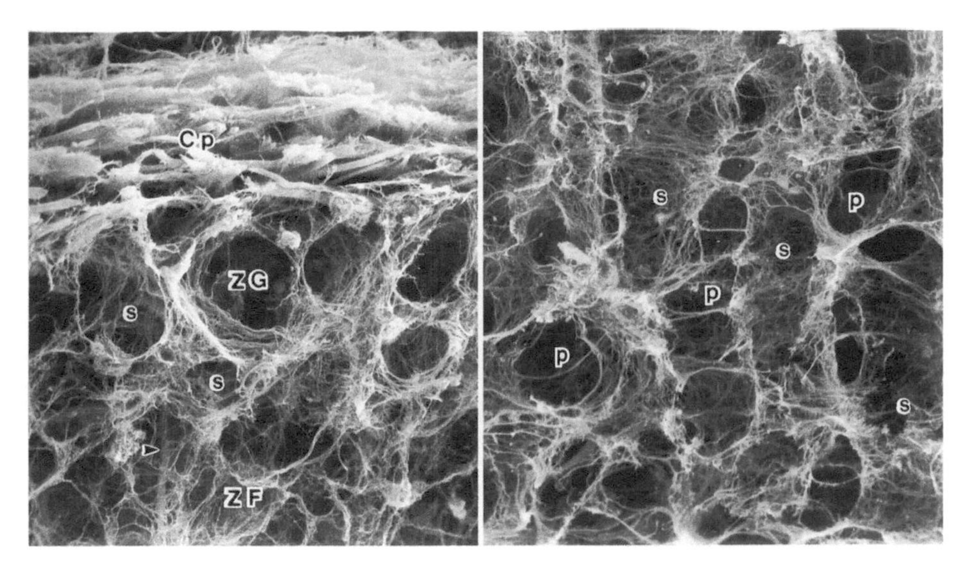

FIGURE N35 These two scanning electron micrographs of the adrenal gland of a rat prepared using an alkali-water maceration method. *Left*: the cut surface of the collagen networks of the capsule (Cp) and zona glomerulosa (ZG). The capsular network is composed of a layered plexus of thick bundles of collagen fibrils. Round spaces for the cortical parenchymal cells are formed by collagenous fiber sheaths. A bundle of collagen fibrils (*arrowhead*) runs straight down into the zona fascicula (ZF). Cortical capillaries are accommodated in the spaces marked *s*, 1000 ×. *Right*: Collagenous networks in the zona fasciculata reinforce the cortical capillary endothelial cells (s). Perivascular sheaths (p) enclose parenchymal cells, 1000 ×.

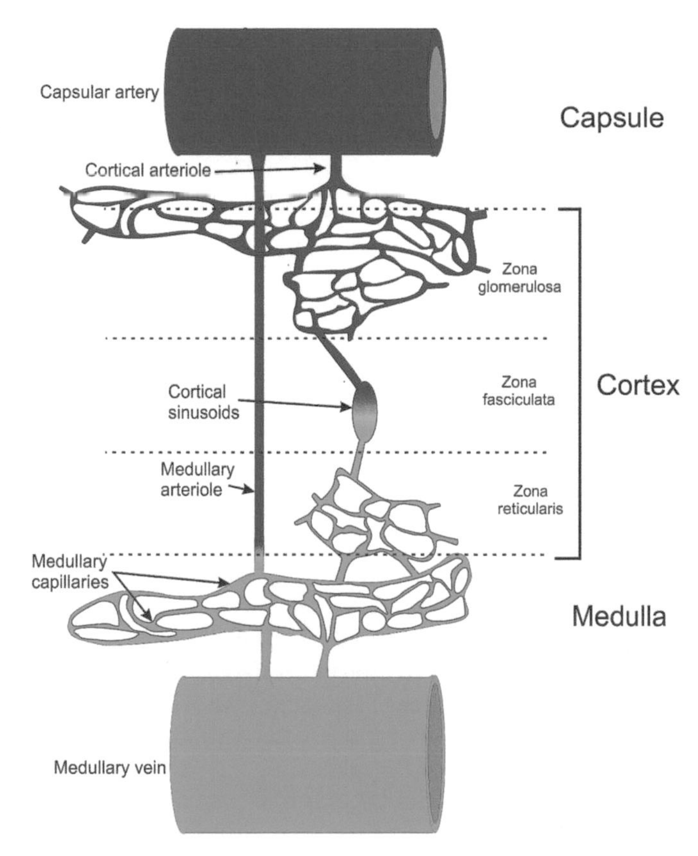

FIGURE N36a The arterial plexus over the surface of the adrenal gland brings blood to separate arterioles supplying the cortex and medulla. The cortical arterioles supply sinusoids in the cortex. At the junction of the cortex and medulla, the sinusoids merge and communicate with the capillaries of the medulla. In addition, the medulla is supplied directly from medullary arterioles that pass through the cortex. Thus the medullary capillary bed receives blood from two sources thereby allowing hormones produced in the cortex to influence the activity of the medullary cells. Blood from the vascular beds of both the cortex and medulla is drained by the medullary veins. Note that the circulation of the adrenal gland is a portal system. *Authors.*

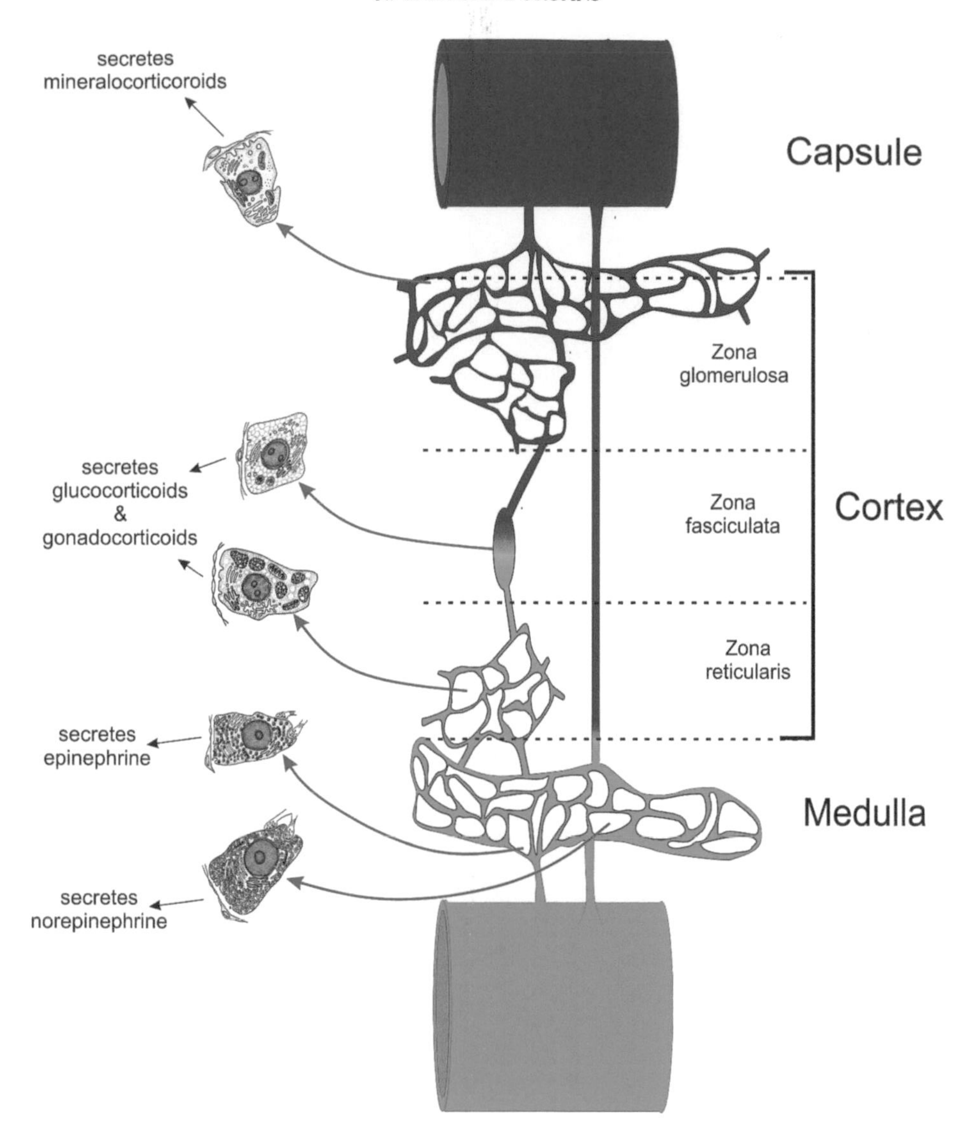

FIGURE N36b Diagram to illustrate the organization of cells in the adrenal gland and their relationship to the blood vessels. *Authors.*

autofluorescent and may be distinguished from those that secrete adrenalin by examining them with ultraviolet light in a fluorescence microscope.

An extensive arterial plexus over the surface of the gland brings blood to separate arterioles supplying the cortex and medulla (Figs. N35 and N36a–N36c). The cortical arterioles supply the sinusoids of the cortex. At the junction of the cortex and medulla the sinusoids merge and communicate with the capillaries of the medulla. In addition, the medulla is supplied directly from medullary arterioles that pass through the cortex. Thus the medullary capillary bed receives blood from two sources. This allows hormones produced in the cortex to influence the activity of the medullary cells. Blood from the vascular beds of both the cortex and medulla is drained by the medullary veins (Fig. N36c).

The ultrastructure of the cells of the zona fasciculata and the zona reticularis is typical of cells active in the secretion of steroids. The numerous mitochondria are packed with characteristic vesicular or tubular cristae and the abundant endoplasmic reticulum is entirely agranular. Cells of the zona glomerulosa contain a more usual

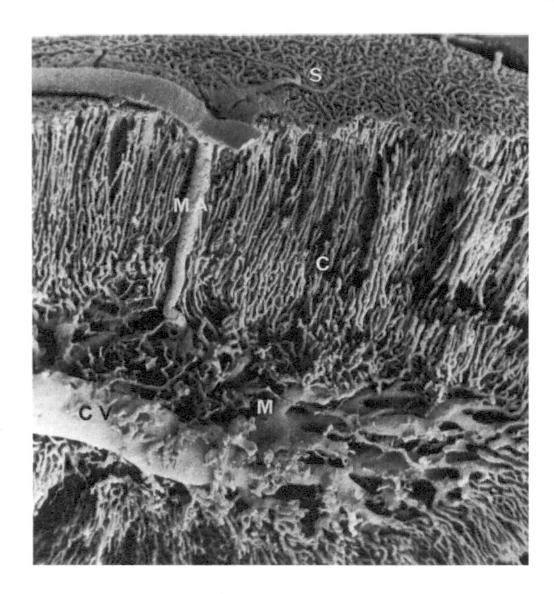

FIGURE N36c Scanning electron micrograph of a vascular cast of the blood vessels in the cortex and medulla of the adrenal gland of a rat. *C*, cortex; *CV*, central vein; *M*, medulla; *MA*, medullary artery; *S*, subcapsular capillary plexus, 80 ×.

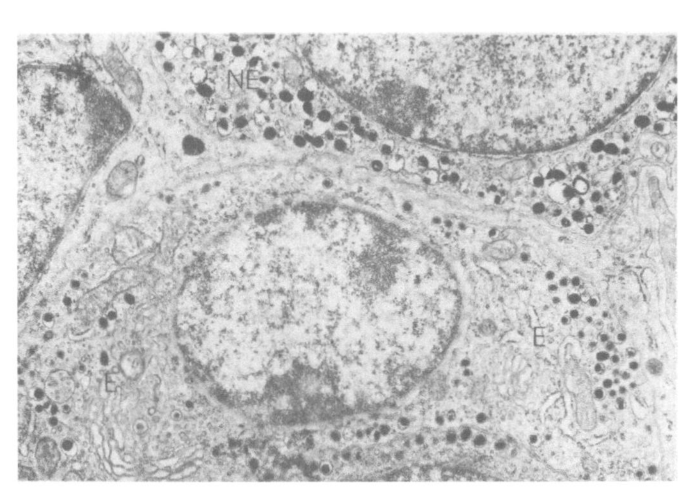

FIGURE N36d Transmission electron micrograph of a section through secretory cells in the adrenal medulla of a mouse. Two types of medullary cells are present. While not definitive, noradrenalin secreting cells (NE) typically show granules with dense cores and adrenalin secreting cells (E) contain less intensely staining granules. 5,000 ×. *Petrovic-Kosanovic D, et al. 2012. Effect of acute heat stress on rat adrenal medulla—a morphological and ultrastructural study. Cent. Eur. J. Biol. 7(4):603–610.*

complement of organelles. Chromaffin cells are packed with dense, membrane-bound granules but the organelles are undistinguished. The granules of the cells that secrete noradrenalin are denser than those of the cells that secrete adrenalin.

C H A P T E R

O

Genital Systems

The ovary and testis are unique exocrine glands in that they produce cells: they are CYTOGENOUS GLANDS. They also secrete sex hormones and are therefore endocrine glands as well. They arise in close association with the kidney from paired GENITAL RIDGES: thickenings of the coelomic epithelium along the dorsal surface of the body cavity on either side of the dorsal mesentery of the gut (Fig. O1). Early development is so similar that sexual differences cannot be recognized (Figs. O2a–O2d). In most vertebrates the ducts that carry gametes to the outside also develop in close association with the kidneys and their ducts.

FEMALE

The frog ovary typifies the general scheme of the vertebrate ovary—it is a mass of richly vascular connective tissue invested with a layer of visceral peritoneum and containing OVARIAN FOLLICLES often in various stages of development (Figs. O3a and O4). Each follicle consists of an OOCYTE surrounded by a layer of FOLLICULAR CELLS (Figs. O3b and O3c). Oocytes develop from cells of the GERMINAL EPITHELIUM surrounding the ovary or from SEX CORDS that grow into the substance of the ovary from the germinal epithelium (Figs. O2a–O2d). Meiotic division occurs so that haploid oocytes are produced. As the oocyte matures it becomes separated from the follicular cells by a homogeneous acidophilic layer, the ZONA PELLUCIDA (Fig. O3c) that is produced by both the egg and the follicular cell. In some follicles, radial striations may be seen traversing this layer. Electron micrographs reveal that these striations represent channels containing microvilli extending outward from the oocyte and inward from the follicular cells. The follicle is enclosed in a sheath of connective tissue, the THECA FOLLICULI, containing abundant capillaries and, in some, smooth muscle.

When the oocyte is mature, the follicle ruptures and the egg is usually cast free in the body cavity (Fig. O4). In several vertebrates the follicular cells round up after ovulation to form an endocrine CORPUS LUTEUM that secretes hormones that delay further ovulation. This is still a matter of dispute in vertebrates other than mammals.

ATRESIA, the degeneration and resorption of oocytes, is a fairly constant feature at all stages of development. Thus eggless follicles (which may have an endocrine function) result either from ovulation or atresia and we can speak of POSTOVULATORY FOLLICLES and ATRETIC FOLLICLES.

Frog Ovary

Sections of a frog ovary provide an example of a fairly unspecialized chordate type (Figs. O3a, O3b, O8a, O8b, and O9). The frog produces large numbers of eggs and immediately before spawning the ovary fills most of the body cavity. The ovary is a thin-walled sac surrounded by peritoneum. The wall is thin and transparent and the ovary may be so distended with eggs that they may erroneously appear to be lying free in the body cavity.

Each OOCYTE is surrounded by a layer of cuboidal or squamous FOLLICULAR CELLS that transmit yolk material from the blood to the cytoplasm of the developing egg (yolk proteins are elaborated in the liver and carried to the ovary in the blood). Surrounding the follicle is a THECA FOLLICULI of connective tissue, small blood vessels, and a few smooth muscle fibers. When the oocyte is mature, the follicle ruptures and the egg passes through these layers, perforates the peritoneum, and arrives in the coelom. It is drawn by ciliated cells into the oviduct where it is stored until the female is clasped by the male in amplexus.

An Atlas of Comparative Vertebrate Histology.
DOI: https://doi.org/10.1016/B978-0-12-410424-2.00019-6

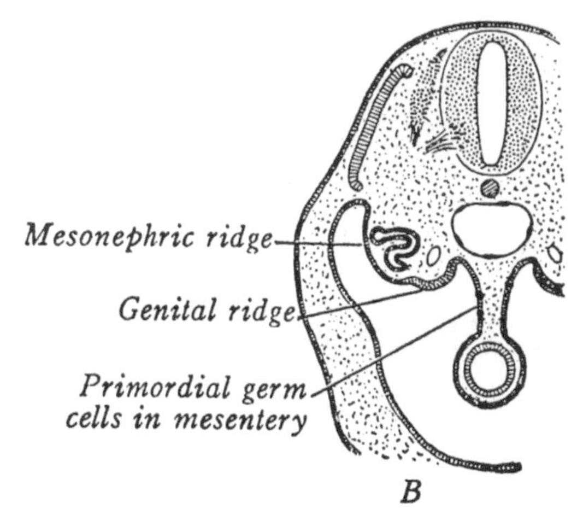

FIGURE O1 The ovary and testis arise in close association with the kidney from paired GENITAL RIDGES: thickenings of the coelomic epithelium along the dorsal surface of the body cavity on either side of the dorsal mesentery of the gut. Early development is so similar that sexes cannot be differentiated.

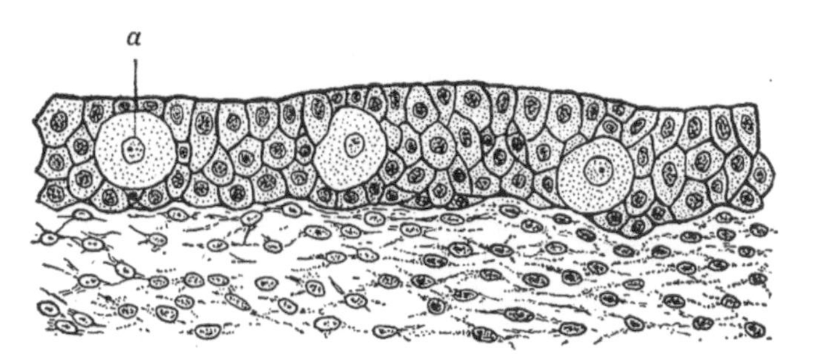

FIGURE O2a The GERMINAL EPITHELIUM of the ridge is continuous with the mesodermal lining of the rest of the coelom. PRIMARY GERM CELLS (a) are scattered throughout. The origin of these cells is enigmatic: it has been suggested that they are endodermal being derived from the lining of the gut or yolk sac and migrating—either through the intervening tissues by amoeboid motion or in the blood—to the genital ridge.

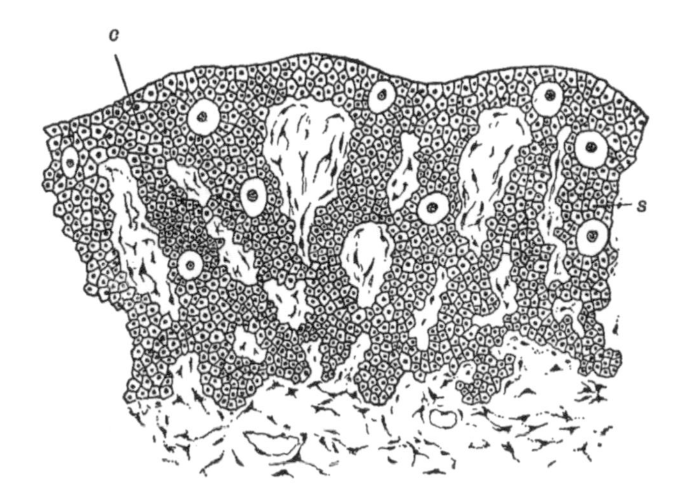

FIGURE O2b Mesenchyme bearing vascular connective tissue and INTERSTITIAL TISSUES invades the epithelial mass (c) from below molding these cells into PRIMARY SEX CORDS (s) which grow inward into the substance of the gonad. These cords contain germ cells as well as supportive elements.

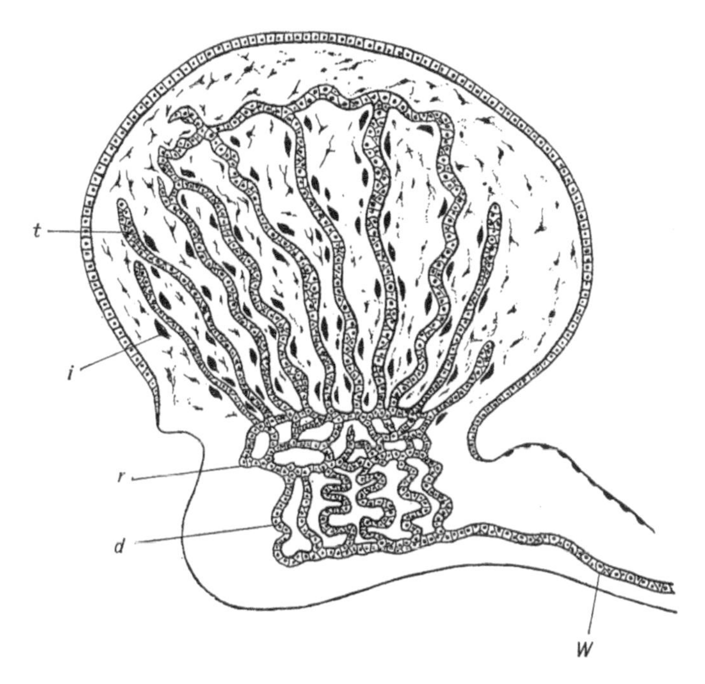

FIGURE O2c The sex cords, containing germ cells, develop into seminiferous tubules (t) of the TESTIS. Germ cells in the walls of the tubules divide repeatedly, eventually producing haploid spermatozoa by meiotic division—the production of sperm is SPERMATOGENESIS. Invading mesenchyme produces the vascular connective tissue between the tubules. Endocrine cells (i), that secrete male sex hormones, are scattered throughout this connective tissue. The rete testis (r) and efferent ducts (d) carry sperm to the vas deferens (W).

Note the intense basophilia and prominent nucleoli of the least mature oocytes. Although ergastoplasm persists in maturing eggs, its basophilia becomes overwhelmed by accumulations of large masses of acidophilic yolk. As the oocyte matures, it acquires a hyaline ZONA PELLUCIDA immediately outside its plasmalemma—especially prominent in the images of lamprey and rock bass developing ova (Figs. O5e and O7c).

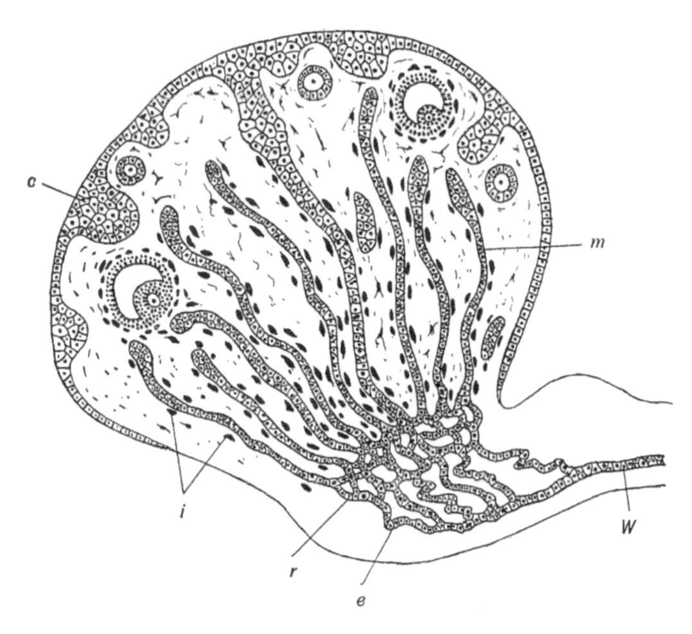

FIGURE O2d In the development of the OVARY, secondary sex cords (c) proliferate inward from the germinal epithelium. The germ cells contained in these cords divide repeatedly and develop eventually into definitive haploid egg cells by *the process of* OOGENESIS. The primary cords, already formed, degenerate in the female. *i*, interstitial cells; *r*, rete ovary; *W*, mesonephric duct.

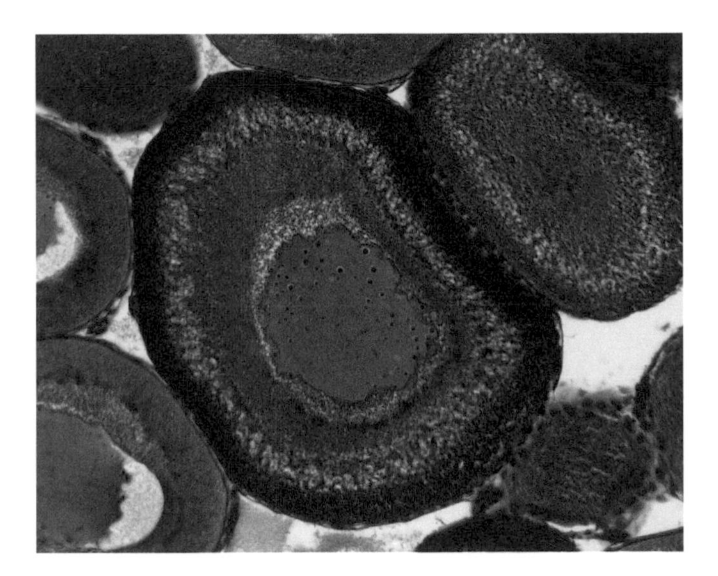

FIGURE O3b Oocytes become increasingly acidophilic as yolk develops and the egg enlarges. Note the prominent nucleoli. Oocytes are actively engaged in the production of proteins and often display several nucleoli in their nuclei, 20 ×.

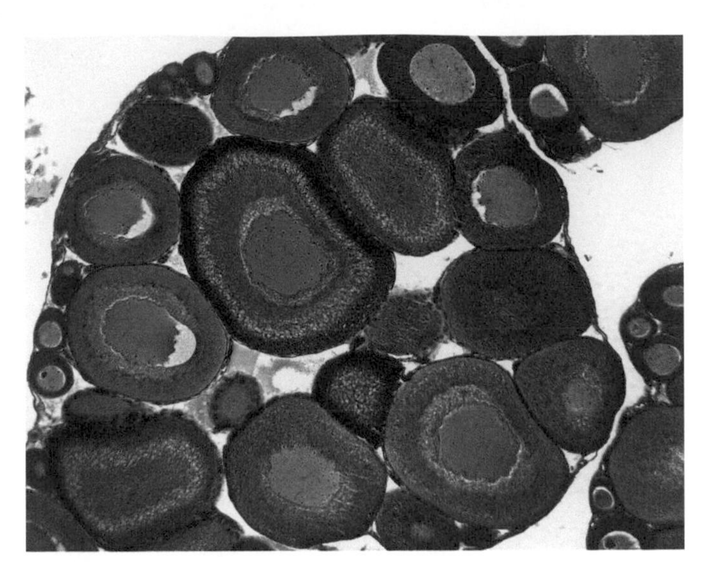

FIGURE O3a Frogs and toads produce large numbers of eggs and, immediately before spawning, the ovary fills most of the body cavity. The ovary is a thin-walled sac enclosed by PERITONEUM. The peritoneum forms a thin sheath that may be seen between the eggs in this section. A portion of the kidney appears in the upper right corner. The OVARIAN FOLLICLES occur in LOBES, each surrounded by thin peritoneum. Two distinct lobes occur at the center of this micrograph; portions of two others are seen at the top center and lower right. The small, deeply basophilic oocytes will not be spawned immediately; the larger, more acidophilic eggs are almost ready for ovulation, 5 ×.

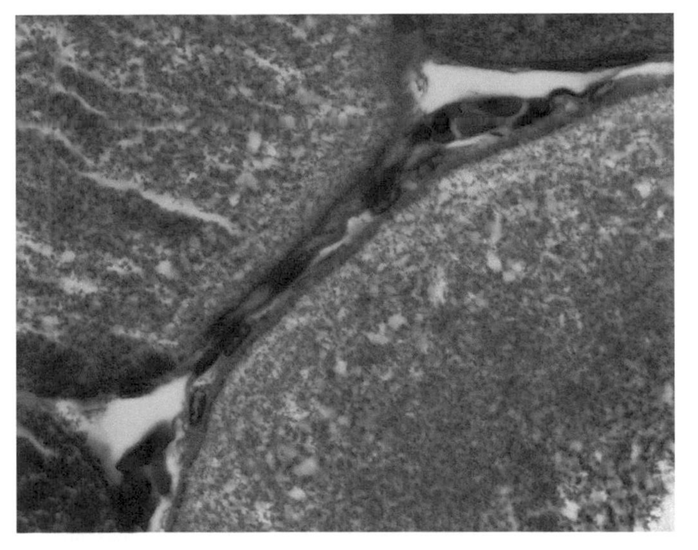

FIGURE O3c These oocytes are surrounded by a squamous layer of FOLLICULAR CELLS that transmit the raw materials for yolk production from the blood to the developing oocyte. Outside the follicular layer a few scattered cells of the THECA FOLLICULI may be discerned. Capillaries, containing brilliant red erythrocytes, course through the connective tissue between the follicles. These mature oocytes have developed a thick, strong membranous ZONA PELLUCIDA outside the OOLEMMA. The delicate cell membrane of such a huge cell, the oolemma, would be inadequate to protect it from the rigors of life after spawning, and this strong membrane fulfills its need. The zona pellucida does not block the transfer of materials into and out of the oocyte—it is perforated by many PORES containing microvilli that extend both from the follicular cells and from the oolemma. These pores give the zona pellucida a striated appearance so that it is sometimes referred to as the zona radiata, 100 ×.

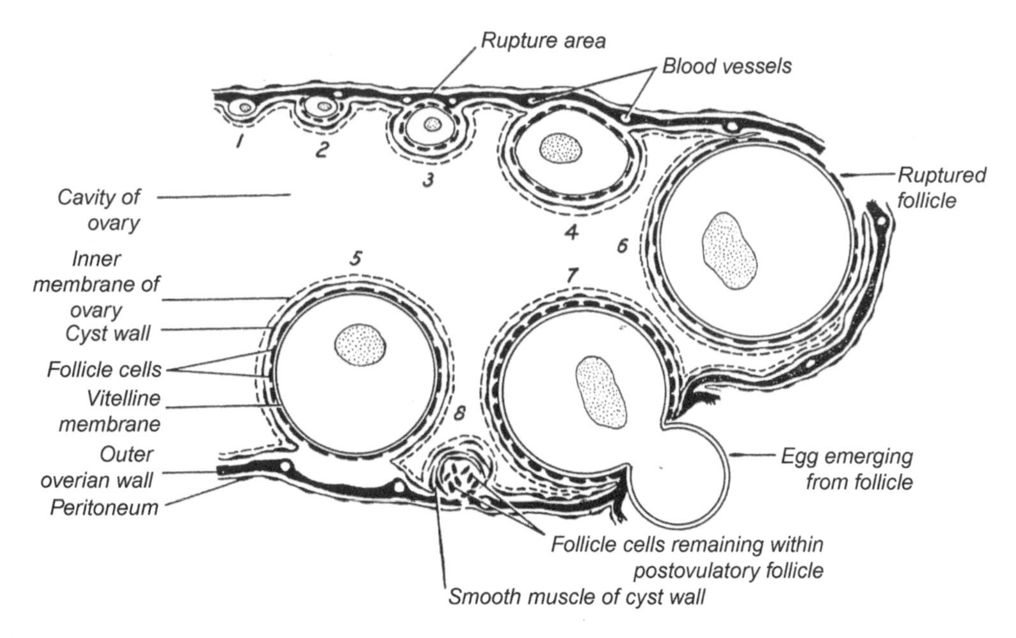

FIGURE O4 When the oocyte is mature, the follicle ruptures and the egg passes through the surrounding layers. It perforates the peritoneum and arrives in the coelom to be drawn by ciliated cells into the oviduct where it is stored until the female is clasped by the male in AMPLEXUS.

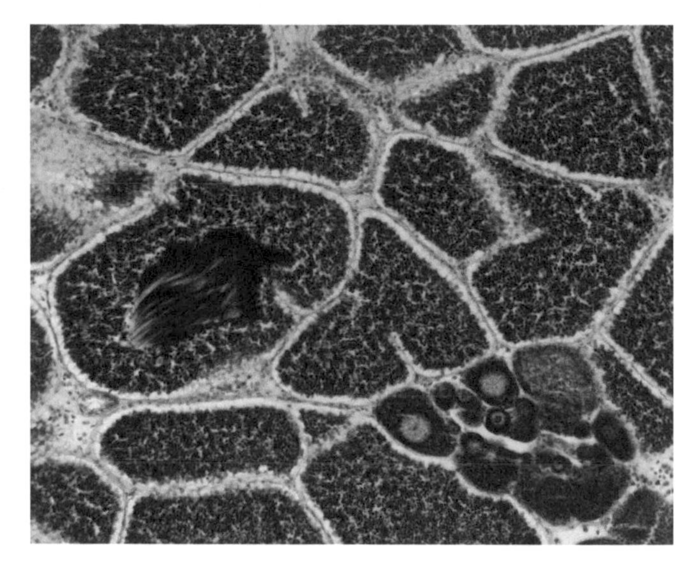

FIGURE O5a Ovary of the sea potato, *Boltenia*. Although *Boltenia* is not a vertebrate, it *is* a chordate. The sea potato is a large sessile tunicate. Its reproductive organ is an OVOTESTIS, containing both developing sperm and oocytes. In spite of this the ovarian portions show the characteristics noted previously. 20×.

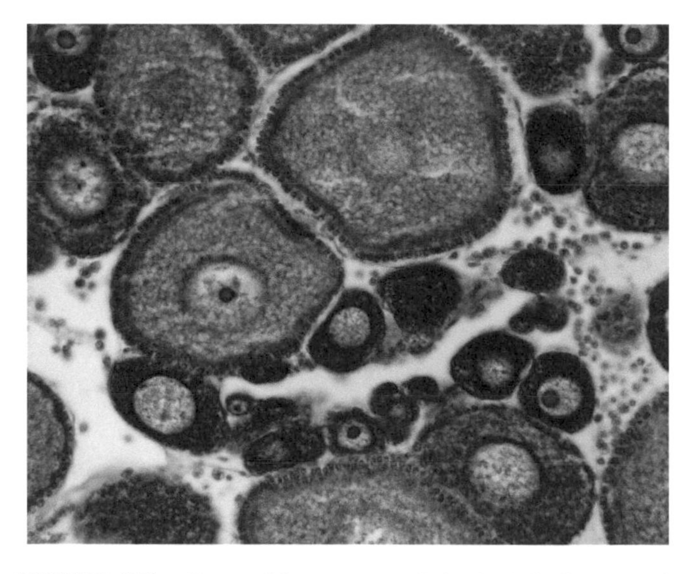

FIGURE O5b Ovary of the sea potato, *Boltenia*. At higher magnification several follicles are seen in various stages of development: some contain small, basophilic oocytes while acidophilic yolk droplets crowd the ooplasm of the larger oocytes. A follicular layer and theca can be seen surrounding the oocytes. A prominent nucleolus is seen in the nuclei of some, 40×.

Ovary of Other Vertebrates

The ovary of many other chordates is similar to this basic plan (Figs. O5a and O5b). Variations can be attributed to differing life habits. Some animals, such as the reptiles and birds, lay fewer eggs than the frog and these eggs are more richly supplied with yolk (Figs. O10, O11a, and O11b). This is reflected in the structure of the ovary where fewer, much larger eggs are seen than in the frog. Mammals produce relatively few offspring and these develop for fairly long periods in the uterus. The mammalian follicle has become a specialized endocrine organ producing the hormones required before and during pregnancy. Animals that breed several times over a short period show a succession of ova in varying degrees of maturity.

Although not vertebrates, as well as being monoecious, even the common sea potato (*Boltenia* a chordate) displays the appearance of a fairly typical ovary (Figs. O5a and O5b).

Hagfish are among the few vertebrates that are monoecious. The single median gonad is attached to the dorsal body wall by the mesovarium. The anterior part produces oocytes, the posterior spermatozoa, but usually at different times. The other cyclostomes (i.e., lampreys) are dioecious and have a single median gonad (Figs. O5c and O5d). Cyclostomes have no oviducts and the eggs leave the body cavity by a GONOPORE that develops in the body wall just prior to spawning. Since the lamprey spawns only once during its lifetime, development of all oocytes in the ovary is synchronous.

Eggs of elasmobranchs are the largest of the fishes and are extremely yolky (Fig. O6). These animals are oviparous or viviparous and only a small number of eggs mature at intervals during the breeding season. Hemopoietic activity may be seen in the connective tissue stroma of the elasmobranch ovary.

By contrast, teleosts have the smallest and most numerous eggs of the an amniotes (Figs. O7a–O7c). In certain teleosts the two ovaries fuse. The oocytes in the ovary of the rock bass (*Ambloplites* sp.) are at various stages of development indicating that there will be several spawnings over the course of the season. The smallest, most basophilic oocytes are preparing to produce proteinaceous yolk. This is indicated by the presence of several nucleoli and the deep basophilia of their ooplasm. The oocytes in the ovary of the rock bass are at various stages of a sign of the presence of granular endoplasmic reticulum which is also associated with protein production. This basophilia is diluted, as acidophilic droplets of yolk are produced. The nuclei of early, deeply basophilic oocytes contain several nucleoli, an indication of active protein production. The protein takes the form of the acidophilic yolk droplets at the lower left and right (Figs. O7a–O7c).

The striated nature of the ZONA PELLUCIDA is obvious in Fig. O7d, the ovary of the rock bass. The striations indicate the presence of PORE CANALS that penetrate the zona pellucida and contain microvilli extending either from the oolemma or the plasmalemma of follicular cells. Outside the zona pellucida is a single layer of squamous FOLLICULAR CELLS, the basal lamina of this epithelium (which you may not be able to discern), and the cells of the THECA. Blood capillaries course through the connective tissue between the follicles.

Several sections of ovaries of Amphibia are provided (Figs. O8a, O8b, and O9) to emphasize the unity of plan of vertebrate ovaries and to indicate that much can be determined of the life style of an animal from its deviation from this simple plan.

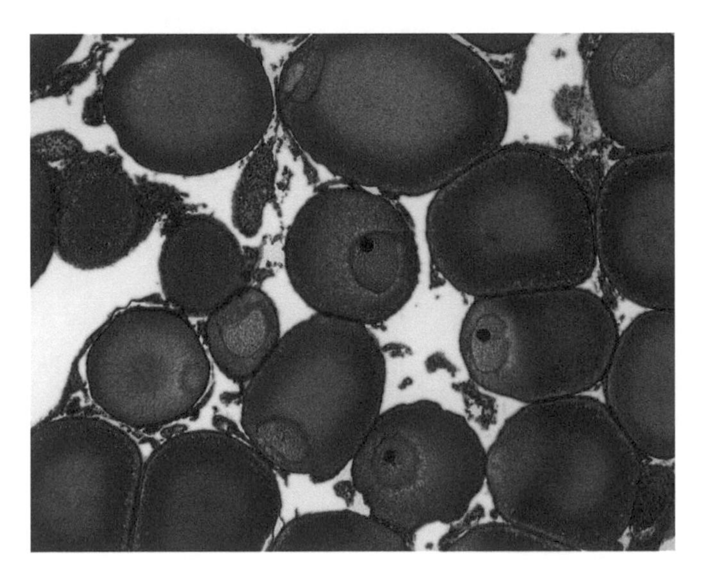

FIGURE O5c Most of the eggs in the ovary of the lamprey are at the same stage of development because the lamprey spawns only once and dies shortly after. Some eggs erroneously appear smaller because they are cut tangentially—note that their ooplasm is similarly acidophilic. Nucleoli are prominent in several nuclei. The connective tissue stroma is richly vascular, 10 ×.

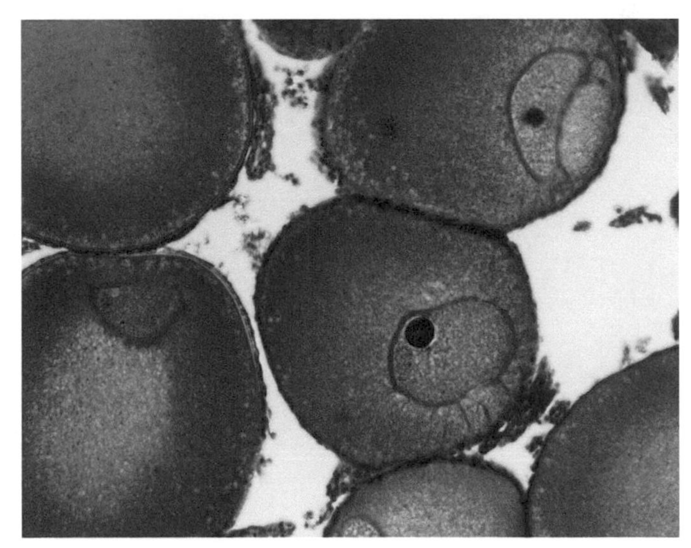

FIGURE O5d Most of the eggs in the ovary of the lamprey are at the same stage of development because the lamprey spawns only once and dies shortly after. Some eggs erroneously appear smaller because they are cut tangentially—note that their ooplasm is similarly acidophilic. Nucleoli are prominent in several nuclei. The connective tissue stroma is richly vascular, 20 ×.

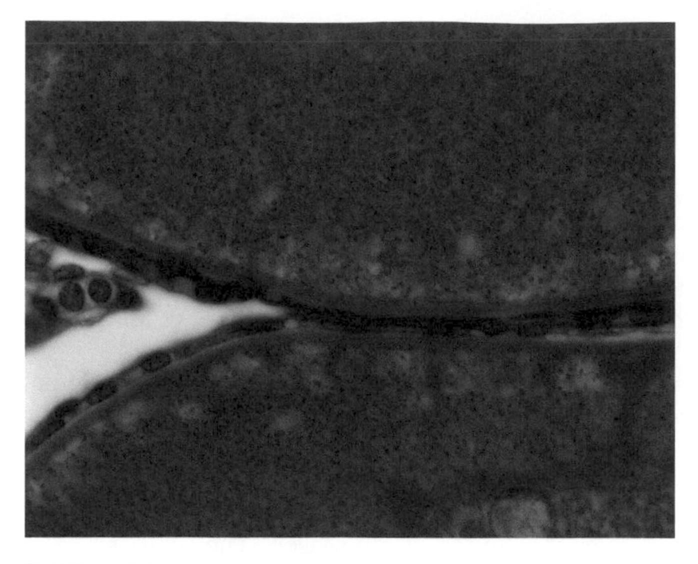

FIGURE O5e Yolk platelets abound in the ooplasm of these two eggs. Striations can be detected in the ZONA PELLUCIDA. Outside the zona pellucida are the squamous cells of the follicular layer. A few cells of the THECA FOLLICULA and some blood vessels enclose the follicles, 100 ×.

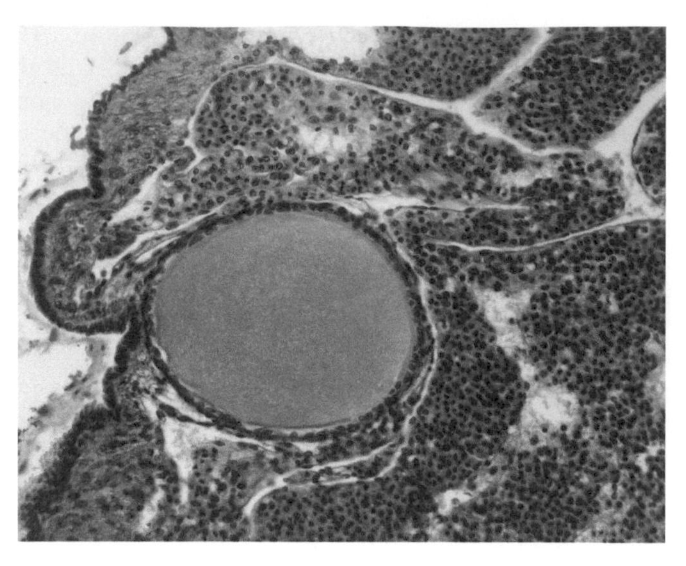

FIGURE O6 Following ovulation in the dogfish, spent follicles remain in the ovary along with a few atretic ovules. Abundant hemopoisis is taking place, 20 ×.

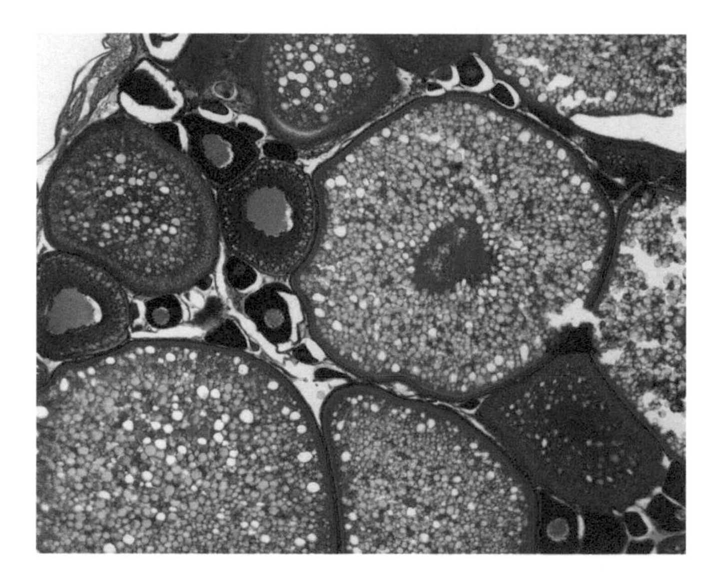

FIGURE O7a The oocytes in the ovary of the rock bass (*Ambloplites* sp.) are at various stages of development indicating that there will be several spawnings over the course of the season. The smallest, most basophilic oocytes are preparing to produce proteinaceous yolk. 10 ×.

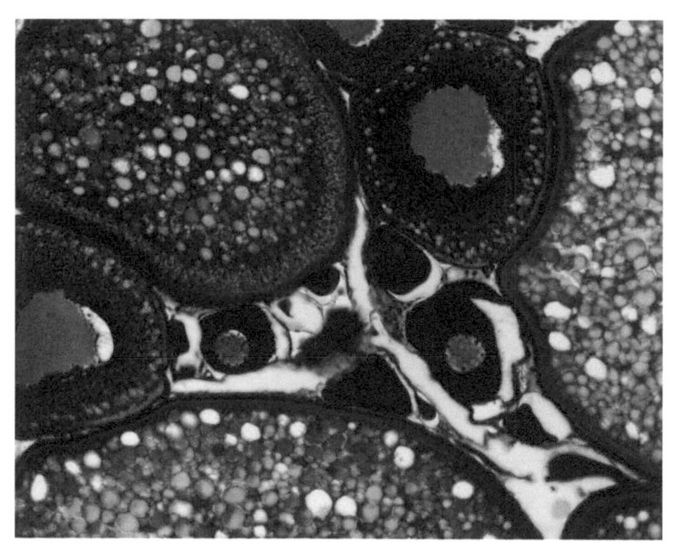

FIGURE O7b The oocytes in the ovary of the rock bass (*Ambloplites* sp.) at various stages of development. The smallest, most basophilic oocytes are preparing to produce proteinaceous yolk. This is indicated by the presence of several nucleoli and the deep basophilia of their ooplasm, a sign of the presence of granular endoplasmic reticulum which is also associated with protein production. 20 ×.

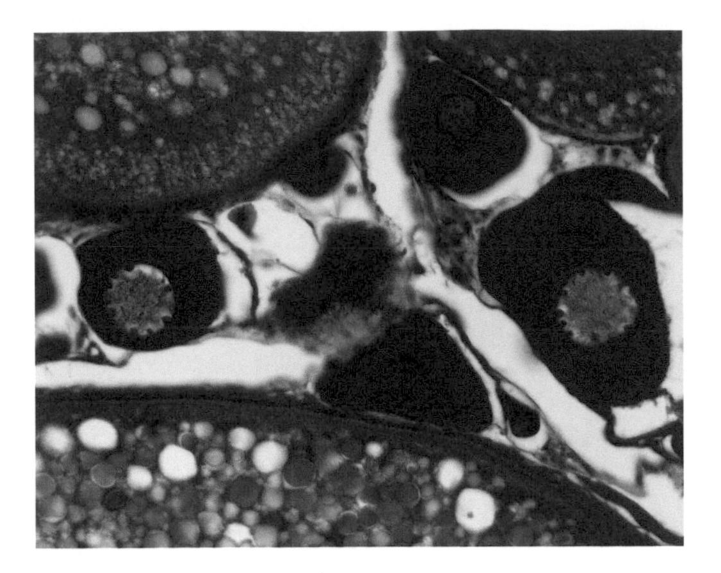

FIGURE O7c The oocytes in the ovary of the rock bass (*Ambloplites* sp.). The basophilia is diluted as acidophili droplets of yolk are produced. The nuclei of early, deeply basophilic oocytes contain several nucleoli, an indication of active protein production. The protein takes the form of the acidophilic yolk droplets as seen in the lower portion of this image. 40 ×.

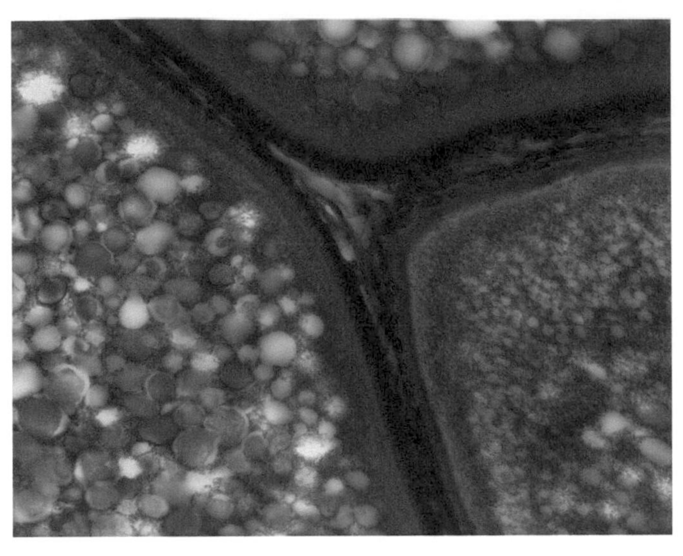

FIGURE O7d The striated nature of the ZONA PELLUCIDA is obvious in these sections. The striations indicate the presence of PORE CANALS that penetrate the zona pellucida and contain microvilli extending either from the oolemma or the plasmalemma of follicular cells. Outside the zona pellucida is a single layer of squamous FOLLICULAR CELLS, the basal lamina of this epithelium (which you may not be able to discern), and the cells of the THECA. Blood capillaries course through the connective tissue between the follicles, 100 ×.

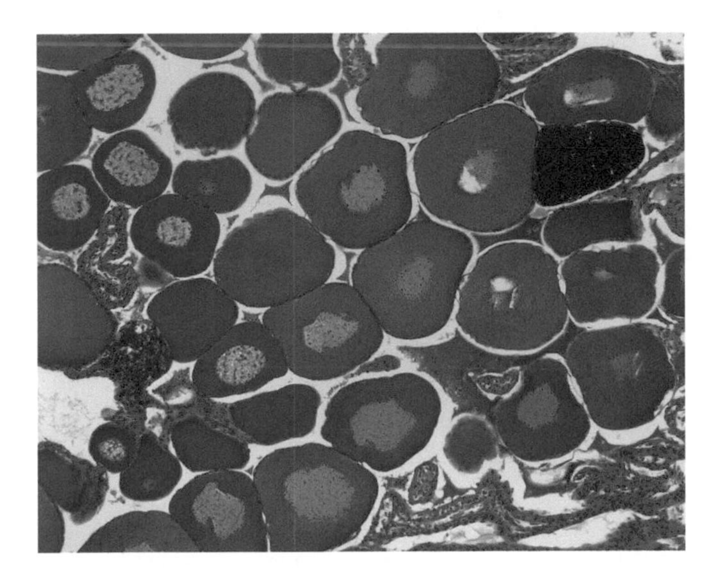

FIGURE O8a Several immature follicles, enclosed with peritoneum, form a part of a single lobe in the ovary of the frog. The black patch in the upper right is probably pigment left from the atresia of a mature, pigmented egg of a previous generation. There is also a small amount of pigment at the lower left, 10 ×.

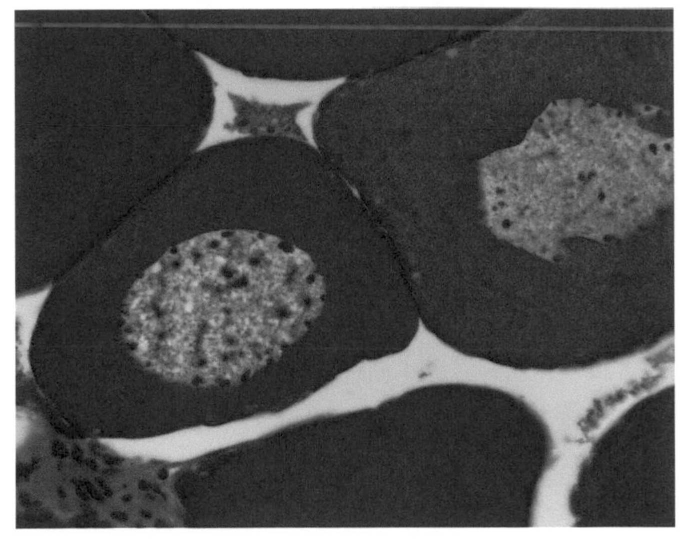

FIGURE O8b A thin, squamous follicular layer surrounds each oocyte. The thecal layer is not obvious. Several nucleoli are scattered within the nuclei. Capillaries occur in the connective tissue between the oocytes, 40 ×.

The ovaries of birds and reptiles produce relatively few large eggs with abundant yolk (Figs. O10, O11a, and O11b). At any one time during the breeding season, the ovary contains a series of "eggs in waiting" which accumulate yolk day by day. During its final day in the ovary, there is a rapid accumulation of yolk and the egg is ovulated. As it passes through the oviduct, the complex and architecturally beautiful shell is added. Several

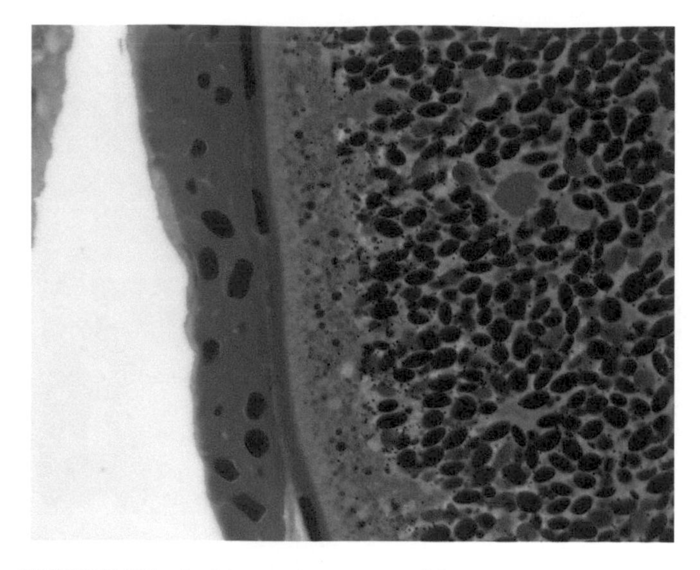

FIGURE O9 In this semithin section of frog ovary, several details can be seen that were only imagined in the above micrographs. Beginning at the left is a thick THECAL LAYER, separated from the FOLLICULAR EPITHELIUM by the thin BASAL LAMINA of the follicular epithelium. This epithelium rests on the ZONA PELLUCIA whose radial striations are discernible. The OOLEMMA separating the zona pellucida from the OOPLASM cannot be seen. The myriad droplets and platelets of YOLK are scattered throughout the ooplasm, 100 ×.

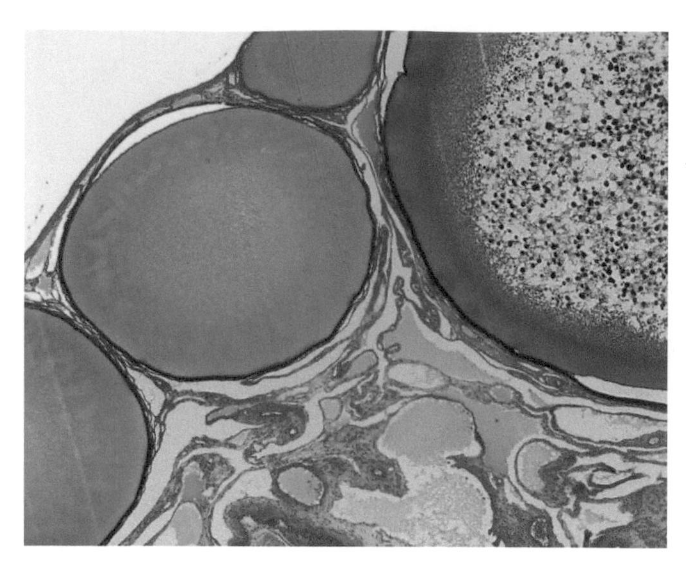

FIGURE O10 The eggs of the turtle are large and yolky. It would be difficult, physically and nutritionally, for a clutch of several eggs to be laid at one time. Instead, she develops oocytes in a graduated series, as can be seen in this micrograph. The yolk complement is completed in one egg at a time and that egg is ovulated. The egg at the upper right is almost ready for ovulation. Ovulation has occurred in many follicles, and the follicular remnants can be seen at the bottom. The leathery shell of the egg is added as it passes through the oviduct, 5 ×.

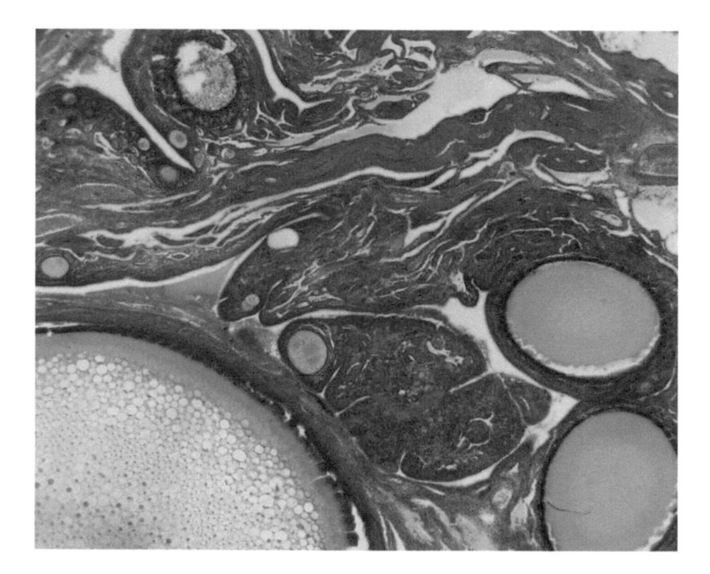

FIGURE O11a Like the turtle, birds produce small clutches of relatively large eggs, usually one egg per day.

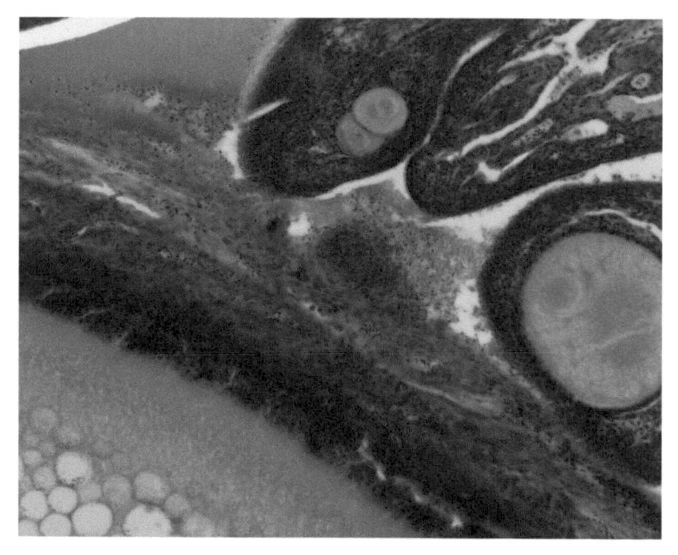

FIGURE O11b A few developing oocytes are scattered throughout this photomicrograph of the ovary of a chicken. The follicular layer and theca are poorly defined in the a mature oocyte at the bottom left, 20 ×.

stages in the maturation of the oocytes can be seen in the micrograph of the ovary of a hen, and one is almost ready for ovulation (Fig. O11a). Eggs are shed at intervals during the breeding season. Egg shells and albuminous coats are added by specialized glands in the oviducts. The ovaries of reptiles are paired but in birds the right ovary is rudimentary and only the left is functional.

Mammalian Ovary

The HILUS may be visible in sections of the mammalian ovary (Fig. O12). This is the point of attachment of the mesovarium where blood vessels and nerves enter the ovary. The broad peripheral layer containing developing

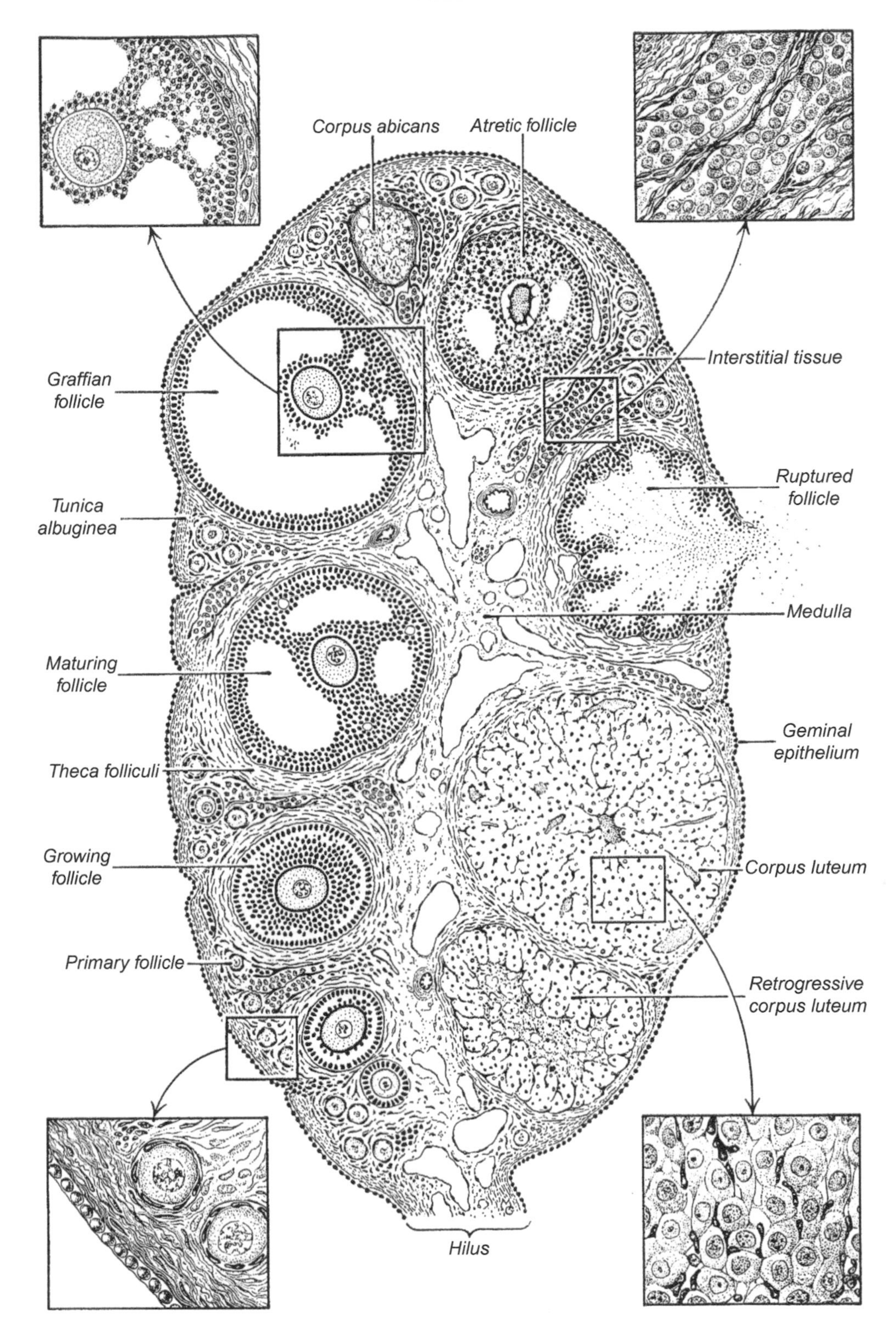

FIGURE O12 The development of oocytes in the mammal—oogenesis—is more complex than we have seen in the lower vertebrates. The development of ovarian follicles is summarized in this figure. Not only is the ovary an ENDOCRINE GLAND, producing hormones in highly synchronized sequences that control the sexual cycles of females, it is also a CYTOCRINE GLAND providing a haven for developing ovarian follicles.

FOLLICLES is the CORTEX (Figs. O13a and O13b). Inside is the MEDULLA with abundant large blood vessels in a STROMA of areolar connective tissue; the medulla is usually devoid of follicles. Smooth muscle fibers in the stroma surround the follicles. In a number of mammals, especially those that have large litters, scattered endocrine cells in the cortex collectively constitute the INTERSTITIAL GLAND.

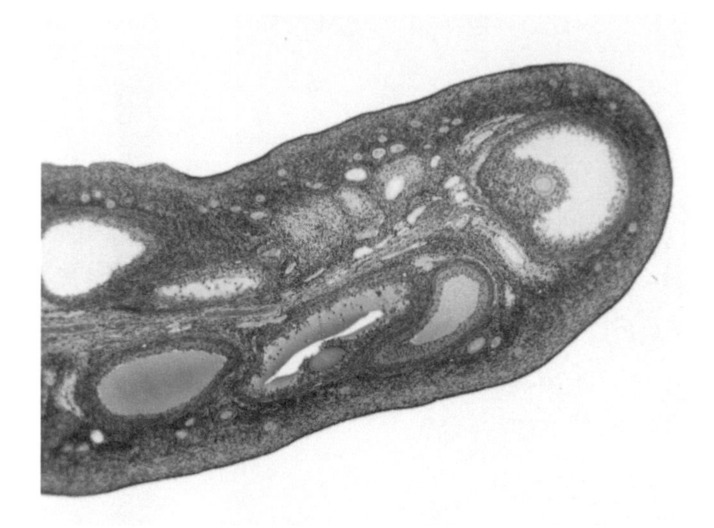

FIGURE O13a The mammalian ovary is attached to the MESOVARIUM at the HILUS (which is not visible in this section), the point when blood vessels enter and leave. Follicles in various stages of development are scattered in the CORTEX at the periphery of the ovary. There is a MATURE FOLLICLE (Graafian follicle) at the right. At the center is the STROMA of connective tissue of the MEDULLA, which contains the large blood vessels from the hilus, 5 ×.

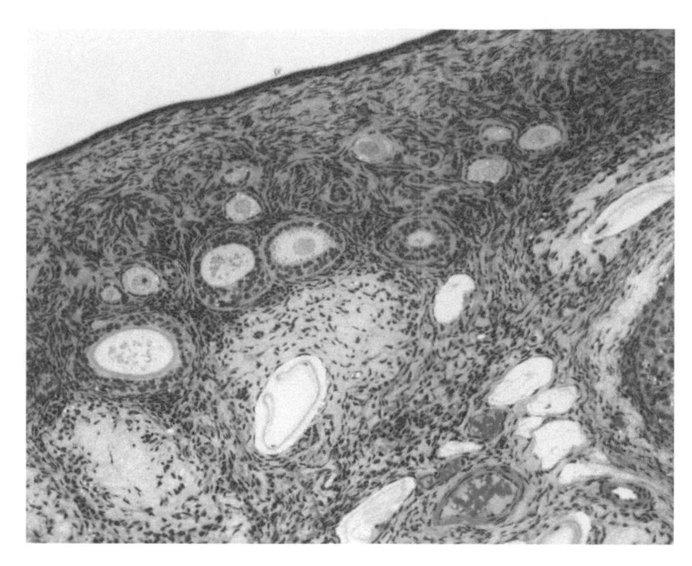

FIGURE O13b Surrounding the ovary (beneath the peritoneum) is the GERMINAL EPIHELIUM of squamous or cuboidal cells. Inside is the TUNICA ALBUGINEA of dense fibrous connective tissue. The rest of the cortex consists of the stroma of loose connective tissue in which the follicles develop, 20 ×.

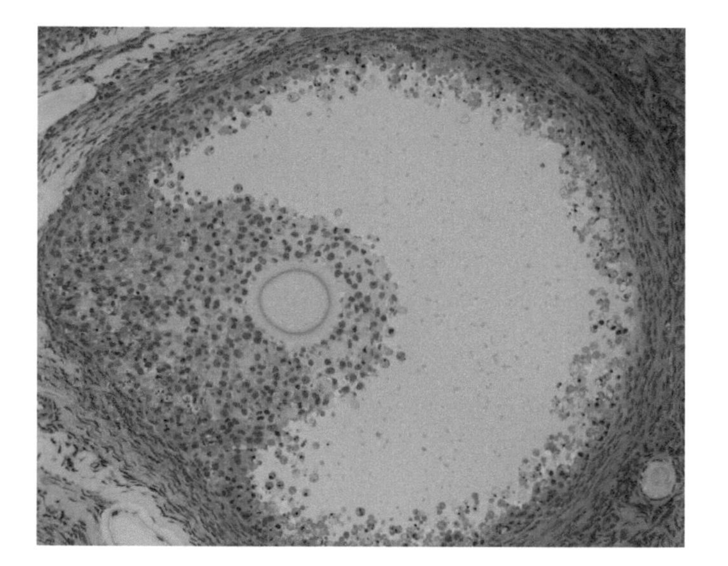

FIGURE O13c This MATURE FOLLICLE is surrounded by a double sheath, the THECA FOLLICULI, consisting of the THECA INTERNA, a highly vascularized layer of secretory cells and the THECA EXTERNA, a coat of connective tissue and larger blood vessels. A loose layer of GRANULOSA CELLS is applied to the inside of the theca interna, 20 ×.

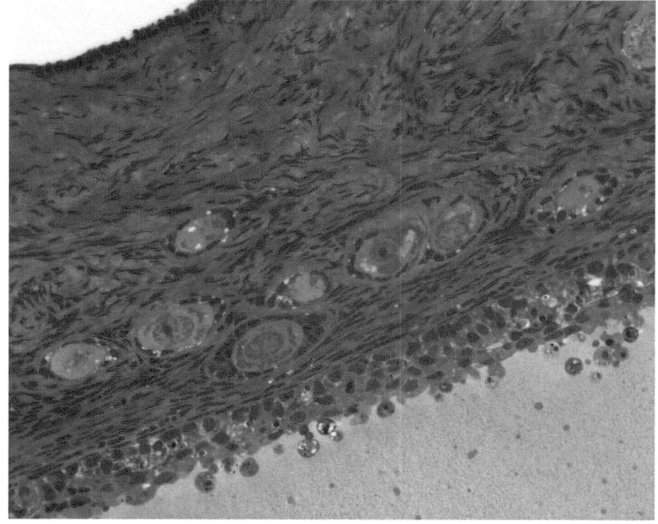

FIGURE O14 The cuboidal cells of the GERMINAL EPITHELIUM are seen at the top of this micrograph. Primordial follicles, surrounded by connective tissue of the theca folliculi, of a mature follicle are at the bottom. A few primordial follicles are lodged in the connective tissue of the cortex, 40 ×.

Surrounding the ovary (under the peritoneum) is the GERMINAL EPITHELIUM of squamous or cuboidal cells (Figs. O13b, O14, O15a–O15c, O16c). Inside is the TUNICA ALBUGINEA of dense fibrous connective tissue. The remainder of the cortex consists of a STROMA of loose connective tissue in which the follicles develop. If ovulation had occurred recently or if the animal was pregnant, CORPORA LUTEA (singular: corpus luteum) may be seen in the cortex (Fig. O22c). These are large endocrine bodies containing cords of pale, liver-like cells. CORPORA ALBICANTIA (*singular*: corpus albicans) (Figs. O22a, O23a and O23b), may also be seen in the cortex. These aggregations of scar tissue are all that remain when the corpora lutea cease to function.

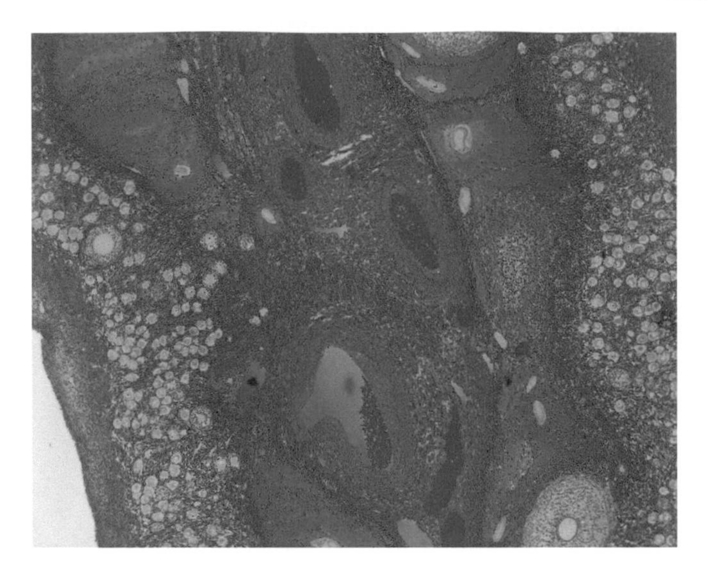

FIGURE O15a Primordial follicles are lodged in the connective tissue of the ovarian cortex. The oocytes are surrounded by a single layer of squamous FOLLICULAR CELLS. Nucleoli are visible in two of the nuclei. CHROMOSOMES, visible within the nuclei, are undergoing meiosis, 5×.

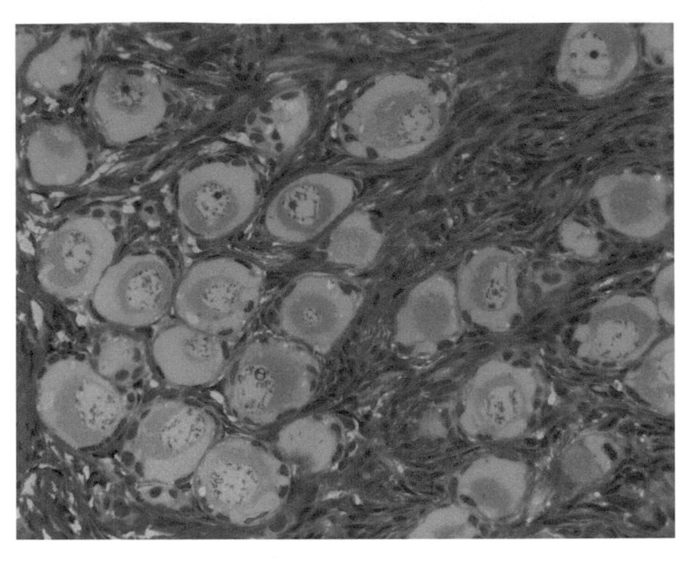

FIGURE O15b Primordial follicles are lodged in the connective tissue of the ovarian cortex. The oocytes are surrounded by a single layer of squamous FOLLICULAR CELLS. Nucleoli are visible in two of the nuclei. CHROMOSOMES, visible within the nuclei, are undergoing meiosis, 40×.

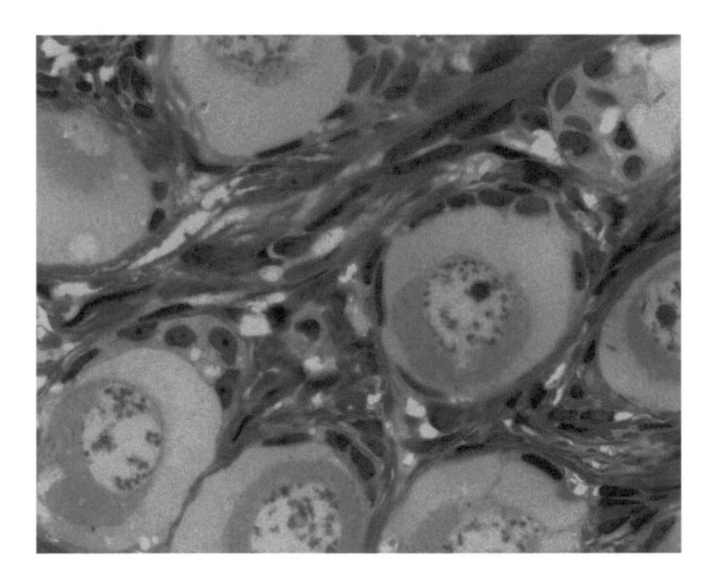

FIGURE O15c Primordial follicles are lodged in the connective tissue of the ovarian cortex. The oocytes are surrounded by a single layer of squamous FOLLICULAR CELLS. Nucleoli are visible in two of the nuclei. CHROMOSOMES, visible within the nuclei, are undergoing meiosis, 100×.

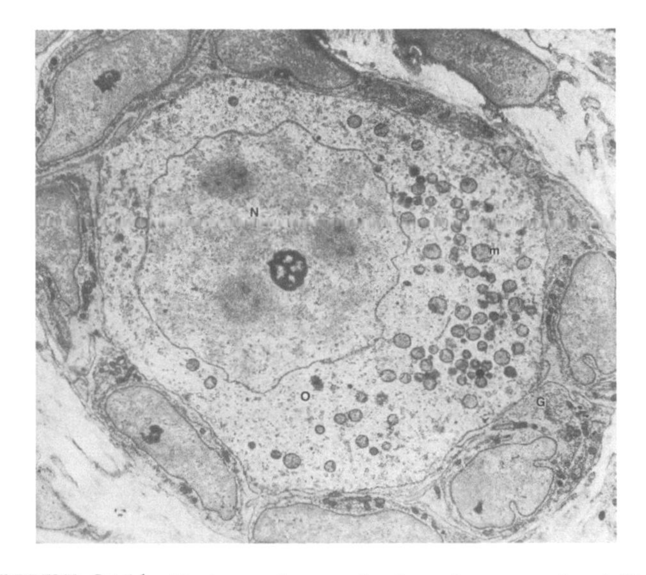

FIGURE O15d Electron micrograph of a primary ovarian follicle in a guinea pig. The nucleus (N) of the primary oocyte (O) is in meiotic arrest during prophase of the first maturation division. Mitochondria (m) contain few cristae. The plasma membrane of the oocyte is apposed to the surface of the surrounding single epithelial layer of granulosa cells (G).

The stages in the development of the ovarian follicle may be followed. PRIMORDIAL FOLLICLES occur at the edge of the section (Figs. O14 and O15a–O15d) each oocyte is surrounded by a single layer of flattened FOLLICULAR CELLS. As a follicle begins growth it is termed a PRIMARY FOLLICLE. It moves deeper in the stroma and the follicular cells become cuboidal or columnar (Figs. O16a–O16d). At some point in this development the refractile deeply staining ZONA PELLUCIDA appears around the oocyte (Fig. O17). It is formed by both the follicular cells and the ovum itself and gradually thickens. Through rapid mitotic proliferation the single layer of follicular cells forms a stratified epithelium, the STRATUM GRANULOSUM, surrounding the oocyte (Figs. O16a–O16d, O18a, O18b, O19a, and O19b). Examining these images allows us to identify the BASEMENT MEMBRANE of this epithelium. Just outside the

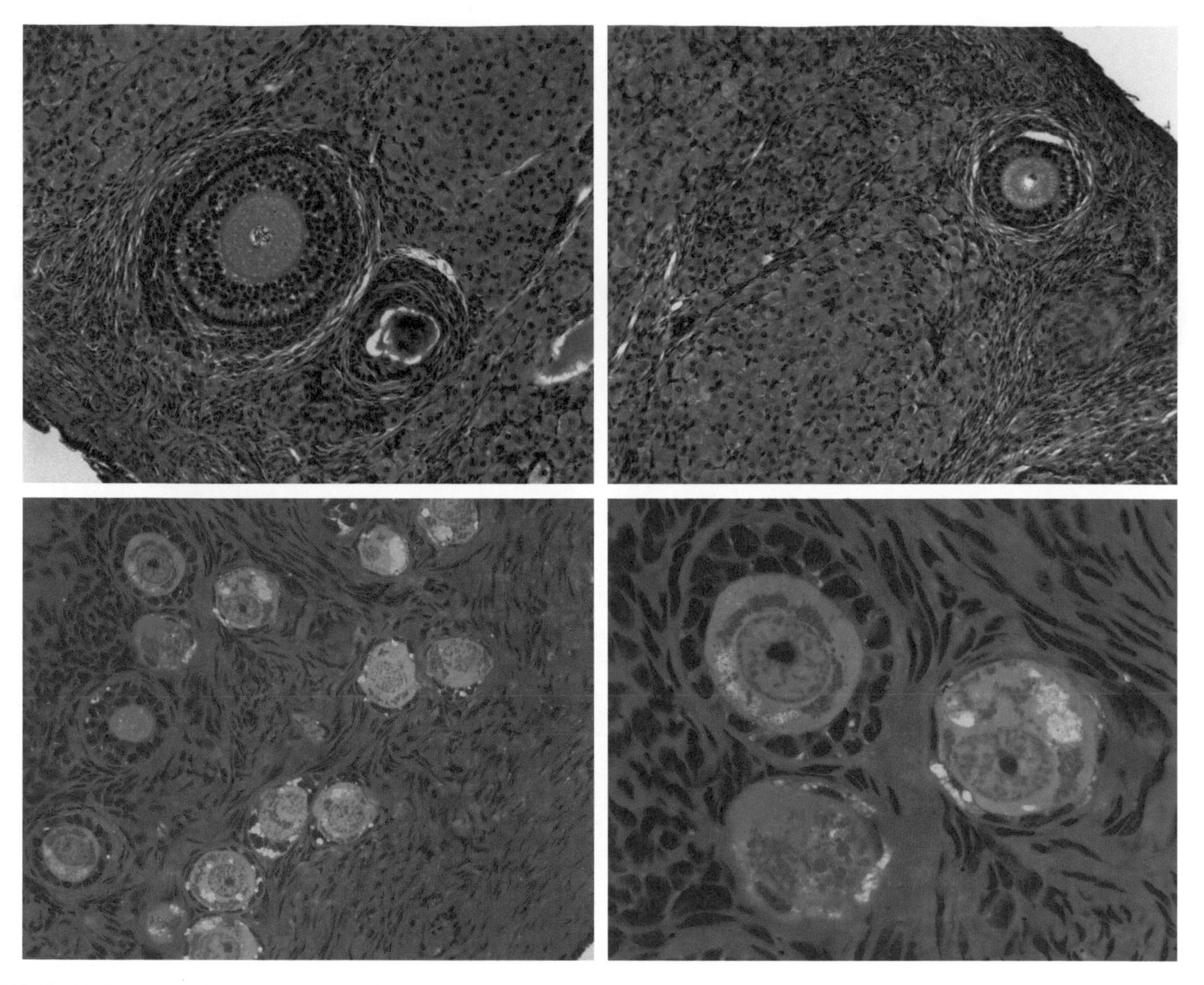

FIGURES O16a, O16b, O16c, and O16d Through rapid mitotic proliferation the single layer of follicular cells forms a stratified epithelium, the STRATUM GRANULOSUM surrounding the oocyte, 20 ×, 40 ×, 100 ×.

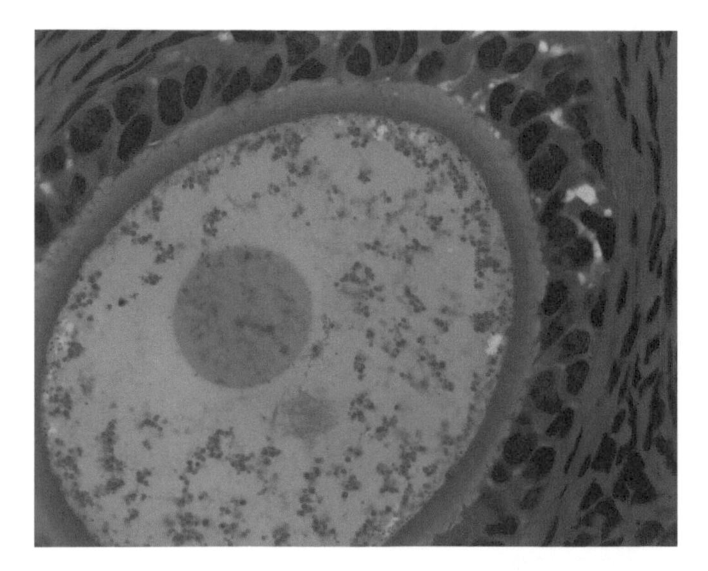

FIGURE O17 As the follicle begins its growth, it moves deeper into the stroma and the follicular cells become cuboidal or columnar, 100 ×.

basement membrane, connective tissue cells of the stroma form a double sheath, the THECA FOLLICULI, consisting of the THECA INTERNA, a highly vascularized layer of secretory cells, and THECA EXTERNA, a coat of connective tissue and larger blood vessels.

As the follicular cells continue to proliferate, fluid-filled cavities appear among the granulosa cells; these coalesce to form a single space, the ANTRUM, filled with LIQUOR FOLLICULI (Fig. O18b). The follicle is now a SECONDARY follicle. The cavity enlarges as the follicle matures. Figs. O19a, O19b, and O20b are the mature follicles where the oocyte and its nucleus show in the section. (Look for the two nucleoli in the upper left quadrant of the oocyte in Figs. O19a and O19b). The oocyte is embedded at one side of the follicle in a "little hill" of follicular cells, the CUMULUS OOPHORUS. As the follicle matures the cumulus becomes more elevated, finally forming a slender stalk bearing the oocyte.

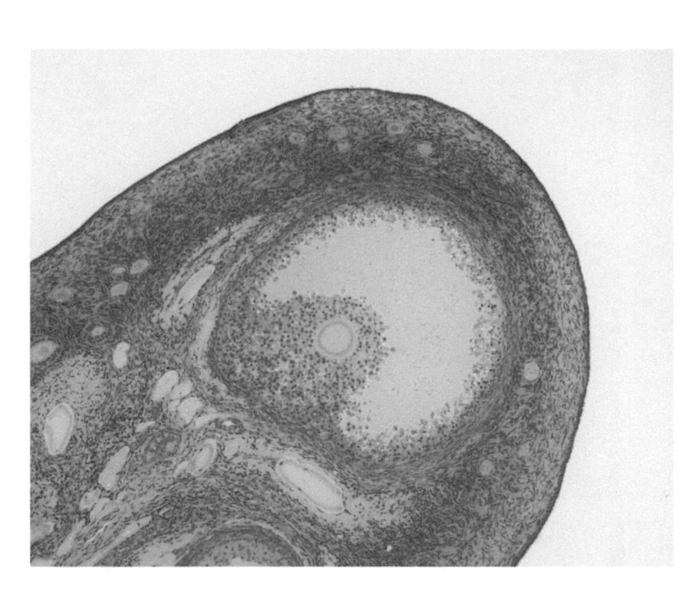

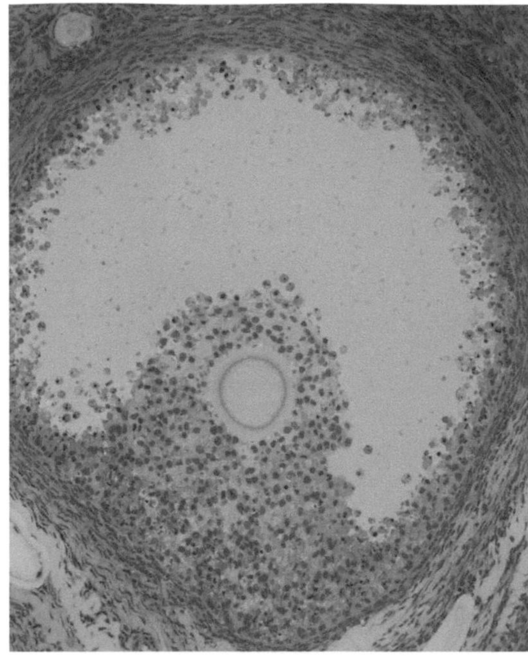

FIGURES O18a and O18b As the follicle cells continue to proliferate, fluid-filled cavities appear among the GRANULOSA CELLS; these coalesce to form a single space, the ANTRUM, filled with LIQUOR FOLLICULI. The follicle is now a SECONDARY FOLLICLE and is seen at the right of this section. The oocyte is held aloft on a "little hill" of follicular cells, the CUMULUS OOPHORUS. As the follicle matures the cumulus becomes more elevated, finally forming a slender stalk bearing the oocyte. The oocyte is surrounded by the ZONA PELLUCIDA, 5×, 10×, 20×.

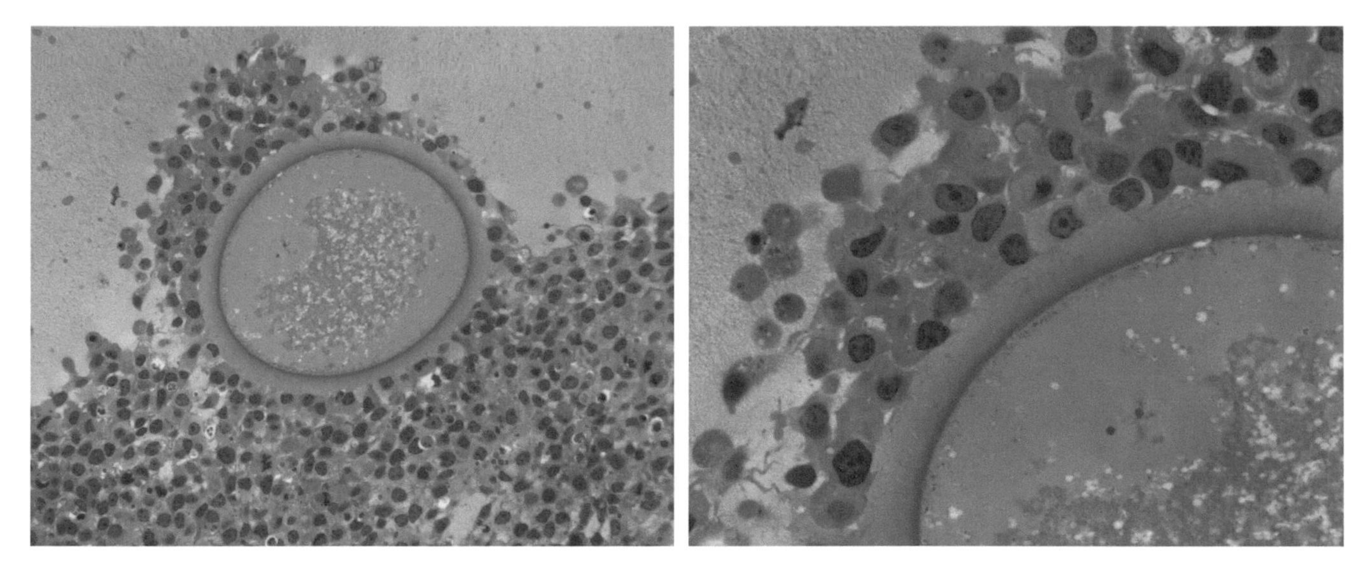

FIGURES O19a and O19b The MATURE FOLLICLE (Graafian follicle) bulges from the surface of the ovary. Cells of the cumulus immediately surrounding the zona pellucida of the oocyte form a single layer of columnar cells, the CORONA RADIATA.

The MATURE FOLLICLE (Graafian follicle) bulges from the surface of the ovary (Figs. O13a, O13c, O18b). Cumulus cells immediately surrounding the zona pellucida of the oocyte form a single layer of columnar cells, the CORONA RADIATA (Figs. O17, O19a, O19b, O20b) ovulation, the follicle ruptures and the oocyte, together with the corona radiata and some cells of the cumulus, is cast into the coelom where it is drawn into the ostium of the oviduct by ciliary currents. The empty follicle fills with clotted blood and follicular fluid and exists for a short time as a CORPUS HEMORRHAGICUM (Fig. O21); corpora hemorrhagica are occasionally visible in sections and are filled with acidophilic granular material. The range of follicular development is shown in Fig. O20a.

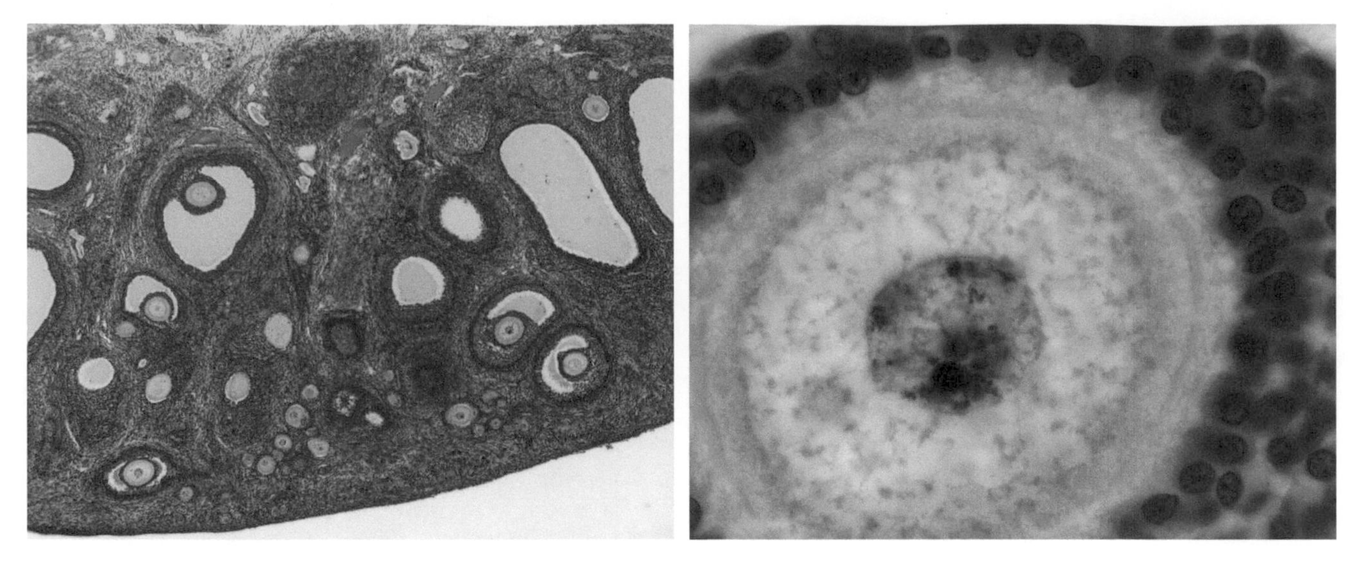

FIGURES O20a and O20b The section in O20a shows various stages in the proliferation of the GRANULOSA CELLS and enlargement of the ANTRUM in the maturing mammalian follicle. The ovum in this mature follicle (O20b) is surrounded by cells of the CUMULUS OOPHORUS. They form a single layer of columnar cells, the CORONA RADIATA. The nucleus contains a large NUCLEOLUS, 5 ×, 100 ×.

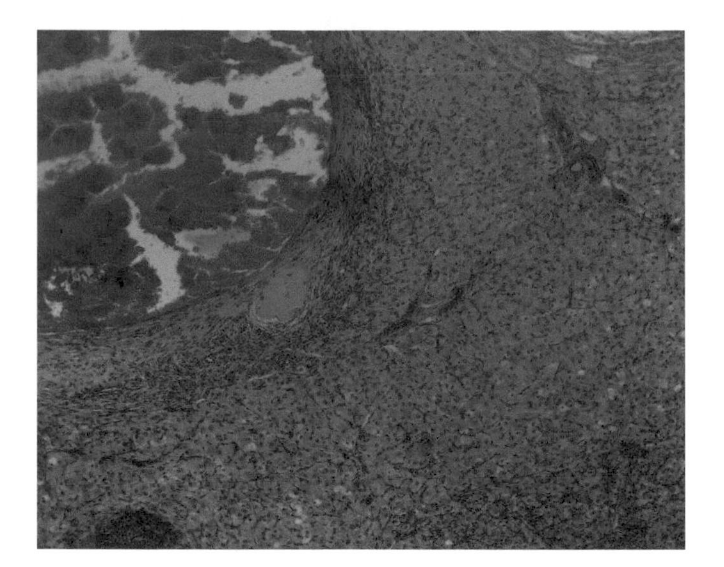

FIGURE O21 At ovulation the follicle ruptures and the oocyte, together with the corona radiata and some cells of the cumulus, is cast into the coelom where it is drawn into the ostium of the oviduct by ciliary currents. The empty follicle fills with clotted blood and follicular fluid and exists for a short time as a CORPUS HEMORRHAGICUM. The remainder of the micrograph consists of a large corpus luteum, 10 ×.

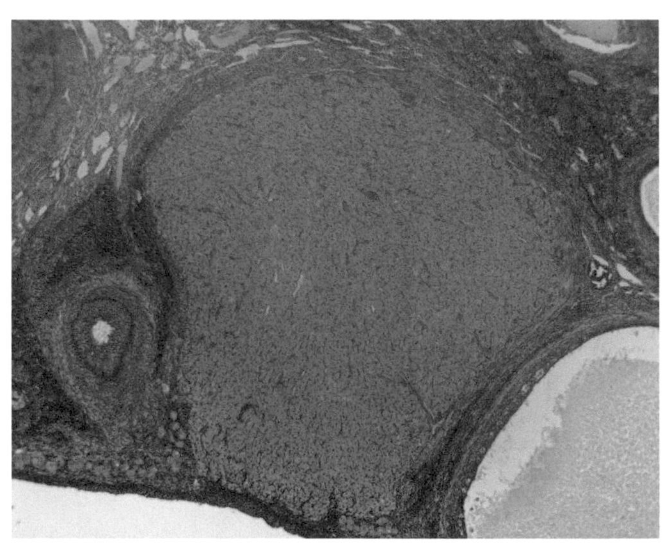

FIGURE O22a After ovulation, extravasated blood of the corpus hemorrhagicum (bloody body) is cleared away by phagocytic invaders. The follicular cells continue to multiply and, together with thecal cells and vascular connective tissue from the storma, actively invade and fill the antrum with cords of pale-staining cells to produce the endocrine CORPUS LUTEUM. If the egg is fertilized, the corpus luteum continues to enlarge, becomes vascular, and remains for the duration of pregnancy, 5 ×.

After ovulation the follicular cells continue to multiply and fill the antrum with cords of pale-staining cells to produce the endocrine CORPUS LUTEUM (Figs. O22a–O22c, O24a, and O24b). If the egg is fertilized, the corpus luteum continues to enlarge, becomes vascular, and remains for the duration of pregnancy. If the ovum is not fertilized, or when pregnancy ends, the corpus luteum degenerates, becomes white scar tissue, the CORPUS ALBICANS, and finally disappear (Figs. O23a and O23b). During anestrus the quiescent ovary of the dog displays no mature follicles, corpora lutea, or corpora albicantia. But many immature follicles are present at the periphery of the ovary (Figs. O25a and O25b).

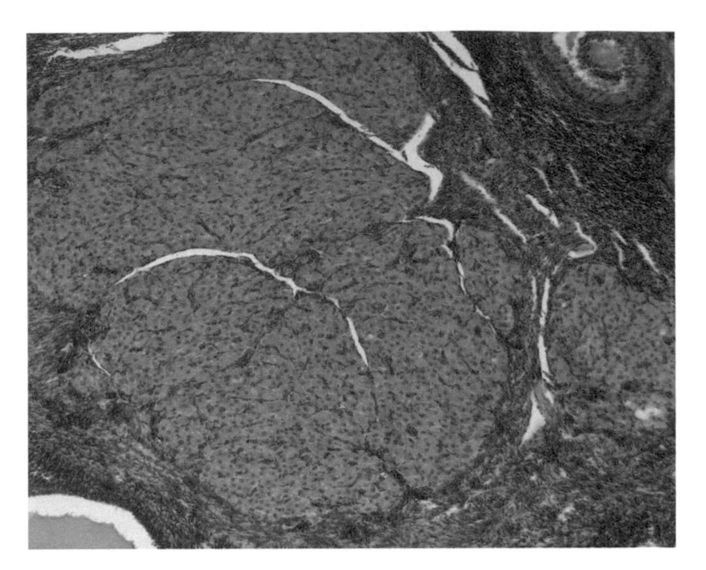

FIGURE O22b A closer look at a CORPUS LUTEUM. If the egg is fertilized, the corpus luteum continues to enlarge, becomes vascular, and remains for the duration of pregnancy, 10 ×.

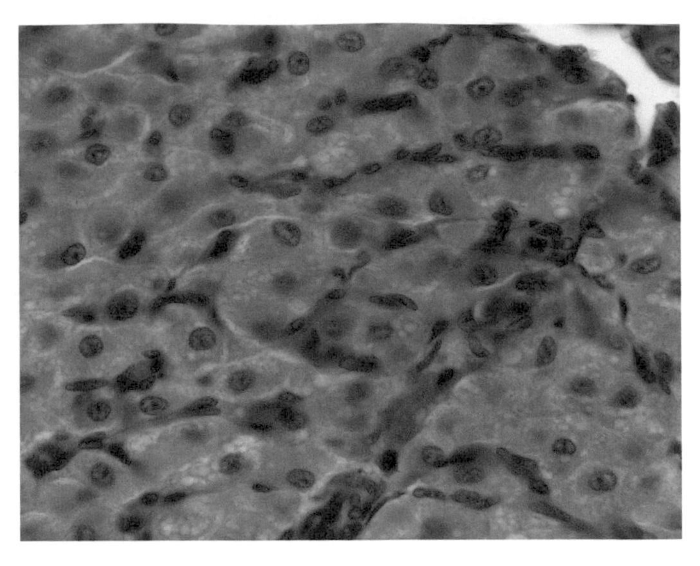

FIGURE O22c The corpus luteum (yellow body) consists of vascularized cords of LUTEIN CELLS irrigated by blood sinusoids—a few endothelial cells may be distinguished in this section. Lipid droplets pack the lutein cells. (Lutein pigment accumulates within the lutein cells and imparts a yellow color to the corpus luteum.) 100 ×.

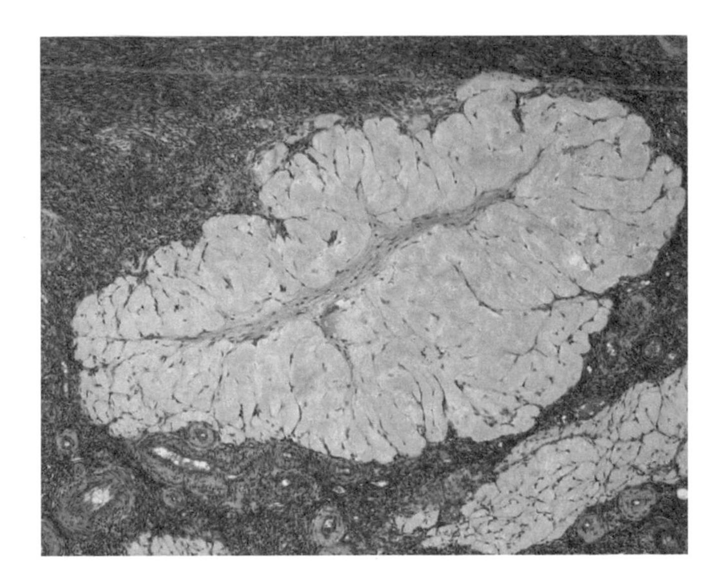

FIGURE O23a If the ovum is not fertilized or, when pregnancy ends, the corpus luteum degenerates, becomes white scar tissue, the CORPUS ALBICANS (white body), and finally disappears, 10 ×.

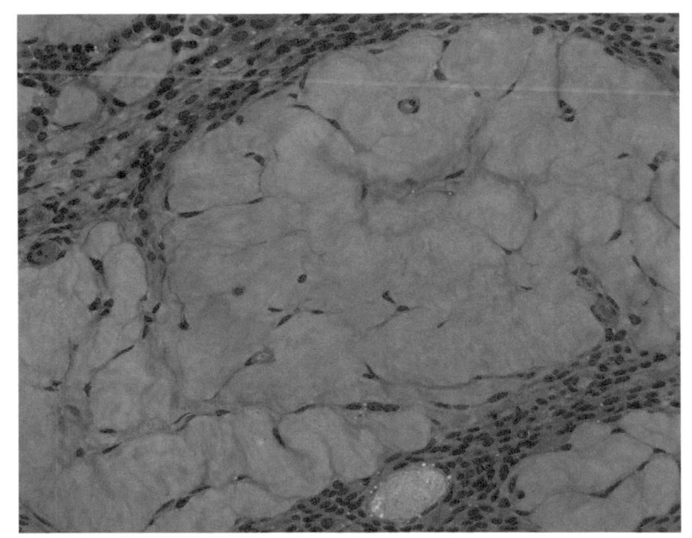

FIGURE O23b The CORPUS ALBICANS (white body) in Fig. O23a at higher power. 40 ×.

Oviduct

There is no oviduct in cyclostomes and the ova leave the body cavity by the gonopore that develops in the body wall just prior to spawning. In teleosts the anterior end of the oviduct is actually a fold of peritoneum that surrounds the ovary and then narrows to form a tubular portion. Eggs are discharged into this peritoneal sac and carried by it to the cloaca. There is a danger that the large number of eggs might choke the coelom if they were shed into it in the usual way.

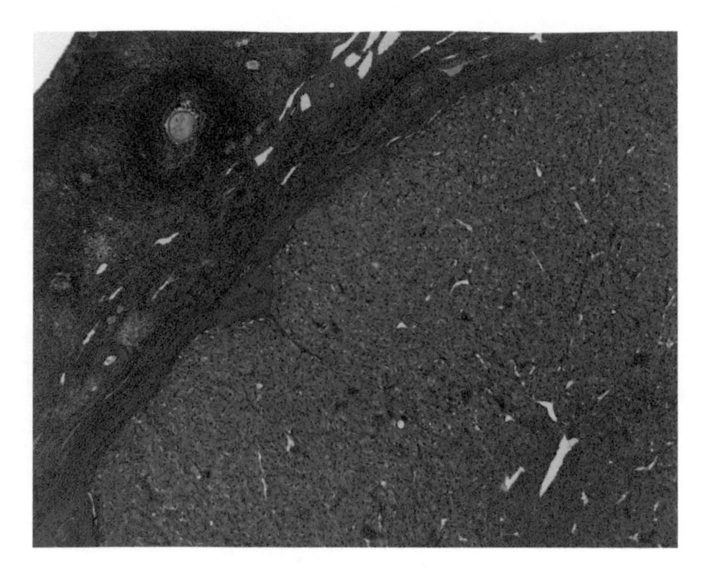

FIGURE O24a The ovary of a dog "in heat" demonstrates an enormous development of corpora lutea, 5×.

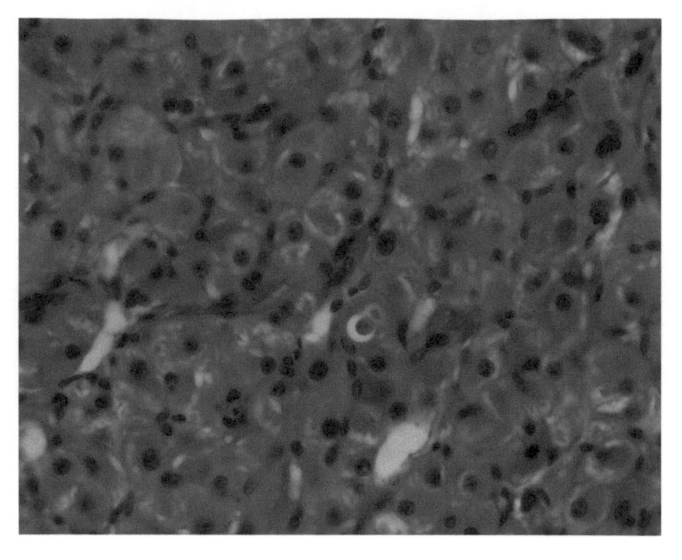

FIGURE O24b The ovary of a dog "in heat" at a higher magnification demonstrating enormous development of corpora lutea, 40×.

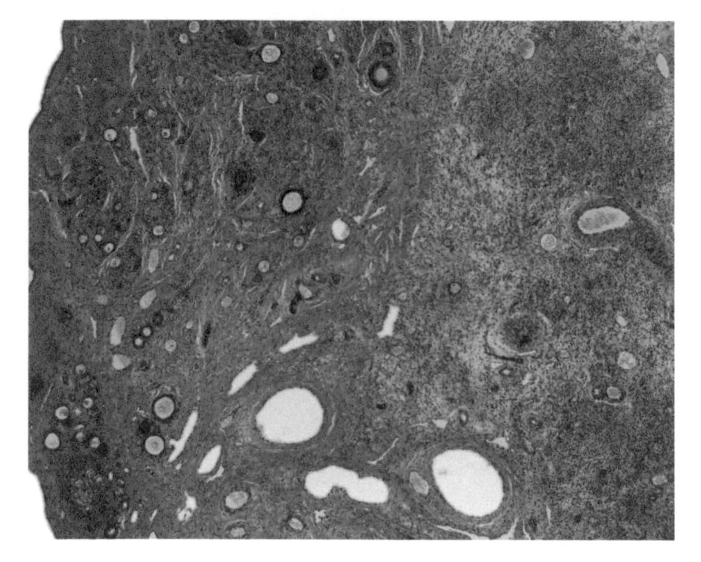

FIGURE O25a There are no mature follicles, corpora lutea, or corpora albicantia visible in this quiescent ovary of the dog during anestrus. Large numbers of immature follicles are present at the periphery of the ovary, 5×.

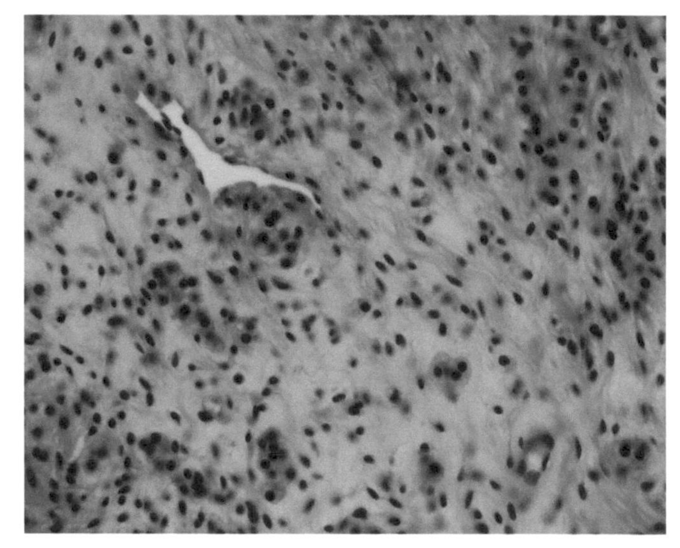

FIGURE O25b There are no mature follicles, corpora lutea, or corpora albicantia visible in this quiescent ovary of the dog during anestrus. 40x.

In all other vertebrates, eggs are shed into the coelom and swept up into a ciliated funnel, the OSTIUM, that carries them to the OVIDUCT. In general, oviducts have a TUNICA MUCOSA consisting of a ciliated columnar epithelium and simple tubular glands that vary with the season, a TELA SUBMUCOSA of loose fibroelastic connective tissue, a two-layered TUNICA MUSCULARIS of smooth muscle, and a TUNICA SEROSA. The tunica mucosa is often thrown up into more or less elaborate folds with a core formed by the tela submucosa. The inner circular layer of the tunica muscularis is often separated from the outer longitudinal layer by a sheet of vascular fibroelastic connective tissue.

Oviducts show modifications depending upon the type of egg and its future development. There may be secretory areas where layers of jelly, albumen, or shell are added; storage areas where eggs are retained until spawning; or areas where embryonic development occurs.

Oviducts of elasmobranchs (Figs. O26a–O26g) fuse at the anterior end so that there is only one ostium. A SHELL GLAND develops in the lower parts of the oviduct. Secretory cells of this compound tubular gland contain an

FIGURES O26(a–g) There is no oviduct in cyclostomes. The ova are shed into the coelom and leave by way of a GONOPORE that develops in the body wall just prior to spawning. In teleosts the anterior end of the oviduct is a fold of peritoneum that surrounds the ovary and then narrows to form a tubular portion. Eggs are discharged into this in the usual way. In all other vertebrates, eggs are shed into the coelom and swept up into a ciliated funnel, the OSTIUM, that carries them to the OVIDUCT.

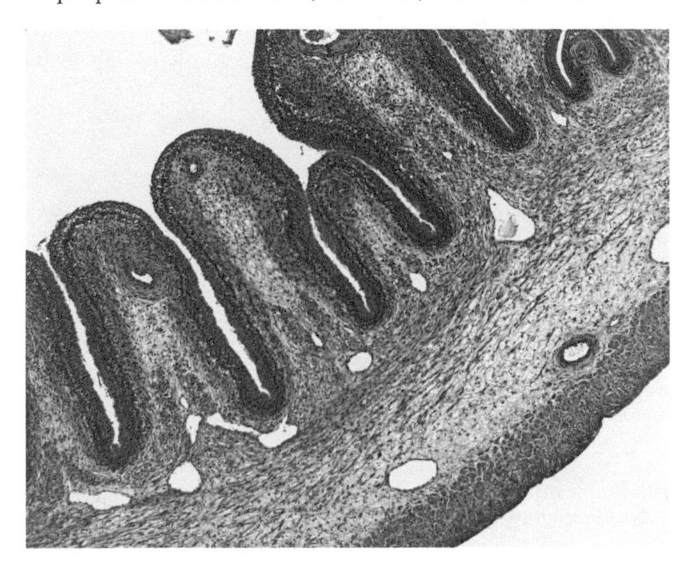

FIGURE O26a In this section of the oviduct of a dogfish, there are no cilia on the mucosal epithelium. The tunica mucosa is thrown up into stout folds with a lining formed by the tela submucosa. Small arteries run through the tela submucisa near the tips of the folds. They supply capillary beds beneath the epithelium. Many thin-walled veins and lymphatic vessels course through the tela submucosa. The layers of the tunica muscularis run longitudinally; they are enclosed by the thin tunica serosa, 5 ×.

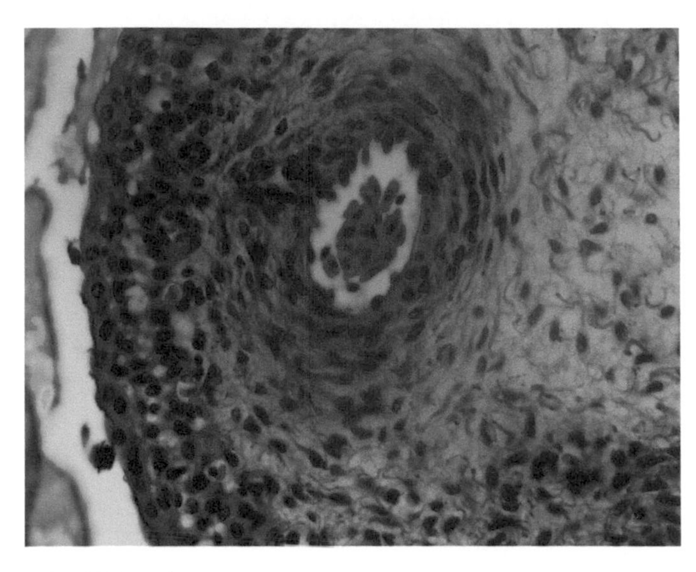

FIGURE O26b A small artery in the connective tissue of the tela submucosa of the fold lining the oviduct supplies the capillaries that run along the outer limit of the tela submucosa, just beneath the epithelium and are visible in this section, 40 ×.

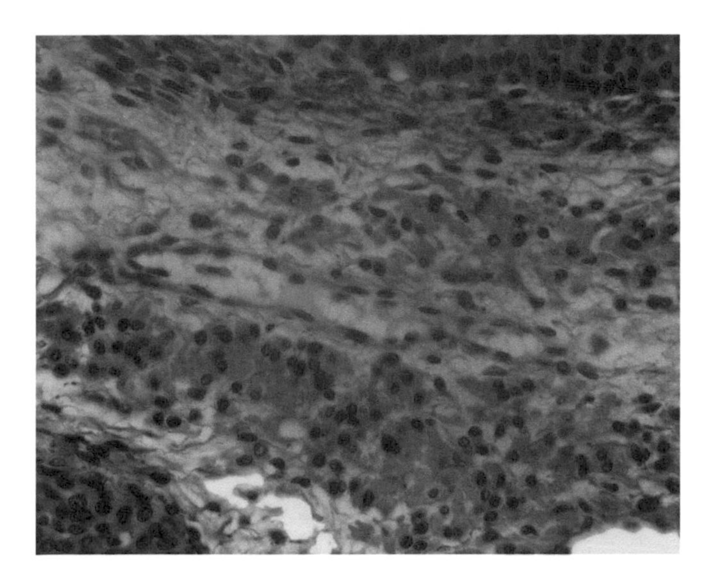

FIGURE O26c These vessels are drained by veins in the deeper tela submucosa.

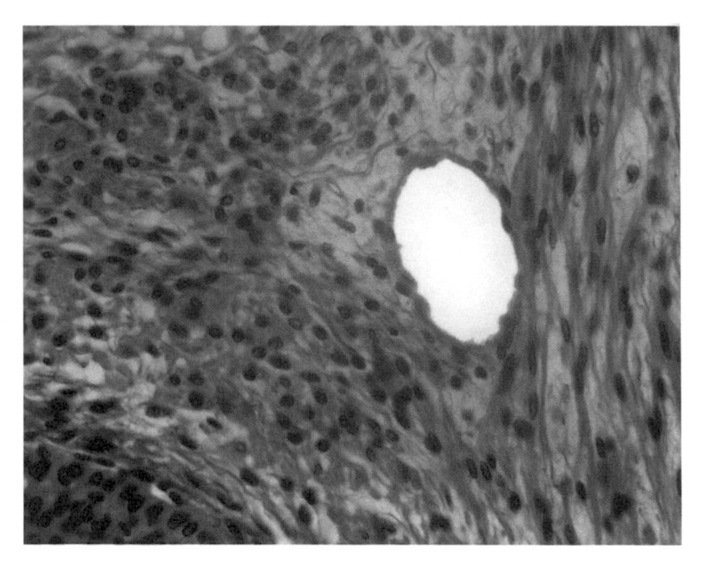

FIGURE O26d Excess tissue fluid is drained from the connective tissue by abundant lymphatics.

acidophilic granular substance. Skates produce an elaborate egg case and these glands are large. They are simpler in dogfish where they produce a temporary shell for the embryo developing in the oviduct. The lower part of the oviduct of dogfish is expanded as an OVISAC where the young, minus the shell, are nourished and develop. The tunica mucosa of the ovisac is lined with a stratified cuboidal epithelium thrown up into vascular ridges containing a core of tela submucosa.

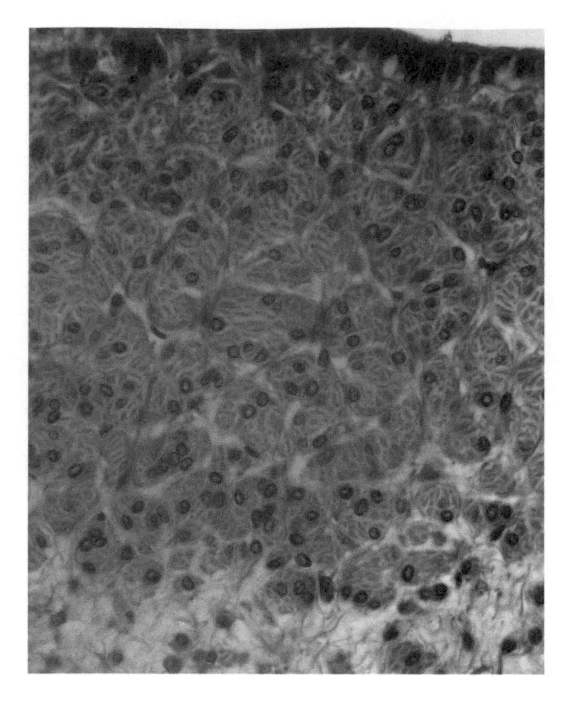

FIGURE O26e Bands of smooth muscle run longitudinally in this section; there appears to be no circular muscle.

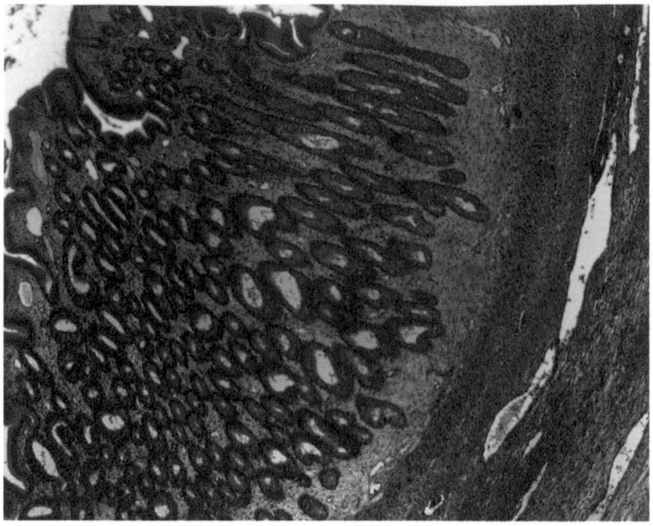

FIGURE O26f Some regions of the oviduct are richly endowed with simple tubular glands. These could be producing gelatinous secretions around the egg or the tough proteins of the egg capsule, 5 ×.

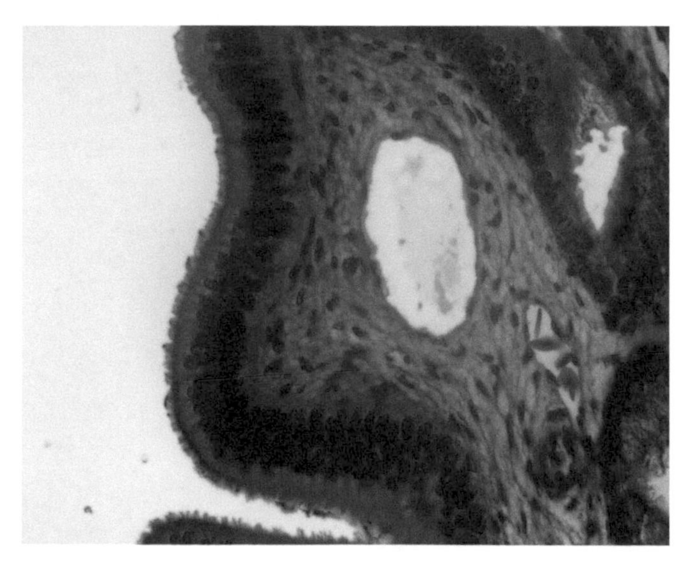

FIGURE O26g The epithelium of this region of the oviduct is richly supplied with cilia. The tela submucosa contains an artery at the lower right, a vein nearby, and a lymph vessel near the center, 40 ×.

The long, contorted oviducts of amphibians are specialized for the application of a jelly coat to the eggs and for storing the eggs until spawning (Figs. O27a–O27c). Between seasons the ducts are thin-walled but in the breeding season they show a great increase in size and convolutions. Upper parts of the oviduct have a ciliated lining with a few goblet cells. Cilia spin the eggs down the oviducts while jelly coats are added by the tubular glands in the long middle region. Eggs are stored in the dilated posterior part of the oviduct; this region is not glandular but is lined with columnar cells.

Oviducts of reptiles and birds are similar to this basic pattern. In birds only the left ovary and oviduct usually develop and the right structures are rudimentary. Albumen and shell layers are added by the oviducts in both groups. The anterior portion of the duct is lined with ciliated columnar cells and goblet cells. The middle region is

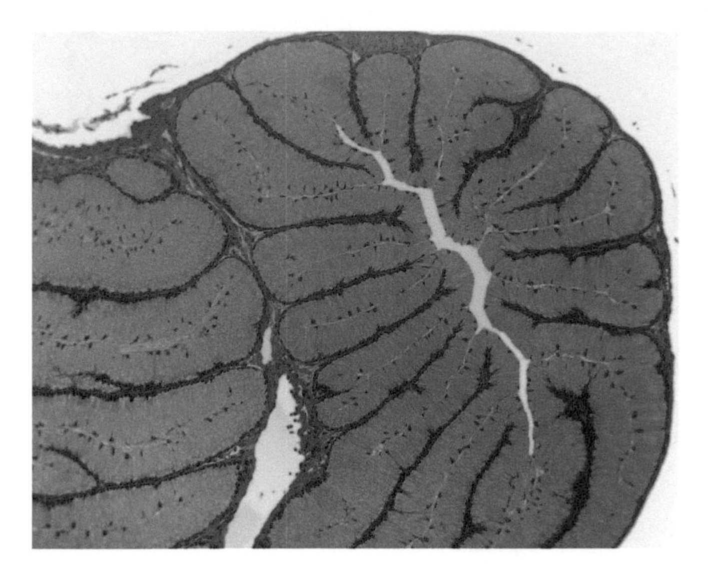

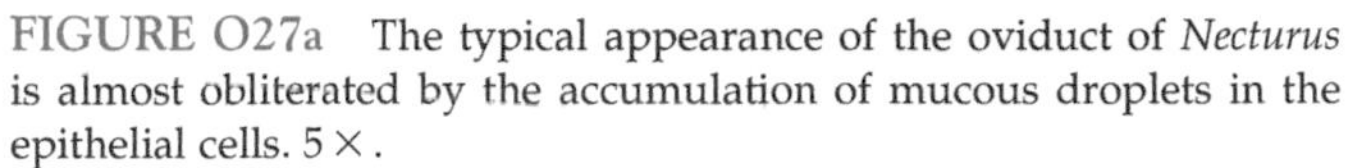

FIGURE O27a The typical appearance of the oviduct of *Necturus* is almost obliterated by the accumulation of mucous droplets in the epithelial cells. 5 ×.

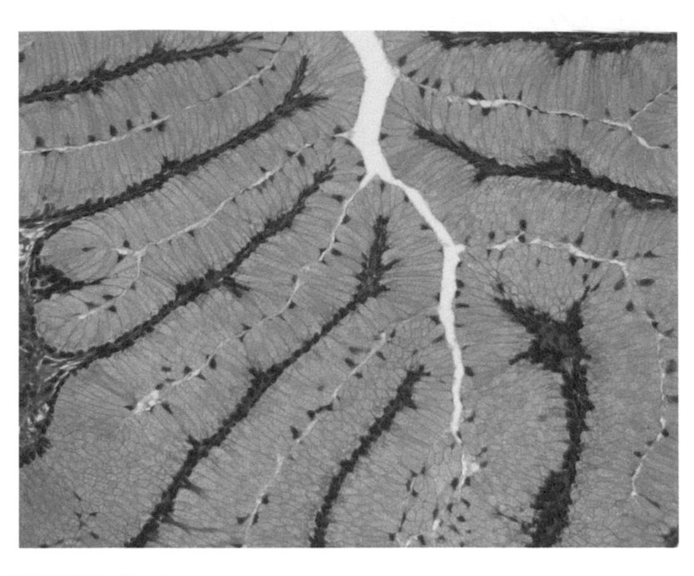

FIGURE O27b The oviduct of *Necturus*. The bloated epithelial cells form simple tubular glands. 10 ×.

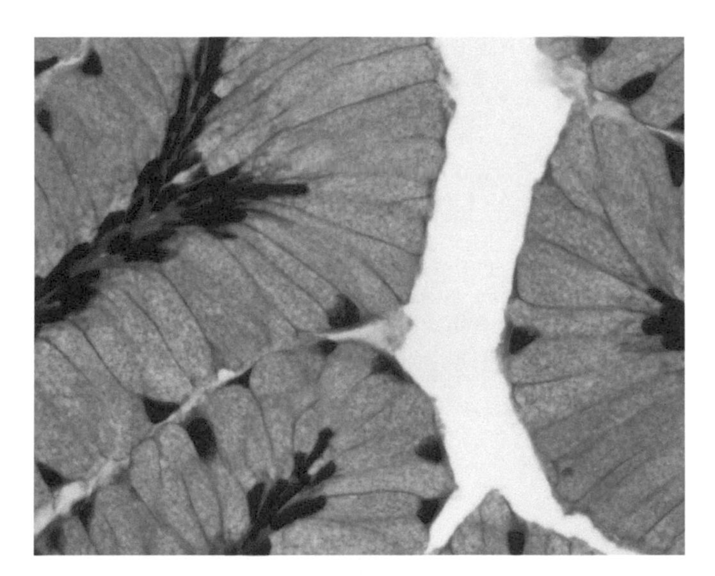

FIGURE O27c The oviduct of *Necturus*. The epithelial cells in these simple tubular glands are packed with mucous droplets that will form the jelly of the egg mass, 40 ×.

a ciliated columnar epithelium with numerous tubular glands containing cells filled with acidophilic granules. These glands secrete albumen (except in snakes and lizards). The posterior region possesses similar tubular glands that secrete the shell. The end of the oviduct is free of tubular glands but contains goblet cells in its epithelium.

The oviduct of mammals is a short, narrow, convoluted duct of undistinguished appearance (Figs. O28a–O28d, O29a, and O29b). There is considered to be no tela submucosa. The TUNICA MUCOSA is thrown into folds that are especially elaborate near the ostium. The folds decrease posteriorly and the wall becomes thicker. The tunica mucosa is composed of a simple columnar epithelium resting on a LAMINA PROPRIA MUCOSAE of loose connective tissue. Both CILIATED CELLS and NONCILIATED CELLS occur in the epithelium; the nonciliated cells are secretory and may produce nutritive substances for the oocyte. Beneath the tunica mucosa is the TUNICA MUSCULARIS of smooth muscle; it consists of a layer of circular fibers enclosed by a layer of longitudinal fibers. It is

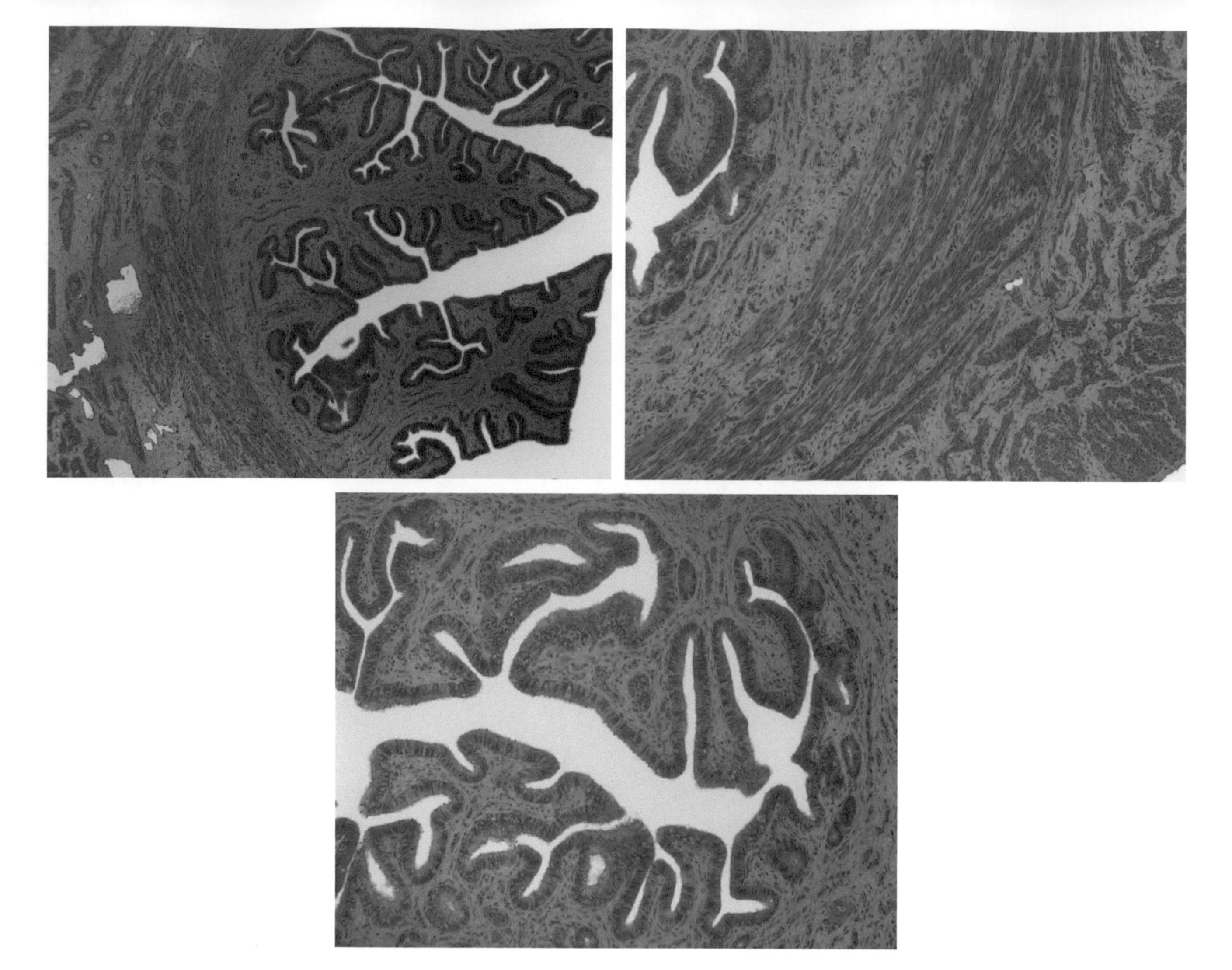

FIGURES O28a–O28c This series is from the oviduct of a rabbit. The tunica mucosa is thrown up into folds that are especially elaborate near the ostium. Although it is said that there is no tela submucosa, this section shows a mingling of connective tissue (which could be considered submucosal) with the circular and longitudinal smooth muscle. A small portion of the tunica serosa appears at the lower right of Fig. O28b, 5×, 10×, 10×.

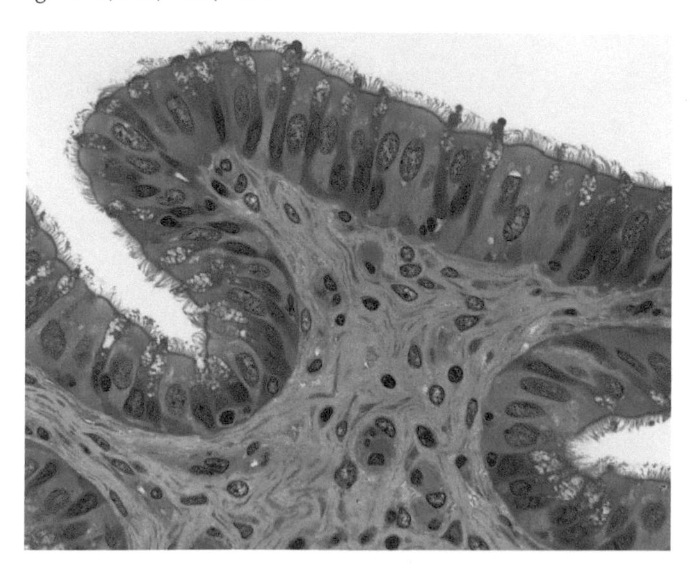

FIGURE O28d The simple columnar epithelium of the oviduct of the rabbit is a mixture of goblet cells and ciliated cells. Note the neat line of basal bodies of the cilia just below the epithelial surface. How could the core of connective tissue of these folds not be considered tela submucosa? It is suggested that the egg is propelled down the oviduct by the combined actions of the cilia and peristalsis of the smooth muscle, 63×.

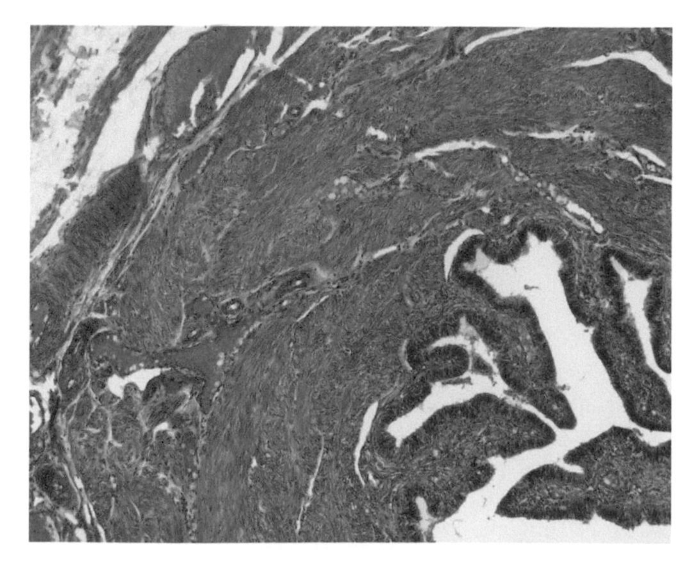

FIGURE O29a The circular layer of smooth muscle surrounding the human oviduct is thick; it contains some of the major blood vessels supplying the organ. An irregular layer of longitudinal muscle is seen at the left, 10×.

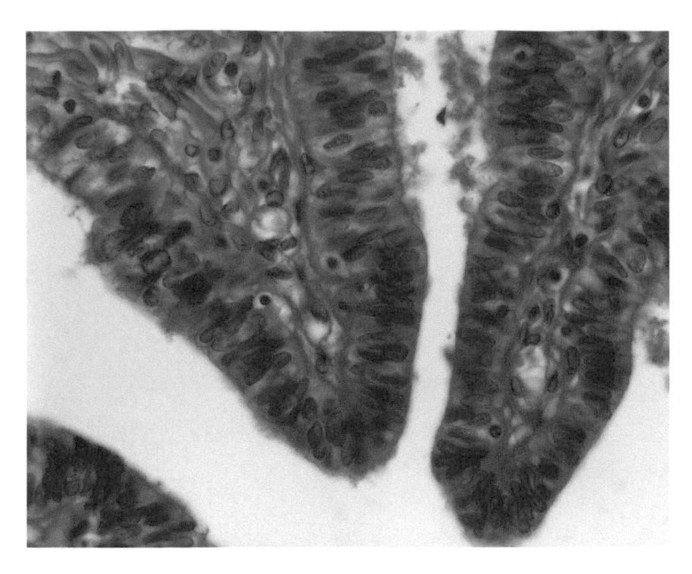

FIGURE O29b One of the folds in the tunica mucosa has a core of vascular loose connective tissue. Some of the columnar cells of the epithelium are ciliated; these are interspersed with secretory nonciliated cells, 63 ×.

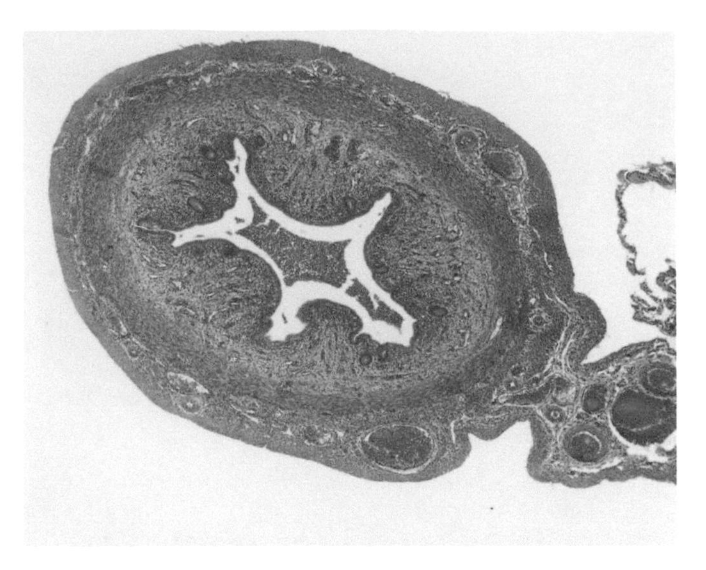

FIGURE O30a The posterior portion of the mammalian oviduct is expanded into a UTERUS. In this section of the resting uterus of a cat, note that the same three layers are present in the oviduct but that they are greatly thickened. The tunica mucosa is called the ENDOMETRIUM and is a nutritive layer. The tunica muscularis is the MYOMETRIUM. Large blood vessels can be seen within its peritoneal sling, the BROAD LIGAMENT, at the lower right. This peritoneum encloses the uterus in a tunica serosa or PERIMETRIUM, 5 ×.

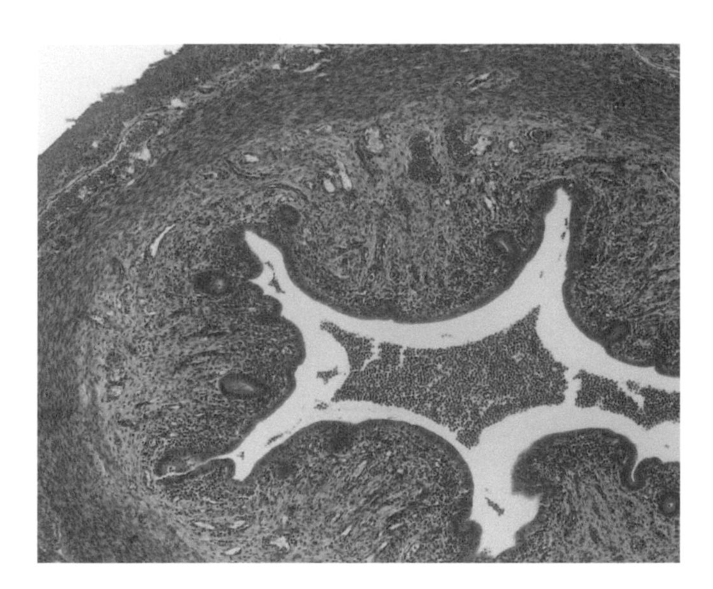

FIGURE O30b The ENDOMETRIUM in the resting uterus of a cat consists of a partially ciliated columnar epithelium with long tubular glands. The epithelium and glands are supported by a LAMINA PROPRIA MUCOSAE of loose fibrous, richly vascular, connective tissue. 10 ×.

covered by a TUNICA SEROSA. Cilia diminish posteriorly and muscle layers increase; movement of the ovum down the duct is mainly by ciliary action at first, mainly by muscular action later.

Uterus

The posterior portion of the mammalian oviduct is expanded into a UTERUS (Figs. O30a–O30c). It is not known whether this is homologous with the ovisac of the lower forms. In a section of uterus, note that the same three layers are present as occurred in the oviduct but they are greatly thickened. The tunica mucosa is called the ENDOMETRIUM and consists of a partially ciliated columnar epithelium with long tubular glands. The epithelium and glands are supported by a LAMINA PROPRIA MUCOSAE of loose fibrous, richly vascular, connective tissue. The tunica muscularis is the MYOMETRIUM and consists of bundles of smooth muscle fibers separated by connective tissue. The tunica serosa or PERIMETRIUM envelops the uterus. In the connective tissue framework of the myometrium, there are many types of connective tissue cells including mesenchymal cells that are capable of forming smooth muscle cells when pregnancy occurs.

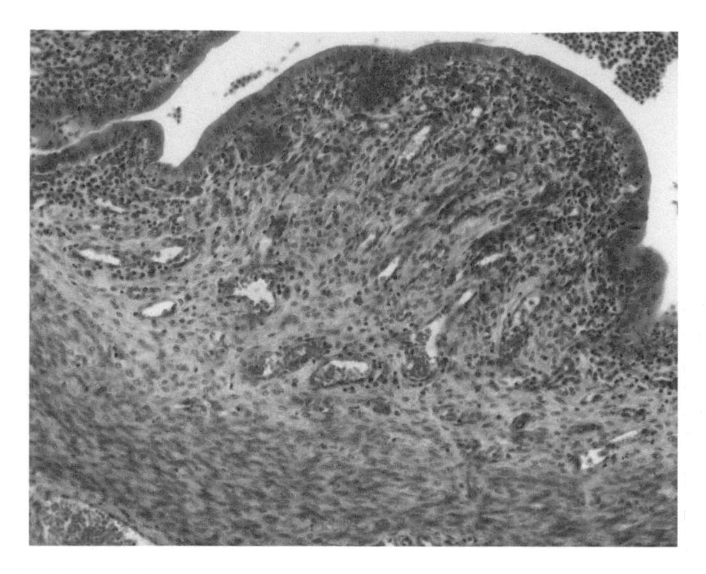

FIGURE O30c The ENDOMETRIUM in the resting uterus of a cat consists of a partially ciliated columnar epithelium with long tubular glands. At this magnification the richly vascular nature of the lamina propria mucosae is obvious. In the myometrium bundles of smooth muscle fibers are separated by connective tissue. Within the connective tissue framework of the myometrium are many types of connective tissue cells, including mesenchymal cells that are capable of forming smooth muscle cells when pregnancy occurs, 20 ×.

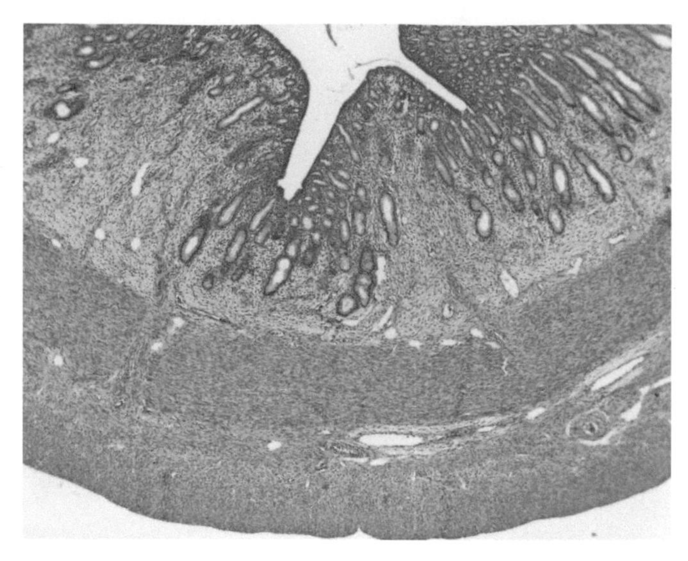

FIGURE O31a In estrus and in pregnancy, the endometrium thickens and becomes more glandular. The blood supply in the lamina propria mucosae is increased and the myometrium thickens, not only by an increase in the size of the smooth muscle cells but also by an increase in their numbers. This section is from the uterus of a cat during estrus, 5 ×.

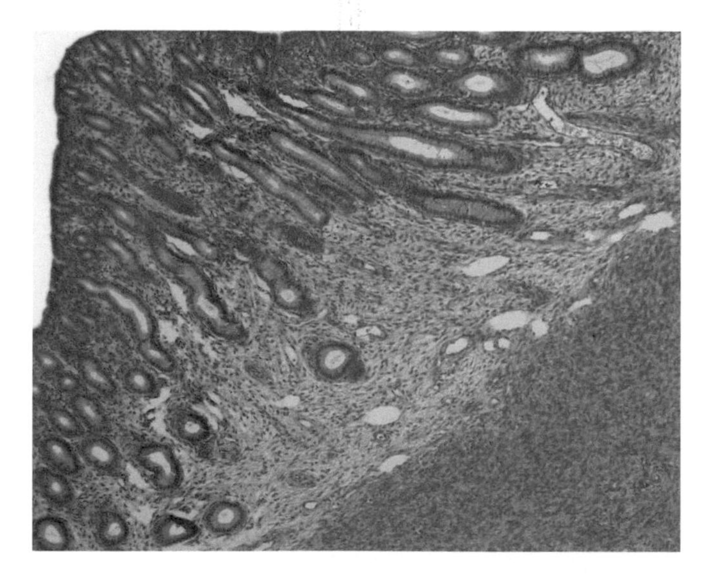

FIGURE O31b This section from the uterus of a cat during estrus clearly demonstrates the endometrial thickening and the abundance of glands. The myometerium has thickened as well - as we see in the lower right, 10 ×.

Female mammals experience a recurrent ESTRUS CYCLE marked by cellular changes in the uterus and vagina and by differences in behavior (Figs. O31a, O31b, O32a–O32c, O33a, and O33b). Successive stages of the cycle are ANESTRUS (quiescent period), PROESTRUS (preparation for mating), ESTRUS, or "heat" (acceptance of the male), and METESTRUS (regressive changes). Ovulation usually occurs late in estrus or soon after. In many mammals, periods of estrus are separated by long periods of anestrus when the reproductive system is quiescent and the female will not mate. In other mammals, estrus periods follow each other in regular cycles, either throughout the year

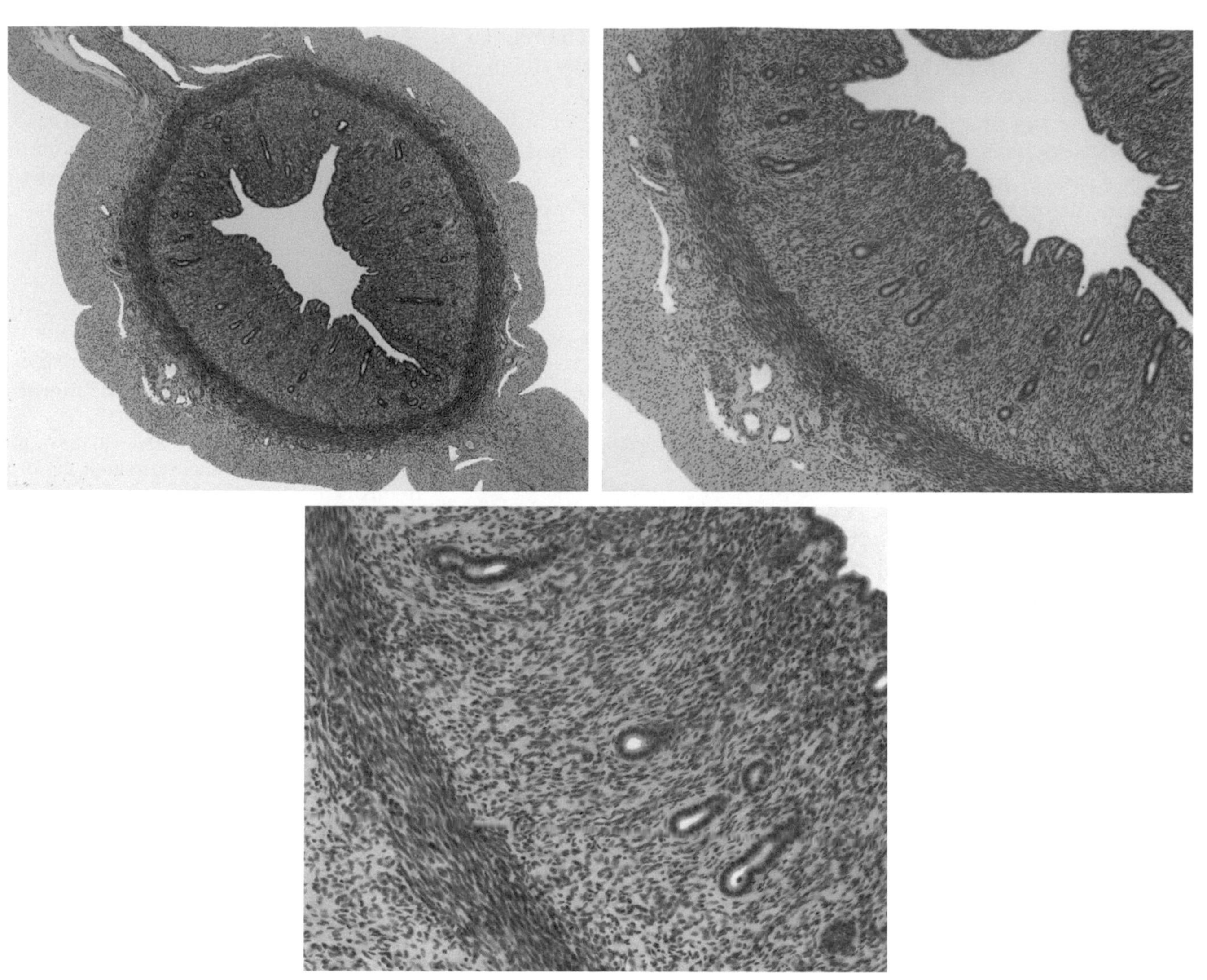

FIGURES O32a–O32c A comparison of the uterus of a dog during anestrus and estrus. They show the same development in the endometrium, vascularization, and myometrium as was seen in the uterus of the cat. Figs. O32a–O32c: Anestrus, 5×, 10×, 20×.

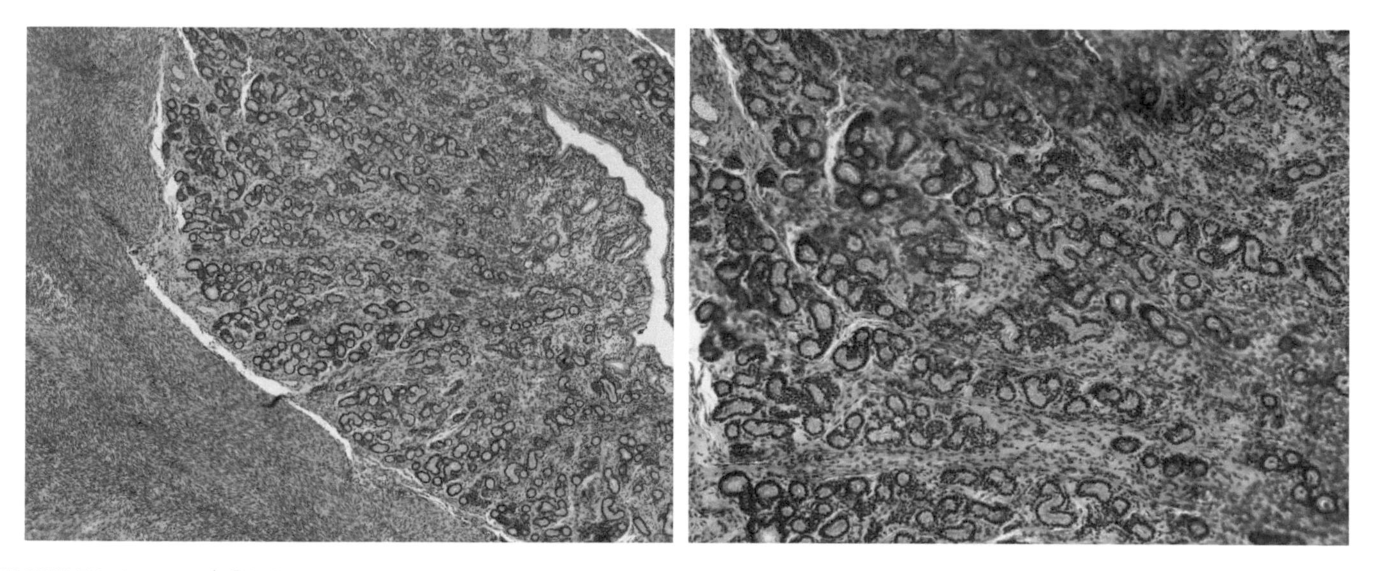

FIGURES O33a and O33b A comparison of the uterus of a dog during anestrus and estrus. They show the same development in the endometrium, vascularization, and myometrium as was seen in the uterus of the cat. Figs. O33 and O33b: Estrus, 5×, 10×.

(human), or during the breeding season (many rodents). The regular MENSTRUAL RHYTHM of a woman is a series of estrus periods in which the thickened endometrium breaks down, producing menstrual flow. This is accompanied by breakdown of blood vessels and loss of blood.

In estrus and in pregnancy the endometrium thickens, not only by an increase in cell size but by an increase in cell numbers as well. Compare sections of uterus during anestrus with uterus during estrus. Also examine a series of slides of human uterine wall from preadolescence to old age, relating differences in the endometrium and myometrium with uterine activity during the phases represented.

Vagina

The mammalian vagina has numerous transverse folds, or RUGAE and is lined with a stratified squamous epithelium that undergoes cyclic changes during the estrus cycle (Figs. O34a–O34c). Beneath is a thick layer of connective tissue, the LAMINA PROPRIA MUCOSAE. It contains numerous leukocytes that are commonly seen migrating

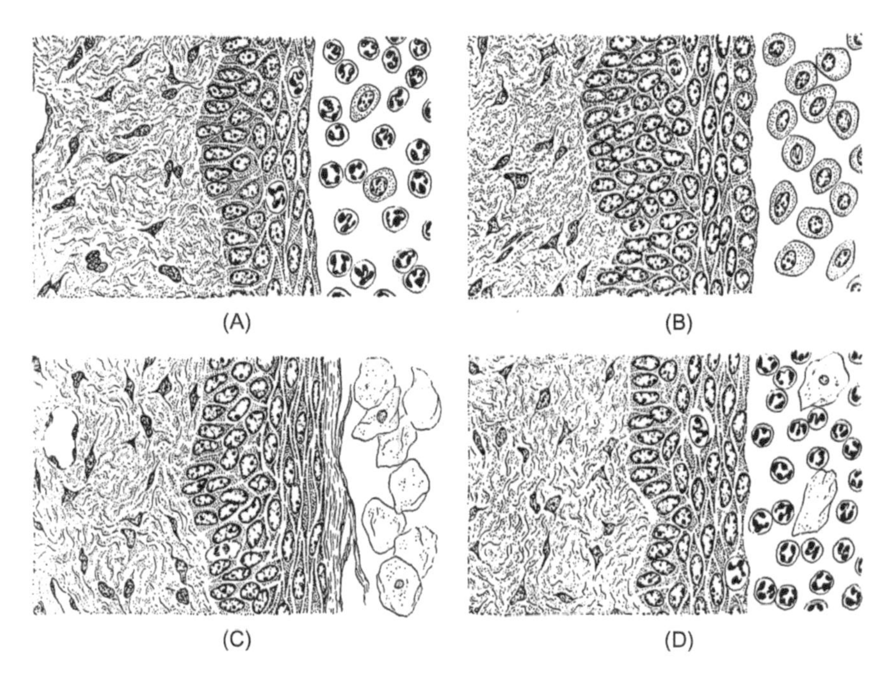

FIGURE O34a Sections through the wall of the vagina of rats during different stages of the estrus cycle with the cells (at the right) which characterize vaginal smears at each of the corresponding stages.

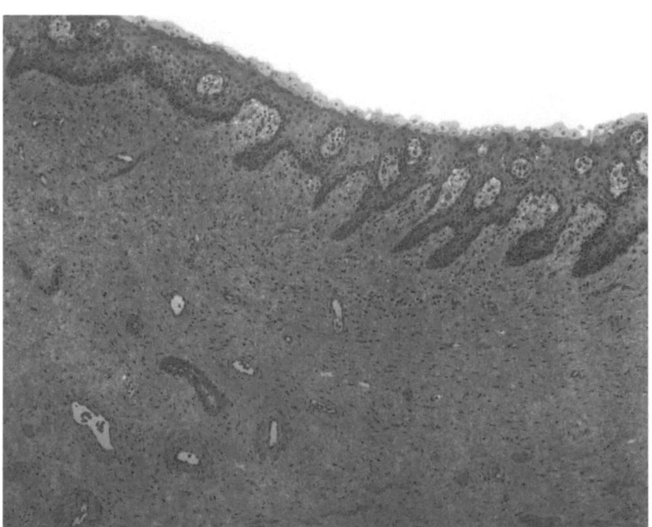

FIGURE O34b The mammalian vagina is lined with a stratified squamous epithelium that undergoes cyclic changes during the estrus cycle. It is these cells that are shed and seen in vaginal smears, 10×.

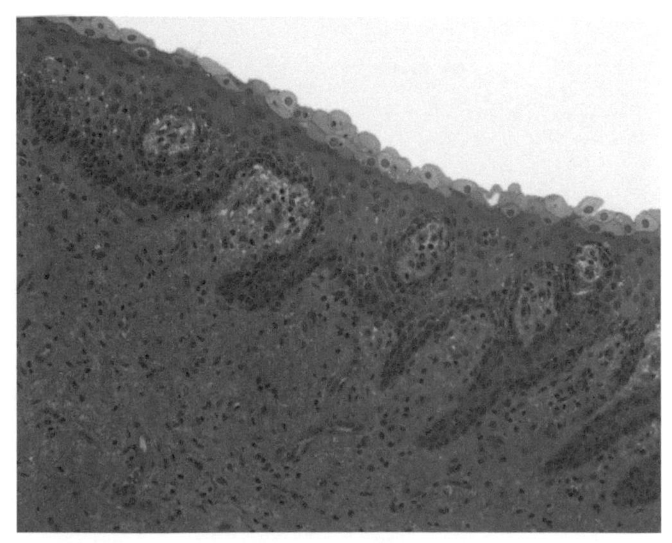

FIGURE O34c The stratified squamous epithelium, obvious here, are shed and seen in vaginal smears, 20×.

into the epithelium. Smooth muscle of the TUNICA MUSCULARIS is organized into two poorly defined layers: an outer longitudinal and an inner circular layer. The TUNICA ADVENTITIA is a layer of dense connective tissue continuous with that of adjacent structures.

Sexual Cycles in Female Mammals

Vertebrates are either CONTINUOUS BREEDERS, breeding continuously throughout the year, or SEASONAL BREEDERS that restrict reproduction to a particular season. With the exception of primates, most mammals are sexually receptive in recurring periods called ESTRUS or "heat." MENSTRUAL CYCLES are characteristic of primates and do not occur in other vertebrate groups. The length of the cycle is variable but 28 days is typical for women. Ovulation occurs in both estrus and menstrual cycles at a time in the cycle when the endometrium has begun to thicken and has become more extensively vascularized, preparing the uterus for possible implantation of an embryo. If pregnancy does not occur, the endometrium is reabsorbed by the uterus and no bleeding occurs. In the menstrual cycle, however, bleeding occurs as the endometrium is shed from the uterus through the cervix and vagina.

During the estrus cycle, the tunica mucosa of the vagina proliferates as it prepares for copulation. Vaginal smears of laboratory rodents are useful for following characteristics to diagnose the stages in the estrus cycle. PROESTRUS: the surface epithelium is underlain by newly cornifying layers and the smear shows nucleated, squamous surface cells. ESTRUS: the nucleated surface has been sloughed and detaching cells appear in the smear as nonnucleated, cornified squamous cells. METESTRUS: the superficial squamous cells are shed entirely and leukocytes that invade the epithelium are apparent in the smear along with nonnucleated squamous cells. DIESTRUS: the detaching cells are nucleated and are shed along with leukocytes in a variable amount of mucus. (Diestrus is the period of sexual quiescence between metestrus and proestrus in mammals with little time between successive estrus cycles.)

MALE

Testes of mammals develop early in embryonic life beneath the peritoneum on the dorsal wall of the abdominal cavity. Because the development of sperm cannot proceed at the normal body temperature of warm-blooded animals, means have developed for the testes to descend into the cooler environment of the SCROTUM (Fig. O35). In birds the testes are located beneath the thin dorsal body wall.

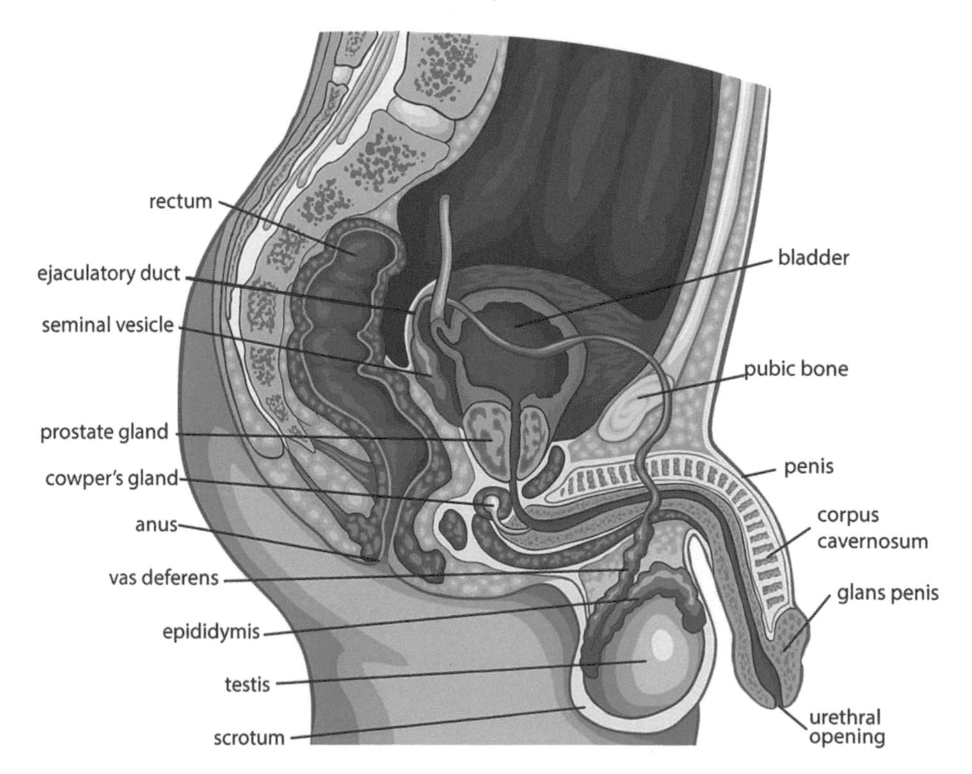

FIGURE O35 Drawing of sagittal section of the human male reproductive system. *Tsaitgaist Own work CC BY-SA3.0 commons.wikimedia.org/wiki/File:Male_anatomy en.svg.*

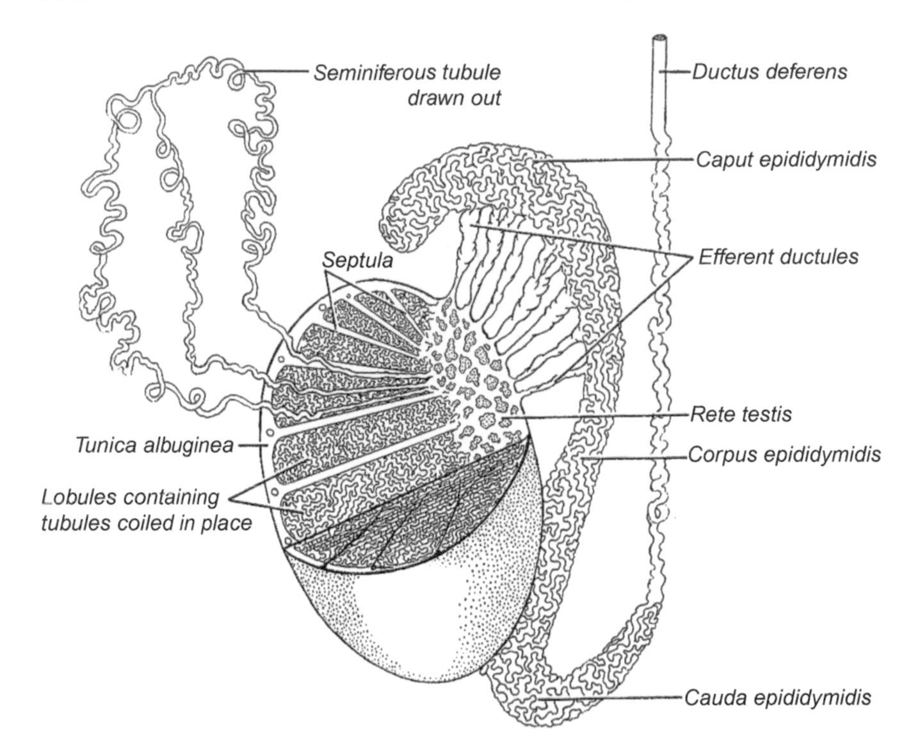

FIGURE O36a A diagram of a mammalian testis teased apart to show the relationships of the tubular system: seminiferous tubules, epididymis, and ductus deferens. One seminiferous tubule is drawn out in order to indicate its extent; the others remain in normal positions within the lobules. Source: *Slightly modified after Corner.*

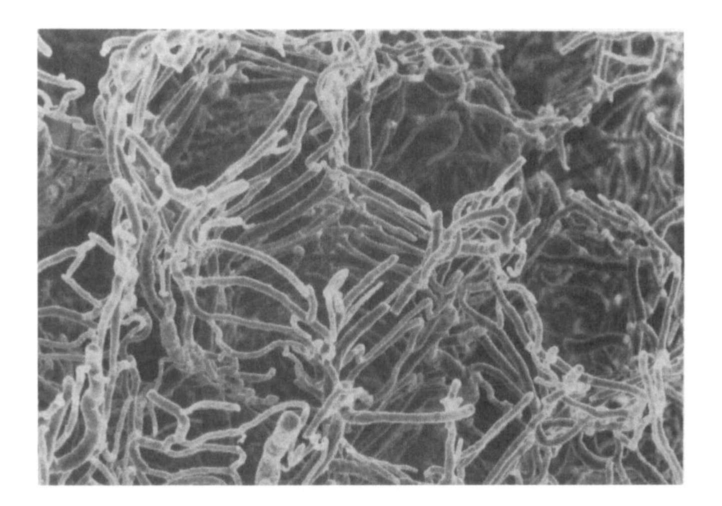

FIGURE O36b Scanning electron micrograph of the microvasculature surrounding several seminiferous tubules that have been removed by enzymatic digestion.

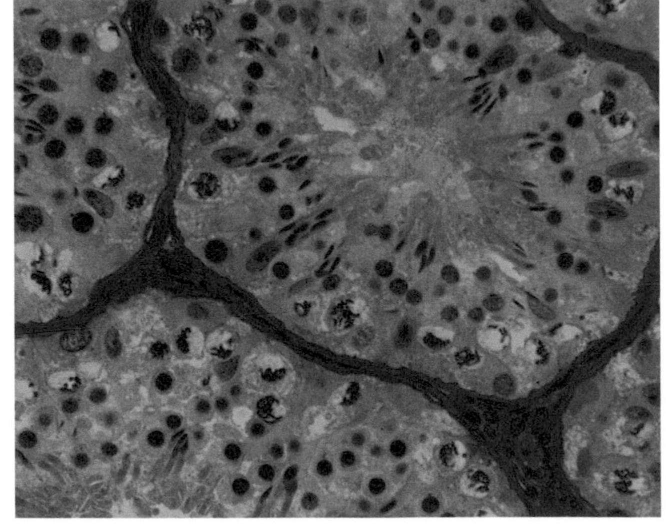

FIGURE O37a Photomicrograph of a semithin section of a mammalian testis. Germinal epithelium lines the tubules. A few INTERSTITIAL CELLS (Leydig cells) occupy the triangular space at the lower right. These cells secrete male sex hormones, 63 ×.

Testes of all vertebrates are similar, consisting of a number of either SEMINIFEROUS TUBULES or AMPULLAE supported by loose vascular connective tissue and contained within a connective tissue CAPSULE (Fig. O36b). The elongated tubules of amniotes are lined with a GERMINAL EPITHELIUM consisting of developing SPERMATOGENIC CELLS (Figs. O36a, O37a, and O37b) and supportive SUSTENTACULAR CELLS (Sertoli cells) (Figs. O38a and O38b). Tubules of an amniotes lack a germinal epithelium and it appears, at least in some forms, that sustentacular cells extend processes through the tubular walls to bring back immature sex cells from the loose connective tissue (Fig. O39).

Seminiferous tubules are derived from sex cords (Fig. O2c) that are homologous with those seen in the development of the ovary. At first they are solid masses of cells but later they develop lumina. In the connective tissue

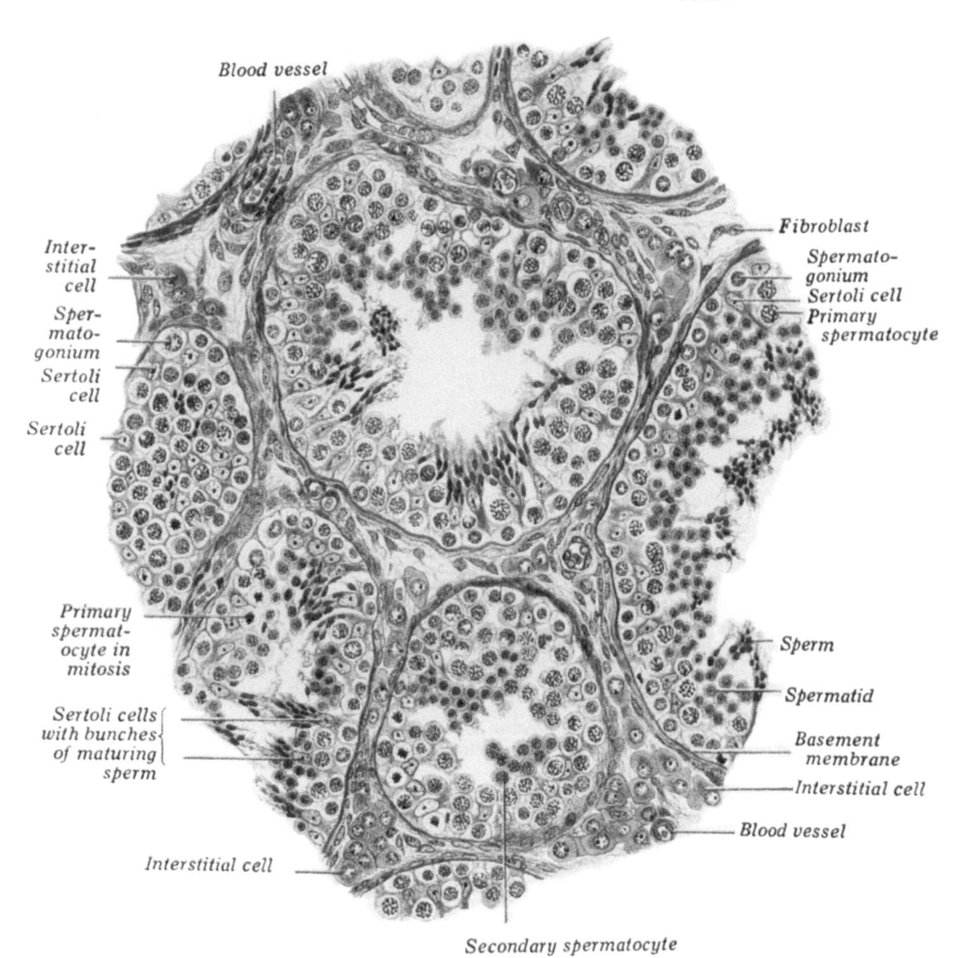

FIGURE O37b Drawing of a cross-section of a small area of a human testis for comparison with Fig. O37a.

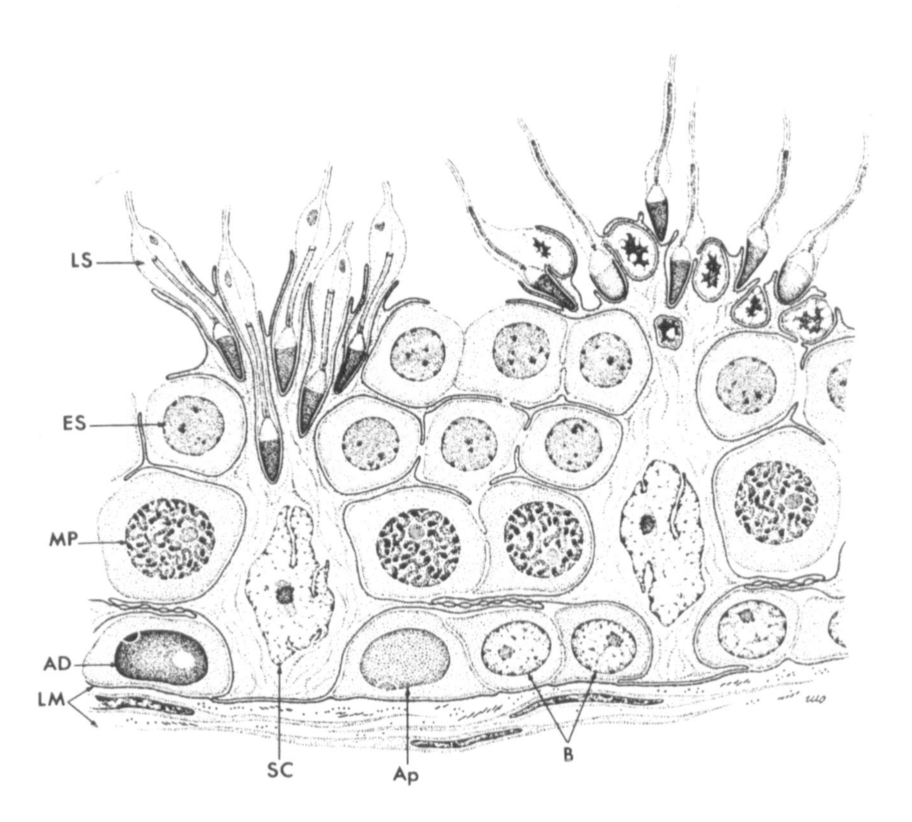

FIGURE O38a A drawing of a seminiferous epithelium showing the relationship of germ cells to sustentacular cells. The germinal cells are engulfed by folds of the SUSTENTACULAR CELLS. *AD*, dark spermatogonium; *Ap*, pale spermatogonium; *B*, spermatogonia; *ES*, early spermatids; *LM*, limiting membrane; *LS*, late spermatids; *SC*, sustentacular cell.

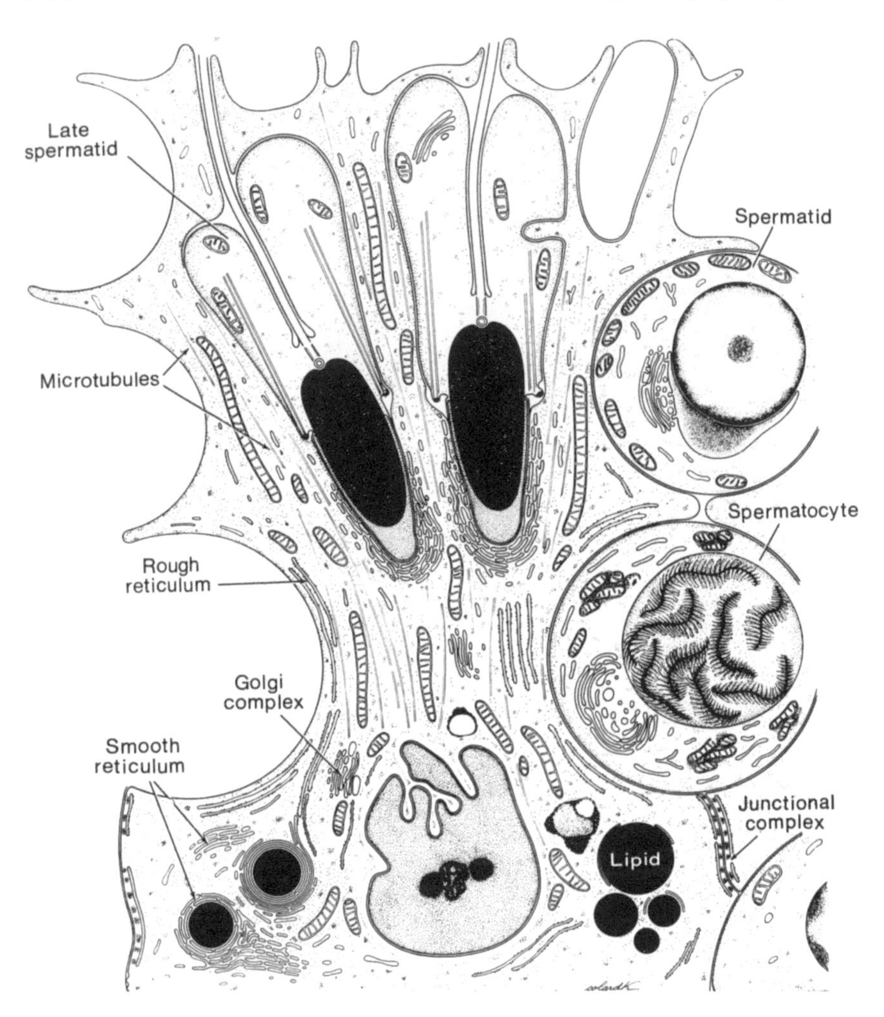

FIGURE O38b Drawing to show the elaborate shape and ultrastructure of the sustentacular cell and its relationship to germ cells that are wrapped in folds of the large sustentacular cells. Spermatids and spermatocytes are engulfed in folds of the sustentacular cell.

between tubules of the testes of mammals and birds, there are endocrine INTERSTITIAL CELLS (Leydig cells) that produce the male sex hormone, TESTOSTERONE (Fig. O40). Although testes of lower forms produce testosterone, interstitial cells appear to be lacking and the site of its formation is still a mystery.

Sperm from the seminiferous tubules is removed from the testes by the DUCTULI EFFERENTES (Fig. O36a). In anamniotes, these join collecting tubules that pass through the opisthonephros to unite with the archinephric duct and therefore serve a combined seminal and urinary function. In amniotes, where the opisthonephros has been replaced by a metanephros, the ductuli efferentes are derived from embryonic kidney tubules and empty their spermatozoa into the EPIDIDYMIS. The epididymis joins the original archinephric duct, which now conducts only spermatozoa and is called the DUCTUS DEFERENS.

In vertebrates with an annual breeding season the size of the testes fluctuates, being largest just prior to the breeding period. After the spermatozoa have been discharged the testes shrink to a small fraction of their former size.

Mammalian Testis

The section of mammalian testis may include a portion of the coiled epididymis (Fig. O36a). It is incompletely surrounded by a peritoneal fold, the TUNICA VAGINALIS, whose visceral layer adheres to the CAPSULE of the testis and parietal layer lines the inner surface of the scrotum. The outer layer of the capsule is the TUNICA ALBUGINEA of dense collagenous tissue and the inner is the TUNICA VASCULOSA of vascular areolar tissue. The parenchyma is divided into several pyramidal LOBULES by SEPTA which extend inward from the capsule to a central mass of connective tissue, the MEDIASTINUM. Each lobule contains one to four convoluted SEMINIFEROUS TUBULES that are cut at various angles in sections. Between them is the vascular STROMA of INTERSTITIAL CONNECTIVE TISSUE that contains the large triangular endocrine INTERSTITIAL CELLS (Leydig cells).

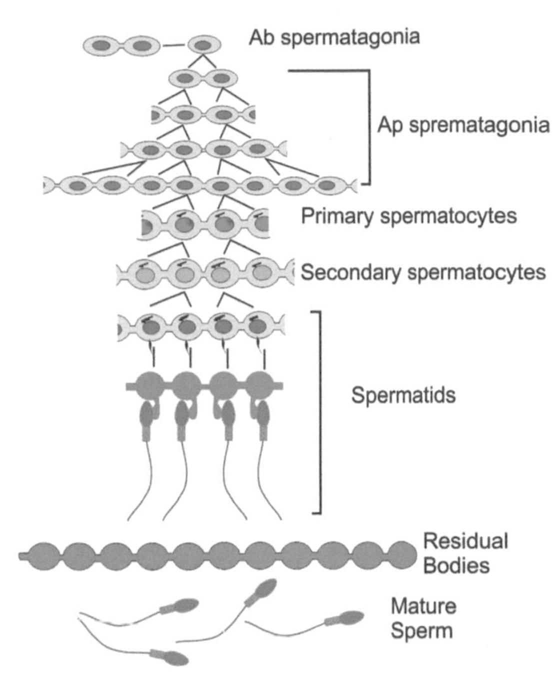

FIGURE O38c Repeated divisions of the progeny of one spermatogonium give rise to a large number of germ cells by meiotic and mitotic division. The cells do not separate completely and remain attached to their fellows by intercellular bridges. After many divisions the maturing germ cells lose most of their cytoplasm, the intercellular bridges are lost and the sperm swim free. Before the sperm is released, each developing germ cell has a nucleus and the cytoplasm of the sustentacular cell is interconnected by cytoplasmic bridges, a SYNCYTIUM is formed. *Authors.*

The germinal epithelium of the seminiferous tubules is composed of two cell types (Figs. O37a and O37b): SPERMATOGENIC CELLS and SUSTENTACULAR CELLS (Sertoli cells). Spermatogenic cells are a proliferating population composed of many stages in the development of the spermatozoa (Fig. O38c). Sustentacular cells are of one type and do not proliferate. This epithelium rests on a thin basement membrane and is enclosed by a BOUNDARY ZONE or LAMINA PROPRIA of fibrous connective tissue containing FIBROCYTES and peritubular contractile MYOFIBROBLASTS or MYOID CELLS. The peripheral layer of spermatogenic cells consists of rounded SPERMATOGONIA. These diploid cells give rise, by mitosis, to more spermatogonia and to the diploid, larger, PRIMARY SPERMATOCYTES that form the next layer of cells toward the lumen. Primary spermatocytes undergo meiosis and produce the next layer of cells, the haploid SECONDARY SPERMATOCYTES, similar in appearance but smaller than their progenitors. Secondary spermatocytes promptly divide, each forming two daughter SPERMATIDS. While this nuclear division of germinal cells is taking place, the daughter cells remain connected by intercellular bridges to be separated only when the spermatozoa develop—thus a CLONE is formed. Since secondary spermatocytes have a short life, they are rarely seen in the germinal epithelium. Each spermatid metamorphoses into a SPERMATOZOON by the process of spermiogenesis; this involves loss of cytoplasm so that the HEAD is almost a naked nucleus, the migration of cytoplasm and mitochondria to form a stalk-like MIDDLE PIECE extending from the head, and the addition of a FLAGELLUM in the TAIL PIECE (Fig. O41a).

The irregular tall SUSTENTACULAR CELLS (Sertoli cells) stand on the basement membrane and extend into the lumen (Figs. O38a and O38b). They envelop all developmental stages of the germinal cells within folds of their ample bodies. (There is, therefore, a centripetal shift of the folds as the germinal cells mature and migrate toward the lumen.) Although large, these cells are difficult to distinguish with the light microscope because of their pale-staining cytoplasm and irregular outlines (Fig. O38d). Their pale ovoid nuclei lie about one-fourth of the distance, or less, between the basement lamina and the lumen. The nucleus may be indented and its long axis is usually radially directed. It may often be recognized by its characteristic large nucleolus whose acidophilic center is flanked by two DNA-containing basophilic CHROMATIN BODIES. When spermatozoa are mature, they leave the sustentacular cells and pass into the lumen.

Although the germinal epithelium may seem disorganized, charts have been prepared that show otherwise (Fig. O38e). They illustrate the SPERMATOGENIC CYCLE, that is, the maturation of a group of germ cells in the seminiferous epithelium that is associated with a single sustentacular cell.

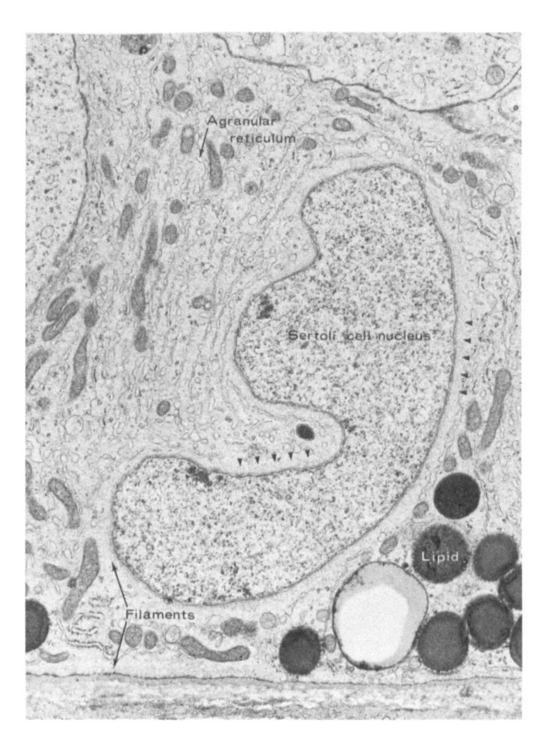

FIGURE O38d Electron micrograph of a section of a sustentacular cell from the testis of a monkey. Note the filamentous zone surrounding the nucleus (*arrowheads*) and the abundant agranular endoplasmic reticulum, 8190 ×.

From the twisted and coiled seminiferous tubules the spermatozoa pass through straight tubules into the RETE TESTIS, a network of fine tubules in the mediastinum (Fig. O36a). Both types of tubules have a simple cuboidal epithelium. Spermatozoa are carried from the rete testis to the epididymis in the EFFERENT DUCTULES that have alternating groups of columnar and cuboidal cells in their wall, giving the lumen an irregular outline. Ciliated and nonciliated cells are present; nonciliated cells have microvilli and display structural manifestations of endocytotic activity. The ductuli efferentes are invested by a thin layer of smooth muscle.

Fig. O38d is an electron micrograph of a SUSTENTACULAR CELL from the germinal epithelium of a mammalian testis. Its nucleus appears indented. Sustentacular cells may contain the characteristic PRINCIPAL NUCLEOLUS flanked by two KARYOSOMES. Tubular agranular endoplasmic reticulum is abundant in the cytoplasm of these cells.

Unique junctional complexes join adjacent sustentacular cells at the level of the primary spermatocytes (Figs. O39a and O39b). These tight junctions form a BLOOD/TESTIS BARRIER between the compartment containing the developing germ cells and the tissue fluid bathing the basal layer of the cellular wall thereby providing an ideal environment for the development of sperm and preventing an autoimmune response when spermatozoa are produced after puberty. The zonulae occludentes are paralleled by flattened cisternae of endoplasmic reticulum and hexagonally packed bundles of actin filaments are interposed between the plasma membrane and these cisternae.

SPERMATOGONIA and SPERMATOCYTES are undifferentiated cells containing abundant free ribosomes but few membranous structures. Cell division is incomplete during mitosis and meiosis so that INTERCELLULAR BRIDGES remain to connect the daughter cells and may be seen at all stages from spermatogonia to late spermatid thereby forming a multinucleate mass of undivided cytoplasm (Fig. O38c). SPERMIOGENESIS is the metamorphosis of a spermatid into a spermatozoon (Fig. O41a). The spermatid develops an extensive Golgi complex at the apical pole of its central nucleus. Numerous mitochondria and two centrioles are distributed throughout the cytoplasm. Several small granules within the Golgi complex coalesce to form a single large ACROSOMAL GRANULE within the membranous ACROSOMAL VESICLE which is derived from part of the Golgi vesicles. The acrosomal vesicle flattens itself over about half the nucleus and shrinks to form a HEAD CAP (Fig. O41b). The remainder of the Golgi complex migrates to the caudal end of the cell. The centrioles come to lie on the caudal side of the nucleus and a FLAGELLUM grows out from one of them within a helical filamentous CAUDAL SHEATH. The other centriole forms a RING or ANNULUS around the flagellum. Fig. O44 (O44a–O44h) shows mature sperm smears from several species. Note the nucleus

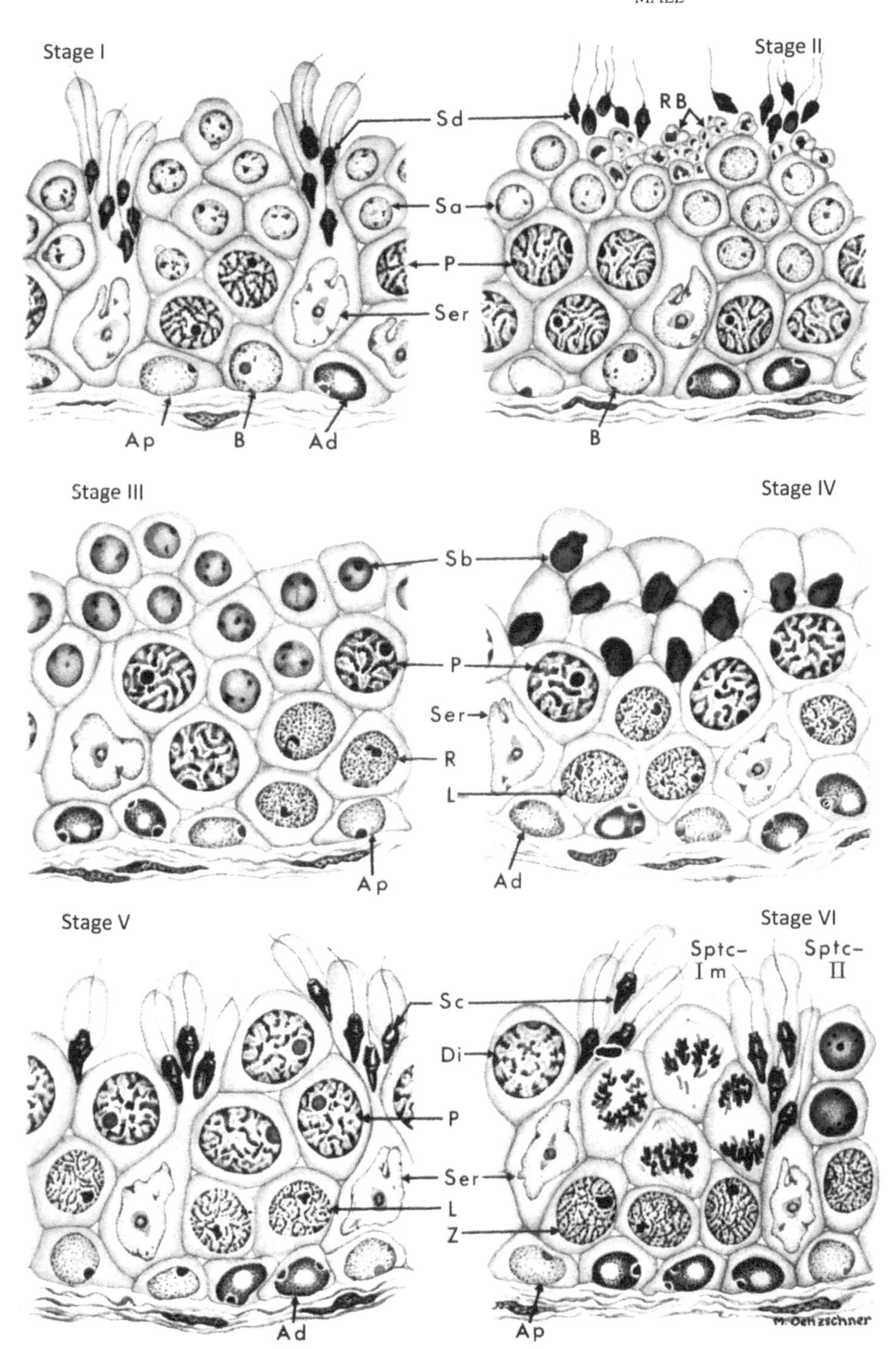

FIGURE O38e Drawings of six stages of spermatogenesis in the cycle of the human seminiferous epithelium. *Ad and Ap*, Type B spermatogonia; *B*, preleptotene spermatocyte; *Im*, secondary spermatocyte; *L*, zygotene spermatocyte; *P*, primary spermatocyte in division; *Pl*, l, eptotene spermatocyte; *Z*, pachytene spermatocyte; *II*, spermatids in various stages of differentiation: *Sa*, *Sb*, *Sc*, and *Sd*; *RB*, residual bodies.

becomes compressed and shaped to form a sperm head that is characteristic of the species. It is also notable that in mammals sperm are all approximately the same size despite great variation in the body size. The bulk of the cytoplasm shifts caudally and the mitochondria form a spiral array around the flagellum. This lies within the caudal tube between the centriole and ring and forms the MITOCHONDRIAL SHEATH of the MIDDLE PIECE of the spermatozoon (Fig. O41c). Most of the cytoplasm is shed as the RESIDUAL BODY, leaving only a thin layer covering the nucleus, middle piece, and tail piece.

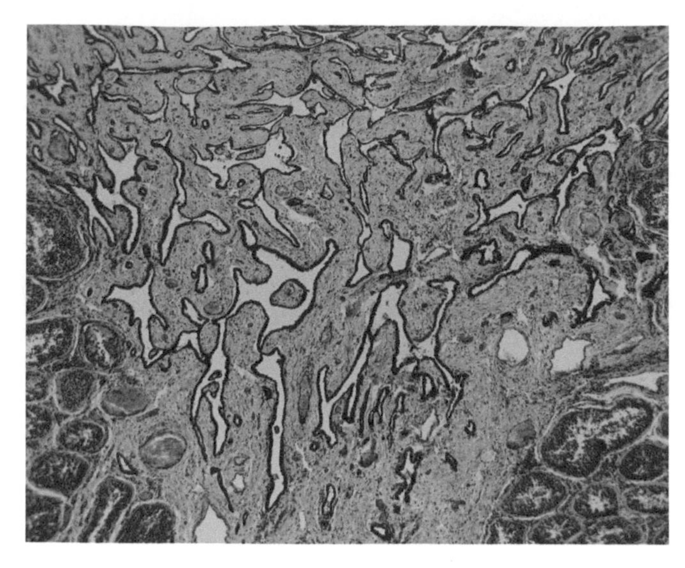

FIGURE O39a Photomicrograph of a section of the rete testis of a man. Sperm that passes to this region from the efferent ductules are not capable of fertilization and they complete some of their maturation here. A few seminiferous tubules can be seen at the left, 5 ×.

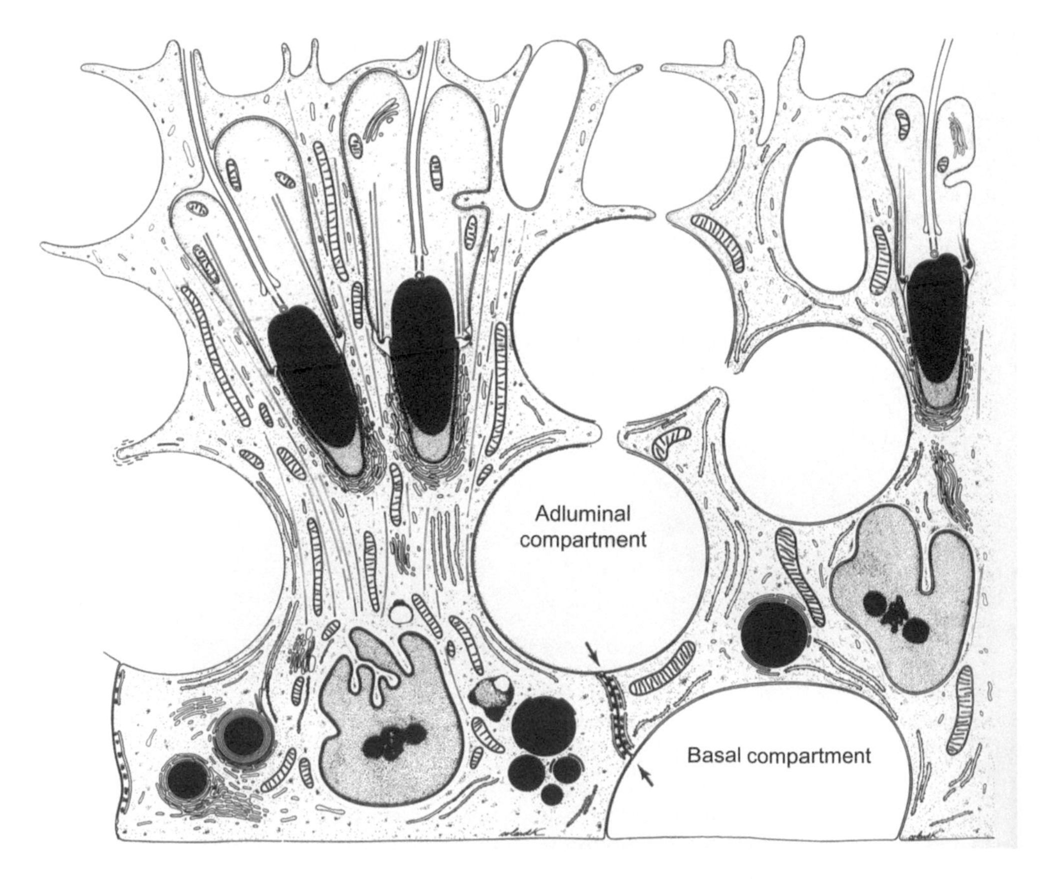

FIGURE O39b Drawing illustrating the blood–testis barrier of occluding junctions (arrows) between adjacent sustentacular cells.

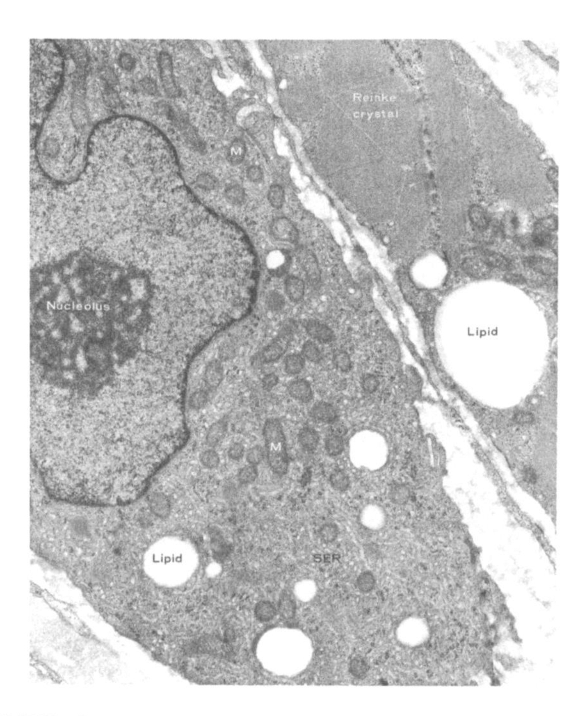

FIGURE O40 Transmission electron micrograph of a section through the human testis. There is abundant agranular endoplasmic reticulum (SER) and numerous mitochondria (M) in the interstitial cell at the left. Although the characteristic of interstitial cells of the testis, the origin and function of the crystals (Reinke Crystals) the cell in the upper right, are unclear, 17,610 ×

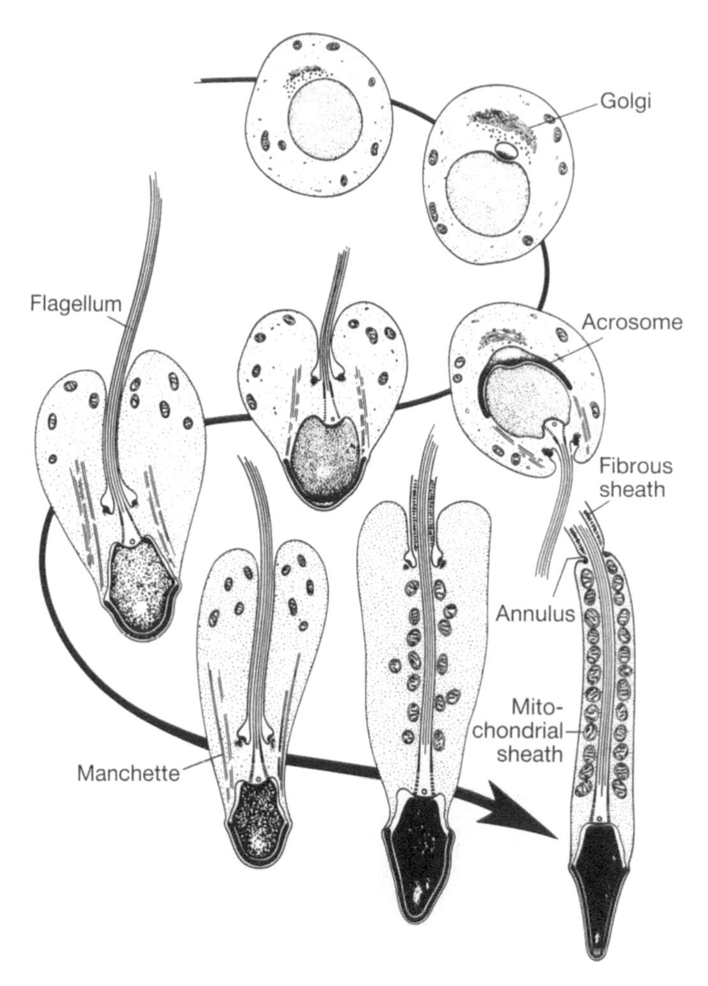

FIGURE O41a Diagram of successive stages of spermiogenesis in the human testis.

The most conspicuous features in electron micrographs of the endocrine INTERSTITIAL CELLS are their densely packed tubular agranular endoplasmic reticulum and the tubular cristae of their mitochondria, both features of cells that secrete steroids (Figs. O42a–O42c).

Testes of Other Vertebrate

Spermatozoa of anamniotes develop within CYSTS. These cysts are readily visible in sections of testis taken during breeding season in frogs (Fig. O43). The seminiferous tubules lack a germinal epithelium and their walls consist only of the BOUNDARY ZONE of a thin layer of fibrous connective tissue and myoid cells. It appears, at least in some anamniotes, that sustentacular cells in the lumen extend part of their massive bulk between the cells of the tubular wall into the interstitium where they engulf several spermatogonia. They withdraw back into the lumen and the spermatogonia divide and metamorphose, as described earlier, and eventually become spermatozoa. All of these stages occur within the engulfing folds of the sustentacular cells. A single sustentacular cell with its complement of developing germinal cells constitutes a cyst. Development of the germinal cells within a cyst is synchronized so that all are at the same stage of development. When the spermatozoa are mature, they are released into the lumen of the tubule and the sustentacular cell is available to engulf another group of spermatogonia. Fig. O44 is a collection of sperm smears from various vertebrates. Spermatozoa survive within the female genital tract for long periods in many vertebrates other than mammals (except bats).

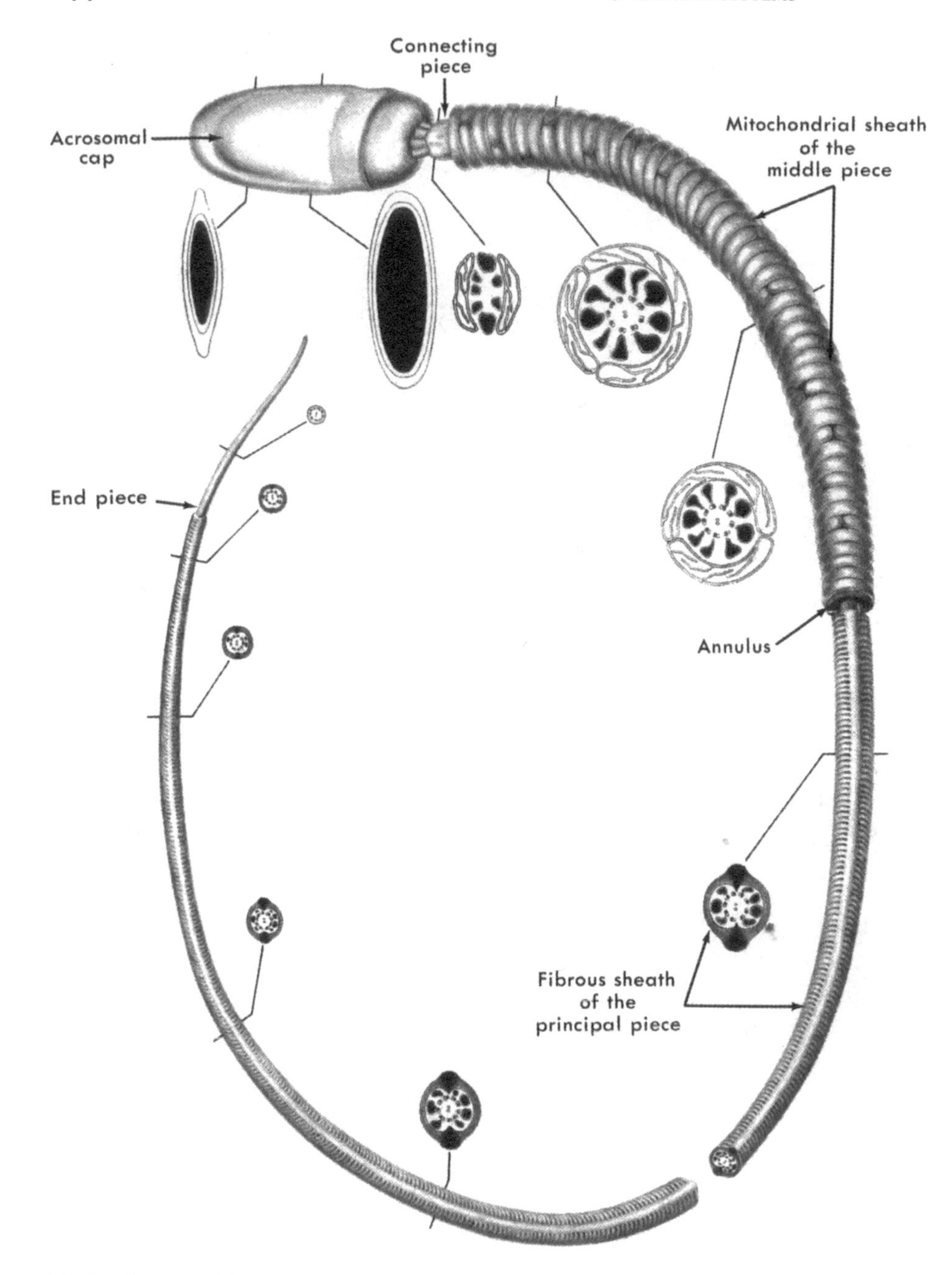

FIGURE O41b Drawing summarizing the ultrastructure of a mammalian spermatozoon. The plasma membranes have been omitted.

Male Genital Ducts

The ducts that carry spermatozoa to the outside of the body in all vertebrates, except the cyclostomes and teleosts, are developed from the archinephric ducts. Like the female the male cyclostome has no genital ducts and the sperm are set free in the body to be released through the GONOPORE that appears just before spawning occurs. Male and female teleosts are similar in that their genital ducts developed from peritoneal folds. In the male the folds enclose the testis and function as its duct. They may open either into the archinephric duct or by a separate pore.

In all other vertebrates the gap is bridged between the seminiferous tubules and the tubules of the mesonephros so that sperm pass through the kidney tubules into the archinephric duct. These anterior kidney tubules may or may not be lost to urinary function. The mesonephric portion of the kidney disappears in amniotes but some of its tubules remain as the DUCTULI EFFERENTES (Fig. O36a) and, with the advent of the separate ureter, the archinephric duct functions exclusively as a genital duct, the DUCTUS DEFERENS. The upper portion of this duct may become extremely coiled, forming the epididymis. (A similar epididymis forms in elasmobranchs.) Along the mammalian tract, there are various glands (seminal vesicle, prostate gland, bulbourethral gland) that secrete SEMINAL FLUID, a vehicle for the spermatozoa.

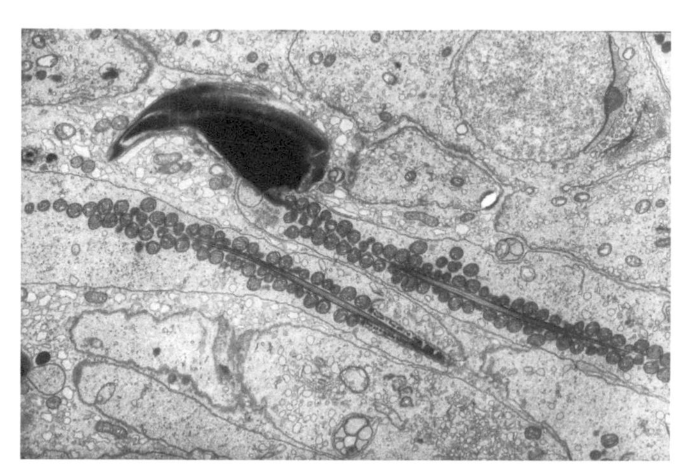

FIGURE O41c Transmission electron micrograph of a developing spermatozoon and its surroundings in the testis of a mouse. The beginning and end of the developmental sequence of spermiogenesis are shown in this micrograph: a spermatid at the lower left and an almost-mature spermatozoon slicing across the micrograph from the upper left to the lower right. The round nucleus of a spermatid is at the lower left corner of the micrograph; its most conspicuous feature is the proacrosomal granule on its left side. The granule participates in the formation of the acrosomal cap of the spermatozoon. A Golgi complex and endoplasmic reticulum are nearby. Directly above the spermatid is a cytoplasmic bridge joining the developing germ cells.
A centriole initiates the formation of a bundle of filaments that will form the core of the middle piece and tail of the spermatozoon. The core comprises the familiar 9 + 2 array of microtubules seen in a cilium. In the middle piece the core of a flagellum that is spirally wrapped by a sheath of mitochondria.
In the final stages of spermiogenesis the spermatids become embedded in the folds of the sustentacular cells, 13,500 ×

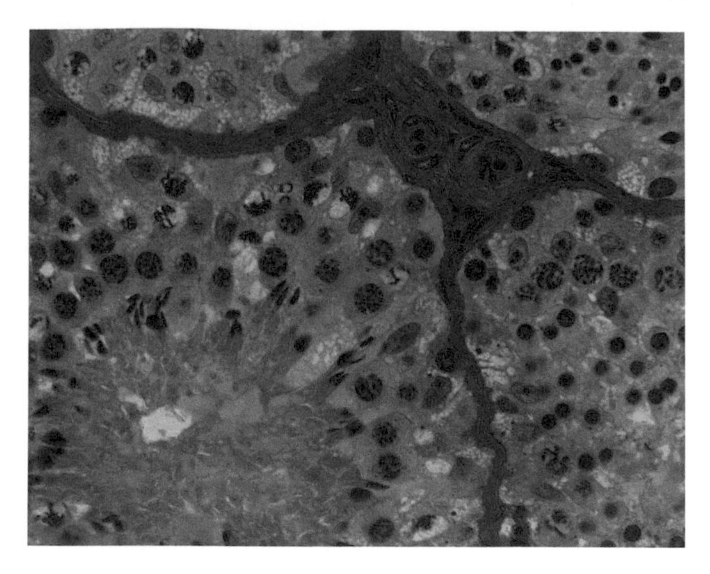

FIGURE O42a Photomicrograph of a semithin section of a mammalian testis. The interstitial space (*top center*) is packed with endocrine interstitial cells, 63 × .

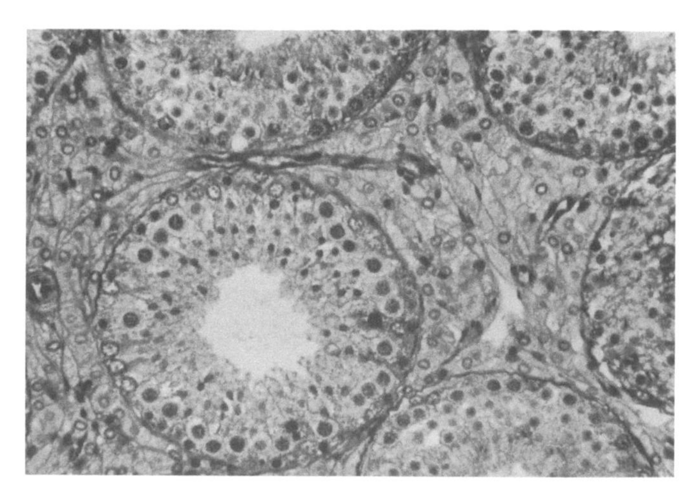

FIGURE O42b Photomicrograph of a section of the testis of an opossum showing large interstitial spaces packed with endocrine interstitial cells.

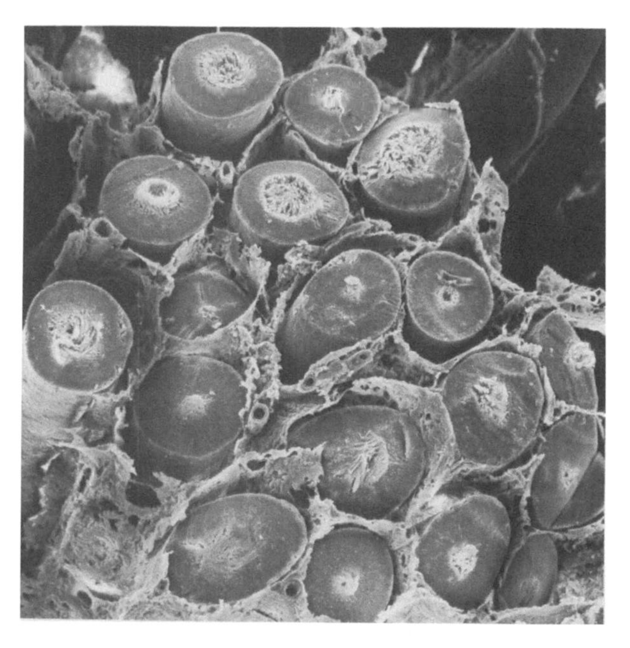

FIGURE O42c Scanning electron micrograph of a portion of the testis of a rat showing several seminiferous tubules in cross-section. Sex hormone-secreting endocrine interstitial cells fill the spaces between the tubules, 33 × .

The EPIDIDYMIS consists of a single twisted, much coiled tubule, the DUCTUS EPIDIDYMIDIS, surrounded by a sheath of dense fibrous connective tissue similar to the tunica albuginea of the testis (Fig. O36a). The tubule is lined by a pseudostratified columnar epithelium consisting of two primary cell types: PRINCIPAL CELLS and BASAL CELLS. The principal cells are columnar with long, nonmotile STEREOCILIA extending into the lumen (Figs. O45a and O45b). The stereocilia are longer than microvilli and branch repeatedly near their bases. In electron micrographs the

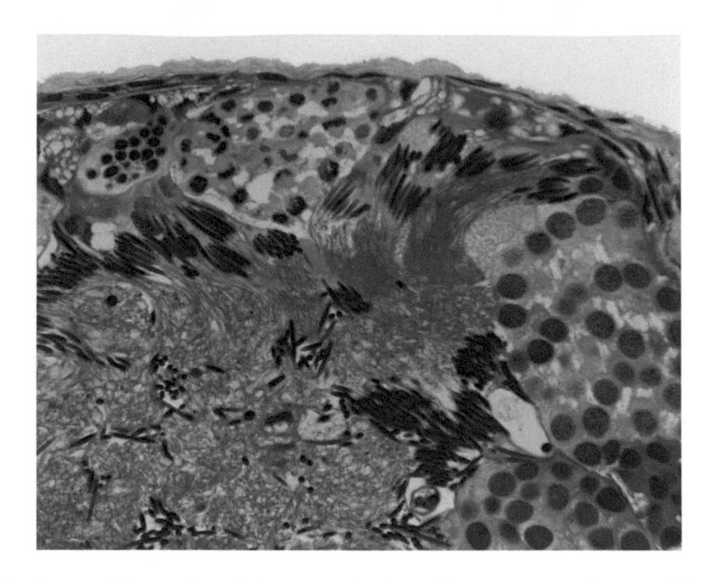

FIGURE O43 Photomicrograph of a semithin section of the testis of a frog. There are several cysts of developing sperm in this photograph. All of the ells in a cyst are at the same stage of development, 63 ×.

FIGURE O44 Photomicrographs of sperm smears of various vertebrates showing the characteristic head shape associated with each species. Body size is not indicative of the size of the mature sperm.

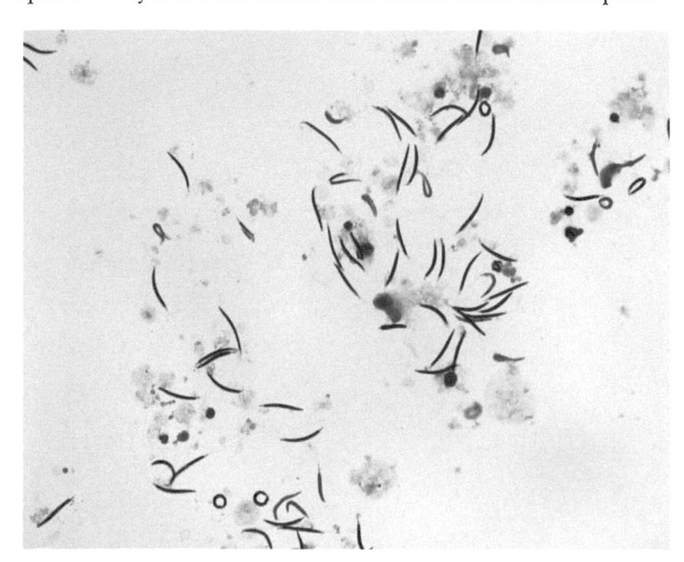

FIGURE O44a Frog, 63 ×.

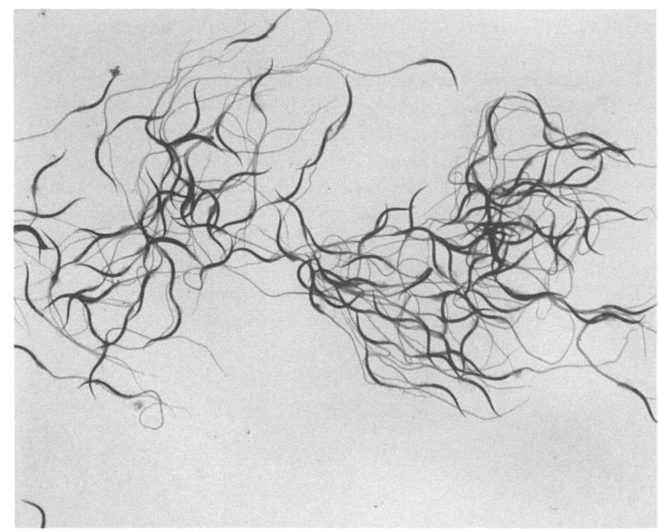

FIGURE O44b Turtle, 63 ×.

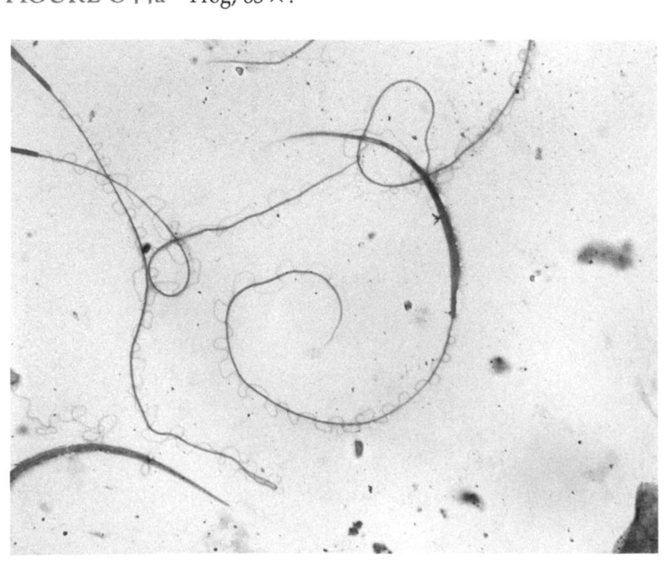

FIGURE O44c Chicken, 63 ×.

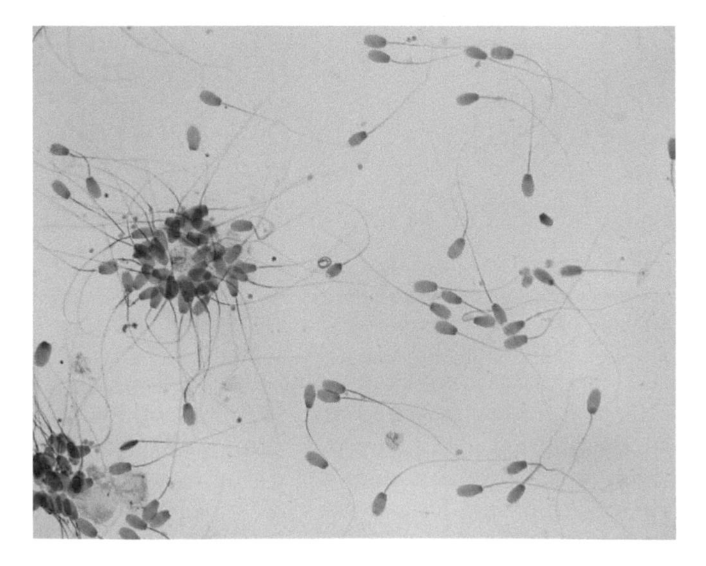

FIGURE O44d Rabbit, 63 ×.

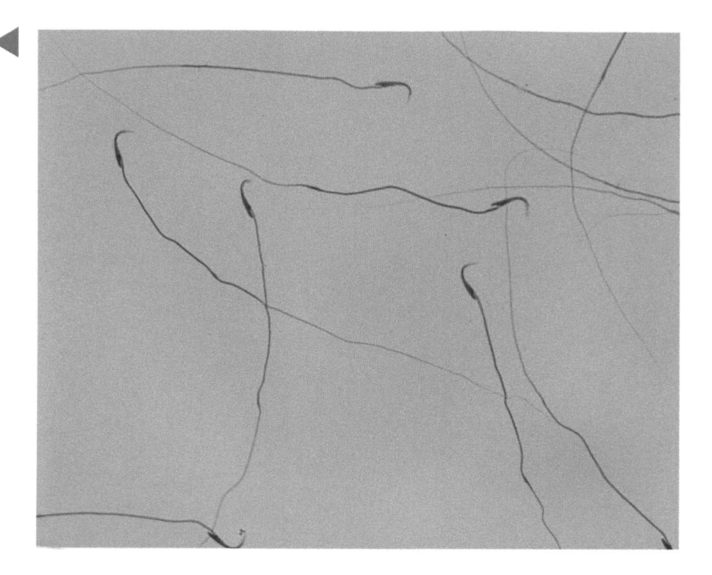

FIGURE O44e Guinea pig, 63 ×.

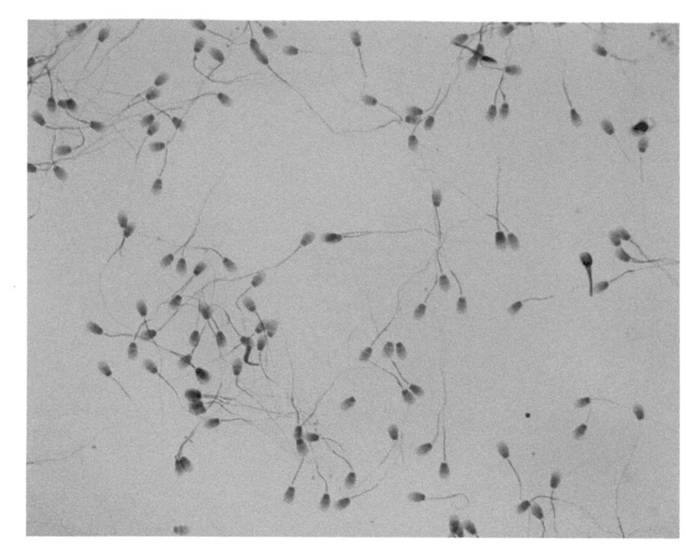

FIGURE O44f Dog, 63 ×.

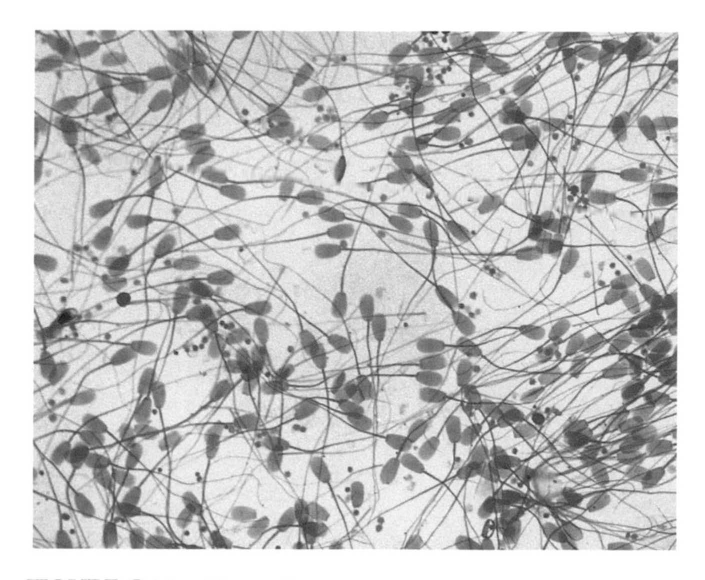

FIGURE O44g Horse, 63 ×.

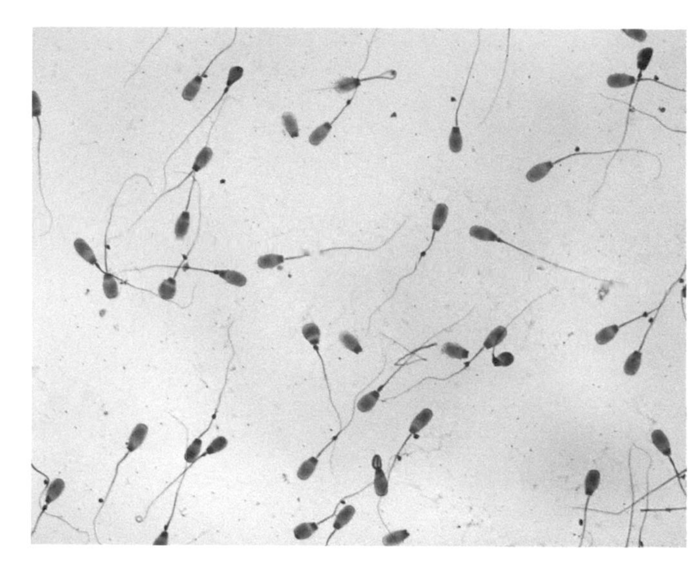

FIGURE O44h Sheep, 63 ×.

principal cells have the appearance of absorptive cells. The basal cells are simple cells located at the base of the epithelium. Surrounding the basement lamina is a dense capillary network and a layer of circular smooth muscle. The coils of the ductus epididymidis are supported by a stroma of loose fibrous connective tissue. Fluids are reabsorbed in the epididymis from the secretion of the testis. This aids in withdrawing the spermatozoa from the seminferous tubules.

The DUCTUS (VAS) DEFERENS is a continuation of the ductus epididymidis (Figs. O46a and O46b). It has a thick muscular wall and a small lumen. The epithelium varies from pseudostratified columnar with stereocilia proximally, to simple columnar distally and rests on a well-developed LAMINA PROPRIA MUCOSAE of fibroelastic connective tissue. The TUNICA MUSCULARIS consists of a thick circular layer of smooth muscle sandwiched between two thinner layers of longitudinal fibers.

The TUNICA MUCOSA of the SEMINAL VESICLE is secretory; it is so elaborately folded that it may appear in section as if there are glandular follicles (Fig. O47). The epithelium is pseudostratified or stratified columnar. There is an inner circular layer and an outer longitudinal layer of smooth muscle.

The PROSTATE GLAND is a much branched tubuloalveolar gland in close association with or even surrounding the urethra into which it opens by numerous ducts (Fig. O48). It is lacking in monotremes, edentates, and cetaceans.

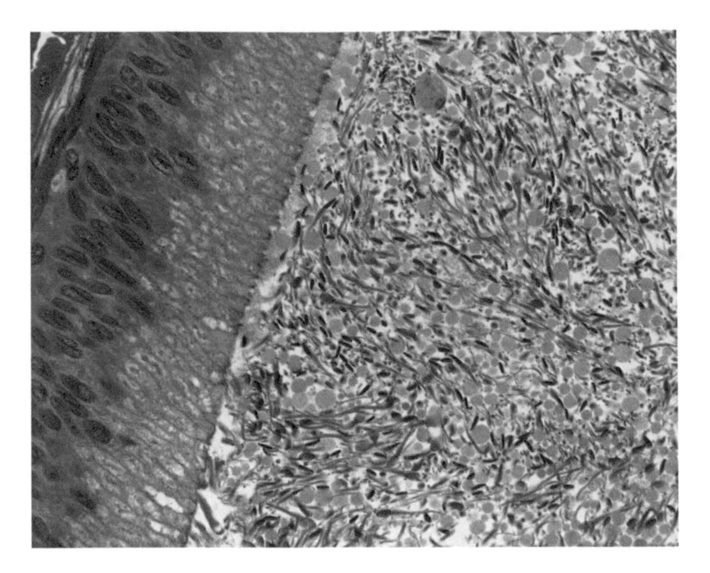

FIGURE O45a Photomicrograph of a semithin section of a mammalian epididymis teeming with spermatozoa. The spermatozoa complete part of their maturation in the epididymis. Close inspection reveals stereocilia and droplets among the spermatozoa, 63×.

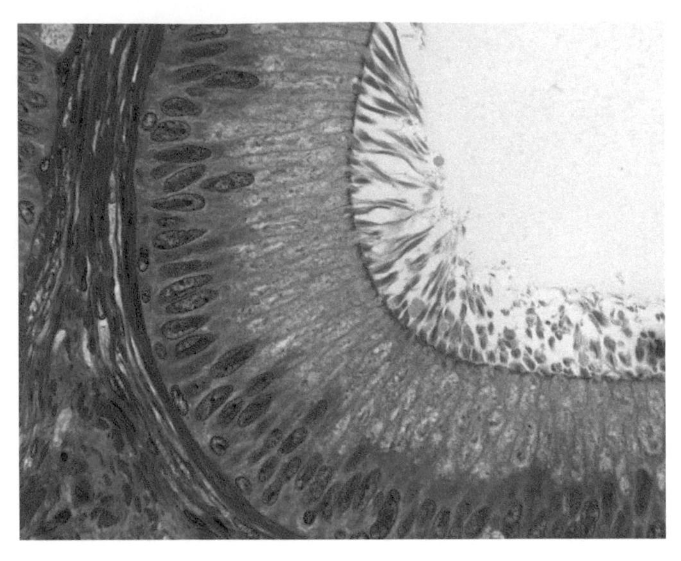

FIGURE O45b Photomicrograph of a semithin section of the mammalian epididymis showing its pseudostratified columnar epithelium. There are few spermatozoa in the lumen and the stereocilia are readily visible, 63×.

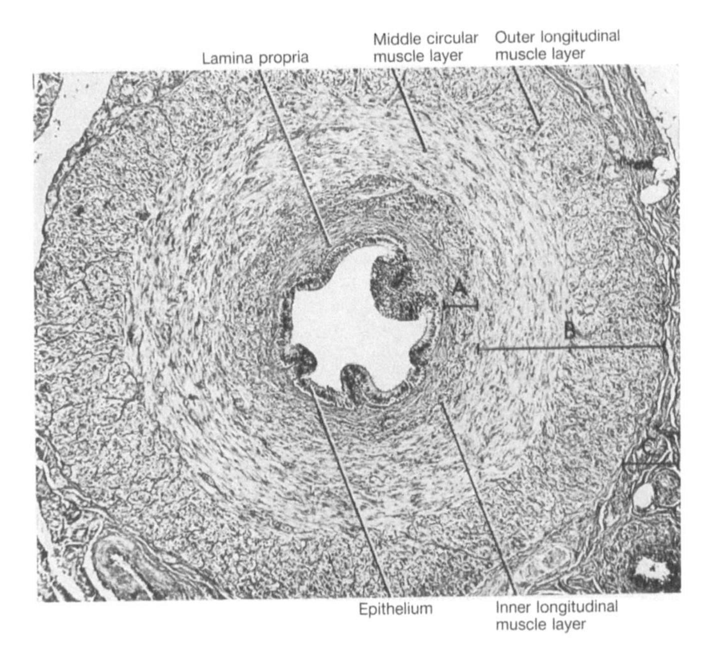

FIGURE O46a Photomicrograph of a cross-section of the human ductus deferens. *A*, muscularis mucosae; *B*, tunica muscularis; *C*, tunica adventitia, 50×.

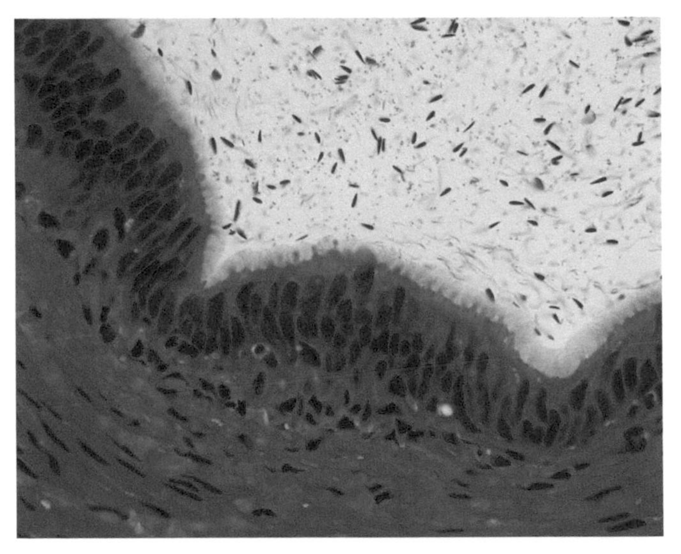

FIGURE O46b Photomicrograph of a cross-section of the vas deferens of a monkey. The epididymis is the site of storage and maturation of spermatozoa. There are two types of cells in the pseudostratified epithelium of the epididymis: tall *principal cells* bearing stereocilia and small, rounded *basal cells* lodged between the bases of the principal cells. These cells appear to have a generative function in the epithelium but there is little evidence to support this idea, 63×.

The secretory portions are surrounded by smooth muscle and are lined by a much folded pseudostratified columnar epithelium. The alveoli have large irregular lumina and may contain acidophilic PROSTATIC CONCRETIONS of condensed secretory substance. The LAMINA PROPRIA MUCOSAE consists of fibrous connective tissue. The CAPSULE is also fibrous connective tissue with smooth muscle.

The paired BULBOURETHRAL GLANDS (Cowper's glands) are compound tubuloalveolar glands that resemble mucous glands and whose ducts open into the base of the urethra (Fig. O49). There is a thin capsule of fibrous connective tissue and smooth and striated muscle fibers that extends SEPTA into the parenchyma to divide the gland into LOBULES.

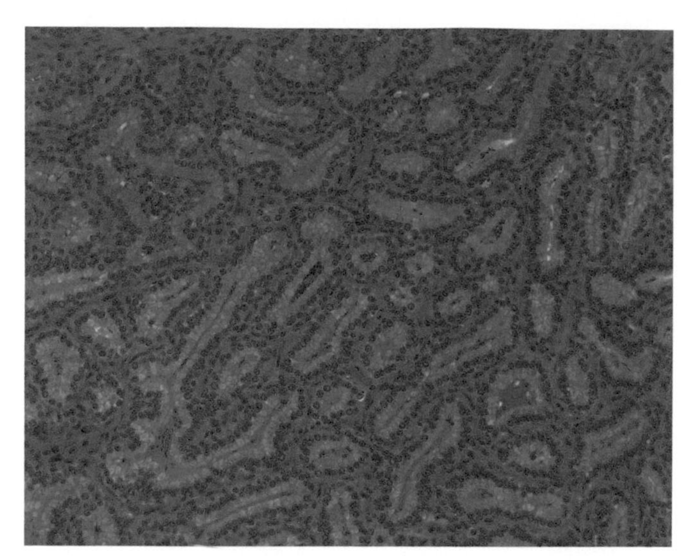

FIGURE O47 Photomicrograph of a section of the seminal vesicle of a monkey. In spite of appearances, the seminal vesicle is actually a long, convoluted tube with many secretory diverticula coming off the tube forming a lumen of "labyrinthine complexity." It contributes to the seminal fluid, 20 ×.

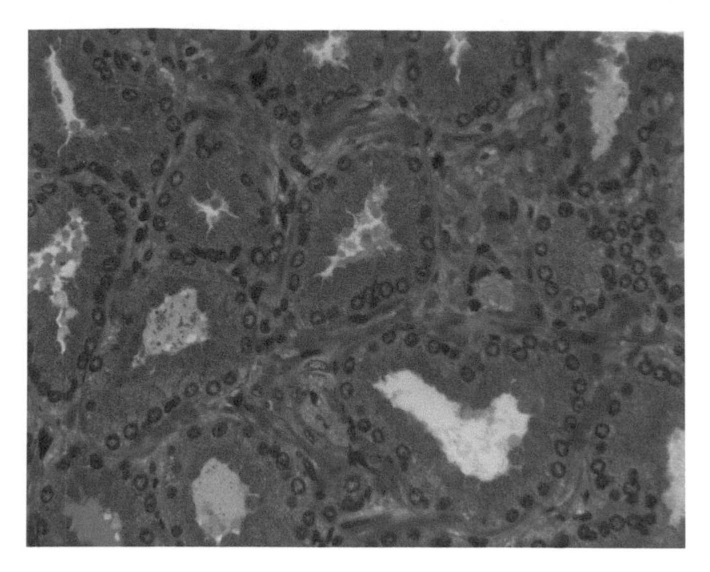

FIGURE O48 Photomicrograph of a section of the prostate gland of a monkey. Together with the seminal vesicle, this tubuloacinar gland contributes to the production of seminal fluid. Prostatic concretions (corpora amylacea) may be seen in the lumina of some of the acini, 40 ×.

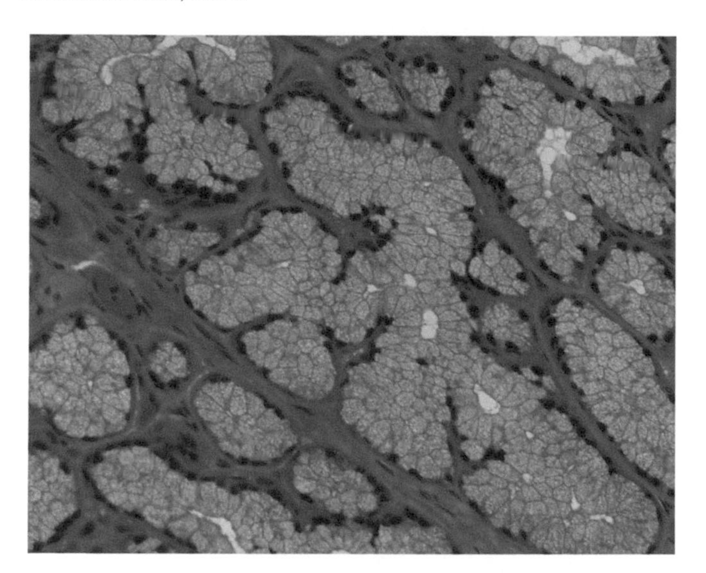

FIGURE O49 The bulbourethral glands (Cowper's glands) are tubuloalveolar glands about the size of a pea. They resemble mucus secretory glands and contribute a small amount of fluid at the onset of ejaculation, 63 ×.

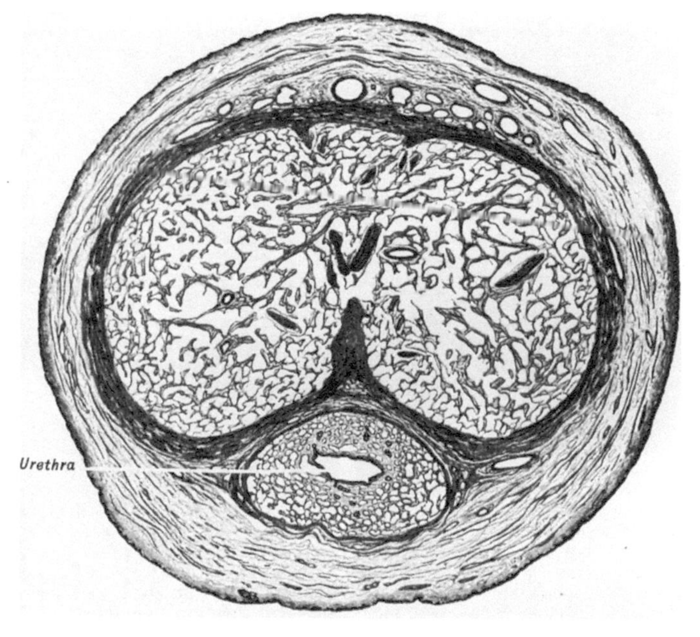

FIGURE O50a Drawing of a cross-section of the penis of a 21-year-old man. The two large *corpora cavernosa* are at the top separated by the *medial septum*. The *corpus cavernosum urethrae* surrounds the urethra.

Intromittent Organs

Elasmobranchs, some teleosts, and amniotes either produce shelled eggs or bear living young and an intromittent organ is used for inseminating the female. This is formed from modified pelvic fins in elasmobranches and from modified anal fins in some teleosts.

In most amniotes the intromittent organ contains erectile tissue that renders it turgid when gorged with blood. HEMIPENES of lizards and snakes are paired eversible sacs attached to the skin adjacent to the cloacal opening. Crocodiles, turtles, and some birds have a single trough-like PHALLUS homologous with the penis of mammals.

In the human penis (Fig. O50a), two large masses of spongy erectile tissue containing large numbers of anastomosing blood vessels are the CORPORA CAVERNOSA (Fig. O50b). A third, ventral CORPUS SPONGIOSUM surrounds the

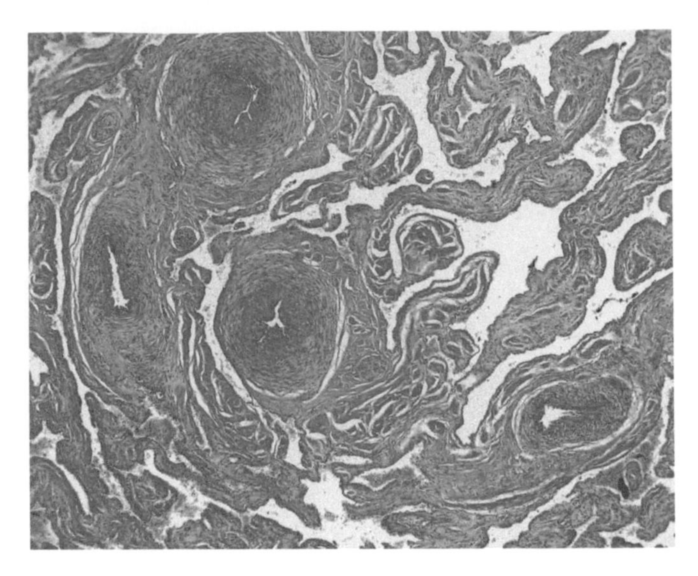

FIGURE O50b Photomicrograph of a cross-section of the penis of a monkey. Blood comes to the penis by way of the muscular arteries; at times of arousal, blood flow from the arteries increases and fills the corpora cavernosa overcoming the ability of the veins, at the periphery of the penis, to drain it away. Erection ensues, 5 ×.

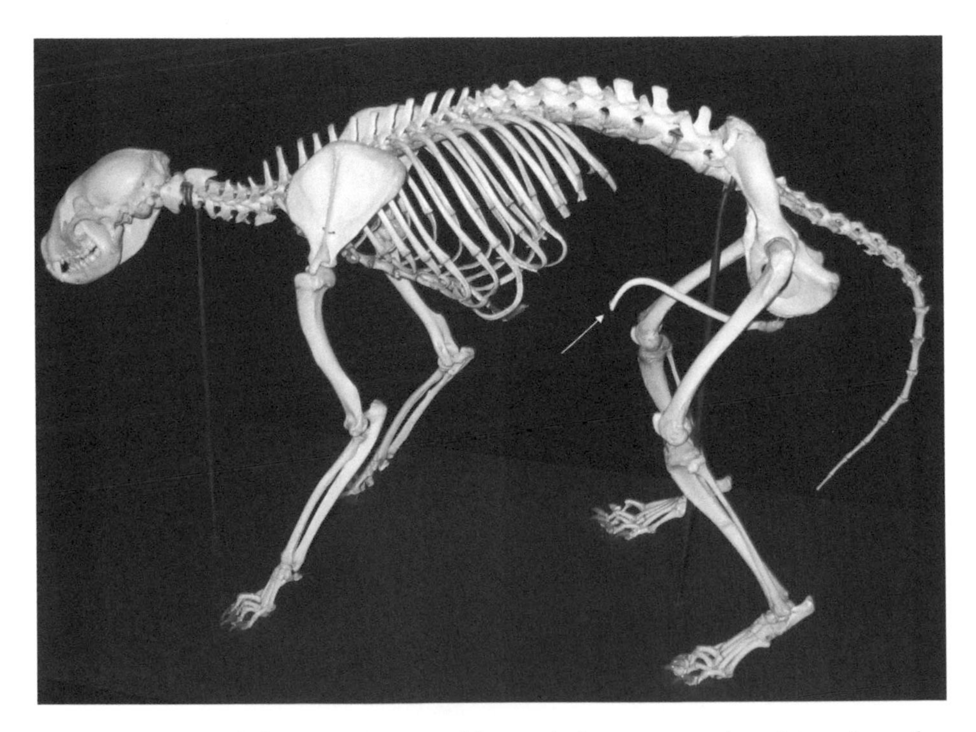

FIGURE O51a As in most male mammals (but not primates and humans), the raccoon penis contains a bone, the *penial bone* (arrow), that provides some measure of support.

urethra. The cavernous bodies are enclosed in a thick fibrous TUNICA ALBUGINEA. The corpus cavernosum may be an unpaired U-shaped mass in other mammals and the corpus spongiosum may be lacking. Blood flows into the center of the cavernous bodies from several arteries. The veins are peripheral and leave at an oblique angle. Sexual excitement causes an increased flow of blood into the tissue and distension of the central spaces compresses the oblique veins. This obstructs the outflow of blood and produces rigidity of the penis. In the flaccid condition, incoming blood pressure is less and the openings of the veins remain free.

A PENIAL BONE (os priapri) is present dorsal to the urethra in many mammals (Fig. O51a). The cavernous bodies are supported by fibrous connective tissue and the penis is covered by thin skin (Fig. O51b).

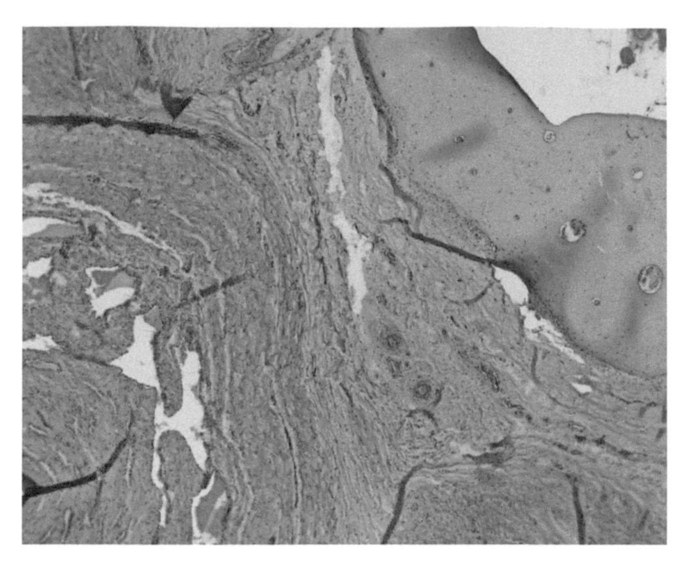

FIGURE O51b The penial bone is a compact bone and is here seen embedded in the penis of a dog (upper right), 5 ×.

The URETHRA is lined by transitional epithelium when it leaves the bladder. This changes to stratified columnar when the genital ducts enter and become stratified squamous at the opening of the penis. The GLANS PENIS is highly vascular and richly supplied with sensory endings (lamellated and tactile corpuscles). Its stratified squamous epithelium is continuous with the lining of the urethra and skin of the prepuce.

CHAPTER

P

Sense Organs

We have already considered simple sense organs: tactile and lamellar corpuscles of touch and pressure. There are other sensory endings, largely in the skin, for touch, pain, heat, and cold, consisting essentially of naked nerve endings, often branched, that pass among the epithelial cells. We have also examined proprioceptors, or organs of muscle sense that might also be included here, although we are not conscious of their activity. In this section we will consider the most complex sense organs: the eye, the lateral line system, and the ear.

EYE

Fig. P1a, a labeled median vertical section of a monkey eye, *and* Fig. P1b, a diagrammatic representation of the mammalian eye, are provided to orient the reader with various major landmarks in the eye. The eyeball is surrounded by three coats: the peripheral TUNICA FIBROSA, middle TUNICA VASCULOSA (uvea), and inner TUNICA INTERNA or RETINA (Fig. P1c). Only the tunica fibrosa surrounds the eyeball completely.

Tunica Fibrosa

The tough outer covering protects the eyeball and gives it form. The anterior portion of the tunica fibrosa is the transparent CORNEA and the posterior opaque part is the SCLERA, the white of the eye.

The CORNEA is nonvascular and consists of a thick layer of transparent fibrous connective tissue, the SUBSTANTIA PROPRIA, sandwiched between two transparent epithelia (Fig. P2a). The outer epithelium is stratified squamous with many free nerve endings and lymphoid elements among its cells. This layer has great regenerative power. The inner epithelium has, by tradition, been termed the "corneal endothelium"; it is a single layer of large squamous cells (Fig. P2b). The substantia propria constitutes about seven-eighths of the thickness of the cornea and consists of lamellae of collagenous fibers and occasional elastic fibers impregnated with a glycoprotein cement. Electron micrographs show that all fibers in a lamella are parallel and cross the fibers in other lamellae, similar to the layers of plywood (Figs. P3a and P3b). Some fibers run between the lamellae. Flat fibroblasts with branching processes and lymphoid cells occur between the collagenous fibers (Fig. P4a). The substantia propria is separated from the epithelia on each side by thin layers of interlacing collagenous fibers.

The opaque SCLERA consists of flat bundles of collagenous fibers and a few elastic fibers running parallel to the surface (Fig. P4b). Tendons of the eye muscles attach to the outer surface (Fig. P5a). The portion of the sclera under the eyelids is covered with a thin transparent stratified columnar epithelium, the BULBAR CONJUNCTIVA. It is continuous with the epithelium covering the cornea and with the PALPEBRAL CONJUNCTIVA lining the eyelids (Figs. P1a and P1b).

Tunica Vasculosa (Uvea)

The tunica vasculosa is the vascular central layer lining the entire eyeball except under the cornea (Fig. P1c). The largest portion is the CHOROID LAYER in the posterior four-fifths of the eyeball. Also derived from the tunica vasculosa are the CILIARY BODY and the stroma of the IRIS.

An Atlas of Comparative Vertebrate Histology.
DOI: https://doi.org/10.1016/B978-0-12-410424-2.00022-6

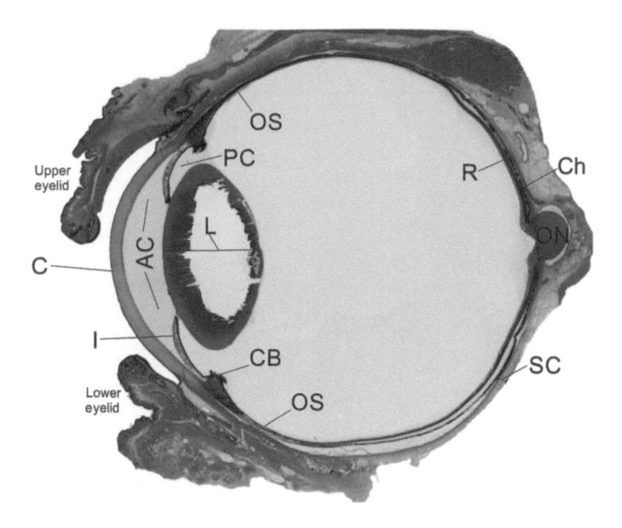

FIGURE P1a Photomicrographic montage of a median section through the of the eye of a monkey. This illustration is useful to orient yourself in the study of the eye. The lens is difficult to section and often appear broken in histological preparations; this lens has been hollowed out during preparation. *C* cornea; *AC*, ciliary body; *Ch*, choroid coat; *I*, iris; *L*, lens; *ON*, optic nerve; *OS*, ora serrata; *PC*, posterior chamber of the eye; *R*, retina; *S*, sclera.

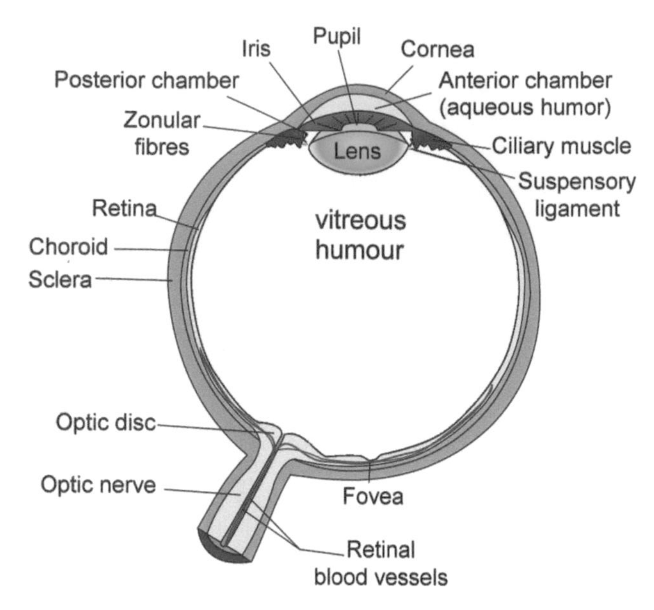

FIGURE P1b Diagram illustrating the major internal structures of the human eye. *By Soerfm - Own work, CC BY SA 3.0, commons.wikimedia.org/w/index.php?curid=29385884.*

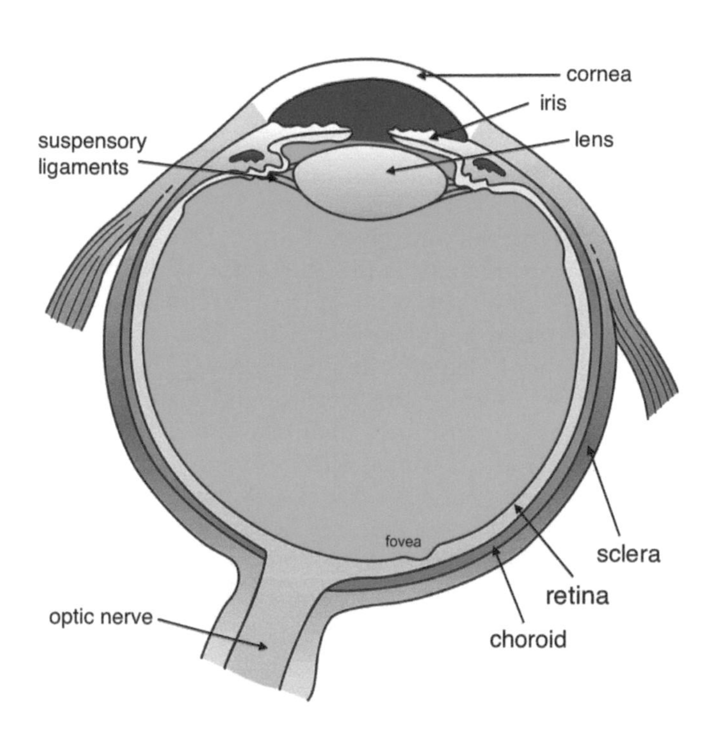

FIGURE P1c Diagram illustrating the three layers of the vertebrate eye: The supportive layer comprising the CORNEA and SCLERA; the vascular coat, the UVEA and the inner photosensitive layer, the RETINA. *Holly Fischer own work CC BY SA 3.0 http://open.umich.edu/education/med/resources/second-look-series/materials. Authors.*

The CHOROID LAYER contains arteries, veins, and, more centrally, a rich vascular layer, the CHORIOCAPILLARIS, consisting of capillaries in a stroma of loose fibrous connective tissue rich in melanocytes (Figs. P4b and P6). The choriocapillaris supplies the outer layers of the retina but does not continue as far forward as the ciliary body. The endothelial walls of capillaries in the choriocapillaris are extremely thin and fenestrated facing the retina while the opposite sides are thick and devoid of fenestrations. The choriocapillaris is separated from the retina by the

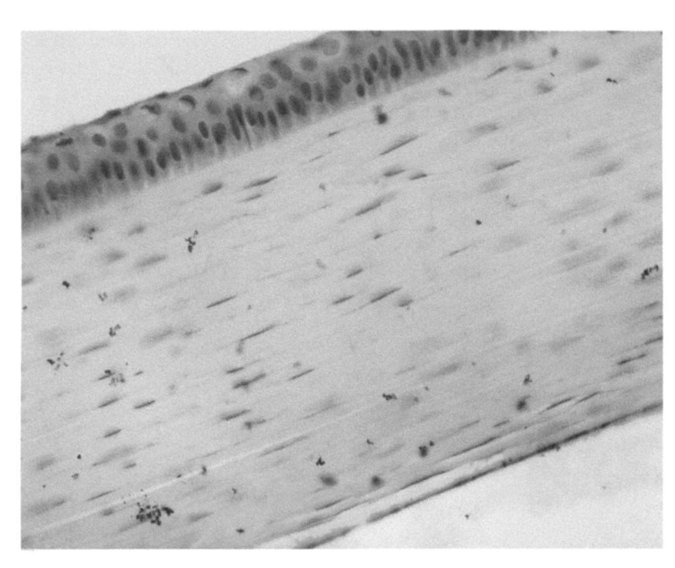

FIGURE P2a Photomicrograph of a section of the cornea of a rabbit. Most of the cornea consists of collagenous fibers (the SUBSTANTIA PROPRIA), arranged in regular sheets—like plywood. It is sandwiched between a stratified squamous EPITHELIUM on the outside (top) and the ENDOTHELIUM on the inside (bottom). The cells of the epithelium do not keratinize and their nuclei can be seen in the section. 50×.

FIGURE P2b Scanning electron micrograph of the simple squamous endothelium that lines the cornea on the posterior surface of the cornea on the eye of an ostrich. *Pigatto, J.A.T, et al., 2009. Scanning electron microscopy of the corneal endothelium of ostrich. Cienc Rural 39(3). doi: 10.1590/S0103-84782009005000001 Creative Commons Attribution License.*

LAMINA VITREA (Bruch's membrane), consisting of the basement membrane of the endothelial cells, layers of collagenous and elastic fibers, and the basement membrane of the retinal pigment epithelium

A thickened ring of the tunica vasculosa, the CILIARY BODY, extends between the lens and sclera and contains connections of the lens (Figs. P5b–P5d). The ciliary body appears roughly triangular in section and its inner surface is lined by the nonnervous portion of the retina. Most of the ciliary body consists of CILIARY MUSCLES that change the shape of the lens. There are three groups of smooth muscle fibers: MERIDIONAL, RADIAL, and CIRCULAR, connected to the lens by ZONULAR FIBERS of the CILIARY ZONULE (suspensory ligament) (Fig. P10d). Immediately behind the origin of the iris are irregular vascular ridges of the CILIARY PROCESSES covering the inner surface of the ciliary muscle. They have a core of loose connective tissue rich in capillaries that is a continuation of the vascular layer of the choroid; they secrete the aqueous humor. Ciliary processes are covered with a pigmented epithelium derived from the nonnervous portion of the retina, Fig. P7.

Excess aqueous humor is drained by a ring-like vein without valves the VENOUS SINUS of the sclera (canal of Schlemm), girdling the eyeball at the junction of the cornea and ciliary body (Figs. P5b, P5c, P8). The venous sinus is embedded in the inner portions of the sclera and extends anastomosing branches into the surrounding sclera. Its wall consists of a single layer of nonfenestrated endothelial cells resting on a discontinuous basement membrane.

The IRIS is largely a circular sheet of pigmented, highly vascular, fibrous connective tissue surrounding the PUPIL and attached to the ciliary body by its outer margin (Figs. P9a and P9b). The quantity of pigment contained in the chromatophores of this connective tissue largely determines the color of the iris. The anterior surface is covered by an epithelium continuous with that of the posterior surface of the cornea. The posterior surface of the iris is covered by a double layer of deeply pigmented epithelium derived from the retina. In the connective tissue surrounding the margin of the pupil is a ring of smooth muscle, the PUPILLARY SPHINCTER. When this muscle contracts, it reduces the size of the pupil. A set of radial muscles, the PUPILLARY DILATOR, forms a sheet between the margin of the pupil and the periphery of the iris and dilates the pupil when its fibers contract.

Lens

The lens, suspended by zonular fibers of the ciliary body, lies directly behind the pupil, separating the aqueous humor from the vitreous body, Figs. P1a, P5b, P5c. Zonular fibers are difficult to see with the light

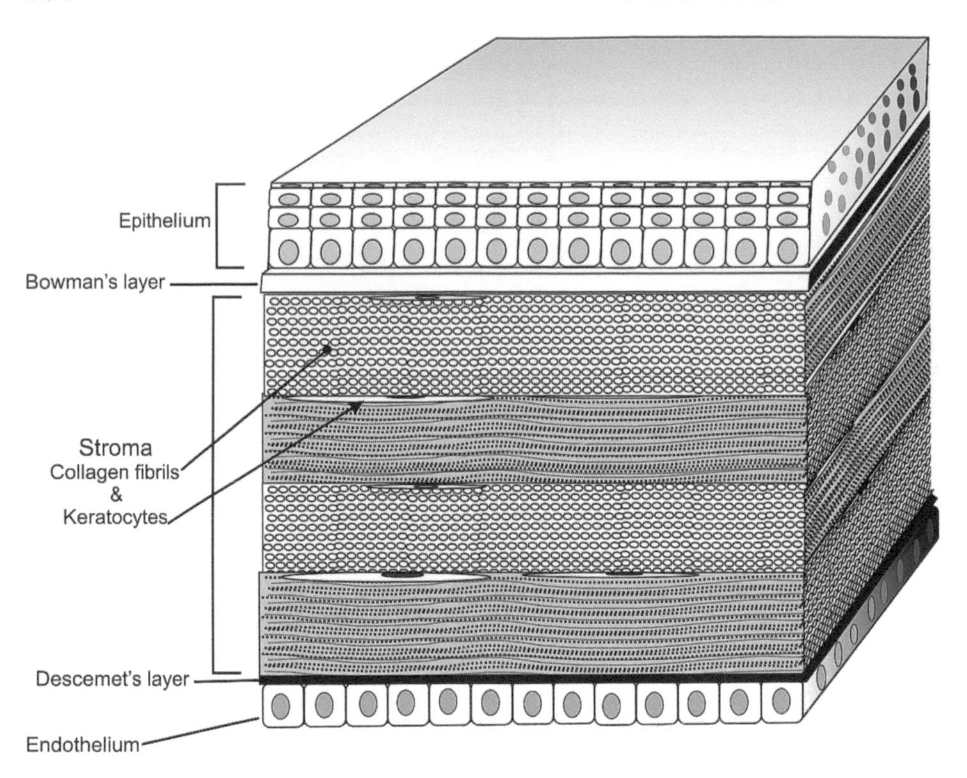

FIGURE P3a Diagrammatic view of the plywood-like construction of the cornea. Flattened fibroblasts sit between the layers of collagenous fibers. *Authors.*

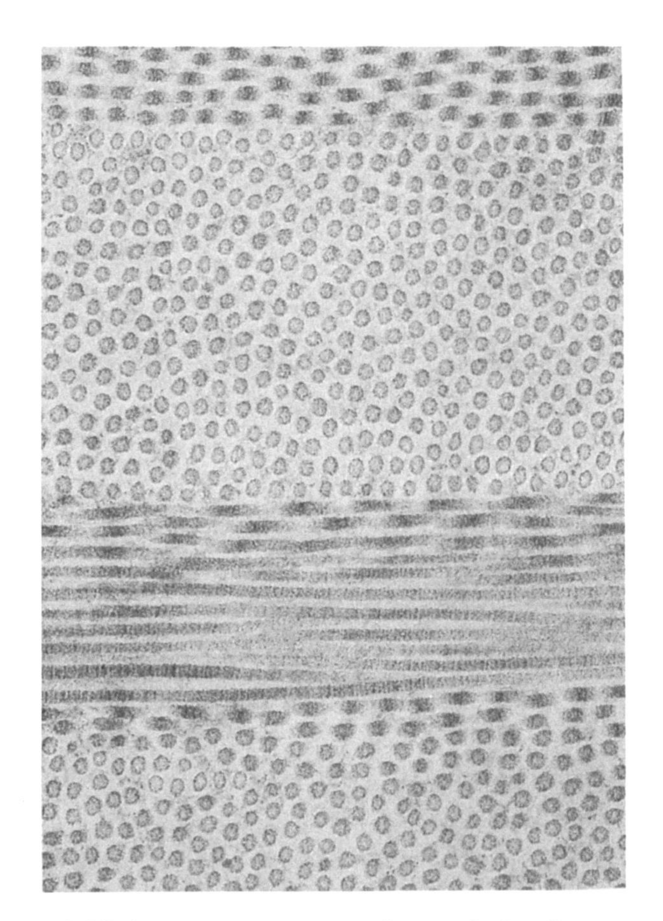

FIGURE P3b Electron micrograph showing the lamellar arrangement of collagen fibers in the vertebrate cornea. The fibers are arranged parallel to the surface of the cornea.

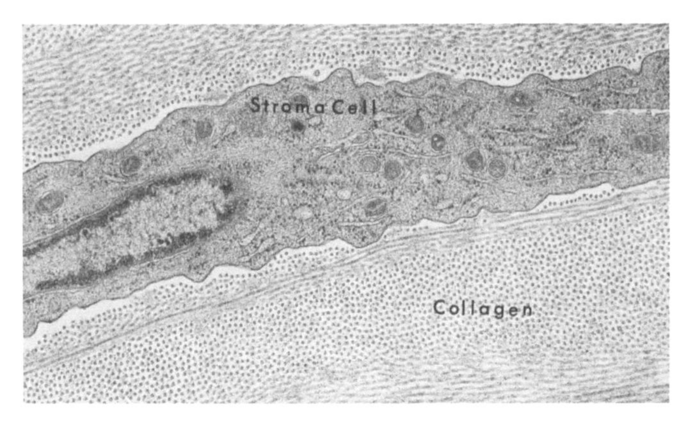

FIGURE P4a Transmission electron micrograph of a portion of a stromal cell sandwiched between collagenous laminae in the vertebrate cornea.

microscope, Figs. P10c and P10d, but show to advantage in scanning electron micrographs (Fig. P5d). The lens is less important than the cornea for refraction of light but it can change its curvature and accommodate the eye for near and far vision. Most of the lens is composed of ribbon-like cells, the LENS FIBERS, in the form of six-sided prisms, forming a homogeneous transparent mass (Figs. P10a and P10b). These cells may attain lengths up to 10 mm in humans. Scanning electron micrographs show their "ball-and-socket" interdigitations. Lens fibers lie parallel to the surface of the lens; toward the interior of the lens, they lose their nuclei and organelles and their cytoplasm becomes

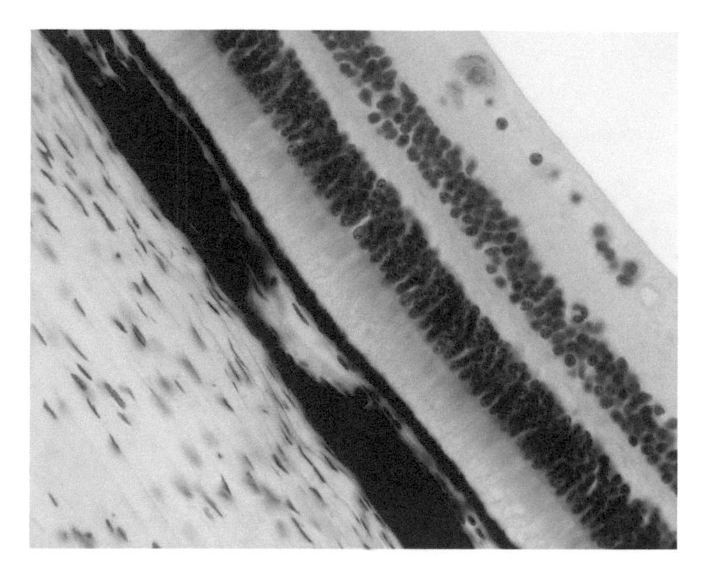

FIGURE P4b Photomicrograph of the photosensory portion of the eyeball of a rabbit. Light travels from the pupil to the photosensory layers of the eye from the upper right and passes through photosensory layers to stimulate the rods and cones.
Just under the photosensitive layer is the sharply defined layer of the pigment epithelium. Below that is the heavily pigmented choroid, which accommodates the blood vessels of the choriocapillaris. The heavily pigmented choroid is sandwiched between the photosensory layer and the sclera. The tough, supportive sclera is at the lower left. 40 ×.

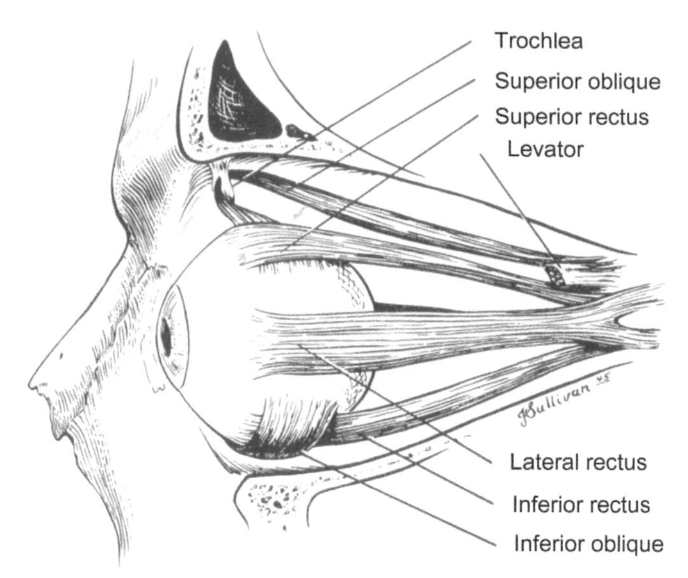

FIGURE P5a Tendons of the eye muscles attach to the outer surface of the sclera.

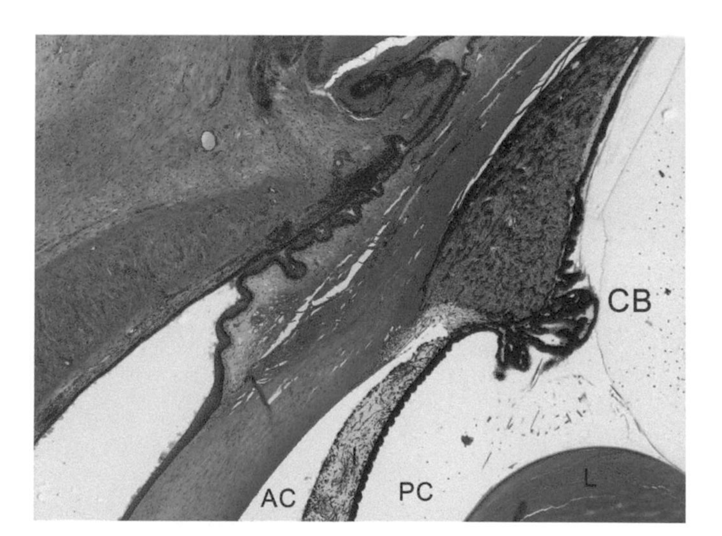

FIGURE P5b Photomicrograph of the ciliary muscles and surrounding structures in the eye of a monkey. Only a hint is visible of the suspensory ligaments that extend between ciliary body (*CB*) and the lens (*L*)—lower right. The iris (*I*—lower left) separates the anterior chamber (AC) of the eye and the posterior chamber (PC), lower right). 2.5 ×.

condensed, giving rise to the NUCLEUS of the lens (Fig. P10e). Lens fibers newly added to the nucleus undergo such strong condensation that the lens grows little during life. The lens is surrounded by a thick homogeneous CAPSULE of cement substance and lamellae of fine collagenous fibers. This capsule is actually the basement membrane that has surrounded the lens since its development from an epithelial lens vesicle (Figs. P12a and P10e). The anterior wall of the lens vesicle forms a single layer of cuboidal cells, the SUBCAPSULAR EPITHELIUM, on the anterior surface of the lens between the capsule and the prismatic lens fibers. Lens fibers are derived from the posterior wall of the lens vesicle.

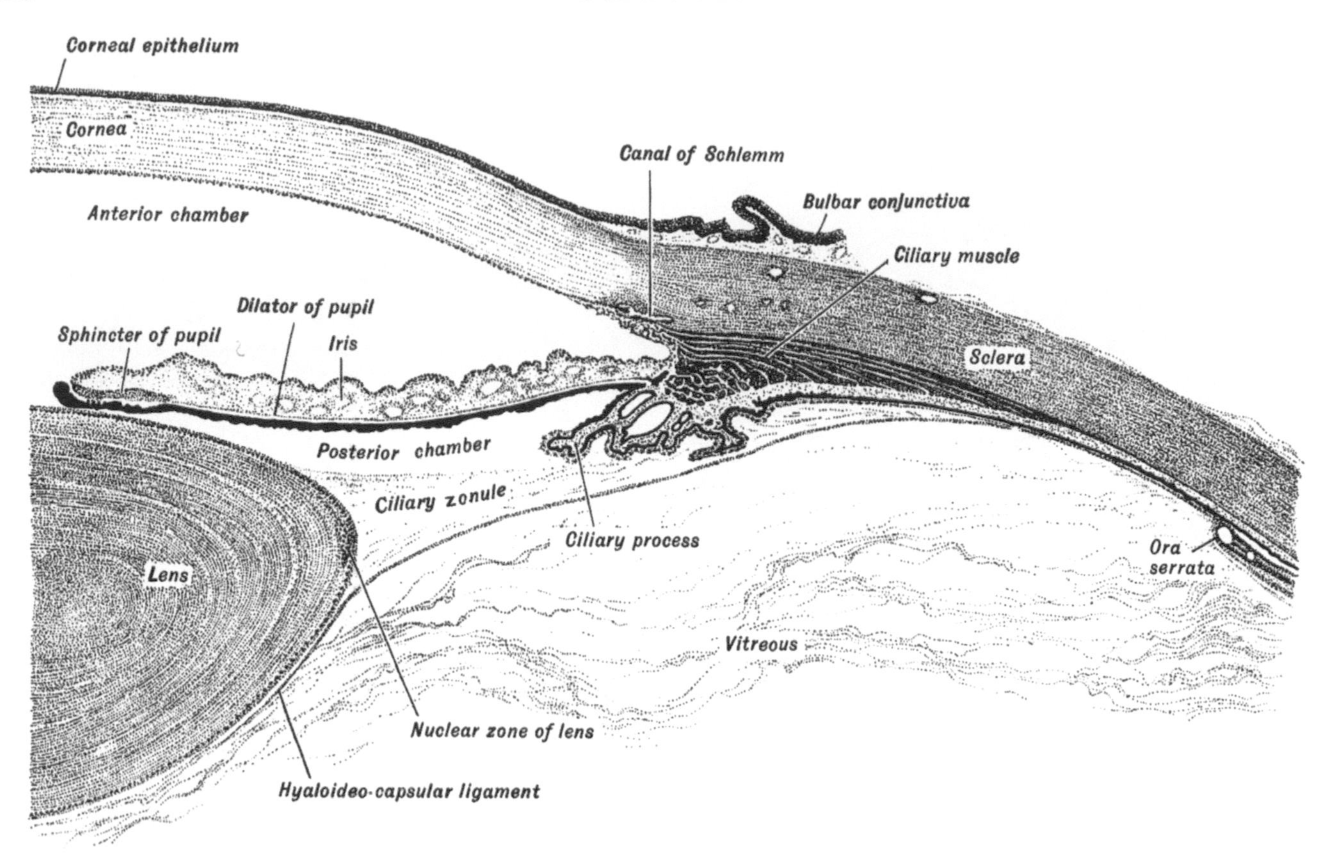

FIGURE P5c Diagram of the ciliary body in the human eye and its surroundings.

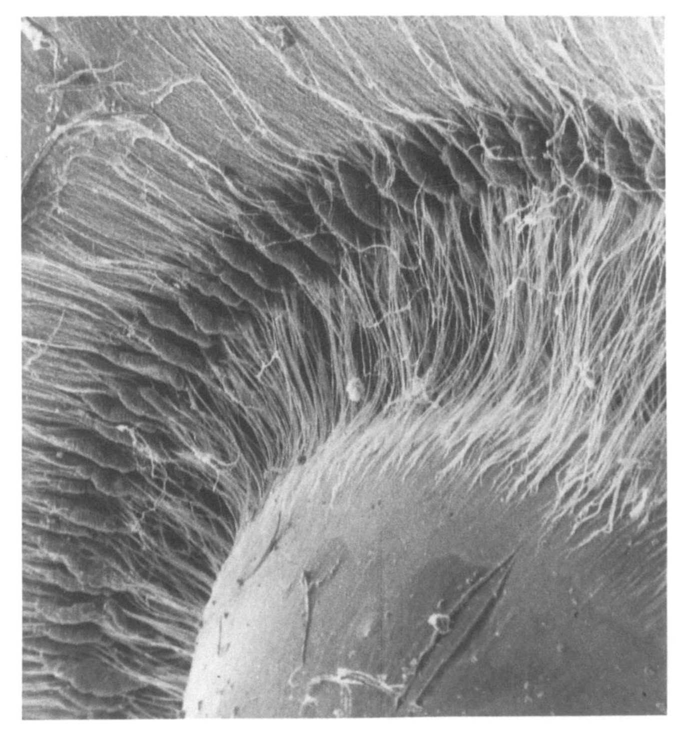

FIGURE P5d Scanning electron micrograph of the annular fibers of a primate ciliary body. These fibers are clearly seen attached to the lens (lower right). 46 ×.

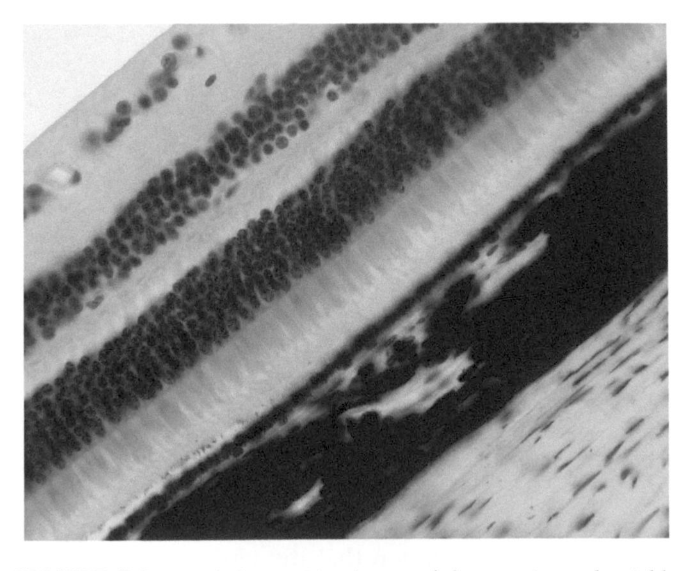

FIGURE P6 The light-sensitive layers of the eye sit on the richly vascularized and heavily pigmented choroid layer and its choriocapillaris. This layer is composed of capillaries in a stroma of loose fibrous connective tissue rich in melanocytes sandwiched between the photosensitive layers above and the dense tissue of the sclera below. 40 ×.

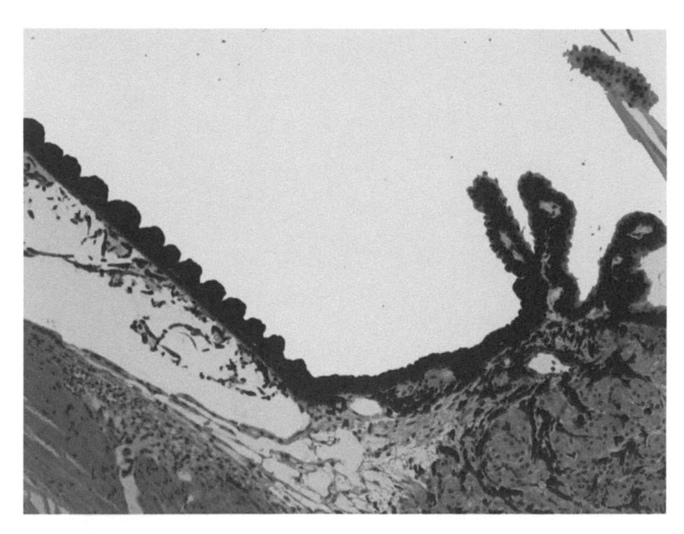

FIGURE P7 The ciliary processes replenish the vitreous humor. Excess fluids pass through trabecular meshwork and are drained by the venous sinus of the sclera (canal of Schlemm). 10 ×.

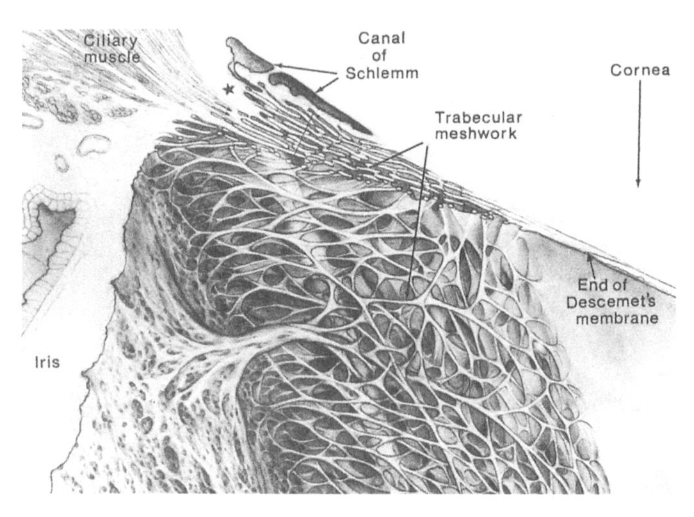

FIGURE P8 Drawing of the outflow system of the aqueous humor showing the venous sinus of the sclera (canal of Schlemm). Aqueous humor is constantly being produced by the ciliary processes. Excess fluid is drained through the trabecular meshwork to the venous sinus of the sclera (canal of Schlemm). Interruptions of this drainage may lead to increased pressure on the eyeball and glaucoma.

FIGURE P9a Photomicrograph of the lens (below) and the iris (above). The pupil is the opening in the iris. 5 ×.

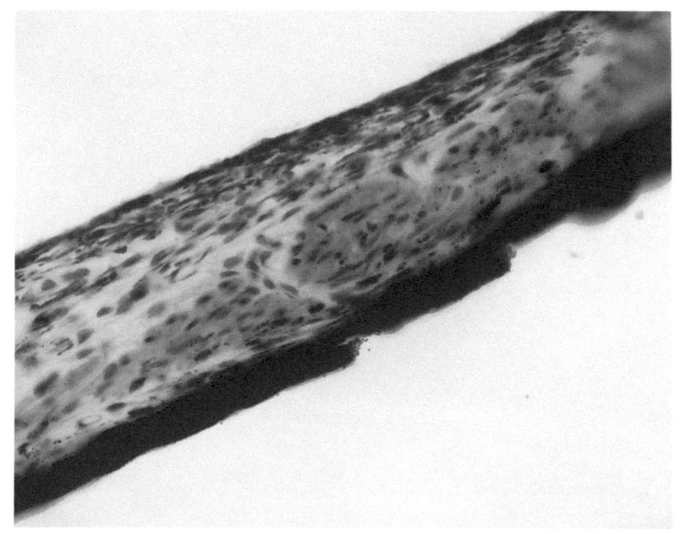

FIGURE P9b A section through the iris of a rabbit. The anterior surface is heavily pigmented. Wisps of the smooth muscle of the pupillary sphincter are scattered throughout the stroma of the iris. 40 ×.

The lens is elastic and tends to round up. It is restrained from this by tension exerted on the ciliary zonule when the ciliary muscles are relaxed. The relaxed eye, then, is accommodated for far vision. When the circular ciliary muscles contract, they oppose the tension of the ciliary zonule, the lens rounds up, and the eye focusses for near vision. Contraction of the meriodonal and radial fibers has the opposite effect.

Retina

The retina is the innermost tunic of the eyeball and lines the entire tunica vasculosa. The largest portion is the nervous OPTIC REGION that lines about two-thirds of the posterior portion of the eyeball as far forward as the ORA SERRATA, the serrated anterior margin of the optic region (Fig. P4b). The nonnervous portion of the retina (Fig. P10d)

FIGURE P10a Photomicrograph of a section of the lens of a rabbit. The lens capsule forms the clear envelope at the upper left; beneath is the anterior epithelium. Lens fibers arranged with parallel precision, form most of the substance of the lens (right). 40 ×.

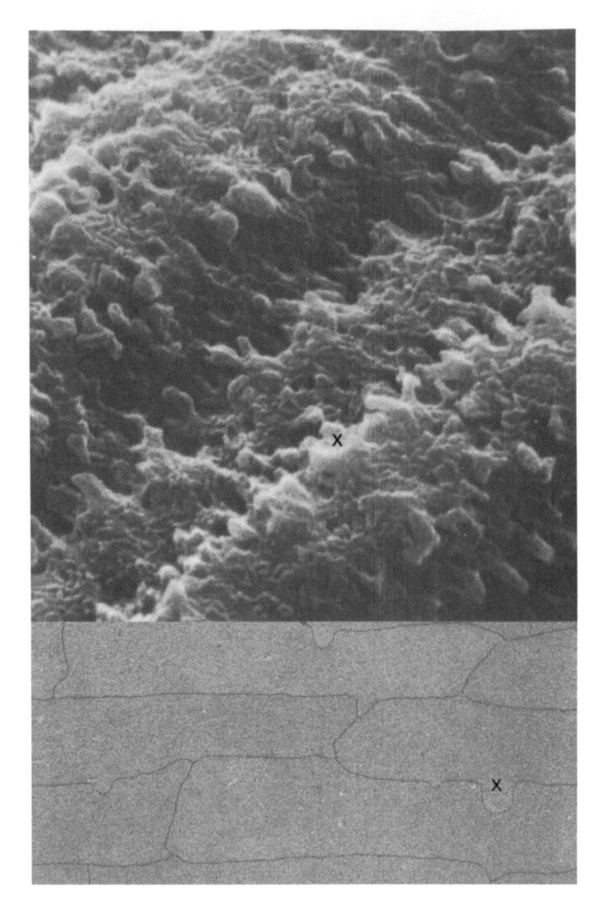

FIGURE P10b Scanning electron micrograph of the fractured cortex of a mammalian lens (upper). The surface of the lens fibers shows prominent undulations (e.g., X) which, in life, intermesh like zippers, with other such structures thereby providing great strength to the integrity of the lens. The lower image is a transmission electron micrograph of a similar area in cross-section showing the interdigitations (e.g., X) between lens cells. *Weiss, L. (ed.), 1988 Cell and Tissue Biology: A Textbook of Histology, 6th ed. Baltimore: Urban & Schwarzenberg.*

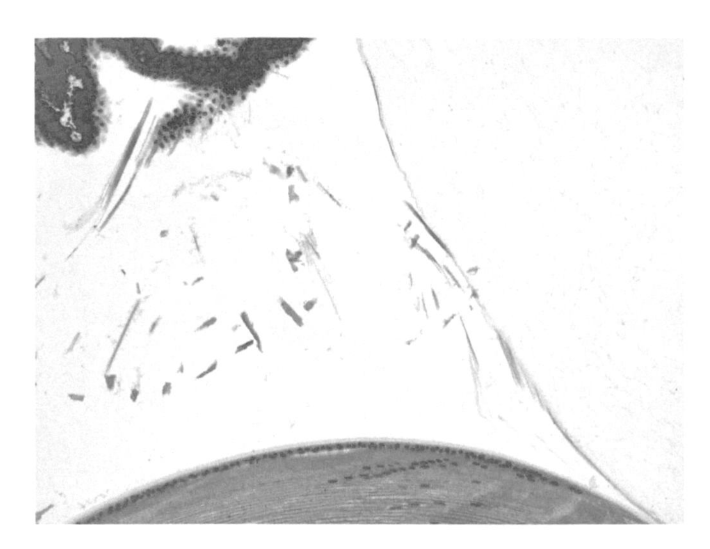

FIGURE P10c The lens is supported by zonular fibers, which extend from the ciliary body (top) to the surface of the lens (bottom). These fibers are difficult to demonstrate with the light microscope but show up well in scanning electron micrographs see Fig. P5d. 10 ×.

continues forward to cover the posterior margins of the ciliary body (CILIARY REGION of the retina) and the posterior portion of the iris (IRIDIAL REGION) (Fig. P5c).

The OPTIC REGION (Fig. P1a) of the retina is the light-sensitive part of the eye on which images are focused by the lens (Fig. P12b). Only the small BLIND SPOT, marking the point of attachment of the optic nerve, is insensitive to light (Fig. P11b). The FOVEA CENTRALIS is the site of keenest vision and is a shallow depression surrounded by a yellowish area, the MACULA LUTEA (Figure P11a and Figure P11c). Light must pass through several layers of

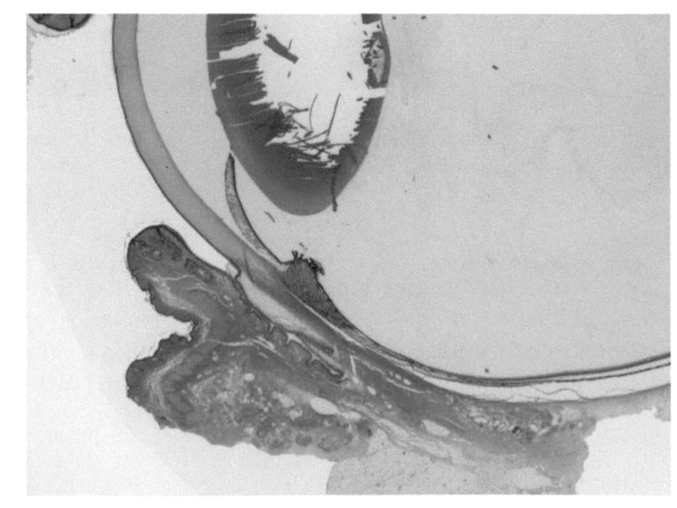

FIGURE P10d The suspensory ligament extends from the lens to the ciliary body but is often not seen with the light microscope. A section of the iris arises from the ciliary body and is seen at the lower left. 6.4 ×.

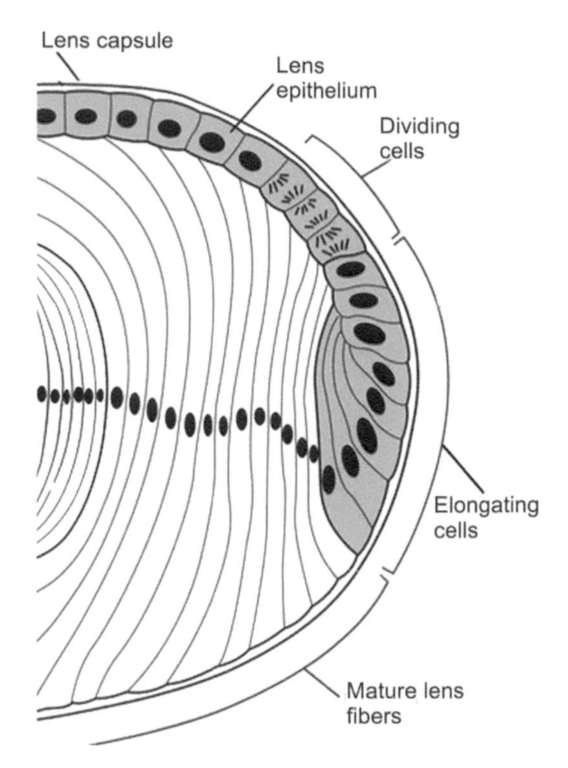

FIGURE P10e This diagram demonstrates the structured arrangement of elongated epithelium cells in the lens. During growth epithelium cells, secondary fibers, at the periphery gradually straighten, elongate and lose their organelles to become the more central primary fibers in the central zone of the lens. *Authors.*

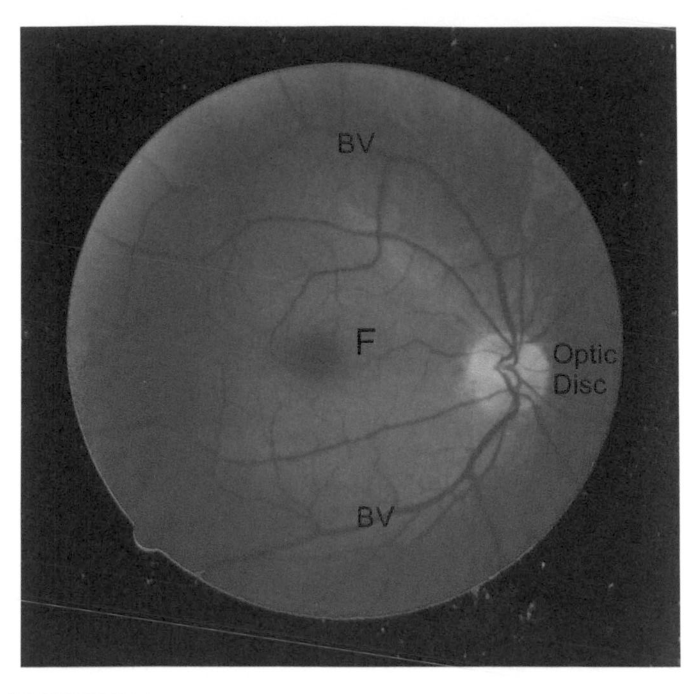

FIGURE P11a Ophthalmoscope photograph of the retinal surface of the senior author's right eye. The optic disc represents the beginning of the optic nerve and is the point where the axons of retinal ganglion cells come together. There are no sensory (rods or cones) cells here hence the name Blind Spot. The fovea (*F*) is surrounded by the macula lutea, the dark spot near the center. Blood vessels (*BV*) can be seen coursing across the retina.

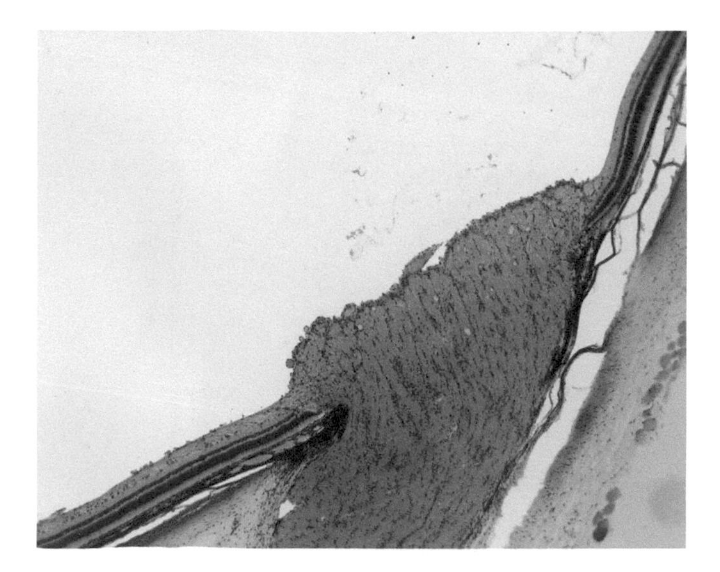

FIGURE P11b The optic disc and blind spot mark the point of entry for blood vessels into the retina and the head of the optic nerve. As seen in this micrograph there are no sensory cells here. 5 ×.

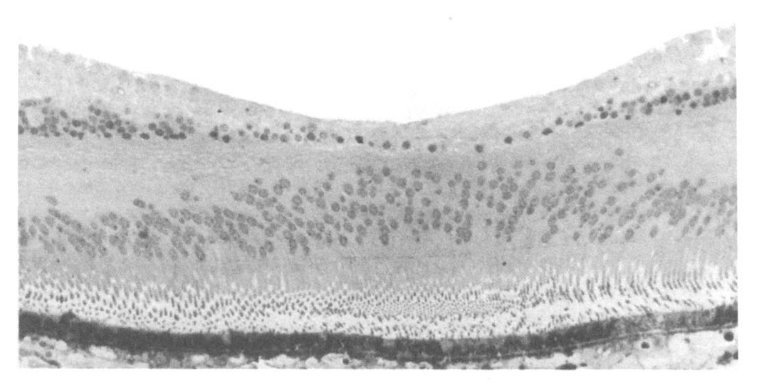

FIGURE P11c A cross section through the macula—this region consists of the central pit (fossa) in which all layers of the retina, except for rods and cones have been displaced toward the outer edge.

the retina to arrive at the light-sensitive RODS and CONES that are adjacent to the choroid layer. Impulses set up by the rods and cones pass forward through the retinal layers to reach the optic nerve. The retina, having arisen from the double optic cup, consists of two epithelial layers, the outer PIGMENT EPITHELIUM and the thicker inner NEURAL RETINA. Ten layers are recognized in the optic region of the retina, beginning at the choroid membrane (Fig. P12c).

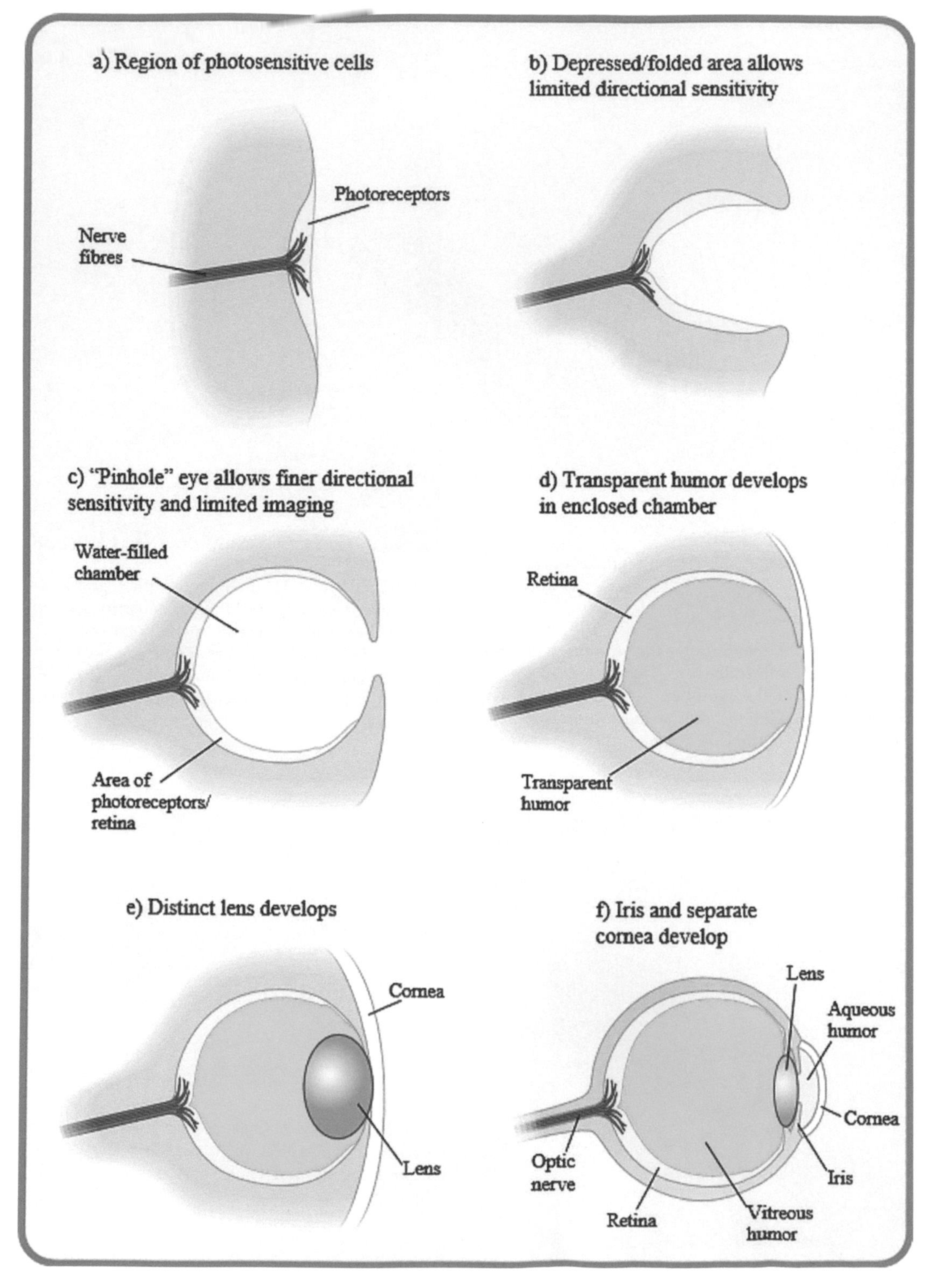

FIGURE P12a Schematic diagram of the stages in the development of the vertebrate eye. *Matticus78 own work CC BY SA 3.0, commons.wikimedia.org/w/index.php?curid=2748615.*

1. The *pigment epithelium* is a single layer of pigmented, cuboidal cells with spherical nuclei at their bases. Pigmented cytoplasmic processes from these cells extend between the rods and cones to enhance visual acuity at high illumination; the pigment granules withdraw from the processes at low light levels. The apical cytoplasm consists of residual bodies filled with lamellar debris, the partially digested residue of the phagocytosed tips of

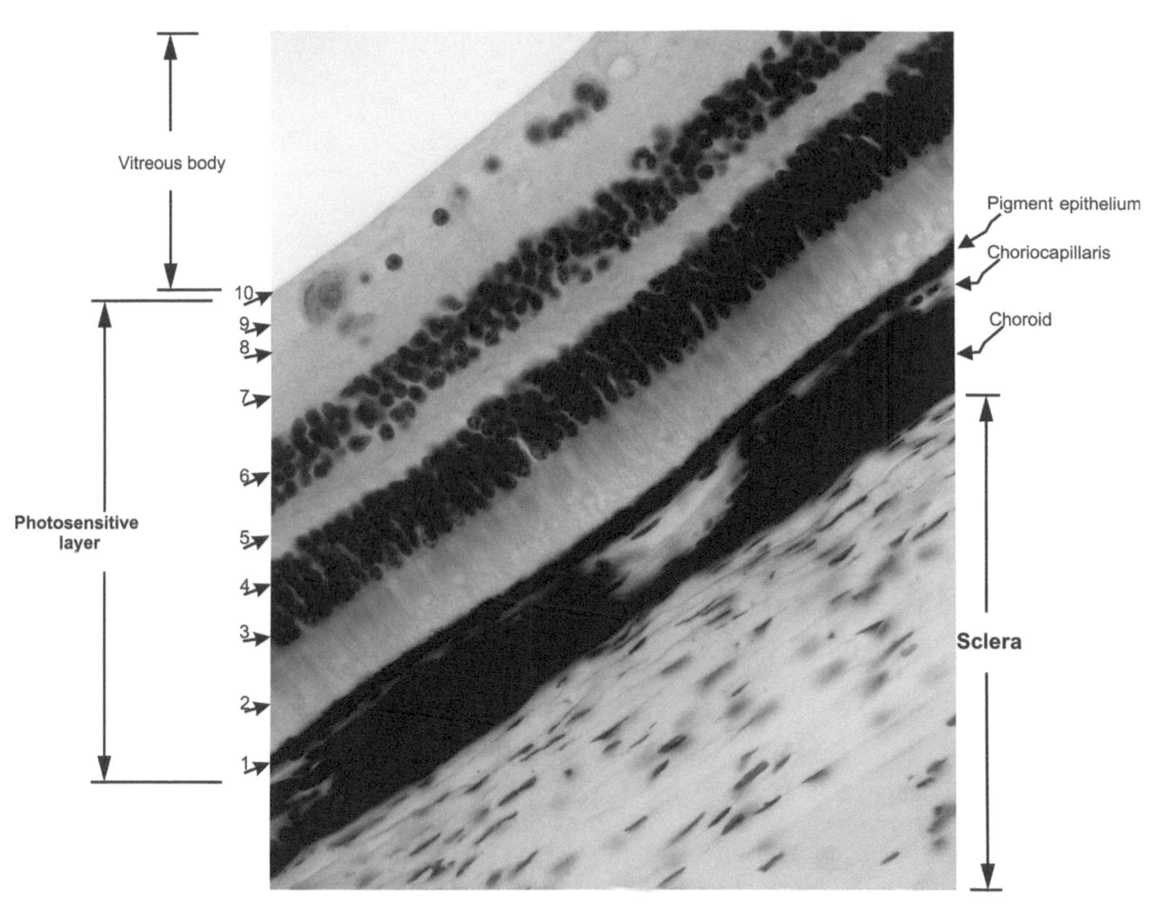

FIGURE P12b A section through the wall of an eye of a monkey. All 10 component layers of the photosensitive layer are readily visible: (*1*) Pigment epithelium, (*2*) Rods and cones, (*3*) Outer limiting membrane, (*4*) Outer nuclear layer, (*5*) Outer plexiform layer, (*6*) Inner nuclear layer, (*7*) Inner plexiform layer, (*8*) Ganglion cell layer, (*9*) Optic nerve fibers, (*10*) Inner limiting membrane.

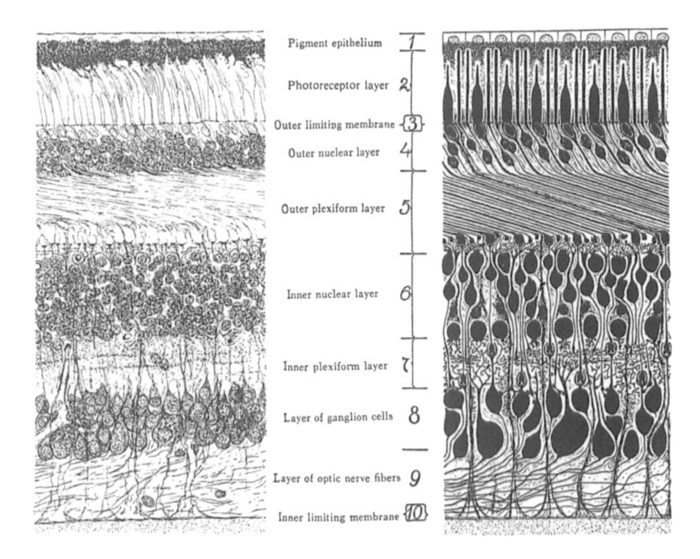

FIGURE P12c Drawings of the layers of the human retina. The left half has been stained routinely; the right half is a reconstruction from sections stained with Golgi's method. Layer identification is made clear here in these drawings.

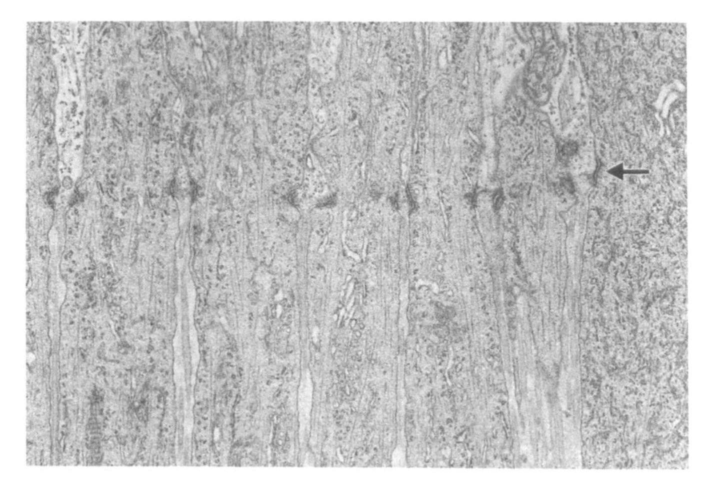

FIGURE P12d The outer limiting membrane (layer 3) is a chain of numerous junctional complexes (arrow) between Müller cells and photoreceptors.

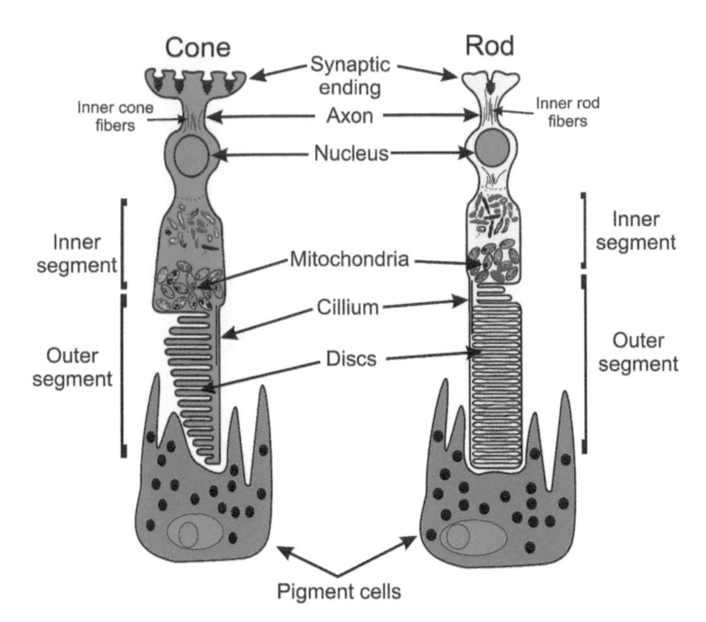

FIGURE P12e Schematic diagrams of rod (left) and cone (right) cells. The outer segments of the rods and cones are closely associated with the adjacent pigment epithelium. *Authors.*

the rod and cone outer segments. The pigment epithelium is separated from the choroid by the LAMINA VITREA (Bruch's membrane). Adjacent cells of the pigment epithelium are connected by junctional complexes consisting of gap junctions and elaborate zonulae occludentes and adhaerentes. The junctional complexes are the site of the *blood/retina barrier*, which prevents the free diffusion of blood solutes into retinal tissue.

2. The *photoreceptor layer* contains closely packed dendritic cytoplasmic processes of the photoreceptors. The CONES are long and conical, considerably wider than the RODS (Fig. P12e). Rods are adapted for perception at low light intensities and for seeing black and white while cones are chiefly concerned with color vision and visual acuity. In the human eye there are about 20 rods for each cone.
3. The *outer limiting membrane* appears with the light microscope as a sieve-like membrane supporting the processes of the photoreceptors. It is actually a row of zonulae adhaerentes between NEUROGLIAL FIBERS (Müller's fibers) that course between the neural elements and support all the layers of the retina from the outer limiting membrane to the inner limiting membrane, the 10th layer (Fig. P12d).
4. The *outer nuclear layer* consists of the cell bodies of the rod and cone cells and contains their nuclei. This is an exceptional situation where cell bodies of afferent neurons are not found in cerebrospinal ganglia.
5. In the *outer plexiform layer* the axons of the rods and cones synapse with dendrites of bipolar nerve cells. Several rods and cones may synapse with one bipolar cell.
6. The *inner nuclear layer* contains the cell bodies and nuclei of the bipolar nerve cells. The nuclei of the neuroglial fibers occur in the outer part of this layer; these cells have more cytoplasm than the nerve cells.
7. In the *inner plexiform layer* axons of the bipolar cells synapse with dendrites of the ganglion cells of the next layer.
8. The *ganglion cell layer* consists of large multipolar nerve cells arranged in a single row with a number of neuroglial cells.
9. Axons of the ganglion cells run parallel to the surface of the retina in the *nerve fiber layer* and plunge outward to form the optic nerve at the BLIND SPOT.
10. The *internal limiting membrane* is formed by feet of the neuroglial fibers. It is a thin membrane that lines the vitreous body.

Note that there are three layers of neurons in the retina: PHOTORECEPTORS (rods and cones), BIPOLAR NEURONS, and GANGLION CELLS (Fig. P12c). These are involved in the *vertical* transmission of impulses. Other neurons are present that transmit impulses *horizontally* and probably have an integrative function. The dendrites of the large pale HORIZONTAL CELLS at the outer edge of the inner nuclear layer dendrites form baskets enclosing the pedicels or end feet of the rods and cones in the outer plexiform layer. AMACRINE CELLS lie at the inner edge of the inner nuclear layer and may be recognized by their large, round, indented nuclei. Their processes synapse in the inner plexiform layer with axons of rod bipolar cells as well as with processes and cell bodies of the ganglion cells. The

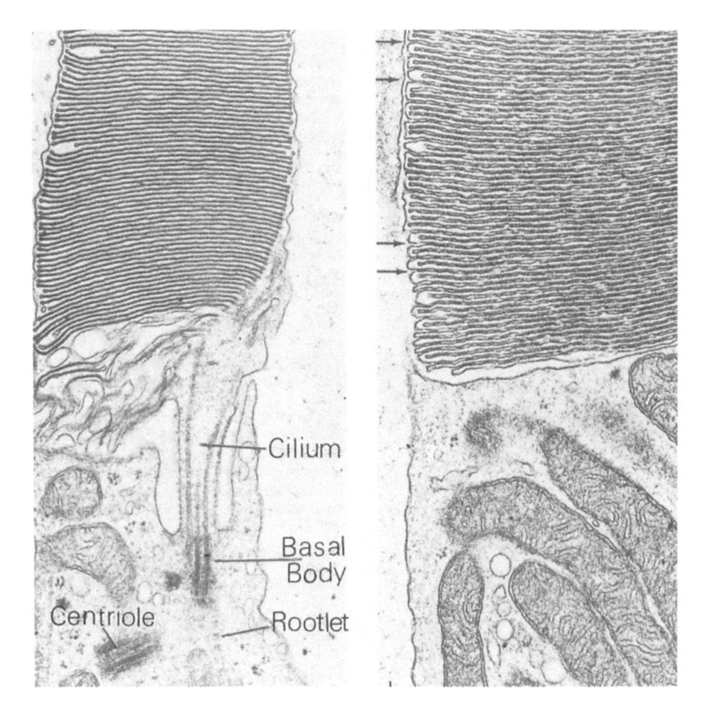

FIGURE P12f Transmission electron micrographs of portions of the outer and inner segments of a rod (left) and a cone (right). In the rod a modified cilium connects the two parts. Some of the discs in the cone are open to the extracellular space (arrows).

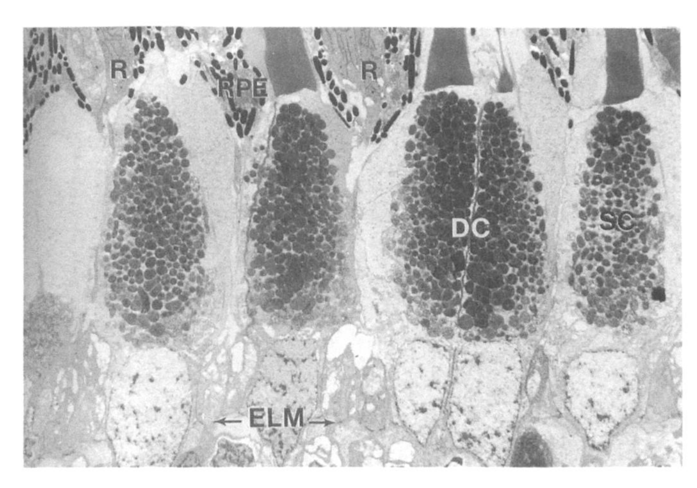

FIGURE P12g Electron micrograph of the photoreceptors in light-adapted rods of tilapia (*Oreochromis nilotica*). A double cone (*DC*) and three single cones *(SC)* are shown. *RPE*, retinal pigment epithelium; *ELM*, external outer limiting membrane; *R*, rod. 4000 ×.

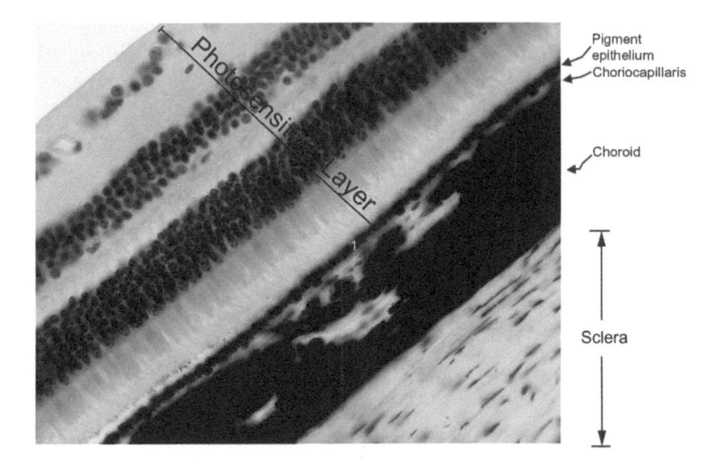

FIGURE P13a Section through the wall of an eye of the monkey showing the choriocapillaris, the main supplier of nutrients to the retina in vertebrates. Its relationship to the photosensitive layer is clear in this image. 40 ×.

retina develops as an outgrowth of the diencephalon and shows some characteristics of the central nervous system such as the complexity of synapse formation in its layers and the possession of neuroglial cells.

The retina is supplied with blood from two sources. Blood vessels from the CHORIOCAPILLARIS penetrate as far as the outer half of the outer plexiform layer (*retinal layer 5*) (Fig. P13a). Branches of the CENTRAL ARTERY OF THE RETINA, running immediately below the inner limiting membrane (*retinal layer 10*), form a capillary network in the inner nuclear layer (*retinal layer 6*). Venous blood is collected by postcapillary venules that join at right angles with veins in the nerve fiber layer (*retinal layer 9*) (Fig. P11a).

Cones are concentrated in the FOVEA CENTRALIS, the area of keenest vision (Fig. P11c). There are no blood vessels and fewer nerve fibers to impede vision here. The cell bodies of the cone cells are clustered around the periphery of the fovea in the MACULA LUTEA. A single cone cell of the fovea synapses with one bipolar cell.

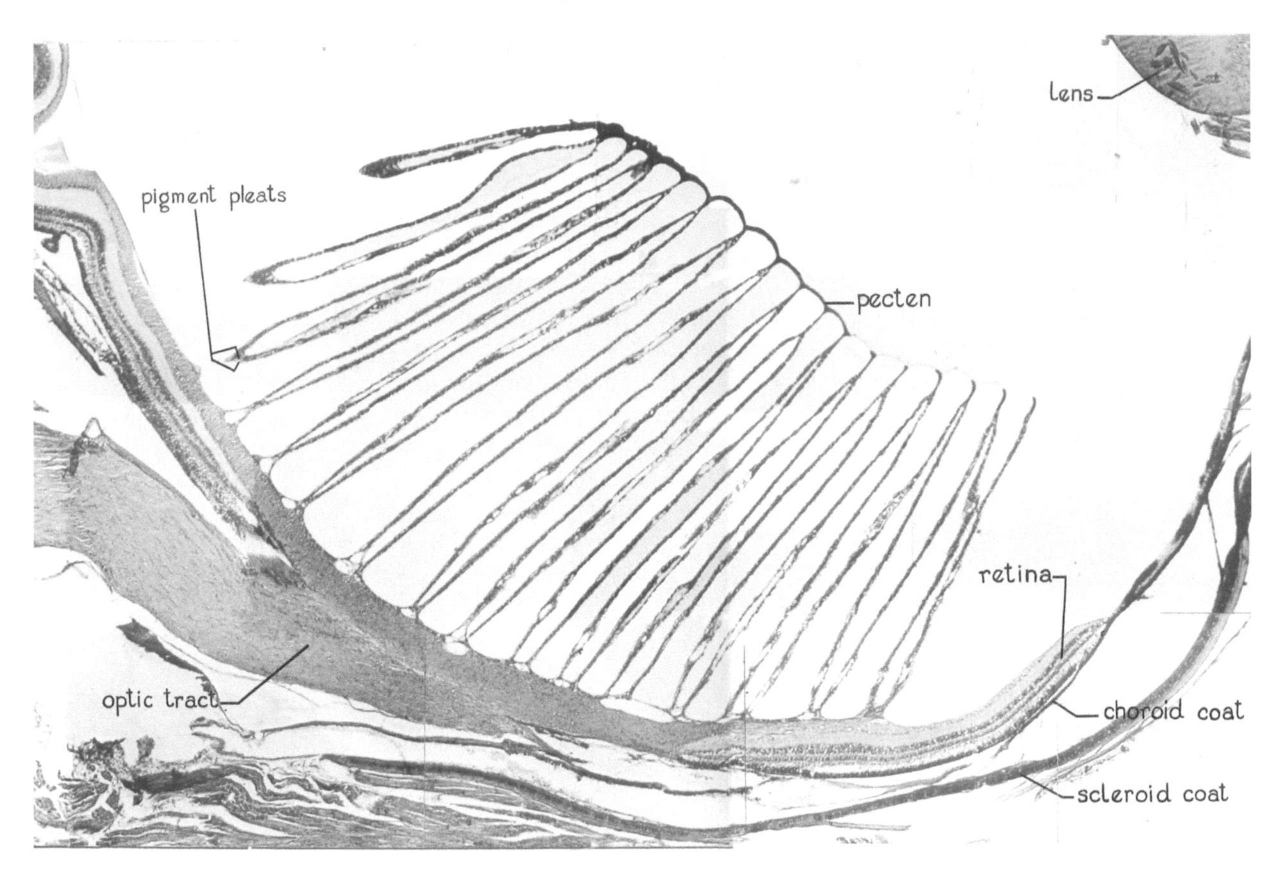

FIGURE P13b The retina of all vertebrates is nourished by the choriocapillaris. In most an ancillary or SUPPLEMENTAL NUTRITIVE DEVICE brings a second vascular system to supply the inner retina. This photomicrographic montage was pieced together by a patient graduate student from 45 photomicrographs. It demonstrates the extensive network of nutritive capillaries of the pecten in the eye of a cowbird. Source: *Assembled by K. Doreen Larsen—used with permission.*

In electron micrographs of rods and cones it is seen that the OUTER SEGMENTS are packed with a stack of flattened, saccular, membranous discs (Fig. P12e). In cones, but not in rods, the edges of the discs are sometimes continuous with the plasmalemma so that the cleft within a disc is open to the extracellular space (Fig. P12f). The outer and inner segments of rods and cones are connected by a constricted CILIARY STALK that contains nine longitudinally oriented fibrils terminating in a centriole in the inner segment; thus the outer segment is a specialized cilium that differs from other cilia in the lack of a central pair of microtubules and in the massive development of the surrounding cell membrane to form the flattened discs. The INNER SEGMENT consists of an outer ELLIPSOID filled with elongate mitochondria and an inner MYOID containing other organelles; the myoid is often rich in glycogen. The remainder of the sensory cell is made up of the INNER and OUTER FIBERS that bear between them an expansion containing the nucleus. A junctional complex may be seen at the point where the inner segment tapers into the outer fiber. This junction is formed between the sensory cell and a neuroglial cell of the outer limiting membrane. At the base of the sensory cell is an expanded PEDICEL that synapses with axons of a bipolar neuron.

There is a constant shedding of the outermost discs of the photoreceptor cells and the cast-off membranes are phagocytosed by the pigment epithelial cells. Lost proteins and their membranes are replaced from the inner segment by way of the ciliary stalk. New discs formed by infolding of the cell membrane displace the older discs outward.

Supplemental Retinal Circulation

In all vertebrates the outer retina (retinal pigment epithelium and photoreceptors) is nourished by the fenestrated capillaries of the CHORIOCAPILLARIS (Fig. P13a). In most vertebrates a supplementary vascular system supplies the inner retina. This can take various forms; in birds it is seen as the *pecten oculi* (Fig. P13b). The pecten projects from the head of the optic nerve into the vitreous chamber and usually takes the form of a folded fan. It is rich in elaborate capillaries festooned on both the inner and outer surfaces with elaborate microfolds. Pericytes are present. The adjacent connective tissue is rich in pigment cells.

Development of the Eye

The retina develops from two sources. An outgrowth of the forebrain forms a vesicle that invaginates to become the cup-shaped retina of two layers while the lens is derived from an adjacent thickening of the outer ectoderm that becomes the LENS VESICLE (Fig. P12a). The invaginated interior wall of the optic cup becomes the NERVOUS LAYER of the retina and the thinner, outer wall becomes the PIGMENTED LAYER of the retina. The optic stalk that connects the optic vesicle to the forebrain becomes trough shaped and encloses blood vessels, which supply the retina and eventually the optic nerve. The mesenchyme surrounding the optic vesicle gives rise to the sclera and choroid coats of the eye as well as the dense connective tissue of the cornea, the substantia propria. This "upside-down" development has the consequence of positioning the retina, so that light rays must pass through several layers before reaching the sensory neuroreceptors, the rods and cones.

The lens results from the unequal development of the anterior and posterior walls of the lens vesicle that has pinched off from the lens ectoderm. The anterior wall remains a thin layer of cells while the cells of the posterior wall elongate to form LENS FIBERS, which constitute most of the substance of the lens. The lumen of the lens vesicle is obliterated. Nuclei of the lens fibers around the equator of the lens continue to divide slowly throughout life. Nuclei and organelles of the central lens fibers fade away and do not distort vision.

Comparative Histology of the Eye

Chordate eyes appear for the first time in cyclostomes and show a great deal of variation, even in members of one class. The account that follows is necessarily general and includes only the most obvious trends. Progress is not a steady upward climb from the cyclostomes to the birds or mammals but rather a series of ups and downs with both well- and poorly-developed eyes to be found in all classes. In general, however, the higher the vertebrate, the more advanced is its eye.

Cartilage occurs in the sclera of all vertebrates except cyclostomes, some teleosts, some urodele amphibians, snakes, and mammals other than monotremes. In reptiles and birds there are bony plates, the SCLERAL OSSICLES, embedded in the sclera. These can sometimes be seen in preparations of the skull.

The cornea of terrestrial vertebrates is the major refractive structure in the eye. In aquatic forms, however, where the refractive index of the water is close to that of the cornea, the lens assumes most of the refractive functions. The spherical lens of the *lamprey* eye is fixed and the pupil is a fixed size. A limited amount of focusing is achieved by flattening the cornea by a special corneal muscle. There is no ciliary apparatus. A later development is a movable lens that can be drawn back and forth in focusing. *Elasmobranchs* use this method and some also have a few muscle fibers in the iris. A ciliary body has appeared that holds the lens in place. Ciliary and iris muscles are lacking in *teleosts* and the lens changes position in focusing. Some *amphibia* retain the movable lens but also have muscles in the ciliary body that change the curvature of the lens slightly. *Reptiles* and *birds* have more advanced ciliary muscles that alter the shape of both lens and cornea by a peripheral squeezing action. In both classes, ciliary and iris muscles are striated. There are no eyelids in *snakes* and the cornea is covered by a transparent window of skin, the SPECTACLE.

Most vertebrates have both rods and cones but the relative differences in their numbers may vary greatly. Most strictly diurnal animals have only cones or else cones greatly outnumber rods; nocturnal animals tend to stress the number of rods. Rods are well developed in animals that live in dark surroundings; cones are almost lacking in deepsea *fishes* and rods are long and numerous.

Twin cones occur in some fishes, especially those exposed to bright light, and appear as cones occurring close together (Fig. P12g). *Amphibia* have developed two kinds of rods, distinguished as red and green rods by the visual pigment in them. They also have single and twin cones. More cones are found in *reptiles* than in lower groups and a pure cone retina seems to be the primitive type in this class. Some nocturnal reptiles appear to have regressed to a rod-rich or rod-pure type but these "rods" are likely transformed cones. Cones are well developed in diurnal *birds* and rods in nocturnal birds. Single and twin cones are also found in birds.

An area of more concentrated cones, the AREA CENTRALIS, occurs in the optical axis of many vertebrates from the elasmobranchs on. It is similar to the macula lutea except that it is not yellow. A fovea centralis is found in some *teleosts* and in all higher classes. Many *birds* with lateral eyes have two fovea: one, in a posterior position on the retina, serves when the bird is flying and looking ahead and one, near the center of the retina, is used for keenest vision at other times.

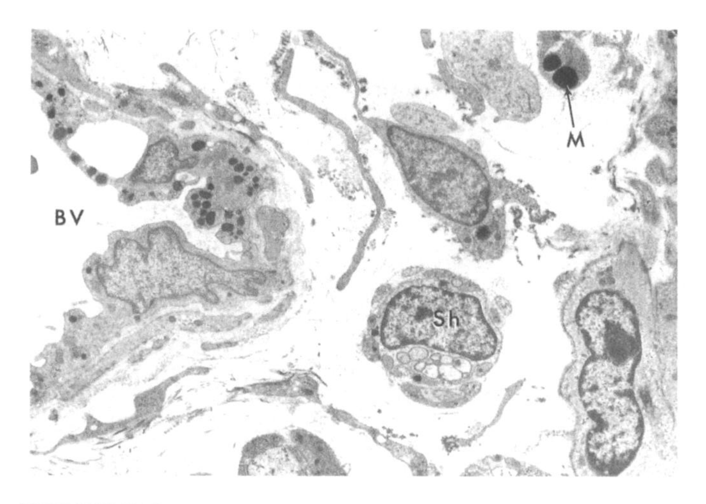

FIGURE P13c Electron micrograph of the conus papillaris of the goanna lizard. Note the loose arrangement of connective tissue elements and the ground substance, A blood vessel (*BV*), a melanocyte (*M*), and a neurilemma cell, (*Sh*) are indicated. 8,600 ×.

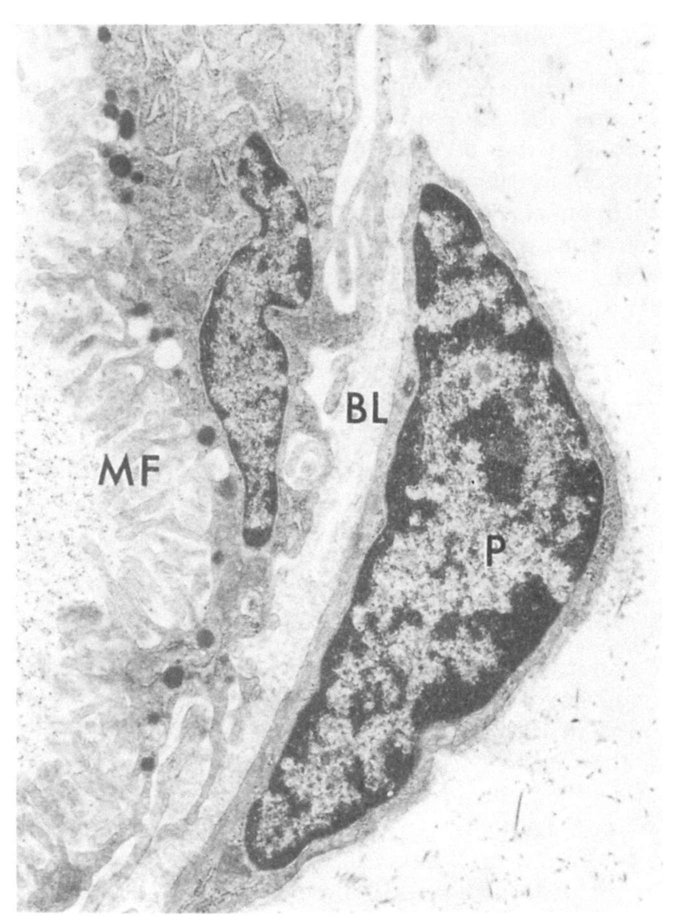

FIGURE P13d Electron micrograph of a capillary with abundant luminal microfolds (*MF*) in the conus papillaris of the goanna but few abluminal microfolds. A pericyte (*P*) and the basal lamina (*BL*) are indicated 12,900 ×.

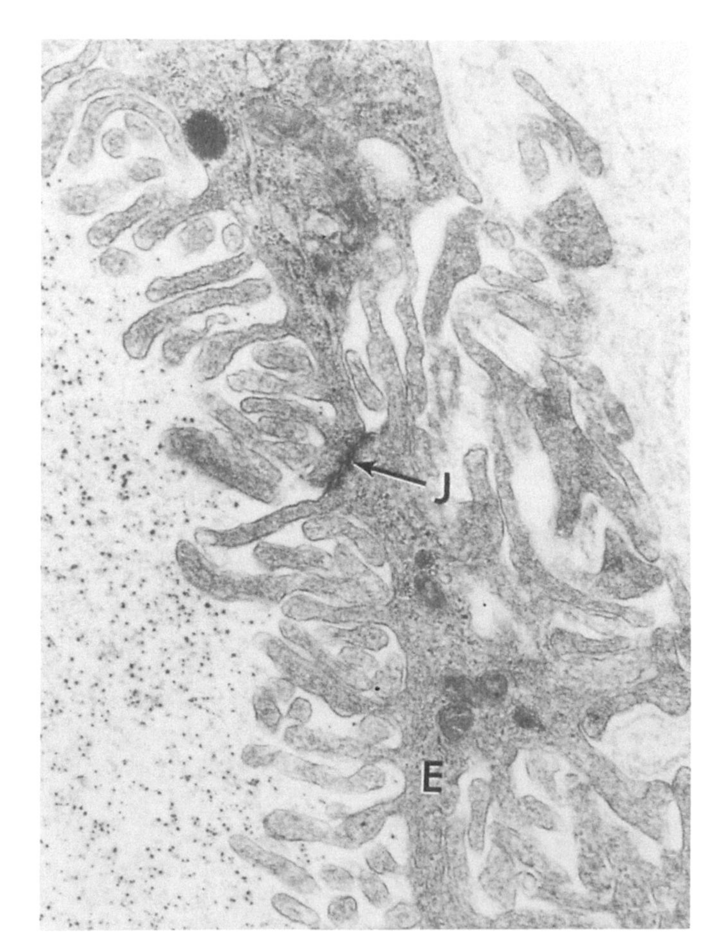

FIGURE P13e Electron micrograph of a capillary with copious luminal and abluminal processes. A cell junction (*J*) is indicated. The endothelial cell body (*E*) is thin. 25,000 ×.

Supplemental Nutritive Devices

Extending from the blind spot in many *birds* and *reptiles* is a projection of richly vascular neuroglial tissue. In reptiles the *conus papillaris* (the PAPILLA) is a highly vascularized finger-like projection jutting into the vitreous chamber of the eye. Within the conus papillaris are numerous capillaries and larger blood vessels. Melanocytes and the occasional mast cell have been described. In transmission electron micrographs capillaries are arrayed within loose connective tissue elements (Fig. P13c) and display extensive and prominent luminal and abluminal microfolds (Figs. P13d and P13e). The conus papillaris is thought to nourish the retina by diffusion of nutrients through the vitreous humor. In birds a more elaborate structure, homologous to the papilla, is the fan-shaped PECTEN (Fig. P13b). It is especially well developed in diurnal birds. Nutrients diffusing from the pecten probably nourish the avascular retina. Transmission electron micrographs of sections of capillaries of the pecten

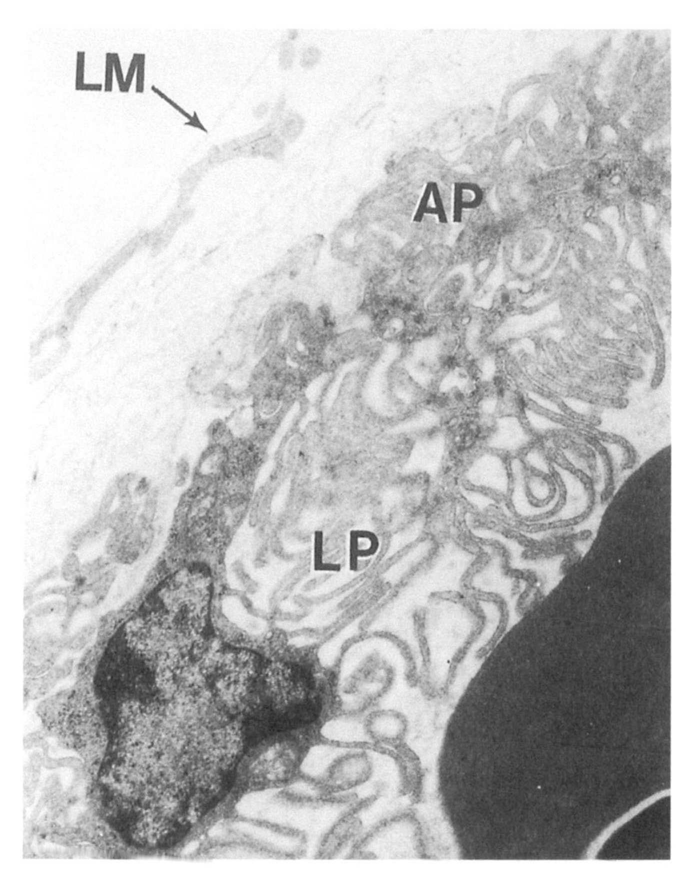

FIGURE P13f Electron micrograph of a capillary (a red blood cell occupies the lower right corner) from the pigment epithelium sporting a forest of luminal (*LP*) and abluminal processes (*AP*). The endothelial cell body is thin. The limiting membrane (*LM*) of the pecten is indicated 25,000 ×.

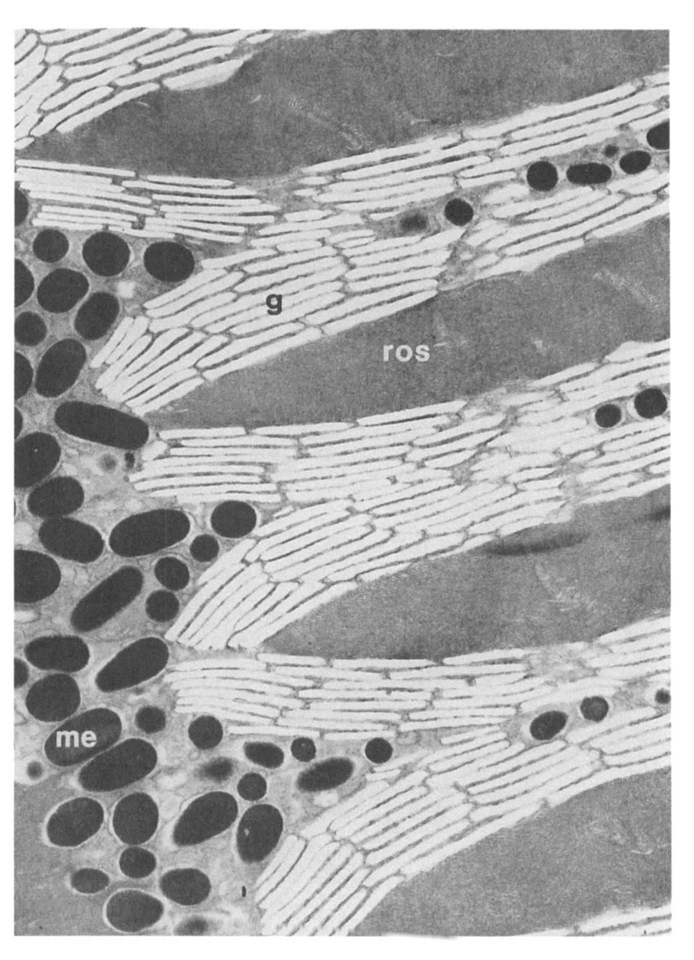

FIGURE P14a Electron micrograph of a tapetum lucidum in the eye of a young anchovy (*Anchoa mitchilli*). Note the pigment epithelial cell processes, extending between the outer segments of the rods. These are filled with plate-like guanine-rich crystals. In the outer segment the discs run vertically. *g*, guanine platelet; me, melanosome; ros, rod outer segment. 18,600 ×.

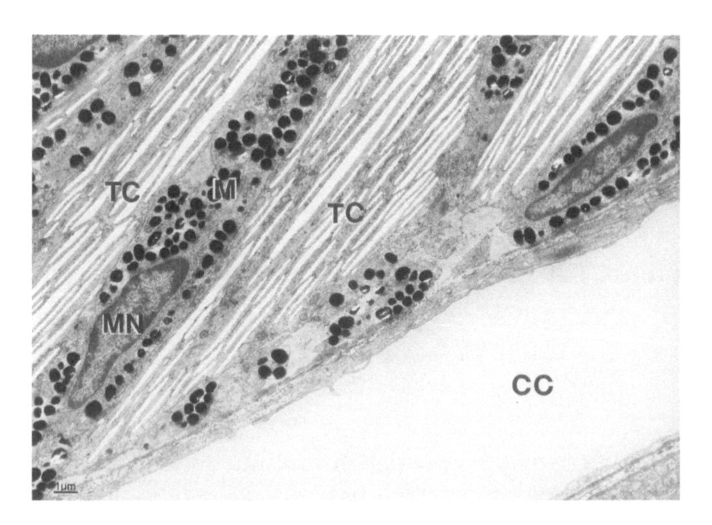

FIGURE P14b Electron micrograph of the retinal side of the tapetum in the eye of a Port Jackson Shark. Tapetal cells (*TC*) alternate with melanocytes (*M*). Note the plate-like structures in the tapetal cells—these are reported to be guanine-rich crystals. *CC*, choriocapillaris; *MN*, melanocyte nucleus are indicated. 6000 ×.

show a thick forest of microfolds on both the apical and basal surfaces of the endothelial cells (Fig. P13f).

Responses to Variability of Ambient Light

Eyes may be adapted for vision in variable intensities of light by the inclusion of reflective materials in the choroid; this constitutes the TAPETUM LUCIDUM. Reflecting light back through the retina with this structure provides the photoreceptors with a second opportunity to be stimulated. The reflectivity may be due to densely packed collagenous fibers (fibrous tapetum), stacks of plate-like cells containing refractile rods in their cytoplasm (cellular tapetum), or to the inclusion of crystals of guanine (or its calcium salt) within the cells of the choroid (guanine-containing tapetum) (Figs. P14a and P14b). In some vertebrates the pigment epithelium may contain reflective elements and be specialized as a RETINAL TAPETUM LUCIDUM; again crystals of

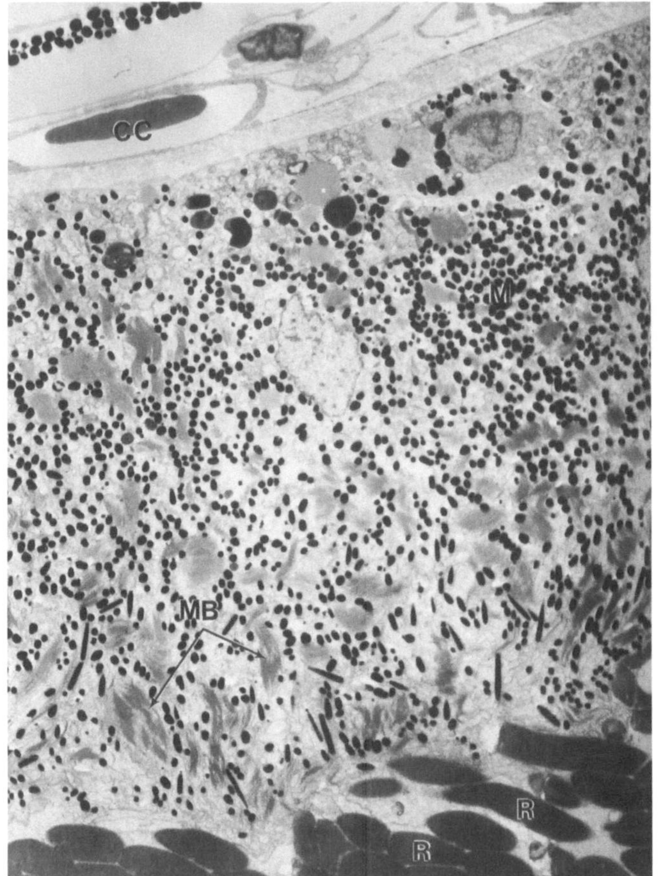

FIGURE P15 Electron micrograph of light-adapted retinal epithelium of a cichlid, *Oreochromis niloticus*. Note the widely scattered melanosomes (*M*) enclosing both rod (*R*) and cone (*C*) photoreceptors. *CC*, choriocapillaris. 3800 ×.

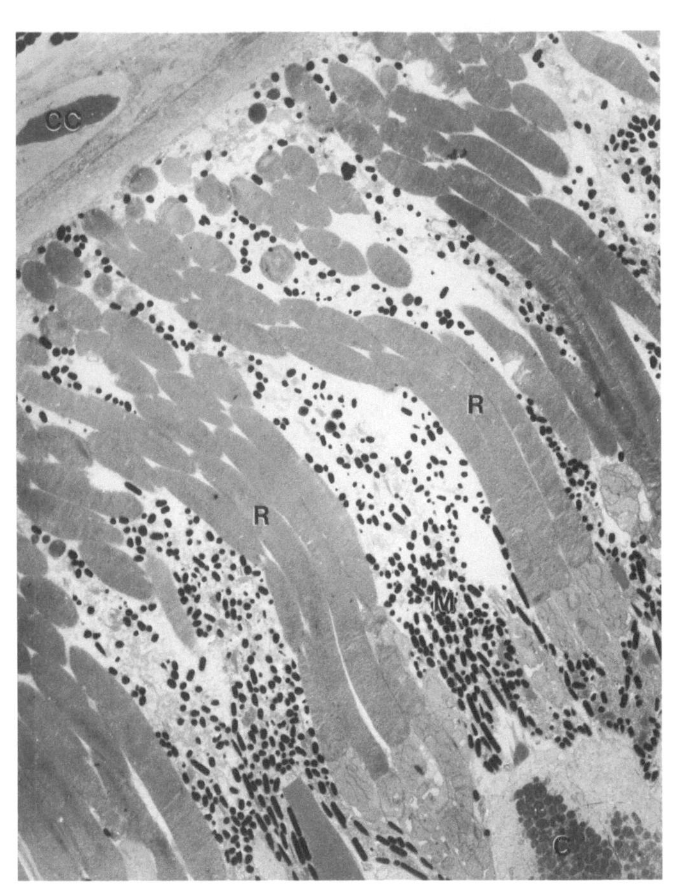

FIGURE P16 Electron micrograph of dark-adapted retinal epithelium of the cyclid, *Oreochromis niloticus*. Melanosomes (*M*) are confined to the cell body and the rod photoreceptors (*R*) have retracted. Note abundant myeloid bodies (*MB*). *CC*, choriocapillaris. 3800 ×.

guanine or a calcium salt of guanine may be the active reflective agent. Alternatively, vertebrate eyes may respond to intense light by means other than closure of an iris. Processes from the pigment epithelium insinuate their way between the photoreceptors, their melanin granules ebbing and flowing as conditions require (Figs. P15 and P16). In periods of bright light, melanin granules roll out reducing the amount of light received by the photoreceptors.

Accessory Structures of the Eye

Terrestrial vertebrates have developed structures that protect the surface of the eye and keep it moist and clean. These are typically eyelids and associated glands.

Eyelids

Eyelids are folds of skin lined on the inside with the PALPEBRAL CONJUNCTIVA, which is continuous with the bulbar conjunctiva covering the sclera of the eyeball (Fig. P17a). The loose connective tissue of the eyelid contains lobules of fat, glandular tissue, and the striated muscles that move the lids (Fig. P17b). A reinforcing band of dense connective tissue in each eyelid is called the TARSAL PLATE. Just below the conjunctiva are the TARSAL GLANDS (Meibomian glands), modified sebaceous glands that pour their oily secretions along the inner margin of the lids. In human, CILIARY GLANDS (glands of Moll), which are modified straight sweat glands, and small SEBACEOUS GLANDS (of Zeis), open into the follicles of the eyelashes (Fig. P17c).

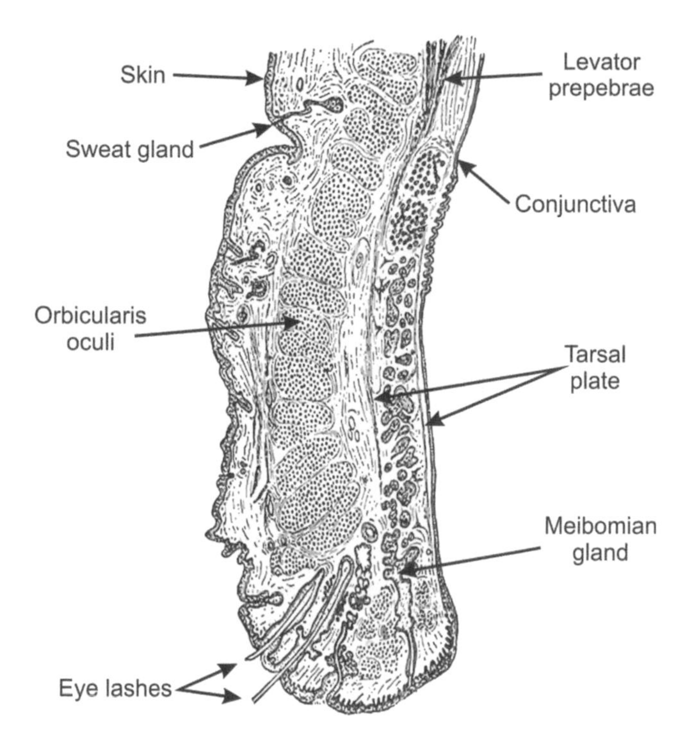

FIGURE P17a Drawing of a section of an upper human eyelid. The surface covering the eye is at the right. The most prominent feature is the muscle that opens the eye. Glands that moisten the surface of the eye are shown. *After H. Gray 1918. Anatomy of the Human Body Lea and Febiger New York & Philadelphia. Public Domain.*

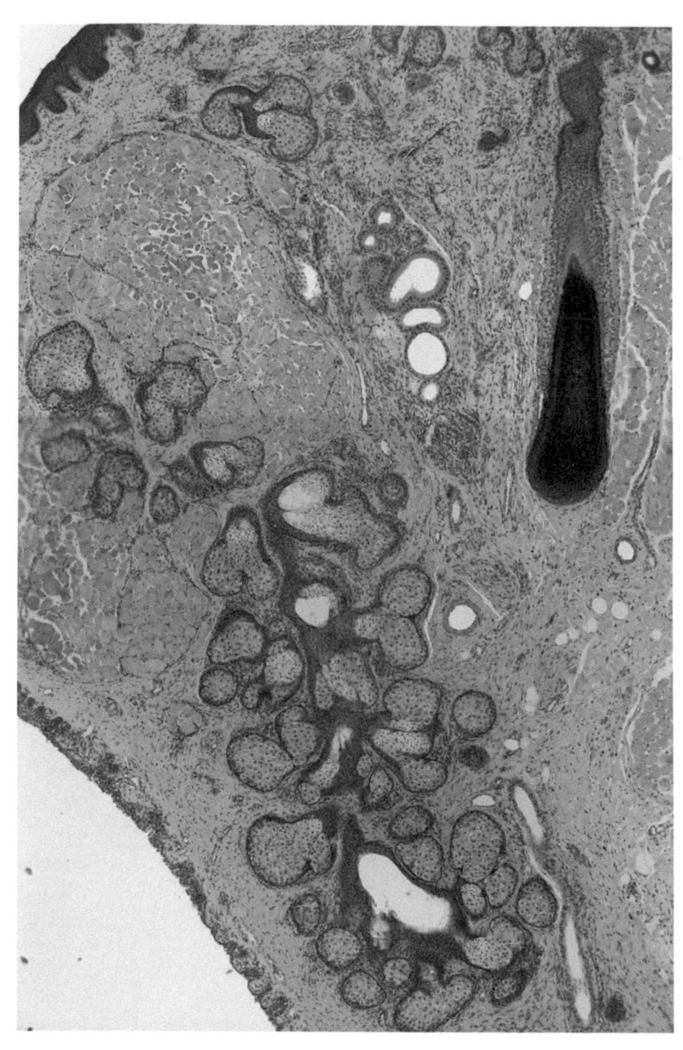

FIGURE P17b Striated muscle and glands fill most of the space in this photomicrographic montage of a section of mammalian eyelid. An eyelash hair follicle and its associated glands are seen in the upper right. 5 ×.

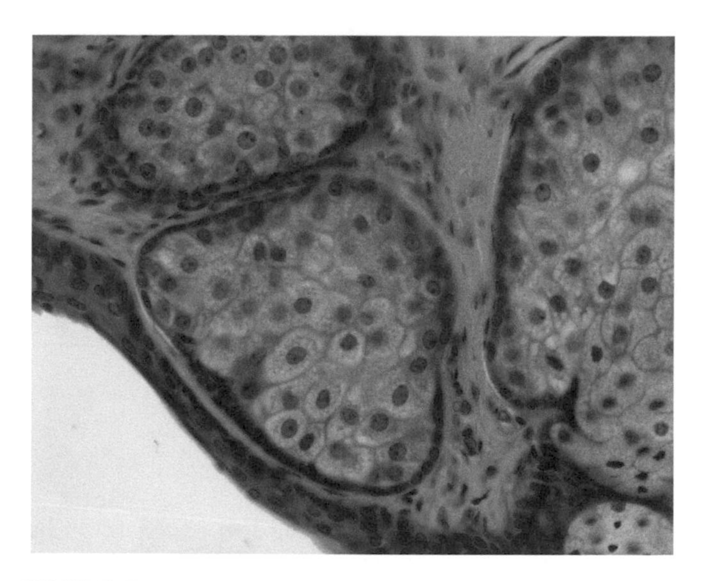

FIGURE P17c Photomicrograph showing glandular tissue in a section of a mammalian eyelid. 40 ×.

In many vertebrates a third eyelid, the NICTITATING MEMBRANE, arises as a vertical fold of the conjunctiva and lies between the other two lids, passing from the inner angle over the surface of the eye. It is usually transparent and consists of a layer of fibrous connective tissue sandwiched between two layers of epithelium. The nictitating membrane is reduced in most mammals; the semilunar fold of mammals may be its homologue. The "nictitating membrane" of amphibians is the transparent upper border of the lower eyelid; when the eye is open, it is folded inside the remaining portion of the lower lid. It is probably not homologous with the nictitating membrane of other vertebrates.

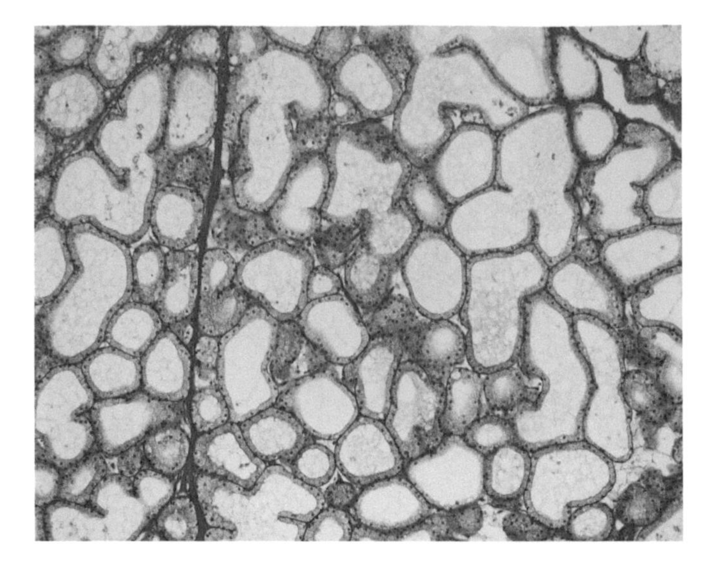

FIGURE P17d Photomicrograph of a section through the lachrimal gland of a rabbit. 10 ×.

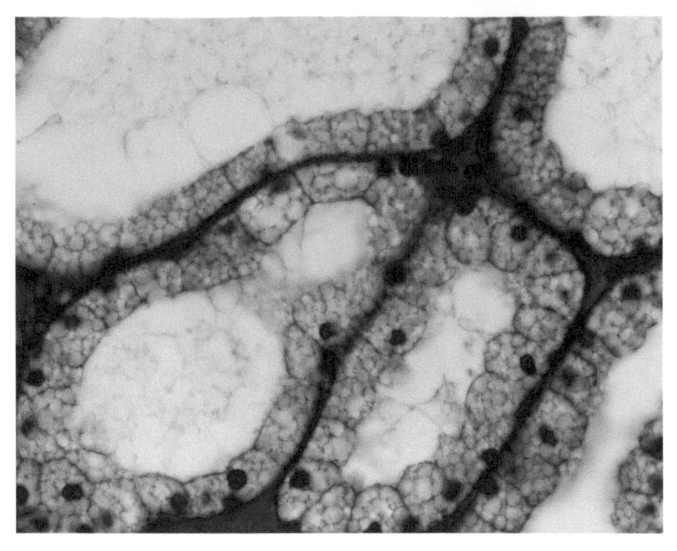

FIGURE P17e At a higher magnification the section through the lachrimal gland above reveals fat droplets in the pyramidal secretory cells of the gland. 63 ×.

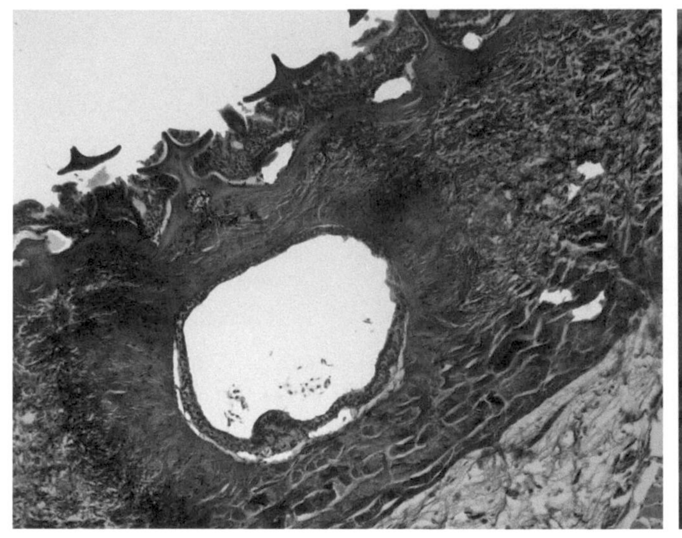

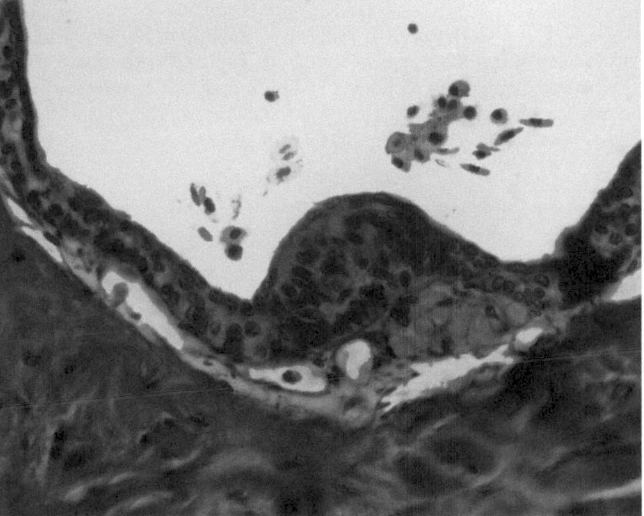

FIGURES P18a and P18b Photomicrographs of neuromasts, consisting of clusters of sensory cells, embedded in the wall of the lateral line canal of a dogfish. (Note the elasmobranch scales at the surface.) 10 ×.

Harderian and Lachrimal Glands

With the assumption of a terrestrial life, two large glandular masses developed in the eye socket of tetrapods, the Harderian gland on the inner angle and lachrimal gland on the outer. They are small in amphibians but well developed in reptiles, birds, and mammals. They are usually lacking or rudimentary in water-living forms, even in the higher classes. In most mammals the Harderian gland degenerates.

The HARDERIAN GLAND is a modified sebaceous gland; it secretes an oily fluid that lubricates the nictitating membrane. The LACHRIMAL GLAND is a compound tubuloacinar serous gland that develops from the conjunctiva (Figs. P17d and P17e). Its pyramidal secretory cells contain fat droplets and secretion granules. The secretory units are surrounded by myoepithelial cells that lie inside the basement membrane; visible in scanning electron micrographs. Recall that the SALT GLANDS of marine turtles are modified lachrimal glands.

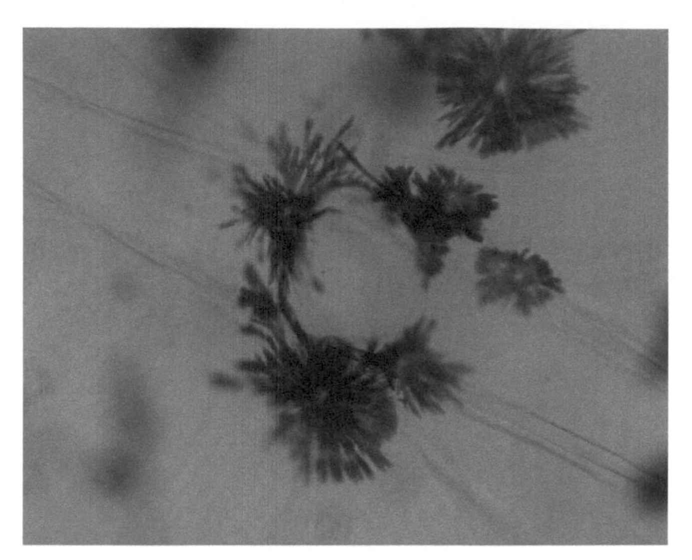

FIGURE P19a Photomicrograph of a whole mount of the skin of a yellow perch (*Perca flavescens*) showing the lateral line canal crossing the field of view. (Note the elaborate chromatophores in the skin.) 20 ×.

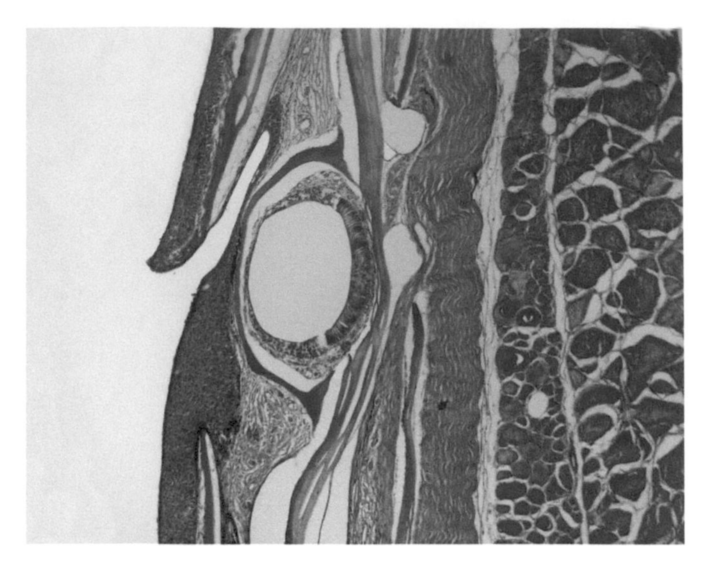

FIGURES P19b Photomicrographs of the skin of a perch showing a cross section of the lateral line canal with the neuromast on the right side of the canal. (Note the bony scales in scale pockets—a characteristic of teleosts.) 10 ×.

LATERAL LINE ORGANS

Unknown in terrestrial vertebrates, lateral line organs are found in fishes and aquatic amphibians. They are thought to function in detecting waves and currents in the water. They are lost in the newt when the animal takes to the land but reappear when it returns to the water during breeding. They do not appear in amniotes, however, even those which are wholly aquatic.

Clusters of sensory cells, the NEUROMASTS, (Figs. P18a and P18b) may occur in isolated patches over the skin or located in a series of grooves or canals on the head and body. These are connected to afferent branches of cranial nerves (VII, IX, or X, depending upon the region of the body). In most forms a closed canal runs the length of the body (Fig. P19a) just below the skin on each side and continues forward to the head where it forms a complex pattern. The canals usually run parallel to the cranial nerves and may be sunk beneath bony plates or scales. They open to the surface at intervals through pores. Grooves are formed instead of canals in some cartilaginous and bony fishes and in cyclostomes neuromasts are located in the skin within isolated pits that show a roughly linear arrangement. Lateral line organs are formed from ectodermal thickenings that sink below the surface of the skin.

In a section through the lateral line showing a NEUROMAST these bundles of SENSORY CELLS are similar in appearance to taste buds (Figs. P19b and P19c). The elongated cells bear sensory hairs whose tips are enclosed in a gelatinous CUPULA secreted by the neuromast cells. The sensory cells are supported by SUSTENTACULAR CELLS and, in some cases, BASAL CELLS. The neuromast is separated from the adjacent epithelial cells by a thin layer of MANTLE CELLS. The afferent nerve from the neuromast is probably a branch of the vagus nerve. The canal is filled with mucus. Currents in the water move the cupula and bend the sensory hairs.

The sensory cells of the lateral line organ are provided with motile CILIA (kinetocilia), STEREOCILIA, and MICROVILLI on their apical surfaces. The tall motile cilium is characteristically placed at one side of a group of stereocilia that often decrease in height in a "steps-and-stairs" fashion.

EAR

The ear of higher vertebrates functions in equilibrium and hearing (Fig. P20a). Its organ of equilibrium is the more fundamental and is found with little variation in all vertebrates. Auditory portions begin to differentiate in the fishes and become more and more complex as the evolutionary scale is ascended. The OUTER and MIDDLE EARS transmit sound and have no function in equilibration. They are totally lacking in cyclostomes and fishes; the

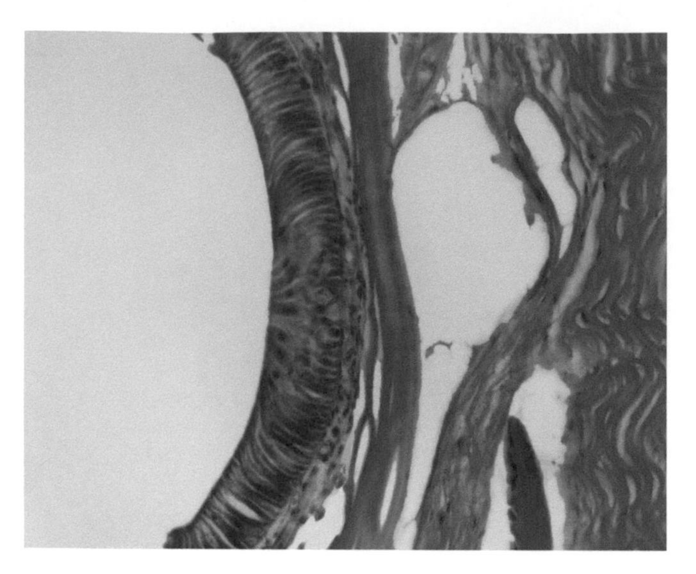

FIGURE P19c Photomicrograph of a neuromast of a perch clearly showing banana-shaped sensory cells. 40 ×.

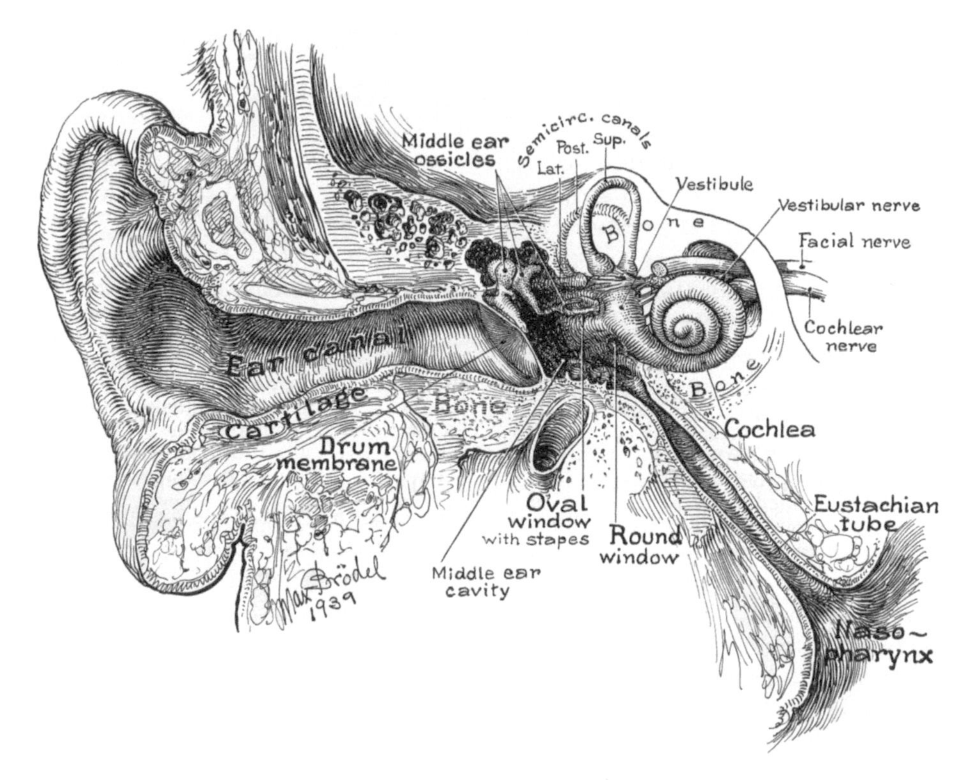

FIGURE P20a Drawing of a dissection of the human ear.

middle ear appears first in amphibians and the outer ear is found only in amniotes. The inner ear is present in all vertebrates and is the location of the sensory receptors (Fig. P20b).

The inner ear arises in the head as a thickening of the superficial ectoderm that forms a pit and later closes off to form a vesicle lying in the mesenchyme near the brain. It differentiates into a complex series of tubes and sacs lined with simple squamous ectodermal epithelium and supported by fibrous connective tissue; this is the MEMBRANOUS LABYRINTH. It consists of two chambers, the upper UTRICULUS and the lower SACCULUS, and the SEMICIRCULAR DUCTS, which communicate with the utriculus. Each semicircular duct bears an enlargement, the AMPULLA, containing sensory cells. With the exception of hagfishes (which have only one) and lampreys (which have two), all living vertebrates have three semicircular ducts. A projection of the ventral wall of the sacculus

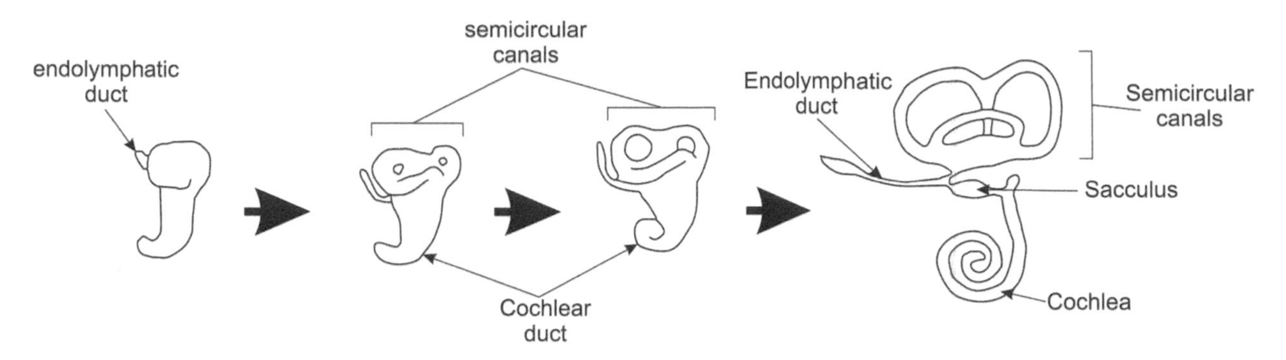

FIGURE P20b Embryological development of the labyrinth and inner ear in a mammal. *Authors.*

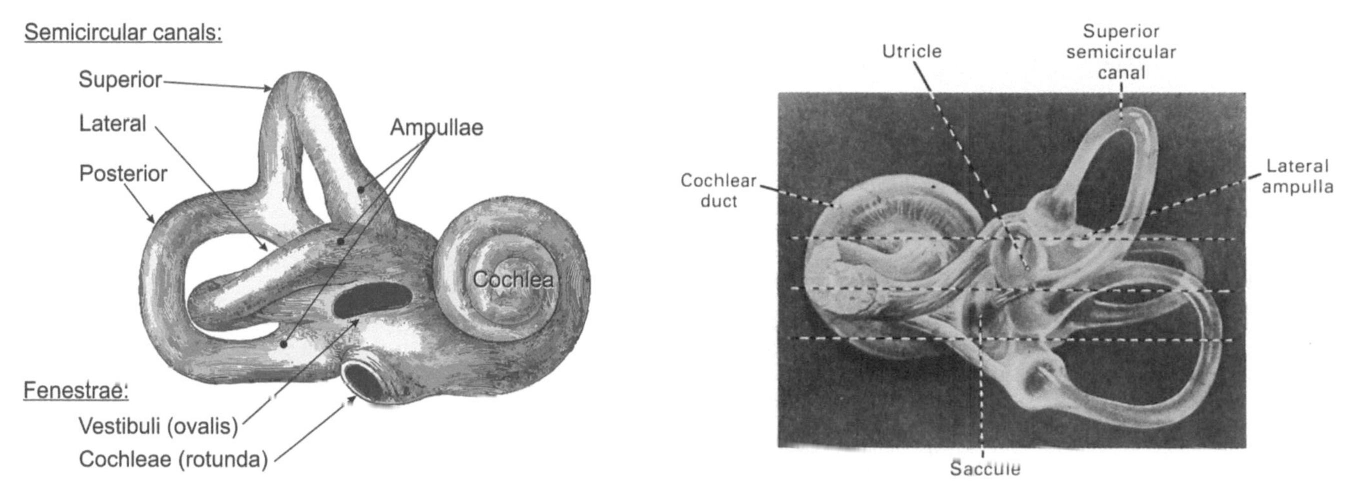

FIGURE P20c Drawing of the bony labyrinth of a mammal. The Fenestrae (oval and round windows) open into the vestibule; the cochlea extends from the vestibule at right and semicircular canals at left. *Authors.*

FIGURE P20d Drawing of the right membranous labyrinth showing relationship between the cochlear duct, utricle, saccule, semicircular canals, and ampullae.

may be present in the lower vertebrates; this is the LAGENA and is a forerunner of the auditory portion of the ear in higher forms. The cavity of the membranous labyrinth is continuous throughout and contains the fluid ENDOLYMPH, which is two or three times as viscous as water.

The membranous labyrinth is suspended by thin strands of connective tissue within form-fitting channels of the bone or cartilage (depending upon the class). These channels constitute the SKELETAL LABYRINTH; they are filled with PERILYMPH and are lined by endosteum or endochondrium. The cavity of the skeletal labyrinth connects to the subarachnoid space of the brain by the perilymphatic duct. So the perilymph surrounding the membranous labyrinth is actually cerebrospinal fluid. In the higher forms the skeletal labyrinth is formed within the petrous portion of the mastoid bone, is lined with endosteum, and constitutes the BONY LABYRINTH (Fig. P20c).

Fig. P20d shows the membranous labyrinth dissected from the cartilaginous skull of an elasmobranch and identifies the sacculus, utriculus, and the semicircular ducts with their ampullae. The membranous labyrinth is lined with simple squamous epithelium and is supported by fibers of connective tissue that cross the perilymphatic space and continue into the fibrous connective tissue of the bony labyrinth.

In the walls of the membranous labyrinth are several patches of sensory cells (Fig. P20e). These are CRISTAE in the ampullae of the semicircular canals and MACULAE in the sacculus, utriculus, and sometimes the lagena. Sensory cells of these areas are comparable to the neuromasts of the lateral line system and their ultrastructure is similar. SENSORY CELLS of the CRISTAE are columnar with long hair-like processes (motile cilia, stereocilia, and microvilli) and are interspersed among SUSTENTACULAR CELLS. The tips of the "hairs" are embedded in a gelatinous mass, the CUPULA, secreted by the sustentacular cells. The cupula may not exist as such in life but may be an artifact resulting from shrinkage of the gelatinous material on fixation. MACULAE are similar but their sensory cells are short and bristly. They too have

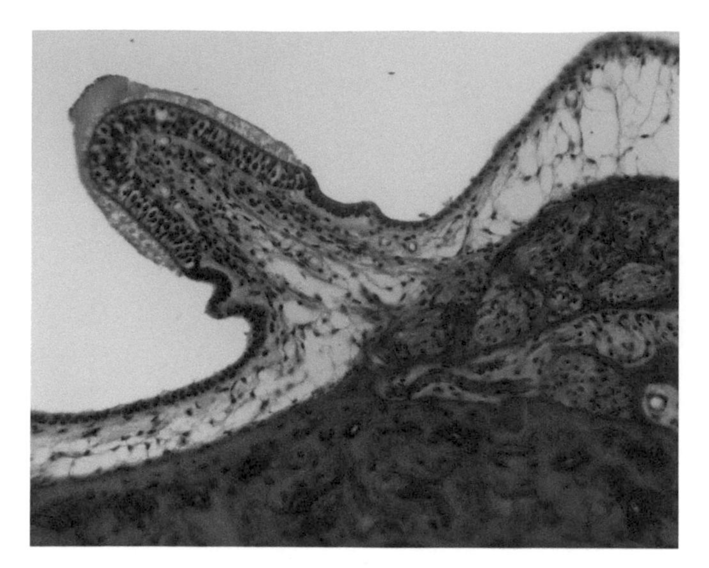

FIGURE P20e Crista ampullaris in a semicircular canal of a mammal.

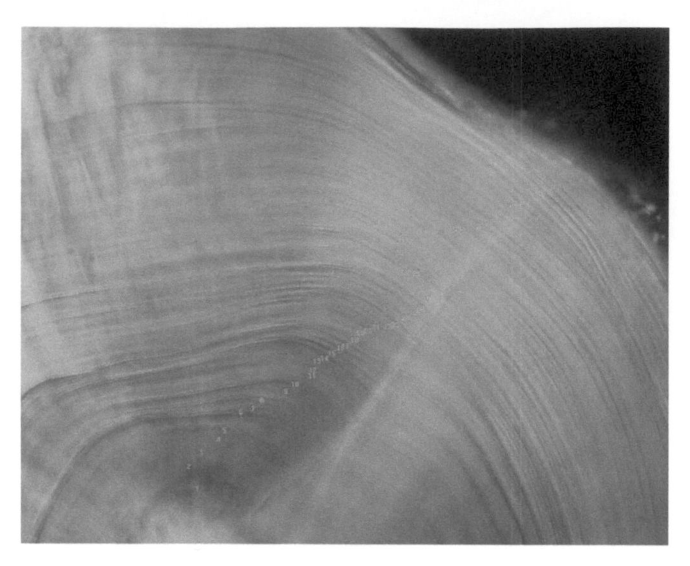

FIGURE P20f Photomicrograph of an otolith from a red horse sucker. The rings seen in otoliths are similar to growth rings in wood, but the correlation of otolith rings to age is not 1:1. This specimen has had its rings numbered for an aging research study. Differential Interference Contrast image. 63 ×.

their sensory hairs embedded in a thickened gelatinous mass, the OTOLITHIC MEMBRANE, enclosing the tufts of these hairs. Small crystals of calcium carbonate and protein, the OTOCONIA, are embedded at the periphery of the membrane and form a mass, the OTOLITH (Fig. P20f). Movements of endolymph in the semicircular canals are believed to stimulate hair cells of the cristae making the animal aware of movement, especially rotation. Changes in position of the otoliths in response to gravity or changes in velocity stimulate sensory cells of the maculae. Afferent branches of the auditory nerve (VIII) connect sensory cells of the cristae and maculae to the brain.

Note the similarities of these structures with the neuromasts: sensory hairs, gelatinous covering of the hairs, and the nervous connections. Their embryological origins as ectodermal thickenings that sink below the skin to form sacs or tubes are similar; the inner ear may be a specialized deeply sunk remnant of the anterior portions of the lateral line system.

In amniotes the lagena elongates to form the auditory organ of the ear, the COCHLEAR DUCT filled with endolymph. Its sensory SPIRAL ORGAN (organ of Corti) runs along its length with a series of SENSORY HAIR CELLS and SUSTENTACULAR CELLS over which forms a gelatinous flap, the TECTORIAL MEMBRANE, corresponding to the cupula of the other neuromast-like structures. As the cochlear duct lengthens it pushes out a U-shaped loop of bony labyrinth, the PERILYMPHATIC DUCT (Fig. P20g). The tips of the U stand on the OVAL and ROUND WINDOWS and the arms form the SCALA VESTIBULI and SCALA TYMPANI, respectively. The cochlear duct lies between the arms. As described earlier, the membranous labyrinth, in this case the cochlear duct, is suspended within the bony labyrinth, here consisting of the scala vestibuli and the scala tympani (Figs. P20h and P20i). In mammals the cochlea forms a spiral but it is straight in reptiles and birds.

As seen in Figs. P20h and P20i the canal of the cochlea makes a few spiral turns around a conical pillar of bone, the MODIOLUS. The canal consists of the membranous labyrinth and the bony labyrinth, which are separated from each other by the perilymphatic space. Blood vessels and branches of the acoustic nerve penetrate through numerous openings in the bony substance of the modiolus. Nerve fibers pass into the SPIRAL GANGLION that extends along the inner wall of the cochlear canal.

The lumen of the cochlear canal is divided along its whole course into an upper and lower section by a shelf, the SPIRAL LAMINA. The inner zone of the spiral lamina contains bone and is called the osseous spiral lamina; the fibrous outer zone is the BASILAR MEMBRANE or membranous spiral lamina. At the point of attachment of the basilar membrane to the outer wall of the cochlea, the periosteum is thickened to form the SPIRAL LIGAMENT. A membrane, the VESTIBULAR MEMBRANE, extends obliquely from the spiral lamina to the outer wall of the osseous cochlea. Thus a cross section of the cochlear canal will show three cavities: the upper SCALA VESTIBULI, the lower SCALA TYMPANI, and the SCALA MEDIA or COCHLEAR DUCT (Fig. P20h).

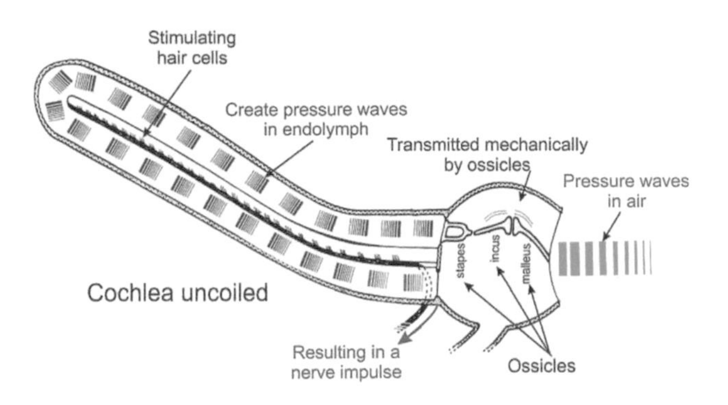

FIGURE P20g Diagram of the mammalian ear showing sound transmission from the eardrum to the auditory nerve. *Authors.*

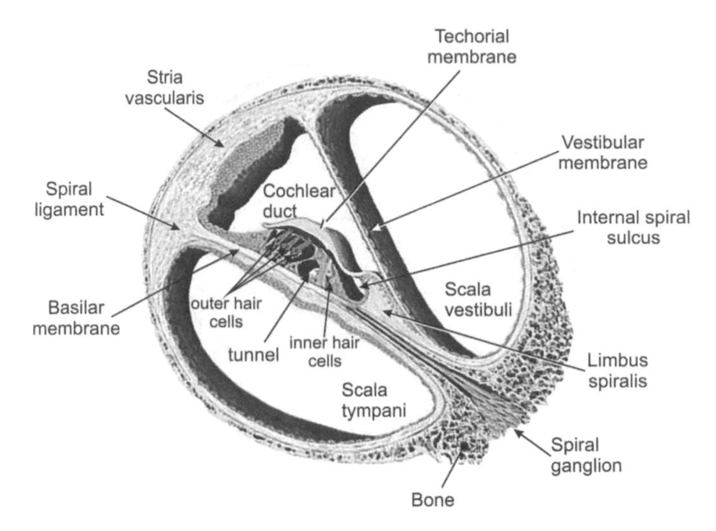

FIGURE P20h A diagram of a section through a turn of the bony cochlea. The anatomical relationships of various components of the inner ear are shown. *Authors.*

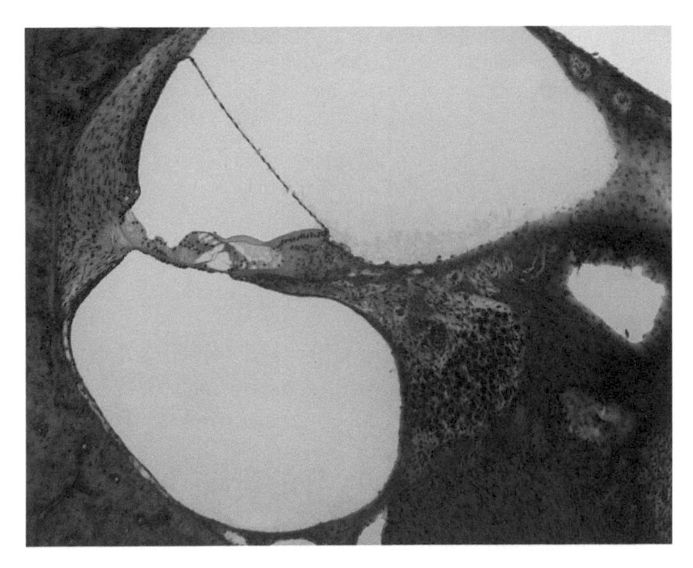

FIGURE P20i Photomicrograph of a section through the bony cochlea. This section is similar to the drawing in Fig. P19h.

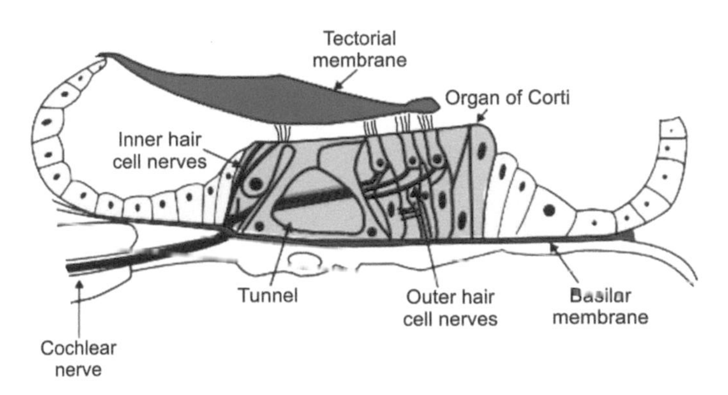

FIGURE P20j This drawing of the organ of Corti of the inner ear shows the relationship of its components. *Authors.*

In the inner corner of the cochlear duct, endosteum of the upper surface of the spiral lamina bulges into the duct as the LIMBUS SPIRALIS. Its edge overhangs a space, the INTERNAL SPIRAL SULCUS with its two margins, the VESTIBULAR and TYMPANIC LIPS. The tympanic lip continues outward into the basilar membrane, which is tightly stretched between it and the spiral ligament. Nerve bundles pass through the tympanic lip to the spiral organ that rests on the basilar membrane and bulges into the cochlear duct. The spiral organ consists of SUSTENTACULAR CELLS and neuroepithelial receptors, the HAIR CELLS. Note that the sensory cells are divided into a single row of INNER HAIR CELLS and three rows of OUTER HAIR CELLS by a deep trench, the INNER TUNNEL (tunnel of Corti) (Fig. P20j). Scanning electron micrographs of the surface of the hair cells clearly show the arrangement of sensory hairs of the inner row are arranged in the form of flattened U's while those of the outer row take the form of V's or W's (Figs. P21a and P21b). The elaborate arrangement of the sustentacular cells also shows to good advantage in scanning electron micrographs (Fig. P21c). Hair cells resemble the sensory cells seen in neuromasts, cristae, and maculae except that, in mammals, the motile cilia degenerate and disappear. The surface of the spiral organ is covered by the ribbon-like gelatinous TECTORIAL MEMBRANE, a continuation of the vestibular lip; it usually becomes

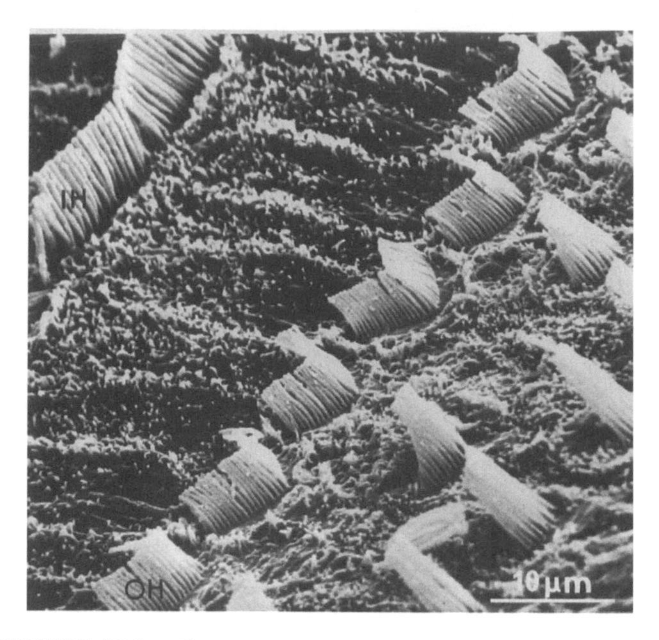

FIGURE P21a The sensory cells (hair cells) of the organ of Corti are arranged into neat rows. The inner hair cells (IH); at the top of this scanning electron micrograph; have stereocilia arranged in a flattened U shape and are spatially separated from the three rows of the outer hair cells. Stereocilia of outer hair cells (OH) have a more angular profile taking on a V or even a W shape. *Weiss, L. (ed.) 1988 Cell and Tissue Biology: A Textbook of Histology, 6th ed. Baltimore: Urban & Schwarzenberg.*

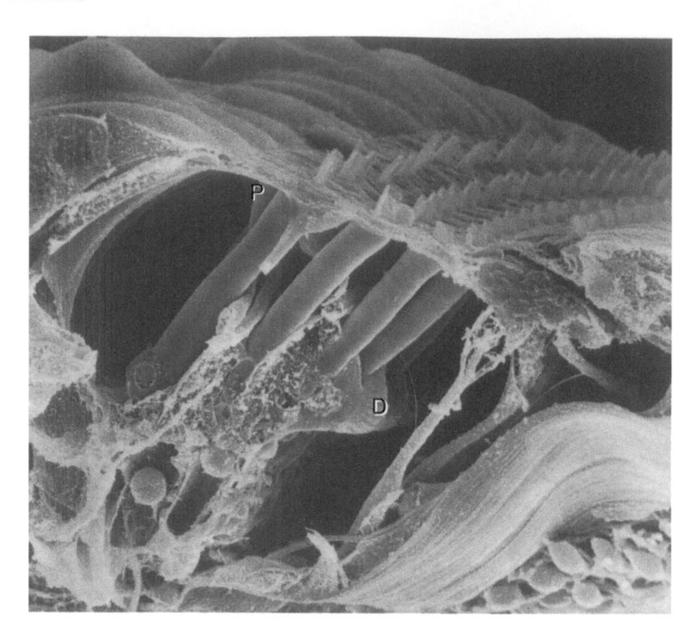

FIGURE P21c A scanning electron micrograph of a fractured organ of Corti shows outer hair cells at the upper right. A portion of the tectorial membrane remains in the upper right corner partially covering some of the hair cells. Hair cells are supported Deiter cells (D) and their phalangeal processes (P). *Weiss, L. (ed.) 1988 Cell and Tissue Biology: A Textbook of Histology, 6th ed. Baltimore: Urban & Schwarzenberg.*

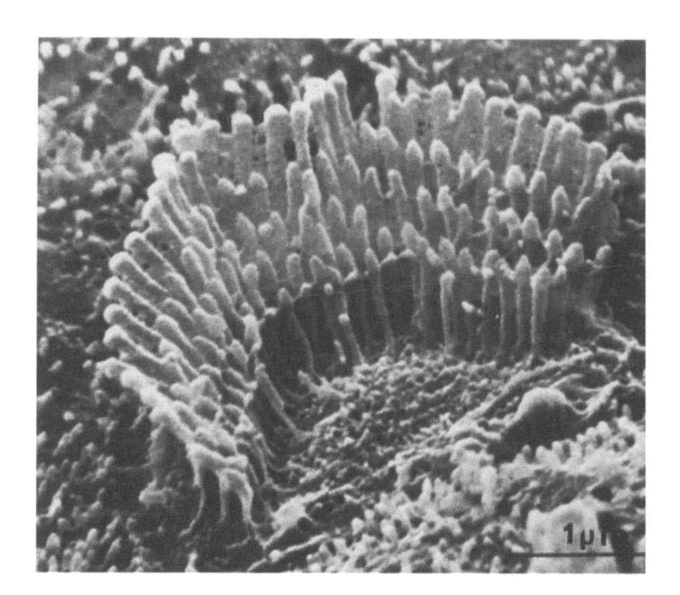

FIGURE P21b A scanning electron micrograph of the typically V-shaped arrangement of stereocilia of an outer hair cell. *Weiss, L. (ed.) 1988 Cell and Tissue Biology: A Textbook of Histology, 6th ed. Baltimore: Urban & Schwarzenberg.*

curled and detached from the epithelium after fixation. It is composed of a homogeneous ground substance with numerous fibrils.

Vibrations are conveyed to the oval window by ossicles in the middle ear moving the perilymph in the scala vestibuli and scala tympani in response. The basilar membrane vibrates and stimulates the hair cells. The basilar membrane seems to be "tuned" to different frequencies over its range so that low frequencies are detected at its apex and high frequencies near its base.

Image and Diagram Acknowledgments

Chapter A	
Figure A1	Wilson, E.B., 1928. The Cell in Development and Heredity, third ed.The Macmillan Company, New York.
Figure A2	Maximow, A.A., Bloom, W., 1930. A Textbook of Histology. W.B. Saunders Company, Philadelphia.
Figure A3	Brachet, J. 1961. The living cell. Sci. Am. 205, (3), 50–61.
Figures A35b, A36, A38, A47, A52, A53, A55, A56, A57, A58, A59, A60	Fawcett, D.W., Bloom, W., Maximow, A.A., 1994. A Textbook of Histology, twelfth ed. Chapman & Hall.New York.
Figure A37	Rogers, A.W., 1983. Cells and Tissues: An Introduction to Histology and Cell Biology. Paris Academic Press, San Diego.
Figures A39, A40, A41, A42, A43, A44, A49, A54, A61, A63	Porter, K.R., Bonneville, M.A., 1973. Fine Structure of Cells and Tissues, fourth ed. Lea & Febiger, Philadelphia.
Figures A64, A65	Krstic, R.V., 1984. Illustrated Encyclopedia of Human Histology. Springer-Verlag, Berlin.
Figure A66	Orci, L., Perrelet, A., 1975. Freeze-etch Histology. A Comparison Between Thin Sections and Freeze-etch Replicas. Springer-Verlag. New York.

Chapter B	
Figure B1b	Maximow, A.A., Bloom, W., 1930. A Textbook of Histology. W.B. Saunders, London, Philadelphia.
Figures B6, B7, B8, B9	Ham, A.W., Cormack, D.H., 1979. Histology. eighth ed. J. B. Lippincott Company, Philadelphia, PA.
Figures B6a, 10a, 10b	Weiss, L. (Ed.), 1988. Cell and Tissue Biology: A Textbook of Histology, sixth ed. Urban & Schwarzenberg, Baltimore, MD.
Figure B11	Fawcett, D.W., Bloom, W., Maximov, A.A., 1994. A Textbook of Histology, twelfth ed. Chapman & Hall, New York.

Chapter C	
Figures C5, C62, C64, C67	Rogers, A.W., 1983. Cells and Tissues: An Introduction to Histology and Cell Biology. Academic Press, San Diego, CA, Paris.
Figures C31 C32 C35, C47, C51a, C56, C60a, C60b	Porter, K.R., Bonneville, M.A., 1973. Fine Structure of Cells and Tissues, fourth ed. Lea & Febiger, Philadelphia.
Figures C33a, C33b, C34a, C69, C74	Ham, A.W., Cormack, D.H., 1979. Histology, eighth ed. J. B. Lippincott Company, Philadelphia.
Figures C34b, C53,C54, C63, C65, C66, C68	Peek, W.D., 1976. Ultrastructural Aspects of Fluid Transport in the Nephron of the Female Garter Snake. PhD Dissertation, Faculty of Graduate Studies Western University, London, ON, CA.
Figures C45, C57, C61	Fawcet, D.W., 1981. The Cell, second ed. W. B. Saunders Co., Philadelphia.

(*Continued*)

(Continued)

Chapter C	
Figures C51b, C51c, C58	Fawcett, D.W., Bloom, W., Maximov, A.A., 1994. A Textbook of Histology, twelfth ed. Chapman & Hall, New York.
Figure C50	Murray, R.G., Murray, A., 1967. Fine structure of taste buds of rabbit foliate papillae. J Ultrastruct. Res. 19, 327–353; Fig. 2.
Figure C42, C51	Weiss, L. (Ed.), 1988. Cell and Tissue Biology: A Textbook of Histology, sixth ed. Urban & Schwarzenberg, Baltimore, MD.
Figure C52	McMillan, D.B., 1966. personal image.
Figure C70	Turner, C.D., 1948. General Endocrinology. W. B. Saunders Co., Philadelphia, PA.
Figure C86	Ross, M.H., Romrell, L.J., Kaye, G.I., 1995. Histology: A Text and Atlas, third ed. Williams & Wilkins, Baltimore, MD.
Figures C88a, C88b	Ishimura, K., Kawamata, S., Fijita, H., 1981. Freeze Fracture Images of Mammary glands of lactating Mice. Anat Embryol. 163, 173–183.

Chapter D	
Figures D4a, D4b, D7, D8, D13,D14	Rhodin, J.A.G., 1974. Histology. A Text and Atlas. Oxford University Press, New York, London, Toronto.
Figures D9, D17, D18a, D25a, D25b, D27, D45, D57, D69	Porter, K.R., Bonneville, M.A., 1973. Fine Structure of Cells and Tissues, fourth ed. Lea & Febiger, Philadelphia, PA.
Figures D10, D40a, D40b	Ross, M.H., Romrell, L.J., Kaye, G.I., 1995. Histology: A Text and Atlas, third ed. Williams & Wilkins, Baltimore, MD.
Figure D15	Fujita, J., Kashimura, M., Adachi, K., 1982. Scanning electron microscopy (SEM) studies of the spleen–normal and pathological. Scan Electron Microsc. 1, 435–444.
Figure D16	Maximow, A.A., Bloom, W., 1942. A Textbook of Histology. W.B. Saunders, London, Philadelphia, PA.
Figure D18b	Motta, P., 1975. Scanning Electron Microscopic observations of mammalian adipose cells. J Micoscopy (Oxford). 1 22, 15–20.
Figures D22a, D22b, D23	Suter, E.R., 1969. The fine structure of brown adipose tissue Pt 1. J Ultrastruct Res. 26, 216–241.
Figure D23b	Krstic, R.V., 1979. Ultrastructure of the Mammalian Cell: An Atlas. Springer-Verlag, Berlin.
Figures D31, 32a	Junqueira, L.C., Raker, E., Porter, K.R., 1974. Studies on pigment migration in the melanophores of the teleost. *Fundulus heteroclitus* (L). Arch. Histol. Jpn. 36 (5), 339–366.
Figures D33, D35	Obika, M., Menter, D.G., Tchen, T.T, Taylor, J.D., 1978. Actin microfilaments in melanophores of *Fundulus heteroclitus*. Their possible involvement in melanosome migration. Cell Tiss. Res. 193 (3), 387–397.
Figure D34	Obika, M., Fukuzawa, T., 1993. Cytoskeletal architecture of dermal chromatophores of the freshwater teleost *Oryzias latipes.* Pigment Cell Res. 6 (6), 417–422.
Figure D36	Obika, M., Meyerrochow, V.B., 1986. Ultrastructure of microtubules in dermal melanophores and spinal nerve of the Antarctic teleost *Pagothenia borcbgrevinki.* Cell Tiss. Res. 244, 339–343.
Figures D38a, D38b	Shinagawa, Y., Shinagawa, Y., 1974. Studies on the collagen fibrils in amphibian skin by means of thin sectioning and densitometric method. J. Elect. Micros. 23 (3), 161–165.
Figures D44a,D44b, D44c	Kikuta, A., Ohtani, O., Murakami, T., 1991. Three-dimensional organization of the collagen fibrillar framework in the rat adrenal gland. Arch. Histol. Cytol. 54 (2), 133–144.
Figure D46	Ham, A.W., Cormack, D.H., 1979. Histology, eighth ed. J. B. Lippincott Company, Philadelphia, PA.
Figures D60, D62a, D62b	Koch, J.C., 1917. The laws of bone architecture. Am J Anat. 21, 177–293.

(*Continued*)

(Continued)

Chapter D	
Figures D61a, D61b	Whitehouse, W.J., 1975. Scanning electron micrographs of cancellous bone from the human sternum. J. Pathol. 116 (4), 213–224.
Figures D67a, D67b	Ejiri, S., Ozawa, H., 1982. Scanning electron microscopic observations of rat tibia using the HCl-collagenase method. Arch. Histol. Jpn. 45 (4), 399–404.
Figure D71	Maximow, A.A., Bloom, W., 1942. A Textbook of Histology. W.B. Saunders, London, Philadelphia, PA.
Figures D78, 81a, 81b	Fawcett, D.W., Bloom, W., Maximow, A.A., 1994. A Textbook of Histology, twelfth ed. Chapman & Hall, New York.
Figure D82	Horton, M.A., Rimmer, E.F., Lewis, D., et al., 1984. Cell surface characterization of the human osteoclast: phenotypic relationship to other bone marrow-derived cell types. J. Pathol. 44 (4), 281–294.
Figure D83	Storer, T.I., Usinger, R.L., 1965. General Zoology, fourth ed. McGraw-Hill, New York, Toronto.
Figure D87	Huettner, A.F., 1949. Fundamentals of Comparative Embryology of the Vertebrates. MacMillan, New York.
Figure D88	Newman, W.A., 1944. Dorland's American Illustrated Medical Dictionary, twentieth ed. Saunders, Philadelphia, PA.
Figure D89	Patten, B.M., 1946. Human Embryology. The Blakiston Company, Philadelphia, Toronto.
Figure D91	Butler, W.F., Fujioka, T., 1972. Fine structure of the nucleus pulposus of the intervertebral disc of the cat. Anat. Anz. 132, 465–475.
Figure D93	Flood, P.R., Guthrie, D.M., Banks, J.R., 1969. Paramyosin muscle in the notochord of Amphioxus. Nature 222, 87–88.
Figure D94	Flood, P.R., 1970. The connection between spinal cord and notochord in Amphioxus (*Branchiostoma lanceolatum*). Z .Zellforsch. 103, 115–128.
Figure D95	Welsch, U., 1968. Über den Feinbau der Chorda dorsalis von *Branchiostoma lanecolatum*. Z. Zellforsch 87, 69–81.

Chapter E	
Figures E4, E11, E34	Porter, K.R., Bonneville, M.A., 1973. Fine Structure of Cells and Tissues, fourth ed. Lea & Febiger, Philadelphia, PA.
Figures E5a, E5b	Orci, L., Perrelet, A., 1975. Freeze-etch Histology. A Comparison Between Thin Sections and Freeze-etch Replicas. Springer-Verlag, New York.
Figures E6, E7	Weiss, L. (Ed.), 1988. Cell and Tissue Biology: A Textbook of Histology, sixth ed. Urban & Schwarzenberg, Baltimore, MD.
Figures E8b, E8c, 10b	Clark, A.W., Schultz, E., 1980. Rattlesnake shaker muscle: II. Fine structure. Tiss. Cell 12, 348.
Figures E9g	Schmalbruch, H., 1974. The sarcolemma of skeletal muscle fibres as demonstrated by a replica technique. Cell Tiss. Res. 150, 377–387.
Figures E10c, E40	Ham, A.W., Cormack, D.H., 1979. Histology, eighth ed. J. B. Lippincott Company, Philadelphia, PA.
Figures E11b, E12b, E12c	Krstic, R.V., 1978. Die Gewebe des Menschen und der Saugetiere. Springer-Verlag, Berlin.
Figures E14, E19, E20, E35	Ross, M.H., Romrell, L.J., Kaye, G.I., 1995. Histology: A Text and Atlas, third ed. Williams & Wilkins, Baltimore, MD.
Figure E16	Trotter, J.A., Samora, A., Baca, J., 1985. Three-dimensional structure of the murine muscle-tendon junction. Anat. Rec. 213, 16–25.
Figure E17	Trotter, J.A., et al., 1985. A morphometric analysis of the muscle-tendon junction. Anat. Rec. 213, 26 – 32.

(Continued)

(Continued)

Chapter E	
Figure E21	Fawcett, D.W., Bloom, W., Maximov, A.A., 1994. A Textbook of Histology, twelfth ed. Chapman & Hall, New York.
Figure E22	Eckert, R., Randall, D.J., Augustine, G. 1988. Animal Physiology: Mechanisms and Adaptations, third ed. W.H.Freeman, New York.
Figure E23	Mellinger, J., Belbenoit, P., Ravaille, M., Szabo, T., 1978. Electric organ development in *Torpedo marmorata. Chondrichthyes.* Dev. Biol. 67 (1), 167–88.
Figures E24a, E24b, E24c	Denizot, J.P., Kirschbaum, F., et al., 1982. On the development of the adult electric organ in the mormyrid fish *Pollimyrus isidori* (with special focus on the innervation). J. Neurocytol. 11 (6), 913–934.
Figure E25	Hoar, W.S., 1966. General and Comparative Physiology. Prentice-Hall, Englewood Cliffs, NJ.
Figures E26, E27	Schwartz, I.R., Pappas, G.D., Bennett, M.V.L., 1975. The fine structure of electrocytes in weakly electric teleosts. J. Neurocytol. 4 (1), 87–114.
Figure E28	Prado-Figueroa, M., Barrantes, F.J., 1989. Ultrastructure of *Discopyge tschudii* electric organ. Microsc Electron. Biol. Cell. 13 (1), 19–37.
Figures E29a, E29b	Allen, T.M., 1975. Scanning electron microscopy of the electric tissue of *Narcine brasiliensis.* Tiss Cell 7 (4), 739–745.

Chapter F	
Figures F1a, F7a, F15a, F15b, F49, F52, F76, F79	Ham, A.W., Cormack, D.H., 1979. Histology, eighth ed. J. B. Lippincott Company, Philadelphia, PA.
Figures F1b,F28, F29, F30, F31, F32, F34	Taylor, N.B., McPhedran, M.G., 1965. Basic Physiology and Anatomy. MacMillan, Toronto.
Figures F10, F13, F14, F17a, F23a	Porter, K.R., Bonneville, M.A., 1973. Fine Structure of Cells and Tissues, fourth ed. Lea & Febiger, Philadelphia, PA.
Figure F11	Rogers, A.W., 1983. Cells and Tissues: An Introduction to Histology and Cell Biology. Academic Press, San Diego, Paris.
Figure F16, F20	Krstic, R.V., 1979. Ultrastructure of the Mammalian Cell: An Atlas. Springer-Verlag, Berlin.
Figures F17b, F19, F20, F45a, F47	Krstic, R.V., 1978. Die Gewebe des Menschen und der Saugetiere. Springer-Verlag, Berlin.
Figure F18	Krstic, R.V., 1984. Illustrated Encyclopedia of Human Histology. Springer-Verlag, Berlin.
Figure F38, F47	Krstic, R.V., 1991. Human Microscopic Anatomy. Springer-Verlag, Berlin.
Figures F39, F40, F41, F64, F68, F69, F80	Fawcett, D.W., Bloom, W., Maximov, A.A., 1994. A Textbook of Histology, twelfth ed. Chapman & Hall, New York.
Figure F42, F43	Denizot, J., Kirschbaum, F., et al., 1982. On the development of the adult electric organ in the mormyrid fish *Pollimyrus isidori* (with special focus on the innervation). J. Neurocytol. 11 (6), 913--934.
Figure F45b	Patten, R.M., Ovalle, W.K., 1991. Muscle spindle ultrastructure revealed by conventional and high-resolution scanning electron microscopy. Anat Rec. 230 (2), 183–198.
Figures F53, F54	Munger, B.L., Yoshida, Y., et al., 1988. A re-evaluation of the cytology of cat Pacinian corpuscles pt.I. The inner core and clefts. Cell Tissue Res. 253 (1), 83–93.
Figures F56, F57, F59, F60, F61, F62, F63	Watanabe, I., Usukura, J., Yamada, E., 1985. Electron microscope study of the grandry and herbst corpuscles in the palatine mucosa, gingival mucosa and beak skin of the duck. Arch Histol Jap. 48 (1), 89–108.
Figure F64	Copenhaver, W.M., Kelly, D.E., Wood, R.L., 1978. Bailey's Textbook of Histology, seventeenth ed. Williams & Wilkins, Baltimore, MD.
Figures F70, F73, F74a	Welsch, U., Storch, V., 1973. Einfuhrung in Cytologie und Histologie de Tiere. Verlag Stuttgart, Gustav Fisher.

Chapter G

Figures G1, G 9, G10, G14, G17, G20a, G20b, G25, G27, G28,G34, G39a, G39b, G43, G45, G47a, G49, G50	Diggs, L.W., Strum, D., Bell, A., 1954. The Morphology of Blood Cells. Pamphlet. Abbott Laboratories, North Chicago, IL.
Figures G2, G3, 11b, G36, G37	Ham, A.W., Cormack, D.H., 1979. Histology, eighth ed. J. B. Lippincott Company, Philadelphia, PA.
Figure G4	Fujita, T., Tanaka, K., Tokunaga, K., 1981. SEM Atlas of Cells and Tissues. Igaku-Shoin, New York.
Figures G8a, G15b, G27b, G35, G41, G46b, G51	Porter, K.R., Bonneville, M.A., 1973. Fine Structure of Cells and Tissues, fourth ed. Lea & Febiger, Philadelphia.
Figure G8b	Campbell, F., 1967. Fine structure of the bone marrow of the chicken and pigeon. J Morphol. 123 (4), 405–440.
Figure G8c	Desser, S.S., Weller, I., 1979. Ultrastructural observations on the erythrocytes and thrombocytes of the tuatara, *Sphenodon punctatus* (gray). Tiss Cell 11 (4), 717–726.
Figure G8d	Vankin, G.L., Brandt, E.M., Dewitt, W., 1970. Ultrastructural studies of red blood cells from thyroxin-treated *Rana catesbeiana* tadpoles. JCB. 47 (3), 767.
Figure G8e	Sekhon, S.S., Maxwell, D.S., 1970. Fine structure of developing hagfish erythrocytes with particular reference to the cytoplasmic organelles. J Morphol. 131, 211–236.
Figures G11a, G13a, G13b, G18b G32, G33, G46a, G47b	Fawcett, D.W., Bloom, W., Maximov, A.A., 1994. A Textbook of Histology, twelfth ed. Chapman & Hall, New York.
Figures G12a, G15a, G18a, G21, G23, G29a	Ross, M.H., Romrell, L.J., Kaye, G.I., 1995. Histology: A Text and Atlas, third ed. Williams & Wilkins, Baltimore, MD.
Figure G12b	Fujita, T., Tanaka, K., Tokunaga, J., 1981. SEM Atlas of Cells and Tissue. Igaku-Shoin Tokyo, New York.
Figures G12d, G16d, G19e	Lester, R.J., Desser, S.S., 1975. Ultrastructural observations on the granulocytic leucocytes of the teleost Catostomus commersoni. Can J Zool 53 (11), 1648–1657.
Figure G12e	Setoguti, T., Fuji, H., Isono, H., 1970. An electron microscopic study on neutrophil leukocytes of the toad. Arch Histol Jpn. 32 (1), 87–94.
Figures G12f, G19a, G31a	Daimon, T., Caxton-Martins, A., 1977. Electron microscopic and enzyme cytochemical studies on granules of mature chicken granular leucocytes. J Anat. 123 (3), 553–562.
Figure G12g	Smith SH, Obenauf SD, Smith DS. 1989. Fine structure of shark leucocytes during chemotactic migration. Tiss. Cell 21 (1), 47–58.
Figures G16a, G19d	Ishizeki, K., Nawa, T., et al., 1984. Hemopoietic sites and development of eosinophil granulocytes in the loach, *Misgurnus anguillicaudatus*. Cell Tiss. Res. 235, 419–426.
Figures G16bi, G16bii, G2a,G2b,G29b	Enbergs, H., Beardi, B., 1971. Zur Feinstruktur der eosinophilen Granulozyten der Hausgans (Anser anser dom.). Z. Zellforsch. 122, 520–527.
Figures G16c, G19b, G26a, G29c, 30a	Maxwell, M.H., Siller, W.G., 1972. The ultrastructural characteristics of the eosinophil granules in six species of domestic bird. J Anat. 112 (pt2), 289–303.
Figure G19c	Kriesten, K., Enbergs, H., 1970. Fine structure of hens basophilic granulocytes. Blut. 20 (5), 282–287.
Figures G20c, G26d, G30d	Nakamura, H., Shimozawa, A., 1984. Light and Electron Microscopic Studies on the leucocytes of the Japanese rice fish (Oryzias latipes). Medaka. 2, 15–21.
Figures G22, G38, G44c, G44d	Weiss, L. (Ed.), 1988. Cell and Tissue Biology: A Textbook of Histology, sixth ed. Urban & Schwarzenberg, Baltimore, MD.
Figure G26b	Zapata, A., Leceta, J., Villena, A., 1981. Reptilian bone marrow. An ultrastructural study in the spanish lizard, *Lacerta hispanica*. J. Morphol. 168 (2), 137–149.

(*Continued*)

(Continued)

Chapter G	
Figures G26c, G30b	Bounous, D.I., Dotson, T.K., et al., 1996. Cytochemical staining and ultrastructural characteristics of peripheral blood leucocytes from the yellow rat snake (*Elaphe obsoleta quadrivitatta*). Comp Haemat Inter. 6 (2), 86–91.
Figure G26e	Cannon, MS., et al., 1980. An ultrastructural study of the leukocytes of the channel catfish, *lctalurus punctatus*. J Morphol. 164, 1–23.
Figures G27c, G40	Leenheer, E.L., 1969. Hemopoiesis in the Kidney of *Oncorhynchus masou* (Brevoort). MSc Dissertation, UWO Faculty of Graduate Studies, London, ON, CA.
Figure G30c	Bielek, E., 1979. Elektronenmikroskopische Untersuchhungen der Blutzellen der Teleostier. Zool Jb Anat. 101, 19–26.
Figure G30d	Hyder, S.L., Cayer, M.L., Petty, C.L., 1983. Cell types in peripheral blood of the nurse shark: An approach to structure and function. Tiss Cell 15 (3), 437–455.
Figures G31a, G31b	Daimon, T., Mizuuhra, V., et al., 1979. The surface connected canalicular system. of carp (*Cyprinus carpio*) thrombocytes: Its fine structure. Cell Tiss. Res. 203, 355–365.
Figure G42	Koury, S.T., Koury, M.J., Bondurant, M.C., 1989. Cytoskeletal distribution and function during the maturation and enucleation of mammalian erythroblasts. J. Cell Biol. 109 (No. 6, Pt. 1), 3005–3013.
Figures G44a, G44b	Campbell, F., 1969. Electron microscopic studies on granulocytopoiesis in slender salamander. Anat Rec. 163 (3), 427–442.

Chapter H	
Figures H1, H4	Storer, T.I., Usinger, R.L., 1965. General Zoology, fourth ed. McGraw-Hill, New York, Toronto.
Figures H2, H3, H11b, H14d, H15c, H19c, H22a	Ham, A.W., Cormack, D.H., 1979. Histology, eighth ed. J. B. Lippincott Company, Philadelphia, PA.
Figure H5a	Original artwork RJH.
Figure H5b	Krstic, R.V., 1991. Human Microscopic Anatomy. Springer-Verlag, Berlin.
Figures H5c, H6d, H10c, H13g, H13h, H16b, H18b	Fawcett, D.W., Bloom, W., Maximov, A.A., 1994. A Textbook of Histology, twelfth ed. Chapman & Hall, New York.
Figures H7a, H7c, H9, H13j, H14c, H14e, H14f, H15d, H18c, H18d, H22c	Weiss, L. (Ed.), 1988. Cell and Tissue Biology: A Textbook of Histology, sixth ed. Urban & Schwarzenberg, Baltimore, MD.
Figures H7b, H12d	Porter, K.R., Bonneville, M.A., 1973. Fine Structure of Cells and Tissues, fourth ed. Lea & Febiger, Philadelphia, PA.
Figures H12c, H22b	Ross, M.H., Romrell, L.J., Kaye, G.I., 1995. Histology: A Text and Atlas, third ed. Williams & Wilkins, Baltimore, MD.
Figures H13f, H14a	Fawcett, D.W., 1966. The Cell An Atlas of Fine Structure. W.B. Saunders Company, Philadelphia and London.
Figure H14b	Orci, L., Perrelet, A., 1975. Freeze-etch Histology. A Comparison Between Thin Sections and Freeze-etch Replicas. Springer-Verlag, New York.

Chapter I	
Figures I1a, I11	Ross, M.H., Romrell, L.J., Kaye, G.I., 1995. Histology: A Text and Atlas, third ed. Williams & Wilkins, Baltimore, MD.
Figure I1b	Fulop, G.M.I., 1979. A Structural Study of Phagocytosis in the Spleen of the Sunfish *Lepomis spp*. MSc. Dissertation – Faculty of Graduate Studies, UWO, London, ON, CA.
Figures I1d, I23d	Forkert, P.G., Thliveris, J.A., Bertalanffy, F.D., 1977. Structure of Sinuses in the Human Lymph Node, Cell Tiss Res. 183, 115–130.

(*Continued*)

(Continued)

Chapter I	
Figure I18a	Langley, L.L., Telford, I.R., Christensen, J.B., 1980. Dynamic Anatomy and Physiology. McGaw-Hill, New York.
Figure I20	Smith, P.E., Copenshaver, W.M., 1948. Bailey's Textbook of Histology. Williams & Wilkins, Baltimore, MD.
Figures I23b, I23c, I23e, I23f	Luk, S.C., Nopajaroonsric, C., Simon, G.T., 1973. The architecture of the normal lymph node and hemolymph node. A scanning and transmission electron microscopic study. Laboratory Invest. 29 (2), 258–265.
Figures I24a, I24b	Fawcett, D.W., Bloom, W., Maximov, A.A., 1994. A Textbook of Histology, twelfth ed. Chapman & Hall, New York.
Figure I24c	Ham, A.W., Cormack, D.H., 1979. Histology, eighth ed. J. B. Lippincott Company, Philadelphia, PA.
Figures I27a,I27b, I27c, I27d, I27e, I27f, I	Fujita, T., 1974. A scanning electron microscope study of the human spleen. Arch Histol Japon. 37 (3), 187–216.

Chapter J	
Figures J4, J5a, J5b, J5c, J6a, J6b, J7a, J7b, J44a, J45,	Fawcett, D.W., Bloom, W., Maximov, A.A., 1994. A Textbook of Histology, twelfth ed. Chapman & Hall, New York.
Figures J5d, J40c, J44	Weiss, L. (Ed.), 1988. Cell and Tissue Biology: A Textbook of Histology, sixth ed. Urban & Schwarzenberg, Baltimore, MD.
Figures J24b, J24c, J47a	Ham, A.W., Cormack, D.H., 1979. Histology, eighth ed. J. B. Lippincott Company, Philadelphia, PA.
Figures J31b	Downing, S.W., Spitzer, R.H., Koch, E.A., Salo, W., 1984. The hagfish slime gland thread cell I. A unique cellular system for the study of intermediate filaments and intermediate filament-microtubule interactions. JCB. 98, 653–669.

Chapter K	
Figures K1a, K43, K59b, K61h, K67c, K98b	Ham, A.W., Cormack, D.H., 1979. Histology, eighth ed. J. B. Lippincott Company, Philadelphia,PA.
Figure K1c	Turner, C.D., 1948. General Endocrinology. W. B. Saunders Co., Philadelphia, PA.
Figures K4b, K11c, K12c, K22, K25b, K25c, K27a, K40, K42b, K42c, K44, K65b, K74d, K101d	Fawcett, D.W., Bloom, W., Maximov, A.A., 1994. A Textbook of Histology, twelfth ed. Chapman & Hall, New York.
Figures K12a, K12b, K19, K27b, K42a, K45g, K45i, K61f, K61g, K62f, K82b, K82d, K97d, K99g, K99h, K-122	Weiss, L. (Ed.), 1988. Cell and Tissue Biology: A Textbook of Histology, sixth ed. Urban & Schwarzenberg, Baltimore, MD.
Figure K15a	Holmes, S.J., 1927. The Biology of the Frog, fourth ed. MacMillan and Co Ltd., Toronto.
Figures K17c, K17d	Takami, S., Hirosawa, K., 1990. Electron microscopic observations on the vomeronasal sensory epithelium of a crotaline snake, *Trimeresurus flavoviridis*. J. Morphol. 205 (1), 45–61.
Figure K25d	Kallenbach, E., 1976. Fine structure of differentiating ameloblasts in the kitten. Am J Anat. 145 (3), 283–317.
Figure K26	Matthiessen, M.E., Romert, P., 1980. Ultrastructure of the human enamel organ. I. External enamel epithelium, stellate reticulum, and stratum intermedium. Cell Tiss. Res. 205, 361–370.
Figures K33d, K50a, K70d, K78	Weichert, C.K., 1967. Elements of Chordate Anatomy, third ed. McGraw-Hill, New York.
Figures K45h, K61e, K62e, K101c, K101e	Porter, K.R., Bonneville, M.A., 1973. Fine Structure of Cells and Tissues, fourth ed. Lea & Febiger, Philadelphia, PA.
Figure K57b	Ross, M.H., Romrell, L.J., Kaye, G.I., 1995. Histology: A Text and Atlas, third ed. Williams & Wilkins, Baltimore, MD.

(*Continued*)

(Continued)

Chapter K	
Figure K74a	Taylor, N.B., McPhedran, M.G., 1965. Basic Physiology and Anatomy. MacMillan Co Ltd., Toronto; Putnam and Sons, New York.
Figure K83e	Tandler, B., 1976. Ultrastructure of baboon parotid gland. Anat Rec. 184 (1), 115–131.
Figures K83g, K83h	Tandler, B., 1986. Ultrastructure of the retrolingual salivary gland of the European hedgehog. J Submicros Cytol Pathol. 18 (2), 249–260.
Figure K83f	Riva, A., Serra, G.P., Proto, E., et al., 1992. The Myoepithelial and Basal Cells of Ducts of Human Major Salivary Glands: A SEM Study. Arch Histol Cytolo. 55 (Suppl), 115–124.
Figure K97a	Storer, T.I., Usinger, R.L., 1965. General Zoology, fourth ed. McGraw-Hill, New York, Toronto.
Figure K99c	Fujita, T., Tanaka, K., Tokunaga, K., 1981. SEM Atlas of Cells and Tissues. Igaku-Shoin, New York.
Figures K101a, K101b	Orci, L., Perrelet, A., 1975. Freeze-etch Histology. A Comparison Between Thin Sections and Freeze-etch Replicas. Springer-Verlag, New York.
Figure K114	Best, C.H., Taylor, N.B., 1963. The Human Body, Its Anatomy and Physiology, fourth ed. Holt, Rinehart and Winston, New York.

Chapter L	
Figures L1c, L2b, L7a, L25a	Storer, T.I., Usinger, R.L., 1965. General Zoology, fourth ed. McGraw-Hill, New York, Toronto.
Figure L3a	Weichert, C.K., 1967. Elements of Chordate Anatomy, third ed. McGraw-Hill Book Co., New York, Toronto.
Figure L5a	Young, J.Z., 1981. The Life of Vertebrates, second ed. Oxford University Press, New York.
Figures L7b, L10d, L10e	Hossler, F.E., Ruby, J.R., Mcilwain, T.D., 1979. The gill arch of the mullet, *Mugil cephalus*. I. Surface ultrastructure. J. Exp. Zool. 208 (3), 379–398
Figures L8c, L9c, L9e	Hughes, G.M., Weibel, E.R., 1972. Similarity of supporting tissue in fish gills and the mammalian reticuloendothelium. J Ultrastr Res. 39 (1-2), 106–114.
Figures L8d, L14	Hughes, G.M., Gray I.A., 1972. Dimensions and ultrastructure of toadfish gills. Biol Bull. 143 (1), 150–161.
Figure L9b	Wright, D.E., 1973. The structure of the gills of the elasmobranch, *Scyliorhinus canicula* (L.). Z. Zellforsch 144, 489–509.
Figures L9d, L11d, L11e	Youson, J.H., Freeman, P.A., 1976. Morphology of the gills of larval and parasitic adult sea lamprey, *Petromyzon marinus* (L). J. Morphol. 149 (1), 73–103.
Figures L10a, L11b	Dunel-Erb, S., Laurent P. 1980. Ultrastructure of marine teleost gill epithelia: SEM and TEM study of the chloride cell apical membrane. J Mophol. 165, 175–186.
Figure L10b	Laurent, P., et al., 1995. Gill structure of a fish from an alkaline lake: Effect of short-term exposure to neutral conditions. Can. J. Zool. 73(6), 1170–1181.
Figure L10c	Paurent, P., Perry, S.F., 1991. Environmental effects on fish gill morphology. Physiol Zool. 64 (1), 4–25.
Figure L11a	Bonga, S.E.W., Flik, G., et al., 1990. The ultrastructure of chloride cells in the gills of the teleost *Oreochromis mossambicus* during exposure to acidified water. Cell Tissue Res. 259, 575–585.
Figures L11c, L11g	Bartels, H., Potter, I.C., 1991. Structural changes in the zonulae occludentes of the chloride cells of young adult lampreys following acclimation to seawater. Cell Tiss. Res. 265, 447–457.
Figure L12	Karlsson, L., 1983. Gill morphology in the zebrafish, *Brachydanio rerio*. J. Fish Biol. 23 (5), 511–524.

(Continued)

(Continued)

Chapter L	
Figures L13a, L13b	Dunel-Erb, S., Bailey, Y., Laurent, P., 1989. Neurons controlling the gill vasculature in five species of teleosts. Cell Tiss. Res. 255, 567–573.
Figures L15a, L15b, L15c	Hossler, F.E., Merchant, L.H., 1983. Morphology of taste buds on the gill arches of the mullet *Mugil cephalus*, and the killifish *Fundulus heteroclitus*. Am. J. Anat. 166 (3), 299–312.
Figures L16a, L17a, L20	Taylor, N.B., McPhedren, M.G., 1965. Basic Physiology and Anatomy. MacMillan Co., Toronto.
Figures L17d	Fawcett, D.W., Bloom, W., Maximov, A.A., 1994. A Textbook of Histology, twelfth ed. Chapman & Hall, New York.
Figures L17e, L23b, L23c, L29h, L29i	Porter, K.R., Bonneville, M.A., 1973. Fine Structure of Cells and Tissues, fourth ed. Lea & Febiger, Philadelphia, PA.
Figures L33a, L33b, L33c, L33d, L33e	Schmidt-Nielsen, K., 1990. Animal Physiology: Adaptation and Environment, fourth ed. Cambridge University Press, Cambridge.

Chapter M	
Figure M1	Weichert, C.K., 1967. Elements of Chordate Anatomy, third ed. McGraw-Hill Book Co., New York, Toronto.
Figures M4, M5a, M5b, M9e	Heath-Eves, M.J., 1970. The morphology of the kidney of the Atlantic hagfish, *Myxine glutinosa* (L.) Faculty of Graduate Studies UWO London ON, MSc Dissertation.
Figure M5c	Romer, A.S., Parsons, T.S., 1977. The Vertebrate Body, fifth ed. Saunders Co., London, Toronto.
Figures M5d, M23	Krstic, R.V., 1991. Human Microscopic Anatomy. Springer-Verlag, Berlin.
Figure M5e	Original to authors.
Figure M5f	Taylor, N.B., McPhedren, M.G., 1965. Basic Physiology and Anatomy. MacMillan Co., Toronto.
Figures M9g, M9h, M9i, M9j, M9k, M9l, M9m, M9n	Bowen, P.C., 1969. The Cytology of the Pronephros of Lampreys. D.Phil Dissertation, University of Oxford, New College.
Figures M11g, M11h	Richter, S., Splechtna, H., 1996. The structure of Anuran podocyte – determined by ecology? Acta Zool (Stockholm) 77 (4), 335–348.
Figure M11i	Bishop, J.E., 1959. A histological and histochemical study of the kidney tubule of the common garter snake (*Thamnophis sirtalis sirtalis*). J. Morph. 104 (2), 307–358.
Figures M26a, M26h	Porter, K.R., Bonneville, M.A., 1973. Fine Structure of Cells and Tissues, fourth ed. Lea & Febiger, Philadelphia, PA.
Figure M26b	Ham, A.W., Cormack, D.H., 1979. Histology, eighth ed. J. B. Lippincott Company, Philadelphia,PA.
Figures M26c, M26d, M26e, M26f, M26f	Fawcett, D.W., Bloom, W., Maximov, A.A., 1994. A Textbook of Histology, twelfth ed. Chapman & Hall, New York.
Figure M28b	Ross, M.H., Romrell, L.J., Kaye, G.I., 1995. Histology: A Text and Atlas, third ed. Williams & Wilkins, Baltimore, MD.
Figure M28c	Barajas, L., Wang, P., et al., 1976. The renal sympathetic system and juxtaglomerular cells in experimental renovascular hypertension. Lab Investig. 35 (6), 574.
Figures M28d M28g	Lamers, A.P.M., van Dongen, W.J., van Kemenade J.A.M., 1974. An ultrastructural study of the juxtaglomerular apparatus in the toad, *Bufo bufo*. Cell Tiss. Res. 153 (4), 449–464.
Figures M28e, M28f	Lamers, A.P.M., van Dongen, W.J., van Kemenade J.A.M., 1973. The morphology of the juxtaglomerular apparatus in the toad, *Bufo bufo*. A light microscopic study. Z. Zellforsch 138, 545–555.
Figure M28h	Barajas, L., 1970. The ultrastructure of the juxtaglomerular apparatus as disclosed by three-dimensional reconstructions from serial sections. The anatomical relationship between the tubular and vascular components. J. Ultrastruct. Res. 33, 116–147.

(*Continued*)

(Continued)

Chapter M	
Figure M44a, M45a	Bulger, R.E., 1963. Fine structure of the rectal gland of the spiny dogfish, *Squalus acanthias*. Anat. Rec. 147 (1), 95.
Figures M45b, M45c, M45d	Lamers, A.P.M., van Dongen, W.J., van Kemenade, J.A.M., 1973. The morphology of the juxtaglomerular apparatus in the toad, *Bufo bufo*. Z. Zellforsch. 138, 545–555.
Figures M46a, M46b, M46c	Fange, R., Schmidt-Niesen, K., Osaki, H., 1958. The salt gland of the herring gull. Biol Bull. 115, 162–171.
Figures M46g, M46h	Butt, M.M., Johnston, H.S., Scothorne, R.J., 1985. Electron microscopic morphometric studies on the resting and secreting nasal (salt) glands of the domestic duck. I. Standardisation of the fixation procedure. J. Anat. 141, 321–239.

Chapter N	
Figures N1a, N2, N9a, N11	Ham, A.W., Cormack, D.H., 1979. Histology, eighth ed. J. B. Lippincott Company, Philadelphia, PA.
Figure N1b	Campbell, N.A., 1993. Biology, third ed. Benjamin/Cummings, Redwood City, CA.
Figures N3, 9b, N10a, N12b	Fawcett, D.W., Bloom, W., Maximov, A.A., 1994. A Textbook of Histology, twelfth ed. Chapman & Hall, New York.
Figures N10b, 14a, N33a	Turner, C.D., 1966. General Endocrinology, fourth ed. W. B. Saunders Co., Philadelphia, PA.
Figures N12a, N24	Porter, K.R., Bonneville, M.A., 1973. Fine Structure of Cells and Tissues, fourth ed. Lea & Febiger, Philadelphia, PA.
Figure N13	Takei Y., Seyama S., et al., 1980. Ultrastructural study of the human neurohypophysis. II. .... Cell Tissue Res. 205 (2), 273–287.
Figure N14b	Bern H.A., Nishioka R.S., et al., 1974. The relationship between nerve fibers and adenohypophysial cell types in the cichlid teleost *Tilapia mossambica*. In: Arvy, L. (Ed.), Recherches Biologiques Contemporaires, CNRS, pp. 179–194.
Figure N15d	Ichikawa T., et al. 1986. The caudal neurosecretory system of fishes. Zool. Sci. 3 (4), 585–598.
Figue N16a	Baumgarten, H.G., Falck, B., Wartenberg, H., 1970 Adrenergic neurons in the spinal cord of the pike (*Esox lucius*) and their relation to the caudal neurosecretory system. Z. Zellforsch. 107 (4), 479–498.
Figure N16b	Oka, S., Chiba A., Honma, Y., 1997. Structures Immunoreactive with porcine NPY in the caudal neurosecretory system of several fishes and cyclostomes. Zool. Sci. 14 (4), 665–669.
Figure N16c	Larson, B., Bern, H., 1987. A double sequential immuno-fluorescence method demonstrating the co-localization of urotensins I and II in the caudal neurosecretory system of the teleost, *Gillichthys* mirabilis. Cell Tiss. Res. 247 (2), 233–239.
Figure N17a	Withers, P.C., 1992. Comparative Animal Physiology. Saunders College Pub, Fort Worth, Toronto.
Figure N17b	Nielsen, J.T., Moller, M., 1978. Innervation of the pineal gland in the Mongolian gerbil (*Meriones unguiculatus*): A fluorescence microscopical study. Cell Tiss. Res. 187 (2), 235–250.
Figure N17d	Furuoka, E., Omura, Y., 1997. Scanning and transmission electron microscopic study on a multilayered basement membrane in the pineal organ of the ayu *Plecoglossus altivelis*. Arch. Histol. Cytol. 60 (5), 511–517.
Figure N18a	Welsh, M.G., Reiter, R.J., 1978. The pineal gland of the gerbil, *Meriones unguiculatus*: I. An ultrastructural study. Cell Tiss. Res. 193 (2), 323–336.
Figure N18b	Mcnilty, J.A., 1981. A quantitative morphological study of the pineal organ in the goldfish, *Carassius auratus*. Can. J. Zool. 59 (7), 1312–1325.
Figure N18c	Vollrath, L., Huss, H., 1973. The synaptic ribbons of the guinea-pig pineal gland under normal and experimental conditions. Z. Zellforsch. 139 (3), 417–429.

(Continued)

(Continued)

Chapter N	
Figures N19c, N25a, N25b	Huettner, A.F., 1949. Fundamentals of Comparative Embryology of the Vertebrates, Rev. ed. MacMillan, New York.
Figure N20d	Gorbman, A., et al., 1983. Comparative Endocrinology. Wiley, New York, Toronto.
Figures N27a, N27b	Weiss, L. (Ed.), 1988. Cell and Tissue Biology: A Textbook of Histology, sixth ed. Urban & Schwarzenberg, Baltimore, MD.
Figure N28a	Jollie, M., 1962. Chordate Morphology. Reinhold Book Corporation, New York.
Figure N28b	Clark, N.B., 1971. The ultimobranchial body of reptiles. J. Exp. Zool. 178 (1), 115–124.
Figure N28c	Shinohara–Ohtani, Y., Sasayama, Y., 1998. Unpaired ultimobranchial glands of the african lungfish, *Protopterus dolloi*. Zool. Sci. 15 (4), 581–588.
Figure N28d	Khairallah, L.H., Clark, N.B., 1971. Ultrastructure and histochemistry of the ultimobranchial body of fresh-water turtles. Z. Zellforsch. 113 (3), 311–321.
Figure N31	Ballard, W.W., 1964. Comparative Anatomy and Embryology. Ronald Press Co., New York.
Figures N33b, N35	Kikuta, A., Ohtani, O., Murakami, T., 1991. Three-dimensional organization of the collagen fibrillar framework in the rat adrenal gland. Arch Histol Cytol. 54 (2), 133–144.
Figures N36a, N36b, N36d	Ross, M.H., Romrell, L.J., Kaye, G.I., 1995. Histology: A Text and Atlas, third ed. Williams & Wilkins, Baltimore, MD.
Figure N36c	Kikuta, A., Muarkami, T., 1982. Microcirculation of the rat adrenal gland: A scanning electron microscope study of vascular casts. Am. J. Anat. 164 (1),19–28.

Chapter O	
Figure O1	Romer, A.S., 1970. The Vertebrate Body, fourth ed.: Saunders & Co., Philadelphia, PA.
Figures O2a, O2b, O2c, O2d	Maximow, A.A., Bloom, W., 1942. A Textbook of Histology. W. B. Saunders Co., London. Philadelphia, PA.
Figures O4, O12, O34a, O36a	Turner, C.D., 1948. General Endocrinology. W. B. Saunders Co., Philadelphia, PA.
Figure O35	Ham, A.W., Cormack, D.H., 1979. Histology, eighth ed. J. B. Lippincott Company, Philadelphia,PA.
Figures O36b, O37b, O38b, O38e, O39b, O41b, O42b, O50a	Fawcett, D.W., Bloom, W., Maximov, A.A., 1994. A Textbook of Histology, twelfth ed. Chapman & Hall, New York.
Figures O38a, O38d, O40, O41a, O42c, O46a	Weiss, L. (Ed.), 1988. Cell and Tissue Biology: A Textbook of Histology, sixth ed. Urban & Schwarzenberg, Baltimore, MD.
Figure O38c	Ross, M.H., Romrell, L.J., Kaye, G.I., 1995. Histology: A Text and Atlas, third ed. Williams & Wilkins, Baltimore, MD.
Figure O41c	Porter, K.R., Bonneville, M.A., 1973. Fine Structure of Cells and Tissues, fourth ed. Lea & Febiger, Philadelphia, PA.

Chapter P	
Figures P1b, P1c, P12e	Ross, M.H., Romrell, L.J., Kaye, G.I., 1995. Histology: A Text and Atlas, third ed. Williams & Wilkins, Baltimore, MD.
Figures P2b, P10e, P12e, P17a, P20c, P20h, P20k	Ham, A.W., Cormack, D.H., 1979. Histology, eighth ed. J. B. Lippincott Company, Philadelphia, PA.
Figure P3a	Elias, H., Pauly, J.E., 1966. Human Microanatomy, third ed. FA Davis Company, Philadelphia.
Figures P3b, P4a, P5a, P11c, P12d	Weiss, L. (Ed.), 1988. Cell and Tissue Biology: A Textbook of Histology, sixth ed. Urban & Schwarzenberg, Baltimore, MD.

(*Continued*)

(Continued)

Chapter P	
Figures P5c, P8, P12c, P12f, P20a	Fawcett, D.W., Bloom, W., Maximov, A.A., 1994. A Textbook of Histology, twelfth ed. Chapman & Hall, New York.
Figure P5d	Courtesy of D. Howard Dixon, Emeritus Professor, Department Anatomy, Dalhousie University Halifax NS, CA.
Figure P10b	Fujita, T., Tanaka, K., Tokunaga, K., 1981. SEM Atlas of Cells and Tissues. Igaku-Shoin, New York.
Figure P12g	Brakevelt, C.J., Smith, S.A., Smith, B.G., 1998. Photoreceptor fine structure in *Oreochromis niloticus* L. (Cichlidae; Teleostei) in light- and dark-adaptation. Anat. Rec. 252 (3), 453–461.
Figure P13b	Courtesy of Doreen Larsen-Riedel.
Figures P13c, P13d, P13e	Braekevelt, C.R., 1989. Fine structure of the conus papillaris in the bobtail goanna (*Tiliqua rugosa*). Histol. Histopath. 4 (3), 287–293.
Figure P13f	Braekevelt, C.R., 1998. Fine structure of the pecten oculi of the emu (*Dromaius novaehollandiae*). Tissue Cell 30 (2), 157–165.
Figure P14a	Nicol J.A.C., Arnott H.J., Best A.C.G., 1973 Tapeta lucida in bony fishes (*Actinopterygii*): A survey. Can. J. Zool. 51 (1), 69–86.
Figure P14b	Braekevelt, C.R., 1994. Fine structure of the choroidal tapetum lucidum in the Port Jackson shark (*Heterodontus phillipl*). Anat Embryol. 190, 591–596.
Figures P15, P16	Brakevelt, C.J., Smith, S.A., Smith, B.G., 1998. Fine structure of the retinal pigment epithelium of *Oreochromis niloticus* (L.) (Cichlidae; Teleostei) in light- and dark-adaptation. Anat. Rec. 252 (3), 444–452.
Figure P20b	Huettner, A.F., 1949. Fundamentals of Comparative Embryology of the Vertebrates, Rev. ed. MacMillan, New York.
Figure P20g	Storer, T.I., Usinger, R.L., 1965. General Zoology, fourth ed. McGraw-Hill, New York, Toronto.
Figures P21a, P21b, P21c	Fujita, T., Tanaka, K., Tokunaga, K., 1981. SEM Atlas of Cells and Tissues. Igaku-Shoin, New York.

Index

Note: Page numbers followed by "*f*" refer to figures.

D

E

M

N

T